PRINCIPLES OF MICROBIOLOGY

PRINCIPLES OF MICROBIOLOGY

ALICE LORRAINE SMITH
A.B., M.D., F.C.A.P., F.A.C.P.

Professor of Pathology, The University of Texas Health Science
Center at Dallas, Texas; formerly Assistant Professor of Microbiology,
Department of Nursing, Dominican College and St. Joseph's Hospital,
Houston, Texas

TENTH EDITION

With 475 illustrations and 4 color plates

TIMES MIRROR/MOSBY
COLLEGE PUBLISHING

ST. LOUIS   TORONTO   SANTA CLARA   1985

Editor   Susan Dust Schapper
Editorial Assistant   Jacqueline Ohlgart Yaiser
Manuscript Editors   Susan K. Hume and April Nauman
Design   Diane M. Beasley
Production   Margaret B. Bridenbaugh

Cover   Photomicrograph of tooth bacteria.
       Manfred Kage, Peter Arnold, Inc.

TENTH EDITION

Library of Congress Cataloging in Publication Data

Smith, Alice Lorraine, 1920-
   Principles of microbiology.

   Includes bibliographies and index.
   1. Medical microbiology.   I. Title.   [DNLM:
1. Microbiology.   QW 4 S642p]
QR46.C35   1985   616'.01   84-25517
ISBN 0-8016-4685-5

To
Adele

PREFACE

I keep six honest serving-men
(They taught me all I knew):
Their names are What and Why and When
and How and Where and Who.
Rudyard Kipling

The scientific answers to the What and When and How and Who are coming so thick and fast in these exciting days that, to no one's surprise, it is time for a revision of *Principles of Microbiology*. Here then is the book, carefully reworked to give contemporary coverage in microbiology and designed to be assimilated readily by students in health science training programs.

What's New in this Edition

Organization

The basic pattern of prior editions continues, comprising six modules (units) in biologic orbit around microorganisms. The text begins with basic concepts of "microbe biology" in the first module, includes laboratory methods for the study of microbes in the second, and outlines methods for their control in the third. The events of microbial injury are developed and the human body's restraints emphasized in the fourth module. Culprits—the pathogens and parasites—are indicted in the fifth module, and the sixth mentions certain benefits that microbes confer. At the request of many loyal readers, the second chapter of the book once again features milestones in the history of microbiology.

Also in response to comments from reviewers and teachers, certain changes in the placement and order of the modules and of chapters within some modules have been made for a more logical and workable sequence for the students. The module on sterilization and disinfection now is placed right after laboratory methods and precedes the module on infection and immunity. Two chapters on immunization encompassing both materials and methods have been placed after the chapters on immunity. The chapter on the classification of bacteria has been moved to the end of the general discussion on the nature of microbes.

New Topics

This book in frame and substance must remain relevant to the here-and-now in health careers. It has been edited and updated to keep it so. Topics new to this edition include the following: AIDS, hybridomas, monoclonal antibodies, human T cell leukemia virus, retroviruses, toxic shock syndrome, the bacterial cause for cat-scratch disease, oncogenes, and a new classification of hepadnaviruses.

Classification of Bacteria

The classification of bacteria throughout this text is that of *Bergey's Manual of Systematic Bacteriology*, volume 1 (1984), with the notable exception of the arrangement of enteric bacteria according to the scheme of W.H. Ewing. For viruses, which are not included in *Bergey's Manual*, the most modern classification is still that based on their properties.

Learning Aids

Some valuable learning aids have been added to this tenth edition. Chapter outlines provide the student with a quick overview of the contents. Internal section and paragraph headings make it easier for the student to grasp not only the key concepts but also the correlations important for mastery of subject matter. New to this edition are boxed essays in many chapters, consisting of tidbits of information to stimulate the student's interest in the various applications of microbiology. Summary statements for each chapter reinforce the key points. As with the previous edition, each chapter contains a set of thought-provoking review questions. Suggested readings for students are gathered at the end of each chapter.

Following each of the six modules is a set of laboratory exercises. A module evaluation also appears here and forms a self-test for the student.

Illustrations and Tables

The illustrations have been selected to enrich the meaning of the prose. For instance, when the subject of tuberculin testing is presented, the photograph of positive skin reactions adds impact. Many new illustrations are found in this edition. Ten color photographs highlight different approaches to the laboratory study of microbes. Every instructor knows that tables dramatize and give quick access to information. Immunization schedules, sterilization maneuvers, incubation periods, comparative sizes, metric equivalents, differential features and biologic properties of microbes, and the contrasting mechanisms in immunity are thus arranged.

Overview of Each Unit

The fund of knowledge in all areas of microbiology is expanding with an unbelievable force. All parts of the book were carefully inspected and edited to ensure that information is accurate and up to date. Much new material has been added.

Module One

With the focus on the expressed needs of the reviewers and other teachers, many changes were made in Module One. Three chapters—two of which are new—have greatly amplified the coverage of basic bacterial physiology compared with previous editions. One of the new chapters covers current concepts in bacterial genetics; another covers microbial metabolism. The chapter on the bacterial cell more carefully defines biologic qualities of form and chemical composition and includes an extended discussion of nucleic acids.

Module Two

Module Two begins with a discussion of the compound light microscope, stressing the guidelines for its care and use. Laboratory technics are indicated for the isolation and identification of bacteria (and other microbes) and the demonstration of their biologic properties. Fine points of microbiologic technic are spotlighted, and precautions for maximal laboratory safety are enumerated. Ground rules for specimen collection in clinical medicine are specified.

Module Three

Various physical and chemical agents that destroy or impede microbes are categorized in a small and fairly compact Module Three, and their applications in the health care field are stated. The action of important chemical agents is explained, and unfavorable side effects associated with their use are stressed. Chemical formulas are given for the various ones, and in some instances chemical equations show their action. Technics for sterilization are discussed and compared, for example, gas sterilization versus steam sterilization. With increasing use and availability of commercial prepackaged, sterile disposable units of all kinds, drastic changes have occurred in concepts of sterilization. However, certain standard, long-reliable measures for practical disinfection and sterilization still merit consideration. For instance, hands are not yet disposable.

Module Four

What follows from the contact of microbes with living cells of the human body is a theme permeating this book. Module Four is fully devoted to it, concentrating not only on the ways microbes produce disease but also on the defenses inherent in the human host. This module deals with basic concepts in immunology. It covers the macrophage system (formerly the reticuloendothelial system), the lymphoid system, mechanisms of immunity, the dual nature of the immune response, and key laboratory reactions for immunologic testing. A new chapter, "Immunologic Disorders, Including Allergy," classifies the major disturbances in immunity and concisely characterizes the resultant disorders. The discussion of allergic concepts and diseases occurs here. Because of the general public's concern over this disease, a section discusses AIDS.

With the rearrangement of chapter order, Module Four now contains the two chapters paired to focus on the best information on immunization available from the U.S. Public Health Service, the American Academy of Pediatrics, the U.S. armed forces, the World Health Organization, and the Centers for Disease Control. The companion chapters sort out modern biologic products, outline technics in passive immunization, tabulate latest schedules for active immunization, give crucial guidelines for administration of biologic products, and provide health information for the traveler. New material includes a discussion of the hepatitis B vaccine with the recommended schedule of administration and a discussion of the pneumococcal polysaccharide vaccine.

Module Five

The largest unit of the book, Module Five contains the roster of significant pathogens and parasites, stressing their identity and the nature of the injuries they cause. Infections are matched to agents and carefully correlated with respective anatomic areas of the body. Material pertinent to the development and progression of a given infection is related to the same causative agent.

Module Six

An unusual unit and the last one, Module Six relates the student to the microbial life of our environment. A survey, yet a practical unit, it accommodates items such as measures to safeguard food, making water safe for drinking, and the consequences of viruses in sewage.

Instructor's Manual

The companion instructor's manual for this text contains extended summary outlines of the chapters and answers to the module evaluations. Also featured is a set of transparency masters of selected line drawings from the text.

Reviewers

The following persons read, criticized, and made valuable suggestions for improving the revision of this text. My sincere thanks to:

Gary Alderson, Palomar College
John Fahey, Bennington College
Philip LoVerde, State University of New York at Buffalo
J.T. Prince, University of Minnesota
Paulette Royt, George Mason University
Frank Varga, University of Cincinnati, Raymond Walters College

Other Acknowledgments

This revision would not have been possible without the counsel and technical know-how of certain talents at the University of Texas Health Science Center at Dallas. In the Department of Pathology, I gratefully acknowledge the kindness of Dr. V.A. Stembridge, Chairman; Drs. R.C. Reynolds, R.G. Freeman, Paul Southern, Jr., Mary Lipscomb, William Welch, and C.S. Petty, professors; Mr. Gale Spring and Mrs. Linda Bolding, medical photographers; and Mrs. Phyllis Kitterman, secretary. The Director of the Library, Mrs. Jean Miller, and able members of her staff, Ms. Jean Gionas, medical photographer, and Mrs. Elinor Reinmiller, formerly Reference Librarian, have been most helpful. Mrs. Earline Kutscher, Chief Technologist, and her staff in the Microbiology Laboratory of Parkland Memorial Hospital, Dallas, have given invaluable assistance. I am especially indebted to Mrs. Mary L. Nelson, illustrator, for sharing her talent.

The making of a book is not possible without the expertise of the publisher. The capable people at Times Mirror/Mosby College Publishing in St. Louis generously blend a dedication to the task, an attention to detail, and considerable skill with a spirit of cooperation and concern.

Now a special word of appreciation to teachers and students whose ever-welcome responses have guided me: may I voice a heartfelt thanks to the many of you who have used this text and who carefully consider the new edition.

Alice Lorraine Smith

CONTENTS IN BRIEF

DETAILED CONTENTS

MODULE TWO

MICROBES: PROCEDURES FOR STUDY

MODULE THREE

MICROBES: PRECLUSION OF DISEASE

MODULE FOUR

MICROBES: PRODUCTION OF INFECTION

MODULE FIVE

MICROBES: PATHOGENS AND PARASITES

MODULE SIX

MICROBES: PUBLIC WELFARE

PRINCIPLES OF MICROBIOLOGY

MICROBIOLOGY
PRELUDE AND PRIMER

DEFINITION AND DIMENSIONS OF MICROBIOLOGY

Take interest, I implore you, in those sacred dwellings which one designates by the expressive term: laboratories. Demand that they be multiplied, that they be adorned. These are the temples of the future—temples of well-being and of happiness. There it is that humanity grows greater, stronger, better.

Louis Pasteur

DEFINITION

Microbiology is the branch of biology (the science of living things) that deals with **microbes**—living, minute organisms, or **microorganisms,** usually one cell and of necessity studied with a microscope. Microbiology considers the occurrence in nature of the microscopic forms of life, their reproduction and physiology, their participation in the processes of nature, their helpful or harmful relationships with other living things, and their significance in science and industry.

Conventionally, within the province of microbiology is placed the study of certain well-known microbes: bacteria (bacteriology), fungi (mycology), protozoa (protozoology), and viruses (virology).

Although human beings have lived with microorganisms from time immemorial and have used certain of their activities such as fermentation to their advantage, the science of microbiology is a product of only the last 150 years. The studies of Anton van Leeuwenhoek in the seventeenth century showed the existence of microscopic forms of life, but it was not until the work of Louis Pasteur toward the end of the nineteenth century (some 200 years later) that the science of microbiology really took shape. The new science stated the germ theory of disease, demonstrated patterns of communicable disease, and gave human beings a measure of protection they had not previously known in their struggle against the injurious forces in the biologic environment. In its time this very young science has influenced practically every phase of human endeavor.

DIMENSIONS
Biologic Classification

All living things may be sorted out in a scheme wherein categories represent successively dependent and related groups. This is done through **taxonomy,** the science of classifying biologic organisms according to their related or differentiating features. The highest possible levels are designated **kingdoms.** Never more than a few of these exist. The basic terms within a scheme used to gather the related groups together in ascending rank are as follows:

1. **Species:** organisms sharing a set of biologic traits and reproducing only their exact kind (Species is the fundamental unit in taxonomy. All other units are defined in terms of each other.)
   a. **Strain:** organisms within the species varying in a given quality
   b. **Type:** organisms within the species varying immunologically
2. **Genus** (*pl.* genera): closely related species
3. **Family:** closely related genera
4. **Order:** closely related families
5. **Class:** closely related orders
6. **Phylum** (*pl.* phyla): related classes (In plant biology the term *division* may be used instead of phylum. In Table 7-1 [p. 117] please note the use of the term *division* for either of the two major classifications in the kingdom Prokaryotae.)

As displayed in the higher and more complex forms of life, differences between the plant and animal kingdoms are dramatic, but for the microbes and the lower forms of life, classification as either plant or animal has always been difficult, since these forms incorporate features of both plants and animals. As bacteria and certain other microbes that had long been relegated to the plant kingdom were studied more closely, the inconsistencies appeared even greater. Because of this, a third biologic compartment with an equivalent rank to the plant and animal kingdoms was advocated as early as 1866 by Ernst Haeckel to sift out the simpler units, designating them as *protists* in a kingdom **Protista.** The term *protist* was coined to mean a one-cell unit that remains so throughout its life history. Even if protists pile cells up in large plantlike masses, their component cells remain the same and do not differentiate.

Subsequently, modern technologic advances coupled with scientific investigation revealed that, out of a major structural difference in the form of the genetic material within the cell, two basic types of living cells stand out. The consequence is two distinct taxonomic categories.

In a cell of a higher organism there is a nucleus, a completely defined one with an expected number of chromosomes and a mitotic apparatus. A double membrane, the nuclear membrane, gathers the chromosomes together in an obvious body and separates it from the cytoplasm. The term **eukaryote** (*adj.* eukaryotic) (Greek *eu*, well, plus *karyon*, nucleus) describes this cell or the organism composed of such cells.

In the lower forms of life an alternate arrangement for nuclear function exists. Within the cell the genetic material present is devoid of any membrane to set it apart from the cytoplasm. Such a cell (or organism) is termed a **prokaryote** (*adj.* prokaryotic).

Other important differences that separate eukaryotes and prokaryotes are given in Table 1-1. Generally, eukaryotic cells are larger than prokaryotic ones, and as would be expected, multicellular organisms are composed of eukaryotic cells. The intracellular organelles—the formed bodies in the cytoplasm that carry out particular biologic functions for the cell—are more numerous and more complex in eukaryotic than in prokaryotic cells. Organelles of eukaryotes are usually encased within membranes (membrane bound), whereas prokaryotes do not contain even membrane-bound organelles such as mitochondria. Prokaryotic cytoplasm does contain small ribosomes on which protein molecules are assembled, and the prokaryote builds its own cell wall outside the cell membrane.

The prokaryotes, the smallest of cellular organisms, are distinctive. Certain components in their cell walls are unique to them, and they display remarkable capabilities with regard to carbon storage, nitrogen fixation, and derivation of energy from oxidation of inorganic compounds. Defined species of prokaryotes number 2700.

Table 1-1 Key Points of Difference Between Prokaryotes and Eukaryotes

Characteristic	Prokaryotes (Especially Bacteria)	Eukaryotes (Fungi, Protozoa)
Nucleus	Not defined	Well defined
Nuclear membrane	Absent: genetic material not enclosed within a nuclear envelope	Present: chromosomes within a nuclear envelope
Genetic recombination	May occur as result of conjugation, transduction, or transformation	Occurs during sexual processes
Ribosomes	70S (small) ribosomes	80S ribosomes in cytoplasm 70S ribosomes in mitochondria
Membrane-bound cytoplasmic organelles	None	Present, variably complex
Mitochondria	Absent	Present
Endoplasmic reticulum	Absent, therefore ribosomes must be distributed in cytoplasm	Present
Golgi apparatus	Absent	Present
Lysosomes	Absent	Present
Cell or plasma membrane	Sterols not present except in *Mycoplasma* species	Sterols present
Cell wall	Present in bacteria; usually contains peptidoglycan	Present in plant but not in animal cells; no peptidoglycan
Photosynthesis	Found in some prokaryotes; absent in most	Present in plants; absent in fungi and animals
Nitrogen fixation (nitrification)	Present in certain bacteria (Chapter 38)	Unknown in any eukaryotic species
Reproduction	Simple transverse fission; no mitoses	Sexual processes; mitoses occur
Nervous system	None	Varies from none in plants, to primitive mechanisms in protozoa for conduction of stimuli, to complex setup in animals
Metabolic mechanisms		
Anaerobic energy–yielding reactions	Present: many bacteria obligate anaerobes	Absent: only a few protozoa are anaerobic
Glycolysis	Present	Present: energy-yielding mechanism utilized
Adaptation of plasma membrane to transfer of large molecules (import or export of soluble particulates)	Not well developed (transforming DNA fragments about only example)	Well developed
Exocytosis	Absent	Present
Endocytosis: phagocytosis, pinocytosis	Absent	Present

Nutritional patterns, as well as structural ones, provide guidelines for classification of microorganisms. If biologic organisms must go to outside sources of organic matter to meet their energy needs and their requirements for biosynthesis, they are termed **heterotrophs** (Greek *hetero*, other, plus *trophē*, nutrition). On the other hand, the ones termed **autotrophs** (Greek *autos*, self) can make their necessary energy-rich organic compounds from the simple inorganic substances available to them. **Phototrophs** (Greek *phōs*, *phōt*, light) directly utilize the sun as their energy source in the process of photosynthesis (the conversion of carbon dioxide from the air to organic molecules for cellular use). Most autotrophs are also phototrophs. A **chemoautotroph** can capture the energy released by specific chemical inorganic reactions.

Using concepts of cellular organization and nutritional patterns, in 1969 H.R. Whittaker of Cornell University presented a classification scheme that is widely accepted. He grouped all biologic organisms into five kingdoms: Monera, Protista, Fungi, Plantae, and Animalia. Later, in 1978, Whittaker and L. Margulis revised somewhat the earlier classification scheme of five kingdoms, placing the emphasis on whether the organism is a prokaryote or eukaryote. This later classification is as follows:

SUPERKINGDOM PROKARYOTAE
    Kingdom Monera
SUPERKINGDOM EUKARYOTAE
    Kingdom Protista
        Branch Protophyta (plantlike protists)
        Branch Protomycota (funguslike protists)
        Branch Protozoa (animal-like protists)
    Kingdom Fungi
    Kingdom Plantae
    Kingdom Animalia

The kingdom **Monera** comprises bacteria and cyanobacteria (all prokaryotes). Bacteria are small prokaryotes but quite complex, and their system of genetic transfer is unique among microorganisms. The classification of bacteria used in this text is given in Chapter 7. Further discussion of bacteria makes up a sizable portion of this book, beginning in Chapter 3.

The **cyanobacteria,** formerly called the blue-green algae, are a group of photosynthetic prokaryotes doing well in alkaline habitats and displaying a gliding kind of motility. Many species fix nitrogen from the air.

The kingdom **Protista** encompasses the algae (Protophyta), slime molds (Protomycota), and protozoa. All are eukaryotic and, for the most part, unicellular, but in a few instances they are found as relatively simple multicellular organisms. Protists may be phototrophs, heterotrophs, photosynthetic autotrophs, or heterotrophic phototrophs, but in every instance each cell is a self-sufficient unit capable of meeting its own biologic needs.

Alga (*pl.* algae) is a Latin term meaning seaweed. Chiefly aquatic in nature, algae form pond scums and seaweeds and float near surfaces of oceans and inland waters. The term *algae*, however, is a broad one applied to single-celled eukaryotes and to some simple multicellular ones that exhibit plantlike photosynthesis. Some of the more primitive ones are classifed here in the kingdom Protista, branch Protophyta, and represent a group of considerable size—around 26,000 species have been described. Complex multicellular types are placed in the plant kingdom. Many algae are microscopic and shaped like bacteria, and they may live in colonies. Flagella, variable in number and position, are found. Algae reveal various biochemical patterns, especially in regard to pigmentation, the nature of their reserve food, and storage products. Although, generally, cell walls have a cellulose matrix, cell wall components differ among the species.

Dr. Wendell Meredith Stanley (1904-1971), biochemist and virologist, defined viruses as follows: "The virus is one of the great riddles of biology. We do not know whether it is alive or dead, because it seems to occupy a place midway between the inert chemical molecule and the living organism."

From Alvarez, W.: Mod. Med. 35:78, Jan. 30, 1967.

Slime molds (some 576 species are known), sometimes referred to as "lower fungi," are classified in the kingdom Protista, branch Protomycota. They are unusual heterotrophs that do resemble fungi in certain ways, yet are distinct. They live in cool, shady, moist places in nature—on decaying wood, dead leaves, or other damp organic matter.

Protozoa are unicellular heterotrophs, typically animal-like in that they can move and display a feeding mechanism, and they are either free living or parasitic. They reproduce asexually but may also have a sexual cycle. Probably the most complex of all cells, protozoa possess an amazing organization of highly specialized structures and functions contained within one cell (Chapter 36).

Fungi (*sing.* fungus) is the term ordinarily used to mean molds, yeasts, and certain related microorganisms. They do not contain chlorophyl and are probably degenerate descendants of chlorophyl-bearing ancestors, most likely the algae. Now placed in a separate kingdom, fungi were formerly classified in Thallophyta, one of the four divisions of the plant kingdom. Thallophytes, or thallus plants (Greek *thallos*, young shoot or branch), are simple forms of plant life that do not differentiate into true roots, stems, or leaves. For current concepts regarding fungi, see Chapter 35.

Members of the animal and plant kingdoms are multicellular eukaryotes but with dramatic differences. In the animal kingdom, organisms can move about and eat and their body parts grow to a certain size and then stop. Plants do not move about or eat in any sense comparable to animals, but they can grow "indefinitely." Plants (autotrophs) meet their energy needs through photosynthesis. Animals (heterotrophs) must ingest organic matter through some kind of mouth opening in a feeding process for it to be utilized. In the plant kingdom are found some algae, all mosses, ferns, conifers, and flowering plants. The animal kingdom encompasses sponges, worms, insects, and animals with backbones (vertebrates).

Microorganisms usually surveyed in a treatise of microbiology include not only unicellular prokaryotes (bacteria) and protozoa but also fungi and a restricted number of metazoan animals (Chapter 37). A special category of living things that is usually incorporated into a text of microbiology is the viruses, although undoubtedly, among living things, viruses stand alone. Viruses are submicroscopic agents, the smallest known living bodies. They demonstrate peculiar properties of their own, and their unique structures and life cycles are quite different from those of other microbes (Chapter 33). Bacterial viruses (bacteriophages) are discussed in Chapter 4 on microbial genetics (see box above).

Naming of Microbes

The scientific name of a living organism is usually made up of two words that, although Latin or Greek in origin, are Latinized in the form used. They are italicized. The first name begins with a capital letter and denotes the genus. The second name begins with a small letter and denotes the species. Either the genus or the species name may be derived from the proper name of a person or place or from a term describing some feature of the organism. The proper name may be that of the scientific investigator or the related geographic area. Biologic characteristics indicated include color, location in

The American Type Culture Collection, located outside Washington, D.C.,*
maintains and distributes authentic cultures of practically all known microorganisms for research or other purposes. This nonprofit, independent agency
grew out of the Bacteriological Collection and Bureau for the Distribution of
Bacterial Cultures, established in 1911 at the American Museum of Natural
History in New York City. There are over 30,000 holdings of bacteria, fungi,
algae, protozoa, viruses, and animal cell lines in this collection. Each year it
supplies thousands of cultures of bacteria, fungi, and viruses to industrial and
educational institutions such as medical schools, breweries, wineries, oil companies, pharmaceutical houses, and food processors.

*American Type Culture Collection, 12301 Parklawn Dr., Rockville, MD 20852.

nature, disease produced, and presence of certain enzymes. For example, *Staphylococcus aureus* is the scientific name for bacteria of genus *Staphylococcus* (Greek *staphylē*,
bunch of grapes, plus *kokkus*, berry) and species *aureus* (Latin *aureus*, golden). It
indicates that the bacteria grow in typical clusters and produce a golden pigment. Honoring Sir David Bruce who discovered it, *Brucella melitensis* (pertaining to the island
of Malta) indicates by its name its disease—Malta fever, or undulant fever—and the
geographic area where it was first recognized.

SUMMARY

1 Microbiology (a product of the last 150 years) deals with the many facets of living,
minute organisms (microorganisms), usually structured as one cell and of necessity
studied with a microscope.

2 Taxonomy is the scientific classification of biologic organisms in successively dependent and related groups according to the distinguishing features. Basic terms ascend
as follows: species, genus, family, order, class, phylum, and kingdom.

3 Modern technology delineates two basic types of living cells out of the difference in
the form of genetic material. *Eukaryotes* have a true nucleus; *prokaryotes* do not—
their genome is not so set apart from the cytoplasm.

4 Eukaryotes are larger than prokaryotes, display more numerous and complex cell
organelles, and make up multicellular organisms.

5 The membrane organelles mitochondria are not found in prokaryotes.

6 Unique biologic capabilities of prokaryotes are related to nitrogen fixation, obligate anaerobiosis, and the derivation of energy for oxidation of inorganic compounds.

7 Twenty-seven hundred species of prokaryotes have been defined.

8 In their classification of organisms as prokaryotes or eukaryotes, Whittaker and Margulis emphasized five kingdoms: Monera for prokaryotes and Protista, Fungi, Plantae, and Animalia for eukaryotes.

9 Bacteria in the kingdom Monera are small but quite complex prokaryotes with
unique systems of genetic transfer.

10 The algae, slime molds, and protozoa of the kingdom Protista are single-cell units,
sometimes simple multicellular ones, that are biologically self-sufficient.

11 Probably the most complexly organized of biologic units, protozoa can move about,
feed, reproduce sexually or asexually, and exist either as parasites or free-living
cells.

12 Probably degenerate descendants of chlorophyl-bearing ancestors, fungi do not now
contain chlorophyl.

13 The smallest known living bodies, the viruses demonstrate peculiar properties and biologically unique structures and life cycles.

14 The American Type Culture Collection in Rockville, Maryland, with over 30,000 holdings, maintains and distributes cultures of practically all known microbes.

QUESTIONS FOR REVIEW

1 What is microbiology? What are microbes? Their importance?

2 Briefly explain these terms: protists, thallus plants, algae, fungi, slime molds, cyanobacteria, eukaryotes, prokaryotes, species, strain, type, *karyon*, taxonomy.

3 What is the role of the nucleus in differentiating prokaryotes from eukaryotes?

4 List the basic units used to classify living organisms.

5 How is the scientific name of a microbe derived?

6 What is the difference between thallophytes and other plants? Do all forms of life contain chlorophyl?

7 What are the basic differences between plants and animals?

8 Give eight major differences between prokaryotes and eukaryotes.

9 What is the American Type Culture Collection? What role does it play in scientific circles?

SUGGESTED READINGS

Curtis, H.: Biology, New York, 1983, Worth Publishers, Inc.

Joklik, W.A.: Contributions of microbiology to present-day biology, ASM News 46:280, 1980.

Noland, G.B.: General biology, St. Louis, 1983, The C.V. Mosby Co.

Singleton, P., and Sainsbury, D.: Dictionary of microbiology, Chichester, England, 1978, John Wiley & Sons, Ltd.

MILESTONES OF PROGRESS

Microbes were probably the first living things to appear on the earth. The study of fossil remains indicates that microbial infections and infectious diseases existed thousands of years ago. However, microbes were not seen until some three centuries ago because lenses of sufficient power to render them visible were not perfected until that time. Even after microbes were discovered, almost 200 years elapsed without any great progress in their study.

Important developments or discoveries in the history of microbiology are presented in this chapter as milestones of progress. A complete listing of all the landmarks in microbiology is not possible, so certain ones with far-reaching effects have been selected. The event or the sequence of events, in italics, is followed by a short discussion.

1546: *Girolamo Fracastoro (1483-1553) wrote a treatise* (De Contagione et Contagiosis Morbis) *which stated that disease was caused by minute "seeds" and was spread from person to person.* The theories of Fracastoro, a scholar of northern Italy, were formulated from his extensive study of infectious diseases prevalent in his day. He defined the word *contagion* as an infection passing from one individual to another in various ways. Although he spoke of "seeds" or "germs" of disease, he compared contagion to exhalation of an onion causing shedding of tears. He realized that all contagions do not behave alike; some attack one organ and some another. For example, he believed that consumption (tuberculosis) was infectious and that the lungs were involved. His work represents a great, although isolated, landmark in the doctrine of infectious disease.

1590-1675: *The microscope emerged as a scientific instrument.* The magnifying glass was known to the ancients. Not quite 100 years after the birth of Christ, Seneca, the Roman philosopher-statesman, recorded the magnifying effects of a glass bulb filled with water. In 1590 Johannes and Zacharias Janssen, father and son spectacle makers in Holland, found that two convex lenses in sequence magnified an image, and if the lenses were arranged in a metal tube with sliding barrels, the magnified image could be focused. In 1624 Galileo, the Italian astronomer better known for his telescopes, also made a microscope.

The first person to engage in what might be called medical microscopy, however, was Athanasius Kircher (1601-1680) (Fig. 2-1), a Jesuit priest and learned man of his day. In 1658 he published a treatise on microscopy in which he recorded seven of his experiments on the nature of putrefaction.

In 1675 Anton van Leeuwenhoek (1632-1723) of Delft, Holland, first described microbes (bacteria and protozoa) under the microscope. Because he first accurately described the different shapes of bacteria (coccal, bacillary, and spiral) and pictured their arrangement in infected material (1683), he is known as the father of bacteriology. Also known as the father of scientific microscopy, he made his own microscopes, which were superior to any of that time. He fashioned about 250, grinding most of the lenses himself. Leeuwenhoek was obsessed with his microscopic findings and examined everything he could think of. In addition to contributions to microbiology, Leeuwenhoek gave the first complete account of the red blood cell, discovered spermatozoa, and demonstrated capillary connections between arteries and veins.

Kircher's microscopes magnified only 32 times. The better scopes fashioned by Leeuwenhoek magnified about 270 times. The compound light microscope of today magnifies 1000 times; the electron microscope, 100,000 times.

1765: *Abbé Lazzaro Spallanzani (1729-1799) conducted experiments to disprove the doctrine of spontaneous generation.* Accelerating the development of microbiology were the arguments that went on for years concerning spontaneous generation. The proponents believed that living forms sprang from nonliving matter, and the opponents believed that every living thing descended from parents like itself. The older advocates

FIG. 2-1   Father Athanasius Kircher.

believed that eels originated from mud, and they gave formulas for producing mice from decaying rags and cheese. With time, both sides conceded that spontaneous generation did not occur in the higher forms of life. Much experimentation was done on both sides, and the controversy continued until Pasteur (p. 13) completely disproved the theory of spontaneous generation.

The great Italian master of experiment, Spallanzani performed several experiments to disprove the doctrine of spontaneous generation. He concluded that what he called "animalcules," carried by air into infusions of organic material, were the explanation. John Needham (1713-1781), a Welshman, criticized much of Spallanzani's work. His observations in 1748 on infusions of organic material in bottles closed with cork stoppers led him to a firm belief in spontaneous generation. Spallanzani retaliated with a series of experiments wherein he boiled his infusions, removed the air from the flasks, and hermetically sealed the containers. He observed that, after a flask of infusion had remained barren a long time, only a small crack in the neck was sufficient to allow development of animalcules in the infusion. Spallanzani's careful work might have ended the discussion on spontaneous generation had his precautions been observed by subsequent workers.

1796: *Edward Jenner (1749-1823), an English physician, introduced the modern method of vaccination to prevent smallpox.* One of the world's epochal contributions to preventive medicine was made on May 14, 1796, when Jenner (Fig. 2-2), having been impressed by the countryside tradition in Gloucestershire, England, that milkmaids who contracted cowpox while milking became immune to smallpox, transferred material from a cowpox pustule on the hand of a milkmaid (Sarah Nelmes) to the arm of a small boy (James Phipps). Six weeks later the boy was inoculated with smallpox and failed to develop the disease. In 1798 Jenner published his results in 23 cases. By 1800 about 6000 persons had been inoculated with cowpox to prevent smallpox, that is, vaccinated, and the scientific basis of the method was established. The word **vaccination** (Latin *vacca*, cow), originally coined to refer to the injection of the cowpox material to prevent smallpox, was first used by Pasteur out of deference to Jenner.

FIG. 2-2   Edward Jenner.

1837: *Theodor Schwann (1810-1882) proved that yeasts were living things.* Schwann may be regarded as the founder of the germ theory of putrefaction and fermentation. In 1837 he related processes of putrefaction and fermentation to living agents that obtained their sustenance from fermentable or putrescible material. In so doing, Schwann discovered and accurately described yeasts and their reproduction by budding.

1838-1839: *The cell theory was advanced.* In 1665 the Englishman Robert Hooke (1635-1703) first used the word **cell.** With his simple and imperfect microscope, he studied thin slices of a cork bottle stopper, noting the "little empty rooms," or cells, each with its own distinct walls. His word *cell* (Latin *cella,* storeroom) has persisted to designate a basic biologic unit. The concept of the cell as the structural unit for plant and animal life did not come until after the scientific studies of Matthias Jakob Schleiden (1804-1881), a German sea lawyer and botanist, and Theodor Schwann, who voiced their conclusions in 1838 and 1839.

1843, 1861: *The contagiousness of childbed fever (puerperal sepsis) was demonstrated.* Oliver Wendell Holmes (1809-1894), also a well-known physician, published in 1843 an article on the contagiousness of puerperal fever in a New England medical journal. In this article he indicated that this disease might be spread by the hand of the nurse or doctor. Some time later Ignaz Philipp Semmelweis (1818-1865), a Hungarian obstetrician in Vienna, noted that the death rate seemed to be higher in certain clinics and wards to which the medical students and physicians came directly from the morgue or autopsy room. He observed in the postmortem examination of a pathologist who had died of an infection complicating a dissection wound that the disease changes were similar to those present in women who had died of puerperal fever. Semmelweis rightly concluded that puerperal fever was infectious. On the wards that he supervised, all hands had to be cleansed carefully before a patient was examined, and the rooms were kept scrupulously clean. The mortality in his areas dropped immediately. In 1847 he stated to the Vienna Medical Society that the cause of puerperal fever could be found in "blood poisoning," and in 1861 he published his views. However, his theories, as those of Holmes earlier, met with strong opposition and were not accepted for 20 years, long after the death of Semmelweis.

1850: *Casimir Joseph Davaine (1812-1882) observed minute infusoria (anthrax bacilli) in the blood of dead sheep and transmitted the disease anthrax by inoculating this blood into other animals.* Davaine, a French pathologist, parasitologist, and experimenter, along with Pierre François Rayer (1793-1867), first saw the bacillus that causes anthrax. A physician in general practice, Davaine had to keep his experimental animals in the garden of a friend.

1854, 1873: *The role of drinking water and milk in the spread of disease was demonstrated.* John Snow (1813-1858), an anesthetist, was the first to recognize that contaminated water spread cholera. In 1854, with "a notebook, a map, and his five senses," he proved epidemiologically that the Broad Street Pump in London was a source of infection in a cholera epidemic killing some 11,000 persons. The causative agent of cholera was discovered by Robert Koch (p. 14) in 1883 in epidemics in Egypt and India.

In 1873, in a treatise on typhoid fever, William Budd (1811-1880) pointed to the role of milk and water in the transmission of this disease. His accurate study of typhoid fever outbreaks in 1856 led him to believe that the disease was infectious and that the causative agent was excreted in the patient's feces. He suspected that contaminated milk and water played an important part in the spread. In 1872 he related an outbreak in Lausanne, Switzerland, to contaminated water. The discovery of the causative organism of typhoid fever by Karl Joseph Eberth (1835-1926) did not come until 1880. In 1884 George Gaffky (1850-1918), one of Koch's co-workers, was the first to cultivate the organism.

1861-1885: *The work of Louis Pasteur (1822-1895) began the era of modern bacteri-*

ology. The great French scientist Louis Pasteur (Fig. 2-3) was born the son of a tanner in Dole, France, and was not a physician but a chemist. At 30 years of age Pasteur, already a distinguished chemist, became interested in the process of fermentation, and his subsequent studies of this process disproved the theory of spontaneous generation.

From 1863 to 1865 Pasteur devised the process of destroying bacteria known as **pasteurization.** He proved that "diseases of wine" could be prevented without altering the flavor by heating the wine for a short time to a temperature (55° to 60° C) a little more than halfway between its freezing and boiling points. This pasteurization process is used throughout the civilized world today to preserve milk and other perishable foods.

Pasteur confirmed Koch's studies on anthrax, the first animal disease proved to be caused by bacteria, and in 1881 produced his anthrax vaccine. Pasteur discovered the staphylococcus, described the streptococcus, and was codiscoverer of one of the microorganisms causing gas gangrene.

Among Pasteur's many scientific achievements, the one for which he is best known is the development of the Pasteur treatment for rabies. He gave the first such preventive treatment in 1885 to an Alsatian boy who had been bitten by a rabid dog. Although today the method of preparing rabies vaccine is completely different from the original method of Pasteur, its administration is still referred to as the *Pasteur treatment.* It is fitting that his name be thus remembered.

1865: *Lord Joseph Lister (1827-1912) applied antiseptic treatment to the prevention and care of wound infections.* The famous English surgeon Lister (Fig. 2-4) was so impressed with the similarity between certain fermentative and putrefactive changes, which Pasteur had proved to be caused by microorganisms and wound infections, that

FIG. 2-3 Louis Pasteur in his laboratory. Pasteur was known as a bold, dedicated, and tireless laboratory scientist, never one for armchair theories. Roux, who adored Pasteur, said that in the hands of his master the scientific approach was one "which resolves each difficulty by an easily interpreted experiment, delightful to the mind, and at the same time so decisive that it is as satisfying as a geometrical demonstration."

he concluded that microbes also caused wound infections. With this idea in mind, he protected wounds with dressings saturated with a solution of carbolic acid (phenol) and devised operating room procedures calculated to destroy microorganisms. The establishment of these methods was so far reaching in effect that Lister will always be known as the father of antiseptic surgery.

1876-1884: *Robert Koch (1843-1910) made important contributions to modern bacteriology.* The great German bacteriologist Koch made his first contribution in 1876 by isolating the anthrax bacillus in pure culture and proving its infectiousness. (That this agent alone caused anthrax had been confirmed by Pasteur.) In 1882 Koch discovered *Mycobacterium tuberculosis*, the microorganism causing tuberculosis, and in 1890 he first prepared and studied tuberculin (p. 600).

Koch is remembered for his discovery of important disease-producing organisms and for his fundamental contributions to bacteriologic technic. He was the first to stain bacterial smears as is done today. He devised a liquefiable solid culture medium (gelatin) and worked out pure culture technics (triggered by chance observations on a boiled potato left out in his laboratory one day). The hanging drop method of studying bacteria used today is a product of his genius.

In 1884 Koch first expounded certain principles related to the germ theory of disease with such clarity that they are known as **Koch's postulates** and remain to this day the basis of the experimental investigation of infectious diseases.

1880-1898: *The cause and transmission of malaria were discovered.* In 1880 Charles Louis Alphonse Laveran (1845-1922), a French army surgeon, discovered the parasite of malaria, and from 1895 to 1898 Sir Ronald Ross (1857-1932) (Fig. 2-5), an English pathologist and parasitologist, worked out its transmission by the mosquito. Ross also devised suitable methods for the widespread elimination of mosquitoes.

FIG. 2-4   Lord Joseph Lister, about the time he discovered antisepsis.

1884-1907: *Controversial theories of immunity and key discoveries marked the beginning of the science of immunology.* With the development of the science of immunology in the late 1800s, two schools arose, and the controversy that raged between them did much to put the study of immunology on a scientific basis, since concepts from both theories have been shown to play a role in immunity. In 1884 Elie Metchnikoff (1845-1916), a zoologist and native of the Ukraine, proposed the phagocytic or cellular theory of immunity, and in 1898 Paul Ehrlich (1854-1915) expounded his "side-chain" theory of immunity.

The French school headed by Metchnikoff, another of Pasteur's pupils, believed the whole process of immunity to be centered on the ingestion and destruction of invading disease-producing agents by the leukocytes and certain body cells. (In some of his earliest experiments on phagocytosis Metchnikoff studied the ingestion of thorns from a rosebush by the larval form of a starfish.)

The second school of immunologists, the German school, was led by Paul Ehrlich (Fig. 2-6), an outstanding German scientist who did important research in immunology from 1890 to 1900. Although he had drawn the outlines of his side-chain theory of immunity more than 10 years before, it was not until the turn of the century that he made his first exposition of it. In his studies Ehrlich conceived the structure of protein molecules to be similar to a benzene ring with unstable side chains. These could act as chemoreceptors that would combine with and thus neutralize harmful materials and then be released in excess after the need had been met. The German school believed that the establishment of immunity stemmed from the development in body fluids, especially in the blood, of certain substances known as *antibodies* or *immune bodies* with the capacity to destroy invading disease-producing agents. This is the humoral or chemical theory of immunity.

In 1896 Max von Gruber (1853-1927) and Herbert Edward Durham (1866-1945) discovered the phenomenon of agglutination, the practical application of which has resulted in agglutination tests for diseases such as typhoid fever, typhus fever, and tularemia. In 1901 the phenomenon of complement fixation was described and developed by Jules Jean Baptiste Bordet (1870-1961) and Octave Gengou (1875-1957), and in 1906 August von Wassermann (1866-1925) applied it in the Wassermann test, the first reliable serologic test for syphilis. In 1907 Leonor Michaelis (1875-1949), a German physical chemist, described precipitation, which is now used in tests such as the Kline, Kahn, Mazzini, VDRL, and Hinton tests for syphilis.

1888-1923: *The demonstration of the toxin-antitoxin relationship was an early landmark in the development of modern immunology.* In 1888 diphtheria toxin was discov-

FIG. 2-5   Sir Ronald Ross.

ered by Pierre Paul Émile Roux (1853-1933) and Alexandre Yersin (1862-1943). (The same year Ludwig Brieger [1849-1919] discovered tetanus toxin.) Roux, a research associate of Pasteur, was considered the greatest French bacteriologist after Pasteur. Together with Yersin, a Swiss bacteriologist in Paris, Roux made important contributions to the bacteriology of diphtheria, one of which was the demonstration of a toxic substance in filtrates of broth cultures of the diphtheria organism.

In 1889 Shibasaburo Kitasato (1852-1931), a Japanese bacteriologist and a pupil and colleague of Koch, cultivated the tetanus bacillus and proved that tetanus (lockjaw) was the result of the tetanus toxin. He showed it to be a chemical intoxication and not the consequence of bacterial invasion.

In 1890 Kitasato and Emil von Behring (1854-1917), also an assistant in Koch's laboratory, found that experimental animals treated with tetanus toxin developed a transferable protection (immunity). The cell-free serum from such animals rendered harmless the toxin elaborated by the bacillus, and this protective serum could be given to other laboratory animals. Normal serum did not affect the tetanus toxin. The fact that such a neutralizing effect was transferable from one animal to another has formed the basis of modern serum therapy and administration of antitoxin. Kitasato and von Behring had discovered passive immunization (p. 419). They published their results, and in a footnote in their article the word *antitoxic* was used. From this word **antitoxin** became firmly established. Only a week after the discovery of tetanus antitoxin was announced, von Behring published an article on passive immunization against diphtheria.

In 1913 von Behring was the first to use toxin-antitoxin mixtures to produce a per-

FIG. 2-6 Paul Ehrlich in his laboratory. (Courtesy National Library of Medicine, Bethesda, Md.)

manent, active immunity against disease in human beings. That same year Béla Schick (1877-1967), a Hungarian pediatrician in Vienna who eventually migrated to New York City, introduced a skin test for susceptibility to diphtheria—the Schick test—which demonstrated immunity to diphtheria by the presence or absence of a reaction at the skin site of the diphtheria toxin injection. Subsequently many other diagnostic skin tests were developed.

Before the end of the nineteenth century William Hallock Park (1863-1939) organized in New York City the first municipal bacteriologic laboratory in the United States. As dynamic head of this laboratory, he was the first to apply the best-known methods of the time to protect the children of the city against diphtheria. For his widespread immunization program he used von Behring's toxin-antitoxin combination.

By 1923 Gaston Leon Ramon (1886-1963), a French bacteriologist and veterinarian, had prepared diphtheria toxoid, an agent that, since it was devoid of the dangers of toxin-antitoxin and demonstrated superior immunizing capacity, quickly supplanted toxin-antitoxin.

1894 to present: *Women microbiologists made important contributions.* Women have played a considerable part in the development of modern microbiology. One of the first such women was Anna Wessels Williams (1863-1954), who in 1894 was appointed assistant bacteriologist of a New York Board of Health laboratory just organized by William Hallock Park. For 40 years she was closely associated with this outstanding health officer during a period of remarkable achievement in the field of preventive medicine in New York City. During the first part of this century the Chicago bacteriologist Ruth Tunnicliff (1876-1946) studied streptococci, anaerobic organisms, measles, and meningitis. Pearl L. Kendrick (1890-1980) has added much to our knowledge of whooping cough. Eleanor A. Bliss (1899-   ) is known for her work on streptococci and the sulfonamide drugs. Senior Bacteriologist at the National Institutes of Health for 27 years, Alice Catherine Evans (1881-1975) was a notable forerunner of women working in investigative laboratories. From 1917 to 1923 she investigated undulant fever.

From 1921 to 1927 George and Gladys Dick, a husband and wife team in Chicago, gained important information about the bacteriology and serology of scarlet fever. In 1923 they causally linked the disease to hemolytic streptococci and in 1924 devised the Dick test. From 1928 to 1933 Rebecca Craighill Lancefield (1895-1981) developed the serologic classification of human and other groups of streptococci in wide use today.

Edna Steinhardt and her associates at Columbia University, who, in 1913, propagated cowpox virus in bits of cornea from the eye of a guinea pig, have been credited as the first to cultivate virus by technics of cell (tissue) culture. The news in 1931 that viruses could be grown in embryonated hen's eggs had a major impact on the science of microbiology. A physiologist, another member of a husband and wife team, Alice Miles Woodruff, working with fowlpox virus, was first able to do this in 1931. The work was reported from the Pathology Laboratory at Vanderbilt University School of Medicine under the direction of Ernest William Goodpasture (1886-1960). Since virus cultivation in fertile eggs and in cell cultures produces an ample supply of virus for study or for vaccine production, the introduction and standardization of these two technics have contributed in large measure to the explosive surge of knowledge in the field of virology. In fact, this period of the last several decades has been called the "golden age" of virology.

Women scientists figure prominently in the field of cancer research. A noted one with a background as a microbiologist is Sarah Stewart (1906-   ), who at the National Institutes of Health worked on tumors induced by viruses. In 1953 her first report came out on the salivary gland tumors in mice induced by the polyoma virus, a fairly common virus infecting many laboratory animals. With the help of Bernice Eddy and co-workers, she subsequently propagated the virus and showed that it could produce a great variety of cancers in the animals studied.

1898: *Theobald Smith (1859-1934) differentiated the human and bovine (cattle) forms of Mycobacterium tuberculosis.* America's foremost bacteriologist, Smith, of Albany, New York, demonstrated the parasitic cause of Texas fever, showed anaphylaxis from bacterial products, and made the first clear differentiation of tubercle bacilli. He laid the groundwork for the identification of the other, nontuberculous mycobacteria.

1898: *Martinus Willem Beijerinck (1851-1931) recognized the first virus (the virus of tobacco mosaic disease).* In 1892 in Russia Dmitri Iwanowski reported the first experiments that pointed to the presence of an infectious agent much smaller than any yet known. The subject of these experiments was the virus of tobacco mosaic disease. This work was reported and extended in 1898 by Beijerinck, a Dutch botanist. Iwanowski doubted his own findings, but Beijerinck realized that a new type of infectious agent had been discovered. In addition to being the first virus discovered, the virus of tobacco mosaic was the first to be purified in a crystalline form. (About 1 ton of infected tobacco must be processed for 1 tablespoon of virus crystals.) This task was accomplished in 1935 by Wendell Meredith Stanley (1904-1971), a biochemist and virologist. In 1955, in Stanley's laboratory at the University of California, the poliomyelitis virus was crystallized, the first virus affecting animals or humans to be so purified.

1901, 1940: *Discoveries were made relating to the human blood groups.* In 1901 Karl Landsteiner (1868-1943), an Austrian-American pathologist, discovered the basic human blood groups. Landsteiner had been associated with Max von Gruber in Vienna. When he came to America, he held an important post with Rockefeller University, New York City. He found that, in the four main groups of normal human blood, blood of some of these groups destroys blood belonging to other groups. This discovery laid the foundation for modern blood transfusion. In 1940 Landsteiner and Alexander S. Wiener (1907-   ) identified the Rh factor of blood. Landsteiner believed blood to be as unique to an individual as that person's fingerprints.

1909: *The concept of chemotherapy was introduced.* In 1909 Paul Ehrlich introduced salvarsan ("the great sterilizing dose") as a treatment for syphilis, and a new branch of medicine was created, dealing with the treatment of disease by chemical agents designed to affect unfavorably the disease-producing agent but not to harm the patient. Previously Ehrlich had begun to synthesize many new drugs of slightly varying chemical formulas and to test each experimentally until a maximal therapeutic effect was obtained. Collaborating with a Japanese physician, Sahachiro Hata (1872-1938), Ehrlich found, in the six-hundred-and-sixth compound of a series tested, salvarsan (arsphenamine), a drug capable of controlling the manifestations of syphilis over a long period.

Ehrlich was a genius of extraordinary activity who added much to our knowledge of medical science, and many technical laboratory methods he proposed are now in daily use. We are indebted to him for the basic procedures in staining, since he was one of the first to apply aniline dye solutions to cells and tissues. He founded and was the greatest exponent of the science of hematology.

1910 to present: *The relationship of viruses to neoplasms began to unfold.* At the turn of the century the suggestion was made that viruses might cause cancer. In 1910 Peyton Rous (1879-1970), a pathologist and medical researcher of the Rockefeller University, used material from a large tumor on the breast of a Plymouth Rock hen to show that cell-free filtrates of it induced the same tumor in hens, that the induced tumor behaved as did the original one, and that the properties of the agent in his experiments were those of a filterable virus. This important landmark in cancer research was years ahead of its time. Rous did not receive the Nobel Prize for his work until 1966 (56 years later). Until the early 1950s, nearly half a century, establishment of the viral etiology of cancer made slow headway, although it was substantiated by many other scientific observations (Table 33-5), and for years it took considerable courage to hold to such a theory.

In 1958 Denis Burkitt (1911-   ) defined Burkitt's lymphoma epidemiologically and postulated a transmissible viral agent. Burkitt defined this jaw tumor in Central African children and also mapped out its geographic distibution. He described it as a cancer, but much about it impressed him as an infectious disease. Partly because it is confined to the yellow fever and mosquito belt of Central Africa, he believed at first that it might be caused by a virus carried in a mosquito; later he dismissed the idea of an insect vector. Today Burkitt's lymphoma is considered by most observers to be caused, at least in part, by a virus, currently the Epstein-Barr virus.

In 1970 Howard Temin (1934-   ) and Satoshi Mitzutani of the University of Wisconsin and David Baltimore (1938-   ) at the Massachusetts Institute of Technology separately but almost simultaneously discovered reverse transcriptase. Reverse transcriptase is a specific viral enzyme in RNA viruses, such as the Rous sarcoma virus, that uses viral RNA as a template to produce DNA, which as genetic information is inserted into the genes of the host cells. When it was found in 1940 that growth and heredity were controlled by DNA, it was believed that the genetic sequence could only go from DNA to RNA. The dramatic nature of the discovery of an RNA-directed enzyme lies in the fact that it showed that the genetic sequence could be reversed—information could go from RNA to DNA. The discovery has been most productive for viral cancer research and has led to the identification of RNA tumor virus–like particles in a variety of tumors (p. 695).

In 1978 the human T cell leukemia-lymphoma virus was isolated. In numerous animal species naturally occurring leukemias or lymphomas are caused by retroviruses (RNA tumor viruses). Since the identification of mammalian retroviruses in the 1950s, considerable effort has been expended in the search for a true *human* retrovirus. The isolation and identification of a true human retrovirus (type C virus), the human T cell leukemia-lymphoma virus, stemmed from scientific discoveries made half a world apart and represent a scientific tale of two cities, one, Bethesda, Maryland, and the other, Kyoto, Japan, where the molecular and clinical studies were done. The relation of human neoplasms to viruses is one of the most challenging research problems today.

1935 to present: *Crucial events occurred in the development of the antimicrobial drugs.* In 1935 to 1936 the remarkable value of Prontosil (the forerunner of the sulfonamide compounds) was established in streptococcal infections, and from 1938 to the present many important sulfonamide compounds have been developed. Ehrlich, from his work in chemotherapy, had the idea that dyes might act as selective poisons, and other workers had found that certain azo dyes had some bactericidal power. In 1927 the German pathologist Gerhard Domagk (1895-1964) began an eventful collaboration with certain organic chemists at the Elberfeld Laboratories of the German firm I.G. Farbenindustrie, who prepared a number of azo dyes for testing. In 1932 Domagk found that one of them, a red dye containing a grouping of atoms that chemists call the **sulfonamide** group, had a definite curative value when injected into mice infected with streptococci. This red dye was given the trade name Prontosil. In 1935 Domagk published his experimental data demonstrating the antibacterial powers of Prontosil, and about the same time several clinical reports in the German literature confirmed his findings. Shortly thereafter French investigators showed that the compound Prontosil was broken down within the body to form the antibacterial compound sulfanilamide. Soon the English-speaking world became interested in bacterial chemotherapy and in these new sulfonamide compounds. Since the introduction of Prontosil more than 5000 sulfonamides have been prepared. However, of these, only a few show antibacterial activity comparable to or greater than that of the parent drug.

Tyrothricin was the first member of the group of antimicrobial agents known as **antibiotics** to be studied extensively and the first to be produced in a pure form. It was discovered in 1939 by René Jules Dubos (1910-1982), later of Rockefeller University, who demonstrated that microorganisms isolated from the soil can produce, under certain conditions, substances possessing antibacterial properties.

In 1943 streptomycin was discovered by Selman Abraham Waksman (1888-1973). The development of streptomycin was the result of a planned, careful search for an antibiotic effective for gram-negative bacteria and of minimum toxicity. Soils, composts, and other natural substrates harboring large number of organisms capable of producing antibiotic substances were carefully studied. Within 5 years nearly 10,000 cultures were examined before streptomycin was obtained by Waksman (Fig. 2-7) and his co-workers at Rutgers University. This antibiotic agent was isolated from two strains of *Streptomyces* identified as *Streptomyces griseus* (a soil organism). Less than a year after the isolation of streptomycin, its action against the tubercle bacillus was demonstrated. For the first time in history an effective drug for the treatment of tuberculosis had been found.

Practical application of the observations of the soil microbiologist, plant pathologist, and medical bacteriologist has resulted in the development of many important antibiotics since World War II. Many diseases that formerly progressed unchecked are now easily controlled.

Sir Alexander Fleming (1881-1955) made his monumental discovery of penicillin in 1928, but his work went practically unnoticed for more than a decade. On a hot August day, Fleming, a Scottish bacteriologist, returned from a week's vacation, noted that a patch of green mold had fallen haphazardly on one of his cultures, and was struck by its antibacterial action. In 1929 he reported that this mold, *Penicillium notatum*, elaborated during its growth a substance that inhibited the development of certain bacteria. He called it **penicillin.** Fleming's own life at one time was saved from an overwhelming pneumonia by the very agent he had discovered. To his great delight, he made a speedy and dramatic recovery.

In 1938 Sir Howard Florey (1898-1968) and Ernst Boris Chain (1906-1979) at Oxford University began a systematic study of Fleming's findings, and by 1940 the value of penicillin as a therapeutic agent was determined in England. Not only did Florey and Chain develop suitable methods for producing penicillin and showing its usefulness, but they were also responsible for introducing it to the American medical world and American industry. American industrial organizations have made possible the production and availability of penicillin on a wide scale.

1938-1963: *The dramatic development of poliomyelitis vaccine took place.* When the National Foundation for Infantile Paralysis (the National Foundation of today) began its support of scientific research in 1938, not much was known about poliomyelitis. Before a successful program of immunization could be launched, more basic facts had to be obtained about the disease. The end of the so-called monkey era in poliomyelitis research came when John F. Enders (1897-   ) and his associates at Harvard University cultivated the poliomyelitis virus in cultures of living cells of nonnervous tissue. Through cell cultures extremely large amounts of virus, free of injurious impurities found in nervous tissue of experimental monkeys, could be produced cheaply, and hundreds of investigations as to the nature of the virus could be easily carried out.

In 1954 the National Foundation launched a field trial using the experimental vaccine that Jonas E. Salk (1914-   ) and associates at the University of Pittsburgh had prepared by formaldehyde inactivation of poliovirus. The events of this field trial were unique in medical history. It was a project of universal interest that involved the talents of research teams, the methods of commerce and industry, the facilities of universities and government, and the resources of the publicly subscribed National Foundation, meanwhile engrossing the press of the world. The success of the field trial was formally announced by the Poliomyelitis Evaluation Center of the University of Michigan on April 12, 1955, the anniversary of the death of Franklin D. Roosevelt (a victim of the crippling effects of polio).

To evaluate the inactivated poliovirus vaccine, the live virus vaccine had to be field tested. Its development started with a series of small, well-controlled trial runs in the United States, from which it jumped to a mass immunization program of global extent and astronomic proportions. Experimental vaccines for the large-scale tests were developed by three major investigators: Hilary Koprowski (1916-   ), Herald R. Cox (1907-   ), and Albert B. Sabin (1906-   ). In 1960 the U.S. Public Health Service reported that more than 60 million persons throughout the world had been fed attenuated live poliovirus vaccines with no ill effects and gave its approval for the commercial production under specified regulations of the oral vaccine prepared by Sabin. Although it had already been used abroad, the Sabin oral vaccine, prepared in 1956, was first used in the United States in 1962. After the three strains of the Sabin oral vaccine were licensed, community-wide immunization programs were carried out to ensure as complete protection of the American public as possible. In the history of infectious disease the conquest of poliomyelitis is a dramatic chapter. Once a major threat, this crippling disease has almost disappeared.

1941: *Albert H. Coons (1912-   ) developed the fluorescent antibody test.* The fluorescent antibody technic represents an important example of modern advances in the field of immunology. It was developed by an immunologist, Coons, and his associates at Harvard Medical School, who demonstrated that fluorescein, a fluorescent dye, could be bound to an antibody without changing the behavior of the antibody. The attachment of the fluorescent dye, however, would cause the antibody to become luminous if it were viewed under ultraviolet light. These workers were the first to tag antibodies with neon lights.

1942: *Dr. Norman Gregg, an Australian ophthalmologist, published his observations*

FIG. 2-7   Selman Abraham Waksman *(left)* and Sir Alexander Fleming *(right)*. (Courtesy Wide World Photos.)

relating congenital cataracts in infants to a severe epidemic of rubella (German measles).
The epidemic had followed a 17-year period of minimal incidence of the disease in
Australia, and consequently many young adults were infected. Gregg noted that practically all mothers of affected infants had had rubella early in gestation, some even
before they realized they were pregnant. His correlations showed that malformations
may result from extrinsic agents, even infectious ones such as viruses, rather than being
exclusively bound to inherent genetic or development mechanisms, a concept that has
had wide-ranging effects.

1944-1953: *The genetic role of DNA, the molecule of heredity, was identified.* In
1944 Oswald T. Avery (1877-1955), Colin MacLeod (1909-1972), and Maclyn Mc-
Carty discovered the genetic role of DNA, and in 1953 James D. Watson (1928-   )
and F.H.C. Crick (1916-   ) demonstrated the structure of DNA. The story of deoxyribonucleic acid (DNA) goes back nearly 100 years, but the dramatic chapters in the
elucidation of its biologic significance did not unfold until well into the twentieth century. An acid substance referred to as nucleic acid, something of its chemical composition, its relation to a carbohydrate, and its association with the chromosomes of the
nucleus were known before Frederick Griffith, a British pathologist, made a remarkable
discovery in 1928. When he inoculated mice (animals peculiarly susceptible to pneumococci) with a mixture of pneumococci, alive but weakened, and pneumococci,
heat-killed but virulent, the mice developed a pneumonia produced by virulent organisms. A mysterious transformation had occurred whereby harmless microbes had
changed their character. In the blood of the sick mice Griffith found a fully encapsulated type of pneumococcus. Since the encapsulated bacteria used for injection had
been killed, the origin of the encapsulated forms could only be explained by the assumption that some genetic material from the killed organisms was able to enter the
living *non*encapsulated cells and change them into living and encapsulated ones. This
work constituted a prologue to what has been called a revolution in biology. Subsequently (as discussed later) it was discovered that the substance transferring the genetic
capability from one bacterium to another resides in DNA alone.

In the following years researchers at the Rockefeller University, because of their
interest in the disease-producing properties of pneumococci, turned their attention to
Griffith's heat-killed pneumococci and repeated his work. Over 10 years they slowly
eliminated, one by one, all genetic factors possibly involved in the transformation of
the pneumococci, including protein of the nucleus (long thought to be the carrier of
genetic information) and the capsule of the microbes, until nothing was left except
DNA, the nucleic acid of the nucleus. When DNA was removed from the heat-killed
virulent pneumococci, it became apparant that it was indeed the transforming principle. By 1944 Avery, MacLeod, and McCarty reported that for the first time they had
the proof that DNA is the stuff of which heredity is made. Three biophysicists, one
American and two British, were responsible for the elucidation of the molecular structure of DNA. In 1953 Watson and Crick reported their construction of the double-helix
model of the DNA molecule, for which M.H.F. Wilkins (1916-   ) had laid the
groundwork by his x-ray diffraction studies.

1953: *Barbara McClintock (1902-   ) reported her discovery of "jumping genes."* In
the early 1950s McClintock, a geneticist, noted that, contrary to the long-held view of
their being stationary, genes can unexpectedly wander from one position on a chromosome to another. As molecular biology continued to evolve into a major scientific
discipline, the importance of her work became increasingly apparent, and several decades after her first report it has been called a great discovery, second only to the
discovery that genes are strands of chemical arrangements in a double helix that can
separate and transmit hereditary traits. McClintock received the Nobel Prize for her
work in 1984.

1957: *Interferon was discovered.* The phenomenon of viral interference whereby infection with one virus can sometimes protect an animal against exposure to another has

been known since 1935. In 1957, in the course of their study of this principle, two workers found interferon. Dr. Alick Isaacs (1921-1967), working with his associate Jean Lindenmann (1924-   ) in London's National Institute for Medical Research, discovered that embryonic chick cells treated with inactivated influenza virus released something into the culture medium that blocked replication of live virus in a fresh batch of cells, apparently with no ill effect to the cells. Since it interfered with the growth of virus, the protein was named **interferon.** In 1962 Kurt Paucker (1924-   ) discovered that interferon also slowed the growth of cancer cells with an antiproliferative effect independent of its antiviral properties.

   1964-1980: *The discovery of the Australia antigen paved the way for the development of the hepatitis B vaccine.* In 1964 at the Institute for Cancer Research in Philadelphia, Baruch S. Blumberg (1925-   ), a population geneticist, found a foreign substance in the blood of an Australian aborigine that he labeled the **Australia antigen.** In 1968 Alfred M. Prince (1928-   ), a virologist of the New York City Blood Center, found the same substance in the blood of patients with viral hepatitis type B, and subsequently the Australia antigen was specifically linked to the disease. In 1970 the complete hepatitis B virus was identified by D.S. Dane of the Bland-Sutton Institute in London. The fact that Blumberg made a remarkable discovery from research of a "basic" type, not goal-oriented at all, reemphasizes the ongoing need for the basic approach to scientific endeavor.

   Since hepatitits B virus has not yet been grown in the laboratory, Blumberg and Irving Millman (1923-   ) suggested that excess antigen found in the blood of carriers of the virus, if separated and purified, could be used as vaccine. (As many as 10% of hepatitis B patients in the United States and as many as 200 million over the world become chronic carriers.) Saul Krugman (1911-   ) and colleagues at the New York University Medical Center proved the feasibility of the concept by injecting a crude, heat-inactivated virus vaccine into children scheduled for admittance to Willowbrook State Institute in New York, where hepatitis type B is endemic. They reported in 1970 and in 1971 that their vaccine protected some 70% of these children. By 1980 W. Szmuness and co-workers reported the existence of a safe and effective vaccine for hepatitis B.

   In a 1981 article H.J. Alter summarized the events from the discovery of the Australia antigen to the development of the vaccine*:

> Within a span of 15 years, an antigen of initially unknown origin and significance has served to elucidate the structure, mechanism of production and transmission pattern of hepatitis B virus, has served to eradicate almost hepatitis B as a determinant in post-transfusion hepatitis and has evolved into an enormously effective vaccine for the prevention of hepatitis B, and possibly for the prevention of hepatocellular carcinoma.

Since strong evidence links primary carcinoma of the liver to chronic hepatitis, the hepatitis B vaccine may be the first successful vaccine against cancer.

   1975: *César Milstein (1927-   ) and Georges Köhler first described the technic of producing monoclonal antibodies in continuously proliferating cells (hybridomas).* At the Medical Research Council Laboratory in Cambridge, England, Milstein and Köhler developed a method for routinely producing monoclonal antibodies in quantities. A hybridoma (p. 345) resulted in which a plasma cell contributed the capacity to produce specific antibody and a myeloma cell contributed the long life in cell culture. The application of monoclonal antibodies already is everywhere. Monoclonal antibodies represent the only viable approach in some instances, as for the analysis and study of biologic materials such as viruses and their genetic variants and newly discovered cell surface antigens.

---

*From Alter, H.J.: Hepatitis B: tribute to nondirected medical research, Semin. Liv. Dis. 1:2, 1981.

SUMMARY

1 The germ theory of disease was stated as early as 1546 but was not accepted until the time of Louis Pasteur, some 300 years later. The validation of the theory rested on Leeuwenhoek's applications of microscopy, Spallanzani's experiments on spontaneous generation, Jenner's discovery of vaccination, and the observations of Semmelweis, Davaine, Snow, Budd, and others.

2 The era of modern bacteriology began with the work of Louis Pasteur from 1861 to 1885. He disproved the theory of spontaneous generation, devised pasteurization, and not only developed the Pasteur treatment for rabies but also was the first to administer it.

3 Robert Koch discovered tubercle bacilli in 1882 and expounded his postulates in 1884. He introduced certain basic bacteriologic technics still in force today, for instance, solid culture technics.

4 The controversy that raged toward the end of the nineteenth century between the proponents of the phagocytic theory of Metchnikoff and those of the humoral theory of Ehrlich did much to put the study of immunology on a scientific basis.

5 Although Iwanowski in Russia had pointed to the presence of such an agent in 1892, it remained for Beijerinck in 1898 to identify the first virus, that of tobacco mosaic disease.

6 When the remarkable genius Paul Ehrlich introduced in 1909 the chemical salvarsan as therapeutic for syphilis, he instituted a new branch of medicine. For chemotherapy, drugs used must strike the infectious agent but in so doing must not harm the patient.

7 The discovery of reverse transcriptase (a specific viral enzyme), the definition of Burkitt's lymphoma, and the isolation of the human T cell leukemia-lymphoma virus are events of the middle part of the twentieth century that further elucidate the relationship of viruses to cancer demonstrated by Peyton Rous at the start of the century.

8 Prontosil in 1932 began the story of the antimicrobial drugs. Antibiotics such as penicillin and streptomycin dramatized it. Penicillin was an accidental finding. Streptomycin and many of the modern antibiotics resulted from planned, careful searches.

9 The elucidation of the genetic role of DNA began with the demonstration in 1928 that genetic material may be passed among bacteria. In 1944 Avery, MacLeod, and McCarty pinpointed this material as DNA, and in 1953 Watson and Crick worked out its structure.

10 A fallout from a basic type of research, the Australia antigen was subsequently linked to viral hepatitis B. In 1970 the complete virus particle was identified, and by 1980 a safe and effective vaccine had been made.

11 Other important milestones of this century include the discovery of jumping genes, interferon, and technics for producing monoclonal antibodies.

QUESTIONS FOR REVIEW

1 Compare the magnification of Kircher's microscopes with that of the scopes fashioned by Leeuwenhoek and with that of the compound light microscope of today.

2 On what observation did Jenner base his approach to vaccination against smallpox?

3 How did Semmelweis drastically reduce the incidence of puerperal fever on his medical wards?

4 What is John Snow's contribution to epidemiology? What role did the Broad Street Pump in London play?

5 Name three important scientific achievements of Louis Pasteur.

6 Name three major contributions of Robert Koch to modern bacteriology.

7 Who first linked the transmission of malaria to the mosquito?
8 Name the proponent of the phagocytic theory of immunity in the nineteenth century and the proponent of the humoral theory held about the same time.
9 Cite three far-reaching contributions of Paul Ehrlich to modern medicine.
10 How did Norman Gregg relate congenital eye defects in infants to rubella in their mothers?
11 Is toxin-antitoxin used today for immunization? If not, why?
12 Who discovered the first virus? What virus was it? What was the first virus to be purified in a crystalline form?
13 Characterize Burkitt's lymphoma. Who first described it?
14 Broadly outline the events in the elucidation of the biologic role of DNA. Name key scientists participating.
15 Name the two scientists who developed the method for producing monoclonal antibodies.
16 Name one important contribution to the science of microbiology made by each of the following persons:

Lazzaro Spallanzani	Anton van Leeuwenhoek
William Budd	Peyton Rous
Emil von Behring	Emile Roux
James Phipps	Alick Isaacs
Barbara McClintock	Karl Landsteiner
Baruch Blumberg	Rebecca Lancefield

17 Explain briefly what is meant by spontaneous generation, Australia antigen, chemotherapy, Prontosil, Lister's application of antisepsis, the word *cell* as used by Robert Hooke, Ehrlich's side-chain theory, Schick test, the human T cell leukemia-lymphoma virus, and antitoxin.

SUGGESTED READINGS

Ackermann, H.W., and others: Felix d'Hérelle: his life and work and the foundation of a Bacteriophage Reference Center, ASM News 48:346, 1982.

Barclay, W.R.: The conquest of smallpox (editorial), J.A.M.A. 240:1991, 1978.

Bardell, D.: Is there a place for Nikolai Gamaleia in the discovery of bacterial viruses? ASM News 45:194, 1979.

Baxby, D.: Jenner's smallpox vaccine: the riddle of vaccinia and its origin, London, 1981, William Heinemann, Ltd.

Bibel, D.J.: Centennial of the rise of cellular immunology: Metchnikoff's discovery at Messina, ASM News 48:558, 1982.

Centennial: Koch's discovery of the tubercle bacillus, Morbid. Mortal. Weekly Rep. 31:121, 1983.

Demain, A.L., and Solomon, N.A.: Industrial microbiology, Sci. Am. 245:66, Sept. 1981.

Doty, R.B.: Microbiology on stamps: the conquest of yellow fever, ASM News 39:701, 1973.

Doty, R.B.: Microbiology on stamps: the battle against tuberculosis, ASM News 40:775, 1974.

Doty, R.B.: Microbiology on stamps: the Pasteur Institute, ASM News 40:613, 1974.

Doty, R.B.: Microbiology on stamps: the pioneers and some historic landmarks, ASM News 40:329, 1974.

Doty, R.B.: Microbiology on stamps: such good friends, ASM News 39:182, 1973.

Dubos, R.: Pasteur's dilemma—the road not taken, ASM News 40:703, 1974.

Goodpasture, E.W.: Use of embryo chick in investigation of certain pathological problems, South. Med. J. 26:418, 1933.

Gröschel, D.H.M.: 100 years of agar use in microbiology, ASM News 47:391, 1981.

Gröschel, D.H.M.: The etiology of tuberculosis: a tribute to Robert Koch on the occasion of the centenary of his discovery of the tubercle bacillus, ASM News 48:248, 1982.

Harrison, G.: Mosquitoes, malaria, and man: a history of the hostilities since 1880, New York, 1978, E.P. Dutton, Inc.

Howard, D.H.: Friedrich Loeffler and his history of bacteriology, ASM News **48:**297, 1982.

Jones, C.D.: Disease and medicine in the Middle Ages, Lab. Med. **10:**54, Jan. 1979.

Keswani, S.G.: Eradication of smallpox by the World Health Organization, J. Am. Med. Wom. Assoc. **32:**334, 1977.

Krause, R.M.: The restless tide: the persistent challenge of the microbial world, Washington, D.C., 1981, National Foundation for Infectious Diseases.

Proter, J.R.: Louis Pasteur sesquicentennial (1822-1972), Science **178:**1249, 1972.

Stetten, D.W., Jr.: Victory over variola, ASM News **44:**639, 1978.

Widmann, F.K.: Notes on the early history of blood transfusion, Lab. Med. **1:**9, Feb. 1970.

CHAPTER 3

THE BACTERIAL CELL
Form and Substance

OVERVIEW
Definition

Bacteria (*sing.* bacterium) are unicellular microorganisms, the smallest ones, having all the necessary protoplasmic equipment for growth and self-multiplication at the expense of available foodstuffs. They can start with rather simple substances and synthesize them into complicated organic moieties. They utilize food materials only in solution and excrete waste products in fluids that must diffuse outward. There is no special structure for intake of solids for digestion or for release of solid particles to the environment.

Classification (Chapter 7)

In the early days of microbiology, bacteria were considered a part of the animal kingdom, but then for a period of years it seemed more appropriate to classify them with plants. Unlike plants, bacteria ordinarily do not contain chlorophyl and may be able to move about independently in their environment. Today we recognize them as prokaryotes, neither as plant nor animal. Bacteria are morphologically simpler than the eukaryotic cells of the higher organisms, and as is true for prokaryotes, they lack an organized nucleus. In spite of their relative simplicity, they have an elaborate and complicated life history. Table 1-1 presents salient differences between prokaryotes and eukaryotes. The characteristics given there for the prokaryotes (especially bacteria) are discussed in this chapter and other sections of the book.

Distribution

Bacteria are widely distributed in nature and have adapted to every conceivable habitat. They are found within and on our bodies and in the food we eat, the water we drink, and the air we breathe. Our skin has a large bacterial population, and bacteria make up a generous portion of the contents of the alimentary tract. They are plentiful in the upper layers of the soil,* and no place on earth, except possibly the peaks of snow-

---

*Take a pinch of dirt. Hold it between your thumb and forefinger. You may be holding as many as 200 million bacteria. In 1 g (½₈ ounce) of fertile soil may be found as many as 2.5 billion bacteria.

capped mountains, is free of them. They are found in frozen Antarctica* and in the hot water of the geysers in Yellowstone National Park. They can be found in the stratosphere at a height of 20 km and in ocean sediments at a depth of 11,000 m; they can grow in environments from 5° to −2° C,† the temperature of 90% of the world's oceans.

Several thousand species of bacteria are defined; of this number, about 100 produce human disease. The ratio is given as 30,000 nondisease-producing bacteria to one disease producer. Some of the bacteria that produce disease in humans also produce disease in lower animals. Others produce disease in lower animals only, and still others attack only plants. The majority, however, do not attack human beings, lower animals, or living plants and either do not affect animals and plants at all or are actually helpful to them. In fact, if the activities of bacteria were to cease, all plant and animal life would soon become extinct.

Bacteria that cause disease are spoken of as **pathogenic;** those that do not cause disease are **nonpathogenic.**

MORPHOLOGY
Size and Shape

Dimensions

Bacteria are so small (no larger than 1/50,000 of an inch‡) that the highest magnification of the ordinary compound microscope must be used to study them (p. 143). Cocci range from 0.4 to 2 μm (μ) in diameter. The smallest bacillus is about 0.5 μm in length and 0.2 μm in diameter. The largest pathogenic bacilli are seldom greater than 1 μm in diameter and 3 μm in length; the average diameter and length of pathogenic bacilli are 0.5 and 2 μm, respectively. Nonpathogenic bacilli may be larger, reaching a diameter of 4 μm and a length of 20 μm. The spirilla are usually narrow organisms from 1 to 14 μm in length. Different species of bacteria show great variation in size, and some variation exists within a species, but as a rule, the size of each species is fairly constant.

Forms

Bacteria assume three well-known shapes (Figs. 3-1 and 3-2)§:
1. Spherical—coccus
2. Rod shaped—bacterium or bacillus
3. Spiral shaped or curved—spirillum, spirochete, or vibrio

Variations exist within these three shapes. Cocci are not necessarily perfectly round but may be somewhat elongated, oval, or flattened on one side. Rod-shaped bacilli may be long and slender or short and plump. Short, thick, oval bacilli resembling

---

*The planet Mars has an environment of extremes like those found in Antarctica. The Viking landers in 1976 searched Martian soil for signs of organic molecules or microbial life but found none.

†The temperature scale called *centigrade* in America has been known in many other countries as *Celsius* after Anders Celsius (1701-1744), the Swede who originated it in 1742. According to the Eleventh General Conference on Weights and Measures (1960), the term *centigrade* is inexact, and the name *Celsius* replaces it, with the capital C retained.

‡A cubic inch would hold 10 trillion medium-sized bacteria, or as many as there are stars in 100,000 galaxies.

§The following are the singular and plural forms:

Singular	Plural
Coccus	Cocci
Bacillus	Bacilli
Spirillum	Spirilla
Bacterium	Bacteria
Medium	Media

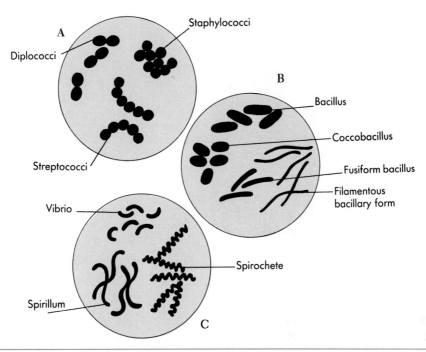

FIG. 3-1   Basic shapes of bacteria. **A,** The coccus is the spherical one: a pair of cocci, *diplococci*; a chain, *streptococci*; and a grapelike cluster, *staphylococci*. **B,** The bacillus is the rod-shaped form. Note variations. **C,** The vibrio, spirillum, and spirochete are spiral shaped.

cocci are known as **coccobacilli.** The ends of bacilli are usually rounded but may be square or concave. The spiral-shaped bacteria are each distinct. Their appearance relates to their movements. In the **spirillum** the long axis remains rigid when it is in motion, whereas in the **spirochete** the long axis bends when it is in motion. The **vibrio** is a curved rod shaped like a comma.

When bacteria, especially cocci, divide, the manner in which they do so and their tendency to cling together often give them a distinct arrangement. Cocci that divide to form pairs are known as **diplococci.** The opposing sides of diplococci may be flattened (for example, gonococci and meningococci). Cocci that divide and cling end to end to form chains are known as **streptococci.** Those that divide in an irregular manner to form grapelike clusters or broad sheets are known as **staphylococci.** Other patterns for cocci are in groups of four *(tetrads)* and cubic packets of eight *(sarcinae)*. No pathogenic cocci are found in the latter group. Bacilli that occur in pairs are known as **diplobacilli,** and those that occur in chains as **streptobacilli.** The diplobacillus and streptobacillus formations are unusual. When some bacilli divide, they bend at the point of division to give two organisms arranged in the form of a V. This is known as **snapping.** In other cases they tend to place themselves side by side. This is known as **slipping.**

Structure

Bacteria, always unicellular, are so tiny and transparent (about the density of water), so slightly refractile that, unless stained with dyes, they are difficult to see even with the compound light microscope. When stained, they appear homogeneous or slightly granular. With the electron microscope, however, microbiologists can visualize minute details of bacterial structure (Fig. 3-3).

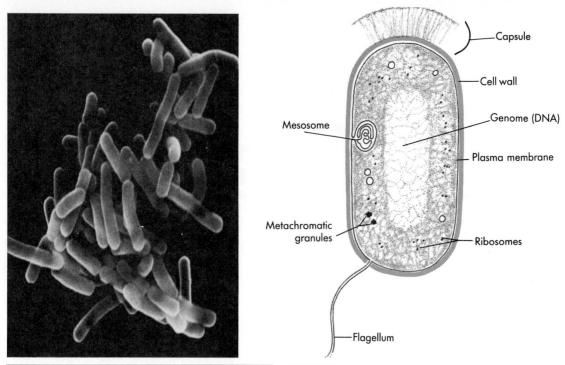

FIG. 3-2   Bacilli (*Lactobacillus* species), scanning electron micrograph. Note three-dimensional effect with this instrument. (Courtesy Dr. R.C. Reynolds, The University of Texas Health Science Center at Dallas.)

FIG. 3-3   Ultrastructure of a bacterial cell.

Cell Wall

The shape of the bacterial cell is maintained by a rigid cell wall, the cell wall being the part of the bacterial cell external to the plasma membrane (p. 32). (The cell envelope refers to the cell or plasma membrane plus all structures, such as the cell wall and capsule, that are external to it.) The protoplasmic substance of bacteria exerts such a high osmotic pressure, equivalent to that of a 10% to 20% solution of sucrose, that in ordinary environments, were it not for the high tensile strength of the cell wall, the bacterial cell would burst. (The internal osmotic pressure is from 5 to 20 atm.)

Peptidoglycan. The stability of the cell wall resides in a single, giant, complex molecule of peptidoglycan (murein, mucopeptide) (see Fig. 5-12). Peptidoglycan (a hard, strong polymer) is composed chemically of a backbone of the alternating amino sugars N-acetylglucosamine and N-acetylmuramic acid (a compound unique to bacteria), a set of identical tetrapeptide side chains attached to the N-acetylmuramic acid, and a set of identical peptide cross-bridges (p. 74). For all bacterial species the backbone is the same, but the side chains and the cross-bridges vary from species to species.

Cell wall variations. In the two major groups of bacteria—the gram-positive bacteria and the gram-negative ones (p. 151)—the chemical structure of the cell wall varies. The difference is reflected in their diverse gram-staining reactions, that is, the cell wall factor is responsible for the gram-positive or the gram-negative staining reaction. One major variation in cell wall makeup between the two bacterial groups lies in the peptidoglycan layer (Fig. 3-4). In gram-positive bacteria this layer consists of concentric sheets cross-linked chemically in three dimensions. In gram-negative ones the peptidoglycan layer is thinner and forms only a two-dimensional monolayer; it is not as fully cross-

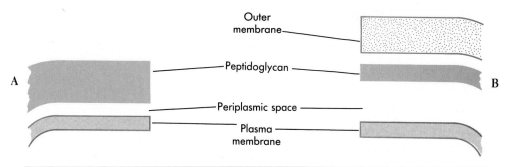

FIG. 3-4   The cell wall of the gram-positive bacterium (**A**) compared with that of the gram-negative one (**B**) in sketch. Note the thickness of the peptidoglycan layer, the presence of the outer membrane, and the extent of the periplasmic space.

FIG. 3-5   Chemical structure of teichoic acid. An accessory polymer in the cell wall of gram-positive bacteria, teichoic acid covers the peptidoglycan.

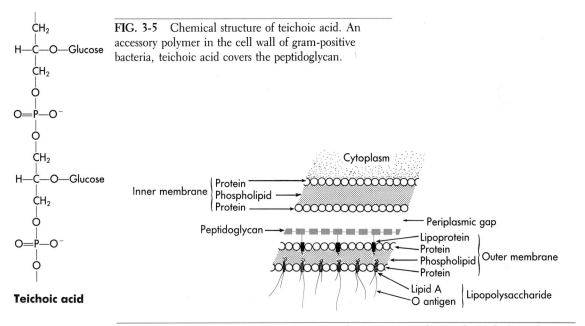

FIG. 3-6   Cell wall structure in gram-negative bacteria. Note relation of peptidoglycan layer and outer membrane to inner, or plasma, membrane.

linked. Still another variation is that most gram-positive cell walls contain teichoic acids (polyol phosphate polymers) in large amounts, even as much as 10% of the dry weight of the total cell. Gram-negative cell walls contain none. Teichoic acids (Fig. 3-5) are related to the outer surface of the peptidoglycan layer and possibly stick out of this layer. They participate in the supply of magnesium to the cell by binding magnesium ions. In bacteria where they are found, they constitute major surface antigens.

The gram-positive cell wall is 20 to 80 nm thick; the gram-negative cell wall, less than 10 nm.

The thick layer of peptidoglycan and the plasma (cytoplasmic) membrane define the cell envelope of gram-positive bacteria (Fig. 3-4). A well-defined outer membrane, a thin layer of peptidoglycan in the periplasmic space, and the inner or cytoplasmic (plasma) membrane make up the cell envelope of gram-negative bacteria (Fig. 3-6).

Gram-negative cell wall. In the gram-negative cell wall (Fig. 3-6) outside the peptidoglycan layer are three polymers: lipoprotein, the phospholipid-protein bilayer of the outer membrane, and lipopolysaccharide. The lipoprotein molecules cross-link the outer membrane with the peptidoglycan layer. The phospholipid-protein bilayer provides a typical structure for the outer membrane, which is a membrane relatively permeable to small molecules but one that hinders the penetration of larger ones, such as penicillin and lysozyme. The outer membrane is not the major permeability barrier in gram-negative bacteria, but the discriminatory capacity in the outer membrane is perhaps the explanation for the relative resistance of gram-negative organisms to certain antibiotics. For molecules of penicillin or lysozyme, access to their target site of action—the peptidoglycan layer—is blocked.

Lipopolysaccharide is the endotoxin of gram-negative bacteria. It is very toxic to animals and is responsible for the many ill effects of gram-negative organisms in human disease. Lipopolysaccharide, embedded in the outer membrane, is tightly bound to the cell surface and released only with lysis of the bacterial cell. It consists of a complex lipid (lipid A) to which is attached a polysaccharide. The lipid is in the outer membrane. The polysaccharide projects out, forming the **O antigen,** which determines the virulence of the bacterial cell. The polysaccharide is made up of a constant chemical core and a terminal series of repeat units, a repeat unit being unique to each gram-negative species. Thus polysaccharide as the O antigen represents a major surface antigen of the gram-negative bacterial cells.

Effect of antibiotics. The viability of bacteria directly depends on the integrity of the cell wall. The complex chemical sequence of events for the biosynthesis of the cell wall provides several possibilities for interference by antimicrobial compounds. Any such compound that inhibits any step can cause the wall to be weakened and, consequently, the cell to be lysed. The sites of action are well known for certain antibiotics that inhibit cell wall synthesis. For instance, in actively growing bacterial cells, penicillin hinders the enzymatic incorporation of the cross-bridges into that complex molecule, peptidoglycan. A defective, no longer rigid cell wall results. With continued growth, the now vulnerable bacterial cells soon are lysed. (See also Chapter 14.)

L-Forms

In certain species of bacteria (for example, *Proteus, Bacteroides, Pseudomonas,* and some of the coliforms), a normal cell may swell to a large entity only to disintegrate into numerous particles approximately 0.2 μm in diameter, known as *L-forms.* This kind of change may occur spontaneously, without a known stimulus, or be induced with a well-defined stimulus, such as a drug. L-forms that grow and divide may be produced experimentally from bacteria treated with lysozyme or penicillin to remove the cell wall. If at the same time the external milieu is maintained at a high osmotic pressure, the bacterial cells so treated are not lysed. If the osmotically fragile bodies that result are from gram-positive bacteria, they are called **protoplasts.** If they are from gram-negative bacteria, they are referred to as **spheroplasts.** Protoplasts are completely devoid of a cell wall, but spheroplasts retain their outer membrane.

Although these forms possess distinguishing features, they not only are akin to the parent cell but also can revert to it. The true importance of L-forms in the production of bacterial infections remains to be established.

Plasma Membrane

The cell wall is so narrow that it cannot be seen with the ordinary compound light microscope. In ultrathin sections it is revealed by electron microscopy as a well-defined structure surrounding a distinct layer, the cell (inner) membrane, or *plasma membrane* (Fig. 3-7), which separates it from the cytoplasm of the bacterial cell. Composed of phospholipids and proteins but no sterols except in bacteria of the genus *Mycoplasma,* the plasma membrane is the site of important enzyme systems, including the respiratory enzyme system (cytochrome enzymes). In fact, in bacteria it corresponds to the inner membrane of the mitochondria of higher organisms (eukaryotes). In regulating the passage of food materials and metabolic by-products between the interior of the cell (where metabolic activities are carried on) and the surroundings, it functions osmotically as the true permeability barrier for bacteria, and in blocking the entry of certain substances and catalyzing the active transport of others into the cell, it is the site of the active transport system.

Bacteria do not contain mitochondria. Since mitochondria in the higher organisms contain their own small set of genes, ribosomes, and the genetic apparatus to fabricate protein, many observers believe that mitochondria may have developed out of a successful and ongoing relationship between a primitive bacterium, originally free living, and a progenitor of the eukaryotic cell.

The **periplasmic gap** (Fig. 3-6) between the outer and inner, or plasma, membrane is an enzyme-containing zone. Between the thin peptidoglycan layer (in gram-negative bacteria) also occupying this gap and the outer membrane, hydrolytic enzymes accumulate.

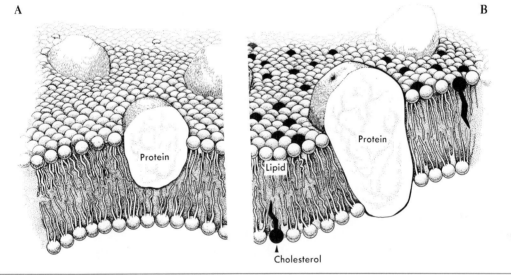

FIG. 3-7   Plasma membrane of prokaryotic and eukaryotic cells. **A,** Plasma membrane of bacteria. Note globular (potato-like) proteins embedded in typical phospholipid bilayer. Proteins make up "active sites" in membrane. **B,** Plasma membrane of eukaryotes. Note that this membrane incorporates sterols (cholesterol). Mycoplasmas (bacteria) incorporate sterol in their plasma membrane if it is available in the media in which they are grown.

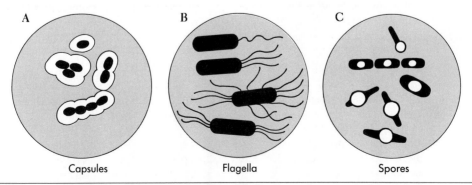

FIG. 3-8   Special features of bacteria. **A,** Capsules. **B,** Flagella. Note placement on bacterial body. **C,** Spores, which variably distort the bacterial cell.

Capsule

Surrounding many bacteria (outside the cell wall) is a mucilaginous envelope or capsule (Fig. 3-8, A). Indistinct in most bacteria because of the small amount of the material, it is well developed in a few (for example, *Streptococcus pneumoniae*, *Clostridium perfringens*, and *Klebsiella pneumoniae*). The capsule is formed by an accumulation of slime excreted by the bacterium. This material is usually a complex polysaccharide and is highly antigenic.

Capsule formation is most prominent in organisms taken directly from the animal body, since when grown on artificial media, the same organisms often lose their ability to form capsules. A capsule does not stain with the ordinary bacteriologic dyes but may appear as a clear halo around the bacterium even two or three times broader than the bacterium. It is stained by special methods. The presence of a capsule appears to enhance the virulence of an organism by protecting it against phagocytosis, and in some cases the capsule gives the organism its specific immunologic nature. For instance, relative to the nature of their capsules, pneumococci are divided into more than 86 types. The specific antigenic nature of a capsule depends on its carbohydrate content.

Flagella

Many bacilli, all spirilla and spirochetes, and most vibrios are motile when suspended in a suitable liquid at the proper temperature. (See p. 147 for the laboratory demonstration of motility.) True motility* is seldom observed in cocci.

The organ of locomotion is the **flagellum** (*pl.* flagella). Flagella are fine, hairlike surface appendages (little whips), 3 to 20 μm long and 12 to 30 nm in diameter, that

---

*True motility, in which the organism changes its position in relation to its neighbors, should not be confused with **brownian motion**, a peculiar dancing motion possessed by all finely divided particles in a liquid.

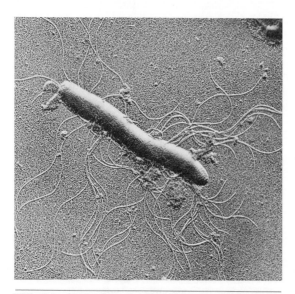

FIG. 3-9 Peritrichous bacillus from mouth, electron micrograph. Note flagella—number and arrangement. (Preparation specimen shadowed with gold, × 28,000.) (Courtesy Dr. J. Swafford, Arizona State University, Tempe, Arizona; from Brown, W.V., and Bertke, E.M.: Textbook of cytology, ed. 2, St. Louis, 1974, The C.V. Mosby Co.)

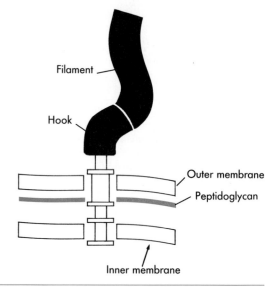

FIG. 3-10 Model to show how a flagellum is anchored into a bacterial cell wall.

are anchored to the bacterial cell in such a way as to cause it to move along by their wavelike, rhythmic contractions (Figs. 3-8, *B*, and 3-9). A flagellum is made up of a cylindric assembly of protein subunits with a hollow core terminating in a hooklike structure and a specialized series of rings that insert and secure it in the cell wall and plasma membrane (Fig. 3-10). Flagella are not revealed by the ordinary bacteriologic dyes, since they are too fine to be visible in the bright-field microscope, but have to be delineated with special staining methods.

Some spirochetes aid the action of their flagella with a sinuous motion of the cell body. A bacterium may have one polar flagellum (monotrichous) or a few or many flagella in a tuft at one pole of the cell (lophotrichous); the flagella may be attached to one end, both ends (amphitrichous), or all around the organism (peritrichous) (Fig. 3-9); in the latter case a few or as many as 100 to 1000 may be present.

Motility. Bacteria may be motile when grown on one medium and nonmotile when grown on another. Also, they may be motile at one temperature and nonmotile at another. Different organisms travel at different speeds. It is reported that the rod-shaped organism that causes typhoid fever (*Salmonella typhi*) is able to progress at a rate of 2000 times its own length per hour. Certain bacteria average from 25 to 55 μm per second, rates that are comparable to automobiles at 40 to 80 miles per hour.

Flagella and chemotaxis. The motility of flagellated bacteria is controlled by certain proteins (receptors) bound to the plasma membrane that recognize specific chemical molecules in the bacterial environment. Motility comes from rotation of the flagella, and the information the bacterium gains from the specialized chemoreceptors regulates either the clockwise or counterclockwise direction. Therefore alternate movements exist. A motile bacterium may be observed to move in a straight line for a second or so and then to tumble aimlessly for a short instant before setting off again in a straight path.

The phenomenon of chemotaxis refers to the response of an organism to a chemical stimulus that it moves toward or away from. Chemoreceptors on the bacterial membrane exist for sugar, amino acids, and other nutrients (attractants) and also for acids, alkalies, and other harmful substances (repellants). Flagellated bacteria can be demonstrated to move in linear runs toward attractants and to tumble away from noxious repellants. The linear run is longer in the direction of an attractant and shorter toward a repellant. How information is transmitted to the flagella or how information from various chemoreceptors is integrated to produce rotation of flagella is not known.

Pili

Pili (Latin, hairs), also referred to as attachment organelles, are filamentous surface projections like flagella found in certain gram-negative bacteria (see Fig. 6-14). However, they are shorter and finer and do not propel the cell. Pili are of two different types. **Common pili** are usually numerous over the cell; **sex** or **conjugal pili** are limited to one to four per cell. Bacteria with common pili can adhere firmly to mucous membranes and other cell surfaces in a susceptible host, a property correlated with their virulence. Sex pili are part of the attachment of bacterial cells in conjugation (p. 96). Pili are also called **fimbriae,** and the two terms are often used interchangeably, but many investigators are now restricting the term *fimbriae* to common pili and the term *pili* to sex pili.

Endospores

Under certain poorly understood conditions some species of bacteria form within their cytoplasm bodies known as *spores (endospores)* that are resistant to influences adverse to bacterial growth (Fig. 3-8, *C*). Spore formation seems to be a trait of bacilli, being exceedingly rare in cocci; it is seen in gram-positive cocci of the genus *Sporosarcina*. Most spore-forming bacteria, about 150 species, belong to the two genera *Bacillus* and

Clostridium. Most of them are nonpathogenic. The important pathogenic sporulating bacteria are those that cause tetanus, gas gangrene, botulism, and anthrax.

Spores seem to form when conditions for bacterial growth are unfavorable, and spore formation is a mechanism for survival, but in the life of certain species of bacteria it does seem to be a normal phase. Bacteria that do not form spores and spore-bearing bacteria in which spores are not forming are known as **vegetative** bacteria. Those in which spores are forming are **sporulating** bacteria. When conditions suitable for bacterial growth are established, the spore converts itself back to the actively multiplying form (germinates). The germinating spore becomes the vegetative form of the bacterium; one spore becomes one bacterium.

There is a temperature for each species at which spore formation is most active, and spore formation is preceded by a period of active vegetative reproduction. Some species form spores only in the presence of oxygen; others form spores only in its absence. Spore formation is not a reproductive phenomenon, since a spore forms only one bacterium and a bacterium forms only one spore.

Sporulation means that the bacterial cell forms within its substance a round or oval, highly refractile body surrounded by a coat (Fig. 3-11). This body increases in size until it is broad or broader than the cell. The portion of the cell that remains gradually disintegrates, leaving only the spore. Spores from which the remainder of the cell has disappeared are **free** spores. In the electron microscope a spore has a structural pattern: a coat, a cortex, and a nuclear core of chromatinic material.

Spores do not take the ordinary bacteriologic dyes, since such dyes cannot penetrate the almost impermeable keratin-like protein coat of the spore. They must be stained by special methods. Ordinary stains of sporulating organisms may show the spore as a clear, unstained area situated in the end of the bacterium (terminal), near its center (central), or in an intermediate position (subterminal). Spores may be spherical, ellipsoidal, or cylindric. In anaerobic bacteria the spore is broader than the remainder of the bacterial cell, causing the bacterium to assume a spindle shape if the spore is central and a club shape if it is terminal or subterminal. The shape and position of the spores help to identify certain bacteria.

Although spores are very resistant to heat, light, chemicals, and drying, the vegetative forms are no more resistant to these or other adverse influences than are nonspore-bearing bacteria. An idea of the protection afforded by spores to bacteria can be drawn from the observation that some spores withstand boiling for hours, whereas the temperature of boiling water kills vegetative bacteria within 5 minutes. On the one hand, spores resist tremendous pressures, but on the other hand, they can persist in a vacuum approaching the emptiness of space. Spores have also been known to resist the temperature of liquid nitrogen ($-190°$ C) for 6 months. This resistance is in large measure the result of the impervious spore coat and the concentrated water-free nature of their substance. Their heat resistance is also correlated with a unique feature of endospores: their content of calcium dipicolinate, a compound rarely found elsewhere in the biologic world but present in large amounts (5% to 15% of the dry weight of a spore) within spores. In fact, the vegetative cell in which the endospore was formed does not contain even detectable amounts of dipicolinic acid.

Although only a few species of bacteria produce spores, these spores are everywhere and can persist for years. This means that *spore-killing* methods in bacteriologic and surgical sterilization and in the canning industry are absolutely necessary.

Cytoplasmic Granules

Within the diphtheria bacillus, at the ordinary magnification of the light microscope, are seen granules, known as **metachromatic granules** (Fig. 3-3), that stain more deeply than the remainder of the cell. They are enzymatically active reserves of inorganic phosphate stored as polymerized metaphosphate (volutin). Metachromatic granules may be arranged irregularly within the bacterial cell or located in one or both ends of the cell, where they are known as **polar bodies.**

Sulfur-oxidizing bacteria convert excess hydrogen sulfide from the environment into intracellular granules of elemental sulfur. At times reserves of carbon source materials are stored in the cytoplasm as insoluble, osmotically inactive granules, available if needed.

No relation exists between the various cytoplasmic granules, including the spherules and submicroscopic particles revealed by electron microscopy, and the ability of the bacteria containing them to produce disease.

Ribosomes

Electron microscopy reveals a dense packing of ribosomes in bacterial cytoplasm. Ribosomes are tiny (around 20 nm* in diameter), uniform, beadlike granules that are

*nm, Nanometer. See Table 8-2.

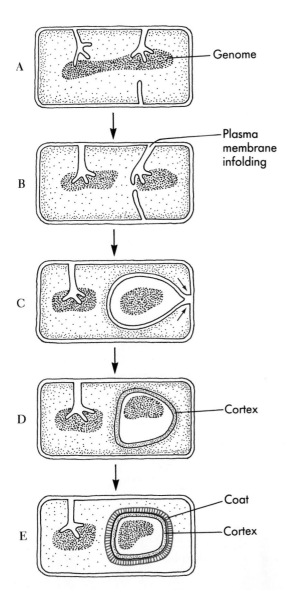

FIG. 3-11   Stages in sporulation. **A,** Vegetative stage of a spore-forming bacterium. **B,** Plasma membrane infolds. **C,** An area of intact genome is cut off. **D,** Cortex of spore formed with material laid down between two membranes. **E,** Coat of spore formed in final stage.

found free in the cytoplasm of a cell and are rich in ribonucleic acid. **Ribonucleic acid (RNA)** is one of the two nucleic acids of physiologic significance, the other being deoxyribonucleic acid (DNA) (as discussed later). Generally, RNA is similar to DNA chemically except that its sugar is a different pentose (ribose), one of its nucleotide bases is uracil instead of thymine, and its chemical pattern is laid out as a single coil, not a double one (Fig. 3-12). Ribosomes are about two-thirds RNA and one-third protein. Within the cell, more ribosomes are present than any other cytoplasmic particle; the rapidly proliferating *Escherichia coli*, a bacterium, contains approximately 15,000 ribosomes, about half the total mass of the bacterial cell.

Ribosomes are distinguished by their sedimentation coefficients expressed in **Svedberg units** (S). The Svedberg unit, a unit for measuring the sedimentation velocity in

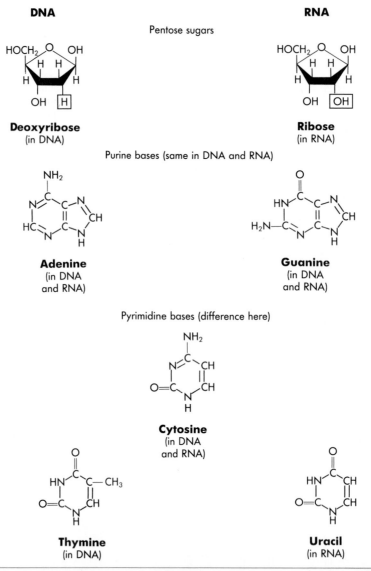

FIG. 3-12   DNA compared chemically with RNA. A difference is found in the pentose sugars and in one pyrimidine base.

the ultracentrifuge, is named for the Swede, Theodor Svedberg, who developed the instrument, a centrifuge with an exceedingly high rate of rotation designed to facilitate the separation of large molecules. Ribosomes in the cytoplasm of bacteria (prokaryotes) sediment at approximately 70S; the cytoplasmic ribosomes of eukaryotes, at approximately 80S.

At low-magnesium concentrations ribosomes can be dissociated into 30S and 50S subunits. Approximately, the content of either subunit is two-thirds RNA. The 30S one contains one 16S RNA molecule with 1600 nucleotides. The 50S subunit contains 2 RNA species—a 23S RNA containing 3200 nucleotides and a 5S RNA with 120 nucleotides. Most of the nucleotide sequences of these molecules are known, and around 60% of the bases are paired. The remainder of the mass (about one third) of the ribosomal subunits is protein. The 30S subunit contains 21 different proteins, and the 50S one contains 34 proteins. These proteins have been identified and purified, and many of their amino acid sequences have been determined.

Protein molecules are chains of amino acids put together in a specific sequence. Ribosomes are the sites in the cell that mediate the complex process of protein assembly. Protein synthesis is directed by the unique conformations of the ribosomal subunits and influenced as well by ionic concentrations, pH, temperature, and other factors in the environment.

Types of RNA. Based on composition, size, function, and location within the cell, three types of RNA are defined (Table 3-1): (1) ribosomal RNA, (2) transfer RNA, and (3) messenger RNA.

Ribosomal RNA (rRNA), the major component of the ribosomes, is the most abundant of the three types. Ribosomes in some bacteria may contain as much as 80% of the total RNA of the cell. rRNA plays a major role in protein synthesis.

Transfer RNA (tRNA), the smallest of the RNA molecules, makes up approximately 10% to 20% of cellular RNA. One molecule consists of about 75 nucleotides. As its designation suggests, tRNA presents activated amino acids to the ribosome for peptide bond formation, and at least one specific tRNA functions for each of the 20 amino acids. Composed largely of the four main ribonucleotides (adenylic, guanylic, cytidylic, and uridylic acids), tRNA also contains small amounts of other nucleotides.

Messenger RNA (mRNA) consists of a single strand of variable length, which may contain more than 5000 nucleotides. It makes up only 5% of the cell's total RNA. It is known to be the template for protein synthesis; molecules of mRNA are the information-carrying intermediaries. Since existing mRNA corresponds to each gene or group of genes being expressed in the genetic material of bacteria, this class of molecules of necessity would be heterogeneous.

Table 3-1 Comparison of the Types of RNA in a Bacterial Cell

Type	Sedimentation Coefficient	Composition (Number of Nucleotides)	Percent of Total Cell RNA	Function	Stability
rRNA	23S	3200	80	Major role in protein synthesis in ribosome	Stable
	16S	1600			
	5S	120			
tRNA	4S	75+	10 to 20	Presents activated amino acids to ribosome (adaptor)	Stable
mRNA		5000+	5	Messenger—a template for protein synthesis	Unstable

Genetic (Nuclear) Material

As special staining methods designed to reveal the chemical substances making up the nucleus in the higher forms of life were applied to the study of bacteria, it became evident that bacteria contained genetic (nuclear) material. This was seen in electron micrographs as a distinct and relatively transparent structure of rounded proportions with no detectable (nuclear) envelope.

The bacterial "chromosome" (genome). Genetic (nuclear) material in bacteria (prokaryotes) consists essentially of a single, continuous, tightly coiled, giant circular molecule of DNA. This structural unit, the genome, is often incorrectly referred to as the chromosome. In prokaryotes, DNA exists naturally in a highly compacted form within the genome. The tight coil is necessary because of the length of the DNA molecule relative to the long axis of the microbe. For instance, the DNA of the bacterium *E. coli* is about 400 times longer than the long axis of this microbe, and the DNA of bacteriophage T2 (a virus) is 500 times longer than the viral particle. If the circular molecule of DNA were unfolded, it would stretch a little over 1 mm. It is hooked at one point to an infolding of the plasma membrane known as a **mesosome** (Fig. 3-3). (The mesosome is involved in DNA replication.)

Within eukaryotic cells are **histones,** chemical compounds with a high concentration of the basic amino acids lysine and arginine that can effectively neutralize phosphate groups in DNA. Because of this, they contribute to the condensation of DNA found within eukaryotic chromosomes. Histones are not found in prokaryotes, but a comparable mechanism is believed to exist, probably effected with low molecular weight polyamines and magnesium ions.

Many bacteria possess one giant molecule of DNA, but two or more may be seen in young, actively dividing cells because cell division does not keep up with DNA replication. As it is for all cells, the genetic (nuclear) material is precisely the governing force for the bacterial cell, and vital activities cannot be carried on in a bacterial cell without it. (The terms *nucleoid* and *nuclear body* are sometimes used today to designate the nuclear or genetic material in bacteria.)

DNA. A very long, threadlike macromolecule, DNA (as well as RNA) is a linear polymer of nucleotides that is formed by phosphodiester linkage between the 5' phosphate of one nucleotide and the 3' hydroxyl group of the sugar of the adjacent one. Three kinds of small chemical molecules of consequence in DNA—(1) nucleotide bases (the four incorporated are guanine, cytosine, thymine, and adenine), (2) a pentose sugar (deoxyribose), and (3) phosphoric acid—are fastened together in a characteristic spiral pattern to give the well-known double helix. Thus two sugar-phosphate ladders—two identical, unbranched, rigid, spiral chains—with a great many paired nucleotide "rungs" are formed. These are at least 1500 times as long as they are wide and are twisted in opposite directions about a central shaft (or axis).

The sugar-phosphate backbone of each strand, the structural component, is external; the purine and pyrimidine bases, which carry the genetic information, are internal in the double helix. Hydrogen bonds between pairs of bases hold the two chains together. Adenine (A) is always paired with thymine (T), and guanine (G) is always matched to cytosine (C). Adenine and guanine are purines; cytosine and thymine, pyrimidines.

In a DNA segment of known length the number of different ways the four bases can be arranged can be calculated. The number of different nucleotide sequences that theoretically may form a stretch 1000 nucleotides long is 10^{600}. (A reasonable size for one gene is 4^{1000}.) The enormous variability of nucleotide sequences in the two polynucleotide chains of the double helix is more than enough to account for the number of different genes found in different genomes (chromosomes in eukaryotes). The DNA of the genome of *E. coli* consists of 4 million base pairs. An average-sized bacterium is said to contain enough DNA to specify about 3000 genes.

Genetic language. Every living organism must possess somewhere in its makeup a master plan that formulates all aspects of its appearance and behavior. Even in simple organisms such a master plan encompasses vast amounts of biologic information that, to be stored efficiently, must be converted to some sort of code. Because of its vantage point in the cell, DNA is just right to do this. Its chemical building blocks are linked in such a way that the four nucleotide bases (adenine, thymine, cytosine, and guanine) can serve as letters in a four-character code alphabet. (The Morse, or International,

FIG. 3-13 Watson-Crick model DNA molecule, sketched to show relation of purine and pyrimidine bases (G, C, T, and A) to helical structure. Steps of a spiral staircase are formed by base pairs bound chemically by hydrogen bonds.

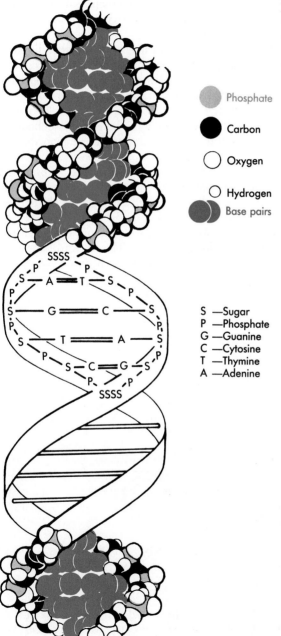

Phosphate

Carbon

Oxygen

Hydrogen

Base pairs

S —Sugar
P —Phosphate
G —Guanine
C —Cytosine
T —Thymine
A —Adenine

telegraphic code, composed of the dot and dash, is a two-character code.) Words can be formed in a biochemical and genetic language. In the language of life the four characters are expressed in linkages of three, called **triplet codes.** There are 64 such triplet codes with so many possible arrangements that they easily describe more than 4 billion human beings populating the world. Think of the chromosome then as a biologic document. From it can be transcribed biologic messages in code, over and over again, even the same message.

Watson-Crick model of DNA. The double helix model of DNA defined by Watson and Crick (Fig. 3-13) demonstrates how all the required genetic information can be built into or encoded in the molecular sequence of chemical structures—the nucleotides in the two strands of DNA—and indicates the biologic mechanism by which faithful copying of the structures can be brought about.

An understanding of the structure of the double helix came from the recognition of the rules of base pairing. With strict complementarity, the nucleotide sequence of one strand is determined by that of the other. Each DNA strand can serve as template for the elaboration of a new strand, and the nucleotide sequence is exact.

An unusual property of the DNA molecule is its ability to reproduce itself, that is, to replicate itself to form two exact copies. Because of this, the hereditary biochemical pattern, or **genetic code,** for the given organism may be passed from one generation to the next.

Forms of DNA. Four forms of DNA are being studied today. The A-form is a double-helical DNA containing 11 base pairs per turn of the helix. The planes of the pairs are tilted 20 degrees away from perpendicular to the helix axis. This right-handed helix is formed by the dehydration of the B-form of DNA (Fig. 3-14).

The B-form of DNA, the one most commonly found in solution and in vivo, is the classic Watson-Crick model containing 10 residues per turn. The planes of the base pairs in this right-handed helix are perpendicular to the axis of the helix.

The C-form of DNA is much like the B-form with nine base pairs per helical turn. The Z-form of DNA is a *left*-handed helix with about 12 residues per turn. Unlike the B-form DNA helix, the repeating unit is a *di*nucleotide. This results in a zigzag course of the sugar-phosphate backbone.

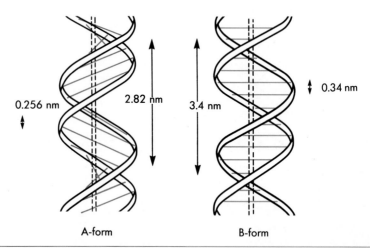

0.256 nm   2.82 nm   3.4 nm   0.34 nm

A-form   B-form

FIG. 3-14   Sketch of two different forms of DNA. About the vertical axis (*dotted lines*) are seen the sugar-phosphate chains with bonding between the base pairs indicated by horizontal rods.

CHEMICAL COMPOSITION

Even in a unit such as the bacterial cell, 500 times smaller than the average plant or animal cell, the chemical composition is exceedingly complex. To gain an idea of the chemical makeup of any bacterial cell, let us look at one that, since it is easily grown and manipulated in the laboratory, is undergoing almost as intensive study today as a human being. The "workhorse" is the colon bacillus *E. coli*.

Biochemically this microbe contains perhaps 3000 to 6000 different types of molecules (Table 3-2). Of these, specific proteins account for 2000 to 3000 kinds. The amount of DNA present is postulated to be that required to code for the necessary amino acid sequences in these proteins. About the DNA are 20,000 to 30,000 spherical ribosomes composed of protein (40%) and RNA (60%). Water, water-soluble enzymes, and a large number of various small and less complex molecules are associated with these nucleic acids.

REPRODUCTION: CELL DIVISION

The typical mode of bacterial reproduction, an asexual process, is by simple **transverse division (binary fission)** (Figs. 3-15 and 3-16). Bacteria do not divide by mitosis—there is no mitotic spindle; bacteria are haploid—they lack the paired homologous chromosomes of higher organisms, which are diploid. Their genetic information is stored in their one circular molecule of DNA, which is always attached to the cell membrane.

Transverse Division

In preparation for division the DNA molecule of the dividing cell replicates, thereby providing duplicate copies of the genetic information. The cell membrane plays an

Table 3-2 Chemical Makeup of a Single, Young, Actively Dividing Colon Bacillus *(Escherichia coli)*

Component	Average Molecular Weight	Estimated Number of Molecules	Number of Different Kinds of Molecules	Percent of Total Cell Weight
Proteins	40,000	1,000,000	2000 to 3000	15
Carbohydrates and precursors	150	200,000,000	200	3
Lipids and precursors	750	25,000,000	50	2
Nucleic acids				
DNA	2,500,000,000	4	1	1
RNA	25,000 to 1,000,000	461,000	1063	6
Amino acids and precursors	120	30,000,000	100	0.4
Nucleotides and precursors	300	12,000,000	200	0.4
Other small molecules (such as breakdown products of food molecules)	150	15,000,000	250	0.2
Inorganic ions (Na^+, K^+, Mg^{++}, Ca^{++}, Fe^{++}, Cl^-, PO_4^{4-}, SO_4^{--})	40	250,000,000	20	1
Water	18	40,000,000,000	1	70

From Watson, J.D.: Molecular biology of the gene, New York, 1976, W.A. Benjamin, Inc.

FIG. 3-15   Simple transverse division (binary fission). Note that the bacterium on the left becomes full size and divides in two. Two new bacteria are seen on the right.

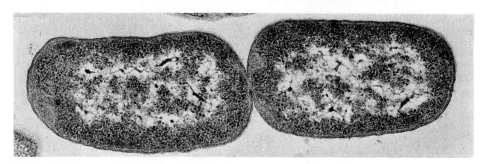

FIG. 3-16   Transverse fission of normal bacterium *(E. coli)*, electron micrograph. Two microorganisms are still attached along line of division. (From Morgan, C., and others: J. Bacteriol. **93:**1987, 1967.)

important role in DNA replication, since, for one thing, it contains a DNA polymerase (p. 82) that is important in DNA replication.

Replication initiates active membrane synthesis at the periphery of the bacterial cell, and a transverse membrane is formed that moves into the interior of the cell. The membrane, along with a new cell wall, constricts the bacterium along its short axis, and partly because of the presence of the mesosome (Fig. 3-3), it segregates the DNA into each of two daughter cells formed by the deepening constriction. Electron microscopy has shown that the segregation of the newly formed DNA molecules occurs when the mesosome is split by the growth of the cell membrane. The cell membrane functions somewhat like the mitotic spindle of higher cells in its influence on segregating genetic material.

Each new cell soon elongates to full size and then in turn divides. The process of bacterial division is very rapid, requiring from 20 to 30 minutes for a newly formed bacterium to reach full size and then divide. Reproduction is always specific; for example, staphylococci always reproduce staphylococci.

A Clone

The progeny (or population of cells) initially resulting from the normal transverse division (asexual reproduction) of a single bacterial cell are genetically identical and therefore constitute a clone (Greek *klōn*, young shoot or twig).

SUMMARY

1 Bacteria are the smallest cells with the necessary protoplasmic equipment for survival. Of several hundred thousand species defined, only about 100 produce disease in humans.

2 Bacteria may be spherical, rod shaped, or spiral or curved.

3 The bacterial cell wall, because of its content of the polymer peptidoglycan, can maintain its shape and integrity. The gram-positive wall is 20 to 80 nm thick; the gram-negative one is less than 10 nm because peptidoglycan is thicker and more complex in gram-positive bacteria.

4 The lipopolysaccharide embedded in the outer membrane of the gram-negative bacterial cell wall accounts for many of the ill effects in disease.

5 The plasma membrane in bacteria corresponds to the mitochondria of higher organisms. It controls the entrance and exit of substances.

6 Outside the cell wall in many bacteria is a mucilaginous capsule that, by protecting the bacterium against phagocytosis, enhances virulence.

7 Flagella are organs of locomotion. Pili are finer, nonpropelling surface projections that help attach certain gram-negative bacteria to cell surfaces.

8 Highly resistant spores are formed when some bacteria are subjected to adverse conditions. About 150 such species are found in the genera *Bacillus* and *Clostridium*. Tetanus, gas gangrene, botulism, and anthrax are serious diseases caused by spore formers.

9 The endospores of the few bacteria that sporulate are everywhere, making mandatory adequate spore-killing methods in sterilization.

10 Ribosomes, small beadlike granules rich in RNA, are the sites in bacterial cytoplasm where amino acids are assembled into proteins.

11 RNA and DNA differ only in their pentose sugar, in one of their nucleotide bases (uracil instead of thymine in RNA), and in the chemical pattern of either a single coil (RNA) or a double one (DNA).

12 The bacterial genome is a single, continuous, tightly coiled, giant circular molecule of DNA, which if unfolded stretches a little over 1 mm.

13 The sugar-phosphate backbone of DNA, which is external, is the structural support. The purine and pyrimidine bases carry the genetic information and are internal. Hydrogen bonds between base pairs hold the two chains together.

14 DNA stores instructions for all cellular life processes in a code based on varying arrangements of its four nucleotide bases, which serve as letters in a four-character code alphabet that is expressed in linkages of three (triplet codes).

15 Bacteria are exceedingly intricate chemical assemblages. *E. coli*, the "workhorse" for many scientific studies, contains 3000 to 6000 different types of molecules. Chemically it is 70% water and 15% protein. The remaining 15% is made up of varied, small, and less complex organic molecules and inorganic ions.

16 A mitotic spindle is nonexistent in bacteria. Bacteria reproduce asexually by simple transverse division (binary fission).

QUESTIONS FOR REVIEW

1 Define bacteria. Indicate their distribution.

2 Discuss the function of the cell wall. Of what is it composed?

3 Name and describe the three forms of bacteria.

4 What is the function of the plasma membrane?

5 Give two characteristics of bacteria at least partly dependent on the capsule.

6 How do bacteria move about in a fluid medium?

7 What are spores? What is their purpose and practical importance?

8 Give the unit of measurement for bacteria.

9 List significant contributions of the electron microscope to our knowledge of the bacterial cell. (Consult outside sources.)

10 Briefly define L-forms, pili, O antigen, peptidoglycan, nanometer (nm), metachromatic granules, mesosome, prokaryote, eukaryote, pathogenic, nonpathogenic, periplasmic gap, nucleoid, nuclear body, triplet codes, and clone.

11 Briefly indicate the importance of the teichoic acids, the lipopolysaccharide in the gram-negative cell wall, pili, the three kinds of RNA, and ribosomes.

12 Briefly discuss the bacterial genome.

13 Name the two nucleic acids. How are they different?
14 Discuss the biochemical structure of DNA. List the different forms.
15 How do bacteria reproduce?

SUGGESTED READINGS

Dickerson, R.E., and others: The anatomy of A-, B-, and Z-DNA, Science **216**:475, 1982.

Hopkins, R.C.: Deoxyribonucleic acid structure: a new model, Science **211**:289, 1981.

Jannasch, H.W., and others: Deep-sea bacteria; isolation in the absence of decompression, Science **216**:1315, 1982.

Jawetz, E., and others: Review of medical microbiology, Los Altos, Calif., 1984, Lange Medical Publications.

Kessel, R.G., and Shih, C.Y.: Scanning electron microscopy in biology: a students' atlas on biological organization, New York, 1976, Springer-Verlag New York, Inc.

Lodish, H.F., and Rothman, J.E.: The assembly of cell membranes, Sci. Am. **240**:48, Jan. 1979.

Schwartz, R.M., and Dayhoff, M.O.: Origins of prokaryotes, eukaryotes, mitochondria, and chloroplasts, Science **199**:395, 1978.

Stanier, R.Y., and others: The microbial world, ed. 4, Englewood Cliffs, N.J., 1976, Prentice-Hall, Inc.

Wang, A.H.J., and others: Left-handed double helical DNA: variations in the backbone conformation, Science **211**:171, 1981.

Watson, J.D., Molecular biology of the gene, New York, 1976, W.A. Benjamin, Inc.

CHAPTER 4

BIOLOGIC ATTRIBUTES OF BACTERIA
Sustenance and Growth

BACTERIAL NEEDS: ENVIRONMENTAL FACTORS

For bacteria to grow and multiply rapidly, certain requirements must be met: (1) sufficient nutrients of the proper kind must be present, (2) the oxygen requirements of the species must be met, (3) moisture must be available, (4) the temperature must be most suitable for the species, (5) the proper degree of alkalinity or acidity must be present, (6) light must be partially or completely excluded, and (7) by-products of bacterial growth must not accumulate in great amounts. Significant departure from any of these conditions modifies bacterial growth, although bacteria generally possess a greater degree of resistance to unfavorable conditions in their environment than do plants and animals.

Food materials prepared for the growth of bacteria in the laboratory are known as **culture media.** Some bacteria will grow on practically any properly prepared culture medium. Others grow only on specially nutritious ones, and a few will not grow on any artificial medium.

Nutrition

The protoplasm of the bacterial cell is composed of numerous organic compounds, including proteins, fats, and carbohydrates, as well as various inorganic components containing sulfur, phosphorus, calcium, magnesium, potassium, and iron. Proteins make up about 50% of the dry weight of the cell (each species has a type of protein peculiar to itself), and bacterial nitrogen makes up 10%. In some species carbohydrates are plentiful, and important traits of the species depend on these compounds.

Nutrition is the provision of food materials (that is, chemical substances) to bacteria so that they can grow, maintain their constituents, and multiply. For their nourishment bacteria require sources of carbon and nitrogen, growth factors, certain mineral salts, and sources of energy. With the exception of some saprophytic species, all bacteria derive their carbon and nitrogen from organic matter.

A number of minerals are required, the most important of the salts being those of calcium, phosphorus, iron, magnesium, potassium, and sodium. Because environmental iron often occurs in highly insoluble compounds, the iron necessary for microbial growth may be present in low and inadequate concentrations. To obtain this iron, many microbes elaborate **siderophores,** iron-chelating molecules that function as iron carriers. Certain minerals are prerequisite to the activation of enzymes.

Many microorganisms can synthesize all the organic compounds of their complex makeup if supplied with the basic nutrients. Many cannot, however, since they require vitamins and certain organic growth factors (a growth factor is utilized as the intact substance) for their activities. In this respect, microorganisms resemble rather closely higher forms of life. In fact, vitamins such as nicotinic acid, pantothenic acid, para-aminobenzoic acid, biotin, and folic acid—requirements for animal nutrition—were first studied and identified as substances necessary for the growth of microorganisms. Growth factor requirements may be determined by the use of chemically defined culture media.

Kinds of Organisms

Organisms that obtain their nourishment from nonliving organic material are known as **saprophytes.** Those that depend on living matter for their sustenance are **parasites. Facultative saprophytes** usually obtain nourishment from living matter but may obtain it from dead organic matter. **Facultative parasites** usually obtain nourishment from dead organic matter but may obtain it from living matter. Some pathogenic bacteria can exist only on living material (for example, the spirochete of syphilis cannot be grown outside a living organism). Most, however, can lead either a parasitic or a saprophytic existence. A few pathogenic bacteria, usually saprophytic, may adapt themselves to a parasitic existence (for example, bacteria that cause gas gangrene). The organism on which a parasite lives is known as a **host.**

Parasites that subsist and produce their effects within cells are **intracellular parasites;** those that do so outside cells are **extracellular parasites. A facultative intracellular parasite** spends only a short time within a parasitized cell, but an **obligate intracellular parasite** depends on the cellular machinery of the host and therefore thrives only within the parasitized cell.

Bacteria that obtain their nutriments by building the organic compounds in their protoplasm from simpler inorganic substances are **autotrophic (lithotrophic).** Autotrophs derive energy from oxidation of inorganic matter; they are important in nature, not in medicine. Organisms that obtain their nutriments by breaking down organic matter into simpler chemical substances are **heterotrophic (organotrophic).** All pathogenic bacteria and many nonpathogenic ones are heterotrophic. Nutrition is generally not a limiting factor for human pathogens because the human body is a rich source of all essential substances for bacterial growth.

Oxygen

Two simple equations help to explain the relationship between bacterial growth and oxygen. When bacteria grow in the presence of molecular oxygen, two toxic products are formed: (1) the superoxide ion (O_2^-) and (2) hydrogen peroxide (H_2O_2). If present, the enzyme superoxide dismutase (SOD) detoxifies the superoxide ion, as indicated in a reaction wherein more H_2O_2 may be produced:

$$2O_2 + 2H^+ \rightarrow H_2O_2 + O_2$$
$$\text{Hydrogen}$$
$$\text{peroxide}$$

The enzyme catalase detoxifies hydrogen peroxide:

$$2H_2O_2 \rightarrow 2H_2O + O_2$$

Organisms that almost always possess the enzymes catalase and superoxide dismutase and therefore can grow in the presence of free atmospheric oxygen are known as **aerobes.** An **obligate,** or **strict, aerobe** cannot develop at all in the *absence* of free oxygen.

Those that possess neither catalase nor superoxide dismutase and therefore cannot grow in the presence of free oxygen but must utilize oxygen-containing compounds (inorganic sulfates, nitrates, and carbonates or certain organic compounds) are **anaerobes. Obligate,** or **strict, anaerobes** are vulnerable to even traces of free oxygen; their enzyme systems are inactivated by atmospheric oxygen because strict anaerobes do not contain even traces of the detoxifying enzymes (p. 571).

Aerobes do not participate in fermentation; anaerobes are fermentative only (p. 73).

Organisms adaptable either to the presence of atmospheric oxygen or to the absence thereof are **facultative anaerobes.** Facultative organisms usually possess both catalase and superoxide dismutase and can switch between oxidative and fermentative metabolic pathways. Those microorganisms growing best in an amount of oxygen less than that contained in the air are **microaerophiles.** Microaerophiles often possess superoxide dismutase but not catalase. (They are usually fermentative.) Certain microaerophilic streptococci prefer anaerobic conditions but can grow in small amounts of atmospheric oxygen.

Microorganisms that need a 3% to 10% increase in carbon dioxide in the environment to initiate growth are designated **capnophiles.** Although most bacteria require carbon dioxide, usually that amount produced endogenously or present in normal atmosphere is enough to satisfy their need.

Moisture

Water is necessary for the growth and multiplication of bacteria. Not only is water a major component of cytoplasm (on an average, 75% to 80% of the bacterial cell is water), but it also dissolves the food materials in the environment of the bacterial cell so that they can be absorbed.

Desiccation is highly detrimental to bacterial growth. Delicate bacteria such as the gonococcus resist drying only a few hours, and even highly resistant bacteria such as the tubercle bacillus succumb to drying within a few days. Spores, however, may resist desiccation for years. As a rule, bacteria with capsules are more resistant in this respect than those with none.

Temperature

For each species of bacteria, a minimum, optimum, and maximum temperature exists, meaning, respectively, the lowest temperature at which the species will grow, the temperature at which it grows best, and the highest temperature at which growth is possible (Figs. 4-1 and 4-2). The optimum temperature for a species corresponds to the average temperature of its usual habitat. For instance, nonpathogenic bacteria that naturally live in the human body or pathogens that attack the human body do best at 37° C (normal body temperature). The lowest temperature at which any of the species continues to multiply is around 20° C, and the highest is from 42° to 45° C. In this temperature range, bacteria are **mesophiles.**

Many bacteria will not grow at a temperature more than a few degrees above or below their optimum. Some pathogens die off rapidly at only 38° C. Most saprophytic bacteria (mesophiles) grow best between 25° and 40° C, but some **thermophiles,** or heat-loving species, grow best at a temperature of about 55° and as high as 70° C. One species, *Sulfolobus acidocaldarius*, obtains its maximum growth at 85° C. Studies on thermophiles done at Yellowstone National Park and other geothermal areas in the world have demonstrated that such heat-loving bacteria are not only there but thriving in waters over 90° C, and bacteria recovered from volcanic vents on the Pacific Ocean floor south of Baja, California, reportedly do well at temperatures as high as 250° C. A few **psychrophiles,** or **cryophiles** (cold-loving species), grow at temperatures just above the freezing point (range 0° to 25° C, best at 10° to 15° C). Since psychrophiles proliferate slowly in the refrigerator, they can be a hazard to the biologic products (vaccines, parenteral medications) that are regularly stored there. Many psychrophilic microbes have red pigments. Where they grow on the surface of ice and snow, they color it to give red snow.

Cold retards or stops bacterial growth, but when the bacteria are later exposed to a temperature favorable for their growth, multiplication resumes. Refrigeration (4° to

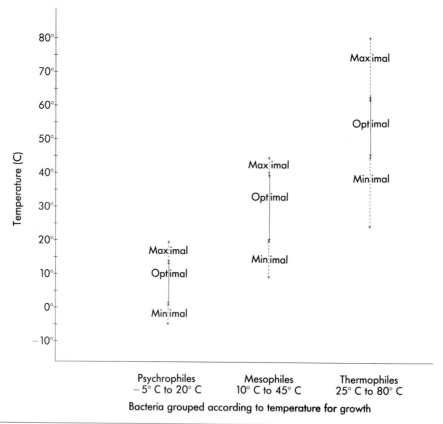

FIG. 4-1   Celsius temperature ranges for bacterial growth.

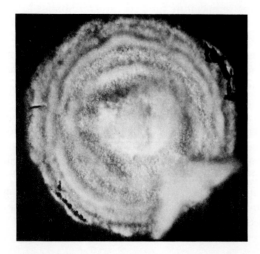

FIG. 4-2   Variation in growth rate from changes in temperature. Note rings of rapid and slow growth in giant colony of fungus *(Histoplasma capsulatum)*. One sector of mutant growth is seen in lower right-hand corner. (Courtesy Dr. R.H. Musgnug, Haddonfield, N.J.)

Table 4-1 Limits of the Hydrogen Ion Concentration (pH) for Two Kinds of Microbes

Microbe	Lowest pH Compatible with Growth	Optimal pH	Highest pH Compatible with Growth
Bacteria	3 to 5	6.5 to 7.5	8 to 10
Fungi	1 to 3	4.5 to 5.5	7 to 8
Molds	1 to 2		
Yeasts	2 to 3		

6° C) is one of the best methods of preserving bacterial cultures, since bacteria are generally resistant to low temperatures and even to freezing. Prolonged freezing, however, destroys them. A few pathogens that can grow slowly at low temperatures in the refrigerator may contaminate the biologic products stored there.

High temperatures are much more injurious to bacteria than low ones and are used effectively in practical situations where bacteria and their spores must be destroyed (Chapter 12).

Hydrogen Ion Concentration (pH)

For each species of bacteria, growth is most rapid at a certain degree of alkalinity or acidity (a certain pH) (Table 4-1). The reaction of culture media must be carefully adjusted to the desired hydrogen ion concentration. The best growth of most microorganisms is found in a narrow pH range of not less than 6 nor more than 8. Most pathogens grow best in a neutral or slightly alkaline medium (like that of the human body). Regardless of the influence of the environment, the reaction of the interior of the cell is just at the neutral point (pH 7).

Light

Violet, ultraviolet,* and blue lights are highly destructive to bacteria, green light is much less so, and red and yellow lights have little bactericidal action. Because of its content of ultraviolet light, direct sunlight kills most bacteria within a few hours. Bright daylight has an effect similar to that of sunlight but one of less potency.

A few species of saprophytic bacteria containing chlorophyl can utilize sunlight to build up the compounds of which they are composed. This bacterial chlorophyl is scattered throughout the cytoplasm, unlike the chlorophyl of plant cells, which is contained in their chloroplasts.

By-products of Bacterial Growth

Were it not for inhibitory influences, bacteria would completely submerge the world. It is estimated that the progeny from the unrestricted growth of a single bacterium would be 280 trillion at the end of only 24 hours. In cultures, bacterial reproduction is so rapid that bacteria soon exhaust their food supply and release products that inhibit further bacterial growth. Notable are organic acids from carbohydrate metabolism that inhibit growth by changing the reaction of the medium. The practical application of this is found in the pickling industry, where the acid medium is used to prevent bacterial contamination.

Electricity and Radiant Energy

By itself electricity does not destroy bacteria, but it causes heat and changes that may be lethal in the medium in which the bacteria are growing. Electric light inhibits bacterial growth, but the inhibition is an effect of light, not of electricity.

*The use of ultraviolet light in sterilization is discussed on p. 214.

The effects of roentgen rays are generally harmful to most bacteria, being about 100 times more effective in their action against bacteria than in ultraviolet light. There are, however, notable exceptions. Microbial species exist that can tolerate an amount of radiation 10,000 times greater than the amount fatal to a human being.

Chemicals

Certain chemicals destroy bacteria, and others inhibit their growth (p. 218). Some attract bacteria (**positive chemotaxis**), whereas others repel them (**negative chemotaxis**).

Osmotic Pressure

The bacterial cell is encased in a membrane said to be **semipermeable** because it allows water to pass freely in and out of the cell but gives a varying degree of resistance to dissolved substances in the fluid medium in which the cell is suspended. This makes the cell a small osmotic unit responsive to changes in its fluid environment (Fig. 4-3).

Under normal conditions there is a higher concentration of dissolved substances within the cell than without it. The greater osmotic pressure inside the cell keeps the protoplasm of the cell firmly against the cell wall, and the cell is said to be **turgid**. If a bacterial cell is placed in solutions having varying concentrations of dissolved substances, changes take place. In a solution with a high concentration of dissolved substances (**hypertonic** solution), water leaves the interior of the cell and the cell begins to shrink. If the difference in concentration between the interior and the exterior of the cell is not too great, the cell may be able to adjust to the hypertonic solution, regain its turgor, and continue its growth. If not, the cell continues to shrink and finally dies (**plasmolysis**).

If a cell is placed in a solution with a low concentration of dissolved substances or in distilled water (**hypotonic** solution), water passes into the cell, the cell swells, and it may burst (**plasmoptysis**). A solution containing that concentration of dissolved substances in which the cell neither swells nor shrinks is said to be **isotonic**.

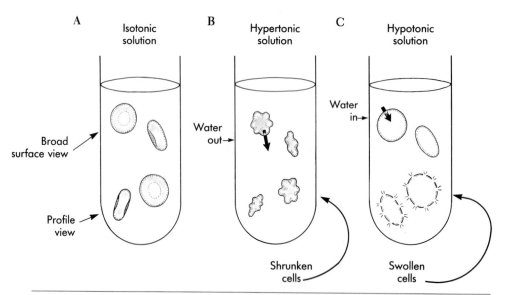

FIG. 4-3   Effect of osmotic pressure on cells. **A,** In isotonic solution, red cell does not change size or shape. **B,** In hypertonic solution, it shrinks. **C,** In hypotonic solution, it swells and bursts.

Most bacteria resist small changes in osmotic pressure but are killed or inhibited by high concentrations either of salt (as used in brines) or of sugar. This fact is used in the preservation of foods such as syrups and jellies and in the preservation of meats in brine. That some microorganisms can adjust to a high concentration of sugar is seen in molds on jellies. To preserve foods safely, higher concentrations of sugar must be used than of salt.

A small amount of various salts in the fluid medium for bacteria is beneficial to bacterial growth. Traces of such salts are furnished by natural foods such as meat extracts. Most bacteria do well in physiologic saline. However, the ability of certain key species to grow in 4.5% to 6% salt solution constitutes one of their distinguishing features. That a few species of bacteria (**osmophiles**) do well in a hypertonic solution is seen from the bacterial life of the oceans. Even the Dead Sea with its high salt content supports a bacterial population of halophiles. **Halophiles** (salt lovers) require even abnormally high concentrations of salt for survival. Other dense waters found in the Great Salt Lake, drying tidal basins, and pickling brines support, in addition to halophiles, other atypical, even strange forms of life existing in the extreme environment.

BACTERIAL GROWTH: BACTERIAL POPULATIONS

Because of the microscopic size of bacteria, aspects of their growth are more readily evaluated not in a single cell but in a collection or community of cells that demonstrates genetic and cultural consistency through several generations of cell division. The large number of bacteria in that community makes up the bacterial population called the *culture* in the laboratory. From the study of the bacterial population has come knowledge of factors related to the growth of bacteria or to the cessation of growth and loss of viability. This knowledge is crucial to understanding bacterial diseases and how these are best managed and controlled, since numbers of bacteria, not single cells, produce disease.

Exponential Growth

The number of cells in a given bacterial population increases with time as a geometric progression, that is, exponentially, because the two daughter cells that result from the division (binary fission) of a single bacterium are capable of growing and dividing at the same rate as the parent cell. Each time the bacteria divide, the population doubles. The time it takes for the population to double, which is the rate of exponential growth, is usually expressed as the **generation,** or **doubling, time.** Since the rate of growth of the population is directly proportional to the number of cells present at a specified time, the generation time can be calculated from the record of the number of bacteria in samples taken from a suitable community of bacteria at various intervals.

Growth rates vary widely among unicellular microbes, and obviously the generation times are shortest with the best conditions for growth. The fastest doubling times, found in only a few bacteria, are around 10 minutes. Much longer generation times of 24 hours or more are found in slow-growing protozoa. In an ideal situation most bacteria double in 30 to 60 minutes or slightly less. At this rate the progeny of but a single bacterium is in the billions within 24 hours. This is why visible bacterial growth appears on the laboratory culture within 24 to 48 hours and why in most instances 24 to 48 hours is the standard incubation period.

Measurement of Bacterial Growth

A bacterial population may be quantitated from a fluid culture or from a colony on a solid medium. Various approaches may be used, but three important measurements are the cell count, cell density, and cell mass.

Bacterial cell counts give very useful information. The two used are the total count and the viable count.

A **total count** is derived from the direct microscopic enumeration of bacteria in a given volume of test material spread out on a microslide or placed in an appropriate counting chamber. From the actual count the total number of organisms per the unit volume can be calculated. A total count, as would be expected, counts all bacteria present and does not distinguish between the dead and the viable ones.

The **viable count** is based on the premise that visible colonies result when viable bacteria are placed on suitable solid culture media and that viability of a bacterial cell is best shown in this way. The colonies are counted with the assumption that each represents a single viable cell in the original inoculum, and the viable count per unit volume of material is calculated. Plate counting techniques usually require preliminary dilution of the specimen for analysis to reduce the number of cells therein to a workable fraction (around 30 to 300) before the plate is inoculated. The colony count is then multiplied by the dilution factor to yield the total number of viable bacteria per unit volume of the original culture. Alternatively, for a viable count the specimen for analysis may be filtered through a plastic membrane filter that retains bacteria. The filter is then inverted over an agar surface, and after incubation the colonies are counted.

Bacterial cell counts are standard procedures in public health laboratories in the analysis of water and milk.

Although generally a method that demonstrates directly and specifically an increase in the number of bacteria is more practical, growth of bacterial cells in a liquid culture may be correlated with an increase in the turbidity (cloudiness) or optical density of the fluid medium into which they were inoculated. Cell density is measured in terms of changes in turbidity by an instrument, the spectrophotometer, that records them in units designated *optical density units.*

The cell mass can be determined from the total measured weight of a population of bacterial cells or from the weight of some significant cellular component—protein or RNA—that is directly related to cell mass, but the procedure is largely a research maneuver.

The Typical Growth Curve

A standard growth curve may be obtained in a prescribed way. A known concentration of bacteria is inoculated into a suitable liquid medium, and at stated intervals of incu-

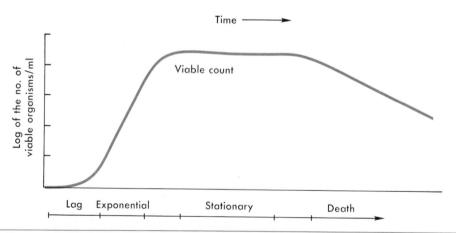

FIG. 4-4   Standard growth curve for a population of bacteria. Note the four distinct phases of growth in relation to viable count.

bation time, viable bacterial counts are performed. Results are plotted on semilog paper, that is, the logarithms of the numbers of bacteria are plotted against incubation time on an arithmetic scale. A typical growth curve (Fig. 4-4) so shaped displays a nearly symmetric curve. Points reflecting multiplication of bacteria at an exponential rate fall on a straight line, generation time is expressed as the slope of the straight-line portion of the curve, and errors in counts tend to be minimized.

When this curve is examined, four distinct phases of bacterial growth stand out: (1) the lag phase, (2) the log, or exponential, phase, (3) the stationary phase, and (4) the death phase or phase of decline.

Lag Phase

Early on, bacterial cells must adapt to the milieu in which they have been placed. Thus a period of time elapses in which no increase in numbers of bacterial cells occurs. After adaptation, cells divide until they reach an optimum rate. Variable factors control this stage, and it lasts a variable period of time.

Log Phase

The log, or logarithmic, phase, so-called because of the logarithmic increase in numbers, is the stage when bacterial growth is at a maximum. Bacteria are doubling at a constant rate.

The generation time is highly variable among different bacteria. Around ½ hour is about as fast as can be demonstrated,* and often hours to days are needed. Among disease-producing bacteria the more rapidly growing ones tend to produce a more acute illness, whereas the slowly growing ones with long generation times tend to produce a slowly developing, chronic illness.

In the log phase the cells are large and uniform in staining, and because they are active metabolically (rich in ribosomes), they are generally most sensitive to antimicrobials, disinfectants, heat, and other toxic influences. At least two factors limit the course of the log phase: the exhaustion of nutrients and the accumulation of toxic by-products that inhibit growth. For instance, the supply of oxygen is a limiting factor for aerobes. If the diffusion of oxygen beyond a certain cellular concentration cannot meet the aerobes' need for it, their growth progressively slows. The study of the log phase gives important information as to the best conditions for growth of a given organism, and the diagnostic features of the organism are most evident at this time.

Stationary Phase

In the stationary phase, bacterial growth slows and eventually ceases, since in a closed system a bacterial population cannot grow at an exponential rate indefinitely. A leveling off of growth may take place to the point where cell division just equals cell death. When this occurs, the total bacterial count slowly increases but the viable count remains the same. With the decline of the metabolic rate in the bacterial cells because of the depletion of nutrients, the buildup of undesirable metabolites, and the pH change, bacterial cells become variable in staining and most resistant to toxic agents. Because of environmental deficits, some bacteria form spores at this time.

Death or Decline Phase

Net cell death comes next. This may continue until all cells in the medium are dead, or it may stop just short of that point. After most bacteria have succumbed, the death

---

Escherichia coli can double in 20 minutes if cultured in a rich nutrient medium.

rate may drop sharply in a way to indicate that some do persist, even for prolonged periods, years in some instances, perhaps because they subsist on essential substances released from cells present that die and are lysed.

Applications

Applications of the physiologic studies of microbial growth are many. In the evaluation of antimicrobial agents, growth patterns help to demonstrate either the killing effect of the bacteri*cidal* drug or the inhibitory, not necessarily lethal, effect of the bacterio*static* drug. In the production of a vaccine using a live agent, careful analysis of the features of its growth is a prime factor in the determination of the safety of that vaccine.

BACTERIAL VARIATION

Living organisms, or the aggregate of living organisms, are seldom if ever exactly the same. Microbes are no exception. Deviation from the parent form in bacteria of the same species growing under different or identical conditions is known as *variation*.

Causes of Variation

Variation in bacteria may be caused by external or internal influences. It may result from factors inherent in bacteria, particularly genetic ones. It may be caused by external factors in the environment, such as the nutrients available, the temperature of growth, the length of time grown in vitro, and exposure to chemical or physical agents (for example, x rays). Although the properties of a bacterium may change with certain external conditions, the range of possible variants is limited and is determined genetically.

Observed Variations

In bacteria, variation may affect all biologic properties—size, shape, biochemical nature, colonial morphology, physiologic activities, and capacity for disease production. It may be temporary or permanent.

A special kind of bacterial variation in which there is a change in the colonial growth on a semisolid culture medium is called **microbial dissociation.** In fact a new type of growth appears. Although a pure culture of the bacteria is used to streak the surface of a culture medium (p. 162), the resultant colonies present contrasts. Some are smooth or mucoid colonies, regular in outline, round, moist, and glistening—the S-type colonies. Other are rough colonies, large, irregular in outline, indented, and wrinkled—the R-type colonies. Between the two are intermediate types. Organisms forming S-type colonies generally seem to be vigorous disease producers, whereas those forming R-type colonies are not harmful. By proper laboratory manipulation bacteria forming S-type colonies can be made to form R-type, and vice versa.

Organisms that are surrounded by a capsule when grown in the animal body often lose that capsule when grown in vitro. Capsule formation may be restored by returning the organisms to a susceptible animal. Capsule formation is pronounced in anthrax bacilli infecting an animal, but from their growth on the artificial media of the laboratory, hardly a sign of a capsule appears for the very same microbe. Bacteria with capsules have a smooth form of growth; those without a capsule produce a rough type of growth.

Bacteria of the same species, growing under ideal conditions, may vary considerably in size, shape, and appearance. This is termed **pleomorphism.** Grown under unfavorable conditions (for example, in old and aging cultures), the members of some species assume irregular, bizarre shapes and stain irregularly. These swollen, shrunken, or granular variants are known as **involution** or **degenerative** forms. In this instance variation in morphology probably reflects the presence of many dying and dead cells in the medium and the injurious effect of the accumulated metabolic by-products.

Adaptation

Variations or changes in bacterial makeup that in its environment improve the bacterium's chances for survival and of leaving descendants are designated by the term *adaptation*. One of the attributes of living cells is their power to adapt themselves to their surroundings. Since the environment on which bacteria depend is changing constantly, this must be true for bacteria. For instance, certain bacteria that require specifically prepared nutrients to sustain their continued growth when first isolated from the animal body gradually acquire the ability to grow on culture media devoid of these growth-promoting and enriching materials. They may also be grown artificially in a gradually changing environment until at last they are able to grow under conditions (food supply, temperature, moisture, and oxygen supply) far different from those in which they originally grew best.

Adaptation may be genetic or nongenetic. Those microbes in a given population genetically endowed to survive do thrive and become dominant. A change in bacterial properties may be independent of any change in genotype. Nongenetic adaptation, also termed *phenotypic* or *physiologic adaptation*, represents the response to a change in the environment of all the cells of a culture and the fact that all the cells adapt physiologically. A change takes place in metabolic activity. The adapted response is reversible but not heritable.

INTERRELATIONSHIPS OF MICROBES

Symbiosis

When two dissimilar organisms (two microbes or a microbe and a host) live together or are closely associated, each is known as a **symbiont,** and they have entered into an association called **symbiosis.** If the association benefits both, it is referred to as **mutualism.** For example, a beneficial relation is maintained between the leguminous plants and the nitrogen-fixing bacteria living in the root nodules of these plants (p. 888). If it benefits one but does not disturb the other, it is **commensalism.** The organisms making up the normal flora of the different areas of the human body are generally considered to be commensals. If the association benefits one and at the same time harms the other, it is **parasitism.** There are many examples of this relationship. If the association harms one but does not disturb the other, it is **amensalism.** If the association is harmful to both, it is referred to as **synnecrosis.**

Staphylococci and influenza bacilli multiply more rapidly when grown together than either does when grown alone. The combined effect of these microorganisms growing together is greater than the algebraic sum of their individual efforts. This is **synergism.**

Antagonism

Sometimes the presence of certain species of microbes inhibits the growth of others. For instance, growth of the gonococcus is inhibited by the presence of almost any other species of bacteria. This is antagonism. Theories advanced to explain antagonism are (1) that one organism secretes a substance toxic to the growth of the other and (2) that one organism promotes a defense mechanism of the animal body against the other. The appearance of certain infections after the administration of antibiotics may be explained in terms of an antagonistic relationship between two organisms, only one of which succumbs to the action of the antibiotic. Released by the antibiotic, the other microorganism then becomes aggressive.

SUMMARY

1 Sources of carbon and nitrogen, growth factors, certain mineral salts, and sources of energy are required for bacterial growth and multiplication. The human body is a rich source of all essential substances for growth.

2 Autotrophs, important in nature but not in medicine, build their organic compounds from simpler inorganic substances, deriving energy from oxidation of inorganic matter. All pathogens and many nonpathogenic bacteria, being heterotrophs, break down organic matter into simple substances.

3 Aerobes can grow in free atmospheric oxygen, since they possess the enzymes to detoxify toxic products of their growth in the presence of molecular oxygen. Anaerobes must utilize oxygen-containing compounds instead. Adaptable microbes are facultative.

4 The best temperature for bacterial growth is the average value of their usual habitat. Human pathogens do well at human body temperature. Most pathogenic bacteria grow best in a neutral or slightly alkaline medium like that of the human body.

5 Most bacteria resist small changes in osmotic pressure but are inhibited by high concentrations of salt or sugar.

6 The fastest time for a bacterial population to double is about 10 minutes; a much longer time of 24 hours is found for slow-growing protozoa. Most bacteria double in 30 to 60 minutes; therefore the progeny of a single bacterium is in the billions within 24 hours and visible on artificial growth medium within 24 to 48 hours.

7 Progeny from the growth of a single bacterium would number 280 trillion at the end of 24 hours were it not for inhibitory influences, such as the by-products of bacterial growth.

8 More rapidly growing bacteria tend to produce a more acute illness, whereas slow-growing ones with long generation times tend to produce a slowly developing chronic illness.

9 Two dissimilar organisms living together or closely associated have entered into symbiosis. If the association benefits both, it is mutualism. If it benefits one but does not disturb the other, it is commensalism. If it benefits one and at the same time harms the other, it is parasitism.

10 Synergism occurs when the combined effect of two microbes growing together is greater than the algebraic sum of their individual efforts.

11 Variation in bacterial makeup representing physiologic adjustment to the environment is designated adaptation. Since the environment on which bacteria depend is changing constantly, bacteria are probably better able to adapt than are many other cell types.

QUESTIONS FOR REVIEW

1 State at least five requirements for bacteria to grow and multiply.

2 Classify bacteria with regard to food requirements and oxygen requirements.

3 How are bacteria affected by extremes of temperature?

4 What is the difference between the following?
Aerobic and anaerobic
Bactericidal and bacteriostatic
Parasitic and saprophytic
Pathogenic and nonpathogenic
Facultative and obligate
Thermophile and psychrophile
Organotrophic and lithotrophic
Organic and inorganic
Total bacterial count and viable bacterial count

Log phase and lag phase
Hypertonic and hypotonic
Plasmoptysis and plasmolysis
Autotrophic and heterotrophic
S-type colonies and R-type colonies

5 How may bacterial growth be measured?

6 Explain exponential growth of bacteria. Name four factors influencing it.

7 Give the four distinct phases of bacterial growth defined in the standard growth curve.

8 Briefly define psychrophiles, halophiles, mesophiles, osmophiles, commensalism, symbiosis, antagonism, capnophiles, microaerophiles, semipermeable, isotonic, synergism, generation time, siderophores, and adaptation.

9 What is the significance of superoxide dismutase?

10 What is bacterial variation? What causes it? How is it measured?

SUGGESTED READINGS

Baron, S., editor: Medical microbiology, Menlo Park, Calif., 1982, Addison-Wesley Publishing Co., Inc.

Bellamy, W.D.: Role of thermophiles in cellulose recycling, ASM News **45**:326, 1979.

Fridovich, I.: Superoxide dismutases, Ann. Rev. Biochem. **44**:147, 1975.

Rogers H.J.: Bacterial iron metabolism and host resistance. In Schlessinger, D., editor: Microbiology—1974, Washington, D.C., 1975, American Society for Microbiology.

Stanier, R.Y., Adelberg, E.A., and Ingraham, J.L.: The microbial world, Englewood Cliffs, N.J., 1976, Prentice-Hall, Inc.

Weinberg, E.D.: Iron and susceptibility to infectious disease, Science **184**:952, 1974.

CHAPTER 5

BIOLOGIC ACTIVITIES OF BACTERIA
Metabolism

OVERVIEW

Bacteria engage in a complex of biologic activities of varying importance. In fact, they first drew attention to their existence in biochemical phenomena such as putrefaction, decay, and soil fertility. Of prime consideration are those activities maintaining the growth and integrity of the microorganisms. Those resulting in the elaboration of toxins and substances harmful to other living cells (Chapter 16) have medical significance in the production or aggravation of disease.

Certain bacterial activities yield biochemical end-products that are easily incorporated into diagnostic tests of great value in the laboratory identification of bacteria. Many highly specific biochemical tests based on the biochemical analysis of the metabolic activity of bacteria and other microbes have been developed and are widely used (Chapter 10).

Much of the general information concerning metabolism in creatures more complex than bacteria has come from studies of comparable processes in bacteria and other microbes. Because bacteria multiply rapidly, many of their physiologic processes can be observed within 2 days. Bacteria are easily manipulated and thereby provide a useful means for investigation of metabolic pathways. To obtain like information from human beings would require 200 years or more.

MAJOR EVENTS IN METABOLISM

The bacterial cell must satisfy two basic needs from its biologic environment. For one, it must obtain from the environment the chemical ingredients with which to build its protoplasm, maintain its viability and function, and continue its reproduction. Building blocks, or subunits, for biosynthesis must be assembled for this purpose. Concurrently, the bacterial cell must derive from its milieu the energy necessary to do the work required. Two energy-related aspects of bacterial metabolism that are geared to this second need are (1) the generation of adenosine triphosphate (ATP) from adenosine diphosphate (ADP) and inorganic phosphate (P_i) and (2) the generation of reducing power, as from nicotinamide adenine dinucleotide (NAD) or from nicotinamide adenine dinucleotide phosphate (NADP).

Metabolism (Greek *metaballein*, to change) encompasses all the biochemical reactions occurring within the bacterial cell whereby these two fundamental needs are met.

These processes are divided into **anabolic,** that is, synthetic, pathways, and **catabolic,** that is, degradative, ones.

The Basic Biochemical Reactions

Role of Enzymes

Enzymes, efficient ingredients in the complex maze of metabolic activity, are proteins serving as metabolic catalysts. A catalyst accelerates a chemical reaction that otherwise would proceed either slowly and ineffectively or not at all. In the reaction the catalyst is not permanently altered, contributing nothing to the end-product and furnishing no energy. Enzymes increase the rate of chemical reactions by bringing reacting molecules closer to one another and by weakening existing chemical bonds, making it easier for new ones to form.

Practically all chemical activity in cells involves enzymes, and life may not be possible without them. Interestingly, in space exploration the enzyme phosphatase, common to most forms of life on this planet, is considered a good indicator of the presence of life or precursors to life if detected in even trace amounts. Of the multitude of proteins built by living cells, most function as enzymes. Approximately 2000 different enzymes have been identified. Bacterial enzymes have been shown to be similar to those found in the higher forms of life.

Enzymes are large, globular, complex proteins consisting of one or more polypeptide chains that are folded to form the **active site,** a groove (Fig. 5-1) into which fits the substrate (the target molecule on which the enzyme acts). Enzymatic reactions are strictly controlled by the cell. Various factors, such as temperature and pH, affect the enzymatic reaction. Enzymes permit cells to carry out chemical reactions at enormous speeds. An enzyme-catalyzed reaction may move 100 million times faster than the uncatalyzed one.

Enzymatic activity does not depend solely on the physical and chemical relationships between the few amino acids of the active site and the substrate but requires the participation of additional nonprotein, low molecular weight, *inorganic* substances called **cofactors.** Certain ions are cofactors. For example, the magnesium ion contributes to

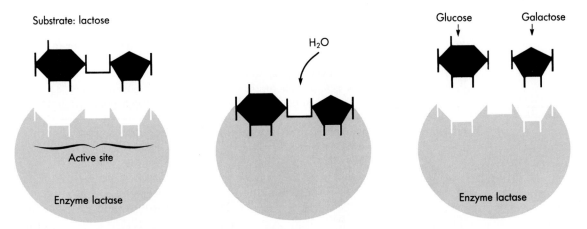

FIG. 5-1   Diagram of enzymatic action. The specificity of the enzyme lactase for the substrate lactose is shown here by the close fit of the substrate lactose at the active site of the enzyme lactase. Enzymatically, glucose and galactose are formed from the hydrolysis of lactose. Lactase is not permanently altered in the reaction but is ready to be used with the next molecule of lactose.

all enzymatic reactions involving the transfer of a phosphate group from one molecule to another.

Coenzymes. Coenzymes are also crucial to certain enzymatic reactions. These are nonprotein *organic* molecules usually containing phosphorus and a vitamin. A coenzyme may be changed during the catalytic reaction, but it reappears at the end of the reaction in its original form, that is, it is recycled. A coenzyme may serve as an electron acceptor. Within any given cell the different kinds of coenzymes present are each arranged to hold the electrons at a slightly different energy level. Examples of coenzymes are coenzyme A (CoA), NAD, and NADP.

Specific enzymes. Many important enzymes participate in metabolic processes. Table 5-1 presents a classification of enzymes in accordance with the recommendations of the International Union of Biochemistry.

Since the biochemical reactions occurring in bacteria are complicated and interrelated, bacterial enzymes may function in a group as an **enzyme system,** with several enzymes being responsible for closely related chemical changes.

Oxidation-Reduction

Oxidation-reduction reactions are very important in bacterial metabolism. In a chemical reaction, energy stored in chemical bonds is transferred to other newly formed bonds by means of electrons shifting from a lower energy level to a higher energy level as they pass from one molecule (or atom) to the next. The loss of an electron(s) is called **oxidation.** Oxygen, with its strong affinity for electrons, is most often the electron acceptor. The chemical compound losing the electron(s) is said to be *oxidized*. The gain of an electron(s) is **reduction.** By accepting the electron(s) lost by the oxidized chemical,

Table 5-1 General Categories of Enzymes

Group	Catalytic Action	Examples
Oxidoreductases	Catalyze redox reactions (hydrogen donor the substrate)	Oxidases Dehydrogenases Reductases Catalase (p. 48) Peroxidases Superoxide dismutase (p. 48)
Transferases	Catalyze reactions (other than redox) in which group with carbon, nitrogen, phosphorus, or sulfur is transferred from one substrate to another	Transferases Kinases DNA polymerases (p. 82) Reverse transcriptase (p. 695)
Hydrolases	Catalyze hydrolytic cleavage or reversal of C—O, C—N, C—C, and other bonds	Nucleases (p. 82) Lactase Urease (p. 174) Coagulase (p. 481)
Lyases	Catalyze cleavage of C—C, C—O, C—N bonds without hydrolysis or redox	Enolase Decarboxylases
Isomerases	Catalyze intramolecular rearrangements	Mutase
Ligases	Catalyze joining together of two molecules with accompanying hydrolysis of high-energy bond	DNA ligase (p. 82)

the molecule (or atom) is *reduced*. Oxidation and reduction thus always occur at the same time. This process is referred to as **redox.**

An electron often travels with a hydrogen atom (a proton). Oxidation is the removal, whereas reduction is the acquisition, of hydrogen atoms. For example, when glucose is completely oxidized, hydrogen atoms are lost from the glucose molecule and gained by oxygen. Electrons move to a lower energy level, and energy is released as follows:

$$C_6H_{12}O_6 + 6O_2 \rightarrow 6CO_2 + 6H_2O + Energy$$

$$Glucose + Oxygen \rightarrow Carbon\ dioxide + Water + Energy$$

To reverse the reaction, since electrons would be moving to a higher energy level, an input of energy is necessary.

Energy stored in particular chemical bonds in a cell can be released in small amounts, as indicated through the chemical mechanism of an ordered sequence of stepwise (consecutive) reactions, some of which are redox reactions. This sequence is referred to as a **pathway.**

In the chemical sequence each step of a pathway has evolved to obtain a small amount of energy. However, this makes available only a small amount of energy. The energy change occurring in the entire pathway may be considerable. If that entire amount were to be released all at once, it would be wasted in the form of heat in amounts that would be destructive to biologic processes.

Some metabolic pathways share steps, for example, those concerned with the synthesis of different amino acids or the various nitrogenous bases. Some converge, for example, the one by which fats are broken down to yield energy also leads to the pathway by which glucose is broken down to yield energy.

Important redox reactions in microbiology are (1) energy-capturing photosynthesis and (2) energy-releasing glycolysis (p. 67) and respiration (p. 72). The reactions of glycolysis and respiration are virtually universal for living systems.

ATP

Importance of ATP. Much of the energy required for bacterial metabolism comes from a single molecule, ATP, which serves as an energy carrier in most of the biochemical sequences in the bacterial cell. Through its formation, energy is made available to the cell. In fact, the study of energy metabolism is the study of ATP.

Chemically, ATP is adenine, the five-carbon sugar ribose, and three phosphate groups (Fig. 5-2). The three phosphate groups are covalently bonded to one another. These covalent bonds are easily broken and then release an amount of energy adequate for essential metabolic reactions. That energy arises both from the movement of bonding electrons to lower energy levels and also from a rearrangement of electrons in other orbitals of the molecule. The phosphate groups each carry negative charges and so repel

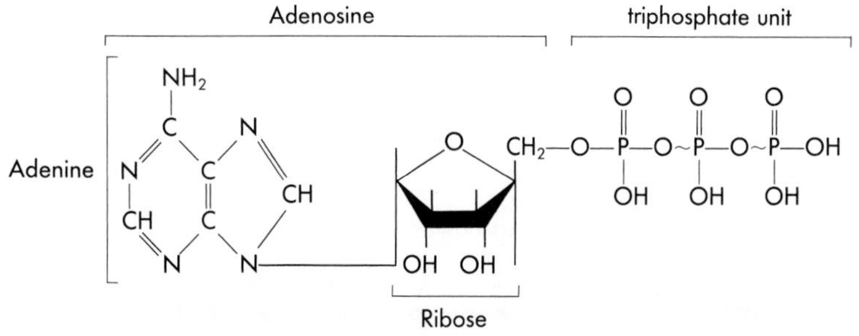

FIG. 5-2   Chemical structure of ATP. (Note that adenosine is adenine linked to ribose.)

each other. When the phosphate group is removed from a molecule such as ATP, the molecule as a whole undergoes a change in electron configuration with a net loss in energy (Fig. 5-3).

Direct hydrolysis of ATP removes its third phosphate. The reaction yields adenosine diphosphate (ADP), an inorganic phosphate (P_i), and energy, as indicated:

$$ATP + H_2O \rightarrow ADP + P_i + Energy$$

Likewise, the removal of a phosphate from ADP yields adenosine monophosphate (AMP), an inorganic phosphate, and energy:

$$ADP + H_2O \rightarrow AMP + P_i + Energy$$

In biologic sequences some of the energy from the terminal phosphate group of ATP is conserved through transfer to another molecule during the process of phosphorylation. A phosphorylated compound is thus energized and made ready for an ensuing biochemical reaction. The ATP molecule will be "recharged" with energy released from catabolism, as occurs in the breakdown of glucose. Energy must be trapped in some chemical form, or else it degrades into heat. Cells would not tolerate such a dissipation of heat.

The ATP-ADP system can be said to be a universal energy-exchange system, since it is an important part of energy-releasing and energy-requiring reactions. With proper enzymes present these compounds can be used as energy sources for biosynthesis of complex organic molecules essential to the bacterial cell, for movement, and for transport of substances across the cell membrane. In biosynthetic reactions, ATP reacts with precursors to give intermediates that form the new carbon-carbon and carbon-nitrogen bonds of the building blocks of the macromolecules.

Generation of ATP. Two important mechanisms for generation of ATP in nonphotosynthetic microbes are (1) substrate (level) phosphorylation and (2) oxidative phosphorylation. In both instances bond energy from a metabolic intermediate is used to elaborate a molecule of ATP from ADP and inorganic phosphate.

In **substrate (level) phosphorylation** a part of the energy that is released is initially conserved in energy-rich compounds formed in dehydrogenase reactions, and then it is later transferred to the ATP-ADP system. Substrate phosphorylation makes up the two steps in the Embden-Meyerhof-Parnas glycolytic pathway (p. 67), in which an energy-rich phosphate bond is created and transferred to ADP.

For a discussion of *oxidative phosphorylation*, see p. 71.

Generation of reducing power. ATP and reduced pyridine nucleotides (the important example, NAD) are the currency of chemical energy, since all pathways of energy generation lead *to* ATP and a reduced pyridine nucleotide, and all pathways of energy utilization lead *from* ATP and the reduced nucleotide. Many of the oxidative steps of catabolism in which ATP is generated are coupled with the reduction of NAD or sometimes of NADP.

NAD. The source of electrons for many biosynthetic reactions, NAD is chemically two ribose units (five-carbon sugars) linked by two phosphate groups. One of the sugars is attached to the nitrogenous base adenine; the other ribose, to the nitrogenous base

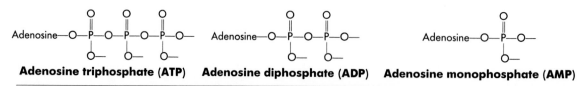

FIG. 5-3   Chemical structures of ATP, ADP, and AMP.

nicotinamide. The nicotinamide ring is the part of the molecule that accepts electrons (Fig. 5-4).

Like other coenzymes, NAD is recycled, that is, NAD$^+$ (oxidized) is regenerated when NADH (reduced) passes its electrons to another electron acceptor. It is the coenzyme used in most reactions in which energy of substrate oxidation is conserved for subsequent ATP synthesis.

Sources of Energy

The plant cell possessing chlorophyl obtains energy from the environment in the form of light (photosynthesis), but microbes with no chlorophyl must gain their energy from the chemical alteration of the substances at hand, a process termed **chemosynthesis.** The microbes are **chemotrophs.**

Chemolithotrophs, or *chemosynthetic autotrophs,* can grow in a simple medium of water, inorganic salts, and carbon dioxide. They can use carbon dioxide as their sole source of carbon, and they can gain energy from the inorganic substances present, such as ferrous compounds, reduced sulfur compounds (sulfide and thiosulfate), ammonia and nitrite, and hydrogen, through the redox reactions involving the inorganic electron donors: hydrogen, hydrogen sulfide, sulfur, and ammonia.

Chemoorganotrophs, or *heterotrophs,* require organic compounds, often complex, as energy sources (electron donors). They cannot use carbon dioxide as their sole source of carbon but require the carbon to be supplied in an organic form. For the heterotrophs, a part of the organic compound that serves as an energy source invariably provides building blocks, since in many instances the pathways used for energy generation channel the required small molecules into biosynthetic reactions.

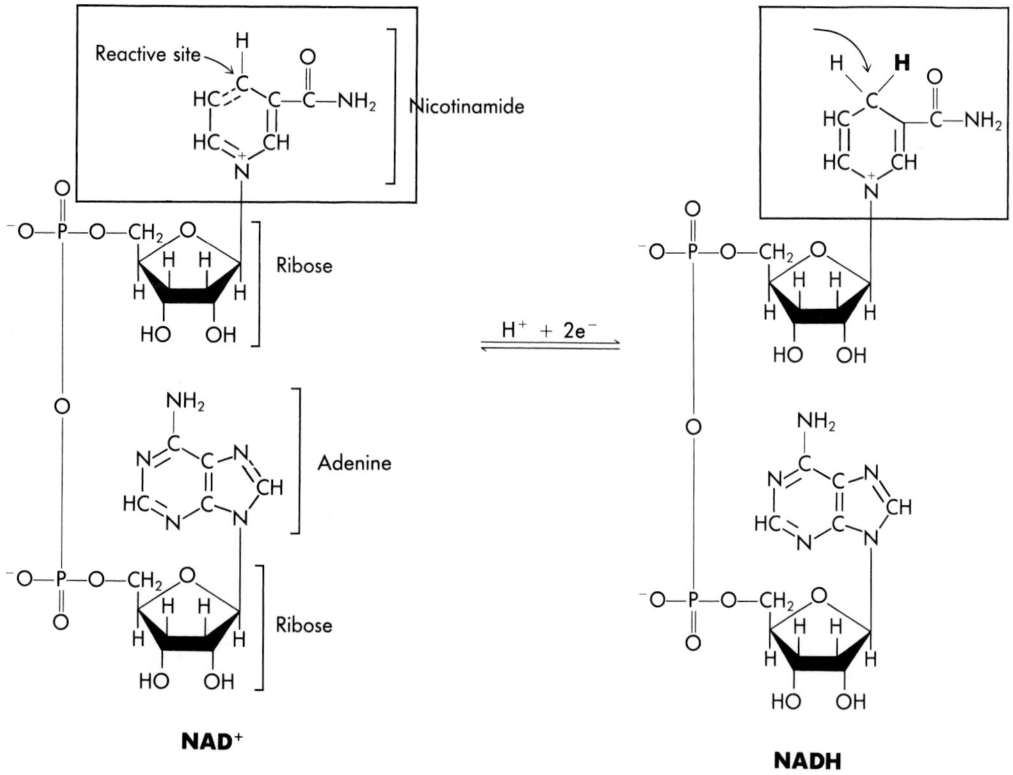

FIG. 5-4   Chemical structures. **A,** NAD$^+$ (NAD oxidized). **B,** NADH (NAD reduced). Nicotinamide is the vitamin niacin.

Practically all proteins, fats, carbohydrates, and organic compounds of whatever nature can serve as energy sources for the metabolically proficient heterotrophs. Because of a tremendous capacity for utilizing an almost limitless variety of organic energy sources, various microbes play an important role in cycling elements of the natural food chains (Chapter 38).

Individual bacterial groups, however, do have special metabolic patterns with special metabolic requirements. Whereas many different bacteria grow on diverse carbohydrates, not all do. These then must derive their carbon and energy from other substances, such as amino acids. Some bacteria are limited enzymatically, but others are not, and over 100 substrates may be utilized by certain ones. In fact, there always seems to be a bacterium available for something extraordinary like an oil spill on the ocean. Pseudomonads are known that grow on vinyl and that seemingly thrive in standard antibacterial solutions, like the quaternary ammonium compounds or the iodophors. If the concentration is low enough, some even grow in phenol.

Most carbohydrates are found in nature, where glucose, as a free sugar, is the most common monosaccharide. Glucose is the sole sugar in cellulose, starch, and glycogen and is a component of oligosaccharides (2 to 10 hydrolyzable monosaccharides) and polysaccharides (more than 10 hydrolyzable monosaccharides). Most disease-producing bacteria grow on glucose.

The general pattern for the use of all carbohydrates is the enzymatic funneling of a few key sugar-phosphates into a few central pathways for catabolism. The enzymes of the various pathways degrade the sugar-phosphates to low molecular weight compounds, mainly pyruvate, acetate, and carbon dioxide.

The carbon source and the energy source may be the same molecule, for example, glucose, or it may be a different molecule, that is, carbon dioxide or ammonia. If the

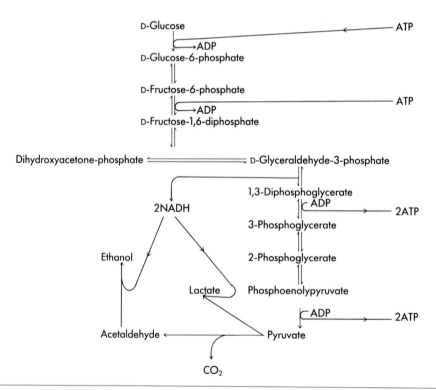

FIG. 5-5   Glycolysis in the Embden-Meyerhof-Parnas metabolic scheme. Small sequential steps, each catalyzed by a specific enzyme, characterize the biochemical process of glycolysis.

cell lacks the biosynthetic pathway, it must obtain the end-product of that pathway from the medium as a growth factor. The medium must also furnish the cell with a terminal hydrogen acceptor so that oxidations occur. For aerobes this is oxygen; for anaerobes it may be an organic compound or an organic by-product from the catabolism of the carbon source.

Glycolysis

Glycolysis is the major route of glucose degradation to pyruvic acid in living systems, *without* the intervention of molecular oxygen and *with* the concomitant production of ATP. It is a strictly anaerobic process. It serves two purposes: (1) to degrade glucose to generate ATP and (2) to provide building blocks for biosynthesis.

Glycolysis occurs in a series of 11 enzymatic reactions as presented in the Embden-Meyerhof-Parnas glycolytic pathway (Fig. 5-5). This is the most commonly occurring sequence for glucose degradation. It can be identified in two phases. Phase I covers the phosphorylation and cleavage of glucose to form glyceraldehyde-3-phosphate. Energy is extracted from the three-carbon units. Phase II covers the conversion of the three-carbon intermediate, glyceraldehyde-3-phosphate, to pyruvic acid in a series of redox reactions that are coupled to the phosphorylation of ADP. Thus the energy originally present in the glucose molecule is conserved.

In summary (Fig. 5-6), the net reaction in glycolysis is the following:

$$\text{Glucose} + 2P_i + 2ADP + 2NAD^+ \rightarrow 2 \text{ Pyruvate} + 2ATP + 2NADH + 2H^+ + 2H_2O$$

Note the two molecules of ATP generated in the process.

Aerobic Respiration

In the presence of oxygen, events following glycolysis are those of aerobic respiration, involving the stepwise oxidation of pyruvic acid (produced during glycolysis) to carbon dioxide and water. Cellular respiration commonly occurs in two sequences that harvest most of the energy contained in glucose. These are (1) the tricarboxylic acid cycle of Krebs (citric acid cycle) and (2) the terminal electron transport (respiratory) chain. These complex metabolic pathways incorporate many chemical reactions that require many enzymes.

Tricarboxylic acid cycle of Krebs. The tricarboxylic acid (TCA), or Krebs, cycle (citric acid cycle) shows how glucose is completely oxidized after the events of glycolysis. It is the most important respiratory mechanism for terminal oxidation and is a unique

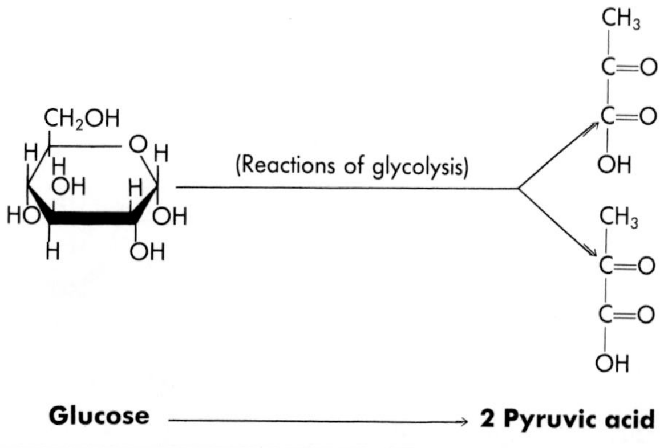

Glucose ⟶ **2 Pyruvic acid**

FIG. 5-6   The six-carbon molecule of glucose is split into 2 three-carbon molecules of pyruvic acid in glycolysis.

setup in being able to provide the cell not only with energy but also with carbon skeletons for cellular structure.

Before entering the TCA cycle, the pyruvate from the glycolytic pathway is enzymatically oxidized (in aerobic organisms) to a two-carbon compound, acetylcoenzyme A (acetyl-CoA), which enters the TCA cycle (Fig. 5-7). Any compound that can generate acetyl-CoA can be oxidized in the TCA cycle. (Lipids and amino acids can be oxidized through pathways leading to pyruvate and on to acetyl-CoA.) The linking reaction between glycolysis and the TCA cycle is as follows:

$$CH_3COCOOH \ + \ NAD^+ \ + \ CoA \rightarrow CH_3{-}\overset{\overset{\displaystyle O}{\|}}{C}{-}S{-}CoA \ + \ CO_2 \ + \ NADH$$

Pyruvic acid (Oxidized)   Acetyl-CoA   (Electron carrier
to electron
transport chain)

In the first reaction of the TCA cycle the acetyl group of acetyl-CoA (two carbons) condensed with oxaloacetate (four carbons) forms citric acid (six carbons) (Fig. 5-8). In one turn of the cycle, so to speak, the chemical reactions effect the decarboxylation and oxidation of citric acid to regenerate the four-carbon oxaloacetate and release two carbon atoms in carbon dixoide molecules. At the same time four pairs of electrons have been enzymatically extracted from metabolic intermediates. A complete cycle consumes one acetyl group and regenerates a molecule of oxaloacetic acid, which is then available for the next turn of the cycle.

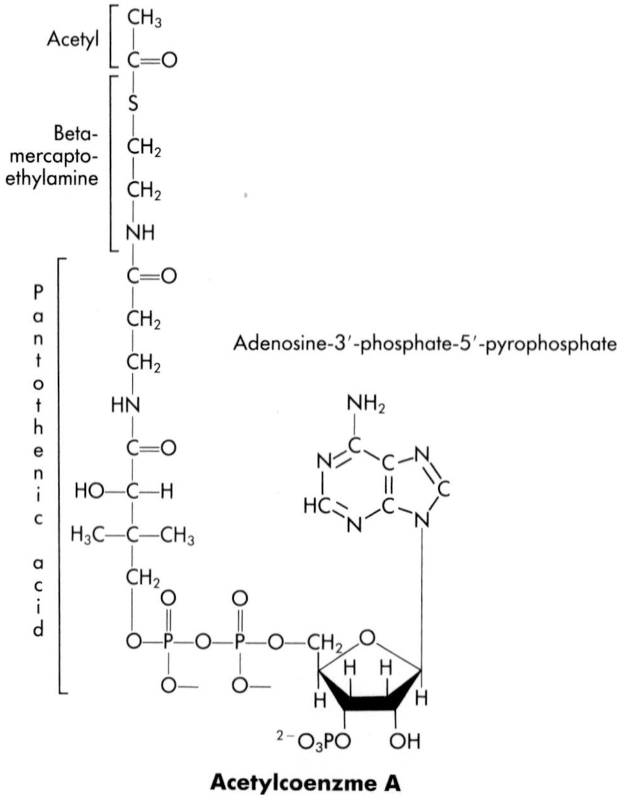

Acetylcoenzme A

FIG. 5-7   Chemical structure of acetylcoenzyme A (acetyl-CoA or acetyl-SCoA). This compound has a nucleotide and pantothenic acid (one of the B complex vitamins) in its makeup.

One molecule of ADP per cycle is converted to ATP by energy released from the oxidation of the carbon-hydrogen and carbon-carbon bonds, and three molecules of NADH (reduced) are produced from three molecules of NAD^+ (oxidized). Electrons and protons removed in the oxidation of carbon are all accepted by NAD^+ or FAD (Fig. 5-9), the other electron carrier (one molecule of $FADH_2$ per cycle). Note the equation:

$$\text{Oxaloacetic acid} + \text{Acetyl-CoA} + \text{ADP} + P_i + 3\text{NAD} + \text{FAD} \rightarrow$$
$$\text{Oxaloacetic acid} + 2CO_2 + \text{CoA} + \text{ATP} + 3\text{NAD} + \text{FADH}_2 + 3H^+ + H_2O$$

Note that the second molecule of oxaloacetic acid is not the exact same molecule as the oxaloacetic acid of the first part of the equation, but another one.

For the two molecules of pyruvate acid produced during glycolysis to be utilized, two complete turns of the TCA cycle are necessary.

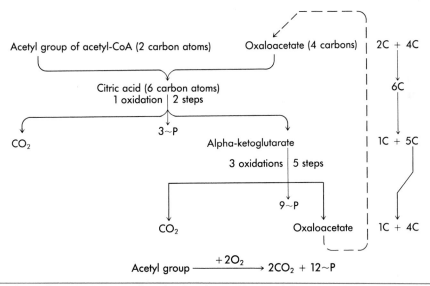

FIG. 5-8   The TCA (citric acid) cycle of Krebs in abbreviated form. It is the sequence of reactions by which acetyl groups are completely oxidized in a way that will cause the formation of ATP. Electrons are removed from acetyl groups and fed into the oxidative phosphorylation pathway with the ultimate transfer to molecular oxygen.

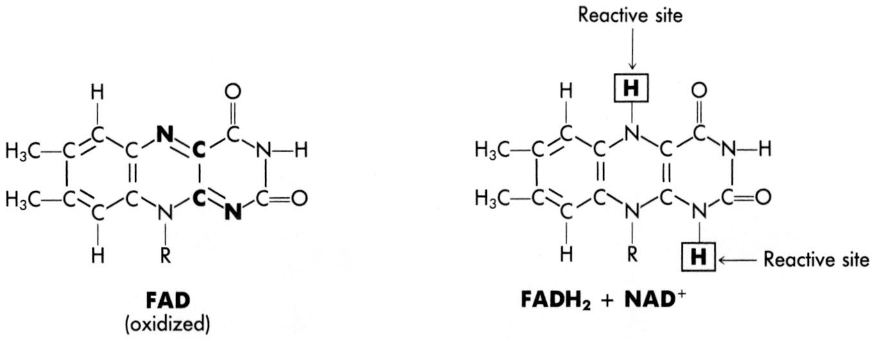

FIG. 5-9   Chemical structures of the reactive sites of FAD and $FADH_2$. Flavin adenine dinucleotide (FAD), a flavoprotein and derivative of riboflavin (vitamin B_2), is a two-electron acceptor. $FADH_2$ is the reduced form.

Electron Transport System

The end-stage of respiration takes place in the bacterial inner membrane and is characterized by an intricate series of electron transfer reactions that occur along with oxidative phosphorylation (to be discussed later).

With each turn of the TCA cycle, electrons in the electron carriers NADH and $FADH_2$ are at high energy levels. In the plasma membrane of a bacterium, NADH and $FADH_2$ are oxidized, and their high–energy level electrons are moved biochemically to successively lower energy levels by passage through a stepwise series of intermediate carrier molecules. These are membrane-bound, chemical compounds that undergo freely reversible redox reactions. The last such carrier participates in a reaction that reduces molecular oxygen to water. This active series of carriers, precisely oriented in the bacterial membrane, is spoken of as an *electron transport chain*.

Important components of the electron transport system are cytochromes (Fig. 5-10),

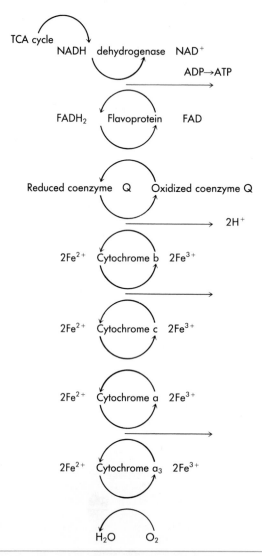

FIG. 5-10   The electron transport system (respiratory chain). Coenzyme Q (ubiquinone) can be reversibly reduced; it functions as a redox agent in prokaryotic electron transport chains. The iron (Fe) atom in the cytochrome combines with the electrons ($Fe^{3+} \rightarrow Fe^{2+}$).

electron transport proteins that contain iron. The protein makeup of the individual cytochromes in a series varies enough to make it possible for these compounds to hold electrons at different energy levels. In each one its iron atom alternately accepts and then releases an electron, passing it to the adjoining cytochrome, which is at a slightly lower energy level. Finally, the electrons, their energy spent, are accepted by molecular oxygen.

The structure for electron transport in bacteria, although comparable to that found in mitochondria of animals, is much more complicated. Some basic types of carriers are present in both, but a wider range of chemical compounds participates and a greater diversity of processes occurs in bacteria.

Oxidative phosphorylation. Oxidative phosphorylation provides the mechanism whereby the abundance of energy liberated during the downgrade passage of electrons in the transport chain can be harnessed to drive the synthesis of ATP (Fig. 5-11). The formation of ATP in oxidative phosphorylation is both a chemical process and a transport phenomenon across a selectively permeable membrane. Bacteria sustain a gradient

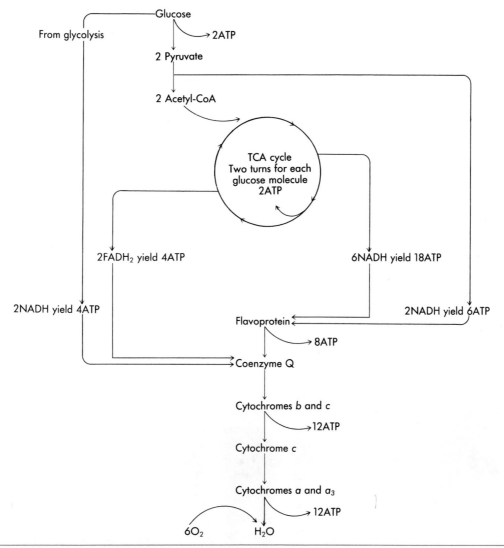

FIG. 5-11   Production of ATP from the catabolism of glucose.

of H^+ ions or protons across their cell membranes, an important fact, since the potential energy coming from this gradient can be used for the active transport of substances into the interior of the bacterium from without. A proton gradient may similarly produce phosphorylation of ADP to ATP. In oxidative phosphorylation the flow of electrons through the chemical complexity of the transport chain generates across the bacterial membrane a proton gradient sufficient to power the formation of ATP.

Biosynthesis of Macromolecules

In a microbial cell, physiologic phenomena are geared primarily for growth, that is, an orderly increase in mass or number of the major cell constituents, most of which are macromolecules. Examples are found in the protein-synthesizing groups of the cytoplasm, in the nucleic acids, in the cell wall, and in the various enzyme systems. A critical metabolic function of the cell is to assemble these macromolecules.

To perform metabolic tasks, the cell needs building blocks (subunits) and available energy. For proteins the building blocks are amino acids; for carbohydrates, simple sugars; for lipids, glycerol (or another alcohol) and fatty acids; for nucleic acids, nucleotides; and for phospholipids, substances such as choline.

Biosynthesis (the alignment of subunits) of macromolecules can occur in one of two ways. It may be either template directed or completely enzymatic. In nucleic acids and proteins, synthesis is template directed. DNA directs its own synthesis (replication) and that of RNA (transcription). Proteins are synthesized from amino acid precursors by a process called *translation* (Chapter 6). By contrast, the synthesis of carbohydrates and lipids does not require a template; polymerization of the subunits is entirely enzymatic.

Whereas some microbes by their metabolic activity can put together the full complement of subunits if provided appropriate sources of carbon, nitrogen, sulfur, and phosphorus in the medium, others cannot and must utilize subunits essential for growth that are found in their environment. When a subunit is not found in the milieu, the bacterial cell must synthesize it by a pathway from a fairly small number of key intermediates that are generated in a limited number of universal metabolic pathways. (Intermediates of metabolism are substances formed in a chemical sequence that are necessary for the formation of the end-products.)

Classes of Biologic Oxidation

In bacteria and certain other microbes, three vital biochemical processes may be demonstrated that transform chemical or radiant energy into a biologically useful form (ATP). These are **photosynthesis** (where light energy is changed to chemical energy), **respiration**, and **fermentation**. Photosynthesis, although common in plants and some algae, is mostly limited in bacteria to the cyanobacteria (blue-green bacteria).

Respiration

Respiration is the process in which the events of biologic oxidation to generate energy culminate with the ultimate and final hydrogen (electron) acceptor being molecular oxygen. It is also designated **aerobic respiration** to separate it from a process termed **anaerobic respiration** in which the ultimate hydrogen acceptor is an inorganic nitrate, sulfate, or carbonate. Anaerobic respiration is found only in bacteria. For example, the sulfate- and carbonate-reducing organisms are obligate anaerobes that rely exclusively on these compounds for energy. The denitrifying bacteria are facultative aerobes that use nitrate only when oxygen is absent.

Fermentation

If both the electron donor and the final hydrogen (electron) acceptor are organic compounds, the processes of biologic oxidation are those of fermentation. A variety of organic compounds can be fermented. Many microbes, including pathogens, live anaerobically so that they can satisfy their energy needs through the fermentation reactions.

Types of fermentation. Most bacteria grow on glucose, utilizing the mechanisms of glycolysis consistently to form pyruvate. At this point, however, metabolic processes diverge widely; some will be fermentative. In this case the different patterns of fermentation become characteristic, even specific, for certain bacteria. As a result, laboratory detection of end-products produced from these reactions is used diagnostically to identify the various microorganisms.

A number of the biochemical pathways in fermentation have been elucidated, and some, such as alcoholic fermentation, are important commercially. The equation here is the following:

$$2 \text{ Pyruvate} \rightarrow 2 \text{ Acetaldehyde} \xrightarrow[\text{(Alcohol dehydrogenase)}]{2\text{NADH}_2 \quad 2\text{NAD}^+} 2 \text{ Ethanol}$$
$$\searrow 2\text{CO}_2$$

Comparison of Aerobic Respiration with Fermentation

Let us compare aerobic respiration with fermentation using microbial metabolism of glucose, since most medically important microbes grow on glucose. In the presence of molecular oxygen, aerobic respiration takes place in all animal cells and in microbes, and glucose is changed in a biochemical sequence to carbon dioxide and water. In the absence of oxygen, microbes ferment the glucose but with different end-products. Among bacteria, incomplete oxidation of this nature is the rule, with a wide variety of end-products accumulating.

Respiration is a much more efficient process than fermentation. Many more intermediate steps mean much more energy can be generated. The useful energy released from complete oxidation of glucose is far greater than that from its conversion to alcohol and carbon dioxide. For instance, aerobic respiration yields a net of 38 molecules of ATP (the energy source), compared with only two for each mole of glucose fermented:

$$C_6H_{12}O_6 + 6O_2 + 38ADP + 38P_i \rightarrow 6CO_2 + 6H_2O + 38ATP$$
$$\text{Glucose } + \text{ Oxygen} \qquad\qquad \text{Carbon } + \text{ Water}$$
$$\text{dioxide}$$

Anaerobic metabolism is restricted in the amount of energy that can be derived from the limited number of substrates serving as energy sources, but it can provide energy quickly and continuously at a high rate as long as the fermentable substance is present.

Microbial fermentation reactions are of great value to humans because of the many by-products of practical worth to home and industry (Chapter 38).

An Event Unique to Bacteria: Cell Wall Synthesis

The cell wall peptidoglycan may be thought of biochemically as a single, enormous, cagelike macromolecule (Fig. 5-12). It is made up of long, parallel polysaccharide chains cross-linked to each other at intervals by short polypeptide chains. The polypeptide chains vary according to the species of bacteria. The synthesis of peptidoglycan in a biochemical pathway represents a metabolic event that is unique to bacteria.

Elaboration of peptidoglycan may be outlined as follows:

1. The sugar-peptide unit is synthesized within the bacterial cytoplasm. The soluble precursors are formed, and the enzymatic, stepwise process begins.
2. Strands of peptidoglycan are formed. Within the bacterial cell membrane, the sugar-peptide unit is transferred to a membrane-bound carrier lipid. Here the sugar-peptide unit is polymerized into strands of peptidoglycan that are not linked at this step.
3. The final step is the biochemical cross-linking of the peptidoglycan strands to form the baglike molecule. This occurs outside the cell membrane at the site of the preexisting cell wall.

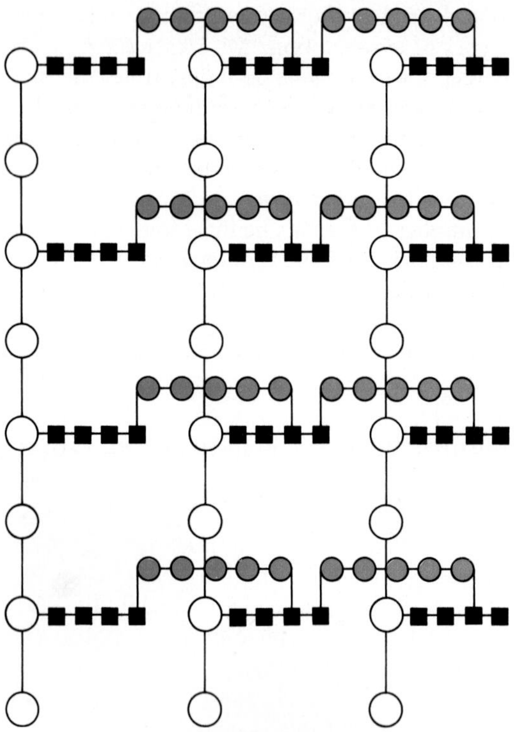

FIG. 5-12   Diagram of peptidoglycan macromolecule (so-called because it is composed of peptide and sugar units), with long parallel sugar chains extensively cross-linked to each other by short peptide chains. The result is a strong, baglike molecule. Sugars are open circles, tetrapeptides are squares, and pentaglycine bridges are colored circles.

Knowledge of the events in this particular pathway is important because certain spots in the sequence provide targets for selective and precise antibiotic action. For instance, penicillin is known to inhibit the enzymatic cross-linkage of the peptidoglycan strands. Because the integrity of the cell wall is crucial to the integrity of the bacterium, this vulnerability of the cell wall has far-reaching implications in the treatment of bacterial disease, notably with the antibiotic compounds.

OTHER EFFECTS OF BACTERIAL ACTIVITY
Pigment Production

So far as is known, pigment production has no relation to disease and no significance other than in identification of pigment-producing organisms. Yellow pigments are most prevalent, but pigments of almost any color may occur. The red and yellow pigments probably belong to the same chemical group as those in turnips, egg yolk, and fruits (carotenoids). Pigments are produced by both parasitic and saprophytic bacteria, but most species of bacteria do not produce any pigment. Common pigment producers are *Staphylococcus aureus*, with a golden pigment, and *Pseudomonas aeruginosa*, with a bluish green pigment. *Serratia marcescens* produces a red pigment seen in high-salt environments: the red salt bacteria in the Dead Sea belong to the species *Halobacterium halobium*. Pigment-producing, or **chromogenic**, bacteria lose that property when grown under unfavorable conditions.

Heat

During the growth of all bacteria, heat is generated but in such small amounts that it can be detected in ordinary cultures only by the most delicate methods. The heating of damp hay or compost in part results from bacterial action.

Light

Some species of bacteria have an extraordinary ability (a curious manifestation of bacterial respiration) to produce light without emitting heat. This process is **bioluminescence,** which results when the flavoprotein luciferin is oxidized by the action of the enzyme luciferase. Most of these bacteria live in salt water where they occupy an unusual symbiotic position, being harbored in specialized light organs of certain marine animals, including fish and squid. Two genera of marine luminescent bacteria (both symbionts) are *Beneckea* and *Photobacterium* (neither pathogenic).

Fox fire, seen on decaying organic material especially in the woods, is from luminescent bacteria. Most luminescent bacteria emit a blue-green light; some emit a yellow light.

Odors

Certain odors are characteristic of some species of bacteria; some arise from the decomposition of the material on which the bacteria are growing.

SUMMARY

1 Bacteria, being easily manipulated, provide a useful tool for the investigation of metabolism in more complex organisms. In 2 days, much of bacterial physiology can be observed; similar information from humans would require 200 years or more.

2 The bacterial cell requires from its milieu the chemical substances with which it can maintain viability and the energy necessary to do so. Metabolism encompasses all the biochemical reactions whereby the bacterium meets these basic needs.

3 Energy in the cell must be trapped chemically. Energy degraded as heat would be intolerable to the cell. Each step of a metabolic pathway makes available only a small amount, although the overall energy change in the pathway may be sizable.

4 Enzymes (proteins) as metabolic catalysts increase the rate of chemical reaction by bringing reacting molecules closer together and by weakening their chemical bonds. Practically all biochemical activity involves enzymes.

5 The study of energy metabolism is the study of ATP, the single molecule that carries the energy in most biochemical sequences in the bacterial cell.

6 In many of the oxidative steps of catabolism, the generation of ATP is coupled with the reduction of NAD. NAD is the source of electrons for many biosynthetic reactions, and like other coenzymes, it is recycled.

7 Various microbes possess a tremendous capacity for using an almost limitless variety of organic energy sources, hence their role in the elemental cycles in nature.

8 Glycolysis (a strictly anaerobic process) is the major route in living systems for degradation of glucose (the most common monosaccharide) to pyruvic acid without the intervention of molecular oxygen and with the concomitant production of ATP. The Embden-Meyerhof-Parnas glycolytic pathway is the chemical sequence most often present.

9 Most bacteria grow on glucose, consistently utilizing mechanisms of glycolysis to form pyruvate. At this point metabolic processes diverge.

10 In the presence of oxygen, events after glycolysis are those of aerobic respiration, involving the stepwise oxidation of pyruvic acid to carbon dioxide and water. The

two complex metabolic sequences harvesting most of the energy contained in glucose are (1) the tricarboxylic acid cycle of Krebs and (2) the terminal electron transport chain.

11 The Krebs cycle provides the cell with carbon skeletons for cellular structure as well as with energy.

12 A chain of electron transfer reactions placed in the bacterial plasma membrane marks the end stage of respiration. In bacteria the process is like that in the mitochondria of eukaryotes but much more complicated.

13 When the events of biologic oxidation culminate with the final electron acceptor being molecular oxygen, the process is aerobic respiration. When the ultimate hydrogen (electron) acceptor is an *in*organic compound, the process, found only in bacteria, is anaerobic respiration. When the final hydrogen (electron) acceptor is an organic compound, it is fermentation.

14 Biosynthesis of macromolecules may be either template directed, as for nucleic acids and proteins, or completely enzymatic, as for carbohydrates and lipids where subunits are polymerized.

15 Many anaerobic microbes satisfy their energy needs through fermentation reactions. Different patterns characterize and even specify these microbes. Such fermentation reactions are the source of many valuable by-products.

QUESTIONS FOR REVIEW

1 Briefly tell what is meant by catalyst, peptide, macromolecule, bioluminescence, chromogenic, fox fire, redox, coenzyme, cofactor, subunit, biosynthesis, intermediate of metabolism, metabolic pathway, electron acceptor, P_i, electron, proton, polymer.

2 List the salient features of enzymes. Name five specific enzymes.

3 Explain oxidation-reduction reactions. What is their importance? List two important redox reactions in microbiology.

4 Give an example of heat production by bacteria.

5 Comment on pigment and light production by bacteria.

6 Define metabolism, photosynthesis, aerobic respiration, anaerobic respiration, and fermentation.

7 What two basic needs must be satisfied by the bacterial cell in its environment?

8 Which is a more efficient process—respiration or fermentation? Why?

9 Discuss the sources of energy for bacteria.

10 Write the summary equation for glycolysis.

11 What is the purpose of the electron transport chain? Where is it located in bacteria? In animal cells?

12 State the importance of the biosynthesis of peptidoglycan. Briefly outline the sequence of events leading to its formation.

13 What is the difference between NAD^+ and NADH? Between FAD and $FADH_2$?

14 What is ATP? Its function in metabolism? Its importance?

15 What are two basic mechanisms for the formation of ATP?

16 State the role of the proton gradient in oxidative phosphorylation.

SUGGESTED READINGS

Baron, S., editor: Medical microbiology, Menlo Park, Calif., 1982, Addison-Wesley Publishing Co., Inc.

Haber, C.L., and others: Methylotrophic bacteria: biochemical diversity and genetics, Science 221:1147, 1983.

Hallas, L.E., and others: Methylation of tin by estuarine microorganisms, Science 215:1505, 1982.

Joklik, W.K., and others, editors: Zinsser microbiology, ed. 17, New York, 1980, Appleton-Century-Crofts.

Leisman, G., and others: Bacterial origin of luminescence in marine animals, Science 208:1271, 1980.

Recommendations of the Nomenclature Committee of the International Union of Biochemistry on the nomenclature and classification of enzymes: Enzyme nomenclature 1978, New York, 1979, Academic Press, Inc.

Rogers, H.J., Perkins, H.R., and Ward, J.B.: Microbial cell walls and membranes, London, 1980, Chapman & Hall.

Smith, E.L., and others: Principles of biochemistry: general aspects, ed. 7, New York, 1983, McGraw-Hill Book Co.

Stryer, L.: Biochemistry, ed. 2, San Francisco, 1981, W.H. Freeman & Co.

BIOLOGIC ATTRIBUTES OF BACTERIA
Genetics

Genetics is the biologic science that encompasses the many facets of heredity. It deals with the characteristics and potentialities that are transmissible from one generation to another, the mechanisms of transmission, and the variation in hereditary traits that are found between ancestors and progeny.

In prokaryotes and eukaryotes, and in many viruses, the molecule of heredity is deoxyribonucleic acid (DNA). In some viruses, however, it is ribonucleic acid (RNA). The best known form of DNA, the double helix of Waston and Crick (see Fig. 3-13), is a double-stranded molecule with a diameter of 2 nm. A few viruses have single-stranded DNA for part of their life cycle; in some viruses the nucleic acid is in fragments (segmented).

The facet of genetics that is concerned with the complex biochemical structure and function, first of DNA and second of RNA, is **molecular genetics.**

Genome is the term used to refer to the complete set of hereditary factors possessed by a given organism. In bacteria it is a single DNA molecule, although other genetic elements known as plasmids are widespread in bacteria. **Genotype** refers to the genetic makeup of an individual with respect to a single trait or set of traits. **Phenotype** designates the measurable features of the organism that are expressed by its genotype in conjunction with environmental factors. These features that can be observed and qualified may be physical, biochemical, or physiologic. For instance, colonial morphology and virulence are qualities that are genetically expressed in bacteria. Phenotype may be applied to the expression of a single trait or to several traits.

Over the last few decades bacteria have proved to be especially valuable models for genetic investigation because of their ease of handling, their measurable attributes, and their rapid rate of reproduction. In a single bacterial cell relatively few genes are present in comparison with other living organisms, and the DNA is not contained within a nucleus. The prime target in the research laboratory, *Escherichia coli*, is today the most thoroughly understood organism in the world.

DNA: THE MOLECULE OF HEREDITY
General Considerations
Description of DNA (p. 40)

In organisms as diverse as viruses and human beings, DNA is the genetic material, and life on earth stems from this fact. DNA stores the genetic information that ultimately determines the biochemical nature of all macromolecules in a cell, and thus it delineates all the structural and functional aspects of the organism. Even minute details in the makeup of a macromolecule are crucial to its biologic function. Since DNA is itself a macromolecule, it includes the instructions for its own replication.

Efficient function of the genetic material involves two basic processes: (1) genetic expression, for instance, the determination of the phenotype of a given organism, and (2) replication of the genetic material.

The structure of DNA can be damaged in its environment by various physical and chemical agents. If damage is not repaired, the result can be a mutation (p. 90) or even lethal damage to the organism.

Molecular Nature of Hereditary

DNA demonstrates two basic properties that account for its efficiency as a molecule directing the storage and transmission of genetic information. The first property is that it is linear. Nucleic acids are long, polymerized chains of nucleotides. (A nucleotide is chemically a sugar and a purine or pyrimidine base combined with phosphoric acid.) The order or sequence of the nucleotide bases along the DNA molecule contains the complete set of genetic instructions, and sequence variation of side chains accommodates enormous amounts of genetic information.

The second property is the complementarity of the two strands in the DNA molecule, which allows the synthesis of nucleic acids to be carried out on a biochemical template. (A **template**, meaning pattern or mold, is neither a catalyst nor a substrate but a source of biochemical information to determine the sequence being fabricated.) As important as the complementary nature of the strands is the inverse polarity of the strands. Each nucleic acid strand has polarity, that is, the ends are not the same, because chemical bonds connect the 3' position of one nucleotide with the 5' position of the next. Each strand terminates with a free 3' position at one end, referred to as the 3' end, and a free 5' position at the other, the 5' end. The nucleotide base se-

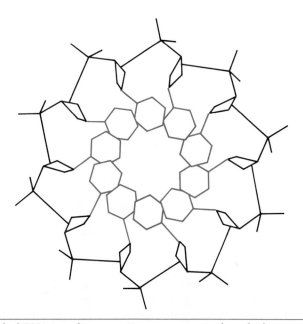

FIG. 6-1   Strand of DNA in a diagrammatic cross section to show the base sequence *(color)* inside with the sugar-phosphate backbone *(black)* outside. Note the tenfold symmetry.

quence is written in the 5'-to-3' direction. **Inverse polarity** means that the strands "run" in opposite directions; they are antiparallel.

In double-stranded nucleic acid, since the genetic information (the base sequence) is inside (Fig. 6-1), near the axis of the helix, the strands must separate for it be "read." An essential feature in the arrangement of genetic information is that it is duplicated in each strand. This redundancy allows not only replication, where each strand determines its complementary strand, but also recombination and repair of a damaged or partly obliterated strand, which can be rebuilt as a complement of the intact strand.

Replication of DNA

The key to the conservation of genetic information for an organism is the exact reproduction of the nucleotide sequence in DNA. This is replication. In this process each strand or chain of the double helix (each half-molecule of DNA) serves as a template for the synthesis of a new strand (Fig. 6-2). In the duplex that is formed, each of the parent strands is conserved because of their complementarity. Because one of the strands of each daughter DNA is newly synthesized but the other continues over from the parent molecule, replication is said to be **semiconservative** (Figs. 6-3 and 6-4).

Replication begins at a growing point on the molecule called a **replication fork** that

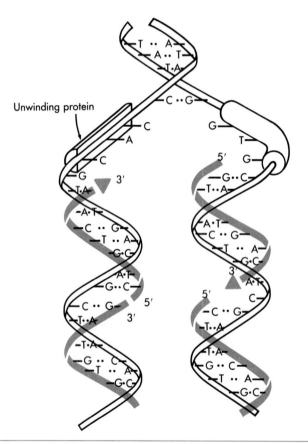

FIG. 6-2   Replication of DNA (diagrammatic simplification). The new chains are formed and grow in the 5'-to-3' direction along both parent strands. Probably with the aid of unwinding proteins, the strands of the parent molecule come apart and separate. Sections of new double helices are formed on each of the parent strands. The daughter strands are stitched together at their ends in the new double helix with DNA ligase.

moves linearly, usually in both directions, with the two strands of DNA unwinding and separating as new chains are formed.

The specificity of base pairing (one member of a base pair being a purine; the other, a pyrimidine) in the DNA molecule is critical to its replication. The structure of the sugar-phosphate backbone of each chain imposes a restriction on the base pairs. The chemical bonds that are attached to a base pair are always 10.85 Å apart. In this space

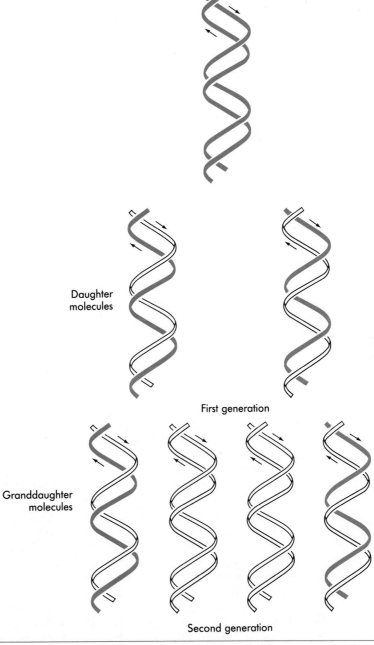

FIG. 6-3   Semiconservative replication of DNA. One strand of the parent molecule is newly synthesized; the other continues over from the parent molecule.

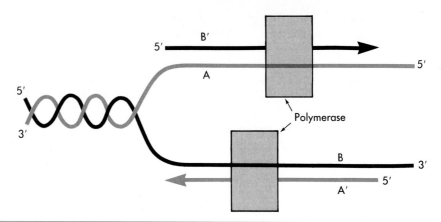

FIG. 6-4 Semiconservative replication of DNA. Complementary strands A and B have their 5′ terminals at opposite ends of the double helix. As a result of the action of DNA polymerase on A and B, new complementary strands, B′ (for strand A) and A′ (for strand B), are laid down. When replication is concluded, the full length of A and that of B will each be matched to the new complementary strand. The new pairs will twist into two molecules of a double helix (each half old and half new).

a purine-pyrimidine unit fits nicely, but not two purines, and although two pyrimidines could fit in, they would be too far apart for hydrogen bonds. The orientation of the hydrogen bonds—two between adenine and thymine and three between guanine and cytosine—is the best arrangement for a strong interaction between the bases.

If the hydrogen bonds between the paired bases are disrupted, the two chains of the double helix of DNA come apart readily. An unwinding of the helix, referred to as **denaturation** or **melting,** can be brought about by heating a solution of DNA. When the temperature is reduced to below the melting temperature, separated complementary strands spontaneously reassociate to reform the double helix. Reassociation of the DNA strands is called **renaturation** or **annealing.**

The number of guanine-cytosine (G-C) base pairs with their three hydrogen bonds is an important factor in the integrity of the DNA duplex. The greater percentage of guanine-cytosine pairs the molecule contains, the higher will be the temperature of melting. With thermal melting, unwinding of the chains begins in regions high in the number of adenine-thymine (A-T) bases and proceeds to regions of increasingly greater guanine-cytosine content. The midpoint melting temperature is determined by guanine-cytosine content.

Enzymatically replication is an intricate process for which a number of enzymatic proteins have been identified. An important one is DNA polymerase, a template-directed enzyme that catalyzes polymerization. Nucleases hydrolyze specifically DNA or RNA. Exonucleases require a free end at which to initiate enzymatic action. Endonucleases do not and therefore may attack one or many specific sites within the DNA molecule. DNA ligase catalyzes the process whereby DNA chains are joined, making possible the reassembly of DNA. Ligase activity is necessary for normal repair and also for splicing of DNA chains in genetic recombination.

Base Ratios

It was noted previously that the purine base adenine (A) always pairs with the pyrimidine base thymine (T) by two hydrogen bonds, and the purine base guanine (G) always pairs with the pyrimidine base cytosine (C) by *three* hydrogen bonds. Base ratios are used to determine the genetic relatedness among microbes. The guanine plus cytosine (G +

C) percentage of total DNA can be determined by direct chemical analysis and a value obtained, as indicated in the following equation:

$$\frac{G + C}{A + T + G + C} \times 100 = \%(G + C)$$

The G + C percent of total DNA is characteristic for each bacterial and viral strain. In bacteria tested the percent G + C (percent GC or GC ratio) ranges from 28 to 30 in the genus *Spirillum* to 70 to 80 in the genus *Mycobacterium*. The more closely related microbes are genetically, the more nearly the same are the base ratios.

The base ratio of microbial DNA may also be given as A + T/G + C. This base ratio varies from below 0.4 in some species of *Micrococcus* to over 2.5 in species of *Clostridium*.

Hybridization of Nucleic Acids

Hybridization is a technical way to determine the homology, or similarity, of two nucleic molecules. (Similar linear sequences of nucleotides are homologous.) For strands to be homologous, they must be derived from the same or closely related organisms. A stringent criterion for close genetic relatedness in microbes is the extent of DNA homology demonstrable.

In this technic a nucleic acid stretch of unknown sequence can be denatured (by heat) with the reference DNA and the two subsequently annealed. In solution the unknown specimen forms hybrid double-stranded molecules with the reference strand in homologous regions. A percentage for the degree of hybridization can be calculated, indicating the percentage relatedness.

DNA, RNA, and Protein Synthesis

Transcription

The genetic expression of DNA requires the synthesis of RNA, which is subsequently used to direct the synthesis of protein (see box below). The nucleotide sequence in DNA is the template not only for DNA replication but also for the elaboration of the complementary, single-stranded RNA molecules. The process that describes this DNA-directed synthesis of RNA is transcription (Fig. 6-5). A complex and massive enzyme, RNA polymerase, catalyzes it.

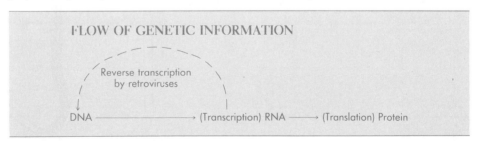

Like DNA, each RNA molecule has a 5' and a 3' end. During transcription RNA polymerase moves along the length of DNA in the 3'-to-5' direction, and nucleotides present in the cell are enzymatically added, one at a time, to the 3' end of the growing RNA chain. Thus sequences of nucleotide bases are converted to their RNA equivalent. The three different types of RNA molecules, which are important in protein synthesis, are so formed: messenger RNA (mRNA), transfer RNA (tRNA), and ribosomal RNA (rRNA) (p. 39).

The Genetic Code

The order of the nucleotides in the DNA molecule (or in its mRNA transcript) is known to determine the order of the amino acids in protein molecules through a mechanism analogous to the deciphering of a code. In the genetic code the unit pattern is three successive nucleotides in DNA (a **triplet**) that together direct the arrangement of, or code for, a specific amino acid in a given protein. The unit pattern with its information is passed from DNA to its RNA transcript. To provide for each of the 20 amino acids (the essential amino acids) required for the fabrication of proteins, a unit pattern of at least three successive nucleotides out of the four possible is necessary. A sequence of only two nucleotides could not code for more than 16 of the essential amino acids, whereas a three-nucleotide sequence (triplet) yields 64 possible combinations (Table 6-1).

FIG. 6-5   RNA transcription (diagrammatic simplification). Where RNA polymerase attaches, DNA opens up and one of its strands becomes the template for the assembly of a single-stranded RNA molecule. Hydrogen bonds between the two DNA strands reform as RNA polymerase moves along the DNA molecule pushing out the newly synthesized RNA.

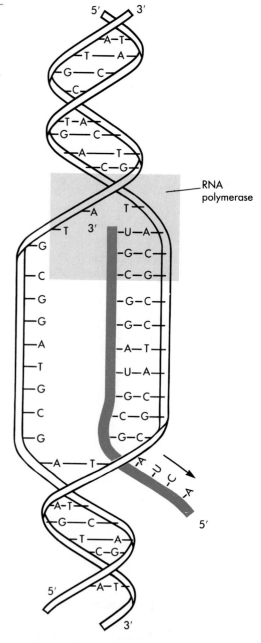

RNA polymerase

Table 6-1 Dictionary of Amino Acid Codons for mRNA* (5'-to-3' Direction)

5' End: First Letter	Second Letter U	Amino Acid	C	Amino Acid	A	Amino Acid	G	Amino Acid	3' End: Third Letter
U	UUU	Phenylalanine	UCU	Serine	UAU	Tyrosine	UGU	Cysteine	U
	UUC	Phenylalanine	UCC	Serine	UAC	Tyrosine	UGC	Cysteine	C
	UUA	Leucine	UCA	Serine	UAA	Stop	UGA	Stop	A
	UUG	Leucine	UCG	Serine	UAG	Stop	UGG	Tryptophan	G
C	CUU	Leucine	CCU	Proline	CAU	Histidine	CGU	Arginine	U
	CUC	Leucine	CCC	Proline	CAC	Histidine	CGC	Arginine	C
	CUA	Leucine	CCA	Proline	CAA	Glutamine	CGA	Arginine	A
	CUG	Leucine	CCG	Proline	CAG	Glutamine	CGG	Arginine	G
A	AUU	Isoleucine	ACU	Threonine	AAU	Asparagine	AGU	Serine	U
	AUC	Isoleucine	ACC	Threonine	AAC	Asparagine	AGC	Serine	C
	AUA	Isoleucine	ACA	Threonine	AAA	Lysine	AGA	Arginine	A
Initiation codon:	AUG	Methionine	ACG	Threonine	AAG	Lysine	AGG	Arginine	G
G	GUU	Valine	GCU	Alanine	GAU	Aspartic acid	GGU	Glycine	U
	GUC	Valine	GCC	Alanine	GAC	Aspartic acid	GGC	Glycine	C
	GUA	Valine	GCA	Alanine	GAA	Glutamic acid	GGA	Glycine	A
Initiation codon:	GUG	Valine	GCG	Alanine	GAG	Glutamic acid	GGG	Glycine	G

*Note that all the amino acids have more than one code word except for methionine and tryptophan.

In DNA and its mRNA transcript the prescribed three-nucleotide sequence is known as a **codon** (code word). Of 64 codons, 61 code for amino acids. Two—AUG and GUG—appear to be initiator, or start, codons. The start codon on the mRNA transcript through an interaction with tRNA triggers the process of polypeptide synthesis. Three codons—UAA, UAG, and UGA—are stop signals; they are also called **nonsense codons.** Each codon specifies only one amino acid and is therefore unambiguous. However, as one might guess, for most amino acids, two or more different codons exist. The amino acid glycine, for example, can be specified by GGU, GGA, GGC, or GGG. Other examples are found in Table 6-2.

The **anticodon** is the triplet of nucleotides in the adapter tRNA molecule (discussed later), which associates (by complementary base pairing) with a specific triplet (or codon) in the mRNA molecule, as indicated:

3' tRNA 5'

Anticodon

—G—U—C—

·   ·   ·

·   ·   ·

—C—A—G—

Codon

5' mRNA 3'

Translation

A much more complicated genetic process than either replication or transcription is translation, the genetic mechanism whereby RNA templates direct synthesis of proteins in a cell. The three types of RNA, as well as special enzymes, and certain soluble factors participate in this intricate process. Actually more than 100 kinds of macromolecules are involved.

The genetic instructions for protein synthesis are carried in the codons of mRNA, which align amino acids for a particular polypeptide chain of a protein. The adapter molecule is tRNA, since each one is specific for a required amino acid (Fig. 6-6). The tRNA molecule possesses several highly specific recognition sites for attachment, such as the one that holds the amino acid and the one that functions as the anticodon.

As the mRNA transcript is being made from DNA, the 5' end of the mRNA strand fastens to a ribosome (Fig. 6-7). At the point of contact of mRNA and ribosome, adapter molecules of tRNA bind temporarily to the mRNA strand, each carrying the precise amino acid called for by the mRNA codon to which the tRNA attaches. As the

Table 6-2 Number of Codons for Some Amino Acids

	Tryptophan	Lysine	Isoleucine	Alanine	Leucine
Number of codons	1	2	3	4	6
Codons	UGG	AAA	AUA	GCU	UUA
		AAG	AUC	GCC	UUG
			AUU	GCA	CUA
				GCG	CUC
					CUG
					CUU

ribosome, with its binding sites for mRNA and tRNA, moves along the mRNA, the anticodon of the tRNA recognizing the codon bonds with it. At the same time the amino acid at the other end of the tRNA links to the protein chain (Fig. 6-8). Now the tRNA is ejected as the ribosome passes on to the next codon. Another tRNA is matched to that codon. Thus in a prescribed fashion amino acids are brought into line one by one, and the polypeptide chain grows until the codon is reached that signals the end.

Kinds of Genes

Traditionally the hereditary unit has been expressed in the word *gene*. Originally a gene was defined as a biologic, self-reproducing unit, located in a fixed position, or **locus**, on a chromosome, influencing specifically the phenotype of the individual, and vulnerable to mutation (p. 90). Currently the emphasis is that a gene represents a partic-

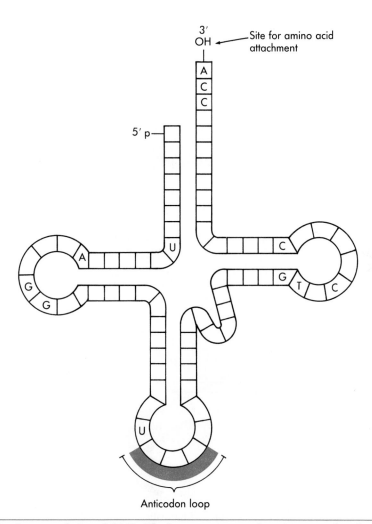

FIG. 6-6   Common design of tRNA. The base sequence can be written in a cloverleaf pattern. The 5′ end is phosphorylated; the 3′ end has a free hydroxyl group. A, Adenine; G, guanine; C, cytosine; T, thymine; U, uracil.

ular sequence of nucleotides in DNA that constitutes a functional unit of inheritance. The term *gene* referring to a particular DNA nucleotide sequence is often used in discussions of transcription and translation.

Genes are designated according to function. Genes that code for polypeptides are **structural** genes. A gene that serves as a starting point for transcription is an **operator** gene. (Operator genes do not specify products.) **Regulator** genes synthesize **repressor** proteins, which interact with the operator gene to block transcription. Genes controlling a series of related biochemical events often are adjacent to each other in a stretch of the DNA molecule referred to as an **operon**. An example is found in the lac operon of *E. coli*. In this microorganism the utilization of lactose as an energy source requires the production of the enzyme beta-galactosidase. When lactose is present, transcription of the structural gene for beta-galactosidase is induced. An interaction of lactose in the bacterial cell with a repressor molecule prevents the repressor from interacting with the operator site (lac operon) on the DNA molecule of *E. coli*.

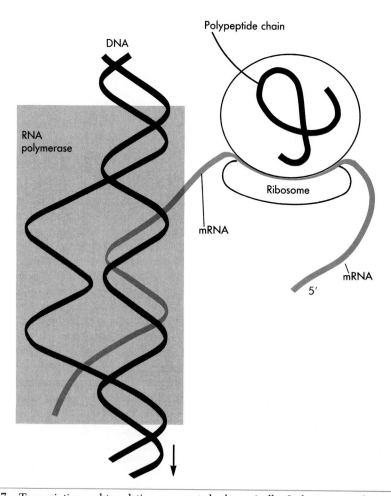

FIG. 6-7 Transcription and translation represented schematically. In bacteria a tight coupling of the two processes exists, since mRNA does not need to be transported from nucleus to cytoplasm. The ribosome can begin to translate mRNA as it is being made (transcribed) by DNA-directed RNA polymerase.

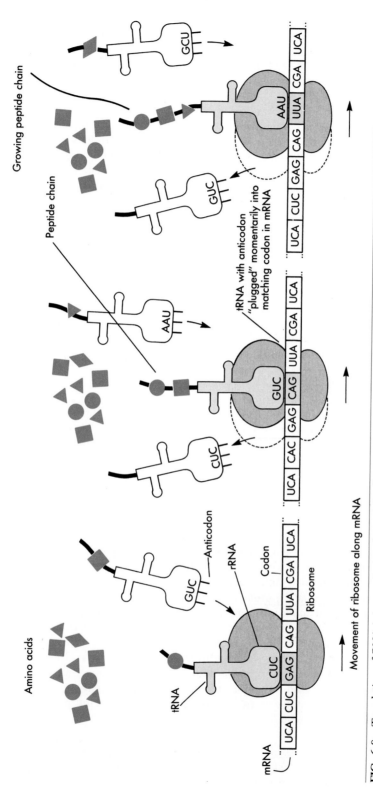

FIG. 6-8 Translation of RNA into protein in a bacterial cell, with sketch showing interactions of the three kinds of RNA—rRNA (the structural component of the ribosome), mRNA (bearer of the genetic instructions), and tRNA (the molecule that hooks an amino acid to a growing peptide chain). Each tRNA is specific to one amino acid, and each amino acid in the growing chain is specified by a codon.

GENETIC PROCESSES IN MICROBES
Mutations

Genetic factors are responsible for well-known variations in properties of bacteria, especially those that produce disease. For instance, whether given bacteria possess a capsule, a measurable property that is genetically controlled, relates directly to their capacity to cause disease in a susceptible host.

Definition of Mutation

A mutation is a permanent, heritable alteration in the genome (genetic apparatus), usually occurring in a single gene. More precisely, a mutation is a heritable change in the nucleotide sequence of DNA. (Remember that alteration of the DNA of the genome is reflected in mRNA and hence in the synthesized protein.)

Mutations are usually identified because of an effect on the phenotype of the individual, although some are **silent,** that is, they have no observable effect on the properties of the organism. For example, mutations are detected in bacteria as a change in nutritional requirements, morphology, susceptibility to antibiotics, or susceptibility to bacteriophages. The individual (bacterial) cell exhibiting or expressing the mutation is the **mutant.**

Mutations in bacteria are analogous to those in higher forms of life (eukaryotes). Since the nature of the genetic apparatus in bacteria is not different from that in other biologic organisms, bacteria have proved to be practical tools for the scientific investigation of heredity and have so provided many valuable correlations. As Koshland said,[*] "A modern molecular biologist might paraphrase the poet Pope by saying, 'The proper study of mankind is the bacterium.'"

Nature of Mutations

Since a specific intracellular enzyme (protein) is regulated biochemically by a specific gene encoded in DNA, the derived structure and function in the cell depend on the integrity of that gene. Any change in the genetic material, such as a loss (deletion), a gain (addition or translocation), or an exchange (transduction), projects an effect on the cell and constitutes the mutation. Rearrangement of the nucleotide sequence of the gene can result from an error in replication or from a breakage of the sugar-phosphate backbone of the DNA molecule. Such a change is often sudden.

Point mutations. Biochemically, the most common mutational change in the gene is a point mutation. Point mutations trace back to a change in a single base or base pair, probably the result of an error in DNA replication or repair or the effect of a mutagen.

Several types of point mutations can occur. In **substitution** mutations one base pair replaces another (or several pairs) in the double helix. A substitution resulting in the replacement of one amino acid by another in a polypeptide chain can alter the properties or function of the protein to a surprising degree.

Frame-shift mutations (phase-shift mutations). These include insertions or deletions of a single nucleotide (or sequences of many, even up to 1000 or more) that alter the reading frame. A gain or a loss of a single nucleotide resulting in an inappropriate sequence of nucleotides shifts the reading frame (Fig. 6-9). During translation the message is read correctly up to the point of loss or addition, but, since the message will continue to be read in triplets, all subsequent codons will necessarily specify the wrong amino acids. Frame-shift mutations often result in proteins that are nonfunctional; thus these mutations tend to be lethal.

---

[*]From Koshland, D.E., Jr.: A response regulator model in a simple sensory system, Science **196:**1055, 1977.

Kinds

Mutations may arise spontaneously or be induced by agents, designated **mutagens,** that directly or indirectly cause alterations in the structure of the DNA molecule.

Spontaneous mutations. Since the machinery of DNA replication is remarkably constant, mutations are relatively rare. When considering bacterial mutation, one must think in terms of populations instead of single cells. A liquid culture can be expected to contain 1 billion bacterial cells per milliliter, and a colony on a solid medium contains from 10 million to 100 million cells. The mutation rate for any given bacterial gene is from 1 in 100,000 to 1 in 1 billion cell divisions. Therefore mutants may well be present in any average population of bacterial cells. Since it is impossible to pinpoint single mutant cells, selective means must be used to isolate them.

Repair mechanisms within the bacterial cell exist to correct potential mutations; therefore the frequency of spontaneous mutation is probably low.

Induced mutations. Two classes of carefully studied mutagens are the physical and chemical agents. Both groups usually exert their mutagenic effect by promoting errors in DNA replication or repair.

Two important physical agents are ultraviolet light and ionizing radiation.

Chemical mutagens are placed in several groups: (1) chemical agents that slip in between pyrimidines or purines to alter base pairs; these include acridine orange, proflavine, and other acridine dyes; (2) agents that interact with DNA and its secondary structure to distort the helix; these include nitrogen and sulfur mustards; and (3) base analogs that, when incorporated into DNA, cause replication errors; 5-bromouracil, an analog of thymine, is the prime example. An **analog** is a chemical with a structure

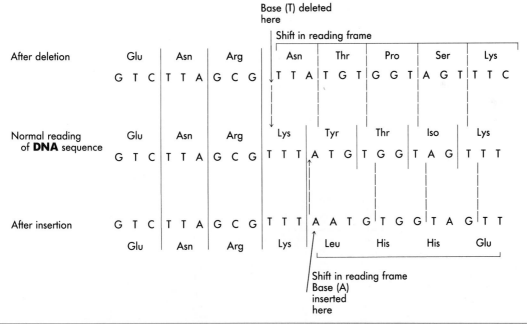

FIG. 6-9 Frame-shift mutations resulting from a single deletion or insertion of a base. The amino acid sequence is completely rearranged beginning with the triplet in which the deletion or insertion occurs. As a consequence, the genetic information intended for a given protein is lost. Most frame-shift mutations are lethal. *Glu,* Glutamine; *Asn,* asparagine; *Arg,* arginine; *Thr,* threonine; *Pro,* proline; *Ser,* serine; *Lys,* lysine; *Tyr,* tyrosine; *Ile,* isoleucine; *His,* histidine.

similar to that of another but different with respect to a certain component and therefore with either a similar or opposite metabolic action.

Interestingly, most mutagens are also cancer producing (carcinogenic) (see box below).

Role of Bacteriophages

Value in Genetic Studies

Since viruses are known to be obligate intracellular parasites, they are often discussed in terms of their host, for example, animal viruses, plant viruses, and bacterial viruses. Given the name of *bacteriophage*, or bacteria-eater (Greek *phagein*, to eat), because of their remarkable effects on bacteria, bacterial viruses have proved especially valuable in genetic studies. Viruses infecting many strains of bacteria have been identified.

Phages are very useful in various kinds of investigative studies both because of their high order of host specificity and because their life cycle is an intricate succession of amazing genetic events. Briefly, the life cycle begins as the viral particle attaches to specific receptors on the cell surface of its specific host (Fig. 6-10). From this vantage point the phage sends its nucleic acid (DNA or RNA) directly into the interior of the host cell. Inside the bacterial cell the viral genome induces the formation of many copies of itself, institutes the buildup of phage-specific proteins, and finally triggers the assembly of these components into many new viruses. When these are released, the host cell is usually lysed.

Kinds

Two major groups of bacteriophages are (1) virulent phages and (2) temperate phages. As would be expected, virulent phages destroy their hosts. Temperate phages are not necessarily destructive (lytic) but can set up within their hosts an ongoing type of infection known as **lysogeny** (Fig. 6-11). The events of their more complex life story are of considerable interest to geneticists. In lysogeny the phage genome within the bacterial cell replicates as a hereditary unit called a **prophage,** and the affected bacterium is said to be *lysogenic*. For example, the prophage of bacteriophage lambda in *E. coli* integrates into the bacterial genome at a specific site and replicates with it, whereas the prophage of bacteriophage P_1 in *E. coli* replicates independently as a plasmid (as discussed later).

Bruce Ames and colleagues from the University of California at Berkeley devised the *Salmonella* mutagenicity test to screen substances from the environment and industry for mutagenicity. The test is applied to the screening of substances for carcinogenicity, since a mutagenic agent has a very high probability of also being carcinogenic and agents that are not carcinogenic are overwhelmingly nonmutagenic.

A microorganism with a genetic defect is required for the test. The one chosen is a mutant of *Salmonella typhimurium* that needs the essential amino acid histidine but cannot make it. However, back-mutation, that is, the return of the capacity to synthesize histidine, may occur, and such back-mutants (revertants) are easily detected, since they grow on a histidine-free medium that contains ammonia as a nitrogen source. One can add a filter paper disk containing the test mutagen to a Petri plate heavily inoculated with 10^9 bacteria and evaluate the ring of colonies around the disk. Since the rate of back-mutation is measurably enhanced by a mutagenic agent, it is possible to determine the relative mutagenicity (and carcinogenicity) of the test substance.

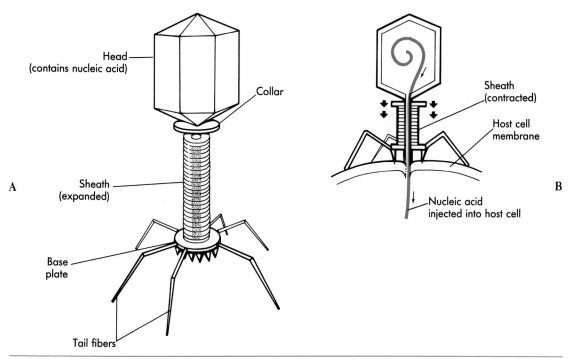

FIG. 6-10   A, Structural model of bacteriophage (p. 691). **B,** Model of bacteriophage to indicate changes with injection of nucleic acid into host cell (one stage in phage life cycle).

Although lysogenic bacteria harbor the genomes of temperate phages as prophages, there are no infectious particles present as such.

Under certain circumstances the DNA of the bacteriophage resident in the lysogenic cell may be induced to replicate and enter the lytic cycle. The term *lysogenic* indicates the potential of the affected cell to lyse should this happen.

Phage Conversion

An important consequence of phage infection is that some temperate phages contain genes that determine bacterial properties. *Phage conversion* is the term used for the phenotypic expression of bacteriophage genes. In medical microbiology, phage conversion is known to be responsible for the production of two poisons directly related to a particular disease—diphtheria toxin in diphtheria and erythrogenic toxin in scarlet fever.

Intermicrobial Gene Transfer Systems

Variations resulting from the passage of DNA from one bacterium to another are easily recognized in alterations of specific properties. Important ones that can be altered in this way include environmental tolerance (ability to survive environmental changes), antigenicity (ability to evade host defense mechanisms, p. 295), toxigenicity (ability to produce toxins, p. 298), pathogenicity (ability to produce disease by mechanisms such as capsule formation, p. 296), and antibiotic susceptibility (ability to induce antibiotic tolerance or resistance, p. 242).

Historical Background

If two bacterial strains, variants in a single species, are incubated together under special circumstances, evidence of genetic transfer is borne out by the appearance in the culture of new organisms displaying qualities of both original strains. The classic example of intermicrobial transfer of genetic material is the experiment performed by the British pathologist Frederick Griffith in 1928 (p. 22).

Mechanisms for Gene Transfer

Three different mechanisms exist by which genes may be passed from one bacterial cell to another (from a donor to a recipient). They are (1) uptake of pieces of DNA (transformation), (2) infection by a virus (transduction), and (3) mating between cells in direct contact (conjugation).

These mechanisms are not necessarily shared by all bacterial species; some bacteria have two or three, some only one. Genes are passed in bacteria most efficiently when participants are of the same species, but at times transfers can cross species barriers.

The various means of gene transfer are at work at all times. Ample opportunities exist in the environment for bacteria in proximity to exchange genetic information.

Transformation. The uptake of naked DNA, or transformation, is a process of direct transfer of nucleic acid from one bacterial cell (donor) to another (recipient). Certain species of bacteria release DNA into the medium in which they are growing, for ex-

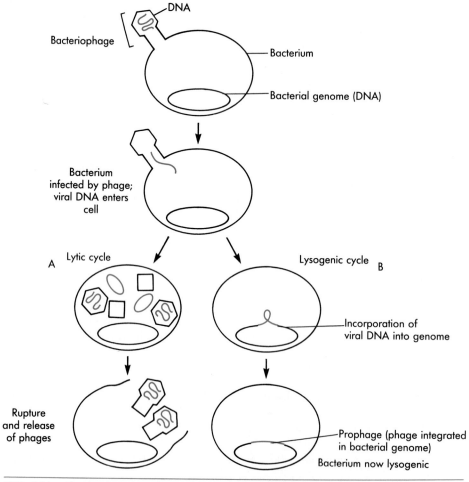

FIG. 6-11 Sketch of infection of a bacterium with a temperate bacteriophage. Viral DNA passed into the bacterium may either set up a lytic cycle (A), destroying the bacterium, or it may initiate a lysogenic cycle (B), in which case it becomes part of the bacterial genome, replicating with it and being passed to bacterial progeny. A bacterium harboring such a virus is *lysogenic*.

ample, certain species of *Neisseria* release DNA in their extracellular slime. A small number of bacteria in a given population can pick up this DNA. If DNA is extracted from a donor cell, it fragments into several hundred pieces, of which perhaps only 10, or 2% to 5% of the total amount, can be taken up by the recipient. *Only double-stranded DNA participates in transformation, and within the recipient bacterium, one strand of that is degraded. The other strand is permanently integrated into and replicates with that cell's genome.* The net effect is that a short region of the recipient genome has been replaced with a new portion of genetic material (Fig. 6-12).

Transformation is widespread in nature. An excellent place for it is the intestinal tract of humans and animals because of the vast numbers of bacteria found there, many of which are dying. As bacteria die and are lysed, their DNA is extruded into the milieu and becomes available. However, the ubiquitous presence in biologic environments of the enzyme DNAse, which degrades DNA, undoubtedly constitutes a physiologic check on such a natural process.

Transduction. Transduction is a special form of indirect transfer of genetic material that is mediated by bacteriophages and other viruses (Fig. 6-13). The bacteriophage carries a small fragment of the genome (host DNA) from the bacterial host cell in which it was produced to the bacterium it now invades and infects. Most phage particles carry phage DNA, but an occasional one carries some host DNA. When this particle infects a cell and releases this DNA into the cell, recombination can occur (p. 99).

Transduction is observed in a wide range of bacteria: coliforms, enteric pathogens, and staphylococci. Transduction is an important method for gene transfer in nature, since phages are universal. The DNAse of the environment has very little effect in

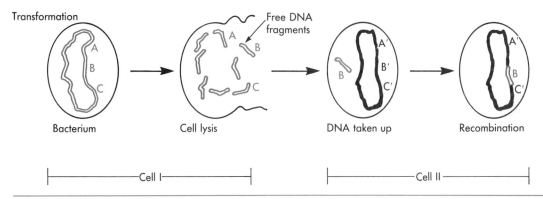

FIG. 6-12   Transformation in bacteria.

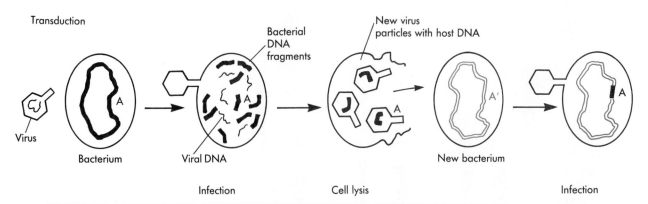

FIG. 6-13   Transduction in bacteria.

regulating the process, since the DNA is protected by the phage coat. The protection of the phage coat also means transduction is easier to perform in the laboratory than is transformation and is more reproducible.

Conjugation. Conjugation is a process effecting passage of genetic material from one bacterial cell to another that requires transient physical contact. In fact, in the electron micrograph (Fig. 6-14) two individual bacteria are seen to unite by means of a sex *pilus*, or conjugal tube, a cytoplasmic bridge between them. Male and female mating types have been defined for the closely related groups in the family Enterobacteriaceae that have been studied. In cells of *E. coli* the two inheritable mating types are called F^+ and F^-. F^+ cells serve as genetic donors (males) and possess the F plasmid (p. 98); F^- cells are recipients (females) and do not possess the F plasmid. Transfer of the plasmid to an F^- cell converts this cell to F^+ (Fig. 6-15). Only the female types must be viable for the process to occur. Conjugation can occur between any two members of the family, not just within a given genus.

The process is not, as its designation suggests, a method of reproduction but a provision for the one-way passage of genetic material. No fusion of cells or new cell results; progeny are produced in the usual way by binary fission of the parent cell. Probably the least common method of genetic exchange in bacteria, conjugation differs from both transformation and transduction in that it may result in the transfer of very large segments, sometimes even the entire genome. Conjugation requires cell contact, which the other processes do not. Unlike transformation, both transduction and conjugation may result in the transfer of autonomously replicating blocks of DNA (viruses, plasmids), which may persist as such in the progeny of the affected cell.

Importance of Gene Transfer

The systems for gene transfer in bacteria are of great medical importance. First, they demonstrate how multiple bacterial resistances to several antibiotics are accomplished rapidly in one step and indicate that the search for new antimicrobials must be continual.

Second, knowledge of the varied aspects of gene transfer has had considerable bearing on recombinant DNA research and the industrial gene-splicing technology that is so exciting today.

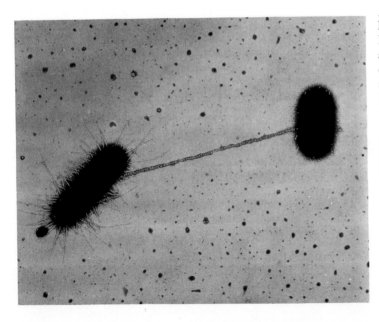

FIG. 6-14 Conjugation in bacteria. Passage of genetic material by means of a long thin filament between bacterium of one species and that of different species (both in family Enterobacteriaceae). A factor inducing resistance to certain antibiotics is known to be so transferred (see also p. 243). Note pili on surface of microbe on left. (Courtesy Dr. C.C. Brinton, Pittsburgh.)

Plasmids

General Considerations

Plasmids represent a large and exceedingly variable group of independent genetic elements ("free floating" in cell cytoplasm) that are found in practically all groups of bacteria. Highly promiscuous plasmids wander between bacteria as diverse as *Neisseria*, *Escherichia*, and *Pseudomonas* species, whereas some of the others have a limited host range. Plasmids are sometimes dubbed "wandering pieces of genetic material."

Plasmids are circular, double-stranded DNA molecules (carrying many genes) that make up to an equivalent of 1% to 3% of the mass of the bacterial genome and are independent, autonomously replicating units in the bacterial cytoplasm. Contributing nothing to the viability of the host, plasmids are not required for normal host metabolism and constitute a kind of excess genetic baggage (an exception is drug-resistance plasmids, p. 243). The host genome must carry all necessary genes for its life processes. However, plasmids may determine bacterial traits crucial to adaptation. In some instances hereditary traits in bacteria are transferred abruptly and almost wholesale by plasmid infection.

Many plasmids carry their own genes for replication and transfer. Some can promote transfer of genetic material from one bacterial cell to another (for example, conjugation, p. 96); others cannot. Various combinations of plasmid genes endow the host bacterium with diverse properties. Some of these properties result in serious consequences to the host. For example, the production of toxins related to serious diseases is often plasmid mediated. The box on p. 98 lists some of the important plasmid-mediated activities in bacteria.

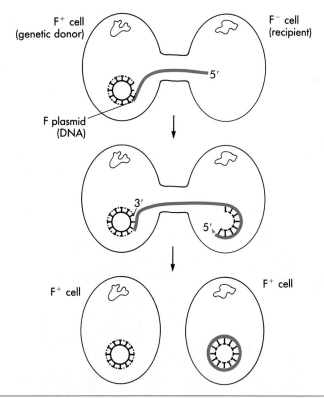

FIG. 6-15   Conjugation in bacteria. An F plasmid (DNA) is transferred by a cell-to-cell contact to an F⁻ cell, converting it to an F⁺ cell.

Comparison with Phages

Since phages and plasmids are alike in many ways (both are autonomously replicating), much that has been learned from the study of phages is applicable to plasmids. Differences exist, however, in the mode of transfer and in the limits to size. The small plasmid circlet of DNA may vary considerably in size as genes are gained or lost, whereas viral DNA is fixed in size by a protein coat.

Unlike viruses, plasmids, as far as we know, have no extracellular existence or phase. Also, their replication being held in check in an unknown way in the cell does not result in the destruction of the cell, as is true for a number of viruses.

Classification

The two main types of plasmids are (1) the large, self-transmissible, or conjugative, plasmids and (2) the small, nonconjugative ones (p. 97). For a plasmid to be conjugative, it must be able to code within one bacterium for an apparatus (pilus) that permits its transfer to another bacterium through a contact between the two bacterial cells. (This entails expression of approximately 20 to 30 genes.) Both large and small plasmids can be transferred by transformation (p. 94).

Examples of large plasmids are the F and R, or resistance factors (drug resistance factors), and certain bacteriocinogens. The F plasmids code for functions that promote mating and transfer of plasmid DNA between bacteria. The classic R factors (p. 243) have two functionally distinct parts. One is the resistance transfer factor, or RTF; it contains genes for autonomous replication and for conjugation. The other is a smaller resistance determinant, or R determinant, which varies in size and gene content. Examples of small plasmids are some bacteriocinogens and some resistance determinants.

Bacteriocinogens and bacteriocins. Bacteriocins are proteins released from enteric bacteria, pseudomonads, and certain other bacteria; they inhibit the growth of a limited number of other strains. Since the ones first discovered were from *E. coli*, they were called *colicins*. A bacteriocin is plasmid mediated, and the corresponding plasmid is called a *bacteriocinogen*.

Importance of Plasmids

There are good reasons to believe that most plasmids benefit their hosts in a relationship that is more like symbiosis than parasitism (see box below). In nature plasmids

---

Some plasmid-mediated activities in bacteria are as follows:
1   Replication, repair, and recombination
2   Restriction and modification
3   Conjugation: some R plasmids provide host cells with the capacity to interact sexually with other bacteria of the same or another species, and thus to transfer R plasmid DNA.
4   Resistance to toxic metals and detergents: some plasmids confer resistance to ultraviolet light; certain ones bestow a degree of resistance to heavy metal ions; survival of the bacteria is thus enhanced.
5   Resistance to bacteriophage
6   Cell adhesion and attachment: certain plasmids influence the synthesis of surface pili needed for the union between organism and surface of host cell.
7   Virulence: production of toxins related to serious diseases (for example, diphtheria and botulism)
8   Resistance to antimicrobial agents: R plasmids determine resistance to a drug or even to multiple drugs.

contribute in a vital way to the geochemical cycles by spreading their genes for the degradation of many complex organic compounds. The spread of resistance genes from plasmids may help investigators trace the epidemiologic spread of disease from one human being to another in an institution or from an outside source in nature to human beings.

The single most visible activity of plasmids is their role in the spread of antibiotic resistance.

Drug resistance (p. 242). A microbial variation that gives an organism an increased tolerance for or resistance to an antimicrobial drug (or drugs) is one with considerable therapeutic significance. Based on genetic mechanisms (rarely, mutation), this tolerance for a given drug or drugs is termed **drug fastness** or **drug resistance.** An organism that becomes resistant to a drug used in the treatment of disease may be said to be *drug fast* or *drug resistant.* Resistance means that the drug is no longer effective in suppressing the growth and multiplication of the organism for which the drug is being given.

R factors or R plasmids. The best known plasmids, perhaps, are the R factor plasmids (R plasmids) and the penicillinase plasmids of staphylococci (p. 244). The genes carried on the R plasmids confer on the host cell resistance to one or more antimicrobial agents, the mechanism, for the most part, being related to the formation of drug-inactivating enzymes. First encountered in members of Enterobacteriaceae, R plasmids recently have been identified in the most prevalent of enterics, the anaerobic genus *Bacteroides.*

Penicillinase plasmids. Penicillinase plasmids endow the staphylococcus with resistance to penicillin, since the genes carried enable that cell to produce the enzyme penicillinase, which inactivates penicillin. Penicillinase plasmids are transmitted from bacterium to bacterium by phage-mediated transduction (p. 95). The dissemination of these plasmids can be an effective means of inducing penicillin resistance, since within a decade of the widespread use of penicillin, 50% to 80% of staphylococci were found to be resistant to it.

Genetic Recombination

General Considerations

Genetic recombination is the process that results in a change in the genetic makeup of the bacterial or other cell when genetic material is passed from one cell (the donor) to another (the recipient). As a consequence of this transfer, the genetic linkage group—the bacterial genome in this case—acquires like information from another such unit. One region of a double-stranded DNA molecule in the recipient is replaced by another from the donor.

When genetic material (DNA) is transferred from one bacterium to another, the recipient cell gains a portion—a single gene or block of genes—from the genome of the donor cell (but not the full complement of the genome). Immediately afterward a reassortment of genetic determinants takes place. The new portion is matched to the corresponding segment of the recipient's genome, and genetic material is exchanged and eliminated. From this rearrangement a newly formed genotype emerges, containing DNA from two bacterial cells. It is the **recombinant genome,** and the process that shapes it is *recombination.* Recombination usually does not occur unless donor and recipient bacterial cells are closely related taxonomically. For the study of this process the phenotypes of the parental bacteria must be clear-cut so that well-defined traits expressed in the recombinant progeny distinguish them from their parents.

Recombination is similar to mutation in that a change is effected in the genetic makeup of the cell, but it is distinct. New biologic traits come from mutation. Methods of recombination permit the dissemination of new (or other) traits to a variety of other bacterial strains so that the selective advantage of a mutation is preserved.

Restriction-Modification Systems

Of particular importance to the modern technics of genetic engineering is the large group of restriction endonucleases, since they make possible a specific kind of cleavage of DNA.

Restriction is the term applied to an enzymatic mechanism existing within a number of bacteria to temper the free interchange of genetic material. Restriction endonucleases ordinarily protect the bacterial cell from intrusion by foreign DNA, since at specific restriction sites in both strands of the DNA helix, they can cut out segments of DNA but in such a way as to permit the strands to realign.

The highly specific restriction system is tied into an equally specific modification system. **Modification** is effected by an enzyme that acts on the same site in the DNA molecule as does the restriction enzyme but alters it so as to prevent cleavage of the DNA.

Restriction enzymes destroy invading foreign DNA from whatever source: phage infection, conjugation, or transformation. The cell's native DNA is not affected because the site recognized by the restriction enzymes is protected by the modification enzymes.

With the aid of restriction enzymes, one can cleave DNA molecules from two different organisms of the same or different species and recombine the fragments to produce a biologically functional hybrid DNA molecule. The hybrid replicates and expresses the foreign DNA within it as though that DNA had always been there.

Protoplast Fusion

Different mechanisms exist by which genetic recombination can occur in bacteria. Important natural ones are the three previously discussed: transformation, transduction, and conjugation. In the laboratory, recombination also may occur by the technic of protoplast fusion. The outer cell wall of two cells—bacterial, fungal, or one of each—is stripped off. With only a cell membrane to contain cytoplasm and to separate them, the target cells in contact fuse together, each bringing its DNA to the fused protoplast. When the new hybrid cell regenerates and divides, exchange of genetic material occurs.

Gene Cloning

For today's burgeoning field of biotechnology, a working definition of genetic recombination might be this. Transformation of a microbial cell is accomplished with the use of a specific DNA molecule that is engineered ("recombined") in the laboratory. This engineered molecule is spliced into a virus or plasmid vector, which is amplified (replicated) in a host organism. Under certain conditions the foreign gene will be expressed more abundantly than the native ones in the host, thus providing an economic source for a given substance. This includes many previously unobtainable substances, such as hormones and growth factors.

Gene cloning encompasses the production of the DNA fragment, uniquely defined for a given product, its insertion into a specific vector, and its subsequent propagation and amplification in a suitable host cell (Fig. 6-16). The desired fragment of DNA or the desired gene describing a single protein may be synthesized by biochemical methods, or it may be possible to extract native DNA from cells, fragment it by restriction enzymes, and isolate the particular gene.

It is technically feasible with recombinant DNA methodology to clone specific DNA fragments from any source into convenient vectors (carriers) for in vivo or in vitro observation. The two commonly used carriers or vectors are bacterial plasmids and the bacterial virus, bacteriophage lambda, both of which replicate in bacteria such as *E. coli*. Plasmids are brought into contact with bacteria in the presence of calcium ions, which increase the permeability of the bacterial membrane and facilitate plasmid penetration of the bacterial host.

Microbiologic Hosts for Recombinant DNA

A mutant of *E. coli* that cannot synthesize the essential amino acid tryptophan is an important host for making recombinant DNA products in the laboratory, since plasmids exist that contain the tryptophan gene. A foreign fragment of DNA can be spliced into these plasmids near the tryptophan gene so that the foreign gene and the tryptophan gene will be expressed at the same time. In the culture of the mutant *E. coli*, plasmids enter the bacterial cells and, by supplying the needed tryptophan, enable them to thrive. Since tryptophan is needed by the bacteria, the gene for it and, correspondingly, the foreign gene will be preferentially expressed. The bacteria are cultured, and subsequently from the culture the amplified gene is isolated and verified. In this way it is possible to mass-produce copies of a single gene with a very high yield of the desired gene product.

Genetic Advantages of Recombination

With the use of the modern methods of genetic recombination, it has been possible to produce in massive amounts pure DNA containing a desired nucleotide sequence. There are at least two advantages to such a development. First, the structure and function of genes and the genetic apparatus can be studied biochemically in ways that would

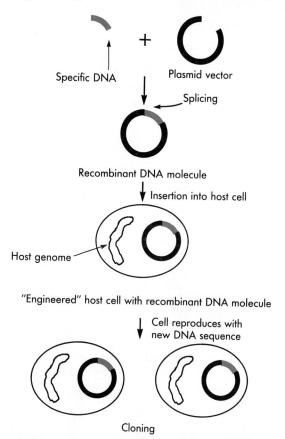

FIG. 6-16   Formation ("engineering") of a recombinant molecule. The desired DNA fragment (for a given product) is inserted into a DNA vector, such as a plasmid. The recombinant DNA molecule is introduced into a suitable host cell (*E. coli*) by transformation or viral infection. The "engineered" host cell reproduces with the new DNA sequence. Expression of the gene foreign to the host cell provides the source of the desired product.

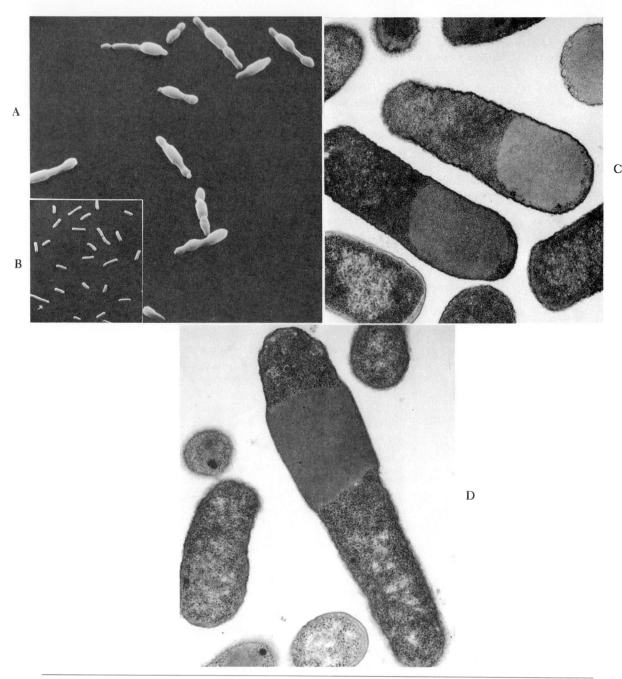

FIG. 6-17   Production of genetically engineered human insulin in the microbiologic host, a plasmid-containing strain of *E. coli*, as observed in electron micrographs. All cells were fixed during the late log phase of growth. **A,** In a scanning electron micrograph, prominent bulges (inclusion bodies) in the bacterial body indicate accumulation of the genetically engineered product, human insulin A-chain chimeric protein. ($\times$10,000.) **B,** For comparison, nonplasmid-containing, nonproductive cells of the same *E. coli* strain. ($\times$5000.) **C,** A transmission electron micrograph of *E. coli* shows the nature and arrangement of the accumulated protein product within the bacterium. ($\times$68,000.) **D,** A higher magnification of the inclusion body. (Courtesy Daniel C. Williams, Ph.D, Lilly Research Laboratories, Indianapolis.)

otherwise be impossible. Second, from the research in recombinant DNA has come the development of technics whereby a gene in a suitable expression vector can be induced not merely to replicate itself but also to express itself as a desired protein.

Genetic Engineering

Recombinant DNA technology provides us with the opportunity to splice together genes from organisms that do not normally exchange genetic material. Technics are available to cause cells to produce molecules they do not normally fabricate, as well as to produce more efficiently molecules that they do. With current technics of genetic engineering, in theory at least, any polypeptide gene product can be synthesized.

The following is a list of genetically engineered products of importance to health science fields:
1. Polypeptide hormones: insulin, interferons, growth hormone, somatostatin, thymosin
2. Blood components: hemoglobin, two clotting factors
3. Vaccines: hepatitis B vaccine, hoof-and-mouth disease vaccine, hog and cattle dysentery vaccine
4. Immune system mediators: antibodies, complement components, lymphokines
5. Enzymes: urokinase
6. Growth factors: platelet-derived growth factor, epidermal growth factor
7. Biologic insecticides: toxin produced by *Bacillus thuringiensis*
8. Certain agricultural products

Human insulin. The first commercial health-care product to come from recombinant DNA biotechnology is human insulin. Its safety and efficacy in the management of human diabetes has been demonstrated.

Insulin is a molecule of 51 amino acids arranged into an A chain of 21 amino acids and a B chain of 30. Each of these two main subunits of insulin is produced separately by bacteria programed to do so with grafts of *human* genes specifying the synthesis of insulin. The subunits produced are later assembled into the insulin molecule by another step in the laboratory.

The specific human genes are integrated into the DNA of the bacterial vector, which may be *E. coli* (Fig. 6-17). Then the small amount of genetic material so introduced is amplified by growing the engineered microbes in culture. From the culture of the microbiologic host the resultant gene product is collected, purified, and tested accordingly.

SUMMARY

1 In prokaryotes, eukaryotes, and many viruses, DNA is the molecule of heredity; in other viruses it is RNA. DNA defines all structure and function of an organism and instructs its own replication.

2 The specificity of base pairing in DNA is critical to its replication. The guanine plus cytosine percent of total DNA, being characteristic for each microbial strain, measures genetic relatedness.

3 Translation is the genetic mechanism whereby RNA templates direct synthesis of proteins in a cell.

4 The unit pattern in the universal genetic code is three successive nucleotides in DNA, referred to as a triplet, or a codon. Through the mRNA transcript, it directs the arrangement of a specific amino acid in a given protein. The triplet of nucleotides in tRNA—the anticodon—associates by complementary base pairing with a specific codon in mRNA.

5 A mutation is a permanent, heritable alteration in the genome, usually of a single gene. The common point mutation results from a change in a single base or a base

pair. A frame-shift mutation results when an insertion or deletion of one or more nucleotides alters the reading frame.

6 Physical and chemical mutagens promote replication errors in DNA. Repair mechanisms exist to correct potential mutations.

7 The machinery of DNA replication is remarkably constant, but mutants are found in an average bacterial population because of its large numbers.

8 Most mutagens appear to cause cancer. The Ames test for mutagenicity applies this fact to the detection of carcinogenic agents.

9 Lysogenic bacteria harbor the genomes of temperate bacteriophages (bacterial viruses) in an ongoing type of infection known as lysogeny. The phage genome replicates therein as a hereditary unit called a prophage.

10 Phage conversion refers to the phenotypic expression of bacteriophage genes. Genes of temperate phages determine bacterial properties, for example, the production of diphtheria toxin in infected diphtheria bacilli.

11 Plasmids are autonomously replicating, circular, double-stranded DNA molecules in the cytoplasm of practically all groups of bacteria. They endow the host bacterium with traits crucial to adaptation and other diverse properties. Bacterial toxins may be plasmid-mediated.

12 The most visible plasmid function is the spread of resistance to one or more antibiotic drugs. Usually plasmids code for drug-inactivating enzymes. The best known example is penicillinase.

13 Genes may be passed from one bacterium to another by (1) uptake of pieces of DNA, (2) viral infection (transduction), and (3) "mating" between cells in direct contact (conjugation).

14 Conjugation effects a one-way passage of genetic material (sometimes the entire genome) from one bacterium to another through a transient physical contact.

15 Genetic recombination means rearrangement of the genetic makeup of the cell when genetic material is passed to it from another cell. The newly formed genotype contains DNA from the two cells. Genetic recombination occurs naturally in bacteria by transformation, transduction, and conjugation.

16 Cloning is the amplification of a specific gene through insertion of a specific DNA fragment into a suitable vector.

17 Genetic engineering (recombinant DNA technology) is a new medical science stemming from bacterial genetics. Genetically engineered products of importance in health sciences are polypeptide hormones, vaccines, and immune system mediators.

QUESTIONS FOR REVIEW

1 List some implications of recombination. (Consult outside sources.)

2 Outline the major steps in protein synthesis.

3 What are the properties of the DNA molecule that account for its efficiency as the molecule of heredity?

4 Name the base pairs for the DNA molecule; for RNA.

5 Discuss plasmids: their nature, genetic importance, classification, and role in genetic engineering.

6 State the biochemical nature of mutations. How do chemical mutagens effect mutagenesis?

7 Describe the process of replication in DNA.

8 Why are bacteriophages important in genetic studies?

9 Briefly discuss the genetic code. What is it? Why is it important? Where is it found?

10 How many codons are functional? Name the ones that make up the so-called nonsense codons.

11 List five classes of biologic products developed as a result of genetic engineering.

12 Briefly define protoplasts, mutation, mutant, transformation, analog, conjugation, transduction, phage conversion, drug fastness, genome, genotype, phenotype, complementarity, polarity, replication, translation, transcription, template, semiconservative, vector, gene amplification, homology, carcinogenic, bacteriocins, hybridization.

13 Comment briefly on the practical importance of penicillinase and R factors.

14 Give the three natural means of intermicrobial gene transfer.

SUGGESTED READINGS

Abelson, J., and Butz, E., editors: Recombinant DNA, Science 209:1317, 1980.

Check, W.A.: Bacterially produced human insulin given therapeutically, J.A.M.A. 245:322, 1981.

Curtis, H.: Biology, New York, 1983, Worth Publishers, Inc.

Dion, A.S., editor: Concepts of the structure and function of DNA, chromatin and chromosomes, Chicago, 1979, Year Book Medical Publishers, Inc.

Dulbecco, R.: Contributions of microbiology to eukaryotic cell biology: new directions for microbiology, Microbiol. Rev. 43:443, 1979.

Johnson, I.S.: Human insulin from recombinant DNA technology, Science 219:632, 1983.

Mainwaring, W.I.P., and others: Nucleic acid biochemistry and molecular biology, London, 1982, Blackwell Scientific Publications, Ltd.

Motulsky, A.G.: Impact of genetic manipulation on society and medicine, Science 219:135, 1983.

Szekely, M.: From DNA to protein, the transfer of genetic information, New York, 1980, John Wiley & Sons, Inc.

Wade, N.: Gene splicing company wows Wall Street, Science 210:506, 1980.

Williams, D.C., and others: Cytoplasmic inclusion bodies in Escherichia coli producing biosynthetic human insulin proteins, Science 215:687, 1982.

CLASSIFICATION OF BACTERIA

The classification of bacteria is difficult with regard to both the separation of bacteria into groups and the placing of certain organisms into the proper group. Biologic classification is based largely on morphology, but the morphologic characteristics of bacteria as a whole are so uniform that they are useful only in dividing bacteria into comparatively large groups. Shape has been an important factor in general classification, but for more exact identification, criteria such as staining reactions, cultural characteristics, biochemical and physiologic behavior patterns, genetic analyses, animal inoculations, and immunologic differences must be used.

In this book we adhere to the scientific classification of bacteria embodied in *Bergey's Manual of Determinative Bacteriology* (1974), a classic source, and *Bergey's Manual of Systematic Bacteriology*, volume 1 (1984). Certain important exceptions will be noted in the text. Bergey's classification is the one that is best known and most generally accepted in the United States. Table 7-1 presents in abbreviated form an overall survey of this classification.

R.G.E. Murray, in 1968, proposed a kingdom for those microbes that are characterized by the possession of nucleoplasm devoid of basic protein and not bounded from cytoplasm by a nuclear membrane. He gave it the name **Procaryotae** (Table 7-1). In the 1984 edition of *Bergey's Manual*, Murray proposed the following breakdown for the higher taxa in the kingdom:

Kingdom *Procaryotae*

    Division I. *Gracilicutes* (Table 7-1)

        Class I. *Scotobacteria* (nonphotosynthetic, gram-negative bacteria)

        Class II. *Anoxyphotobacteria* (light-requiring bacteria not producing oxygen; anaerobic; contain bacteriochlorophyls)

        Class III. *Oxyphotobacteria* (light-requiring bacteria that produce oxygen in the light; contain chlorophyls; gliding motility; aerobic; rigid multilayered wall with inner peptidoglycan layer)

Division II. *Firmicutes* (Table 7-1)

 Class I. *Firmibacteria* (Latin *firmus*, strong, plus Greek *bakterion*, small rod; simple gram-positive bacteria whose name means "strong bacteria")

 Class II. *Thallobacteria* (Greek *thallos*, branch, plus *bakterion*, small rod; gram-positive bacteria with branching whose name means "branching bacteria"; actinomycetes and related microbes here)

Division III. *Tenericutes* (Latin *tener*, soft, plus *cutis*, skin; prokaryotes of pliable, soft nature, since no rigid cell wall) (Table 7-1)

 Class I. *Mollicutes* (distinctive group of diverse prokaryotes with no cell wall) (Table 7-1)

Division IV. *Mendosicutes* (Table 7-1)

 Class I. *Archaeobacteria* (diverse prokaryotes with unusual walls, membrane, lipids, ribosomes, and RNA sequences)

Since currently bacterial classification is in a state of transition, the authors of *Bergey's Manual of Systematic Bacteriology*, volume 1, have used a few readily definable criteria to group bacteria into sections (Table 7-1). The vernacular name of each section reflects the primary basis for the organization. The higher taxa (class, order, family) are not necessarily incorporated, although a section may give the name of a taxon. All accepted genera have been placed in what appears to be the most appropriate section, but some sections mention no taxa at all above the level of genus.

Text continued on page 123.

Table 7-1 Abbreviated Classification of Microbes from *Bergey's Manual of Determinative Bacteriology* (1974) and *Bergey's Manual of Systematic Bacteriology,* volume 1 (1984)

Kingdom Procaryotae

Highest level taxon encompassing microbes wherein nucleoplasm lacks basic protein and is not bounded by nuclear membrane; cells are single or found in simple filamentous, mycelial, or colonial associations; cytoplasm is immobile; nutrients are acquired in molecular form, and enclosure by rigid wall is common but not universal

 Division I. *Gracilicutes* (phototropic or nonphototropic, nonsporulating, usually gram-negative prokaryotes with complex cell wall of gram-negative type; variable in shape, some sheathed or encapsulated; reproduction by binary fission or, in some instances, budding; species are aerobic, anaerobic, or facultatively anaerobic; some obligate intracellular parasites; when present, motility by swimming or gliding)

 Division II. *Firmicutes* (usually gram-positive prokaryotes with cell wall of gram-positive type; chemosynthetic heterotrophs, not photosynthetic; variable in shape with spheres and occasionally branching rods and filaments; reproduction by binary fission; aerobic, anaerobic, or facultatively anaerobic species include asporogenous and sporogenous bacteria and actinomycetes and their relatives)

 Division III. *Tenericutes* (highly pleomorphic [no cell wall] prokaryotes enclosed only by unit membrane; filamentous forms common with branching; wide size range; genome size smaller than for other prokaryotes; reproduction by budding, fragmentation, and/or binary fission; usually nonmotile but, if present, motility by gliding; complex media required for growth; fried-egg colony common, since growth penetrates surface of media; species saprophytic, parasitic, or pathogenic; cause diseases in animals, plants, and tissue cultures)

Based on data from Buchanan, R.E., and Gibbons, N.E., co-editors: Bergey's manual of determinative bacteriology, ed. 8, Baltimore, 1974, The Williams & Wilkins Co.; and Holt, J.G., editor-in-chief, and Krieg, N.R., editor, vol. 1: Bergey's manual of systematic bacteriology, vol. 1, Baltimore, 1984, The Williams & Wilkins Co. Volume 1 covers sections 1 to 11. Volumes 2 (sections 12 to 17), 3 (sections 18 to 23), and 4 (sections 27 to 30) are as yet unpublished.

Continued.

Table 7-1 Abbreviated Classification of Microbes from *Bergey's Manual of Determinative Bacteriology* (1974) and *Bergey's Manual of Systematic Bacteriology*, volume 1 (1984)—cont'd

Division IV. *Mendosicutes* (ecologically and metabolically diverse prokaryotes; nonspore-forming, gram-variable, and variable in shape; most strictly anaerobic, some aerobic; many motile by flagella; most with some form of cell wall but without muramic acid conventional peptidoglycan cannot be present; some cell walls of protein macromolecules, some of heteropolysaccharides; known members, such as methanogens, strict halophiles, and thermoacidophiles, live in extreme environments)

Section 1. The Spirochetes

Order I. *Spirochaetales**
  Family I. *Spirochaetaceae*
    Genus I. *Spirochaeta* (helical cells, motile, free living in H₂S-containing mud and sewage)
    Genus II. *Cristispira* (helical cells with 2 to 10 complete turns; commensal in mollusks)
    Genus III. *Treponema*
      Treponema pallidum (type species; syphilis)
      Treponema pallidum subsp. *pertenue* (yaws)
      Treponema carateum (pinta, a chronic disease of children endemic in South and Central America)
      Treponema denticola [*microdentium*] (oral microflora)
      Treponema [*Borrelia*] *vincentii* (oral microflora)
    Genus IV. *Borrelia*
      Borrelia recurrentis (louse-borne epidemic relapsing fever)
      Borrelia species (tick-borne endemic relapsing fever)
  Family II. *Leptospiraceae*
    Genus I. *Leptospira*
      Leptospira interrogans (type species; leptospirosis)

Section 2. Aerobic/Microaerophilic, Motile, Helical/Vibrioid Gram-Negative Bacteria

  Genus *Aquaspirillum* (small water spirals found in stagnant freshwater environments)
    Species incertae sedis *Spirillum minus* [*minor*] (one type of human rat-bite fever)
  Genus *Spirillum* (small spirals found in stagnant, fresh water; no pathogens described; *Spirillum minus* does not belong here)
  Genus *Campylobacter* (slender, spiral rods found in reproductive and alimentary tracts of humans and animals)
    Campylobacter [*Vibrio*] *fetus* (type species; abortion in sheep and cattle; human infections)

Section 3. Nonmotile (or Rarely Motile), Gram-Negative Curved Bacteria

Aerobic inhabitants of soil, fresh water, and marine water, with characteristic morphology

Section 4. Gram-Negative Aerobic Rods and Cocci

  Family I. *Pseudomonadaceae*
    Genus I. *Pseudomonas*
      Pseudomonas aeruginosa (type species; wound, burn, and urinary tract infections)
      Pseudomonas [*Actinobacillus*] *mallei* (glanders and farcy in horses and donkeys; infection transmissible to humans)
      Pseudomonas pseudomallei (human and animal melioidosis)
      Pseudomonas cepacia (opportunistic human infections)
    Genus II. *Xanthomonas* (plant pathogens)

*Names indicating bacterial orders end consistently in *ales*; those indicating families, in *aceae*.

Table 7-1 Abbreviated Classification of Microbes from *Bergey's Manual of Determinative Bacteriology* (1974) and *Bergey's Manual of Systematic Bacteriology,* volume 1 (1984)—cont'd

Genus III. *Frateuria* (all strains isolated in Japan)

Genus IV. *Zoogloea* (motile rods in natural waters and sewage)

Family II. *Azotobacteraceae* (large, motile, aerobic gram-negative rods fixing atmospheric nitrogen; found in soil and water and on leaf surfaces)

Genus I. *Azotobacter* (nitrogen fixation)

Genus II. *Azomonas* (nitrogen fixation)

Family III. *Rhizobiaceae* (nitrogen fixation; symbionts in root nodules of legumes; cortical overgrowths [galls] in plants)

Genus I. *Rhizobium* (nitrogen fixation)

Genus II. *Bradyrhizobium* (plant galls; no nitrogen fixation)

Genus III. *Agrobacterium* (plant pathogens—tumorigenic phytopathogens; galls on stems of more than 40 plants; free nitrogen fixed; found in soil)

Family IV. *Methylococcaceae* (diverse group of rods, vibrios, and cocci utilizing methane as sole source of carbon and energy under aerobic conditions; widespread in nature)

Family V. *Halobacteriaceae* (high concentration of sodium chloride for growth; found in salterns, salt lakes, Dead Sea, proteinaceous material preserved with solar salt [fish, sausage casings, and hides])

Genus I. *Halobacterium* (salt-loving rods)

Genus II. *Halococcus* (halophilic cocci, red colonies)

Family VI. *Acetobacteraceae*

Genus I. *Acetobacter* (motile aerobic vinegar rods; oxidize ethanol to acetic acid; found on fruits and vegetables, in souring fruit juices, vinegar, alcoholic beverages)

Acetobacter aceti (acetic acid bacteria)

Genus II. *Gluconobacter* (ellipsoids or rods in flowers, souring fruits, vegetables, cider, wine, baker's yeast, garden soil; ropiness in beer and wort)

Family VII. *Legionellaceae* (motile, nonencapsulated rods isolated from surface water, mud, and thermally polluted lakes and streams; human pathogens; no known soil or animal source)

Genus I. *Legionella* (legionellosis)

Legionella pneumophilia (Legionnaire's disease [human pneumonia], Pontiac fever)

Legionella micdadei (Pittsburgh pneumonia)

Family VIII. *Neisseriaceae*

Genus I. *Neisseria* (parasites of mucous membranes of mammals)

Neisseria gonorrhoeae (type species; gonorrhea)

Neisseria meningitidis (epidemic cerebrospinal fever)

Neisseria sicca (human nasopharynx)

Neisseria subflava [*flava*] [*perflava*] (yellowish green pigment; human nasopharynx)

Neisseria flavescens (xanthophil pigment)

Neisseria mucosa (human rhinopharynx)

Neisseria lactamica (nasopharynx of infants and children)

Genus II. *Moraxella* (parasites of mucous membranes of humans and warm-blooded animals)

Subgenus *Moraxella*

Moraxella (Moraxella) lacunata (type species; pink-eye [conjunctivitis])

Subgenus *Branhamella* (parasites of mammalian mucous membranes)

Moraxella (Branhamella) catarrhalis (venereal discharges; catarrhal inflammations)

Genus III. *Acinetobacter* (opportunistic pathogens)

Acinetobacter calcoaceticus [*Herellea vaginicola*, acid-forming] [*Mima polymorpha*, nonacid-forming] (type species, soil and water bacteria)

Continued.

Table 7-1 Abbreviated Classification of Microbes from *Bergey's Manual of Determinative Bacteriology* (1974) and *Bergey's Manual of Systematic Bacteriology,* volume 1 (1984)—cont'd

      Genus IV. *Kingella* (part of normal flora of mucous membranes of upper respiratory tract in humans)
   Other genera (no family assigned)
      Genus *Beijerinckia* (nitrogen fixation)
      Genus *Derxia* (found in tropical soils of Asia, Africa, and South America; fixation of atmospheric nitrogen)
      Genus *Xanthobacter* (free living in soil and water; fixation of atmospheric nitrogen)
      Genus *Thermus* (gram-negative nonmotile rods and filaments; often pigmented; common in hot springs, hot water tanks, and thermally polluted rivers; unusually thermostable enzymes, ribosomes, plasma membrane found here)
      Genus *Flavobacterium* (proteolytic soil and water bacteria producing yellow, orange, or red pigments; found on vegetables and in dairy products; rare human infection by unpigmented organism)
      Genus *Alcaligenes* [*Achromobacter*] (motile aerobic rods or cocci; common saprophytes in intestines of vertebrates in dairy products, rotting eggs, fresh water, soil; important in decomposition and mineralization processes)
        Alcaligenes faecalis (type species; some strains denitrify)
      Genus *Brucella* (brucellosis)
        Brucella melitensis (type species; Malta fever; infection in goats, sheep, and cattle)
        Brucella abortus (abortion in cattle; disease in humans)
        Brucella suis (infection in pigs, other animals, and humans)
      Genus *Bordetella*
        Bordetella pertussis (type species; whooping cough)
        Bordetella parapertussis (whooping cough)
        Bordetella bronchiseptica (found in respiratory tract of animals, sometimes human beings; rodent bronchopneumonia)
      Genus *Francisella*
        Francisella [*Pasteurella*] *tularensis* (type species; tularemia)

Section 5. Facultatively Anaerobic Gram-Negative Rods

   Family I. *Enterobacteriaceae*
   Genus I. *Escherichia*
      Escherichia coli (type species [the colon bacillius]; important opportunistic pathogens)
   Genus II. *Shigella* (shigellosis)
      Shigella dysenteriae (type species; bacillary dysentery plus effects of diffusible neurotoxin)
      Shigella flexneri (bacilliary dysentery)
      Shigella boydii (bacilliary dysentery)
      Shigelli sonnei (one cause of summer diarrhea in young children, milder form of bacillary dysentery in adults)
   Genus III. *Salmonella* (salmonellosis)
     "Subgenus" I
      Salmonella choleraesuis (type species; salmonellosis)
      Salmonella hirschfeldii (paratyphoid C bacilli; enteritis)
      Salmonella typhi (typhoid fever)
      Salmonella paratyphi-A (paratyphoid [enteric] fever)
      Salmonella schottmuelleri (enteritis)
      Salmonella typhimurium (food poisoning in humans)
      Salmonella enteritidis (enteritis)
      Salmonella gallinarum (fowl typhoid)
     "Subgenus" II
      Salmonella salamae (type species of subgenus)

Table 7-1 Abbreviated Classification of Microbes from *Bergey's Manual of Determinative Bacteriology* (1974) and *Bergey's Manual of Systematic Bacteriology,* volume 1 (1984)—cont'd

"Subgenus" III
  Salmonella arizonae (type species of subgenus; isolated from reptiles)
"Subgenus" IV
  Salmonella houtenae (type species of subgenus)
"Subgenus" V
  Salmonella bongor (isolated from a lizard in Chad)
Genus IV. *Citrobacter*
  Citrobacter freundii (type species; found in water, food, and human excreta)
Genus V. *Klebsiella*
  Klebsiella pneumoniae (type species; pneumonia, infections of respiratory and urinary tracts)
  Klebsiella pneumoniae subsp. *ozaenae* (found in ozena and chronic respiratory disease)
  Klebsiella pneumoniae subsp. *rhinoscleromatis* (found in rhinoscleroma, a granulomatous disorder of nose and pharynx associated with nodular induration of tissues)
  Klebsiella oxytoca (opportunistic pathogen in human intestine)
Genus VI. *Enterobacter*
  Enterobacter cloacae (type species; opportunistic pathogens)
  Enterobacter [*Aerobacter*] *aerogenes* (opportunistic pathogens)
Genus VII. *Erwinia* (plant pathogens)
Genus VIII. *Serratia*
  Serratia marcescens (type species; opportunistic pathogens)
Genus IX. *Hafnia* (Hafnia, the old name for Copenhagen; opportunistic pathogens)
Genus X. *Edwardsiella*
  Edwardsiella tarda (type species; found in intestine of snakes, sometimes in humans)
Genus XI. *Proteus* (urinary tract infections, community and hospital acquired)
  Proteus vulgaris (type species; urinary tract and wound infections, rarely peritonitis, meningitis)
  Proteus mirabilis (most frequent species in medical specimens)
Genus XII. *Providencia* (urinary tract, burn, and wound infections)
  Providencia alcalifaciens [*inconstans*] (found in diarrheal stools in children)
  Providencia rettgeri (urinary tract infections)
Genus XIII. *Morganella* (opportunistic pathogens in feces of mammals; respiratory, urinary tract, and wound infections)
  Morganella morganii (opportunistic pathogen in humans; one cause of summer diarrhea)
Genus XIV. *Yersinia*
  Yersinia [*Pasteurella*] *pestis* (type species; plague)
  Yersinia [*Pasteurella*] *pseudotuberculosis* (pseudotuberculosis in animals, usually mesenteric lymphadenitis; human septicemia)
  Yersinia enterocolitica (widespread; has been found in sick and healthy animals and in material likely contaminated by their feces; enterocolitis in young children, mesenteric lymphadenitis, variety of other infections)
Family II. *Vibrionaceae*
Genus I. *Vibrio*
  Vibrio cholerae [*comma*] (type species; cholera)
    Vibrio cholerae biovar *eltor* (El Tor vibrio; cholera)
  Vibrio parahaemolyticus (acute gastroenteritis)
  Vibrio [*Photobacterium*] *fischeri* (luminescent saltwater bacteria)
Genus II. *Photobacterium* (luminescent saltwater bacteria)
Genus III. *Aeromonas* (motile gas-forming rods)
  Aeromonas hydrophilia (type species; nonluminescent fresh water bacteria; infections of

Continued.

Table 7-1 Abbreviated Classification of Microbes from *Bergey's Manual of Determinative Bacteriology* (1974) and *Bergey's Manual of Systematic Bacteriology*, volume 1 (1984)—cont'd

cold-blooded animals—red leg [bacteremia] in frogs and septicemia in snakes; rarely infectious in compromised host)

Genus IV. *Plesiomonas* (motile rods growing mostly on mineral media containing ammonia as sole source of nitrogen and glucose as sole source of carbon; found in feces; infectious gastroenteritis reported in humans)

Family III. *Pasteurellaceae*

Genus I. *Pasteurella*

Pasteurella multocida (type species; chicken cholera; shipping fever of cattle; hemorrhagic septicemia in warm-blooded animals; cat- and dog-bite wound infections in human beings)

Pasteurella pneumotropica (infections in animals; dog-bite wounds in humans)

Pasteurella haemolytica (enzootic pneumonia of sheep and cattle; septicemia of lambs)

Genus II. *Haemophilus*

Haemophilus influenzae (type species; purulent meningitis in young children; acute respiratory infection; acute conjunctivitis)

Haemophilus parasuis [*suis*] (with virus causes swine influenza)

Haemophilus haemolyticus (commensal in upper respiratory tract of humans)

Haemophilus parainfluenzae (found in upper respiratory tract of humans and cats)

Haemophilus parahaemolyticus (found in human upper respiratory tract; associated with acute pharyngitis; pleuropneumonia and septicemia in swine)

Haemophilus aphrophilus (endocarditis and other human infections)

Haemophilus ducreyi (chancroid)

Haemophilus aegyptius (Koch-Weeks bacillus; acute infectious conjunctivitis)

Genus III. *Actinobacillus* (actinobacillosis)

Actinobacillus lignieresii (type species; actinobacillosis of cattle [wooden tongue] and of sheep)

Actinobacillus equuli (actinobacillosis in horses and pigs)

Other genera

Genus *Chromobacterium* (soil and water bacteria producing violet pigment violacein; infections in animals; food spoilage)

Genus *Cardiobacterium*

Cardiobacterium hominis (type species; found in human nose and throat; endocarditis)

Genus *Calymmatobacterium* (pleomorphic encapsulated rods like safety pins)

Calymmatobacterium [*Donovania*] *granulomatis* (type species; granuloma inguinale)

Genus *Gardnerella*

Gardnerella vaginalis (major cause of bacterial nonspecific vaginitis)

Genus *Eikenella* (colonies may appear to corrode agar surface)

Eikenella [*Bacteroides*] *corrodens* (oral and intestinal microflora; cause of serious visceral and wound infections)

Genus *Streptobacillus* [*Haverhillia*] (rods and filaments in chains with filaments showing bulbous swellings, like a string of beads; parasites and pathogens of rats and other mammals)

Streptobacillus moniliformis [*Actinomyces muris ratti*] (type species; necklace-shaped bacteria found in nasopharynx of rats; streptobacillary rat-bite fever)

Section 6. Anaerobic Gram-Negative Straight, Curved, and Helical Rods

Family I. *Bacteroidaceae*

Genus I. *Bacteroides* (bacteroidosis)

Bacteroides fragilis (type species; opportunists in visceral and wound infections; most common anaerobe in soft tissue infections)

Table 7-1 Abbreviated Classification of Microbes from *Bergey's Manual of Determinative Bacteriology* (1974) and *Bergey's Manual of Systematic Bacteriology*, volume 1 (1984)—cont'd

Bacteroides oralis (gingival crevice of humans; oral, upper respiratory, and genital infections)

Bacteroides melaninogenicus (brown to black pigment on blood agar; opportunists in infections of mouth, soft tissue, and respiratory, alimentary, and urogenital tracts)

Genus II. *Fusobacterium* (purulent or gangrenous infections)

Fusobacterium nucleatum (type species; wound and respiratory tract infections)

Fusobacterium varium (wound and serous cavity infections; intestinal contents of roaches, termites)

Fusobacterium necrophorum [*Sphaerophorus necrophorus*] (abscesses of humans and animals)

Fusobacterium mortiferum (visceral abscesses and septicemia in humans)

Genus III. *Leptotrichia* (human oral cavity; not known as pathogens but found in clinical material)

Leptotrichia buccalis (type species)

Genus IV. *Butyrivibrio* (many members, biochemically versatile in rumen of most ruminants and in intestinal tract of other mammals)

Section 7. Dissimilatory Sulfate- or Sulfur-reducing Bacteria

Diverse, strictly anaerobic, gram-negative bacteria utilizing sulfate and other oxidized sulfur compounds as electron acceptors, reducing them to hydrogen sulfide; found in anaerobic mud and sediments of fresh water or brackish water, marine environments, and alimentary tract of humans and animals

Section 8. Anaerobic Gram-Negative Cocci

Family I. *Veillonellaceae* (alimentary tract parasites of humans, ruminants, rodents, and pigs)

Genus I. *Veillonella* (oral microflora; intestinal and respiratory tracts of humans and animals)

Veillonella parvula (type species; oral microflora)

Genus II. *Acidaminococcus* (amino acid cocci—amino acids sole energy source; intestinal tract of humans and pigs)

Genus III. *Megasphaera* (large cocci; rumen of cattle and sheep)

Section 9. The Rickettsias and Chlamydias

Order I. *Rickettsiales* (prokaryotic microbes; parasitic forms in macrophages and vascular endothelial cells or erythrocytes of vertebrates; mutualistic forms in insects; arthropods are vectors or primary hosts; disease in vertebrates and invertebrates)

Family I. *Rickettsiaceae* (intracellular parasites of tissue cells, not erythrocytes; arthropod vectors important)

Tribe I. *Rickettsieae* (adaptation to existence in arthropods but can infect vertebrate hosts; humans usually incidental hosts)

Genus I. *Rickettsia* (human pathogens; growth in cytoplasm, sometimes in nucleus, but not in vacuoles of cells; not cultivated in cell-free media)

Rickettsia prowazekii (type species; typhus fever—humans the reservoir)

Rickettsia typhi (murine typhus)

Rickettsia rickettsii (Rocky Mountain spotted fever)

Rickettsia sibirica (Siberian tick typhus)

Rickettsia conorii (fièvre boutonneuse, tick-bite fever)

Rickettsia australis (Queensland tick typhus)

Rickettsia akari (rickettsialpox)

Rickettsia tsutsugamushi (scrub typhus)

Continued.

Table 7-1 Abbreviated Classification of Microbes from *Bergey's Manual of Determinative Bacteriology* (1974) and *Bergey's Manual of Systematic Bacteriology,* volume 1 (1984)—cont'd

Genus II. *Rochalimaea* (like *Rickettsia* genus but usually in extracellular position in arthropod host; can be cultured in host cell-free media)

Rochalimaea quintana (type species; trench fever—humans the primary hosts)

Genus III. *Coxiella* (growth in vacuoles of host cell; highly resistant outside cells)

Coxiella burnetti (type species; Q fever)

Tribe II. *Ehrlichieae* (minute rickettsia-like organisms; only a few species; adapted to invertebrate existence; pathogens of mammals but not humans)

Genus IV. *Ehrlichia* (tick-borne diseases of dogs, cattle, sheep, goats, and horses)

Genus V. *Cowdria* (heartwater diseases of domestic ruminants)

Genus VI. *Neorickettsia* (helminth-borne disease of dogs, wolves, jackals, and foxes)

Tribe III. *Wolbachieae* (symbionts in arthropods but not in vertebrates; miscellaneous group)

Genus VII. *Wolbachia* (associated with arthropods—seldom pathogenic for hosts)

Genus VIII. *Rickettsiella* (pathogenic for insect larvae and other insect hosts)

Family II. *Bartonellaceae* (intracellular parasites, sometimes extracellular, in or on red blood cells of vertebrates; arthropod transmission; cultivatable on cell-free media)

Genus I. *Bartonella* (occur in or on erythrocytes and within fixed tissue cells; often have flagella; found in humans and sandfly *Phlebotomus*)

Bartonella bacilliformis (type species; Oroya fever, verruga peruana)

Genus II. *Grahamella* (intraerythrocytic; no flagella; no multiplication in fixed tissue cells; not found in humans; grahamellosis of rodents and other mammals)

Family III. *Anaplasmataceae* (very small viruslike particles within or on red blood cells of vertebrates; anemia main feature of disease; arthropod transmission)

Genus I. *Anaplasma* (parasites form inclusions in red cells—no appendages; anaplasmosis of ruminants)

Genus II. *Aegyptianella* (inclusions in red cells; aegyptianellosis in birds)

Genus III. *Haemaobartonella* (parasites within or outside erythrocytes, ring forms rare; pathogens in rodents, dogs, cats)

Genus IV. *Eperythrozoon* (parasites on red cells and in plasma; ring forms common; infections in various animals)

Order II. *Chlamydiales* (gram-negative parasites of vertebrates causing various diseases; obligately intracellular reproduction typical)

Family I. *Chlamydiaceae*

Genus I. *Chlamydia* [*Bedsonia*] (parasites of tissue cells of vertebrates; three ecologic niches: [a] humans—oculourogenital and respiratory diseases; [b] birds—respiratory and generalized diseases; and [c] mammals [not primates]—respiratory, placental, arthritic, and enteric diseases)

Chlamydia trachomatis (type species; trachoma, inclusion conjunctivitis, lymphogranuloma venereum, urethritis, proctitis in humans)

Chlamydia psittaci (ornithosis and psittacosis in birds; pneumonitis in cattle, sheep, goats; polyarthritis in sheep, cattle, and pigs; placentitis in cattle and sheep)

Section 10. The Mycoplasmas

Division Tenericutes

Class I. *Mollicutes* (prokaryotes lacking true cell wall; sometimes ultramicroscopic; saprophytes, parasites, or pathogens for animals; possibly arthropod-borne plant pathogens)

Order I. *Mycoplasmatales*

Family I. *Mycoplasmataceae* (sterol required for growth)

Genus I. *Mycoplasma* (pleuropneumonia group; tiny, very pleomorphic, delicate saprophytes and pathogens)

Mycoplasma mycoides (type species; contagious pleuropneumonia of cattle)

Mycoplasma gallisepticum (diseases in poultry)

Table 7-1 Abbreviated Classification of Microbes from *Bergey's Manual of Determinative Bacteriology* (1974) and *Bergey's Manual of Systematic Bacteriology,* volume 1 (1984)—cont'd

        Mycoplasma neurolyticum (rolling disease in mice and rats, epidemic conjunctivitis in mice; common in healthy as well as in diseased mice)

        Mycoplasma pulmonis (infectious catarrh and pneumonia of mice and rats)

        Mycoplasma canis (inhabitants of upper respiratory and genital tracts of dogs)

        Mycoplasma hyorhinis (arthritis in swine)

        Mycoplasma pneumoniae (cold hemagglutin–associated primary atypical pneumonia of humans)

        Mycoplasma agalactiae (mastitis of animals)

        Mycoplasma arthritidis (purulent polyarthritis of rats)

        Mycoplasma orale [*pharyngis*] (common parasites of human oropharynx)

        Mycoplasma salivarium (human gingival crevice; possible role in periodontal disease)

        Mycoplasma hominis (common parasites of lower genitourinary tract in humans; potential pathogens in postpartum fever and pelvic inflammatory disease)

        Mycoplasma fermentans (possible role in rheumatoid arthritis)

      Genus II. *Ureaplasma*

        Ureaplasma urealyticum ["T mycoplasmas"] (common mucous membrane parasites in urogenital tract of humans and animals)

    Family II. *Acholeplasmataceae* (sterol not required for growth)

      Genus I. *Acholeplasma* (free-living saprophytes; mammalian and avian parasites; wide host range; possible pathogens)

        Acholeplasma laidlawii (type species; saprophytes from sewage as well as parasites in animals)

Section 11. Endosymbionts

Intracellular symbionts of protozoa, insects, fungi, and invertebrates other than arthropods

Section 12. Gram-Positive Cocci

  Family I. *Micrococcaceae* (saprophytes and important pathogens; normal occupants of human skin and associated structures)

    Genus I. *Micrococcus* (small aerobic cocci common in soil and fresh water and on skin of humans and animals)

      Micrococcus luteus [*Sarcina lutea*] (type species; golden yellow pigment produced; pattern of tetrads)

    Genus II. *Stomatococcus* (encapsulated, coagulase-negative, salt-intolerant, and biochemically distinct from genera *Staphylococcus* and *Micrococcus*; probably part of normal flora of mouth and upper respiratory tract in humans)

    Genus III. *Planococcus* (motile cocci in seawater)

      Planococcus citreus (type species; yellowish orange pigment produced)

    Genus IV. *Staphylococcus* (primary relation to skin and mucous membranes of warm-blooded animals; wide host range; important pathogenic strains)

      Staphylococcus aureus (type species; type lesion: the abscess)

      Staphylococcus epidermidis (normal occupants of human skin and mucosae; opportunists in stitch abscesses and other wound infections)

Other organisms

  Family *Streptococcaceae* (saprophytic and pathogenic bacteria in chains)

    Genus *Streptococcus*

      Streptococcus pyogenes (type species; many types of infection, usually of diffuse or spreading nature [cellulitis])

      Streptococcus equisimilis (upper respiratory tract infections in humans and animals; erysipelas and puerperal fever)

Continued.

Table 7-1 Abbreviated Classification of Microbes from *Bergey's Manual of Determinative Bacteriology* (1974) and *Bergey's Manual of Systematic Bacteriology,* volume 1 (1984)—cont'd

Streptococcus zooepidemicus (septicemia of cows, rabbits, and swine; wound infections of horses)

Streptococcus equi (strangles in horses)

Streptococcus dysgalactiae (mastitis in cows; polyarthritis, or joint-ill, in lambs)

Streptococcus [Diplococcus] pneumoniae (lobar pneumonia, bronchopneumonia)

Streptococcus anginosus (varied infections; alpha- and gamma-hemolytic strains formerly known as *Streptococcus* MG relate to primary atypical pneumonia)

Streptococcus agalactiae (mastitis of cows; human infections)

Streptococcus salivarius (tongue, saliva, and feces of humans)

Streptococcus mutans (dental caries)

Streptococcus mitis (human saliva, sputum, feces)

Streptococcus bovis (alimentary canal of cows, sheep, and other ruminants)

Streptococcus equinus (alimentary tract of horses)

Streptococcus thermophilus (in milk and milk products; starter culture for Swiss cheese and yogurt)

Streptococcus faecalis (enterococcus in intestine of humans and warm-blooded animals; urinary tract infections, endocarditis)

Streptococcus lactis (important in dairy industry, some strains starter cultures in manufacture of cheese and cultured milk drinks)

Streptococcus cremoris (raw milk and milk products)

Genus *Leuconostoc* (colorless nostoc; saprophytes found in slimy sugar solutions, on fruits and vegetables, in milk and dairy products)

Leuconostoc mesenteroides (type species; used in industrial fermentation)

Genus *Pediococcus* (microaerophilic saprophytes in fermenting plant material, especially spoiled beer)

Genus *Aerococcus* (microaerophilic saprophytes widely distributed in air, in meat brines, on raw and processed vegetables)

Family *Peptococcaceae* (occupants of alimentary and respiratory tracts of humans and animals and of normal human female genital tract; found in soil and on surface of cereal grains; lesions of human female genitalia)

Genus *Peptococcus* (lesions of viscera and serous cavities; postpartum septicemia; black colonies seen among species)

Peptococcus niger (type species)

Genus *Peptostreptococcus* (puerperal fever, pyogenic infections, septic war wounds, osteomyelitis, pleurisy, gangrene, sinusitis, dental infection, vulvovaginitis)

Genus *Ruminococcus* (in rumen and cecum and colon of animals; important there in fermentation of cellulose)

Genus *Sarcina* (nearly spherical cells in packets of 8 or more; found in soil, mud, surface of cereal seeds, and stomach contents of humans and animals)

Section 13. Endospore-forming Gram-Positive Rods and Cocci

Family *Bacillaceae* (saprophytic gram-positive rods, mostly; endospore formation dominant feature in defining genera)

Genus *Bacillus* (genus of great diversity in properties of members; several species produce antibiotics, some strains more than one; because of pattern of spore formation in nature, few species have distinctive habitats)

Bacillus subtilis (type species; endospores widespread, may even be in heat-treated surgical dressings and canned foods; some strains produce antibiotics)

Bacillus licheniformis (source of bacitracin; spores with high heat tolerance)

Bacillus cereus (found in foods)

Bacillus anthracis (anthrax)

Table 7-1 Abbreviated Classification of Microbes from *Bergey's Manual of Determinative Bacteriology* (1974) and *Bergey's Manual of Systematic Bacteriology,* volume 1 (1984)—cont'd

Bacillus thuringiensis (microbial insecticide)
Bacillus polymyxa (source of polymyxin; participates in retting of flax)
Bacillus stearothermophilus (thermophilic; endospores highly resistant to heat)
Bacillus brevis (source of tyrothricin)
Genus *Sporolactobacillus* (bacteria-like *Lactobacillus* but spore-forming)
Genus *Clostridium* (strict anaerobes, motile rods; soil and water bacteria; occupants of intestinal tract of humans and animals)
Clostridium butyricum (type species; soil, animal feces, cheese, naturally soured milk)
Clostridium beijerinckii (wound infections)
Clostridium sporogenes (wound infections; low-grade bacteremia in debilitated patients)
Clostridium botulinum (potent exotoxin causes botulism)
Clostridium histolyticum (wound infections)
Clostridium novyi (gas gangrene)
Clostridium perfringens [*Bacterium welchii*] (gas gangrene)
Clostridium septicum (gas gangrene)
Clostridium tertium (wound infections, low-grade bacteremia in debilitated patients)
Clostridium tetani (potent exotoxin causes tetanus)
Genus *Desulfotomaculum* (sausage-shaped bacteria reducing sulfur; strict anaerobes; common soil and water saprophytes)
Genus *Sporosarcina* (spore-forming spherical cells, strict aerobes)

Section 14. Regular, Nonsporing, Gram-Positive Rods

Family *Lactobacillaceae* (rods, nonmotile, anaerobic or facultative, highly saccharoclastic; unusual as pathogens; found in fermenting animal and plant products)
Genus *Lactobacillus* (found in dairy products and effluents, grain and meat products, water, sewage, beer, wine, fruits, fruit juices, pickled vegetables, sourdough, mash; some species in tooth decay)
Lactobacillus delbrueckii (type species; fermenting grain and vegetable mashes)
Lactobacillus leichmannii (compressed yeast, grain mash)
Lactobacillus lactis (milk, cheese; starter cultures in manufacture of cheese)
Lactobacillus bulgaricus (sour milk as yogurt)
Lactobacillus acidophilus (feces of infants, mouth and vagina of young human adults)
Lactobacillus casei (milk, cheese, dairy products, sourdough, cow dung, silage, human alimentary tract and vagina)
Lactobacillus brevis (milk, kefir, cheese, sauerkraut, sourdough, soils, ensilage)
Genus *Listeria* (listeriosis)
Listeria monocytogenes (type species; listeriosis in humans and animals)
Genus *Erysipelothrix* (parasites of mammals, birds, fish; widespread in nature)
Erysipelothrix rhusiopathiae [*insidiosa*] (type species; swine erysipelas, erysipeloid in humans)

Section 15. Irregular, Nonsporing, Gram-Positive Rods

Animal and saprophytic corynebacteria
Genus *Corynebacterium*
Section I. Human and animal parasites and pathogens (club bacteria, nonspore-forming, gram-positive, irregular rods; widespread in nature)
Corynebacterium diphtheriae (type species, diphtheria—the effect of a highly lethal exotoxin)
Corynebacterium pseudotuberculosis (pseudotuberculosis; chronic purulent infections in warm-blooded animals, rarely in humans)
Corynebacterium xerosis (harmless occupants of conjunctiva; nontoxigenic)

Continued.

Table 7-1 Abbreviated Classification of Microbes from *Bergey's Manual of Determinative Bacteriology* (1974) and *Bergey's Manual of Systematic Bacteriology*, volume 1 (1984)—cont'd

Corynebacterium kutscheri (parasite and opportunistic pathogens of mice and rats)

Corynebacterium pseudodiphtheriticum (normal in human throat; nontoxigenic)

Corynebacterium equi (pneumonia of horses)

Section II. Plant pathogenic corynebacteria

Section III. Nonpathogenic corynebacteria (soil, water, and air)

Genus *Arthrobacter* (jointed gram-positive rods, strict aerobes; among dominant soil bacteria)

Genus *Cellulomonas* (coryneform bacteria with ability to attack cellulose)

Genus *Brevibacterium* (coryneform bacteria found in water, soil, dairy products, and insects; widespread in nature, sewage; many in industry)

Genus *Microbacterium* (small diphtheroid rods with rounded ends; found in dairy products)

Family *Propionibacteriaceae* (gram-positive, nonspore-forming, anaerobic, pleomorphic rods, saccharolytic members; found in skin and respiratory and alimentary tracts of most animals; some members in soft tissue infections)

Genus *Propionibacterium* (nonmotile, anaerobic to aerotolerant bacteria producing propionic and acetic acids; some pathogens here)

Propionibacterium freudenreichii (type species; raw milk, Swiss cheese, and dairy products)

Propionibacterium [*Corynebacterium*] *acnes* (soft tissue abscesses, wound infections; common laboratory contaminants)

Genus *Eubacterium* (motile or nonmotile, obligately anaerobic rods; found in cavities of humans and animals, plant products, soil; infections of soft tissue)

Eubacterium foedans (type species; dental tartar; varied infections)

Order *Actinomycetales* (gram-positive soil bacteria, mostly aerobic, forming branching filaments, which in some families develop into mycelium; some members acid-alcohol fast; some species fix nitrogen as obligate symbionts in plant root nodules; some members pathogenic to humans, animals, and plants)

Family *Actinomycetaceae* (diphtheroid bacteria; no mycelium, nonacid-fast, usually facultative anaerobes; branching filaments)

Genus *Actinomyces* (actinomycosis)

Actinomyces bovis (type species; lumpy jaw in cattle)

Actinomyces israelii (human actinomycosis)

Actinomyces naeslundii (oral cavity of humans—in tonsillar crypts and dental calculus)

Genus *Arachnia* (branched diphtheroid rods, gram-positive, nonacid-fast pathogens)

Arachnia propionica (type species; human actinomycosis)

Genus *Bifidobacterium* (bifid bacteria; highly variable rods, gram positive, nonacid-fast)

Bifidobacterium bifidum [*Lactobacillus bifidus*] (type species; stools and alimentary tract of breast-fed and bottle-fed infants, and adults)

Genus *Bacterionema* (thread-shaped bacteria, "whip handle" cells, nonacid-fast with filamentous branching morphology; on teeth and in oral cavity, especially in calculus and plaque deposits)

Genus *Rothia* (gram-positive, nonacid-fast aerobes with filamentous branching; common in normal mouth and throat)

Section 16. Mycobacteria

Family *Mycobacteriaceae* (important acid-fast bacteria, parasitic and saprophytic; branching inconspicuous)

Genus *Mycobacterium* (mycobacteriosis, chronic granulomatous disease of varied forms, tuberculosis)

Mycobacterium tuberculosis (type species; human tuberculosis)

Table 7-1 Abbreviated Classification of Microbes from *Bergey's Manual of Determinative Bacteriology* (1974) and *Bergey's Manual of Systematic Bacteriology*, volume 1 (1984)—cont'd

Mycobacterium bovis (tuberculosis in cattle, humans)

Mycobacterium kansasii (yellow bacillus, Group I photochromogens; chronic pulmonary disease similar to tuberculosis)

Mycobacterium marinum (tuberculosis of saltwater fish, swimming pool granulomas in humans)

Mycobacterium scrofulaceum (scrofula scotochromogen; suppurative cervical lymphadenitis in children)

Mycobacterium intracellulare (Battey bacillus; severe chronic pulmonary disease in humans)

Mycobacterium avium (tuberculosis in birds)

Mycobacterium ulcerans (skin ulcers in humans)

Mycobacterium phlei (timothy bacillus, hay bacillus; widespread in nature)

Mycobacterium smegmatis (smegma bacillus)

Mycobacterium fortuitum (mycobacteriosis in animals and in humans; found in soil)

Mycobacterium paratuberculosis (Johne's disease in cattle and sheep)

Mycobacterium leprae (leprosy)

Mycobacterium lepraemurium (rat leprosy)

Section 17. Nocardioforms

Family *Nocardiaceae* (nocardiosis; gram-positive aerobic actinomycetes; mycelium rudimentary or extensive; spore production variable)

Genus *Nocardia* (acid-fast to partially acid-fast, gram-positive, obligate aerobes; some strains pigmented; found in soil)

Nocardia asteroides (pulmonary nocardiosis, chronic subcutaneous abscesses, mycetomas)

Nocardia brasiliensis (mycetoma)

Genus *Pseudonocardia* (false nocardia; soil organisms, nonacid-fast)

Section 18. Gliding, Nonfruiting Bacteria

Order *Cytophagales* (rods or filaments, gram negative, with slow or rapid gliding; no fruiting bodies; chemolithotrophs, chemoorganotrophs, or mixotrophs)

Family *Cytophagaceae* (carotenoid pigments present; motility by gliding)

Genus *Cytophaga* (type genus; cellulose or chitin digested; common in soil and in fresh water and saltwater)

Order *Beggiatoales* (colorless, flexible filaments with cells in chains; motility by gliding; gram negative; mixotrophs or chemoorganotrophs; aerobic or microaerophilic; metabolism respiratory)

Family *Beggiatoaceae* (cells may or may not contain sulfur in presence of H_2S)

Genus *Beggiatoa* (cells contain granules of sulfur when grown in presence of H_2S)

Section 19. Anoxygenic Photosynthetic Bacteria

Order *Rhodospirillales* (mostly water bacteria; gram negative, variably shaped, all with bacteriochlorophyls and carotenoid pigments; purple-violet, purple, red, or orange-brown, brown, or green colors from photopigments in cell suspensions; some can fix nitrogen; purple [sulfur or nonsulfur] and green sulfur groups of bacteria here)

Section 20. Budding and/or Appendaged Bacteria

Soil and water bacteria reproducing by budding; may have excreted appendages and holdfasts; in some a semirigid appendage, the prostheca, proceeds out from the cell [stalk of *Caulobacter* genus] extending the length of the rod and holding a small bit of glue at its tip

Continued.

Table 7-1 Abbreviated Classification of Microbes from *Bergey's Manual of Determinative Bacteriology* (1974) and *Bergey's Manual of Systematic Bacteriology,* volume 1 (1984)—cont'd

Section 21. Arachaeobacteria

Order *Methanobacteriales*

Family *Methanobacteriaceae* (gram-positive or gram-negative rods or cocci, very strict anaerobes, motile or nonmotile, in a highly specialized physiologic group; to gain energy for growth, they reduce carbon dioxide–forming methane [Fig. 7-1] or ferment compounds such as acetate and methanol; widespread in nature in anaerobic habitats—sediments of natural waters, soil, anaerobic sewage digestors, and gastrointestinal tract of animals and humans)

Genus *Methanobacterium* (methane-producing rodlets)

Order *Methanococcales*

Family *Methanococcaceae*

Genus *Methanococcus* (methane cocci)

Order *Methanomicrobiales*

Family *Methanosarcinaceae*

Genus *Methanosarcina* (methane-producing sarcinae—large spherical cells in regular packets)

Order *"Sulfolobales"* (organisms metabolizing sulfur and sulfur compounds)

Family *"Sulfolobaceae"*

Genus *Sulfolobus* (spherical cells with lobes, sulfur-oxidizing, and resembling mycoplasmas, in solfatara areas containing hot acid environments, both soil and water)

Section 22. Sheathed Bacteria

Sheath present, may be encrusted with iron or manganese oxides; single cells; flagella may be found

FIG. 7-1   Methane gas burning out of a hollow increment borer bit drilled 20 cm into a cottonwood growing on shore of Lake Wingra, Wisconsin. (Timed exposure at night.) (From Zeikus, J.G., and Ward, J.C.: Science **184**:1181, 1974. Copyright by the American Association for the Advancement of Science.)

Table 7-1 Abbreviated Classification of Microbes from *Bergey's Manual of Determinative Bacteriology* (1974) and *Bergey's Manual of Systematic Bacteriology*, volume 1 (1984)—cont'd

Genus *Leptothrix* (gram-negative, strictly aerobic straight rods in chains within a sheath; also free-swimming as single cells, in pairs, or in motile short chains; sheaths often impregnated with hydrated ferric or manganic oxides; prevalent in iron-containing, uncontaminated, slow-running, fresh waters

Section 23. Gliding, Fruiting Bacteria

Order *Myxobacterales* (slime bacteria; gram negative, strict aerobes with slow gliding movements; found on soil and decomposing plant and animal matter; no photosynthetic pigments; chemoorganotrophs; energy-yielding mechanism respiratory, never fermentative; fruiting bodies formed from cell aggregates, often brightly colored and macroscopic; bacteriolytic and cellulolytic [attacking cellulose] groups here)

Section 24. Chemolithotrophic Bacteria

Organisms oxidizing ammonia or nitrite
   Family *Nitrobacteraceae* (soil and water bacteria converting ammonia to nitrite, nitrite to nitrate, and fixing carbon dioxide; nitrifying bacteria—one group oxidizes ammonia, the other nitrite)
      Genus *Nitrobacter* (nitrate rods oxidizing nitrite to nitrate)
      Genus *Nitrospina* (nitrate spines oxidizing nitrite to nitrate; straight slender rods found in South Atlantic Ocean)
      Genus *Nitrococcus* (nitrate spheres, yellowish to red, oxidizing nitrite to nitrate; found in South Pacific Ocean)
      Genus *Nitrosomonas* (ellipsoidal, yellowish to red bacteria oxidizing ammonia to nitrite)
      Genus *Nitrosospira* (nitrous spirals, yellowish to red, oxidizing ammonia to nitrite)
      Genus *Nitrosococcus* (nitrous spheres oxidizing ammonia to nitrite)
      Genus *Nitrosolobus* (nitrite-producing lobes; pleomorphic and lobate, yellowish to red bacteria oxidizing ammonia to nitrite; found in soils from South America, Southwest Africa, and Russia)
Organisms metabolizing sulfur and sulfur compounds
      Genus *Thiobacillis* (sulfur rodlets, small rods in soil, seawater, fresh water, acid mine water, sewage, and sulfur springs and near sulfur deposits or where hydrogen sulfide produced)
      Genus *Thiobacterium* (small sulfur rods near surface of sulfurous brackish and marine waters)
      Genus *Macromonas* (large cylindric or bean-shaped cells in fresh waters)
      Genus *Thiovulum* (small sulfur egglike cells with cytoplasm at one end of cell, a large vacuole at the other; found in fresh waters and seawaters where sulfide-containing waters contact oxygen-containing waters)
      Genus *Thiospira* (sulfur spirals in waters overlaying sulfurous muds)
Organisms depositing iron and/or manganese oxides
   Family *Siderocapsaceae* (iron bacteria of iron-bearing waters)
      Genus *Siderocapsa* (iron bacteria of fresh water; may be attached to surface of water plants)
      Genus *Naumanniella* (encapsulated rods widespread in iron-bearing waters; capsule encrusted with iron compounds and manganese oxide)
      Genus *Ochrobium* (bacteria yellowish from iron oxides; widespread in fresh waters bearing iron)
      Genus *Siderococcus* (iron cocci in fresh water and mud)
Magnetotactic bacteria (magnetically-responsive motile; passively align north-south in an applied magnetic field; contain novel, iron-rich, intracellular crystals)

Continued.

Table 7-1 Abbreviated Classification of Microbes from *Bergey's Manual of Determinative Bacteriology* (1974) and *Bergey's Manual of Systematic Bacteriology,* volume 1 (1984)—cont'd

Section 25. Cyanobacteria

Blue-green algae with gliding motility, producing oxygen in light; photosynthetic prokaryotes as single cells or simple or branched chains of cells; photopigments including chorophyl *a*

Section 26.

Order *Prochlorales* (photosynthetic containing both chlorophyls *a* and *b*; unicellular; reproduce by binary fission; classified with oxyphotobacteria)
    Family *Prochloraceae*
      Genus *Prochloron*

Section 27. Actinomycetes that Divide in More than One Plane

    Family *Frankiaceae* (symbiotic filamentous, mycelium-forming bacteria, which induce and live in root nodules of a wide variety of plants; nodules can fix molecular nitrogen)
      Genus *Frankia* (characteristic root nodules formed and inhabited by bacteria; also free state in soil)
    Family *Dermatophilaceae* (nonacid-fast, gram-positive organisms; skin lesions of mammals)
      Genus *Dermatophilus* (characteristic mycelium; skin pathogens)
      Genus *Geodermatophilus* (rudimentary mycelium, tuber-shaped thallus; soil organisms)

Section 28. Sporangiate Actinomycetes

    Family *Actinoplanaceae* (soil bacteria; distinctive mycelium; shape of sporangia and spore structure make division into genera)
      Genus *Actinoplanes* (globose spores)
      Genus *Spirillospora* (spiral spores)
      Genus *Streptosporangium* (spores coiled within sporangia)
      Genus *Amorphosporangium* (irregularly shaped sporangia)
      Genus *Ampullariella* (bottle-shaped sporangia)
      Genus *Pilimelia* (rod-shaped spores, end-to-end in parallel chains; about 1000 per sporangium)
      Genus *Planomonospora* (single, large, motile spore in each sporangium)
      Genus *Planobispora* (longitudinal pair of large spores in sporangium)
      Genus *Dactylosporangium* (finger-shaped sporangia)

Section 29. Streptomyctes and Their Allies

    Family *Streptomycetaceae* (gram-positive, aerobic, primarily soil forms; well-developed branched mycelium and special spores for multiplication; many species important in production of antibiotics)
      Genus *Streptomyces* (about 500 distinctive antibiotics come out of this genus; many members produce one or more antibacterial, antifungal, antialgal, antiviral, antiprotozoan, or antitumor antibiotics [see Table 14-1, p. 239, for examples]; well over 500 species known)
      Genus *Streptoverticillium* (whorled actinomycetes)
      Genus *Sporichthya* (spores motile in water)
      Genus *Microellobosporia* (small spores in a pod)

Section 30. Other Conidiate Genera

    Family *Micromonosporaceae* (saprophytic soil forms)
      Genus *Micromonospora* (well-developed branched septate mycelium, small single spores; nonacid-fast proteolytic and celluloytic organisms; some antibiotics found here)
        Micromonospora purpurea (source of gentamicin complex)

Table 7-1 Abbreviated Classification of Microbes from *Bergey's Manual of Determinative Bacteriology* (1974) and *Bergey's Manual of Systematic Bacteriology,* volume 1 (1984)—cont'd

Genus *Thermactinomyces* (heat-loving ray "fungus")
Genus *Actinobifida* (ray "fungus" with bifurcations)
Genus *Thermomonospora* (heat-loving, single-spored organisms)
Genus *Microbispora* (small two-spored organisms)
Genus *Micropolyspora* (small many-spored organisms)

SUMMARY

1 Various criteria must be used in the classification of bacteria: the morphology, cultural characteristics, physiologic behavior patterns, genetic analyses, and immunologic reactions.

2 The scientific classification of bacteria embodied in the classic *Bergey's Manual of Determinative Bacteriology* is the best known and the most generally accepted one, but it is being updated by *Bergey's Manual of Systematic Bacteriology,* of which only 1 of 4 volumes has been published to date.

3 Since bacterial classification is in a state of transition, the authors of volume 1 of *Bergey's Manual of Systematic Bacteriology* have used a few important criteria to define bacteria in sections. Eleven sections cover spirochetes, gram-negative bacteria, rickettsias, chlamydias, mycoplasmas, and bacterial endosymbionts.

QUESTIONS FOR REVIEW

1 What are the criteria most useful in the classification of bacteria?
2 Name the four bacterial divisions proposed in the kingdom *Procaryotae.*
3 Give the suffix indicating the name of a bacterial order and the suffix indicating the name of a bacterial family.
4 Name a pigmented bacterium.
5 Give the shape of spirochetes.
6 Name four organisms that are known as opportunistic pathogens.
7 Name five disease-producing bacteria and give the disease produced.
8 What is the major difference between oxyphotobacteria and anoxyphotobacteria?
9 Briefly define taxon, phototrophic, halobacteria, chemolithotrophic, cyanobacteria, biovar, endosymbiont, morphology, magnetotactic.

SUGGESTED READINGS

Brenner, D.J.: Impact of modern taxonomy on clinical microbiology, ASM News **49:**58, 1983.

Buchanan, R.E., and Gibbons, N.E., co-editors: Bergey's manual of determinative bacteriology, ed. 8, Baltimore, 1974, The Williams & Wilkins Co.

Holt, J.G., editor-in-chief, and Krieg, N.R., editor vol. 1: Bergey's manual of systematic bacteriology, vol. 1, Baltimore, 1984, The Williams & Wilkins Co.

Lapage, S.P., and others: International code of nomenclature of bacteria, 1975 revision, Washington, D.C., 1976, American Society of Microbiology.

Mandel, M.: New approaches to bacterial taxonomy: perspective and prospects, Ann. Rev. Microbiol. **23:**239, 1969.

Morowitz, H.J.: On first looking into *Bergey's Manual,* Hosp. Pract. **10:**156, April 1975.

Skerman, V.D.B., and others, editors: Approved lists of bacterial names, Intern. J. System. Bacteriol. **30:**225, 1980.

LABORATORY SURVEY OF MODULE ONE

Information's pretty thin stuff unless mixed with experience.
Clarence Day

DESIGN

A laboratory course in microbiology is planned to (1) present the student with the scientific method, (2) impress the fact of microbes as living agents, and (3) demonstrate characteristics of microbes and their effects on the human body.

The following exercises may be varied to meet the needs of the individual student and fit the teaching facilities of the school. Details of technic have been omitted in this type of abbreviated survey because it is felt that the student can be taught technics to which the instructor is accustomed. In all instances the exercises are designed to be carried out with minimum equipment and maximum safety.

The student needs a notebook in which to record accurately each experiment: (1) its purpose, (2) how it is performed, (3) results obtained, and (4) conclusions reached. The notebook should be neat and be corrected frequently by the instructor.

SAFETY FIRST!

RULES FOR YOUR OWN PERSONAL PROTECTION

Avoid any exposure of yourself or others to unnecessary danger. Do not take chances! Use common sense in an emergency. For safety in the laboratory, observe the following precautions:

1. Wash your hands thoroughly with soap and water at the beginning and end of each laboratory period. Disinfect your hands when necessary. Use hand lotion regularly to prevent dryness and cracking of the skin.
2. Wear a laboratory coat or jacket made of material easily cleaned and sterilized. The coat or jacket should be laundered at least once a week (more frequently when indicated).
3. Scrub the working surface of your desk with a cleansing agent and then disinfect that surface at the beginning and end of each laboratory period. A suitable detergent-germicide may be used.
4. Do not put anything in your mouth during the laboratory exercise (pipetting, etc.). Do not, under any circumstances, lick gummed labels or put pencils in your mouth. Do not eat, drink, or smoke in the laboratory during laboratory time.
5. At all times, keep your hands away from your face and mouth. Fingers and the areas under the nails can become contaminated quite easily.
6. Dispose of infectious material, cultures, and contaminated material carefully and *only* in the way prescribed by the instructor. Note the location of the special containers of disinfectant and use regularly.
7. Always flame-sterilize wire inoculating loops and needles carefully, faithfully, and thoroughly before laying them aside.
8. Be very careful to avoid spilling any kind of contaminated material. If infectious material contacts the desk, hands, clothing, or floor, notify the instructor at once.
9. Report even a minor accident to the laboratory instructor *immediately!*

10. Arrange cultures and infectious material in a secure place on the desk. Do not allow unused equipment to accumulate in your work area. Avoid clutter.
11. Keep glassware and other equipment clean and in its proper place. Be clean and orderly at all times.
12. Keep personal items in the places designated and away from the work area.

THE COMPOUND LIGHT MICROSCOPE
Care of the Microscope

"Faith" is a fine invention
When Gentlemen can see
But Microscopes *are prudent*
In an Emergency.
Emily Dickinson

The microscope, one of our most valued scientific instruments, is obviously "prudent" in a wide variety of situations. That its proper care is "prudent" is likewise evident. As you use the microscope, follow these instructions:

1. Keep both eyes open. You can do so with a little practice.
2. Many modern microscopes have built-in base illuminators. If yours does not and you must work with the mirror, avoid direct sunlight. North light is advantageous. Good results are obtained with daylight.
3. When you place a slide on the stage, see that it lies flat against the platform. Adjust the light so that the object is evenly illuminated.
4. Learn to focus with the low-power objectives (such as the 16 mm objective). Use the coarse adjustment to move the low-power objective until it nearly (but *not quite*) touches the cover glass or upper surface of the mounted specimen. Then focus *up* until the object comes plainly into view. Complete focusing with the fine adjustment. Keeping the specimen in focus, use very short back-and-forth strokes of the fine adjustment to produce an illusion of depth in the specimen. When the immersion objective is used, first place a drop of immersion oil on the object so that it can be clearly brought into view. Parfocal objectives are designed so that when switching from one objective to another, the microscopist can keep the specimen essentially in focus; most modern microscopes are equipped with them.
5. Keep the microscope clean, and handle all parts with care. Do not touch the glass parts of the scope with the fingers. Do not allow chemicals to contact the microscope, since they may injure it. Clean the mechanical parts with an application of olive oil on gauze. Wipe the oil off with chamois or lens paper. Keep optical glass parts (ocular lenses, condenser lens system, and nonimmersion objective lenses) of the microscope clean by frequent use of lens paper. Carefully remove immersion oil from the oil-immersion objective lens right after you have finished your microscopic study. At times it may be necessary to remove dried immersion oil with lens paper moistened with xylol. Do this as rapidly as possible to prevent injury to the optical settings of the oil-immersion objective lens system.
6. Clean the microscope thoroughly when you are finished. Leave the lowest power objective in the working position; thus the least expensive objective would be injured should the optical system be jammed down accidentally. Keep the microscope covered when you are not using it.

PROJECT   Use of the Compound Light Microscope

1. *Parts of the microscope*
   a. Place the microscope on a table in the proper working position at a convenient height. Be certain that you are comfortably seated and that you do not have to stretch or that you are not cramped.

      b. Locate all parts shown in Fig. 8-1. Refer to the discussion of the microscope on pp. 141-143.

      c. By means of their numbers, locate the low-power, high dry, and oil-immersion objectives. Explain the use of each in your notebook.

2. *Resolution*

      a. Place a prepared microslide on the stage of the microscope. (A section of tissue may be used.) Be sure that the specimen slide lies flat on the stage.

      b. Elevate the top of the condenser until it is flush with the surface of the stage.

      c. With the aid of the instructor, focus the low-power objective on the specimen.

      d. Observe the changes when the coarse and fine adjustments are manipulated. Note how an image comes in and out of focus.

3. *Illumination*

      a. Observe the changes in lighting brought about with a change in the position of the mirror. (NOTE: A microscope with a built-in lamp base does not have a mirror.) When you can fill the field of observation with light, have the instructor check to see if the lighting can be improved.

      b. Adjust the illumination so that it is even throughout the microscopic field of view.

4. *Contrast*

      a. Observe the changes in the specimen when the iris diaphragm is opened or closed.

      b. Note changes in contrast of the specimen image in the microscope. Note lesser changes in focus.

5. *Magnification*

      a. Focus on a specimen image with the low-power objective. (Complete focusing with the fine adjustment.) Center the specimen image.

      b. Move the high-power objective into position. Complete focusing with the fine adjustment. Center the specimen image. Note change in magnification. NOTE: The constant movement of the fine adjustment is essential in producing a three-dimensional effect on the specimen image in microscopy. Rotate the fine adjustment slightly clockwise and then slightly counterclockwise.

      c. Under the direct supervision of the instructor, carefully focus with the oil-immersion objective; first place a drop of immersion oil on the cover glass of the microslide (or on the upper surface of the specimen preparation). Lower the oil-immersion objective with the coarse adjustment into the drop of oil until the objective engages the drop and spreads it somewhat. The objective does not quite touch the surface of the microslide. Complete focus with the fine adjustment. Note the change in magnification.

      d. Under the supervision of the instructor, clean the microscope and return it to the locker.

6. *Problem*

      a. Compare and calculate the magnification of the different objectives of your microscope with each eyepiece and complete the following table:

	Objective		
Ocular	16 mm 10×	4 mm 43×	1.8 mm 97×
5×			
10×			
15×			

PROJECT   Microscopic Appearance of Bacteria
1. *Study of prepared slides*
   a. Using the oil-immersion objective of the microscope, study the appearance of bacteria in the prepared stained slides carefully. Draw representative fields.
   b. Note the different shapes of bacteria. Draw cocci, bacilli, and spirilla.
   c. Note the arrangements of bacteria. Draw bacteria arranged as staphylococci, diplococci, streptococci, and streptobacilli.
   d. Note the following structures related to bacteria and draw: (1) spores, (2) capsules, (3) metachromatic granules, and (4) flagella.
2. *Demonstration of motility of bacteria*
   a. Observe motility in hanging drop preparations set up as demonstrations.
   b. Contrast true motility with brownian motion. Define brownian motion.
   c. Note the steps in making a *hanging drop preparation* (demonstration and discussion by instructor).
3. *Use of the hanging drop*
   a. *Motile organisms*
      (1) Consult p. 147 for the method.
      (2) Use a broth culture of a motile organism to make a hanging drop preparation. NOTE: Be careful to sterilize the wire loop in the flame each time *before* and *after* it is used.
      (3) Examine the preparation with the high dry objectives of the microscope. The amount of light passing through the substage condenser must be somewhat reduced by partly closing the diaphragm. A properly made preparation shows the cells standing out distinctly against a dimly lit background.
      (4) Carefully discard the preparation according to the method given by your instructor. *Why is this important?*
   b. *Nonmotile organisms*
      (1) Make a hanging drop preparation of a suspension of carmine—note brownian motion.
      (2) Demonstrate similarly nonmotile organisms; note that they do not change their position in relation to each other. Note brownian motion.
4. *Demonstration of the use of the dark-field microscope*
   The instructor demonstrates the use of the dark-field microscope and briefly discusses its application.
5. *Application of negative or relief staining*
   a. Place a loopful or two of a broth culture or watery suspension of bacteria on a slide.
   b. Add an equal amount of commercial india ink that has been diluted with two parts of water.
   c. Mix thoroughly and spread out slightly.
   d. Allow the smear to dry, and examine with the oil-immersion objective. In satisfactory portions of the smear the bacteria stand out as colorless bodies against a gray-brown or black background. With *relief* staining, the background is stained, *not* the bacteria.

PROJECT   Staining of Bacteria
1. *Simple stains*
   a. Make three smears of organisms furnished by the instructor according to directions given on p. 150. *Do not neglect proper sterilization of the wire loop.*
   b. Fix the smears; then pour a few drops of methylene blue on the first, a few drops of gentian violet on the second, and a few drops of carbolfuchsin on the third. Let the stains act for 1 minute.
   c. Wash the stains off with distilled water, and drain the slides. Blot with blotting

paper. Examine with the oil-immersion lens. Make drawings. (Use colored pencils.) What is the color given to the organisms by the dye used in each smear?

2. *Gram stain*
   a. Obtain fixed smears already prepared by the instructor, or prepare such smears from suitable organisms furnished by the instructor.
   b. Refer to the technic of staining outlined on p. 152.
   c. Examine smears with the oil-immersion objective of the microscope. Are the organisms furnished you gram positive or gram negative? How do you know?

3. *Acid-fast stain*
   a. Obtain prepared and fixed slides of sputum from the instructor.
   b. Refer to the technic of acid-fast staining outlined on p. 152.
   c. Examine smears with the oil-immersion objective. What color is *Mycobacterium tuberculosis* when stained by this method? What is the color of nonacid-fast organisms? Sketch acid-fast bacilli. (Use colored pencils.)

4. *Special stains*
   a. Make a methylene blue stain of a spore-forming organism. Note the spores, which appear as unstained areas in the cell body. Make a drawing.
   b. Make a Gram stain of pneumococci. Note the capsule, which stands out as an unstained halo around the cell. Make a drawing.

EVALUATION FOR MODULE ONE

PART I

In the following statements or questions, please circle the number in the column on the right that correctly completes the statement or answers the question.

1. In what order do we arrange the following groups of living organisms, beginning with the largest or most general classification?

   (a) Species
   (b) Family
   (c) Phylum
   (d) Order
   (e) Genus
   (f) Class

   1. c, d, f, b, e, and a
   2. b, c, a, f, e, and d
   3. d, a, f, c, b, and e
   4. c, f, d, b, e, and a
   5. none of above is correct

2. Which of the following bacteria are spiral in shape?

   (a) Diplococci
   (b) Bacilli
   (c) Vibrios
   (d) Spirochetes
   (e) Streptococci

   1. a and e
   2. a
   3. b
   4. c and d
   5. e

3. Which of the structures listed below would be resistant to conditions adverse to bacterial growth?

   (a) Capsules
   (b) Flagella
   (c) Metachromatic granules

   1. a only
   2. d only
   3. a and d

(d) Spores
(e) Polar bodies

4. a, b, c, and d
5. none of these

4. The method by which bacteria reproduce is:
   (a) Budding
   (b) Transverse fission
   (c) Spore formation
   (d) Conjugation
   (e) Capsule formation

1. a only
2. b only
3. c only
4. d only
5. c and e

5. Which of the following maintains the shape of the bacterial cell?
   (a) Spore
   (b) Flagella
   (c) Plasma membrane
   (d) Capsule
   (e) Cell wall

1. e only
2. a and e
3. c and e
4. d and e
5. c only

6. A structure in the bacterial cell rich in enzymatic activity is the:
   (a) Plasma membrane
   (b) Cell wall
   (c) Metachromatic granule
   (d) Capsule
   (e) Flagellum

1. a and b
2. d and e
3. b and c
4. b only
5. a only

7. A protoplast is not:
   (a) A bacterial cell with a spore
   (b) A bacterial cell without a spore
   (c) A bacterial nucleus
   (d) A bacterial spore without a coat
   (e) A bacterial cell without a cell wall

1. a and e
2. a, c, and e
3. a, b, and c
4. d only
5. a, b, c, d, and e

8. Plasmids:
   (a) Can replicate autonomously
   (b) Can be a vector in recombinant DNA tech-
       nology
   (c) Can exist in the cytoplasm
   (d) Can vary in size
   (e) Are the same as phages

1. all of these
2. none of these
3. a, b, d, and e
4. a, b, c, and d
5. b and d

9. In the stationary phase of bacterial growth:
   (a) Cell division and cell death may be in equi-
       librium
   (b) Rate of cell metabolism is maximal
   (c) Bacteria do not form spores
   (d) Morphology and staining characteristics are
       most uniform
   (e) Viable bacterial counts cannot be done

1. all of these
2. a, b, and c
3. a only
4. d only
5. none of these

10. Bacteria are capable of which of the following variations?
    (a) A change from coccus to bacillus
    (b) A change from an encapsulated to a nonen-
        capsulated organism
    (c) A change from a nonvirulent to a virulent or-
        ganism
    (d) A change from a spirillum to a spirochete
    (e) A change from an antibiotic-susceptible to an
        antibiotic-resistant organism

1. all are true
2. all but d
3. all but a and d
4. all but a, b, and e
5. none is true

11. Which of the following organisms exist only on living material?
    (a) Facultative saprophyte
    (b) Strict parasite

1. b
2. b and d

(c) Facultative parasite   3. c
(d) Strict saprophyte   4. a

12. Which terms indicate that two organisms can live together?
    (a) Commensalism   1. d
    (b) Symbiosis   2. b and e
    (c) Angatonism   3. a and b
    (d) Positive chemotaxis   4. c
    (e) Plasmoptysis   5. a, b, and c

13. In which groups would you *not* expect to find concentrated the organisms of greatest medical importance?
    (a) Autotrophs   1. a
    (b) Lithotrophs   2. b
    (c) Psychrophiles   3. all of these
    (d) Obligate halophiles   4. a, b, c, and d
    (e) Thermophiles   5. c and d

14. The process by which the red cell gives up water to a hypertonic plasma is called:
    (a) Plasmoptysis   1. c
    (b) Plasmolysis   2. b
    (c) Osmosis   3. b and c
    (d) Metabolism   4. a and c
    (e) None of the above   5. e

15. The nuclear equivalent for bacteria:
    (a) Undergoes a true mitosis during division 1. a, b, and e
    (b) Is enclosed in a nuclear membrane   2. b only
    (c) Is rich in RNA   3. all of these
    (d) Is rich in DNA   4. b and d
    (e) Contains neither of the nucleic acids   5. d only

16. The lag phase of bacterial growth:
    (a) Is a period of active multiplication   1. a only
    (b) Involves adaptation to a new environment 2. b only
    (c) Occurs between the exponential phase and 3. c only
        the stationary phase   4. d and e
    (d) Is not influenced by environmental factors 5. a and e
    (e) Is constant in duration

17. Public health authorities are concerned about the increasing prevalence of plasmids among enteric bacteria because of the following characteristics of plasmids:
    (a) Relatively high rates of transmissibility 1. a, b, and e
    (b) Transmission across species and genus bar- 2. b, c, and d
        riers   3. e only
    (c) Frequent carriage of multiple antibiotic resis- 4. none of these
        tances   5. all of these
    (d) Determination of bacterial traits crucial to ad-
        aptation
    (e) Widespread occurrence among bacteria

18. Lipopolysaccharide of gram-negative bacteria is found in the:
    (a) Cytoplasmic membrane   1. a, b, and c
    (b) Outer membrane   2. c only
    (c) Inner membrane   3. d and e
    (d) Peptidoglycan   4. b and c
    (e) Teichoic acid polymer   5. b only

19. Extracellular deoxyribonuclease (DNAse from lysed bacteria) should not inhibit the process of:
    (a) Mutation   1. all but a

(b) Transformation
(c) Transduction
(d) Conjugation
(e) Recognition

2. all but b
3. all but c
4. all but d
5. all but e

20. The lipid A portion of lipopolysaccharide is:
   (a) Responsible for the endotoxic activity of gram-positive bacteria
   (b) Involved as a carrier lipid in the biosynthesis of O antigens
   (c) Usually associated with teichoic acids
   (d) Associated with the outer membrane
   (e) Associated with the inner membrane

1. a
2. b
3. c
4. d
5. e

21. The oxidation of nutrients in microbes:
   (a) Always requires oxygen
   (b) Is the means for supplying energy to the microbial cell
   (c) Is usually completed to carbon dioxide and water
   (d) Is balanced by the reduction of substrates and/or intermediates
   (e) Does not involve enzymatic action

1. a, b, and c
2. b, c, and d
3. b and d
4. d and e
5. a and e

22. The phenotype of a bacterial strain:
   (a) Depends on the genotype
   (b) Depends on the environment
   (c) Can vary without changes in the genotype
   (d) Refers to a limited number of physiologic characteristics
   (e) Is not easily measured

1. a, b, and c
2. b, c, and d
3. c, d, and e
4. d and e
5. a and d

23. Bacterial cell walls:
   (a) Contain antigens
   (b) Are synthesized by pathways that are sensitive to the action of certain antimicrobial compounds
   (c) May be removed without destroying the cell viability
   (d) Are reduced or absent in cells of rough colony type
   (e) Do not influence the Gram-staining reaction of the cell

1. a, b, and d
2. a, b, and e
3. a, b, and c
4. b and c
5. all of these

24. Bacterial spores:
   (a) When present cannot be seen to be stained in a routine Gram-stained preparation
   (b) Are more resistant than vegetative cells to most ordinary physical and chemical agents
   (c) Are produced by species of the genus *Bacillus*
   (d) Are rarely seen in cocci
   (e) Represent a mechanism for bacterial survival

1. a and c
2. b and d
3. e only
4. all of these
5. none of these

25. A normal bacterial growth curve includes:
   (a) A lag phase
   (b) A log phase
   (c) A smooth phase
   (d) A rough phase
   (e) A phase contrast

1. a, b, and e
2. a and b
3. c, d, and e
4. e only
5. all of these

26. Flagella are more likely to be found:
    (a) On a rod-shaped bacterium   1. a, b, and c
    (b) On all cocci   2. a and b
    (c) In the Gram-stained smear   3. a and c
    (d) On all curved organisms   4. a and d
    (e) As surface appendages to a bacterial cell   5. a and e
        rather than in other positions

27. The generation time during the logarithmic phase of growth:
    (a) Is not the same for all bacterial species   1. all of these
    (b) Is the time required for the viable cell count   2. none of these
        to double   3. a, b, c, and d
    (c) Is shortest when growth temperature is opti-   4. b, c, d, and e
        mal   5. b and d
    (d) Is related to the time of onset of the disease
        produced by the given bacterium
    (e) Is usually 20 to 30 minutes at a minimum

28. The genetic process of recombination:
    (a) Can be accomplished with cell-free extracts of   1. all of these
        bacteria   2. none of these
    (b) May occur when two genotypes of bacteria   3. a, b, c, and d
        are grown together   4. b, c, d, and e
    (c) Can never involve a change in more than one   5. b and e
        trait
    (d) Is essential for bacterial reproduction
    (e) Usually does not occur unless donor and re-
        cipient bacterial cells are closely related

29. Which of the following observations are accurate with regard to microbes that are
    strict anaerobes?
    (a) They lack the enzyme superoxide dismutase.   1. a and c
    (b) They possess the enzyme superoxide dismu-   2. a and d
        tase.   3. c and e
    (c) They are catalase negative.   4. b and d
    (d) They are catalase positive   5. b and c
    (e) They are of no medical importance.

30. The number of variable bacteria in a suspension can be determined by:
    (a) Measuring the degree of turbidity   1. a
    (b) Counting the number of cells in a measured   2. b
        volume under the microscope   3. c
    (c) Counting the colonies that develop from a   4. d
        measured volume   5. e
    (d) Measuring the amount of bacterial nitrogen
        present
    (e) Measuring the amount of bacterial carbon
        present

31. Conjugation in *Escherichia coli:*
    (a) Requires a transient cell-to-cell contact   1. all of these
    (b) Involves a bidirectional exchange of genetic   2. none of these
        material   3. a, b, and c
    (c) May result in the transfer of very large seg-   4. a, c, and d
        ments of DNA   5. a, d, and e
    (d) Involves inheritable mating types called F^+
        and F^-
    (e) Is a method of bacterial reproduction

32. Bacterial histones are:
    (a) More prevalent in gram-positive than in gram-negative bacteria
    (b) More prevalent in gram-negative than in gram-positive bacteria
    (c) More basic than eukaryotic histones
    (d) Less basic than eukaryotic histones
    (e) None of the above

    1. a and c
    2. e only
    3. a and d
    4. b and d
    5. b and c

33. Strict anaerobes:
    (a) Grow well in a candle jar
    (b) Grow well in highly reduced media
    (c) Include species of spore-forming rods
    (d) Do not tolerate atmospheric oxygen
    (e) Can utilize oxygen-containing compounds (inorganic nitrates, for example)

    1. a, b, and c
    2. all of these
    3. none of these
    4. b and c
    5. e only

34. Virulent bacteriophages:
    (a) Can carry out the lytic cycle only
    (b) Can carry out lysogeny only
    (c) Can carry out lysogeny or the lytic cycle
    (d) Can determine bacterial properties
    (e) Are responsible for the production of diphtheria toxin

    1. e
    2. d
    3. c
    4. b
    5. a

35. The most likely reason that some bacteria require a very rich medium in which to grow is that:
    (a) Their plasmids control the selection of media.
    (b) Their Gram-staining pattern is a factor.
    (c) The internal osmotic pressure is lower in these organisms.
    (d) An inability to synthesize a number of organic compounds needed for growth exists in these bacteria.
    (e) None of the above

    1. a
    2. b
    3. c
    4. d
    5. e

PART II

1 Consider the characteristics listed in the column on the left. Please indicate by a check mark in the appropriate column the relation to eukaryotes or prokaryotes.

	EUKARYOTES	PROKARYOTES
Nucleus present	☐	☐
Nucleus absent	☐	☐
Ribosomes 70S (in cytoplasm)	☐	☐
Ribosomes 80S (in cytoplasm)	☐	☐
Mitochondria absent	☐	☐
Endoplasmic reticulum present	☐	☐
Cell wall usually contains peptido-glycan	☐	☐
Cell membrane present	☐	☐
Nitrogen fixation well known	☐	☐
Mitosis present	☐	☐
Mitosis absent	☐	☐
Simple binary fission noted	☐	☐
Genetic combination occurs	☐	☐

2 Consider the characteristics listed in the column on the left. In the appropriate column, please indicate by a check mark the relation of these to the cell wall of gram-positive bacteria or to that of gram-negative bacteria.

	CELL WALL OF GRAM-POSITIVE BACTERIA	CELL WALL OF GRAM-NEGATIVE BACTERIA
Peptidoglycan present	☐	☐
Large amounts of teichoic acid present	☐	☐
Lipopolysaccharide present	☐	☐
Outer membrane present	☐	☐
O antigen identified	☐	☐
High tensile strength	☐	☐
Barrier to penetration by molecules of penicillin	☐	☐
Vulnerable to action of penicillin	☐	☐
Direct relation to toxicity of microorganisms	☐	☐

3 Consider the characteristic listed in the column on the left. In the appropriate column, please indicate by a check mark the relation of these to DNA or RNA.

	DNA	RNA
Pentose sugar ribose	☐	☐
Pentose sugar deoxyribose	☐	☐
Pentose sugar undetermined	☐	☐
Purine bases adenine and guanine	☐	☐
Pyrimidine base cytosine	☐	☐
Pyrimidine base thymine	☐	☐
Pyrimidine base uracil	☐	☐
Types—ribosomal, transfer, and messenger	☐	☐
Forms B, C, and Z	☐	☐
Molecule of heredity	☐	☐
Best known as single stranded	☐	☐
Best known as double stranded	☐	☐
Found in essentially all living organisms	☐	☐
Found in bacterial cell wall	☐	☐
Found in metachromatic granules	☐	☐

4 Match the item in column Beta with the phrase best defining it in column Alpha. Please indicate the match by placing the related letter in the appropriate space.

COLUMN ALPHA

_____ 1. Important in genetic engineering
_____ 2. Catalyze redox reactions
_____ 3. Hydrolytic activity for nucleic acids
_____ 4. May be exoenzyme or endoenzyme
_____ 5. Catalyze polymerization of nucleic acids
_____ 6. Coenzyme related to substrate oxidation
_____ 7. Remove mismatched DNA residues
_____ 8. Catalyze translation
_____ 9. Effect breakdown of toxic H_2O_2
_____ 10. Join DNA chains

COLUMN BETA

(a) Polymerases
(b) Nucleases
(c) Ligases
(d) Oxidoreductases
(e) Catalase
(f) NAD
(g) Restriction enzymes

PART III

1 Match the item in column Beta with the phrase best defining it in column Alpha. Please indicate the match by placing the related letter in the appropriate space.

COLUMN ALPHA

_____ 1. Enhanced at times by radiant energy
_____ 2. Dissimilar microbes closely associated
_____ 3. Smooth and rough colonies grown from pure culture
_____ 4. Variation in size, shape, and other features
_____ 5. Loss in disease-producing ability of microbe
_____ 6. Direct transfer of DNA into culture medium
_____ 7. Indirect transfer of genetic material by virus
_____ 8. Physiologic adjustment to environment
_____ 9. First observed with pneumococcal capsular types

COLUMN BETA

(a) Adaptation
(b) Pleomorphism
(c) Dissociation
(d) Commensalism
(e) Attenuation
(f) Recombination
(g) Transformation
(h) Mutation
(i) Transduction
(j) Conjugation
(k) Symbiosis

2 Identify the organisms from the following list as being either gram positive or gram negative. Please place the identifying number (before the name of the genus) in the appropriate column.

GENUS OF ORGANISMS
TO BE IDENTIFIED

1. *Neisseria*
2. *Staphylococcus*
3. *Streptococcus*
4. *Shigella*
5. *Escherichia*
6. *Salmonella*
7. *Lactobacillus*
8. *Clostridium*
9. *Haemophilus*
10. *Brucella*
11. *Pasteurella*
12. *Bacillus*
13. *Corynebacterium*
14. *Pseudomonas*
15. *Proteus*

GRAM POSITIVE		GRAM NEGATIVE	
____	____	____	____
____	____	____	____
____	____	____	____
____	____	____	____
____	____	____	____
____	____	____	____
____	____	____	____
____	____	____	____

3 Match the item in column Beta with the phrase best defining it in column Alpha. Please indicate the match by placing the letter related in the appropriate space. (Give the answer you consider to be the one best suited to the phrase.)

COLUMN ALPHA

_____ 1. Renaturation of DNA
_____ 2. DNA-directed synthesis of RNA
_____ 3. Hereditary biochemical pattern for living organisms
_____ 4. Determines genetic relatedness
_____ 5. Denaturation of DNA (unwinding of helix)
_____ 6. Allows synthesis of nucleic acid to be carried out on biochemical template
_____ 7. Reproduction of DNA
_____ 8. May be semiconservative
_____ 9. Template-directed synthesis of DNA
_____ 10. Three-nucleotide sequences in mRNA transcript
_____ 11. Triplet of nucleotides in tRNA
_____ 12. Protective methylation of purine and pyrimidine bases at specific sites on DNA
_____ 13. Gives genetic information for a particular amino acid
_____ 14. Determines homology of nucleic acids
_____ 15. Specificity of base pairs in DNA critical to the process
_____ 16. Mechanism protecting bacterial cell from intrusion by foreign DNA

COLUMN BETA

(a) Transcription
(b) Translation
(c) Genetic code
(d) Base ratio of DNA
(e) Complementarity
(f) Annealing
(g) Melting
(h) Hybridization
(i) Codon
(j) Anticodon
(k) Restriction
(l) Modification
(m) Replication
(n) Genome

PART IV

1 Match the item in column Beta with the phrase best defining it in column Alpha. Please indicate the match by placing the related letter in the appropriate space.

COLUMN ALPHA

_____ 1. Oxidation of glucose completed after glycolysis
_____ 2. End-stage of aerobic respiration in bacterial membrane
_____ 3. Molecular oxygen the final hydrogen acceptor in process of biologic oxidation
_____ 4. Organic compound the final hydrogen acceptor in process of biologic oxidation
_____ 5. Formation of ATP both a chemical and a transport process across bacterial membrane
_____ 6. Important components are cytochromes.
_____ 7. Active series of carriers in bacterial membrane
_____ 8. Any compound that can generate acetyl-CoA can be oxidized here.
_____ 9. Many different end-products from this process
_____ 10. Process can provide energy quickly and continuously as long as substrate present.

COLUMN BETA

(a) Tricarboxylic (citric) acid cycle
(b) Electron transport system
(c) Chemosmosis
(d) Fermentation
(e) Aerobic respiration
(f) Anaerobic respiration

2 True-False. Please circle either the *T* or the *F*.

T   F   1. Certain nonprotein organic molecules are crucial to enzymatic reactions.

T   F   2. Enzymes are not usually protein molecules.

T   F   3. Much of the energy required for bacterial metabolism comes from ATP.

T   F   4. Energy within a cell must be trapped in some chemical form, or else it degrades into heat, which is intolerable to the cell.

T   F   5. Adenosine diphosphate is a metabolic catalyst.

T   F   6. An agent causing an alteration in genetic material of a cell is a carcinogen.

T   F   7. An obligate intracellular parasite important in genetic studies is the bacteriophage.

T   F   8. Bacteriophages are essentially the same as plasmids.

T   F   9. Plasmids are like bacteriophages in that they are always the same size.

T   F   10. Penicillinase plasmids endow the staphylococcus with resistance to penicillins.

T   F   11. R plasmids are not very widespread and do not cross species barriers.

T   F   12. The best known microbiologic vector for recombinant DNA technology is *Escherichia coli*.

T   F   13. Vectors in bioengineering also include plasmids and bacteriophages.

T   F   14. The yeast *Saccharomyces cerevisiae* is used for recombinant DNA technology.

T   F   15. Both bacteriophages and plasmids are incorporated into the host genome.

T   F   16. In the process of lysogeny the affected bacterium harbors a genetic determinant called a prophage.

T   F   17. The loss of an electron is called oxidation in a chemical reaction.

T   F   18. Oxygen has a strong affinity for electrons and is therefore most often the electron acceptor in a chemical reaction.

T   F   19. Redox reactions are of little importance in microbiology.

T   F   20. Chemolithotrophs can use carbon dioxide as their sole source of carbon.

T   F   21. Chemoorganotrophs can also use carbon dioxide as their source of carbon.

T   F   22. The major route of glucose degradation to pyruvic acid in living systems is a strictly anaerobic process known as *glycolysis*.

T   F   23. Synthesis of carbohydrates and lipids in a bacterial cell is template directed.

T   F   24. Certain important coenzymes are recycled in metabolism.

T   F   25. The most commonly occurring sequence for glucose degradation is the Embden-Meyerhof-Parnas pathway.

PART V

1 Match the name of the person with the scientific discovery or work for which he or she is best known. Please indicate the match by placing the related letter from column Beta in the appropriate space in column Alpha

COLUMN ALPHA

_____ 1. Developed the polio vaccine first used in widespread immunization

_____ 2. Developed oral polio vaccine

_____ 3. Laid foundations for germ theory of disease

_____ 4. First identified penicillin

_____ 5. Discovered "jumping genes"

_____ 6. First described microbes under the microscope

_____ 7. Introduced modern method of vaccination against smallpox

_____ 8. Introduced the technic for isolating bacteria in a pure culture

_____ 9. Discovered the Australia antigen

_____ 10. First demonstrated a significant relation between cancer and viruses

COLUMN BETA

(a) Louis Pasteur
(b) Baruch S. Blumberg
(c) Anton van Leeuwenhoek
(d) Edward Jenner
(e) Jonas E. Salk
(f) Sir Alexander Fleming
(g) Robert Koch
(h) Albert B. Sabin
(i) Barbara McClintock
(j) Peyton Rous

2 Unscramble the following sequences for important terms in microbiology:

1. O O M C R M S O H E _____

2. E N S E G _____

3. S I C B M R O E _____

4. S T S I O R T P _____

5. A E P T Y H L O L H T S _____

6. G A A E L _____

7. B R I M O O S S E _____

8. O O M M S S E E _____

9. C N T G E E I O E C D _____

10. I O U A M T N T _____

11. N M N C E R O B A T I I O _____

MODULE TWO

MICROBES
PROCEDURES FOR STUDY

CHAPTER 8

VISUALIZATION OF MICROBES

PLAN OF ACTION FOR LABORATORY STUDY OF MICROBES

The microbiologist studies microbes in many ways, but in general five procedures, each with merits and some limitations, are standard for identifying and evaluating microbes in the microbiologic laboratory. These methods are presented in a sequence of five chapters, but discussions of the laboratory methods for identification and diagnosis are also included in material specifically related to the different microbes throughout the book.

The five methods are:

1. *Direct microscopic examination.* Masses and aggregates of microbial cells (colonies) are visible to the naked eye, but to visualize the individual microbial cells too tiny to be seen with the unaided eye, the microbiologist must use an instrument of precision, the microscope, to magnify them. A test specimen from an indicated source, such as an area of contamination or disease, is prepared for microscopic study. Some of the specimen is applied to, or spread thinly on, a glass microslide so that light rays from the light source on the microscope can pass through the material and allow visualization. Sometimes the microscopic preparation is viewed unstained, but more often the preparation is stained by one of several methods before the microbiologist studies it.

2. *Culture.* The microbiologist may make a culture from the test specimen. After the microorganisms have multiplied sufficiently to form visible growth, the microbiologist studies the physical pattern of that growth. Again the microbes may be studied in unstained or suitably stained microscopic preparations, this time in material taken from the colonies.

3. *Biochemical tests.* The microbiologist may use some of the colonial growth of the microbes (obtained from the original specimen) to determine their biologic properties. The growth is subjected to a series of biochemical tests from which identifying characteristics emerge.

4. *Animal inoculation.* The microbiologist may inject microbes recovered from the test specimen by cultural methods into a suitable laboratory animal and observe the reactions in that animal. Some of the original specimen itself may at times be directly inoculated into the test animal.

5. *Immunologic reactions.* The microbiologist may use antigen-antibody tests for the identification of microorganisms recovered.

Importance of Laboratory Safety

Pathogenic microbes must be handled with extreme caution and in accord with well-known principles of standard conduct and aseptic technic in microbiologic laboratories. *Such disease-producing agents can be very dangerous, and accidental laboratory infections can be fatal.* In one survey of 1300 infections among laboratory workers, 39 ended fatally. In many instances the infections occurred in research workers and highly trained technologists.

Remember that laboratory accidents cause serious infections. A potential hazard for injury to the laboratory worker exists at all times. Whether you are an instructor or a student, stress precautions. For instance, carefully handle contaminated needles and broken glass to prevent cuts and puncture wounds in the skin. Restrain laboratory animals firmly to avoid bites or scratches. Take measures to prevent bites of ectoparasites. Pipette carefully to avoid swallowing infectious material. Manipulate specimen containers gently to avoid the leaks, spills, splatters, and splashes that disperse infectious material to the environment.

TOOLS FOR STUDY

Microscopes

Man is a tool-using animal . . . without tools he is nothing, with tools he is all. . . .
Thomas Carlyle

The microbiologist has many instruments of precision. Some are in constant use; others are needed only in special investigations.

The Compound Light (Bright-field) Microscope

The compound light (bright-field) microscope (Fig. 8-1) is an instrument that magnifies an object suitably mounted and positioned so that it can be seen and studied by the human eye. It consists of a light source and lens systems for resolution and magnification. Condensers and prisms incorporated therein guide the passage of light through the microscope and assist in adjustments required for illumination of the specimen (the object) and for focusing on it.

The compound light microscope is the instrument most often used by the microbiologist and one to be handled with utmost care. Needless to say, its workmanship should be of the highest quality.

The term **resolution** (or resolving power) is important in microscopy. Resolution is the property whereby detail in an object can be seen as clearly distinct with the human eye, microscope, or camera. The limit of resolution, another important concept in microscopy, is given as the minimal distance between two points that allows for their definition as two separate points. The limit of resolution has been exceeded if two given points are seen as one point.

General description. Microscopes are of two kinds: simple and compound. A **simple** microscope is little more than a magnifying lens. A **compound** microscope incorporates two or more lens systems so that the magnification of one system is increased by the other (Fig. 8-1). Practically, a compound microscope consists of two parts: the supporting stand and the optical system. The supporting stand includes (1) a base and pillar, (2) an arm to support the optical system and house the fine adjustment, (3) a platform (stage) on which the object to be examined rests, and (4) a condenser and mirror fitted beneath the stage. The condenser and mirror focus the light either from an external source, such as a special kind of microscope lamp, or from an illuminating system

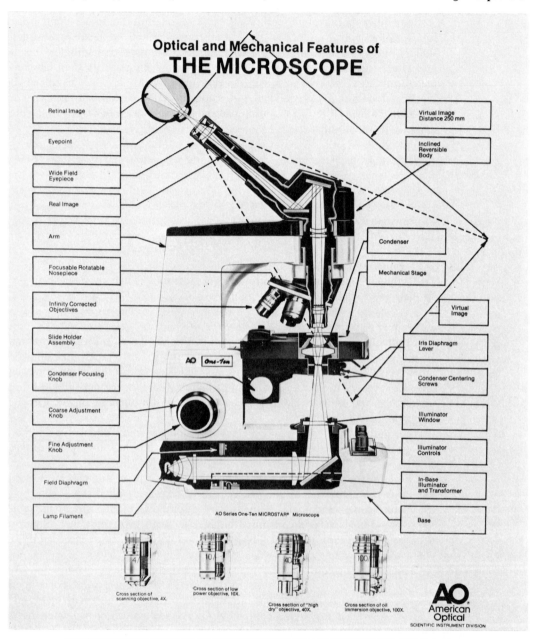

fitted into the base of the scope. Where present, the built-in base illuminator (self-contained substage illuminator) houses the light source, a collecting lens system, mirror, and a condenser lens (Fig. 8-1). An iris field diaphragm, a variable transformer attached to the system, and properly placed filters permit adjustments of the light.

The optical system consists of a body tube that supports the **ocular** lenses in the eyepiece(s) at the top end and the **objective** lenses attached to a revolving **nosepiece** at

FIG. 8-1 Dissection of microscope. Parts are labeled and optical features of monocular microscope are indicated schematically. Note how light rays are reflected from mirror in built-in base illuminator through microscope to eye of observer. This modern microscope is fitted with movable stage—specimen may be focused to objective and observer need not change eye level. (Courtesy AO Reichert, Scientific Instruments.)

the other end. The optical system is connected to the arm of the supporting stand by an **intermediate slide,** which moves up and down on the arm in response to movement of the **fine adjustment.** The intermediate slide contains the rack and pinion for the **coarse adjustment,** which acts directly on the tube of the optical system. The platform of the microscope is usually equipped with a **mechanical stage** to hold the microslide firmly in place. The object is mounted on the microslide so that it can be moved from place to place by set screws. The advantages of this device are that the specimen can be examined systematically and, unless moved, the specimen remains in a fixed position.

The objectives of a microscope are given $4\times$, $10\times$, $45\times$, and similar stipulations to indicate the magnifications of the system of lenses contained therein. American-made microscopes generally are fitted with $4\times$, $10\times$, $45\times$, and $100\times$ objectives as depicted in Fig. 8-1. For best results with the $100\times$ objective, a liquid (oil or water) must be placed between the objective and the object. Thus this objective is an **immersion** objective, and since only rarely would the lens system be immersed in water, it is best known as an **oil-immersion** objective. Most modern immersion oil is highly refined mineral oil having the same refractive index as glass (1.52).

The oculars (or eyepieces) of a microscope are given $5\times$, $10\times$, and similar designations to indicate that they increase the magnification of the objective, 5, 10, or more times, respectively. To obtain the magnification of any combination of ocular and objective lenses, multiply the magnification of objective by that of the ocular (Table 8-1). Remember that magnification refers to both the length and the width of an object; that is, a magnification of 100 means that the object is made to appear 100 times as long and 100 times as wide.

Microscopic units of measurement. For living objects of such small dimensions that they must be viewed with the microscope, a special unit of measurement is needed. For a long time this was the *micromillimeter* or *micron* designated by the Greek letter mu (μ). A micron is 1/1000 of a millimeter, or 1/25,400 of an inch, long, and a millimicron (mμ) is 1/1000 of a micron. (The paper on which this book is printed is about 100 μ thick.) Today both micron and millimicron are superseded by the equivalent and preferred terms, **micrometer** (μm) for micron and **nanometer** (nm) for millimicron. A nanometer is 1/1000 of a micrometer, or 10 angstroms (Å) (see Table 8-2 for the metric equivalents).

Ocular micrometer. In the light microscope microbes can be measured microscopically by means of a device known as a micrometer. The simplest type is the ocular micrometer, which consists of a scale on a glass disk that fits (scale side down) between the lenses of the eyepiece. The spaces between the lines on this scale do not represent true measurements but arbitrary units. Real values are obtained by calibration of the ocular micrometer against a stage micrometer. This is a glass slide with a true measurement scale on it—the lines on the scale are either exactly 10 or 100 μm apart. By

Table 8-1 Magnifications Attainable with Compound Light Microscope*

Oculars (Eyepieces)	Objectives						
	$2.5\times$	$4\times$	$10\times$	$20\times$	$45\times$	$50\times$	$100\times$
$5\times$	$12.5\times$	$20\times$	$50\times$	$100\times$	$225\times$	$250\times$	$500\times$
$10\times$	$25\times$	$40\times$	$100\times$	$200\times$	$450\times$	$500\times$	$1000\times$
$15\times$	$37.5\times$	$60\times$	$150\times$	$300\times$	$675\times$	$750\times$	$1500\times$
$20\times$	$50\times$	$80\times$	$200\times$	$400\times$	$900\times$	$1000\times$	$2000\times$

*The magnification of the given ocular times the magnification of the objective equals the total magnification.

placing the stage micrometer in the position occupied by a smear of cells and by looking through the ocular of the microscope, one may superimpose the scale of the ocular on that of the stage micrometer. One can then determine the actual unit length that the distance between the lines of the ocular micrometer represents.

The Dark-field Microscope

The dark-field microscope is essentially the bright-field microscope fitted with a dark-field condenser for certain applications in microbiology.

In the dark-field microscope an object appears bright against a dark background (the reverse of what is seen in the bright-field microscope). Such an optical effect may be produced by an opaque stop that is either inserted below or built into the condenser (the dark-field condenser) to permit only peripheral rays of light to enter the condenser. Since these rays pass through an object at an angle, they do not fill the microscopic field with light and the field appears black. However, because particles in the field do reflect the light, they stand out brightly against the dark background. In some instances, in contrast to the effect of the bright-field scope, light objects against a black background are more easily seen by the human eye. Therefore dark-field microscopy is used to visualize objects poorly defined by the bright-field and phase-contrast scopes; for example, the difficult-to-stain treponeme of syphilis. Dark-field examination is performed with living organisms in a wet mount. Motility is readily demonstrated.

The Phase-contrast Microscope

Phase-contrast illumination changes the path of light through the light microscope. The condenser used has an annular diaphragm, and it is paired to an objective with a phase plate or annular grove. The principle of phase contrast is simple. With light from the source the phase condenser produces a halo around the points where the object is being examined meets the milieu in which it is being viewed. Rays of light pass through its microscopic structure with parts varying in refractive index and emerge out of phase to display a pattern of bright and dark relief to the observer. Flagella, granular cytoplasmic inclusions, and sometimes unstained bacteria are examined more effectively with the phase-contrast microscope.

Table 8-2 Equivalents in the Metric System

	Meter (m)	Centimeter (cm)	Millimeter (mm)	Micrometer* (μm)	Nanometer* (nm)	Tenth Nanometer* (0.1 nm)
Meter (m)	1	100 10^2	1000 10^3	1,000,000 10^6	1,000,000,000 10^9	10,000,000,000 10^{10}
Centimeter (cm)	0.01 10^{-2}	1	10 10^1	10,000 10^4	10,000,000 10^7	100,000,000 10^8
Millimeter (mm)	0.001 10^{-3}	0.1 10^{-1}	1	1000 10^3	1,000,000 10^6	10,000,000 10^7
Micron (μ)	0.000001 10^{-6}	0.0001 10^{-4}	0.001 10^{-3}	1	1000 10^3	10,000 10^4
Millimicron (mμ)	0.000000001 10^{-9}	0.0000001 10^{-7}	0.000001 10^{-6}	0.001 10^{-3}	1	10 10^1
Angstrom (Å)	0.0000000001 10^{-10}	0.00000001 10^{-8}	0.0000001 10^{-7}	0.0001 10^{-4}	0.1 10^{-1}	1

*Preferred term.

The Electron Microscope

The information gained through the study of cells by the compound light microscope in just over 100 years contributed significantly to the development of modern medicine. However, the nature of light is such that it is not possible to get a clear image in the light microscope when the magnification of the object is greater than 2000 times. The limit of resolution of the compound light microscope, even with the best of lenses, is about half the wavelength (200 nm or 0.2 μm) of the visible light used, which has a minimum wavelength of about 400 nm. Hence many minute (submicroscopic) structures within cells were not visualized until the electron microscope was used.

The electron microscope, one of the most powerful research tools, differs from the compound microscope. In the latter, rays of ordinary light pass from the object being examined to the eye to be focused there, whereas in the electron microscope electrons pass through the specimen to be focused on a viewing screen from which a photograph is made—an **electron micrograph.** In the most modern electron microscopes the dimensions of the object under observation may be magnified 1 million times. If careful photographic technics are used, a further enlargement can be accomplished, so that today a potential magnification of 2 million or more times exists. (With this magnification a human hair viewed in its entirety in an electron micrograph would appear twice the size of a California redwood tree.)

The scanning electron microscope provides a three-dimensional capability for the study of specimens at magnifications up to 100,000 times the actual size. It sweeps a very narrow beam of electrons back and forth across a specimen, revealing its surface features rather than its internal structure.

A unit even smaller than the micrometer must be used to designate the size of objects too small to be visualized in the light microscope. This unit is the *nanometer* (nm). (*Angstrom,* although not the preferred term, is still used extensively.) In Table 8-3 the sizes of *microscopic* and *ultramicroscopic* objects are compared. An ostrich egg and the mature human ovum are added to illustrate the magnitude of the size discrepancy.

Table 8-3 Comparative Sizes of Biologic Objects

Limit of Resolution	Biologic Object	Diameter	
		Micrometer (μm)	Nanometer (nm)
Human eye (100 μm)	Ostrich egg	200,000	200,000,000
	Mature human ovum	120	120,000
Light microscope (0.2 μm)	Erythrocyte (red blood cell)	7.5	7500
	Serratia marcescens (bacterium)	0.75	750
	Rickettsia species	0.475	475
	Chlamydia psittaci	0.27	270
Electron microscope (0.001 μm*)	*Mycoplasma* species	0.15	150
	Influenza virus	0.085	85
	Genetic unit (Muller's estimation of largest size of a gene)	0.02×0.125†	20×125†
	Poliomyelitis virus	0.027	27
	Tobacco necrosis (plant virus)	0.016	16
	Egg albumin molecule (protein molecule)	0.0025×0.01†	2.5×10†
	Hydrogen molecule	0.0001	0.1

*For most biologic specimens the limit of resolution of the electron microscope is 0.001 μm, or 1 nm. Under special conditions, resolution may be obtained at 2 or 3 Å.
†Width × length.

The Ultraviolet Microscope

Greater resolution than can be achieved with white light can be obtained with ultraviolet light, with magnification two or three times higher, because of the shorter wavelength of ultraviolet light. Since this kind of light is invisible, a photographic plate must be used to record to image of the object. Glass is impervious to ultraviolet light; therefore quartz lenses must be used in the instrument. This kind of microscope is used for fluorescence microscopy.

The Fluorescence Microscope

A fluorescent substance, such as a fluorescent dye or fluorochrome, absorbs ultraviolet light and then, after it has absorbed energy, emits visible light. Fluorescence microscopy (p. 402) is designed to visualize the microscopic preparation stained or in some way tagged with fluorescent material. For this purpose the ultraviolet microscope is appropriate, with a source of ultraviolet light, various filters (an integral part of the system), and a dark-field condenser (Fig. 8-2).

   For illumination, the commonly used source of ultraviolet light is either a high-pressure mercury lamp, contained within its own housing unit, or a halogen quartz

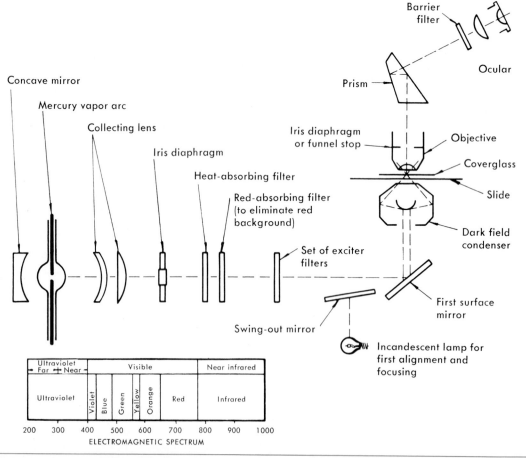

FIG. 8-2   Schematic representation of equipment for fluorescence microscopy. (Prepared with suggestions by Dr. Peter Bartals, E. Leitz, Inc., New York, N.Y.) (From Finegold, S.M., and Martin, W.J.: Bailey and Scott's diagnostic microbiology, ed. 6, St. Louis, 1982, The C.V. Mosby Co.)

lamp. Both light sources require an exciter filter to select the indicated wavelength for exciting the fluorescent material, and both require the right barrier filters to allow the longer fluorescent wavelengths to pass while eliminating the shorter wavelengths. The dark-field condenser spreads the exciting light at such a great angle that negligible amounts enter the objective, and only the fluorescent light off the specimen is viewed (against a dark background).

The most intense fluorescent dyes for practical application are fluorescein with a yellow-green fluorescence and rhodamine with a reddish orange fluorescence (Color plate 1, A). Of the two, fluorescein is more useful because the human eye is more sensitive to its yellow-green color than to the reddish orange color of rhodamine. (Also, in nature red autofluorescence is more prevalent than green autofluorescence.) An example of the application of fluorescence microscopy to the visualization of bacteria is fluorochrome staining of the tubercle bacillus *(Mycobacterium tuberculosis)*. When a specimen containing tubercle bacilli, such as sputum, is stained with the fluorescent dye auramine, the tubercle bacilli readily take up the fluorochrome, whereas most other bacteria present do not. When the specimen is observed in the fluorescence microscope, the tubercle bacilli are nicely identified as brightly fluorescent objects against the black background.

EXAMINATION OF UNSTAINED BACTERIA

It may be desirable to examine bacteria that are unstained to determine their biologic grouping, motility, and reaction to chemicals or specific serums. These features may be determined in a **hanging drop** preparation or in a wet mount. A few species of bacteria that cannot be stained by the methods to be discussed are often examined by dark-field illumination.

Hanging Drop Preparations

To examine bacteria microscopically in a hanging drop, one must use an inoculating loop for transferring the material to be examined to a cover glass to fit over a hanging drop slide.

The inoculating loop is a piece of fine wire about 3 inches long. One end is fastened in a handle, and the other end is fashioned into a loop about 1/16 inch in diameter (Fig. 8-3). Platinum, Nichrome V, and tungsten alloy are used for making inoculating loops because repeatedly heating these metals in a flame to sterilize them does not destroy them, and the wire loop cools quickly after being heated.

A hanging drop slide is a thick glass slide with a circular concavity or depression in the center. A cover glass (coverslip) is a piece of very thin glass about 7/8-inch square.

To make the preparation, first spread a small amount of petroleum jelly around the concavity of the slide. If the specimen to be examined is a culture growing on a solid

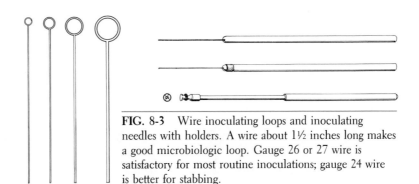

FIG. 8-3 Wire inoculating loops and inoculating needles with holders. A wire about 1½ inches long makes a good microbiologic loop. Gauge 26 or 27 wire is satisfactory for most routine inoculations; gauge 24 wire is better for stabbing.

medium or material such as thick pus, take up a loopful of specimen with the wire loop and mix thoroughly with a drop of sterile isotonic salt solution placed in the center of the cover glass. If bacteria growing in a liquid medium are to be examined, transfer a drop of the fluid to the cover glass by means of the wire loop. Place the hanging drop slide over the cover glass so that the center of the depression lies over the drop. The petroleum jelly seals the cover glass to the slide, holds it in place, and prevents evaporation. Invert the slide now so that the drop to be examined hangs from the bottom of the cover glass but does not touch the surface of the concavity at any point. The preparation is ready for microscopic examination. Examine with the 4 mm (high dry) lens, and reduce the amount of light passing through it by partly closing the diaphragm of the substage condenser of the microscope. Examine all parts of the drop, but remember that the best areas for microscopic study are usually near the edges where cells in a single layer are more evenly dispersed in the fluid medium. When hanging drop preparations are observed, brownian motion and flowing of organisms in currents should not be mistaken for true motility. Care must be taken when one is viewing a hanging drop with the oil-immersion objective that the contamination from the specimen is not spread and that the objective is not soiled from either the specimen or the ring of petroleum jelly.

The inoculating loop must be sterilized immediately before and after each transfer of material containing bacteria (Fig. 8-4). Since hanging drop preparations contain living bacteria, discard the slide and cover glass into a suitable container of disinfectant after the examination is completed.

Wet Mount

The wet mount is similar to the hanging drop preparation, except that an ordinary microslide is used instead of the thick hanging drop slide with its central depression, and the fluid specimen spreads to fill the narrow space between cover glass and microslide (Fig. 8-5). Many of the applications are the same (p. 759).

Dark-field Illumination

Dark-field illumination is used to examine certain delicate bacteria that are invisible in the living state in the light microscope, that cannot be stained by the standard methods, or that are so distorted by staining that they lose their identifying characteristics. Its greatest usefulness is in the demonstration of *Treponema pallidum* in chancres and other

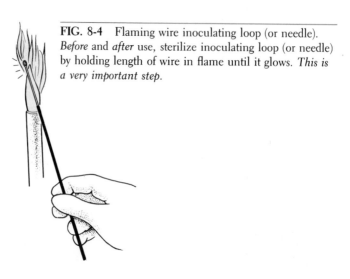

FIG. 8-4   Flaming wire inoculating loop (or needle). *Before* and *after* use, sterilize inoculating loop (or needle) by holding length of wire in flame until it glows. *This is a very important step.*

syphilitic lesions, but it is of value in the examination of many other organisms as well (Fig. 8-6).

The suitably prepared specimen to be examined is placed on a microslide and covered with a cover glass. Sealing the cover glass to the slide with a ring of melted paraffin prevents slipping of the cover glass and accidental infection of the fingers. Dark-field illumination depends on the use of a substage condensor constructed so that the light rays do not pass directly through the object being examined, as is the case with an ordinary condenser, but strike it from the sides at almost a right angle to the objective of the microscope. The microscopic field becomes a dark background against which bacterial or other particles appear as bright silvery objects (Fig. 8-6). A similar effect is seen when a beam of light enters a darkened room and renders visible particles of dust that cannot be seen in a better-lighted room.

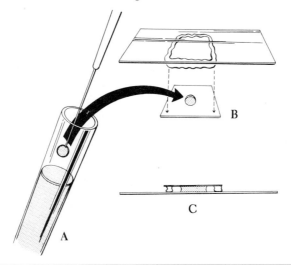

FIG. 8-5   Wet mount preparation (not only simple but adaptable). Rim clean cover glass with petroleum jelly (or make ring with petroleum jelly on clean microslide). *Be sure to flame wire loop before you proceed as shown in figure.* **A,** Use loop to take drop of fluid specimen (either test fluid or test material in sterile isotonic saline). **B,** Place drop on cover glass. Invert coverslip onto microslide, or invert microslide to contact rim of coverslip, and then return microslide with attached cover glass to upright position shown in **C.** Allow the fluid specimen to spread (by capillary action) and fill narrow space between cover glass and microslide.

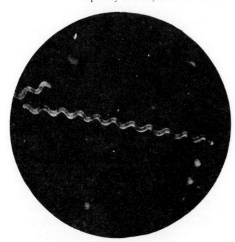

FIG. 8-6   Dark-field microscopy. Large spiral organism is white against black background. It is *Spirulina jenneri,* one of the algae. The dark-field microscope is an important tool in the identification of various kinds of spiral microbes.

EXAMINATION OF STAINED BACTERIA
Staining

Bacterial cells are so small that, when examined in hanging drop preparations, little of their finer structure can be made out; to be studied more closely, they must be colored with some dye. This process is called *staining*. The dyes most often used are aniline dyes, derivatives of the coal tar product aniline. (Dye-impregnated paper strips for staining bacteria are available commercially.)

To make a preparation of stained bacteria, place a small amount of the material to be examined on a perfectly clean glass slide and spread out into a thin film with a wire loop or swab (Fig. 8-7, A). Subsequent flame sterilization of the inoculating loop is sketched in Fig. 8-4. The film is known as a **smear**. Let the film dry in the air; then slowly pass the slide, smear side up, through the flame two or three times (Fig. 8-7, B). **Flaming** kills the bacteria in the smear and causes them to stick to the slide. The smear is thus fixed; the process is called **fixing**. Other methods of fixing, such as immersion in methyl alcohol or in Zenker's solution, are sometimes used, but heat fixing is the most suitable for routine work. Apply the desired stain to the fixed smear (Fig. 8-7, C). Wash off excess stain with water. Blot the slide dry between sheets of absorbent paper. Observe stained smears with the microscope.

Three classes of stains are used in bacteriology: (1) simple stains, (2) differential stains, and (3) special stains. There is also the process of **negative staining**.

Simple Stains

A simple stain is usually an aqueous or alcoholic solution of a single dye. It is applied to the fixed smear for 1 to 5 minutes and washed off. The stained preparation is then ready for microscopic examination. Widely used simple stains are Löffler's alkaline methylene blue, carbolfuchsin, gentian violet, and safranin. The length of time that the stain remains in the smear depends on the avidity with which it acts. Sometimes a chemical is added to the solution to make it stain more intensely. Such a chemical is called a **mordant**.

Most bacteria stain easily and quickly with simple stains, but some do not stain so readily and a few do not stain at all. Capsules and spores are not stained with simple

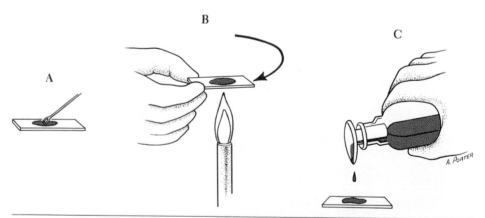

FIG. 8-7   Preparation of stained bacterial smear. **A,** Spread loopful of test specimen thinly on clean glass microslide. Allow to dry. **B,** Fix specimen onto glass slide (held between fingers) by passing it through flame. **C,** Apply drops of stain from dropper bottle to slide (according to technic indicated). Be sure to flame-sterilize wire loop before and after smear is made.

stains but may give contrast as clear, *unstained* structures. Flagella cannot be demonstrated at all in this way.

Differential Stains

More complex staining methods divide bacteria into groups according to their reaction to the chemicals used for staining. Of these, the **Gram stain** and the **acid-fast stain** are most often used.

The method of staining introduced in 1884 by Hans Christian Jochim Gram (1853-1938), a Danish physician working in Berlin, is an important method of differential staining that remains essentially unaltered in use today. It divides bacteria into two great groups: those that are **gram positive** and those that are **gram negative.** This method depends on the fact that when bacteria are stained with either crystal violet or gentian violet and the smear is then treated with a weak solution of iodine as a mordant, some bacterial cells combine with the dye and iodine to produce a color that cannot be removed easily by alcohol, acetone, or aniline, whereas the color is readily removed from certain other bacteria by these solvents. In the last step of the staining procedure a counterstain, such as safranin, is applied to give a contrasting color to the decolorized bacterial cells. Stains used to give contrast in color are **counterstains.** The ones most often used in the Gram stain are safranin and dilute carbolfuchsin, both of which give a red color, and Bismarck brown, which, as its name implies, gives a brown color. Many modifications have been devised for Gram's original method.

Bacteria from which the blue color of the stain–iodine–bacterial cell complex cannot be removed are spoken of as being *gram positive,* and those from which it can be removed and which are stained with the counterstain, as *gram negative.* Gram-positive bacteria do not stain with the counterstain because they are already completely stained. A few bacteria that sometimes retain the stain and other times do not are referred to as **gram variable.** Table 8-4 gives the reactions of the Gram stain for important pathogenic bacteria.

Gram-positive and gram-negative bacteria take up the initial dye in equal amounts, but the difference in the chemical composition of the cell wall between the two groups explains the opposite reaction of the two groups to the action of acetone, alcohol, or other organic solvent. The organic solvent used dehydrates the thick gram-positive cell wall and reduces its porosity. In so doing it fixes the dye-iodine complex within the gram-positive cell. In gram-negative cells, however, the cell wall is thinner, and the solvent readily extracts the dye-iodine complex.

Certain physiologic differences are generally correlated with Gram staining. Gram-

Table 8-4 Gram-Stain Reactions for Some Pathogenic Bacteria

Gram Positive (Dark Violet or Purple)		Gram Negative (Color of Counterstain—Red if Safranin)	
Cocci	Bacilli	Cocci	Bacilli
Staphylococcus aureus	*Bacillus anthracis*	*Neisseria gonorrhoeae*	*Bacteroides fragilis*
Streptococcus pneumoniae	*Clostridium* species	*Neisseria meningitidis*	*Bordetella pertussis*
Streptococcus pyogenes	*Corynebacterium diphtheriae*		*Escherichia coli*
			Haemophilus species
			Proteus vulgaris
			Pseudomonas aeruginosa
			Salmonella species
			Shigella species

positive bacteria tend to be more resistant to the action of oxidizing agents, alkalis, and proteolytic enzymes than are gram-negative ones. They are more susceptible to the acids, detergents, sulfonamides, and antibiotics such as penicillin than are gram-negative ones.

The method of Gram staining outlined below is often used. The explanations appended will apply to any technic used:

1. Make a thin smear of material for study and fix in a flame. Let cool.
2. Stain with crystal violet or gentian violet* for 1 minute.
3. Blot thoroughly to take up excess stain.
4. Cover the smear with Gram's iodine solution,† the mordant, for 1 minute.
5. Drain and blot dry. Both gram-positive and gram-negative bacteria are now stained a dark violet or purple color.
6. Drop acetone on the smear until no more color flows off (about 5 to 10 seconds). Blot dry. All gram-negative bacteria are completely decolorized. The gram-positive ones are not affected.
7. Cover the smear with a stain‡ that gives a contrast in color (counterstain) for 1 minute.
8. Wash with water, blot dry, and examine.

Another method of differential staining is the acid-fast or Ziehl-Neelsen stain, for which several standard modifications exist. The credit for staining tubercle bacilli with the acid-fast stain belongs to Paul Ehrlich (1854-1915), whose method is used in slightly modified form today. The modifications were made by Franz Ziehl (1857-1926) and Friederich Neelsen (1854-1894), whose names were given to the staining method. When most bacteria and related forms are stained with carbolfuchsin, they stain easily, but when the smear is treated with acid-alcohol, they are completely decolorized. It is relatively difficult to stain certain other microbes with carbolfuchsin, but once stained, they retain the dye even when treated with acid-alcohol. Those that retain the stain are spoken of as being *acid fast*. The property of being acid fast derives from complex lipids, fatty acids, and waxes that are not possessed by cell walls for nonacid-fast bacteria.

The technic of acid-fast staining is as follows:

1. Make a smear on the slide and fix. Let cool.
2. Flood the slide with carbolfuchsin§ (red) and gently steam over a flame (do not boil) for 3 to 5 minutes. Keep the slide covered with dye.
3. Allow the slide to cool. Wash off the excess stain with water (all bacteria are now red).
4. Dip slide repeatedly in acid-alcohol‖ until all red color is removed from smear. Wash with water. (At this step the acid has removed the red color from all bacteria that are not acid fast. Acid-fast organisms are unaffected and remain stained bright red.)
5. Apply a counterstain for 1 minute to give contrast. Löffler's alkaline methylene blue** is frequently used. (Since acid-fast organisms are completely saturated with

---

*Mix a 10% solution of crystal violet in 95% ethyl alcohol (solution A) with a 1% solution of ammonium oxalate in distilled water (solution B) in the ratio of 1 part of solution A to 4 parts of solution B (Hepler's ratio 1:8).

†Dissolve 2 g of potassium iodide and 1 g of iodine crystals in 300 ml of distilled water; solution should be stored in a brown glass bottle.

‡Mix 10 ml of a 2.5% alcoholic solution of safranin with 90 ml of distilled water.

§The carbolfuchsin solution for the acid-fast stain is made by combining 1 part of a saturated solution of basic fuchsin in 95% ethyl alcohol with 9 parts of a 5% aqueous solution of phenol.

‖Acid-alcohol for decolorizing is 3% hydrochloric acid (concentrated) in 95% ethyl alcohol.

**Löffler's alkaline methylene blue is made by mixing 30 ml of a saturated solution of the dye methylene blue chloride in 95% alcohol (1.5% filtered) with 100 ml of a weak (0.01%) aqueous solution of potassium hydroxide. Modern samples of the dye have been considerably purified, and the addition of alkali may not be necessary, distilled water alone giving a satisfactory preparation for staining.

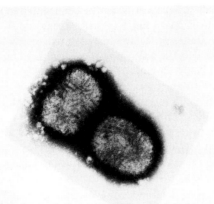

FIG. 8-8   Negative staining applied to study of viruses. Example here is virus of smallpox. Phosphotungstic acid provides background opaque to electrons when specimen is examined in electron microscope and one against which surface structures of viral particle are defined by contrast. (×120,000.) (Courtesy Dr. Derrick Baxby, University of Liverpool, England.)

the red carbolfuchsin, they do not take any of the counterstain. Nonacid-fast organisms, having had all their stain removed by the acid, stain a deep blue.)

Important examples of acid-fast organisms encountered in medicine (some nonpathogenic) are *Mycobacterium tuberculosis*, *Mycobacterium leprae*, other *Mycobacterium* species, and *Nocardia asteroides* (Color plate 1, B).

Special Stains

Important special stains are those for capsules, spores, flagella, and metachromatic granules. Stains primarily designed to demonstrate metachromatic granules are especially valuable in identifying *Corynebacterium diphtheriae* and in differentiating it from related bacteria. Important stains of this type are Albert's stain, using toluidine blue and malachite green, and Neisser's stain, using methylene blue.

Negative (Relief) Staining

Microorganisms such as *Treponema pallidum*, not stained by ordinary dyes, may be made visible by the process known as *negative*, or *relief*, *staining*, in which the background, but *not* the microorganisms, is stained. The microorganisms are mixed with india ink or 10% nigrosin (both of which are black); the mixture is spread out into a thin smear and allowed to dry. The microbes appear as colorless objects against a black background. Negative staining may be done with the dye Congo red; this has been used to display spiral organisms and chlamydiae (bedsoniae). By a technic of negative staining using an electrodense material, such as phosphotungstic acid, viruses are prepared for visualization in the electron microscope (Fig. 8-8).

SUMMARY

1 A potential hazard for injury to the laboratory worker exists at all times in the laboratory. All specimens possibly containing microbes must be handled with great care in accord with well-known principles of standard conduct and aseptic technic.

2 The compound light microscope is the microscope most often used, but the dark-field, the phase-contrast, the fluorescence, and the electron microscopes also have important applications in biology.

3 The dark-field microscope is essentially the bright-field scope fitted with a dark-field condenser so that an object viewed microscopically appears silvery bright against a dark background (the opposite effect of the bright-field scope).

4 The magnification possible in an electron micrograph is so powerful that a human hair viewed in its entirety appears twice the size of a California redwood.

5 The scanning electron microscope reveals surface features rather than internal structure of a specimen. It can magnify up to 100,000 times the actual size.

6 The ultraviolet microscope is used for fluorescence microscopy, which is used to visualize microscopic preparations stained or in some way tagged with fluorescent material.

7 Resolution is the property whereby detail in an object can be seen as clearly distinct by the human eye, microscope, or camera. The limit of resolution is the minimal distance between two points that allows for their definition as two separate points.

8 Biologic grouping, motility, and reaction to certain chemicals may be determined for unstained bacteria in a hanging drop preparation or in a wet mount.

9 The wire inoculating loop must be sterilized immediately before and after each microbiologic transfer of material. The loop is flamed or otherwise heated until it glows.

10 The differences in the chemical nature of the cell wall between gram-positive and gram-negative bacteria explain the opposite reactions in the Gram stain. The organic solvent dehydrates the thick gram-positive cell wall fixing (not extracting) the dye-iodine complex taken up in steps one and two of the Gram stain. Since the gram-negative cell wall is thinner, the solvent readily extracts the dye-iodine complex from it.

11 Certain bacteria contain complex lipids and waxes in their cell wall so that when stained with carbolfuchsin, they retain the dye even when treated with acid-alcohol, that is, they are acid fast. Important acid-fast bacteria cause serious diseases (for example, tuberculosis and leprosy).

12 Organisms not easily stained by ordinary dyes may be viewed with negative, or relief, staining, in which the background is stained. Dyes for this purpose are Congo red, india ink, and 10% nigrosin.

QUESTIONS FOR REVIEW

1 Name the procedures used in the study of microbes.

2 What are the purposes of a hanging drop preparation? How is one made? What is a wet mount?

3 Briefly define fixing, flaming, mordant, counterstain, gram variable, micrometer, and nanometer.

4 What are the three classes of stains used in microbiology? Describe each one and give examples.

5 How is a simple stain made? Do many bacteria stain with simple stains?

6 Name two very important differential stains. Give the underlying principles of each.

7 Name two important gram-positive cocci, three important gram-positive bacilli, two important gram-negative cocci, and three important gram-negative bacilli.

8 Name several important acid-fast bacilli.

9 What is negative, or relief, staining?

10 Briefly describe the bright-field microscope. Name two other kinds of microscopes.

11 Give the steps in the preparation of a bacterial smear.

12 When is dark-field illumination used to study microbes?

13 Give an application of fluorescence microscopy. Name two important fluorochromes.

14 What is the important biochemical difference between gram-positive and gram-negative bacteria that relates to their Gram-staining reaction?

SUGGESTED READINGS

Bish, D.J.: You and your microscope, Lab. Med. **10**:168, 1979.

Bohnhoff, G.L.: Setting up proper illumination on your microscope, Am. J. Med. Technol. **45**:650, 1979.

Goldfogel, G.A., and Sewell, D.L.: Preparation of sputum smears for acid-fast microscopy, J. Clin. Microbiol. **14**:460, 1981.

Murray, P.R., and others: The acid-fast stain: a specific and predictive test for mycobacterial disease, Ann. Intern. Med. **92**:512, 1980.

Siegle, M.D.: Urinoscopy: first the microscope, Lab. Med. **12**:781, 1981.

CULTIVATION

The small size and similarity in appearance and in staining reactions of microbes often make identification by microscopic methods alone practically impossible. One of the most important ways to identify them is to observe their growth on nutrients (for example, beef extract and peptone) prepared in the laboratory. This is cultivation of the microorganisms, or **culturing**. The nutrient substance on which they are grown is a **culture medium** (*pl.* media), and the growth itself is a **culture**. Some 10,000 different kinds of culture media have been prepared.

Culture methods help to identify microbes and also to find organisms in test material of low microbial content. When such material is placed on a suitable culture medium, each organism present multiplies many times. Practically speaking, the smear may contain relatively few (or no) organisms, whereas the culture soon produces thousands of them.

CULTURE MEDIA
General Considerations

For the most satisfactory growth of bacteria and related forms on artificial culture media, the proper temperature, right amount of moisture, required oxygen tension, and proper degree of alkalinity or acidity (pH) must be provided. The culture medium itself must contain the necessary nutrients and growth-promoting factors and be free from contaminating microorganisms; in other words, *it must be sterile.*

Most disease-producing bacteria require complex foods similar in composition to the fluids of the animal body. For this reason, the base of many culture media is a neutral or slightly alkaline infusion of meat containing meat extractives, salts, and peptone, to which various other ingredients may be added.

The basic infusion can be prepared by soaking 500 g of fresh lean ground beef in 1 L of distilled water in the refrigerator. After 24 hours the surface layer of fat is removed with absorbent cotton, the mixture is passed through muslin or gauze, and the meat is discarded. After ingredients related to the organism under study are added and the reaction of the medium adjusted to neutral or slightly alkaline, the infusion is heated to 100° C for 20 minutes to remove coagulated tissue proteins and then filtered through coarse paper. At each step the volume is reconstituted to 1 L.

Culture media are of three physical types: (1) liquid, (2) solid that can be liquefied by heating but returns to the solid state when cooled, and (3) solid that cannot be liquefied. Many times a solid culture medium is preferred because bacteria and related forms can be viewed on the surface. Agar (extract of marine algae) is a solidifying agent widely used. It melts completely at the temperature of boiling water and solidifies when cooled to about 40° C. With minor exceptions, it has no effect on bacterial growth and is not attacked by bacteria growing on it. The low temperature at which agar solidifies is crucial if test material must be inoculated directly into melted media before it solidifies. If the agar began to solidify at a temperature high enough to kill bacteria or fungi (that is, 100° C), this could not be done. Gelatin is a well-known but less frequently used solidifying agent.

Numerous materials are found for various reasons in different culture media—carbohydrates, serum, whole blood, bile, ascitic fluid, and hydrocele fluid. For instance, carbohydrates are added (1) to increase the nutritive value of the medium and (2) to indicate the fermentation reactions of microbes being studied. Serum, whole blood, and ascitic fluid are added to promote the growth of the less hardy organisms.

Dyes added to culture media act as indicators to detect acid formation when fermentation reactions are being tested, or they act as selective growth inhibitors for certain bacteria. An example of an indicator dye is phenol red, which is red in an alkaline or neutral medium but yellow in an acid one. An example of an inhibitory dye is gentian violet, which inhibits the growth of most gram-positive bacteria.

Media Making

In actual practice media making is greatly simplified by the availability of commercially prepared mixtures containing the basic ingredients and any special growth factors in just the right proportions. These preparations are sold in the dehydrated state as a powder or tablet, either of which may be reconstituted in water. (The powder is weighed and dispensed into a given volume of distilled water, and the mixture is heated carefully over a water bath to effect solution.)

Before sterilization, hydrated culture media are poured into suitable test tubes or flasks that are closed with cotton plugs or screw-on plastic caps. After sterilization, some of the tubes of hot liquid media containing agar may be left vertical for the agar to solidify, but some are laid on a flat surface with their mouths raised so that when the agar cools and solidifies there is a slanting surface, or an **agar slant.** Slanting the agar gives a larger surface area for colonial growth. Cotton plugs allow the access of moisture and oxygen but block the entrance of contaminating microorganisms; for the same purpose, screw caps are left loosely attached. The medium remains sterile until used, and the microbes with which it will be inoculated are not contaminated by those from the outside.

A large surface area for bacterial growth is provided when a solid culture medium partly fills a Petri dish. A **Petri dish** is a circular glass (or plastic) dish about 3 inches in diameter with perpendicular sides about ½ inch high. A glass, metal, or plastic cover exactly like it fits over it. The edges of the two are smooth, and a container is formed that prevents both the entrance and exit of bacteria. The sterile Petri dish may be filled with a sterile agar medium still warm from the sterilization process, or a tube of solidified agar may be melted and poured into the dish. Until the agar has completely cooled, the lid is slightly raised on the Petri dish. When the agar is solid, the lid is lowered.

Sterilization

Most media are sterilized by autoclaving (pp. 210 and 212). Those that contain carbohydrates may have to be sterilized by the fractional method because many carbohydrates, as well as some amino acids and vitamins, do not withstand the high temperature of the autoclave. Enriching materials that are injured by even moderate heat, such as serum and whole blood, are collected in a manner to be kept sterile and mixed with the medium only after it has been sterilized and allowed to cool. With agar media, these substances are added when the medium reaches a range of 42° to 45° C, because when the temperature falls slightly lower than this, the agar begins to solidify.

Applications

Almost every species of microbe has some medium on which it grows best. A few grow only on media specially prepared for them, but most pathogens grow more or less luxuriantly on certain routine or standard media (Fig. 9-1). Table 9-1 lists media suitable for the isolation and growth of important pathogenic microorganisms.

A good example of a **general-purpose nutrient medium** is trypticase, or tryptic soy broth, a pancreatic digest of casein and soybean peptone. It is widely used because it supports the growth of many, even fastidious, microbes. Another example of a general-purpose medium is thioglycollate, which supports the growth of many facultative organisms, microaerophiles, and anaerobes.

An **enriched medium** by definition is the basic nutrient medium to which serum or blood (commonly) or any one of a number of growth factors may be added. The addition of the nutritive supplement enhances the growth and recovery of the desired organism.

The isolation of the few pathogenic bacteria that may be overshadowed by the many commensals of the normal intestinal tract requires the use of an **enrichment medium.** This medium chemically inhibits or even kills the undesired normal flora while encouraging a profuse growth of the desired pathogen. Examples of enrichment media are tetrathionate broth and selenite broth, both used primarily in enteric bacteriology.

Differential Media

Media that by virtue of their ingredients distinguish organisms growing together are differential media. Examples are eosin–methylene blue (EMB) agar and MacConkey agar (Color plate 1, C), used in the differentiation of gram-negative bacteria from the intestinal tract. Incorporation of lactose into differential media makes possible a sharp

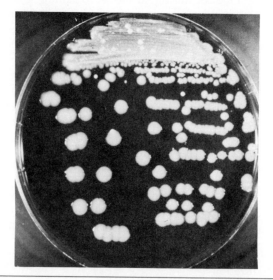

FIG. 9-1   Use of blood agar in Petri dish. Bacterial colonies are those of *Escherichia coli*, the colon bacillus.

Table 9-1 Microbes Related to Culture Media

Organism	Media Most Suitable for Growth
Actinomyces bovis	Brain-heart infusion glucose broth and agar; thioglycollate broth medium (Brewer modified) (anaerobic)
Bacillus anthracis	Growth on almost all media
Bacteroides species	Blood agar (anaerobic system)
Blastomyces dermatitidis	Blood agar; beef infusion glucose agar; Sabouraud glucose agar (antibiotics added to media used)
Bordetella pertussis	Glycerin-potato-blood agar (Bordet-Gengou agar)
Brucella species	Trypticase soy broth and agar
Candida albicans	Sabouraud glucose agar; brain-heart infusion blood agar; corn meal agar; rice infusion agar
Clostridia of gas gangrene	Thioglycollate broth medium (Brewer modified); anaerobic blood agar; cooked meat medium under petroleum; Clostrisel agar
Clostridium tetani	Same as for clostridia of gas gangrene
Coccidioides immitis	Sabouraud glucose agar
Corynebacterium diphtheriae	Löffler blood serum medium; cystine-tellurite-blood agar
Cryptococcus neoformans	Blood agar; beef infusion glucose agar; Sabouraud glucose agar
Entamoeba histolytica	*Entamoeba* medium (Difco) with dilute horse serum and rice powder
Escherichia coli and coliforms	Eosin–methylene blue (EMB) agar; Endo agar; MacConkey agar; deoxycholate agar; blood infusion agar (growth on almost any medium)
Francisella tularensis	Cystine glucose blood agar
Fungi, including dermatophytes	Sabouraud glucose agar with antibiotics; Littman oxgall agar; corn meal agar
Haemophilus influenzae	Chocolate agar with yeast extract; rabbit blood agar with yeast extract
Histoplasma capsulatum	Brain-heart infusion glucose blood agar with antibiotics; Sabouraud glucose agar with antibiotics; brain-heart infusion glucose broth
Mycobacterium tuberculosis	Slow growth on special media such as Petragnani medium, Löwenstein-Jensen medium, or Dubos oleic agar; Dorset egg medium
Neisseria gonorrhoeae	Chocolate agar
Neisseria meningitidis	Chocolate agar
Nocardia asteroides	Beef infusion blood agar; beef infusion glucose agar; Sabouraud glucose agar; Czapek agar; Dorset egg medium
Proteus vulgaris	EMB agar; Endo agar; MacConkey agar; deoxycholate agar; blood infusion agar (growth on almost any medium)
Pseudomonas aeruginosa	Blood agar; EMB agar; Endo agar; MacConkey agar; deoxycholate agar (growth on almost any medium)
Rickettsias (including chlamydiae)	Chick embryo; cell culture
Salmonella species	Deoxycholate agar; deoxycholate-citrate agar; *Salmonella-Shigella* (SS) agar; MacConkey agar; tetrathionate broth; selenite-F enrichment medium
Salmonella typhi	Bismuth sulfite agar; SS agar; deoxycholate agar; deoxycholate-citrate agar; MacConkey agar; tetrathionate broth; selenite-F enrichment medium
Shigella species	SS agar; deoxycholate agar; deoxycholate-citrate agar; MacConkey agar; tetrathionate broth; selenite-F enrichment medium
Staphylococcus species	Growth on almost any medium
Streptococcus species	Blood infusion agar; trypticase soy broth; tryptose phosphate broth; brain-heart infusion media
Viruses	Chick embryo; cell culture

separation in colonial characteristics between the organisms that ferment lactose and those that do not. Colonies of the lactose fermenters are deep red or possess a metallic sheen; colonies of the nonfermenters are colorless. This is important because generally pathogens in the intestinal tract, such as *Salmonella* and *Shigella* species, do *not* ferment lactose, whereas certain of the normal inhabitants, the coliforms, do.

Selective Media

Media that by virtue of their complexity of composition promote the growth of one organism and retard the growth of others are selective media. Their inhibitory component is highly selective in many instances. Selective media may also *differentiate* among certain genera of microorganisms. Examples of selective media are deoxycholate-citrate agar and Löwenstein-Jensen medium. Deoxycholate-citrate agar inhibits the growth of most coliforms, along with many strains of *Proteus*, while favoring the selective isolation of intestinal pathogens, especially those of the *Salmonella* and *Shigella* species. The coliform colonies that do appear on the medium are readily separated from the nonfermenters of lactose (the pathogens) by their color and opacity. Löwenstein-Jensen medium (another example, Petragnani medium) promotes the growth of tubercle bacilli but retards the growth of any other organism present. Although EMB agar and MacConkey agar are primarily thought of as differential media, they display a selective function in that they allow the growth of gram-negative bacteria but not gram-positive bacteria.

Selective and differential media are of great value in the diagnosis of infections in areas of the body such as the respiratory and intestinal tracts, where normally a variety of organisms reside and the presence of pathogenic bacteria may not disrupt the normal bacterial pattern for the area (the normal flora). In test specimens taken from these areas the pathogens tend to be mixed with many other microorganisms, and only with the use of selective and differential media is it possible to isolate and identify the disease producers.

Synthetic Media

The culture media just discussed are made up of components of variable composition such as meat infusions, serum, or other body fluids. Such media are *nonsynthetic (complex) media*. Synthetic (chemically defined) media, on the other hand, contain ingredients of definite chemical composition. Used primarily in research work, these media have the advantage that, wherever prepared, their composition is the same. Therefore the results of microbial action on these media in one laboratory are strictly comparable with those in other laboratories thousands of miles away.

Transport Media

A transport medium, often a liquid, maintains the viability and infectivity of microorganisms while the specimen in it is carried to the laboratory where it can be properly handled. The pH of such a medium must be right, and the specimen in it must not be allowed to dry out. Its chemical composition must prevent oxidation or enzymatic deterioration of any pathogen present, and anaerobes must be sheltered from atmospheric oxygen. A number of transport media have been devised for delicate organisms, anaerobes, and viruses. Examples are Stuart's transport medium, Hanks' balanced salt solution with protein supplement, and for anaerobes, double-stoppered tubes or vials that have been gassed out with oxygen-free carbon dioxide or nitrogen.

CULTURE METHODS
Inoculation

A culture is made or inoculated (Fig. 9-2) when the specimen to be cultured, such as sputum, urine, or pus, is placed into a fluid medium or rubbed gently over the surface of a solid medium with either a sterile swab or a flame-sterilized inoculating wire loop (see Fig. 8-4). The inoculated medium is then incubated for a period, routinely 24 to 48 hours.

Standard aseptic bacteriologic technics are implied in the discussion to follow.

Bacteriologic Incubator

A bacteriologic incubator consists of an insulated cabinet fitted with an electric heating element and a thermoregulator to maintain the temperature at a set point. A good thermoregulator in a properly constructed incubator maintains the temperature constant from day to day to within 0.5° C of the set temperature. Most specimens are incubated at 35° C for the isolation of pathogenic microbes, but in some instances the temperature may be as low as 32° C or as high as 42° C.

The incubator must be properly ventilated. It may be constructed so that circulation of air is brought about by the combined effects of gravity and the difference in weight of warm and cold air. Incubators are also ventilated with blowers or fans. All are fitted with perforated shelves and most with sets of double doors. While keeping out cold air, the inner glass doors allow the contents of the incubator to be viewed.

The relative humidity in an incubator should be about 70%. If a regulatory mechanism for this purpose is not incorporated into the incubator, a pan of water placed on the lowest shelf may suffice. The water level should be checked daily to ensure that it is kept constant.

Incubators may be fitted with mechanisms to maintain a controlled environment within the incubator, such as a desired carbon dioxide tension at constant temperature and humidity, thereby facilitating the growth of carbon dioxide–dependent and other fastidious bacteria with special requirements for growth.

To support the growth of anaerobes, the anaerobic incubator is constructed so that

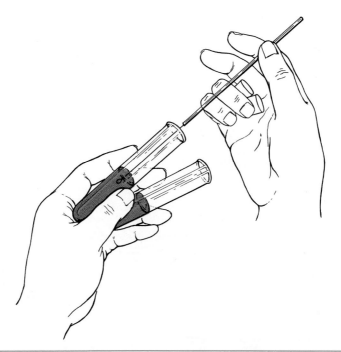

FIG. 9-2   Inoculation of liquid culture medium. Note slanting position of test tubes during maneuver; thus, contaminating organisms from air are less likely to drop into open tubes. Fingers of right hand are lifted away from handle of wire loop to show how caps are held.

after a vacuum has been pulled for the interior, the environment of the evacuated chamber may be charged with whatever gas is desired—nitrogen, hydrogen, or carbon dioxide, alone or in various mixtures.

Inspection of Cultures

Each microbe in the medium inoculated and incubated multiplies rapidly, and in a few hours there are many more microorganisms in the culture than there were in the original sample. Consequently bacteria or fungi are readily identified in cultures, whereas in smears of the same test sample they may be found with difficulty or not at all. Diphtheria bacilli can be found in throat cultures twice as often as in throat smears.

When a culture is made on a solid medium, all the bacteria that grow from each individual bacterium deposited on the medium cling together to form a colony, a mass visible to the naked eye. The colony has certain characteristics, such as texture, size, shape, color, and elevation (Fig. 9-3). These are fairly constant for each species and are valuable in differentiating one species from another. Theoretically, each organism should give rise to one colony, but two or more may cling together and, when planted on a medium, give rise to a single colony. If organisms that cling together are different, the colony will contain the different kinds. As a rule, with *good laboratory technic* a colony contains only one kind of microorganism.

Petri dish cultures permit good observation of colonies, since the large surface area favors separation of individual microorganisms. Petri dish cultures are usually called *plates*, and the process of making such cultures is referred to as **plating** or **streaking**.

Pure Cultures

An **axenic culture** (Greek *xenos*, guestfriend, stranger; meaning "not contaminated by or associated with any foreign organism") contains only one kind of bacteria; a **mixed culture** contains two or more kinds. As a rule, infectious material contains more than one kind of bacteria. So that one kind alone can be studied, it must be separated from all others and grown alone as a pure culture. Pure cultures are usually obtained by the pour plate or streak plate method. The pour plate method is as follows (Fig. 9-4):

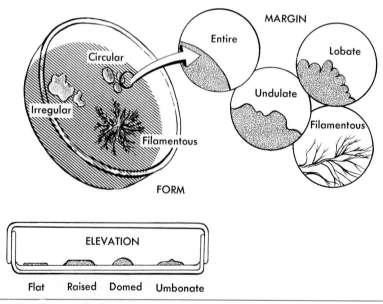

FIG. 9-3   Examination of colonies.

1. Melt at least three (or more if needed) tubes of agar in boiling water, then allow to cool to about 42° C. If an enriching material injured by heat, such as serum or whole blood, is needed, add it to the medium at this time.
2. Transfer a loopful of test material from which you expect to obtain a pure culture to one tube. Replace the cotton plug and roll the tube between the palms of the hands to distribute the bacteria throughout the medium.
3. Flame-sterilize the wire loop thoroughly. Transfer three loopfuls of the contents of the inoculated tube to the second tube.
4. Mix the contents of the second tube with the inoculum. Sterilize the inoculating loop. Transfer five loopfuls of the mixture to the third tube and mix.
5. Pour the contents of each tube of medium into separate Petri dishes and allow to solidify.
6. Incubate the Petri dishes for 24 to 48 hours. Observe the colonies.

Successive dilution of the specimen in the three tubes reduces the number of bacteria and disperses them in the medium so that the colonies in the Petri dish cultures are more likely to be distinctly separate from each other. Those to be studied further are removed from the Petri dish with an inoculating needle and rubbed over the surface of one or more slants of suitable media. This maneuver is known as **fishing** (Figs. 9-5 and 9-6). If the colonies are not separated from each other, it is impossible to fish one colony without touching other colonies. Thus other types of bacteria are mixed in during the transfer.

The inoculated slants (subcultures) are allowed to incubate for 24 to 48 hours and then are studied further (Fig. 9-7). In the majority of cases all bacteria growing on a slant will be alike because they all grew from the members of a single colony on the

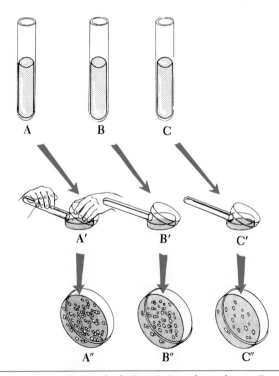

FIG. 9-4   Pure culture, pour plate method. *Step 1:* Inoculate tubes **A**, **B**, and **C**, as indicated by arrows. *Step 2:* Transfer contents of each tube to Petri dishes **A′**, **B′**, and **C′**, respectively. Incubate plates. *Step 3:* Inspect plates (**A″**, **B″**, and **C″**) for number and distribution of colonies. Fish representative colony for further study.

plate, and these in turn grew from a single bacterium in the original matter. If, as sometimes happens, the original colony on the plate contains two or more kinds of bacteria, the same two or more kinds of bacteria will grow on the slant. Subsequent separation is made by suspending in sterile salt solution some of the growth from the slant and then replating it.

Streak Plates

To prepare a streak plate, streak a single loopful of infectious material over the surface of the solid medium (agar) in a Petri dish. The pattern illustrated in Figs. 9-8 and 9-9 is widely used and gives good separation of colonies, which can then be used for subculturing.

Bacterial Colony (Plate) Count

The accuracy of the bacterial colony (plate) count rests on the premise that, when material containing bacteria is cultured, every bacterium present develops into a colony. This statement is not strictly true, because some bacteria may fail to multiply and two or more bacteria may cling together to form a single colony. Nevertheless, the method is of decided value in the examination of water and milk (pp. 899 and 910).

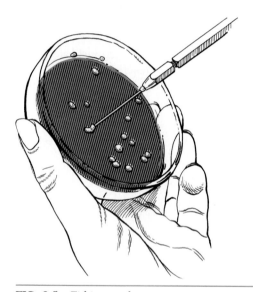

FIG. 9-5   Fishing a colony.

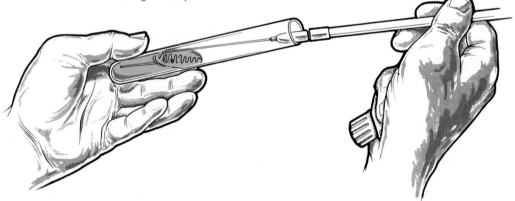

FIG. 9-6   Streaking an agar slant.

Although varying numbers and volumes of dilution blanks may be used, depending on the nature of the test sample, one method of performing a bacterial colony count uses four tubes containing 9 ml of sterile distilled water (dilution blanks) as follows.

Use a sterile 1 ml pipette to add 1 ml of the test sample (water or milk, for example) to the first dilution tube, and mix. Transfer 1 ml from this tube (using another sterile pipette) to the second dilution tube and mix again. From the second tube transfer 1 ml

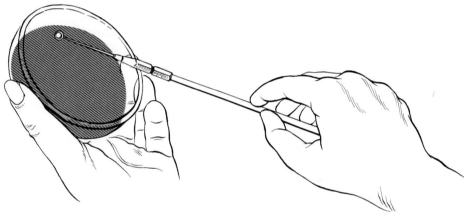

FIG. 9-7   Pure culture *(Staphylococcus aureus)* on agar slant. (Courtesy Ayerst Laboratories, New York, N.Y.)

FIG. 9-8   Streaking an agar plate.

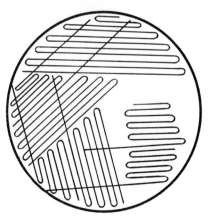

FIG. 9-9   Pattern for streaking an agar plate. Sections of plate are streaked successively. After each section is streaked, wire loop is flame sterilized. A bit of inoculum for next section in turn is obtained when loop is passed back onto or through section just inoculated. (From Smith A.L.: Microbiology: laboratory manual and workbook, ed. 5, St. Louis, 1981, The C.V. Mosby Co.)

The bacterial colony (plate) count is a bacteriologic technic that may be used to diagnose clinical infection of the urinary tract, since enumeration of the bacteria present in the urine is a valid way of telling whether they represent disease-producers or contaminants. For this purpose the urine is a first-morning, mid-stream, "clean-catch" specimen collected after the external genitalia have been cleansed. It is not centrifuged. A measured amount of the specimen is handled as indicated, and a count is determined. A positive test result is considered to be a count of more than 100,000 colonies per cubic millimeter of urine. In this event the same specimen of urine may next be cultured and antibiotic sensitivity determined in microbes isolated. Generally, in the absence of therapy, when the count is less than 10,000 colonies from 1 ml of specimen, the microorganisms demonstrated are considered to represent members of the normal flora of the urethra or contaminants.

to the third tube (using another sterile pipette) and mix again. From the third tube transfer 1 ml to the fourth tube (with another sterile pipette). Again mix. You now have in the four dilution tubes 1:10, 1:100, 1:1000, and 1:10,000 dilutions, respectively, of the original material. With a sterile pipette transfer exactly 1 ml from each tube to a Petri dish and add sufficient melted agar (about 20 ml) cooled to 42° C. Mix the contents of each dish by rotating and allow the mixture to solidify. After incubation for 24 to 48 hours, count the colonies that have grown. If the second plate shows, for example, 200 colonies, this means that at least 200 bacteria were present in the 1 ml of material placed in the tube, and since this was a 1:100 dilution, the original test material must have contained at least 20,000 bacteria per milliliter. More bacteria may have been introduced into the dish, since some may have failed to grow and some may have clung together in groups of two or more that grew as single colonies. The number of bacteria indicated by the count is certainly present, and more may be also (see box above for application to urinary tract infection).

Cultures of Anaerobic Bacteria

To grow some anaerobic bacteria in a routine laboratory, one may use a tube of suitable culture medium about 8 inches long and about half full of medium, which is heated in a boiling water bath for several minutes to expel oxygen present in the medium. The tube is quickly cooled and inoculated; then enough sterile melted petroleum jelly to

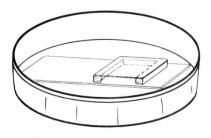

FIG. 9-10   Slide culture. To contain indicated culture medium, a small three-sided compartment on a sterile microslide may be constructed with use of quick-drying plastic cement. A small inoculum to center of media in chamber is covered with a sterile cover glass. A wet strip of filter paper provides moisture for culture in covered Petri dish, which is incubated at desired temperature. When culture of fungus is made, growth is inspected with low magnification at stated time intervals so that early sporulation may be detected. After fungal growth is exposed to formalin vapor, it is more closely inspected and studied with higher microscopic magnification. To make a permanent mount of slide culture, seal off all sides of chamber with an adhesive.

make a layer about ¾-inch thick is poured onto the top of the medium. Thus the culture is sealed off from the air. If agar is used, inoculation is made when the heated medium cools to a temperature of about 42° C.

Thioglycollate broth, a special medium containing thioglycollic acid, supports the growth of anaerobes in the depths of a partly filled tube without a special seal. Methylene blue indicator colors the upper layers of the fluid medium as oxidation occurs.

Cultures may be made in the usual way and placed in a specially constructed anaerobic jar (for example, a GasPak jar), a tightly sealed chamber from which the oxygen is either removed by some chemical reaction that is made to take place in the chamber or replaced by a gas, such as hydrogen, that does not affect the growth of the bacteria (see Figs. 24-2 and 27-2). Cultures may be placed in an anaerobic incubator. Better results are obtained with the more sophisticated methods indicated on p. 573.

Slide Cultures

Slide cultures (Fig. 9-10) used to study fungi may be placed on the stage of the microscope and actual growth observed from time to time. The depression of a sterile culture slide is filled with suitable medium, inoculated, and covered with a sterile cover glass. Liquid cultures can be made by inoculating a drop of medium on a sterile cover glass and proceeding as for a hanging drop preparation.

Cultures in Embryonated Hen's Egg

Important sites of growth for viruses and rickettsias, including chlamydiae (bedsoniae), are the yolk sac and embryonic membranes of the developing chick embryo (p. 527). Bacteria other than rickettsias are occasionally grown in this way.

Cultivation of Microbes in Cell Cultures

Cell (tissue) cultures are composed of animal cells in a suitable substrate in which they multiply and grow. The cells may be supported on a solid or semisolid substrate, such as fibrin, agar, and cellulose, or they may be suspended in a liquid. The cells used are taken from mammalian or fowl embryos or selected adult tissues such as the cornea, kidney, and lung, and minced tissue of the cell source is inoculated. Pathogenic agents, such as rickettsias (also chlamydiae) and viruses, that do not grow in lifeless media multiply with ease when incorporated into cell (tissue) cultures.

ANTIBIOTIC SUSCEPTIBILITY TESTING

The susceptibility (or resistance) of bacteria to different antibiotics may be determined by suitable laboratory procedures, usually designated as susceptibility tests. Two are used: (1) the disk diffusion test (Fig. 9-11) and (2) the tube dilution method. The disk diffusion test gives qualitative results, whereas the tube dilution test gives quantitative ones.

The first and more common procedure employs disks of filter paper impregnated with precise amounts of different antibiotics. A Petri dish is heavily inoculated with test bacteria (plated out as a lawn), generally from a pure culture. The disks are dropped onto the freshly inoculated surface, and the plate is inoculated for 24 hours. Meanwhile, as the bacteria grow, the antibiotic diffuses throughout the culture medium. If the bacteria are susceptible to the antibiotic, a zone of inhibition (no growth) encircles the disk, and the area of inhibition around the disk increases in size with increasing concentration of the antibiotic. If the bacteria are unaffected by the drug (that is, resistant), growth will cover the area around the disk. Several antibiotics can be tested on one plate, and the method is simple and rapid. Results correlate well with the effectiveness of clinical treatment.

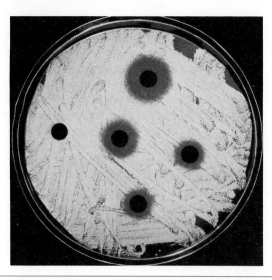

FIG. 9-11   Antibiotic susceptibility testing: single paper disk diffusion method. Petri dish of agar is heavily inoculated with test organisms. Disks of filter paper impregnated with different antibiotics are dropped onto freshly seeded surface. Note lawn of white bacterial growth with zones of inhibition around four disks and none around one. A zone of inhibition indicates that growth of organisms would probably be limited in a patient receiving that antibiotic.

The U.S. Food and Drug Administration has set its stamp of approval on the standardized disk diffusion test as precisely defined step by step in the *Federal Register*.* The test is a modification of the well-known Kirby-Bauer method introduced in 1966. In the official test Mueller-Hinton agar (beef infusion, peptone, starch, and agar at a pH of approximately 7.4) is used for culture of test microbes, since it is low in interfering substances and supports the growth of most pathogens for which evaluation is needed.

The second method of susceptibility testing employs a series of tubes of culture medium containing both a known quantity of test bacteria and known dilutions in increasing concentrations of a given antibiotic. The tubes are incubated, and turbidity in each is evaluated as a measure of bacterial growth. The lowest concentration of antibiotic that inhibits the growth of a given microorganism in the series of tubes is the **minimal inhibitory concentration (MIC)**. Next a check for viability of test bacteria is made. An appropriate aliquot is taken from each of the tubes and inoculated onto a plate of suitable medium referenced to the antibiotic concentration used. Plates are incubated and observed for growth. The minimal concentration of antibiotic that has caused the death of test bacteria is the **minimal bactericidal concentration (MBC)**. The minimal bactericidal concentration is usually greater than the minimal inhibitory concentration when bactericidal agents are used and invariably so with bacteriostatic agents, which kill bacteria only at very large doses.

With this sort of testing, microbes are reported as being susceptible or resistant to a given antibiotic. There are also technics whereby the amount of certain antibiotics can be determined in the blood. Automated methods of susceptibility testing are available that claim to speed up the process to approximately 3 hours while good correlation with the Kirby-Bauer method is maintained.

*Federal Register **37**(191):20527-20529, 1972.

SUMMARY

1 Culture methods are valuable in the isolation and identification of organisms as small and similar in appearance as microbes. These methods can demonstrate microorganisms in test material of low microbial content.

2 Because most pathogenic bacteria require complex foods similar in composition to the fluids of the animal body, the base for many culture media is a neutral or slightly alkaline infusion of meat containing meat extractives, salts, peptone, and various other items. Agar is the usual solidifying agent.

3 Carbohydrates in culture media increase the nutritive value and indicate fermentation reactions. Dyes act as indicators to detect acid formation, or they selectively inhibit growth of certain microbes.

4 Thioglycollate medium, a good general-purpose medium, supports the growth of many facultative organisms, microaerophiles, and anaerobes. The anaerobes survive in its depths without a special seal.

5 An enriched medium is the basic nutrient medium to which suitable growth factors are added to enhance growth and recovery of the desired organism.

6 An enrichment medium (for example, tetrathionate broth used in enteric bacteriology) chemically inhibits or even kills undesired normal flora while encouraging profuse growth of the sought for pathogen.

7 Differential media distinguish organisms growing together by virtue of ingredients such as lactose. Eosin–methylene blue (EMB) agar and MacConkey agar are used to differentiate gram-negative bacteria cultured from the intestine.

8 Selective media promote the growth of one and retard the growth of another organism by virtue of their complex composition. Deoxycholate-citrate agar inhibits the growth of most coliforms while favoring the selective isolation of intestinal pathogens, notably *Salmonella* and *Shigella* species.

9 Transport media (often liquid) maintain viability and infectivity of microbes while the specimen stored therein is carried to the laboratory. An example is Hanks' balanced salt solution with protein.

10 A colony, the composite growth from an individual bacterium deposited on the solid medium, displays a number of descriptive features that are fairly constant for each microbial species.

11 The two major technics for antibiotic susceptibility testing are the disk diffusion test and the tube dilution method. The disk diffusion test is qualitative; the tube dilution, quantitative. Susceptibility in both methods is reflected in an absence of growth when the test bacteria contact the antibiotic. Microbes are reported as either susceptible or resistant.

12 The minimal bactericidal concentration (MBC) is usually greater than the minimal inhibitory concentration (MIC) when bacteri*cidal* agents are used and invariably so with bacterio*static* agents, which kill bacteria at very large doses only.

QUESTIONS FOR REVIEW

1 Name important ingredients of culture media.

2 Why is agar the most commonly used solidifying agent?

3 What purposes do dyes serve in culture media?

4 What advantages are there to culture by comparison with other methods for studying microbes?

5 Explain differential media and selective media. Give examples.

6 What is the practicality of the Petri dish in microbiologic culture?

7 Outline the steps in making a pure culture. What is meant by a pure culture? What purpose does it serve?

8 Briefly describe how and under what circumstances a bacterial count is made.

9 Briefly explain slide culture, culture on synthetic media, culture of anaerobic bacteria, cell (tissue) culture, fishing a colony, streaking, plating, inoculation of a culture, axenic, subculture, dilution blank, and enriched media.

10 How is antibiotic susceptibility testing carried out? What is its purpose? Briefly describe two such procedures.

11 What are the special requirements for transport media? What is the purpose of this kind of medium?

SUGGESTED READINGS

Bish, D.J.: You and your microscope, Lab. Med. 10:168, 1979.

Blazevic, D.J., and others: Cumitech 3: practical quality control procedures for the clinical microbiology laboratory, Washington, D.C., 1976, American Society for Microbiology.

Bohnhoff, G.L.: Setting up proper illumination on your microscope, Am. J. Med. Technol. 45:650, 1979.

Dalton, H.P.: State of the art of antimicrobial agent susceptibility testing: a clinical microbiologist's view, ASM News 48:513, 1982.

Finegold, S.M., and Martin, W.J.: Bailey and Scott's diagnostic microbiology, ed. 6, St. Louis, 1982, The C.V. Mosby Co.

Gerhardt, P., editor: Manual of methods for general bacteriology, Washington, D.C., 1981, American Society for Microbiology.

Goldfogel, G.A., and Sewell, D.L.: Preparation of sputum smears for acid-fast microscopy, J. Clin. Microbiol. 14:460, 1981.

Isenberg, H.D., and others: Cumitech 9: collection and processing of bacteriological specimens, Washington, D.C., 1979, American Society for Microbiology.

Jorgensen, J.H., and others: Comparison of automated and rapid manual methods for the same-day identification of Enterobacteriaceae, Am. J. Clin. Pathol. 79:683, 1983.

Knezek, K.H., and others: Is your incubator really calibrated? Lab. Med. 14:366, 1983.

Mascart, G., and others: Evaluation of the Replireader™ for identification and sensitivity determination of gram-negative bacilli, Lab. Med. 14:563, 1983.

Murray, P.R.: Antibiotic susceptibility testing, part I, Lab. Med. 14:345, 1983.

Murray, P.R.: Antibiotic susceptibility testing, part II, Lab. Med. 14:417, 1983.

Murray, P.R., and Jorgensen, J.H.: Quantitative susceptibility test methods in major United States medical centers, Antimicrob. Agents Chemother. 20:66, 1981.

Murray, P.R., and others: The acid-fast stain: a specific and predictive test for mycobacterial disease, Ann. Intern. Med. 92:512, 1980.

Siegle, M.D.: Urinoscopy: first the microscope, Lab. Med. 12:781, 1981.

Thornsberry, C.: NCCLS standards for antimicrobial susceptibility tests, Lab. Med. 14:549, 1983.

Thornsberry, C., and others: Cumitech 6: new developments in antimicrobial agent susceptibility testing, Washington, D.C., 1977, American Society for Microbiology.

Tilton, R.C.: The drug/disc dilemma, Diagn. Med. 5:64, March/April 1982.

Tilton, R.C., and Ryan, R.W.: Micro-Media Systems' bacterial identification and MIC panels, Lab. World 31:38, April 1980.

CHAPTER 10

LABORATORY IDENTIFICATION

BIOCHEMICAL REACTIONS

Biochemical tests demonstrate the presence of enzyme systems in the microbial cell, such as those responsible for fermentation of carbohydrates or decomposing of proteins, and are done so that individual species of microbes may be identified. By noting the presence of specific enzymes in a pure culture of bacteria being studied, one can usually make its identification.

Miniaturization of microbiologic technics refers to the use of commercially prepackaged test units wherein basic reagents for a given biochemical reaction are premeasured, standardized, and compacted into a tablet or onto a paper disk or filter paper strip. The test unit is applied to cultures in liquid or solid media and is chemically designed so that enzymatic action of the test organism usually results in a color change, or end point, a quick and easy observation. There are many varieties of such tests available.

Fermentation of Sugars

Microbes ferment many organic compounds, including carbohydrates, generating in the process energy-rich chemical bonds. Through fermentation, simple sugars can serve as the main source of energy for many different kinds of microbes. The end-products of fermentation depend on the substrate, enzymes present, and conditions under which the reaction proceeds. For instance, the yeast enzymes that break down glucose into carbon dioxide and alcohol have no effect on sucrose (cane sugar), whereas enzymes produced by pneumococci and other streptococci ferment glucose to lactic acid. Products of bacterial fermentation are lactic acid, formic acid, acetic acid, butyric acid, butyl alcohol, acetone, ethyl alcohol, and the gases carbon dioxide and hydrogen.

Fermentation reactions, though varying among species of microorganisms, are constant and therefore of great value in characterizing the species. From their specific action on a given sugar in the laboratory, bacteria (and other microbes) are classified as (1) those that do not ferment the sugar, (2) those that ferment it with the production of acid only, and (3) those that ferment it with the production of both acid and gas. Sugar-containing media are inoculated with test microbes and observed for gas and acid formation. Gas production in liquid media is detected by an accumulation of gas that displaces the fluid medium contained in the closed arm of a special tube, as in a Smith fermentation tube. Gas may be detected in a small tube (one end of which is sealed off) placed in an inverted position inside a larger tube of liquid with the sealed end projecting slightly above the level of the fluid. As gas is formed in the inoculated liquid, it collects in the small tube in the depths of the culture and rises toward the sealed end (Fig. 10-1) of the smaller tube, where it is trapped.

One inoculates solid media for fermentation studies by plunging a straight needle carrying some of the test microbes (bacteria or other) deep into the medium, a maneuver known as **stabbing** (Fig. 10-2). The solid medium may also be melted, cooled to 42° C, and then inoculated.

In solid media gas production is seen as "bubbles" of gas disrupting the medium. Acid formation in both liquid and solid media is measured by color changes in pH indicators incorporated therein.

Many carbohydrates are used in routine fermentation studies. Important sugars are glucose, lactose, maltose, sucrose, xylose, mannitol, and salicin. The fermentation of lactose is crucial in the identification of bacteria recovered from the intestinal tract since it makes a working division between the ordinarily nonpathogenic coliform organisms, which rapidly ferment lactose, and the pathogenic species, such as *Salmonella* and *Shigella*, which cannot ferment lactose. There are exceptions, however.

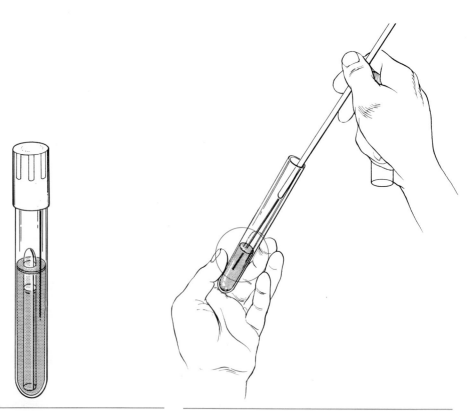

FIG. 10-1   Durham fermentation tube. Note gas collected in tip of smaller inverted tube.

FIG. 10-2   Stabbing tube of solid culture medium.

Hydrolysis of Starch

When microbes are grown on starch agar, a clear zone develops around the colonies of those that digest starch. If the agar is covered with a weak solution of iodine, the *undigested* starch assumes a blue color.

Liquefaction of Gelatin

A gelatin medium is stabbed with an inoculating wire with bacteria on it, incubated at 20° C, and observed for liquefaction. It may be incubated at 37° C (gelatin is liquid at this temperature) and then placed in the refrigerator, where the unliquefied gelatin solidifies. Incubation should be continued for at least 2 weeks unless liquefaction occurs before then, and a tube of uninoculated gelatin should be incubated as a control.

The proteolytic enzymes (protein-splitting enzymes, proteinases), produced by bacteria, split complex proteins into proteans, peptones, polypeptides, amino acids, ammonia, and free nitrogen. Protein decomposition is known as **putrefaction.** Some authorities restrict the term *putrefaction* to the decomposition of proteins by anaerobic bacteria, which results in the formation of hydrogen sulfide and other foul-smelling decomposition products, and they use the term *decay* for the decomposition of proteins by aerobic bacteria. The latter does not result in the malodorous decomposition products.

Citrate Utilization

Utilization of citrate by bacteria is determined by inoculation of Simmons citrate agar, a commercially prepared test medium, colored green, that contains ammonium phosphate as a nitrogen source and sodium citrate as the sole source of carbon. If the test microbe possesses the enzyme *permease* to transport citrate into the cell, the medium becomes alkaline, turning from its original green to a deep Prussian blue. The test is useful in the differentiation of certain coliforms, such as *Enterobacter* and *Klebsiella* species, which can use citrate as their sole carbon source, from *Escherichia coli*, which cannot.

Indole Production

Production of indole (from the amino acid tryptophan) is determined by culture of test organisms in a medium containing tryptophan. The reaction is mediated by *tryptophanase*, the complete system of enzymes responsible. A strip of filter paper soaked with a saturated solution of oxalic acid may be hung over the culture and held in place either by the cotton plug or the screw cap of the culture tube. A pink color on the paper indicates indole production. When Kovacs reagent (paradimethyl aminobenzaldehyde dissolved in amyl alcohol and concentrated hydrochloric acid) is added to the broth, a red color indicates indole production. Notable indole producers are *Escherichia coli* and *Proteus* species, which avidly decompose proteins.

Nitrate Reduction (Nitrite Test)

Nitrate reduction means the removal of oxygen from the nitrate radical (NO_3) to convert it to nitrite (NO_2). The reaction is mediated by the enzyme *nitratase*, produced by a wide variety of microorganisms. Test microbes are incubated in broth containing 0.1% potassium nitrate, and the broth is tested for nitrite with sulfanilic acid and alpha-naphthylamine reagents. A red color means a positive test.

Deoxyribonuclease Elaboration

Formation of deoxyribonuclease (DNAse) is demonstrated by culture of test microbes on an agar surface into which deoxyribonucleic acid (DNA) has been incorporated. After 24 hours' incubation, the surface is flooded either with 1N hydrochloric acid to highlight the clear zones around the DNAse-positive growth or with 0.1% toluidine blue to display a bright rose-pink color that as an end point marks the presence of DNAse. Staphylococci produce this enzyme.

Hydrogen Sulfide Production

A stab culture in an agar that contains basic lead acetate or iron acetate is incubated for 1 to 4 days. The production of hydrogen sulfide from the sulfur-containing amino acids of the medium is indicated by the appearance of a black compound, lead sulfide (or iron sulfide), formed from the combination of the hydrogen sulfide with the lead (or iron) acetate. Instead of being incorporated into the medium, the basic lead acetate may be impregnated on sterile filter paper strips suspended over the medium as in the test for indole production. The production of hydrogen sulfide facilitates the identification of *Brucella* species and enteric bacilli.

Splitting of Urea

Certain bacteria (for example, *Proteus* species) convert urea to ammonia by producing *urease*. With phenol red as the indicator, the presence of urease is seen by the appearance of the red color. This test helps to separate species of the genus *Proteus* from the urease-negative *Salmonella* and *Shigella* species.

Digestion of Milk

Bacterial growth in sterile milk may be alkaline or acid, with or without curdling. Curdling may or may not be followed by liquefaction of the casein curd. Excessive gas production in milk produced by *Clostridium perfringens* is called "stormy fermentation."

Oxidase Reaction

The enzyme *oxidase* is produced by *Neisseria* species, and its detection is of great value in the identification of *Neisseria gonorrhoeae*. The oxidase reagent (*N*,*N*-dimethyl-*p*-phenylenediamine monohydrochloride) colors the positive colonies pink to red to black.

Niacin Test

The niacin test is used to distinguish *Mycobacterium tuberculosis* from other species of mycobacteria. A 4% alcoholic aniline solution and a 10% aqueous solution of cyanogen bromide (NOTE: *toxic to humans*) are added to 1 or 2 ml of emulsified bacterial growth. A complex yellow compound is formed when niacin or nicotinic acid, formed by the tubercle bacillus but not by other mycobacteria, reacts with the cyanogen bromide and a primary amine.

Optochin Growth-inhibition Test

The Optochin growth-inhibition test is performed by placement of a 6 mm absorbent paper disk containing Optochin (ethylhydrocupreine hydrochloride) in contact with a heavy inoculum of test streptococci in a pure culture on a blood agar plate, which is incubated overnight. A zone of inhibition greater than 18 mm identifies the growth as that of pneumococci, whereas zones around other streptococci, if present, are much smaller. Currently this is the most widely used test for differentiating pneumococci from other alpha-hemolytic streptococci.

Catalase Test

The catalase test is performed with hydrogen peroxide poured over the growth of a heavily inoculated, 18- to 24-hour agar slant that is then placed in an inclined position to demonstrate the rapid evolution of oxygen gas bubbles, the positive reaction. In microbes the transfer of hydrogen to oxygen during respiration may result in the production of hydrogen peroxide, which is toxic to the microbial cell. Most aerobic (but not anaerobic) bacteria produce an enzyme, *catalase*, that oxidizes the hydrogen per-oxide to water and oxygen. Staphylococci are catalase positive; streptococci are catalase negative. Acid-fast bacilli produce catalase, and the detection of catalase activity is useful in the laboratory evaluation of mycobacteria.

Demonstration of Specific Enzymes

Reagent systems for the identification of bacterial enzymes are commercially impreg-nated onto easy-to-use strips of paper suitable for application to bacterial cultures. Three enzymes of practical importance are packaged this way: phenylalanine deaminase, cy-tochrome oxidase, and lysine decarboxylase.

Phenylalanine deaminase catalyzes the metabolism of the amino acid phenylalanine to phenylpyruvic acid, which in turn reacts with ferric ions present to give a brownish or gray-black color. The presence of this enzyme in cultures of *Proteus* species is re-sponsible for the positive reaction obtained with them.

$$\text{Phenylalanine} \xrightarrow{\text{Phenylalanine deaminase}} \text{Phenylpyruvic acid } + \text{ Ferric ions} \rightarrow \text{Color}$$

Cytochrome oxidase is produced by *Pseudomonas*, *Alcaligenes*, *Neisseria*, *Vibrio*, *Brucella*, and some *Halobacterium* species. This enzyme catalyzes a coupling reaction to give a blue color (the positive reaction) as follows:

$$\text{Dimethyl-}p\text{-phenylenediamine} + \text{Alpha naphthol} + \text{Oxygen} \xrightarrow{\text{Cytochrome oxidase}}$$
$$\text{Indophenol blue}$$

Lysine decarboxylase, which is formed by most of the *Salmonella* species, catalyzes the conversion of lysine to cadaverine, a more alkaline compound than lysine. In one commercial, prepackaged test kit the Prussian blue reaction is used to indicate the positive test.

$$\text{Lysine} \xrightarrow{\text{Lysine decarboxylase}} \text{Cadaverine } + \text{ CO}_2 \rightarrow$$
$$\text{Bromthymol blue (yellow) changing to blue as pH rises}$$

Rapid Identification of Enteric Bacilli

At least six commercially packaged systems are in current use for the rapid identification of enteric bacilli. An example of one such multitest system is Enterotube System.* Enterotube is made with a number of small compartments in sequence, each of which is filled with a different test medium, all pertinent to the identification of enteric bacilli. The arrangement provides for various diagnostic biochemical reactions to be carried out simultaneously on the same microorganism, and the battery of tests is calculated to be sufficient to permit the identification of most enteric bacilli without additional testing. (Enterotube II with 12 compartments can give results for 15 tests.) After growth from the clinical specimen has been obtained on suitable media, such as MacConkey or EMB agar, a discrete colony is selected for multitesting. The inoculating needle sup-plied with the diagnostic set is touched to that colony and then pulled through all of the compartments in Enterotube in one motion. Results are read after 18 to 24 hours of incubation.

Limulus Test for Endotoxemia

Serum or cerebrospinal fluid (or suitable test material) from a patient that possibly con-tains bacterial endotoxin is incubated with lysates derived from the blood cells (amebo-cytes) of the horseshoe crab (*Limulus polyphemus*). (The horseshoe crab is not a crab at all and is not even closely related to any other living thing in the sea. Its blood is blue

---

*Roche Diagnostics, Division of Hoffman-LaRoche, Inc., Nutley, N.J.

from the copper-containing molecule therein that carries oxygen.) In the presence of even minute amounts of bacterial endotoxin there is a greatly increased viscosity of the test medium, the *amebocytic lysate*. This becomes the gelation reaction, which can progress to a solid gel. The limulus test is the most sensitive and practical in vitro assay for endotoxin.

Automated Identification of Microbes

Reliable methods for the automated identification of microbes came into being largely as the result of the search for life on other planets. Today several automated and semi-automated technics in existence promise to revolutionize the microbiology laboratory. The advantages are considerable. Automated technics shorten the work time required for characterization of microbes from 2 to 3 days (or more) to less than 1 day. They economize the laboratory operation, reducing the amounts of test materials such as media and substrates needed, and requiring perhaps fewer work hours from the laboratory personnel. When automated tests give accurate, reproducible, and unequivocal results, they afford an important means of providing uniform standards to all laboratories, regardless of size.

Applications of automated instruments for microbiology have been largely confined to testing in the detection of bacteria in the urine (bacteriuria), determination of antibiotic susceptibility, and identification of members of Enterobacteriaceae.

Pyrolysis

Highly sensitive technics result when sequential combinations of standard chemical procedures are automated; an example is the automation of pyrolysis (the carefully controlled heat decomposition of organic substances) linked to gas-liquid chromatography or mass spectrometry for the analysis of the decomposition products. In microbiology this approach is used for the identification of certain microbes.

Suitable fragments for analysis of given bacteria are obtained by heating a sample very quickly in the absence of air, usually in less than 1 second, to a temperature between 300° and 800° C. As bacterial compounds are degraded by heat, fragments from proteins, lipids, nucleic acids, and other molecules provide a **pyrogram**, a chemical fingerprint, which apparently is peculiar to each species of microorganism and can be used to distinguish between even closely related strains. In fact it is possible in this way to separate bacterial strains differing by only one gene. The pyrolytic fragments released are analyzed by the technic of either gas-liquid chromatography or mass spectrometry, and thus identification of the test microorganisms is made.

ANIMAL INOCULATION

The inoculation of suitable laboratory animals, such as mice, is an essential part of the study of many microbes. After an animal has been inoculated, it is observed for effects produced by the microbes. In some cases the animal is killed after a certain period and examined for evidence of disease. Smears and cultures are made, and gross and microscopic changes in the organs are noted. In other cases the animal is not killed, but blood and body fluids are examined at given intervals.

Well-known laboratory animals used for inoculation are guinea pigs, white mice, white rats, hamsters, and rabbits. Inoculations or injections (Figs. 10-3 and 10-4) may be given with syringe and needle either subcutaneously (beneath the skin), intradermally (between the layers of skin), intravenously (into a vein), intraperitoneally (into the peritoneal cavity), subdurally (beneath the dura of the brain), or intracerebrally (into the brain).

Advantages of animal inoculation in recovery and identification of microbes are as follows:

1. Some microbes are most easily detected by animal inoculation (for example, *Francisella tularensis* and, under some conditions, *Mycobacterium tuberculosis*).
2. Microbes readily demonstrate their virulence when inoculated into animals (for example, *Corynebacterium diphtheriae*).
3. Some microbes are most easily obtained in pure culture by animal inoculation (for example, *Streptococcus pneumoniae*).
4. Microorganisms that cannot be cultured at all on artificial media can be readily recovered from suitable laboratory animals (for example, spiral bacteria, rickettsias, and most viruses). A few microbes can be propagated in no other way.
5. Animal inoculation is often necessary to determine the action of drugs on microbes.
6. Animal inoculation is a part of the experimental determination of the action of pathogenic and nonpathogenic agents.
7. Animal inoculation is basic to the manufacture of antitoxins and other antiserums.

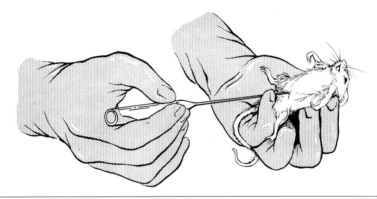

FIG. 10-3 Intraperitoneal inoculation of white mouse. Note point of capillary pipette directed into peritoneal cavity as mouse is held firmly by nape of neck.

FIG. 10-4 Rabbit restrained for intravenous inoculation of marginal vein in ear. (Courtesy Plas Labs, Lansing, Mich.)

A good example of animal inoculation as the best method of detecting microbes is the ease with which tuberculosis of the kidney is proved in a patient whose urine shows no bacteria on direct microscopic examination but whose urine produces classic disease in a guinea pig. A good example of animal inoculation being used to obtain a pure culture is seen in pneumococcal typing. Some of the sputum in which the type of pneumococcus is to be determined is injected into the peritoneal cavity of a white mouse, and within a few hours the growth of pneumococci has outstripped that of all other organisms to such an extent that the peritoneal content consists of practically a pure culture of pneumococci ready to be typed.

A laboratory animal must be handled carefully to prevent its spreading the disease under study to the laboratory staff. It can also bite or scratch the person handling it. Because of many considerations inherent in the maintenance of animal quarters and because improved culture media and new technics are facilitating the recovery and identification of pathogenic microorganisms, many laboratories find routine diagnostic animal inoculations impractical.

REVIEW OF REACTIONS: IDENTIFICATION OF BACTERIA

An overview of the various methods of studying bacteria shows that bacteria can be classified in comparatively large groups as follows:

I. Direct microscopic examination
   A. Staining reactions
      1. Reaction to Gram stain
         a. Gram positive
         b. Gram negative
      2. Reaction to acid-fast stain
         a. Acid fast
         b. Nonacid fast
   B. Size
   C. Shape
      1. Coccus—spherical
      2. Bacillus—rod shaped
      3. Vibrio, spirillum, spirochete—spiral shaped
   D. Presence of spores
      1. Spore-formers
      2. Nonspore-formers
   E. Capsule formation
      1. Encapsulated
      2. Nonencapsulated
   F. Flagellar motility
      1. Motile
      2. Nonmotile
II. Culture
   A. Food requirements
      1. Saprophytes—growth on dead organic matter
      2. Parasites—growth on living matter
      3. Autotrophs—growth on carbon dioxide and inorganic salts
   B. Media most suitable for growth
      1. Growth on chemically simple media
      2. Growth only on special media
      3. No growth on any media
   C. Appearance of growth on different media

    D. Oxygen requirements
       1. Obligate aerobes—growth only in the presence of free oxygen
       2. Obligate anaerobes—growth only in the absence of free oxygen
       3. Facultative anaerobes—growth in the presence or absence of free oxygen
       4. Microaerophiles—growth in the presence of small amounts of free oxygen
    E. Optimum temperatures (and range of temperatures) for growth
    F. Characteristics of the pure (axenic) culture
       1. Size, shape, texture of colonies
       2. Pigment production in colonies
         a. Water soluble
         b. Water insoluble
    G. Production of hemolysins on blood agar
       1. Beta-hemolytic—complete hemolysis of red blood cells
       2. Alpha-hemolytic—partial hemolysis of red blood cells
       3. Gamma-hemolytic—nonhemolytic
 III. Biochemical reactions (A great variety of biochemical reactions are used; only a few are mentioned.)
    A. Fermentation of carbohydrates (for example, the sugar lactose)
       1. Fermentation with acid and gas (lactose fermenter)
       2. Fermentation with acid only; no gas (lactose fermenter)
       3. No fermentation (nonlactose fermenter)
    B. Hydrolysis of urea
       1. Urease positive
       2. Urease negative
    C. Reduction of nitrate
       1. Reduction of nitrate (nitratase positive)
       2. No reduction of nitrate (nitratase negative)
    D. Formation of indole from tryptophan
       1. Formation of indole (tryptophanase positive)
       2. No formation of indole (tryptophanase negative)
 IV. Animal inoculation
    A. Disease production (virulence tests)
       1. Pathogenic—disease produced
       2. Nonpathogenic
    B. Toxin production
       1. Exotoxin producers
       2. Endotoxin producers
  V. Immunologic reactions (Chapter 20)

SUMMARY

1 Since biochemical tests demonstrate the presence of specific enzyme systems within microbial cells, they are valuable in the characterization of microbes.

2 Many enzymes may be tested for in the microbiologic laboratory. Examples of important ones and the procedures used are permease (citrate utilization), tryptophanase (production of indole), urease (splitting of urea), oxidase (oxidase reaction), and catalase (catalase test).

3 Citrate utilization is used to differentiate certain coliforms utilizing citrate as their sole source of carbon from *E. coli*, which cannot.

4 *Proteus* species and *E. coli*, which avidly decompose proteins, are notable producers of indole.

5 Certain bacteria (the best-known example is the genus *Proteus*) convert urea to am-

monia by producing urease. Urease-negative *Salmonella* and *Shigella* species cannot do so.

6 *Neisseria* species, including the gonococcus, produce oxidase. The detection of oxidase is valuable in their identification.

7 Most aerobic (but not anaerobic) bacteria produce catalase. Staphylococci are catalase positive; streptococci are catalase negative. Since acid-fact bacilli produce catalase, the detection of this activity is useful in the laboratory evaluation of mycobacteria.

8 Constant patterns of sugar fermentations greatly facilitate differentiation of microbial species. Bacteria may not ferment the given sugar, they may ferment it with the production of acid only, or they may ferment it with the production of both acid and gas.

9 Excessive gas production in milk by *Clostridium perfringens* is called "stormy fermentation."

10 Many laboratories do not make routine diagnostic animal inoculations because of the impracticalities of animal care. Some of the advantages of this method of studying microbes are that some microbes are most easily detected and best display their virulence in animals.

11 Automated methods for the identification of microbes shorten the working time for the test, reduce the amounts of needed test materials, and provide accurate and uniform standards to laboratories regardless of size.

QUESTIONS FOR REVIEW

1 How are biochemical reactions used in the study of microbes?
2 List five sugars important in routine fermentation studies.
3 What two observable changes may occur in the culture medium when carbohydrates are fermented?
4 What is the importance of lactose fermentation?
5 List the laboratory animals routinely used for inoculation.
6 What are the advantages of animal inoculation?
7 What is meant by putrefaction? Decay?
8 Name two bacteria that produce indole.
9 What bacterial genus produces urease?
10 Briefly outline the niacin test.
11 How may specific enzymes be demonstrated? Name three of diagnostic value.
12 Summarize methods used to classify and categorize bacteria.
13 Explain briefly miniaturization of microbiologic technics, citrate utilization, demonstration of deoxyribonuclease, permease, urease, catalase, one multitest system for identification of enteric bacilli, and pyrolysis.
14 Briefly discuss automation in the microbiologic laboratory (consult outside sources).

SUGGESTED READINGS

Balows, A.: CDC reference and diagnostic services: a new look, ASM News **49**:371, 1983.

Edberg, S.C.: Gas-liquid chromatography in microbiology, Lab. Manage. **18**:31, June 1980.

Freeman, D.H.: Liquid chromatography in 1982, Science **218**:235, 1982.

Goel, M.C.: Why a color change? Biochemical tests for Enterobacteriaceae, Lab. World **31**:22, April 1980.

Leighton, P.M., and Little, J.A.: Clinical comparison of the Enterotube II and API 20E systems for bacterial identification, Am. J. Clin. Pathol. **79**:367, 1983.

Merchant, D.J.: The future of reference and diagnostic services in microbiology, ASM News **49**:374, 1983.

Peterson W.C., and others: An evaluation of the practicality of the spot-indole test for the identification of *Escherichia coli* in the clinical microbiology laboratory, Am. J. Clin. Pathol. **78**:755, 1982.

Romeo, J.: A dichotomous key for the identification of miscellaneous gram-negative bacteria, Lab. Med. **10**:547, 1979.

CHAPTER 11

SPECIMEN COLLECTION

Ground Rules
Clinical Specimens
Pathogens Related to Specimens

To the extent that the microbiologic investigation defines the diagnosis and nature of disease, it determines the mode of treatment and outlook for the patient affected. In the practice of scientific medicine, proper specimen collection is a crucial first step on which the validity of each succeeding step rests.

GROUND RULES

The following admonitions apply to the collection of specimens for microbiologic study:

1. Collect the specimen from the actual site of disease and do not contaminate it with microbes from nearby areas. For example, in making smears and cultures from ulcers in the throat, be very careful to take the specimen from the actual site of ulceration and do not contaminate it unduly with secretions of the mouth.
2. Always use sterile equipment and materials to collect the specimen. Place it in a sterile container.
3. Sterilize material used to collect the specimen as soon as possible.
4. Collect adequate amounts of specimen. For instance, when pus is to be examined, collect several milliliters, if possible. If it is necessary to collect the material on swabs, use more than one swab.
5. Insofar as possible, do not send specimens to the laboratory on swabs (a swab can take up a small amount of material, dry quickly, and enmesh in its fibers important cells and organisms that fail to be transferred to smears or culture media).
   NOTE: If swabs must be used, specially devised transport media are available (for example, Transgrow medium for detection of gonococci).
6. Collect the specimen in such a manner that it does not endanger others. Sputum or other excreta must *not* soil the outside of the container.
7. Take great care in handling specimens collected in cotton-plugged tubes. Cotton plugs can entirely soak up a small specimen. Also, microbes from the environment pass through wet plugs and contaminate the specimen.
8. Whenever possible, make smears directly from the original specimen.
9. Do not add any preservative or antiseptic to the specimen. If possible, take the specimen *before* the patient has received any antimicrobial drug or *before* the wound has had local treatment. If the patient has already received such a drug, notify the laboratory.
10. Label and identify the specimen properly as indicated in the medical setting. Give the patient's name, hospital or clinic identification number, source of specimen, date, time of day specimen was taken, physician's name, and tentative diagnosis.
11. Deliver the specimen to the laboratory promptly.

Sometimes the person delivering the specimen to the laboratory may have to care for the specimen if laboratory personnel are temporarily off duty. In this event specimens already inoculated onto culture media are placed in the incubator, and those not on culture media are placed in the refrigerator. When cultures must be made without delay, the medium chosen should be that most likely to grow the suggested organisms as well as other microbes that might be present. Since it supports the growth of microorganisms better than most other culture media found in the usual microbiologic laboratory, blood agar is generally used. Blood agar plates are better than tubed media except when the cultures must be shipped a distance to the diagnostic laboratory (that is, to a state-operated public health laboratory or comparable facility). Plates are not easily transported.

CLINICAL SPECIMENS

Urine for microbiologic examination may be carefully collected as a clean-voided specimen. The collection of a urine specimen with a sterile catheter has long been recommended, especially in women and girls, but many physicians now believe that catheterization is unnecessary. No matter how carefully done, the maneuver is traumatic and pushes infectious organisms into the urinary tract. For most purposes a distinction can be made between infectious and contaminating microorganisms in a *carefully collected*, clean-voided specimen. The meatus (opening) of the urethra is gently cleaned with soap (or detergent) and water. In both voided and catheterized specimens the first portion should be discarded, and the last portion should be received in a sterile container. A "clean-catch" (or midstream) urine specimen is one obtained during the midpart or toward the end of urinating; it is received into a sterile receptacle. The sample should be promptly cultured or refrigerated if laboratory examination is delayed. A healthy person's urine has relatively few microorganisms.

Blood for cultures must be collected with special care because microbes, especially staphylococci, are present on the surface of the skin and within its superficial layers. A blood culture is taken by venipuncture (Fig. 11-1) as follows:
1. Paint the skin over the veins in the bend of the elbow with tincture of iodine or other suitable disinfectant (70% isopropyl alcohol may be used).
   WARNING: *Some persons are allergic to iodine.*
2. Remove disinfectant with 70% alcohol (or make a second application of 70% alcohol).
3. Place a 70% alcohol compress on the area.
   NOTE: Clean but unsterile cotton balls can be used with isopropyl alcohol to disinfect the skin, but they cannot be used with quarternary ammonium disinfectants ("quats") because of the danger of skin surface contamination with *Pseudomonas* species.
4. Secure a tourniquet, not too tightly, around the arm just above the elbow and ask the patient to close and open the hand several times.
5. Puncture a prominent vein with a 20- or 21-gauge needle to which may be attached either a 20 ml syringe or Vacutainer, a specially designed holder for a vacuum tube (Fig. 11-1). Remove 10 to 15 ml of blood (or whatever amount is indicated for the particular laboratory test).
6. Add the blood directly to a culture medium or place it in a sterile bottle containing a suitable anticoagulant (to prevent coagulation), such as sodium citrate, and carry it to the laboratory, where the specimen can be distributed to proper culture media. To prevent contamination, remove the needle from the syringe into which blood was received before the blood is expelled from it.
7. Remember that the venipuncture site is a wound, although a small one. After the bleeding has stopped, cleanse the area as indicated and apply a small sterile covering.

Bacteremia is often transient. If the blood culture is not collected at the proper time, the etiologic organisms of the disease may not be found. Since the advent of antimicrobial drugs, positive blood cultures are fewer. Even one dose of an antimicrobial drug before the culture is drawn may mask the infection. The bloodstream in health is free of microorganisms.

Blood for serologic examinations (such as the precipitation tests for syphilis and various agglutination tests for continued or persistent fevers) may be collected from a vein in the bend of the elbow. The technic is the same as that for collecting blood for culture, except that (1) the preliminary sterilization usually consists of scrubbing the area with 70% alcohol only and (2) the blood is placed in a chemically clean test tube and allowed to clot. Five milliliters is usually sufficient.

NOTE: Hepatitis can be conveyed to the person from whom the blood is drawn by the use of a clean *but unsterile* syringe. Use sterile *disposable needles* and preferably sterile *disposable syringes* to prevent hepatitis. (Otherwise syringes and needles must be carefully sterilized by autoclaving.)

Sputum is often collected improperly, and samples tend to be too small. Many specimens consist of secretions from the nose, mouth, and throat and contain no sputum at all, only saliva. Sputum should represent a true pulmonary secretion and should be expelled after deep coughing. Instruct the patient to bring the specimen up from the lower part of the respiratory tract. If necessary, nebulized aerosols can be used. The

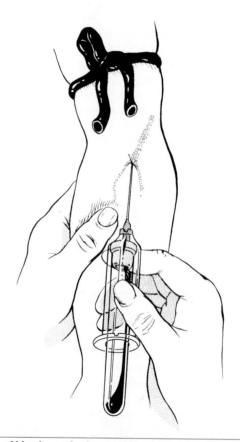

FIG. 11-1   Collection of blood samples by aseptic venipuncture. Note use of evacuated blood collection tube in plastic holder. (Syringe can also be used.)

teeth should be carefully cleaned with toothpaste and the mouth rinsed with sterile water before the specimen is taken. Usually more sputum is raised in the morning; this sputum may contain *Mycobacterium tuberculosis*, whereas specimens taken later in the day do not. Generally, sputum specimens for bacteriologic study are collected daily for at least 3 days. The ideal container for sputum is a sterile 6-ounce widemouthed bottle (or plastic receptacle) with a tight-fitting stopper or cover.

Children swallow all their sputum. Adults swallow sputum in their sleep. Generally, the only effective way to collect sputum from a child is to aspirate the stomach contents. When it is impossible to obtain a satisfactory cough specimen from an adult, the stomach contents may be examined. The stomach should be aspirated early in the morning before any food or water is taken.

Bronchial secretions are secured through the bronchoscope. After bronchoscopy, sputum specimens should be collected again for a 3-day period.

Cultures from the nose and throat are made many times, since sore throats and nasal discharges are common disorders and are often part of systemic illness. Taking the culture just after an antiseptic has been used defeats its purpose because the antiseptic retards the growth of bacteria present. (This applies to other cultures also.) To take a throat culture, use a good light and insofar as possible allow the swab to touch only the diseased area. Use at least two swabs. When cultures are to be taken from the nose, pass a small, tightly wound swab directly back through the nose. Avoid large loose swabs because they may slip off the applicator and lodge in the nose.

When diphtheria is suspected, cultures made from both nose and throat instead of from the throat alone improve the detection rate. The microbes causing diphtheria are often found abundantly in cultures when they have not been seen on direct smears. Smears should be made routinely when diphtheria-like lesions of the throat are encountered, since the exudate in lesions of Vincent's angina, the etiologic agents of which are detected only in smears, often closely resembles the membrane of diphtheria.

Cultures from the nasopharynx (the upper part of the throat behind the soft palate) are needed for the detection of meningococcal carriers and are used in infants and children suspected of having whooping cough. To collect specimens from this area, pass a thin swab (made of Dacron, cotton, or calcium alginate) through the nose into the nasopharyngeal area, gently rotate it, remove it, and place it in a suitable transport medium. Expedite transport to the laboratory. A satisfactory swab may be made by wrapping cotton around the end of a piece of wire bent at a right angle about 1 inch from the end. When specimens from the throat and nasopharynx are to be obtained, take care to avoid contamination with saliva.

Feces for routine microbiologic examination (see also p. 642) may be collected directly by the patient in a sterilized cardboard carton or other suitable container. At times it may be desirable for the patient to pass a stool into a larger, previously sterilized container, with a sterile tongue blade or spoon used to remove a small amount to a sterile widemouthed bottle. The stool in the latter case should be examined superficially. Portions of the fecal material with mucus are preferable for microbiologic study.

Rectal swabs are very satisfactory for most microbiologic purposes and are easily obtained from both adults and children. If a disease process involves the lining of the terminal portion of the rectum, these swabs are likely to obtain material from within the focus of disease and are therefore more likely to contain disease-producing agents than the stool of that person. Swabs are of particular value in testing for carriers of bacteria causing typhoid fever, dysentery, or cholera. To obtain the rectal swab, clean the skin around the anus with soap, water, and 70% alcohol. Introduce a sterile cotton swab moistened with either sterile isotonic solution or sterile broth through the anus and gently rotate it around the circumference of the lower rectum to contact a large area of the mucosal lining. A fecal specimen may also be obtained from the physician's glove after digital examination of the rectum.

In all cases the stool specimen should be sent to the laboratory at once because nonpathogenic intestinal bacteria quickly overgrow the pathogenic ones, and pathogenic amebas rapidly lose their motility if the fecal material containing them cools. If acute amebiasis is suspected, the warm feces should be examined immediately for pathogenic amebas showing their typical, diagnostic movement. If some delay is anticipated, refrigerate the fecal sample or preserve it in polyvinyl alcohol fixative. Never incubate a stool specimen before laboratory manipulation. Examination for typhoid bacilli is facilitated by collection of feces in a special brilliant green–bile medium; bile facilitates the growth of typhoid bacilli, and brilliant green retards the growth of many other intestinal organisms. Enrichment media other than brilliant green–bile medium used for the recovery of enteric bacterial pathogens include selenite-F enrichment medium and tetrathionate broth.

Pus from abscesses and boils may be obtained after drainage. The abscess or boil is painted with suitable disinfectant, the area is allowed to dry, and an incision is made with a sterile scalpel. Some of the contents are obtained by means of sterile swabs. If the lesion is opened widely, as much of the superficial portion should be removed as possible and the specimen taken from the deeper part. When this is not done, the specimen is usually contaminated with surface microorganisms. Protect all specimens taken on swabs from drying by placing the swabs in a sterile test tube containing sterile physiologic saline (a drop or two suffices), a nutrient broth that may be directly inoculated with the swabs, or another suitable transport medium. The swabs, which are longer than the test tube, should be placed in the test tube so that the cotton pledget is just above the salt solution and does not touch it. A cotton plug is inserted to hold the swabs in place.

Smears of gonococci (p. 511) in pubertal girls or adult women should be taken from the meatus of the urethra, the cervix uteri, and the rectum. In prepubertal and younger girls suspected of having gonorrheal vulvovaginitis, smears and cultures are made from the vagina; the vagina is primarily attacked. In acute gonorrhea in men and boys smears

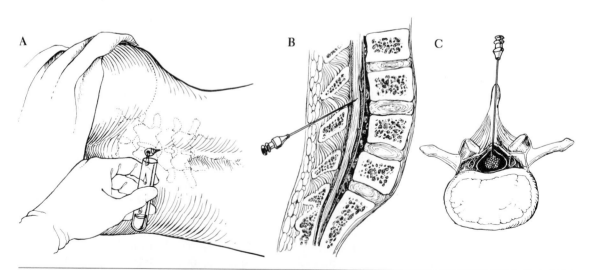

FIG. 11-2 Collection of cerebrospinal fluid by lumbar puncture. **A,** Note spinal needle in place in lower back. **B** and **C,** Inserts indicate anatomic position of needle in subarachnoid space.

are obtained from the urethral discharge. In chronic cases in males a specimen may be obtained from the physician's massage of the prostate gland and seminal vesicles.

Cerebrospinal fluid is obtained by a procedure known as *lumbar puncture* (Fig. 11-2) done by the physician. After the overlying skin of the lower back has been disinfected and anesthetized, a special needle with stylet is introduced slightly to one side of the midline between the third and fourth lumbar vertebrae and passed into the spinal sub-arachnoid space. Strictest asepsis must be observed, and a sterile dressing is placed over the site of puncture. An infant or child seated with head and arms resting on a folded sheet, blanket, or pillow set against the abdomen may be held securely while the physician performs the lumbar puncture. Normal cerebrospinal fluid is sterile.

Pleural and peritoneal fluids are obtained through the lumen of a trochar or specially designed needle inserted through the wall of the chest or abdomen. With the trochar in place, fluid is aspirated with a syringe or freely drained into a sterile container. Strictest asepsis should be observed during this procedure, which is done by the physician.

Smears and cultures from the conjunctiva are made with swabs wet with sterile physiologic solution. The swabs should be handled carefully to prevent physical injury and spread of infection to adjacent parts of the eye or to the other eye.

Specimens for anaerobic culture are collected as carefully as possible to (1) avoid contamination of the anaerobes with other organisms of the given area and (2) minimize the exposure of the specimen to the oxygen of the air. Direct aspiration (with needle and syringe) of a given focus of disease gives a workable preparation. For example, in pulmonary infection, direct lung puncture or transtracheal aspiration is done to obtain pulmonary secretions or tissue, and in urinary tract infection percutaneous suprapubic bladder puncture is done to obtain urine.

Specimens for the viral and rickettsial laboratory, such as blood, cerebrospinal fluid, respiratory secretions, feces, and urine, should be collected with strict aseptic technic and placed in sterile containers. Avoid further contamination of throat swabs, nasal washings, stools, and specimens already containing bacteria. Swabs may be received in tubes of sterile physiologic saline. The addition of a suitable antibiotic to the saline may be indicated.

Rush all specimens to the viral diagnostic laboratory *immediately after collection*. Often this means transporting them a distance to a medical center, large hospital, or state health laboratory. When this is the case, freeze the specimen (unless it is blood) immediately and keep it frozen until shipment can be made. Note that some viruses, such as respiratory syncytial virus, parainfluenza viruses, and cytomegaloviruses, are very sensitive to freezing. Refrigerate blood samples, let them clot, and separate serum from clot. Transit is best made with the specimen container packed in dry ice. Shipment is expedited by air carrier with provisions for rushing the specimen to its destination. A frozen shipment is likely to be satisfactory provided the ice has not thawed by the time it gets there. (Small packages of dry ice are likely to thaw in 8 to 12 hours or less.) To send blood samples on dry ice, place the clot and serum in separate tubes. Although cold shipment of fresh material is preferable most of the time, specimens may be preserved in sterile 50% glycerin solution or in buffered glycerol transport medium (but *never* in formalin).

Paired samples of blood for serologic examination in viral and rickettsial diseases (p. 687) are to be collected aseptically, the blood allowed to clot, and the serum separated from the clot. The serum is sent at once to the diagnostic laboratory by the quickest method (direct delivery or air carrier, depending on the location of the laboratory).

Remember that proper labels on the outside of the package help prevent accidental infection of laboratory personnel receiving the specimen. The diagnostic specimens discussed here should always be packaged so that they do not leak in transit. Warn mailhandlers of the potential health hazard.

PATHOGENS RELATED TO SPECIMENS

This chapter has considered body fluids and anatomic sites in the body from which specimens for microbiologic examination are regularly taken. Since the emphasis here is on specimen collection for diagnosis of disease, Table 11-1 is given to show the distribution of disease-producing agents in routine specimens taken from patients with infectious diseases.

Table 11-1 Medical Specimen and Associated Pathogens

Specimen	Important Pathogens	Specimen	Important Pathogens
Blood	Anaerobic cocci	Urine—cont'd	*Mycobacterium tuberculosis*
	Arboviruses		Polioviruses
	Bacteroides species and related anaerobes		*Proteus* species
	Brucella species		*Pseudomonas* species
	Campylobacter species		*Salmonella* species
	Chlamydia psittaci		*Staphylococcus* species
	Clostridia		*Streptococcus pyogenes* and other streptococci
	Coxiella burnetii		
	Escherichia coli and other coliforms	Nose and throat secretions	Adenoviruses
	Francisella tularensis		Anaerobic cocci
	Haemophilus influenzae		Arboviruses
	Hepatitis viruses		*Bacteroides* species
	Legionella species		*Bordetella pertussis*
	Leptospira species		*Candida albicans*
	Listeria monocytogenes		*Chlamydia psittaci*
	Neisseria gonorrhoeae		Coliforms
	Opportunistic fungi		*Corynebacterium diphtheriae*
	Plasmodium species		Enteroviruses
	Polioviruses		Fungi
	Proteus species		*Haemophilus influenzae*
	Pseudomonas species		Measles viruses
	Rickettsias of spotted fevers		Mumps virus (saliva)
	Rubeola and rubella viruses (measles viruses)		Mycobacteria
	Salmonella species		*Mycoplasma pneumoniae*
	Spirillum minor		Myxoviruses
	Staphylococci		*Neisseria* species
	Streptobacillus moniliformis		Polioviruses
	Streptococcus pneumoniae		Rabies virus
	Streptococcus pyogenes and other streptococci		Respiratory viruses
Urine	*Alcaligenes* species		*Staphylococcus aureus*
	Arboviruses		*Streptococcus pyogenes* and other streptococci
	Brucella species	Sputum	*Actinomyces israelii*
	Candida albicans		Anaerobic streptococci
	Escherichia coli and other coliforms		*Blastomyces dermatitidis*
	Gardnerella vaginalis		*Candida albicans*
	Measles virus		*Coccidioides immitis*
	Mumps virus		Coliform bacilli
			Cryptococcus neoformans
			Haemophilus influenzae

Table 11-1 Medical Specimen and Associated Pathogens—cont'd

Specimen	Important Pathogens	Specimen	Important Pathogens
Sputum—cont'd	*Histoplasma capsulatum*	Fluid from conjunctiva	Anaerobes
	Klebsiella pneumoniae	of eye—cont'd	*Chlamydia trachomatis*
	Legionella species		Fungi
	Mycobacterium tuberculosis		*Haemophilus* species
	and other mycobacteria		Measles viruses
	Mycoplasma species		*Moraxella lacunata*
	Nocardia asteroides		*Neisseria gonorrhoeae*
	Proteus species		*Pseudomonas* species
	Pseudomonas species		Rabies virus
	Staphylococcus aureus		*Staphylococcus aureus*
	Streptococcus pneumoniae		*Streptococcus pneumoniae*
	Streptococcus pyogenes and		*Streptococcus pyogenes* and
	other streptococci		other streptococci
Stool (feces)	Adenoviruses		*Toxoplasma* species
	Anaerobic cocci	Pleural fluid	*Haemophilus influenzae*
	Bacteroides species		*Mycobacterium tuberculo-*
	Candida albicans		*sis*
	Clostridia		*Staphylococcus aureus*
	Enteroviruses		*Streptococcus pneumoniae*
	Hepatitis viruses	Peritoneal fluid	Coliforms
	Certain metazoa (*Taenia*,		Enterococci
	etc.)		*Mycobacterium tuberculo-*
	Polioviruses		*sis*
	Proteus species	Pus, exudates, wound	*Actinomyces* species
	Certain protozoa (*Enta-*	drainages, and the	Anaerobic cocci (*Pepto-*
	moeba histolytica,	like	*streptococcus, Peptococcus*
	etc.)		species)
	Pseudomonas species		*Bacillus anthracis*
	Salmonella species		*Bacteroides* species and re-
	Shigella species		lated anaerobic rods
	Staphylococcus aureus		*Blastomyces* and other sys-
	Vibrio cholerae		temic fungi
Cerebrospinal fluid	Adenoviruses		*Clostridium* species
	Arboviruses		Coliforms
	Bacteroides species		*Corynebacterium diphthe-*
	Coliforms		*riae*
	Cryptococcus neoformans		Enterococci
	Enteroviruses		*Francisella tularensis*
	Haemophilus influenzae,		Mycobacteria
	type b		*Mycobacterium tuberculo-*
	Leptospira species		*sis*
	Listeria monocytogenes		*Neisseria gonorrhoeae*
	Mycobacterium tuberculosis		*Nocardia* species
	Neisseria meningitidis		*Pasteurella multocida*
	Polioviruses		*Proteus* species
	Rabies virus		*Pseudomonas* species
	Staphylococcus aureus		*Staphylococcus aureus*
	Streptococcus pneumoniae		*Streptococcus pyogenes* and
	Streptococcus pyogenes		other streptococci
Fluid from conjunctiva	*Adenoviruses*	Genital secretions and	See Table 24-2, p. 516
of eye		exudates	

SUMMARY

1 Microbiologic investigation can define the diagnosis and nature of infectious disease (disease caused by a living agent).

2 Microbiologic investigation depends on careful collection of specimens for reliable results.

3 The person collecting any specimen for microbiologic examination should adhere to the ground rules carefully; this is particularly true with collection of specimens for the viral diagnostic laboratory.

4 Proper labels on the outside of the packages containing specimens to be shipped help prevent accidental infection of laboratory personnel receiving the shipments.

5 Certain clinical specimens are routinely tested by microbiologic methods: urine, blood, respiratory secretions, nose and throat discharges, feces, drainages from foci of disease, cerebrospinal fluid, and serous fluids.

QUESTIONS FOR REVIEW

1 Why is careful specimen collection stressed?

2 State the ground rules for the collection of any specimen for microbiologic examination.

3 What precautions should be taken in collecting urine specimens for microbiologic examination? Is it always necessary to catheterize the human female?

4 How should the skin be prepared for collection of a blood culture? Why is this necessary?

5 What vein is commonly used as a collection site for blood specimens?

6 What are the precautions for collecting sputum specimens? For collecting stool specimens?

7 How would you handle a specimen of pus that is to be cultured?

8 Briefly discuss the collection of cerebrospinal fluid.

9 How are specimens handled that must be shipped to an out-of-town laboratory for diagnosis of viral disease?

SUGGESTED READINGS

Barry, A.L., and others: Cumitech 2: laboratory diagnosis of urinary tract infections, Washington, D.C., 1975, American Society for Microbiology.

Bartlet, R.C.: Making optimum use of the microbiology laboratory. II. Urine, respiratory, wound, and cervicovaginal exudate, J.A.M.A. 247:1336, 1982.

Brook, I.: Anaerobic bacteria in pediatric respiratory infection: progress in diagnosis and treatment, South. Med. J. 74:719, 1981.

Cade, R., and others: New method for obtaining uncontaminated urine from women, South. Med. J. 71:1536, 1978.

Crates, R.H.: Improper microbiological specimens: a shocking deficit in health care, Lab. Med. 10:234, 1979.

Fuchs, P.C.: Focus on clinical relevance in microbiology, Lab. Med. 10:521, 1979.

Huntoon, C.F., and others: Recovery of viruses from three transport media incorporated into culturettes, Arch. Pathol. Lab. Med. 105:436, 1981.

Isenberg, H.D., and others: Cumitech 9: collection and processing of bacteriological specimens, Washington D.C., 1979, American Society for Microbiology.

Lewis, J.F., and Alexander, J.J.: Overnight refrigeration of urine specimens for culture, South. Med. J. 73:351, 1980.

Slockbower, J.M.: Venipuncture procedures, Lab. Med. 10:747, 1979.

LABORATORY SURVEY OF MODULE TWO

PROJECT   Growth of Bacteria in the Laboratory—An Overview

Part A—Demonstration by the Instructor with Discussion

1. Materials and equipment for culture
   a. *Petri dish*—what it is, how it is used, how it is manipulated with sterile technic, how culture medium is poured into the dish, why it is inverted in the incubator, how it is handled on the laboratory bench
   b. *Culture tube*—how it is used, purpose of cotton plug or screw cap, care and handling of the cotton plug, how tube is manipulated with sterile technic
   c. *Culture media*—solid and liquid, various kinds, various ways media is dispensed, incorporation of specific substances for specific reasons as, for example, indicators, test reagents, and materials to enhance growth
   d. *Wire loop and wire inoculating needle*—manipulation, importance of flame sterilization
   e. *Microbiologic incubator*—its mechanism, how it is used
2. Precautions to avoid contamination, species variation from inconsistencies in technic, and spread of infection or undesired bacterial growth
3. Examination of bacterial growth—how to study it in the laboratory
   a. Enumerate and stress descriptive features (refer to Figs. 9-1, 9-3, and 9-7). List workable adjectives, useful terms, and the like.
   b. Comment on variations.
   c. Note how identification of bacteria is indicated from gross morphology.

Part B—Student Participation: Studying Bacterial Growth

NOTE: It is suggested that for the following exercise students work in groups of four or more. Each group is furnished with a set of prepared cultures. Each member of the group carefully inspects each culture and records a description of its pattern of growth in a notebook. Twelve cultures have been selected. Although these make up a practical and workable set, they can easily be varied.

1. Study the following prepared cultures assigned to your group:
   a. Three Petri dish cultures (streak plates on trypticase soy agar) in each of which growth of one organism on one side is separate from growth of another organism on the opposite side
      (1) The first plate displays *Escherichia coli* and *Bacillus subtilis* (incubated 24 hours at 37° C).
      (2) The second plate shows *Micrococcus luteus* and *Serratia marcescens* (grown at room temperature for 24 hours).
      (3) The third plate shows *Streptococcus faecalis* and *Pseudomonas aeruginosa* (grown at room temperature for 24 hours).
   b. Two pour plates (trypticase soy agar incubated 24 hours at 37° C), one showing growth of *Serratia marcescens* and the other of *Pseudomonas aeruginosa*
   c. Four agar slant cultures (trypticase soy agar incubated at room temperature), one each for *Escherichia coli*, *Serratia marcescens*, *Streptococcus faecalis*, and *Pseudomonas aeruginosa*

        d. Three broth cultures (trypticase soy broth grown at room temperature for 48 hours), one each for *Escherichia coli*, *Serratia marcescens*, and *Pseudomonas aeruginosa*

2. Note colonial morphology and gross features of bacterial growth in the laboratory setting.

    a. Inspect growth found on the prepared Petri dishes (three), pour plates (two), agar slants (three), and broth cultures (four).

    b. Record observations for each in your notebook.

       (1) Note appearance of discrete colonies on agar plates. Wherever a bacterium contacts the surface of the medium, a colony develops. Count the number of colonies for each organism. Do colonies vary for a given organism? For different organisms? Are the edges of the colonies smooth or irregular? What might this mean? Is the growth on the agar slant discrete or confluent? Why?

       (2) List terms and phrases that help describe bacterial growth in colonies.

       (3) Note location and nature of growth in broth culture. Why are both solid and liquid media used?

    c. Compare the following:

       (1) Growth patterns for the same microbe in different media

       (2) Growth patterns of one microbe with those of another in different media, in the same medium

       (3) Bacterial growth in solid media with that in fluid media

    d. Keep the cultures in a place designated by the laboratory instructor so that you may observe any changes in them after a few days and after a week. Note the differences.

    e. Answer the following in your notebook:

       (1) Is it possible to identify a bacterium by the nature of its growth under specified conditions in the laboratory?

       (2) Must requirements for culture of bacteria in the laboratory be specific and constant if reproducible results are desired?

PROJECT   Conditions Affecting Growth of Bacteria

1. Influence of temperature on bacterial growth

    a. The instructor will furnish you three broth cultures just inoculated with bacteria. Note that all are clear because bacterial growth has not occurred yet.

    b. Place one in the refrigerator, another at room temperature, and another in the microbiologic incubator.

    c. Observe after 24 hours.

    d. Record results. In which tube is the growth, indicated by cloudiness, most abundant? What does this prove?

	Refrigerator Temperature 4° C (24 hr)	Room Temperature 20° C (24 hr)	Incubator Temperature 37° C (24 hr)
Bacterial growth			

2. Effect of reaction of medium (pH) on bacterial growth

    a. The instructor will furnish you five recently inoculated tubes of broth. Make additions of alkali or acid to four of them as follows: (1) 0.2 ml 1N NaOH, (2) 0.5 ml 1N NaOH, (3) 0.2 ml 1N HCl, and (4) 0.5 ml 1N HCl. The fifth serves as a control.

    b. Incubate at 37° C for 24 hours. Compare the multiplication of bacteria in the different tubes as indicated by clouding of the fluid medium.

c. Record results.

	Alkaline pH		Acid pH		
	0.2 ml 1N NaOH	0.5 ml 1N NaOH	0.2 ml 1N HCl	0.5 ml 1N HCl	Control
Bacterial growth					

3. Effects of sunlight on bacterial growth
   a. The instructor will furnish you a recently inoculated Petri dish. Invert and cover half the Petri dish in such a manner that direct sunlight is not allowed to strike the medium. A piece of black paper may be used.
   b. Expose to sunlight for several hours.
   c. Incubate at room temperature until colonies appear. On which half of the plate are the colonies more numerous? What does this prove?

PROJECT   Special Activities of Bacteria
           Part A—Fermentation
1. Fermentation of sugars
   a. The instructor will furnish each group of students three Smith fermentation tubes containing sterile sugar solutions. The first contains 10% dextrose (grape sugar) solution; the second contains 10% lactose (milk sugar) solution; and the third contains a 10% solution of sucrose (cane sugar).
   b. Rub a small piece of yeast cake in water and add a portion to each tube.
   c. Set in a warm place and observe after 24 hours.
   d. Chart results. What differences are noted? Why?

	Dextrose, 10% (Grape Sugar)	Lactose, 10% (Milk Sugar)	Sucrose, 10% (Cane Sugar)
Changes observed after inoculation with yeast			

2. Prove that gas formed in no. 1 is carbon dioxide as follows:
   a. Measure the length of the gas column in the arm of the fermentation tube.
   b. Remove a small amount of material from the bulb of the tube with a pipette and reserve for no. 3.
   c. Fill bulb of tube with 10% potassium hydroxide (KOH).
   d. Place thumb over mouth of tube and mix by inverting so that gas in arm of tube comes in contact with KOH.
   e. Collect all remaining gas in arm of tube. Reduction in amount of gas proves that CO_2 was present and was absorbed according to the following chemical equation: $2KOH + CO_2 = K_2CO_3 + H_2O$.
3. Prove that the fermentation in no. 1 produced alcohol as follows:
   a. Filter the liquid removed in no. 2.
   b. To the filtrate add a few drops of weak iodine solution.
   c. Add enough 10% sodium hydroxide solution to change the color from brown to a distinct yellow.
   d. Warm slightly. A distinct odor of iodoform will be obtained, and a yellow precipitate, appearing under the microscope as small hexagonal crystals, will be formed. If the amount of alcohol is small, the crystals may not appear until the solution has cooled or stood some time.

4. Fermentation with acid and gas formation
   a. The instructor will furnish you a phenol red dextrose broth fermentation tube recently inoculated with *Escherichia coli*.
   b. Incubate 24 to 48 hours.
   c. Note multiplication of bacteria as indicated by turbidity. What other two changes have taken place? What does each indicate? What is the purpose of the phenol red?
5. Fermentation with reduction of pigment
   a. Nearly fill a long narrow test tube with milk.
   b. Add enough weak aqueous solution of methylene blue to give the milk a distinct blue color.
   c. Add a little milk that has soured naturally.
   d. Mix and incubate at 36° C for several days.
   e. Observe. What change has taken place?

Part B—Pigment Production by Bacteria

1. The instructor will give you three agar slants. One has been inoculated with *Pseudomonas aeruginosa*, one with *Serratia marcescens*, and one with *Staphylococcus aureus*.
   a. Incubate at 37° C until pigment production is well developed.
   b. Observe. What color is produced by each? Is the color confined to the growth of bacteria or does it diffuse into the medium?
   c. Record colors.

	Pseudomonas aeruginosa	*Serratia marcescens*	*Staphylococcus aureus*
Color produced			

2. Examine cultures of *Escherichia coli* and *Proteus vulgaris* prepared on eosin–methylene blue (EMB) agar or MacConkey agar (p. 158).
   a. Note pigmentation. Record findings. What color is produced by *Escherichia coli*? By *Proteus vulgaris*? What is the color of the culture medium?
   b. Explain and interpret the findings.

Color Produced	*Escherichia coli*	*Proteus vulgaris*
EMB agar		
MacConkey agar		

PROJECT   Preparation of Culture Media

Part A—Demonstration of Bacteriologic Technics by the Instructor with Discussion

1. Flaming the mouth of a container to preserve sterility of contents
2. Holding tubes of culture media in a slanting position during inoculations
3. Removing and inserting cotton plugs or other stoppers during inoculations
4. Transferring bacterial growth
   a. Broth–to–agar slant cultures
   b. Agar slant–to–broth cultures
5. Pouring melted agar into a Petri dish

Part B—Preparation of Nutrient Broth

The composition of 1 L of nutrient (beef extract) broth is as follows:

Meat extract	3 g	Sodium chloride (may be omitted)	5 g
Peptone	5 g	Distilled water	1 L

Commercially, one can obtain Difco (or equivalent product*)—a dehydrated medium that contains the meat extract, peptone, and buffer substance in such proportions that the finished product has a pH of 6.8. It is not necessary to adjust the pH of this product. Using the commercial, dehydrated medium to make nutrient broth, proceed as follows:

1. Dissolve 4 g of dehydrated nutrient broth (containing 1.5 g of meat extract and 2.5 g of peptone) in 500 ml of distilled water contained in a liter Erlenmeyer flask. Add 2.5 g of sodium chloride if desired. (For the subsequent experiments, 500 ml amounts of nutrient broth will be sufficient.)
2. After solution is complete, pour 10 to 15 ml amounts in each of 25 culture tubes.
3. Plug tubes with cotton plugs and autoclave at 121° C for 15 minutes. During the process of plugging and when handling the culture medium, do not let the liquid contact the cotton plugs. Under no circumstances can the pressure of the autoclave be rapidly released when steam pressure sterilization is finished, because rapid release of pressure causes the liquid in the tubes to boil over and wet the plugs. What are some of the objections to wetting of cotton plugs?

Part C—Preparation of Nutrient Agar

The composition of 1 L of nutrient (beef extract) agar is as follows:

Meat extract	3 g	Agar	15 g
Peptone	5 g	Distilled water	1 L
Sodium chloride (may be omitted)	5 g		

The Difco dehydrated medium (or equivalent commercial product) contains the meat extract, peptone, and agar in a dehydrated condition buffered so that the final pH of the medium is 6.8. No further adjustment of pH is necessary.

To make nutrient (beef extract) agar, proceed as follows:

1. To 500 ml of cold distilled water contained in a liter flask, add 11.5 g dehydrated nutrient agar (Difco).
2. Place in pan of water or on boiling water bath and heat until solution is effected. The material may be boiled over a free flame, but there is danger of scorching.
3. To have sufficient culture medium for the experiments to follow in this and the next unit, put up the following:
   a. Six tubes containing 5 ml of agar (these are slanted to solidify after sterilization)
   b. Six test tubes containing 25 ml of agar
   c. Four Erlenmeyer flasks (150 ml) containing about 89 ml of agar
4. Plug containers with cotton plugs and autoclave at 121° C for 15 minutes.
5. After pressure in autoclave has returned to normal, slant agar tubes to solidify as previously indicated.

PROJECT   Cultivation of Bacteria

1. Taking a throat culture
   a. Demonstration by the instructor with discussion
      The instructor will show how a sterile cotton swab can be used to take a throat culture. The swab, after it has been rubbed over the back part of the throat of a

---

*Commercial products are usually supplied with precise directions for use.

given individual, is then rubbed over the surface of a suitable culture medium. No attempt in this exercise is made to separate bacterial organisms. Media such as Löffler serum medium, blood agar, trypticase soy agar, or nutrient (beef extract) agar may be used.

    b. Student participation

        (1) Use a sterile swab to take a throat culture. (Students may work in pairs.) Apply swab to the surface of a suitable culture medium.

        (2) Incubate cultures and observe after 24 hours.

        (3) Make Gram-stained smears of the bacterial growth. Make drawings to show representative microorganisms.

2. Transferring bacterial growth

    a. Demonstration by the instructor with discussion

      The instructor reemphasizes the proper technics of sterile transfers, indicates the differences in ability of microorganisms to grow in various laboratory media, and explains the usefulness of cultural transfers.

    b. Student participation

        (1) Obtain a culture (tube or plate) showing growth of a nonpathogenic organism.

        (2) Make the transfer of organisms from your culture to a different kind of culture medium

        (3) Incubate the culture for 24 to 48 hours. Observe and describe growth.

PROJECT   Isolation of Pure Cultures

    1. Pour plate method

      NOTE: The students, working in small groups, will use tubes of culture media that they have prepared, sterile Petri dishes, and a mixed culture furnished by the instructor.

      a. Refer to the pour plate method as outlined on p. 163.

      b. Use this method to obtain pure cultures of organisms in the mixed culture provided to you. Incubate the Petri dish cultures for 24 to 48 hours.

    2. Streak plate method

      NOTE: The students are each given three Petri dishes of sterile culture medium, and mixed cultures are laid out in the laboratory.

      a. Refer to the discussion of streak plates on p. 164 and to the illustration of the method in Fig. 9-9. Consult the instructor for assistance when you carry out this technic for the first time.

      b. Make a streak plate using one of the mixed cultures available in the laboratory. Incubate for 24 to 48 hours.

      c. Observe colonial growth as obtained by the two methods of isolation. How do the distributions of the colonies on the streak and pour plates differ?

      d. Note the importance of making Gram-stained smears for microscopic study of different types of colonies obtained. How can a Gram-stained smear be used to check a pure culture? Make Gram-stained smears for representative colonies that you have obtained from your cultures.

    3. Fishing of colonies

      a. Demonstration by the instructor with discussion

      The instructor should demonstrate how given colonies are transferred from Petri dishes to agar slants. Why is it often necessary to do this?

PROJECT   Bacterial Colony (Plate) Count

    1. Work in small groups of two to four.

    2. Obtain necessary materials for this experiment as follows:

      a. A small amount of milk

      b. Four tubes containing 9 ml sterile distilled water

      c. Five sterile 1 ml pipettes

      d. Four sterile Petri dishes

3. Place a flask of agar in a pan of water and heat water until agar melts. Let cool to about 40° C. This may be roughly determined by considering a temperature of 40° C as that temperature at which the flask may comfortably be held against the back of the hand. Keep at this temperature.

4. Proceed as follows:

      a. Transfer 1 ml of milk to a tube of 9 ml sterile distilled water and mix. With a fresh pipette transfer 1 ml of the mixture to a second tube of 9 ml sterile distilled water. Continue in this manner until the fourth tube is mixed, being careful to use a fresh sterile pipette to make each mixture. You now have 1:10, 1:100, 1:1000, and 1:10,000 dilutions of milk.

      b. With a sterile pipette transfer 1 ml of each dilution to a Petri dish. Begin with the highest dilution and go to the lowest when making transfers. The material is transferred to the Petri dishes by slightly raising the lid and depositing the material in the middle of the plate. Gently raise the lid and add about 15 ml of melted agar to each dish. Mix contents of each dish by gently rotating. Let agar solidify; invert and incubate 48 hours.

      c. Select a plate showing well-distributed, easily counted colonies and count the colonies present. Multiply the number of colonies counted by the times the milk in that plate was diluted. The result is the number of bacteria in 1 ml of the milk used. Why was a new pipette used for making each dilution of the milk? Why can transfers from the tubes to Petri dishes be made with a single pipette by beginning with the highest dilution and proceeding to the lowest?

PROJECT   Animal Inoculation

1. Demonstration by instructor with discussion

      a. Technic of animal inoculation

      b. Value of animal inoculation

   The organisms suspected and the results obtained should be explained.

2. Study of preserved organs from autopsies of guinea pigs inoculated with *Mycobacterium tuberculosis* and *Francisella tularensis*.

   Compare diseased animal organs with normal ones.

EVALUATION FOR MODULE TWO

1 In the following statements or questions, use the letters A to G to indicate the response that *best* completes the statement or answers the question. Please place the appropriate letter in the space to the left, as follows:

A if entries 1, 3, and 5 are best answers   E if only entry 3 is correct
B if entries 1 and 5 are best answers   F if no entry is correct
C if entries 2 and 3 are best answers   G if all entries apply correctly
D if entries 2 and 4 are best answers

_____ 1. Why is refrigeration a good method for preserving bacterial cultures?
(1) Cold is necessary for bacterial growth.
(2) Freezing may destroy bacteria.
(3) Cold retards bacterial growth.
(4) Cold excludes oxygen.
(5) Cold protects bacteria from sunlight.

_____ 2. A spirochete can best be studied by which of the following preparations?
(1) Gram stain
(2) Hanging drop preparation
(3) Dark-field illumination
(4) Acid-fast stain
(5) Culture on artificial medium

_____ 3. Strict anaerobes grow well:
(1) Under 5% to 10% increase of carbon dioxide tension
(2) In broth beneath a layer of petroleum jelly
(3) In thioglycollate broth
(4) In atmospheric oxygen
(5) On routine blood agar

_____ 4. Testing the antibiotic susceptibility of a bacterial culture by the disk sensitivity method has:
(1) Become unnecessary since discovery of broad-spectrum drugs
(2) Little application to subsequent treatment of a patient
(3) Been used only with penicillin
(4) Been too time consuming to be practical
(5) Interfered with routine blood cultures

_____ 5. The bacterial plate count is important in the microbiologic evaluation of urine. Generally the microbes recovered per milliliter of specimen are not considered significant until the count exceeds:
(1) 100/ml of urine
(2) 1000/ml of urine
(3) 10,000/ml of urine
(4) 100,000/ml of urine
(5) 1,000,000/ml of urine

_____ 6. The most common complication that follows urethral catheterization is:
(1) Urinary tract closure

      (2) Rupture of urethra

      (3) Urinary tract infection

      (4) Bleeding from bladder

      (5) Loss of the patient's ability to void naturally

_____ 7. The number of *viable* bacteria in a suspension can be determined by:

      (1) Counting the number of bacterial cells in a measured volume of suspension

      (2) Counting the number of bacterial cells retained on a plastic membrane filter

      (3) Counting the number of bacterial colonies that develop from a measured volume of suspension

      (4) Measuring the amount of bacterial nitrogen present

      (5) Measuring the turbidity of the culture medium

_____ 8. Body fluids that are appropriate for anaerobic culture include:

      (1) Blood

      (2) Sputum

      (3) Joint fluid

      (4) Voided urine

      (5) Cerebrospinal fluid

_____ 9. From a patient with a lung abscess, the following specimens may be collected for anaerobic culture:

      (1) Material suctioned from the nose and throat

      (2) Saline washings taken at the time of bronchoscopy

      (3) Material aspirated transtracheally

      (4) Saliva

      (5) Gastric washings

_____ 10. Animal inoculation as a method of studying microbes is used for:

      (1) The demonstration of the virulence of a given microbe

      (2) The routine isolation of microbes in the laboratory

      (3) The cultivation of microbes that cannot be cultured on artificial media

      (4) The determination of antibiotic susceptibility

      (5) The demonstration of the action of drugs on given microbes

_____ 11. The catalase test performed in the laboratory:

      (1) Identifies most anaerobic organisms

      (2) Is an important characteristic of acid-fast bacilli

      (3) Utilizes hydrogen peroxide, a product not toxic to bacterial cells

      (4) Catalyzes the oxidation of hydrogen peroxide to water and oxygen

      (5) Utilizes oxygen gas

_____ 12. Bacterial fermentation of sugars:

      (1) Is always associated with gas formation

      (2) Is generally constant for a given bacterial species

      (3) Is never associated with gas formation

      (4) Constitutes one factor in bacterial classification

      (5) Is not dependent on enzyme action

_____ 13. MacConkey agar:

      (1) Is selective for gram-positive bacteria

      (2) Is used primarily for the detection of bacteria from the intestinal tract

      (3) Permits the growth of intestinal anaerobes under aerobic conditions

      (4) Differentiates lactose fermenting from nonfermenting colonies

      (5) Is an example of an enrichment medium

_____ 14. A pigmented bacterium is isolated from a clinical specimen. It is negative for oxidase but positive for DNAse. It is likely to be:
   (1) *Serratia marcescens*
   (2) *Pseudomonas aeruginosa*
   (3) *Staphylococcus aureus*
   (4) A tubercle bacillus
   (5) *Escherichia coli*

_____ 15. Which of the following make good dyes for negative staining?
   (1) Nigrosin
   (2) Crystal violet
   (3) Carbolfuchsin
   (4) Acetone
   (5) Congo red

_____ 16. When cell cultures are intended for demonstration of a virus, bacterial contamination can usually be avoided:
   (1) Because most bacteria do not grow in the presence of living cells
   (2) Because the inoculum can be passed through an ordinary bacteria-retaining filter
   (3) Because a bacteriostatic concentration of antibiotics can be added to the cell culture medium
   (4) Because the inoculum can be heated to a temperature that destroys bacteria but not the contained virus
   (5) Because a small amount of inoculum can be used

2 Fill in the blanks.

1. Name four kinds of microscopes.

   (1) _____

   (2) _____

   (3) _____

   (4) _____

   Which one is most widely used? _____

2. Which microscopic preparation is especially well-suited for the determination of motility in bacteria? _____

3. Write the scientific name of a bacterium that is an example of the following:

   (1) Gram-positive coccus _____

   (2) Gram-negative coccus _____

   (3) Gram-positive bacillus _____

   (4) Gram-negative bacillus _____

   (5) Acid-fast mycobacterium _____

4. Name three sugars important in routine fermentation studies.

   (1) _____

   (2) _____

   (3) _____

5. Name two laboratory animals used in the identification and study of microbes.

(1) _____

(2) _____

PART II

1 Before each principle, please place the letter from column Beta in the appropriate space to indicate the procedure based on that principle.

COLUMN ALPHA—Principles

_____ 1. Bacteria can be partially classified by their characteristic action on media containing different carbohydrates.

_____ 2. A single species of bacteria can be isolated from infectious material by making a dilute mixture of the medium and the infectious material. Then, when the inoculated medium is allowed to harden in a Petri dish, the colonies will be widely separated.

_____ 3. We can assume that for each colony appearing on a pour plate, there was only one bacterium originally present in the medium.

_____ 4. Oxygen can be removed from an airtight container by bringing about some chemical reaction in the container that uses up the oxygen.

COLUMN BETA—Procedures

(a) Pure culture, pour place method
(b) Culture of anaerobic bacteria
(c) Pure culture, streak plate method
(d) Bacterial count, plate method
(e) Determination of fermentation reactions

2 Comparisons: match the item in column Beta to the most descriptive word or phrase in column Alpha. Please place the related letter in the appropriate space.

COLUMN ALPHA
Respiration (Aerobic Oxidation) with Fermentation

_____ 1. Yields more energy for same amount of substrate
_____ 2. End-products carbon dioxide and water
_____ 3. End-product ammonia
_____ 4. End-product pyruvate
_____ 5. Carried out by facultative organisms
_____ 6. Carried out by strict anaerobes
_____ 7. Important laboratory application in identification of microbes

COLUMN BETA

(a) Aerobic oxidation of glucose
(b) Fermentation of glucose
(c) Both
(d) Neither

Differential Culture Media with Selective Media

_____ 8. Retard growth of microbes in certain instances

_____ 9. Important in culture of feces

_____ 10. Petragnani for growth of tubercle bacilli, an example

_____ 11. EMB agar for growth of gram-negative bacilli, an example

_____ 12. Lactose may be incorporated

_____ 13. Dye added to indicate fermentation in certain instances

_____ 14. Dye added to suppress growth in certain instances

_____ 15. Selected embryonic tissues incorporated in certain instances

(a) Selective culture media
(b) Differential media
(c) Both
(d) Neither

Rough Bacterial Colonies with Smooth Ones

_____ 16. Expression of bacterial pleomorphism

_____ 17. Expression of microbial dissociation

_____ 18. Association with encapsulated bacteria

_____ 19. Association with nonencapsulated microbes

_____ 20. Association with virulence of bacteria forming them

_____ 21. Association with nonvirulent bacteria

_____ 22. Association with presence of O antigen

(a) Rough colonies
(b) Smooth colonies
(c) Both
(d) Neither

Nutrient Agar with Blood Agar

_____ 23. Useful for demonstration of pigment produced by bacteria

_____ 24. Useful for maintenance of bacterial cultures

_____ 25. Differential medium for intestinal pathogens

_____ 26. Selective medium for tubercle bacilli

_____ 27. Preferred medium for most pathogenic bacteria

_____ 28. Useful as a base for more complex media

(a) Nutrient agar
(b) Blood agar
(c) Both
(d) Neither

**Disk Diffusion Antibiotic Susceptibility Testing
with Tube Dilution Testing**

_____ 29. Multiple antibiotics tested at one time

_____ 30. Single antibiotic tested in the technic

_____ 31. Multiple strengths of a given antibiotic tested simultaneously in the technic

_____ 32. Utilizes Mueller-Hinton agar

_____ 33. Utilizes filter paper impregnated with antibiotic

_____ 34. Modification of Kirby-Bauer method

_____ 35. Utilizes bacterial growth plated out as a lawn

_____ 36. Requires check for viability of bacteria

(a) Disk diffusion method
(b) Tube dilution method
(c) Both
(d) Neither

3 Match the item in column Beta to the most closely related phrase in column Alpha. Please place the proper letter in the appropriate space.

COLUMN ALPHA

_____ 1. Important in identification of _Neisseria_ species

_____ 2. Important in identification of _Proteus_ species

_____ 3. Demonstration of microbes with tryptophanase activity

_____ 4. Demonstration of bacterial endotoxin

_____ 5. Demonstration of organisms with permease activity

_____ 6. Separation of certain pathogens from nonpathogens in the intestinal tract

_____ 7. Demonstration with appearance of black compound in agar containing basic lead acetate

_____ 8. Rapid evolution of gas from surface of culture medium

COLUMN BETA

(a) Niacin test
(b) Limulus test
(c) Hydrogen sulfide production
(d) Citrate utilization
(e) Fermentation of lactose
(f) Indole production
(g) Oxidase reaction
(h) Catalase test
(i) Splitting of urea
(j) Elaboration of DNAse

PART III

Complete the following statements by filling in the blanks to the left. Please be sure that the number of the answer corresponds with that in the statement. Note vocabulary for reference.

Vocabulary

Absorption	Colony	Lipase	Penetration
Active transport	Commensalism	Macromolecule	Pigment
Aerobic	Contaminated	Microaerophilic	Pili
Agar	Decolorizer	Mixed	Positive
Amino acid	Dye	Mordant	Pure
Amylase	Endoenzyme	Mutualism	Routine
Anaerobic	Enzyme	Negative	Spore
Antagonism	Exoenzyme	Niacin	Stormy fermentation
Biologic oxidation	Fatty acid	Nicotinic acid	Substrate
Chemosynthesis	Fermentation	Optimal	Symbiosis
Chlorophyl	Kinase	Organotrophic	Synthetic
Coagulase	Leukocidin	Oxidase	Urease
Coenzyme	Light		

1. _____ Certain bacteria can form a __1__ when surrounding conditions are unfavorable for growth.

2. _____ A visible mass formed by rapid reproduction of an organism on culture media is called a __2__.

3. _____ When the organism sought is found in a culture, we designate the culture a __3__ one.

4. _____ Iodine is used as a __4__ in the Gram-staining procedure.

5. _____ When the presence of two species of microbes in the same environment favors the development of both, the condition is called __5__.

6. _____ A growth of one kind of bacteria is called a __6__ culture.

7. _____ Chromogenic bacteria produce __7__.

8. _____ Microbes without chlorophyl must obtain their energy by __8__.

9. _____ __9__ are important structural constituents of cells.

10. _____ Excessive gas formation in milk produced by *Clostridium perfringens* is __10__.

11. _____ The enzyme __11__ is produced by *Neisseria* species.

12. _____ The __12__ test is important in the differentiation of tubercle bacilli from atypical mycobacteria.

13. _____ If a culture contains two or more kinds of microbes, it is termed a __13__ culture.

14. _____ Culture media of known or definite chemical composition are __14__ media.

15. _____ The enzyme __15__ is responsible for conversion of urea to ammonia.

16. _____ A nonprotein compound necessary for enzyme action is __16__.

17. _____ The temperature most suitable for growth of a given microbe is termed the __17__ one.

18. _____ Organisms growing best in an amount of oxygen less than that of the air are __18__.

19. _____ The organic catalyst of the body is the __19__.

20. _____ A protein may be composed of several hundred __20__ building blocks.

21. _____ The material acted on by a given enzyme is termed the __21__.

22. _____ An enzyme released from a given cell into the extracellar medium is called __22__.

23. _____ Microbes utilizing organic matter as a source of energy are termed __23__.

24. _____ The process whereby a cell works actively to absorb molecules is __24__.

25. _____ __25__ destroys polymorphonuclear neutrophilic leukocytes.

26. _____ __26__ accelerates clotting of blood.

MICROBES
PRECLUSION OF DISEASE

PHYSICAL AGENTS IN STERILIZATION

Sterilization is the process of killing or removing all forms of life, especially microorganisms, associated with a given object or present in a given area. This includes bacteria and their spores, fungi (molds, yeasts), and viruses (which must be either destroyed or inactivated). An object on and within which all microbes are killed or removed is said to be **sterile.** The length of time an object remains sterile depends on how well it is protected from microorganisms after it is sterilized. For instance, the outside of a tube of culture medium soon becomes contaminated because it directly contacts microorganisms in the air, whereas inside the tube the culture medium, protected from microbes by the cotton plug or screw cap, remains sterile indefinitely. Bacteria that have been killed cannot multiply, but their bodies are not necessarily completely destroyed. The dead bodies of certain bacteria retain their shape and staining qualities and even promote the production of antibodies when introduced into the bodies of humans or lower animals.

Sterilization may be accomplished with the aid of mechanical means, heat (moist or dry), chemicals (Chapter 13), and certain other physical agents.

MECHANICAL MEANS

Three key methods for removing microbes mechanically are (1) scrubbing, (2) filtration, and (3) sedimentation.

Scrubbing

Scrubbing is usually done with water to which some chemical agent such as soap, detergent, or sodium carbonate has been added. The process is both mechanical and chemical. Scrubbing, by itself, removes many microorganisms mechanically while the incorporated compound acts on them chemically. Scrubbing with soap (or detergent) and water is a process basic to sterilization because the removal of dirt, debris, and extraneous matter from an area or object is crucial to the effective removal of microbes by any method. Hands and person, floors, walls, woodwork, furniture, utensils of all kinds, glassware, linens, clothing, instruments, thermometers—all must be clean! Note that mechanical cleansing cannot *sterilize* the hands and skin surfaces of the human body, since bacteria within the depths of the skin may not be reached by the scrubbing process.

Filtration

In bacterial filtration, liquid containing bacteria is passed through a material with pores so small that the bacteria are held back. The mechanical removal of the bacteria from the fluid sterilizes it. In the laboratory this process is used for sterilizing liquids and culture media that cannot be heated and for separating toxins, enzymes, and proteins from the bacteria that produced them. Certain pharmaceutical preparations are sterilized in this way. Materials most often used for bacterial filtration are unglazed porcelain, diatomaceous earth, compressed asbestos, sintered glass, and cellulose membranes. (The finest mesh filter paper of the best quality will not hold back bacteria, but the pore size of a cellulose membrane filter may be reduced so that even certain viruses are retained; viruses passing such filters are said to be **filterable**.) Bacterial filters are constructed so that the material to be filtered passes through a disk or the wall of a hollow tube made of the filtering material.

Filtration is an important step in the purification of a city water supply (p. 901). Bacterial filtration by a plastic membrane technic is widely used as a laboratory procedure in sanitary microbiology (p. 899).

Sedimentation

Sedimentation is the process by which suspended particles settle to the bottom of a liquid, an important event in the purification of water by natural or artificial means. In nature, large particles and suspended bacteria sink to the bottom of lakes, ponds, and flowing streams. In the water purification plant, sedimentation plays a significant part in the artificial purification of the community's water supply (p. 901).

MOIST HEAT

The most widely applicable and effective sterilizing agent is heat. It is also the most economic and easily controlled. Aside from burning (really a chemical process), heat is applied as *moist* or *dry heat*. Moist heat may be applied as hot water or steam and is the method of choice in sterilization except for items altered or damaged by it. See Table 12-1 for applications of heat sterilization.

The temperature that kills a 24-hour liquid culture of a certain species of bacteria at a pH of 7 (neutral reaction) in 10 minutes is known as the **thermal death point** of that species. Since this represents the temperature at which all bacteria are killed, it is obvious that many are destroyed before this temperature is reached. In fact, most are destroyed within the first few minutes. For standardization, bacteria must be in a neutral medium when their thermal death point is determined because in either a highly acid or a highly alkaline medium they are more susceptible to heat. The **thermal death time** is the time required to kill all bacteria in a given suspension at a given temperature.

Boiling

Boiling is a commonly employed, although incompletely effective, method of sterilizing by moist heat. Boiling kills vegetative forms of pathogenic bacteria, fungi, and most viruses in a matter of minutes. Hepatitis viruses are probably destroyed at the end of 30 minutes, but for elimination of hepatitis viruses, boiling is *not* recommended. Spores are less readily destroyed. Although most of the spores of pathogenic bacteria can be destroyed in a boiling time of 30 minutes, boiling is not a reliable method when materials are likely to contain spores. Certain heat-loving saprophytes can survive at high temperatures, and their spores resist *prolonged* boiling for many hours.

Completely immerse the object in boiling water, and continue boiling long enough to ensure even distribution of heat through the object being sterilized. Remember that microbes cannot be eliminated from the interior of materials boiled until heat has penetrated there. Prolong boiling time 5 minutes for each 1000 feet above sea level.

The addition of sodium carbonate to make a 2% solution in boiling water hastens destruction of spores and helps prevent rusting of instruments. Surgical instruments, needles, and syringes that are boiled must be clean and free of organic material.

Sterilization by Steam

Steam yields heat by condensing back into water. For instance, when a bundle containing fabrics is sterilized by steam, the steam contacts the outer layer, where a portion of it condenses into water and gives up heat. The steam then penetrates to a second layer, where another portion condenses into water and gives up heat. The steam thus moves centrally, layer after layer, until the whole package is sterilized.

Steam may be applied as free-flowing steam or as steam under pressure. Free-flowing steam has about the same sterilizing action as boiling water. Steam under pressure is the most powerful method of sterilizing that we possess and is the preferred one unless the material being sterilized is injured by heat or moisture. The process is carried out in the **pressure steam sterilizer,** familiarly known as the **autoclave,** a square sterilizing chamber surrounded by a steam jacket, the outside of which is insulated and covered. The chamber is loaded with supplies to be sterilized (the load) through a door that closes the front end of the sterilizing chamber. This is a safety steam-locked door made tight against a flexible heat-resistant gasket. The design is such that steam can be admitted to the closed chamber under pressure. The source of the steam varies; it may come from the central boiler supply of a large hospital or from an electrically heated boiler on the instrument itself. Valves on the autoclave control the flow and exhaust of steam. On some of the pressure steam sterilizers these valves are operated by hand, but in more modern versions the entire operation of the instrument is automatically controlled, and the most modern sterilizers use a microcomputer to control all phases of each sterilization cycle.

Steam under pressure is hotter than free-flowing steam, and the higher the pressure, the higher the temperature. The temperature of free-flowing steam (atmospheric pressure, sea level) is 100° C. At 15 pounds' pressure in the autoclave (atmospheric pressure, sea level) the temperature of steam is 121° C, and at 20 pounds' pressure it is 126° C. *Sterilization by steam under pressure is the result of the heat of the moist steam under pressure and not of the pressure itself.* Steam under a chamber pressure of 15 or 20 pounds will kill all organisms and spores in 15 to 45 minutes (depending on the materials involved). To maintain these temperatures at higher altitudes, the pressure must be increased 1 pound for each 2000 feet of increase in altitude.

Operation of an Autoclave

Steam sterilization in an autoclave (Fig. 12-1) for a long time has used the downward displacement gravity system. (Steam is admitted to the sterilizing chamber in such a way as to drive air down and out, and steam, being lighter than air, displaces it downward.) In the operation of an autoclave, pressure is first generated in the steam jacket. The connection to the sterilizing chamber is kept closed until jacket pressure is constant at 15 to 17 pounds. (The pressure within the steam jacket is kept constant during the procedure to keep the walls of the chamber heated and dry.) The load is placed in the chamber and the door secured. Then steam is admitted to the sterilizing chamber and the load heated to the temperature of the steam in the chamber. At the time steam enters the chamber, the load and the chamber are both filled with air, which, if not evacuated, would reduce the moisture content of the autoclave and lessen its sterilizing capacity. Pressure steam sterilizers are vented so that the air can escape to the atmosphere as the temperature is raised. When all the air has been evacuated, steam contacts the thermostatic valve, and it closes. The moisture that condenses on the door or back part of the sterilizing chamber drains downward from behind a steam deflector plate to the bottom part of the chamber and is then discharged to the waste line. After the end of the sterilizing cycle the steam is exhausted from the chamber but not from the jacket. At this point a drying cycle is effected before the door is opened. This is done by creation of a partial vacuum in the chamber, with steam from the jacket passing through an ejector tube on the autoclave at the same time that air is admitted through a bacteria-retentive filter.

Since sterilization by steam under pressure is primarily a matter of temperature and the increase in pressure serves only to increase the temperature, the height to which the thermometer rises, rather than the reading on the pressure gauge, should be the guiding factor in the operation of the autoclave. This is particularly true because many pressure gauges are inaccurate, generally reading too high. (Autoclaves usually have two pressure gauges, one to indicate the pressure in the steam jacket and the other to indicate the pressure in the sterilizing chamber.) A thermometer or the sensing element of a thermometer placed at the bottom of the sterilizer is a better indicator of the efficiency of the sterilizing process than one placed at the top because, if any part of the autoclave fails to receive full benefit of the steam, it is the lower part. The ther-

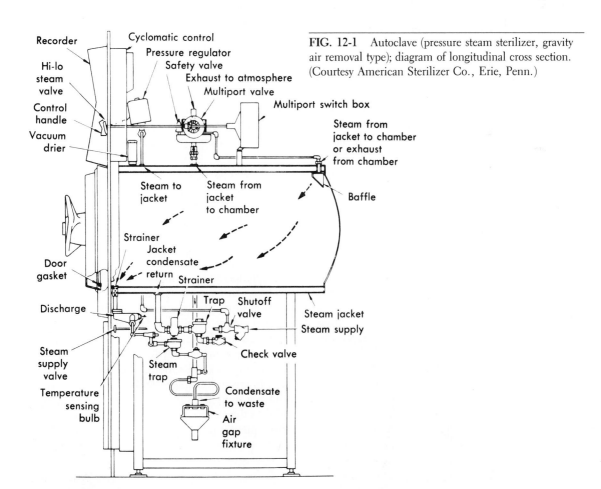

FIG. 12-1 Autoclave (pressure steam sterilizer, gravity air removal type); diagram of longitudinal cross section. (Courtesy American Sterilizer Co., Erie, Penn.)

Recorder
Hi-lo steam valve
Control handle
Vacuum drier
Door gasket
Discharge
Steam supply valve
Temperature sensing bulb

Cyclomatic control
Pressure regulator
Safety valve
Exhaust to atmosphere
Multiport valve
Multiport switch box
Steam from jacket to chamber or exhaust from chamber
Steam to jacket
Steam from jacket to chamber
Baffle
Strainer
Jacket condensate return
Strainer
Trap   Shutoff valve
Steam jacket
Steam supply
Check valve
Steam trap
Condensate to waste
Air gap fixture

mometer in the discharge path of air and moisture coming from the sterilizing chamber can never indicate less than the lowest temperature in the system.

Modern autoclaves are equipped with a number of controls to increase the efficiency of sterilization and to remove, insofar as possible, the human factor. The **recording thermometer** is a clock-thermometer mechanism that indicates (1) the time at which the material being sterilized reaches the desired temperature, (2) whether the temperature remains stable, (3) how long the exposure lasts, and (4) how many times the autoclave is in operation during the day. The **indicating potentiometer** is an instrument for actually measuring the temperature of material in the autoclave. The **automatic time-temperature control** (1) operates the autoclave at the time and temperature for which it is set, (2) exhausts the steam from the chamber, (3) regulates drying, and (4) sounds an alarm indicating that the operation is complete.

Indicators are placed in with the load to be sterilized. An indicator predictably changes its physicochemical properties or biologic nature when the prescribed temperature for sterilization has been reached. Indicators used are strips of paper impregnated with biologic material such as the dried spores of *Bacillus stearothermophilus*, the thermal death time of which is known. For steam sterilizers, *B. stearothermophilus* is the monitor (or challenge) microorganism of choice; it grows readily and is more resistant to heat than the microbes usually found on the material being sterilized. For dry heat and ethylene oxide sterilizers (p. 229) *Bacillus subtilis* (var. *globigii*) is preferred.

Ethylene oxide and dry heat sterilizers should be monitored with every cycle. Steam sterilizers should be biologically monitored at least once a week, preferably once a day.

High-Prevacuum Sterilizer

The basic principles for sterilization are the same in both the gravity air-removal type of sterilizer (just described) and the high prevacuum one. An improved pressure steam sterilizer, the high-prevacuum sterilizer is in wide use. With a vacuum system incorporated into the sterilizer unit, a precisely controlled vacuum is pulled at the beginning and end of the sterilizing cycle. Saturated steam at a temperature of 132° to 134° C (270° to 274° F) (under a pressure of 28 to 30 pounds) enters the preevacuated chamber and instantly penetrates the load to be sterilized. Microbes present are killed within a few minutes. Under these high-temperature, high-pressure conditions of steam sterilization, there is a considerable shortening of the sterilizing (exposure) time (sometimes only 3 minutes). The vacuum pulled at the end of the sterilizing cycle dries the load. There is less damage to fabrics and to items made of rubber because of reduced exposure.

The recommended time for prevacuum steam sterilization at 132° to 134° C is 4 minutes.

Fractional (Intermittent) Sterilization

When something that cannot withstand the temperature of an autoclave has to be sterilized, fractional, or intermittent, sterilization can be done. This procedure consists of exposing the material to free-flowing steam at atmospheric pressure for 30 minutes on 3 successive days; between times it is stored under conditions favorable for bacterial growth. With the first application of heat, all vegetative bacteria are killed, but the spores are not affected. Under conditions suitable for growth the spores develop into vegetative bacteria, and the second application of heat kills them. The second incubation and third application of heat are added to ensure complete sterilization. Best results depend on the material being sterilized having such a nature as to promote the germination of spores. Therefore it is most useful in the sterilization of culture media. Sometimes referred to as **tyndallization**, it is infrequently used.

The low-temperature method of sterilizing vacines (p. 277) may be applied to biologic products that cannot withstand a temperature of 100° C. These may be sterilized by heating to a temperature of 55° to 60° C for 1 hour on 5 or 6 successive days.

Pasteurization

All nonspore-bearing disease-producing bacteria and most nonspore-bearing nonpathogenic bacteria are killed when exposed in a watery liquid to a temperature of 60° C for 30 minutes. This is the basis of pasteurization (p. 693), a special method of heating milk or other liquids for a short time to destroy undesirable microorganisms without changing the composition and food value of the material itself.

DRY HEAT

Dry heat (hot air) sterilization means baking the item to be sterilized in a suitable oven. Dry heat at a given temperature is not nearly as effective a sterilizing agent as moist heat of the same temperature. Under controlled conditions of dry heat a temperature of 120° to 130° C kills all vegetative bacteria within 1½ hours, and one of 160° C kills all spores within 1 hour; but with moist heat a temperature of 120° C kills all vegetative bacteria and most spores within 15 to 20 minutes. Whereas moist heat sterilization is primarily a process of protein coagulation, dry heat sterilization is one of protein oxidation, and that oxidation goes on more slowly than coagulation. Moist heat also has greater penetrating power than does dry heat.

An advantage of dry heat is that it does not dull cutting edges, but for most fabrics even a moderate degree of dry heat is injurious. A temperature of more than 200° C causes cotton and cloth to turn brown. Therefore hot air is used mostly to sterilize glassware, metal objects, and articles injured by moisture or items such as petrolatum (Vaseline), oils, and fats that resist penetration by steam or water. In this form of sterilization the temperature should be slowly raised, and after sterilization is complete, the oven should be allowed to cool slowly to prevent breakage of glassware. *Instruments to be sterilized must be clean and free of oil or grease films.* For temperatures and times in practical dry heat sterilization, see Table 12-1.

There are two causes of ineffective dry heat sterilization: (1) the materials to be sterilized are too closely packed, and (2) the temperature is not uniform within the sterilizing ovens. An attempt to overcome the uneven distribution of heat has been made in sterilizers and sterilizing ovens constructed in such a manner that either gravity aids in the circulation of hot air through the sterilizer (gravity convection) or circulation is

Table 12-1 Heat (Physical) Sterilization of Reusable Instruments and Supplies

Method	Administration		Application
	Temperature	Time*	
Autoclave	121° to 123° C (250° to 254° F) 15 to 17 lb pressure†	30 min	Gloves, drapes, towels, gauze pads, instruments, glassware, and metalware
Dry heat	170° C (340° F)	1 hr	Glassware, metalware, and dull instruments (any temperature listed)
	160° C (320° F)	2 hr	Small quantities of powders, petrolatum (Vaseline), anhydrous oils, and petrolatum gauze
	150° C (302° F)	3 hr	Sharp instruments and metal-tip syringes
	121° C (250° F)	6 hr or longer	
Boiling	100° C (212° F)	30 min†	Method not recommended when dry heat and autoclave sterilization available

*With a high-prevacuum sterilizer, sterilizing (exposure) times are shorter; at a temperature of 132° to 134° C (270° to 274° F), sterilizing time is 4 minutes.
†Atmospheric pressure, sea level.

carried on by means of blowers (mechanical convection). Mechanical convection is more satisfactory than gravity convection.

Burning (Incineration)

Burning is a form of intense dry heat very effective in removing infectious materials of various kinds and generally applicable when materials and supplies are disposable or expendable. The inoculating wire loop used to inoculate cultures is repeatedly and quickly sterilized by **flaming**—heating the wire in an open flame until it glows.

All contaminated objects that are of no value or cannot be used again are best burned!

OTHER PHYSICAL AGENTS
Natural Methods

If a culture of bacteria is dried, the majority of the bacteria are quickly killed but some live for quite a while. Spores and encysted protozoa resist drying for a long time. Although drying is an important natural method of removing or destroying microbes, it is not applicable to "artificial" sterilization, except that sterile dressings and similar objects should be kept dry.

Sunlight has an inhibitory and destructive action on microbes. It will kill *Mycobacterium tuberculosis* within a few hours and many other bacteria in a shorter time. Sunlight is nature's great sterilizing agent, but its presence is so irregular that one cannot depend on its action. The antimicrobial action of both drying and sunlight is applied to advantage in the home drying of food.

Ultraviolet Radiation

The purity of the air in the "wide-open spaces" has long been recognized, and it is well known that the sterilizing effect of sunlight there comes from the ultraviolet rays present. This fact has been applied to the construction of ultraviolet lamps in wide use to prevent the airborne spread of disease-producing agents, especially in public places, hospitals (operating rooms, treatment rooms, nurseries), microbiologic laboratories, and quarters used to house animals.

Ultraviolet irradiation is especially effective in killing organisms contained in the minute dried respiratory droplets that tend to disperse rapidly through the atmosphere of a building or hospital. It is not as active against dust-borne agents and microorganisms on surfaces. It inactivates certain viruses. Its bactericidal effect drops sharply when the humidity of the area exceeds 55% to 60%. Even in appropriate amounts it damages the skin or conjunctivae.

The effect of ultraviolet light in sterilization reflects its heavy absorption by the nucleic acids and, to a lesser extent, by the proteins of microbial or other cells. Since the cellular function most sensitive to ultraviolet irradiation is synthesis of DNA, the activation of that macromolecule produced by the energy of ultraviolet light results in chemical changes in the molecule that are either destructive in themselves or that inhibit its replication.

Indirectly, ultraviolet irradiation affects the growth of microbes by inducing unfavorable changes in their environment. It causes the formation of compounds toxic or inhibitory to them, such as ozone in air, hydrogen peroxide in water, and organic peroxides in media where various organic compounds are present.

X Rays and Other Ionizing Radiations

X rays and other ionizing radiations are known to be lethal to microbes and to living cells as well, but there is no practical application for their use in routine sterilization. The damage done to microbes and other cells by ionizing radiation, the energy of which is many times greater than that of ultraviolet radiation, results from the ionization or activation of the component atoms (not molecules) that absorb it. This means that the atoms are made to eject very high–energy electrons, which in turn ionize each of many atoms of whatever kind are found in their path within the cell, regardless of the chemical molecules that the atoms make up. Since the energies involved would be great when ionizing radiation is applied to sterilization, many molecules would be made to undergo extensive chemical changes of a destructive nature. An indirect effect of radiation of this kind is that exerted on the water within the cell, leading to the formation of products in the intracellular milieu that adversely affect the essential molecules therein.

In industry, beta-rays or electrons sterilize prepackaged materials such as sutures and plastic tubing. High-energy electrons have been proposed for the treatment of sewage and wastewater.

One interesting application is the combination of heat and gamma-radiation for the sterilization of spacecraft. If heat alone is used, the spaceship is subjected to a temperature of 125° C (257° F) for 60 hours. The temperature has to be controlled carefully to prevent heat damage to and failure of components in items such as silver-zinc batteries and tantalum capacitors. If the spaceship with its equipment is sprayed with 150,000 rad of gamma-radiation, the time for sterilization at the temperature of 125° C can be cut down to 2 hours.

Lasers

Recent investigation suggests the use of a laser to sterilize medical instruments, clear the air in operating rooms, and pick organisms off a wound surface. The technical problem is that the laser beam must reach all parts of the item to be sterilized. If feasible, laser sterilization would be a split-second procedure.

Ultrasonics

Sound waves are mechanical vibrations. In the range of vibration (18,000 cycles/second or more) where they are no longer heard as sound (supersonic or ultrasonic), these waves have been demonstrated to coagulate protein solutions and to destroy bacteria. **Cold boiling** results from the passage of ultrasonic pressure waves through a cleaning solution. Very tiny empty spaces in the liquid form and collapse thousands of times a second. This type of scrubbing action can blast material from the surface of objects made of metal, glass, or plastic. The use of such vibrations (pitched too high to be heard) is not widely practical, but the principle has been incorporated into a commercially available dishwasher. Cleaning medical instruments is a common application. Because it can effect disruption of viruses, bacteria, and chemicals such as phosphates and nitrogen-containing compounds, "noiseless sound" is also used experimentally to treat sewage water.

Action of Fluorescent Dyes

By photodynamic sensitization, certain dyes with the property of fluorescence (such as methylene blue, rose bengal, and eosin) are lethal to bacteria and viruses if in contact with these microbes in strong visible light, which can be made almost as active as ultraviolet light if passed through a dye that causes it to fluoresce. In the process of photodynamic sensitization, biologic molecules of the microorganism are oxidized in the presence of oxygen or a suitable hydrogen donor, the dye, and visible light activated by the dye.

The Microwave Oven

The microwave oven found in many homes today *does not kill microbes*. The medical attendant caring for patients outside the hospital would find it a convenience because of its availability and simplicity of operation, but the use of the oven in sterilization is ineffective and unsatisfactory. Not only are medical items not sterilized in the microwave oven, but certain ones that are transparent to microwaves are not even heated, and plastic or rubber parts are melted. Also, it is not possible to set up and maintain any kind of uniform standards or apply quality control methods to individual microwave ovens placed under widely variable conditions in diverse living and work areas.

SUMMARY

1 Sterilization (an absolute term) is the process of killing or removing from a given object or area all forms of life—bacteria and their spores, fungi, and viruses, which must be either destroyed or inactivated.

2 Sterilization may be accomplished by moist or dry heat, chemicals, or certain other physical agents (for example, ionizing radiation and fluorescent dyes).

3 Scrubbing, filtration, and sedimentation are key methods for mechanically removing microbes.

4 The most effective, economic, easily controlled, and widely applicable sterilizing agent is heat.

5 Boiling kills vegetative forms of pathogenic bacteria, fungi, and most viruses within minutes, but it is not a reliable method for eliminating hepatitis viruses and spores.

6 Steam yields heat by condensing back into water, and steam under pressure is a powerful method of sterilizing. It is the method of choice if the items being sterilized are not injured by heat or moisture.

7 Steam under pressure is hotter than free-flowing steam; the higher the pressure, the higher is the temperature. Sterilization by steam under pressure is a matter of temperature; the increase in pressure serves only to increase the temperature.

8 Dry heat means baking the item to be sterilized in a suitable oven. Since it is not nearly as effective as moist heat of the same temperature, longer exposure times are needed.

9 Dry heat sterilization does not dull cutting edges, but it does injure most fabrics. Its best use is for glassware, metal objects, articles injured by moisture, or items such as petrolatum that resist penetration by steam.

10 Burning is a form of intense dry heat effective in removing infectious material. All contaminated objects that are of no value or not reusable are best burned.

11 Other physical agents useful in removing microbes include ultraviolet irradiation applied in ultraviolet lamps to prevent airborne spread of microbes and ultrasonic sound waves incorporated into dishwashers.

12 The microwave oven is ineffective and unsatisfactory in practical sterilization.

QUESTIONS FOR REVIEW

1 Name three ways in which sterilization may be accomplished.

2 Define sterilization, bacterial filtration, sedimentation, thermal death point, pasteurization, cold boiling, and indicator.

3 Name and describe briefly three mechanical means of removing or destroying microbes.

4 Why is moist heat more effective than dry heat as a sterilizing agent?

5 What is the effect of pressure on steam sterilization?

6 Explain intermittent, or fractional, sterilization.

7 What is an autoclave? Indicate basic principles of its operation.

8 Briefly describe the high-prevacuum sterilizer. State its advantage.

9 How is dry heat applied for sterilization? Cite examples of items that must be sterilized in this way.

10 Tabulate all physical agents used for sterilization.
11 Briefly discuss the use of biologic indicators to test for efficacy of sterilization. Cite
two examples of microbes used. Indicate recommended frequency of testing.

SUGGESTED READINGS

Alshire, R.L., and Dunton, H.: Resistance of selected strains of *Pseudomonas aeruginosa* to low-intensity ultraviolet radiation, Appl. Environ. Microbiol. **41**:1419, 1981.

Christensen, E.A., and Kristensen, H.: Biological indicators for the control of ethylene oxide sterilization, Acta Pathol. Microbiol. Scand. **87**:147, 1979.

Insist on clean environment to protect the patient, AORN J. **25**:1375, 1977.

McInnes, M.E.: Essentials of communicable disease, ed. 2, St. Louis, 1975, The C.V. Mosby Co.

Meers, P.D., and Yeo, G.A.: Shedding of bacteria and skin squames after handwashing, J. Hyg. **81**:99, 1978.

Perkins, J.J.: Principles and methods of sterilization in health sciences, Springfield, Ill., 1973, Charles C Thomas, Publisher.

Simmons, B.P., and others: Guidelines for hospital environmental control, Atlanta, 1981, Centers for Disease Control, Department of Health and Human Services.

CHAPTER 13

CHEMICALS AS ANTIMICROBIAL AGENTS

EFFECTS OF CHEMICAL AGENTS ON MICROBES
Definitions

Certain definitions are necessary for the discussions that follow. As we have learned, **sterilization** is an absolute term referring to the destruction or removal of all microorganisms present under given conditions. **Disinfection,** on the other hand, means death to disease-producing organisms and the destruction of their products, usually with chemical agents known as **disinfectants.** (A more practical definition might indicate that disinfection *halts* the progress of undesirable microorganisms by inducing structural or metabolic derangements in them.) Disinfection does not refer directly to saprophytic organisms present in a given setting that may or may not be killed. The terms *disinfection* and *disinfectant* are applied to procedures and chemical agents used to destroy microbes associated with inanimate objects. The term **antiseptic** is usually applied to an agent that acts on microorganisms associated with the living body to prevent their multiplication but not necessarily kill them. We should disinfect the excretions from a sick person but apply an antiseptic to his wounds. However, the terms are often interchanged.

Germicides are chemical that kill microbes (not necessarily their spores). **Bactericides** kill bacteria, **virucides** destroy or inactivate viruses, **fungicides** destroy fungi, and **amebicides** destroy amebas, especially the protozoan *Entamoeba histolytica*. **Asepsis** means the absence of pathogenic microbes from a given object or area. In aseptic surgery, to avoid infection of the patient, the field of operation, instruments, and dressings are rendered free of microorganisms by sterilization, and the operation is conducted in such a manner that it is kept as free of microbes as possible.

Bacteriostasis is that condition in which multiplication is blocked but bacteria are in no other way affected. *Agents causing bacteriostasis are known as* **bacteriostatic agents.** Examples are low temperatures, weak antiseptics, and dyes. *Antiseptics* and *chemical bacteriostatic agents* are synonymous terms. When used to prevent the deterioration of foods, serums, vaccines, and other biologic products, such agents are called **preservatives.**

Two terms with practical applications are **degerm** and **sanitize.** To degerm is to remove bacteria from the skin by mechanical cleaning or application of antiseptics. To sanitize is to reduce the number of bacteria to a safe level as judged by public health requirements. It refers to the daily control of the microbial population on utensils and equipment used in dairies and establishments where food and drink are served. In short, sanitization refers to a "good cleaning" and is basic to technics of sterilization. As used, the term *sterilization* implies a mechanically clean item. The word **decontamination** applies to the process of killing all microbes from an item known to be mechanically dirty and containing a heavy growth of microorganisms.

Fumigation is the liberation of fumes or gases to destroy insects or small animals. **Deodorants** are substances that destroy or mask offensive odors. They may have neither disinfectant nor antiseptic action and may generally tend to obscure infectious material rather than destroy it.

An **astringent** is a locally acting drug that precipitates proteins but has so little penetrability that only the surface of the cell is affected. The permeability of the cell membrane is greatly reduced, but the cell itself remains viable.

Qualities of a Good Disinfectant

Certain qualities specify the ideal disinfectant for general use. Unfortunately, at present no chemical agent possesses all of them, and the selection of a disinfectant is often a workable compromise. The more of the following qualities that a disinfectant has, the more nearly it qualifies as ideal. It should:

1. Attack all types of microorganisms
2. Be rapid in its action
3. Not destroy body tissues or act as a poison if taken internally
4. Not be retarded in its action by organic matter
5. Penetrate material being disinfected
6. Dissolve easily in or mix with water to form a stable solution or emulsion
7. Not decompose when exposed to heat, light, or unfavorable weather conditions
8. Not damage materials being disinfected, such as instruments or fabrics
9. Not have an unpleasant odor or discolor the material being disinfected
10. Be easily obtained at a comparatively low cost and be readily transported

The most important feature of a disinfectant is its ability to form lethal combinations with microbial cells. *Remember that different species of microbes, especially bacteria, show much greater variation in their susceptibility to disinfectants than they do to sterilization by physical agents.*

Action of Antiseptics and Disinfectants

Antiseptics and disinfectants act by (1) oxidation of the microbial cell, (2) hydrolysis, (3) combination with microbial proteins to form salts, (4) coagulation of proteins, (5) inactivation of vital enzymes by chemical interference with certain parts of an enzyme necessary for its specific reaction, (6) modification of the permeability of the microbial plasma membrane, or (7) disruption of the cell.

A chemical agent may be destructive because it can coagulate cell proteins or disrupt the cell membrane (p. 32). Many chemical agents interfere with enzymatic reactions. Cellular enzymes and coenzymes containing cysteine possess side chains that terminate in sulfhydryl (—SH) groups. An example is coenzyme A (p. 68). The sulfhydryl group must remain free and reduced for proper function of the enzyme and coenzyme. An oxidizing agent or heavy metal compound can link neighboring sulfhydryl groups, as indicated:

$$\text{R—HS + HS—R} \xrightarrow{-2\text{H}} \text{R—S—S—R}$$

(R represents the rest of the protein molecule to which the sulfhydryl group is attached.)

Since many sulfhydryl groups are part of the enzymatic makeup of the cell, such agents can do much damage.

Factors Influencing the Action of Disinfectants

Factors influencing the action of disinfectants are (1) qualities of the disinfectant, (2) nature of the material to be disinfected, (3) concentration of the disinfectant, and (4) manner of application. A chemical in a solution of one strength may be a disinfectant, whereas in a weaker solution it may act as an antiseptic, and in certain weak solutions it may actually stimulate microbial growth.

The item for disinfection is evaluated as to (1) kind and number of microbes present, (2) presence of vegetative forms or spores, (3) distribution of microbes in clumps or uniformly, and (4) presence of organic compounds or other chemicals that inactivate the disinfectant.

Most chemical disinfectants in common use are *germicidal but not absolutely sporicidal*; that is, they do not kill all spores present. As a rule, the process of disinfection is gradual, and a few microbes survive longer than the majority. To be effective, the disinfectant must be applied for a length of time sufficient to destroy *all* microorganisms. Many chemical disinfectants must be used for a long time to obtain the maximal effect; this often means 18 to 24 hours. An important factor relating to the disinfectant is the temperature at which it is applied. The higher the temperature, the more active it is. Remember that a disinfectant must penetrate all parts of the material being disinfected because it must contact microbes to destroy them. An article being disinfected must be *completely submerged* in the working solution. Also, a disinfectant should be properly chosen in accord with the physicochemical nature of the material to be disinfected.

Surface Tension in Disinfection

The molecules lying below the surface of a liquid are pulled in all directions by the cohesive forces of neighboring molecules. Those at the surface are pulled downward and sideways only but not upward because there are no molecules above the surface to attract them. This phenomenon is surface tension, which is nontechnically defined as *that property stemming from molecular forces by which the surface film of all liquids tends to bring the contained volume into a shape having the least superficial area*. Since a sphere has the least area for a given volume, surface tension causes drops of liquid to become spherical.

Surface tension is important in disinfection because liquids of low surface tension spread over a greater area and contact cells more intimately than do liquids of high surface tension. Low–surface tension liquids are often spoken of as **wetting agents.** A good wetting agent spreads over a surface rapidly and remains in a thin film. Chemicals that reduce the surface tension of water when dissolved in it are more concentrated on the surface of cells than they are throughout the solution, and thus they are in proximity to the cell membrane, an essential but delicate structure. The interface between the lipid-containing microbial membrane and the surrounding aqueous milieu provides a peculiarly vulnerable site for the chemical action of such an agent. Some wetting agents, as a consequence, may also be effective antiseptics.

Wetting agents regarded primarily as surface-cleaning agents are *detergents*, a diverse group, and many detergents are important for their combined action as cleansers and as antiseptics (inhibitors of microbial growth). The action of different kinds of detergents varies in degree, but certain ones can be demonstrated to dissolve lipids and to denature proteins. Concentrated at the unit membrane of a microbe, such a detergent may combine with the lipid of the membrane, alter its permeability, allow leakage of crucial cellular constituents, and then penetrate the cell's interior to denature its proteins.

The classic example of a wetting agent or detergent is soap, although soap has only a mildly antiseptic action (also related to an effect on the permeability of the cellular membrane). Many synthetic organic detergents (liquids, granules, and such) on the market are sometimes spoken of as soapless soaps or nonsoap cleaners. Some detergents, such as Tween 80, do not have an antiseptic action but promote bacterial growth. Tween 80 provides a source of oleic acid (a fatty acid) necessary for the growth of the tubercle bacillus.

Standardization

Various procedures have been designed to evaluate the antimicrobial activity of a given chemical agent, as well as to indicate its toxicity for tissues. Rideal and Walker in 1903 devised the original phenol coefficient test, which compared the disinfectant or antiseptic activity of a given compound with that of phenol under standard conditions. A phenol coefficient of greater than 1 indicated a stronger agent than phenol; a coefficient less than 1, a weaker one. Despite the number of methods existing today, inadequacies still remain and the tests fail to give all the information needed.

COMMON DISINFECTANTS AND ANTISEPTICS

Myriads of cleaning and disinfecting chemicals and combinations exist. With exceptions, there seems to be no general consensus today as to which of these is best for application in any given situation, and the use of such chemical agents varies considerably, even in the same community. Table 13-1 (p. 230) compares the antimicrobial activity of some of the more important ones.

Surface-active Compounds

Soap (bar, liquid, granule, leaflet, or soap-impregnated tissue) is our most important cleaning agent. Although its utility as a disinfectant is limited, it is generally used before one is applied. The major action of soap is to aid the mechanical removal of microbes, primarily through scrubbing. In cleaning, soap and water separate particulate contamination of whatever kind from a given area, for example, the skin surface of the human body. This is an area constantly accumulating dead cells, oily secretions, dust, dried sweat, dirt, soot, and microorganisms. Soap breaks up the grease film into tiny droplets; water and soap acting together lift up the emulsified oily materials and dirt particles and float them away as the lather is washed off (Fig. 13-1).

Although the term *detergent* means any cleaning agent, even water, it is used to distinguish the synthetic compounds from soap, both of which lower the surface tension of water. Soap is made from fats and lye; detergents are made from fats and oils by complicated chemical processes, and most contain a biodegradable linear sulfonate derivative of petroleum. Soaps depend for their cleaning action on their content of alkali, which suspends the grime in water from the surface of an object so it can be washed off. The detergent ionizes in water; its electrically charged ions attach themselves to the dirt. The washing action releases the ions, which carry the dirt away. Detergents dissolve readily in cold water and completely in even the hardest water. Soap combines with the calcium and magnesium salts in hard water to form an insoluble scum.

Because of its sodium and alkali content, soap is germicidal for pneumococci, streptococci, gonococci, meningococci, spirochetes of syphilis, and influenza viruses, but it is too mildly antiseptic to remove most bacteria effectively. Therefore scrubbing with soap must be followed by the application of a suitable disinfectant.

Since soap and water constitute a colloidal solution, washing with water usually does not remove all soap, but soap is completely soluble in 70% alcohol. *If the use of soap is to be followed by the application of a germicide, the soap must be thoroughly washed off with 70% alcohol before the germicide is applied. If this is not done, the residual soap and germicide might combine to form an inert compound.* So-called germicidal soaps generally have little or no advantage over ordinary soaps. If soap is not properly handled and dispensed, it may itself become a source of infection.

Most pathogenic bacteria and viruses are removed or chemically killed by the soaps and detergents ordinarily used in the commercial self-service laundry machine. The temperature of the dryer usually is high enough to eliminate any remaining bacteria. If the clothes are ironed, the heat of the hot iron destroys any microbes that possibly could have survived.

Enzyme detergent refers to a laundry presoak product in which certain proteolytic enzymes from bacteria are incorporated (in amounts up to 1.0% active enzyme). Enzymes are obtained through a fermentation process from the widely distributed nonpathogenic soil organism *Bacillus subtilis*. Enzyme detergent dissolves organic (protein) stains such as blood, feces, and meat juices without harming the fabric or user. Enzyme activity is quickly dissipated during the washing process and is inactivated by chlorine bleach.

Benzalkonium (Zephiran) chloride, a mixture of high molecular weight alkyl dimethylbenzylammonium chlorides, is a well-known member of the surface-active chemical disinfectants known as **quaternary ammonium disinfectants,** or quats. Quats possess a characteristic chemical structure that is built around a nitrogen atom of five valence bonds (Fig. 13-2). The efficacy of benzalkonium chloride has recently been questioned. In reported instances where this agent has been used to prevent the growth of the disease-producing bacteria, it has instead been demonstrated to support it. Since one cannot rely on complete disinfectant action, the use of this disinfectant is not recommended.

Cetylpyridinium chloride (Ceepryn, Cēpacol) is another quaternary ammonium compound (Fig. 13-2). A commercially available 0.5% solution is recommended as a mouthwash or gargle; it combines a foaming detergent cleaning action with antibacterial activity against certain pathogenic organisms of the mouth and throat.

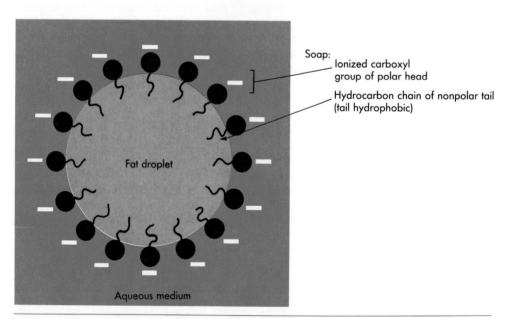

FIG. 13-1   Action of soap. Soap forms an emulsion with fat droplets. Around each droplet a shell is formed of hydrophilic, highly polar carboxylate groups. The negatively charged carboxylate groups are electrically balanced by an equal number of positive ions, such as sodium ions.

Heavy Metal Compounds

In extremely high dilutions, heavy metals, such as mercury, copper, and silver, destroy microbes. They exert an inhibitory effect on the function of vital intracellular enzymes. Certain enzymatic reactions proceed only in the presence of free and reduced sulfhydryl groups, integral parts of the enzymes. Heavy metals combine with these groups.

Organic mercury compounds were developed to avoid the toxicity of inorganic mercurials and yet retain the disinfecting qualities of mercury. They are soluble in water and body fluids with a low toxicity for tissues—an advantage. They kill pathogenic bacteria that do not form spores, other than tubercle bacilli. *Organic mercurials have no action against spores.* Among them are merbromin (Mercurochrome), nitromersol (Metaphen), thimerosal (Merthiolate), and Mercresin, a mixture instead of a single compound.

Merbromin is a combination of mercury and a derivative of fluorescein (Fig. 13-3). A 1% solution is tolerated by the urinary bladder and kidney pelvis; a 2% solution is used to disinfect wounds, and stronger solutions may be used for skin antisepsis. *Nitromersol* is used for sterilization of instruments, skin antisepsis, and irrigation of the urethra. In strengths ranging from 1:10,000 to 1:1000, it is said to be comparatively nontoxic, nonirritating, and nondestructive to metallic instruments and nonmetallic materials. *Thimerosal* is used for disinfection of instruments, skin, and mucous membranes and as a biologic preservative for vaccines, serums, and blood cells. Aqueous solutions are considered to be fungistatic also.

Mercresin combines the germicidal action of the mercurials with that of the phenolic derivatives, giving maximal disinfection with minimal tissue injury. Its action is not inhibited by the presence of serum, and it is widely used for local skin antisepsis.

Silver nitrate ($AgNO_3$), a caustic, antiseptic, and astringent, is the inorganic silver salt most often used. Silver nitrate pencils cauterize wounds. Silver nitrate in a 1:10,000 solution inhibits the growth of bacteria, and increasing the strength of the solution, even to 10% in certain instances, enhances germicidal activity. It has a selective action for gonococci, and a 1% solution instilled in the eyes of newborn babies prevents ophthalmia neonatorum (p. 500). A 0.5% solution used in the continuously

FIG. 13-2 The quaternary ammonium compound. **A,** General chemical formula. R_1 to R_4 are alkyl groups, alike or different; for maximal antibaterial activity, one of the radicals should have 8 to 18 carbon atoms. X is usually a halogen. **B,** Example of a quat, cetylpyridinium chloride. **C,** Benzalkonium chloride, another example.

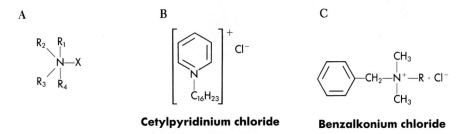

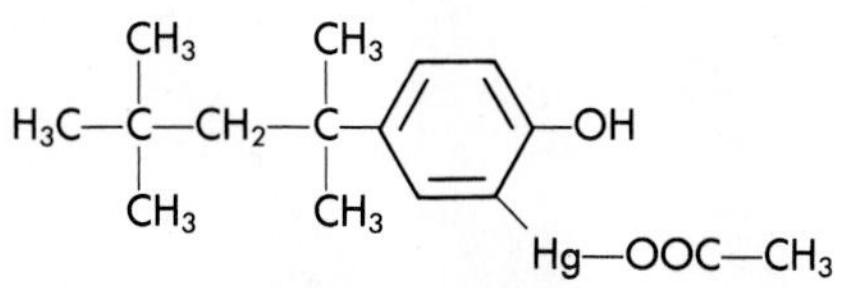

Merbromin

FIG. 13-3 Chemical formula of a heavy metal compound, merbromin (Mercurochrome).

soaked dressings applied to burns reduces infection, but pseudomonads can still grow beneath the dry black eschar that results. The action of silver nitrate is retarded by chlorides, iodides, bromides, sulfates, and organic matter. It is reduced on exposure to light and, because of its many incompatabilities, *should be used only with distilled water*.

Zinc salts are mild antiseptics and astringents. *Medicinal Zinc Peroxide* is a mixture of zinc peroxide, zinc carbonate, and zinc hydroxide. The commercial powder should be sterilized at 140° C dry heat for 4 hours. In water suspensions and ointments it is of special value in the control of infections caused by anaerobic and microaerophilic organisms in injuries such as gunshot wounds, bites, and deep puncture wounds. *Calamine Lotion* is mostly zinc oxide with some ferric oxide.

Copper sulfate ($CuSO_4 \cdot 5H_2O$) is valuable chiefly for its destructive action on the green algae that often grow in reservoirs and render water obnoxious. A copper sulfate concentration of 1 part per million parts (ppm) of water kills algae if the water does not contain an excess of organic matter. One part of copper sulfate added to 400,000 parts of water destroys typhoid bacilli in 24 hours. For a short time it is not harmful to drink water that contains this amount of copper sulfate. Copper sulfate is also an important ingredient of sprays used to combat fungal diseases of plants (see box below).

Alcohols and Aldehydes

Ethyl alcohol (C_2H_5OH) is one of the most widely used disinfectants and one of the best. For alcohol to coagulate proteins (its disinfectant action), water must be present. Because of this fact, 70% has long been considered the critical dilution of ethyl alcohol. However, there is good reason to think that against microorganisms in a moist environment alcohol acts over a range of 70% to 95%. Alcohol in a 70% dilution may be used to disinfect certain delicate surgical instruments, although it tends to rust instruments and dissolves the cement from around the lights of endoscopes. It does not kill spores. Alcohol kills tubercle bacilli rapidly and is a tuberculocidal disinfectant of choice.

Isopropyl alcohol ($CH_3 \cdot CHOH \cdot CH_3$) is slightly superior to ethyl alcohol as a disinfectant. It is also cheaper, and its sale is not subject to legal regulations. Like ethyl alcohol, it acts against vegetative bacilli (not spores) and tubercle bacilli. Recent evidence indicates that this agent also is an effective disinfectant in dilutions stronger than the conventional 70% and that the most effective one may well be full strength (99%).

Formaldehyde (Fig. 13-4), a gas, is contained in a watery 37% solution known as *formalin*. In addition to its disinfecting properties, formaldehyde deodorizes, preserves tissues, and converts toxins into toxoids. Its action depends on the presence of moisture, concentration of gas, temperature, and condition of the object to be sterilized. The presence of 1% of the gas in a room or chamber destroys all nonsporulating pathogenic bacteria. Cystoscopes and certain specialized instruments that would be damaged by heat are sometimes disinfected in airtight sterilizing cabinets equipped with electric terminals so that formaldehyde pastils can be vaporized in them. The instruments are left in the cabinet and exposed to the formaldehyde fumes for 24 hours.

Mixtures of formalin with alcohol and hexachlorophene make active germicidal so-

1 ppm = 1 inch in 16 miles
      = 1 minute in 2 years
      = 1 penny in $10,000
      = 1 needle (weight, 1 g) in a 1-ton haystack
      = 1 mouthful of food when compared with the food a person will eat in a lifetime

lutions for sterilizing surgical instruments. Spores, tubercle bacilli, other bacteria, and many viruses are speedily killed therein. One example of such a mixture is *Bard-Parker Germicide*, a commercial solution of formaldehyde, isopropanol, methanol, and hexachlorophene. It has been especially useful in the sterilization of knife blades and suture needles.

Glutaraldehyde (Fig. 13-4), a bactericidal aldehyde that is acidic in water, is one of the latest developments and one of the best. If alkali is added, the bactericidal action is enhanced and speeded up. It is marketed as *activated* (alkalinized) *glutaraldehyde solution (Cidex)*, an alkaline solution in 70% isopropyl alcohol, which is bactericidal, virucidal, and sporicidal in short exposure times. It is especially suitable for sterilization of anesthetic equipment, the intermittent positive-pressure breathing apparatus, and instruments containing optical lenses.

Phenol and Derivatives

Phenol (carbolic acid), a corrosive poison, in 0.2% solution inhibits the growth of bacteria (Fig. 13-5). A 5% solution kills all vegetative bacteria and the less resistant spores in a short time. Contact with alcohol, ether, or soap decreases its action. The addition of 5% to 10% hydrochloric acid increases its efficiency. Since the action of phenol is not greatly retarded by the presence of organic matter, it is an excellent disinfectant for feces, blood, pus, sputum, and proteinaceous materials. It does not injure metals, fabrics, or painted surfaces. The crude acid may be used for woodwork because it is cheaper and more effective than the pure substance. The crystals or strong solutions should not touch the skin. If this happens, alcohol should be applied at once. Dilute solutions should not contact the skin or mucous membranes for more than 30 minutes to 1 hour because they injure tissues.

Cresol has a higher germicidal power than does phenol and is less poisonous. *Saponated cresol solution (Lysol)*, an alkaline solution of cresol in soap, in a 2.5% solution makes a good disinfectant for feces and sputum. Lysol, a widely used and popular proprietary disinfectant, is most important in the disinfection of inanimate objects, including instruments, furniture, table surfaces, floors, walls, rubber goods, rectal thermometers, and contaminated objects of varied description, especially when these have been contaminated by *Mycobacterium tuberculosis*. Phenol derivatives such as Lysol can be used to disinfect excreta and contaminated secretions and excretions from patients with infectious diseases, especially tuberculosis. They are of little value in an-

Formaldehyde   **Glutaraldehyde**

FIG. 13-4   Two aldehydes used in disinfection.

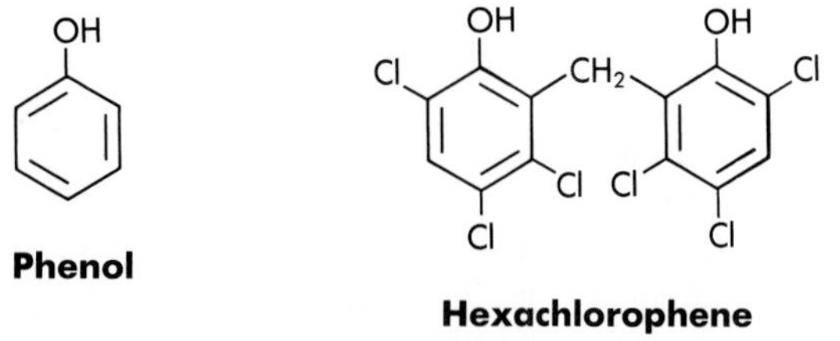

Phenol

Hexachlorophene

FIG. 13-5   Phenol and phenolic derivative in disinfection.

tisepsis of the skin because in concentrations that would not injure the skin they possess little bactericidal activity. *They do not destroy all spores present.*

Amphyl, a modern and greatly improved phenolic disinfectant, has wide and varied applications as a germicide. (Chemically it is a mixture of orthophenylphenol and paratertiary amylphenol with potassium ricinoleate in propylene glycol and alcohol.) Nontoxic, noncorrosive to metals, and nonirritating to skin and mucous membranes, Amphyl can be mixed with soap and certain other antiseptics. Because it is an agent with low surface tension, it spreads and penetrates materials. Surfaces treated with it tend to retain an antimicrobial action for several days. It is effective in dilutions of 0.25% to 3% for a range of routine disinfecting procedures involving the skin, mucous membranes, floors, walls, furniture, dishes, utensils, and surgical instruments. Fungi, bacteria (including the tubercle bacillus), and viruses, but *not* spores, are destroyed by its action; heating increases its germicidal properties.

Staphene, a phenolic disinfectant related to Amphyl, is a mixture of four synthetic phenols (paratertiary amylphenol, orthobenzylparachlorophenol, orthophenylphenol, and 2,2'-methylene-bis[3,4,6-trichlorophenol]). NOTE: Some of the phenolic detergent germicides widely used in cleaning solutions can cause depigmentation of the user's skin.

O-syl, an antiseptic, germicide, and fungicide, is considered to be especially effective against the causative agent of tuberculosis. Chemically it is one synthetic phenol (orthophenylphenol).

Hexachlorophene (G-11) is a diphenol (Fig. 13-5). In concentrations from 1% to 3% it has been incorporated into soaps and combined with detergents used for the preoperative hand scrubs of the surgical team and for the preoperative and postoperative preparation of the patient's skin. Hexachlorophene is long retained on the skin, from which it can be recovered more than 48 hours after use. Its benefit is attributed to the bactericidal film left on the skin after repeated application; the concentration on the skin is cumulative up to 2 to 4 days. (NOTE: The skin cannot be completely sterilized.) Mechanical cleaning plus the use of germicidal agents removes the superficial growth of bacteria from the skin surface, which fortunately includes most of the pathogens. Hexachlorophene has been widely used because of its bacteriostatic action against grampositive microbes, notably the staphylococcus. Bathing infants with a detergent containing hexachlorophene has been shown to reduce considerably their incidence of staphylococcal infections.

Generally, hexachlorophene is not irritating to the skin, although it is sometimes associated with allergic manifestations. It is readily absorbed through the normal skin, more easily so through abraded or burned skin. Its use has been blamed for brain seizures developing in young burn victims who had been washed with it. Animal experiments relate the uptake through the skin over a period of days to subsequent brain damage and even paralysis.

The U.S. Food and Drug Administration (FDA) warns against total body bathing of infants and adults with products containing 2% or 3% concentrations of hexachlorophene and bans over-the-counter sale of most products containing it.

Hexachlorophene has been perhaps the most universal of the antibacterial agents; it has been incorporated into a wide variety and number of consumer products: soaps, shampoos, toothpastes, deodorants, lotions, powders, ointments, cosmetics, and medicinal cleaners. According to the FDA, the ingredient replacing hexachlorophene in many antibacterial products—tribromsalan—should also be eliminated.

Currently, the Committee on Fetus and Newborn of the American Academy of Pediatrics recommends that only with a serious outbreak of staphylococcal infections may one resort to total body bathing of newborns with hexachlorophene and then only if (1) a solution of not more than 3% is used, (2) it is applied to full-term infants only,

(3) it is washed off thoroughly after each application, and (4) it is applied no more than two times to a given infant.

Available today only by prescription, it is still used in hospitals and similar institutions because of its effective antibacterial action.

pHisoderm, a detergent cream, is not a single compound but a proprietary mixture of wool fat, cholesterol, lactic acid, and sulfonated petrolatum. It cleans faster and more effectively than does soap and is used for degerming the hands and other skin areas. It may be used alone, but it is most often combined with hexachlorophene.

pHisoHex is a mixture of the detergent base pHisoderm and 3% hexachlorophene. It is still popular for surgical hand scrubs because regular use gives maximal bactericidal effect. It is not irritating, only small amounts need to be used, and the time required for the surgical scrubbing is much shorter than with soap preparations. Its emulsifying action on oily material and its ability to form suds are desirable features.

Halogen Compounds

The halogens, such as chlorine and iodine, are strong oxidizing agents that form halogenated derivatives of many of the nitrogenous compounds of protoplasm in reactions that are destructive to microbes.

Iodine, one of the best known and most widely used disinfectants, is a potent amebicide, wide-spectrum bactericide, good tuberculocide, fungicide, and virucide. Elemental iodine is lethal to microbes; its germicidal properties are attributable to certain alterations produced in proteins by direct iodination. For example, iodination of the amino acid tyrosine biologically inactivates the protein of which it is a constituent. The ubiquitous *tincture of iodine* containing 2% iodine is one of the best disinfectants for small areas of skin, minor cuts, abrasions, and wounds. The strong tincture of iodine (7% in alcohol) is too toxic for most purposes. Iodine becomes freely soluble in water in the presence of soluble iodides such as those of sodium and potassium. *Iodine solution* (aqueous) containing 2% iodine is as effective as the alcoholic solutions (tinctures) but is less irritating to abraded tissue. Very strong and very old solutions of iodine burn the skin. *Iodine should not be applied under a bandage.* Although in some persons iodine compounds produce allergic skin rashes, iodine is less toxic than other routinely used germicides. Iodine has been used for disinfection of water in swimming pools, and the tincture of iodine may be employed to render contaminated water safe for drinking (p. 901). The use of iodine as an antiseptic dates back to 1839, and during the U.S. Civil War it was applied to battle wounds.

Iodophors are compounds in which iodine is carried by a surface-active solvent. The germ-killing action results from release of free iodine when the compound is diluted with water; only the free iodine has the disinfectant effect. An iodophor enhances the bactericidal action of iodine and reduces odor. It does not stain the skin or items on which it is placed as does the tincture. The proprietary preparation *Wescodyne*, an iodine detergent germicide (also a good tuberculocide), is an example, as are *Hi-Sine*, *Iosan*, and *Betadine* (povidone-iodine complex). The widely used povidone-iodine is a complex of iodine with the solubilizing agent, polyvinylpyrrolidone. The latter effects the liberation of the elemental and bactericidal iodine. Povidone-iodine complex is fungicidal, amebicidal, and virucidal, but it must be used in concentrations no lower than 3% to 5%. Recently contamination of certain lots of this product with *Pseudomonas cepacia* has been reported.

Chlorine, one of the most effective and widely used of all chemical disinfectants, is potent both as elemental chlorine and as the strong oxidizing agent hypochlorous acid, which results from the combination of chlorine with water. It is used in the disinfection of drinking water, purification of swimming pools, and treatment of sewage. It is applied with the release of gas from cylinders or with the use of compounds that release free chlorine. The chlorine from chlorine-liberating compounds is spoken of as "available" chlorine. For effective disinfection the chlorine content of water must reach a concentration of 0.5 to 1 ppm.

With no organic material present a concentration of 0.1 ppm of chlorine in water destroys the poliovirus, but in a strength of 1 ppm, chlorine has no effect on the cercariae that represent one stage in the development of flukes (p. 828) (nor are these minute worms removed by sand filtration or aluminum sulfate clarification of the water).

Cysts of *Giardia lamblia*, now a recognized pathogen, can survive cool or tepid water for as long as 3 months but quickly succumb to a temperature of 50° C (122° F). They can survive in 0.5% chlorinated water for 2 or 3 days. However, they may be killed by iodine compounds in doses recommended for water purification.

Cysts of *Entamoeba histolytica* (p. 794), the cause of amebic dysentery, can live as long as 1 month in water. The usual chlorination treatment for drinking water does not kill them; water containing an adequate amount of chlorine to do so would not be fit to drink. The World Health Organization recommends that commercial airlines filter their water supply and overchlorinate it to attain 8 to 10 ppm of free available chlorine in order to kill cysts of *Entamoeba histolytica* and viruses of infectious hepatitis. The water must then stand for at least 30 minutes. It is dechlorinated to eliminate the disagreeable taste and smell.

Sodium hypochlorite (NaOCl), made by the reaction of chlorine on sodium hydroxide, is a powerful oxidizing agent. It cannot be prepared as a powder but is manufactured in solutions of varying strength. The stronger solutions are used as bleaches by laundries and other establishments. The weaker solutions (for example, Clorox) are used as household bleaches and for the bactericidal treatment of food-handling equipment. Hypochlorite solutions *prepared fresh daily* are active against viruses if they contain 5000 to 10,000 ppm of available free chlorine (0.5% to 1% solutions) and may be used in hemodialysis units, laboratories, and blood banks for disinfection of nonmetal equipment and surfaces likely to contain hepatitis viruses. (Surfaces contaminated with blood are disinfected by cleaning them with a solution of 1 part of household bleach containing 5.24% sodium hypochlorite and 9 parts of water.) In the clinical laboratory, sodium hypochlorite is an effective and economic agent, but it is corrosive to metal. *Dakin's solution* and modifications are weak neutral solutions of sodium hypochlorite, liberating from 0.5% to 5% available chlorine. The official *Sodium Hypochlorite Solution* (5%) is a valuable agent in dental surgery.

Chloramines are organic chlorine compounds that slowly decompose and liberate chlorine. They are inferior when speed is required, since their value lies in their prolonged action. They may be used to sanitize glassware and eating utensils and to treat dairy and food-processing equipment, but their practicality is mainly for emergency disinfection of drinking water, particularly of small amounts, and *Halazone* is the one applied (Fig. 13-6).

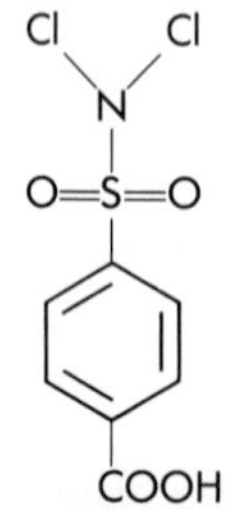

Halazone

FIG. 13-6   An organic chlorine compound (Halazone) used as a disinfectant.

Acids

Boric acid (H$_3$BO$_3$) is a weak antiseptic most often used as an eyewash. However, when taken internally, boric acid is highly toxic. Because of the risk of accidental poisoning, its use in hospitals is condemned. It has no place in the pediatric division or in any nursery. If even a dilute solution of boric acid mistaken for distilled water should be used in an infant formula or in a parenteral fluid, the results could be disastrous. Dusting the powder repeatedly over the diaper area of a baby to remedy diaper rash can cause death. A so-called bland ointment applied to abraded areas of the skin can produce an unfavorable reaction.

Fuming nitric acid (HNO$_3$) is the best agent for cauterizing the wounds inflicted by rabid animals.

Benzoic acid (C$_6$H$_5$ · COOH) and *salicylic acid* (OH · C$_6$H$_4$ · COOH) are fungistatic agents. A mixture of 6% benzoic acid and 3% salicylic acid *(Whitfield's ointment)* is used to treat fungal infections of the feet. Benzoic acid is an important food preservative (p. 918).

Undecylenic acid (CH=CH(CH$_2$)$_8$COOH) *(Desenex)* is used for athlete's foot and other fungal infections of the skin.

Oxidizing Agents

Hydrogen peroxide (H$_2$O$_2$) owes its disinfecting qualities to the free oxygen that it liberates. Its spectacular effervescence mechanically produces a weak cleaning effect in certain wounds. Not an overly reliable antiseptic, it deteriorates rapidly.

Potassium permanganate (KMnO$_4$) owes its antiseptic and antifungal effectiveness to strong oxidizing qualities. However, its action is weakened by the presence of organic matter, and it can irritate tissues if the concentration of the solution exceeds 1:5000.

Dyes

Dyes such as *gentian violet* and *crystal violet* in suitable dilutions inhibit the growth of gram-positive bacteria and fungi but have little effect on gram-negative bacteria, whereas dyes such as *acriflavine* and *proflavine* inhibit the action of gram-negative bacteria but have little effect on gram-positive bacteria. Dyes are used to obtain pure cultures in the laboratory. Gentian violet, 1%, is sometimes used to treat fungal infections. With the advent of the sulfonamide drugs, the need for dyes as therapeutic agents no longer exists.

Miscellaneous Agents

Ethylene oxide (ETO) is a toxic, chemically active gas with a broad range of antimicrobial activity that has been used for years as a germicide in the processing and packaging of pharmaceuticals, cosmetics, and food and as a fumigant (Fig. 13-7). It is an odorless, poisonous, and explosive gas that can be easily kept as a liquid. Since its vapor is highly inflammable in air, it is transported commercially in a noninflammable mixture with carbon dioxide. This gas is used especially to treat items that are damaged by heat, water, or chemical solutions and is a potent agent if properly used. The article to be disinfected must be exposed to the gas in an appropriately designed sterilizing chamber in which the temperature can be raised and from which the air must be evacuated. Polyethylene and paper are suitable for packaging. A prolonged period of exposure in the sterilizing chamber is required—at least 8 hours in one large commercial model.

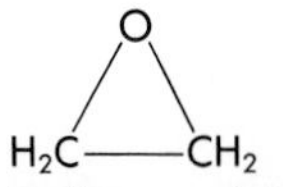

Ethylene oxide

FIG. 13-7 Ethylene oxide, important and very effective in gaseous sterilization.

Bacteria (including tubercle bacilli and staphylococci) and spores are killed; hepatitis viruses are destroyed. Ethylene oxide properly used has an irreversible chemical reaction with all organisms, including their spores; this reaction is speeded by heat. This form of **cold** sterilization is widely used for delicate surgical instruments with optical lenses and for the tubing and heat-susceptible plastic parts of the heart-lung machine. Many prepackaged commercial items, such as rubber goods and plastic tubes, are sterilized with ethylene oxide. Because of the penetration of the gas, this form of sterilization is effective for blankets, pillows, mattresses, and bulky objects.

Ethylene oxide sterilization is generally considered safe and is effective, provided adequate aeration time is allowed for the gas to elude from items so sterilized. (Certain products of cloth or rubber, however, do continue to emit significant levels of ethylene oxide even after extensive ventilation.) This is important, since the gas and its by-products are highly irritating to skin and mucous membranes, including those of the eye. Recently in isolated instances ethylene oxide has been cited as a mutagen and suspected carcinogen, but the Occupational Safety and Health Administration's (OSHA) current standard for workplace exposure to this gas of 50 ppm as an 8-hour time-weighted average (TWA) has not changed. Many hospitals have now instituted monitoring programs to limit the exposure, and efforts are being made to decrease OSHA's limit.

Lime is one of the most common and, when properly used, most effective germicidal agents. Limestone (calcium carbonate [$CaCO_3$]), which occurs plentifully in nature, is converted by heating into *quicklime* (calcium oxide [CaO]), with release of carbon dioxide. When quicklime is treated with half its weight of water, *slaked lime* (calcium hydroxide [$Ca(OH)_2$]) is formed. Slaked lime, a powerful disinfectant, is used as *milk of lime*, prepared by mixing 1 part of freshly slaked lime with 4 parts of water. Milk of lime is especially useful as a disinfectant for feces. A more dilute suspension of slaked lime is known as *whitewash*. Lime preparations must not be exposed to the air because they combine with the carbon dioxide of the air to form calcium carbonate, which has no antiseptic action.

Chlorinated lime (chloride of lime, bleaching powder, calcium hypochlorite), made by passing chlorine through freshly slaked lime, is one of the most important of the chlorine-liberating compounds used as disinfectants. It is an unstable compound whose

Table 13-1 Antimicrobial Activity of Commonly Used Cold Sterilants

Agent	Destructive Action Against				
	Bacteria	Tubercle Bacilli	Spores	Fungi	Viruses
Alcohol, ethyl (70% to 95%)	+ *	+	0	+	±
Alcohol, isopropyl (70% to 90%)	+ +	+	0	+	±
Alcohol-iodine (2%)	+ +	+	±	+	+
Formalin (37%)	+	+	+	+	+
Glutaraldehyde (buffered, 2%) (Cidex)	+ +	+	+ +	+	+
Iodine (2% to 5% aqueous)	+ +	+	±	+	+
Iodophors (1%) (povidone-iodine complex)	+	+	±	±	+
Mercurials (Merthiolate)	±	0	0	+	±
Phenolic derivatives (0.5% to 3%)	+	+	0	+	±
Quats (benzalkonium chloride, 1:750 to 1:1000)	+ +	0	0	+	0

After DiPalma, J.R., editor: Drill's pharmacology in medicine, New York, 1971, McGraw-Hill Book Co.
* + +, Very good; +, good; ± fair (greater concentration or more time needed); 0, no activity.

antiseptic action results from the release of hypochlorous acid, which is toxic to the bacterial cell. A 0.5% to 1% solution kills most bacteria in 1 to 5 minutes. A 1:100,000 solution destroys typhoid bacilli in 24 hours. Chlorinated lime bleaches and destroys fabrics and decomposes when exposed to the air. As a means of disinfecting excreta, chlorinated lime is probably without rival.

Ferrous sulfate (copperas [$FeSO_4 \cdot 7H_2O$]) is used in the form of the impure commercial salt. Besides being a good disinfectant, it is an effective deodorant because it combines with both ammonia and hydrogen sulfide. It is an ideal disinfectant for use in damp musty places and around houses.

Chlorophyl in the form of water-soluble components has been used in the local treatment of wounds. It cleans the wound, stops pus formation, destroys odors, and promotes healing.

See Table 13-1 for several cold sterilants and their antimicrobial activity.

SUMMARY

1 *Disinfection* means destruction on inanimate objects of disease-producing organisms and their products, with chemical agents known as *disinfectants*.

2 Antisepsis prevents multiplication of microbes associated with the living body but does not necessarily kill them.

3 Ideally, a good disinfectant attacks all types of microbes rapidly, is not slowed by organic material, and does not destroy body tissues. It should form a stable mixture with water, not injure materials to be sterilized, be easily obtainable at a reasonable cost, and be readily transported.

4 Different species of microbes, especially bacteria, vary more in their susceptibility to chemicals than to the physical agents used for sterilization.

5 A chemical agent may destroy microbes by coagulating cell proteins, by disrupting the cell membrane, or by inactivating enzymes.

6 Factors that influence disinfection are the nature and contamination of the material to be disinfected, the nature and concentration of the chemical agent, and the way it is applied.

7 The kind and number of microbes contaminating the item to be sterilized, whether they are clumped or dispersed, and whether they are vegetative forms or spores are important.

8 Surface tension is important in disinfection, since liquids of low surface tension spread over a greater area and contact the cell more intimately than do liquids of high surface tension.

9 Commonly used disinfectants and antiseptics are classified as surface-active compounds, heavy metal compounds, alcohols and aldehydes, phenols and derivatives, halogen compounds, acids, oxidizing agents, and dyes.

10 Soap, a surface-active compound, facilitates the mechanical removal of microbes through the scrubbing process. Because of its sodium and alkali content, soap is germicidal for pneumococci, streptococci, gonococci, meningococci, spirochetes of syphilis, and influenza viruses, but it is too mildly antiseptic for general use.

11 Since silver nitrate selectively affects gonococci, a 1% solution is instilled into the eyes of newborn babies to prevent the serious complications of ophthalmia neonatorum.

12 Ethyl alcohol is one of the best disinfectants, acting over a range from 70% to 95%. It does not kill spores but is a tuberculocidal disinfectant of choice. The cheaper and comparable isopropyl alcohol is slightly superior.

13 The bactericidal aldehyde, glutaraldehyde, is virucidal and sporicidal in short exposure times.

14 Phenol and the modern phenolic derivatives are excellent for objects contaminated

by tubercle bacilli. These disinfectants are good for woodwork, floors, and furniture, as well as for feces, blood, pus, and other proteinaceous material.

15 Halogen compounds such as chlorine and iodine are strong oxidizing agents that form halogenated derivatives of many of the nitrogenous compounds of protoplasm in reactions that destroy microbes. Chlorine is used in the disinfection of drinking and swimming pool water and in the treatment of sewage.

16 Ethylene oxide has a broad range of antimicrobial activity for which a prolonged period of exposure is required in an appropriately designed sterilizing chamber. This form of so-called cold sterilization must be monitored carefully.

QUESTIONS FOR REVIEW

1 Define disinfection, germicide, bactericide, microbicide, antiseptic, asepsis, fumigation, bacteriostasis, preservative, iodophor, detergent, enzyme detergent, Tween 80, astringent, degerm, sanitize, decontamination, hydrolysis, wetting agent, and surface-active compound.

2 How do chemical disinfectants act?

3 Broadly classify antiseptics and disinfectants. Give examples.

4 Discuss ethylene oxide as a sterilizing agent.

5 Explain *cold* sterilization.

6 What agents are most effective in the disinfection of articles contaminated with *Mycobacterium tuberculosis?*

7 List the qualities of a good disinfectant. Does any presently available agent have all of them?

8 Why is surface tension a factor in disinfection?

9 What is the action of soap in disinfection?

10 What special usefulness does copper sulfate have in water treatment?

SUGGESTED READINGS

Bryan, R.M., and Bland, L.A.: Occupational exposure to ethylene oxide; its effect and control, J. Environ. Health **43:**254, 1981.

Glaser, R.: Special occupational hazard review with control recommendations for the use of ethylene oxide as a sterilant in medical facilities, NIOSH Publication no. 77-200, Rockville, Md., 1977, National Institute for Occupational Safety and Health.

Pseudomonas aeruginosa peritonitis attributed to a contaminated iodophor solution—Georgia, Morbid. Mortal. Weekly Rep. **31:**197, 1982.

Vanell, L.D.: On-site monitoring of ethylene oxide sterilizers, Am. Lab. **11:**70, Dec. 1981.

CHAPTER 14

CHEMICAL AGENTS THERAPEUTIC FOR MICROBIAL DISEASES

CHEMOTHERAPEUTIC AGENTS

Although it is possible to treat many local infections with chemicals, finding a chemical that, taken orally or parenterally, *selectively* destroys microorganisms without injuring body cells is not easy (p. 241). Examples of such chemicals are quinine, against the parasites of malaria, and emetine (the active principle of ipecac), against *Entamoeba histolytica*, the organism causing amebic dysentery. These are known as *chemotherapeutic agents*. The treatment of disease with chemotherapeutic agents is **chemotherapy**. The **chemotherapeutic index** compares the toxicity of a drug for the body with its toxicity for a disease-producing agent. It is obtained by dividing the maximal tolerated dose per kilogram of body weight by the minimal curative dose per kilogram of body weight. In recent years there have been phenomenal developments in the field of chemotherapy, beginning with the discovery of the sulfonamide compounds.

Sulfonamides

Sulfonamide compounds contain the group —$SO_2N<$. Many of them are derived from the compound known as *sulfanilamide*, which was among the first of the sulfonamides to be developed as a chemotherapeutic agent. Sulfanilamide was derived from a dye, known by the trade name Prontosil, that had been found effective in the treatment of streptococcal infections (Fig. 14-1). Prontosil was used for only a short time, and sulfanilamide itself was soon replaced by less toxic drugs.

Action

The sulfonamide compounds are bacteriostatic in their mode of action because they act as antimetabolites in a special way. They interfere with certain enzymes in the targeted bacterial cells. Certain microbes must synthesize their own folic acid, an essential factor for growth (a vitamin to these microbes), and a part of the chemical process is the incorporation of paraaminobenzoic acid (PABA) into the folic acid molecule. Sulfon-amides closely resemble PABA structurally, and because of this chemical similarity, sulfonamides can prevent the utilization of PABA by the sensitive microorganisms, replacing it or, as it were, competing with it for participation in the chemical sequences involved (Fig. 14-2). Thereby their normal production of folic acid is blocked. This kind of activity is termed **competitive antagonism**. Since mammalian cells do not synthesize folic acid but require it as a vitamin from outside the cell, sulfonamides do not interfere with their metabolism in this way.

Effectiveness

Sulfonamide compounds are not self-sterilizing, and their action is inhibited by pus. Sulfonamides are effective against gram-positive bacteria, some gram-negative diplococci and bacilli, chlamydiae, and actinomycetes. Generally, sulfonamides have been replaced by antibiotics that are not as toxic and that are more certain and rapid in their action. Where an antibiotic and a sulfonamide are of equal value, it is customary to give the antibiotic. Sulfonamides have been used for treatment of epidemic meningitis because they pass into the cerebrospinal fluid more easily than do antibiotics. The highly soluble sulfonamides are important in the treatment of some infections of the urinary tract as well. Certain sulfonamides for intestinal use suppress the growth of the intestinal flora, an important measure in the preparation of the patient for surgery of the large bowel.

A patient receiving sulfonamide therapy should be continuously monitored for early toxic symptoms. If such appear, the drug must be discontinued.

Specific Compounds

Today there are many sulfonamide compounds with varying actions on bacteria and varying toxic effects on the human body (Fig. 14-3). The ones most readily available are sulfadiazine, sulfamerazine, sulfamethazine, sulfacetamide (Sulamyd), sulfadimethoxine (Madribon), sulfisoxazole (Gantrisin), sulfamethoxypyridazine (Kynex), sulfachloropyridazine (Sonilyn), sulfisomidine (Elkosin), sulfamethizole (Thiosulfil), sulfa-

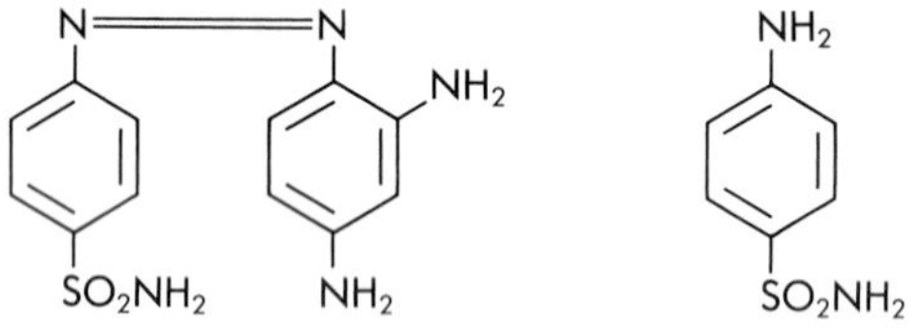

Prontosil   **Sulfanilamide**

FIG. 14-1   Two early sulfonamide compounds.

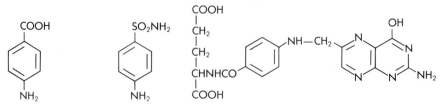

Paraaminobenzoic acid   Sulfanilamide   **Folic acid**

FIG. 14-2   Because of the close chemical similarity of sulfanilamide to paraaminobenzoic acid (PABA), sulfonamides exert an antimicrobial activity termed *competitive antagonism*. The incorporation of PABA into folic acid is blocked.

cytine (Renoquid), and sulfaethidole, all for systemic use. Other compounds include sulfaguanidine, succinylsulfathiazole (Sulfasuxidine), sulfasalazine (Azulfidine), and phthalylsulfathiazole (Sulfathalidine) for intestinal use.

Co-trimoxazole

Co-trimoxazole (Septra), an important urinary tract antiseptic, is an oral synthetic antibacterial drug combining sulfamethoxazole with trimethoprim, a diaminopyrimidine (Fig. 14-4). The combination results in a true synergistic action against common urinary tract pathogens with the exception of *Pseudomonas* species. This action is to block at one crucial point two consecutive steps in the biosynthesis of nucleic acids and certain proteins essential to many bacteria. The drug combination has been applied with some success to the treatment of toxoplasmosis and pneumocystosis. Both drugs of the combination are toxic to the bone marrow.

Tuberculostatic Drugs

Certain chemotherapeutic drugs (Fig. 14-5) specifically inhibit the growth of the tubercle bacillus and have proved invaluable in the treatment of tuberculosis (p. 606).

Isoniazid (isonicotinylhydrazine) (INH) is a remarkably potent bacteriostatic agent against *Mycobacterium tuberculosis* but has little effect on most other microbes. Its mode of action is not known at present. INH is an effective drug not only in treating tuberculosis but also in preventing tuberculous infection from becoming active disease. The American Thoracic Society, American Lung Association, and Centers for Disease

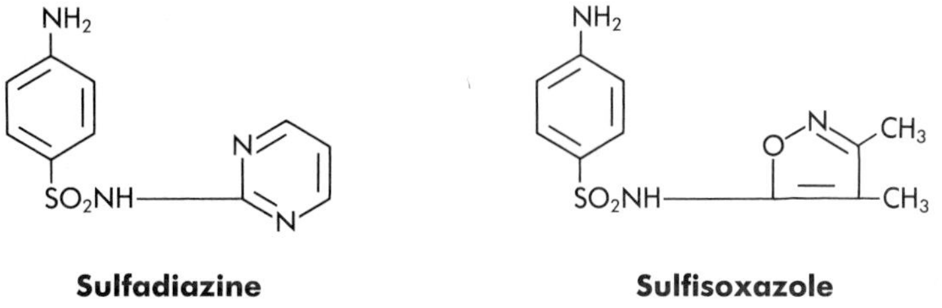

Sulfadiazine   **Sulfisoxazole**

FIG. 14-3   Two sulfonamide compounds in use.

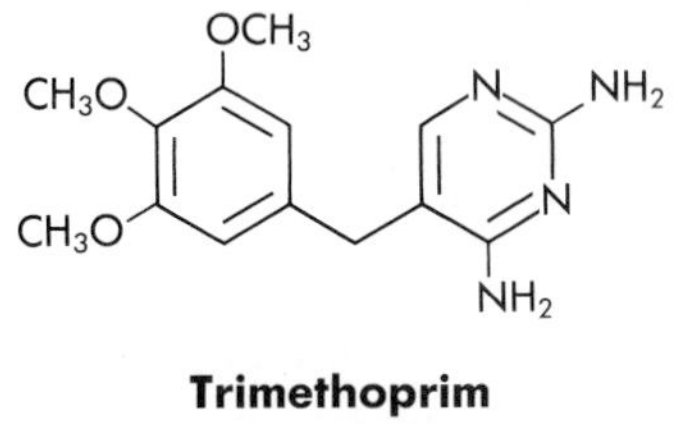

Trimethoprim

FIG. 14-4   The combination of trimethoprim with a sulfonamide is more effective than either drug given alone (an example of drug synergism).

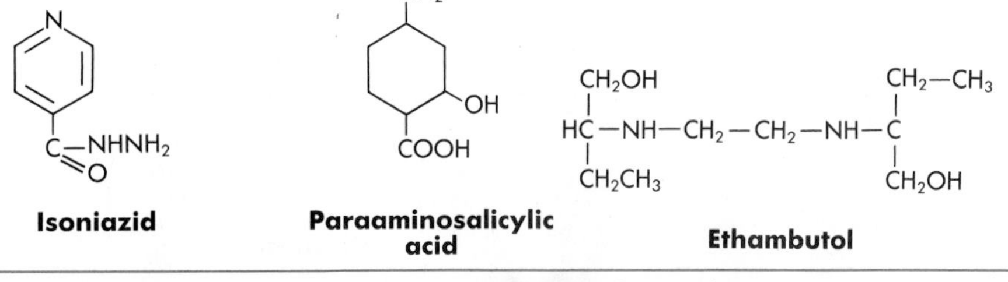

Isoniazid   **Paraaminosalicylic acid**   **Ethambutol**

FIG. 14-5   Chemical structures of the important tuberculostatic drugs.

Control recommend ideally that all tuberculin-positive individuals receive a year of INH prophylaxis (p. 606). Generally, there have been few unfavorable reactions, but instances of drug toxicity have appeared. Some 10% of individuals receiving INH prophylactically may develop evidence of liver dysfunction, rarely severe hepatitis. Hepatitis usually appears early in the course of therapy; nearly half the cases are reported during the first 12 weeks. Many authorities state that the possibility of hepatotoxicity is no reason to discontinue the regimen but rather an indication to screen the recipients initially and, once they are receiving the drug, to check them at monthly intervals for any sign of liver disorder.

Paraaminosalicylic acid (PAS) is another chemotherapeutic agent highly specific for the tubercle bacillus and not effective against other microorganisms (Fig. 14-5). It is often used in combination with INH and streptomycin in the treatment of tuberculosis. Its antimetabolic activity provides another example of competitive antagonism, being comparable to that of the sulfonamides, since PAS is structurally similar to PABA and its presence interferes with normal use of PABA by microorganisms in the synthesis of folic acid.

Ethambutol (2,2'-[ethylenediimino]-di-1-butanol dihydrochloride) is often used together with INH for the treatment of tuberculosis (Fig. 14-5). It is valuable in the treatment of disease caused by tubercle bacilli resistant to other tuberculostatic drugs and is effective against certain atypical mycobacteria. Its mechanism of action is unknown.

Nitrofurans

A group of synthetic antimicrobial drugs with activity against gram-positive and gram-negative bacteria, fungi, and protozoa but little toxicity for tissues are the nitrofurans. The mechanism of the antimicrobial activity of the nitrofurans is unknown, but it is presumed that the compounds interfere with enzymatic processes essential to carbohydrate metabolism in the bacterial cell. An important one is *nitrofurazone* (Furacin), applied topically to prevent infections of wounds, burns, ulcers, and skin grafts. *Nitrofurantoin* (Furadantin) is used in bacterial infections of the urinary tract (Fig. 14-6). *Furazolidone* (Furoxone) is effective against *Giardia lamblia* in intestinal infections; in a mixture with *nifuroxime* (Micofur) it is used in the treatment of vaginitis caused by bacteria, by *Trichomonas vaginalis*, or by *Candida albicans*.

ANTIBIOTICS

Experiments and observations from past decades have proved that numerous organisms can produce substances with the power to inhibit the multiplication of other organisms or even to kill them. Such substances are known as *antibiotics*. This information is not as new as it might seem: Pasteur in 1877 demonstrated that the growth of anthrax bacilli was retarded by the presence of common soil organisms, and several antibiotic-like substances (for example, pyocyanase) were prepared before 1900. The ancient Egyptians painstakingly listed, on papyri, concoctions with activity corresponding to that of the

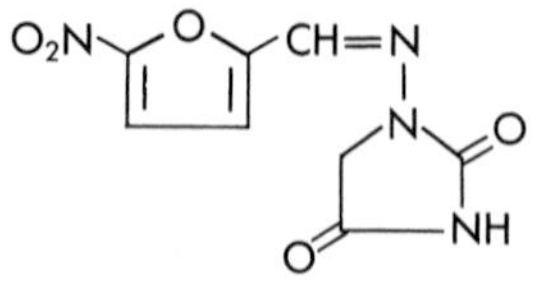

Nitrofurantoin

FIG. 14-6   A nitrofuran.

The term antibiosis was introduced in 1889 by Vuillemin, who said, "No one considers the lion which leaps on its prey to be a parasite nor the snake which injects venom into the wounds of its victim before eating it. Here there is nothing equivocal; one creature destroys the life of another to preserve its own; the first is completely active, the second completely passive. The one is in absolute opposition to the life of the other. The conception is so simple that no one has ever thought of giving it a name. This condition, instead of being examined in isolation, can appear as a factor in more complex phenomena. In order to simplify words we will call it antibiosis. . . ."

Cited by Florey, H.: Antibiotics, Fifty-second Robert Boyle Lecture presented to the Oxford University Scientific Club, Springfield, Ill., 1951, Charles C Thomas, Publisher.

modern antibiotics; one item they felt to be most effective was "moldy wheaten loaf" (see box above).

In 1941 the term *antibiotic* was defined by Waksman as follows: "An antibiotic is a chemical substance produced by microorganisms which has the capacity to inhibit the growth of bacteria and even destroy bacteria and other microorganisms in dilute solution."*

Antibiotics are biosynthesized by bacteria, fungi, and plants, including certain flowering ones. Some organisms produce more than one antibiotic, and certain antibiotics are produced by more than one organism (Table 14-1). Although thousands of species of microorganisms elaborate antibiotics and hundreds of antibiotics are known, only a few are practically useful in the treatment of disease. We consider them collectively, yet their physical, chemical, and pharmacologic properties are divergent.

Spectrum of Activity

Some antibiotics affect mainly gram-positive bacteria, with little or no effect on gram-negative bacteria. Others seem to affect only certain species of bacteria, regardless of whether they are gram negative or gram positive. A few are active against rickettsias; a few attack fungi.

The susceptibility (or resistance) of bacteria to different antibiotics may be determined by suitable laboratory procedures (p. 167) designated **susceptibility tests** (see Fig. 9-11, p. 168), and there are also satisfactory ways to determine the amount of certain antibiotics in the blood.

The **spectrum** of a compound is its range of antimicrobial activity. A **broad-spectrum** antibiotic indicates a wide variety of microorganisms affected, including usually both gram-positive and gram-negative bacteria. A **narrow-spectrum** antibiotic limits only a few. Table 14-2 lists antibiotics in both categories.

Mode of Action of Antimicrobial Drugs

Several mechanisms explain in part the harmful or lethal effects that antimicrobial drugs have on microbes. Most of the time the mechanisms of injury are subtle, usually reflecting an inhibitory effect on some aspect of microbial physiology. The first mechanism studied was the role of the sulfonamides in competition for a place in the metabolic activities of the cell (metabolic antagonism). A drug utilized by mistake because of a close chemical resemblance to a vital cytoplasmic substance can halt microbial growth and function.

*From Waksman, S.A.: Bull. N.Y. Acad. Med. **42:**623, 1966.

Table 14-1 Sources of Antibiotics in Microbes

Microbe*	Antibiotic
Bacteria: Gram-Positive Rods	
Bacillus subtilis (group *licheniformis*)	Bacitracin
Bacillus polymyxa (var. *colistinus*)	Colistin
Bacillus polymyxa	Polymyxin
Bacillus brevis	Tyrothricin
Streptomycetes	
Streptomyces parvullus	Actinomycin D (Dactinomycin) (antineoplastic)
Streptomyces nodosus	Amphotericin B (Fungizone) (antifungal)
Streptomyces tenebrarius	Apramycin (Amyblan)
Streptomyces fragilis	Azaserine (antineoplastic)
Streptomyces verticillus	Bleomycin (antineoplastic)
Streptomyces griseus	Candicidin (antifungal)
Streptomyces capreolus	Capreomycin (Capastat)
Streptomyces venezuelae	Chloramphenicol (Chloromycetin)
Streptomyces aureofaciens	Chlortetracycline (Aureomycin)
Streptomyces bellus (var. *cirolerosis*, var. *nova*)	Cirolemycin (antineoplastic)
Streptomyces noursei	Cycloheximide (Actidione)
Streptomyces peucetius	Daunomycin (antineoplastic)
Streptomyces aureofaciens (mutant)	Demeclocycline (Declomycin)
Streptomyces peucetius (var. *caesius*)	Doxorubicin (Adriamycin) (antineoplastic)
Streptomyces erythraeus	Erythromycin (Ilotycin)
Streptomyces fradiae	Fosfomycin (antibiotic)
Streptomyces fradiae	Fradicin (antifungal)
Streptomyces tanashiensis	Kalafungin (antibiotic)
Streptomyces kanamyceticus	Kanamycin (Kantrex)
Streptomyces lincolnensis	Lincomycin (Lincocin)
Streptomyces melanogenes	Melanomycin
Streptomyces plicatus	Mithramycin (antineoplastic)
Streptomyces caespitosus	Mitomycin
Streptomyces carzinostaticus (variant)	Neocarzinostatin (antineoplastic)
Streptomyces fradiae	Neomycin
Streptomyces noursei	Nystatin (Mycostatin)
Streptomyces antibioticus	Oleandomycin
Streptomyces rimosus	Oxytetracycline (Terramycin)
Streptomyces rimosus	Paromomycin (Humatin)
Streptomyces lincolnensis	Ranimycin (antibiotic)
Streptomyces mediterranei	Rifampin (Rimactane)
Streptomyces spectabilis (variant)	Spectinomycin (Trobicin)
Streptomyces griseus	Streptomycin
Streptomyces lavendulae	Streptothricin
Streptomyces achromogenes	Streptozotocin
Streptomyces tenebrarius	Tobramycin
Streptomyces orientalis	Vancomycin

*From members of the genus *Streptomyces* are derived the majority of the 150 antibiotics marketed for the treatment of human infections and over 60% of the 7000 known naturally occurring antibiotics. An additional 15% are formed by members of the related genera *Nocardia* and *Micromonospora*. (More than 30,000 semisynthetically derived antibiotics have been defined.)

Table 14-1 Sources of Antibiotics in Microbes—cont'd

Microbe	Antibiotic
Streptomycetes—cont'd	
Streptomyces antibioticus	Vidarabine (adenine arabinoside, ara-A) (antiviral)
Streptomyces puniceus	Viomycin
Streptomyces bikiniensis (variant)	Zorbamycin
Actinomycetes	
Micromonospora purpurea	Gentamicin
Nocardia lurida	Ristocetin (Spontin)
Micromonospora inyoensis	Sisomicin
Fungi	
Arachniotus aureus	Aranotin (antiviral)
Cephalosporium acremonium	Cephaloridine (Loridine)
Cephalosporium species	Cephalothin (Keflin)
Aspergillus fumigatus	Fumagillin
Penicillium griseofulvum dierckx and *Penicillium janczewski*	Griseofulvin (Fulvicin)
Aspergillus giganteus	Nifungin (antifungal)
Penicillium notatum and *chrysogenum*	Penicillin
Penicillium stoloniferum	Statolon (antiviral in animals)

Table 14-2 Antibiotics—Range of Activity

Narrow-spectrum Antibiotics	Broad-spectrum Antibiotics
Penicillin	Chloramphenicol
Streptomycin	Chlortetracycline
Dihydrostreptomycin	Demeclocycline
Erythromycin	Oxytetracycline
Lincomycin	Tetracycline and derivatives
Polymyxin B	Kanamycin
Colistin	Ampicillin
Vancomycin	Cephalosporins
Nystatin	Rifampin
Spectinomycin	Gentamicin
	Tobramycin
	Paromomycin

The physiologic integrity of the microbial cell depends on key macromolecules and the processes in which they are involved. Certain of those vulnerable to antimicrobials include the following four major modes of action.

Cell Wall Synthesis

The rigid cell wall jackets the bacterial cell cytoplasm with its uniquely high osmotic pressure. If the cell wall is faulty in construction and weakened because of the action of the antibiotic, the cell membrane cannot withstand the internal pressure. In a medium of ordinary tonicity the bacterial cell explodes.

Cell wall synthesis takes place essentially in three stages (p. 73), the first two occurring in the bacterial cell cytoplasm and the third outside the cell membrane. Bacitracin and vancomycin inhibit the second stage, at which time strands of peptidoglycan are formed within the cell membrane. Vancomycin also inhibits spheroplasts, probably by an additional effect on the cell membrane. Penicillins and cephalosporins inhibit the cross-linking of the peptidoglycan strands that is done in stage 3 (see Fig. 5-12, p. 74).

Nucleic Acid Synthesis

Antibiotics can interfere with the synthesis of the nucleic acids within the bacterial cell by inhibiting key enzymes or by mimicking a necessary ingredient. Some antibiotics can complex with DNA and block the formation of the messenger RNA (mRNA).

A poorly understood relationship exists between the structure of griseofulvin as an analog of a purine nucleotide bound to cell lipids and the fact that DNA synthesis is slowed (as are also growth and metabolism) in drug-treated fungal cells. The drug interferes with cell division and induces the formation of faulty DNA in susceptible fungi.

Protein Synthesis

Other antibiotics inhibit the ribosomes (rRNA) directly, thereby materially affecting the elaboration of proteins within the cell. Since the biochemical activities relating to both nucleic acid and protein synthesis are complicated, the possibilities for associated injury to the microbial cell are many. In fact, the complex events of protein synthesis represent the most frequent sites of antibiotic action.

Aminoglycosides block protein synthesis effectively by interfering with the function of mRNA, causing mRNA to misread transfer information, and leading it to direct an assembly of amino acids in an *incorrect* sequence in the polypeptide. A **nonsense protein,** a chemical mix-up, is the result, which does not support the viability of the organism. Aminoglycosides also interfere with protein synthesis by binding to 30S and 50S ribosomal subunits. Tetracyclines inhibit protein synthesis, it is believed, by attaching to the 30S ribosomal subunit and preventing the amino acid–transfer RNA (tRNA) complex from binding to the ribosome. Tetracyclines also inhibit certain bacterial enzymes and can act as efficient chelators of heavy metals. Chloramphenicol interferes with the synthesis and function of mRNA. Erythromycin and spectinomycin

Table 14-3 Cellular Disturbances (Mode of Action) Related to Antimicrobial Drugs

| | | Inhibitory Effect on | | | |
| | | Protein Synthesis | | | Metabolic Antagonism (Competitive Interference in Cell Metabolism) |
Cell Wall Synthesis	Nucleic Acid Synthesis	Ribosome Function	Other	Cell Membrane Function	
Bacitracin*	Actinomycin D	Chloramphenicol	Chlortetracycline	Amphotericin B*	Isoniazid*
Cephalosporins*	Griseofulvin*	Dihydrostreptomycin*	Clindamycin	Colistin*	Nitrofurans
Penicillins*	Idoxuridine	Gentamicin*	Erythromycin	Nystatin*	Paraaminosalicyclic
Vancomycin*	Rifampin*	Kanamycin*	Lincomycin	Polymyxins*	acid
		Neomycin*	Methacycline		Sulfonamides
		Paromomycin*	Oxytetrcycline		
		Streptomycin*	Tetracycline		

*Bactericidal antibiotic.

bind to the ribosome to inhibit polypeptide formation. Lincomycin and clindamycin bind to the ribosome and inhibit formation of peptide bonds for polypeptide molecules.

Plasma Membrane Function

Some antibiotics alter the permeability or the integrity of the plasma membrane. For an antibiotic to exert a deleterious effect through the plasma membrane, the organism does not have to be dividing actively.

The antifungal antibiotics amphotericin and nystatin bind to sterols in the cell membrane of the susceptible fungus, destroying membrane integrity. Cell contents leak out, and the cell dies. Polymyxins act like detergents on the cell membrane, since they are surface-active agents with both lipophilic and lipophobic groups in the same molecule. After the molecules of polymyxin have penetrated the cell wall of a bacterium and entered the cell membrane, they become arranged with their lipid tails embedded in the lipid layer of the membrane and their water-soluble heads buried in the protein layer. The cell membrane thus altered is no longer an effective osmotic barrier and soon ruptures to destroy the cell. Colistin's action is similar to that of the polymyxins. The disinfectant benzalkonium chloride disrupts the bacterial cell membrane by a detergent action also.

Table 14-3 lists antimicrobial compounds according to category of action.

Selective Toxicity

An important factor in any consideration of the mode of action of antimicrobial compounds is selective toxicity. The agent, to be useful, must kill the microbial cells without damaging the cells of the animal host. An antimicrobial agent can act selectively if it can modify a biochemical sequence vital to target microbes but one that is not necessarily comparable and that may not even be present in the host organism. The best example is the action of penicillin on the synthesis of the bacterial cell wall. A structure comparable to the cell wall is not found in the animal host cell. Agents such as disinfectants that coagulate proteins in *all* cells are **nonselective.**

Side Effects

There are four serious complications of antibiotic therapy: (1) hypersensitivity or allergy, (2) toxicity, (3) effects of the replacement flora, and (4) induction of bacterial resistance.

Hypersensitivity

The turnover of drugs by the human body involves a series of complex enzymatically mediated reactions. It is increasingly apparent that there can be vast individual differences. Signs of hypersensitivity to the antibiotics vary in different persons but include

Table 14-4 Some Examples of Pathologic Complications to Antibiotic Therapy

Antibiotic	Therapeutic Background	Physiologic Derangement	Pathologic Lesion
Neomycin	Prolonged administration	Changes in lining of small intestine (jejunum) with interference in absorption of glucose, iron, vitamin B_{12}	Malabsorption syndrome (deficiency states, weight loss, skin pigmentation, abdominal distention, and the like)
Tetracycline	Administration during pregnancy or to infants	Deposition of antibiotic in tooth enamel	Dental defects—pigmentation
Tetracycline	Large doses given intravenously	Toxic effect on liver	Liver disease

skin rashes, joint pains, lymph node enlargement, lowered white blood cell count, and sometimes hemorrhages.

Toxicity

Examples of toxic reactions to antibiotic therapy include the aplastic anemia induced by chloramphenicol and the deafness sometimes produced by streptomycin. Table 14-4 gives additional examples with the background and lesion indicated.

Some antibiotics in the doses ordinarily given are associated with psychiatric disturbance. Depression and hallucinations have been related to certain long-acting sulfonamides. Psychosis, especially in the elderly, can be precipitated by the antiviral agent amantadine. A buoyant state of well-being and elation has been noticed in INH-treated patients with tuberculosis.

Replacement Flora

The appearance of a replacement microbial flora is troublesome at times. Not being affected by the drug themselves, such organisms tend to increase during the period of therapy, being no longer held back by the organisms that succumb to the antibiotic. They can then multiply in an uncontrolled way, and, if potentially pathogenic, they may start a disease process difficult to treat. The replacement flora may be very resistant to standard antibiotics. Such secondary infections are often caused by *Proteus* or *Pseudomonas* species, and infections with *Candida albicans* easily follow prolonged and extensive antibiotic therapy, especially with broad-spectrum preparations. **Superinfection** as a complication is more likely with the broad-spectrum antibiotics than with the narrow-spectrum ones.

A change in the pattern of the normal microbial flora of a given area as a result of antibiotic administration may lead to an enhancement of the effects of the pathogen, for example, with *Salmonella* species. Antibiotics may be so much more effective against the normal flora than against salmonellas that the growth and persistence of the pathogen into a carrier state are considerably enhanced. Generally, the normal flora constitute a better defense against *Salmonella* species than does the antibiotic.

Drug Resistance (p. 99)

Strains of bacteria are always emerging that are resistant to one or more antibiotics, a resistance often acquired when clinical infectious disease has been inadequately treated with the given antibiotic. Resistance is always a possibility when a bacterio*static* drug is given, since such a drug merely limits the activities of the given microbes—body mechanisms dispose of them. Many antimicrobials are only bacteriostatic within the human body. If only *bactericidal* (lethal) antibiotics could be given, resistance would cause less concern. Bacteria vary in their ability to develop resistance to a given drug. They sometimes may become tolerant to more than one antimicrobial drug, that is, display a **cross-resistance,** which is noted especially in the case of closely related antibiotics.

A chromosomal mutation or other genetic change in bacteria can be a factor leading to some forms of drug resistance. From changes in the DNA come metabolic rearrangements in bacterial cells that can provide the necessary biochemical mechanisms. Such genetic mechanisms may enable bacteria to block penetration by an antimicrobial drug, to destroy it more effectively, or alter the structural target to site of action, to bypass the particular metabolic reaction inhibited, or to produce an enzyme that is still metabolically efficient although less vulnerable to drug interference.

Antibiotics do not induce mutation. A mutation is a spontaneous event resulting from chemical damage to DNA. An antibiotic, however, may provide the environment for the rare resistant mutant to survive. By definition, resistance resulting from a mutation involves a single bacterial species descendant from a single parent. This type of change in a chromosome does not easily pass between different species or genera.

Therefore resistance is not disseminated widely and is probably of minor epidemiologic significance. Also, under normal growth conditions, mutationally altered bacteria tend to be at a metabolic disadvantage in comparison with their wild-type associates, but if both types of microorganisms are growing in the milieu of the given antibiotic drug, the mutants may be selectively maintained and then can certainly threaten the well-being of the patient under treatment.

The systems for gene transfer in bacteria are of great medical importance because these demonstrate how bacterial resistance to antibiotics is accomplished rapidly in one step and indicate that the search for new antibiotics is to be continual.

One of the most interesting mechanisms for drug resistance yet described concerns the **R plasmids,** which are by far the most common determinants of drug resistance in bacteria. The origins of plasmids is not known. Previously unsuspected, they were first detected in *Shigella* species in Japan in 1959. Characteristically, R plasmids are disseminated widely among bacterial populations. They are transmitted in several ways, certainly by transduction, transformation, and conjugation, and sometimes rapidly from one strain to another, even from one species to another, and occasionally from one genus to another.

When R plasmids were first defined genetically, they were found to encode resistance for as many as six or seven antimicrobial agents that were not necessarily related. R plasmid resistance is commonly associated with cross-resistance patterns to drugs that are both related and unrelated structurally.

Bacteria transferring R plasmids have been studied, especially in members of the genera *Enterobacter, Escherichia, Klebsiella, Salmonella, Shigella, Serratia, Proteus,* and *Citrobacter;* all of Bergey's family Enterobacteriaceae; the genus *Pseudomonas* from Bergey's family Pseudomonadaceae; and the genus *Vibrio* of Bergey's family Vibrionaceae. For a time it was presumed that antibiotic resistance transmitted by plasmid-mediated conjugation occurred only in gram-negative bacteria. The transfer of resistance to several antibiotics by plasmid-mediated conjugation also has been demonstrated in gram-*positive* bacteria that were taken from dental plaques.

With drug resistance being the effect of an R plasmid, a wide range of possibilities for expression exists because of the genetic material brought into the bacterial cell by the plasmid. A number of mechanisms for drug resistance exclusive to plasmid-coded function have been studied (Table 14-5). Plasmid-mediated resistance can provide the microbial cell with many new enzymatic capabilities. The formation of varied drug-inactivating enzymes is given frequently as the explanation for plasmid-mediated resistance. The antibiotic can be chemically inactivated or destroyed in other ways. Its penetration into the cell can be stopped. A metabolic step at which the antibiotic is active

Table 14-5 Different Expressions of Plasmid-Mediated Resistance to Antibiotic Drugs

R Plasmid–Coded Activity in Microbial Cell	Antibiotic Effect Nullified
Blocks entry of drug into cell with substances formed in inner or outer membranes to interfere with transport system or to cover membrane pores	Antibiotic not able to enter cell
Modifies site of drug action (ribosome) by enzymatic action	Drug binding site not available
Inactivates or destroys antibiotic by enzymatic modification or hydrolysis (intracellular enzymes, extracellular enzymes, or both involved)	Antibiotic inactivated or destroyed chemically
Replaces an essential enzyme that is antibiotic sensitive with one that is antibiotic resistant	Substitute enzyme allows cell growth; antibiotic unable to compete

can be bypassed, and a target site for drug action can be altered biochemically to render the antibiotic ineffective.

That the genetic events in bacteria are well suited for the rapid transfer of antibiotic resistance genes throughout the world is evidenced by the fact that 60% to 90% of all antibiotic resistance in gram-negative bacterial pathogens is carried on plasmid R factors.

Transposition is widely believed to be the basis in nature for the construction of resistance plasmids from various genetic sources. In this process certain resistance determinants seem literally to "jump" from one plasmid to another plasmid, to the bacterial genome, or to a bacteriophage. The transposable element creates its own site of insertion in each instance. This mechanism is important for understanding certain forms of drug resistance.

The "jumping genes," properly called **transposons,** are bits of linear DNA that carry some of the genes of a plasmid. In the plasmid DNA in which they originate, they form a segment of the DNA circlet. Under appropriate conditions they break loose from their plasmid and are incorporated into the DNA of another plasmid, into a virus, or into the host bacterial genome, carrying with them the traits specified by their DNA sequence. Repeating DNA sequences located at either end of the transposon apparently enable it to insert into certain common but definite sites in bacterial or plasmid DNA. A transposon cannot replicate independently and can be passed from one organism to another only when it is part of a plasmid, virus, or bacterial genome. Once inside the new bacterial cell, the transposon may again "jump" from its plasmid into another plasmid or into the bacterial genome.

Penicillinase is the adaptive enzyme that inactivates penicillin G. Its synthesis by disease-producing staphylococci is perhaps the best-known mechanism of drug resistance (p. 99). Penicillinase is a beta-lactamase, one of several such enzymes that inactivate penicillins and cephalosporins by breaking the beta-lactam linkage in the chemical nucleus of these compounds (Fig. 14-8). The elaboration of a beta-lactamase may be directed by the bacterial genome (rarely), by an R factor, or by a transposon. In pathogenic staphylococci, penicillinase production is determined by a penicillinase plasmid. Penicillinase-producing strains of *Haemophilus*, gonococci, and many enterobacilli contain all or part of a certain transposon that carries the gene for a particular penicillinase widespread among gram-negative bacteria.

Table 14-6 Some Examples of Antibiotics Used as Feed Additives

Antibiotic	Cattle	Sheep	Hogs	Chickens
Bacitracin	1,2,3*		1,2,3	1,2,3,4
Chlortetracycline†	1,2,3	2,3	1,2,3	1,2,3,4,5
Erythromycin	1,2		1,2	1,2,3,4
Lincomycin			3	1,2,3
Oxytetracycline	1,2,3	2,3	1,2,3	1,2,3,4,5
Penicillin G			1,2	1,2,3

*1, Antibacterial agent added (at low level) to enhance growth indirectly by improving general health of animal

2, Agent that stimulates growth by direct action on metabolism of animal (mechanism unknown)

3, Antimicrobial agent added for therapeutic reasons

4, Agent to improve egg production

5, Agent added to combat coccidiosis, a parasitic disease caused by protozoa in genus *Eimeria*, which infect intestinal epithelium

†Chlortetracycline, the first antibiotic discovered to enhance growth in animals, is still the most widely used today.

The use of *antibiotics in animal feed* is a controversial subject today. The benefits to livestock of a few grams of antibiotic in a ton of feed grew out of the discovery of chlortetracycline in 1948. The fermentation process required for chlortetracycline also yielded a needed animal food supplement, vitamin B_{12}. The small amount of chlortetracycline in the crude residues given to farm animals for the content of vitamin B_{12} was found to enhance remarkably the growth and well-being of the animals. By 1951 the practice of adding antibiotics to livestock feed began to spread rapidly over the world. Today some 16 antibiotics, especially penicillin and tetracycline, are used as additives, and nearly 50% of antibacterial products manufactured in the United States are used in animal feeds or for other nonhuman purposes. Most poultry, 90% of hogs and veal calves, and 60% of cattle receive small amounts of antibiotics and other drugs (Table 14-6). Farmers also rely on antibacterial drugs to treat infections such as mastitis in milk cows. In industrial food processing, antimicrobials help control spoilage. All these factors probably contribute significantly to the problem of bacterial resistance. The Food and Drug Administration has concluded that they do and that the potential for breeding resistant strains of bacteria in the barnyard is a threat to human health and should be controlled.

The Roster of Antibiotics

Tyrocidins: Cyclic Decapeptide Antibiotics

Tyrocidins are cyclic decapeptide antibiotics, the chemical structure of which contains the one or more free amino groups necessary for the bactericidal effect. Tyrocidins are produced by *Bacillus brevis*.

Tyrothricin, isolated out of the soil from the aerobic, motile, nonpathogenic *B. brevis*, became one of the first antibiotics to be studied carefully. From it have been obtained two polypeptides: *gramicidin* (named in honor of the bacteriologist Gram) and *tyrocidine*, both highly antagonistic to gram-positive cocci but very toxic (Fig. 14-7). Tyrothricin is without effect if given orally and very dangerous if given parenterally. It therefore must be applied locally. It prevents bacterial growth and lyses bacteria already present

The antibacterial activity of tyrothricin comes from the destruction of the bacterial membrane. The loss of the osmotic barrier allows leakage of essential substances from within the bacterial cell with a resultant arrest of metabolic activity.

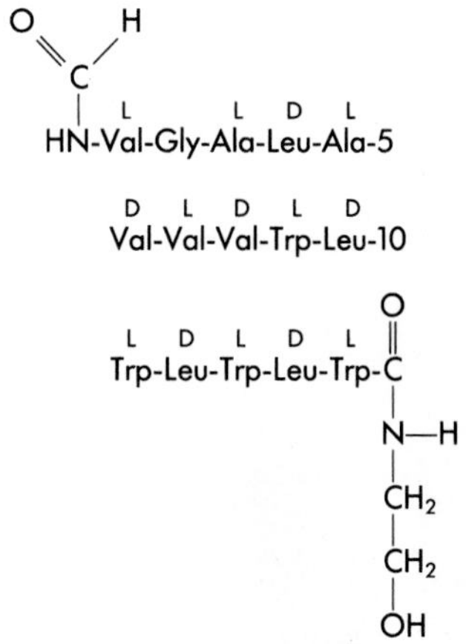

Gramicidin A

FIG. 14-7 Chemical structure of a cyclic decapeptide antibiotic. *Val*, Valine; *Gly*, glycine; *Ala*, alanine; *Leu*, leucine; *Trp*, tryptophan.

Beta-Lactam Antibiotics

Beta-lactam antibiotics are so designated because of the presence of the beta-lactam, a four-membered ring in which a carbonyl and a nitrogen are joined in an amide linkage (Fig. 14-8). The beta-lactam antibiotics are the penicillins and cephalosporins, both bactericidal agents with a wide safety/toxicity ratio. Beta-lactams are highly specific for bacteria.

Penicillin, an organic acid and a beta-lactam antibiotic, is isolated from certain molds, the most important of which are *Penicillium notatum* and *Penicillium chrysogenum* (Fig. 14-9). The first antibiotic to come into general use, it is still in many ways the best. Penicillin is an inexpensive drug. The chances are that a course of penicillin will cost only a fraction of that with another antibiotic.

Penicillin may be administered intravenously, subcutaneously, into body cavities, orally, and locally, and in contrast to many antibiotics that are bacteriostatic, it is also bactericidal. *It is one of the most important antiinfectives because it is bactericidal!* Penicillin interferes with synthesis of the bacterial cell wall of actively growing bacteria. It impedes the formation of cross-links in the peptidoglycan of the cell wall. Instead of

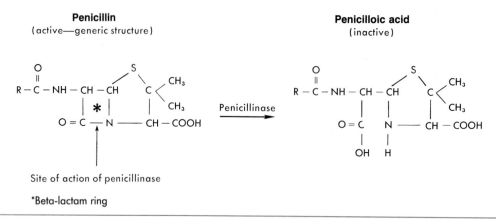

Penicillin
(active—generic structure)

Penicilloic acid
(inactive)

Penicillinase →

Site of action of penicillinase

*Beta-lactam ring

FIG. 14-8 Inactivation of penicillin by penicillinase. Note that antibiotic activity of penicillin can be destroyed by one simple change in its molecular structure, since when the beta-lactam ring is split, penicilloic acid, an inactive compound (without antibiotic capacity), is formed.

FIG. 14-9 *Penicillium chrysogenum,* giant colony. From this mutant form of the green mold almost all the world's supply of penicillin is obtained. (From Arnett, R.H., Jr., and Bazinet, G.F.: Plant biology: a concise introduction, ed. 4, St. Louis, 1977, The C.V. Mosby Co.; courtesy Chas. Pfizer & Co., Inc., New York.)

a layer of high tensile strength being formed, a structurally weak layer results that cannot hold back osmotic lysis of the cell. The peptidoglycan of the cell wall in gram-negative bacteria is relatively inaccessible, probably the reason for the lower penicillin sensitivity of gram-negative microbes.

In the individual not allergic to penicillin its toxicity for human tissue is almost nonexistent. It is an effective agent, with some exceptions, against staphylococci, streptococci (group A), pneumococci, and meningococci. Penicillin is still the unquestioned drug of choice in the nonallergic individual for infections caused by group A beta-hemolytic streptococci. In over 40 years of its use, group A streptococci have not been reported to be resistant to it. Its use revolutionized the treatment of syphilis. It does not have an effect on most gram-negative bacilli with the usual doses given. Most of the microorganisms originally sensitive to penicillin still are, with the notable exceptions coming from the penicillinase-producing staphylococci and gonococci.

The enzyme penicillinase (beta-lactamase), accounting for penicillin resistance in both gram-positive and gram-negative microorganisms, inactivates penicillin by opening the beta-lactam ring (Fig. 14-8). Penicillinase can destroy 200,000 to 400,000 units of penicillin a day. The term *penicillinase* is widely used, but since such enzymes have been found to act against the beta-lactam ring of both the penicillins and the cephalosporins, the more appropriate term is *beta-lactamase*.

Penicillin has been isolated in six closely related forms called F, G, X, K, O, and V. Penicillin G is the most satisfactory type to manufacture and use and constitutes about 90% of commercial penicillin. The sodium, potassium, and procaine salts are the common ones.

The strength of penicillin is designated in **units**. A unit (International Unit) corresponds to the activity of 0.6 μg (0.000006 g) of pure sodium penicillin G.

The generic name *penicillin* includes all penicillins, both natural and semisynthetic (Fig. 14-10). The *natural* penicillins are extracted from cultures.

Semisynthetic penicillins are the sequel to a basic observation on the chemical structure of penicillin made in Great Britain in 1959. When chemists learn how the molecule can be modified, they can change it to give the new product properties more desirable than those of the natural substance. Consequently, more than 500 new semisynthetic penicillins have been made. Phenethicillin potassium (Syncillin), an oral preparation, was the first one to be dispensed by prescription.

Penicillin resistance is most important from a pathogenic staphylococcus. Generally the newer penicillins are indicated for pathogenic staphylococci resistant to penicillin G; in other cases the potency of these new drugs is less than that of the original penicillin. Compounds designed to counteract staphylococcal penicillinase (penicillinase-resistant penicillins) are methicillin (Staphcillin), oxacillin (Prostaphlin), nafcillin (Unipen), cloxacillin (Tegopen), and dicloxacillin (Pathocil).

Ampicillin (Polycillin) is a broad-spectrum semisynthetic penicillin with activity against a number of gram-positive and gram-negative bacteria, including *Salmonella*, *Shigella*, and *Proteus* species and *Escherichia coli*. If disease of the biliary tract is not present, ampicillin can be used to treat the typhoid carrier, and it has value in the treatment of intestinal amebiasis. It is ineffective in the presence of penicillinase. A close chemical relative to ampicillin, *amoxicillin*, is also a broad-spectrum semisynthetic penicillin. With improved oral absorption over ampicillin, better blood levels are obtained with this compound.

Carbenicillin (Pyopen) is a British semisynthetic penicillin with pronounced activity in high doses against *Pseudomonas* species. *Disodium carbenicillin* (Geopen) is active against the gram-negative *Pseudomonas* and *Proteus* species. *Ticarcillin* (Ticar), similar to carbenicillin, is said to be more active against *Pseudomonas* species than is carbenicillin and to have fewer side effects because of the lower dose that can be used.

Piperacillin and *mezlocillin* are two new semisynthetic penicillins with the broadest spectrum of activity of any current penicillin. The newer penicillins are inactivated by beta-lactamases. They appear to be relatively safe, but none has greater inherent activity than penicillin G against penicillin-sensitive microbes.

Cross-allergenicity exists among the penicillins; that is, a person sensitive to penicillin G will be sensitive to the semisynthetic penicillins. It is estimated that at least 5% to 6% of the population is allergic to penicillin.

Cephalosporins, also beta-lactam antibiotics and closely related chemically to penicillin, represent a heterogeneous group of antibiotics of an entirely new class (Fig.

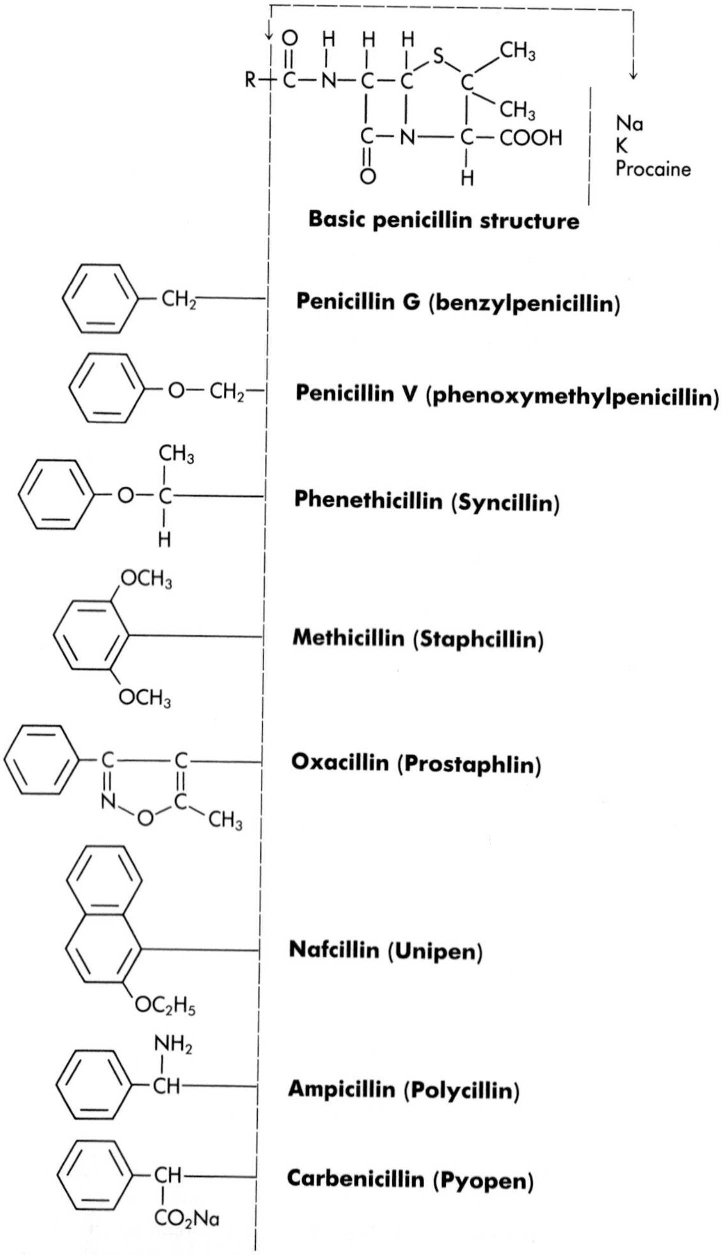

FIG. 14-10   Chemical structure of the penicillins.

14-11). The cephalosporins were developed following a discovery in 1945 by Professor Guiseppe Brotzu in Sardinia of a fungus apparently of the genus *Cephalosporium* that secreted material inhibiting the growth of bacteria. Cephalosporins as we know them were first used in 1965, and the number of compounds in this group has been increasing steadily ever since. These compounds are semisynthetic derivatives prepared from the fermentation products of the fungus *Cephalosporium* (Fig. 14-12) (now *Acremonium* since sexual spores were identified), whose natural habitat is the seacoast of Sardinia, Italy. They are bactericidal antibiotics that inhibit critical enzymatic reactions necessary for intact bacterial cell walls. Specifically, in binding to receptor proteins in the cell wall, they suppress the formation of peptidoglycan cross-bridges in newly synthesized cell walls. They are less vulnerable to beta-lactam inactivation by gram-negative organisms than are pencillins.

Cephalosporins in current use include *cephalothin* (Fig. 14-11), *cephaloridine, cephaloglycin,* and *cephalexin.* Cephalothin (Keflin) is a broad-spectrum antibiotic bactericidal against gram-positive and gram-negative bacteria. It must be given parenterally; it is unaffected by penicillinase and is of low toxicity. Cephaloridine (Loridine) is similar to cephalothin. Cephaloglycin (Kafocin) and cephalexin (Keflex) can be given orally. The newer cephalosporins *cefamandole nafate* (Mandol) and *cefoxitin* (Mefoxin) are more effective against certain gram-negative bacilli but are believed to be less so against penicillinase-producing staphylococci. Of the cephalosporins, cefoxitin is the most active against *Bacteroides fragilis.* Unlike previous compounds, cefoxitin is derived not from a fungus (*Cephalosporium*) but from a bacterium, *Streptomyces lactamdurans* and is therefore a cepha*mycin.* Another new compound in this class, *moxalactam,* is unlike the others in being totally synthetic.

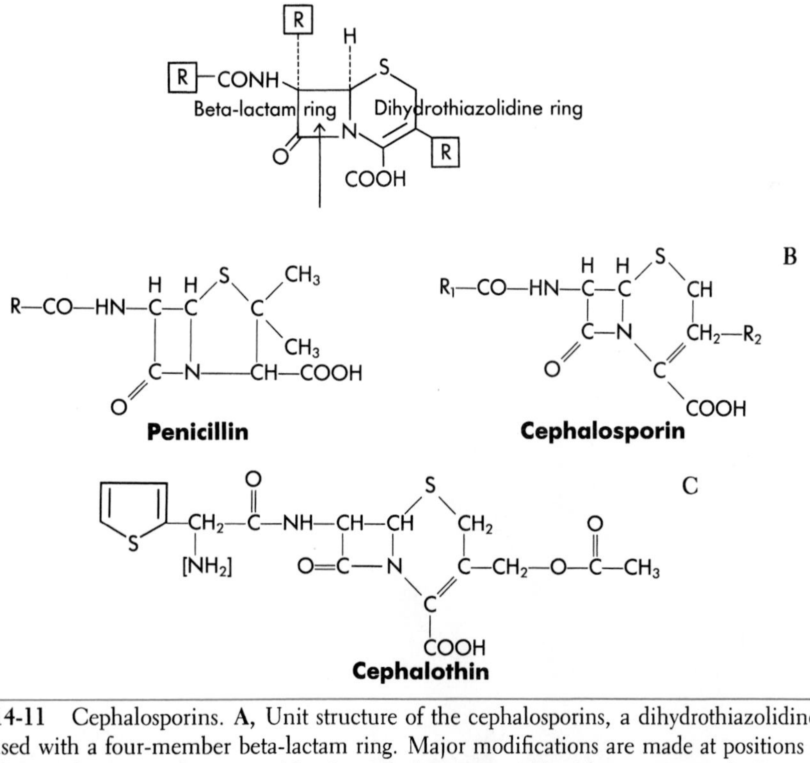

FIG. 14-11   Cephalosporins. **A,** Unit structure of the cephalosporins, a dihydrothiazolidine ring fused with a four-member beta-lactam ring. Major modifications are made at positions marked R, and many such are possible. Arrow indicates site of beta-lactamase attack. **B,** Comparison of chemical structures of penicillin and cephalosporin. **C,** Cephalothin, an important cephalosporin.

An advantage of the cephalosporins is the general lack of cross-allergenicity with penicillin. However, the opinions expressed today are that cephalosporins should not be given to a patient with known penicillin allergy. Allergic reactions can be expected in 2% to 5% of persons receiving cephalosporins but in 5% to 16% of persons with known penicillin allergy.

Many new cephalosporins (as well as penicillins) have been developed, and many compounds are in testing stages. There are several reasons for the proliferation of cephalosporins. First, gram-negative bacteria are increasingly resistant to available drugs. Second, good antimicrobial drugs in other respects (for example, the aminoglycosides) have undesirable toxic effects. Finally, advances in chemical technics have allowed the production of new compounds with good in vitro activity. Important differences among cephalosporins result from changes in the side chains attached to the molecular nucleus, and the basic nucleus can be modified. The antibacterial spectrum of these compounds can be expanded, yet the safety of the early ones can be retained.

Aminoglycoside Antibiotics

Aminoglycoside antibiotics are a large group of compounds sharing chemical, pharmacologic, antimicrobial, and toxic features. With the exception of streptomycin, they all have the same chemical nucleus (Fig. 14-13). Their particular chemical configuration, a predominantly carbohydrate structure, contains amino sugars in a glycosidic linkage. They are polycations, and their polarity is the cause of their pharmacologic properties. Their antimicrobial action is interference with protein synthesis. In the target cell they induce a "misreading" of the genetic code at the level of the ribosomes, which results in the formation of faulty bacterial proteins. Since they are bactericidal with a broad antibacterial spectrum, aminoglycoside antibiotics represent some of the most effective antimicrobial weapons known against serious *aerobic* gram-negative bacillary infections. Obligate anaerobes, lacking oxidative transport pathways necessary for the assimilation of such drugs, are universally resistant to the aminoglycosides. Most are derived from various species of *Streptomyces.* Streptomycin was the first to be developed, but discovery of the others came also as the result of a deliberate search.

Streptomycin, an organic base, is an antibiotic that is active against gram-negative bacilli. However, its action against acid-fast bacilli is much more important. Although seldom given intravenously, it may be administered by many routes. There are three disadvantages to streptomycin: (1) toxicity—its use may induce injury to the auditory

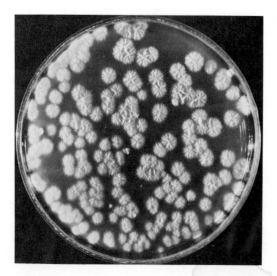

FIG. 14-12. *Cephalosporium* colonies on agar plate. Important antibiotics are derived from this fungus. (From Profile of an antibiotic, Indianapolis, 1966, Eli Lilly & Co.)

portion of the eighth cranial nerve, resulting in deafness; (2) drug resistance—bacteria sensitive to it can readily become resistant during the course of therapy; and (3) poor oral absorption—if the drug is given by mouth, it is not significantly absorbed. A unit of streptomycin corresponds to the activity of 1 μg of pure crystalline streptomycin base.

Dihydrostreptomycin, a derivative of streptomycin, is much like it. Its use also may be followed by neurotoxic symptoms, but as a rule, it is less toxic than streptomycin.

Neomycin and *viomycin*, also aminoglycosides, are similar to streptomycin. Neomycin can damage the kidney. Viomycin is sometimes used in the treatment of tuberculosis, although it is also nephrotoxic.

Kanamycin (Kantrex), a broad-spectrum aminoglycoside antibiotic, must be given parenterally. Its activity is directed against many aerobic gram-positive and gram-negative bacteria, including most strains of staphylococci and many strains of *Proteus*. The major toxic manifestation of this drug is partial or complete deafness resulting from its action on the auditory portion of the eighth cranial nerve. It is also nephrotoxic. Similar to kanamycin is *amikacin*, a semisynthetic aminoglycoside produced by acylation of kanamycin A. Its antimicrobial scope is much greater than that of the parent compound, encompassing a wide range of gram-negative rods, even *Pseudomonas aeruginosa*.

Aminoglycoside antibiotics are derived from organisms classified in the order Actinomycetales. In the name given to the antibiotic, the suffix *mycin* or *micin* indicates this origin. This part of the name of the antibiotic also refers to the parent genus in the order. Those antibiotics derived from species in genus *Streptomyces* were given the suffix

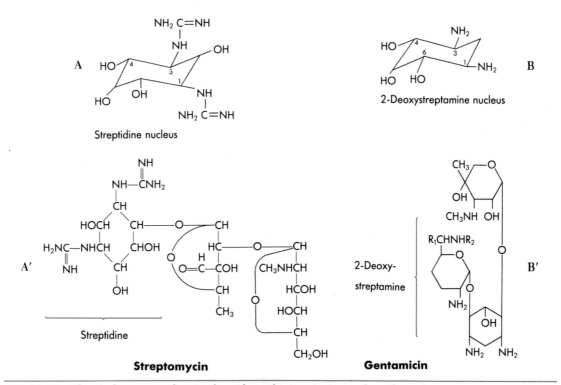

FIG. 14-13 Chemical structure of aminoglycoside antibiotics. **A,** Aminoglycoside nucleus streptidine found only in streptomycin and dihydrostreptomycin. **A′,** Chemical formula for streptomycin. Note position of chemical nucleus streptidine. **B,** Aminoglycoside nucleus 2-deoxystreptamine in all other available aminoglycosides (for example, gentamicin). **B′,** Chemical formula for gentamicin. Note position of 2-deoxystreptamine.

mycin, with the *y* (for example, kanamycin, streptomycin), whereas those obtained from genus *Micromonospora*, an actinomycete rather than a streptomycete, were given the suffix *micin*, with the *i* (for example, gentamicin).

Gentamicin (Garamycin), another broad-spectrum aminoglycoside antibiotic and chemically related to neomycin and kanamycin, can be given parenterally (Fig. 14-13). It is used in the treatment of serious systemic infections caused by gram-negative bacteria. Like the other aminoglycoside antibiotics, it is nephrotoxic and ototoxic. Loss of hair, including that of the eyebrows, has been reported with its use.

Tobramycin (Nebcin), a basic water-soluble aminoglycoside given parenterally, has a wide range of antibacterial activity and is bactericidal against staphylococci and gram-negative bacteria, including *Pseudomonas* species. It acts synergistically with cephalosporins and some penicillins and is useful in infections caused by organisms resistant to kanamycin and gentamicin. Ototoxicity, both vestibular and auditory, and nephrotoxicity are the complications.

Spectinomycin (Trobicin), another member of the streptomycin-kanamycin family, is designed for the one-dose intramuscular treatment of gonorrhea. It is also being used for uncomplicated anogenital infections of gonococcal origin. Although an aminoglycoside antibiotic, it is not considered to be nephrotoxic or ototoxic, and it also is not bactericidal.

Paromomycin (Humantin) is an aminoglycoside antibiotic that is an amebicide. Be-

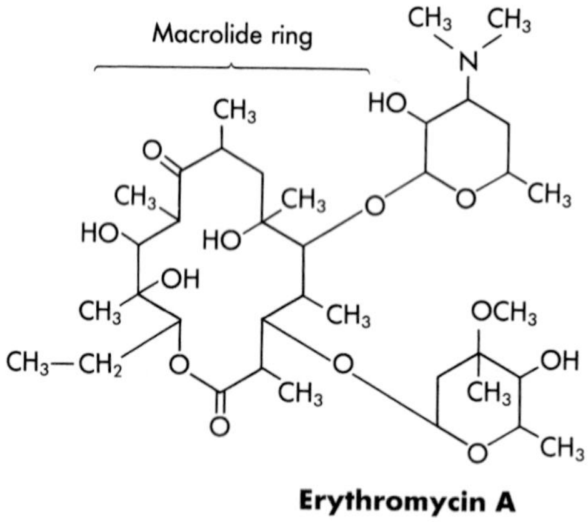

Erythromycin A

FIG. 14-14   A macrolide antibiotic.

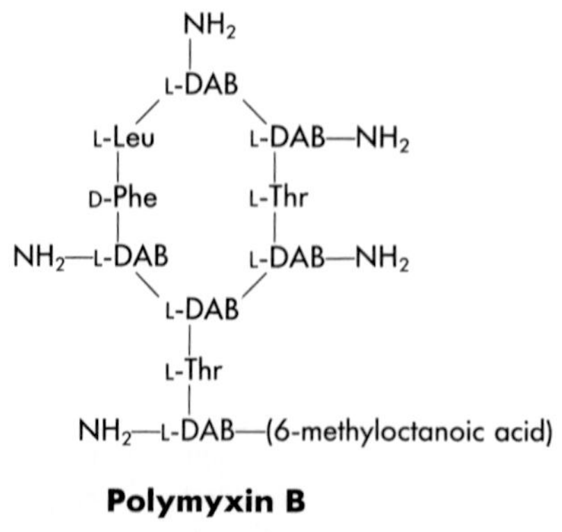

Polymyxin B

FIG. 14-15   A polypeptide antibiotic. *DAB*, α,γ-Diaminobutyric acid; *Leu*, leucine; *Thr*, threonine; *Phe*, phenylalanine.

cause of its nephrotoxicity, it is given not by injection but orally, although it is poorly absorbed from the gastrointestinal tract. It is effective against *Entamoeba histolytica* and holds back the secondary infection of the bowel that complicates amebiasis. In addition to paromomycin, the antibiotics with amebicidal activity are neomycin, kanamycin, and the tetracyclines.

Macrolide Antibiotics

A macrolide antibiotic is one containing a many-membered lactone ring to which are attached one or more deoxy sugars (Fig. 14-14). Macrolides inhibit protein synthesis by binding to specific sites on microbial ribosomes.

Erythromycin (Ilotycin), a macrolide, is an effective antibiotic, when given orally, against gram-positive and some gram-negative bacteria. It is especially recommended for the treatment of infections caused by organisms that have become resistant to penicillin. Its use is seldom followed by toxicity.

Polypeptide Antibiotics

The polypeptide antibiotics listed here are generally reserved for special purposes and used only rarely as systemic agents (Fig. 14-15). The three, polymyxin, colistin, and bacitracin, are derived from species in genus *Bacillus* and are appreciably nephrotoxic.

Polymyxin is a general name for a number of polypeptide antibiotics derived from different strains of *Bacillus polymyxa*, known as polymyxin A, B, C, and D. They are bactericidal to both growing and resting cells. The antimicrobial activity of polymyxin comes from the strong but devastating union it makes with a cell membrane, apparently by way of phosphate groups. The permeability of the membrane is so altered that the membrane can no longer function as an osmotic barrier for the cell. Polymyxin is toxic, especially to the kidney and brain. It is used locally and as an intestinal antiseptic. It is important in topical preparations, but its systemic use is hazardous. Polymyxin B is usually effective against infections with *Pseudomonas aeruginosa*; however, *Proteus* species resists it.

Colistin (Coly-Mycin), a polypeptide antibiotic isolated in Japan, is chemically similar to polymyxin. It is effective against gram-negative bacteria in general and, like polymyxin B, against *P. aeruginosa* in particular. Less toxic than polymyxin B and with fewer side reactions in the central nervous system and kidney, it is of special value in urinary tract infections. It is given intramuscularly.

Bacitracin, a polypeptide antibiotic, inhibits the growth of many gram-positive organisms, including penicillin-resistant staphylococci and the gram-negative gonococci

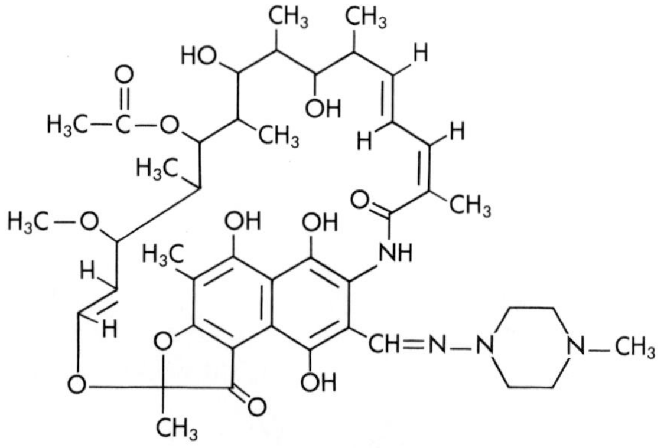

Rifampin

FIG. 14-16   The complex macrocyclic antibiotic rifampin.

and meningococci, but it can only be used locally. Bacitracin interferes with cell wall synthesis in bacteria.

Rifamycins

Rifamycins are a group of related complex macrocyclic antibotics acquired from *Streptomyces mediterranei* (Fig. 14-16). Rifampin is a derivative of one of these, rifamycin B.

Rifampin (Rifadin, Rimactane) is a semisynthetic, broad-spectrum antibiotic taken by mouth, with bactericidal activity against the tubercle bacillus. Rifampin blocks RNA synthesis by the inhibition of DNA-dependent RNA polymerase; fortunately the bacterial enzyme is more susceptible to this action than the mammalian counterpart. This drug is an important contribution to antituberculosis therapy, although resistance does develop to it. To minimize the appearance of resistant strains of tubercle bacilli, combinations of drugs, which include rifampin, are generally recommended. Rifampin eliminates carriers of the meningococcus and eradicates nasopharyngeal carriage in contacts of *Haemophilus influenzae* type b disease. It has been used in the treatment of leprosy. It has an effect against the causative chlamydiae (bedsoniae) of trachoma and inhibits replication of poxvirus by blocking synthesis of viral protein. Adverse reactions are infrequent. The urine, feces, saliva, sputum, tears, and sweat may take on an orange-red color because of therapy with the drug. It has been reported that orange tears stain soft contact lenses.

Overdosage of rifampin produces the "red-man syndrome." Several hours after ingestion the drug and its metabolic products impart a yellow-orange or orange-red to red discoloration of the skin. Since the drug and its metabolites are eliminated by the cutaneous sweat glands, the pigment can readily be removed by washing the skin.

Tetracyclines

Tetracyclines possess a common chemical moiety and, as members of a group, are alike in most respects (Fig. 14-17). They are truly broad-spectrum antibiotics that are usually taken orally. Their discovery was the result of planned experiments on soil microorganisms. Tetracyclines act to suppress protein synthesis but are essentially bacteriostatic drugs. As a group, tetracyclines sometimes irritate the alimentary tract, and they may discolor the teeth. If a tetracycline gains access to the circulation of an unborn baby because the drug is given to the mother, it is deposited in areas of developing bones and teeth. The damage results in staining of tooth enamel (Fig. 14-18). A microbe sensitive (or resistant) to one of these compounds is sensitive (or resistant) to the others. Tetracyclines are given orally to treat intestinal amebiasis.

Tetracycline is the chemical moiety common to the group. It is produced chemically from chlortetracycline but has been isolated from a natural source—a streptomycete found in Texas. Tetracycline affects gram-positive bacteria and certain gram-negative bacteria, and its antibacterial action is notable against rickettsias and chlamydiae (also

Disco jaundice, a side effect, not a disease, is a condition to which young fun-loving patients taking tetracycline for their acne are susceptible. Something in the medication causes the face to fluoresce a bright yellow when the patient is under the lights permeating discotheques. The same kind of light may be used in the doctor's office to determine whether the patient is taking the prescribed medicine. The drug leaves no residual damage, but who wants to dance with a yellow face?

bacteria) but *not* against true viruses or fungi. Adverse side effects are sometimes seen with its use.

Chlortetracycline (Aureomycin), another broad-spectrum antibiotic, is produced commercially from the natural source. It may be given orally or intravenously.

Oxytetracycline (Terramycin) is also produced commercially from the natural source. It is a comparatively nontoxic member of this group.

Demeclocycline (Declomycin), or demethylchlortetracycline, was first prepared chemically but was later found in a mutant strain of *Streptomyces aureofaciens*. In general, its antibacterial activity is similar to that of the other tetracyclines, perhaps somewhat prolonged. Cases of photosensitization (sensisivity to light) have followed the use of this particular tetracycline.

Methacycline (Rondomycin), *minocycline* (Minocin), and *doxycycline* (Vibramycin) are semisynthetic derviatives of tetracycline and comparable to it pharmacologically. Phototoxicity is also reported for doxycycline, but it is of a lesser degree than with

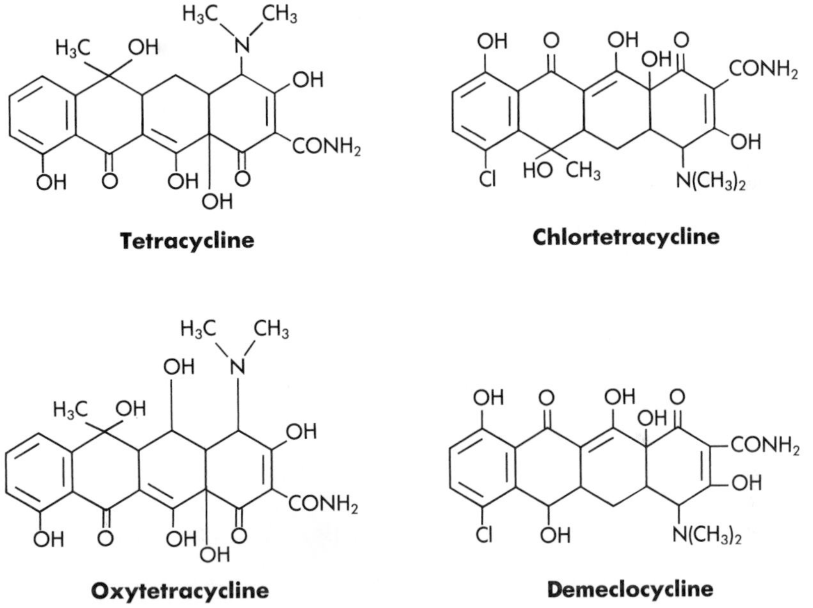

Tetracycline

Chlortetracycline

Oxytetracycline

Demeclocycline

FIG. 14-17   Tetracyclines. Note chemical kinship.

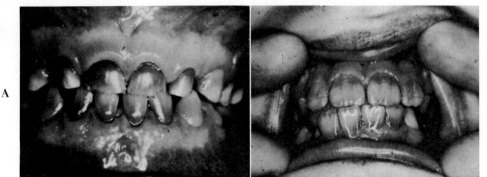

FIG. 14-18   Discoloration of teeth caused by tetracycline. **A,** Deciduous teeth (brownish gray). **B,** Permanent teeth (yellowish gray). (From Gorlin, R.J., and Goldman, H.M.: Thoma's oral pathology, ed. 6, St. Louis, 1970, The C.V. Mosby Co.)

demeclocycline. The metabolism of minocycline in the thyroid gland may yield a non-fluorescent pigment that blackens the gland.

Polyene Antibiotics

Polyene antibiotics, a functionally allied group of macrolide antibiotics, are so called because chemically they contain a large lactone ring with a conjugated double-bond system. They are synthesized by microbes in the family Streptomycetaceae. Polyene antibiotics interact with sterols in the plasma membrane of sensitive cells to alter the permeability of the membrane and to allow leakage of small molecules from the interior of the cell. Because of this affinity, they can be active against most fungi (particularly those with yeast forms), protozoa, and algae and also, unfortunately, against the cells of the mammalian host. Hemolytic anemia may be a serious side effect of therapy because of the affinity of these drugs for the cholesterol of the red cell membrane. They do not affect bacteria; their membrane, except for that of mycoplasma, does not contain sterol.

Nystatin (Mycostatin), an antifungal polyene antibiotic, is a large, conjugated, double-bond ring system linked to an amino acid sugar, mycosamine. It is used with favorable results in infections caused by *Candida albicans*. It is often given with a broad-spectrum antibiotic to suppress the growth of the resistant intestinal fungi (replacement flora) that might produce lesions in the bowel after prolonged antibiotic therapy as a superinfection. The antifungal activity of nystatin depends on its being bound to a sterol moiety in the cell membrane of sensitive fungi. Nystatin was named after the New York State Health Department, where it was discovered.

Amphotericin B (Fungizone), a broad-spectrum antifungal drug with a mechanism for antifungal activity the same as that of nystatin, is a polyene macrolide (Fig. 14-19) that binds to sterols in eukaryotic cell membranes to disrupt them. Its avidity has been demonstrated for ergosterol, the primary sterol of the fungal cell membrane, rather than for cholesterol, the primary sterol of the animal cell membrane. It is relatively specific for fungi but without effect on bacteria. The toxicity of this drug is severe, but it is the only polyene antibiotic with a level of toxicity that would permit its use in the treatment of systemic and deep-seated fungal infections, especially those involving internal organs and those widespread within the body. It should be given intravenously, and the initial infusion is regularly accompanied by a chill-fever reaction.

Other Antibiotics

Other antibiotics that do not fall into one of the above well-known categories include the following.

Lincomycin (Lincocin), a medium-spectrum antibiotic that can be administered or-

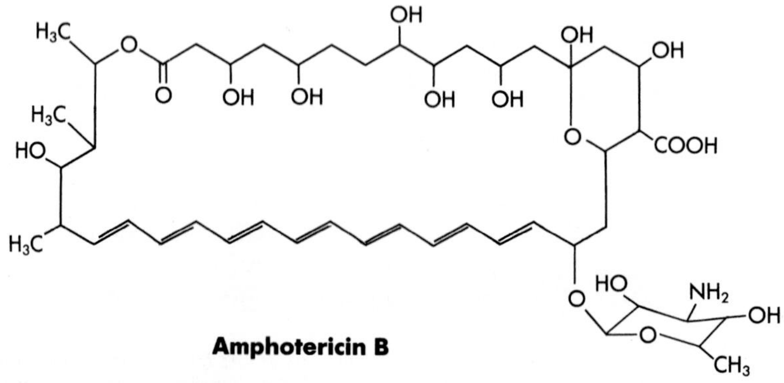

Amphotericin B

FIG. 14-19 A polyene antibiotic.

ally or by injection, inhibits the growth of gram-positive cocci, including staphylococci that produce penicillinase. It does so by suppressing protein synthesis. It binds to a specific site on the bacterial ribosome that is the precise site of action for two other antibiotics, erythromycin and chloramphenicol. No one of these three is structurally related to the other two; in fact, lincomycin differs chemically from other antibiotics in general. Although its action (not its structure) is similar to that of erythromycin, erythromycin is antagonistic to it, as are also kaolin-pectin preparations and the cyclamate artificial sweeteners formerly used in various low-calorie preparations. Lincomycin is a parent drug to clindamycin.

Clindamycin (Cleocin) may be the first antibiotic that is an antimalarial drug. An oral agent, it is particularly useful in the treatment of anaerobic infections, especially

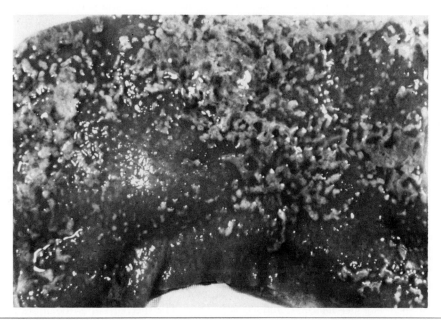

FIG. 14-20   Pseudomembranous colitis as a complication of clindamycin therapy. Note denuded mucosal surface with loss of normal markings; surface is partly covered by ragged grayish membranous exudate.

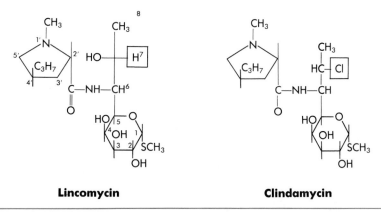

Lincomycin   **Clindamycin**

FIG. 14-21   Comparison of the chemical structure of lincomycin with clindamycin. The replacement of the 7-hydroxy group by chlorine (in clindamycin) gives a more active compound that is better absorbed.

those of *Bacteroides* species. A serious intestinal disorder, called pseudomembranous colitis (Fig. 14-20), may complicate lincomycin and clindamycin therapy. This antibiotic-associated colon lesion of characteristic appearance is secondary to the overgrowth of a clostridial species, probably *Clostridium difficile*. The cytotoxic agent produced by that species most likely relates to the lesion. Clindamycin has almost completely supplanted lincomycin because of an excellent oral absorption and the fact that absorption of clindamycin is not modified by antidiarrheal preparations (Fig. 14-21).

Chloramphenicol (Chloromycetin), the first of the broad-spectrum antibiotics, was isolated from *Streptomyces venezuelae*, found in the soil of Venezuela, but it is now crystallized (Fig. 14-22). Derived from nitrobenzene, chloramphenicol has a chemical structure unique for natural compounds. It is a bacteriostatic drug primarily; it interferes with protein synthesis. It is effective against certain aerobic and anaerobic bacteria, including rickettsias and *Bacteroides fragilis*, and is the best antibiotic for the treatment of typhoid fever. It is usually given orally, but it may be administered intramuscularly or intravenously. Two peculiar toxic disturbances of a very serious nature occasionally complicate chloramphenicol therapy. One is hematologic: blood cell formation ceases in the bone marrow, and an aplastic anemia ensues. (This occurs in approximately 1 in 40,000 patients treated.) The other is called **gray baby syndrome** to describe the cardiovascular collapse seen in babies 2 months of age and younger. Since chloramphenicol is toxic, it should not be used in those cases where tetracyclines are sufficiently potent.

Vancomycin (Vancocin), a glycopeptide antibiotic, must be given intravenously. It is a bactericidal drug promoted for the therapy of infections caused by antibiotic-resistant gram-positive bacteria, particularly by methicillin-resistant staphylococci. Its mode of action is complex; most important, it does inhibit cell wall synthesis. Ototoxicity, especially, and nephrotoxicity are complications to the use of this drug.

Capreomycin, a cyclic peptide and a new antibiotic, is related to viomycin. It is mentioned because of its activity against atypical mycobacteria and, in combination with other drugs, against tubercle bacilli resistant to standard antituberculosis drugs.

Griseofulvin (Fulvicin) is an oral antibiotic obtained from at least four species of *Penicillium* (Fig. 14-23). A *fungistatic* drug, not a fungicidal one, it has a wide range of antifungal activity against practically all the dermatophytic fungi *(Microsporum, Epidermophyton,* and *Trichophyton)*. The mechanism of the antifungal activity is unclear, although it is known that this drug is bound to cellular lipids. Young, actively metabolizing cells are vulnerable. Although it may be administered systemically, it does not

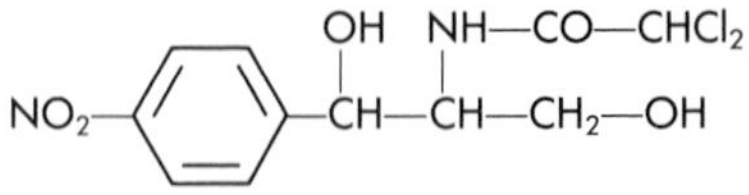

Chloramphenicol

FIG. 14-22.   Chemical structure of chloramphenicol.

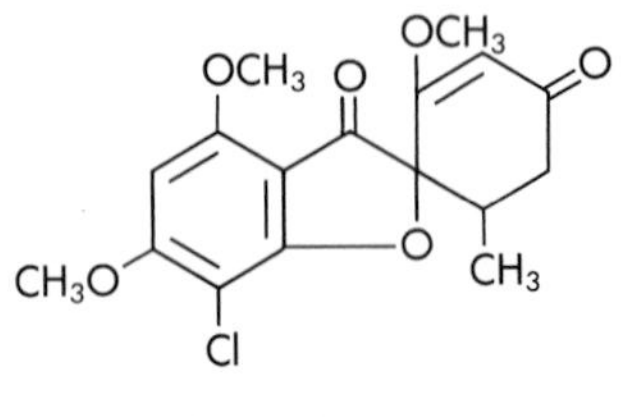

Griseofulvin

FIG. 14-23   A fungistatic antibiotic.

act on *Candida albicans*, the deep systemic fungi, or bacteria. The drug is concentrated in keratin-containing tissue of the body and can be demonstrated to attack the advancing hyphal threads of invading fungi in skin, nails, and hair, causing these segments to shrivel and curl. Toxicity seems to be low; in anticipation of an effect on the blood-forming organs, periodic white blood cell counts are advised.

ANTIVIRALS

Even after several decades, antiviral chemotherapy is still in a period of development, and antiviral agents are still to a certain extent investigational. An increased understanding of the events unique to the viral life cycle has speeded up research activity, since a potential (or actual) site of drug action may be established. Currently licensed antiviral compounds primarily inhibit DNA synthesis.

Virally coded enzymes seem to be unique in comparison with the corresponding enzymes in noninfected cells. The fabrication of compounds that affect viral enzymes is gaining momentum. Because of the enzyme specificity, this focus on viral biosynthesis represents an exquisitely selective approach.

Ideally, an antiviral agent should be stable—metabolically, thermally, and chemically. In the human body it must be soluble in the physiologic milieu and readily transported to the site of infection. (In a number of serious viral diseases this is the brain.) Acute toxicity or an immunosuppressive effect is undesirable, as are the activation of latent viruses or the emergence of drug resistance. The antiviral agent should not be integrated into the DNA of a noninfected cell. It must not be teratogenic, mutagenic, or carcinogenic. Finally, it must be cost effective.

Generally, antiviral compounds are most effective when the virus load is limited and treatment is begun early in the course of the disease. Many of the agents are topical rather than designed for internal use. This is changing, however, as compounds are being developed for use parenterally, orally, or as aerosols. Unfortunately resistant strains to the various antiviral compounds are being found.

There is a great deal of interest in antivirals, with many being studied and some being released for use. About a dozen or so have demonstrated a definite clinical usefulness. Currently four categories stand out: (1) thiosemicarbazone, (2) purine and pyrimidine analogs, (3) amantadine, and (4) interferon.

Thiosemicarbazone

In thiosemicarbazones the mechanism of antiviral action is unclear. *Methisazone*, a synthetic compound, has been shown to prevent smallpox in susceptible persons exposed in the epidemics studied in India. In vitro this drug can be a potent inhibitor of poxvirus replication. It is well absorbed when taken orally, but since the eradication of smallpox it is not used.

Purine and Pyrimidine Analogs

Purine and pyrimidine analogs, since they are incorporated into viral DNA, result in transcription errors. These analogs are "misread." They can also inhibit viral DNA polymerase.

Idoxuridine, an antimetabolite and pyrimidine analog (Fig. 14-24) that inhibits normal cellular DNA synthesis, has been found effective against eye infection caused by the herpes simplex virus. Idoxuridine represents a spectacular achievement in the treatment of herpetic keratitis.

Vidarabine, also known as adenine arabinoside or *ara*-A, is a purine nucleoside analog (Fig. 14-24) from *Streptomyces antibioticus* and a first-generation antiviral agent. It is a nonspecific inhibitor of viral replication with a wide range of activity against important DNA viruses (herpes simplex and varicella-zoster viruses) and certain animal

RNA viruses that have an RNA-dependent DNA polymerase (reverse transcriptase). Although it is said to have no acute toxicity, side effects do exist and treatment failures may occur in immunosuppressed patients. In animals this antiviral agent has been shown to be mutagenic, teratogenic, and carcinogenic.

Cytarabine (Cytosar), also known as cytosine arabinoside or *ara-C*, is a pyrimidine analog. As it was studied as an antileukemic agent in the laboratory, it was found to be an antagonist of DNA synthesis with potent antiviral properties. Both cytarabine and idoxuridine are rapidly inactivated in vivo, and both can suppress bone marrow function.

Ribavirin (Virazole) (Fig. 14-24), a synthetic triazole nucleoside, is a topical viru-*static*, not viru*cidal*, agent that in tissue culture blocks replication in a number of DNA and RNA viruses, including influenza viruses A and B. An aerosol preparation shows good activity against influenza viruses A and B and also against respiratory syncytial virus. A number of mechanisms may explain its effect on RNA synthesis, particularly those altering mRNA formation. The drug has no effect against viruses whose genomes act directly as mRNA. Ribavirin does not induce interferon, it does not affect viral attachment or penetration into host cells, and its action may not be confined to infected cells.

Acyclovir, or *acycloguanosine* (Zovirax), one of the most promising of the new antivirals with specificty for herpesviruses, was discovered in 1974 during a search for analogs of guanine that would selectively impede DNA replication. Acyclovir is a guanosine derivative containing a substitution for a sugar moiety in the guanine base (Fig. 14-24).

Virus-infected cells selectively take in this purine. It interferes with the action of the viral enzyme thymidine kinase in the infected cell, thus blocking viral replication. The degree of human toxicity is low. Acyclovir is most potent against *Herpesvirus* types 1 and 2, as well as against the varicella-zoster virus. To a lesser degree this compound is active against herpesviruses lacking thymidine kinase, such as the Epstein-Barr virus. It can be given intravenously.

A major disadvantage is that acyclovir does not eradicate dormant or nonreplicating

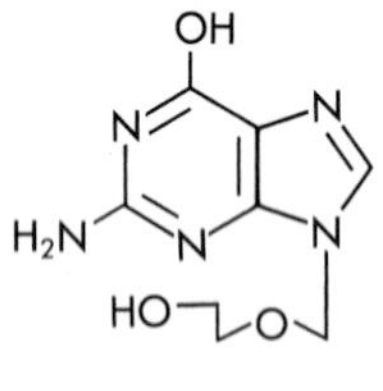

Acyclovir (purine analog) 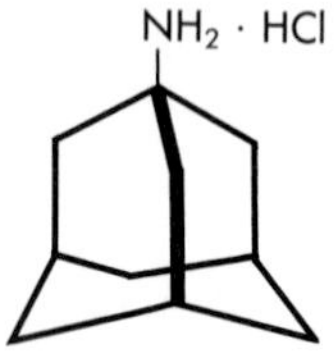   **Amantadine** (adamantine cage carboxylic derivative)

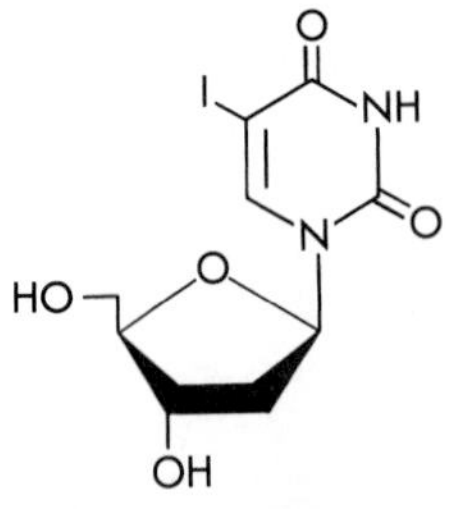

Idoxuridine (pyrimidine analog) 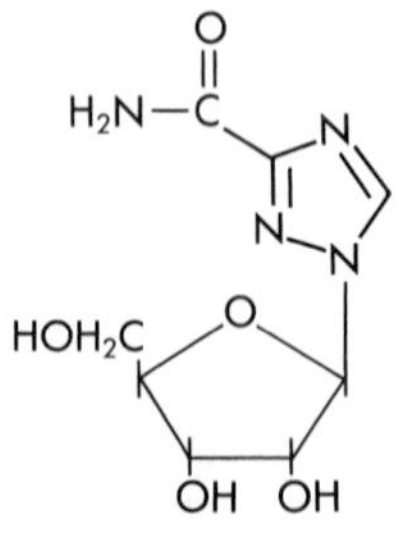   **Ribavirin** 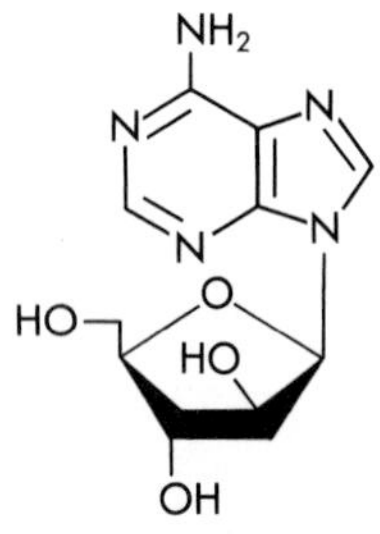   **Vidarabine** (purine analog)

FIG. 14-24   Antiviral agents.

virus, so recurrences are likely. Reports of drug resistance have indicated two possible mutations: one involving the viral enzyme thymidine kinase and the other involving an altered viral DNA polymerase.

Amantadine

The synthetic antiviral agents in the amantadine category in structure and in properties are unlike any of the other antimicrobial agents. Both compounds, amantadine and rimantadine, probably prevent infection by stopping viral activity in the early stage when the virus penetrates the cell.

Amantadine hydrochloride (Symmetrel), first reported in 1964, is a relatively safe synthetic tricyclic amine (l-adamantanamine) with a selective antiviral action (Fig. 14-24). It is licensed for systemic use. Although readily absorbed when given orally, it is not metabolized in the human body. Fully 90% of the amount of drug given can be recovered unchanged from the urine of the patient. Because of its broad-spectrum action against influenza viruses type A, even to suppress new antigenic variants, amantadine is of value in the prevention and symptomatic treatment of infections with this virus only. It has no appreciable effect against the type B viruses. It stimulates the central nervous system and in large doses can cause convulsions.

Rimantadine, an amantadine analog, is a promising new drug being used against influenza virus type A and against other RNA viruses as well.

Interferon

Interferon, a naturally occurring nonimmune protein elaborated in living virus-infected cells, was the first substance to be associated with an antiviral action and today is the antiviral agent on which most attention is focused.

Interferon (actually a class of rather small, slightly different glycoproteins) interferes with the multiplication of many viruses (hence the name) in ways that have not been fully elucidated. Over the years interferon has been made from human white blood cells, from fibroblasts, and from lymphoblastoids (transformed lymphocytes). To produce interferon from white blood cells, the leukocytes from buffy coats of blood bank blood are induced with Sendai virus. Four milliliters of donor blood yields 1 to 2 million units of interferon, but for clinical trials with humans, from 1 to 30 million units per day might be needed. For fibroblast interferon one company obtains fibroblasts from the foreskins of newborn babies, an inexpensive source. For lymphoblastoid interferon a large-scale culture of transformed lymphocytes is first stimulated with virus, after which the interferon can be harvested. However, large amounts of pure material, not available from the methods just indicated, are needed for the studies that will establish the role of interferon in the treatment of humans, and such amounts are coming from the new recombinant DNA technology. The manufacture of interferon appears to be one of the first major commercial applications of this new biotechnology. Quantities of pure material adequate for pilot studies and field applications in human disease are now ensured, and as more interferon is thus produced, the cost to treat one patient should drop from between $20,000 and $30,000 to a level of $200 to $300.

To date, in patients studied with various viral infections, interferon is definitely therapeutic against both RNA and DNA viruses. Rhinovirus infections in volunteers are effectively handled. In fact, in the respiratory tract interferon seems to possess a broad spectrum of activity. Of special note are pilot studies with interferon in patients with chronic hepatitis B infection. As would be expected, applications of pure preparations of interferon center around cytomegalovirus and other herpesvirus infections. The pure preparations are also being widely investigated in patients with a variety of cancers, with reportedly modest success at this time. The role of interferon in cancer therapy remains to be defined.

Toxic side effects with interferon appear to be dose dependent. With the high doses given in cancer therapy, side effects are seen that are comparable to those with other cancer medications. These include hair loss, nausea, lethargy, and fever. If interferon is given for antiviral therapy in the much lower doses required, toxic effects are almost nonexistent.

Interferon inducers are certain chemicals that are given to induce formation of interferon in an individual needing protection. One of the best is a synthetic, double-stranded RNA in which all the bases of one strand are inosines and all the bases of the second strand are cytosines. Sometimes called poly(I) · poly(C), the compound polyinosinic acid: polycytidylic acid, if complexed and stabilized with polylysine, forms a potent interferon inducer. The use of interferon inducers in enhancing resistance to viral infection is still largely experimental.

SUMMARY

1 Sulfonamides block normal synthesis of the essential growth factor folic acid in microbes by an antibacterial action termed *metabolic antagonism*. Microbes utilize sulfonamides by mistake because they resemble closely paraaminobenzoic acid, a key substance in the elaboration of folic acid. Mammalian cells are spared, since they must obtain their required folic acid from an external source.

2 Antibiotics are biosynthesized by bacteria, fungi, and certain plants. Thousands of species elaborate them. Hundreds of antibiotics are known, but only a few can be used. Most of the 150 antibiotics marketed today and 60% of the ones known are from the genus *Streptomyces*.

3 Antibiotics affect deleteriously certain key macromolecules and the biologic events in which they are involved: cell wall synthesis, nucleic acid synthesis, protein synthesis, and plasma membrane function.

4 The cell membrane cannot withstand the uniquely high osmotic pressure of the bacterial cell cytoplasm without a rigid cell wall. Penicillins and cephalosporins weaken the cell wall by inhibiting the cross-linking of the peptide strands in the synthesis of the cell wall peptidoglycan.

5 By inhibiting a key enzyme or mimicking a necessary ingredient, antibiotics can interfere with nucleic acid synthesis. Griseofulvin induces the formation of a faulty DNA in susceptible fungi.

6 The most frequent sites of antibiotic action are found in the complex events of protein synthesis. A "nonsense" protein that cannot support the viability of the microbe results when aminoglycosides cause mRNA to misread transfer information in the synthesis of that protein.

7 Polymyxin (a polypeptide antibiotic) embedded in the layers of the plasma membrane destroys it as an osmotic barrier. The altered membrane soon ruptures. Because of their damage to the plasma membrane, polypeptide antibiotics are bactericidal and appreciably toxic.

8 Four serious complications of antibiotic therapy are (1) allergy to the drug, (2) toxicity of the drug, (3) effect of resistant microbes replacing antibiotic-sensitive ones, and (4) induction of bacterial resistance to the antimicrobial agent.

9 Resistance is always a possibility with the administration of the many antimicrobial drugs that are only bacterio*static* in the human body. Body mechanisms must dispose of still viable organisms.

10 Widely disseminated R plasmids carry 60% to 90% of all antibiotic resistance in gram-negative pathogenic bacteria. Plasmids encode resistance to as many as six or seven antimicrobial agents, not necessarily related.

11 "Jumping genes," or transposons, are bits of linear DNA carrying plasmid genes that are so called because they seem to "jump" from one plasmid to another. Transposition is considered the basis in nature for the formation of resistance (R) plasmids.

12 Today nearly 50% of antibacterial products manufactured in the United States are for animal feeds. Most poultry, 90% of hogs and veal calves, and 60% of cattle receive small amounts of antibiotics and other drugs.

13 Important biochemical categories of antibiotics are the cyclic decapeptides, beta-lactam antibiotics, aminoglycosides, macrolides, polypeptides, rifamycins, tetracyclines, and polyenes.

14 Beta-lactamases inactivate the beta-lactam antibiotics (penicillins and cephalosporins) by breaking the beta-lactam linkage in their chemical nucleus. The best known one, penicillinase, destroys 200,000 to 400,000 units of penicillin a day.

15 Penicillin is still the unquestioned drug of choice in the nonallergic person for the treatment of infections caused by group A streptococci.

16 Both the aminoglycosides and the tetracyclines interfere with protein synthesis. The aminoglycosides are bactericidal; however, the tetracyclines are essentially bacteriostatic.

17 Rifampin, derivative of rifamycin, is a broad-spectrum bactericidal agent effective against tubercle bacilli. It blocks RNA synthesis.

18 Polyene antibiotics damage the plasma membrane through their interaction with the sterols therein. Hence they are active against most fungi but for the same reason toxic to the mammalian host cells.

19 Currently licensed antiviral drugs primarily inhibit DNA synthesis; most are still investigational. Thiosemicarbazone, purine and pyrimidine analogs, amantadine, and interferon represent four categories of such agents.

20 Interferon, a naturally occurring nonimmune protein and the first substance recognized for its antiviral action, affects both DNA and RNA viruses, interfering in an unknown way with viral multiplication. The manufacture of interferon appears to be one of the first major commercial applications of the new recombinant DNA technology.

QUESTIONS FOR REVIEW

1 Define bacteriostasis, antibiotic, R plasmids, penicillinase, interferon, cross-allergenicity, toxicity, selective toxicity, hepatotoxicity, competitive antagonism, metabolic antagonism, interferon inducer, cephamycin, pentapeptide, "nonsense" protein, superinfection, "jumping genes," synergistic drugs, semisynthetic drugs, and wild-type microbes.

2 What is an antibiotic susceptibility test?

3 Name five sulfonamide compounds used in medicine.

4 What is meant by *chemotherapy* and *chemotherapeutic agent?*

5 What are broad-spectrum antibiotics? Name five.

6 What is a narrow-spectrum antibiotic? Name three.

7 What antimicrobial agents are used in the treatment of pulmonary tuberculosis?

8 Briefly describe one antifungal agent and one amebicide.

9 Indicate four mechanisms to explain the antimicrobial activity of drugs. Give an example of each.

10 What is the current status of the antiviral agents? Name six such compounds.

11 Briefly discuss harmful side effects of antibiotic therapy. What is meant by a replacement flora?

12 Broadly classify antibiotic drugs. Give an example for each category.

13 Discuss the importance of transposition in relation to drug resistance. What is a transposon?

14 Give ways in which drug resistance in microbes may come about.

15 State the consequences of drug fastness and cross-resistance.

SUGGESTED READINGS

AMA drug evaluations, ed. 5, Littleton, Mass., 1983, Publishing Sciences Group, Inc.

Aminoglycosides: use with caution, Nursing '80 80:86, Feb. 1980.

Baron, S.: The interferon system, ASM News 45:358, 1979.

Burg, R.W.: Fermentation products in animal health, ASM News 48:460, 1982.

Cohen, S.N., and Shapiro, J.A.: Transposable genetic elements, Sci. Am. 242:40, Feb. 1980.

Davies, J.: Mechanisms of antibiotic resistance, Kalamazoo, Mich., 1980, The Upjohn Co.

Galasso, G.J.: Perspectives in antivirals, ASM News 45:353, 1979.

Gilman, A.G., Goodman, J.S., and Gilman, A.: Goodman and Gilman's the pharmacological basis of therapeutics, New York, 1980, Macmillan Publishing Co., Inc.

Goth, A.: Medical pharmacology: principles and concepts, ed. 10, St. Louis, 1981, The C.V. Mosby Co.

Kelly, W.N.: Antibiotics, superdrugs against bacteria, Diagn. Med. 2:35, May 1979.

Medical news: chemotherapeutic agents against RNA viruses: ranks swelling, J.A.M.A. 249:989, 1983.

Merigan, T.C.: Interferon—the first quarter century, J.A.M.A. 248:2513, 1982.

Sun, M.: Interferon: no magic bullet against cancer, Science 212:141, 1981.

Thornsberry, C.: Beta-lactamase-positive organisms: microbiologic methods, Lab. Manage. 75:29, Nov. 1980.

Vournakis, J.N., and Elander, R.P.: Genetic manipulation of antibiotic-producing microorganisms, Science 219:703, 1983.

White, S.J., and Williamson, K.: What to watch for when you give aminoglycosides, R.N. 42:73, Sept. 1979.

White, S.J., and Williamson, K.: What to watch for when you give a cephalosporin, R.N. 42:29, Aug. 1979.

White, S.J., and Williamson, K.: What to watch for when you give penicillin, R.N. 42:21, June 1979.

White, S.J., and Williamson, K.: What to watch for when you give tetracyclines, R.N. 42:31, July 1979.

Wilhelmus, K.R., and Jones, D.B.: What's new: acyclovir for treatment of ocular viral infections, Tex. Med. 79:27, July 1983.

CHAPTER 15

PRACTICAL TECHNICS

SURGICAL DISINFECTION AND STERILIZATION

Many different methods of sterilizing surgical instruments and supplies and of disinfecting wounds, the field of operation, and the hands of persons taking part in surgical operations are used in different hospitals and related medical or health care settings. An overview of the standard methods as evaluated in an experimental and testing laboratory is presented in Table 15-1. In the following paragraphs some of the ones commonly used are given. Note that an increasing number of commercially prepackaged, sterile, disposable items are in widespread use, with the limits being set apparently only by the cost. The advantages of disposables are obvious in most instances, and thus the use of disposables is changing to a remarkable degree the patterns of sterilization in the health care fields.

Steam Sterilization

The method of choice wherever applicable is sterilization by steam under pressure as usually carried out in a pressure steam sterilizer (preferably the modern high-prevacuum steam autoclave). Exceptional circumstances do exist in which chemical disinfection, including sterilization with ethylene oxide gas, has to be substituted for autoclave sterilization. The environment of the pressure steam sterilizer damages certain items, such as surgical instruments with optical lenses, instruments or materials made of heat-susceptible plastics, articles made of rubber, certain delicately constructed surgical instruments, and nonboilable gut sutures.

Generally, the pressure steam sterilizer (gravity air removal type) for most routine purposes of sterilization is maintained at 121° to 123° C (250° to 254° F), 15 to 17 pounds' pressure, for 30 minutes (Table 12-1, p. 213). For certain items of rubber subjected to steam sterilization the time is shorter (15 to 20 minutes). (Note the shorter time with the higher temperatures in the high-prevacuum sterilizer.)

Gas Sterilization

Ethylene oxide is widely applied today in the sterilization of many items that cannot be steam sterilized. Gas sterilization is carried out in a specially devised chamber. Table 15-2 compares this form with that of steam under pressure.

Surgical Instruments and Supplies

In Table 15-3 practical suggestions are given for the sterilization or chemical disinfection of specific instruments and supplies (see also Table 15-4). Although one chemical disinfectant may be indicated in Table 15-3 and for various purposes in this chapter, it does not follow that the one cited is necessarily the best. There are certainly many other good ones. It is imperative that the directions supplied with commercial containers of disinfectant be read carefully and that the instructions for the preparation and application of the working germicide solution be followed meticulously.

Syringes and needles contaminated by even minute traces of serum or plasma may be sterilized by one of the methods recommended in Table 15-4, but *remember, to eliminate hepatitis viruses, chemical disinfection is used only when application of physical methods is impossible.* Disposable needles, blood lancets, and syringes are available commercially, and their use is strongly recommended whenever feasible because one important method of spreading the hepatitis viruses is thereby blocked.

Note that disposable items are designed to be used only once and should never be sterilized and reused. The proper disposition of disposables must be stressed! Disposable

Table 15-1 Overview of Antimicrobial Methods

Method	Temperature Required	Minimum Exposure Time (Minutes)
Sterilization*		
Saturated steam under pressure	140° C (285° F)	0–2 (Instantaneous)
	132° C (270° F)	0–4
	121° C (250° F)	0–12
Ethylene oxide gas (12% ETO, 88% Freon-12)	54° C (130° F)	0–90
	40° C (105° F)	0–150
Hot air	160° C (320° F)	0–120 or more
Chemical sporicide (solution)	Room temperature	0–180 or more
Disinfection		
Steam—free flowing or boiling water	100° C (212° F)	0–10
Chemical germicide (solution)	Room temperature	0–12
Sanitization		
Water, detergents, or chemicals aided by physical means for soil removal	Maximum 93° C (200° F)	0–15

Modified from chart copyrighted 1967, Research and Development Section, American Sterilizer Co., Erie, Penn.
*See pp. 208 and 218 for definitions.

syringes and needles must be kept away from children and drug addicts and must not wound the hand of the garbage collector. Once used, they can be incinerated, deformed in boiling water, or crushed in a machine.

One method for safe discard of disposable syringes and needles is to place the needle inside the syringe, needle point upward toward the piston. Insert the piston into the syringe, and jam it down, locking the needle tightly between the barrel of the syringe and the plunger. The unit may then be thrown into the trash.

It is hard to render instruments with optical lenses completely free of microbes. To kill all microbes present might loosen the cement in the lens system in the process or corrode the metal in some of the instruments. A recently developed heat-resistant cement makes it possible to carry out a type of pasteurization wherein endoscopes are submerged in hot (not boiling) water at a temperature of 85° C for 1 hour. This method kills most of the ordinary bacterial contaminants. Telescopic instruments with improved cements in the lens systems can also be sterilized with ethylene oxide.

Formerly the preparation and care of rubber tubing used for intravenous medication were difficult and time-consuming tasks. The use of plastic disposable units should *completely eliminate* the need for rubber tubing for intravenous medication and blood transfusion.

Dressings and Linens (Reusable, Woven)

Towels, gowns, and other reusable articles of cloth may be sterilized by autoclaving at 15 pounds' pressure for 30 minutes. They may be arranged in packs of suitable size and wrapped in two layers of muslin or one layer of muslin and one or two layers of heavy paper.

Reusable Gloves

When an operation is finished, the surgeon and assistants should wash their gloves in cold water before removing them. After removal the gloves are washed outside and

Table 15-2 Comparison of Steam Sterilization with Gas Sterilization

	Steam	Gas
Medical indications	Syringes, instruments, drapes, gowns, some plastics, rubber	Heat-sensitive plastics, rubber, bulky equipment (ventilators, telescopic instruments*)
Equipment	Automatic chamber (large and bulky)	Automatic chamber (large and bulky); also portable model
Penetration of material to be sterilized	Rapid	Very rapid
Time required	Minutes (15 min or less in some models)	Hours (2 to 12)
Efficiency	Excellent	100%
Technical snags	Removal of air	Prevacuum; correct humidification; temperature over 70° C
Danger	None	Undiluted ethylene oxide explosive†
Hazard	Sterilization inadequate if air not completely removed	Toxic residues if aeration inadequate
Contraindications	Many plastics, sharp instruments, rubbers deteriorate	Previously irradiated plastics such as polyvinyl chloride

Modified from Rendell-Baker, L., and Roberts, R.B.: Med. Surg. Rev. **5:**10 (fourth quarter), 1969.
*Modern cements used on optical lenses of such instruments permit sterilization with ethylene oxide gas.
†The mixture of 12% ethylene oxide (ETO) and 88% Freon-12, which is generally used, is safe.

Table 15-3 Technics of Sterilization (or Disinfection) of Commonly Used Items in Health Care

Items to Be Sterilized	Autoclave (Method of Choice)	Chemical Disinfection (Choice Indicated)*	Other Technics Applicable
Surgical Instruments and Supplies, Mechanically Clean			
Noncutting instruments	Yes	1. Amphyl, 2%, 20 min 2. Activated glutaraldehyde, 10 min to 10 hr 3. Ethylene oxide gas, 2 to 12 hr	Boil completely submerged 30 min
Sharp instruments	Yes, in certain instances	1. Amphyl, 2%, 20 min 2. Activated glutaraldehyde, 10 min to 10 hr 3. Ethylene oxide gas, 2 to 12 hr	1. Bake in hot air sterilizer, 150° C, 3 hr 2. Boil 30 min (cutting part wrapped with cotton; instrument kept from tossing about; water boiling *before* instruments placed in it†)
Surgical needles	Yes, in packages	1. Amphyl, 2%, 20 min 2. Isopropyl alcohol, 70% to 90%, 20 min 3. Ethylene oxide gas, 2 to 12 hr Wash needles in sterile water	Hot air at 160° C, 2 hr
Hypodermic syringes and needles	Yes	Cold sterilization inadequate to remove hepatitis viruses	1. Dry heat, 170° C, 2 hr 2. Boil 30 min
Hinged instruments	Yes, in certain instances	1. Isopropyl alcohol, 70% to 90%, 20 min 2. Activated glutaraldehyde (2% aqueous), 20 min to 10 hr 3. Ethylene oxide gas, 2 to 12 hr	—
Transfer forceps	Yes, for the container	Activated glutaraldehyde (change solution once a week)	—
Endoscopes and instruments with optical lenses (cystoscopes and others)	—	1. Ethylene oxide gas, 2 to 12 hr (method of choice) 2. Aqueous formalin, 20%, 12 hr 3. Activated glutaraldehyde (2% aqueous) 10 hr‡	—

*Sodium nitrite added to alcohols, formalin, formaldehyde-alcohol, quaternary ammonium, and iodophor solutions prevents rusting. Sodium bicarbonate added to phenolic solutions prevents corrosion. Antirust tablets are available commercially.
†Boiling dulls knives and other sharp instruments because iron tends to enter solution in the ferrous state when heated. The site of corrosion—the part that is electropositive—is the sharp edge of the instrument. A "cathodic" sterilizer is equipped electrically so that such instruments are connected with the negative pole of a battery and corrosion prevented. This permits long-continued boiling without dulling.
‡Exposure to nos. 2 or 3 for 15 min eliminates tubercle bacilli and enteroviruses.

Table 15-3 Technics of Sterilization (or Disinfection) of Commonly Used Items in Health Care—cont'd

Items to Be Sterilized	Autoclave (Method of Choice)	Chemical Disinfection (Choice Indicated)*	Other Technics Applicable
Surgical Instruments and Supplies, Mechanically Clean—cont'd			
Polyethylene tubing	—	Immerse tubing carefully filled with disinfectant: 1. Activated glutaraldehyde, 10 hr 2. Ethylene oxide gas, 2 to 12 hr 3. Aqueous formalin, 20%, 12 hr	—
Rubber tubing, silk catheters, and bougies	—	1. Ethylene oxide gas, 2 to 12 hr (method of choice) 2. Iodophor, 500 ppm available iodine, 20 min* 3. Amphyl, 2%, 20 min* 4. Activated glutaraldehyde (2% aqeous), 15 min*	—
Inhalation equipment for anesthesia	—	1. Ethylene oxide gas, 2 to 12 hr 2. Activated glutaraldehyde (2% aqueous), 10 hr	—
Thermometers	—	1. Activated glutaraldehyde, 10 hr 2. Ethylene oxide gas, 2 to 12 hr (cold cycle only) 3. Isopropyl alcohol, 70% to 90%, plus 0.2% iodine, 15 min (to eliminate tubercle bacilli and enteroviruses)	—
Surgical Instruments and Supplies, Contaminated	Yes, wherever possible—as a rule, double exposure time, other conditions the same	Ethylene oxide gas, 2 to 12 hr	—

*To eliminate tubercle bacilli and enteroviruses, tubing must be filled.

inside with tincture of green soap, thoroughly rinsed, and dried. They are then powdered, placed in glove envelopes, and wrapped. To sterilize reusable gloves, (1) autoclave at 15 pounds' pressure for 20 minutes, (2) follow with a vacuum for 5 to 10 minutes, (3) open the door of the sterilizer slightly and allow gloves to dry in the sterilizer for about 5 minutes, and (4) remove and separate the packages to allow rapid cooling (this prevents sticking). CAUTION: Gloves may be ruined if left in the sterilizer too long.

NOTE: In some sterilizer models all these steps will not be necessary.

By putting an iodine solution in the water used to wash off glove powder and by taking advantage of the well-known color change, one can detect holes in gloves. If there is even a tiny hole, a brownish discoloration of the hand is seen after the glove is removed.

The availability and practicality of prepackaged, sterile or unsterile, disposable gloves make their use strongly recommended.

Hand Brushes

To sterilize hand brushes, (1) clean thoroughly, rinse free of soap under running water, and submerge in suitable disinfectant for 30 minutes (leave in the disinfectant until needed) or (2) autoclave (preferably) and store in a sterile dry container. (Note the availability and practicality of prepackaged, sterile, disposable hand brushes.)

Materials of Fat or Wax, Oils, and Petrolatum Gauze

To sterilize these heat-susceptible objects, place them in the sterilizing oven at 160° C for 2 to 2½ hours.

Water and Saline Solutions

The various intravenous solutions regularly used in the health care field are available commercially in sterile, sealed bottles (bags or other receptacles) designed to be used

Table 15-4 Methods of Sterilization to Eliminate Hepatitis Viruses*

Method	Time	Temperature
Physical		
Autoclave†	30 min	121.5° C (15 pounds' pressure)
Dry heat	2 hr	170° C
Boiling (active)	30 min	100° C
Chemical		
Sodium hypochlorite solutions, 0.5% to 1% (5000 to 10,000 ppm available chlorine)	30 min	
Formalin, 40% aqueous (16% aqueous formaldehyde)	12 hr	
Formalin, 20% in 70% alcohol	18 hr	
Aqueous activated glutaraldehyde, 2%	10 hr	
Ethylene oxide gas sterilization	Note manufacturer's recommendations	

*Most human viruses are destroyed at 60° C for 30 minutes, with the exception of the hepatitis viruses. At this temperature a 10-hour period is required for their inactivation.
†The autoclave is definitely the preferred method and strongly advised over the other two methods of physical sterilization. Heat sterilization is the method of choice for instruments and objects that can be so handled.

with a sterile prepackaged unit of tubing in such a way that the likelihood of contamination is minimal. At times water solutions, saline solutions, and the like must be prepared specially. They should be placed in suitable containers and autoclaved at 15 pounds' pressure for 30 minutes. If they are to be given intravenously, great care must be exercised to keep lint fibers and other particles out of the solution.

Surgical Suite

The operative field is exposed to airborne bacteria, both pathogenic and possibly pathogenic. These arise from the noses and throats of the surgical team and the occupants of the surgical suite, and they may be destroyed by ultraviolet radiation.

At the end of the operation after the patient has been removed, the operating room (or theater of action) is disinfected, a maneuver referred to as **terminal disinfection.** A phenolic derivative, iodophor, or suitable germicidal detergent combination compatible with tap water and used properly is satisfactory.

Body Sites

Hands

Many methods exist for the disinfection of the hands. The following one gives good results:
1. Remove jewelry. Wash hands and arms well with soap and water.
2. With soap and a sterile hand brush, scrub nails and knuckles of both hands.
3. Scrub hands; begin with thumbs, and in succession scrub inner and outer surfaces of thumbs and fingers; give several strokes to each area.
4. Scrub palms and backs of the hands and then forearms to the arms 3 inches above elbow for 3 minutes; rinse well with running water.
5. With a sterile orange stick (a pointed stick of orangewood used in a manicure) clean under each nail thoroughly.
6. With a second sterile hand brush, scrub hands and forearms for 7 minutes in the same manner as with the first brush.
7. Thoroughly rinse off all soap with water and then rinse hands in 70% alcohol.
8. With one end of a sterile towel, dry one hand and, with a circular motion, dry forearm to arm above elbow. With the other end of the towel, dry the other hand and forearm in the same manner.

When one is scrubbing the hands for a surgical operation, plenty of soap should be used, and a surgical scrub should be done with running water. When one is using cake soap, the soap is held between the hand brush and the palm until the scrubbing is complete. During the entire procedure the hands should be held higher than the elbows to prevent contamination.

Remember that soap dispensers should be emptied *regularly*, cleaned, and refilled with fresh solution. Cake soap, although not maintaining contamination directly (organisms inoculated on bar soap die quickly), may be left in pools of water supporting growth of microorganisms. If bar soap is desired, small bars that can be changed often and soap racks that allow drainage are recommended. Because of such problems, many health care centers prefer an expendable form of dry soap such as the soap leaflets or soap-impregnated paper tissues (more expensive items, however).

At present the methods of disinfection of the hands are shortened by the habitual use of detergent soaps containing germicidal substances. The following method, known as the Betadine (p. 227) Surgical Scrub regimen, is widely used today. Because some members of the surgical team, as well as some patients, are allergic to iodine, pHisoHex (p. 227) can still be found in a medical setting.
1. Remove jewelry. Clean under nails. (Keep fingernails short and clean.)
2. Wet hands and forearms with water.

3. Apply to hands two portions, about 5 ml, of Betadine Surgical Scrub from a dispenser. (Depress Betadine dispendser foot pedal twice.)
4. Without adding water, wash hands and forearms thoroughly for 5 minutes. Use a hand brush if desired.
5. With a sterile orange stick, clean under nails. Add a little water for copious suds. Rinse thoroughly with running water.
6. Apply 5 ml from the dispenser to the hands. Repeat steps just given. Rinse with running water.

Actual sterilization of the hands is not possible. Microbes live not only on the surface of the skin but also in its deep layers, in ducts of sweat glands, and around hair follicles. The resident microbes are chiefly nonpathogenic staphylococci and diphtheroid bacilli and are plentiful about the nails (see box below).

Site of Operation

Although the deeper portions of the skin cannot be sterilized, enough bacteria can be removed to make infection unlikely. The selection of an antiseptic for preoperative skin preparations has had the attention of surgeons since the earliest times, and many different ones have been used.

The operative site may be prepared several hours ahead of time. It is scrubbed thoroughly with a suitable detergent soap or green soap solution, and the area is shaved. However, many authorities object to the skin shave being done before the patient is taken to the surgical suite. They state that the best time is right before the surgical procedure, in the operating room, after the patient has been put to sleep and positioned. If lather from a dispenser and a new razor blade are used, there is no danger of cross-infection, and the lather traps the hair. When the skin shave is done many hours before surgery, there is a good chance bacteria will grow in the inflammatory reaction about small razor nicks. The incision the surgeon makes is then likely to go through minute abscesses that have formed in the interim.

The following outline using the Betadine Surgical Scrub represents a satisfactory stepwise method of skin preparation. Although it calls for a given iodophor, a suitable

Bacteria on the skin of the human body may be either *transient* microorganisms or the *resident* flora. The transient population found on exposed surfaces may be either free on the skin or loosely bound thereon by fatty materials, such as grease. Transient bacteria are variously and unpredictably disease- or nondisease-producers. Generally, the residents are different, making up a relatively stable population of nondisease-producers. However, they can be modified by prolonged exposure to certain contaminating influences. The origin of the residents is unclear.

The surgical scrub of the hands and skin of the operative site is designed to deal with bacteria, transient or resident, that are found on the skin. Three good reasons exist for doing the surgical scrub: (1) gross dirt and grease must be removed to eliminate as completely as possible the transient microbes; (2) the overall microbial population of the skin must be reduced to the greatest extent possible, in keeping with the principles of aseptic technic; and (3) a residual germicidal action of several hours' duration must be laid down. When the microbial population is reduced greatly and a residual germicidal action is present, the possible deleterious effects of the resident flora are eliminated or minimized.

detergent soap or green soap solution may be used, and the method can be modified as indicated.

1. Shave field of operation. Wet with water.
2. Apply Betadine on a sponge to the umbilicus, if skin of the abdomen is beng scrubbed, and discard sponge.
3. Scrub the indicated line of incision with Betadine on a sponge for 1 minute; use a rotary motion.
4. Carry the same sponge out from the line incision, and scrub outer portions of the surgical area for 1 minute; again use a circular motion and discard the sponge.
5. Repeat this entire manuever five separate times. When skin area is prepared for operation, the indicated line of incision is never painted with the same sponge more than once.
6. Use a little water to make suds. Use wet sterile gauze to rinse surgical area.
7. Pat dry with sterile towel.
8. Paint area with Betadine. Allow to dry. The surgical area is now ready for surgical drapes.

Operating room personnel and those who prepare patients for operations should not let the fact that many surgical infections may be controlled by antibiotics lull them into complacency with careless technics that should be strictly performed antiseptic or aseptic ones.

Other Areas of Skin

For rapid sterilization before routine hypodermic injections and venipunctures for intravenous medication, 70% to 90% ethyl or isopropyl alcohol or 2% iodine in 70% alcohol is effective when applied to a clean area of skin for 30 to 60 seconds. When lumbar puncture is to be done, the overlying skin should be cleaned with soap or detergent, rinsed, dried, and disinfected with a suitable preparation.

Mucous Membranes

Absolute disinfection of mucous membranes is impossible. The mouth and throat may be cleaned with Dobell's solution, hydrogen peroxide, hexylresorcinol, or cetylpyridinium chloride, 1:2000 to 1:10,000.

DISINFECTION OF EXCRETA AND CONTAMINATED MATERIALS FROM INFECTIOUS DISEASES

A patient with a communicable disease is not properly supervised until every avenue by which the infectious agent may be spread from the patient to others has been closed. These avenues are not closed until all excreta and objects that convey infection have been properly disinfected. A patient who is being cared for in such a manner that all avenues by which the infection may spread to others are closed is said to be **isolated**. A ward patient with a conscientious and capable nurse is more strictly isolated from the microbiologic point of view than is that patient in a room with plastered walls and airtight doors who has a careless or incompetent nurse. Isolation refers to avenues of infection, not to walls and doors. *Remember that pathogenic microbes are most virulent when first thrown off from the body.*

Disinfection in infectious diseases is of two types: concurrent and terminal. **Concurrent disinfection** means immediate disinfection of the patient's excreta or of objects that have been contaminated by the patient or the excreta. **Terminal disinfection** means the final disinfection of the room, its contents, and environs after it has been vacated. The final chapter is written when it has been proved that patient and attendants are free of the agent that caused the disease; that is, they have not become carriers. In the following paragraphs are methods for the disinfection of excreta, materials, and objects by which infection is most often transmitted from the sick to the well.

Hands

Unless properly disinfected, the hands of nurses, physicians, or other medical attendants in contact with the patient are almost certain to transfer the infectious agent to themselves or to others. Although disinfection of hands to prevent the spread of communicable diseases is not so time consuming as for a surgical operation, it should be just as conscientiously done. The hands and forearms are submerged in a basin of suitable disinfectant, such as iodophor-containing preparations, and rubbed vigorously for 2 minutes. Ethyl (or isopropyl) alcohol, 70%, can be combined with a quaternary ammonium compound into an emollient mixture, effective and less irritating to the hands. Afterward the hands are washed with soap and running water (and with good mechanical friction). A hand brush is not necessary for scrubbing the hands and forearms, although in some institutions it is required. The clean hands and forearms should be dried thoroughly. Hand lotion is advisable to keep the skin soft and in good condition. The iodophor-containing preparations and other disinfectants tend to dry the skin excessively. The nails should be kept clean at all times. The hands should be cleaned in this way every time they contact the patient for any purpose whatsoever. To allow one's hands to contact a person suffering from a communicable disease and then, without disinfection, to allow them to contact another person is little short of criminal neglect.

Soiled Linens and Clothing

All linens and clothing in contact with the patient should be considered contaminated and kept in the patient's room until ready for final disposal. In the hospital they may be wrapped in sheets or placed in pillowcases and put into a special bag, with care being taken that the outside of the bag does not become contaminated. Before the bag is full, it is closed tightly, marked to indicate that it contains infectious material, and sent to the laundry. On reaching the laundry, the bag and contents are put directly into the washer. The washer is then closed, and live steam is introduced for 15 to 20 minutes. After this the regular laundering procedure is carried out. In the home the linen or bedclothes are bundled and carried to the kitchen or home laundry to be placed in a boiler containing warm water and soap suds and boiled from 10 to 15 minutes. After boiling, the process of laundering may be completed in the home or by a commercial laundry. On occasion the boiler is carried to the patient's room to receive the soiled linen.

Mattresses, pillows, and blankets may be sterilized by heat in special sterilizing chambers, or they may be sterilized in a formaldehyde chamber. They also may be dried in the sunlight for an exposure period of 24 hours. Mattresses and pillows are best covered with impervious plastic covers. Such plastic covers may be laundered readily or cleaned and disinfected with a germicidal detergent.

Shoes

Disinfection of shoes is best accomplished with a pledget of cotton saturated with formalin placed in each shoe; the shoes are then kept in a closed box for 24 hours. Shoes may also be disinfected by being sprayed with formalin from an atomizer for three successive evenings.

Feces and Urine

Where connections exist to modern sewage disposal systems with adequate chlorination procedures, feces and urine may be disposed of in the commode without disinfection. Otherwise, feces and urine, especially from patients with typhoid fever, viral hepatitis, and similar disorders, must be disinfected. Disinfection may be carried out with either 5% phenol or, preferably, chlorinated lime. The volume of disinfectant is three times that of the material to be disinfected, and feces should be thoroughly broken up to ensure even penetration. After the urine and feces are thoroughly mixed with the disinfecting solution, the mixture is allowed to stand for at least 2 hours before disposal. Contaminated colon tubes and such may be disinfected in Amphyl, 2% solution, for 20 minutes.

Discharges from Mouth and Nose

Discharges from the mouth and nose are received on squares of paper tissue and placed in a paper bag pinned to the side of the bed. After the bag is wrapped in several layers of clean newspaper, it is burned in the incinerator.

Sputum

Sputum must not be allowed to dry and thus permit particles to float off in the air. It may be received in a paper cup held in a special metal holder. The paper cup can be discarded and the metal holder sterilized at least twice daily (more often if necessary). One way to discard a paper cup is to remove the cup from the metal holder with forceps and place it on several thicknesses of newspaper. Then fill the cup with sawdust (or like material) to decrease the amount of moisture and make it burn more readily. Wrap paper around the cup, and tie the wrapping securely; next place the packet in the incinerator or in a special receptacle to be carried to the incinerator later. After this has been done, autoclave the metal holder or place it in boiling water for 30 minutes.

Sputum also may be received into a waxed paper cup with a tight-fitting lid that can be secured before the cup is discarded. Ambulatory patients often use collapsible cups.

Clinical Thermometers

A clinical thermometer must be sterile each time it is used. After a thermometer is used and the temperature recorded, the thermometer must be washed with soap or detergent and water (with friction applied). It may be disinfected as indicated in Table 15-3.

After sterilization a thermometer is wiped with sterile cotton and placed in a sterile container for later use. Preferably it is stored in alcohol, in fresh-daily solutions of 0.5% tincture of iodine, or in an aqueous iodophor, 150 ppm available iodine. Thermometers are rinsed in cold water before use.

For each patient with hepatitis there should be a thermometer restricted to use only by that patient, and it should be destroyed when that patient leaves the hospital.

Eating Utensils

Dishes and eating utensils may be collected in a special covered container placed in the sickroom. If possible, the patient should be the one to put such items into that container. Care must be exercised to prevent contamination of the outside of the container. Leftover bits of food may be placed in a paper sack, wrapped well with newspaper, and burned in the incinerator. If this is not practical, the food not eaten by the patient is disinfected in chlorinated lime or phenol solution for 1 hour. Afterward the container with the dishes and eating utensils is carried to the kitchen, where the container and contents are boiled in soapy water for 5 minutes.

Terminal Disinfection

When a patient is removed from a room, the floors, walls, woodwork, and furniture should be scrubbed with soapsuds and thoroughly aired. The phenol derivatives are valuable and widely used in terminal disinfection. Other good housekeeping disinfectants are 1% sodium hypochlorite and Wescodyne or another iodophor, 500 ppm available iodine.

DISPOSAL OF CONTAMINATED MATERIALS IN THE DIAGNOSTIC LABORATORY

In the clinical diagnostic laboratory, in the course of testing for infectious disease, not only the suspect specimen from the patient, but also the laboratory supplies, items of various sorts, parts of equipment, and the like that physically contact that specimen must be carefully handled and disposed of. Table 15-5 gives recommendations for safe disposal of such infectious materials.

DISINFECTION OF ARTICLES FOR PUBLIC USE

The woodwork, floors, sinks, basins, and toilets of public rest rooms should be washed frequently with hot soapsuds or detergent and rinsed thoroughly. Phenolic disinfectants or one of the iodophors are suitable for disinfection.

The woodwork of schoolhouses, churches, theaters, or other places of public assembly should be washed frequently with a germicidal detergent solution.

All parts of public conveyances touched by the hands of passengers should be frequently washed with a germicidal detergent solution.

Telephones in the clinical laboratory where infectious materials are handled and in other comparable areas should be disinfected on a regular basis.

Table 15-5 Disposal Procedures for Disposable Items and Specimens in the Clinical Laboratory

Disposable Item or Specimen	Use Special Cardboard Box for Incineration	Use Plastic or Discarded Glass for Incineration	Flush into Sewer System	Autoclave, then Discard	Incinerate
Needles, razor blades, scalpel blades	X				
Syringes		X			
Cotton balls					X
Gauze and cotton swabs					X
Pipettes		X			
Capillary tubes		X			
Residual reagent or diluent			X		
Other disposable glassware		X			
Counterimmunoelectrophoresis plates, immunodiffusion plates				X	
Paper, trash					X
Culture plates				X	
Sputum				X	
Urine			X		
Stools					X
Blood, serum					X
Specimen tubes		X			
Vomitus				X	
Gastric washes				X	
Bronchial fluid				X	
Sample cups from automated and semiautomated equipment	X				
Cerebrospinal fluid				X	

FUMIGATION OF ROOMS AND DISINFECTION OF AIR

Fumigation

In former years it was an almost universal procedure to "put a sickroom into fumigation" as soon as the room was vacated. Fumigation at that time meant terminal gaseous disinfection with formaldehyde, the gas of choice. This gaseous method of disinfection has been abandoned because it does not kill insects, small animals, or pathogenic microbes. The scrubbing of walls and floors to clean them thoroughly has replaced fumigation in the control of communicable disease; natural drying of surfaces effects the destruction of bacteria.

The term *fumigation* now denotes the destruction of disease-carrying animals, insects, or vermin. The fumigant of choice is hydrocyanic acid, a deadly poison that must be handled with extreme care. Sulfur dioxide is a splendid insecticide, but because of its deleterious effect on metals and household goods, it is seldom used in homes, public buildings, or hospitals.

Air Disinfection

Two methods of destroying bacteria and other disease-producing agents as they are dispersed through the air are ultraviolet radiation and chemical disinfection by means of aerosols.

Aerosols are chemical vapors liberated into the air for disinfection. Two of the most potent are propylene glycol, effective in a dilution of 1 part in 100 million parts of air, and triethylene glycol, effective when diluted to 1 part in 400 million parts of air. Glycols in the vapor state are highly soluble in water. Airborne bacteria, suspended in aqueous droplets that absorb the glycol vapors, are killed when a bactericidal concentration of the glycol is reached in the drop. Propylene glycol and triethylene glycol are also virucidal. Aerosols may be introduced into rooms through ventilating systems, bombs, or special equipment. The importance of thorough cleaning and disinfection of air conditioning units must be kept in mind.

STERILIZATION OF BIOLOGIC PRODUCTS

In the preparation of bacterial vaccines it is necessary to kill the bacteria at as low a temperature as possible, since a temperature of more than 62° C coagulates bacterial protein and renders the vaccine worthless. Any of the ordinary bacteria used in the manufacture of vaccines are killed when exposed for 1 hour to a water bath temperature of 60° C. Bacterial vaccines are sometimes sterilized by the addition of chemicals such as phenol or cresol. Mercury-vapor lamps as a source of ultraviolet radiation are used in the preparation of bacterial as well as viral vaccines, but ultraviolet radiation, ideal in theory, is practically ineffective at times. Ethylene oxide gas is also used for this purpose.

Hepatitis viruses may be eliminated in pooled plasma or serum by exposure to ultraviolet light followed by the use of beta-propiolactone (Fig. 15-1), 1.5 g/liter of material.

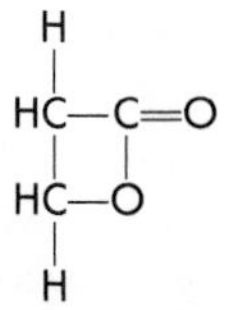

```
         H
         |
       HC—C=O
         |   |
       HC—O
         |
         H
```

Beta-propiolactone

FIG. 15-1.   Beta-propiolactone, an agent used in sterilization of biologic products.

SUMMARY

1 Sterilization by steam under pressure as usually carried out in a pressure steam sterilizer (preferably the modern high-prevacuum steam autoclave) is the preferred method whenever applicable.

2 The expanding use of disposables, limited apparently only by the cost, is changing the patterns of sterilization in health care fields.

3 The use of disposable needles, blood lancets, and syringes is strongly recommended to block one important method of spread of hepatitis viruses.

4 Disposable items are to be used *only once*; they cannot be sterilized and reused.

5 Directions supplied with commercial containers of disinfectant should be read carefully and the instructions followed meticulously.

6 To eliminate hepatitis viruses, chemical disinfection is used only when physical methods cannot be applied.

7 An appropriate method for the safe discard of disposable syringes and needles should be followed. These items can be incinerated or mechanically deformed.

8 The surgical scrub of the hands and skin of the operative site is designed to reduce maximally the microbial population of the skin and to lay down a residual germicidal action of several hours' duration.

9 Even though actual sterilization of the hands and skin is not possible, since microbes live in the deeper, as well as the upper, layers of the skin, technics of scrubbing should always be meticulously followed.

10 Disinfection in infectious disease is concurrent and terminal. *Concurrent* means immediate disinfection of the patient's excreta or of objects contaminated by either the patient or the excreta. *Terminal* means the final disinfection of the patient's room, its contents, and environs after the patient is gone.

11 Disinfection of hands to prevent spread of communicable disease may be accomplished by submerging the hands and forearms in a basin of suitable disinfectant, rubbing briskly for a few minutes, and then washing the hands with soap and running water, again with good mechanical friction.

12 Discharges from a patient are preferably handled so they can be burned.

13 Either 5% phenol or chlorinated lime can be used for the disinfection of feces or urine from a patient with enteric disease.

14 Articles used by the public should be kept clean with the use of germicidal detergent solutions.

15 Procedures for disposal of certain items and specimens in the clinical laboratory include handling in a way that allows incineration, flushing into the sewer system, or autoclaving before discarding.

QUESTIONS FOR REVIEW

1 How may knives be sterilized?

2 What is the preferred method of sterilization for endoscopes? What is meant by pasteurization of endoscopes?

3 List three surgical items easily damaged by heat.

4 Give a method of preparing the skin for operation.

5 How may the air be kept free of pathogenic bacteria during an operation?

6 What is the difference between physical and microbiologic isolation?

7 How would you dispose of the feces and urine of a patient with typhoid fever?

8 Discuss the drugs used in wound disinfection.

9 Explain what is meant by concurrent disinfection; terminal disinfection.

10 What precautions should be taken in handling sputum from a patient with tuberculosis? How is sputum sterilized?

11 How are clinical thermometers sterilized?

12 Briefly discuss the use of hydrocyanic acid as a fumigant.
13 How may hepatitis viruses be eliminated?
14 How are biologic products sterilized?
15 Compare steam sterilization with gas sterilization.
16 Briefly discuss the disposal of contaminated materials in the diagnostic laboratory.

SUGGESTED READINGS

Barrett-Connor, E., and others: Epidemiology for the infection control nurse, St. Louis, 1978, The C.V. Mosby Co.

Beck, W.C.: Places to get information, AORN J. **25**:1392, 1977.

Borick, P.M., editor: Chemical sterilization, Stroudsburg, Penn., 1973, Dowden, Hutchinson & Ross, Inc.

Decontamination of CPR training mannequins, Morbid. Mortal. Weekly Rep. **27**:132, 1978.

Duckworth, J.K.: Clinical laboratory precautions against viral hepatitis, Pathologist **30**:412, 1976.

Fox, M.K., and others: How good are hand washing practices? Am. J. Nurs. **74**:1676, 1974.

Lauer, J.L., and others: Decontaminating infectious laboratory waste by autoclaving, Appl. Environ. Microbiol. **44**:690, 1982.

Pavlech, H.M.: Reusable laboratory glassware or disposable glassware, ASM News **40**:275, 1974.

Perspectives on the control of viral hepatitis, type B, Prepared by the Committee on Viral Hepatitis of the National Academy of Sciences–National Research Council and the Public Health Service Advisory Committee on Immunization Practices, Pathologist **30**:405, 1976.

Price, P.B.: The bacteriology of normal skin: a new quantitative test applied to a study of the bacterial flora and the disinfectant action of mechanical cleansing, J. Infect. Dis. **63**:301, 1938.

Rhodes, M.J., and others: Alexander's care of the patient in surgery, ed. 6, St. Louis, 1978, The C.V. Mosby Co.

Simmons, B.P.: Guidelines for the prevention and control of nosocomial infections—antiseptics and handwashing, Washington, D.C., 1981, U.S. Department of Health and Human Services.

LABORATORY SURVEY
OF MODULE THREE

PROJECT Effects of Physical Agents on Bacteria

Part A—Sterilization of Filtration with Demonstration by Instructor and Discussion

1. Method of filtration using either the Berkefeld or Seitz filter
2. Filtration of a test solution, which can be a broth culture of some nonpathogenic organism diluted 1:20 with sterile isotonic saline solution
3. Cultures taken before and after filtration of test material
4. Observations made on cultures after incubation period of 24 hours

Part B—Effects of Heat on Bacteria

1. Prepare five suspensions each of *Bacillus subtilis* and *Escherichia coli* by placing 1 or 2 ml of the respective bacterial suspension furnished by the instructor in sterile cotton-plugged test tubes.
2. Treat a properly labeled tube of each suspension as follows:
    a. Autoclave at 15 pounds' pressure for 15 minutes.
    b. Stand in a pan of boiling water for 30 minutes.
    c. Expose to free-flowing steam in an Arnold sterilizer for 1 hour.
    d. Stand in a pan of water at 60° C for 1 hour.
    e. Use the fifth pair as a control.
3. Transfer a loopful of the bacterial suspension from each tube to a culture tube of broth and incubate at 37° C for 24 hours.
4. Observe growth in cultures made. Record growth obtained.

Exposure to Heat	Growth Obtained	
	Bacillus subtilis	*Escherichia coli*
Autoclave, 15 pounds' pressure, 15 min		
Boiling water, 30 min		
Free-flowing steam, 1 hr		
60° C, 1 hr		
Control		

5. Which tubes of the heated suspension of *Bacillus subtilis* show growth? What does this indicate? What is the difference between the growth of *Bacillus subtilis* and that of *Escherichia coli*? What does this indicate? What part does spore formation play in this experiment?

Part C—Effects of Boiling

1. Obtain test specimens of bacteria.
    a. Dip three sterile cotton swabs in a 5-day broth culture of *Bacillus subtilis*.
    b. Dip three sterile cotton swabs into a 24-hour broth culture of *Escherichia coli*.

2. Expose test specimens to the effects of boiling water for 1 minute.
   a. Hold a swab from each suspension in a separate beaker of boiling water for 1 minute, and then dip into a tube of nutrient broth.
   b. With the other swabs, repeat the procedure, using periods of 5 and 10 minutes.
   c. Incubate the six tubes of broth 24 to 48 hours.
   d. Indicate growth obtained in the following table.

Organism	Period of Boiling		
	1 Min	5 Min	10 Min
Bacillus subtilis			
Escherichia coli			

   e. What do the foregoing findings indicate?

Part D—Operation of Autoclave with Demonstration by Instructor and Discussion

1. Discussion of the autoclave or pressure steam sterilizer given on pp. 210-212
2. Demonstration of the parts of the instrument and its operation
3. Direct supervision of students as they operate the autoclave
   CAUTION: An autoclave or pressure steam sterilizer is filled with steam under pressure; therefore, if the pressure gets too high, there is always danger of an explosion.

PROJECT   Effects of Chemical Agents on Bacteria

Part A—Effects of Chemical Disinfectants

1. Obtain from the instructor four sterile cotton-plugged tubes containing the following:
   a. 5 ml of 70% isopropyl alcohol
   b. 5 ml of 5% phenol or phenolic disinfectant
   c. 5 ml of some widely advertised antiseptic (one or several may be used)
   d. Broth suspension of staphylococci
2. Proceed as follows:
   a. To each tube of disinfectant, add 0.5 ml of the bacterial suspension. Do not let the broth run down the side of tube. Mix by gently tapping the bottom of the tube.
   b. Place in water bath at 20° C.
   c. At the end of 5, 10, and 15 minutes, transfer a loopful of material from each tube of disinfectant to a tube of sterile broth.
   d. Label the tube, noting the organism and the time held in water bath, and incubate 48 hours.
   e. Record growth on the accompanying chart, using a plus sign to indicate growth and a minus sign to indicate absence of growth.
3. Repeat the preceding experiment using *Bacillus subtilis* or another spore-forming organism instead of staphylococci. Record on chart.

Disinfectant	*Staphylococcus*			Spore-former		
	5 Min	10 Min	15 Min	5 Min	10 Min	15 Min
Isopropyl alcohol, 70%						
Phenol or phenolic derivatives, 5%						
Commercial disinfectant						

Part B—Susceptibility of Bacteria to Penicillin

NOTE: Students may work in groups of three or four for this experiment.

1. Obtain from instructor the following:
   a. Petri dish of trypticase soy agar that has just been inoculated with an organism susceptible to the action of penicillin
   b. Petri dish of trypticase soy agar that has just been inoculated with an organism *not* susceptible to the action of penicillin
2. Use sterile forceps to drop filter paper disks* impregnated with different strengths of penicillin onto the inoculated surface of the agar plates. Place the disks about 2 cm apart.
3. Invert plates, and incubate for 24 to 48 hours.
4. Note the presence or absence of bacterial growth around the filter paper disks.

Part C—Susceptibility of Bacteria to Other Antibiotics

NOTE: Students may work in groups of three or four.

1. Obtain from the instructor the following:
   a. Petri dish of trypticase soy agar that has just been inoculated with bacteria susceptible to some of the antibiotics to be used and *not* susceptible to other test antibiotics
   b. Disks of filter paper* impregnated with different antibiotics to be used.
2. Use sterile forceps to place the antibiotic disks on the surface of the agar about 2 cm apart.
3. Invert and incubate at 37° C for 24 to 48 hours.
4. Examine and record which antibiotics inhibited bacterial growth and which did not.
5. Perform a susceptibility test on an agar plate inoculated with *Pseudomonas aeruginosa*, using a number of different antibiotics.
6. Note the resistance exhibited by this organism.

PROJECT   **Testing the Efficiency of Sterilization**

NOTE: Students may work in groups of two or more to test the efficiency of some of the sterilizing processes outlined in this book.

Part A—Test of the Effectiveness of Methods for Disinfecting the Hands

1. Lightly touch the tips of the fingers of the recently disinfected hand to the surface of a Petri dish of nutrient agar.
2. Invert the dish and incubate. Examine the growth at the end of 24 hours.
3. Set up a control. Rub the palm of one hand before washing or disinfection with a moistened sterile cotton swab. Inoculate the swab onto sterile nutrient agar. Incubate for 24 hours and observe growth. Make comparisons.

Part B—Test of the Sterility of Recently Sterilized Solutions

1. Use a sterile pipette to remove a few drops of isotonic saline solution from a recently sterilized flask. (Sterile distilled water or commercially packaged solutions may be used.)
2. Add the test solution to a tube of nutrient broth. Incubate for 24 hours and examine for growth.

Part C—Test of the Sterility of a Clinical Thermometer

1. Dip the tip of a recently disinfected thermometer into a tube of nutrient broth.
2. Incubate the broth for 24 hours and examine for growth.

---

*Disks impregnated with graded amounts of penicillin or other antibiotics may be obtained from commercial sources.

EVALUATION FOR MODULE THREE

PART I

In the following exercise please place in the blank space the letter from column Beta that represents the one preferred method of sterilizing or disinfecting each item named in column Alpha.

COLUMN ALPHA—Task

_____ 1. Transfer forceps
_____ 2. Dressings and linens
_____ 3. Silk catheters and bougies
_____ 4. Cystoscopes
_____ 5. Noncutting instruments
_____ 6. Hypodermic syringes and needles
_____ 7. Hand brushes
_____ 8. Sharp cutting instruments
_____ 9. Inhalation equipment
_____ 10. Water and saline solutions
_____ 11. Shoes
_____ 12. Sputum from patient with tuberculosis
_____ 13. Feces and urine
_____ 14. Air of nursery for newborn infants
_____ 15. Rectal thermometers
_____ 16. Hands of communicable disease nurse
_____ 17. Bacterial vaccines
_____ 18. Glassware
_____ 19. Inoculating wire loop
_____ 20. Blood-smeared tabletop
_____ 21. Petrolatum gauze
_____ 22. Furniture
_____ 23. Floors, walls, and woodwork
_____ 24. Endoscopes
_____ 25. Mucous membranes
_____ 26. Plastic parts of heart-lung machine
_____ 27. Polyethylene tubing
_____ 28. Suture needles
_____ 29. Eating utensils
_____ 30. Skin site of operation
_____ 31. Powders
_____ 32. Hinged instruments
_____ 33. Contaminated surgical instruments
_____ 34. Plastic Petri dishes

COLUMN BETA—Procedure

(a) Pressure steam sterilization (in an autoclave)
(b) Chemica disinfection
(c) Dry heat sterilization
(d) Incineration
(e) Seitz filtration
(f) Gaseous chamber sterilizations (as with ethylene oxide)
(g) Boiling
(h) Ultraviolet radiation

283

PART II

Please place the letter from column Beta in front of the corresponding definition in column Alpha.

COLUMN ALPHA	COLUMN BETA
_____ 1. Destroys or masks offensive odors (gas)	(a) Fumigant
_____ 2. Prevents growth of bacteria, but does not destroy them	(b) Disinfectant
_____ 3. Prevents deterioration of food	(c) Preservative
_____ 4. Destroys disease-producing agents and their products	(d) Deodorant
_____ 5. Destroys insects and small animals (gases or fumes)	(e) Germicide
_____ 6. Prevents multiplication of bacteria	(f) Antiseptic
_____ 7. Destroys bacteria (chemical)	(g) Bacteriostatic agent

PART III

Please circle the *one* item that best completes the introductory statement.

1. The most efficient sterilization procedure involving heat is:
   (a) Pasteurization at 62° C for 30 minutes
   (b) Dry heat at 100° C for 30 minutes
   (c) Steam under pressure at 121° C for 15 minutes
   (d) Freezing at 20° C for 24 hours
   (e) Boiling at 121° C for 10 minutes
2. In using a germicidal agent it is best to know:
   (a) The volatility of the agent
   (b) The concentration needed and the time of contact
   (c) The exact number of bacterial spores present
   (d) The exact number of bacteria present
3. In the management of infection by an organism that has become highly resistant to tetracycline it would be best to:
   (a) Withdraw all antibiotic treatment
   (b) Test the causative microorganism for susceptibility to other antibiotics
   (c) Increase the dose
   (d) Switch to penicillin
   (e) Do nothing
4. The development of resistance to an antibiotic by a microorganism comes from an alteration of:
   (a) Cell wall permeability
   (b) Plasma membrane
   (c) Capsule
   (d) Mitochondria
   (e) None of these
5. The phenol coefficient of a germicide is determined by comparison of that germicide with phenol on the basis of:
   (a) The concentration that kills the same number of bacteria
   (b) Penetrating power of the agent
   (c) Proportion of bacteria killed
   (d) Zone of inhibition of bacteria
   (e) None of these
6. If 0.1 ml of serum is added to 0.9 ml of saline solution and mixed, and if half the mixture is transferred to a second tube containing 0.5 ml of saline solution, the dilution in the second tube (after mixing) would be:
   (a) 1:5
   (b) 1:10

    (c) 1:20

    (d) 1:40

    (e) 1:100

7. Sulfanilamide is effective as a chemotherapeutic agent against:

    (a) Organisms that synthesize folic acid

    (b) Organisms that require but do not synthesize folic acid

    (c) Organisms that cannot utilize folic acid

    (d) All the above

    (e) None of the above

8. Which of the following antimicrobial agents interferes with cell wall synthesis?

    (a) Penicillin

    (b) Colistin

    (c) Isoniazid

    (d) Tetracycline

    (e) Actinomycin D

9. Which of the following antimicrobial agents does *not* inhibit protein synthesis?

    (a) Chloramphenicol

    (b) Tetracycline

    (c) Kanamycin

    (d) Polymyxin

    (e) Streptomycin

10. The antimicrobial action of the polymyxins and colistin is related primarily to:

    (a) Interference in cell membrane function

    (b) Inhibition of cell wall synthesis

    (c) Inhibition of nucleic acid synthesis

    (d) An unknown mechanism

    (e) Inhibition of protein synthesis

11. Which of the following interferes with bacterial ribosome function?

    (a) Chloramphenicol

    (b) Tetracycline

    (c) Erythromycin

    (d) All the above

    (e) None of the above

12. Idoxuridine, an antiviral agent, competes with thymidine and thus prevents synthesis of:

    (a) DNA

    (b) Penicillinase

    (c) Catalase

    (d) Interferon

    (e) Penicilloic acid

13. The most frequent cause of operating room infection is:

    (a) Nasopharnygeal infections of the staff or patient

    (b) Normal flora of the patient's skin and upper respiratory tract

    (c) Inadequate cleaning of the operating room suite

    (d) Lack of sterility of the operating room air

    (e) Lack of sterility of instruments and drapes

14. Which of the following agents is of value in the prevention of influenza type A?

    (a) Methisazone

    (b) Cytarabine

    (c) Idoxuridine

    (d) Amantadine

    (e) Acyclovir

15. Drugs that are effective in the treatment of tuberculosis include:
    (a) Rifampin
    (b) Isoniazid
    (c) Ethambutol
    (d) Streptomycin
    (e) All the above

16. Antibiotic-associated colitis is associated with which of the following microbes that is postulated to secrete an enterotoxin damaging the intestinal mucosa?
    (a) *Staphylococcus* species
    (b) Tubercle bacillus
    (c) *Clostridium difficile*
    (d) *Escherichia coli*
    (e) *Clostridium perfringens*

17. Which of the following statements about chloramphenicol is true?
    (a) It inhibits protein synthesis in susceptible bacteria.
    (b) Some resistant bacteria enzymatically inactivate it.
    (c) Aplastic anemia is an infrequent but serious side effect.
    (d) It is bacteriostatic for susceptible organisms.
    (e) All the above

18. All the following antibiotics are bacteriostatic except:
    (a) Tetracycline
    (b) Lincomycin
    (c) Erythromycin
    (d) Sulfanilamide
    (e) Gentamicin

19. For a chemotherapeutic agent to show selective toxicity it has to:
    (a) Be able to inhibit the growth of all bacteria
    (b) Inhibit a pathway unique to bacteria (one not found in the host)
    (c) Inhibit a pathway present both in host and in bacteria
    (d) Selectively inhibit the growth of pathogenic bacteria but not the growth of non-pathogens
    (e) None of the above

20. Rifamycin, an important tuberculocidal drug:
    (a) Binds to bacterial DNA
    (b) Inhibits bacterial DNA polymerase
    (c) Inhibits eukaryotic DNA polymerase
    (d) Is not a broad-spectrum drug
    (e) Is not associated with any pigmentary changes

21. Resistance to streptomycin can be due to:
    (a) An altered protein in the 30S ribosomal subunit
    (b) An altered protein in the 50S ribosomal subunit
    (c) An alteration in the 23S ribosomal RNA
    (d) The production of a streptomycin adenyl transferase
    (e) None of the above

22. Penicillin treatment of a patient with a bacterial infection
    (a) Is ineffective if the pathogen produces beta-lactamase
    (b) Directs the formation of penicillin-resistant mutants
    (c) Is always useless because of the selection of spontaneously occurring penicillin-resistant mutants
    (d) Is ineffective because of cross-resistance with aminoglycosides
    (e) None of the above

23. Drug resistance may be the effect of:
    (a) An R plasmid

(b) An independently replicating transposon
(c) A lethal mutation
(d) Plasmid-mediated conjugation in gram-negative bacteria only
(e) Adaptive enzymes of various kinds

PART IV

1 When the paired statements are compared in a quantitative sense, one may be found
to exert or imply a greater effect or value. If so, please indicate which is the stronger
statement in the following way:
A if the first statement is of a greater degree or the one with the preferred action; it
  is "more"
B if the second statement represents a greater degree
C if the two statements are equal or very nearly so

_____ 1. Effectiveness of chemical disinfectant in saline solution
         Effectiveness of chemical disinfectant in serum
_____ 2. Lethal effect of temperatures 10° C above maximal growth temperature
         Lethal effect of temperatures 10° C below minimal growth temperature
_____ 3. Effect of chemical disinfectant at pH 7
         Effect of chemical disinfectant at pH 4
_____ 4. Time required for sterilization in autoclave
         Time required for sterilization in hot air sterilizer
_____ 5. Time required for sterilization in pressure steam sterilizer
         Time required for sterilization in high-prevacuum steam sterilizer
_____ 6. Time required for sterilization in pressure steam sterilizer
         Time required for sterilization in ethylene oxide gas sterilizer
_____ 7. Sterilizing action of hot air at 170° C for 15 minutes
         Sterilizing action of free-flowing steam for 15 minutes
_____ 8. Disinfectant action of ethyl alcohol against tubercle bacilli
         Disinfectant action of mercurials against tubercle bacilli
_____ 9. Disinfectant action of iodophors against spores
         Disinfectant action of glutaraldehyde against spores
_____ 10. Likelihood of drug resistance in tuberculosis if streptomycin is used alone
          Likelihood of drug resistance in tuberculosis if drug combinations are
          used
_____ 11. Sensitivity of vegetative cells to chemical treatment
          Sensitivity of spores to chemical treatment
_____ 12. Genetic material transferred in sexual crosses
          Genetic material transferred in transformation
_____ 13. Temperature for sterilization by autoclaving (in pressure steam sterilizer)
          Temperature for sterilization by dry heat (in hot air sterilizer)
_____ 14. Variability in Gram-staining reaction of gram-positive organisms
          Variability in Gram-staining reaction of gram-negative organisms
_____ 15. Optimal growth temperature of a psychrophile
          Optimal growth temperature of a mesophile
_____ 16. Spectrum of microbes sensitive to penicillin
          Spectrum of microbes sensitive to tetracycline
_____ 17. Capacity of an autotrophic organism for biosynthesis
          Capacity of an obligate parasite for biosynthesis
_____ 18. Generation time of bacterial cells in lag phase of growth
          Generation time of bacterial cells in log phase of growth
_____ 19. Ease of staining of vegetative bacterial cells
          Ease of staining of spores from the same bacterial cells

_____ 20. Death rate of cells in the stationary phase of growth

Multiplication rate of cells in the stationary phase of growth

2 Match the item in column Alpha to the item in column Beta that indicates the major action of the group of antibiotics. Please place the letter in the appropriate place.

COLUM ALPHA

_____ 1. Beta-lactam antibiotics

_____ 2. Aminoglycosides

_____ 3. Macrolides

_____ 4. Polypeptide antibiotics

_____ 5. Rifamycins

_____ 6. Tetracyclines

_____ 7. Polyene antibiotics

COLUMN BETA

(a) Interference with cell wall synthesis

(b) Interference with nucleic acid synthesis

(c) Interference with protein synthesis

(d) Interference with plasma membrane function

MICROBES
PRODUCTION OF INFECTION

ROLE IN DISEASE

INFECTION

When microbes or certain other living agents enter the body of a human being or animal, multiply, and produce a reaction there, an *infection* results. The host is *infected*, and the disease is an *infectious* one. The reaction of the body may or may not be accompanied by outward signs of disease. Infection is distinguished from **contamination,** which refers to the mere presence of infectious material (no reaction produced). Superficial wounds in skin and mucous membranes are often the sites of large microbial populations. As long as these microorganisms do not invade the deeper tissues and induce a reaction there, they are contaminants instead of agents of infection.

RESIDENT POPULATION: MICROBES NORMALLY PRESENT

The mere presence of microbes in the body does not mean infection because microorganisms normally inhabit many parts of the body without invading the deeper tissues to cause disease. Some body fluids, such as blood and urine, are normally sterile, that is, free of the presence of microorganisms. In parts of the body other than the circulatory system and urinary tract the secretions or excretions* are normally in contact with a resident population of microorganisms, consistently present in varying

---

*Secretion is the term for a specific, useful product resulting from the activity of a gland in the body.
Excretion is "waste matter," such as sweat or urine, not useful, to be eliminated from the body.

proportions. These microorganisms constitute the **normal flora** of the body (Fig. 16-1). This pattern of growth, conspicuously bacterial, is associated with the well-being of the person and in an area such as the intestinal tract is even necessary to that person's health. Although it is true that viruses may be recovered from the throats of otherwise healthy children, viruses are not generally believed to be part of the normal flora in humans. Table 16-1 gives the normal habitats in the human body for such microorganisms.

Microbes of the normal flora contribute to the well-being of their host primarily in two ways: (1) by satisfying certain nutritional needs and (2) by functioning as a barrier to infection (Chapter 18). Resident microorganisms in the intestine can synthesize certain vitamins—K, folic acid, B_{12}, pyridoxine, biotin, pantothenic acid, and riboflavin—in excess of their needs. Some of these are probably absorbed by humans. Most of the time resident microbes compete successfully with pathogens for anatomic space and by varied mechanisms can inhibit would-be troublemakers.

Among the members of the normal flora, the **resident flora** constitute those microbes that one expects to find in a given host at a given age at a given anatomic site. These include commensals ("freeloaders") that derive benefit from the host but give nothing in exchange (they do not harm, however) and symbionts ("paying guests") that interact with the host for the mutual good of both host and organism. **Transient flora** from the environment are present for a limited time only—hours, days, or weeks. Since they tend not to colonize at the site, they are mostly without significant effect as long as a healthy normal flora is maintained. Their potential to injure comes only if they are able to take advantage of the host because of some unfavorable alteration in the host's condition.

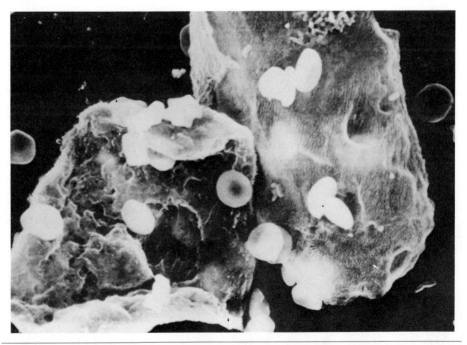

FIG. 16-1   Two oral squamous epithelial cells with cocci (dense whitish objects) of normal mouth flora attached, scanning electron micrograph. (Approximately ×1100.) (Courtesy Dr. R.C. Reynolds, The University of Texas Health Science Center at Dallas.)

Table 16-1 Distribution of Resident Population

Anatomic Site	Important Resident Microbes	Anatomic Site	Important Resident Microbes
Skin	*Bacillus* species Coliforms Diphtheroids (aerobic and anaerobic) Enterococci Fungi (lipophilic) Mycobacteria Neisseriae *Peptostreptococcus* species *Propionibacterium acnes* *Proteus* species *Pseudomonas* species Staphylococci Streptococci	Respiratory tract (primarily the upper part)— cont'd	*Candida albicans* and other fungi Diphtheroids Enterococci *Fusobacterium* species Hemophilic bacilli Mycobacteria Mycoplasmas Neisseriae *Peptostreptococcus* species Pneumococci Spirochetes Staphylococci Streptococci (aerobic and anaerobic) Veillonellae
Mouth (including gingival crevice)	*Actinomyces* species Amebas *Bacteroides* species *Bifidobacterium* species Borrelias *Candida* species and other fungi Clostridia Diphtheroids *Fusobacterium nucleatum* Hemophilic bacilli Lactobacilli *Leptotrichia buccalis* Mycoplasmas Neisseriae Peptococci *Peptostreptococcus* species Pneumococci Spirochetes Staphylococci Streptococci *Trichomonas hominis* Veillonellae *Vibrio* species	Gastrointestinal tract* (mostly in lower ileum and colon)	*Actinomyces* species *Aeromonas* species *Alcaligenes faecalis* *Bacteroides* species *Bifidobacterium* species *Candida* species *Clostridium* species Coliforms Diphtheroids *Entamoeba coli* Enterococci Enteroviruses *Eubacterium* species Fusobacteria Lactobacilli† Mycobacteria Mycoplasmas *Penicillium* species *Peptococcus* species *Peptostreptococcus* species *Proteus* species *Pseudomonas aeruginosa* Spirochetes Staphylococci
Respiratory tract (primarily the upper part)	*Acinetobacter* species Actinomycetes *Bacteroides* species		

*The stomach is normally free of microbial growth because of its acid content. Microorganisms are much more plentiful in the lower ileum (small intestine) and large bowel because the pH and motility of the duodenum and jejunum (upper small intestine) are factors that act to reduce bacterial concentration in these areas.

†A baby is born with a sterile intestinal tract, but before or with the first feeding, bacteria are introduced. If the child is breast fed, the predominant organism is said to be *Bifidobacterium bifidum* (formerly *Lactobacillus bifidus*). If the child is bottle fed, *Lactobacillus acidophilus* (bacteria of genus *Lactobacillus* convert carbohydrates into lactic acid) predominates. In addition, other bacteria are present. Development of the normal flora in the healthy term infant is usually complete by the first 36 to 48 hours of life.

Table 16-1 Distribution of Resident Population—cont'd

Anatomic Site	Important Resident Microbes	Anatomic Site	Important Resident Microbes
Gastrointestinal tract (mostly in lower ileum and colon)—cont'd	Streptococci (aerobic and anaerobic)	Genital tract—cont'd	Streptococci (aerobic and anaerobic)
	Yeasts		Veillonellae
	Vibrio species	Vagina	
Genital tract	*Bacteroides* species	Before puberty	Coliforms
	Bifidobacterium species		Diphtheroids
	Clostridia		Streptococci
	Coliforms	During childbearing period	Döderlein's bacilli (*Lactobacillus* species)
	Diphtheroids	After childbearing period	Coliforms
	Enterococci		Diphtheroids
	Fusobacterium species		Streptococci
	Haemophilus vaginalis	Organs of special sense	
	Lactobacilli	Eye	Alpha-streptococci
	Mycobacterium smegmatis		Diphtheroids
	Mycoplasmas		Hemophilic bacilli
	Neisseriae		Pneumococci
	Peptococci		Staphylococci
	Peptostreptococcus species	External ear	*Bacillus* species
	Proteus species		Diphtheroids
	Saprophytic yeasts		Nonpathogenic acid-fast organisms
	Spirochetes		Staphylococci
	Staphylococci		

INTRUSION OF MICROBES
Source of Microbes Causing Infections

The microbes that cause infections fall into two classes. The first covers those that can cause disease in healthy persons and reach such persons directly or indirectly from animals, persons ill with the disease, or carriers. Such microbes are considered pathogens and cause communicable diseases. The second covers those that attack either when the host is more susceptible because of some factor lowering host resistance or when they themselves are increased in virulence. To the second group belong both microorganisms that normally inhabit the body (normal flora) but produce disease only under given conditions and certain ones that are inadvertently introduced into the body by wounds, injuries, and the like. Ordinarily these are saprophytes that can be opportunists.

How Microbes Reach the Body

According to the manner in which the causative agent reaches the body, infectious diseases may be classified as communicable and noncommunicable.

A **communicable disease** is one whose agent is directly or indirectly transmitted from host to host. Examples are diphtheria and tuberculosis. The host from which the infection is spread is usually of the same species as the recipient, but not necessarily. For instance, cattle may transmit tuberculosis and undulant fever to human beings.

A **noncommunicable disease** is one whose agent either normally inhabits the body, only occasionally producing disease, or resides outside it, producing disease only when introduced into the body. For example, tetanus bacilli, inhabitants of the soil, produce disease only when introduced into abrasions or wounds. Although *not* communicable, tetanus is infectious. The term **contagious** is applied to diseases that are easily spread directly from person to person.

Infectious diseases are also classified as exogenous and endogenous. **Exogenous** infections are those in which the causative agent reaches the body from the outside and enters through one of the portals of entry. An **endogenous** infection is one from organisms normally present in the body. Endogenous infections occur when the defensive powers of the host are weakened or, for some reason, virulence of the microorganism is increased.

How Microbes Enter the Body

Microorganisms invade the body by several avenues, and each species has its own favored one. The way of access to the body is known as the **portal of entry,** as indicated in the following areas.

Skin

Most pathogens do not penetrate unbroken skin or mucous membrane, but some do. Staphylococci and some fungi can penetrate the hair follicles and cause disease in the deeper tissues of the skin. The bacteria of tularemia pass through the unbroken skin, as do hookworm larvae. In their penetration of intact skin, malarial parasites have the help of a biting insect. Many microbes are found in the superficial layers of the skin but under normal conditions are not able to penetrate the underlying area. As soon as the superficial barrier is broken, infection is easily accomplished.

Respiratory Apparatus

Pulmonary tuberculosis, pneumonia, and influenza are contracted by way of the respiratory tract, and the viruses causing measles, smallpox, and German measles enter the body this way.

Alimentary Tract

Some of the most important pathogens enter the body through the digestive tract, for example, the dysentery and typhoid bacilli, cholera vibrios, and amebas of dysentery. In many cases food and drink are the vehicles. The great majority of pathogenic microorganisms invade the body through the respiratory system or the digestive tract.

Genitourinary System

Certain infections are acquired chiefly through the genitourinary system, notably the venereal, or sexually transmitted, diseases (STDs).

Placenta

Most microbes do not cross the placenta, but the spirochete of syphilis and the smallpox virus may.

EVENT OF INFECTION
Koch's Postulates

The fact that microbes have entered the body in no manner indicates that infection has occurred. Traditionally the proof that an organism causes a given disease rests on the fulfillment of certain requirements, known as Koch's postulates:
1. The organism must be observed in every case of the disease.
2. The organism must be isolated and grown in pure culture.
3. The organism must, when inoculated into a susceptible animal, cause the disease.
4. The organism must be recovered from the experimental animal and its identity confirmed.

These postulates have adequately proved the cause of many important bacterial diseases in the past. However, they cannot always be met. Some disease-producing organisms, including all viruses, cannot be grown on artificial media, and some microbes cause disease only in humans. In exceptional instances an organism has been accepted as the cause of a disease although all these requirements have not been met.

Factors Influencing the Occurrence of Infection

Whether an infection ensues when microbes are in contact with the human body depends on (1) the portal of entry, (2) virulence of the organisms, (3) their number, and (4) defensive powers of the host (Chapter 17).

Most pathogenic bacteria have definite portals by which they enter the body, and they can fail to produce disease when introduced into the body by another route. For instance, typhoid bacilli produce typhoid fever when swallowed but only a slight local inflammation when rubbed on the abraded skin, whereas streptococci rubbed into the skin produce intense inflammation (cellulitis) but are generally without effect when swallowed. If streptococci are breathed into the lungs, they may cause pneumonia. A few bacteria, such as *Francisella tularensis*, the cause of tularemia, can enter the body by several different routes and produce disease in each area. Usually the route of entry for a given organism determines the disease process induced.

Pathogenicity means the ability of microbes to overcome the defensive powers of the host and to induce disease. **Virulence** refers to the degree or intensity of disease produced. A difference in virulence is found not only among pathogenic species but also among members of the same species. As a rule, microbes are most virulent when freshly discharged from the patient in whom they have caused disease. Organisms harbored by carriers are generally less virulent.

Virulence may be increased by rapid transfer of organisms through a series of susceptible animals. As each animal becomes ill, the organisms are isolated from its excreta and transferred to a well animal (**animal passage**). Herein are explained epidemics. The agent of the disease, by repeated passage from person to person, becomes so virulent that everyone whom it contacts is made ill.

An organism that is highly virulent for one species of animals and less virulent for another may, with repeated passage through the animal for which it is less virulent, show a **transposal of virulence;** that is, it becomes less virulent for the animal for which it was originally highly virulent and highly virulent for the animal for which it was originally less so.

Attenuation indicates a loss in disease-producing ability; an organism whose virulence is decreased is **attenuated** (weakened). A highly pathogenic organism may be rendered temporarily or permanently nonpathogenic if repeatedly subcultured on artificial laboratory media. Virulence, although artificially eliminated, may often be restored by animal passage.

The *number* of microbes is crucial to infection. If only a few enter the body, they may well be overcome by the local defenses of the host even though they are highly virulent. Therefore, if infection is to occur, enough microorganisms to overcome the local defenses of the host must invade the body.

How Microbes Cause Disease

When pathogens enter the body, two opposing forces are set in motion. The organisms strive to invade the tissues and colonize there. **Invasiveness,** an important factor in the determination of the virulence of an organism, refers to its ability to attach to tissues, to penetrate them, to multiply soon after, and in certain instances to spread. The phenomenon of **attachment** often precedes invasion; it is brought about usually by special bacterial structures that connect to specific receptor sites on the cells of the host. The body, utilizing its defensive powers, strives to block the onslaught of the microbes, destroy them, and cast them off. If the body wins the contest, the microbes are destroyed, and the body suffers no ill effects. If the microbes prevail, infection occurs.

Colonization of Surfaces

A delicate balance exists between the normal resident flora and the host. It is influenced by various factors, such as diet, nutritional status, occupation, presence of disease or debilitating state, and the administration of antibiotics to the host. The latter factor results in the most drastic change in the host's normal flora, and such a change may result in serious bacterial disease.

Viral infection may disturb that balance between host and indigenous flora. The bacterial pneumonia complicating influenza, for instance, is more feared than the viral infection itself. Apparently the influenza virus and certain other respiratory viruses can promote colonization of the epithelium of the respiratory tract by pathogens. Their antigens on the surfaces of infected cells may be binding sites for bacteria, or viral infection of a cell may damage the cell membrane, thus making it easier for the pathogen to attach.

For microbes to colonize a surface lined by an epithelium they must be able not only to adhere to the surface but also to survive in the environment of the surface. Normal flora and pathogens alike must overcome certain inherent obstacles found on epithelial surfaces if they are to establish themselves there. The peristaltic activity of the gastrointestinal tract and the mucociliary clearance system of the respiratory tract are powerful forces. As a result of the smooth muscle contractions in the bowel or the beating effect of the cilia of the respiratory tract, bacteria along with luminal contents are readily moved from the area.

However, bacteria anchored to epithelial cells can persist in the area, and their growth may be even stimulated, since nutrients and certain macromolecules tend to concentrate at solid-liquid interfaces.

Microbes possess a remarkable degree of specificity for particular cell types to which they adhere. Specific microbial **adhesins** are one reason. These are proteinaceous, filamentous projections on the surface of bacteria that bind to specific receptors on the cell membrane.

Invasiveness

Generally speaking, the more rapidly a microbe invades a tissue and multiplies therein, the greater the potential for serious disease it possesses. This capability varies among different microorganisms, and invasiveness is not necessarily synonymous with disease. Some viruses may be widely distributed in the body without causing overt illness.

A number of mechanisms enable disease-producing organisms to invade tissues. With some, penetration of tissue spaces and cells is facilitated by enzymes that dissolve blood clots or weaken tissue components. Some possess lysosomal enzymes that damage cell membranes. Certain microbes, since they are able to survive within the cytoplasm of phagocytic cells, are protected from host defenses. Safe within the cytoplasm, they may be carried deep into areas of the body.

A constituent of the bacterial cell wall may be an injurious factor. A part of the microbial cell such as its capsule, although not harmful in itself, may so protect the microbe from phagocytosis as to enhance its virulence. The carbohydrate (polysaccharide) capsules of pneumococci invading the lower tract protect them from the defensive cells of the lungs, thereby facilitating the development of pneumonia.

Mechanical Effects

Microbes cause disease in various ways. In some cases the mechanical effects of microorganisms operate, for example, when the organisms of falciparum malaria occlude the capillaries to the brain. Microbes that possess independent motility may be at an advantage over the host.

Harmful Metabolic Products

The ability of pathogenic microorganisms to produce disease is closely tied in with certain complex chemical substances that the organisms either elaborate and release or that are a vital part of their makeup. A soluble exotoxin is an example of a product released into body fluids and even into the bloodstream of the host. A number of enzymatic substances are not directly toxic but are related significantly to disease.

Hemolysins are substances that cause the lysis (dissolution) of red blood cells. There are many types, among which are immune hemolysins, hemolysins of certain vegetables, hemolysins contained in venoms, and bacterial hemolysins. The bacterial hemolysins are of two types: (1) those that may be separated from the bacterial cells by filtration (filterable hemolysins) and (2) those that are demonstrated around the bacterial colony on a culture medium containing red blood cells. The filterable hemolysins are named after the bacteria that give rise to them (for example, **staphylolysin,** derived from staphylococci, and **streptolysin,** derived from streptococci). The filterable hemolysins have a certain degree of specificity; that is, a given hemolysin may act on the red blood cells of one animal species but not on those of another. These are protein in nature and inactivated by exposure to a temperature of 55° C for 30 minutes. In the body of an animal they induce the formation of specific antibodies.

Staphylococci, streptococci, pneumococci, and *Clostridium perfringens* are important producers of these hemolysins. An organism may have more than one filterable hemolysin; for instance, streptococci produce two, streptolysin O and streptolysin S.

The hemolysins detected when bacteria are grown on a culture medium containing blood are of two types, **alpha** and **beta.** Beta hemolysins give rise to a clear, colorless zone of hemolysis around the bacterial colony. Alpha hemolysins give rise to a greenish zone of partial hemolysis. The relation between filterable hemolysins and those detected by growing the bacteria on a culture medium containing blood is not known. The relation between hemolysin production and virulence is obscure. As a rule, the hemolytic strains of pathogenic bacteria have greater capacities to cause disease than the nonhemolytic ones. On the other hand, certain nonpathogenic bacteria produce hemolysins. Hemolysins probably contribute to the invasive capacity of some bacteria.

Leukocidins destroy polymorphonuclear neutrophilic leukocytes. The leukocytes, or white blood cells, take an active part in the battle of the body against infection. Leukocidins are formed by pneumococci, streptococci, and staphylococci. It is likely that they enhance virulence. Certain hemolysins and leukocidins appear to be identical.

Coagulase, elaborated by bacteria to accelerate the coagulation of blood, under suitable laboratory conditions causes oxalated or citrated blood to clot. Coagulase formation appears to be confined to the staphylococci. The **coagulase test,** performed by mixing a culture of staphylococci with a suitable amount of oxalated or citrated blood under specified conditions, differentiates pathogenic from nonpathogenic staphylococci. Because the former cause the blood to coagulate, they are said to be **coagulase positive.** Coagulase protects organisms against the destructive action of leukocytes (phagocytosis). The coagulum from plasma that has leaked into the tissues forms a barrier between the bacteria and the white blood cells.

Bacterial kinases act on certain components of the blood to liquefy fibrin. Kinases interfere with coagulation of blood, since a blood clot is made up of red blood cells enmeshed in interlacing strands of fibrin, and liquefy clots already formed. Kinases enhance virulence because they destroy blood or fibrin clots that form around any infection site to wall it off. The most important one is **streptokinase,** also known as **fibrinolysin,** elaborated by many hemolytic streptococci. Staphylococci and certain other bacteria also produce kinases. Strepotokinase has been used in certain locations in the body to dissolve clots and to prevent the formation of adhesions that would be laid down on the fibrin precipitated in the body cavities.

Hyaluronidase is secreted by certain bacteria to make the tissues more permeable to the bacteria elaborating it. It hydrolyzes hyaluronic acid, a constituent of the intercellular ground substance of many tissues that helps to hold the cells of the tissue together. Hyaluronidase was formerly spoken of as the "spreading factor." It is produced by pneumococci and streptococci.

Toxins

Many species of bacteria produce poisonous substances known as toxins (Table 16-2). The capacity to elaborate toxins is **toxigenicity,** and the resultant toxin may be almost entirely responsible for the specific action of the toxigenic bacteria. Toxigenicity is another important factor in the determination of an organism.

Toxins are of two types: **exotoxins,** diffused by the bacterial cell into the surrounding medium, and **endotoxins,** liberated only when the bacterial cell is destroyed. Some bacterial exotoxins are more deadly than any mineral or vegetable poison.

Among the gram-positive bacteria noted for their exotoxin production are *Corynebacterium diphtheriae*, *Clostridium tetani*, and *Clostridium botulinum*. In disease caused by *Corynebacterium diphtheriae* (diphtheria) and *Clostridium tetani* (tetanus) the bacteria grow restricted to a superficial area in the body. Within themselves these organisms produce little effect, but the soluble exotoxins elaborated at the site of growth are absorbed into the body to cause serious and often fatal illness. Such illness is better considered chemical poisoning (toxemia) rather than disease induced by growth of bacteria. It is interesting to note that some bacteria are harmful only when they are diseased. The aforementioned organisms make exotoxin only because they are infected

Table 16-2 Distinguishing Features of Toxins

Differential Point	Exotoxins	Endotoxins
Location	Gram-positive and gram-negative bacteria	Gram-negative bacteria
Relation to cell	Extracellular (in fluid medium)	Closely bound to cell wall (release with destruction of cell)
Toxicity	Great	Weak
Febrile reaction	Minimal	Pronounced
Tissue affinity	Specific (for example, nerves—tetanus toxin; adrenals, heart muscle—diphtheria toxin)	Nonspecific (local reaction—injection site; systemic reaction—fever, shock)
Chemical nature	Polypeptides mostly (molecular weight 10,000 to 900,000)	Lipopolysaccharide complexes
Stability	Usually unstable (denatured with heat, ultraviolet light)	Stable (relatively)
Antigenicity	High (neutralizing antibodies)	Weak
Conversion to toxoid	Yes	No
Diseases where action important	Diphtheria	Salmonellosis
	Tetanus	Brucellosis
	Gas gangrene	
	Staphylococcal food poisoning (heat stable)	
	Botulism	
	Bacillary dysentery	
	Cholera	

with bacterial virus (bacteriophage). The virus initiates (codes for) the production of toxin (p. 93).

Some gram-negative enteric bacteria elaborate an exotoxin referred to as an **entero-toxin** because it exerts its effects in the intestine. These include *Escherichia coli* and *Vibrio cholerae* (cause of cholera). Another gram-negative microbe, *Pseudomonas aeru-ginosa*, produces an exotoxin that behaves biologically like the diphtheria exotoxin.

Exotoxins are protein in nature, polypeptides mostly, and hence easily heat-inactivated. They are antigenic and specific. When a small amount of a toxin is injected into an animal, it stimulates the production of **antitoxin.** The **specificity** of exotoxins refers to the fact that diphtheria bacilli elaborate a toxin that causes diphtheria and nothing else and tetanus bacilli elaborate a toxin that causes tetanus and nothing else. When an exotoxin is inactivated with formaldehyde, it no longer causes the disease but still can produce an immunity to the disease. Such modified toxins are known as **ana-toxins** or **toxoids.** Toxoids are used to produce permanent immunity to diphtheria and tetanus.

Substances contained in the seeds of certain plants and some of the secretions of animals resemble exotoxins. Among these are crotin, derived from the seed of the croton plant; ricin, from the castor bean; and the venoms of snakes, spiders, and scorpions.

Endotoxins are complex lipopolysaccharides that are an integral part of the cell wall of most gram-negative bacteria. Endotoxins are not found in the surrounding medium when bacteria are grown in a liquid. They do not effectively promote the formation of antitoxins, do not possess the specificity of exotoxins, and cannot be converted into toxoids. The lipid fraction (lipid A) of the complex is related to the numerous biologic disturbances of endotoxin—fever, circulatory disturbances, and other toxic effects (Table 16-2). Noteworthy for their production of endotoxins are *Salmonella typhi*, the cause of typhoid fever, and *Neisseria meningitidis*, the cause of epidemic meningitis. Endotoxic activity in *Neisseria* species may also come from blebs released from the outer membrane that contain lipopolysaccharide.

Biochemical Effects

Biochemical substances implicated in pathogenicity of microbes are many, and their action is not always well understood. Their major effects are summarized as follows:
1. They interfere with mechanical blocks to the spread of infection set up in the body (effect of bacterial kinases).
2. They slow or stop the ingestion of microbes by the phagocytic white blood cells (action of leukocidins).
3. They destroy body tissues (effect of hemolysins, necrotizing exotoxins, lethal factor).
4. They cause generalized unfavorable reactions in the host, resulting in fever, discomfort, and aching (effect of endotoxins). Such features are usually collected together under the term **toxicity.**

Elective Localization

Most disease-producing microbes prefer a given part of the body; they have their favored site for involvement. This is elective localization. For instance, dysentery bacilli attack the intestine, pneumococci attack the lungs, and meningococci localize in the lepto-meninges. Toxic products released by microbes also have a tissue affinity; the toxin of tetanus acts on the central nervous system, and the toxin of diphtheria affects not only the central nervous system but also the heart.

Local Effects of Infection

Local effects refers to changes produced in the tissues in which given microbes are multiplying, summed up in the process of **inflammation** (Chapter 17). Inflammation is the body's answer to injury. Its design is to halt the invasion and destroy the invaders. The features of the inflammatory process vary greatly with the different causative organisms and are significantly related to the disease-producing capacity of the microbe.

General Effects of Infection

Certain host reactions are found in nearly all infectious diseases. In addition, many diseases have their own peculiar features. Among the general effects are fever (p. 313), increased pulse rate (tachycardia), increased metabolic rate, and signs of toxicity. The degree of fever approximates the severity of the infection. Anemia is a result of prolonged and severe infections.

A common effect of infection is a change in the total number of circulating leukocytes (white blood cells) and in the relative proportion of the different kinds. This is why total and differential white cell counts are so important in the diagnosis of disease. In most infections the total number of leukocytes is increased (**leukocytosis**). In some the number is decreased (**leukopenia**), and in a few it is unchanged. If fever or leukocytosis fails to occur in an infection where either ordinarily does, a severe infection or decreased host resistance is indicated.

An important consequence of infection is the development of immunity (Chapter 18).

How Disease-Producing Agents Leave the Body

Just as they have definite avenues of entry, pathogenic agents have definite routes of discharge from the body, known as **portals of exit,** which to a great extent depend on the part of the body that is diseased. The following list gives the important ones with examples:
1. *Feces*—bacteria of salmonellosis, bacillary dysentery, and cholera; protozoa of dysentery; viruses of poliomyelitis and hepatitis type A
2. *Urine*—bacteria of typhoid fever, tuberculosis (when affecting the genitourinary tract), and undulant fever
3. *Discharges from the mouth, nose, and respiratory passages*—bacteria of tuberculosis, whooping cough, pneumonia, scarlet fever, and epidemic meningitis; viruses of measles, smallpox, mumps, poliomyelitis, influenza, and epidemic encephalitis
4. *Saliva*—virus of rabies
5. *Blood (removed by biting insects)*—protozoa of malaria; bacteria of tularemia; rickettsias of typhus fever and Rocky Mountain spotted fever; virus of yellow fever

PATTERN OF INFECTION
Course of Infectious Disease

The course of many infectious diseases extends over the following:
1. *Period of incubation*—interval between the time the infection is received and the appearance of disease (Table 16-3).

In some diseases its length is constant. In others it varies greatly. The length of the incubation period depends on (a) the nature of the agent (for example, the incubation period of diphtheria is less than that of rabies), (b) its virulence, (c) resistance of the host, (d) distance from site of entrance to focus of action (for instance, the incubation period of rabies, which affects the brain, is shorter when the innoculation site is on the face), and (e) number of infectious agents invading the body.

The incubation period cannot be equivalent to the lag phase of bacterial growth. The growth curve of a given bacterium in an infected host is similar to its in vitro curve (p. 548), but the doubling time is nowhere near what can be achieved in the laboratory. If it were, the bacterial mass would exceed the weight of the host.

2. *Period of prodromal symptoms*—short interval (prodrome) that sometimes follows

Table 16-3 Incubation Periods of Important Infectious Diseases*

Disease	Usual Incubation Period
Adenovirus infections	5 to 6 days
Amebiasis	2 weeks (varies)
Ascariasis	8 to 10 weeks
Brucellosis	5 to 30 days (varies)
Cat-scratch fever	3 to 10 days (varies)
Chickenpox	14 to 16 days
Cholera	1 to 3 days
Coccidioidomycosis, primary infection	1 to 3 weeks
Coxsackievirus infections	2 to 14 days
Diphtheria	2 to 6 days
Echovirus infections	3 to 5 days
Encephalitis	2 to 21 days (varies as to type)
Escherichia coli diarrhea	2 to 4 days
Gas gangrene	1 to 5 days
Gonorrhea	3 to 5 days
Hepatitis, viral, type A	15 to 50 days
Hepatitis, viral, type B	6 weeks to 6 months
Herpesvirus infections	4 days
Histoplasmosis	5 to 18 days
Hookworm disease	6 weeks
Impetigo contagiosa	2 to 5 days
Infectious mononucleosis	2 to 6 weeks
Influenza	1 to 3 days
Leprosy	3 months to 20 years (?)
Leptospirosis	2 to 20 days
Measles (rubeola)	10 to 12 days
Meningitis, acute bacterial	1 to 7 days
Molluscum contagiosum	2 to 8 weeks
Mumps	14 to 21 days
Mycoplasmal pneumonia (primary atypical)	7 to 21 days
Pertussis	5 to 21 days
Pinworm infection	2 to 6 weeks
Plague	2 to 6 days
Pneumonic	2 to 3 days
Bubonic or septicemic	2 to 7 days
Poliomyelitis	7 to 14 days
Psittacosis	4 to 15 days
Rabies†	2 to 6 weeks (to 1 year)
Rocky Mountain spotted fever	3 to 12 days
Rubella	14 to 21 days
Salmonelloses	
Food poisoning (intraluminal)	6 to 72 hours
Enteric fever (extraluminal)	1 to 10 days
Typhoid fever	7 to 21 days
Shigellosis (bacillary dysentery)	1 to 7 days
Smallpox	12 days
Streptococcal (group A) infections (scarlet fever, etc.)	2 to 5 days

*In this table and throughout the book the length of the incubation period is usually that given in the Red Book of the American Academy of Pediatrics.
†Rabies incubation in dogs is 21 to 60 days.

Continued.

Table 16-3 Incubation Periods of Important Infectious Diseases—cont'd

Disease	Usual Incubation Period
Syphilis, primary lesion	10 to 90 days
Tetanus	3 days to 3 weeks
Trichinosis	2 to 28 days
Tuberculosis, primary lesion	2 to 10 weeks
Tularemia	1 to 10 days
Yellow fever	3 to 6 days

the incubation period, described by such ill-defined symptoms as headache and malaise.

3. *Period of invasion*—time that disease develops fully to maximum intensity.

Invasion may be rapid (a few hours, as in pneumonia) or insidious (a few days, as in typhoid fever). With bacteria, such as pneumococci, that seem to grow without restrictions in a human being, serious manifestations develop with a short time.

At the onset of acute infectious disease, rigors and chills often precede the temperature rise. The sensation of cold is difficult to explain but probably results from a difference between the temperatures of the deep and superficial tissues of the body. To conserve heat, the superficial vessels of the skin constrict, and sweating stops. The skin is pale and dry. As heat loss is decreased, the temperature rises rapidly.

4. *Fastigium or acme*—disease is at its height.

5. *Period of defervescence or decline*—stage during which manifestations subside.

During this state profuse sweating occurs. Heat loss soon exceeds heat production. As the temperature falls, the normal hue of the body returns, and sweating ceases. Defervescence may be by **crisis** (within 24 hours) or by **lysis** (within several days, with the temperature going down a little each day until it returns to normal). Fever that begins abruptly usually ends by crisis. During the stage of **convalescence** the patient regains lost strength.

Many diseases are **self-limited,** which means that under ordinary conditions of host resistance and microbial virulence the disease will last a certain, rather definite length of time, and recovery will take place.

Types of Infections

In a **localized** infection the microbes remain confined to a particular anatomic spot (for example, boils and abscesses). In a **generalized** infection microorganisms or their products are spread generally over the body by the bloodstream or lymphatics. A **mixed** infection is caused by two or more organisms. If a person infected with a given organism becomes infected with still another, a **primary** infection is complicated by a **secondary** one. Secondary infections of the skin and respiratory tract are common, and in some cases the secondary infection is more dangerous than the primary, for example, the group A streptococcal bronchopneumonia that often follows measles, influenza, or whooping cough. Lowered body resistance resulting from the primary infection facilitates the development of the secondary infection.

A **focal** infection is one confined to a restricted area from which infectious material spreads to other parts of the body. Examples are infections of the teeth, sinuses, and prostate gland. An infection that does not cause any detectable manifestations is an **inapparent** or **subclinical** one. An infection held in check by the defensive forces of the body but activated when the body resistance is reduced is a **latent** infection. An infection from the accidental or surgical penetration of the skin or mucous membranes is sometimes spoken of as an **inoculation** infection. Patients with chronic wasting diseases such as cancer often die from the immediate effects of some bacterial infection, especially group A streptococcal and pneumococcal infections. These are **terminal** infections.

When bacteria enter the bloodstream but do not multiply, the condition is **bacteremia.** If bacteria enter the bloodstream and multiply, causing infection of the bloodstream itself, the condition is **septicemia** ("blood poisoning"). When pyogenic bacteria (pus formers) in the bloodstream spread to different parts of the body to lodge and set up new foci of disease, the condition is called **pyemia.** When toxins liberated by bacteria enter the bloodstream and cause disease, the condition is **toxemia.** (Diphtheria is a good example of a toxemia.) Saprophytic bacteria may grow on dead tissue such as a retained placenta or a gangrenous limb and produce poisons that cause disease when absorbed into the body. This is **sapremia.**

SPREAD OF INFECTION
Transmission of Communicable Diseases

The etiologic agents of communicable diseases may be transmitted from the source of infection to the recipient by (1) direct contact, (2) indirect contact, or (3) arthropod carriers.

Direct contact is the term applied when an infection is spread more or less directly from person to person or from a lower animal to a human. It does not necessarily mean actual bodily contact, except with STDs (p. 515), but does indicate a rather close association. **Droplet infection,** infection by microbes cast off in the fine spray from the mouth and nose during coughing, talking, or laughing, is a form of direct contact, as is placental transmission. By direct contact, blood transfusion transmits malaria and viral hepatitis. Diseases that spread directly from person to person include tuberculosis, diphtheria, measles, pneumonia, scarlet fever, the common cold, smallpox, syphilis, gonorrhea, and epidemic meningitis, and from lower animals to humans, rabies and tularemia. (Some of these are transmitted also by indirect contact.)

Indirect contact refers to the spread of the etiologic agent of a disease by conveyers such as milk, other foods, water, air, contaminated hands, and inanimate objects. Infections arising from indirect contact are usually more widely separated in space and time than those arising by direct contact. The diseases ordinarily spread by indirect contact are those in which the infectious material enters the body through the mouth.

Food, including milk, may convey infection. Salmonellosis, bacillary dysentery, and cholera may be spread by contaminated food. Botulism is caused by the consumption of canned foods insufficiently heated. Trichinosis and tapeworm infection are contracted from ingestion of improperly cooked pork.

Air is established in the spread of infection; particles of dried secretion from the mouth or nose (**droplet nuclei**) may float in the air for a considerable time and be carried a rather long distance. Naturally, airborne transfer operates in respiratory infections.

Contaminated fingers are important conveyors of infection. Persons with contaminated fingers may pass infection to other people or to different parts of their own bodies. Fingers easily soil food and drink.

Fomites are inanimate objects that spread infection. The most important are items such as handkerchiefs, towels, blankets, bed sheets, diapers, pencils, and drinking cups. Money is grossly contaminated, especially the small unit coins and paper bills that constantly change hands. Potential pathogens can be cultured from 13% of coins and 42% of bills.

Filth is responsible for disease, and certain diseases are primarily ones of filth. Of these, typhus fever, as it is seen in parts of the world, is an example. However, filth plays little or no part in the spread of typhus fever in the United States, where it is rare. Persons who are unclean in their personal habits are more likely to contract an infectious disease than are those who are clean, and disease is more difficult to control in an unsanitary community.

Arthropods (p. 822), including insects, convey disease mechanically or biologically. In the **mechanical** transfer of disease arthropods merely hold the pathogens on their feet or other parts of their body. Flies can so transfer the agents of typhoid fever and bacillary dysentery from the feces of patients to food to infect the consumer of that food. Flies act as more or less important carriers of 30 diseases. The distance that flies may carry infectious agents is surprising—they have been known to travel miles in search of food.

In the **biologic** transfer of disease the arthropod bites a person or animal ill with a disease or a carrier and ingests some of the infected blood. The microbes of the blood meal undergo a cycle of developmernt within the body of the insect or other arthropod within a specified time. The insect or other arthropod can then transfer the infection to a well person, usually by biting that person. Spread in this manner are malaria, yellow fever, and encephalitis by mosquitoes (Fig. 16-2); typhus fever by lice; Texas fever (a disease of cattle) by ticks; African sleeping sickness by tsetse flies (Fig. 16-3); and plague by fleas. For diseases that are biologically transferred to be prevalent in a community, both the transmitting insect or arthopod vector and hosts to harbor the infection must be present. As a rule, a particular infection is spread by only one species, and a given insect or other arthropod is able to spread only one type of infection. There are, however, important exceptions.

Chapter 37 considers arthropods briefly because of their importance as vectors of disease. Table 37-1 is presented near the end of the book to give an overview of all the diseases so spread that have been discussed in the text.

Sources of Infection in Communicable Diseases

In only a few instances does the infectious agent live in lifeless surroundings outside the body long enough to maintain the source of infection. For the continuous existence of a disease there must be a **reservoir of infection.** Reservoirs of infection and the sources of most communicable diseases are found in practically all communities as (1) human beings or animals with typical disease, (2) human beings or animals with unrecognized

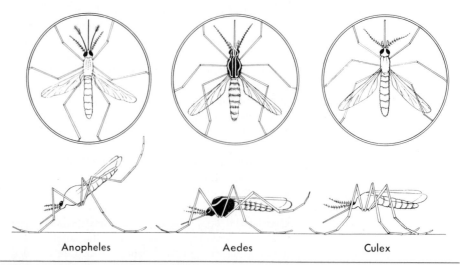

Anopheles   Aedes   Culex

FIG. 16-2   Three mosquitoes as biologic vectors in spread of disease. Typical resting positions aid in identification of the mosquito. Mosquitoes of genus *Anopheles* transmit malaria; members of genus *Aedes* transmit yellow fever and dengue fever; common house mosquito (genus *Culex*) is vector for arboviral encephalitis and filariasis.

disease, and (3) human or animal carriers. Plants may be the reservoir in some fungal infections.

Human carriers are persons who harbor pathogenic agents in their bodies but show no signs of illness. The carrier is infected but asymptomatic. **Convalescent carriers** harbor an organism during recovery from the related illness. **Active carriers** harbor an organism for a long time *after* recovery. **Passive carriers** shelter a pathogen without having had its disease. **Intestinal** and **urinary carriers** discharge it from the body in the feces and urine, respectively. **Oral carriers** discharge infectious material from the mouth. **Intermittent carriers** discharge organisms only at intervals. (Intestinal carriers are often intermittent carriers.) Human carriers play a significant role in the spread of diphtheria, epidemic meningitis, salmonellosis, amebic dysentery, bacillary dysentery, streptococcal infections, and pneumonia.

As a rule, an actual case of a disease is more likely to spread infection to others than a carrier. Carriers and persons with unrecognized disease keep epidemic diseases in existence during interepidemic periods; the number of carriers increases just before the epidemic (for example, diphtheria and acute bacterial meningitis). The prominence of carriers in the spread of a disease depends on the frequency with which people become carriers and the length of the carrier state.

Animals spread disease to humans in a number of ways. A human being may so acquire disease from (1) direct contact with an infected animal, (2) contamination of food by discharges of the animal, (3) arthropod or rodent vectors, (4) contaminated air or water, and (5) consumption of animal products such as milk or eggs.

Rabies is acquired by the bite of rabid dogs, cats, or other animals. Children may contract tuberculosis by consumption of the raw milk of infected cows. Bubonic plague is primarily a disease of rats, transmitted from rat to a human by the flea. Undulant fever may be transmitted to humans by milk, or it may be acquired when a human handles the meat of infected animals. Psittacosis is an infectious disease of parrots in which transmission may be airborne from the sick parrot to a human being. Those han-

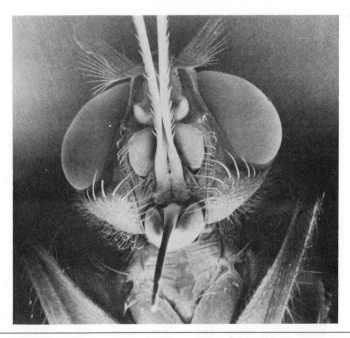

FIG. 16-3   Head of tsetse fly, scanning electron micrograph. (×60.) (Courtesy Eastman Kodak Co., Rochester, N.Y.)

dling the hides of anthrax-infected animals may contract anthrax. Tetanus may be directly contracted from horses because the tetanus bacillus is a normal inhabitant of the intestinal canal of the horse. This is why wounds contaminated by barnyard dirt are likely to be followed by tetanus. Shellfish such as oysters may transmit hepatitis and enteric infections if human excreta contaminate the water in which the shellfish are grown.

All told, about 181 different infections are known to be transmitted naturally from *animals to human beings.* Rats and mice carry more than 15 different diseases, and the casualties from rodent-borne diseases are greater than those from all of history's wars. (Rats are increasing in number in the underdeveloped areas of the world. In parts of Africa and Asia rats may outnumber humans 10:1.)

Animals found in the home as pets may be significant sources of infection (Table 16-4), since many persons live closely associated with their pets. The two most common household pets are dogs and cats (more dogs than cats). An estimated 110 million of them may be in the United States today. In North America the dog can transmit 24 diseases to humans and the cat can transmit 12. Exotic animals that are increasingly popular as household pets may present real hazards. For example, monkeys in the home, pet shop, or even some zoos may spread diseases to humans such as shigellosis, hepatitis, salmonellosis, and tuberculosis.

The most prevalent of the **zoonoses** (infections of animals secondarily transmissible to humans) are cat-scratch disease, salmonellosis, and fungal infections.

Epidemiologic Approach to Disease

Epidemiology is the science that presents the pattern of disease in a given community and that focuses on factors and mechanisms influencing its presence or absence. In this discipline the epidemiologist critically examines interrelationships involving a given pathogenic agent, the environment, and a population group of relevant hosts. Therefore three considerations are basic to the epidemiologic study of disease: (1) the virulence of the given pathogenic agent, (2) the herd immunity, and (3) the environment.

Herd immunity refers to the **collective resistance** to the disease displayed by the community in its environmental setting. An important feature of herd immunity is the ratio of the number of resistant persons to the number of susceptible ones. The application of vaccination, the use of quarantine measures, and the effect of carriers are variables affecting the level of herd immunity.

Environmental factors favor or impede the transmission of a disease-producing agent. These include standards of personal hygiene, the quality of community water supplies, the general state of sanitation, and the like.

Two concepts of importance in epidemiologic studies are the incidence and the prevalence with reference to disease. The **incidence** of disease is the number of new cases per block of population in a specific time period. The **prevalence** of the disease is the number of cases in existence at any given time in that population. An occasional case only in a community is **sporadic** disease. An increase in the number of susceptibles in a population may lead from an **endemic,** wherein a disease is one that constantly is present to a greater or lesser extent, to an **epidemic.** An epidemic is said to exist when a disease attacks a larger number of persons in the community within a short time. It is said that the incidence of diphtheria among schoolchildren would reach epidemic levels if the number of susceptibles in that group would be permitted to exceed 30%. When a disease becomes epidemic in a great number of countries at the same time, it is **pandemic.**

In epidemiologic investigation three basic study plans are used: (1) retrospective (looking back) case-control study, (2) prospective (looking forward) study, and (3) clinical trial. The first two plans are usually done as preliminary investigations before the randomized controlled clinical trial is undertaken. Data are collected from personal inter-

Table 16-4   Sources of Infection in Household Pets

Disease	Dogs	Cats	Farm Animals	Caged Birds (Pigeons, Parakeets, Parrots, Myna Birds, Canaries, and Such)	Poultry	Rodents (Mice, Rabbits, Rats, Hamsters, and Such)	Nonhuman Primates	Reptiles (Snakes, Turtles, Lizards, and Such)
Viral								
Encephalitides			X	X	X	X	X	
Infectious hepatitis							X	
Rabies	X	X	X					
Bacterial								
Anthrax	X		X					
Brucellosis	X		X					
Chlamydial								
Cat-scratch disease	X	X						
Psittacosis-ornithosis				X	X			
Leptospirosis	X		X	X		X		
Melioidosis			X			X	X	
Rickettsial—Q fever			X		X		X	
Salmonellosis	X	X	X	X	X	X	X	X
Tuberculosis	X	X	X		X		X	
Tularemia			X			X		
Fungal								
Ringworm	X	X	X	X	X	X	X	
Protozoan								
Amebiasis	X						X	
Giardiasis	X							
Toxoplasmosis		X	X			X	X	
Metazoan								
Roundworm infection	X	X	X			X		
Scabies	X	X						
Tapeworm infection	X	X	X			X		

views, medical records, vital statistics, and other readily available sources. The controlled clinical trial is the actual experiment. It is usually the strongest study design, but it is also the most expensive and difficult to do.

Spread of Disease in the Jet Age

This is an age of travel at high speeds. More and more people are and will be traveling over the world aboard the jumbo jets or supersonic airliners and consequently exposing themselves to the infections of the geographic localities they visit. The problems in public health are enormous. First, disease is so easily and quickly spread as to jeopardize public health controls. A traveler can bring back a disease into a country that had eliminated it. Second, since no two points on the earth are separated by more than 48 hours' travel time, the tourist can return home during the incubation period (feeling well), to come down with the disease days or weeks later. By this time, the diagnosis may be far from obvious. Third, the world sightseer may fail to take adequate precautions against a disease that person inevitably contacts. The best example of this is malaria, the major risk. It is a widespread disease, but nowhere is chemoprophylaxis required. Travelers must find out for themselves what to do for protection. Even then, they may fail to take the suppressive drugs as required or may discontinue the medication after they have left an infected area, believing suppressive medication is no longer needed. Fourth, still another problem has to do with the establishment of international standards and guidelines for sanitary facilities, which are required for tourists in all geographic areas. Maintenance of sanitary facilities is mandatory aboard the big jets, in airport terminals, and in the accommodations associated therewith.

SUMMARY

1 When a microbe enters the human body, multiples there, injures tissue cells, and induces a reaction to its presence, an infection occurs. Outward signs of infection are sometimes absent.

2 Known microbes normally reside in many anatomic areas of the human body, and the integrity of the normal flora of the area is essential to the well-being of the host.

3 Infectious microbes follow reasonably consistent anatomic patterns of entry into and exit from the human host.

4 Absolute proof that an organism causes a given disease rests on the fulfillment of Koch's postulates. The organism observed in every case of the disease must be isolated, be grown in pure culture, cause the given disease in a susceptible animal, be recovered from the experimental animal, and have its identity confirmed.

5 Factors influencing the event of infection include the portal of entry, virulence of the microbe, number of invaders, and defensive powers of the host.

6 Important to disease-producing microbes is their ability to adhere to and colonize body surfaces, to invade directly body tissues, and to elaborate substances poisonous to body tissues.

7 Harmful bacterial products include hemolysins (several kinds), toxins (exotoxins and endotoxins), leukocidins, coagulase, and bacterial kinases.

8 General features of most infections are fever, increased pulse rate, increased metabolic rate, and signs of toxicity.

9 An infection often displays the following stages: (1) incubation, (2) prodrome, (3) invasion, (4) acme, and (5) defervescence or decline.

10 The etiologic agents of communicable diseases may be spread from the source of infection to the recipient by (1) direct contact, (2) indirect contact, and (3) arthropod carriers.

11 Sources of most communicable diseases found in most communities are (1) humans

with typical disease, (2) humans or animals with unrecognized disease, and (3) human or animal carriers.

12 Some 181 different infections are known to be transmitted naturally from animals to human beings. Rats and mice carry more than 15 different diseases. The casualties from rodent-borne diseases are greater than those from all of history's wars.

13 Epidemiology is the science that presents the pattern of disease in a given community and that focuses on factors and mechanisms determining its presence or absence.

QUESTIONS FOR REVIEW

1 Explain infection and contamination.
2 Give the routes by which microbes enter the body. Name two diseases whose causative agent enters the body by each route.
3 What is virulence? How do microbes produce disease?
4 What is one important local effect of bacterial invasion?
5 List some of the general effects of bacterial invasion.
6 What is the difference between bacteremia and septicemia?
7 Give the routes by which disease-producing agents leave the body. Name two agents eliminated by each route.
8 List five diseases spread by animals to humans.
9 Define carrier. List the different kinds.
10 Name three diseases spread by carriers.
11 List important arthropod vectors and diseases so spread.
12 State Koch's postulates.
13 What are toxins? Name three species of bacteria important as exotoxin producers and two producers of endotoxins.
14 What are toxoids? For what are they used?
15 Define hemolysin, leukocidin, coagulase, and streptokinase. Name a microbe producing each. What is the application of the coagulase test?
16 What is hyaluronidase? Why is it sometimes called the "spreading factor"? In what infections with what microbes is it important?

SUGGESTED READINGS

Abrams, H.L.: Special report: medical problems of survivors of nuclear war, infection and the spread of communicable disease, N. Engl. J. Med. 305:1226, 1981.

Anderson, R.M., and May, R.M.: Infectious diseases and population cycles of forest insects, Science 210:658, 1980.

Keusch, G.T.: Recognition mechanisms in infectious disease, Hosp. Pract. 14:33, Aug. 1979.

Thacker, S.B., and others: The surveillance of infectious diseases, J.A.M.A 249:1181, 1983.

THE BODY'S DEFENSE

PROTECTIVE MECHANISMS

If infection occurred each time infectious and injurious agents entered our bodies, we would be constantly ill. When microorganisms attempt to invade the body, they must first overcome certain mechanical, physiologic, and chemical barriers existing on the body surface or in the body cavity at the site of entry. The body not only reacts to the event of invasion but is also endowed with a measure of protection against it.

The very important *immunologic* mechanisms for defense and protection of the human body are discussed in Chapter 18. Here we focus on certain inherent, nonspecific, and nonimmunologic factors.

Anatomic Barriers

During their agelong struggle for existence, human beings have developed certain mechanisms that enable them to overcome many agents of potential harm in their environment. In the first place nature has provided them with special senses that act as watchdogs against danger. Human beings protect themselves against major injuries from moving objects by batting their eyes and jumping aside. Although some of these acts are voluntary, they are executed as reflex movements and therefore may be considered protective natural mechanisms.

The body's first line of defense is the epithelium that covers the exterior of the body and lines its internal surfaces. A condensation of resistant cells along the most superficial part of the skin offers a barrier against physcial and chemical injuries or microbial invasion. Epithelium may become quite thick at a site of irritation (for example, a callus or corn in the skin). The lining cells of the mucous membranes opening on body surfaces also afford a barrier, but these cells are softer and more vulnerable to injury than are the surface cells of the skin. The hairs in the anterior nares protect the respiratory tract by filtering bacteria and larger particles from the inspired air. The respiratory passages are lined with epithelial cells from the surface of which spring hair-like appendages, known as cilia, that sweep overlying material from the deeper portion of the tract to the upper portion from where it may be discharged to the environment.

Closure of the glottis during swallowing prevents food from entering the respiratory tract. Coughing and sneezing serve to expel mechanically any irritating materials from the respiratory tract. Peristalsis mechanically rids the intestinal tract of irritants. Urine, in the act of micturition, flushes out bacteria from the urinary tract.

Chemical Factors

Body cavities opening on the surface are protected by secretions that wash away bacteria and foreign materials. The physical consistency of the mucus of the respiratory and alimentary tracts helps to entrap as well as eliminate mechanically microbes and other foreign particles. Body fluids such as saliva, tears, gastric juice, and bile exert an antiseptic action that reduces microbial invasion. For instance, the acid gastric juice destroys bacteria and almost all important bacterial toxins except that of *Clostridium botulinum*. Because of their acid reaction and content of fatty acids, sweat and sebum (sebaceous gland secretion) on the skin have antimicrobial properties. Fatty acids are toxic to gram-negative bacteria. Fatty acids result from the combined action of *Staphylococcus epidermidis* and *Corynebacterium acnes*, both gram-positive residents of the sebaceous glands, on the lipids of sebum. Most microbes require iron for growth. During an infection the body can withhold iron, thus inducing a decreased concentration of iron in the plasma (hypoferremia). The result is inhibitory to the intruder, since there is less iron available to that microbe.

Lysozyme

The antimicrobial action of mucus and tears is partly related to the presence of lysozyme (muramidase), found in white blood cells and body secretions from organs exposed to airborne bacteria. Lysozyme, a substance dissolving the cell wall of certain gram-positive bacteria, is a first line of defense against a gram-positive microorganism. It works to hydrolyze linkages in the peptidoglycan molecule. It has no antiviral activity. An enzyme, not an antibody, lysozyme was discovered by Fleming in 1922— before he discovered penicillin.

Interferon

Another internal defense is a naturally occurring substance, interferon (not an antibody) (pp. 261-262). Its target action is against viruses to block viral infection. Many cells form it quickly if threatened by viral attack, and the substance diffuses to nearby, uninfected cells, protecting them not just against the challenging virus but also against any virus. (Interferon can be produced in all cells and almost anywhere in the body.) Although it is a natural product, it is produced in minute quantities. To extract but a few grams of crude natural interferon, about a ton of leukocytes has to be treated with virus.

Interferon is not a single substance but a family of low molecular weight glycoproteins secreted in response to viral and mitogenic stimuli. All thus far characterized are 165 or 166 amino acids long. At least 15 interferons, coded for by multiple structural genes, are classified into three distinct types: alpha, beta, and gamma (Table 17-1). Alpha-interferon is elaborated by leukocytes in response to viral infection or stimulation with double-stranded RNA. A number of genetic variations are known. Beta-interferon is made by fibroblasts in response to the same stimulus as alpha-interferon; it comes in two genetic varieties. Only one kind of gamma-interferon is known. It is made by lymphocytes after mitogenic stimulation. (Phytoagglutinin is an example of a mitogen, an agent that induces mitosis and consequently cell division.)

Although interferon is well known for antiviral properties, the underlying mechanism of its action is not well defined. Interferon does act on the cell surface, not in the interior, its activity involving a receptor site on the exterior of the plasma membrane. It affects cell function only after it has been excreted into the extracellular fluid. It has been demonstrated to inhibit DNA synthesis and thereby cell division. This important property has obvious implications for anticancer therapy. Interferon is present before mechanisms of humoral and cell-mediated immunity (Chapter 18) are developed. Because effects of interferon on immunologic function have been demonstrated, investigators have suggested that immunoregulation may be one of its important actions.

Normal Flora in Host Defense

An important factor in host defense is the presence of the normal flora, that characteristic assortment of bacteria and other microbes found in a given anatomic area in contact with the environment (p. 312). There, conditions have been established that are favorable to the growth of the residents: the oxygen tension, the level of nutrients, the particular pH, and the provision for the accumulation of toxic substances associated with bacterial growth. The same situation may not be at all acceptable for the growth of a newcomer and possible pathogen.

A number of ways have been demonstrated whereby normal flora block potential pathogens from gaining a foothold. Some examples follow:

1. Bacteriocins (p. 98) produced by resident bacteria prevent colonization by pathogens in the alimentary and lower urinary tracts. For instance, bacteriocins produced by *Streptococcus viridans* in the throat provide a barrier there to the intrusion of pathogenic pneumococci, streptococci, and gram-negative bacilli.
2. Normal flora can suppress the adherence of would-be pathogens to an epithelial surface, particularly when they completely carpet that surface.
3. Normal flora influence certain clearance mechanisms already operating to rid their area of undesirables. For instance, they may stimulate the peristaltic movements of the intestinal tract.

Table 17-1 Comparison of the Three Types of Interferon

Current Terminology	Former Name	Produced Primarily by	Induced by	Molecular Weight	Genetic Types
IFN-alpha	Le (leukocyte), type I	B lymphocytes	Viruses Polynucleotides Bacterial products Foreign cells Tumor cells	18,000	14 or more
IFN-beta	F (fibroblast), Fi, type I	Body's solid tissue cells	Viruses Polynucleotides Bacterial products	38,000	2
IFN-gamma	IIF (immune), type II, T	T lymphocytes	Antigens Antigen-antibody complexes Mitogens Antilymphocyte serum	30,000 to 100,000	1

Physiologic Reserves

The blood supply of many parts of the body is protected by a series of **anastomoses** (connections) among the branches of the supply vessels. If a vessel is occluded, the blood is detoured around the obstruction by way of the anastomoses. A **collateral** or **compensatory circulation** is set up.

When the body becomes chilled, the superficial vessels of the skin contract to prevent heat from being dissipated at the surface, and sweating stops to slow cooling caused by evaporation. To aid in this conservation of heat, the involuntary muscles of the skin contract (gooseflesh). In animals having hair or feathers, the hair or feathers are raised to enclose in their meshes a thick layer of air, which is a poor conductor of heat. The muscular activity associated with rigor and shivering increases heat production. When the body becomes too hot, the superficial vessels dilate and sweating occurs. Heat is dissipated from the surface, and evaporation has a cooling effect.

The functional capacities of the vital organs (organs necessary for life) extend far beyond the normal demands of life. We have much more liver, pancreas, adrenal gland, and parathyroid gland tissue than we ordinarily use. That we are able to lead a normal life after one kidney has been removed or after one lung has collapsed illustrates the abundance of reserve possessed by our vital organs.

Vital organs (brain, heart, lungs) are enclosed within protective bony cases, and their surfaces are bathed with a watery fluid. The fluid surrounding the brain serves as a water bed to prevent undue jarring, and the fluid that bathes the surface of the heart and lungs acts as a lubricant.

Fever

Fever is a condition characterized by an increase in body temperature. Fever may be related to many factors that incite an inflammatory process and may be associated with a variety of events ranging from mechanical to microbial injury, but perhaps most regularly with bacterial infection. Although manifestations of fever may be disagreeable, it is primarily beneficial and generally in an infectious illness is a good prognostic sign. If the temperature is not too high, fever accelerates the destruction of injurious agents by an increase in phagocytosis and the induction of immune mechanisms (Chapter 18). A degree or two rise in body temperature slows the growth of some organisms, for example, the spirochete of syphilis.

In health, human body temperature remains remarkably constant and, as determined by a thermometer placed in the mouth, ranges from 96.7° F (35.94° C) to 99° F (37.22° C) with an average of about 98.6° F (37° C). The rectal temperature is about 1° F higher and the temperature taken in the armpit is about 1° F lower than the oral temperature. The rectal temperature is the most dependable because both oral and axillary readings are influenced by outside factors. The temperature of the internal organs is 2° or 3° F higher than that of the skin.

The body's thermostat is a thermoregulatory center located in the hypothalamus, which integrates the responses from various thermal receptors throughout the body. The physiologic thermostat is affected by a substance released into the bloodstream when a macrophage engulfs an invading microbe. With fever the center still regulates the body temperature but at a higher level than normal. (The setting on the thermostat is raised.)

There is a close relation between pulse rate and temperature. Ordinarily an increase of around eight pulse beats per minute occurs for each 1.8° F increase in temperature. Certain exceptions to this rule are diagnostically significant. For instance, the pulse in typhoid fever, malaria, miliary tuberculosis, and yellow fever is comparatively slow.

Cellular Resources

Erythrocytes (red blood cells), leukocytes (white blood cells), and platelets are three groups of formed elements found in the blood at all times. The red blood cells participate in the physiologic exchange of oxygen and carbon dioxide that takes place in the respiratory passages of the body, and platelets are essential to the blood-clotting process. Major components of the body's basic, ongoing defense plan are the white blood cells. White blood cells both in the bloodstream and in the body tissues are constantly on guard. They function to maintain the normal state, and in the event of injury to the body their numbers increase and their activities intensify. In an encounter with an injurious agent white blood cells are referred to as inflammatory cells, and the subsequent sequence of events is the process of inflammation (p. 323).

Kinds of Cells

Unlike red blood cells, which in their mature state are a single type, there are three kinds of white blood cells. According to their appearance and activities, they are designated monocytes (part of the macrophage system), granulocytes, and lymphocytes (part of the lymphoid system).

Role of Bone Marrow

Blood cells, including those important in the body's defense, are formed in the bone marrow, a dispersed tissue that functions as an organ. Although it is distributed in certain bones of the body, it has a mass comparable to that of the liver. In the adult human active red marrow is found in the trunk and skull bones and in the upper end of each extremity.

Bone marrow is a soft, pulpy, highly vascular, sinusoidal tissue. Within the sinusoids are primitive, nonfunctional but multipotential cells, designated **stem cells**. These are the ultimate ancestors of the formed elements of the blood (the red cells, white cells, and platelets). From these very early cells, the development of the different blood cell lines proceeds in an orderly and defined sequence (Fig. 17-1).

THE MACROPHAGE SYSTEM (RETICULOENDOTHELIAL SYSTEM)

The system of macrophages (reticuloendothelial [RE] system) with its widely dispersed cells constitutes one of the main organs of the body designed primarily for a human being's defense against both the living and the nonliving agents in the environment that harm that human being. The macrophages not only can directly fight the invader,

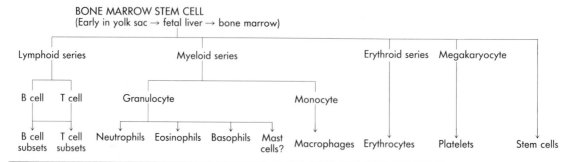

FIG. 17-1   Blood cell formation begins in the early yolk sac of the embryo and continues in the liver of the fetus. After birth, this function is taken over by the bone marrow, where it remains throughout life. The marrow stem cell maintains the formed elements of the blood— red blood cells, white blood cells, and platelets. It seeds other organs with these cells and renews itself through the elaboration of more stem cells. Immature granulocytes (the precursor cells) in the bone marrow are often called myeloid cells; myeloid series is the designation for the cell line.

which they do by phagocytosis, but can also clean up the debris and eliminate the cellular and metabolic breakdown products incident to the encounter. In short, they are also scavengers.

Phagocytosis

The ingestion of microbes or other particulate matter by cells is known as phagocytosis, and the cells that ingest such materials are **phagocytes.** Phagocytosis resembles closely the feeding process of unicellular organisms and is a universal response on the part of the body to penetration or invasion by microbes, alien cells, or other foreign particles. It is an essential protective mechanism against infection and probably plays an important part in natural immunity. Phagocytosis is a general process because the few kinds of body cells having this power can ingest many different kinds of particulate matter (bacteria, dead body cells, mineral particles, dusts, and pigments).

Study of the process has revealed that phagocytosis takes place in three phases (Fig. 17-2). First, the particle (for example, a microbe) to be engulfed is isolated and incorporated into an invagination of the cell membrane of the phagocyte. The result is a **phagosome,** a vacuole (within the cell) surrounded by cytoplasm. Next, a burst of metabolic activity ensues within the cell. The phagosome is moved into the interior of

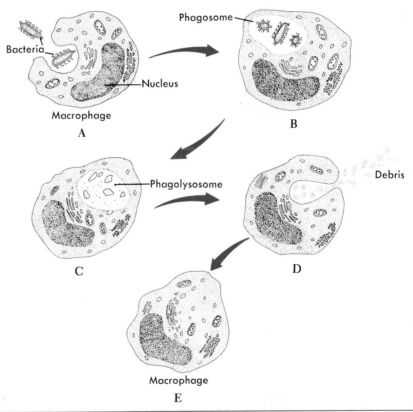

FIG. 17-2 Phagocytosis by a macrophage, sketch. **A,** Opsonized bacteria (p. 317) engulfed by phagocyte (macrophage). **B,** Phagosome formed. **C,** Phagosome becomes phagolysosome; bacteria digested. (To this point, process of phagocytosis is comparable in either a macrophage or neutrophil, not shown.) **D,** Debris is egested. (Neutrophil would succumb here.) **E,** Macrophage returns to resting state. (From Smith, A.L.: Microbiology and pathology, ed. 12, St. Louis, 1980, The C.V. Mosby Co.)

the cell, contacting its lysosomes. When hydrolytic enzymes are released into the phagosome, it becomes a **phagolysosome.** A special kind of microbicidal system is activated
in preparation for the second phase, in which occurs the killing or inactivation of the
microbe, so that in the third phase the microbe can be digested.

If, after ingesting bacteria, phagocytes fail to destroy them, the phagocytes themselves may be destroyed, with liberation of the ingested organisms. Relatively harmless
microbes are usually completely destroyed. Sometimes the microbes persist and even
continue to multiply within the cytoplasm of the phagocyte. Virulent bacteria, especially those with capsules, resist phagocytosis.

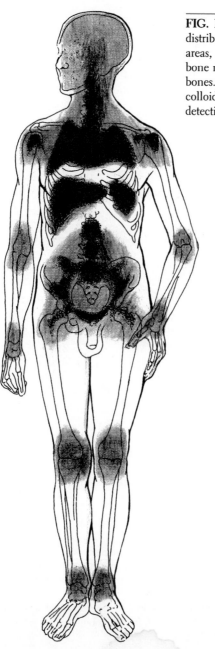

FIG. 17-3   The macrophage system. Note anatomic
distribution of maximal activity as indicated by black
areas, including definition of liver, spleen, and active
bone marrow in axial skeleton and proximal parts of long
bones. To produce such an image, certain radioactive
colloidal particles are given to subject, and with radiation
detection technics, tissue uptake is delineated.

The student who wishes to see bacteria undergoing phagocytosis should stain a smear from an ordinary boil, whereon staphylococci can be found easily within the cytoplasm of the leukocytes. Smears from gonorrheal exudates and the cerebrospinal fluid in epidemic meningitis usually show organisms undergoing phagocytosis inside leukocytes. Different bacteria vary in their susceptibility to phagocytosis. For instance, gonococci undergo phagocytosis readily, whereas tubercle bacilli are quite resistant.

Phagocytosis is not as great in the first 3 years of life as it is later on.

Opsonization

The process of making microbes more susceptible to destruction by phagocytes is known as opsonization. It involves both immune and nonimmune proteins (complement components), since these proteins provide a bond on the surface of the microbe to which phagocytes can attach and initiate the sequence of events in phagocytosis. Substances that act on certain bacteria, fungi, and other particles to increase their palatability to phagocytic cells are called **opsonins**. An organism such as *Escherichia coli* can resist opsonization because its surface contains a large amount of sialic acid.

Phagocytic Cells

Three major types of phagocytic cells with antimicrobial properties follow: (1) polymorphonuclear neutrophilic leukocyte—a cell that is most important in the host defense against acute bacterial infection; it is responsible for many manifestations of acute inflammation because of the nature of its contents and metabolic products; (2) monocyte—a cell that becomes the macrophage or histiocyte in tissues; the macrophage is peculiarly efficient in killing microbial pathogens that survive inside other cells; and (3) eosinophil—a cell that seems to function in host defense against multicellular parasitic invasion and also as a modulator of the immediate immune response (p. 380); its phagocytic expression is poorly understood. All three of these cells are related to processes of immunity discussed in Chapters 18 to 20.

Anatomy of the System of Macrophages (RE System)

The macrophage system comprises one group of cells in the human body especially endowed with phagocytic powers—both the cells themselves and those to which they give rise. Although widespread throughout the body (Fig. 17-3), these cells are concentrated in lymph nodes, spleen, bone marrow, and liver, and although scattered, these uniform cells are spoken of collectively as the **macrophage system**. There are two main cell types: (1) the stationary or fixed (littoral) cells and (2) the free or wandering cells. The wandering cells in tissues are known as **macrophages**; in the bloodstream, as **monocytes**.

The Principal Cell

The macrophage, because it moves about in body fluids and because its observed properties of phagocytosis are dramatic, receives a lot of attention. A variety of names—histiocyte, epithelioid cell, clasmatocyte, multinucleated giant cell (if confluence of several cells)—relate this cell type to different circumstances and reflect an active participation in diverse roles.

Cellular forebears of the macrophage are identified in the bone marrow as **monoblasts** and **promonocytes,** which progress there into monocytes. At this level of development the monocyte is released into the peripheral blood. After circulating there for only a few hours, the monocyte leaves the bloodstream, never to return, to migrate into the tissues where it completes its development into a macrophage and may live thereafter for months. The anatomic area selected modifies somewhat the structural and functional changes that ensue. The cell becomes larger in size (up to 15 to 25 μm) and may even become multinucleated, and the number of its cytoplasmic lysosomes and mitochondria increases. When the monocyte has acquired the property of phagocytosis, the capacity to make protein, and surface receptors for immunoglobulin G and certain complement components, it is a fully mature macrophage.

The macrophage is well equipped for the phagocytic functions it performs with an abundance of lysosomes (cytovesicles) filled with varied enzymes—proteases, nucleases, lipases, and phosphatases. It is designed to collect and remove cellular detritus, tissue breakdown products, and extravasated blood (hemorrhage). It stands as the first line of defense against facultative and obligate intracellular parasites such as the tubercle bacillus (*Mycobacterium tuberculosis*). Interacting with lymphocytes in the immune response (Chapter 18), it processes—binds, ingests, and degrades—antigen.

Table 17-2 summarizes the prominent features of the macrophage.

Granulomatous Inflammation

In granulomatous inflammation the disease-producing agent is dealt with directly by the macrophages (the main reacting cells). The formation of tiny nodules encompassing an ovoid arrangement of macrophages is characteristic. These nodules are called **granulomas** and are usually walled off from surrounding tissues by fibrous tissue. Granulomas are associated with chronic diseases such as tuberculosis, syphilis, leprosy, and certain fungal infections. The granuloma of tuberculosis is the **tubercle**; of syphilis, the **gumma**; of leprosy, the **leproma**.

GRANULOCYTES

Granulocytes are cells so called because of distinctive granules within their cytoplasm. When granulocytes are viewed microscopically on Wright-stained blood smears, three kinds are seen: the **basophil** (p. 379) (with an affinity for basic dyes), containing coarse bluish black granules; the **eosinophil** (p. 380) (with an affinity for acid dyes), containing coarse round reddish granules; and the **neutrophil** (which stains with neutral dyes), containing many small lavender granules within its cytoplasm.

The Neutrophil

The most important of the granulocytes is the **polymorphonuclear neutrophilic leukocyte,** a cell especially designed to phagocytize bacteria, that is, a microphage. (A *mi-crophage* ingests small particles, notably bacteria; a *macrophage*, larger ones.) The term

---

Table 17-2 Characterization of the Cell of the Macrophage System

Formation in	Bone marrow
Maturation in	Bone marrow and bloodstream
Concentration in	Liver, spleen (sinusoids), lymph nodes (medullary sinuses), lungs, peritoneum, bloodstream
Life span	Months to years
Primary products	Hydrolytic enzymes, complement components, interferon
Functions	Antigen processing, presentation; phagocytosis (most active of body's phagocytic cells) of pathogenic cells and inert particles; participation in inflammatory process
Various forms	Monocytes in bloodstream, Kupffer cells lining sinusoids in liver, histiocytes in connective tissues, Langerhans' cells in skin, peritoneal macrophages in peritoneal cavity, alveolar macrophages or dust cells in lungs, microglial cells in nervous system, giant cells and epithelioid cells in certain inflammatory processes

polymorphonuclear refers to the several large segments or lobes usually displayed by the nucleus of this granulocyte. This is a well-known cell and familiarly called the "poly," "PMN," and "pus cell" (because of its association with pus formation).

In the development of the neutrophil in the bone marrow, certain metabolic changes must take place in the immature myeloid cells to equip the mature neutrophils for the type of milieu in which they often have to express their biologic role, such as the oxygen-poor medium of an inflammatory site. When maturation is complete, the neutrophil is released into the bloodstream; in fact, billions of them leave the marrow each day to circulate throughout the body. Whereas the maturation sequence in the bone marrow extends over several days, the life span of the neutrophil in the peripheral blood is but several hours.

With the majority of bacterial inflammations, an increase in the number of these cells occurs at the site of injury in the body and in the bloodstream. An increase of circulating leukocytes is termed **leukocytosis.** In some infections, particularly those caused by viruses or those complicating severe gram-negative bacteremia, a decrease in number, a **leukopenia,** may be present. In conditions that produce leukocytosis, the phagocytic power of the leukocytes is increased in patients who are doing well. The phenomenon whereby neutrophils (motile cells) are specifically drawn or attracted to an inflammatory reaction is **chemotaxis,** and their accelerated formation in the bone marrow and rapid accumulation at the site follow the release of a variety of chemical (chemotactic) factors from the anatomic area of injury. Neutrophils are rich in proteolytic enzymes, known as **leukoproteases,** contained within their lysosomes. Leukoproteases participate in the intracellular digestion of phagocytized particles.

Call it what you will, the polymorphonuclear neutrophilic leukocyte is the major circulating white blood cell and has long been known for its crucial role in the body's defense.

LYMPHOID SYSTEM (IMMUNOLOGIC SYSTEM*)
Lymphoid Tissue

Closely interwoven with the macrophage system and encompassed by it is the lymphoid system (the immune system), which comprises the lymphoid tissues and organs of the body. Lymphoid tissue is widely distributed in the body; it is concentrated in lymph nodes, spleen, thymus, Peyer's patches (in the ileum of the bowel), small nodules in the bone marrow, and nodules scattered in the lining of the alimentary tract. Its major constituent cells are lymphocytes, which stand out in its mixed population of plasma cells, macrophages, and respective precursors (forebears). Lymphocytes are small round cells with a relatively large and darkly staining nucleus, and their constant companions, the plasma cells, are slightly larger cells with an eccentric nucleus of unique appearance.

In lymphoid tissue the component cells are packed more or less densely onto a spongelike support provided by a special type of stroma, and collections of lymphocytes, including precursors and progeny, in a well-defined, usually rounded aggregate constitute a **lymphoid nodule.** A nodule may be solitary, or several nodules may be grouped together, as in Peyer's patches of the lower small intestine. What is referred to as gut-associated lymphoid tissue includes the solitary nodule of the lower bowel (Fig. 17-4), Peyer's patches, and the diffuse arrangement of lymphoid tissue that makes up the lymphoid stroma (lamina propria) of the mucous membrane of stomach and intestine.

Lymphoid tissue organized into a small bean-shaped organ well enscribed by a connective tissue capsule becomes a **lymph node** (Fig. 17-5). The outer part is referred to

---

*Immunologic function is an important property of lymphoid tissue. Mechanisms are discussed in Chapter 18.

as the **cortex;** the inner part is referred to as the **medulla.** In the cortex there is an orderly alignment of lymphoid follicles. Lymph nodes are found throughout the body, strategically placed in all major organ systems and body areas. Lymphocytes circulate in the bloodstream as single cells and account for one fifth of the white blood cells present.

FIG. 17-4   Solitary lymphoid nodules of lower ileum (small bowel). Mucosal surface of bowel showing thereon numerous small pearly gray foci of lymphoid tissue.

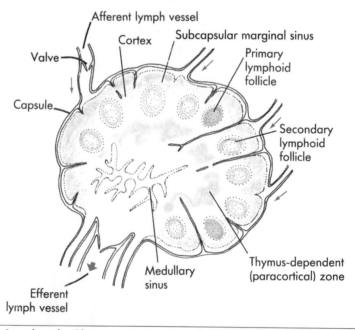

FIG. 17-5   Lymph node. Observe anatomic pattern for immunologically important areas in this sketch. Arrows indicate flow of lymph.

Lymphatic Circulation

In the exchange of nutrients for waste products that goes on continuously between the cells and the blood of the capillaries, excess fluid (lymph) leaks into the tissue spaces. From here it moves into the lymphatic capillaries, the many small thin-walled tubes of the lymphatic circulation draining the vast intercellular region of the body. The fluid circulates as lymph into successively larger vessels, and ultimately through the main lymphatic trunks, the right lymphatic duct and the thoracic duct (on the left), it pours back into the bloodstream. Along the course of the lymphatic channels at definite in-

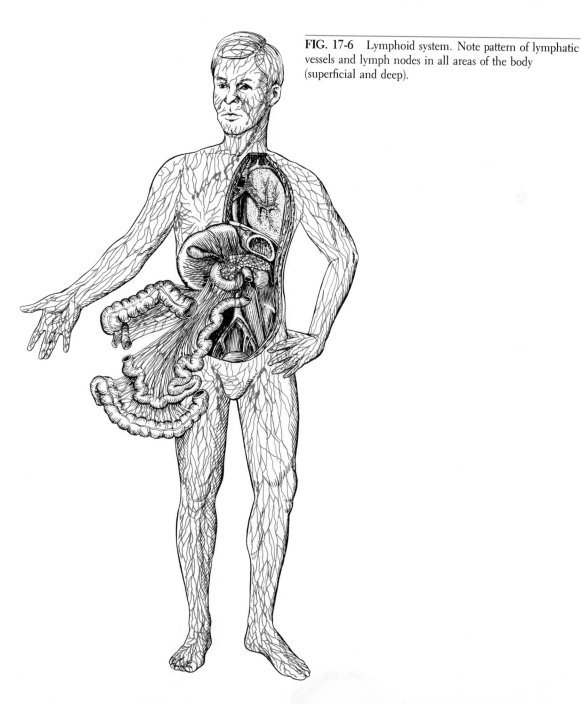

FIG. 17-6   Lymphoid system. Note pattern of lymphatic vessels and lymph nodes in all areas of the body (superficial and deep).

tervals are placed the lymph nodes in chain formations (Fig. 17-6). All the lymph of the body is filtered and strained through these chains.

Spleen

The largest mass of lymphoid tissue in the body is the spleen, found in the upper left abdomen as a large, encapsulated lymphoid organ adapted to a peculiar circulation of blood. Like the lymph nodes it is a filter, but unlike them it strains the blood not the lymph stream. It functions as an important member of the macrophage system in the filtration of foreign particles from the blood, including living organisms. Fixed and free cells of the system are found here.

Thymus

The thymus, a key organ in immunologic processes, is situated in the front part of the chest just behind the breast bone. It is a lobular, partly epithelial and partly lymphoid organ that is present at birth and continues to enlarge until puberty. After puberty it shrinks (atrophies) and is soon replaced by fatty tissue in the adult (Fig. 17-7). (This sequence of events has always suggested that the thymus plays a significant role in the person's development.) At or about the time of birth it processes certain lymphocytes, which then migrate out to colonize the spleen, lymph nodes, and other areas of the body. These in turn give rise to cells that maintain cell-mediated immunity (Chapter 18).

Lymphatic Vessels in Infection

When an agent gains access to the tissues, it may sometimes be carried away from the site of entry in the lymphatic circulation and be deposited in the nearest lymph nodes. (In the lungs of persons residing in coal-burning regions, coal dust is thus deposited in drainage nodes.) Similarly, bacterial invaders may be borne from the site of disease to the regional nodes, where their presence stimulates phagocytic cells normally present

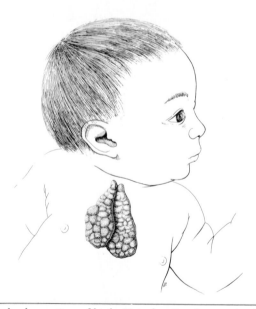

FIG. 17-7 Thymus gland near time of birth. Note functional state, size (larger than baby's fist), and position in upper chest. By contrast thymus gland of adult is a functionless, shriveled, and ill-defined mass of tissue.

to increased activity. Ideally, these phagocytes dispose of the invaders. Should they fail to do so, the bacteria flow with the lymph stream through successive nodes in the chains into the thoracic duct, from there to be emptied into the bloodstream (bacteremia) and disseminated throughout the body. Infection of the bloodstream (septicemia) before the days of antibiotic therapy was a rapidly fatal condition.

Responses in Lymphoid Organs.

An intimate relationship exists between elements designated as lymphoid and those designated as phagocytic. If an infection is localized, the macrophages of the regional lymph nodes proliferate and the nodes enlarge as a consequence (lymphadenopathy). Regional lymph nodes are an important second line of defense. Many injurious agents carried there are destroyed. If an infection becomes generalized, the macrophage system throughout the body responds. One notable feature of this is the enlargement of the spleen (splenomegaly) seen with such acute infectious processes as malaria, typhoid fever, and septicemia.

INFLAMMATION
Definition

Inflammation is the sum of the reactions in the body incited by an injury. Within itself, inflammation is not a pathologic condition (although usually discussed as one), but rather it is an exaggeration of physiologic processes set in motion by an irritant. The initial purpose of inflammation is to destroy the irritating and injurious agent and to remove it and its related by-products from the body. If this is not possible, the inflammatory process serves to limit its extension through and effects on the body. Finally, inflammation is the mechanism within the body for the repair and replacement of tissues damaged or destroyed by the offending agent.

The inflammatory process consists of three parts: (1) localized vascular and cellular responses at the site of injury, (2) general body reactions (fever, leukocytosis, formation of antibodies), and (3) events designed to repair the injury done and restore the part to normal. Reparative processes are so intimately associated with inflammation that authorities consider them to be a definite part of it. Reparative processes come into play almost with the beginning of the inflammatory reaction.

Causes

The injury in inflammation may be produced by a great number of agents, either living or nonliving. Inflammation produced by a living agent is called **infection.** The most important living agents are bacteria. Also important, however, are viruses, fungi, protozoan parasites, metazoan parasites, arthropods, and higher plants and animals. Inflammation produced by immunologic reactions is a special category accompanying allergic and autoimmune states (Chapter 19), and the immunologic injury to cells is responsible for the manifestations of disease.

Agents causing inflammation are diverse, but basic pathogenetic mechanisms are essentially the same. The common denominator seems to be injury to body cells, with death required of a baseline number, for dying cells release at the inflammtory site certain chemical substances called **mediators.** Although the pathogenesis of inflammation is as yet incompletely understood, it can be shown experimentally that mediators do trigger key vascular and cellular components of the reaction proper. Some of the ones best studied include histamine; serotonin; the polypeptides leukotaxine, bradykinin, and kallidin; and the proteases plasmin, kallikrein, and globulin permeability factor. Mediators not only figure in local responses, but if they enter the bloodstream, they also induce certain systemic reactions.

Local Changes

After an irritant has gained access to a tissue and injured some of the cells, the vascular response is triggered. There is a brief transient period of vasoconstriction, after which the small vessels and capillaries dilate and become filled with blood as the speed of the blood current increases. The increased velocity of blood that occurs with the dilation of the small vessels results from a more direct transmission of arterial pressure to these vessels. The current of blood then slows because of an increased viscosity of the blood and swelling of the endothelial cells lining the small vessels. At about this time, the walls of the vessels have begun to leak the cellular and liquid constituents of the blood.

When the blood is flowing normally, the red cells and leukocytes travel along in an axial stream surrounded by a zone of plasma, but when the current slows, the leukocytes fall out peripherally into the plasma zone (margination) and adhere to the vessel wall. By their ameboid activity the polymorphonuclear neutrophilic leukocytes pass between the poorly cemented endothelial cells to initiate the cellular response (emigration). Once outside the vessel, they travel in the tissue spaces toward the injurious agent. Red blood cells, platelets, and plasma also escape from the vessels. One of the plasma proteins, fibrinogen, precipitates from the plasma to form fibrin. An **inflammatory exudate** soon surrounds the irritant agent. It is formed by the accumulation of cells and fluid from the lumen of blood vessels of the injured area, and each component plays its own specific part in the ensuing battle. The polymorphonuclear neutrophils remove bacteria, cellular debris, and solid particles by phagocytosis. The plasma (known as serum after fibrin has formed) brings antibodies to the scene, dilutes toxins that microbes produce, and floats away particles of debris. Wandering macrophages attracted to the site and certain leukocytes other than polymorphonuclear cells clear the ground for repair; fibrin forms a restraining wall around bacteria and acts as a framework for the repair of the destroyed tissues.

At the beginning of the inflammatory reaction the flow of lymph away from the site of inflammation is increased but is later decreased. The lymph vessels may bear microbes or other inflammatory agents to the regional lymph nodes, where they set up inflammatory reactions. Lay persons often refer to these swollen lymph nodes as *kernels*.

Timing in Inflammation

As to duration, inflammation may be classified as acute, subacute, or chronic. An **acute** inflammation lasts for only a few days to a few weeks. A **chronic** inflammation lasts for many months to many years, indicating that the causative agent is able to persist partly unchecked by the body for an indefinite period of time. A **subacute** inflammation is intermediate between the two forms, being a few weeks to a few months in duration. An acute inflammation may become chronic, or an inflammation may be chronic from the beginning.

The polymorphonuclear neutrophilic leukocyte is the characteristic cell of the exudate in acute inflammation, whereas in chronic inflammation the cells of the exudate are chiefly lymphocytes and plasma cells, with little fluid and no fibrin associated. Proliferation of the connective tissue cells in the vicinity of the inflammatory process is a prominent feature of chronic inflammation, whereas it is a negligible feature in acute inflammation. The exudate of subacute inflammations is intermediate in character between that of acute and chronic.

Kinds of Exudates

The series of vascular changes is designed so that an inflammatory exudate is formed as quickly as possible. The composition of an exudate may vary. Depending on the constituent that predominates, inflammatory exudates may be classified as serous, fibrinous, purulent, catarrhal, pseudomembranous, or hemorrhagic (sanguineous). If two components of the exudate are prominent, the inflammatory exudate is so designated (for example, serofibrinous, fibrinopurulent).

A **serous** exudate is one composed chiefly of the liquid portion (serum) of the blood, with few cells and little fibrin. **Fibrinous** exudates are characterized by the presence of a large amount of fibrin. They occur most frequently on serous surfaces of body cavities (pleural, peritoneal, pericardial) and often lead to permanent fibrous adhesions, since fibrin once deposited in a serous cavity is poorly reabsorbed by the body. A **purulent** exudate is one composed chiefly of polymorphonuclear neutrophilic leukocytes (pus cells). Purulent exudates are most often caused by pyogenic organisms (streptococci, staphylococci, pneumococci, meningococci, and gonococci). A **hemorrhagic** exudate contains many red blood cells. When there is necrosis of an epithelium, when fibrin is deposited on the epithelial surface, and when many leukocytes, dead epithelial cells, and bacteria are enmeshed in the fibrin threads, the exudate is described as **pseudo-membranous**. A good example of this is the pseudomembrane of diphtheria, which usually forms in the throat. Inflammations of mucous surfaces accompanied by a great outpouring of mucus, as in a cold, are spoken of as **catarrhal** inflammations. (The microbe causing an infection plays an important part in determining the nature of the exudate.)

The inflammatory exudates described as being serous, fibrinous, purulent, and hemorrhagic are associated with acute inflammations, the exudate being better defined in acute than in chronic inflammations. A chronic inflammation may be associated with pus formation, in which case it is described as a chronic **suppurative** process. The terms **productive** and **fibrous** in connection with chronic inflammations emphasize the proliferation of connective tissue cells, an unsuccessful attempt on the part of the body to resolve and heal.

Resolution

What happens in inflamed tissue after the initial events depends on whether the protective powers of the exudate prevail against the destructive agent. If the exudate quickly overcomes the injurious agent and too much tissue has not been destroyed, the rate of blood flow returns to normal, the caliber of the vessels is restored to normal, the fluid of the exudate is absorbed, and the fibrin, red blood cells, and dead tissue cells are removed by leukocytes entering the lymph stream. The area has returned to its natural state. This type of repair is known as resolution.

Suppuration

If the injurious agent gains the upper hand, there is death of the tissue, with or without suppuration (pus formation). Suppuration, which means a measurable degree of tissue necrosis, contributes to the progress of the injurious agent, which may gain access to the lymphatics or bloodstream and spread to distant parts of the body.

The suppurative focus consists initially of a central mass of dead tissue cells and dead leukocytes surrounded by a zone of active leukocytes, macrophages, and proliferating connective tissue. The surrounding leukocytes and macrophages attempt to separate the dead cells from the living; the dead tissue cells liberate enzymes, as do leukocytes, that liquefy dead tissue cells. Because of the action of these enzymes, the inflammatory focus undergoes liquefaction. The liquefied material is known as **pus**, and the chief cell of pus is the polymorphonuclear neutrophilic leukocyte. In addition to leukocytes, dead tissue cells, and red blood cells, bacteria are present in pyogenic infections.

Chronic Inflammation

When neither the injurious agent nor the bodily defenses can completely gain the ascendancy, then what started out as an acute inflammation merges into one that is chronic.

If the agent can persist in causing pus to form, then a chronic suppurative focus has been set up. Pus formation can be progressive but still effectively limited by the body mechanisms responsible for the proliferation of a connective tissue wall and barrier around the focus. Such a lesion is simultaneously characterized by progressive destruction and by productive reactions that are attempts at healing. A peculiar kind of balance has been struck, and in the case of certain chronic inflammations this state of affairs may persist for many years.

Inflammatory Lesions

An **abscess** is a circumscribed collection of pus surrounded by a wall of inflammatory tissue. If an abscess is located near the surface, the presence of pus is indicated by a yellow or green area on the surface near the center. As abscesses enlarge, they attempt to open onto a surface or into a cavity by extending (pointing) in the line of least resistance. In this way they may travel great distances, such as along muscle sheaths. Abscesses may reach the surface by the formation of narrow tracts known as **sinuses.** **Boils** or **furuncles** are abscesses located in the deeper layers of the skin and the subcutaneous tissue. The condition of generalized boils is known as **furunculosis.**

An **ulcer** is a circumscribed area of necrosis of the epithelium of the skin or mucous membrane that is often, but not always, caused by infection. A tuberculous ulcer has thin undermined edges; a syphilitic ulcer has a punched-out appearance.

Repair

When the injurious agent has been overcome and dead tissue cells, dead leukocytes, and other casualties are being removed from the scene, reparative processes, although active before, assume the place of greatest prominence. If the walls of the injured area are not too far apart, the gap is filled in with elements of the exudate, chiefly plasma and interlacing strands of fibrin. Small capillaries grow into the fibrin, become filled with blood, and form a network between the walls. The connective tissue of the walls proliferates and forms young cells known as fibroblasts that grow into and completely replace the framework of fibrin, which is absorbed. Because of the reddish granular appearance, this youthful tissue, composed of capillaries and fibroblasts, is known as **granulation tissue.** * This method of repair is known as **primary union.**

If the wound is so large that the gap cannot be filled with exudate, the capillaries grow into the exudate formed on the sides and bottom of the wound. Failing to find anchorage beyond the surface of the exudate, the capillaries form branches that join similar branches from other capillaries to form arclike structures extending from capillary to capillary. The areas are filled with blood. Fibroblasts lay down fibrous tissue as usual, and the exudate is converted into granulation tissue. The layers of granulation grow inward and upward until they fuse. This is known as **secondary union** and is most common in wounds complicated by bacterial infection. When a certain point is reached in the healing of surface wounds by either primary or secondary union, the epithelium at the edges of the wound proliferates and covers over the gap. If a scab forms in the early part of the healing process, it should not be removed, since it protects the underlying epithelium.

Regardless of location or type of healing, there is a time when the fibroblasts contract, the capillaries are absorbed, and a white, bloodless glistening scar (cicatrix) remains. Scars do not contain hair follicles, sweat glands, sebaceous glands, or sensory nerve endings.

---

*Granulation tissue is a very delicate tissue, and it may be destroyed or its purpose defeated by vigorous cleaning, strong antiseptics, or frequent dressings. Overtreatment, infection, or any unhealthy condition of the tissue may stimulate the granulation tissue to such activity that it overfills the wound and protrudes beyond the surface as exuberant granulations, or proud flesh.

Regeneration

The complete restoration of destroyed tissue to normal depends on a number of factors, the most important of which is the ability of the tissue to regenerate, that is, to reproduce tissue of its exact kind. The regeneration of cells destroyed by the natural wear and tear of life is known as **physiologic regeneration;** the regeneration of tissues destroyed by disease or injury is **pathologic regeneration.** When tissues that have little or no regenerative capacity are destroyed, the defect is repaired by proliferation of nature's omnipresent repair material—connective tissue. The same kind of repair usually takes place when there is *extensive* destruction of tissue having well-developed regenerative capacity. Naturally, in a large wound several different kinds of tissues having different powers of regeneration are involved.

In humans connective tissue and capillary endothelium, which are the body's repair tissues, regenerate with extreme ease. Next to these, surface epithelium possesses the greatest regenerative capacity. This applies not only to the epithelium covering the surface of the body but also to that covering mucous membranes such as the lining of the gastrointestinal, respiratory, urinary, and genital tracts.

Periosteum and bone regenerate well, or healing of fractures would be impossible. Muscle regenerates poorly, and wounds of muscle are usually repaired by the formation of scar tissue. Nerve cells never regenerate when destroyed, but their processes may regenerate, provided that the cell body has not been injured. Regeneration is more active in childhood than in old age.

Table 17-3 The Hallmark Lesion of Some Important Infectious Agents

Microbe	Type Lesion	Disease
Bacteria		
Staphylococcus	Abscess (predilection for skin, also systemic)	Variety
Streptococcus	Spreading inflammation	Scarlet fever, puerperal sepsis, erysipelas
Pneumococcus	Exudate rich in fibrin	Lobar pneumonia, bronchopneumonia
Gonococcus	Catarrhal inflammation	Gonorrhea
Meningococcus	Purulent inflammation	Epidemic meningitis
Diphtheria bacillus	Pseudomembrane; toxemia from exotoxin	Diphtheria
Typhoid bacillus	Granulomatous inflammation	Typhoid fever
Shigella species	Pseudomembrane	Bacillary dysentery
Tubercle bacillus	Tubercle	Tuberculosis
Leprosy bacillus	Granuloma	Leprosy
Pseudomonas species	Necrotizing inflammation	Variety
Spirochete of syphilis	Gumma, fibrosis, vascular changes	Syphilis
Rickettsia	Vasculitis, widespread	Variety
Viruses	Intracellular parasitism; little tissue reaction	Great variety
Fungi		
Systemic fungus	Granuloma plus suppuration; presence of etiologic agent important	Great variety
Protozoa		
Plasmodium of malaria	Hemolysis	Malaria

Patterns in Infection

Since variability is a characteristic of living organisms, the infectious diseases they produce vary considerably. This may make difficult at times the recognition and treatment of an infectious disease.

Agents of infectious diseases are commonly linked to the area of the animal body that they injure. There are those that single out an organ or a system. The bacillus of dysentery primarily attacks the intestinal tract, and the pneumococcus primarily attacks the lungs. In infections in which the attack is directed primarily to a single organ, a generalized intoxication from some product released by the agent can contribute to disability and even cause death. On the other hand, certain organisms can and regularly do produce infection in a variety of tissues in widespread anatomic locations, for example, the microbes of wound infections, staphylococci, streptococci, and coliforms.

Table 17-3 presents the pathologic patterns for certain infectious disease-producing agents.

SUMMARY

1 If infection occurred each time a person encountered an infectious agent, that person would be constantly ill. Certain mechanical, physiologic, and chemical barriers in the human body afford a constant measure of protection to the body.

2 Anatomic barriers constitute the first line of defense; for instance, an epithelial cell lining covers and protects body surfaces.

3 Chemical factors operate in body fluids. Such a factor is the antimicrobial substance lysozyme.

4 Interferon, another natural substance in the body's internal defense, is a family of low molecular weight glycoproteins secreted in response to viral and mitogenic stimuli. Three distinct types are alpha, beta, and gamma.

5 The normal flora protects the host. The normal flora can suppress adherence of would-be pathogens to an epithelial surface, and they influence clearance mechanisms already in force; for instance, they may enhance peristalsis in the alimentary tract.

6 In health, normal body temperature taken orally is usually given as 98.6° F (37° C). Rectal temperature, the most dependable, is usually about a degree higher and the axillary temperature about a degree lower than this.

7 A close relationship exists between pulse rate and body temperature. An increase of some eight pulse beats per minute occurs for each rise of 1.8° F in the body temperature.

8 The widely dispersed system of macrophages (the reticuloendothelial system) constitutes one of the main organs for the body's defense against both living and nonliving harmful agents. Although ubiquitous, macrophages are concentrated in the lymph nodes, spleen, liver, and bone marrow.

9 Macrophages ingest many different kinds of particulate matter—bacteria, dead body cells, mineral particles, dusts, and pigments.

10 Opsonization is the process of making microbes more susceptible to phagocytosis. Opsonins carry out this action.

11 The three major phagocytic cells are (1) polymorphonuclear neutrophilic leukocytes (against bacteria), (2) eosinophils (against certain parasites), and (3) monocytes that become macrophages in the body tissues (against larger particles and intracellular pathogenic microbes).

12 An increase in the number of circulating white blood cells is leukocytosis; a decrease, leukopenia.

13 The lymphoid (immunologic) system comprises the lymphoid tissues and organs of the body: the lymph nodes, spleen, thymus gland, Peyer's patches, the solitary

lymphoid follicles of the alimentary tract, and the small nodules in the bone marrow. Its major constituent cells are the lymphocytes.

14 The spleen, the largest mass of lymphoid tissue, is an organ adapted to filter blood similar to the way that the lymph node filters lymph.

15 The thymus, partly epithelial and partly lymphoid, processes certain lymphocytes for release to lymphoid tissues in the body. It shrinks after puberty.

16 Inflammation is the sum of the reactions in the body incited by injury to cells. Inflammation is designed to destroy and remove the injurious agent and its byproducts, if possible, or if not, to limit the extension of the agent and minimize its effects. Finally, inflammation is the mechanism within the body for repair and replacement of tissues damaged or destroyed.

QUESTIONS FOR REVIEW

1 List mechanisms that serve to protect the body from disease or injury, and explain how each functions.

2 What is lysozyme? Who discovered it?

3 What are cilia? What is their function in the respiratory passages?

4 What is a vital organ? Name three. How are the vital organs protected in the body?

5 Define fever.

6 What is considered normal body temperature?

7 What is the relation between pulse rate and temperature?

8 Briefly describe the system of macrophages. What is its chief function?

9 Briefly discuss phagocytosis. What are the three phases? Why is this an important process?

10 Outline the anatomic distribution of the macrophage system.

11 Give the component cells and organs of the lymphoid system.

12 What is meant by lymphadenopathy? By splenomegaly? With what are they associated?

13 Briefly define leukocytosis, leukopenia, granulocyte, anastomoses, collateral circulation, sebum, inflammation, granuloma, phagosome, monocyte, chemotaxis, lymphoid nodule, microphage, lymph, and leukoproteases.

14 Define inflammation.

15 What are the three parts of the inflammatory process?

16 What is the purpose of the inflammatory exudate?

17 Describe the different kinds of inflammatory exudates.

18 What are the differences between acute and chronic inflammation?

19 Define abscess, boil, pseudomembrane, furuncle, ulcer, resolution, and regeneration.

SUGGESTED READINGS

Jabs, D., and others: Assaying of human neutrophil function, Lab. Manage. 18:37, Feb. 1980.

Platt, R.: Editorial: infection after splenectomy, J.A.M.A. 248:2316, 1982.

Snyderman, R., and Goetzl, E.J.: Molecular and cellular mechanisms of leukocyte chemotaxis, Science 213:830, 1981.

Tatsumi, N., and others: Neutrophil evaluation: hidden clues to the body's defenses, Diagn. Med. 2:30, Nov./Dec. 1979.

Whicher, J.T.: The role of the acute phase proteins, Diagn. Med. 4:52, May/June 1981.

CHAPTER 18

IMMUNOLOGIC CONCEPTS

Meaning of Immunity
Immunity and Infection
  Kinds of Immunity
  Level of Immunity

Immune System
  Humoral Immunity
  Cell-Mediated Immunity
  Résumé of Features of the Immune
    System

MEANING OF IMMUNITY

Immunology is the division of biology concerned with the study of **immunity**. With the steady progression of knowledge in the field of immunology and the rapidly expanding application of the word immunity, a working concept of immunity becomes hard to define. In a literal sense *immunity* refers to a state of being exempt from injury or harmful influence, and *immunology* would then refer to the study of the immunity of living organisms to harmful agents, regardless of their nature. However, the body's arrangement of cells and their products for the implementation of immunity, although it does indeed protect human beings from invading microorganisms and probably assists them in fighting cancer, is a defense system with the potential to reverse its purpose at times and do harm to its host. It can be responsible for the development of an autoimmune disease or an allergic disorder.

The scheme of things in the body that function in this discriminatory way is the machinery of immunity, and the elements making it up are collectively the **immune system.** The term **immunobiology** surveys the subject of immunity and its many ramifications in modern medicine.

Susceptibility is the reverse of immunity and the result of suppression of factors that produce immunity.

IMMUNITY AND INFECTION

Infectious agents such as bacteria and other microorganisms are easily "foreign" to the body (nonself), their effect detrimental, and the cause-effect relationship clear-cut. The applications of immunity with infectious processes have been dramatic. The classic definition that immunity is a highly developed state of resistance has referred directly to infectious disease.

Kinds of Immunity

Traditionally immunity has been classified as shown in Fig. 18-1.

A **natural** immunity is a more or less permanent one with which a person or lower animal is born; that is, it is a natural heritage. It may be the heritage of a species, race, or individual; it is also known as **innate** or **genetic** immunity. **Species** immunity is that peculiar to a species (for example, a human being does not have distemper, nor do dogs have measles). A **racial** immunity is one possessed by a race (for example, ordinary sheep are susceptible to anthrax, whereas Algerian sheep seldom contract the disease). Species and racial immunities are more highly developed in plants than in animals. **Individual** immunity is a rare condition, and most so-called cases stem from unrecognized infections.

An **acquired** immunity is one in which protection *must be obtained*. It is never the heritage of a species, race, or individual. It may be **naturally** acquired when a mother transmits antibodies (protective substances) to her fetus by way of the placenta or when a person has an attack of a disease. When a person has a disease such as measles, the disease is rather severe for a time, after which recovery occurs. The person will not then contract the disease again although repeatedly exposed; that person has developed a permanent immunity.

An immunity may be **artificially** acquired by a vaccination or by the administration of an immune serum.

According to the part played by the body cells of the animal or person being immunized, acquired immunity (natural or artificial) is classified as **active** and **passive.** If a person has an attack of measles or is vaccinated against it, that person's cells respond to the presence of the agent (or its products) with the production of antibodies that destroy the causative agent should it again gain access to the body. When certain body cells react to the agent or its products in this way, the naturally acquired immunity is an active one. When a horse is given frequent injections of diphtheria toxin of graded strength, the first of which is a very small dose, the horse becomes immune to the toxin. As doses of increased amounts of toxin are given, the horse becomes able eventually to withstand an amount of diphtheria toxin thousands of times greater than that required to kill an untreated, or nonimmune, animal.

Immunity transferred by way of the placenta from the mother to her child is a passive immunity, which the child has naturally *acquired*. When the serum of an actively immunized animal is injected into a nonimmune animal, the latter becomes temporarily immune. This artificially acquired immunity is passive because the immunity-producing principle is introduced in the serum, and the cells of the recipient animal take no part in the process. For the first 4 to 6 months of life an infant may have a passive immunity to measles, smallpox, diphtheria, mumps, tetanus, influenza, and certain staphylococcal and streptococcal infections because of the transfer of immune bodies from the blood of the mother to the child across the placenta. The immunity that the mother has passed to the child still in her womb may be enhanced by the content of protective substances that she passes to her child in her milk, especially in the colostrum. Naturally, the child will not receive any immunity if the mother is not immune. Except in newborn babies, passive immunities are established by the administration of an immune serum (the serum of a subject [human or animal]) that contains antibodies because that subject has been actively immunized.

Active immunity artificially acquired does not last so long as when it is naturally

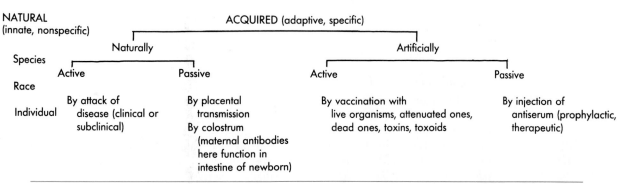

FIG. 18-1   Kinds of immunity related to disease prevention.

acquired, but active immunity always lasts longer than passive immunity. One should bear in mind that an active immunity can be established only with an attack of a given disease or with vaccination against it and that active immunity is slowly established (days or weeks) but of long duration (months or years). On the other hand, passive immunity is ordinarily established at once by the injection of an immune serum but is of short duration (1 or 2 weeks). From these considerations the principle emerges that active immunity, which is desirable, should be produced when no immediate danger from the disease exists and that passive immunity should be established only when danger from the disease is imminent or the patient is already ill.

Level of Immunity

It seems that, when an infectious disease attacks a population that has never been exposed, the attack rate is very high, and many succumb to its ravages. On the other hand, when a disease endemic in a population for years suddenly becomes epidemic, the number of persons attacked and the percentage of deaths are not so great. People develop a degree of inherent resistance to a given disease through exposure for generation after generation. There is little doubt that syphilis was at one time a far more virulent disease and more often fatal than it is now. The susceptibility of aboriginal people to tuberculosis is much greater than that of people in communities in which the disease has been prevalent for a long time.

In 1951 measles was introduced into Greenland by a recently arrived visitor who attended a dance during the onset of his symptoms. Within 3 months more than 4000 cases were reported, with 72 deaths. The attack rate reached the unprecedented figure of 999 cases per 1000 people. When measles was first introduced into the Fiji Islands in 1875, 30% of the population died.

IMMUNE SYSTEM

Immunity is important as a defense against infection, but in light of current achievements it is more. Broadly speaking, immunity comprises the complexity of things that help human beings to maintain their structural and functional integrity and to ward off certain kinds of injury. Collectively the immune mechanisms of their immune system furnish to vertebrate organisms a most remarkable and powerful defense arrangement.

Some of the factors in immunity are hard to delineate. One cannot easily measure the basic mechanisms of resistance built into all living organisms and expressed as the inflammatory process, the presence of anatomic barriers, phagocytosis of bacteria by selected cells, and the effect of naturally occurring antimicrobial substances. Therefore such factors are collected loosely into the concept of **innate,** or **nonspecific, immunity,** but another approach to immunity emphasizes the study of the highly developed, precisely specialized physiologic mechanisms, the topics of **specific,** or **adaptive, immunity.**

Three phenomena are at the center of any understanding of adaptive immunity. One involves the recognition of nonself by the body. A second provides for recall of immunologic experience, or memory, and the third is the property of specificity.

Recognition of Nonself

Basic to an understanding of the complexity of physiologic mechanisms encompassed by immunity are the concepts of self and nonself, or foreign. For a given individual to preserve biologic integrity (self), that person must be able to recognize and deal with threatening factors in the environment. On the one hand, there is self, which must remain intact. On the other, there are the elements it contacts, not self, although perhaps of comparable makeup and potentially harmful. Because of the many biologic parallels in nature, the distinction is sometimes a fine one. We require an arrangement whereby the cells and cell substances of our bodies are marked off as "ours" and the substances that appear on the scene are accosted as "intruders."

Memory

The first time the immune system contacts an infectious agent, a **primary immune response** occurs. Information concerning the event is retained and stored in the immune system, and a population of memory cells is established to prepare the body for future encounters with the same agent. When the second (or subsequent) encounter occurs, the immune system being primed the first time, the memory of the first evokes the events of the **secondary immune response.**

This phenomenon may be demonstrated graphically. A standard bacterial product is injected into a suitable animal that has not been previously exposed. Several days elapse before antibodies are detected in the animal's blood. When the concentration of antibodies is plotted against time (Fig. 18-2), a peak in the level is seen and then a decline. At this point the animal is allowed to rest. Then a second injection of the same bacterial product can be given. When the concentration of antibodies is plotted against time, this time the curve is quite different. Within a shorter time, the antibody level in the blood rises sharply to a much higher peak than before. The early and explosive type of antibody production is the desired result of a process such as vaccination, and with vaccination the first exposure to the antigen can be carefully controlled.

Specificity

Specificity is the property whereby the body can differentiate precisely between two different entities inducing an immune response. Therefore the protection established against one agent does not extend to another unrelated agent.

Role of the Lymphoid Tissues

In humans the lymphoid system (Table 18-1) is the basic biologic defense network, with the lymphoid tissue being responsible for the events of specific immunity. The cells (predominantly lymphocytes) in the different areas of lymphoid tissue look very

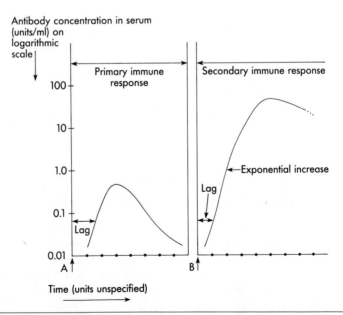

FIG. 18-2 Comparison of the primary with the secondary immune response. First dose of injectable immunogen given at A. Second dose given at B.

much alike when viewed under the microscope, but in the immunologic setting they behave quite differently. At least two distinct cell populations of lymphocytes, the T and B cells, have been identified. Both have a common origin in bone marrow stem cells, but they diverge in development and function.

T cells. One population is called a **thymus-dependent** (processed) **system** of cells because its early development is influenced and its destiny fixed by the thymus gland. These are the T cells or T lymphocytes. The thymus gland may act directly on these cells or effect their maturation through secretion of a hormone (or hormones) such as **thymosin.** During the maturation process in the thymus, the surface membrane of the T cells changes and takes on a unique group of protein fractions designated surface antigens (p. 409). After the T cells leave the thymus, they circulate in the blood and lymph. In fact most of the circulating lymphocytes are T lymphocytes. They selectively concentrate in the pericortical regions of lymph nodes and in the periarterial lymphoid sheaths of the spleen (Table 18-1). T cells have a very long life expectancy—many years, even a decade or so.

Subpopulations of both types of lymphocytes exist with specialized functions, but they are especially important with T cells (Table 18-2). Differentiation of their surface antigens has allowed for further characterization of T cells into functionally distinct

Table 18-1 Distribution of T and B Cells in the Lymphoid System

	T Lymphocytes	B Lymphocytes
Site of origin	Bone marrow from a common stem cell	Bone marrow from a common stem cell
Site of maturation	Primary lymphoid organ—thymus	Primary lymphoid organ—bone marrow or bursa equivalent (exact site unknown)
Site of antigen recognition in host	Peripheral lymphoid organ—lymph nodes, spleen, gut-associated lymphoid areas	Peripheral lymphoid organ—lymph nodes, spleen, gut-associated lymphoid areas
Site of temporary or permanent residence	Peripheral lymphoid organ—lymph nodes: pericortical regions; spleen: periarterial lymphoid sheaths	Peripheral lymphoid organ—lymph nodes: follicles and medullary cords; spleen: white pulp

Table 18-2 Subpopulations of T Lymphocytes (T Cells)

Subset	Immunologic Role
T helpers and/or T amplifiers	Assist antigenic activation of B cell; aid T effector ("killer") cell
T suppressors	Block induction and/or activity of T helper cells
T regulators	Develop into T helper or T suppressor cells; effect balance in cellular responses to antigen
T_{DTH}	Mediate inflammation and nonspecific increased resistance to many infectious agents associated with activated macrophages in **delayed-type hypersensitivity** (p. 382)
T killers, cytotoxic T lymphocytes	Lyse target cells

subsets. T cells may be effector cells—"killer" or cytotoxic cells that specifically attack and lyse a targeted cell, for example, certain virus-infected cells. T cells also have regulatory properties. "Helper" T cells initiate the functions of the effector T cells and are necessary for certain antibody responses. "Suppressor" T cells have an opposite effect; they block the activity of the helper cells.

B cells. The other population of lymphoid cells, the B cells or B lymphocytes (Fig. 18-3), is a thymus-*in*dependent system that owes its differentiation to a site not yet identified with certainty in human beings. Some observers consider the bone marrow to be the likely site of maturation for these cells, but others state that the fetal liver, spleen, and gut-associated lymphoid tissue (Peyer's patches, Fig. 17-4) are the regions involved. In the course of their development B cells also get a unique set of surface antigens (distinct from those of the T cells). They selectively concentrate in the follicles and medullary cords of lymph nodes and in the follicles of the splenic white pulp (Table 18-1). The small lymphocytes of this cell population have a short life expectancy—only a week or two.

Dual Nature of the Immune Response

Generally, the immune system can be compartmentalized into the B cell system and the T cell system, and the two systems can be sorted out for practical purposes in humans and animals, although the distinct functions expressed by the principal cells nevertheless interact and overlap *to an extraordinary degree.* The B lymphocytes are responsible for the production of chemically mediated (humoral) immunity, best seen in the neutralization of toxins and viruses and in the body's main defense against bacterial infections with virulent encapsulated pathogens, such as those caused by pneumococci, streptococci, or meningococci. T lymphocytes regulate the cell-mediated immune responses (those resulting from the action of certain cells) in graft (solid tissue allograft) rejection, delayed hypersensitivity, immunologic surveillance against cancer, and infections from facultative intracellular bacteria (for example, acid-fast bacteria), fungi, and many viruses.

In adaptive immunity, then, two fundamental processes are found, both related to cells in the lymphoid system. One involves the elaboration of molecules—this is **humoral immunity. (Immunochemistry** is the study of the complex chemical reactions

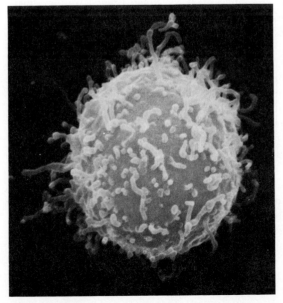

FIG. 18-3   B lymphocyte or B cell, scanning electron micrograph. (×16,000.) Numerous fingerlike protrusions over the surface of the cell are surface markers. (Courtesy Dr. Peter Andrews, The University of Texas Health Science Center at Dallas.)

in immunity.) The other focuses on activities of certain cells—that is **cell-mediated,** or **cellular, immunity.**

If a lymph node (see Fig. 17-15) is studied during the course of an immune response, definite changes may be seen within its cortex. With hormonally mediated responses, the follicles of the cortex become very active and greatly enlarged; numerous plasma cells appear within them. With cell-mediated immune responses, the cortical follicles appear to be spared, but pronounced changes are found in the lymphoid tissue adjoining them. Within these **paracortical** areas numerous mononuclear cells containing increased amounts of RNA accumulate, later to become small lymphocytes ready for release into the circulation.

Reflecting the dual nature of the immune response, our discussion will take the twofold approach with, first, a consideration of humoral immunity and, second, a discussion of the cell-mediated kind.

Humoral Immunity

Humoral (antibody-mediated) immunity is that related to certain molecules called antibodies.

Antigens

When a person or animal becomes immune to a disease, the immunity largely results from the development within the body of substances capable of destroying or inactivating the causative agent of the disease when it gains access to the body. These substances, known as **antibodies,** are produced by the body in response to a specific stimulus. Microbes and their products may constitute such a stimulus, as may certain vegetable poisons, snake venoms, and, from an animal of a different species, red blood cells, serum, and other proteins. A substance that elicits the production of antibodies against itself and can react specifically with them is an **antigen** or **immunogen.** (*Antigen* and *immunogen* are used interchangeably, although *immunogen* is better used to indicate induction of specific immunity to infection or to toxin.) To act as an antigen, a substance *must be* foreign to the body of a given individual, that is, the body must *not* recognize it as "self." Antigens are usually macromolecules with a molecular weight over 10,000. Most are proteins, but in some instances complex carbohydrates (polysaccharides), or combinations such as glycoproteins or nucleoproteins, may act as antigens. Enzymes and many hormones are antigenic. It is estimated that 1 million different antigens exist to which the human immune system can react directly.

Although the antigenic molecule is a reasonably large one, the antibodies formed in the response to the antigen react with only a small part of the antigenic molecule; the part that specifically combines with an antibody is the determinant group for that particular antigen. The antigenic determinant by itself is unable to induce production of antibodies. When certain body proteins (for example, thyroglobulin and lens protein from the eye) of one animal are injected into another animal of the same species, antibodies against the proteins are produced. These antigens are called **isoantigens** (*iso* meaning from an individual of the same species), and the antibodies to them are **isoantibodies.** Naturally occurring isoantigens are those residing on the red blood cells and making up the blood groups.

Haptens, or **partial antigens,** are comparatively simple chemical substances (certain low molecular weight lipids and carbohydrates) that are too small to induce antibody synthesis but, when combined with a large carrier protein, may become one of the antigenic determinant groups toward which antibody synthesis is directed. Antibodies formed are specific for the small hapten molecules of the protein-hapten complex, the complete antigen. When separated from the protein of the antigen, haptens do *not* elicit the formation of antibodies, but they can combine with antibodies already produced against the complete antigen (the protein-hapten complex).

Certain chemicals that within themselves are not antigenic increase the potency of antigens. They are known as **adjuvants.** Among these are alum and aluminum hydroxide, used in preparation of toxoids. In addition to increasing the potency of the antigen, they concentrate it.

Antibodies

Plasma is the liquid portion of the blood in which the red corpuscles and other formed elements are suspended. It is composed chiefly of water (90% to 92%), proteins (6% to 7%), and various mineral constituents. The plasma proteins are albumin, globulin, and fibrinogen. By electrophoresis it is shown that the globulin is made up of these main factors—alpha-1, alpha-2, and beta, and gamma globulins. Most antibodies migrate electrophoretically in the gamma and beta regions, a few with the alpha globulins. Largely associated with gamma globulin, antibodies represent about 20% of total serum protein. This is why this fraction (immune globulin) can be used in the prevention of diseases such as infectious hepatitis.

Serum. The fluid plasma of the blood minus blood cells and fibrin, which is derived

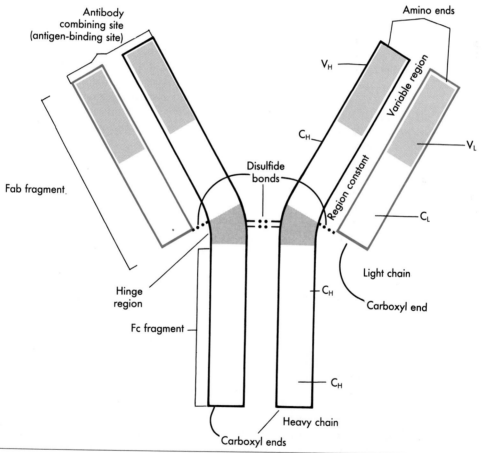

FIG. 18-4   Basic immunoglobulin (IgG class) structure in stick diagram. Two identical heavy (H) chains and two identical light (L) chains are united by disulfide bonds. The L and H chains are divided into domains. The L chain has two domains—the variable, V_L, and the constant, C_L. The H chain has one variable domain, V_H, and several constant ones, C_H. V_L and V_H form the **antibody combining site,** or the **antigen-binding site.**

from the plasma protein fibrinogen, is serum. To collect serum, allow a sample of blood to clot, with the fibrin formed in the clotting process entrapping the blood cells. When the clot shrinks, the fluid expressed therefrom is the serum. Practically, serum is an important source of antibodies, and in the laboratory **serology** is the study of antigen-antibody reactions in a specimen of serum.

Nature. Antibodies, the specialized globulins in serum and body fluids that react precisely with the antigen inducing their formation, are better designated as **immunoglobulins** (Ig). Antibodies or immunoglobulins segregate into a family of related proteins with distinctive properties when studied by electrophoresis (see Fig. 20-5). They are very large and heterogeneous, multichain proteins and the most complex group known.

Chemically, immunoglobulins adhere to a basic pattern on which significant variations are superimposed (Fig. 18-4). The basic unit is structured of four polypeptide chains in two pairs. Two of the paired chains are of greater molecular weight—heavy, or H, chains—and two are of lower molecular weight—light, or L, chains. The light chains are linked to the heavy ones by single disulfide bonds. Light chains are 214 amino acids long; heavy chains are from 450 to 700 amino acids long. The heavy chains together with the light chains spread out in the shape of a Y. Antigen binding occurs at the extremity of the two limbs of the Y, where the two "Fab" fragments ("Fragment, antigen-binding)" are formed. The opposite ends of the heavy chains, at the base of the stalk of the Y, form the other main part of the molecule called the "Fc" fragment ("Fragment, crystalline"), which is involved in complement fixation and other biologic processes.

Antigen-antibody combining site. Light chains display a **variable** region in the Fab fragment, one of highly variable amino acid sequence, on the amino terminal half of the chain. The other end containing the terminal carboxyl group is a **constant** region in light chains of the same immunoglobulin class. A similar variable region exists in the same part of the heavy chain; the remaining three fourths is a constant region (but different from that in the light chain).

Each immunoglobulin molecule has a unique combining site that corresponds to the structure of an antigen, making it fit to one antigen but not to another (Fig. 18-5). This is possible because of the variable regions in the chains.

Classification. In different antibody (immunoglobulin) molecules, light chains are composed of the same two kinds of protein, but according to the protein of which they are composed, heavy chains vary in at least five ways; thus classification of immunoglobulins according to the protein of the heavy chain is possible. Five structurally and

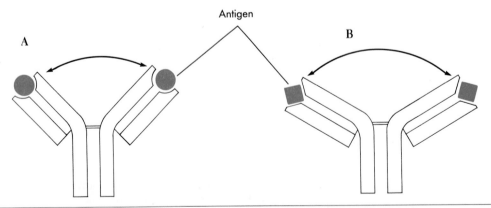

FIG. 18-5   Diagrams (**A** and **B**) to show antigen binding sites in clefts in the variable part of heavy and light chains. (Variable part varies with specificity of antibody; constant part does not.) **B,** Antibody molecule is flexible, adapting to spacing of antigen.

antigenically distinct classes of immunoglobulin molecules, different in physicochemical and biologic properties, are enumerated as immunoglobulin G (IgG), immunoglobulin A (IgA), immunoglobulin M (IgM), immunoglobulin D (IgD), and immunoglobulin E (IgE). Subclasses of IgG, IgA, and IgM (a macroglobulin, Fig. 18-6) are recognized, but the biologic properties do not vary so much within the subclass. IgG is the predominant globulin of plasma and is relatively abundant extravascularly as well. The major classes of immunoglobulins as defined by the World Health Organization Committee on Nomenclature of Immunoglobulins are given in Table 18-3 with their distinguishing features; Table 18-4 indicates their immunologic capacity.

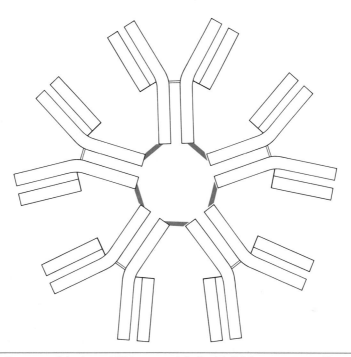

FIG. 18-6   Diagram of the large 19S immunoglobulin M (IgM), a pentamer. Five of the basic four-chain units are bound together.

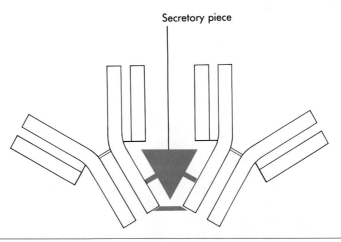

FIG. 18-7   Diagram of secretory immunoglobulin A (IgA), a dimer—four heavy, four light chains, linked by secretory piece.

Table 18-3 Analysis of Major Classes of Immunoglobulins (Antibodies)

WHO Term*	Where Found in Body	Sedimentation Coefficient (Ultracentrifugal Analysis†)	Molecular Weight	Chemical Composition Known	Half Life in Days
IgG	Serum (40% intravascular)	7S	150,000	4 polypeptide chains—2 light, 2 heavy	25 to 35
IgA	Serum (40% intravascular) External secretions (secretory IgA): saliva, parotid gland, gastrointestinal, respiratory tract, colostrum	7S (monomer) 11S	160,000 (monomer) or 400,000 (secretory IgA a dimer)	4 polypeptide chains—2 light, 2 heavy	5
IgM	Serum (80% intravascular)	19S	900,000	20 polypeptide molecules of five basic 4-chain units—2 light, 2 heavy (Fig. 18-6)	9 to 11
IgD	Serum (75% intravascular)	7S	180,000	4 polypeptide chains—2 light, 2 heavy	2 to 3
IgE	Serum (mostly extravascular)	8S	190,000	4 polypeptide chains—2 light, 2 heavy	2.5

*Designation recommended by the World Health Organization Committee on Nomenclature of Immunoglobulins.
†The ultracentrifuge is an important tool in the identification of antibodies. By means of the exceedingly high gravitational forces in Svedberg units (S).

One form of IgA in Table 18-3 is mentioned as secretory IgA. By nature, this primary immunoglobulin of the secretory immune system is well adapted to provide humoral defense needed in external body secretions that bathe mucous membranes of the body (Table 18-3). Note that secretory IgA is a dimer (four heavy, four light chains)—basically two immunoglobulins linked by a digestion-resistant protein called the secretory (or transport) piece (Fig. 18-7). Secretory piece is not like antibody and is formed not by plasma cells of the local mucosa, as is IgA, but by regional epithelium. The strong affinity for mucus of the secretory piece prolongs the retention of immunoglobulin on lining surfaces, which is especially important to the secretory IgA coating on the intestinal tract. Secretory IgA binds antigens mechanically, such as bacterial and viral antigens and certain macromolecules, and thus blocks penetration and colonization of mucous surfaces by potential pathogens. Its concentration is high in breast milk, for the local immune protection on the mucosal surfaces of the very young person is especially critical.

Immunoglobulins responding to but a single given antigenic stimulus usually make up a heterogeneous population, the composition of which is variable. The classes represented and their proportionate distribution in plasma are influenced by factors such as the age, species, and genetic makeup of the subject immunized; the kind of antigen and adjuvant used; the route and duration of immunization; and the tissue site of antibody formation.

Formation. Antibodies (immunoglobulins) are formed in the lymphoid tissues of the body in response to a particular antigen within the body. On the surface of an antigen there is a special pattern of atoms designated as the **antigenic determinant** or **marker.** The immunoglobulin resulting from the stimulus of a given antigen must be structured to fit neatly into the unique arrangement of the antigenic determinant. The main anatomic sites involved with synthesis of antibodies are the spleen, gut-associated lymphoid tissue, lymph nodes, and bone marrow, but antibody synthesis may be demonstrated

Estimated Percent Ig in Population	Serum Level (mg/100 ml)	Crosses Placenta	Binding of Complement by	Percent Carbohydrate	Electrophoretic Mobility (Principal)
75% to 85%	1000 to 1500	Yes	Classic pathway	4%	γ
5% to 15%	150 to 400	No	Alternative pathway	10%	Slow β, fast γ
5% to 10%	60 to 170	No	Classic pathway	12%	Between γ and β
0.2%	3	No	Alternative pathway	13%	Fast γ
0.5%	0.01 to 0.03	No	No	12%	γ

possible in this instrument, serum protein fractions may be separated according to their molecular weights. Results are quantitated

wherever there is lymphoid tissue—*with the exception of the thymus!* The lymphoid tissue directly associated with the gastrointestinal tract, an amount approximately that contained in the spleen, most likely synthesizes more immunoglobulin than any other lymphoid organ.

The principal cells involved are lymphocytes—B cells primarily, but T cells participate, as well as plasma cells (derivative cells of B lymphocytes) and macrophages (mobile and sessile cells able to ingest fairly large particles of foreign matter). T cells provide regulatory mechanisms within the immune system with powers to assist antibody formation as helper cells or inhibit it as suppressor cells (Fig. 18-8). We speak of lymphocytes as **immunologically competent** cells, meaning cells that can undertake an immunologic response when engaged by an antigen. During a lifetime, because of the many and varied antigens of our environments, each of us produces tens of thousands of different kinds of antibodies.

Although B cells and T cells look alike under the light microscope, important differences on the cell surfaces can be demonstrated by special technics. Bound to the plasma membrane of the B cell are numerous (estimated 100,000) specific antigen receptors shown to be molecules of IgM (the monomer) and IgD. (The purpose of IgD on the cell membrane is not known.) Among these are also receptors for the third factor of complement (C3) and the Epstein-Barr virus (p. 718). By contrast, T lymphocytes show considerably fewer surface markers—no identifiable immunoglobulin, but they do possess receptors for normal sheep red blood cells (p. 409) and the thymic hormone thymosin.

Each B cell and its clone synthesizes only one combination of variable regions from the heavy and light chains of a given immunoglobulin molecule to make up an antibody combining site. The exact identity of this site in the Fab fragment is determined during fetal development. According to the **clonal selection theory of Burnet,** clones of lymphocytes are present in the human body that are *predetermined* to respond to a

Table 18-4 Immunologic Activities of Major Classes of Immunoglobulins

Immunoglobulin	Agglutination	Precipitation	Complement Fixation	Lysis	Neutralization	
					Viruses	Toxins (and Enzymes)
IgG (Gamma-G, γG)	Weak	Strong	Strong	Weak	+	+

General remarks:
1. Ones best studied—in this category the bulk of long-range antibody activity
2. Responsible for passive immunity of newborn
3. Identified here:
   a. Certain Rh antibodies
   b. Bacterial agglutinins
   c. LE cell factor*
   d. Antitoxins
   e. Antiviral antibodies

IgA (Gamma-A, γA)	+	±	+	−	+	?

General remarks:
1. Known to be made by plasma cells
2. Chief Ig in external secretions—plays major role in secretory immune system†
3. Protective function on mucosal surfaces exposed to environment—transported across mucous membrane with secretions
4. Identified here:
   a. Diphtheria antitoxin
   b. Blood group antibodies
   c. Antibodies against *Brucella* species and *Escherichia coli*
   d. Antibodies against respiratory viruses
   e. Antinuclear factors in collagen disease*
   f. Certain other autoantibodies* (against insulin in diabetes, thyroglobulin in chronic thyroiditis)

IgM (Gamma-M, γM)	Strong	Variable	Avid	Strong	+	−

General remarks:
1. Rapid protection here—first antibodies noted after antigen injection in an adult; first capacity for Ig synthesis in newborn that of IgM
2. Powerful agglutinins and hemolysins here (700 to 1000 times stronger than those of IgG in agglutinating red cells or bacteria)
3. Identified here:
   a. Blood group antibodies (ABO)
   b. Antibodies against somatic O factors of gram-negative bacteria
   c. Human anti-A isoantibody
   d. Isohemagglutinins, cold agglutinins, rheumatoid factor

IgD	?	?	+	?	?	?

General remarks:
1. Minor Ig in serum—biologic function undefined
2. On surface of B lymphocytes serves as surface receptors—reaction with antigen influences lymphocyte activity

IgE (Reagin)	−	−	−	+	−	−

General remarks:
1. Role in allergy—governs responses of immediate-type hypersensitivity (mast cell fixation)
2. Carries skin-sensitizing antibody
3. Prominent in external secretions
4. Protective role in certain parasitic infections

*For discussion of autoimmunity and autoimmune diseases, see pp. 362 and 363.
†Most infectious agents enter the body through the alimentary, respiratory, and genitourinary tracts. A local immune response at the portal of entry primarily involves IgA.

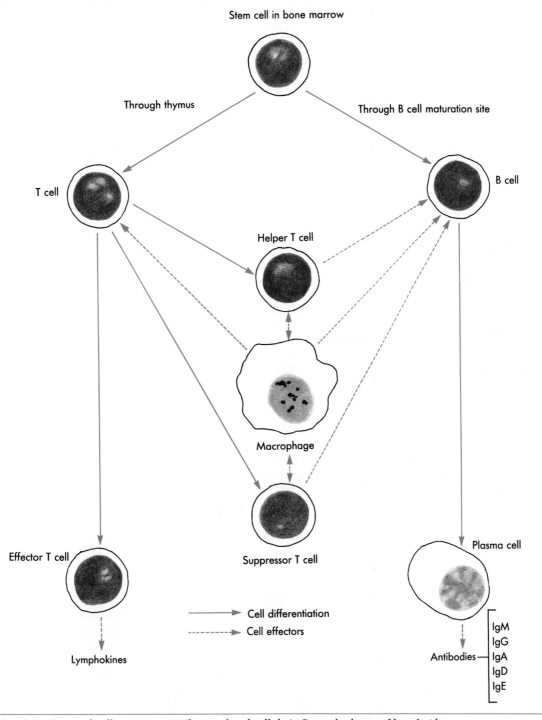

Stem cell in bone marrow

Through thymus

Through B cell maturation site

T cell

B cell

Helper T cell

Macrophage

Effector T cell

Suppressor T cell

Plasma cell

Cell differentiation
Cell effectors

Lymphokines

Antibodies —— IgM
IgG
IgA
IgD
IgE

FIG. 18-8 Principal cells in immunity (humoral and cellular). General scheme of lymphoid cell differentiation, interaction, and function. In humoral immunity one set of thymus-derived lymphocytes (T cells) functions as helper T cells in the maturation of B lymphocytes to the antibody-forming plasma cells. Another, the suppressor T cells, regulates immunoglobulin synthesis by inhibiting the maturation of B cells to plasma cells. In cellular immunity the T cells are the main reacting cells. Macrophages, because of their special properties, are intimately involved in both kinds of immunity.

particular antigen. When the human being is so exposed, that antigen finds or "selects" the particular clone that has been programmed specifically to meet it (and no other). That clone then preferentially proliferates and responds immunologically.

When an antigen (immunogen) gains access to an appropriate lymphoid area within the body, a sequence of events ensues wherein the antigen is processed, recognized, and bound to an antigen-sensitive lymphocyte. For most antigens to attain the maximum immunologic response in the production of immunoglobulin, they must induce an interaction between B and T cells in which macrophages are involved. The macrophage participates by processing (chemically altering) the antigen, a complex process in itself. This results in changes in the antigen but not the loss of its antigenic marker. With the collaboration of the T cell, the B cell receives the altered immunogen presented to it at the specific receptor site on its plasma membrane. The engagement of the immunogen with the membrane receptor is the signal for the B lymphocyte to become immunologically active.

Some antigens are T cell independent and therefore for an immune response need only to affect B lymphocytes. However, most antigens are T cell dependent; that is, the interaction of the T cell as a "helper" cell is necessary to activate the B cell and to start it on a course of immunologic activity. When a T cell recognizes an antigen, it releases an antigen-specific factor, which in the case of a B cell response turns on B cell activity. The same antigen also binds to the appropriate B cell; at the same time the suppressor T cell for this particular B cell function is activated. The balance between the helper T cell and the suppressor T cell influences the net expression of B cell activity.

When B cell receptors on the cell surface recognize and bind specific antigen, such bound antigen acts as a mitogenic factor, thereby stimulating the proliferation of the resting B cells and their transformation into plasma cells (the cells elaborating antibodies). With B cell activation by a T cell–independent antigen, the resultant plasma cells produce almost exclusively IgM and few or no memory cells (p. 333) are formed. By comparison with T cell–independent antigens, T cell–dependent ones induce in given instances the formation of IgM, IgG, and IgA, and memory cells as well. T cell–dependent antigens affect three cell types—T lymphocytes, macrophages, and B lymphocytes—whereas T cell–independent antigens involve only the one—the B lymphocytes.

When the surface IgM of the B cell combines with the specific antigenic determinant for which it is designed, that B cell to which it is attached is activated. As a result, the B cell proliferates and soon generates a clone of differentiated and immunologically competent cells. A small number of the progeny become plasma cells, cells that synthesize and secrete IgM **polymer.** This is a primary immune response. Some five to eight divisions of a B cell are required for its transformation into a plasma cell—an end cell incapable of further mitotic activity. Plasma cells are found in the germinal centers of lymph nodes, spleen, and diffuse lymphoid tissue of the alimentary and respiratory tracts. They have long been known to synthesize and secrete antibody globulin, which they do very efficiently. The antibody secreted has the very same specificity as the surface membrane immunoglobulin of the B lymphocyte from which the given plasma cell is derived. After a certain amount has been released, the plasma cell dies.

The remaining progeny of that activated B cell become "memory" cells, cells that remember the antigenic determinant to which their parent B cell was exposed. These cells are disseminated through blood, lymph, and tissues for extended periods and establish a reservoir of antigen-sensitive cells that are primed with a memory factor for immunoglobulin production. When the same immunogen reappears on the scene, these memory cells will proliferate rapidly and synthesize a large quantity of antibody. The secondary immune response will be expedited and may be augmented even as much as a hundred times or more. If the antigen persists with stimuli from helper T cells, some of that progeny of the original activated B cell will respond by forming a large number of memory B cells and plasma cells. A switchover from a primary to a secondary immune response will occur. The plasma cells will synthesize IgG (instead of IgM), IgA, or IgE.

Many B cell clones have already been activated by antigens fairly early in life. Thus the predominant immunoglobulin in the serum is IgG, since the primary immune response of these clones has switched over to the secondary immune response. Because multitudes of cell clones with individual specificities produce immunoglobulins of a given class (IgM, IgD, or other), the resultant immunoglobulins or antibodies are known as **polyclonal** antibodies. Many clones secreting IgG yield the typical serum concentration of 10 g/L; however, the amount of IgG from any single clone of B cells—a **monoclonal** antibody—is less than 1% of the total serum level.

Genetic background. Of the 1 million or so genes in the genome of an animal, only a small fraction specifies antibodies. Since animals contact unbelievably large numbers of antigens (in the millions) to which they must form specific antibodies, a genetic mechanism allowing for such scope must exist. The mechanism is one of gene construction from component parts or separate segments of DNA. Because genes specifying the structure of antibodies are not present in their final form in germ cells or even in embryonic cells, the B cell making the antibodies is not endowed with a definitive set. With the "bits and pieces" it does possess, the B cell is able to rearrange and to recombine the genetic material to make possible an almost limitless variety of antibodies. The number given is 18 billion.

Monoclonal antibodies. In 1975 the discovery was made that a pure, defined molecule with a single specificity to a given antigen—a monospecific or monoclonal antibody—could be produced by certain hybrid cells.

HYBRIDOMAS. Certain hybrid cells (hybridomas) can be structured for this purpose. The two cells first successfully fused were (1) an antibody-producing cell from the mouse—the B lymphocyte—and (2) a cancer cell from the same species—the myeloma plasma cell. The myeloma cell was chosen because it can make huge quantities of immunoglobulin and lives indefinitely in cell culture. As a result of the laboratory manipulations, a hybridoma (hybrid cell line) emerged with desired properties. This hybrid cell line expresses antibodies of the same single specificity as the parent B cell, and because of its cancerous origin in a myeloma (cancer) cell, it is "immortalized." (The term *immortal* is used when a cell line continues to be viable and to secrete a product as long as it is properly cared for.) Such a hybridoma is a long-standing source of specific antibody.

PRODUCTION. The steps in the production of monoclonal antibodies follow:

1. A mouse is immunized with a selected antigen. An advantage to this technic is that the antigen does not have to be pure.
2. An immune response occurs, and B lymphocytes may be harvested from the animal's spleen.
3. The B cells recovered are chemically fused with myeloma plasma cells from the same animal species.
4. The fused mixture goes into a selective cell culture medium where neither the B cell nor the myeloma cell can survive alone (unfused). Whatever grows out has to be hybridized.
5. The surviving hybrid cells are next cultured to obtain cell lines grown from only one cell (clones), thus making it easy to identify an antibody. A clone can then be selected because of its specificity.
6. To amplify production, the hybridoma is injected into a mouse. An ascites tumor results, a tumor associated with accumulation of fluid in the abdominal cavity. The tumor generates large amounts of the particular monoclonal antibody and releases it in a sizable accumulation of the ascitic fluid.

CHARACTERISTICS. The monoclonal antibody is specific for a single antigenic determinant. Since all the cells of the clone producing the antibody are progeny of a single hybrid, it is absolutely homogeneous. This order of specificity is not possible with conventional polyclonal sera. The discriminatory quality of even the best preparations of conventional polyclonal sera is considerably less.

Monoclonal antibodies are more concentrated and more easily purified than polyclonal antibodies generated in large animals. Titers for monoclonal antibodies in cell culture supernatants are high—100 to 1000 times higher than the titer from the serum of an animal—and often number over 1 million in ascitic fluid from the mouse.

APPLICATIONS. The applications of monoclonal antibodies in the health care setting are twofold (Table 18-5). First, they are highly specific reagents for use in diagnostic tests. They are already crucial reagents in many of the modern immunologic binding assays, such as radioimmunoassay and fluorescence microscopy (Table 18-6). They do not form precipitation complexes with antigen, since they are homogeneous. Direct diagnostic tests utilizing monoclonal antibodies require only 15 to 20 minutes to perform, whereas 3 to 6 days may be required with conventional technics.

Second, monoclonal antibodies provide a highly specific immunologic agent for the treatment of certain disorders. A good example is the application of antibodies to the detection and possible deletion of tumor-specific antigens. Antibody-producing hybridomas have already been developed against many human serum components, blood groups, transplantation antigens (p. 369), and many microbes.

VALUE.. The essential value of hybridoma is its ability to yield a stable, well-defined chemical agent against virtually any known antigen and to manufacture it in large quantities whenever needed.

Think of the hybridoma as an immunoglobulin factory, operating in a space about one fifth the diameter of a human hair. Each hybrid cell can make 1000 antibody molecules per

---

Table 18-5 Usefulness of Monoclonal Antibodies

Present	Future
Diagnosis of infectious diseases (gonorrhea, chlamydial infections, herpesvirus infections, and others)	Identification of autoantibodies
	Passive immunization against infectious agents, allergens
Epidemiology of infectious diseases	Localization of tumor cells
Identification of lymphocyte subsets	Manipulation of immune response
Identification of tumor antigens	Therapy of autoimmune diseases
Tissue and blood typing	Provision of graft protection
Purification of macromolecules	Potentiation of tumor rejection

---

Table 18-6 Some Monoclonal Antibodies Approved for In Vitro Diagnostic Testing

Testing for	Trade Name	Assay System
Chlamydia species (p. 655)	Cultureset	Immunofluorescence or immunoperoxidase
Gentamicin (TDM) (p. 252)	Optimate	Fluorescent immunoassay
HBsAg (p. 739)	NML	Radioimmunoassay
Specific IgE	IgE FAST test	Fluorescence
T and B cells	Cytotag	Fluorescence

second, and 10 million hybrid cells can be packed into the volume of 1 ml.

Currently sales of monoclonal antibodies are around $50 million per year for human immunodiagnostic test kits and $8 million annually for research agents. Predictions are that in the early 1990s the annual research market will reach $500 million and that for diagnostic materials $4 *billion*.

Antigen-Antibody Reaction

The antigen-antibody reaction is specific: a given antigen promotes the production of antibodies only against itself, and a given antibody acts only against the antigen promoting its development. The hemolysin in the serum of a rabbit that has been repeatedly injected with the red blood cells of a sheep dissolves the red blood cells of the sheep but not those of other animals.

Within the antibody molecule there is a pattern of small areas specially designed for a close-fitting union with the antigen (like a key fitted into a lock). Most immunoglobulins are bivalent; that is, their structure allows combination with the antigen molecule at two areas. IgM (a macromolecule) is pentavalent and perhaps decavalent.

If the union of antigen and antibody occurs in such a way that the aggregation is manifest as a visible phenomenon, the antibody is said to be a **complete antibody** (Fig. 18-9). When union of an antibody occurs with specific antigen but the antibody lacks the capacity to change the surface properties of the antigen and therefore a visible reaction does not take place, that antibody is referred to as an **incomplete** or **blocking antibody**.

Detection. Further discussion of the antigen-antibody reaction and the detection of antibodies is found in Chapter 20.

Complement. The widely used term *complement* is a system of proteins (components) that contributes, because of the bioactivity of the individual components, an essential function to normal host defenses against infection. Although complement (alexin)* is not antibody, its presence is necessary for the complete action of certain **lytic** antibodies, notably hemolysins, bacteriolysins, and cytolysins, and complement (the first four components) enhances the action of opsonins. The complement system participates in the antigen-antibody reaction but only after the antigen has been sensitized by the antibody.

The bioactivity of complement is also a part of the process of inflammation, and complement plays a crucial role in the mediation of immunologic tissue injury. In fact removal of circulating complement has been demonstrated to prevent such injury. A deficiency in one or the other of the complement proteins predisposes the person to autoimmune disease. Diseases in which mediation by complement is implicated as a significant factor include glomerulonephritis, rheumatoid arthritis, and certain acute viral diseases such as dengue hemorrhagic fever.

Complement is present in freshly drawn serum (10% of globulin fraction) and is not increased by immunization. It is destroyed by exposure to room temperature for a few hours or to a temperature of 56° C for 30 minutes. The freshly drawn serum of a guinea pig is regularly used as a source of complement because its content shows little variation in different animals and is comparatively high. Complement has been best studied in guinea pigs and in humans. It may be preserved by various chemical methods or by drying of the fresh frozen guinea pig serum to a powder in a vacuum.

Complement components together with inhibitors make up a complex enzymatic system of 20 antigenically distinct, interacting beta-globulins that constitute 20% of the nonimmunoglobulin protein in the serum. Each component is numbered. C3, the third component, appears to play the most significant role in disease.

---

*The term *complement* originated from the fact that it was demonstrated to "complement" certain immunologic reactions.

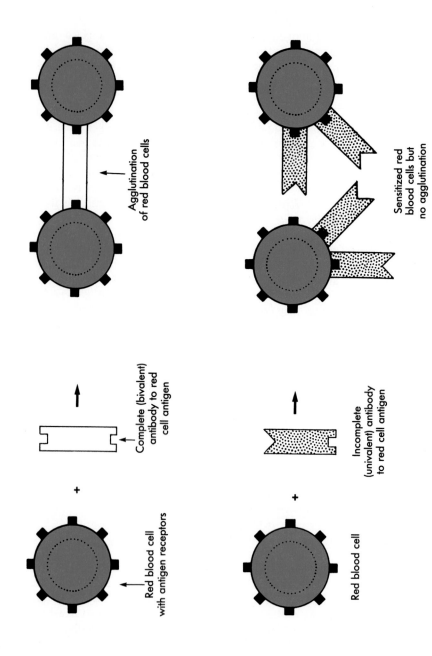

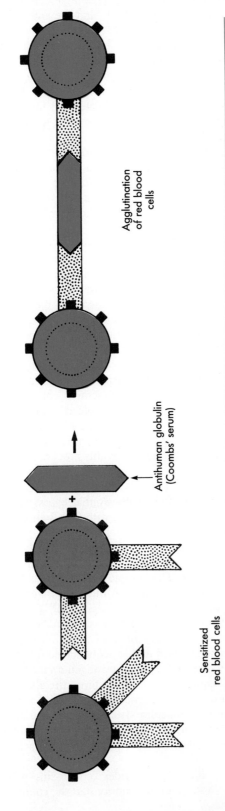

Agglutination
of red blood
cells

Antihuman globulin
(Coombs' serum)

Sensitized
red blood cells

FIG. 18-9 Antigen-antibody reaction. Note three reactions diagrammed horizontally between antigens on erythrocytes and antibody. *Top,* Agglutination (visible clumping) results from bonding of antigens and antibody. *Middle,* Antigen-antibody complex, but without clumping. Red blood cells here are altered (sensitized). *Bottom,* In contact with antiglobulin serum (a special complete antibody, p. 415) sensitized red cells are clumped. (Redrawn from Miale, J.B.: Laboratory medicine: hematology, ed. 5, St. Louis, 1977, The C.V. Mosby Co.)

On contact with activators of the system, the complement proteins align themselves to form two pathways of activation—the classic and the alternative—and the common terminal pathway of membrane attack (Fig. 18-10).

THE CLASSIC COMPLEMENT PATHWAY. The biochemical sequence of the classic pathway usually begins when an antigen and antibody combine to form an immune complex (immunoglobulin bound to cell membrane) and the immune complex interacts with ("fixes") the first component of complement (C1). Along the classic pathway the components react and interact in a complicated but orderly sequence, referred to as the complement cascade, and thereby produce certain immune phenomena, notably cytolysis (dissolution of cells).

In the classic pathway the proteins of the complement system, according to their participation in the events of the complement cascade, fall into three categories: the recognition, activation, and membrane attack units. The recognition unit is C1, which is actually three discrete proteins. One of these has the unique ability to bind the immune complex, thus triggering the activation of the complete C1 unit that is necessary for the assembly of the activation unit.

The activation unit of proteins appears to prepare the target cell surface for eventual lysis. Several complement components are involved, notably C4, C2, C3, and C5. They participate in a series of reactions in which various complexes and fragments attach to sites on the cell surface, thus marking the cell for attack. The splitting of C3 into a and b fragments is a key step. (C3a released into the serum acts as mediator of anaphylaxis.) C3b fragments, by coating the target cell surface, play a role in immune adherence; phagocytes and lymphocytes are drawn to these cells. C3b also participates in the splitting of C5. (The C5a fragment is released in serum to take part in the inflammatory process.)

The activation of C5 begins the final sequence of events in the classic pathway—the formation of the membrane attack unit. Complement components C6, C7, C8, and C9 are important at this stage. C6 and C7 complexed with the C5b fragment attach firmly to the lipid layer of the cell membrane. This in turn activates C8. Activated C8 begins cell lysis by opening a channel in the cell membrane between the inside and outside of the cell. Activation of C9 augments this process, and complete cell dissolution comes quickly.

THE ALTERNATIVE (PROPERDIN) PATHWAY. Activation of complement may also occur by a bypass or alternative chemical pathway that bypasses the first components of complement and starts the activation sequence with C3. Both pathways lead to cleavage of C3, a crucial step in the generation of biologic activity.

In the alternative pathway Factor 1, a constituent of the target cell membrane and an initiating component, is altered so that it can combine with serum C3 and two other proteins, Factor B and Factor D. The complex formed splits C3 enzymatically into a and b fragments, with C3b being deposited on the cell surface. The binding of C3b to the cell membrane is the meeting point of the classic and alternative pathways. Bound C3b fragments combine with Factor B, and the combination complexes with properdin, which acts as a stabilizer. These events lead to the cleavage of C5, which means the formation of the membrane attack unit, a sequence common to both pathways.

The alternative pathway does not require the antigen-antibody reaction for activation. Since antibody formation requires an interval of time, the alternative pathway provides an early defense mechanism against agents that may be harmful in the absence of antibodies. It does involve the properdin system, which consists of the unique serum protein properdin, a C3 activator, and certain other serum proteins. This system enhances resistance to gram-negative infections but also is a factor in the mediation of immunologic injury.

Certain polysaccharides, lipopolysaccharides (endotoxins without antibody), IgA, and some immunoglobulin aggregates that cannot start the complement sequence of the

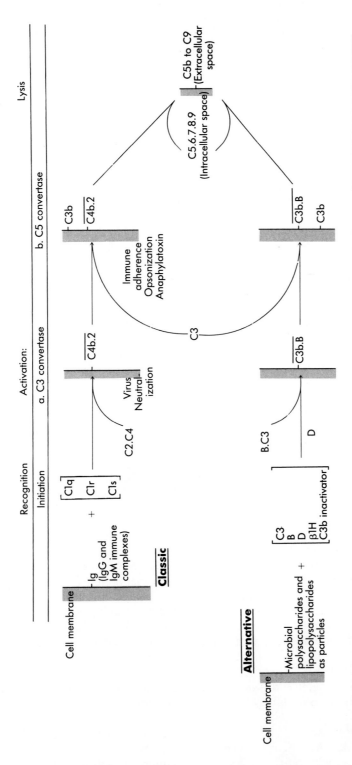

FIG. 18-10   Schema of the classic and alternative pathways of complement activation and of the common terminal pathway of membrane attack (lysis). Shaded areas correspond to biologic membranes on the surface of which the sequential reactions proceed. Certain biologically active by-products are not included. (A bar above the compound indicates that it is enzymatically active.)

classic pathway may participate early in the events of the alternative pathway, since they activate properdin.

COMPLEMENT FIXATION. During the process of antigen-antibody-complement union, bound (activated) complement is designated as fixed. Applicable in the laboratory to the diagnosis of disease, this phenomenon is complement fixation. The cytolytic effect of complement activity directed to cell membranes provides the working basis of a standard laboratory procedure.

Cell-Mediated Immunity

In acute bacterial infections with gram-positive cocci the mechanisms of humoral immunity in the normal individual are efficient. Persons with agammaglobulinemia, who cannot make measurable amounts of antibody globulin, do not fare well with the cocci, but they do adequately resist infections caused by viruses, fungi, protozoa, and certain other bacteria. Such resistance must stem from an immunologic mechanism other than the classic antigen-antibody combination.

To the host, infection with these latter agents presents a peculiar problem in defense, since these microbes colonize the interior of host cells. There they obtain shelter from antimicrobial elements of blood and tissue fluids. Some are in danger if they pass from cell to cell, but others seem to be unaffected in transit. The patterns of cell-mediated immunity are designed to meet this situation.

Although less is known of immunity related to the activities of certain cells than is known of that related to antibodies, cell-mediated (cellular) immunity, because of its practical implications for tissue transplantation and rejection, is a subject of considerable interest. The notion that, without demonstrable antibody, cells can bring about immune responses was hard to accept for a long time.

Comparisons

Like humoral immunity, cell-mediated immunity is an important part of the body's defense. Unlike it, no clear-cut antibody is demonstrable. For this reason no equivalent of passive immunization exists with cellular immunity—it cannot be passed in serum. However, transfer can be made in a less predictable fashion with lymphoid cells from a sensitized host.

For the study of cellular immunity there are fewer, less well-standardized immunologic procedures to match the extended list of well-standardized tests used to describe antibodies. For instance, the traditional skin test, important as it is in its application to cell-mediated immunity, is considerably influenced by local tissue factors and variably dependent on the subject's prior contacts with antigens. Another point of difference between cell-mediated and humoral immunity is illustrated by the skin test. The immune responses of cellular type emerge more slowly—within days—than those mediated by antibodies, which can evolve in a matter of hours.

In both kinds of immunity, contact with antigen triggers the sequence of immune events. With cell-mediated immunity, concurrence with presence of antigen is crucial. If antigen is destroyed, the immune process subsides. In tuberculin testing, for example, the previously sensitized person reacts until all injected tuberculoprotein is gone.

The Antigen

Antigens that induce the sequence of events in cellular immunity are usually protein. The antigenic protein may be quite small or more complex.

Principal Cells

The primary reacting cells of cellular immunity are (1) small, thymus-dependent lymphocytes, the T cells, which provide immunologic specificity, and (2) macrophages, which are secondary participants because of their property of phagocytosis.

The initial function of the macrophage in cellular immunity is to prepare and present the antigen for T cell recognition, since the receptors on the T cell surface do not see the antigen as "foreign" by itself but only after it has been processed by the macrophage. The macrophage in so doing greatly increases the immunogenicity of the antigen, which it retains on its plasma membrane. Here the recognition response takes place in a cell-to-cell contact between lymphocyte and antigen-macrophage complex. T lymphocytes possess their unique specificity for reaction with certain processed antigens as a result of the prior influence of the thymus gland. When contact with that particular antigen is made as indicated, the cells become sensitized, are activated, and respond immunologically. T cells recognize antigens in the processes of both cellular and humoral immunity.

Events

The first step, the recognition of the antigen, results in the immunologic activation of a population of T lymphocytes. Then ensues **blastogenesis,** the transformation and enlargement of the T cells into rapidly dividing blast cells (cells with an immature appearance). The immunologically active T cells elaborate and release certain soluble factors called **lymphokines,** which are the chemical mediators of the immune response to follow. Lymphokines are designed to activate macrophages, that is, to summon macrophages into the anatomic area and to program them for an attack. Thus affected, the macrophages congregate in large numbers at the site, primed to sequester and destroy the "foreign" element that does not belong there. Although the alteration of the macrophage as an immunologic phenomenon is incompletely explained, it can be shown that the lethal capabilities of the macrophage are thus enhanced and that a larger and more effective phagocytic cell is the result. To summarize, it might be said that cellular immunity involves an antigenic stimulus (for example, an infectious agent) to a specific population of cells (T cells). These cells can then activate phagocytic monocytes so that these phagocytes can better destroy the particular antigen.

Lymphokines

Lymphokines are regulatory proteins that transmit growth and differentiation signals among cells of the macrophage system (Fig. 18-11). (They may affect cancer cells directly or indirectly.) Similar products (monokines) may be produced by macrophages.

As suggested by the names given early on to these soluble factors, lymphokines as a group display a variety of immunologic properties and activities. Some of these biologically active substances have been designated as follows:

1. **Blastogenic** or **lymphocyte mitogenic factor**—induces blast transformation of lymphocytes
2. **Macrophage chemotactic factor**—recruits and draws macrophages to the anatomic area
3. **Macrophage migration inhibitory factor**—blocks movement of macrophages out of the area
4. **Macrophage activating factor**—alters macrophages immunologically
5. **Skin reactive factors**—relate manifestations of cell-mediated immunity to the skin
6. **Lymphotoxin**—damages or kills cells foreign to the body (microbial, cancer, and transplanted)

One important substance released in cellular immunity is **transfer factor,** recoverable as a cell-free white blood cell extract that transmits immunologic information from one person to another; the passage of sensitivity in this way is measured by skin test results.

Lymphokines also participate in responses of humoral immunity in that they amplify antibody production and the rate at which antigen is bound by antibody (helper T cell activity). They may impede antibody production (suppressor T cell activity). By enhancing antibody formation, they heighten the activation of the complement cascade.

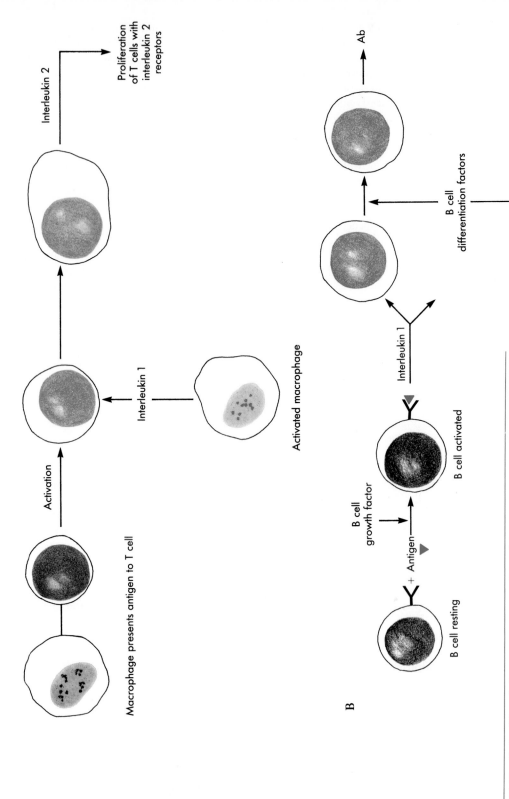

FIG. 18-11 Role of lymphokines in activation of T and B cells. **A,** T cell activation: T cell receives processed antigen from macrophage. Activated macrophage releases interleukin 1. Production of interleukin 2 stimulated in the targeted T cell. Interleukin 2 stimulates proliferation of T cells bearing interleukin 2 receptors. **B,** B cell activation: B cell receives antigen. B cell growth factor needed to activate B cell. Interleukin 1 stimulates clonal expansion to antibody-producing cells, with contribution from differentiation factors. *Ab,* Antibody.

Recently much investigative activity has focused on lymphokines—their role and their cell of origin—particularly because of their potential in biotechnology. Progress is being made. Table 18-7 presents some of the best-defined lymphokines to date.

In 1979 the term *interleukin* was coined for substances made by white blood cells *(leuko)* that regulate other leukocytes *(inter)*. Currently three lymphokines are so designated. **Interleukin 1** is needed to initiate and enhance T cell immune responses and to induce production of soluble factors (Fig. 18-8). **Interleukin 2** is a growth factor for T cells. **Interleukin 3** is one of several lymphokines involved in the development of effector or cytotoxic T cells.

Interferon. Since more is known about the molecular biology of interferons than of other lymphokines, interferon-gamma (or immune interferon) is viewed as the prototype lymphokine. Interleukin 2 stimulates the production of interferon-gamma in T cells (one of the best sources). Included in its many effects is the augmentation of the cytotoxic activities of T cells, macrophages, and natural killer cells.

Immune Result

The cytotoxic effects of the immunologically active T lymphocytes coupled with the efforts of immunologically active macrophages result in localization and destruction of antigen or antigen carrier. Some of the progeny of the T lymphocytes resume the small lymphocyte status to recirculate as memory cells.

Manifestations

Cell-mediated immunity is a major component of the body's resistance to viruses, fungi, and certain bacteria. Classic expressions of it are found in delayed-type hypersensitivity (once known as bacterial hypersensitivity), allograft rejection, and graft-versus-host reactions. It is also considered to be the capacity possessed by the body to resist most forms of cancer.

Tuberculin test (p. 605). A well-known example of cell-mediated immunity or de-

Table 18-7 Some of the Best-Defined Lymphokines

Lymphokine: Current Name	Lymphokine: Early Name	Cell of Origin	Target Cell	Immunologic Capacity
Interleukin 1	Lymphocyte activating factor	Monocytes or macrophages	T cells	T cell activation
Interleukin 2	T cell growth factor	T cells	Cytotoxic T lymphocytes	Required for continued T cell growth
Interleukin 3		T cells	Certain cytotoxic T cells	Required for T cell growth
Interferon-gamma	Immune interferon	T cells		Regulation of certain humoral and cell-mediated responses
B cell growth factor		T cells	B cells	Stimulates growth of B cells
Tumor necrosis factor		Macrophages	Tumor cells	Lysis of tumor cells
Lymphotoxin		T cells	Tumor cells	Lysis of tumor cells (in the laboratory)

Table 18-8 Dual Nature of the Immune System

Component	Humoral Axis (Antibody-Producing Limb)	Cellular Axis (Cell Mediated)
Specific cells	B lymphocytes	T lymphocytes
Specific molecules	Antibody	Lymphokines
Effector cells	Polymorphonuclear neutrophilic leukocytes (PMNs)	Macrophages
Amplifying mechanism	Complement system	Lymphokines
Interactions	Antibody with effector cell in opsonization	Lymphokines activate macrophages (effector cells) to stimulate phagocytosis and eradication of infectious agent
Response rate (in sensitized person)	Minutes to hours	Hours to days
Protective immunity in	Opsonization and lysis of bacteria Neutralization of viruses and toxins	Control of intracellular pathogens Tumor surveillance

layed-type hypersensitivity is the tuberculin reaction in the skin. When a purified preparation of tuberculin (p. 600), such as purified protein derivative (PPD), is injected into the skin of a subject in whom a previous tuberculous infection has occurred, a characteristic reaction occurs that will not be present in the nonexposed and hence nonimmune person. A first-time infection with tubercle bacilli induces a state of cell-mediated immunity, which can be detected subsequently by skin testing with antigen. Once the test is positive, it remains so for life. The first contact with tubercle bacilli sensitizes the T lymphocytes of the subject to the tuberculin. Thereafter the sensitized lymphocytes on contact with the antigen will be stimulated by macrophages to release lymphokines, which will mediate the reaction.

The changes in the skin of a person with a positive tuberculin reaction appear several hours (are delayed) after injection of the tuberculin and become full blown in 24 to 48 hours, thereafter subsiding. The site is reddened, swollen, and hardened (indurated) (see Fig. 29-4). With a pronounced reaction, the skin site may be hemorrhagic or even necrotic. The redness and swelling tend to clear readily, but the diagnostic feature, the induration, can be felt for days or weeks.

RÉSUMÉ OF FEATURES OF THE IMMUNE SYSTEM

Table 18-8 summarizes and emphasizes further the salient features of the two limbs of the immune system that have been presented in this chapter. Table 18-9 indicates that different patterns of immune response are the important ones in different infections. It presents a classification that refers to the dual nature of the immune system as characterized in Table 18-8.

SUMMARY

1 Immunology, the division of biology concerned with the study of immunity, comprises the complexity of factors that help humans to maintain their structural and functional integrity and to ward off certain kinds of injury. Susceptibility, the reverse, is the suppression of factors that produce immunity.

Table 18-9 Immune Mechanisms Operative in Infectious Disease

Intracellular Infections—Cellular Immunity: Lymphocytes and Macrophages Decisive in	Extracellular Infections			
	Humoral Immunity: Opsonins and Neutrophils Decisive to Recovery in Infections Caused by	Humoral Immunity: Antibody Decisive to Prevention or Recovery in		Both Humoral and Cellular Immunity: Immune Collaboration Decisive in Host Defense in
		Antibody Action	Disease	
Tuberculosis	Pneumococci	Neutralizes toxin	Diphtheria	Cryptococcosis
Leprosy	Meningococci		Tetanus	Syphilis
Histoplasmosis	Influenza bacilli, type b		Botulism	Salmonellosis
Coccidioidomycosis	Gonococci		Cholera	Candidiasis
Brucellosis	Group A streptococci	Blocks epithelial attachment	Cholera	Listeriosis
Tularemia	*Klebsiella pneumoniae*		Shigellosis	
	Pseudomonas aeruginosa		Salmonellosis	
	Plague bacilli	Participates in complement-mediated bacteriolysis	Cholera	
	Anthrax bacilli		Meningococcal disease	
		Neutralizes virus	Poliomyelitis	
			Influenza	
			Adenovirus infection	

2   Three basic phenomena in immunity are (1) recognition of nonself; (2) memory—recall of prior immunologic experience; and (3) specificity.

3   The important role of the lymphoid system in immunity is expressed in two distinct populations of lymphocytes, the T lymphocytes and the B lymphocytes.

4   T lymphocytes are influenced by the thymus gland early in their development. They circulate in the peripheral blood and lymph for many years. Subpopulations of T cells are effector or cytotoxic, helper, and suppressor cells.

5   B lymphocytes are short-lived, thymus-independent cells.

6   Although the functions of the principal cells overlap, the immune system can be compartmentalized into the B cell and the T cell systems. B lymphocytes are responsible for the production of humoral immunity, that is, immunity tied in with the elaboration of chemical molecules called antibodies or immunoglobulins. T lymphocytes regulate cell-mediated immune responses. This kind of immunity focuses on the activities of certain cells.

7   An antigen is a substance foreign to the body, usually protein, that elicits the production of specific antibodies against itself. The part of the antigenic molecule that specifically combines with antibody is the antigenic determinant.

8   Antibodies or immunoglobulins, the specialized globulins that react precisely with specific antigens, segregate into a family of distinct and related proteins. They are very large and heterogeneous, multichain proteins, and the most complex group known. Largely associated with gamma globulin of the blood, antibodies represent about 20% of total serum protein.

9   Immunoglobulins are classified as immunoglobulin G (IgG), immunoglobulin A (IgA), immunoglobulin M (IgM), immunoglobulin D (IgD), and immunoglobulin E (IgE). The predominant globulin of plasma is IgG, which is also abundant extravascularly.

10 Secretory IgA provides humoral defense in external secretions bathing mucous membranes in the body. It binds antigens mechanically, thus blocking penetration and colonization of surfaces by potential pathogens.

11 Immunoglobulins are formed in the lymphoid tissues of the body in response to a given antigen; the largest source is the gut-associated lymphoid tissue.

12 The principal cells in antibody synthesis are lymphocytes, primarily B cells; plasma cells derived from B cells; and macrophages. T lymphocytes regulate and assist the process.

13 According to the clonal selection theory of Burnet, clones of lymphocytes already present in the human body are preset and programmed specifically to respond to one particular antigen and none other.

14 An antigen in the body is processed by a macrophage and with the collaboration of a T lymphocyte is bound to the antigen-sensitive B cell at the specific receptor site. The B cell is thus activated immunologically.

15 Proliferating B cells become transformed into plasma cells, the antibody-producing cells. The antibody secreted by the plasma cell has the same specificity as the surface membrane immunoglobulin of the B lymphocyte from which it is derived.

16 Memory cells, progeny of an activated B cell, remember the antigenic determinant to which their parent B cell was exposed. Disseminated through blood and tissue fluids, they proliferate rapidly, synthesizing a large quantity of specific antibody when the same immunogen reappears.

17 In 1975 a pure, defined molecule with a single specificity to a given antigen—a monospecific or monoclonal antibody—was produced by certain hybrid cells.

18 The term *complement* refers to a complex system of proteins contributing an essential function to normal host defenses against infection. Complement is needed for the complete action of certain lytic antibodies, enhances the action of opsonins, but also may play a role in immunologic tissue injury.

19 The immune responses of the cellular type emerge more slowly—within days—than those mediated by antibodies, which can evolve in a matter of hours. The principal reacting cells are the T lymphocytes and macrophages. The macrophage initially prepares and presents the antigen in cellular immunity for T cell recognition, which occurs in a cell-to-cell contact between the macrophage and the T cell. The T cell then responds immunologically.

20 T cells recognize antigens in the processes of both humoral and cellular immunity.

21 Blastogenesis is the transformation of T cells into large, rapidly dividing cells that elaborate and release lymphokines, the chemical mediators of the cellular immune response.

22 The cytotoxic effects of the immunologically active T lymphocytes coupled with the efforts of immunologically active macrophages result in localization and destruction of the antigen or antigen carrier. Some of the T cell progeny become memory cells.

23 Cell-mediated immunity is a major component of the body's resistance to viruses, fungi, and certain facultatively intracellular bacteria.

QUESTIONS FOR REVIEW

1 Fill out the table below, indicating whether each example of acquired immunity (a) is obtained naturally or artificially and (b) is active or passive.

2 Classify immunity. Give examples.

3 What type of immunity is of the longest duration? The shortest? Explain.

Cause of Immunity	Acquired			
	Naturally	Artificially	Active	Passive
Attack of measles				
Attack of smallpox				
Vaccination against poliomyelitis				
Vaccination against rubella				
Pasteur treatment				
Diphtheria antitoxin				
Diphtheria toxoid				
Tetanus antitoxin				
Tetanus toxoid				
Immune globulin				
Bacterial vaccine				
Placental transmission of antibodies				
Breast feeding				
Viral vaccine				

4 What are immunoglobulins? How may they be classified?

5 Compare humoral immunity with cellular immunity; B cells with T cells.

6 What is the role of the lymphoid system in immunity? Role of thymus gland?

7 Comment on the immunologically competent cell.

8 What is meant by the memory phenomenon in immunity? What is its importance?

9 Compare the primary immune response with the secondary immune response.

10 How does the property of specificity apply to the immune system?

11 What is the role of the different T cell subsets? What does a "killer" T cell do?

12 What is the difference between serum and plasma? Fab and Fc fragments in an immunoglobulin molecule? Variable and constant regions in an Ig molecule? Polyclonal and monoclonal antibodies? Classic and alternative pathways of complement activation?

13 Briefly describe and indicate the biologic function of the following: secretory IgA, the plasma cell, the macrophage, the complement cascade, lymphokines, blastogenesis, haptens or partial antigens, the T cell, HLA, and gut-associated lymphoid tissue.

14 Define or explain immunity, immunology, immunopathology, immunoglobulin, immunochemistry, immunobiology, antigenic determinant, isoantigen, adjuvant, immune system, susceptibility, opsonins, serology, dimer, marker, activation, and cross-match.

SUGGESTED READINGS

Abruzzo, L.V., and Rowley, D.A.: Homeostasis of the antibody response: immunoregulation by NK cells, Science 222:581, 1983.

Barrett, J.T.: Textbook of immunology, St. Louis, 1983, The C.V. Mosby Co.

Braun, W.E.: The HLA antigens in diagnostic medicine, Lab. Manage. 19:45, Sept. 1981.

de Macario, E.C., and Macario, A.J.L.: Monoclonal antibodies for bacterial identification and taxonomy, ASM News 49:1, 1983.

Gordon, D.S.: Fast track for monoclonal antibodies, MLO 15:53, 1983.

Grouse, L.D.: Bone marrow transplantation, a lifesaving applied art, an interview with E. Donnall Thomas, MD, J.A.M.A. 249:2528, 1983.

Kahan, B.D.: What's new in renal transplantation, Texas Med. 78:44, Aug. 1982.

Kolata, G.: Drug transforms transplant medicine, Science 221:40, 1983.

Macek, C.: Medical news: cyclosporine's acceptance heralds new era in immunopharmacology, J.A.M.A. 250:449, 1983.

Marx, J.L.: A closer look at the genes of the MHC, Science 220:937, 1983.

Nowinski, R.C., and others: Monoclonal antibodies for diagnosis of infectious diseases in humans, Science 219:637, 1983.

Peter, J.B., and Hawkins, B.R.: HLA antigens and disease, what the associations signify for diagnosis, Diagn. Med. 4:54, Jan./Feb. 1981.

Polin, R.A., and Wasserman, R.L.: Monoclonal antibodies: progress and promise, Lab. Manage. 21:33, Oct. 1983.

Roitt, I.M.: Essential immunology, London, 1980, Blackwell Scientific Publications, Ltd.

Wade, N.: Inventor of hybridoma technology failed to file for patent, Science 208:693, 1980.

Wade, N.: Hybridomas: the making of a revolution, Science 215:1073, 1982.

IMMUNOLOGIC DISORDERS, INCLUDING ALLERGY

Unfortunately the immunologic response is not always favorable in the body and may even be quite harmful. **Immunopathology** is the study of tissue injury consequent to the immune reaction.

Considerable attention is being given today to the elucidation of both normal and abnormal mechanisms in immunity. As a result, immunologic disorders are better understood. Some have only recently been defined.

DISTURBANCES IN IMMUNITY

The following listing cites the broad categories of immunologic disturbances or abnormalities:

1. Immunologic accident—best example, reaction complicating mismatched blood transfusion (p. 413)
2. Immunologic depression (immunosuppression)—administration of immunosuppressive agents materially interfering with or inhibiting the immune system (humoral or cellular); such are the steroid hormones in huge doses, irradiation, cytotoxic chemicals (antimetabolite drugs and alkylating agents), and antilymphocytic serum (p. 372)
3. Immunologic deficiency—failure or impairment of normal development of immunologically competent cells of the lymphoid system of the body, both antibody deficiencies and defects of cell-mediated immunity included

    An absence of lymphocytes and plasma cells and hence of all immunoglobulins is termed **lymphoid aplasia**. Abnormal development of the thymus gland (**thymic dysplasia**) or a congenital absence of the thymus will result in severe impairment of immunologic mechanisms. Death comes early in infancy.

    An absence of plasma cells with very low levels of gamma globulin is either **hy-**

pogammaglobulinemia (antibodies in greatly decreased amounts) or **agammaglobu-linemia** (no demonstrable gamma globulin in the blood). An absence of or a defect in some of the immunoglobulin components of serum is **dysgammaglobulinemia.**

Certain patients with depressed cellular immunity are peculiarly vulnerable to infections caused by *Mycobacterium tuberculosis, Pseudomonas aeruginosa,* and *Pneumocystis carinii.*

4. Immunologic effect of nonimmunologic diseases—infections with bacteria, viruses, and protozoa producing striking elevations in serum globulin levels (**hyperglobuli-nemia**)

5. Immunologic aberration secondary to malignancies of the lymphoid system—as would be expected, associated with striking abnormalities

The malignant prototype of the lymphoid cell can yield an abnormal type of globulin, often in large quantities. Examples of frankly abnormal globulins synthe-sized by malignant cells are the myeloma proteins, the macroglobulins of Walden-ström, and the Bence Jones proteins. The malignancies include the leukemias, lymphomas, or other neoplastic cell proliferations arising in the tissues occupied by the macrophage system.

6. Allergic states or hypersensitivity reactions (p. 376)

7. Autoimmunity and autoimmune diseases—defects of immunologic tolerance

Immunologic tolerance, immunologic paralysis, and **immunologic unresponsive-ness** are interchangeable terms. They are used for a state of nonreactivity to a given, specific antigen, an antigen that is usually an effective one. Immunologic tolerance is an acquired quality; a prior exposure to the given antigen has occurred, and the antigen must persist for tolerance to be maintained.

Because of the specificity, immunologic tolerance is unlike immunosuppression, which is a state of *non*specific *non*reactivity to all antigens.

In the prenatal period it is believed that the contact of potential antigens with the immunologically immature lymphoid cells induces tolerance that in some way prevents any future response coming from the contacts of the mature lymphoid cells with the same specific antigens. In this way the body is considered to establish a necessary tol-erance for its own constituents (self). Its immune system is geared to discriminate be-tween what belongs, the self, which it tolerates, and what does not, the nonself, which it attacks.

One might reasonably expect that on this basis immunologic paralysis could be built up artificially. The foreign cell for which tolerance might be desired at a future date could be introduced into the body at the time in life when the immunologic system is not fully functional. Tolerance is induced less readily in adult life and generally then with much higher doses of antigen.

Autoimmunization

Generally, the immunologic mechanisms in the body are protective. Microbes or sub-stances that would enter and injure are attacked. Because the body recognizes its own cells and tissues, it differentiates between the protein of the foreign invader and that which belongs to itself. "Unfortunately," as William Boyd said, "the immune system is a two-edged sword which can be turned against the body in that biological paradox nicknamed autoimmunity. The same forces that normally reject foreign material act in reverse and reject the cells and tissues of the body itself with unpleasant conse-quences."*

Many potentially antigenic substances exist in or on the surface of a person's own cells, but before these can induce an immune response, they must surely be altered in

---

*From Boyd, W.: A textbook of pathology, Philadelphia, 1970, Lea & Febiger.

some way, possibly by the action of bacteria, viruses, chemicals, or drugs. Once that change has been made, these antigens, now **autoantigens,** may stimulate the body of the person in whom they occur, to form the corresponding antibodies, or **autoantibodies.** The process by which antibodies are made by the body against its own cellular constituents is referred to as autoimmunization. The importance of autoimmunization lies in the unfavorable and disease-producing effects of certain immune responses to the autoantigen. The category of human disease resulting is that of the **autoimmune diseases.**

Autoimmune Diseases

Autoantibodies can exist without apparent effect, but under certain conditions the reaction of autoantibodies (or sensitized immunocytes of the lymphoid series) with the specific tissues (the autoantigens) does result in disease. Autoantibodies may cause disease by reacting with tissue antigens to effect tissue injury directly, or they may form circulating antigen-antibody complexes that deposit in and thus injure tissue (p. 381). The disease resulting may be one of widespread involvement or may be limited to one organ. The overall clinical and pathologic picture is determined by the particular tissue attacked immunologically, its distribution in the body, and the extent of the damage done by the reaction. Autoimmunity is important in disorders of the thyroid gland, brain, eye, skin, joints, kidneys, and blood vessels.

Six features are always emphasized in discussions of autoimmune disease: (1) elevation of serum globulin, (2) presence of autoantibodies in the serum, (3) deposition of denatured protein in tissues, (4) infiltration of lymphocytes and plasma cells in damaged areas, (5) benefit from treatment with corticosteroids, and (6) coexistence of multiple autoimmune disorders.

In some autoimmune disorders the autoantigens involved are known (Table 19-1), and circulating autoantibodies can be demonstrated in the patient's serum, especially when there is an associated elevation of gamma globulin. As a group these diseases often give a history of preceding infection, tend to run in families, and are relieved by steroid hormones (ACTH and cortisone), and the patient's serum gives a false positive test for syphilis. Virus particles have been associated with autoimmune diseases, but the relationship is unclear at this time.

Disseminated (Systemic) Lupus Erythematosus

One of the most interesting of the autoimmune diseases is disseminated lupus erythematosus (Latin *lupus*, wolf). A disease primarily of women, it runs an acute or subacute course interspersed with periods of improvement. Although the mechanisms are ob-

Table 19-1 Examples of Disease and Antigen in Autoimmunity

Autoimmune Disease	Autoantigen
Acquired hemolytic anemia	Antigens on surface of red blood cells
Idiopathic thrombocytopenic purpura	Platelet antigens
Rheumatoid arthritis	Denatured gamma globulin
Systemic lupus erythematosus	Nuclear material including DNA; certain cytoplasmic substances
Sympathetic ophthalmia	Uveal (eye) pigment
Postvaccinal and postinfectious encephalomyelitis	Myelin (lipid-forming sheath around certain nerve fibers)
Rheumatic fever	Streptococcal antigen
Glomerulonephritis (kidney disease)	Streptococcal antigen

FIG. 19-1   Lupus erythematosus (LE) cell. This distinctive cell is found in blood and bone marrow of patient with lupus. It is a mature neutrophil that has phagocytized a homogeneous mass of nuclear material (degraded DNA) in the presence of antinuclear autoantibody in serum.

scure, immunologic injury is implicated in the pathogenesis. It is postulated that disease-producing complexes are formed when the autoantibodies, the antinuclear globulins, react with breakdown products of nuclei in the normal wear and tear of cells (Fig. 19-1). Anatomically, connective tissue, a tissue widely distributed over the body, is destroyed. Therefore lesions can occur anywhere but classically select the kidney, heart, spleen, and skin (butterfly rash on face). With the connective tissue injury come joint pains, fever, malaise, and anemia.

Idiopathic Thrombocytopenic Purpura (Autoimmune Thrombocytopenia)

Idiopathic thrombocytopenic purpura is an autoimmune disorder affecting platelets, one of the formed elements in the blood, important for their role in normal blood clotting. It is called idiopathic because the exact mechanism of its production is obscure. The salient feature here is an accelerated destruction of platelets as a result of the action of known autoantibodies against platelets. Hence the number of platelets circulating in the bloodstream is considerably depressed below the normal level. (This lack of platelets is thrombocytopenia.) The bone marrow, the site of platelet formation, appears normal, and the megakaryocytes from which platelets are derived are present in normal or even in increased numbers. The disorder primarily affects children and young adults and in most cases occurs some 3 weeks after an infectious process, such as a viral respiratory illness or one of the childhood diseases. The onset is acute, and spontaneous bleeding sites (no history of trauma) in skin, mucous membranes, joints, and the gastrointestinal tract are the manifestations. Fortunately, clinical hemorrhages are usually limited, although the thrombocytopenia may be pronounced. The rare hemorrhage into the cranial cavity can be serious. In 80% or more of the patients, removal of the spleen is curative, suggesting an immunologic role for the spleen in disease production. Either the spleen is the major site of removal of sensitized platelets, or it is the major source of the autoantibodies.

Deficiencies of Complement Components

A number of diseases have a clear relationship with deficiencies in the complement system (Table 19-2). Deficiencies may be acquired or congenital. Acquired deficiencies are often associated with disease in which the immune complexes (p. 381) formed depress the serum level of the components participating early in the sequence of complement activation. Examples of such diseases are the autoimmune diseases, agammaglobulinemia, and serum sickness. In certain instances the serum level returns to normal when the disease abates.

Another feature of complement system deficiencies is an increased susceptibility to infection, the most severe possibility being a lack of C3. Since C3 is at the junction of the two pathways of complement activation, the products of its activation are crucial in both. As a consequence, the serum of patients without C3 demonstrates little in the way of hemolytic, chemotactic, opsonic, or bactericidal activity.

Table 19-2 Some Diseases Associated with Deficiencies of Complement Components

Component Deficient	Disease Related
C1	Systemic lupus erythematosus (SLE), severe recurrent infections, glomerulonephritis (p. 387)
C4	Systemic lupus erythematosus
C2	Recurrent infections, systemic lupus erythematosus
C3	Severe recurrent infections
C5	Systemic lupus erythematosus, recurrent infections
C6	Meningococcal and gonococcal infections (p. 507)
C7	Renal disease
C8	Systemic lupus erythematosus, glomerulonephritis, gonococcemia

Acquired Immunodeficiency Syndrome (AIDS)

General Features

First described in 1979, acquired immunodeficiency syndrome (AIDS) is by all criteria a new disease with no evidence for its existence before this date. As a new disease, AIDS has generated more interest in the scientific community and frightened more of the general public than has any disease for a long time. The number of cases is still small, but the infection rate is doubling every 6 months with no sign of abating. In the number of deaths it has taken the highest toll of any American epidemic since the swine flu outbreak of 1918-1919. Two years after the onset of disease, the case fatality rate exceeds 90%.

Nature of the Disorder

As its name indicates, AIDS means loss of normal immune function, and the loss is a profound one. The severe immunodeficiency renders the patient especially vulnerable to a number of opportunistic infections (Table 19-3), coming either sequentially or simultaneously, and to a rare malignancy, Kaposi's sarcoma. AIDS is an extremely virulent disease, a killer. No known cure exists, and none of its victims regains normal immunity.

The disease is predominantly one of homosexual or bisexual men living in large cities—75% of cases. Intravenous drug abusers account for 17% of cases, Haitian immigrants for 5%, and hemophiliacs for a smaller number. AIDS is rare in other persons or groups, although infants born to high-risk mothers may have the disease. Why Haitians are susceptible is one of the mysteries of the disease.

Cause

The cause is unknown. No one knows why the immune system breaks down in the way that it does. The disease behaves as an infectious one, and a transmissible agent is postulated, most likely a virus, but no known virus produces the kind of immune dysfunction seen in AIDS. Presumably any causative virus is a new one, a mutant form of an old one, or maybe an animal virus recently introduced into humans. Such a virus, if it exists, is contagious in much the same way as the hepatitis B virus, but it is definitely not the same virus.

Spread

The major route of transmission appears to be through homosexual contacts. Among homosexuals, one sexual contact with an infected person can transmit the disease.

Table 19-3 **Opportunistic Infections in AIDS Patients**

Agent	Infection(s)
Viral	
Cytomegalovirus	Generalized
	Retinitis
	Pneumonia
	Encephalitis
Herpes simplex	Progressive
Herpes zoster	Skin vesicles
Fungal	
Candida albicans	Thrush
	Esophagitis
	Disseminated
Cryptococcus neoformans	Meningitis
Histoplasma capsulatum	Generalized
Aspergillus species	Pulmonary
Protozoan	
Pneumocystis carinii	Pneumonia
Toxoplasma gondii	Encephalitis
Cryptosporidium species (intestinal coccidian rarely seen in humans)	Enteritis
Mycobacterial (Other Bacterial Infections Infrequent)	
Mycobacterium avium–intracellulare	Disseminated
Mycobacterium tuberculosis	Generalized

However, there is no reason to believe that it can be acquired from a casual physical contact. In drug abusers the sharing of infected needles is relevant. Except for hemophiliacs, spread through blood and blood products is the least likely way. Why the disease favors such an unlikely combination of sexually active homosexuals, drug addicts, and Haitians remains an enigma.

The Lack of Immunity

In AIDS T cell–mediated immunity is selectively impaired (Table 19-4). The humoral (B cell) immune response may also be altered but not to the extent that the cell-mediated one is. (Bacterial infections do not so regularly complicate AIDS as do infections where cell-mediated mechanisms operate to a greater degree.) Humoral mechanisms may appear to be intact as demonstrated both by in vitro testing and by the presence of a normal antibody titer.

Lymphopenia is present. T lymphocytes are sparse. Most striking is the imbalance between the two subpopulations of T lymphocytes—the helper T cells and the suppressor T cells. Usually twice as many helper T cells are found as suppressor/cytotoxic T lymphocytes. The ratio of these two cells is popularly referred to as the T_4/T_8 ratio because the helper T cells are identified by a monoclonal antibody, one brand of which is called OKT_4, and the suppressors with a different monoclonal antibody, OKT_8. In AIDS the number of helper T cells is dramatically decreased, whereas the number of suppressor lymphocytes is definitely increased. Reversal of the T_4/T_8 ratio is seen transiently following many common bacterial and viral infections, but the reversal is not so

Table 19-4 Immune Changes in AIDS

State of Immunity	Laboratory Findings
Cellular	Lymphopenia—total number of T cells reduced
	Percentage and numbers of T helper cells decreased
	Percentage and numbers (variable) of T suppressor cells increased
	Ratio of T helper to T suppressor cells reduced
	T cell proliferative responses to mitogen and antigen diminished
	Anergy with delayed hypersensitivity skin testing demonstrated
Humoral	B cells normal in numbers
	Serum IgA and IgG increased in patients with opportunistic infections
	Levels of circulating immune complexes increased in patients with opportunistic infections

severe nor so prolonged as in AIDS. Reflecting the severity of the immunodepression, infections are more extensive and varied than in any other immunosuppressed group of persons.

One of the reasons that the cytomegalovirus has been implicated in AIDS is that it is known to cause transient immunosuppression in normal persons. (Note that the immunosuppression is not transient here.)

Role of human T cell leukemia virus. Infection with a recently discovered human retrovirus—the human T cell leukemia virus—may be involved in AIDS (p. 698). This virus infects T helper cells and causes them to exhibit suppressor activity in vitro. As many as 30% of AIDS patients reportedly carry serum reactivity to the viral antigens, and the virus itself has been isolated from a few AIDS patients. In one group studied this virus was found in a third of blood samples tested.

Clinical Picture

The disease is probably transmitted before symptoms appear, since the diagnosis is usually made with the occurrence of opportunistic disease predictive of cellular immune deficiency (most commonly Kaposi's sarcoma or *Pneumocystis carinii* pneumonia). The incubation period, considered to be from 9 to 22 months, is at least 6 to 8 months. Because of a protracted incubation, asymptomatic persons probably spread disease for many months before signs of clinical illness appear.

The manifestations of the prodrome are persistent generalized lymphadenopathy, lymphopenia, and impaired T cell function. Most AIDS patients appear to be partially immunosuppressed before the disease becomes full blown. Notable among the presenting features of the disease are (1) fatigue, (2) low-grade fever with night sweats, (3) enlarging, hardening, painful lymph nodes, (4) persistent diarrhea, (5) a thick, whitish coating of the tongue or throat, and (6) recent, slowly enlarging, purplish, discolored nodules or plaques in the skin.

An important complication of AIDS is Kaposi's sarcoma, a cancerous disorder distinctive for the red-purple, flattened lesions found especially in the skin of the feet and legs. Previously Kaposi's sarcoma had been thought of chiefly as an indolent tumor of the skin among elderly men of Jewish and Mediterranean origin. It had been seen in renal transplant patients, but it would resolve after cessation of the immunosuppressive therapy. Now it is an aggressive disease in 30- to 40-year-old homosexual men who have lesions anywhere in the skin, on mucous membranes, in lymph nodes, or in the internal organs.

Laboratory Diagnosis

No diagnostic test is available. The well-defined T cell subset abnormalities can be demonstrated, and AIDS patients are anergic when tested by a battery of skin tests.

Importance in Blood Banking

Over the last 3 to 4 years approximately 10 million or more persons have received blood transfusions. Only 15 of the cases of AIDS reported to the Centers for Disease Control (CDC) in Atlanta in that period of time were suspected of being related to a blood transfusion. The risk of getting AIDS through a blood transfusion is about 1 in 1 million.

Because of its concern that this disorder might be transmitted through blood donated by an AIDS victim, the CDC has recommended that (1) blood clotting factors used by hemophiliacs be pasteurized, (2) patients scheduled for an elective surgical procedure be encouraged to donate their own blood ahead of time for their own use if a transfusion is necessary, (3) measures to discourage persons at high risk from donating blood be instituted, and (4) blood donors be as tightly screened as is possible.

Prevention

The CDC of the U.S. Public Health Service has published its recommended precautions for hospital personnel providing medical care to possible AIDS patients, for persons performing laboratory tests or studies on potentially infectious material from known or suspected AIDS patients, and for persons involved in studies in which experimental animals are inoculated with tissues or potentially infectious material from persons with known or suspected AIDS.*

TISSUE TRANSPLANTATION

Many times in medicine it would be desirable to take a normal tissue (or organ) from a healthy person (donor) and transplant it into the body of a patient (recipient) whose corresponding tissue (or organ) is severely diseased. It is true that a piece of skin can be relocated as a graft from one site on the same body to another (**autograft**) and that a transplant of living tissue from one identical, *not* fraternal, twin will take in the other (**syngraft, isograft, isogeneric** transplant) (Table 19-5). (**Xenografts** or **heterografts** are those between members of different species.) In plastic surgery pieces of nonviable bone, cartilage, and blood vessel from one person can be used in another as a framework for new growth in certain areas (**homostatic graft**). But immunologically each of us is a rugged individual! In the natural course of things what would be lifesaving grafts

---

*See Morbid: Mortal. Weekly Rep. **31**:577, 1982; and **32**:61, 1983.

Table 19-5 Types of Grafts

New Term	Former Term	Adjective	Genetic Relationships
Allograft	Homograft	Allogeneic	Genetically different individuals within one species are involved.
Syngraft	Isograft	Syngeneic	Genetically identical but separate individuals are involved.
Autograft	Autograft	Autogeneic	Donor and recipient are same individual.
Xenograft	Heterograft	Xenogeneic	Individuals from different species are involved.

of viable skin, bone marrow, kidney, and so on are cast off and rejected, because very subtle antigenic differences exist among even closely related human beings.

Allograft Rejection

A graft from one member to another within a noninbred species is a **homograft** or **allograft.** Without intervention of any kind this graft is characteristically rejected. For about the first week every indication is present that the graft has taken. After that time the situation changes. The circulation of blood established in the graft is cut down; the graft is infiltrated by mononuclear cells, loses its viability, and is soon cast off (Fig. 19-2).

What has happened immunologically is an expression of cell-mediated immunity. (Although antibodies may also play a part, they are not primarily implicated.) Small thymus-dependent lymphocytes detected the foreign antigen, became transformed, were sensitized in the regional lymph nodes, and were released back into the circulation as "activated" lymphocytes. They returned to the graft site to set the rejection process in motion. Although they initiated it, the actual demolition of the graft was carried out by macrophages, with the help of a few polymorphonuclear leukocytes, that the lymphocytes secured to the site.

In allograft rejection (or acceptance) interplay among three T cell populations exists. Cytotoxic T cells destroy allografts, helper T cells accelerate cellular T lymphocyte responses, and suppressor T cells terminate the response once the offending agent has been eliminated.

Transplantation Antigens

Histocompatibility refers to the degree of compatibility, or lack of it, between two given tissues. Of some eight to ten genetically independent systems in the body affecting

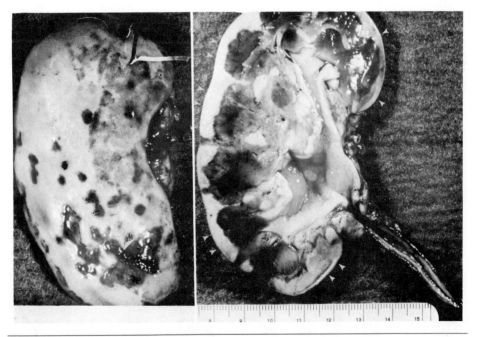

FIG. 19-2   Graft rejection of transplanted kidney (acute). Outer surface and cross section of kidney shown. Diffuse death and destruction of tissue seen as whitish areas on outer and sectioned surfaces. Mottling is produced by hemorrhage. (From Smith, A.L.: Microbiology and pathology, ed. 12, St. Louis, 1980, The C.V. Mosby Co.)

graft survival or rejection, there are two that are of practical importance.

The first, a very important human histocompatibility factor, is the ABO blood group system, which is present on red blood cells and practically all tissues.

The second is the major histocompatibility complex (MHC) known in humans as the HLA (human leukocyte antigens) complex.

HLA Complex

The major histocompatibility complex contains genes that govern the important biologic functions of tissue graft rejection, immune responsiveness to varied antigens, resistance to certain infectious agents, provision of complement components, and resistance to tumors. It is determined by histocompatibility genes of the intricate HLA region of the short arm of the sixth chromosome in humans. In this genetic region a large number of genes are located in close proximity to each other. To date, four gene loci have been defined (Fig. 19-3). These are A, B, C, and D/DR (DR for D-related) in order of their discovery, not in order of position on the chromosome (Table 19-6). Each locus has from 8 to 39 codominant alleles (alternative genes). The number of possible gene combinations is high, easily several million.

The HLA complex controls allograft (homograft) immunity because it contains a group of antigens (glycoproteins) associated with cell membranes that are found on all nucleated cells of the body, especially in lymphoid and blood-forming tissues (not on red blood cells, however). This complex system is most easily identified in leukocytes, hence the name. Therefore tissue typing is commonly done with white blood cells. Practically speaking, the antigens usually referred to as **transplantation antigens** are

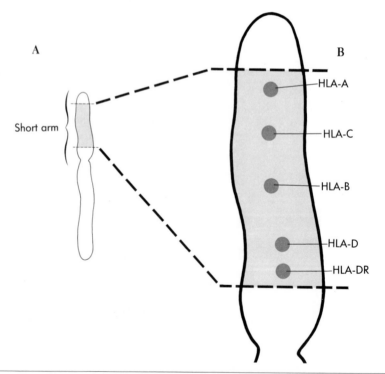

FIG. 19-3   The gene loci—A, B, C, and D/DR—have been identified within the major histocompatibility complex, which in human beings is located on the short arm of chromosome 6. **A,** Chromosome 6. **B,** Close-up of major histocompatibility complex.

those of the HLA system. Part of the graft, they are the ones that mark the graft as "trespasser" to the recipient host.

HLA-Associated Diseases

The HLA complex was first probed because of its relation to survival of organ transplants. Now it is used not only to select donors for transplants (the most important application) but also in the identification of genetic predisposition to certain diseases. More than 50 diseases have been associated with or linked to HLA (Table 19-7). The various diseases that have been described with a preponderance of certain HLA antigens generally share the following features: chronicity, obscure etiology, high familial incidence, and immunologic abnormalities. Examples are ankylosing spondylitis (vertebral column involved), juvenile rheumatoid arthritis (joints), Graves' disease (thyroid gland), insulin-dependent diabetes mellitus, and multiple sclerosis (central nervous system). The strongest association between the HLA complex and disease is found between HLA-B27 and ankylosing spondylitis. In a patient with this kind of arthritis, bony

Table 19-6 Major Histocompatibility Complex and Related Genetic Regions in Humans

Designation	Genetic Region for
HLA	Major histocompatibility complex (MHC)
HLA-B	Classic, serologically defined antigens (Class I molecules)
HLA-A	Gene products ubiquitous on almost any cell surface; serve as targets when cell infected by virus or covered with hapten
HLA-D	Lymphocyte-stimulating determinants (Class II molecules) Gene products serve as regulators between various cell subgroups involved in immune response; found on surface of an immunocompetent and certain specialized cell only
HLA-DR	B cell antigens
HLA-C (located between HLA-B and HLA-A)	Other histocompatibility antigens
C2, C4	Complement components (Class III molecules) Regulatory and structural genes coding for C2, C4, and Factor B in complement system

Table 19-7 Some Examples of HLA Disease Associations

Disease	HLA Antigen	Relative Risk*
Ankylosing spondylitis (arthritis of spine)	B27	126
Reiter's syndrome (p. 660)	B27	36
Subacute thyroiditis (inflammation of thyroid gland)	Bw35	12
Insulin-dependent diabetes	DR3/DR4	6
Multiple sclerosis (chronic destructive disease of central nervous system; lesions multiple, irregularly distributed)	DR2	4
Chronic active hepatitis (p. 738)	DR3	3

*A value assigned for the relative risk indicates the number of times a person with a particular antigen is more likely to have a particular disease than is a person without the antigen. For instance, a relative risk of 1 indicates that a person with the antigen is no more at risk than someone without the antigen.

changes in the ligaments and fusion of adjacent vertebrae produce stiffness and fixation of the spine.

Testing for Histocompatibility

When tissue transplantation is anticipated, typing and cross-matching of blood specimens from the postulated donor and patient-recipient are done as though for blood transfusion. Any incompabitility in the ABO blood groups is an *absolute* contraindication to the use of the donor's tissue.

Tissue typing for transplantation, done with white blood cells or platelets, is partly analogous to red blood cell typing and cross-matching for clinical blood transfusion. Of the various technics for tissue typing, lymphoagglutination and lymphocytotoxicity are widely used. **Lymphoagglutination** compares the clumping reactions of would-be donor and recipient lymphocytes when tested with a panel of selected antiserums. **Lymphocytotoxicity** measures the lethal effect of plasma membrane injury to presumed donor and recipient lymphocytes under similar conditions. Recipient serum can be mixed with the would-be donor's lymphocytes as a "cross-match." Reaction in this indicates probability of immediate graft rejection. In evaluating the degrees of donor-recipient incompatibility for HLA types determined from tissue typing, there are no absolute matches (except with identical twins). A "good" match still means differences, although small.

Immunosuppression

Natural regulatory mechanisms are a part of the immunologic system and consequently modify the immune process as indicated when the need for the immune protection generated no longer exists. Two important ones are (1) a negative feedback by an antibody that reduces further antibody production and (2) suppression of T cell functions by the suppressor T cell subset.

Immunosuppression is the suppression, that is, the nonspecific interference with either the induction or the expression, of an immunologic response by means of chemical, physical, or biologic mechanisms. **Immune tolerance** refers to the state wherein an immune response cannot be mounted to a given antigen that is normally effective. Tolerance differs from other forms of immunosuppression in its restriction to one antigen. Immunosuppression is an unlimited and nonspecific blockade against all antigens.

Various clinical and experimental technics have been designed to suppress or manipulate artificially the body's normal immune responses so that tissue transplantation would be possible. For instance, a potent antiproliferative drug such as 6-mercaptopurine (6-MP) that selectively interferes with either DNA or RNA synthesis may be used singly or combined with steroid hormones and high-dose irradiation. Depletion of lymphocytes has been attempted by drainage of the thoracic duct. Antilymphocyte globulin has been administered to eliminate selectively the cellular population mediating the immune responses and perhaps to spare the sorely needed humoral ones. Splenectomy is done.

The disadvantage of immunosuppressive agents and regimens is that they concurrently impair normal host resistance. The risk for serious viral and bacterial infections is a significant one. Such an individual with either subnormal or practically nonexistent immune responses is spoken of as the **compromised host.** An effort is being made experimentally to induce tolerance to the graft under conditions that do not compromise the body's normal defenses when they are needed elsewhere in the body.

Cyclosporin A

A new compound, the fungal antimetabolite cyclosporin A, may revolutionize the field of organ transplantation. First discovered and studied as a natural product in two strains of fungi (*Trichoderma polysporum* Rifae and *Cylindrocapon lucidum* Booth), it can now be synthesized. A cyclic molecule, it consists of 11 amino acids, one of which has never been seen before in nature (Fig. 19-4).

The excitement over cyclosporin A comes from the fact that its action is uniquely selective and highly specific. It suppresses cell-mediated immunity without killing cells and presumably does not affect B cells and antibody formation, and its action is reversible. The mechanism of drug action is believed to be as follows: this drug allows T cells to be primed to divide but blocks them from doing so, since its action prevents T helper cells from synthesizing the needed differentiation factor, interleukin 2. The drug also interferes with the synthesis of certain other T cell lymphokines, such as interleukin 3, that recruit and activate macrophages. The fully developed cytotoxic mechanisms are not touched by this agent, but their precursors are. The suppressor T cells are resistant.

Complications to the use of the drug have been encountered, the most serious of which is the nephrotoxicity. At present these are not considered to be contraindications to its use. The drug has no reported myelotoxicity, which is an important feature.

Cyclosporin appears to be of value in the treatment of autoimmune disorders and has demonstrated an unexpected effect on two human parasites—the blood fluke (all three species) (p. 826) and the malarial plasmodium (p. 806). It starves the blood fluke by inhibiting one of the worm's enzymes that degrades hemoglobin; it kills the malarial parasite in a similar way.

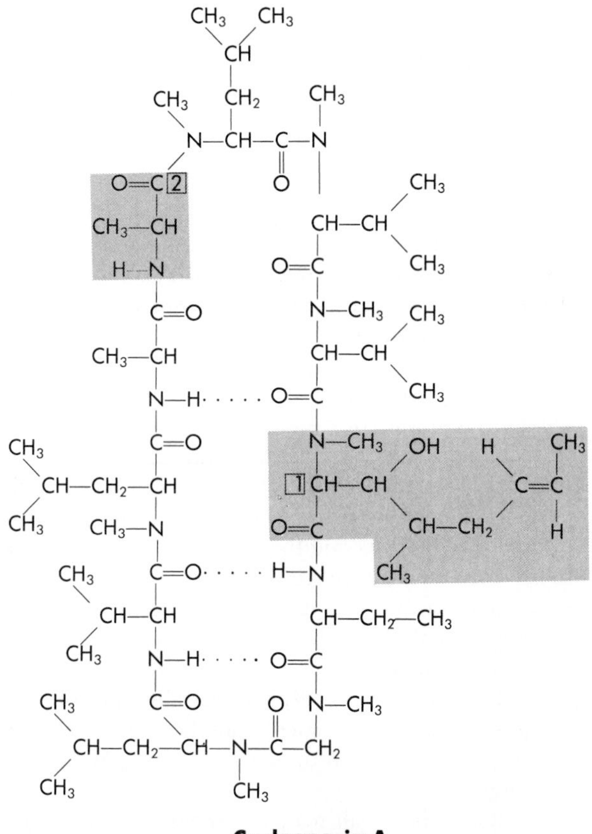

Cyclosporin A

FIG. 19-4   Chemical structure of cyclosporin A, metabolite of fungus *Trichoderma polysporum*, that was discovered in 1972. Of 11 amino acids making up this cyclic polypeptide, 1 (*1* in the formula) has not been observed in other organisms. Alanine, the other (*2*), exists here in the *dextro* form rather than in the usually encountered *levo* form.

Graft-Versus-Host Reaction

It has been observed in animals and a few humans that at times, instead of the host rejecting the graft, the graft has "rejected" the host. This has been found especially with transplantation of bone marrow and lymphoid tissue. In the immunologically depressed host the graft survives. The immunologically competent lymphoid cells of the graft recognize the "foreignness" of the environment, and the stage is set for rejection. The end result of it—runt disease in animals, secondary disease in humans—is characterized by wasting, diarrhea, dermatitis, infection, and death.

Present Status

Much has been accomplished in tissue transplantation. Transplantation of the kidney is established; worldwide more than 30,000 recipients of a kidney transplant have survived.

If the kidney was transplanted from a living donor, case studies show that 94% of the recipients return to full-time activity (89% of those who receive a kidney from a cadaver). By contrast, less than 30% of patients on dialysis are employed full-time; 17% only work part-time.

Although transplantation of other organs is still experimental, a reasonable expectancy of success holds with transplant of the heart, liver, pancreas, lung, and even en bloc heart and lung. Around 500 heart transplants have been done, and more than 200 segmental transplants of the pancreas have been done.

Unexpected benefits have been considerable in terms of new knowledge in basic disciplines, control of infection, and a breakthrough in cancer immunology. It was discovered that with immunosuppression there is also a greater likelihood for the development of certain cancers. Problems are formidable still, but optimism and enthusiasm prevail.

IMMUNITY AND CANCER

Significant advances in oncology (the study of cancerous growth) today focus on the immunologic aspects of neoplasms and on **immunotherapy,** their possible application to the treatment of cancer. Immunotherapy is based on the premise that immune responses normally operating for prevention and control of cancer can be manipulated or adapted to enhance host resistance to the cancer. The immunology of cancer is very complex, and a lot of work, largely experimental, is being done in this field.

In humans and animals the existence of a relationship between the individual's immunologic system and that individual's cancer is generally accepted today. Convincing evidence supporting present concepts of tumor immunology stems from an abundance of experimental work and the many interesting observations made on patients with organ transplants. The incidence of cancer after organ transplant, especially of the lymphoid system and in the skin, is about 100 times that observed in the same age group of the general population. (Immune mechanisms must be suppressed dramatically to allow for organ transplantation.) The increased risk for the development of cancer in the transplant recipient appears within 1 year after transplantation and continues over the next 5 or more years. Malignancies* complicate immunosuppression in many clinical settings, and in congenital immunologic deficiencies as well, a high incidence of cancer is encountered.

Most and maybe all neoplastic cells contain antigens that are new and foreign to the host. These neoantigens, called **tumor-associated antigens,** are absent on normal cells and probably result from the initial change of the normal cell into the cancerous one. Tumor-associated antigens are tumor specific and provoke an immune response (cellular

---

*Malignancies, neoplasms, and tumors are scientific terms used for cancer. Oncogenic and carcinogenic refer to the ability to induce cancer, and an oncogen is such an agent.

and humoral) either in the original host or in the recipient host for the transplanted tumor. Such a response may be of consequence, for, in certain instances, immunocompetent cells of the host do recognize these foreign antigens and halt progression of the neoplastic process. Perhaps certain early, incipient cancers, vulnerable to the host's immune defenses, are indeed quietly and completely destroyed.

Immune surveillance is considered an ongoing function of T lymphocytes and natural killer (NK) cells (discussed below). In keeping with this concept, a mechanism is postulated to exist at all times in humans and animals whereby cytotoxic T cells distinguish the antigens unique to cancer cells as foreign to the host and destroy them in a reaction comparable to that of graft rejection. These T cells may be important in the elimination of early leukemia, lymphoma, and other tumors.

As one would expect, different kinds and degrees of tumor immunity have been identified. A special kind is found with tumors induced by viruses. Surface antigens appear as a direct effect of viral activity, and these are consistent for the same virus in other circumstances. Not so if the oncogenic mechanism is *non*viral. For instance, a chemically induced neoplasm has its own distinct antigens, it is true, but even though the same chemical carcinogen is applied in exactly the same way to other anatomic areas in the very same animal, the same tumor antigens are not repeated. Consequently, many antigenic types become possible with varying immunologic potential.

In view of the existence of immune mechanisms, how does one explain the prevalence of clinical cancer? It is certainly well known that often tumor grows in spite of neoantigens and an apparent host immune response, cellular, humoral, or both.

Various theories are given to account for the failure of immunologic controls over neoplasms, if such do exist. One possibility is that tumor antigens sometimes elicit the production of **enhancing** antibodies, antibodies that offer protection against the sensitized cytotoxic T cells, perhaps by combining with antigen to form a complex that in some way blocks the activity of the T cell. Another theory is that neoplasm may trigger immunologic mechanisms that produce chromosomal abnormalities, which in turn favor the development of tumors. Also, a cancer-causing agent may possibly suppress immune reactions. Still another theory is that early in the development of a neoplasm, immune mechanisms being weak may serve to stimulate rather than inhibit growth of neoplastic cells—this would be **immunostimulation** of tumor growth.

Considerable emphasis has been given to the postulate that suppressor T cells of the host and factors associated with their activity stifle effective antitumor responses. The net result enhances tumor growth. In specific instances suppressor T cells inhibit the antitumor activity of cytotoxic T cells.

Fetal Antigens

Many human tumors express antigens (proteins, glycoproteins, or polysaccharides) that are normally present during the very early phases of embryonic and fetal life but are depressed to trace amounts in adults. The two best-defined fetal antigens are **carcinoembryonic antigen (CEA)**, related especially to cancer of the colon, and **alpha-fetoprotein (AFP)**, related to cancer of the liver. Testing for fetal antigens is of value in the medical management and follow-up of patients known to have these cancers.

Natural Killer (NK) Cells

Certain cells have been found that mediate a **nonspecific** cytotoxic reaction against other cells; that is, they can kill cells in the absence of a prior sensitization process or an antibody being involved, hence the term *natural* killer cells. The cell lineage of these natural killer cells is unknown. They look like lymphocytes and possibly are a subset of T cells. In animal studies natural killer cells are plainly involved in recovery from certain viral infections and in tumor surveillance. They seem to play a major role in the immune surveillance against the transformation of noncancerous cells into cancerous ones. Interferon regulates their activity and can heighten it.

ALLERGIC DISORDERS
What Is Meant by Allergy (Hypersensitivity)

In a previous chapter we saw that immunity is a state wherein by prior contact a subject gains protection against an agent that would otherwise harm that subject. The reaction is specific for the particular agent, and a benefit comes from the immune setup. Unfortunately the complex immunologic response of the body is not always favorable and in fact may be quite harmful. Consider another situation comparable immunologically but one in which the immune mechanisms injure instead of protect. This state is allergy (a term first used by von Pirquet), or hypersensitivity.

Both *allergy* and *hypersensitivity* are terms also used to mean conditions such as emotional upsets, nonallergic food reactions, toxic responses to drugs, and others that are nonimmunogenic. Since hypersensitivity is more often so used than allergy, some authorities believe that the term *allergy* has priority in this context.

Allergy, or hypersensitivity, then, is a state of altered reactivity directly related to the operation of the immune system. Broadly defined, either of the terms refers to a large group of untoward physiologic responses to normally innocuous substances, the responses being mediated by immunologic mechanisms. Except in the United States, *allergy* is a term so used. In the United States the word *allergy* is mostly restricted in its use to disease resulting from IgE-mediated reactions. The word *hypersensitivity* is not limited in its applications.

The study of the effects and changes in body tissues consequent to the adverse immunologic mechanism is included in **immunopathology.**

General Characteristics

A person may exhibit unusual and heightened manifestations on coming into contact with certain substances that are often, but not necessarily, protein and that do not affect the average or another individual. For example, such a subject may have hay fever on contact with plant pollens or asthma on contact with horse dander. (Dander is animal

Table 19-8 Classification of Hypersensitivity Reactions

Type	Mediated by	Mechanism
Disorders of Humoral Immunity (Antibody-Mediated)		
I—Immediate or anaphylactic	IgE (reagin)	Specific IgE combined with allergen on mast cell surface releases pharmacologic mediators.
II—Cytotoxic	IgG, IgM	Specific antibody combined with antigen fixed to target cell surface activates complement system; thereby leads to cytolysis of target cell.
III—Immune complex or toxic complex	IgG, IgM	Soluble complex of soluble antigen with antibody activates complement system (antigen present in excess); deposition of immune complex in blood vessel wall results in tissue injury.
Disorder of Cell-Mediated Immunity		
IV—Delayed	T lymphocytes	Activation of T lymphocytes

dandruff—the composite of castoff epithelial cells of the skin, sebum, and other body oils.)

An antigenic substance that can trigger the allergic state is an **allergen.** It may be a protein or frequently a nonprotein of low molecular weight. Allergens reach the body in several ways: by the respiratory tract or the digestive tract, by direct contact with skin or mucous membranes, and by mechanical injection. Of these routes, the respiratory pathway is most consistent. Presumably unchanged protein must pass through the intestinal wall to the bloodstream for sensitization through the intestinal tract to occur. This can occur in children and in adults with digestive disturbances. A passive sensitization may be transmitted from the mother to her child in utero across the placenta. This type of sensitization is short lived, as is passive immunity in the newborn infant.

An allergic person is often sensitive to several different allergens. At least 10% to 20% of persons in the United States develop allergic disorders under the natural conditions of life.

Kinds

As you would expect from our discussion of immunity, the reactions of allergy or hypersensitivity are separated into two main divisions: (1) those in which the allergic responses result from the presence of humoral antibody and (2) those in which the responses are not mediated by circulating antibody but by specifically sensitized cells; that is, the responses are cell mediated. When emphasis is placed on certain medical considerations, the classification of four basic types of hypersensitivity reactions is made (Table 19-8). Three types are subdivisions of the first category of antibody-mediated disorders. The four patterns identified with immune injury to tissue and the consequent host response are type I—immediate, anaphylactic, or IgE-mediated hypersensitivity; type II—cytotoxic, cytolytic, or necrotizing reactions; type III—immune complex, toxic complex, or antigen-antibody complex disease; and type IV—cell-mediated, delayed, tuberculin, or bacterial hypersensitivity.

Effectors	Special Features	Time Interval (Usual)	Examples
Histamine, eosinophilic chemotactic factor, slow-reacting substance, and others	Antibody on cell surface; antigen in circulation	Short—within seconds after exposure to allergen	Urticaria, allergic rhinitis, allergic asthma, insect sting hypersensitivity, some drug allergies, anaphylaxis
Complement cascade	Antigen on cell surface; antibody free in circulation	Short—within minutes	Transfusion reaction, some drug allergies, hemolytic disease of newborn
Complement cascade	Antigen and antibody both in circulation; circulating immune complex deposited at various anatomic sites	Short—within hours	Arthus phenomenon, serum sickness, acute poststreptococcal glomerulonephritis, some drug allergies
Lymphokines	Phenomena related to cells	Long—within days	Contact dermatitis (poison ivy), tuberculin or bacterial hypersensitivity, autograft rejection, graft-versus-host disease, some drug allergies

Mechanisms

Mechanisms of allergy or hypersensitivity are intricate and incompletely defined. Only a sketch is given of some of the background factors.

Immediate-type allergy. The sequence in the development of the immediate hypersensitivity reaction starts with the first exposure to an allergen (Fig. 19-5). This event initiates biosynthesis of the humoral antibody immunoglobulin E (IgE). Before the identity of the humoral antibody was known, it was called **reagin.**

IgE. Of 5000 different immunoglobulin molecules found in normal persons, only one seems to be an IgE. Most people have some IgE in their serum, but in allergic persons it is usually present in increased amounts. Like all immunoglobulins, IgE is produced by plasma cells. It is formed mainly in tonsillar tissues, Peyer's patches, and generally in the lymphoid tissue near the mucosal surfaces of the gastrointestinal and respiratory passages as well as in the lymph nodes regional to these areas. It is not formed to any considerable extent in the spleen and in most other lymph nodes, however.

People vary in their production of IgE, but it is highly specific for the allergen

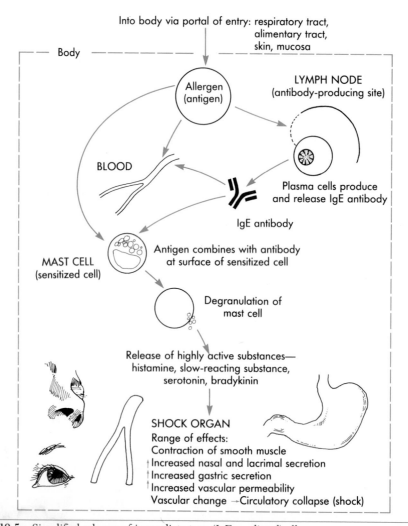

FIG. 19-5 Simplified schema of immediate-type (IgE-mediated) allergy.

inducing its formation, and it possesses a unique immunologic affinity for basophils of peripheral blood and mast cells of the tissues. From the first exposure to allergen, the IgE formed is attached by its Fc part to specific receptor sites on these cells, and thus it sensitizes them. (The Fab fragment of IgE recognizes the allergen.)

MAST CELLS AND BASOPHILS. The mast cell (Fig. 19-6) is a tissue-based cell near blood vessels and most numerous in the places where IgE is formed—in the lining areas of the respiratory and gastrointestinal tracts and under the skin. A cell about 15 μm wide, it is known for its distinct appearance. Some 300 to 500 large, densely placed, metachromatically staining granules found in its cytoplasm contain high concentrations of pharmacologically potent compounds such as histamine, heparin, proteolytic enzymes, and chemotactic factors for neutrophils and eosinophils. Degranulation or loss of these unique granules releases their contained substances into tissue fluids. Nonimmune as well as immune factors can cause degranulation. (Degranulation of basophils occurs not only in allergic reactions mediated by antibody [IgE] but also in cell-mediated responses.)

The basophil circulating in the bloodstream is a look-alike to the mast cell and

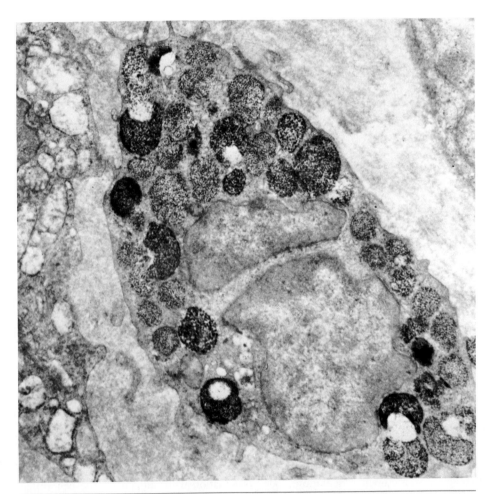

FIG. 19-6 Human mast cell (basophil of peripheral blood is a look-alike), electron micrograph. Note large dark granular masses (large dark blue cytoplasmic granules of Wright-stained blood smears). They store pharmacologically potent compounds. (From Barrett, J.T.: Textbook of immunology, ed. 4, St. Louis, 1983, The C.V. Mosby Co.)

mimics its responses in allergy. The basophil is smaller than the mast cell and has fewer granules and a larger nucleus. The mast cell and the basophil are the only cells to which IgE is affixed and the only two cells in the body containing histamine.

EVENTS. With the second and subsequent exposure to the allergen, the stage is set for the release of the mediators of allergy, that is, those compounds with the chemical potential to induce the manifestations of immediate-type allergy. The allergen reaches the sensitized cell, complexes there with its specific IgE, and is locked into the receptor site. The allergen must bridge two adjacent molecules of surface IgE on the mast cell. This combination at the surface of the sensitized cell sets in motion an active, energy-requiring process along the plasma membrane, which causes the cell to degranulate. In so doing, the cell releases histamine, the eosinophilic chemotactic factor of anaphylaxis, slow-reacting substance of anaphylaxis, and the other mediators that profoundly influence vascular permeability and smooth muscles. IgE bound to the sensitized cell has an extremely high order of biologic activity, and even very low concentrations of allergen can induce reactions ranging from a mild allergy to fatal anaphylactic shock.

The spectrum of effects in immediate-type allergy includes contraction of smooth muscle, increase of gastric secretion, increase of nasal and lacrimal secretions, increase of vascular permeability, and vascular changes that can lead to circulatory collapse (shock). Of the chemical mediators of allergy, histamine has been the most extensively studied, and this allergic reaction has even been said to constitute a response to too much histamine.

EOSINOPHILS. Eosinophilic chemotactic factor (tetrapeptides) attracts eosinophils into the anatomic site of the allergic reaction. The eosinophil, one of the granular leukocytes normally present in small numbers in the circulating blood, is often called the cell of allergic inflammation. A cell just larger than the polymorphonuclear neutrophilic leukocyte, it is distinct for large, round, bright red-orange granules within its cytoplasm demonstrated in Wright-stained blood smears. The eosinophil releases substances (secondary mediators) that have been shown to be antagonistic to certain of the chemical mediators from the basophils and mast cells, and it also has been demonstrated to ingest antigen-antibody complexes. Therefore it is suggested that the eosinophil modulates or limits the allergic response.

The highly visible eosinophil also responds immunologically in tissues and in the circulating blood to the presence of helminths or worms. It is not generally seen with protozoan infections. Eosinophils are not well known for their phagocytic capacity or as killer cells, but at times they may be seen to inflict their damage on the surface of a worm. The eosinophil binds through IgG-Fc receptor and through C3b on the surface of the parasite.

EFFECTS. In immediate-type allergy the consequences of the antigen-antibody combination follow shortly thereafter, even within a minute or two, and the reaction reaches a maximum within a few hours. The responses occur in vascularized tissues and relate primarily to smooth muscle, blood vessels, and connective tissues. A specific shock tissue (or organ) is important; that is, a target tissue or organ that, because of its anatomic makeup, shows maximal effects. Manifestations are acute and short lived. Epinephrine, antihistaminic drugs, and related compounds are of benefit. This type of reactivity can be transferred passively to a nonsensitized person by antibody-containing serum.

Cytotoxic reaction. The cytotoxic reaction causes cell and tissue death. As a consequence of the antibody combination with antigen on the cell surface, the cell is destroyed. Complement often participates in the reaction, but its action is not always required. Target cells are usually either the formed blood cells—the red blood cells, white blood cells, or platelets—or a special cell type within a given area, such as in the skin. The transfusion reaction (p. 414) characterized by the destruction of large numbers of red blood cells is an important example of this type of abnormal immunologic response.

Immune complex disease. The immune complex reaction indicates an abnormal state in which soluble antigen that has been injected combines with the circulating antibody that results. The specific antibody is usually a precipitating IgG. When a certain ratio of the number of antibodies to the number of antigens inducing them is attained with a moderate excess of antigen, soluble antigen-antibody aggregates called **immune complexes** appear. These attract and activate complement and become depos-

---

The sequence of events in tissue injury from immune complexes is as follows:

Ag-Ab combined → Complement fixed → Mediators of inflammation generated and released → Neutrophils drawn to site → Offending complexes phagocytized; Neutrophils degranulate → Lysosomal enzymes released from neutrophils → Blood vessel wall damaged, disrupted, or partly dissolved

---

Table 19-9 Antibody-Mediated Allergy of the Immediate Type Compared with Cell-Mediated Allergy of the Delayed Type

	Immediate-type Reactions (Anaphylactic Type)	Delayed-type Reactions (Bacterial or Tuberculin Type)

Clinical state	Hay fever
	Asthma
	Urticaria
	Allergic skin conditions
	Anaphylactic shock
		Drug allergies
		Infectious allergies
		  Tuberculosis
		  Tularemia
		  Brucellosis
		  Rheumatic fever
		  Smallpox
		  Histoplasmosis
		  Coccidioidomycosis
		  Blastomycosis
		  Trichinosis
		Contact dermatitis

Onset	Immediate	Delayed
Duration	Short—hours	Prolonged—days or longer

Allergens	Pollen
	Molds
	House dust
	Danders
	Drugs
	Antibiotics
	Horse serum
	Soluble proteins and
	  carbohydrates
	Foods
		Drugs
		Antibiotics
		Microbes—bacteria, viruses,
		  fungi, animal parasites
		Poison ivy and plant oils
		Plastics and other chemicals
		Fabrics, furs
		Cosmetics

Passive transfer of sensitivity	With serum	With cells or cell fractions of lymphoid series

ited in the wall of a blood vessel. To cause disease, the immune complexes must penetrate vessel walls; the hallmark of this immunologic disorder is vasculitis, the inflammatory involvement of the blood vessel wall that results. Once deposition occurs, the immune complex triggers a number of interrelated events that culminate in tissue damage, inflammation at the anatomic site ensues, and the disease state appears. Classic type III reactions include the Arthus phenomenon and serum sickness; other examples are acute poststreptococcal glomerulonephritis and certain parasitic infections—leishmaniasis, trypanosomiasis, quartan malaria, and schistosomiasis (Chapters 36 and 37). In infections such as leprosy, typhoid fever, syphilis, and viral hepatitis, deposits of circulating immune complexes are an important feature.

Delayed-type allergy. In delayed-type allergy, by contrast, no relation exists to conventional antibody. The phenomenon is closely associated with cells—T lymphocytes. The responses are set in motion by T lymphocytes specifically modified or sensitized so that they can react to the specific allergen deposited at a given site. The main feature of the reaction is the direct destruction of target tissue containing the antigen. When antigen-activated T cells not only dispose of allergen but also damage cells in the process, their activity results in disease. One important manifestation of T cell activity is dermatitis (inflammation of the skin), which is related to mediators—skin reactive factors and blastogenic factors—released from rich stores within the T cells. Delayed-type allergy may be passively transferred to a nonsensitized person if immunologically competent cells (lymphoid cells) are used.

In allergic reactions of this type a much longer period of time elapses between the presentation of the antigen and manifestations of the disorder. The period of incubation ranges from 1 to several days. Reactions do not favor any particular tissue but may occur anywhere. There is no specific shock organ or tissue as with immediate-type allergy. Many cells in the body are vulnerable. Manifestations tend to be slow in onset and prolonged. No histamine effect is demonstrated as in immediate-type reactions, and antihistamines are without effect. The steroid hormones seem to help in the treatment of some allergic disorders of this kind and in some of the immediate type as well.

Table 19-9 compares salient features of antibody-mediated allergy with those of cell-mediated allergy. Immediate-type reactions contrasted with delayed-type reactions emphasize certain differences between the two major forms of immunologic disorders.

Immediate-type Allergic Reactions

Anaphylaxis

In immediate-type allergy with sudden release of chemical mediators, the response is acute (within seconds or minutes). The acute reaction is anaphylaxis, the prototype of immediate-type hypersensitivity. Anaphylaxis is spoken of as a symptom complex because it refers to a grouping of acute diverse manifestations that include any or all of the following: skin changes, itching of the skin, nausea, vomiting, crampy abdominal pain, diarrhea, and asthma. Anaphylaxis may occur with or without cardiovascular collapse. When it is associated with hypotension (low blood pressure), it is **anaphylactic shock.**

Systemic anaphylaxis is a generalized reaction induced in a sensitized animal on contact with adequate amounts of antigen administered so that rapid release and dissemination of mediators occurs (as intravascularly).

If a small amount of foreign protein (horse serum or egg white), in itself nonpoisonous, is injected into a suitable animal such as a guinea pig, the **first** and **sensitizing dose** will be without noticeable effect. When the second and larger injection, the **provocative dose,** is given after an interval of from 10 to 14 days, severe manifestations of anaphylaxis ensue. The guinea pig becomes restless and breathes with increasing difficulty within *only a minute or two!* A period of frantic activity precedes death of the animal in respiratory failure.

Anaphylaxis is specific; that is, for anaphylaxis to occur, the sensitizing and provocative doses must be the same substance. If the serum of a sensitized animal is injected into a normal animal and if after an interval of 6 to 8 hours the normal animal is injected with some of the material to which the first animal is sensitive, signs of anaphylaxis appear in the normal animal. This is **passive anaphylaxis.** The passive transfer of the hypersensitive state assumes considerable importance when it is realized that the use of an allergic donor for blood transfusion may render the recipient hypersensitive. One interesting case is reported in which, after receiving blood from a donor sensitive to horse dander, the recipient had an attack of asthma on contact with a horse. If an actively sensitized animal survives an anaphylactic attack, it is desensitized for a few days but eventually becomes sensitive again.

Role of smooth muscle. The dramatic manifestations of anaphylaxis result from the contraction of smooth muscle fibers, and the part of the body primarily attacked in a given animal depends on the distribution of smooth muscle fibers in the species. In the guinea pig, smooth muscle fibers are plentiful in the lungs; contraction of these closes bronchioles and bronchi and leads to death from respiratory tract obstruction. In the rabbit, smooth muscle is plentiful in the pulmonary arteries. If the rabbit is used as test animal, these arteries contract, and a circulatory burden that leads to failure is thrown on the right side of the heart.

Anaphylaxis in humans. The phenomenon of anaphylaxis, generally speaking, is influenced by the route of administration of the inciting agent, the dose, and the degree of host susceptibility. But remember that while anaphylaxis in human beings may present only mild, annoying manifestations, *it can be immediate, severe, and fatal, running its course within minutes or even seconds* (Table 19-10). If the person is sensitive to the given allergen, a very small amount can precipitate the reaction. Anaphylaxis in humans is most likely to occur after intravenous injection of the allergen.

When anaphylaxis was first described at the turn of the century, one of the chief causes then was a sting by a member of the insect order Hymenoptera. Included here are bees, yellow jackets, wasps, and hornets. This is still true. It is said today that far more deaths occur from insect stings ("insect anaphylaxis") than from snakebites. A person with no history of allergy can have a fatal reaction to an insect bite.

In all areas of our complex society the number of different substances to which

Table 19-10 Makeup of the Anaphylactic Set*

Item	Number Needed	Item	Number Needed
Ampules of 1:1000 solution of epi-nephrine, 1 ml	2	Gauze sponges	
		Syringes, 2 ml	2
Hypodermic needles	2	Needles (1 and 4 inches long)	2
Ampules of aminophylline (0.24 g each)	2	Bottle of 5% solution of dextrose in distilled water, 1000 ml	1
Intravenous set	1	Ampule of diphenhydramine hydro-chloride, 10 ml (10 mg/ml)	1
Bottle of hydrocortisone (dilution to 2 ml gives 50 mg/ml)	1	Ampule of sterile water	1
Scalpel	1	Hemostat	1
Ampule of absorbable surgical suture (catgut) with needle	1	Syringe, 20 ml	1
Tongue depressors		Alcohol, 70% (disinfectant)	

*This kind of emergency must be anticipated. At the Mayo Clinic, Rochester, Minn., "anaphylactic sets" are placed in strategic areas. They contain the above items, quickly available, to be used in the event of such a reaction.

people may become sensitized is astonishing. This is especially true in the field of medicine, where a variety of chemicals are used for many reasons. Most of the anaphylactic reactions today are the result of sensitivity to drugs that have been used in one way or another for diagnosis and treatment of disease. Reactions of this type are referred to as **iatrogenic** (treatment-induced). Table 19-11 lists diverse agents known to have caused systemic anaphylaxis in human beings.

Contacts with insects cause a number of injuries, a few fatal, each year, especially in the seasons when people like to "get outside." The greatest danger from an encounter with an insect is the possibility of an allergic reaction to a sting or bite. The insects most likely to trigger this are bees, wasps, and yellow jackets.

Insects produce poisonous substances or venoms that when injected into a human subject may induce one of the following pathologic responses:

1. Blisters (superficial skin lesions) may be formed by stinging caterpillars, centipedes, and blister beetles.
2. Damage to the deeper layers of the skin may be done by fire ants, wheel bugs, brown recluse spiders, mites, scorpions, chiggers, bees, wasps, yellow jackets, and hornets.
3. Defects in clotting of the blood may be produced by fleas, lice, mites, biting flies, and true bugs.
4. Manifestations of central nervous system disturbance may be produced by scorpions, black widow spiders, bees, wasps, yellow jackets, and hornets.

A given person's response to insect injury is not necessarily a predictable one. Because human beings differ widely in immunologic makeup, allergic reactions to insect bites will vary greatly. The range of severity is indicated in the following classification:

1. Slight reaction—the skin is inflamed and probably itches.
2. Moderate reaction—the bite area swells, and the person stung becomes nauseated and experiences pain in the abdomen.
3. Severe reaction—the person bitten begins to breathe with difficulty, becomes hoarse, and may be confused. This subject can go quickly into shock with a drop in blood pressure and circulatory collapse. Unconsciousness follows.

Needless to say, when the reaction is moderate or severe, call the doctor. If possible, capture and hold the offending insect.

Take every precaution to avoid an unfortunate contact with a venomous insect. If a stinging insect is near, remain still. Brush it off if it attacks, but do not slap it to prevent a bite. If attacked by a swarm of wasps, yellow jackets, hornets, or bees, leave the area at once, meanwhile protecting the face with hands and arms. When bees or wasps are active, avoid exposure. Postpone mowing the lawn or working the flower beds. Remember that insects are less active in the early morning hours. Avoid activities or circumstances that attract insects. Since insects are attracted to the odor of ripened fruit, watermelons, soft drink cans, and other sweet smelling substances, keep these covered.

If one has had a prior reaction, then every precaution in areas of contact is crucial. Wear the necessary protective clothing. If need be, carry an emergency kit of appropriate medication. Insect-sting kits are available commercially.

Atopic Allergy

Atopic allergy is the designation for a group of well-known human allergies to naturally occurring allergens with chronic manifestations of immediate-type allergy (IgE mediated). Here are found hay fever and asthma. By contrast with systemic anaphylaxis, where larger doses of antigen are given intravenously, small doses of antigen repeatedly contact mucous membranes.

Allergic diseases are not inherited, but the special tendency for the individual to

Table 19-11 Systemic Anaphylaxis in Humans—Documented Agents

Chemical Class	Route of Inoculation	Chemical Class	Route of Inoculation
Proteins		*Food*	
Antiserum from horse (in passive immunization)*	Parenteral	Egg white*	Intradermal
Antirabies serum		Buckwheat*	Intradermal
Diphtheria antitoxin		Cotton seed*	Intradermal
Tetanus antitoxin		Nuts	
Snake venom antitoxin		Seafoods (especially shellfish)	
Bivalent botulism antitoxin		Fruits, especially citrus and	
Antilymphocyte globulin	Parenteral	strawberries	
Blood products	Parenteral	Seeds	
Whole blood		Certain legumes—chickpeas,	
Plasma		soybeans	
Platelets			
Gamma globulin		**Polysaccharides**	
Cryoprecipitates		*Acacia* (emulsifier)	Intravenous
Other clotting factors		*Dextran* (plasma expander)	Intravenous
Other blood components			
Hormones		**Others: Action as Haptens**	
Insulin	Subcutaneous	*Antibiotics*	
Corticotropin	Intravenous	Penicillin*	Oral and parenteral
Enzymes		Cephalothin	Parenteral
Chymotrypsin	Intramuscular	Demethylchloretracycline hydrochloride	Oral
Trypsin	Intramuscular		
Penicillinase	Intramuscular	Nitrofurantoin	Oral
Sting of Hymenoptera (bees, wasps, hornets, fire ants, and yellow jackets)*	Subcutaneous (insect bite)	Streptomycin	Intramuscular
		Other medications	
		Sodium dehydrocholate*	Intravenous
Pollen		Thiamine*	Subcutaneous and intravenous
Bermuda grass*	Intradermal		
Ragweed*	Intradermal		
Allergenic extracts (skin test agents, and such)	Intradermal	Salicylates (aspirin)	Oral
		Procaine*	Parenteral
*Glue**	Intradermal	*Diagnostic agents (halogenated compounds)*	
		Sulfobromophthalein*	Intravenous
		Iodinated organic contrast agents*	Oral and intravenous

Adapted from Austen, K.F.: Systemic anaphylaxis in man, J.A.M.A. **192**:108, 1965.
*Deaths reported.

develop the state of altered reactivity in tissues is. Atopic allergy seems to be present more in some families than in others; however, all the members of an affected family do not have the same condition. One may have hay fever, another asthma, and still another some type of skin eruption. Seldom will all the members of a family manifest the hypersensitive state. High levels of IgE are present in the serum of persons with atopy.

Asthma. Asthma is a recurrent type of breathlessness coming in acute episodes and described by certain types of respiratory movements and wheezing. It is an allergic condition most often caused by animal hair, feathers or dander, house dust, foods, microbes, and certain cosmetics. Important asthma-producing foods are milk, milk products, eggs, meat, fish, and cereals. A person may become sensitive to the bacteria normally inhabiting the upper respiratory tract and thereby become a victim of asthmatic attacks. This type of asthma is spoken of as **endogenous** asthma. An estimated 4% of the population has asthma.

In asthma, as would be expected, contact of an allergen with a sensitized mast cell triggers an allergic reaction (Fig. 19-5). Although the exact mechanisms that are responsible for asthma are not worked out, it is believed that effects from certain chemical mediators released from the mast cell are operative. For instance, in asthma, histamine has been found in particularly high concentrations in the lungs. Histamine, for one, and slow-reacting substance of anaphylaxis, for another, both induce the smooth muscle of the bronchus to contract or even to go into spasm. Histamine also stimulates the secretion of mucus. These chemically mediated changes reflect two pathophysiologic disturbances in asthma—bronchial spasm and hypersecretion of mucus.

Hay fever. Hay fever, with its well-known nasal distress and seasonal occurrence, results from sensitivity to pollens (Fig. 19-7), and the period of attack corresponds to the time of the year that the offending plant or plants pollinates. (In season, plant pollens usually shed in the morning hours between 6 AM and 10 AM.) Hay fever was first described in 1819 as a disease that came from exposure to the "effluvium of new hay." It is neither precipitated by exposure to vapors nor associated, except rarely, with fever. Pollen, the airborne allergen, has a high molecular weight and contains much protein. To be important as a cause of hay fever, a plant must produce a light dry pollen easily carried a long distance by the wind. This excuses both goldenrod and roses, long accorded an unearned distinction as causes of hay fever. Early spring hay fever is usually caused by the pollen of trees. Late spring and early summer hay fever is most often from grass pollens, and more than 80% of the cases of fall hay fever result

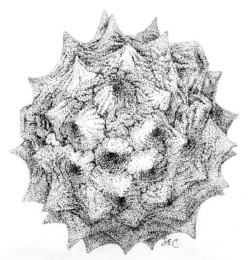

FIG. 19-7 Windborne pollen grain of ragweed looks like battered golf ball in scanning electron micrograph. (×2000.) Wind-pollinated plants produce pollen in huge quantities. A single ragweed plant may expel 1 million pollen grains in 1 day, 1 billion in one season. More than 250,000 tons of ragweed pollen may be blown hundreds of miles across this nation during a fall season.

from ragweed pollen. Ragweed is found only in the United States, parts of Canada, and Mexico. The importance of individual trees, grasses, or other plants depends on geographic location. Ten percent of the population has hay fever.

Urticaria and Allergic Skin Eruptions

Allergic skin conditions may come from foods, drugs, chemicals, and various other allergens. Urticaria, known sometimes as nettle rash, is an allergic disorder of the skin depicted by the presence of wheals (whitish swellings) or hives (lesions of a strikingly transitory nature at times).

Cytotoxic Reactions: Hemolytic Transfusion Reactions

For a discussion of transfusion reactions, see p. 413.

Immune Complex Diseases

Serum Sickness

After the administration of an immune serum or of any biologic product in an appreciable amount, a reaction known as *serum sickness* may appear in subjects who have never had a previous injection of that product (for example, horse serum) and who, insofar as is known, are not sensitive to the foreign protein (horse or some other). Serum sickness is an expression of a basic type of hypersensitivity, the immune complex reaction, and it is a reaction that can occur within a relatively short time.

The manifestations of serum sickness, while unpleasant, seldom threaten life. It usually begins some days (perhaps 8 to 12) after the single-dose injection of an immune serum, most likely an antiserum used for passive immunization. Such a preparation contains a large quantity of foreign antigen. The disease is typified by localized swelling at the injection site; skin rash; fever; swollen, painful, and stiff joints; enlarged lymph nodes (lymphadenopathy); leukopenia (decreased number of circulating leukocytes); and decreased coagulability of the blood. All these manifestations come from the localization in the tissues of the soluble immune complexes. **Local serum disease** may occur if the reaction appears only around the injection site.

CAUTION: An immune serum or any biologic product should be administered with extreme caution to an asthmatic patient or to any person who has had a previous injection, especially of horse serum. Before an immune serum is given, tests (p. 434) to detect sensitivity to that product should always be done, because the foreign protein in the product is the factor responsible for untoward reactions. The fall in blood pressure, the drop in body temperature, and respiratory difficulty resulting from the injection of an immune serum are most likely to appear when the second injection is given 2 or 3 weeks after the first, but severe reactions may occur when a second injection is given months or even years after the first. Rarely, manifestations may lead to immediate collapse and death.

Acute Poststreptococcal Glomerulonephritis

Acute poststreptococcal glomerulonephritis is a complication of a sore throat or skin infection, such as impetigo, produced by certain nephritogenic strains of group A beta-hemolytic streptococci (Chapter 23). **Nephritogenic** refers to the propensity of these strains to induce inflammation of the kidney. From 1 to 3 weeks after the acute infection, a child or a young adult develops blood in the urine (hematuria), edema (swelling), and high blood pressure—three manifestations characteristic of glomerulonephritis, an inflammation of the tuft of capillaries making up the glomerulus, the filtration unit of the kidney. In such a patient the urine is dark and smoky and contains red blood cells, red cell casts, and protein. The level of serum complement (especially C3) is low, and serologically there is evidence of the recent streptococcal infection. Fortunately, the outcome is favorable in 80% to 90% of children and in 50% to 70% of adults.

Acute poststreptococcal glomerulonephritis is among the diseases in which immune complex deposits in a target organ initiate the disorder. By immunofluorescent technics, lumpy deposits of IgG, IgA, or IgM and C3 can be seen in the glomerulus indicating the presence there of antigen-antibody complexed with complement. The antigen is assumed to be streptococcal and nonglomerular in origin. Such an antigen has been detected in the lesion, and it is known that nephritogenic strains of streptococci possess a lipoprotein that is serologically cross-reactive with kidney tissue. Most likely, circulating streptococcal antigen-antibody complexes filter out in the glomerulus, fix complement, and trigger the reactions. The normal endothelial fenestrations found in glomerular capillaries favor leakage of such complexes and help to explain the unique susceptibility of the glomerulus to this condition.

The pathologic changes seen in the glomerulus result both from the damaging deposits and from the activation of complement. Activation of complement releases factors that enhance the disorder. Capillary permeability is increased, and neutrophils are attracted to the site of injury. Neutrophils release lysosomes, which cause further damage in the area (see box on p. 381). The eventual outcome of this disorder may be correlated with the degree of glomerular inflammation and destruction noted in the renal biopsy. With the extensive damage seen in a small percentage of cases, the process is not reversible.

Arthus Phenomenon

The Arthus (toxic complex) phenomenon shows us how dramatically immunologic mechanisms may damage tissues (Fig. 19-8). When material (an antigen) to which an individual is sensitized is reintroduced into the body, the Arthus phenomenon may develop in a localized area of skin around the site of intradermal injection; an ulcer characteristically appears after a delay of several hours. Skin involvement results from the intricate sequence of events set in motion by the formation of the immune complex, the activation of complement, and the precipitation of the immune complex in the subendothelial layer of the small blood vessels. As a net result, the area of skin becomes swollen, reddened, hemorrhagic, and even necrotic (dead).

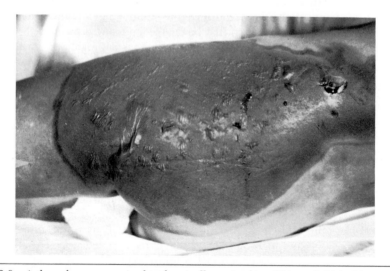

FIG. 19-8 Arthus phenomenon in sharply set-off area in skin of hip and thigh. Note discoloration associated with hemorrhage and damaged tissue. (From Top, F.H., Sr.: Communicable and infectious diseases, ed. 6, St. Louis, 1968, The C.V. Mosby Co.)

Delayed-type Allergic Reactions
Sensitivity to Drugs*

Persons are not hard to find who exhibit drug allergy, that is, an untoward reaction to a certain drug or drugs. Drugs represent one of the commonest causes of hypersensitivity reactions (antibiotics are high on the list of drugs implicated). When a particular drug is taken regularly into the body, a chemical combination probably occurs between that drug and certain body proteins; the result can be a complete antigen in a hapten-protein conjugate that is foreign to the body and therefore one against which the mechanisms responsible for allergic manifestations can be directed. Skin lesions of varied nature are often associated with drug reactions, especially to the antibiotics, but the manifestations of allergy depend on host factors, as well as on drug characteristics. They may be systemic or local according to the route of administration of the drug.

True drug allergies may be associated with the humoral antibodies of types I, II, and III hypersensitivity or with the cell-mediated immune responses of type IV reactions. Multiple reactions such as anaphylaxis (type I), urticaria (type I), and serum sickness (type III) may occur with one drug.

An estimated 500 drugs can bring about the allergic state. The ones most commonly responsible for a hypersensitive state in the person using them are opiates (morphine, codeine), salicylates (aspirin), barbiturates, iodides, bromides, arsenicals, sulfonamides, and antibiotics. Among the drugs inducing allergy, penicillin is the worst offender. Most modern anaphylactic reactions come from penicillin and its related compounds; in this country at least several hundred deaths each year are attributed to penicillin anaphylaxis. Penicillin is also the most common cause of other types of drug reactions. These may appear hours, days, or even weeks after the administration of the drug. An allergy is said to exist in 10% of the instances in which penicillin is given. For other drugs, fortunately, the incidence is much lower. Because of the likelihood of sensitivity after topical use of sulfonamide drugs and penicillin as well, this type of therapy has almost completely disappeared.

Penicillin hypersensitivity. Penicillin as such does not induce hypersensitivity. A small molecule with a molecular weight of 400, it is too small to be immunogenic in itself. Its degradation products, however, can become haptens; and the immune response can occur to any one of several such metabolic products. A skin test done with one benzylpenicilloyl-polylysine (Pre-Pen), indicates the probable risk of an allergic reaction.

If a prior reaction to penicillin has taken place or if there is reason to suspect that the patient is hypersensitive to the drug, it should *not* be given. Remember that penicillin is said to have caused more deaths than any other drug. No reliance should be placed on immunotherapy (desensitization). (Apart from allergy, penicillin is one of the least toxic of drugs and is still the drug of choice, when it can be used, for many bacterial infections.)

Infectious Allergy (Hypersensitivity to Infection)

Repeated or chronic infections may sensitize a patient to the etiologic microbes. This is well illustrated in the allergy to *Mycobacterium tuberculosis* and its products. When an extract of *Mycobacterium tuberculosis* is applied to the skin of the person who has or who has had tuberculosis, a reaction ensues, and the tuberculin test is said to be positive. Patients may likewise be hypersensitive to other organisms, such as *Brucella*

---

*__Drug intolerance__ is the state in which a person receiving the drug reacts in a characteristic way to unusually small doses of the drug, doses with no physiologic effect. __Drug idiosyncrasy__ refers to the situation in which the recipient of the drug reacts in an unusual way, such as when a dose of a barbiturate (a sedative) produces excitement.

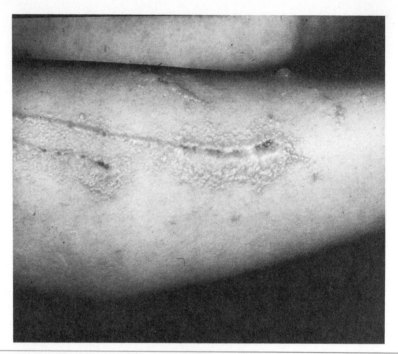

FIG. 19-9 Rhus dermatitis (poison ivy), well-known example of contact dermatitis. The linearity of the many, varying-sized vesicles (blisters) of the acute phase is seen here.

(the causative agents of undulant fever). In fact the manifestations of acute rheumatic fever are considered an allergic reaction to streptococci.

Other infections in which hypersensitivity to the causative agent is measured are leprosy, chancroid, lymphogranuloma venereum, coccidioidomycosis, blastomycosis, and histoplasmosis. In veterinary practice, hypersensitivity of infection is encountered in glanders, Bang's disease (undulant fever), and Johne's disease. In some of these, diagnostic skin tests routinely detect sensitivity (see Table 20-3).

Contact Dermatitis

Contact dermatitis (eczema or eczematous dermatitis) localizes in the skin, especially in areas exposed to direct physical contact with the irritant substance. As would be expected, this inflammatory condition often involves the hands. The first time that the skin contacts the offending substance no visible signs appear, but with sensitization subsequent contacts result in skin changes. The long and varied list of offenders includes the well-known poisons of poison ivy and poison oak. Poison ivy, poison oak, and poison sumac belong to the *Rhus* genus of trees and shrubs; the skin lesions may be called rhus dermatitis (Fig. 19-9). The *Rhus* plants are responsible for more allergic contact dermatitis than all other allergens combined. The list also contains a vast host of substances related to trade and industry, such as dyes, soaps, lacquers, plastics, woods, fabrics, furs, formalin, cosmetics, drugs, chemicals, metals, and explosives.

LABORATORY TESTS TO DETECT ALLERGY

The clinical laboratory provides several routes to the diagnosis and evaluation of allergy (or hypersensitivity).

Identification of the eosinophil, the cell of allergic inflammation, in blood and other body fluids is a simple way of testing for the allergic state. **Eosinophilia,** an increase in the number of eosinophils, is typically associated with allergic reactions and also parasitic infections. In both these conditions a mild to moderate blood eosinophilia is usually present, and in the area of involvement a fairly intense tissue eosinophilia is seen. For instance, in allergic rhinitis from 30% to 90% of cells viewed microscopically in a stained smear of nasal secretions consistently are eosinophils, and they are also prominent in the sputum of asthmatics.

Basophil counts may be used to study allergies, as with the in vitro **basophil degranulation test.** A preparation of fresh basophils from either a human being or a rabbit (a good source) is mixed with a patient's serum and stained supravitally. Specific antigen is added, and on the premise that in the presence of the specific antigen-antibody reaction degranulation occurs, the extent of degranulation in the preparation is measured by the reduction in the number of intact basophils. In a positive test over 30% of them have been degranulated.

The levels of IgE in the serum of normal, nonallergic subjects are very low but do tend to be much higher in the serum of allergic ones. Most IgE is fixed in the target organs in which the allergic reaction occurs; therefore the serum levels tend to be low. Also, certain classes of allergens are more efficient in bringing about synthesis of IgE. For instance, weeds and grass are responsible for higher levels than are dusts, molds, or danders. The IgE level in serum can be measured by radioimmunoassay, and the increase with allergic and parasitic diseases can be so demonstrated. Normal values of this test are stated to be from 50 to 500 units/ml of serum.

The **radioallergosorbent test** (RAST) applies the technic of radioimmunoassay to the detection of specific IgE against a variety of antigens in patients with atopic allergy. Forty-five different allergens used include Bermuda grass, ragweed, oak, cedar, elm, house dust, and house mites.

The **skin test** is of prime value. If some of the antigen to which a person is sensitive is rubbed into a scratch on the skin or if dilute antigen is injected between the layers of the skin, a wheal and flare 0.5 to 2 cm in diameter will occur at the site within 20 to 30 minutes (cutaneous anaphylaxis). If the person is not sensitive to the antigen, no reaction appears.

The **ophthalmic test** helps to detect sensitization to foreign protein such as horse serum. A drop of the *diluted* serum is instilled into the conjunctival sac; if the patient is sensitive to it, redness of the conjunctiva and a watery discharge will appear within 10 to 20 minutes. The reaction may be controlled by epinephrine.

The **patch test** is useful in detecting the cause of *contact dermatitis*. In this test suspect material is applied directly to the skin and held in place by means of adhesive tape from 1 to 4 days. A positive reaction reproduces the lesion from which the patient is suffering, with blister and papule formation.

IMMUNOTHERAPY (DESENSITIZATION)

In some cases desensitization (or hyposensitization) may be accomplished by the administration of repeated injections of very small amounts of the antigen to which the patient is sensitive. A patient hypersensitive to horse serum may sometimes be given an immune serum if the administration is preceded by several injections of very small amounts at 30-minute intervals. The desensitizing doses are graded by the reaction of the patient, and the administration of an immune serum to a hypersensitive person should be undertaken only by those experienced in immunotherapy. A similar but much prolonged method of immunotherapy is used in the treatment of hay fever and selected patients with asthma and urticaria. With this method, small amounts of an antigen, especially a high molecular weight polymerized antigen such as ragweed, are administered at weekly intervals to a person hypersensitive to that antigen. The intent is to induce the formation of IgG antibodies that will bind the allergen before it binds to a mast cell receptor. Thus one hopes to prevent the immediate-type hypersensitivity, that is, hay fever.

Desensitization, unlike sensitization, is of relatively short duration. In some forms of allergy it is accomplished with great difficulty.

SUMMARY

1 Immunologic responses may be quite harmful. Immunopathology is the study of tissue injury consequent to the immune reaction. Seven broad categories are noted.

2 Immunosuppression is an immunologic disturbance relative to the administration of immunosuppressive agents—steroid hormones, cytotoxic chemicals, or forms of radiant energy. This is a state of nonspecific nonreactivity to all antigens.

3 Immunologic deficiency is the developmental failure of the normal immune mechanisms. An absence of plasma cells results in very low levels of gamma globulin in the blood.

4 Immunologic aberrations may occur secondary to lymphoid cancers. The malignant plasma cell can yield an abnormal immunoglobulin in large quantities.

5 Immunologic tolerance is an acquired state of nonreactivity to a given, specific antigen, one that is ordinarily effective.

6 Generally, the normal immunologic mechanisms protect the body, but at times the immune system seems to act in reverse, making antibodies against the body's own constituents. The result is autoimmunity. Autoimmune diseases thus represent defects of immunologic tolerance.

7 Autoantibodies may react with specific autoantigens to cause disease directly, or they may form circulating antigen-antibody complexes depositing in and injuring tissues. Susceptible sites are the thyroid gland, brain, eyes, skin, joints, kidneys, and blood vessels.

8 Complement system deficiencies are associated with recurrent infections, with the most severe deficiency being that of C3.

9 The victim of acquired immunodeficiency syndrome (AIDS) is vulnerable to opportunistic infections, notably pneumocystosis and Kaposi's sarcoma.

10 T cell–mediated immunty is selectively impaired in AIDS. T helper cells are depressed; T suppressor cells are increased.

11 The CDC has spelled out precautions to be taken in the event of exposure to a patient with AIDS.

12 The ABO blood group system is an important histocompatibility factor.

13 The major histocompatibility complex on the short arm of the sixth human chromosome contains genes governing tissue graft rejection and immune responsiveness to varied antigens.

14 The HLA complex is used both to select donors for organ transplants and to identify genetic predisposition to more than 50 diseases. Specific examples of HLA-associated diseases are ankylosing spondylitis and multiple sclerosis.

15 The use of cyclosporin may revolutionize the field of organ transplantation. Cyclosporin A in a unique way selectively and specifically suppresses cell-mediated immunity without killing cells. It probably allows T cells to be primed to divide but blocks them from doing so.

16 The existence of a relationship between a person's immunologic system and that person's cancer is generally accepted today.

17 The two best-known fetal antigens that are expressed by some human cancers are carcinoembryonic antigen (CEA) and alpha-fetoprotein (AFP).

18 Allergy, or hypersensitivity, is a state of altered reactivity directly related to the operation of the immune system. Usually innocuous substances trigger the allergic state.

19 Examples of the immediate, IgE-mediated type of allergy are asthma, hay fever, and anaphylaxis from an insect sting.

20 The best example of the cytotoxic, cytolytic, or necrotizing reaction of hypersensitivity is the hemolytic transfusion reaction.

21 Examples of immune complex disease are serum sickness, acute poststreptococcal glomerulonephritis, and the Arthus phenomenon.

22 Sensitivity to penicillin, infectious allergy (the tuberculin reaction), and contact dermititis are examples of cell-mediated, delayed-type, tuberculin or bacterial hypersensitivity reactions.

23 Laboratory tests to detect allergy include the identification of the eosinophil, basophil counts, determination of serum IgE levels, the RAST test, the skin test, the eye test, and the patch test.

24 Desensitization of short duration to an allergen may be accomplished in a person allergic to it by repeated injections of small amounts of the given allergen. In so doing one hopes to induce the formation of IgG that will bind the allergen before it can bind to a mast cell receptor.

QUESTIONS FOR REVIEW

1 Categorize in general terms the disturbances in immunity.
2 Cite examples of immunosuppressive therapy.
3 What is meant by graft-versus-host reaction? What is the probable cause?
4 State the importance of the transplantation antigens.
5 Indicate the position of the HLA complex on the sixth human chromosome. Locate the four defined gene loci.
6 Briefly discuss immunity and cancer.
7 Classify and compare the major categories of allergic disorders.
8 Briefly discuss anaphylaxis.
9 Briefly describe the most important allergic diseases.
10 What is serum sickness? Why is it less common today than in the past?
11 Comment on the chemical mediators of allergy.
12 Outline the laboratory diagnosis of allergy.
13 How is desensitization accomplished? What is the underlying principle?
14 Explain the role in allergy of IgE.
15 Explain the relation of the basophil to allergic disorders.
16 List five agents known to have caused anaphylaxis in humans.
17 What are the infectious allergies? How can they be detected?
18 Define or briefly explain hypersensitivity, allergy, sensitizing dose, provocative dose, allergen, immunopathology, iatrogenic, endogenous, drug intolerance, drug idiosyncrasy, passive anaphylaxis, cutaneous anaphylaxis, immune complex, vasculitis, penicillin hypersensitivity, and RAST.
19 Give salient features of an autoimmune disease, the Arthus phenomenon, complement component C3 deficiency, idiopathic thrombocytopenic purpura, ankylosing spondylitis, AIDS, and Kaposi's sarcoma.
20 What is the current importance of cyclosporin A? Give its specific effect on the immune system.

SUGGESTED READINGS

Acquired immune deficiency (AIDS): precautions for clinical and laboratory staffs, Morbid. Mortal. Weekly Rep. **31**:577, 1982.

Acquired immunodeficiency syndrome (AIDS): precautions for health-care workers and allied professionals, Morbid. Mortal. Weely Rep. **32**:450, 1983.

Aledort, L.M.: AIDS: an update, Hosp. Pract. **18**:159, Sept. 1983.

Austen, K.F.: Tissue mast cells in immediate hypersensitivity, Hosp. Pract. **17**:98, Nov. 1982.

Check, W.A.: Medical News: A new age in asthma research: study of chemical mediators, J.A.M.A. **244**:745, 1980.

Check, W.A.: Medical News: Treatment improved for ragweed, penicillin allergy, J.A.M.A. **243**:1703, 1980.

Diffley, S.A.: Performing safe phlebotomy on AIDS patients, MLO **15**:75, Nov. 1983.

Drew, W.L., and others: Cytomegalovirus and Kaposi's sarcoma in young homosexual men, Lancet **2**:125, 1982.

Guarda, L.A.: Acquired immune deficiency syndrome (AIDS): new biohazard in the laboratory, Lab. Med. **14**:664, 1983.

Lichtenstein, L.M., and others: Once stung, twice shy: when should insect sting allergy be treated? J.A.M.A. **244**:1683, 1980.

Marx, J.L.: Human T-cell leukemia virus linked to AIDS, Science **220**:806, 1983.

Modlin, R.L., and others: T-lymphocyte subsets in lymph nodes from homosexual men, J.A.M.A. **250**:1302, 1983.

Oettgen, H.F.: Immunologic aspects of cancer, Hosp. Pract. **16**:85, July 1981.

Prevention of acquired immune deficiency syndrome (AIDS): report of inter-agency recommendations, Morbid. Mortal. Weekly Rep. **32**:101, 1983.

Reinherz, E.L., and Rosen, F.S.: New concepts of immunodeficiency, Am. J. Med. **71**:511, 1981.

Toewe, C.H., II: Bug bites and stings, Am. Family Pract. **21**:90, May 1980.

CHAPTER 20

IMMUNOLOGIC TESTING

REACTIONS FOR HUMORAL IMMUNITY

Because of the specificity of the antigen-antibody reaction, if either the antigen or antibody is known, it is possible to identify the other and, in many instances, to obtain reliable information as to how much of the other is present. Since antigen-antibody reactions are commonly measured in serum, they are often called **serologic reactions.***

In the clinical laboratory **serology** is the study of the antigen-antibody reaction in a specimen of serum. With a positive serologic test, the unit that measures the number of antibodies present is the **titer.** Many technics are available. Some of the important ones are discussed here.

Detection of Antibodies

The type of antigen-antibody reaction that applies to any given situation depends largely on the physical state of available antigen. Different types of reactions may be obtained with the same antigen and antibody; one reaction may be more efficient than another.

Practically speaking, antibodies (immunoglobulins) are detected by what they do to produce a *visible* (or measurable) reaction with a specific antigen under specified conditions (Table 20-1). In the hospital and clinical laboratory, antibodies are ordinarily classed as (1) complement-fixing antibodies, (2) antitoxins (neutralizing), (3) precipitins, (4) cytolysins, (5) agglutinins, (6) opsonins, (7) virus-neutralizing antibodies, and (8) fluorescent antibodies. In Table 20-1 this list of antibodies is tied in with antigens involved, and nature of the antigen-antibody reaction is briefly indicated.

Serum may be treated in such a way that the class of antibodies reacting with the antigen may be determined. Thus a primary immune response may be separated from a secondary one. For instance, immunoglobulin may be inhibited in serum by certain reducing agents. Fractionation of serum may be accomplished by special technics such as ultracentrifugation.

---

*To collect serum, allow a sample of blood to clot. The fibrin formed in the process entraps the blood cells. When the clot shrinks, the fluid expressed is the serum, an important practical source of antibodies.

Table 20-1 Detection of Important Antigen-Antibody Combinations

Positive Reaction	Antigen	Antibody (in Serum)	Nature of Reaction
Complement fixation	Microbes	Complement-fixing antibodies	Binding of complement detected by hemolytic indicator system; *absence* of hemolysis a positive test
Flocculation	Exotoxins	Antitoxins	Flocculent precipitate
Precipitation	Microbes Animal proteins	Precipitins	Fine precipitate from clear solutions

Cytolysis
  Bacteriolysis
  Hemolysis
	Intact cells
	  Bacteria
	  Red blood cells
		Cytolysins
		  Bacteriolysins
		  Hemolysins
			Cells dissolved

Agglutination	Agglutinogens
	  Bacteria
	    Proteus bacilli
	
	  Human red blood cells
	  Sheep red blood cells
		Agglutinins
		
		  Heterophil antibodies
		  Isoantibodies
		
		  Heterophil antibodies
			Gross clumping of cells

Inhibition of viral hemagglutination	Human type O red cells coated with viral antigen	Inhibiting antibodies	Clumping of red cells *blocked*
Opsonization	Bacteria	Opsonins	Bacteria phagocytized in greater numbers by leukocytes
Neutralization or protection	Viruses	Virus-neutralizing antibodies	Protection of an animal or cell culture from harmful effects of agent
Fluorescence under ultraviolet light	Microbes	Fluorescein-tagged antibodies	Reaction tagged with marker

A listing of antigens for such important tests as complement fixation, precipitation, or passive hemagglutination includes crude or purified extracts of organisms, capsular polysaccharides, protein derivatives, endotoxins, and exotoxins. Generally, antigens used today are highly purified, especially in the highly sophisticated fluorescent immunoassay, radioimmunoassay, and enzyme-linked immunoassay. Intact microbes are often used, and a formalin or heated-fixed suspension of bacteria is still standard.

Kinds of Reactions

Complement Fixation

The complement-fixation test, designed to detect complement-fixing antibodies in a serologic test, is based on this fact: when serum containing the complement-fixing antibodies against the etiologic agent of a disease, specific microbial antigen (perhaps the microbe itself), and complement are mixed in suitable proportions and incubated, the three enter into a combination whereby the complement is bound (fixed) and is not available for cell lysis. Since this combination is not accompanied by any visible change, an indicator system must be used to detect it. Because hemolysis, or lysis of red blood cells, is dramatic, a hemolytic system is a good indicator. In this system the combination of complement, **hemolysin (amboceptor),** and sensitized red cells (of the species of animal against which hemolysin was prepared) effects **hemolysis,** or dissolution of the red cells, as an end point.

If a sample of serum from a patient is mixed with complement and a test bacterial

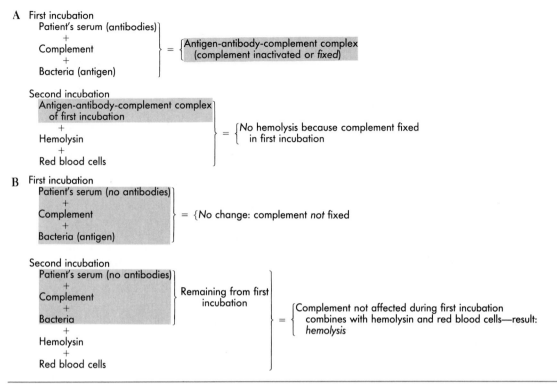

A First incubation
Patient's serum (antibodies)
+
Complement
+
Bacteria (antigen)
= { Antigen-antibody-complement complex
    (complement inactivated or *fixed*)

Second incubation
Antigen-antibody-complement complex
    of first incubation
+
Hemolysin
+
Red blood cells
= { No hemolysis because complement fixed
    in first incubation

B First incubation
Patient's serum (*no antibodies*)
+
Complement
+
Bacteria (antigen)
= { No change: complement *not* fixed

Second incubation
Patient's serum (*no antibodies*)
+
Complement
+
Bacteria
+
Hemolysin
+
Red blood cells

Remaining from first incubation

= { Complement not affected during first incubation
    combines with hemolysin and red blood cells—result:
    hemolysis

FIG. 20-1   A, Positive complement-fixation test. **B,** Negative complement-fixation test.

suspension and incubated, complement will be fixed when complement-fixing antibodies to the given bacteria are present in that serum. If suitable proportions of hemolysin and red blood cells are then added and the mixture incubated a second time, hemolysis *cannot* occur because complement is not available to effect lysis of red blood cells. This is a positive test, graphically illustrated in Fig. 20-1, A. If, on the other hand, the patient's serum does *not* contain the specific complement-fixing antibodies, complement is *not* fixed during the first incubation, and hemolysis can occur in the second because this time complement is available to effect lysis in the indicator system. This is a negative test, with the colorful results of red cell lysis (Fig. 20-1, *B*).

Complement-fixation tests are available for most viral diseases, syphilis, rickettsial diseases, fungal infections, and a few protozoan and metazoan infections.

Wassermann test. The Wassermann test for syphilis applies the principle of complement fixation. It has always been impossible to grow the spirochetes of syphilis artificially. When the Wassermann test first came into use, extracts of the livers of syphilitic fetuses were used as antigens because such a syphilitic liver was known to contain myriad spirochetes and such extracts were the nearest approach to extracts of the organisms. Today certain lipid fractions of normal organs, such as a lipid extract of beef heart, are efficient antigens that give clinically reliable and consistent results. Of course the term *antigen* used in this context is scientifically incorrect. In diseases other than syphilis, the antigens used in complement-fixation tests are derived directly from the causative organisms.

Assay of complement. The complement system itself can be measured by the pattern of complement fixation outlined previously. The level of complement is altered in many disease states, and the level of complement component C3 is quite low in important diseases. Assay of complement with a hemolytic system can indicate both amount of complement (quantity) and its functional activity (quality) (p. 347).

More recently the method of radial immunodiffusion has been used for assay of C3 (the third component of complement) and when specific serum is available for quantitation of other complement components as well.

Toxin-Antitoxin Flocculation

Antitoxins are antibodies that neutralize toxins, a combination demonstrable by flocculation. **Flocculation** refers to the presence of aggregates visible microscopically or macroscopically with certain antigen-antibody reactions. Toxin-antitoxin flocculations are similar to precipitin reactions (see below) but show a very sharp end point, the reaction being blocked by either too much antigen or too much antibody.

Precipitation

When a suitable animal is immunized with certain bacteria, the animal's serum acquires the ability to cause a fine, powdery precipitate from the clear filtrate of the bacterial growth (**soluble** bacterial antigen is present). The antibodies formed are **precipitins.** The same phenomenon is noted when animals are immunized with soluble antigens of certain animal and vegetable proteins. Precipitins are specific; bacterial precipitins react only with the filtrates of the bacteria that produced them, and precipitins formed by immunization of an animal with the serum of an animal of a different species react only with the proteins of the species used for immunization. Precipitins do not form regularly in infectious disease, but where they are present, they are nicely demonstrated by technics of immunodiffusion (p. 404).

In the precipitin test the antigens detected are usually soluble macromolecules, such as proteins or polysaccharides. They are multivalent, since they display three or more antigenic determinants with binding sites for precipitating antibodies. Precipitins are obtained by injecting an animal with specific antigen or from patients responding to infection or immunization. Precipitins are bivalent immunoglobulins, each with two combining sites for antigen. Where the amount of specific antigen is balanced to specific antibody in the reaction, a large and complex aggregate precipitates visibly in a liquid or semisolid medium. Certain factors control the rate of precipitation, such as the proportions of the reactants. Salt is needed, and the pH must be near neutral. The rate of reaction proceeds faster at higher temperatures, but the maximum amount precipitates in the cold. The reaction is fastest in the zone of **equivalence,** the point where, because optimal proportions exist between antigen and antibody, both antigen and antibody are completely precipitated. (At this point there is also maximal complement fixation.) The amount of the precipitate relates to the proportions of the reactants. Generally speaking, the complexes generated in antigen excess tend to be soluble, whereas those with antibody excess tend to be *in*soluble.

Precipitin reactions may be demonstrated by layering two solutions. With reaction, a precipitate forms a ring at the interface. If precipitation takes place in a semisolid medium, such as soft agar or the purer polysaccharide agarose, distinct bands of precipitate form on contact of antigen with antibody. A variety of technics that have been developed depend on the visualization of the precipitate for the quantitation of soluble antigens and antibodies. Because the precipitin reaction can be accurately quantitated, it has been studied in great detail.

The precipitin reaction has a wide medicolegal application in the determination of whether blood stains are of human origin and in the detection of the adulteration of one kind of meat with another. It is the test on which Lancefield's classification of streptococci is based. Flocculation tests for syphilis, such as the Kline, Kahn, and VDRL, are modified precipitin tests.

C-reactive protein. A biologic coincidence occurs in a number of inflammatory conditions of either infectious or noninfectious nature. A peculiar protein (a beta-globulin) that can form a precipitate when in contact with the somatic C-polysaccharide of pneumococci appears in the blood of the affected person. For this reason the protein is spoken of as C-reactive protein; it is *not* an antibody. (Its structure is different, as is the organ where it is formed.) The serum of an animal immunized to this protein is used in a precipitation test to detect C-reactive protein in serum of persons suspected of having one of the diseases in which the protein appears. Among such diseases, both infectious and noninfectious, are acute rheumatic fever, subacute bacterial endocarditis, staphylococcal infections, infections with enteric bacteria, and neoplasms.

C-reactive protein functions as an acute phase protein, which means that the level of this protein in the serum shoots up within hours of the onset of an inflammatory process, peaks during acute stage, and then decreases with resolution of that process. The mechanism for this acute phase response is not worked out, but the C-reactive protein level is believed to be a more reliable and sensitive indicator of inflammation than is the widely used sedimentation rate. Like other acute phase proteins, it is synthesized primarily in the liver.

Recently C-reactive protein has been shown in vitro to influence certain reactions of both cellular and humoral immunity. It binds to lymphocytes and may affect their functioning. It stimulates phagocytosis both directly and indirectly through complement-mediated activity. It appears in vitro to be as much a part of certain immunologic processes as is an immunoglobulin.

Cytolysis

Antibodies that dissolve or lyse cells are known as **cytolysins** (or **amboceptors**). **Bacteriolysins,** cytolysins that lyse bacteria, are produced or at least increased by bacterial infection. Cytolysis requires both cytolytic antibodies and complement; if only one is present, cytolysis does not occur. A cytolysin of note is **hemolysin,** which dissolves red blood cells. It is prepared by the administration to an animal (most often a rabbit) of a series of injections of the washed red blood cells from an animal of a different species (most often a human or sheep). Hemolysin develops in the serum of the recipient animal, specific only for the red blood cells of the species used. It does *not* affect the red blood cells of other species.

Agglutination

If the serum of a person who has had a certain disease such as salmonellosis is mixed with a suspension of the bacteria responsible, the bacteria adhere to one another and, in the test tube, form easily visible clumps that sink to the bottom of the tube. Likewise, if an animal receives several injections of a given microbe, its serum will acquire the ability to clump, or **agglutinate,** those microbes. This kind of clumping is *agglutination* (Fig. 20-2), and the antibodies that induce it are **agglutinins.** Particulate substances (antigens) that, when injected into an animal, induce the formation of agglutinins are known as **agglutinogens.** Agglutinogens may be microorganisms, red blood cells, or latex spheres coated with adsorbed antigen. In each instance, particles must be present for this kind of clumping to occur.

Motile bacteria lose their motility before agglutination. Bacteria are not necessarily killed by agglutination, and dead bacteria are agglutinated as easily as are viable ones. Agglutinated microorganisms are more readily ingested by phagocytic cells.

Kinds of agglutinins. Normally occurring agglutinins exist in the blood of some persons. For instance, human serum sometimes shows a weak agglutinin content against typhoid bacilli although the person has never been knowingly so infected. The agglutinin content is so low, however, that when the serum is diluted 20 or 40 times and then mixed with a suspension of the bacilli, agglutination does not occur.

Immune agglutinins are agglutinins brought about by infection or artificial immu-

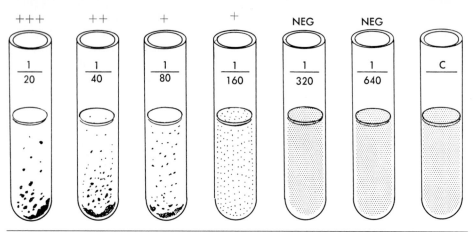

FIG. 20-2 Bacterial agglutination, test tube method. Dilute 0.2 ml of patient's serum serially in isotonic saline solution through first *six* tubes (left to right). Add 1 ml of suspension of bacteria to each tube through all seven. Final dilutions indicated by numbers on each tube (as 1 part serum in 20 parts, etc., suspension). Control tube on far right contains only bacterial suspension and saline solution. Note maximal agglutination of bacteria (large masses) in first three tubes on left, some in fourth, and none in fifth and sixth. Agglutination test positive; titer of serum 1:160.

nization. Such agglutinins can occur plentifully in serum, and such serum causes agglutination although diluted 500 to 1000 times. Agglutinin formation is usually specific; that is, when an organism is introduced into the body, the body forms agglutinins against that organism only or, in some cases, against very closely related organisms. Agglutinins for closely related organisms are known as **group agglutinins.** However, the agglutinin content of the serum is much higher for the organisms directly inducing the production of the agglutinins than it is for the closely related ones. Usually, in the first 7 to 10 days of an infection, the human (or animal) body has not manufactured enough agglutinins for positive identification.

Comparatively few organisms induce appreciable agglutinin formation with infection or artificial immunization. Important ones that do are *Salmonella typhi*, other *Salmonella* species, *Brucella* species, and *Francisella tularensis*. In infections with these (salmonellosis, undulant fever, tularemia) the agglutination test (Fig. 20-2) may become a diagnostic procedure. For example, in typhoid fever, the serum agglutinins that appear during the course of disease may be measured by the Widal test. This is an agglutination test employing test tube dilutions of the patient's serum with known cell suspensions of *Salmonella typhi*. In the lower animals the agglutination test is employed to detect Bang's disease (contagious abortion of cattle) caused by *Brucella abortus* and, in chickens, bacillary white dysentery caused by *Salmonella gallinarum*.

Sometimes **nonspecific** agglutinins are produced against organisms with apparently no relation to the disease. Nonspecific agglutination is of special diagnostic value in typhus fever and other rickettsial diseases. In typhus fever nonspecific agglutinins develop against certain members of the *Proteus* genus of bacteria; typhus fever is caused by rickettsias, *not* by *Proteus* bacilli. Many such agglutinins are probably formed as a result of the introduction into the body of **heterophil** antigens (*hetero,* from an individual of a *different* species). (The heterophil antigen is an exception to the tenet of antigen specificity.) Heterophil antigens cause heterophil antibodies to be formed not only against themselves but against other antigens as well. Such antigens are common in the lower orders of life. In typhus fever the causative agent of the disease, a rickettsia, seems to act as a heterophil antigen, and the antibodies formed react not only to the rickettsias but to certain *Proteus* bacilli as well. The agglutination of certain

Proteus types by the serum of a patient with typhus fever constitutes the **Weil-Felix reaction.**

The serum of human patients with infectious mononucleosis agglutinates the red blood cells of the sheep. This antibody is a **heterophil antibody,** since it agglutinates an antigen (in the red cells) of a different species (the sheep). This diagnostic test is the **heterophil-antibody test,** or the **Paul-Bunnell test.**

For the laboratory detection of serum agglutinins, a battery of typing serums may be formulated by the artificial immunization of a group of animals of the same species. Each individual animal is immunized artificially with a specified one (but a different one) of a chosen group of known bacterial types. After immunization, the animal is bled and its serum is separated and preserved for use. Such typing serums retain their potency for months. For the agglutination test, a suspension of unknown bacteria from a subject is mixed in turn with each of the serums of the battery. If one specific serum agglutinates the test suspension, the unknown bacteria must be the same as those against which the serum was prepared.

In the application of the agglutination test to the identification of bacteria, the antibody is known and the antigen is unknown. In the application of the agglutination test to the diagnosis of disease, a known antigen is used to identify unknown antibody. In serum from a patient laboratory agglutination implicates the agent responsible for the production of serum agglutinins demonstrated as the one responsible for the disease.

Hemagglutination. Many types of antigens can be absorbed onto the surface of a red blood cell. A red cell properly prepared with a test antigen can serve as a reagent for the detection of antibodies in the serum of a patient by the technic of hemagglutination. The **titer** of antibody is the dilution of serum in which no further agglutination of red cells takes place.

Hemagglutination-inhibition. Red cell agglutination is used to identify certain viruses (p. 688). The action of a virus to clump red cells (viral hemagglutination) can be blocked by specific antibody to the virus. This is hemagglutination-inhibition, a valuable test in a number of viral infections, including the following:

Rubella	Vaccinia	Dengue
Influenza	St. Louis encephalitis	Adenovirus infections
Mumps	Western equine encephalitis	Reovirus infections
Measles	Japanese B encephalitis	Some enterovirus infections

Passive (indirect) hemagglutination. Agglutination forms the basis of many useful laboratory tests, some of which are among the most effective laboratory procedures done today for the detection and identification of pathogenic bacteria. With agglutination, bacteria serve as both carriers and indicators of the antigen-antibody reaction. Because of the great sensitivity of agglutination reactions, many soluble antigens have been placed on inert particles. Latex particles, fixed staphylococcal cells, bentonite (potter's clay), and treated red blood cells become "passive" carriers and indicators. Passively coated particles agglutinate specifically. (Reagents for the detection of C-reactive protein are latex particles coated with an anti–C-reactive protein immunoglobulin.) Technics of indirect agglutination or of passive hemagglutination are sensitive, detecting as little as 10 ng of antibody per milliliter. (Precipitin tests detect in the range of 20 μg/ml.)

The following microorganisms may be identified by latex particle agglutination or passive hemagglutination tests:

Haemophilus influenza type b	Streptococci groups, A, B, C, D, F, and G
Neisseria meningitidis	*Neisseria gonorrhoeae*
Streptococcus pneumoniae	*Staphylococcus aureus*

Bacterial microagglutination. A bacterial microagglutination test for the agents of tularemia and brucellosis is carried out in microwell plates with stained bacteria. One twentieth the amount of antigen used in the standard tube test is required here for the diagnosis of tularemia; one tenth the amount, for brucellosis.

Opsonization

Opsonins are antibodies or other components of serum that act on bacteria, other microbial cells, or particulate antigens in such a manner as to render them more easily ingested and digested by phagocytes. The binding process, whereby a microbial cell, for example, becomes more susceptible to phagocytosis by combination with an opsonin, is opsonization (Greek *opsonein*, to prepare food for). If the microbes causing a certain disease are mixed with the white blood cells of a patient (human or animal) who might have the disease, the degree of phagocytosis resulting is of diagnostic significance and can be evaluated in the **opsonocytophagic test.** This test has been used in the diagnosis of undulant fever and tularemia.

Phagocytosis (p. 315) is not an activity of cells alone but depends on the action of opsonins in the serum. Opsonins are to some degree present in normal serum but are present in an increased amount in immune serum, and their action is enhanced by complement. C3 on a cell surface is a powerful opsonic agent.

Macrophages have receptors on their surface membrane for the Fc part of antibody and for the C3 component of complement. The possession of these receptors greatly facilitates their phagocytosis of opsonized or antibody-coated particles.

Nitroblue tetrazolium test. When normal neutrophils (microphages) in an individual with a normally functioning immune system are confronted by bacterial pathogens, certain metabolic changes consequent to their phagocytic activity take place, probably related to the cell membrane; opsonins are not pinpointed. Because of these changes, the neutrophils can be shown to reduce in appreciable amounts the supravital dye pale yellow nitroblue tetrazolium to blue-black formazan crystals. An in vitro test, the nitroblue tetrazolium test, has been devised to display precisely in human blood the ability of the neutrophils to do so. This test helps to separate certain bacterial infections from nonbacterial illness in patients in whom clinical manifestations are confusing and to monitor patients with increased vulnerability to infection (because of tissue transplant, immunosuppressive therapy, and the like).

Neutralization or Protection Tests

Neutralization tests are important in a number of infections caused by viruses and a few caused by bacteria. Antibodies that give immunity to viruses are known as **virus-neutralizing antibodies.** They can block infectivity of a given viral agent under test conditions in a susceptible host (p. 688).

A crucial first step in viral infection is the binding of the virus to a specific receptor on the surface of a target cell. Virus neutralization is accomplished when specific antibodies block that attachment and thus prevent disease. In the immune host, lymphocytes have been primed to recognize viral antigen on the viral surface. When the receptors on one of these cells encounter specific virus, the immune process is set in motion. Specific antibodies are secreted that bind to viral receptors, coat the surface of the virus, and effectively stop viral action.

Virus neutralization may be demonstrated in the following diseases:

Influenza	Mumps	Poliomyelitis
Parainfluenzal infections	Measles	Rubella
Respiratory syncytial virus infection	Varicella	Cytomegalovirus infection
Common cold	Herpesvirus infections	

Immunofluorescence

When antibodies are chemically tied to the fluorescent dye fluorescein, they demonstrate no change in basic properties as antibodies and react with specific antigens in the usual way. The combination of such tagged antibodies, or **fluorescent antibodies,** with the specific antigens forms a precipitate that, because of the accompanying fluorescein, can be seen as a luminous area if viewed under ultraviolet light. A special type of microscope has been devised with which one can do just this (see Fig. 8-2 and Color plates 1, A, and 3, C).

In the **direct** fluorescent antibody test fluorescent antibodies are used to detect and identify specific antigens in cultures or smears containing test microbes or in tissue cells of the human or animal body. Specimens for testing with fluorescent antiserums may be taken from various body fluids and various anatomic sites of disease.

The direct test is applicable to infection with the following agents:

Actinomyces	Enteroviruses	*Neisseria gonorrhoeae*
Blastomyces dermatitidis	Group A streptococci	*Neisseria meningitidis*
Bordetella pertussis	*Haemophilus influenzae*	Parainfluenza viruses
Brucella	Herpesvirus	Rabies virus
Candida albicans	*Histoplasma capsulatum*	*Shigella*
Coccidioides immitis	Influenza viruses	*Sporothrix schenckii*
Cryptococcus neoformans	*Legionella pneumophilia*	*Staphylococcus aureus*
Enteropathogenic *Escherichia coli*	*Listeria monocytogenes*	*Streptococcus (Diplococcus) pneumoniae*

In the **indirect test** *non*fluorescent antibody is bound to its antigen as the first part of the test. In the second part the combination is made visible after application of a second antibody (antihuman serum), this time a fluorescent one. The test is set up so that the unconjugated serum acts as antibody in the first part and as antigen when exposed to fluorescent serum in the second part. The indirect test has been used with infections caused by the following:

Aspergillus	Epstein-Barr virus	Rubella virus
Brucella	*Haemophilus influenzae*	*Schistosoma*
Chlamydia trachomatis	Herpesvirus	*Toxoplasma gondii*
Cryptococcus neoformans	*Leptospira*	*Treponema pallidum*
Cytomegaloviruses	*Mycoplasma pneumoniae*	*Trichinella spiralis*
Echinococcus granulosus	*Plasmodium* of malaria	*Trypanosoma*
Entamoeba histolytica	*Pneumocystis carinii*	

Because of the rapid results with this technic, a wide variety of applications exist for the identification of antigens related to bacteria (including rickettsias), viruses, fungi, and protozoa. Pathogenic amebas can be quickly distinguished from nonpathogenic ones, streptococci may be detected quickly in throat swabs, and rabies virus can be tagged in animals. Complement (especially C3) can be assayed. Autoantibodies can be measured. In tissues, fluorescent staining helps to localize the precise anatomic sites of antigen-antibody combination in diseases such as lupus erythematosus, rheumatoid arthritis, and certain hypersensitivity reactions.

Special Technics

There are several other laboratory tests contributing much to the study of the structure and nature of antibodies (immunoglobulins) and to the elucidation of immunologic disorders. Measurement of immunoglobulin in serum and body fluids is said to be the commonest assay in clinical immunology.

Electrophoresis

Electrophoresis refers to the movement in an electrical field of electrically charged particles suspended in a suitable medium. When proteins are thus placed in an electrical field at a given pH, they move in distinct paths according to their own specific electrical charges. The pattern of migration for different proteins can be visualized under pre-scribed test conditions on a suitable strip, such as paper, stained with dye (Figs. 20-3 and 20-4). **Zone electrophoresis** refers to the technic of using a stabilizing medium to trap migrating proteins into more or less separate areas or zones that can then be stained and identified later. Routinely used solid support media include paper, starch, agar, and cellulose acetate. With zone electrophoresis, antibodies tend to show up as an undifferentiated group. They can be sorted, however, by technics indicated below.

Radial Immunodiffusion

Under the appropriate conditions, antigen-antibody combinations form a visible precip-itate. **Gel diffusion** is a term used for the antigen-antibody precipitin reaction detected in semisolid media. If protein (antigen) is placed in a specially designed hole or well in an agarose gel medium (no electrical charge), that protein diffuses concentrically out of the well. Since specific antibody precipitates antigen in gels, radial immunodiffusion can be shown nicely in the clear gel if specific antibody has been impregnated into it. A ring of precipitate forms at the points of contact of antigen with specific antibody, and the reaction can be quantitated easily for amounts of protein (or antigen). Known also as **single gel diffusion,** this is an important method for analysis of immunoglobu-lins.

Electroimmunodiffusion is single gel diffusion in an electric field.

Double Gel Diffusion

In this method antibody is not incorporated throughout the agarose gel as in single gel diffusion but is placed in a trench (or other reservoir) cut in the medium. The arrange-ment for double gel diffusion is such that both soluble protein (antigen) and antibody diffuse out into the agar to form a band or line of precipitate at points of contact. If a mixture of proteins (or antigens) is analyzed, each specific antigen-antibody combina-tion tends to present as a separate band at a definite position in the medium between the reservoir for the antigen and that for the antibody, since in soft agar different anti-gens and antibodies are likely to diffuse at different rates. As a consequence, optimal proportions for precipitation occur at different sites in the agar and distinct bands form. This method is a quantitative one.

Agar diffusion methods based on this principle, such as the Ouchterlony technic,

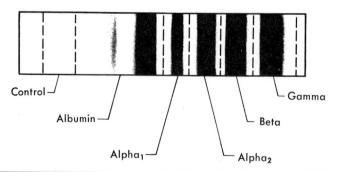

FIG. 20-3 Electrophoresis of normal human serum. Serum sample is placed near one end of cellulose acetate strip moistened with buffer. An electric potential is applied. Serum proteins migrate along strip at different speeds. Afterward, strip is dried and proteins are fixed and stained (bromphenol blue). Black bands correspond to separated and stained proteins, depth of color in each band proportional to amount of that protein present. Strip may be cut crosswise as indicated and protein fractions analyzed further. (From Bauer, J.D., and others: Clinical laboratory methods, ed. 9, St. Louis, 1982, The C.V. Mosby Co.)

make possible the detection of the number of antigens in mixtures and also the identification and recognition of the diversity of antigens interacting with a single antibody.

Ouchterlony technic. In the Ouchterlony technic for double diffusion a suitable diffusion medium, such as a 2% solution of hot agar in isotonic saline, is poured onto a microslide or into a Petri dish to a depth of 1 or 2 mm. With a special punch, various patterns of circular holes are cut into the agar layer after it has cooled. A workable pattern is a hexagon of antigen wells around a central reservoir holding serum—human or animal. Test specimens containing soluble antigens are placed in the peripheral walls. When fully loaded, the dish or microslide is incubated in a moist chamber for 24 to 48 hours. Since both antiserum and corresponding antigen in solution diffuse through the agar to meet and combine, specific reactions can be read from precipitation lines.

The relationship between two antigens can be determined by this method. Two immunologically identical antigens form a continuous band, a complete chevron, between the antigen wells and the antiserum reservoir. This is the **reaction of identity.** Two unrelated antigens reacting with the same antiserum produce two lines completely crossing each other, the **reaction of nonidentity.** If two antigens share some antigenic

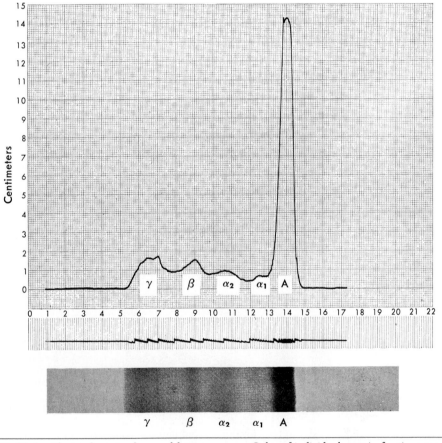

FIG. 20-4   Electrophoresis of normal human serum. Color of individual protein fractions may be measured directly on cellulose acetate strip with recording densitometer and curve obtained as above. (From Bauer, J.D., and others: Clinical laboratory methods, ed. 9, St. Louis, 1982, The C.V. Mosby Co.)

determinants but one reacts more completely with antibodies present than the other, a spur is formed, the **reaction of partial identity.**

When two reactants in this technic are present in balanced proportions, the line formed will be concave to the well containing the reactant of higher molecular weight, because larger molecules move more slowly.

Precipitin-in-agar technics (of which this is one) are simple, fairly rapid screening methods of value in the identification of various microbes.

Immunoelectrophoresis

If the technics of electrophoresis and double gel diffusion are combined, we have the process of immunoelectrophoresis, a valuable method for separating complex mixtures of proteins (antigens), particularly in body fluids (Fig. 20-5). In the test specimen in an electric field the unknown proteins are separated and spread out in a series of differently charged masses. After electrophoresis, specific antibody is placed in a trough parallel to the line of migration of the proteins in the gel and allowed to diffuse inward in an incoming wave. As specific antibody meets antigen, precipitation occurs in a series of arcs (Fig. 20-6). The immunoprecipitin bands are readily seen, and the position and shape are consistent and stable for known proteins. Deviations represent abnormal constituents (Fig. 20-7).

Counterimmunoelectrophoresis

Counterimmunoelectrophoresis (CIE) is a modification of immunodiffusion whereby both antigen and antibody are brought together in an electric field; that is, they are "driven" toward each other electrophoretically. It is only applicable under special circumstances, but it can be used to detect a number of bacterial antigens in serum, urine, cerebrospinal fluid, and sputum.

FIG. 20-5   In two-dimensional immunoelectrophoresis of normal serum in agarose, immunoglobulins are seen as flat lines on left separated from prominent peak of albumin on right by numerous alpha and beta globulins. (From Thompson, R.A.: The practice of clinical immunology, ed. 2. In Current topics in immunology, London, 1978, Edward Arnold [Publishers] Ltd.)

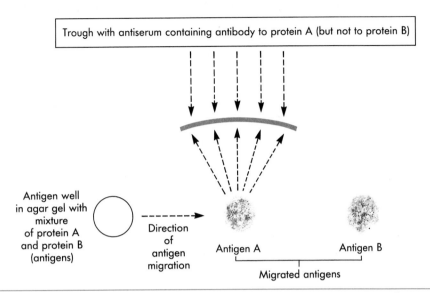

FIG. 20-6 Principle of immunoelectrophoresis in sketch. From a mixture in agar gel, antigens migrate in an electric field to positions as indicated. After the current is cut off, specific antibody is placed in the trough. Antigen diffuses radially from a point source, and antibody diffuses from the trough along a plane front. They contact in such a way (in optimal proportions) that precipitation occurs along an observable arc. The arc is closest to the trough where antigen is most concentrated.

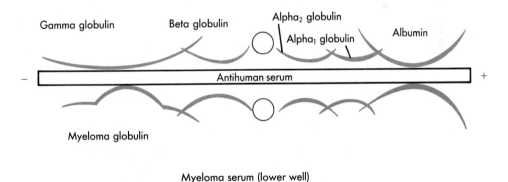

FIG. 20-7 Immunoelectrophoresis used to compare a normal serum with an abnormal one. The abnormal serum is from a patient with multiple myeloma. *Step 1.* Electrophoresis of antigens in agar gel: test sera (antigens) are applied to wells punched out of agar slides. Agar slide is placed in electric field. Proteins of the test sera migrate. (Albumin to the anode [+], gamma globulin to the cathode [−].) Note that the abnormal myeloma globulin is restricted in its mobility. *Step 2.* Formation of precipitin arcs: antisera to five human serum proteins are applied to central trough. Antibodies diffuse into agar toward serum proteins of test material, which also diffuse in agar. Where serum proteins and homologous antibodies meet, precipitation arcs appear. Remember that an arc is characteristic for a given protein (antigen). Normal IgG forms a long sweeping arc, but the arc formed by the abnormal IgG of the myeloma serum is more of a blister.

Radioimmunoassay (RIA) (Competitive Binding Analysis)

This method is most sensitive for quantitating antigens that can be radioactively labeled. It is based on competition under test conditions for a limited number of binding sites in a fixed amount of specific antibody between (1) the known, a fixed level of purified antigen labeled with radioactive iodine, and (2) the unknown, an unlabeled test sample of antigen. The antigen-antibody complexes formed are separated, and the amount of radioactivity is determined. The concentration of the unknown, unlabeled antigen is calculated by comparison with reference standards.

Enzyme-Linked Immunosorbent Assay (ELISA)

An enzyme immunoassay technic known as ELISA is similar in design to radioimmunoassay but uses enzyme-labeled rather than radioactively labeled reagents, a solid phase on which to conduct the test, and the production of a color to indicate the concentration of the unknown. For linkage to either an antigen or antibody, a number of enzymes such as peroxidase, glucose oxidase, beta-galactosidase, or alkaline phosphatase may be used. For the solid phase, polyacrylamide or solid materials such as polystyrene or polyvinyl disks, tubes, microtitration plates, and beads are employed. For the appraisal of the concentration of the unknown in the test, the production of color rather than the level of radioactivity is monitored. Concentrations of either antigen or antibody in the sample are measured by a color change.

PROCEDURES FOR CELLULAR IMMUNITY

Cellular immunity may be measured by immunologic yardsticks although the technics are not so well established as those with which we evaluate humoral immunity.

Lymphocytes in Peripheral Blood

The presence, number, and kind of lymphocytes circulating in the peripheral blood can be determined. This is information of a general nature; nonetheless it can be useful. If the total number of lymphocytes is decreased (lymphopenia, a significant finding), this is reflected in the total lymphocyte count (sometimes done sequentially). The critical level for lymphocytes is 1200 per cubic millimeter. The morphology of the lymphocytes may be studied in peripheral blood smears and differential counts made. Thymus-derived T cells are easily recognized: they are small, mostly nucleus with little cytoplasm. Approximately 70% of the peripheral blood lymphocytes are T cells; some 25% are B cells. (The small percentage remaining, called null cells, are indeterminate.)

Battery of Skin Tests

Skin testing with recall antigens is fairly straightforward for the presence or absence of cellular immunity. A battery of antigens is selected with the idea that most normal adults should have a positive skin test to at least one of them. The antigens used routinely include one for *Candida* species, one for tetanus, a mumps antigen, streptokinase-streptodornase, trichophytin, and purified protein derivative (PPD) (tuberculin). Children, however, even under normal circumstances may show negative skin test results with all of these. In a child, therefore, sensitivity may be artificially induced with dinitrochlorobenzene and challenge made 2 weeks later.

Immunologic Competency In Vitro

There are tests based on the premise that cell-mediated immunity requires for its expression a direct interaction between the host lymphoid cell and the agent (microbe, foreign cell, and the like) that carries the antigen into the area of contact. When the natural situation is simulated, the immunologic behavior of the lymphocytes can be observed.

Rather precise assays are being developed along these lines. One such assay starts with peripheral lymphocytes isolated in cell culture. A mitogen such as phytohemagglutinin (PHA) is added to stimulate increased DNA and RNA synthesis (in T cells only), a response of the lymphocytes measurable in two ways. One method quantitates uptake of radioactive thymidine. A second method makes a differential count of the cells. Small T lymphocytes exposed to PHA become larger cells (blast transformation). The nucleolus swells, and cytoplasmic organelles develop. After suitable staining, one can record the number of transformed blast cells.

Another essay starts with a mixed white cell culture, one in which test lymphocytes from a given person are cultured with lymphocytes from an unrelated donor. Competent lymphocytes can recognize the cells that do not belong, the nonself. This they do because of the histocompatability antigens foreign to them on the donor cells. In severe immunologic deficiency disorders involving cell-mediated immunity the ability to recognize and respond to HLA antigens is lost.

Migration Inhibitory Factor Assay

Several methods have been devised to assay factors released after the interaction of sensitized lymphocytes with specific antigen. One such method is the assay of the migration inhibitory factor (MIF). For this test cellular suspensions are prepared from peritoneal exudates of guinea pigs sensitized to antigen(s). In this material there are macrophages (70%), polymorphonuclear leukocytes (only a few), and lymphocytes, specifically sensitized. The exudative material is placed into capillary tubes, and, *in the absence of antigen*, the macrophages are observed as they migrate out of the tubes onto glass coverslips in culture chambers. When antigen is added, migration ceases. It takes only 1 sensitized lymphocyte to block the movement of 99 macrophages. With this assay, the patient is studied to determine first whether sensitized lymphocytes are circulating and then what sensitivities to a given variety of tissue antigens are present.

Lymphocyte Surface Markers

Surface markers on human lymphocytes make possible their identification as T or B cells (Table 20-2). Because of their surface receptors for sheep red blood cells, T lymphocytes can be identified by the **E (erythrocyte) rosette** test. When human T lymphocytes from peripheral blood and sheep red blood cells are incubated together for a short time, the T cells bind the sheep cells to form rosettes (Fig. 20-8). B lymphocytes in human peripheral blood can be identified by the **B cell immunofluorescence test** in which fluorescent antiimmunoglobulin serum demonstrates the surface immunoglobulin on B cells. Receptors for complement on the surface of some B cells are surface markers also demonstrable by a rosette test. Red blood cells of an ox, *not* a sheep, are coated with anti–red cell antibody (IgM antibody to red cells) and complement and then incubated with separated lymphocytes. Rosettes form around the B cells.

Table 20-2 Identification of Certain Immunocompetent Cells

Procedure	Lymphocytes		Monocytes/ Macrophages
	T	B	
Rosettes formed with sheep red blood cells	Yes	No	No
Surface membrane IgM and IgD present	No	Yes	No
Membrane Fc receptor (IgG) present	Yes (in some)	Yes	Yes
Surface receptor for complement present	No	Yes	Yes

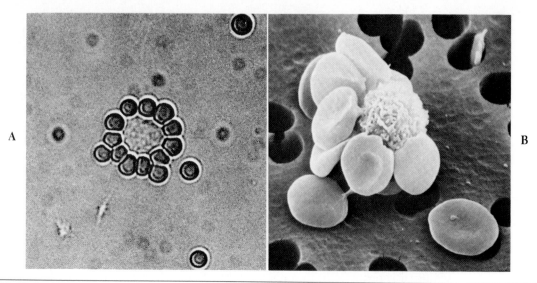

FIG. 20-8   Human T cell rosettes. **A,** Human T lymphocyte encircled by red blood cells from sheep as seen in a hemocytometer (instrument for counting red blood cells). **B,** Human T lymphocyte surrounded by red blood cells from sheep as seen in scanning electron micrograph. (Reproduced with permission of E.M. Block, Ph.D., The Upjohn Co., Kalamazoo, MI.)

Table 20-3 Immunologic Approach to Disease

Test(s)	Application	Test(s)	Application
Serologic		**Intradermal (Skin)**	
1. Complement fixation	Diagnosis of syphilis, viral infections, and certain bacterial infections	1. Cell-mediated immunity to infection with:	Diagnosis of infection, past or present, caused by microorganisms of:
2. Flocculation	Diagnosis of syphilis	a. Bacteria	
3. Precipitation	Diagnosis of syphilis	Lepromin	Leprosy
Protein precipitation	Detection of origin of bloodstains	Mallein	Glanders
	Detection of meat adulteration	Tuberculin	Tuberculosis
4. Agglutination	Identification (serologic typing) of bacteria; diagnosis of undulant fever, tularemia, and salmonelloses	b. Fungi	
		Blastomycin	Blastomycosis
a. Widal	Diagnosis of typhoid fever	Coccidioidin	Coccidioidomycosis
b. Weil-Felix	Diagnosis of typhus fever and Rocky Mountain spotted fever	Cryptococcin	Cryptococcosis
		Histoplasmin	Histoplasmosis
c. Paul-Bunnell (heterophil antibody)	Diagnosis of infectious mononucleosis	Spherulin	Coccidioidomycosis
		Sporotrichin	Sporotrichosis
		Trichophytin	Epidermophytosis
d. Viral hemagglutination-inhibition	Diagnosis of viral infections—rubella, rubeola, influenza, parainfluenzal infection	c. Protozoa	
		Leishmanin	Leishmaniasis
		d. Viruses	
	Diagnosis of viral infections	Mumps	Mumps
5. Virus neutralization	Identification of protozoa, fungi, bacteria, and viruses; diagnosis of rabies, syphilis, amebiasis	2. Susceptibility	
6. Fluorescent antibody		Schick	Susceptibility to diphtheria
		Dick	Susceptibility to scarlet fever

APPLICATION OF IMMUNOLOGIC METHODS

Immunologic methods are important in the diagnosis and management of bacterial infections, but in the identification of viral, rickettsial, and many parasitic infections the value of immunoassays of all kinds cannot be matched.

Some of the more frequently used immunologic methods for the identification and study of microbes and for the definition of their role in disease production are given in Table 20-3.

IMMUNOLOGY OF RED BLOOD CELL

The red blood cell (Fig. 20-9) is composed of a compact stromal protein that supports molecules of lipid and hemoglobin. Antigens of the blood groups are contained in both the stroma and on the surface of the red cells. On the surface, exposed antigens are ready to react with specific antibodies. The spatial features of the relatively small antigenic groups out on the cell surface are such that each red cell can have a large number and variety of them.

Agglutinins of the Blood

The blood of every person falls into one of four ABO blood groups, depending on the ability of that person's serum to agglutinate the cells of persons whose blood is in another group (Fig. 20-10). This is the result of the distribution of two agglutinins (antibodies) in the serum and two agglutinogens (antigens) on the red blood cells. The

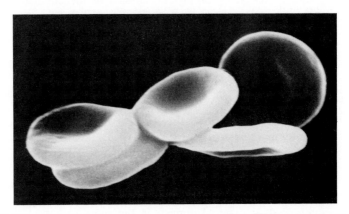

FIG. 20-9   Red blood cells (erythrocytes) are seen as biconcave disks in scanning electron micrograph. ($\times$9000.)

FIG. 20-10   Technic of blood grouping (typing). Make suspension in isotonic saline solution of red blood cells from blood to be typed. Mark out two circles (wells) on microslide; label A and B. Place drop of test suspension in each. In well labeled B, mix in drop of serum from person of blood group A (anti-B serum); in well labeled A, mix in drop of serum from person of blood group B (anti-A serum). Macroscopic agglutination of red blood cells in drop of typing serum seen in A. No agglutination in B. Well A contains anti-A typing serum and well B, anti-B serum. What type of blood do we have?

agglutinins are known as **anti-A** and **anti-B**. The agglutinogens are designated A and B. Agglutinin anti-A causes cells containing agglutinogen A to clump. Agglutinin anti-B causes cells containing agglutinogen B to clump. The blood groups and their agglutinin and agglutinogen content follow:

Group AB	Serum contains no agglutinin; cells contain agglutinogens A and B
Group A	Serum contains agglutinin anti-B; cells contain agglutinogen A
Group B	Serum contains agglutinin anti-A; cells contain agglutinogen B
Group O	Serum contains agglutinins anti-A and anti-B; cells contain no agglutinogen (Table 20-4)

If a person whose blood belongs to one group receives a transfusion of blood from a donor of another group, a hemolytic transfusion reaction is likely to occur. The reason is that the serum of the recipient may agglutinate the cells of the donor, or the serum of the donor may agglutinate the cells of the recipient. Reactions to transfusion of mismatched blood are associated with agglutination of red cells in the circulation, hemolysis of red cells, and liberation of free hemoglobin into the plasma; this leads to hemoglobinuria, fever, prostration, failure of kidney function, and in some cases death. A reaction is much more likely to occur if the recipient's serum agglutinates the donor's cells than if the donor's serum agglutinates the recipient's cells because the donor's serum is considerably diluted in the circulation of the recipient. This effect tends to minimize the agglutinating capacity of the donor's serum. For kinds of transfusion reactions, see Table 20-5.

The Rh Factor

An agglutinogen with no relation to the ABO blood groups is the Rh factor. When a guinea pig is given repeated injections of the red blood cells of a rhesus monkey, its serum acquires the ability to agglutinate the red cells of rhesus monkeys and also to agglutinate the red cells of most human beings (85% of the white race; more in others). The red blood cells that are agglutinated possess the Rh factor, and the person from whom the cells were removed is said to be **Rh positive**. Persons who do not have the Rh factor on their red blood cells are **Rh negative**. Actually the Rh factor encompasses a system of many antigens. The six most commonly detected are designated by capital and lower case letters as follows: C, D, E, c, d, e. The D antigen is the most potent of these and is synonymous with the Rh antigen or factor when only one is referred to. (An Rh-positive person possesses this blood antigen; an Rh-negative person lacks it.)

The plasma of Rh-negative persons does not carry agglutinins against the Rh factor, but such agglutinins may develop if blood transfusions of Rh-positive blood are given to

Table 20-4 Determination of Blood Type from Agglutination of Red Blood Cells in Specific Typing Serum

If Agglutination of Red Cells Occurs in		Then Blood Type Is	How Many People in the United States with It	
Anti-A Serum	Anti-B Serum		If It Is Rh Positive	If It Is Rh Negative
−	−	O	1 in 3	1 in 15
+	−	A	1 in 3	1 in 16
−	+	B	1 in 12	1 in 67
+	+	AB	1 in 29	1 in 167

these people. Such incompatible blood transfusions, if continued, may lead to serious reactions.

If an Rh-negative woman and an Rh-positive man have children, one or more of the children will probably be Rh positive. The Rh-positive cells of the fetus can sometimes get into the bloodstream of the mother during gestation but usually in such small amounts that they do not induce antibody formation. The Rh-negative mother likely has anti-D antibodies against the Rh factor during a given pregnancy because she was immunized in a previous one. When the Rh-positive cells of the baby enter the mother's circulation at time of delivery, they can be most effective in stimulating antibody formation. Antibodies formed against the Rh-positive cells begin to appear about 6 weeks later.

When antibodies against Rh factor are present in the plasma of the Rh-negative mother, they freely pass through the placenta to attack the red blood cells of the fetus and destroy them. As a consequence, the child may be born with **erythroblastosis fetalis,** or **hemolytic disease of the newborn.** The disease afflicting the fetus before it is born or within the first few days of postnatal life is manifested by anemia, jaundice, edema, and enlargement of the infant's spleen and liver. As would be expected, erythroblastosis is rare in the first pregnancy, but once it happens, the condition recurs in about 80% of subsequent pregnancies.

Designated the Rh "vaccine," an immunizing agent Rh_O (D) immune globulin (human) (trade name RhoGAM) is available to prevent a susceptible Rh-negative mother

Table 20-5 Outline of Transfusion Reactions

Kind	Cause	Clinical Features
Hemolytic	Mismatched transfusion (for example, blood type A given to patient with blood type O)	Severe chill; lumbar pain; nausea and vomiting; fever; suppression of urine (urine may be reddish brown); jaundice; death
Pyrogenic	Fever-producing substance in blood (sterile chemical contaminants, bacterial toxic products, antibodies to white blood cells)	Occurs 30 to 60 minutes after transfusion; flushing; nausea and vomiting; headache; muscular aches and pains; chills and fever
Contaminated blood	Break in aseptic technic; bacteria (usually gram-negative rods) introduced at time blood collected or with use of unsterile equipment	Chills, fever; generalized aching and pain; skin quite red; drop in blood pressure; shock
Allergic	Occurs in patients with history of allergy; cause unknown	Occurs within 1 or 2 hours of transfusion; itching of skin; hives or definite rash; swelling of face and lips; sometimes asthma
Sensitivity to donor white blood cells, platelets, or plasma	Multiple transfusions favor development of leukoagglutinins (those to white cells) and similar substances	Chill and fever; headache; malaise, serum may agglutinate white blood cell suspension from donor
Circulatory overload	Blood transfused too rapidly into person with failing circulation; likely to occur in elderly, debilitated, or cardiac patients, also in children	Difficulty in breathing; cyanosis; cough with frothy, blood-tinged sputum; heart failure
Embolic	Infusion of air: (1) transfusion under pressure, (2) tubing not completely filled before venipuncture	Sudden onset of cough, cyanosis, syncope, convulsions

from developing anti-Rh antibodies. It is human gamma-globulin containing anti-Rh-antibodies obtained from sensitized Rh-negative mothers who have given birth to erythroblastotic babies. Injected as a single dose within 72 hours of delivery, it cancels out the antigenic effect of the Rh-positive cells that have entered the circulation. Immune mechanisms for the production of Rh antibodies are blocked. In the event of future pregnancy with an Rh-positive fetus, the immunologic basis for hemolytic disease of the newborn has been eliminated.

Blood Grouping (Typing)

Human blood is routinely grouped (or typed) in modern blood bank laboratories into one of the four major blood groups, and the presence or absence of the Rh factor is determined. Blood is typed as O, A, B, or AB and as either Rh positive or Rh negative. However, the red blood cell is a very complex structure and posesses many antigens or factors (at least 60), and new ones are constantly being discovered. As antigens are studied, an attempt is made to classify them into categories containing related antigens, designated **systems.** At least 15 well-known systems of blood groups include (1) ABO, (2) MNS, (3) Rh, (4) P, (5) Lewis, (6) Kell, (7) Lutheran, (8) Duffy, (9) Kidd, (10) Sutter, (11) Vel, (12) Diego, (13) I, (14) Auberger, and (15) Xg (the only sex-linked blood group).

Cross-Matching

In routine blood grouping or typing for blood transfusion, the presence or absence of antigens in all of these systems is not determined. But it is important to know whether blood from a donor will be safe to give to a recipient, most likely a patient with some medical or surgical condition. Certain technics have been devised to determine the relative safety of such a procedure and to prevent a blood transfusion reaction. The first step is called the **major cross-match,** and in this the red blood cells of the donor are mixed with the serum of the person to receive the blood. The mixture is carefully observed for any sign of clumping of the red blood cells, with any agglutination meaning that the donor blood is incompatible with the blood of the recipient. If the recipient were to receive the blood even in small amounts, a transfusion reaction could easily occur.

The second step is the **minor cross-match,** in which red blood cells, this time from the recipient, are admixed with serum from the donor. Again the mixture is carefully observed for any sign of clumping of the red blood cells, which would mean an incompatibility of the two bloods. Agglutination at this step indicates the possibility of a transfusion reaction should the donor blood be transfused into the recipient, but the reaction would probably not be so severe as an incompatibility picked up in the major cross-match.

A third step is further indicated to detect unusual or uncommon antibodies that for some reason are not demonstrated in the major or minor cross-match. In many blood bank laboratories this is accomplished by the use of the **Coombs' test** (also known as the antiglobulin test). If we remember that antibodies are primarily related to the globulin fraction of the plasma proteins, and if we know that the protein globulin itself can function as an antigen to stimulate the production of antibodies, then it should be easy to understand that an antihuman globulin serum (Coombs' serum) can be prepared by the successive injection of human serum into a suitable animal such as the rabbit or goat. The serum of this immunized animal will then contain an antibody against the globulin of human serum. Since this fraction contains the antibodies, it means that Coombs' serum will also have an action against antibodies, an action, as one can see, of a general or nonspecific nature. The Coombs' test does not in any way identify specific antibodies; its chief use lies in the fact that it can detect their presence (Fig. 20-11). It helps indicate the presence of antibodies even when an antigen-antibody combination has not resulted in an observable reaction. Such a test is of immeasurable value in blood typing and, on the whole, is easily carried out. If this test is negative when donor red blood cells and recipient serum are used in the procedure and if the major and minor cross-matches show no incompatibility, it is then considered safe for the donor blood to be given as a transfusion to the recipient in question.

The direct Coombs' or direct antiglobulin test (DAT) detects antibodies. The indirect Coombs' or indirect antiglobulin test is performed to detect either blood group antibodies in serum of patients or to determine certain blood group antigens.

Blood Component Therapy

Whole blood is a complex mixture of cellular and fluid constituents that may be safely and efficiently separated in the modern blood bank. Since in many instances the physiologic role of each component is known, it may be administered to a patient to fill a specific need. This is often a more efficient use of a precious commodity. Not only is

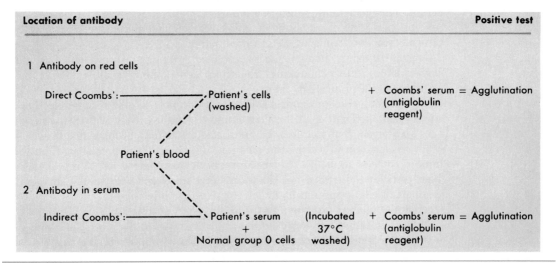

FIG. 20-11   Coombs' antiglobulin test (direct and indirect), outline. The antiglobulin test (patient's cells plus Coombs' serum) indicating presence of antibodies on red blood cells is *direct* Coombs' test. *Indirect* Coombs' test (patient's serum plus O cells plus Coombs' serum) identifies presence of antibodies in patient's serum. No agglutination is negative test. (Modified from Bove, J.R.: J.A.M.A. **200**:459, 1967.)

Table 20-6 Important Blood Components

Name	Content	Indication for Use
Platelet concentrate	Platelets, few white blood cells, some plasma	Bleeding because of platelet deficiency
Fresh frozen plasma	Plasma, all factors required for blood to clot; no platelets	Treatment of bleeding disorders
Cryoprecipitate	Certain blood clotting factors (I, VIII, XIII)	Treatment of certain bleeding diseases (for example, one kind of hemophilia)
Leukocyte concentrate	White blood cells, red blood cells, few platelets	Agranulocytosis (marked reduction of white blood cells)
Packed red blood cells	Red blood cells, some plasma, variable numbers of white blood cells	Replacement of red cell mass (to correct anemia)

component therapy useful in filling a specific need but also it allows for a concentration of the therapeutic effect not possible with whole blood transfusion. A good example of this is the treatment in a bleeding episode of the hemophiliac who needs antihemophilic globulin. The amount it takes to stop bleeding may have to be harvested from many units of whole blood; on occasions over 100 units have been required. This much antihemophilic globulin could not be supplied to such a bleeder in whole blood units. The technic of removing plasma from a unit of blood and immediately returning the red cells to the donor is **plasmapheresis.** Table 20-6 outlines applications of blood component therapy.

SUMMARY

1 Because of the specificity of the antigen-antibody reaction, if either specific antigen or antibody is known, the other can be identified and usually quantitated. Serology is the study in the clinical laboratory of antigen-antibody reactions in a specimen of serum. The quantitating unit is the titer.

2 The physical state of the available antigen largely determines the type of antigen-antibody reaction. Antigens are of varied nature.

3 Complement-fixing, neutralizing or antitoxic, precipitating, cytolytic, agglutinating, opsonic, virus-neutralizing, and fluorescent antibodies are most commonly tested for in the clinical laboratory.

4 In precipitation reactions antigens detected are usually soluble macromolecules and multivalent. Precipitins are bivalent immunoglobulins, each with two combining sites for antigen, that are nicely demonstrated by technics of immunodiffusion.

5 With precipitation in a semisolid medium, distinct bands of precipitate form where antigen contacts antibody. This reaction can be quantitated.

6 Inert particles and treated red blood cells that are coated with soluble antigens agglutinate or clump specifically to detect pathogenic microbes.

7 Technics of indirect or passive agglutination detect as little as 10 ng of antibody per milliliter. Precipitin tests detect in the range of 20 μg/ml.

8 The combination of fluorescein-tagged antibodies with their specific antigens forms a precipitate that can be viewed as a luminous area under ultraviolet light.

9 Electroimmunodiffusion is single gel diffusion (the antigen-antibody precipitin reaction in semisolid media) in an electric field. In double gel diffusion antibody is placed in a trench cut in the agarose gel, not incorporated throughout the medium. Both soluble antigen and antibody diffuse out in the agar to precipitate in an arc at the point of contact. From a mixture of antigens, each specific combination tends to be a separate band at a definite position.

10 The Ouchterlony technic for double diffusion detects the number of antigens in mixtures and the diversity of antigens interacting with a single antibody.

11 Radioimmunoassary (RIA) is based on competition under test conditions for a limited number of binding sites in a fixed amount of specific antibody between a fixed level of purified antigen labeled with radioactive iodine and an unknown, unlabeled test sample of antigen. The antigen-antibody complexes are separated and the amount of radioactivity determined. The concentration of the unknown antigen is then calculated.

12 An enzyme immunoassay known as ELISA is similar in design to RIA but uses enzyme-labeled reagents rather than radioactive ones and relies on color production to indicate test results.

13 Skin testing with recall antigens helps to determine the level of cellular immunity, since most normal adults have positive skin reactions for certain standard tests antigens.

14 Surface markers on human lymphocytes make possible the identification of T or B

cells. Approximately 70% of peripheral blood lymphocytes are T cells; some 25% are B cells. A small number of the lymphocytes are undefined as either T or B cells.

15 There are four blood groups. Persons with group AB blood contain agglutinogens (antigens) A and B on their red blood cells and no agglutinin (antibody) in their serum. Persons with blood group O contain no agglutinogens on their red cells but contain anti-A and anti-B agglutinins in their serum. Persons with blood group A contain agglutinogen A on their red cells and agglutinin anti-B in their serum. Those with blood group B contain agglutinogen B on their red cells and agglutinin anti-A in their serum.

16 A person receiving mismatched blood as a transfusion is at life-threatening risk for a hemolytic transfusion reaction.

17 The D antigen in the Rh system is the one that is most often referred to when a person is designated as Rh positive (with the D antigen) or Rh negative (without the D antigen).

18 The Rh vaccine prevents a susceptible Rh-negative mother from developing Rh antibodies. A single dose within 72 hours of delivery cancels out the antigenic effect of any Rh-positive cells that enter her circulation.

19 Antihuman globulin serum (Coombs' serum) in the direct antiglobulin test (Coombs' test) detects antibodies even when the antigen-antibody combination may not be visible.

QUESTIONS FOR REVIEW

1 What is an agglutination test? Name diseases in which it is used as a diagnostic procedure.

2 Briefly define antitoxin, cytolysin, agglutinin, precipitation, precipitin, serology, heterophil antibody, hemolysis, hemolysin, amboceptor, serum, hemagglutination, virus neutralization, flocculation, migration inhibitory factor, lymphopenia, blast formation, rosette, electrophoresis, fluorescence, and phytohemagglutinin.

3 How may antibodies be detected?

4 What is meant by the titer of a serum?

5 What are the advantages of the immunofluorescent technics?

6 Explain the difference between the direct and indirect fluorescent antibody test.

7 Briefly discuss immunoelectrophoresis and its applications.

8 Diagram a negative complement-fixation test. How is complement fixed in a *positive* test?

9 State ways in which cell-mediated immunity may be evaluated.

10 Discuss the applications of the precipitin reaction in semisolid media.

11 Describe the Ouchterlony technic.

12 Briefly discuss C-reactive protein as an acute phase protein.

13 What is meant by indirect hemagglutination? What is the role of latex particles or of bentonite clay in agglutination reactions?

14 Characterize an opsonin.

15 Compare B with T cells with respect to surface markers that may be used in identification of the two cells.

16 What practical application is made of the Coombs' test?

17 In the following diagram, place arrows pointing from the serum of each blood group to the cells agglutinated by that serum.

18 Why is a transfusion reaction more likely to occur when the donor's cells are agglutinated by the recipient's serum than when the recipient's cells are agglutinated by the donor's serum?

19 List the different kinds of transfusion reactions. Briefly explain.

SUGGESTED READINGS

Anhalt, J.P., and others: Detection of microbial antigens by counterimmunoelectrophoresis, Washington, D.C., 1978, American Society for Microbiology.

Deodhar, S.D., and Valenzuela, R.: C-reactive protein: new findings and specific applications, Lab. Manage. **19**:47, June 1981.

Detrick-Hooks, B., and Bernard, A.: T and B lymphocytes: assays in malignant and nonmalignant disease, Lab. Manage. **19**:41, May 1981.

Friedman, H., and Specter, S.: Microbiology and serology: progress in automation, Lab. Manage. **19**:37, Sept. 1981.

Guttmann, R.D., and others: Immunology, a Scope publication, Kalamazoo, Mich., 1981, The Upjohn Company.

Hamilton, R.G., and Adkinson, M.F.: Clinical laboratory methods in allergic disease, Lab. Manage. **21**:37, Dec. 1983.

Harmening, D.M.: Blood preservation: a look to the future, Lab. Med. **9**:6, Dec. 1978.

Jabs, D., and others: Identifying human lymphocyte subpopulations, Lab. Manage. **18**:33, Jan. 1980.

Klein, T., and others: Newer immunoassays in clinical bacteriology, Lab. Manage. **21**:37, Nov. 1983.

Leder, P.: Genetic control of immunoglobulin production, Hosp. Pract. **18**:73, Feb. 1983.

Marx, J.L.: Chemical signals in the immune system, Science **221**:1362, 1983.

Marx, J.L.: The T cell receptor—at hand at last, Science **221**:444, 1983.

Myhre, B.A., and Van Antwerp, R.: Diagnosis of unexpected reactions to transfusion, Lab. Med. **14**:153, 1983.

Normansell, D.E.: Medical laboratory immunology: future trends, ASM News **49**:217, 1983.

Waldman, A.A., and Stryker, M.H.: Blood components: preparation and quality control, Lab. Manage. **19**:43, Feb. 1981.

IMMUNIZING BIOLOGICALS

The two terms *vaccine* and *immune serum* are often confused, any product of this nature being spoken of as a "serum." Vaccines and immune serums differ in fundamental properties, method of production, and resultant type of immunity.

A **vaccine** is the causative agent of a disease (toxin or microbe such as bacterium or virus) so modified as to be incapable of producing the disease yet at the same time so little changed that it is able, when introduced into the body, to elicit production of specific antibodies against the disease. Vaccines are always antigens; therefore they always induce active immunity and are most useful in the *prevention* of disease.

An **immune serum** is the serum of a human or other animal that has been immunized against a given infectious disease. The salient feature of an immune serum pertains to the content of antibodies. Immune serums confer passive immunity; the antibodies therein do *not* stimulate further antibody production.

Both vaccines and immune serums are specific in their action; that is, they induce immunity to no disease other than the one for which they are prepared.

In this chapter are presented the biologic products that have been standardized and are available. The administration of certain of these for the production of passive immunity is given. In the next chapter are recommended schedules for the production of active immunity.

IMMUNE SERUMS (PASSIVE IMMUNIZATION)

Passive immunization induced by the administration of an immune serum makes for short-term immunity when active, long-term immunity either has not been induced before exposure to the disease or is unavailable. Remember that passive immunization is of variable effectiveness and of variable duration (1 to 6 weeks) and, when the immune product is not of human origin, can be associated with highly *un*desirable reactions. Since animal immune serum represents proteins foreign to the human being, a significant risk goes with its use; hence, wherever feasible, human immune serum has replaced the animal one. Animal serums must still be used in the following situations:

(1) for passive immunization with diphtheria, botulism, and gas gangrene; (2) for passive immunization with tetanus and rabies when special human gamma-globulin is not at hand; (3) for immune protection after snake and spider bites; and (4) for immunosuppression in selected instances (for example, a disorder in which antilymphocyte serum is indicated).

Preparations used for passive immunization include animal antitoxins or antiserums, special human immune globulin for a specific illness, and standard human serum globulin (immune globulin) for general use.

Antitoxins

Antitoxins are immune serums that neutralize toxins. They may be prepared artificially, and they also develop in the body as a result of repeated slight infections. This is why some adults are immune to diphtheria. Antitoxins have no action on the bacteria that produce the toxins. For example, diphtheria antitoxin neutralizes diphtheria toxin in the tissues and body fluids but has no effect on the diphtheria bacilli growing in the throat and producing the toxin. Neutralization of the circulating toxin in the body does favor the defense mechanisms operating to eliminate the membrane formed in the throat.

Preparation

Antitoxins can be successfully prepared only against exotoxins (extracellular toxins). The antitoxins that have been used longest and saved the most lives are those against diphtheria and tetanus toxins. Both are prepared in the same general way. The bacteria are grown in liquid culture until a large amount of exotoxin has been released. The bacilli are then filtered from the medium and its toxin. It is necessary to determine the strength of each lot of toxin or antitoxin because two lots prepared exactly alike seldom have the same strength.

The unit of toxin is measured as the **minimum lethal dose (MLD),** the amount of toxin that when injected into a test animal will kill it in a prescribed time. For diphtheria toxin it is the amount that will kill a guinea pig weighing 250 g in 4 days, and for tetanus toxin, the amount that will kill a 350 g guinea pig in the same length of time. Tetanus toxin is so potent that 1 ml of a broth culture of tetanus bacilli may contain enough toxin to kill 75,000 guinea pigs. To determine the MLD, one begins with a very small dose and gives increasing amounts of toxin to a series of guinea pigs. The animals receiving small doses may not be affected or else become but slightly ill and recover; those receiving larger doses become ill, and some may die, but it is more than 4 days before death occurs. Finally, an animal receiving a still larger dose dies on the fourth day. The amount of toxin given this animal contains one MLD.

When the toxin of suitable strength is found, horses (or sometimes cattle) may be immunized with it or with toxoid (modified toxin, p. 429). First, a small dose is used; it is increased at each of several successive injections. The first dose of toxin may be preceded by an injection of antitoxin. At the time that experience has shown antitoxin production to be at its height, the horse is bled and the antitoxin strength of its serum tested. If the serum is found to contain sufficient antitoxin, it is further refined and purified for use. Refining serves to (1) concentrate the antitoxin and (2) eliminate some of the horse serum protein. Horse proteins can sensitize the recipient of the antitoxin. If an immune serum is given the second time to a sensitized person, anaphylactic shock may result. It is *not* the antibodies in antitoxins and other immune serums that lead to allergic manifestations but the **serum protein** of the animal used in preparing the immune serum.

Standardization

The following definitions pertaining to the standardization of antitoxins are given for reference:

1. *Standard antitoxin*—an antitoxin of known strength prepared, stored, and distributed according to precise specifications
2. *Unit of antitoxin*—an amount of antitoxin equivalent to 1 unit of standard antitoxin
    This definition is used instead of the original definition of Ehrlich—the amount of antitoxin required to neutralize 100 MLD of toxin—since toxin also contains variable quantities of toxoid. Although toxoid has no disease-producing capacity, it can combine with antitoxin. Different lots of antitoxin tested against different lots of toxin therefore have different strengths. The standard antitoxin unit used in this country contains sufficient antitoxin to neutralize 100 MLD of the particular toxin that Ehrlich used originally in establishing his unit of antitoxin.
3. *L+ dose of diphtheria toxin*—an amount of toxin that, when combined with 1 unit of antitoxin, causes the death of a 250 g guinea pig on the fourth day

Three methods are used to standardize antitoxins: (1) animal protection tests, (2) skin reactions, and (3) flocculation tests. In 1896 Paul Ehrlich introduced methods of standardizing toxins and antitoxins. To him goes the credit for the concept of MLD.

Diphtheria antitoxin is an example of an antitoxin that may be standardized by the determination of its protective action against diphtheria toxin in a susceptible animal, in the following way. Different amounts of diphtheria toxin are mixed with 1 unit of standard diphtheria antitoxin, and the mixtures are tested in 250 g guinea pigs. The L+ dose of toxin as thus determined is mixed with different amounts of the antitoxin being tested. These are injected into 250 g guinea pigs, and the mixture that causes the death of a guinea pig on the fourth day obviously contains 1 unit of antitoxin. Tetanus antitoxin is standardized in much the same manner as diphtheria antitoxin, but different amounts of antitoxin and toxin are used. Botulism and gas gangrene antitoxins are standardized similarly, but the mouse is the test animal.

Standardization of antitoxins by skin reactions is based on the same principles as standardization by animal protection tests. The toxin and antitoxin are injected into the skin of rabbits or guinea pigs, and the production of a skin reaction has the same significance as the death of a guinea pig in the guinea pig antitoxin-toxin injection method.

The flocculation method depends on perceptible flocculation in a test tube when antitoxin and toxin are brought together in certain proportions.

Since antitoxin can be successfully prepared only against the few organisms producing extracellular toxins, the number of antitoxins is limited. Generally speaking, all antitoxins should be given *early* in the disease and *in sufficient amount* because they cannot repair injury already done.

Listing of Antitoxins

Botulinal Equine Antitoxin, types A, B, and E, is a refined and concentrated preparation distributed and stored by the Centers for Disease Control in Atlanta (pp. 580-581). It is given to prevent damage from toxin not already taken up by the central nervous system. Therefore it is of value when given before manifestations have appeared but of little effect afterward. Antitoxin against one type of *Clostridium botulinum* is ineffective against the toxin of the other types. Therefore, to be therapeutically expedient, antitoxin must be available against toxins of the A, B, and E types of bacilli, the ones causing human disease.

Before receiving the antitoxin, a person must be **skin tested**—about 15% of persons receiving the antitoxin show allergic reactions. For the treatment of botulism it is best to give large doses early, with the first one given intravenously and an additional dose intramuscularly. Most antitoxin is given in 2 to 4 hours if indicated. For prophylaxis in a person who has consumed suspect food, antitoxin is given intramuscularly, and the person is watched for signs of botulism.

Diphtheria equine antitoxin (purified, concentrated globulin, equine) is a therapeutic agent that has given possibly more brilliant results than any other. By its use, much suffering has been prevented, and many lives have been saved. In the production of diphtheria antitoxin in horses, toxoid has replaced diphtheria toxin as the immunizing agent. Refinements in manufacture give a diphtheria antitoxin that contains more than 5000 units/ml and is less likely to trigger serum reactions than previous antitoxins, since in production a very high proportion of the horse serum protein is removed.

Diphtheria antitoxin is *not* effective against toxin that has combined with the body cells, and no amount of antitoxin given late in the disease can repair injury already incurred. For best results, the disease must be recognized early, and sufficient antitoxin must be given at once. After subcutaneous administration, the antitoxin content of the blood does not reach its maximum for 72 hours. Injections must therefore be made intravenously (or intramuscularly).

The Committee on Infectious Diseases of the American Academy of Pediatrics advises that antitoxin be given as soon as the clinical diagnosis of diphtheria is made, with *no delay* even in waiting for bacteriologic results. The site of the membrane, the degree of toxicity, and the length of illness are much more reliable guidelines for determining the dose of diphtheria antitoxin than the weight and age of the patient. (Children and adults receive the same doses.) The Committee's schedule for the administration of diphtheria antitoxin is given in Table 21-1.

NOTE: When diphtheria antitoxin is given, *remember that a serum reaction is a possibility*, especially when the antitoxin is given intravenously. Preliminary testing for serum hypersensitivity is mandatory.

The symptom-free, nonimmunized contacts of the diphtheria patient need not receive antitoxin, provided that daily surveillance of these persons can be maintained, active immunization started, and prophylactic erythromycin or penicillin given to them.

Although penicillin G and erythromycin have no effect on the toxin of *Corynebacterium diphtheriae*, they do affect the microbe. The administration of these antibiotics is a valuable adjunct to serum therapy, never a substitute. These antimicrobials reduce the number of secondary invaders, decrease the severity and length of illness, and help to prevent the carrier state.

Tetanus antitoxin (TAT) or **antitetanus serum** (ATS) is a product usually obtained from horses, but it may also be produced in cattle. Although patients have undoubtedly benefited by its use, to the person allergic to horse serum there is a real hazard. **Tetanus immune globulin (human)** (TIG[H]) is prepared from the blood of persons actively immunized with tetanus toxoid. As a gamma globulin fraction of a hyperimmunized human being, it contains antitoxin without foreign protein. *Equine or bovine antitoxin is never recommended if human tetanus immune globulin is available.* Table 21-2 compares the human antitoxin with that of animal origin.

Table 21-1 Administration of Antitoxin in Diphtheria

	Duration of Illness		
	48 Hours	48 Hours	Over 48 Hours
Lesions	Throat and larynx	Membrane in nasopharynx	Brawny swelling of neck, or extensive disease (3 + days)
Dose of antitoxin (units)	20,000 to 40,000	40,000 to 60,000	80,000 to 120,000
Route of administration*	Intravenously	Intravenously	Intravenously

Recommendations (1982) of the Committee on Infectious Diseases of the American Academy of Pediatrics.
*Preferred route is intravenous (after eye and skin tests for hypersensitivity) to neutralize toxin rapidly. If patient reacts to antitoxin, desensitization is indicated.

If the tetanus-susceptible wound is seen immediately after injury and if it can be adequately cared for by standard surgical technics and antibiotics, most authorities believe that the risk involved in giving equine or bovine tetanus antitoxin is not justified regardless of the immune status of the patient. On the other hand, if the patient delays seeking medical attention for a day or more, or if the wound is one in which adequate surgical care is impossible because of its extent and nature, passive protection must be given to nonimmunized persons and to those patients who have failed to take their tetanus booster in the preceding 5 years.

Gas gangrene antitoxin is prepared against the important causative organisms. Polyvalent antitoxin may be of some value in clostridial infections, particularly with sepsis. Since its value is questioned, however, many surgeons do not give it.

Antivenins

Antivenins against the venoms of snakes and the black widow spider are prepared in the same general way as antitoxins, that is, by immunization of a horse with serial doses of the venom of the snake or spider. Antivenins may be prepared against a single species of snake or against a group of closely related species, depending on the geographic location in which it is to be used. **North American antisnakebite serum** (also known as Antivenin [Crotalidae] Polyvalent and crotaline antitoxin, polyvalent) is effective against rattlesnakes, copperheads, and cottonmouth moccasins, the most common poisonous snakes of North America. It is not effective against the venom of the coral snake. To combat that potent poison, **North American coral snake antivenin** (Antivenin [*Micrurus fulvius*] [Equine origin]) is distributed to state and local health departments through the Centers for Disease Control in Atlanta.

The use of antivenin in snakebite or black widow spider bite should in no way replace first aid and supportive measures.

Antibacterial Serums

Before the discovery of the antimicrobial compounds, antibacterial serums played a crucial role in the treatment of various infections. Among such were those against meningococci (antimeningococcal serum) and pneumococci (antipneumococcal serum). Many were prepared by injection of horses with the given bacteria. Rabbits were employed in the manufacture of antipneumococcal serum, and a specific serum for each

Table 21-2 Evaluation of Biologic Products for Passive Immunization in Tetanus

	Tetanus Antitoxin (TAT)	
Point of Comparison	Equine or Bovine	Human (Tetanus Immune Globulin [Human]—TIG[H])
Efficacy	Less effective	More effective, 10:1
Dosage	Usual—3000 to 10,000 units†	Usual adult dose—250 to 500 units
Duration of protective levels of antitoxin in recipient	5 days	Up to 30 days (detectable levels reported to 14 weeks)
Danger of allergic reaction	Present; estimated 6% to 7% or more	Remote
Cost	Less expensive	More expensive
Evaluation	Not advocated if human product available	Recommended in allergic persons and generally

*The majority of persons sensitive to horse serum are also sensitive to bovine serum.
†Recommendations vary as to dose of tetanus antitoxin; 500 units may be as effective according to some authorities.

known type of pneumococcus was prepared. A serum against *Haemophilus influenzae*, type b, was prepared by injection of rabbits with the bacilli and was formerly used in the treatment of meningitis caused by *Haemophilus influenzae*, type b. Antipertussis serum was prepared by injection of rabbits with *Bordetella pertussis* and its products and was used with success in very young children to prevent the disease in those exposed (when given early enough) and to decrease the severity of the established disease.

Antibacterial serums act to destroy bacteria by combining with a surface antigen, thereby rendering the bacteria more susceptible to leukocytes. A serum that acts on several strains of bacteria is **polyvalent;** one that acts on only one strain, **univalent.** Today the treatment of disease processes with antibacterial serums has largely been supplanted by antimicrobial therapy.

Table 21-3 Clinical Value of Immune Globulin (Gamma Globulin) (IG)

Disease	Indications	Preparation
Antibody deficiency disease (agammaglobulinemia, hypogammaglobuline- mia, dysgammaglobuli- nemia)*	Treatment	Immune globulin (human)
Rh hemolytic disease	Prevention	Rh_O (D) immune globulin (human)†
Tetanus	Prevention—children and adults Treatment	Tetanus immune globulin (human)
Varicella (chickenpox)	Prevention	Varicella immune globulin (human)‡
Viral hepatitis type A	Prevention or modification in children or adults 1. Short-term, moderate risk	Immune globulin (human)
	2. Long-term, intense exposure (also prophylaxis in institu- tions)	
Viral hepatitis type B (serum)	Prevention	Hepatitis B immune globulin (HBIG) (immune globulin can be used)

Compiled from the 1984 Physicians' Desk Reference, Oradell, N.J., 1984, Medical Economics Co., and the 1982 Red Book of available in the United States; (2) rubella—immune globulin does not prevent viremia, and infants with congenital rubella have exposed; in an outbreak, oral poliovirus vaccine should be given to all susceptibles older than 6 weeks of age; and (4) pertussis—
*See also p. 361.
†Rh_O (D) immune globulin (RhoGAM) is discussed on p. 413.
‡Zoster immune globulin (ZIG) and varicella-zoster immune globulin (VZIG), both investigational, are available on a limited
§HBIG titer of antibodies is said to be 50,000 times greater than that of immune globulin (not preselected for anti-HBs content); error in a susceptible person (anti-HBs negative). HBIG prepared from plasma is preselected for a high titer. In the United States

Antiviral Serums

The best-known antiviral serum is **antirabies serum,** which is used to establish an immunity directly after exposure and provide a passive immunity until an active one can be established by vaccination.

Antirabies serum has been prepared by injection of horses with rabies virus, but remember that with any hyperimmune serum of equine origin, the danger of an anaphylactic reaction or severe serum sickness always exists. Therefore the recommended antiserum today for rabies is the effective human one, **rabies immune globulin, human (HRIG),** which is obtained by immunization of human volunteers. (The human preparation replaced the serum prepared in the horse in the mid-1970s.) Table 22-2 presents the recommendations of the U.S. Public Health Service, Centers for Disease Control.

Dose	Schedule	Route	Comments
0.66 ml/kg body weight	One dose every 3 to 4 weeks (double dose at onset of therapy)	Intramuscular	Maximum dose 20 to 30 ml at any one time; gamma globulin deficiency here; adequate levels must be supplied to protect against infections
One vial, usually	One dose	Intramuscular	Give to nonsensitized Rh-negative mothers after delivery of Rh-positive infant or after abortion
250 to 500 units (4 to 5 units/kg body weight)	One dose	Intramuscular	Product from person hyperimmunized with tetanus toxoid; given where tetanus-prone wound in person with uncertain vaccination status and when such wound is more than 24 hours old
3000 to 10,000 units	May be repeated 1 month later		Active immunization should be started with tetanus toxoid
0.6 to 1.2 ml/kg body weight	One dose	Intramuscular	Give within 3 days of exposure; may decrease severity of illness
0.02 to 0.04 ml/kg body weight	One (repeat in 3 to 5 months if necessary)	Intramuscular	Give as soon as possible
0.06 ml/kg body weight	One (repeat in 5 to 6 months if necessary)		Protection for 10 to 12 months
0.06 ml/kg body weight (3 to 5 ml for adult)	One dose within 1 week after exposure; one dose 1 month later	Intramuscular	Give as soon as possible; antibodies persist 2 or more months§

the American Academy of Pediatrics. Recommendations are not listed for (1) measles—measles immune globulin is no longer been born to women given this shortly after exposure; (3) poliomyelitis—immune globulin is no longer recommended for persons immune globulin is no longer recommended by the American Academy of Pediatrics.

basis for patients with serious background disease.

HBIG is used for small-volume exposure to contaminated material, as occurs with needle stick, accidental splash, or pipetting HBIG has an anti-HBs titer greater than 1:100,000 as measured by radioimmunoassay.

Convalescent Serum

Convalescent serum therapy consists of the injection of the whole blood or serum of a person *recently* recovered from a disease into one ill with the disease (as a therapeutic measure) or into one exposed to the disease (as a preventive measure). In this way the blood of the convalescent patient containing antibodies confers a passive immunity on the recipient.

Caution should be observed in giving convalescent serum or any human serum because it may transmit viral hepatitis!

Gamma Globulin

Gamma globulin is the main globulin fraction of blood plasma with which antibodies are associated. For its commercial production, a series of precipitations in pooled normal adult plasma, in venous blood, or in pooled extracts of human placentas is carried out with varying concentrations of alcohol at a low temperature. The gamma globulin fraction removed is more than 40 times richer in antibodies than the original plasma from which it is taken. It is free of hepatitis viruses, and the contained antibodies are concentrated into a small volume for intramuscular injection. Five hundred milliliters of blood yield an average dose.

Gamma globulin, usually dispensed as immune serum (IG) (formerly immune serum globulin, human), is emphasized as indicated in Table 21-3. Gamma globulin is always given intramuscularly—never intravenously (Fig. 21-1). *The consequence can be a severe shocklike state.* As ordinarily given, gamma globulin is one of the benign injectables, being associated with few side reactions.

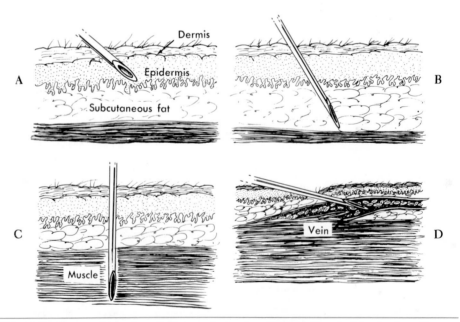

FIG. 21-1   Routes of injection, diagrammatic sketches. Note position and depth of needle. **A,** Intradermal—used for BCG vaccine and skin tests for allergy. **B,** Subcutaneous—used for measles-mumps-rubella (single or combined antigens), influenza, cholera, yellow fever, and typhoid fever vaccines. **C,** Intramuscular—used for plague, pertussis, and hepatitis B vaccines; diphtheria and tetanus toxoids; immune serums; and administration of antibiotics. **D,** Intravenous—used for diphtheria antitoxin.

PLATE 1

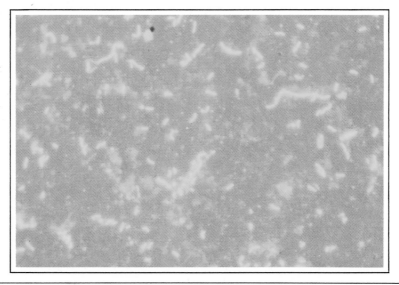

A Bacterial fluorescence in the yellow-green glow from fluorescent antibody staining of *Legionella pneumophilia*, cause of Legionnaires' disease. Background color is that of rhodamine counterstain.

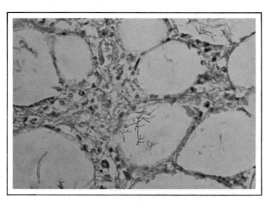

B Laboratory study of microbes. Direct smear: acid-fast stain shows filaments of *Nocardia asteroides* against a background stained with methylene blue.

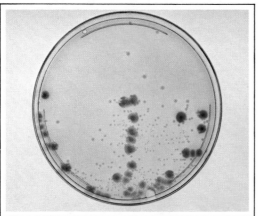

C Laboratory study of microbes. Culture: with the use of the differential medium MacConkey agar, red-purple colonies of lactose-fermenting *Escherichia coli* contrast with colorless colonies of nonlactose-fermenting *Shigella* species.

PLATE 2

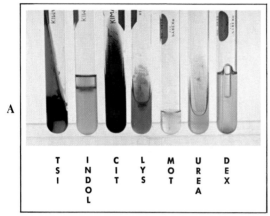

A

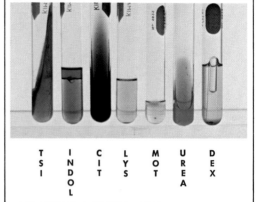

B

Laboratory study of microbes. Biochemical tests: key reactions for one gram-negative enteric microorganism (**A**) are tabulated and contrasted with those for another gram-negative enteric microorganism (**B**).

| Examples | Triple Sugar Iron Agar (TSI) | | | | INDOL | CIT | LYS | MOT | UREA | DEX |
	H₂S	Lactose	Sucrose	Glucose						
Salmonella species (A)	+ (black)	− (red slant)	−	Obscured	−	+ (blue)	+	+	−	Acid, gas (yellow)
Morganella morganii (B)	−	−	−	Beginning fermentation	+	− (green)	− (yellow)	+	+ (red-purple)	Acid, gas (yellow)

TSI, triple sugar iron agar, detects sugar fermentation and hydrogen sulfide production in gram-negative bacilli. TSI incorporates glucose, lactose, and sucrose with ferrous sulfate and phenol red indicator. (Phenol red indicator is yellow in an acid medium, red in an alkaline one.) TSI is prepared as a slant with a deep butt. The test microbe is passed across the slant surface and stabbed deep into the butt. After incubation the medium is inspected for an acid or alkaline reaction along the slant and in the butt, presence of gas, and production of hydrogen sulfide (*H₂S*), which is indicated by the blackening of the medium along the stab line where ferrous sulfide is formed. *INDOL* production is seen in a red color when Kovacs reagent (*p*-dimethylaminobenzaldehyde) is added to tryptophan broth. *CIT*, Simmons citrate agar with bromthymol blue indicator, identifies by a deep Prussian blue color a test microbe utilizing citrate as sole source of carbon. *LYS*, lysine decarboxylase test medium with cresol purple indicator, detects the enzymatic decarboxylation of lysine. By transmitted light, the positive test (an alkaline reaction) is red; the negative test (an acid reaction) is yellow. *MOT*, semisolid nutrient motility agar, demonstrates the growth of motile microbes out from the stab line. *UREA*, urease test medium with phenol red indicator, indicates the conversion of urea to ammonia. A positive test is the red-purple color; no change in the color of the medium is seen in the negative test. *DEX*, dextrose broth with indicator in a Durham fermentation tube, displays the fermentation of sugar by the yellow color (acid formed); gas collects in the inverted tube.

PLATE 3

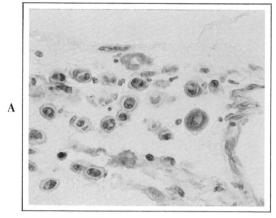

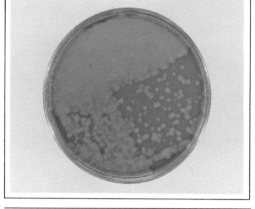

A Immunoperoxidase staining method applied to individual cells. This enzyme-linked immunoassay can identify colorimetrically a particular antigen in a cell. When the peroxidase-antiperoxidase enzyme system with color indicator is bound to the specific monoclonal antibody, the combination of that antibody with its cellular antigen produces a color where the antigen is present in the cell. The brown color of the positive reaction depicted here identifies the given antigen in the cell cytoplasm. (Courtesy Hector Battifora, M.D., Director of Surgical Pathology, City of Hope National Medical Center, Duarte, Calif.)

B Routine blood agar sets off beta-hemolytic *(clear)* zones around pinpoint gray colonies of *Streptococcus pyogenes.*

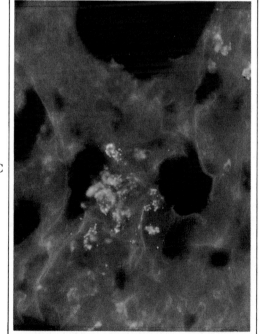

C The green glow of bacterial fluorescence from fluorescent antibody staining identifies *Francisella tularensis,* the cause of tularemia (rabbit fever). (From White, J.D., and McGavran, M.H.: J.A.M.A. **194:**294, 1965.)

D The selective medium of Petragnani favors the dry buff-colored growth of *Mycobacterium tuberculosis.* Background color comes from incorporation of malachite green, a dye inhibitory to other bacteria.

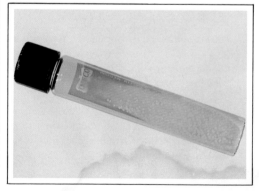

PLATE 4

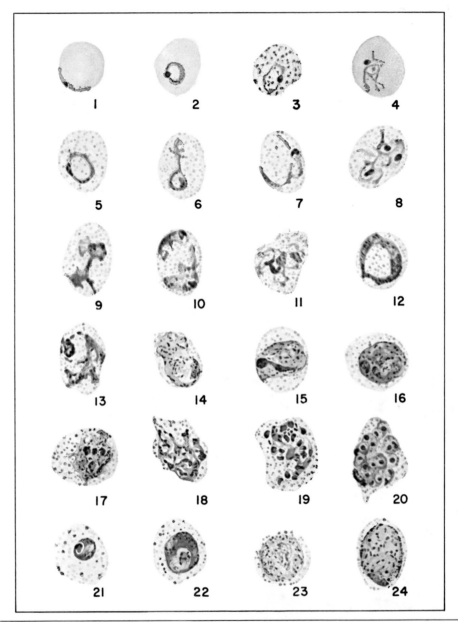

Plasmodium vivax. 1, Normal-sized red cell with ring form at margin. 2, Young ring form (trophozoite) in enlarged red cell. 3, Ring form in red cell showing basophilic stippling. 4, Red cell containing young parasite with pseudopods. 5, Ring form in enlarged cell containing Schüffner's dots. 6 to 14, Successive stages. 15, Mature trophozoite in process of division. 16 to 19, Schizonts showing steps in division (presegmentation). 20, Mature schizont. 21 and 22, Developing gametocytes. 23, Mature microgametocyte. 24, Mature macrogametocyte. (From Wilcox, A.: Manual of the microscopical diagnosis of malaria in man, Washington, D.C., 1960, Department of Health, Education, and Welfare, Public Health Service, U.S. Government Printing Office.)

VACCINES (ACTIVE IMMUNIZATION)
Bacterial Vaccines

Bacterial vaccines are suspensions of killed bacteria in isotonic sodium chloride solution. After culture for 24 to 48 hours, the bacterial growth is emulsified in sterile saline solution and the number of bacteria in each milliliter of the washings determined. Isotonic sodium chloride solution is added until the desired number per milliliter is obtained. The bacteria are then killed at the lowest possible lethal temperature in the shortest possible time. Without heat, they may be killed by formalin, cresol, thiomerosal (Merthiolate), or ultraviolet light. The finished product is cultured to verify sterility. Vaccines are usually prepared so that the number of dead microorganisms given at a single injection ranges from 100 million to 1 billion.

Bacterial vaccines are of two types: **stock,** made from stock cultures maintained in the laboratory, and **autogenous,** made from the organisms of a specific lesion in the patient for whom the vaccine is prepared. A **mixed** vaccine contains bacteria belonging to two or more species. A **polyvalent vaccine** contains several strains of organisms belonging to the same species, for instance, one containing several strains of *Pseudomonas aeruginosa*. Bacterial vaccines show their best results in the prevention of typhoid fever, whooping cough, and plague.

Prompted by the growing resistance of bacteria to standard antibiotics and aided by improvements in laboratory technology, researchers are presently concentrating on the development of bacterial vaccines. Major bacterial antigens, both polysaccharide and protein, have been isolated, purified, and studied, and, of special interest in vaccine production, certain avirulent bacterial hybrids have been formulated.

Listing of Bacterial Vaccines

Bacille Calmette-Guérin (BCG) vaccine (tuberculosis vaccine) is derived from a strain of bovine tubercle bacilli cultivated on artificial media (containing bile) so long that they have completely lost their virulence for humans. There are many BCG vaccines in the world today, quite variable in properties, but all derived from the original strain. The BCG strain of *Mycobacterium bovis* prepared by French scientists Albert Calmette and Camille Guérin was first given to a human in Paris in May 1921.

In the United States freeze-dried BCG vaccines are available in two concentrations, the higher for use in the multiple-puncture technic and the lower for intradermal inoculation. Administered either way, the vaccine is given *only to persons with a negative tuberculin test*. Afterward the tuberculin test remains positive for 2 or 3 years, gradually becoming negative or only weakly reactive over the years. Even though BCG-vaccinated when a child, an adult with a definitely positive tuberculin test should be considered to have had a second contact with tubercle bacilli, the second time an infectious one.

BCG vaccine has been used extensively in European countries to immunize children. The World Health Organization in recent years has vaccinated over 250 million persons with BCG. This has been an especially worthwhile program in parts of the world where the incidence of tuberculosis is high. In the United States, where the prevalence rate of tuberculosis is low, BCG vaccination, a controversial issue for over 50 years, has not been used widely.

Cholera vaccine given to the military personnel of the U.S. armed forces sent to duty in cholera-endemic areas contains 8 billion killed vibrios per milliliter. Cholera vaccine is toxic, of limited usefulness, and must be injected at frequent intervals to maintain immunity.

Since an oral vaccine inducing antibody formation in the intestinal tract at the actual site of disease is feasible, several such oral vaccines for cholera are under study. One, the **Texas Star vaccine,** is a live, chemically mutated version of the pathogen *Vibrio* *cholerae.* This mutant bacterium elaborates the part of the cholera toxin molecule that is responsible for the vibrio's immunogenicity.

Pertussis vaccine is a suspension of the inactivated organisms. Pertussis vaccine is commonly prepared and used in the aluminum phosphate–absorbed form in combination with diphtheria and tetanus toxoids. Severe neurologic reactions rarely complicate its administration.

Newer pertussis vaccines are partially purified and detoxified acellular or extract vaccines, containing antigens associated with the etiologic microorganism.

Plague vaccine of the United States armed forces contains 2 billion killed plague bacilli per milliliter. Currently plague vaccine is prepared from a culture of the etiologic agent *(Yersinia pestis)* in artificial media, inactivatead with formalin, and preserved in phenol. Plague vaccine is recommended for laboratory and field personnel at risk.

Pneumococcal polysaccharide vaccine, licensed in the United States in 1977, combines the purified capsular material, separately extracted, from 14 serotypes of *Streptococcus pneumoniae*—Danish types 1, 2, 3, 4, 6A, 7F, 8, 9N, 12F, 14, 18C, 19F, 23F, and 25. In the United States these serotypes are responsible for 68% of pneumococcal disease that is associated with bacteremia (bacteria in the bloodstream). Cross-reactivity and therefore protection may occur with certain serotypes not in the vaccine that are related immunologically to those that are. The vaccine is given only once to adults, since adverse reactions are seen with a second dose. It is recommended for persons at risk, such as (1) those over 2 years of age with absence of or dysfunction of the spleen, (2) those over 2 years of age with chronic illness likely to be complicated by pneumococcal disease, and (3) persons 50 years of age or older. It is definitely indicated in an immunodeficient patient.

Tularemia vaccine, live, attenuated, is a lyophilized, viable, weakened variant of *Francisella tularensis.* **Foshay's tularemia vaccine** is made of killed organisms and produces an immunity that lasts about 1 year. Tularemia vaccine has also been prepared by the U.S. Army as an inhalant-type vaccine. Large numbers of persons can be vaccinated against this disease simply by their marching through a room with vaccine droplets in the atmosphere. The interest of the military in this disease stems from its possibilities in bacteriologic warfare. Airborne immunization is being evaluated experimentally in other diseases transmitted by droplet infection.

Typhoid vaccine for the U.S. armed forces is an acetone killed-dried strain of *Salmonella typhi* (the AKD vaccine). An oral vaccine containing live attenuated typhoid bacilli from a mutant strain of *Salmonella typhi* is being tested. The U.S. Public Health Service discredits paratyphoid A and B vaccines as immunizing agents and has implicated them in reactions following administration of mixed vaccines.

Rickettsial (bacterial) vaccines in use at the present time are prepared from the culture of the respective rickettsias in the yolk sac of the developing chick embryo. A vaccine so prepared is known as a **Cox vaccine.** Cox vaccines have been made against epidemic typhus fever, murine typhus fever, Rocky Mountain spotted fever, and Q fever. An effective Rocky Mountain spotted fever vaccine has also been made from the pulverized bodies of infected ticks. In a new and improved Rocky Mountain spotted fever vaccine, prepared from formalin-killed rickettsias of a chick embryo culture, the egg yolk proteins and lipids of the older vaccine have been eliminated.

Vaccination against the rickettsial diseases is advised for persons likely to contact the infectious agent because of living conditions or occupation. Protection against epidemic typhus is strongly urged for those who have been in contact with a patient. Although typhus immunization is not required by any country in the world as a condition of entry, it is nevertheless recommended to travelers when travel plans include a geographic area that is not only infected but one in which living conditions are generally poor.

Toxoids

In 1923 Gatson Leon Ramon (1885-1963), French bacteriologist and veterinarian, prepared diphtheria toxoid. This agent soon supplanted the toxin-antitoxin combination used in immunization to that date, since it was not only devoid of the dangers of toxin-antitoxin but also was of superior immunizing capacity.

Diphtheria toxoid is diphtheria toxin detoxified so that it cannot cause disease but can induce formation of specific antitoxin. Formalin, 0.2% to 0.4%, is added to diphtheria toxin and the mixture incubated at 37° C until detoxication is complete (several weeks). This treatment, used with other bacterial toxins as well, reduces toxicity but preserves the antigenic qualities of toxin. After purification to remove inert protein, the preparation is available as **plain** or **fluid toxoid.** (Fluid toxoids are used in the United Kingdom and Canada.)

Alum added to diphtheria toxoid precipitates the antigenic portion. The precipitate, after being washed and suspended in sterile physiologic saline solution, is **diphtheria toxoid, alum precipitated.** Its advantage is that the alum is not absorbed but remains at the injection site. The toxoid slowly separates from it to give a prolonged antigenic stimulation. If aluminum hydroxide or aluminum phosphate is added to the liquid toxoid, the antigenic portion adheres to the particles of the aluminum compound. This process is known as **adsorption,** and diphtheria toxoid so treated is known as **aluminum hydroxide–** or **aluminum phosphate–adsorbed diphtheria toxoid** (depot toxoid or antigen). Its properties are similar to those of the alum-precipitated toxoid. Diphtheria toxoid should not be given to a person more than 10 to 12 years of age without preliminary sensitivity tests.

Tetanus toxoid establishes permanent immunity to tetanus. It is manufactured in the liquid (or fluid) and aluminum phosphate–adsorbed forms.

Most manufacturers market mixtures of diphtheria and tetanus toxoids (DT) or mixtures of diphtheria and tetanus toxoids and pertussis vaccine (DTP). Combinations are available in the unconcentrated and aluminum phosphate–adsorbed forms. They have proved to be very satisfactory.

Botulinum toxoid, an effective pentavalent preparation available for active immunization against botulism, incorporates aluminum phosphate–adsorbed toxoids for the antitoxins of *Clostridium botulinum,* types A, B, C, D, and E. It is given to laboratory workers at risk in three spaced injections followed by a booster.

Viral Vaccines

With the exception of the antibiotics, nothing has done more to protect us against infection than the unborn chick, because in 1931 a physiologist, Alice Miles Woodruff, working with fowlpox virus, showed that virus could be grown in embryonated hen's eggs. Her work was reported from the pathology laboratory at Vanderbilt University School of Medicine under the direction of E.W. Goodpasture.

Viral vaccines are possible because of the development of modern technics for cultivation of viruses in embryonated hen's eggs and also in cell cultures, thus furnishing the large supply of virus essential to the production of vaccines. In chick embryos or in cell cultures, cultivation of a virus involving multiple transfers from one medium to another (serial passage) can alter the organism's pathogenicity. A normally virulent virus can thus be **attenuated** (weakened) or domesticated. Fortunately, with loss of virulence there is *no* loss of antigenicity. This makes possible the production of very effective vaccines.

In a **killed** or **inactivated** virus vaccine the infectivity of the virus and its ability to reproduce have been destroyed physically or chemically, but its capacity to induce antibody formation has been preserved. Formalin is a standard inactivating agent. Killed virus vaccines must be injected in several doses, but live virus vaccines can be taken orally or inhaled (as an aerosol).

Virologists have long preferred living agents to killed or inactivated viruses for vaccines. A living agent continues to multiply in the body of the animal or human to which it is given, thus exerting a prolonged and increasingly strong stimulation to the host to make antibodies. The immunity so produced is stronger and longer lasting because the presence of the live virus stimulates actual infection. The possibility that the virus will revert to its original level of pathogenicity in the vaccinated person is an obvious disadvantage to the use of the living agent. Live viral vaccines are well known in veterinary medicine. The two best known in human medicine are the measles vaccine and the live poliovirus vaccine, Sabin-type strains.

Listing of Viral Vaccines

Adenovirus vaccine has been prepared by the formalin and ultraviolet inactivation of viral types grown on monkey kidney cells. Because of the discovery that some of the inactivated adenovirus serotypes produced tumors in experimental animals, active immunization with these agents is suspended. Trials with *live* adenovirus types 4 and 7 vaccine given orally in an enteric-coated tablet are in operation with good results, since live adenoviruses presumably have no capacity to cause cancers.

Hepatitis B vaccine is a suspension of the inactivated, alum-adsorbed surface antigen particles of the hepatitis B virus. The source of the antigen is plasma of asymptomatic persons with chronic hepatitis B, and such plasma is selected to contain high concentrations of noninfectious surface antigen particles but low concentrations of infectious hepatitis B virus particles. The ratio can be as great as 10,000 antigen particles to 1 viral particle. The vaccine is inactivated biophysically with ultracentrifugation and biochemically with digestion and formalin inactivation. The methods used have been shown to inactivate hepatitis B virus and representative viruses from all known virus groups.

Immediate side effects are minimal after administration of the vaccine; soreness at the site is the most common unfavorable reaction. No long-term reactions have been reported. No known cases of hepatitis B or of non-A, non-B hepatitis have been vaccine transmitted, and there is no known occurrence of AIDS associated with it.

Influenza virus vaccine is prepared from virus grown in the fertile hen's egg, inactivated with formalin or ultraviolet irradiation, and concentrated (Fig. 21-2). American flu vaccines (with an equal amount of viral protein) are either "whole-virus" or "split-virus" vaccines. In the processing of split-virus vaccine, in an additional step, an organic solvent or detergent disrupts or splits the viral protein. Some degree of antigenicity is lost from the vaccine, but fewer side reactions (sore arm, low-grade fever) with its use are the benefit, especially for children. In children under 13 years of age, split-virus vaccine is preferred.

The Bureau of Biologics, Food and Drug Administration, regularly reviews the formulation of influenza vaccines, suggesting changes as needed to take in the strains expected to cause trouble during the next flu season. Usually the strains of the recent epidemics are known, and the most practical arrangement from year to year seems to be one wherein two strains recently implicated are tagged for the vaccine. Generally, but not necessarily, a bivalent vaccine with greater amounts of antigen is preferred to a polyvalent one containing smaller amounts of four or five strains, but polyvalent vaccines are used when indicated. In the event that a large-scale epidemic is anticipated, a monovalent vaccine might be advised.

The peculiar ability of influenza viruses to change their antigenic structure from time to time poses a problem to the manufacturer of flu vaccines. A strain against which there is no protection in a current vaccine can easily emerge. Great care must be taken that strains containing a wide pattern of antigenic substances are selected for vaccine production.

The production of active immunity by influenza virus vaccine has met with considerable success. However, the immunity is short-lived, the vaccine sometimes gives fairly severe reactions, and the subject can become sensitized to egg protein, as well as react to a previously existing sensitivity. Highly purified vaccines eliminate most of the egg protein.

New vaccines being tested include an attenuated live virus vaccine, a recombinant administered intranasally (by nose drops), and a vaccine utilizing the subunit proteins hemagglutinin and neuraminidase in the viral envelope to induce antibody formation. Antibody to neuraminidase reduces the amount of virus replicating in the respiratory tract and lessens the likelihood of transmission to contacts.

The **Guillain-Barré syndrome** (polyradiculitis, acute febrile polyneuritis), sometimes called *French polio*, is the only known delayed reaction to influenza vaccination, appearing within 8 weeks after inoculation in one out of 100,000 vaccine recipients. A prominent feature of this acute paralytic, but self-limited, disorder is a symmetric weakness of the extremities with sensory loss. Recovery usually is complete. Events other than flu vaccination may trigger the appearance of this syndrome of unknown etiology, the most common of which is viral infection. Because of the possible link between the swine influenza vaccine and the development of the Guillain-Barré syndrome, the government-sponsored program of influenza immunization of 1976 was ended about 6 weeks after vaccinations were begun.

Measles virus vaccine is a chick embryo cell culture vaccine containing the Edmonston-Schwarz strain of *more attenuated* measles virus—virus attenuated beyond the level of the original Edmonston B strain, which is no longer distributed. Measles virus vaccine is available either as a monovalent vaccine or in combination with mumps and/or rubella vaccines.

An aerosol vaccine (one that is inhaled) has been prepared—a human diploid cell vaccine (HDC) containing the Ikić (Edmonston-Zagreb) strain of more attenuated measles virus.

FIG. 21-2 Influenza vaccine, commercial production. Virus is injected into fertile hen's eggs along an assembly line. (Courtesy Eli Lilly & Co., Indianapolis.)

Remember that live virus vaccines for measles give permanent protection, but inactivated virus vaccines for measles are in disrepute. NOTE: Measles vaccines must be carefully refrigerated.

Mumps virus vaccine, live, is an attenuated live virus vaccine adapted to the chick embryo. It contains egg proteins and a small quantity of neomycin. The duration of protection afforded by the mumps vaccine is unknown, but continued protection has been observed for at least 15 years.

Salk poliomyelitis vaccine, or **poliovirus vaccine, inactivated** (IPV), is a highly effective, formaldehyde-inactivated viral vaccine, prepared from growing all three types of poliovirus in cell cultures of monkey kidney. It must be injected.

Sabin poliomyelitis vaccine, or **poliovirus vaccine, live oral** (OPV), is prepared with the three types of attenuated live poliovirus grown in human (not monkey) cell culture. After the vaccine has been fed to an individual, the virus of the vaccine multiplies in the lining and in the lymphoid tissue of the alimentary tract. A satisfactory immune response from the vaccinated person occurs usually within 7 to 10 days. The implantation of the virus in the bowel blocks further infection by the same type of a wild poliovirus. This materially cuts down on the number of carriers. The fear that competition from other enteroviruses might check growth of poliovirus in the intestinal tract has been largely obliterated by studies of children in the tropics. Their alimentary tracts literally swarm with viruses and yet they show antibody responses indicating growth of poliovirus. To minimize possible interference from other enteric viruses, the oral vaccine should be given during the spring and winter months in temperate climates.

The chief disadvantage of the Sabin vaccine is that it is difficult to preserve. It can be kept frozen for years but only *for 7 days in an ordinary refrigerator* and *3 days at room temperature!* Tap water cannot be used to dilute it, since *the contained chlorine destroys the poliovirus.* Like other vaccines manufactured on kidney cells, the final preparation contains traces of penicillin and streptomycin from the culture medium. The overall effectiveness of the Sabin vaccine is better than 95%.

Rabies vaccines are no longer made from brain tissue but are prepared by propagation of virus in nonnervous culture systems. Duck embryo vaccine (DEV) is made from the growth of fixed rabies virus in the embryonated duck egg. Because of the absence of central nervous system tissue, this vaccine is much less hazardous than the one prepared in years past from the nervous tissues of the rabbit (the Semple vaccine).

Currently a potent, inactivated-virus rabies vaccine, the human diploid cell vaccine (HDCV), is available. It is prepared by the concentration and inactivation of rabies virus harvested from human fetal lung diploid cell (nonneural) culture. The rabies virus used is derived from Louis Pasteur's original strain. The new vaccine is safe, more immunogenic but less allergenic than the duck embryo vaccine. Fewer doses are required to establish protection, antibody response appears to be ten times better, and the reactions at the injection sites have been minor.

A genetically engineered rabies virus vaccine is being developed.

The U.S. Public Health Service recommends that all dogs be immunized against rabies by the age of 3 months. For dogs a vaccine is grown on a series of chick embryos to reduce its virulence. Such a live virus vaccine, given in a single dose, will produce immunity of over 3 years' duration in the dog. Modern vaccines undoubtedly provide 3 years' protection. Since the immunity resulting from some of the older vaccine preparations still used may not extend beyond 15 months, these vaccines should be abandoned. As a precautionary measure, some local governments require that dogs and "exotic animals" kept as pets be vaccinated annually. In areas where vampire and other bats exist it may be necessary to immunize livestock.

According to the Centers for Disease Control in Atlanta, the keeping of exotic or wild animals, including skunks, as pets is ill-advised. In some of these the incubation period for rabies is unpredictable, variable, and sometimes long, and a vaccine for their protection does not exist. In fact no vaccine available today demonstrates any effect in the immunization of wildlife against rabies. Licensed animal vaccines are for use only in dogs, cats, sheep, cattle, horses, and goats.

Rubella virus vaccine, live, the RA 27/3 vaccine, is prepared in a human diploid cell line. It induces secretory IgA nasal antibody, as well as serum antibody. It must be given subcutaneously.

Episodes of arthritic joint pain, found especially in the older age groups, may rarely complicate the use of rubella vaccine. Although arthritis was a disturbing event when first recognized as a consequence to vaccination, it has usually turned out to be mild and self-limited, without sequelae. Hypersensitivity reactions to egg protein and neomycin may occur.

Yellow fever vaccine is prepared from a specified strain of the virus grown in chick embryos. For purposes of international travel, yellow fever vaccine must be approved by the World Health Organization and administered at a designated Yellow Fever Vaccine Center.

PRECAUTIONS FOR ADMINISTRATION OF BIOLOGIC PRODUCTS

To ensure the development of the desired immunity and to prevent, insofar as possible, certain untoward side effects or complications, the following precautions in the administration of vaccines, serums, and other biologic products are strongly advised:

1. *Read carefully the label on the package and the accompanying leaflet when any biologic product is to be given!*
2. Disinfect the skin properly at the injection site. (Note also proper disinfection of the puncture site on the vial of medication.)
3. Use a sterile syringe and needle for each injection, preferably a sterile *disposable* unit. (The use of several needles with the same syringe can be a way to spread viral hepatitis.)
4. When the needle is placed into the subcutaneous or intramuscular area, pull the plunger of the syringe outward to check that the needle has not inadvertently entered a vein. If it has, blood wells up in the syringe.
5. Immunize only well children. Delay any such procedure for sick children until their recovery.
6. *Anaphylactic and allergic accidents in biotherapy are unpredictable—the next one may be yours!* Carefully question the recipient of the injection as to reactions to previous doses of the same or comparable biologic products. Inquire carefully as to previous injection of horse serum, any known allergy to horse serum products, and any history of allergic conditions such as asthma. Did the child have reactions to a previous dose, such as fever or sleeplessness? Was there an area of redness and tenderness about the injection site? If the child had only a mild reaction, the injections may be continued. In some instances the amount of the material must be reduced and the overall number of injections increased. If the child had a severe reaction, *do not repeat* the injection. Consult the physician.
7. Ascertain the presence of allergy to egg or chicken dander if certain viral vaccines prepared in chick embryos (influenza, yellow fever, and measles) are to be given. Practically speaking, if the recipient can eat eggs without event, it is safe to give the vaccine. Some of the newer vaccines may contain neomycin, to which an allergic state may also exist.
8. Note these special situations.
   a. Do *not* immunize persons with altered or deficient immunologic mechanisms.
   b. Do *not* immunize persons receiving steroid hormones.
   c. Unless there is an emergency, do *not* immunize during a poliomyelitis outbreak.

Tests for Hypersensitivity

Preliminary testing to detect hypersensitivity by all means is mandatory.

The following tests for hypersensitivity must be used discreetly, and a syringe containing 1 ml of epinephrine 1:1000 should be handy (pp. 383 and 385).

1. In the **scratch** test, a 1:10 dilution of the biologic product in isotonic saline solution (tetanus antitoxin, for example) is applied to the abraded skin.
2. In the **eye** test, one drop of a 1:10 dilution is placed in the conjunctival sac. If this is negative, the eye test may be repeated with undiluted material.
3. The **intradermal** skin test is usually carried out with 0.02 to 0.03 ml of a 1:10 dilution. *(An intradermal test with 0.1 ml of undiluted tetanus antitoxin may be fatal.)* If indicated, skin tests may be carried out serially with 0.02 to 0.03 ml of 1:10,000, 1:1000, 1:100, and 1:10 dilutions. The positive reaction is a hivelike wheal with redness.
4. In the **intravenous** test, the patient's blood pressure is recorded before the 5-minute interval taken to inject 0.1 ml of biologic product in 10 ml of isotonic saline solution and every 5 minutes thereafter for 30 minutes. If the blood pressure falls 20 points or more, a hypersensitive state is indicated.

Injection Site for Biologic Products

The optimal site for intramuscular injection is the outer lateral aspect of the thigh (Fig. 21-3). By directing the needle downward as it enters the tissues at the junction of the upper third with the lower two thirds, one can deliver serum or other biologic products into the middle third of the thigh. This area is preferred because (1) in the event of an

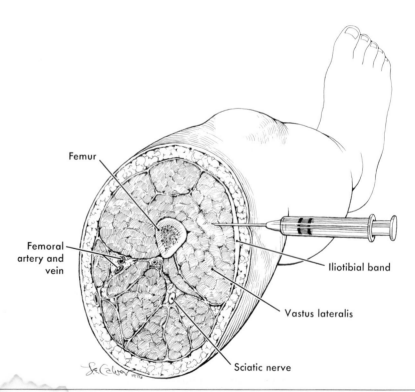

FIG. 21-3 Optimal site for intramuscular injection into outer lateral aspect of thigh. The needle points downward to deliver injectable into middle third of thigh. Cross section shows anatomy of area.

impending serum reaction a tourniquet can be placed on the thigh above the injection site, and the absorption of the biologic product greatly delayed, and (2) here there are no important anatomic structures to be injured by pressure of injected material. If speed of absorption is desired, the fascia lata, a tense band stretched across this area, facilitates it.

STANDARDS FOR BIOLOGIC PRODUCTS
Manufacture

It is so important that serums, vaccines, and other biologic products, including human blood and its derivatives, meet acceptable standards of safety, purity, and potency that, when offered for sale, import, or export in interstate commerce, they must be manufactured under the license and regulations of the federal government. Authority for this is delegated to the Food and Drug Administration of the U.S. Department of Health and Human Services.

Label

Each package must be properly labeled. The name, address, and license number of the manufacturer, the lot number, and expiration date must be shown.

Date of Expiration

On the **date of issue,** the product is placed on the market. This must be within a certain time after manufacture, depending on the kind of product and the temperature of storage. The **expiration date** means the date beyond which the product cannot be expected to exert its full potency. The expiration date of most biologic products is 1 year after manufacture or issue.

SUMMARY

1 An immune serum confers passive immunity for 1 to 6 weeks. Whenever feasible, human immune serum should be used instead of animal serum because the animal serum contains proteins foreign to the human. These can produce highly undesirable reactions.
2 Animals serums must still be used for passive immunization with diphtheria and botulism, after snake and spider bites, and for immunosuppression in selected instances (antilymphocyte serum).
3 Agents for passive immunization include antitoxins, antivenins, antibacterial serums, and human immune globulin.
4 The best-known antiviral serum is the antirabies serum obtained by immunization of human volunteers.
5 Gamma globulin, the main fraction of blood plasma with which antibodies are associated, is always given *intramuscularly*, never intravenously. It is dispensed as immune serum. As ordinarily given, it is a benign injectable, being associated with few side reactions.
6 Agents used for active immunization include bacterial vaccines, toxoids, and viral vaccines. Vaccines are always antigens; they induce active immunity.
7 Important bacterial vaccines include the BCG vaccine (for tuberculosis), pertussis vaccine, pneumococcal polysaccharide vaccine, plague vaccine, tularemia vaccine, typhoid vaccine, and rickettsial vaccines.
8 Toxoids are prepared to prevent diphtheria, tetanus, and botulism.
9 Viral vaccines have been developed from both attenuated (live) virus and from killed or inactivated virus. A living viral agent is preferred, since it allows a prolonged and increasingly strong immunogenic stimulation to the host.

10 Hepatitis B vaccine is a suspension of the inactivated, alum-adsorbed surface antigen particles of the hepatitis B virus.

11 The human diploid cell vaccine, a potent, inactivated-virus rabies vaccine, replaces the duck embryo vaccine for rabies.

12 Guillain-Barré syndrome, a delayed reaction somtimes seen after influenza immunization, is an acute, paralytic, but self-limited disorder. Recovery is usually complete.

13 Before any biologic product is given to a human being, the label on the package and the package insert should be read carefully.

14 Every precaution must be taken to avoid anaphylactic and allergic reactions when biologic products are administered.

15 The optimal site for intramuscular injection of a biologic product is the outer aspect of the thigh.

QUESTIONS FOR REVIEW

1 Define vaccine, immune serum, antitoxin, antivenin, antibacterial serum, convalescent serum, MLD, date of issue (of vaccine), expiration date (of vaccine), autogenous vaccine, immune serum globulin, gamma globulin, whole virus vaccine, and split-virus vaccine.

2 Cite at least four differences between vaccines and immune serums.

3 What is the term used to designate the strength of tetanus and diphtheria antitoxins?

4 Briefly discuss the three methods of standardizing diphtheria antitoxin.

5 List three antitoxins, one antiviral serum, two antibacterial serums, six bacterial vaccines, and seven viral vaccines.

6 List six diseases for which immune globulin may be of benefit. Indicate how gamma globulin is used to modify or prevent disease.

7 How is toxoid prepared? Name three important toxoids.

8 What is BCG? Comment on its usefulness.

9 Compare live virus vaccines with inactivated virus vaccines.

10 Briefly describe the hepatitis vaccine.

11 Briefly discuss rabies vaccines for humans and animals.

12 What is meant by the Guillain-Barré syndrome?

13 Explain intradermal, subcutaneous, intramuscular, and intravenous.

14 Give the optimal site for injection of biologic products.

15 List the tests for hypersensitivity.

16 Give precautions in the administration of biologic products.

SUGGESTED READINGS

Berlin, B.S., and others: Rhesus diploid rabies vaccine (adsorbed) a new rabies vaccine, J.A.M.A. 249:2663, 1983.

Brunell, P.A.: Childhood infectious diseases in adults, Tex. Med. 77:41, July 1981.

Compendium of animal rabies vaccines, 1983, Morbid. Mortal. Weekly Rep. 31:685, 1982.

Ellis, R.J.: Immunobiologic agents and drugs available from the Centers for Disease Control, ed. 2, Atlanta, 1980, Centers for Disease Control.

Francis, D.P.: Hepatitis B virus vaccine, an opportunity for control, J.A.M.A. 250:1891, 1983.

Gerety, R.J., and Tabor, E.: Newly licensed hepatitis B vaccine, known safety and unknown risks, J.A.M.A. 249:745, 1983.

Hadler, A.C., and others: Effect of immunoglobulin on hepatitis A in day-care centers, J.A.M.A. 249:48, 1983.

Lerner, R.A.: Synthetic vaccines, Sci. Am. 248:66, Feb. 1983.

Recommendation of the Immunization Practices Advisory Committee (ACIP): Immune globulins for protection against viral hepatitis, Morbid. Mortal. Weekly Rep. 30:423, 1981.

Recommendation of the Immunization Practices
    Advisory Committee (ACIP): Pneumococcal
    polysaccharide vaccine, Morbid. Mortal. Weekly
    Rep. **30**:33, 1981.
Recommendation of the Immunization Practices
    Advisory Committee (ACIP): Mumps vaccine,
    Morbid. Mortal. Weekly Rep. **31**:617, 1982.
Recommendation of the Immunization Practices
    Advisory Committee (ACIP): Plague vaccine,
    Morbid. Mortal. Weekly Rep. **31**:301, 1982.

Recommendation of the Immunization Practices
    Advisory Committee: Influenza vaccines 1983-
    1984, Morbid. Mortal. Weekly Rep. **32**:333,
    1983.
The safety of hepatitis B vaccine, Morbid. Mortal.
    Weekly Rep. **32**:134, 1983.
Smallpox vaccine no longer available for civilians—
    United States, Morbid. Mortal. Weekly Rep.
    32:387, 1983.

IMMUNIZING SCHEDULES

In this chapter methods of immunization for the prevention of the more important infectious diseases are given. Since opinions from public health physicians and authorizing agencies concerning the most efficacious methods of immunization are not uniform, the student will note variations in procedures endorsed in standard references.

Dosage may be indicated in text discussions, but in actual practice, *one must always carefully check the manufacturer's recommendations and admonitions concerning the product as stated in the package insert and label.*

RECOMMENDATIONS OF THE COMMITTEE ON INFECTIOUS DISEASES OF AMERICAN ACADEMY OF PEDIATRICS*

The Red Book of the American Academy of Pediatrics is widely accepted as a source of the best immunization procedures in both children and adults. Table 22-1 schedules the protection recommended by the Committee on Infectious Diseases for normal infants and children from 2 months through 16 years of age (after which the immunizing procedures are those for adults).

Combined Active Immunization

In the schedule in Table 22-1 diphtheria and tetanus, adsorbed (depot), toxoids combined with pertussis vaccine, adsorbed (DTP), are preferred as primary immunizing agents over the nonadsorbed (fluid or plain) mixtures. A dose is injected deep into the deltoid or midlateral muscles of the thigh or given by intradermal jet injection (Fig. 22-

---

*A copy of the Red Book may be obtained from the American Academy of Pediatrics, Inc., P.O. Box 1034, Evanston, IL 60204. The Academy has developed a personal immunization card, billfold size and plastic, to provide a permanent record.

Table 22-1 Schedule for Active Immunization and Tuberculin Testing in Normal Infants and Children*

| Age | Administration of Immunizing Agents | | | Tuberculin Testing‡ |
	Combined Vaccines	Others†	
2 months	DTP (diphtheria and tetanus toxoids, adsorbed, and pertussis vaccine, adsorbed)	OPV (trivalent oral poliovirus vaccine)	
4 months	DTP	OPV	
6 months	DTP	OPV (optional)	
12 months			Yes
15 months	MMR (measles-mumps-rubella)§		
18 months	DTP	OPV	
4 to 7 years	DTP	OPV	
7 to 16 years	Td (adult-type combined tetanus-diphtheria toxoids for persons more than 7 years of age)		
Over 16 years	Td every 10 years		

*See also p. 438.
†Inactivated polio vaccine instead of OPV can be given with DTP as indicated in this table.
‡Risk of exposure indicates frequency of tuberculin testing.
§Combined viral vaccines (single injection) can be given from age 15 months, *but not before*.

1). The concentration of antigens varies in various products. The package insert supplied with the vaccine should be carefully studied as to volume of dose.

The standard preparations of **combined diphtheria-tetanus toxoids** (DT) used in infants and young children contain 7 to 25 Lf (flocculating units) diphtheria toxoid per dose. The *adult type* of **combined tetanus-diphtheria toxoids** (Td) with adjuvant used in children over 7 years of age, teenagers, and adults contains not more than 2 Lf diphtheria toxoid per dose. Td contains less antigen than DT, since fairly severe reactions in older children and adults may be associated with the increase of diphtheria antigen. Adult-type toxoids are specially purified preparations.

Live oral (attenuated) poliovirus vaccine (OPV) is recommended (Table 22-5) over the killed vaccine, poliovirus vaccine inactivated, because it is more effective as an immunizing agent, antibodies induced persist longer, and it is easier to give. OPV refers now to the preferred trivalent oral poliovirus vaccine (containing virus types 1, 2, and 3).

The immunity for each disease produced by the combined method of immunization is as great as would be obtained by separate immunization against the given disease.

Tetanus

For individuals to whom the combined primary immunizations of early childhood do not apply, the Red Book recommends administration of tetanus toxoid, adsorbed. The total recommended dose of the manufacturer may be given in fractional doses of 0.05 or 0.1 ml. A booster (recall) injection of alum tetanus toxoid should be given 1 year later.

When a child or an older person sustains an injury and the wound remains clean, the Committee on Infectious Diseases advises that in the fully immunized person no booster dose is needed unless more than 10 years have elapsed since the last dose. When the child or older person incurs a contaminated wound likely to be complicated by tetanus, such as a deep puncture wound, dog bite, certain crusted or suppurating lesions, and wounds containing soil or manure, T (tetanus toxoid and adjuvant) or Td should be given regardless of the immunization status of the patient *unless* the patient

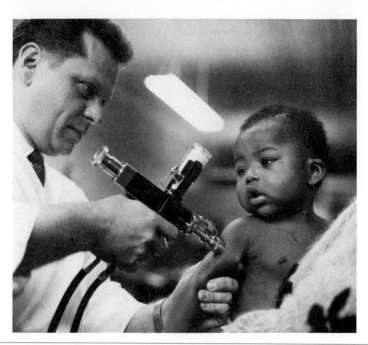

FIG. 22-1   Jet injector (air gun) delivering vaccine to young Chicagoan; it is instant and pain free. This method may be used for certain vaccines mentioned in text. (From Medical News, J.A.M.A. **196:**29, May 23, 1966; courtesy American Medical Association.)

is known to have received Td or T in the past 5 years. In that case the inoculation at this time need not be given. Nonimmunized individuals also require passive immunization (p. 422).

It has been found that the immunity conferred by tetanus toxoid is extremely long lasting and that frequent recall doses seem to provoke reactions. The U.S. Public Health Service recommends that tetanus boosters of adsorbed (*not* fluid) toxoid be given at 10-year intervals.

Pertussis

Combined immunization as indicated in Table 22-1 is preferable, utilizing the antigens combined with adjuvant, especially when immunization is begun under the age of 6 months. Pertussis vaccine ordinarily is not given beyond the age of 7 years.

When the young child has been intimately exposed to whooping cough or there is an epidemic and protective immunity must be developed as soon as possible, pertussis vaccine, adsorbed, should be given intramuscularly in three doses of 4 protective units each, 8 weeks apart. Routine recall injections are given 12 months later and at school age with adsorbed vaccine. Exposure recall injections indicated for children up to the age of 7 years are given with adsorbed vaccine.

Diphtheria

Diphtheria toxoid is rarely given alone. With a severe reaction after the injection of the multiple vaccine DTP, primary immunization against diphtheria in infants and children *less than* 7 years of age may be continued with DT, adsorbed, but *without* pertussis vaccine. For immunization of children older than 7 years of age, adult-type Td, adsorbed, is used.

Full, active immunization against diphtheria prevents the disease but not the acquisition and carriage of *Corynebacterium diphtheriae*. Antibiotics are necessary to eliminate the carrier state.

Smallpox

Smallpox vaccination is now indicated only for laboratory workers directly exposed to smallpox or closely related orthopox viruses. No evidence exists that it has any therapeutic value in the treatment of recurrent herpes infections, warts, or any other disease, and smallpox vaccination in such instances is strongly contraindicated.

For a discussion of the eradication of smallpox, see p. 709.

Mumps

Live, attenuated mumps virus vaccine given in a single 0.5 ml subcutaneous injection or in combination with measles and rubella vaccines (Table 22-1) protects for at least 12 years. Febrile reactions have not been noted. The vaccine is advocated for susceptible children approaching puberty, adolescents, and adults, particularly men with no history of the disease. In special circumstances vaccination may be considered for younger children in closed populations such as special schools, camps, and institutions. It should *not* be given to pregnant women or to children less than 12 months of age.

Rubella

Rubella immunization is carried out with a single *subcutaneous* injection of vaccine (or preferably given in a combined vaccine as indicated in Table 22-1). It is advocated for boys and girls between the ages of 15 months and puberty and especially for those of kindergarten age. Vaccinees will shed virus from their throats for 2 or more weeks afterward, but they are not considered infectious. Most likely, vaccination gives long-range protection, and a booster is not now advised.

Table 22-2 Guidelines for Postexposure Prophylaxis of Rabies

Animal	Condition of Animal at Time Injury Inflicted	Treatment of Exposed Human
Domestic (dog, cat)	Healthy—observe for 10 days	None, unless animal develops rabies
	Rabid or suspected of being rabid	Immune rabies globulin and rabies vaccine
	Unknown (animal not apprehended)	Consult public health officials; if treatment needed—rabies immune globulin and rabies vaccine
Wild (skunk, bat, fox, coyote, raccoon, bobcat, other carnivores)	Regard as rabid unless proved negative by laboratory testing	Consult public health officials; if treatment needed—rabies immune globulin and rabies vaccine
Other (livestock, rodents, rabbits, hares)		Consider each case individually. Consult public health officials. NOTE: Bites of squirrels, hamsters, guinea pigs, gerbils, chipmunks, rats, mice, and rabbits rarely call for antirabies prophylaxis.

Immunization is contraindicated for infants less than 15 months of age, for pregnant women, and for persons with altered immune states (those with leukemia or lymphoid malignancies or recipients of immunosuppressive therapy). It should not be given routinely to adolescent girls (after the twelfth birthday) or women because it might dangerously complicate a pregnancy. It is not given in severe febrile illness.

Rabies

With rabies there is an exception to the rule that there is not enough time between exposure and onset of disease to establish an *active* immunity to protect the person exposed. Since the incubation period of rabies may be much longer than that of most diseases, an active immunity can usually be established before the disease begins. The other disease in which the exposed person can be protected against development of the disease by the production of an active immunity is smallpox. In smallpox an active immunity develops very quickly. In rabies the period of incubation is sometimes so long that an active immunity can be produced before the disease takes effect. The onset of the disease *must* be prevented because there is no treatment; *the disease must be considered to be 100% fatal.*

The amount of postexposure protection needed from hyperimmune serum and vaccine depends on many factors, such as the type of animal, whether is has been vaccinted, location of the bite, extent of injury, protection afforded by clothing, whether rabies is present in the community, and whether observation of the animal is possible. *It is very important that the animal inflicting the bite wound be apprehended and kept for a 10-day period of observation.* Table 22-2 gives guidelines for postexposure prophylaxis formulated by the U.S. Public Health Service Centers for Disease Control.

Postexposure Prophylaxis

The best prophylaxis against rabies after exposure is the use of hyperimmune serum (preferably HRIG) in combination with vaccine (preferrably HDCV). Rabies immune globulin, human (HRIG), is given as soon as possible. The recommended dose is 20 units per kilogram body weight, with at least half the dose being used to infiltrate the wound inflicted by the animal for local antiviral effect. The rest of it is given intramuscularly in the buttocks.

Active immunization with the human diploid cell vaccine (HDCV) is carried out in a series of five doses given intramuscularly. (Other routes of administration have not been fully tested.) The first dose should be given as soon as possible after exposure, and additional ones are given thereafter on days 3, 7, 14, and 28 after the first dose.

If duck embryo vaccine (DEV) (no longer sold in the United States) is used, active immunization is carried out in a series of at least 14 single daily injections of duck embryo vaccine given subcutaneously in the abdomen, lower back, or lateral aspect of the thigh, with rotation of sites suggested. When serum is given with vaccine, active immunization requires a series of 21 daily doses of vaccine. An acceptable alternative is the administration of 14 of the 21 doses during the first 7 days, either as two separate doses per day or as a double daily dose, with the remaining doses of the series given singly each of the next 7 days. A booster is needed 10 days and 20 days after the last dose of the series.

If an exposure to rabies occurs in an immunized person, two doses of HDCV are given, one immediately and one 3 days later. If DEV is used with a nonbite exposure to rabies in an immunized individual, one booster dose is injected. If reexposure is a bite by a rabid animal, five daily doses of vaccine are given, with a booster 20 days later.

Preexposure Prophylaxis

For active preexposure immunization, HDCV is given intramuscularly as one dose on each of days 0, 7, 21, or 28 (total three doses). Boosters of HDCV are recommended every 2 years. Persons at high risk because of regular, direct contact with the rabies virus should have the rabies antibody titer of their serum determined every 6 months and should receive booster doses of vaccine as needed to maintain an adequate titer.

Prophylactic vaccination with DEV consists of two subcutaneous inoculations of 1 ml each into the deltoid area, 1 month apart, followed by a booster dose 6 to 7 months later. Boosters here are recommended every 1 to 3 years.

The U.S. Public Health Service advises prophylactic rabies immunization for persons at high risk. Such persons include veterinarians, veterinary students, animal caretakers, spelunkers, laboratory workers in direct contact with the virus, children living overseas in a highly endemic area, and persons in contact with potentially rabid animals, such as skunks, bats, unvaccinated dogs, and the like. The advantages of such protection are (1) it covers persons at high risk, (2) it is there when prophylaxis is delayed, (3) it provides a basis for a rapid anamnestic response when booster doses of vaccine are given, (4) it eliminates the need for immune globulin, and (5) it reduces the number of doses should postexposure therapy be needed.

Tuberculosis

BCG vaccination against tuberculosis is done as superficially as possible *into the skin (intracutaneously)* of the upper arm over the deltoid or triceps muscle. To the newborn infant 0.05 ml (or half the usual dose) is given; to the older *tuberculin-negative* individual, 0.1 ml. Tuberculin testing—Mantoux test with 5 units of purified protein derivative (PPD) tuberculin—must be done before vacinnation except in infants less than 2 months of age. A preliminary radiograph of the chest may be indicated. Two to three months after vaccination, tuberculin testing is repeated, and if the test is negative, BCG vaccination is repeated.

BCG vaccination is advised for children or those persons traveling in areas of the world where tuberculosis is a major health problem and for those likely to be exposed to adults with the disease. It should be carried out 2 months before exposure. The U.S. Public Health Service stresses vaccination also for health workers at risk from unrecognized cases of tuberculosis and the members of socially disaffiliated groups or those without a regular source of health care, such as alcoholics, drug addicts, and migrants. BCG vaccination is said to be 75% effective with protection lasting for 10 to 15 years. A low incidence of postvaccinal complications is reported.

RECOMMENDATIONS OF THE U.S. PUBLIC HEALTH SERVICE
Influenza

The U.S. Public Health Service recommends annual influenza vaccination for persons over 65 years of age and for those in whom the onset of flu would represent an added health risk, such as individuals of all ages with chronic debilitating diseases (diabetes, cardiovascular ailments, pulmonary disease, and others). There is some indication for

Table 22-3 Recommendations for Administration of Influenza Vaccine

Age	Recommended Vaccine	Dosage (ml)	Number of Doses	Interval (in Weeks)
6 to 35 months	Split virus	0.25	2	More than 4
3 to 12 years	Split virus	0.5	2	More than 4
13 years and over	Whole virus or split virus	0.5	1	—

immunization of persons responsible for furnishing essential public services, such as law enforcement officers, firemen, and many others.

The primary series has been traditionally two doses given subcutaneously (beneath the skin) 6 to 8 weeks apart. Currently, a *single* dose for adults and children over 12 years of age is indicated for either the primary or annual booster vaccination (Table 22-3). In the package insert supplied with the vaccine the manufacturer gives the recommended dose for adults and children and directions for administration. Vaccination should be done by the middle of November.

The status of influenza immunization (both vaccine and schedule) can change rapidly. For this reason the recommendations of its Committee on Immunization Practices are regularly published by the U.S. Public Health Service and should be consulted before any procedure is carried out.

Hepatitis B

The recommended schedule for immunization against hepatitis B is given in Table 22-4. Immunization is urged for those persons at high risk because of their geographic origins, their life-styles, or exposure to the hepatitis B virus at home or at work.

Poliomyelitis

The U.S. Public Health Service is urging regular immunization against poliomyelitis for all children from early infancy. Routine immunization of infants should begin at 6 to 12 weeks of age and be completed against all three types of the virus. The schedule for poliomyelitis vaccination is given in Table 22-5. In the United States inactivated poliovirus vaccine (IPV) is not used except in special instances. A Canadian vaccine licensed for use in the United States is available when needed. Trivalent oral poliovirus vaccine (OPV) has largely replaced the monovalent form (MOPV) in immunization programs.

Although the risk to persons with a normal immune status is slight, vaccine-associated paralytic poliomyelitis rarely complicates vaccination with OPV (not with IPV) in either the recipient of the vaccine or in a contact of the recipient. The current estimated risk of vaccine-associated paralysis with OPV is estimated at one in 8 million doses of vaccine. Of the small number of cases reported, a significant proportion have

Table 22-4 Immunization Schedule for Administration of Hepatitis B Vaccine

Group to Be Immunized	HBsAG Given per Dose (μg/ml)	First Dose (ml)	Second Dose— 1 Month Later (ml)	Third Dose— 6 Months After First Dose (ml)	Comments
Infants and young children (birth to 10 years)	10	0.5	0.5	0.5	Vaccine given intramuscularly
Adults and older children	20	1	1	1	Vaccine given intramuscularly
Immunosuppressed patients and those on dialysis					
Adults	40	2	2	2	Each milliliter given in a separate anatomic site (intramuscularly)
Children	10	0.5	0.5	0.5	

occurred in young white men. Vaccinated children can excrete virus and so inadvertently transmit poliovirus infection to nonimmunized parents, especially to the father (Table 22-5).

Vaccination with oral poliovaccine is contraindicated in patients with severe underlying diseases, such as leukemia or generalized malignancy, and in persons with immune deficiencies or altered immune states.

Table 22-5 Administration of Poliovirus Vaccines

Group	Doses	Poliovirus Vaccine	Time Interval	Comments
Administration of Trivalent Oral Poliovirus Vaccine (OPV)				
Infants	First	OPV (I, II, III)		Primary immunization begun at 6 to 12 weeks of age; may be given with DPT
	Second	OPV	6 to 8 weeks after first dose (no less than 6 weeks)	
	Third	OPV	8 to 12 months after second dose	
	Booster	OPV		Given at time that child enters school; boosters with OPV otherwise not indicated except for children at increased risk
Children and adolescents (to 18 years of age)*	First	OPV		
	Second	OPV	8 weeks after first dose	
	Third	OPV	8 to 12 months after second dose	Third dose given as early as 6 weeks after second dose in unusual circumstances
Administration of Inactivated Trivalent Poliovirus Vaccine (IPV)				
All ages†	First	IPV (I, II, III)		Parenteral administration; immunization series: four doses; for infants 6 to 12 weeks of age, IPV may be given with DPT
	Second	IPV	1 to 2 months after first dose	
	Third	IPV	1 to 2 months after second dose	
	Fourth	IPV	6 to 12 months after third dose	
	Booster	IPV	5 years between doses to age 18 years unless primary series given with OPV	Booster advised at time child enters school

*Routine primary immunization of persons over 18 years of age with OPV is not indicated for most adults residing in the continental United States.

†For primary immunization of adults at risk, IPV carries less risk of vaccine-associated paralysis than does OPV. Unvaccinated parents of a child to be immunized with OPV may for their protection against the very small risk of OPV-associated paralysis receive two doses of IPV a month apart before immunization of the child is begun.

Measles

The U.S. Public Health Service urges immunization for all susceptible children (those not having had vaccine or natural measles). The prime target groups are children no younger than 15 months of age and susceptible children entering nursery school, kindergarten, or elementary school (Fig. 22-1). Protection against measles is especially needed for children in the high-risk groups—children in institutions and those with chronic heart or pulmonary disease and cystic fibrosis.

Active immunization against measles is carried out with a single dose of live measles virus administered preferably in the combined measles-mumps-rubella vaccine (Table 22-1). No booster dose is required.

Revaccination for measles with a current live virus vaccine is a matter of some concern to health authorities today. The groups at risk because of their inadequate vaccination status include the following: (1) persons vaccinated with live measles vaccine before the age of 12 months; (2) those vaccinated at any age with the inactivated vaccine that was available from 1963 through 1967; (3) persons unaware of their age at vaccination; (4) those vaccinated before 1968 with a vaccine of unknown type; (5) those immunized with live, further attenuated measles virus vaccine along with immune serum globulin, regardless of age at the time; and (6) those who received live measles virus vaccine within 3 months of inactivated measles virus vaccine.

Live attenuated vaccine should not be given to pregnant women or patients being treated with steroids, irradiation, antimetabolites, or agents that depress an individual's immunologic capacities. Leukemia, lymphoma, and other generalized malignancies are also contraindications, as is untreated tuberculosis.

Typhoid and Paratyphoid Fevers

The U.S. Public Health Service no longer recommends routine typhoid immunization in the United States, even for individuals in flood disaster areas. Circumstances in which vaccination would be indicated include an intimate exposure to a known carrier, an outbreak of the disease in a community, or travel to an endemic area (for instance, Mexico). The U.S. Public Health Service does *not* advise paratyphoid immunization because there is good evidence that paratyphoid A and B vaccines are worthless. (Schedules for typhoid immunization may be found in the Red Book of the American Academy of Pediatrics.)

REQUIREMENTS FOR U.S. ARMED FORCES

Immunization procedures for the U.S. armed forces are regularly evaluated and updated. So that the requirements may be precisely related to the geographic area of military duty, the world is divided into the following areas:

Area I (routine)—United States, Canada, Greenland, Iceland, Kwajalein Atoll, Guam, Pacific islands east of the 180th meridian, North and South Polar regions including Antarctic continent, Bermuda, Bahama Islands, Baja California, the area in Mexico north of a line 50 miles south of the United States–Mexico border, and the Naval Base at Guantanamo Bay, Cuba

Area II—All areas outside Area I, including Areas IIP and IIY

Area IIP (plague)—Cambodia, Laos, and Vietnam

Area IIY (yellow fever)—Central America southeast of Isthmus of Tehuantepec, Panama, South America, and Africa south of the Sahara

Table 22-6 summarizes the immunization requirements for U.S. military personnel. Military dependents and civilians in the employ of the armed forces or organizations serving the armed forces are not required to take immunizations while in the United States or Canada. Outside the United States or Canada they are required to take essentially the same ones as military personnel.

Table 22-6 Immunizations Required for U.S. Armed Forces

Immunizing Agent	Basic Series*	Required Area I (Excluding Alert Forces)	Required Area II	Reimmunization Area I	Reimmunization Area II
Smallpox vaccine, (freeze-dried)	Vaccination to be read after 6 to 8 days; successful reaction required	X	X	3 years	3 years
Typhoid vaccine (acetone killed-dried)	Two 0.5 ml injections 4 weeks apart	X	X	None	0.5 ml—3 years
Tetanus and diphtheria toxoids, adsorbed (for adult use)	Two 0.5 ml injections 1 to 2 months apart with third dose of 0.1 ml 12 months later	X	X	0.1 ml—10 years; 0.5 ml after injury or burn	0.1 ml—10 years; 0.5 ml after injury or burn
Poliovirus vaccine, live oral, trivalent, types I, II, and III	Two oral doses 6 to 8 weeks apart; third dose 12 months later	X	X	None	None
Yellow fever vaccine	One 0.5 ml injection of concentrated vaccine diluted 1:10		X (Area IIY)		0.5 ml of 1:10 dilution—10 years
Influenza virus vaccine	One 0.5 ml injection	X	X	0.5 ml—1 year	0.5 ml—1 year
Cholera vaccine	0.5 ml injection followed by 1 ml in 1+ weeks		X (Areas IIY and IIP)		0.5 ml—6 months (while in area)
Plague vaccine	Two *intramuscular* injections—0.5 ml followed in 3 months by 0.2 ml		X (Area IIP)		0.2 ml (IM)—6 months (while in area)

*Smallpox vaccination is given either by multiple-puncture technic or by intradermal jet. Plague vaccine must be given intramuscularly. Other injections are subcutaneous or intramuscular. (See Fig. 21-3.)

INTERNATIONAL TRAVEL
Immunizations

The Red Book of the American Academy of Pediatrics also contains immunization information and recommendations for international travel. In Table 22-7, which presents vaccination schedules recommended by this organization, there is a breakdown of special requirements for travel by age groups.

This book also advises suppressive medication for malaria (a major worldwide problem) before entry into an infected area, during sojourn there, and even for a number

Table 22-7 Special Immunization Requirements by Age for International Travel

Disease	Travel Destination	Age	Doses of Vaccine (ml)	Route of Inoculation	Intervals	Recall Needed
Cholera	Asian countries, Middle East	6 months to 4 years	(1) 0.2 (2) 0.2 Booster: 0.2	Subcutaneously or intramuscularly	1 week to 1 month or more apart	6 months (with exposure)
		5 to 10 years	(1) 0.3 (2) 0.3 Booster: 0.3			
		10 years and older (adults also)	(1) 0.5 (2) 0.5 Booster: 0.5			
Yellow fever*	Central and South America, Africa	All ages	0.5 of 1:10 dilution	Subcutaneously		10 years
Plague	High-risk area	Less than 1 year	(1) 0.2 (2) 0.04 (3) 0.04 Booster: 0.04	Intramuscularly only	30 days between 1 and 2; 4 to 12 weeks between 2 and 3	6 to 12 months
		1 to 4 years	(1) 0.4 (2) 0.08 (3) 0.08 Booster: 0.08			
		5 to 10 years	(1) 0.6 (2) 0.12 (3) 0.12 Booster: 0.12			
		10 years and older (adults also)	(1) 1.0 (2) 0.2 (3) 0.2 Booster: 0.2			
Typhoid fever	Developing countries and those with low standards of sanitation	6 months to 10 years 10 years and older (adults also)	(1) 0.25 (2) 0.25 (1) 0.5 (2) 0.5	Subcutaneously	3 + weeks	3 years if exposed (0.1 ml intradermally or 0.25 or 0.5 ml subcutaneously)

Based on data from the Red Book of the American Academy of Pediatrics, 1982.

*Note that yellow fever vaccination for international travel must be given at a designated Yellow Fever Vaccination Center, registered by the World Health Organization. Centers are located in 46 states, the District of Columbia, Puerto Rico, and American Samoa. Yellow fever vaccination certificate is not valid until 10 days after primary immunization.

of weeks thereafter. Remember that outside the United States, Canada, and Europe most of the world is malarious and that malaria is a disease costing pennies to prevent but thousands to cure. (Fatalities have resulted in a few tragic instances of neglect.)

Where to Find a Doctor

The International Association for Medical Assistance to Travelers, Inc. (IAMAT),* supplies without charge a directory of English-speaking physicians available 24 hours a day in the different cities of the world. It also provides a world climate chart that contains information regarding the type of clothing to be worn in the different countries, the availability of food there, and the safety of the water. International SOS Assistance† provides coverage of certain emergency services for travelers for a prescribed fee. A long way from home one may also obtain medical aid from the U.S. embassies and consulates (also the British consulates), the Red Cross, travel agencies, the police, medical associations, hospitals, clinics, and U.S. armed forces bases and installations.

Traveler's Diarrhea

Traveler's diarrhea is known by the many colorful synonyms—Aztec two-step, "turista," Montezuma's revenge, Cromwell's curse, Delhi belly—that mark a widespread geographic distribution, and preventive immunization does not exist for this unpleasant complication of the initial phase of the tourist's trip out of the country. Traveler's diarrhea is described consistently by nausea and vomiting, abdominal cramps, chills, low-grade fever, and diarrhea productive of loose and watery, foul-smelling stools. Manifestations, though distressing and temporarily incapacitating, disappear within a few days. Relapses or sequelae are rare. Microbes found in the stools and thereby implicated include *Shigella* species, *Salmonella* species, enterotoxigenic *Escherichia coli*, and *Giardia lamblia*. Enterotoxigenic *Escherichia coli* has been singled out as a major cause in travelers to Mexico and some other countries. However, the cause is uncertain much of the time, and a cure-all nonexistent.

To minimize the hazard, especially where sanitary practices are substandard, certain precautions are stressed. The tourist can safely eat only foods thoroughly cooked or recently peeled and can drink with impunity only boiled water, carbonated mineral water, or beverages boiled or carbonated in cans. Very hot water out of the tap is not likely to transmit viable enteric bacteria. After it has cooled, it can be used for oral hygiene and various other purposes.

Centers for Disease Control (CDC)

The responsibility for assisting the states in controlling communicable and vector-borne diseases and for participation in international programs of disease eradication rests with an agency of the U.S. Public Health Service—Centers for Disease Control (CDC). The headquarters in Atlanta is a vast complex of specialty laboratories and other disease control facilities, and there are field stations across the United States and Puerto Rico. Valuable advice for the traveler is given in its annual booklet, *Health Information for International Travel.*‡

World Health Organization (WHO)

The World Health Organization in Geneva, or WHO as it is usually called, is an agency of the United Nations set up to provide for the highest level of health for all

---

*350 Fifth Ave., Suite 5620, New York, NY 10001 (also note Intermedic, 777 Third Ave., New York, NY 10017, and World Medical Association, 10 Columbus Circle, New York, NY 10019.
†1420 Walnut Street, Philadelphia, PA 19102.
‡Health information for international travel 1984, Morbid. Mortal. Weekly Rep. **33**(suppl.): Aug. 1984, Department of Health and Human Services Pub. No. (CDC) 84-8280.

peoples of the world. It is composed of 135 member states and two associate members. It works with the United Nations, governments, and special health groups to devise standards, train personnel, carry on research, and improve public health in every way at the international level. To protect the sightseer of the jet age, WHO is concerned with uniform standards of hygiene in aviation. Guidelines are detailed for aspects of sanitation applied to aircraft and international airports, the design to provide every safety factor for the traveler in rapid transit over the globe.

For the traveler, WHO publishes *Vaccination Certificate Requirements for International Travel, Situation as of 1 January, 1983*, and *Health Advice to Travelers*.

SUMMARY

1 The Red Book of the American Academy of Pediatrics is an excellent source of the best immunization procedures for both children and adults.
2 The Red Book provides recommendations and a schedule for active immunization and tuberculin testing in normal infants and children.
3 Smallpox vaccination is indicated only for laboratory workers at risk.
4 The U.S. Public Health Service Centers for Disease Control (CDC) has prepared guidelines for postexposure prophylaxis of rabies.
5 Preexposure prophylaxis of rabies is needed for persons at high risk, such as veterinarians, animal caretakers, spelunkers, laboratory workers in contact with the virus, children overseas in a highly endemic area, and persons in contact with potentially rabid animals—skunks, bats, and the like.
6 The U.S. Public Health Service recommends annual influenza vaccination for persons over 65 years of age and for those in whom the onset of flu would represent an added health risk.
7 Immunization with hepatitis B vaccine is urged for those persons at high risk because of their geographic origin, their life-style, or their exposure to the B virus in their occupation.
8 The Red Book contains immunization information and recommendations for international travel.
9 Outside the United States and Canada, most of the world is malarious. Suppressive medication for malaria is strongly advised for the international traveler. Malaria is a disease costing pennies to prevent but thousands of dollars to cure.
10 Outside the United States medical aid may be obtained from embassies and consulates, the Red Cross, travel agencies, the police, hospitals, clinics, and U.S. armed forces bases and installations.
11 The CDC, headquartered in Atlanta, bears the responsibility for assisting the states in the control of communicable and vector-borne disease and also for participating in international programs of disease eradication.
12 The World Health Organization (WHO) is a 135-member agency of the United Nations set up to provide for the highest level of health for all the peoples of the world. It works with nations and special health groups to devise standards, train personnel, carry on research, and improve public health in every way at the international level. For the traveler, it publishes *Vaccination Certificate Requirements for International Travel, Situation as on 1 January, 1983*, and *Health Advice to Travelers*.
13 The CDC publishes an annual *Health Information for International Travel*.

QUESTIONS FOR REVIEW

1 When should tuberculin testing be done in infants and young children?
2 At what age may active immunization be started in children?

3 Immunization for measles should not be given before the age of 15 months. Why is this so?

4 Beginning at what age must adult-type combined tetanus-diphtheria toxoids be given? Why are adult-type toxoids given from this age on?

5 What is the best prophylaxis against rabies after an exposure?

6 Name five categories of persons at high risk for rabies.

7 How is vaccination against tuberculosis carried out? What is the immunizing agent?

8 Briefly outline the recommendations of the U.S. Public Health Service for immunization against poliomyelitis.

9 Why is prophylaxis against malaria recommended for world travelers?

10 Briefly describe traveler's diarrhea. Name three agents implicated in its cause. State precautions the traveler may take to prevent or minimize this hazard.

11 What are functions of the Centers for Disease Control and of the World Health Organization? Cite their publications for international travelers.

12 Briefly explain prophylaxis, Red Book of the American Academy of Pediatrics, combined active immunization, combined vaccines, an attenuated viral vaccine, OPV, IPV, DPT, booster dose, subcutaneous, intracutaneous, and intramuscular.

SUGGESTED READINGS

Chin, J.: Vaccine requirements and recommendations for international travelers (editorial), J.A.M.A. 248:2163, 1982.

Clemens, J.D., and others: The BCG controversy, J.A.M.A. 249:2362, 1983.

DuPont, H.L., and others: Travelers; diarrhea: can it be eluded? (editorial), J.A.M.A. 249:1193, 1983.

Goodman, R.A., and others: Vaccination and disease prevention for adults, J.A.M.A. 248:1607, 1982.

Hepatitis B virus vaccine safety: report on an Inter-Agency Group, Morbid. Mortal. Weekly Rep. 31:465, 1982.

Influenza vaccines, 1983-1984, Ann. Intern. Med. 99:497, 1983.

Kim-Farley, R., and others: Letters: vaccination and breast-feeding, J.A.M.A. 248:2451, 1982.

Krugman, S.: Hepatitis B immunoprophylaxis, Lab. Med. 14:727, 1983.

Lane, J.M.: Hazards of smallpox vaccination (editorial), J.A.M.A. 247:2709, 1982.

Mann, J.M.: Emporiatric policy and practice: protecting the health of Americans abroad, J.A.M.A. 249:3323, 1983.

Patterson, T., and others: What's new with "turista": causes, treatment and prevention, Tex. Med. 79:40, 1983.

Recommendation of the Immunization Practices Advisory Committee (ACIP): Diphtheria, tetanus, and pertussis: guidelines for vaccine prophylaxis and other preventive measures, Morbid. Mortal. Weekly Rep. 30:392, 1981.

Recommendation of the Immunization Practices Advisory Committee (ACIP): General recommendations on immunization, Morbid. Mortal. Weekly Rep. 32:1, 13, 1983.

Recommendation of the Immunization Practices Advisory Committee (ACIP): Measles prevention, Morbid. Mortal. Weekly Rep. 31:217, 1982.

Recommendation of the Immunization Practices Advisory Committee (ACIP): Poliomyelitis prevention, Morbid. Mortal. Weekly Rep. 31:22, 1982.

Recommendation of the Immunization Practices Advisory Committee (ACIP): Rubella prevention, Morbid. Mortal. Weekly Rep. 30:37, 1981.

Recommendation of the Immunization Practices Advisory Committee (ACIP): Supplementary statement on pre-exposure rabies prophylaxis by the intradermal route, Morbid. Mortal. Weekly Rep. 31:279, 1982.

Rubella vaccination during pregnancy—United States, 1971-1982, Morbid. Mortal. Weekly Rep. 32:429, 1983.

LABORATORY SURVEY
OF MODULE FOUR*

PROJECT   Sources of Infection
Part A—Bacteria on the Human Body
1. *Hands*
    a. Melt a tube of nutrient agar, and allow it to cool until it feels just comfortable to the back of the hand.
    b. Wash your hands in a sterile dish containing 300 ml of sterile water. Do not use soap.
    c. With a sterile pipette place 1 ml of the wash water in a sterile Petri dish.
    d. Pour the cooled agar into the same Petri dish, replace cover, and set the dish on your desk.
    e. Slide the dish *gently* over the desk with a rotary motion until the melted sodium and water are well mixed.
       NOTE: Do not allow the agar to come up the sides of the Petri dish and touch the cover.
    f. Let the medium harden, invert the dish, and incubate at 37° C for 24 to 48 hours.
    g. Observe the growth of colonies. Note the number and different kinds.
2. *Fingertips*
    a. Gently touch the ends of your fingers over the surface of a sterile agar plate.
    b. Incubate plate at 37° C for 24 hours and examine.
3. *Breathing passages*
    a. Normal breathing
       (1) Remove the lid from a sterile nutrient agar plate.
       (2) Hold the plate about 6 inches from your mouth and breathe normally but directly on the medium for about 1 minute.
       (3) Replace the cover, invert, and incubate at 37° C for 24 to 48 hours.
       (4) Observe the growth of colonies. Note the different kinds.
    b. Violent coughing
       (1) Remove the lid from a sterile nutrient agar plate.
       (2) Hold the plate about 1 foot from your mouth and cough violently onto the surface of the agar.
       (3) Replace the cover, invert, and incubate at 37° C for 24 to 48 hours.
    c. Violent coughing—plate at arm's length from the mouth
       (1) Proceed as above.
       (2) Observe the growth of colonies on this plate. Note the different kinds.
       (3) Which plate of the three prepared in a, b, and c of this exercise has the most colonies? Why?

---

*There is no end to the laboratory projects that may be carried out in connection with this module. Some of them require such highly skilled technic and are so time-consuming that it would be unwise to include them. Others yield little information. Our selection is therefore limited to a few—simple to perform and yet applicable to the content of this module.

4. Other body sites
   a. Select body sites such as the nose, throat, or ears. You can gently pass a sterile cotton swab moistened with sterile isotonic saline solution over or into the given area. Use the cotton swab to make a culture of the area chosen.
   b. Use tubes or plates of media that would be likely to grow all organisms present. Blood agar is suitable.
   c. Incubate cultures taken at 37° C for 24 to 48 hours.
   d. Observe the growth. Note the appearance of colonies.
   e. Make smears from colonies and stain by Gram's method. Note the shape of bacteria and whether they are gram positive or gram negative.
   f. Compare the bacterial growth obtained from the body sites selected for culture.

Part B—Bacteria on Insects

1. Prepare a Petri dish of nutrient agar.
2. Allow an insect, such as a fly, to crawl for a time on the surface of the hardened agar under the cover of the dish.
3. Invert the plate after the fly has escaped and incubate at 37° C for 24 to 48 hours.
4. Note the colonies.
5. Make Gram-stained smears and examine.

Part C—Bacteria in the Air

1. Open a Petri dish and expose the surface of sterile nutrient agar to the air in the laboratory for 5 minutes. (Exposure is made by removing the lid from the dish but the lid is not inverted.)
2. Replace the lid, invert the dish, and place in the incubator.
3. Incubate for 24 to 48 hours.
4. Study the colonies obtained.
5. Make Gram-stained smears, examine, and record findings.

Part D—Bacteria at Specific Sites in Our Surroundings

1. Sample areas of the laboratory and its environs—tabletops, doorknobs, cabinet handles, floors, water faucets, water from the faucets, walls, windowsills, and the like. Feel free to investigate your surroundings bacteriologically.
2. Expose the surface of the nutrient agar in a given Petri dish to the given site for 5 minutes. The agar surface may be applied directly to the specified site, or a sterile swab stick moistened with sterile saline solution may be rubbed over the given area and then used to inoculate the surface of the agar plate.
3. Cover the dish, invert, and incubate.
4. Note colonies. Describe them. Do you recognize any?
5. Make Gram-stained smears, examine, and record findings. Do you encounter the same or different microbes as you sample the bacterial population of the above exercises?

PROJECT   Study of Blood Cells

Part A—Cells in the Blood Smear

1. Obtain a specimen of peripheral blood.
   NOTE: Students may work in pairs to obtain finger puncture specimens.
   a. Cleanse side of the tip of the fourth finger of either hand with 70% alcohol; remove alcohol with dry sterile cotton.
   b. Prick the finger with a sterile lancet or needle forcibly enough to get a large drop of blood. Discard the *first* drop.
   c. Gently squeeze finger until a second drop of blood forms.

2. Make the blood smear.
   a. Touch a clean slide to a drop of blood on the finger. The drop should be about 1.5 cm from one end of the microslide.
   b. Place the slide with the level surface with the drop of blood up. With one hand, steady the slide, and with the other hand, place the end of another slide flat against the surface of the blood slide so that an angle of about 45 degrees is formed. The second slide acts as the spreader or pusher.
   c. Draw the spreader slide back until its edge contacts the drop of blood. Let the blood spread along the junction of the two slides. Push spreader slide forward. Keep the 45 degrees of contact to form a film of blood.
   NOTE: The thickness or thinness of the blood smear depends on the acuteness of the angle and the rapidity with which the slide is pushed along. A wider angle and increased speed of spreading tend to thicken the blood film greatly. Since thin films are more easily studied, the angle of contact should be kept small and the speed of spread fairly slow.

3. Stain the blood smear.
   a. Allow the blood film to dry in air.
   b. Cover with Wright's stain* and let stand 4 to 6 minutes.
   c. Do not remove any stain, but add an equal volume of distilled water or preferably phosphate buffer.
   d. Let remain until a metallic scum forms (usually 2 to 3 minutes).
   e. Wash with distilled water and dry between sheets of blotting paper.
   NOTE: A phosphate buffer† used for 5 to 10 minutes after the application of Wright's stain (instead of distilled water) gives nicely stained smears. The staining times for Wright's stain and phosphate buffer are variable, and the actual time for a given stain must be determined by trial intervals.

4. Examine the smear to observe the red blood cells (erythrocytes) and the different kinds of white blood cells (leukocytes).
   a. Use the low-power objective of the microscope first to examine the blood film.
   b. Place a drop of immersion oil on a thinner part of the smear.
   c. Use the oil immersion lens to study the different kinds of cells present.
   NOTE: Red cells appear reddish orange in a well-stained smear. Those from normal adult human blood do not have nuclei. Most white cells have dark purple nuclei with lighter blue or lavender cytoplasm. Some white cells have large conspicuous granules in the cytoplasm. What might these cells be called?
   d. Observe the proportion of red blood cells to nucleated white blood cells.
   e. Observe and identify the different kinds of leukocytes—neutrophils, eosinophils, basophils, lymphocytes, and monocytes.
   f. Sketch at least one representative of each of these leukocytes.
   g. Identify the platelets. Observe their size, their characteristic cytoplasm, their lack of a nucleus, and their distribution on the smear. Make a sketch of a representative group.

---

*Wright's stain and practically all reagents used in the laboratory exercises may be obtained in ready-to-use form—either as single reagents or in kit combinations—from any of several commercial sources. To prepare Wright's stain working solution in the laboratory, grind 0.3 g of commercial powder with a pestle in 3 ml of glycerin and successive small portions of 100 ml of acetone-free methyl alcohol in a mortar. Allow the solution so made to stand, with occasional shaking, for at least 1 week. Filter before use.

†Potassium phosphate, monobasic, 6.63 g, and sodium phosphate, dibasic, 2.56 g, in 1 L of distilled water give a buffer with a final pH of 6.4.

Part B—Neutrophils in Pus

1. Examine microscopically a prepared and stained smear of pus. Use the oil immersion objective.
2. Observe the predominant cells—the pus cells. Sketch a few.
3. Do pus cells resemble certain white blood cells that you have just visualized in the peripheral blood smear? Are they the same cells?

PROJECT   Measure of Immunity

Part A—Demonstration of Serologic Technic by the Instructor with Discussion

1. Technic of pipetting solutions for serologic tests
2. Technic of making serial dilutions for serologic testing
3. Technic of manipulating serologic tubes

Part B—Demonstration of Complement Fixation by the Instructor with Brief Discussion

1. Principles of complement fixation
2. Explanation of results in positive and negative tests
3. Demonstration of positive and negative tests, with ample opportunity for student observation of the exhibits

Part C—Detection of Antigen-Antibody Combinations

1. *Demonstration of phenomenon of agglutination—tube agglutination*
   a. Set up a row of seven serologic tubes in a test tube rack.
   b. To the first tube add 1.8 ml isotonic saline solution, and to each of the remaining six tubes add 1 ml isotonic saline solution.
   c. To the first tube add 0.2 ml agglutinating serum against *Salmonella typhi* (or suitable *Salmonella* species) and mix.
   d. From tube 1 remove 1 ml and transfer to tube 2 and mix. From tube 2 remove 1 ml, transfer to tube 3, and mix. Continue transferring and mixing until tube 6 has been mixed. Discard 1 ml. Tube 7 acts as a control.
   e. To all tubes add 1 ml of a suspension of *killed* typhoid bacilli. Calculate the dilution of agglutinating serum in each tube.
   f. Incubate tubes in the test tube rack at 37° C for 2 hours.
   g. Note clumping of bacteria (agglutination) in the tubes containing the lower dilutions of agglutinating serum.
   h. Relative to this experiment fill out the following table:

	Tube						
	1	2	3	4	5	6	7
Dilution of serum							
Agglutination*							

   *Complete, partial, or none.

   NOTE: Suppose human serum had been used in this test instead of an artificial agglutinating serum and the results had been as indicated in the table. What would be the significance of the test? What would agglutination in tube 7 indicate? What is the name of the agglutination test for typhoid fever?
2. *Agglutination in blood typing (grouping)—slide agglutination*
   NOTE: Students should work in groups of two. The blood type of each student will be determined.

a. Mark a microslide down the middle with a wax pencil. Mark the left end A and the right end B. Make two wax rings, one of each end of the slide (see Fig. 20-10).

b. Place 1 drop of anti-A serum in the wax ring on the A side of the slide and a drop of anti-B serum on the B side.

NOTE: These serums may be obtained commercially. One drop of each is sufficient for the test.

c. Puncture the finger or the earlobe of the person to be tested. Discard the first drop of blood. Transfer a minute drop of blood, by means of a clean applicator, to the drop of anti-A serum, mixing to make a smooth suspension of the cells. Discard applicator. With a fresh applicator, transfer a like drop to the anti-B serum and mix thoroughly. Do *not* use the same applicator for both serums. Why?

d. Allow the slide to stand for 5 minutes, occasionally rolling or tilting it to ensure thorough mixing. Examine under the low power of the microscope.

NOTE: It may be difficult to distinguish between true agglutination and formation of **rouleaux**. Stir with no applicator. Rouleaux will be broken up, but true agglutination will be unaffected.

e. If there is no agglutination at the end of 5 minutes, cover each mixture with a cover glass and examine at 5- to 10-minute intervals, making a final reading at the end of 30 minutes.

NOTE: Sometimes a final reading must be deferred for 60 minutes, especially with group A or AB blood; in this event, ring the preparation with petroleum jelly to prevent evaporation.

f. Determine the blood type as follows:
   (1) If the cells are not agglutinated on either end of the slide, the subject belongs in group O.
   (2) If the cells on both ends are agglutinated, the subject belongs in group AB.
   (3) If the cells on the A end are agglutinated but those on the B end are not, that person belongs in group A.
   (4) If the cells on the B end are agglutinated but those on the A end are not, that person belongs in group B.

3. *Demonstration of the phenomenon of hemolysis*
   a. Prepare serial dilutions of antisheep hemolysin.
      (1) Prepare a 1:100 dilution of antisheep hemolysin (having a titer of about 1:3000) by adding 0.1 ml of hemolysin to 9.9 ml of isotonic sodium chloride solution and mixing.
      (2) Set up a row of 10 clean serologic tubes in a test tube rack.
      (3) To the first tube add 1 ml of the 1:100 hemolysin, and 0.5 ml of the 1:100 hemolysin to the ninth tube.
      (4) To the remaining tubes, except the ninth, add 0.5 ml isotonic sodium chloride solution.
      (5) To make serial dilutions of hemolysin, remove 0.5 ml from the first tube, add to the second tube, and mix. From the second tube remove 0.5 ml, transfer to the third, and mix. Continue in this manner until the contents of tube 7 have been mixed. Discard 0.5 ml of contents of this tube.
      (6) Calculate the dilution of hemolysin so produced in each tube and chart the results in Table A.
   b. Add complement.
      (1) To the first eight tubes add 0.3 ml of a 1:30 dilution of fresh guinea pig serum (complement).
      (2) Note that complement is not added to tubes 9 and 10.

Table A Serial Dilution of Hemolysin

	Tube Number									
	1	2	3	4	5	6	7	8*	9*	10*
Isotonic NaCl	—	0.5 ml	0.5 ml	0.5 ml	0.5 ml	0.5 ml	0.5 ml	0.5 ml	—	0.5 ml
Hemolysin	1.0 ml of 1:100 dilution	0.5 ml of 1:100 dilution No. 1	0.5 ml from mixture of No. 2	0.5 ml from mixture of No. 3	0.5 ml from mixture of No. 4	0.5 ml from mixture of No. 5	0.5 ml from mixture of No. 6; discard 0.5 ml	—	0.5 ml of 1:100 dilution	—
Dilution of hemolysin	1:100								1:100	
Total volume	0.5 ml	0.5 ml	0.5 ml	0.5 ml	0.5 ml	0.5 ml	0.5 ml	0.5 ml	0.5 ml	0.5 ml

*Controls.

Table B Hemolysis of Red Blood Cells

	Tube Number									
	1	2	3	4	5	6	7	8*	9*	10*
Isotonic NaCl	1.7 ml	1.7 ml	1.7 ml	1.7 ml	1.7 ml	1.7 ml	1.7 ml	1.7 ml	2.0 ml	2.5 ml
Hemolysin serial dilution (from Table A)	0.5 ml 1:100	0.5ml	0.5 ml	0.5 ml	0.5 ml	0.5 ml	0.5 ml	—	0.5 ml 1:100	—
Complement 1:30 dilution	0.3 ml	0.3 ml	0.3 ml	0.3 ml	0.3 ml	0.3 ml	0.3 ml	0.3 ml	—	—
Sheep red cell suspension, 2%	0.5 ml	0.5 ml	0.5 ml	0.5 ml	0.5 ml	0.5 ml	0.5 ml	0.5 ml	0.5 ml	0.5 ml
Presence of hemolysis?										
Final volume	3.0 ml	3.0 ml	3.0 ml	3.0 ml	3.0 ml	3.0 ml	3.0 ml	3.0 ml	3.0 ml	3.0 ml

*Controls.

Table C Suggested Combinations of Antigen and Antibody

	Antigen(s)		Antibody(ies)
	A	**B**	**Anti-AB**
Test I	Human serum from "normal" volunteer	Human serum from "normal" volunteer	Rabbit antihuman IgG
Test II	Human serum from "normal" volunteer	Cholera toxin	Mixture of rabbit antihuman IgG and rabbit anticholera toxin
Test III	Goat IgG	Human IgG	Rabbit antihuman IgG

    c. Add isotonic saline solution.
        (1) Add 1.7 ml to the first eight tubes.
        (2) Add 2 ml isotonic sodium chloride to tube 9 and 2.5 ml to tube 10.
        (3) Note that by this part of the exercise all 10 tubes contain isotonic saline solution.
    d. Add to all tubes 0.5 ml of a 2% suspension of sheep red blood cells.
    e. Incubate tubes in the water bath at 37° C for 1 hour.
    f. Observe. Note the solution of cells (hemolysis).
    g. Chart the results using Table B.
    h. Answer the following questions:
        (1) What is the highest dilution of hemolysin showing complete hemolysis?
        (2) In what dilutions is hemolysis partial?
        (3) In what dilution does hemolysis cease?
        (4) Is hemolysis present or absent in the eighth tube? Why?
        (5) Is it present in the ninth tube? Why?
        (6) What would hemolysis in the tenth tube indicate?

4. *Demonstration of precipitin reaction—agar-gel diffusion (Ouchterlony technic)*
    NOTE: Students may participate in small groups, or the instructor may demonstrate and discuss the technic. The procedure for preparing the Ouchterlony plate (Petri dish) is outlined. For the three combinations of antigen and antibody given (Table C), three such plates are needed.
    a. Pour the first of two tubes of warm Noble agar (pH 7 with Merthiolate) into a flat-bottomed Petri dish to make a thin layer about 3 mm or so. Let harden.
    b. Use the diagram below as a guide. Place Petri dish over it. Using sterile forceps, place three templates (or cylinders) over small circles in the diagram, pressing them gently into the agar. (A ¼-inch or size 2 cork borer may be used.)
    c. Pour the second tube of warm agar onto the plate. Let harden. Remove templates, leaving wells or reservoirs in the agar. Gingerly clear out excess agar from the wells. With wax pencil markings on the underside of the plate, identify reservoirs as indicated on diagram at top of p. 459.
    d. Using a sterile Pasteur pipette for each test substance, fill wells slowly with 0.2 ml amounts.
    e. Replace lid on dish. Leave plate at room temperature to be checked at regular intervals for 7 to 10 days.
    f. Draw lines representing precipitin bands on diagram.
        NOTE: Antigen and specific antibody diffuse out in the agar. Where they meet in optimal proportions, white lines of precipitation form, the precipitin bands.

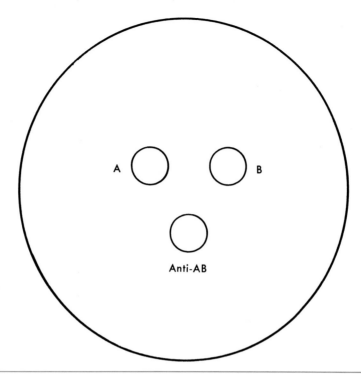

Diagram of Ouchterlony plate. *Anti-AB,* Reference label for well to contain antiserum or antibody source; A, well to contain antigen designated A; B, well to contain antigen B.

g. Interpret results.
   (1) If precipitin band forms a complete chevron, antigens A and B are serologically identical—reaction of identity.

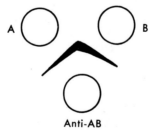

   (2) If the bands form a spur, antigens A and B are partly related serologically but one has reacted more fully with antibodies present than the other—reaction of partial identity.

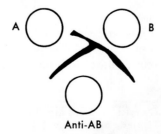

(3) If lines completely cross, anti-AB (antiserum) contains antibodies against both A and B (antigens) but they are *not* related serologically—reaction of nonidentity.

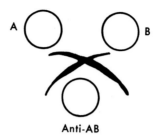

Part D—Phagocytosis

1. Examine stained smears of pus showing phagocytosis by pus cells.
2. Compare with smears of pus in which phatocytosis is absent.
3. What is the pus cell? What is its usual function?

PROJECT   Measure of Allergy

Part A—Demonstration of Anaphylaxis by the Instructor with Discussion*

NOTE: Prepare test guinea pigs 2 weeks beforehand.

1. Administer sensitizing dose—inject *intraperitoneally* 0.1 ml of horse serum (or of 5% egg albumin or other suitable foreign protein) in 2 ml of sterile isotonic saline solution.
2. Inject provocative (or shocking) dose—1.0 ml of horse serum (or of 5% egg albumin or other suitable foreign protein) *into the heart* of the test animal.
3. Observe and describe results. Encourage student comments.

EVALUATION FOR MODULE FOUR

PART I

Select the one number on the right that best represents completion of the statement or answer to the question. Please circle.

1. Members of the normal microflora in a given anatomic area:
   (a) Can become pathogens if the host's defenses are depressed
   (b) Can interfere with colonization and invasion by true pathogens
   (c) Can be confused with the true etiologic agent of a disease

   1. a, b, c, and d
   2. a, c, d, and e
   3. b, c, d, and e
   4. all of these
   5. none of these

---

*A 12-minute color and sound film, "Anaphylaxis in Guinea Pigs," no. 290, is available from the American Society for Microbiology, 1913 I St., N.W., Washington, D.C. 20006.

(d) Can be suppressed with antibiotic treatment
(e) Can immunize the host against pathogens in special circumstances when related or cross reacting antigens are shared with pathogens

2. Which of the following articles from an active case of typhoid fever most likely carry the organisms?

(a) Bedpan	1. c and d
(b) Urinal	2. e
(c) Mouth wipes with nasal secretion	3. c
(d) Eating utensils	4. a and b
(e) Bloody bandage from arm wound	5. none

3. The approximate age at which immunologic maturity occurs in human beings is:

(a) 2 months before birth	1. e
(b) At birth	2. d
(c) 3 weeks after birth	3. c
(d) 6 weeks after birth	4. b
(e) 6 months after birth	5. a

4. The skin is the common portal of entry for the causative agents of all of the following diseases of humans *except*:

(a) Malaria	1. a, b, and c
(b) Tularemia	2. a, b, c, and d
(c) Hookworm disease	3. a and c
(d) Measles	4. d and e
(e) Cholera	5. e

5. In which of the following tests is the serum from a patient mixed with a suspension of bacteria to test its ability to clump bacterial cells?

(a) Agglutination test	1. a
(b) Complement-fixation test	2. b and c
(c) Tuberculin test	3. a and d
(d) Precipitin test	4. c, d, and e
(e) Schick test	5. all but e

6. A patient has type B blood. A person (not type B) of which of the following blood types could be used in *unusual* circumstances as a donor?

(a) Type A	1. b and c
(b) Type AB	2. b and d
(c) Type B	3. d
(d) Type O	4. a
	5. all but a

Which of these types should be given to this patient under ideal conditions?

---

7. The thymus-derived lymphocytes, or T cells, have been shown to:

(a) Represent 70% of lymphocytes circulating in the peripheral blood
(b) Have a shorter life span than B lymphocytes
(c) Demonstrate less detectable immunoglobulin on their surface membranes than B cells
(d) Form characteristic rosettes
(e) Play an important role in delayed-type hypersensitivity

1. a, b, c, and d
2. b, c, d, and e
3. a, c, d, and e
4. all of these
5. none of these

8. Which of the following most nearly states your chances of developing an allergy during your lifetime?
    (a) A tendency to become hypersensitive runs in families to a noticeable degree.
    (b) Allergy is always inherited.
    (c) There is no connection between allergy and heredity.
    (d) Hypersensitivity may be inherited as a constitutional abnormality, or it may develop later in life.
    (e) If hypersensitivity runs in a family, all members will suffer the same type of allergic condition.

    1. a
    2. b
    3. c
    4. d
    5. e

9. Which of the following is an unlikely source of microbes usually responsible for pulmonary infections?
    (a) Soil
    (b) Normal microflora of host
    (c) Bloodborne infectious agents
    (d) Skin diseases
    (e) Inhaled aerosol particles greater than 20 μm in diameter

    1. a, c, and e
    2. a, c, and d
    3. d and e
    4. all of these
    5. none of these

10. The cell directly related to antibody formation is:
    (a) Polymorphonuclear neutrophilic leukocyte
    (b) Red blood cell
    (c) Plasma cell
    (d) Squamous cell
    (e) Eosinophil

    1. a
    2. b
    3. c
    4. d
    5. a and c

11. Normal body temperature is usually given as:
    (a) 37° C
    (b) 35° C
    (c) 99° F
    (d) 97.6° F
    (e) 98.6° F

    1. a
    2. a and d
    3. a and e
    4. d
    5. e

12. Koch's postulates:
    (a) Can always be satisfied
    (b) Are important in establishing that a given organism causes a given disease
    (c) Refer to Koch's procedure for isolating tubercle bacilli
    (d) Are the basis of experimental investigation of infectious disease
    (e) Are only of historical interest

    1. a and e
    2. b, c, and d
    3. e
    4. b and e
    5. b and d

13. All the following diseases are commonly transmitted to humans from animals *except*:
    (a) Rabies
    (b) Bubonic plague
    (c) Typhus fever
    (d) Psittacosis
    (e) Candidiasis

    1. a, b, and c
    2. b, c, and e
    3. e
    4. d
    5. b

14. The cells that produce significant amounts of antibody:
    (a) Are derived from the B lymphocyte population
    (b) Are the same morphologically at the time of initial stimulation with exposure to antigen as when they are synthesizing antibody
    (c) Are able to produce antibody molecules probably of only one specificity per cell
    (d) Are able to function also as killer cells
    (e) Are found in lymphoid tissues of the body

    1. a, b, and c
    2. a, b, c, and d
    3. a, c, and e
    4. a, b, d, and e
    5. all of these

15. Interferon is described as:
    (a) An antibody to a virus   1. b
    (b) A substance in the complement system   2. b and c
    (c) A substance formed by host cells   3. c
    (d) A substance with action against viruses   4. c and d
    (e) A substance comparable to an antigen   5. all of these
16. A tissue transplant from a person to an identical twin is designated:
    (a) Autograft   1. a
    (b) Allograft   2. b
    (c) Syngraft   3. c
    (d) Xenograft   4. d
    (e) Homograft   5. e
17. Complement:
    (a) Is found normally in circulation of humans   1. a
    (b) Is an antibody   2. b
    (c) Is increased by immunization   3. a and b
    (d) Is made up of components that act together in a se-   4. a, d, and e
        quence   5. a, c, and e
    (e) Is activated in combination with antigen and antibody
18. Phagocytic cells:
    (a) Are active producers of antibody   1. a and e
    (b) Are found only in the bloodstream   2. b and c
    (c) Can ingest bacteria in absence of antibody   3. a and d
    (d) Are not related to immune mechanisms   4. d
    (e) Always digest ingested bacteria   5. c
19. A certain serologic test performed on serum samples from 500 normal, presumably
    healthy (disease-free) persons is negative in 485 of the samples and positive in 15
    of them. One may say that the test is:
    (a) 97% specific   1. a
    (b) 3% specific   2. b
    (c) 97% sensitive   3. c
    (d) 3% sensitive   4. d
    (e) 97% specific and sensitive   5. e
20. The immunoglobulin found in the highest concentration in normal human serum
    is of the class:
    (a) M   1. a
    (b) G   2. b
    (c) D   3. c
    (d) E   4. d
    (e) A   5. e
21. The immunoglobulin found most prominently in human breast milk is:
    (a) M   1. a
    (b) G   2. b
    (c) D   3. c
    (d) E   4. d
    (e) A   5. e

22. Systemic anaphylaxis differs from serum sickness in that systemic anaphylaxis:
    (a) Occurs within seconds to minutes after administration   1. a and b
        of the inciting agent   2. a, b, and c
    (b) Results from smooth muscle contraction rather than   3. a, b, c, and d
        from vascular changes   4. a, c, d, and e
    (c) Usually involves a different immunoglobulin   5. a, b, and e
    (d) Results from the presence of immune complexes
    (e) Represents a medical emergency
23. Pollen that is important in inducing an allergic reaction when it is inhaled is usually:
    (a) Produced in small quantities   1. c
    (b) Windborne   2. b
    (c) Produced by brightly colored flowers   3. a and b
    (d) Larger than 500 μm in diameter   4. d
    (e) Insect-borne   5. e
24. The gel diffusion type of precipitin test is applied for:
    (a) The identification of cross reactions between antigens   1. a, b, and c
    (b) The identification of antigens from different sources   2. a only
    (c) The identification of contamination of an antigen   3. b only
        preparation with other antigens   4. c only
    (d) The detection of atopic antibodies   5. d and e
    (e) The detection of delayed hypersensitivity
25. Indirect fluorescent antibody staining involves:
    (a) Fluorescein conjugated antibody to the infectious mi-   1. a
        crobe   2. b
    (b) Fluorescein conjugated antibody to globulin constitut-   3. c
        ing antibody to the infectious microbe isolated from a   4. d
        given patient   5. e
    (c) Fluorescein conjugated antibody to complement from
        the animal species producing the antibody to the infec-
        tious microbe
    (d) Fluorescein conjugated antibody to known stock cul-
        tures of various infectious agents that might be ex-
        pected to cause the patient's disease
    (e) Fluorescein conjugated antibody to tissue agents of ex-
        perimental animals
26. An organism needing another living organism for its survival in nature is called a:
    (a) Saprophyte   1. a
    (b) Neophyte   2. b
    (c) Pathogen   3. c
    (d) Carrier   4. d
    (e) Parasite   5. e
27. The normal flora of the vagina consists of:
    (a) Lactobacilli   1. a
    (b) Herpesviruses   2. b
    (c) A protozoon—*Trichomonas*   3. c
    (d) Yeast   4. d
    (e) Pathogenic streptococci   5. e

28. Measles vaccines are licensed by:
    (a) American Medical Association
    (b) National Institutes of Health
    (c) Children's Bureau
    (d) Food and Drug Administration
    (e) American Public Health Association

    1. a
    2. b
    3. c
    4. d
    5. e

29. Authorities recommend that all children be actively immunized against the following diseases:
    (a) Pertussis
    (b) Diphtheria
    (c) Smallpox
    (d) Rabies
    (e) Scarlet fever

    1. a and c
    2. b and d
    3. c, d, and e
    4. a and b
    5. all of them

30. In which of the following diseases can toxoid be used for active immunization?
    (a) Pertussis
    (b) Typhoid fever
    (c) Diphtheria
    (d) Tetanus
    (e) Whooping cough

    1. c and d
    2. a, c, and d
    3. c and e
    4. a and e
    5. all of these

31. In which of the following is active immunization not universally accepted?
    (a) Poliomyelitis
    (b) Tuberculosis
    (c) Influenza
    (d) Typhoid fever
    (e) Diphtheria

    1. e
    2. d
    3. c
    4. b
    5. a

32. Bites from which of the following animals should be considered as necessitating postexposure prophylaxis for rabies?
    (a) Skunk
    (b) Raccoon
    (c) Coyote
    (d) Bat
    (e) Fox

    1. a only
    2. b and c
    3. a and d
    4. d only
    5. all of these

33. Active immunization against tetanus confers immunity lasting:
    (a) 6 months
    (b) 1 year
    (c) 3 years
    (d) 5 years
    (e) 10 years

    1. a
    2. b
    3. c
    4. d
    5. e

34. The World Health Organization (WHO) serves as:
    (a) An international health agency with power to enforce health and sanitary regulations
    (b) An agency concerned principally with health promotion
    (c) An agency concerned essentially with research in tropical medicine
    (d) An advisory agency that provides consultants and guidance for national health organizations
    (e) An agency concerned primarily with epidemiologic aspects of disease in the world

    1. a, b, and c
    2. a, d, and e
    3. b only
    4. d only
    5. none of these

35. A 28-year-old ranch hand is brought into the emergency room of the hospital with an infected leg wound. The information is readily obtained that this person completed at age 4 a three-dose primary immunization series for tetanus and received one booster 8 years later. A physical examination reveals that in addition to the leg wound, this person is seeing double, talking thickly, has a dry mouth, and is so weak that he is unable to stand. The treatment of this patient may include in addition to wound debridement the following:
    (a) Administration of trivalent botulinal antitoxin
    (b) Administration of botulinal toxoid
    (c) Administration of tetanus toxoid
    (d) Administration of tetanus immunoglobulin
    (e) Administration of rabies immune globulin
    1. a and b
    2. c and d
    3. a and c
    4. d only
    5. none of these

36. A positive tuberculin reaction in a 1-year-old child suggests that:
    (a) Immunity to tuberculosis has been acquired by placental transfer
    (b) Active immunization with BCG vaccine should be instituted
    (c) An infection was acquired in utero
    (d) The child has contacted an active case of tuberculosis
    (e) The child is susceptible to the primary infection
    1. e
    2. d
    3. c
    4. b
    5. a

37. Because immunization with BCG vaccine induces hypersensitivity to tuberculin:
    (a) It should not be given to a person whose Mantoux test is negative
    (b) It limits the diagnostic usefulness of the negative tuberculin reaction
    (c) It increases resistances to infection with the bovine tubercle bacillus only
    (d) It has no value in the prevention of tuberculosis
    (e) It cannot be given to infants and young children
    1. a, b, and c
    2. b and c
    3. a only
    4. all of these
    5. none of these

38. A 3-year-old child is bitten rather severely on the forearm by a vicious dog that otherwise appears to be healthy.
    (a) The dog should be killed at once, and its brain examined for Negri bodies.
    (b) The child should receive rabies vaccine at once.
    (c) The child should be watched for manifestations of rabies and given hyperimmune serum at that time.
    (d) The dog should be confined for 10 days, and then, even if it shows no signs of rabies, its brain should be examined for Negri bodies.
    (e) The dog should be confined, and at the first sign of rabies in the animal, rabies vaccine should be given to the child.
    1. c only
    2. d only
    3. a and b
    4. e only
    5. a and c

39. Reactions aided by or requiring complement demonstrate:
    (a) Antigen-antibody precipitation
    (b) Hemolysis of red cells with specific antibody
    (c) Phagocytosis of bacteria when adequate specific antibody is present
    (d) Fluorescent antibody staining
    (e) None of these
    1. a and b
    2. b and c
    3. c and d
    4. d only
    5. e only

40. Gram-positive bacteria are generally more resistant to lysozyme than gram-negative ones because:
    (a) They possess an outermembrane   1. a
    (b) They possess a more cross-linked peptidoglycan layer   2. b
    (c) They possess endotoxic activity   3. c
    (d) They may possess the O antigen   4. d
    (e) They possess a plasma membrane   5. e
41. Guinea pig serum used in a complement-fixation test must be titrated because:
    (a) Too much complement tends to give a false positive   1. a
        test   2. b
    (b) Too little complement tends to give a false negative test 3. c
    (c) The complementary activity of different lots of serum 4. d
        may vary   5. e
    (d) It contains antibodies for sheep erythrocytes
    (e) It contains antibodies for human erythrocytes
42. The immunoglobulin found in the lowest concentration in normal human serum is:
    (a) IgM   1. a
    (b) IgG   2. b
    (c) IgD   3. c
    (d) IgA   4. d
    (e) IgE   5. e
43. Of the following, which is most likely to cause a transfusion reaction?
    (a) Accidental heating of donor blood   1. a
    (b) Accidental freezing of donor blood   2. b
    (c) Technical error in cross-matching   3. c
    (d) Technical error in typing of donor or recipient   4. d
    (e) Clerical error   5. e
44. Which of the following insects can cause death from the hypersensitivity reaction induced by a bite or sting?
    (a) Honeybee   1. a and b
    (b) Fire ants   2. a, b, and d
    (c) Scorpions   3. a, b, and e
    (d) Hornets   4. b, c, and d
    (e) Black widow spiders   5. c, d, and e
45. Transplantation antibodies are bound primarily to:
    (a) Lymphocytes   1. e
    (b) Erythrocytes   2. d
    (c) Mast cells   3. c
    (d) Basophils   4. b
    (e) Eosinophils   5. a
46. All but one of the following is useful in immunosuppression:
    (a) Irradiation   1. a
    (b) Antilymphocytic globulin   2. b
    (c) Cyclosporin   3. c
    (d) Gamma globulin   4. d
    (e) Steroid hormones   5. e
47. What is the most common cause of death following an organ transplant?
    (a) Infection   1. a
    (b) Chronic rejection   2. b
    (c) Bleeding   3. c
    (d) Graft-versus-host reaction   4. d
    (e) Autoimmune disease   5. e

48. The antibody seen in systemic lupus erythematosus acts selectively against:
    (a) Cell membrane   1. a
    (b) Cell wall   2. b
    (c) Cell nucleus   3. c
    (d) Cell mitochondria   4. d
    (e) Cell cytoplasm   5. e
49. The usual site of deposition of an immune complex in the kidney is in:
    (a) The tubules   1. e
    (b) The supporting tissues   2. c
    (c) The basement membrane of the glomerulus   3. d
    (d) The large blood vessel walls   4. a
    (e) All of the above   5. b
50. B lymphocytes may transform into:
    (a) Basophils   1. a
    (b) Mast cells   2. c
    (c) Eosinophils   3. e
    (d) Plasma cells   4. b
    (e) Polymorphonuclear neutrophils   5. d
51. The cancers most likely to be seen in kidney transplant patients are:
    (a) Cancer of the lung   1. a
    (b) Cancers of the skin and lymphomas   2. b
    (c) Cancers of the urinary bladder and urethra   3. c
    (d) Cancers of the stomach and pancreas   4. d
    (e) Brain tumors   5. e
52. The donor-recipient pair with the greatest likelihood of allograft success is:
    (a) Parent-to-child transplant   1. a
    (b) Child-to-parent transplant   2. b
    (c) Sibling-to-sibling transplant   3. c
    (d) Donor unrelated to recipient   4. d
    (e) Cadaver transplant   5. e
53. Atopic allergic asthma is associated with which immunoglobulin?
    (a) A   1. a
    (b) G   2. b
    (c) M   3. c
    (d) D   4. d
    (e) E   5. e

PART II

1 Before each type of acquired immunity listed in column Alpha, please place the
  number of each method by which it may be conferred as listed in column Beta.

COLUMN ALPHA	COLUMN BETA
_____ Active	1. Injection of a vaccine
_____ Passive	2. Recovery from a given disease
	3. Injection of convalescent serum
	4. Injection of an antiserum
	5. Injection of an antitoxin
	6. Injection of a toxoid

2 Use the letter in front of the appropriate term from column Beta to designate the kind of immunity resulting from the situation indicated in column Alpha. Please place the letter in the appropriate space.

COLUMN ALPHA

_____ 1. Administration of tetanus immune globulin

_____ 2. Administration of Salk poliomyelitis vaccine

_____ 3. Administration of influenza vaccine

_____ 4. An attack of measles

_____ 5. Transfer of antibodies from the mother to her newborn child

_____ 6. An attack of whooping cough

_____ 7. Administration of antivenin

_____ 8. Administration of gamma-globulin

_____ 9. An attack of rubella (German measles)

_____ 10. Administration of duck embryo rabies vaccine

_____ 11. Administration of Sabin live poliomyelitis virus vaccine

_____ 12. Parenteral injection of convalescent serum

_____ 13. The Pasteur treatment

_____ 14. Vaccination with tetanus toxoid

_____ 15. Resistance of human beings to canine distemper

_____ 16. Injection of diphtheria toxoid

_____ 17. Resistance of Algerian sheep to anthrax

COLUMN BETA

(a) Natural immunity
(b) Naturally acquired, passive immunity
(c) Naturally acquired, active immunity
(d) Artificially acquired, passive immunity
(e) Artificially acquired, active immunity

3 On the line at the left, please place the letter of the term from the right side that does *not* bear a significant relationship to the other terms in the aspect mentioned.

_____ 1. Occurrence

(a) Sporadic
(b) Endemic
(c) Pandemic
(d) Epidemic
(e) Congenital

_____ 2. Action

(a) Phagocytes
(b) Precipitins
(c) Agglutinins
(d) Lysins
(e) Antitoxins

_____ 3. Vitamins

(a) Folic acid
(b) Polysaccharide
(c) Biotin
(d) Nicotinic acid
(e) Pantothenic acid

_____ 4. Phagocytes

(a) Scavenger cells
(b) Polys
(c) Macrophages
(d) Lymphocytes
(e) Microphages

_____ 5. Antibodies

(a) Antitoxin
(b) Agglutinin
(c) Bacteriolysin
(d) Opsonin
(e) Thymosin

_____ 6. Immediate-type allergic reactions

(a) Hives
(b) Hay fever
(c) Asthma
(d) Insect anaphylaxis
(e) Contact dermatitis

4 Match the item in column Beta to the one in column Alpha with which it is most closely associated. Please place the letter in the appropriate space.

COLUMN ALPHA

_____ 1. Widal test
_____ 2. Paul-Bunnell test
_____ 3. Fluorescent antibody test
_____ 4. Coombs' test
_____ 5. Weil-Felix reaction
_____ 6. Wassermann test
_____ 7. Histoplasmin test
_____ 8. Neutralization test
_____ 9. Dick test
_____ 10. Schick test
_____ 11. Coccidioidin test
_____ 12. Tuberculin test
_____ 13. Gel diffusion
_____ 14. Patch test

COLUMN BETA

(a) Skin test to detect cause of contact dermatitis
(b) Antiglobulin test
(c) Agglutination test
(d) Heterophil antibody test
(e) Skin test for fungal disease
(f) Skin test for bacterial allergy
(g) Skin test for chlamydial infection
(h) Skin test for susceptibility to scarlet fever
(i) Immunologic reaction requiring special microscope
(j) Precipitin reaction
(k) Serologic test for viruses
(l) Complement-fixation test
(m) Skin test for susceptibility to diphtheria
(n) Serologic test for syphilis

5 Comparisons: Match the item in column Beta to the word or phrase in column Alpha best describing it. Please place the indicated letter in the appropriate space.

COLUMN ALPHA
Humoral Immunity with Cell-Mediated Immunity

_____ 1. Graft rejection an example
_____ 2. Major role in resistance to viruses
_____ 3. Major role in resistance to bacteria
_____ 4. Presence of antigen not necessary
_____ 5. Passive immunization possible
_____ 6. Immune response within hours
_____ 7. Immune response within days
_____ 8. Contact with antigen the trigger mechanism
_____ 9. Antigen reaches lymphoid tissues
_____ 10. Postulated role in cancer immunology
_____ 11. Mediating cells thymus-dependent lymphocytes
_____ 12. Mediating cells thymus-independent lymphoid cells
_____ 13. Part of body's defense against infection
_____ 14. Presence of demonstrable antibody
_____ 15. Presence of penicillinase
_____ 16. Important molecules are antibodies
_____ 17. Important molecules are lymphokines

COLUMN BETA

(a) Humoral immunity
(b) Cell-mediated immunity
(c) Both
(d) Neither

COLUMN ALPHA   COLUMN BETA
Immediate-type Allergic Reaction with Delayed-type Allergic Reaction

_____ 18. Not associated with circulating anti-   (a) Immediate-type allergy
              body   (b) Delayed-type allergy
_____ 19. Similar symptoms induced by hista-   (c) Both
              mine   (d) Neither
_____ 20. Drug allergies
_____ 21. Passive transfer with cells or cell frac-
              tions of lymphoid series
_____ 22. Short duration
_____ 23. Prolonged duration
_____ 24. Contact dermatitis
_____ 25. Anaphylactic shock
_____ 26. Eosinophilia present
_____ 27. Drug idiosyncrasy
_____ 28. Triggered by well-defined allergen
              usually
_____ 29. Emotional upsets
_____ 30. Asthma

PART III

1 Use the letter in front of the appropriate term from column Beta to indicate the nature of the immunizing substance in column Alpha. Please place it in the appropriate space.

COLUMN ALPHA   COLUMN BETA

_____ 1. Antivenin (Crotalidae)   (a) Dead bacteria
_____ 2. BCG vaccine   (b) Attenuated bacteria
_____ 3. Botulinal antitoxin   (c) Inactivated virus
_____ 4. Diphtheria toxoid   (d) Live, attenuated virus
_____ 5. Gamma-globulin   (e) Modified exotoxin
_____ 6. Hepatitis B vaccine   (f) Antibodies (immunoglobulins)
_____ 7. Influenza vaccine   (g) Capsular material from bacterium
_____ 8. Measles vaccine   (h) Inactivated antigen particles from
_____ 9. Pneumococcal polysaccha-   virus
          ride vaccine
_____ 10. Rabies chick embryo vac-
          cine for dogs
_____ 11. Rabies vaccine (human dip-
          loid cell)
_____ 12. RhoGAM
_____ 13. Sabin poliomyelitis vaccine
_____ 14. Salk poliomyelitis vaccine
_____ 15. Zoster immune globulin

2 Consider the characteristics listed in the column on the left. Please indicate by a check mark in the appropriate column the relation to exotoxin or endotoxin.

	EXOTOXIN	ENDOTOXIN
Found in gram-positive bacteria	☐	☐
Found in gram-negative bacteria	☐	☐
Great toxicity	☐	☐
Polypeptide mostly	☐	☐
Lipopolysaccharide complex	☐	☐
Important in diphtheria, tetanus	☐	☐
Weakly antigenic	☐	☐
Conversion to toxoid	☐	☐
Extracellular	☐	☐
Specific tissue affinity observed	☐	☐
Unstable	☐	☐
Important in salmonelloses	☐	☐

3 True-False. Please circle either the *T* or *F*.

T F   1. A given microbe may invade the body by any portal of entry and leave by any portal of exit.

T F   2. Most microbes cross the placenta.

T F   3. Microbes do not need to adhere to surfaces to colonize them.

T F   4. The agent of a communicable disease is spread directly or indirectly from host to host.

T F   5. Members of the same microbial species vary in virulence.

T F   6. Salmonellosis is an example of a zoonosis.

T F   7. The process of ingestion of particles by cells is referred to as phagocytosis.

T F   8. Lymphokines are regulatory proteins in cell-mediated immunity.

T F   9. An example of a lymphokine is an interleukin.

T F   10. Antibiotic therapy may drastically change the pattern of the resident flora in the human body.

T F   11. Viruses may promote colonization of the respiratory tract by pathogenic bacteria.

T F   12. Microbes undergo no cycle of development in a biologic vector.

T F   13. Inanimate objects spreading infection are referred to as fomites.

T F   14. Bacteria in the bloodstream but not multiplying there constitute a bacteremia.

T F   15. Microbes can never survive the environment of the cytoplasm of a macrophage.

T F   16. Microbes may invade an area of the body without producing disease there.

T F   17. Inflammation is the human body's response to injury.

T F   18. Leukocytosis is an increase in the number of leukocytes in the human body.

T F   19. Elective localization indicates that most pathogenic microorganisms can localize anywhere in the body that seems favorable at the time.

T F   20. Malaria and encephalitis are spread by biologic vectors.

T F   21. Infections of animals secondarily transmitted to humans are zoonoses.

T F   22. Three doses of inactivated hepatitis B vaccine will provide, it is believed, lifelong immunity to hepatitis B virus.

T F   23. Immunocompromised adults should be treated with doses of hepatitis B vaccine twice those given to healthy adults.

T F   24. Cyclosporin is of major importance in immunosuppression.

T F   25. Acute poststreptococcal glomerulonephritis is not an immune complex disease.

T F   26. Alpha-fetoprotein is a fetal antigen.

T F   27. More than 50 diseases have been associated with the HLA complex.

T F   28. Complement system deficiencies are characterized by an increased susceptibility to infection.

T F   29. Autoantibodies may exist without apparent disease.

T F   30. Active immunization for prophylaxis of hepatitis B is currently available in the form of an inactivated hepatitis B vaccine.

T F   31. The autoantigens postulated to be involved in autoimmune diseases are in every instance unknown.

T F   32. The HLA complex controls allograft immunity.

PART IV

1 After the initial statement or phrase, please indicate by a checkmark which of the items in the list apply. Otherwise leave the space blank.

The following are harmful bacterial products:

_____ Hemolysins	_____ Kinases	_____ Anatoxins
_____ Leukocidins	_____ Crotin	_____ Oxidase
_____ Opsonins	_____ Toxoids	_____ Interferon
_____ Ricin	_____ Hyaluronidase	_____ Lymphotoxin
_____ Coagulase	_____ Indole	_____ Lymphokines

The following are cells with phagocytic properties:

_____ Eosinophil	_____ Basophil	_____ Kupffer cell
_____ B lymphocyte	_____ Macrophage	_____ Null cell
_____ T lymphocyte	_____ Polymorphonuclear	_____ Monocyte
_____ Plasma cell	neutrophil	

AIDS:

_____ Is a recently described disease

_____ Is characterized by a loss of immune function

_____ Is not associated with opportunistic infections

_____ Is relatively benign

_____ Is a disease predominantly of intravenous drug users

_____ Is a disease of unknown cause

_____ Is characterized by a selective impairment of T cell–mediated immunity

_____ Is characterized by a long incubation period (in months)

_____ Is frequently transmitted by blood transfusions

_____ Is characterized by a lymphopenia and an imbalance between T and B lymphocytes

_____ Is commonly complicated by Kaposi's sarcoma

_____ Is characterized by a normal number of B lymphocytes

Monoclonal antibodies:

_____ Are produced by certain hybrid cells
_____ Are nonspecific antibodies, unlike polyclonal antibodies
_____ Have many applications
_____ Are derived from hybridomas
_____ Are not approved for commercial use
_____ Are derived from a process that fuses an antibody-producing B cell with an
       "immortal" myeloma plasma cell
_____ Are specific for multiple antigenic determinants
_____ Make up highly specific reagents for diagnostic tests
_____ Are absolutely homogeneous
_____ Are more easily concentrated and purified than polyclonal antibodies

2 Match the item in column Beta to the one in column Alpha with which it is most
closely related.

COLUMN ALPHA
Types of Hypersensitivity Reactions

_____ 1. Immediate hypersensitivity
_____ 2. Serum sickness
_____ 3. Cytotoxic reaction
_____ 4. Contact dermatitis
_____ 5. Immune complex disease
_____ 6. Anaphylaxis
_____ 7. Asthma
_____ 8. Tuberculin reaction
_____ 9. Poison ivy
_____ 10. Hemolytic transfusion reaction
_____ 11. Drug allergies
_____ 12. Acute poststreptococcal glomerulone-
      phritis
_____ 13. Delayed hypersensitivity
_____ 14. Hay fever
_____ 15. Arthus phenomenon

COLUMN BETA

(a) IgE
(b) IgG, IgM
(c) T lymphocytes
(d) B lymphocytes
(e) Null cells

Characteristics of Lymphocytes

_____ 1. Processed by the thymus gland
_____ 2. Role in immunity unknown
_____ 3. Killer cells found here
_____ 4. Cell with short life span
_____ 5. Cell with long life span
_____ 6. Most of circulating lymphocytes in
      bloodstream
_____ 7. Transformation into plasma cells
_____ 8. Development primarily in liver
_____ 9. Processed primarily by bone marrow
_____ 10. Concentrated in peripheral lymph or-
      gans
_____ 11. Cells containing no surface markers
_____ 12. Recognition of antigen in peripheral
      lymphoid organs

(a) B lymphocyte population
(b) T lymphocyte population
(c) Neither
(d) Both

Stages of Disease

_____ 1. Interval after infection before appearance of disease

_____ 2. Disease at its height

_____ 3. Manifestations of disease subside

_____ 4. Disease developing to maximum intensity

_____ 5. Ill-defined period wherein manifestations of disease are seen but before disease is full blown

(a) Incubation period
(b) Prodrome
(c) Period of invasion
(d) Fastigium
(e) Defervescence

MICROBES
PATHOGENS AND PARASITES

PYOGENIC COCCI

The pathogenic cocci are often called the *pyogenic* cocci because of their ability to cause pus formation (suppuration). The important gram-positive cocci are the staphylococci (genus *Staphylococcus*), the streptococci, and the pneumococci (genus *Streptococcus*).

The pneumococcus has long been considered a streptococcus by many authorities who have even classified it in the same genus. Recognizing the biologic similarity of the two, the eighth edition of *Bergey's Manual* reclassifies the pneumococcus. Formerly *Diplococcus pneumoniae*, it is now *Streptococcus pneumoniae*, a name long used in some parts of the world. In this chapter the pneumococcus is discussed separately because of certain properties peculiar to it and because of its relation to an important disease.

STAPHYLOCOCCUS SPECIES (THE STAPHYLOCOCCI)
General Considerations
General Characteristics

Staphylococci occur typically in grapelike clusters. Individual cocci, approximately 1 μm in diameter, tend to cluster because their cell division occurs in three planes, and daughter cells so formed tend to remain close by. Under special conditions, staphylococci may occur singly (Fig. 23-1), in pairs, or in short chains. They are gram positive, nonmotile, and nonspore-forming and grow luxuriantly on all culture media, generally forming opaque, smooth, and glistening colonies. Since staphylococci produce the enzyme catalase, the hydrogen peroxide formed as a metabolite under aerobic conditions is not toxic to them, and most grow best in oxygen. They also grow easily in the absence of oxygen, and a few are strict anaerobes. They grow best with temperatures between 25° and 35° C but may grow at a temperature as low as 8° C or as high as 48° C.

When staphylococci are grown aerobically on blood agar, characteristic pigments appear. (No pigment is seen under anaerobic conditions.) The colors range from deep gold to lemon yellow to white. The deep golden color observed with growth was originally responsible for the species name, *aureus*. We now know that there are white variants of the golden staphylococci.

Among bacteria, staphylococci are most resistant to the action of heat, light, drying, extremes of temperature, and chemicals. Although most vegetative bacteria are destroyed when exposed to a temperature of 60° C for 30 minutes, staphylococci frequently resist a temperature of 60° C for 1 hour, and some strains may resist 80° C for 30 minutes. Because of their resistance to drying, they can be carried on dust particles and live for weeks or months in dried pus or sputum. Another mechanism for staphylococcal survival is their tolerance of salt or a salty medium (up to a concentration of 15% sodium chloride), as found in preserved foods. In a medium that would otherwise be satisfactory for food preservation (for example, a ham cured with salt), staphylococci grow and form their enterotoxin. These microbes are resistant to the action of phenols and many other disinfectants, but they are very sensitive to the basic dyes. Staphylococci also tend to become resistant to the sulfonamides and antibiotics. They adapt quickly and easily to such agents. About 80% are penicillin resistant.

There are two species of note for the genus *Staphylococcus*: *Staphylococcus aureus* and *Staphylococcus epidermidis* (formerly *Staphylococcus albus*), two organisms microbiologically separated by a biochemical reaction and the presence of an enzyme. *S. aureus* ferments mannitol, which the other does not, and is coagulase positive. *S. epidermidis* is coagulase negative.

A third species, *Staphylococcus saprophyticus*, deserves mention as a cause of urinary tract infection. Primarily a saprophyte, as its name suggests, it perhaps has a limited invasive capacity. It is coagulase negative.

Toxic Products

Several metabolic products with notable toxic properties are elaborated by staphylococci. Probably no other bacterium produces as many. These include extracellular toxins, hemolysins (staphylolysins), enzymes, and the like, all of which at one time or another have been seriously implicated in the virulence or disease-producing capacity of the organisms. No single strain produces all of these, and many produce none of them. Some of the more important ones are considered here.

Hemolysins are exotoxins released by staphylococci, the activity of which is directed to the cell membrane. Because of this, red blood cells, white blood cells, platelets,

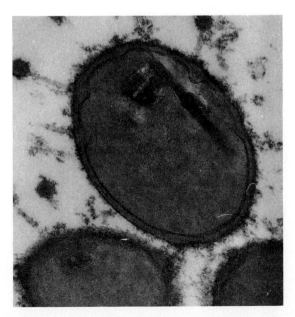

FIG. 23-1 Staphylococcus, electron micrograph. Note attached staphylophages. (Courtesy S. Tyrone, University of Texas Health Science Center, Dallas.)

and macrophages are lysed, and many tissue cells are injured as well. The action of hemolysins probably contributes to all fatal staphylococcal infections.

Colonial growth of the staphylococcus exhibits either the clear zones of beta-type hemolysis or no hemolysis at all. Unlike the growth of the streptococcus, the growth of the staphylococcus is *not* associated with partial or alpha hemolysis (p. 297). (Note that Greek letters are used to designate immunologically distinct types of staphylococcal he-molysins, whereas with streptococci, the Greek letters designate types of hemolysis. For instance, a hemolysin, termed alpha hemolysin, alpha lysin, or alpha toxin, is re-sponsible for the *clear* zones of hemolysis found around some colonies of staphylococci.) One of the most important staphylococcal virulence factors, a cytotoxin, causes aggre-gation of platelets and acts selectively on the smooth muscle of small veins.

Leukocidin, a nonhemolytic exotoxin, destroys white blood cells. It degranulates the polymorphonuclear neutrophilic leukocyte (the pus cell) and the macrophage.

Enterotoxin, an extracellular toxin elaborated by about 50% of coagulase-positive strains of staphylococci, particularly of phage group III (p. 486), is the most common cause of food poisoning in the western world. The production of this toxin, the action of which is directly on the vomiting center in the central nervous system, is mediated by a bacteriophage (a staphylophage), and five types, A, B, C, D, and E, are demon-strable. Preformed enterotoxin in contaminated food is responsible for staphylococcal food poisoning, a gastroenteric syndrome that is not really an infection but rather a toxemia. The presence of toxin in suspect food can be proved immunologically by a precipitin test.

Exfoliatin (exfoliative toxin), found in staphylococci of phage group II, is an exo-toxin, the production of which is mediated by a plasmid. It selectively damages certain cells of the skin so that large sheets of lining cells may be peeled therefrom. Young infants and neonates are peculiarly vulnerable to the action of this toxin. The appear-

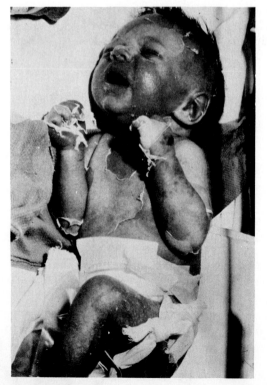

FIG. 23-2   Scalded skin syndrome of an infant. Note peeling of superficial skin layer. (From Melish, M.E., and Glasgow, L.A.: N. Engl. J. Med. **282:**114, 1970.)

ance of the injured skin led to the descriptive designation of Ritter's disease as the **scalded skin syndrome** (Fig. 23-2).

Coagulase, an important extracellular enzyme produced *only* by some staphylococci, causes the plasma of the blood to clot. In the laboratory the coagulase test indicates the presence of coagulase, a finding that is generally accepted as the best single bit of evidence for the pathogenicity of a staphylococcus. Avirulent strains repeatedly fail to show coagulase. Because of this enzyme, a surface layer of fibrin accumulates on an individual coccus and protects it from phagocytic attack, and the formation of coagulase correlates well with the production of other toxic products. Coagulase-positive staphylococci also possess **clumping factor;** a cell-bound but antigenically distinct form of coagulase, it induces a rapid clumping of cells emulsified in a drop of plasma.

Lipases of staphylococci are enzymes that break down lipids in skin structures and lipoproteins in the blood. The fact that staphylococci utilize the metabolites so derived may explain the intense colonization of staphylococci on the skin surface. Lipase formation is correlated with staphylococcal invasion of healthy skin and subcutaneous tissues, as a consequence of which localized abscesses form. Strains without lipase are more likely to be related to generalized lesions.

Other extracellular enzymes of staphylococci include hyaluronidase, nuclease, and staphylokinase. **Hyaluronidase** (the spreading factor, invasin), formed by more than 90% of pathogenic staphylococci, increases the permeability of tissues to the cocci and their toxic substances (hyaluronic acid, one of the substances cementing tissue cells together, is degraded). **Nuclease,** found in 90% to 96% of S. aureus, cleaves DNA and RNA. (The nuclease of S. aureus is thermostabile; the nuclease of other, coagulase-negative staphylococci is not.) **Staphylokinase** dissolves fibrin clots and as a consequence contributes to the spread of a focal, initially confined infection.

Pathogenicity

Some staphylococci are nonpathogenic; others, such as S. aureus, produce severe infections. S. epidermidis, though sometimes responsible for very mild, limited infections, is generally considered nonpathogenic except under unusual medical circumstances. (It has a predilection for therapeutic devices implanted in the body and foreign to it. Currently, S. epidermidis is considered a frequent cause of endocarditis of prosthetic heart valves and infections of prosthetic orthopedic devices and neurosurgical shunts.)

The best known staphylococcal infections are those of the skin and superficial tissues of the body, where staphylococci cause boils, pustules, pimples, furuncles, abscesses, carbuncles, paronychias, impetigo contagiosa (Fig. 23-3), and infections of accidental or surgical wounds. Features of skin infections may be influenced by the age of the patient. For instance, the scalded skin syndrome occurs in the very young (Fig. 23-2).

Staphylococci also produce systemic disease, and all organ systems in the body may be affected. They are one cause of pneumonia, empyema, endocarditis, meningitis, brain abscess, puerperal fever, parotitis, phlebitis, cystitis, and pyelonephritis. Staphylococcal pneumonia has been a fatal complication in epidemic influenza. Staphylococci are the most common cause of osteomyelitis. Systemic staphylococcal disease is often acquired in the hospital, especially in patients already seriously ill, and staphylococcal pneumonia as a superinfection is a threat after large doses of antibiotics have been given.

Staphylococcal septicemia assumes two forms. The first is a fulminating, profound toxemia followed by death a few days later. The second and more frequent form lasts longer, is typically severe with metastatic abscesses that are formed in different parts of the body, but is not inevitably fatal. Staphylococcal septicemia may be primary, but mostly it results from secondary invasion of the bloodstream by organisms from a localized site of infection in the skin (often a trivial one). It can come from an infected wound, a dental abscess, pneumonia, or an infected intravenous catheter site. "Main line" drug addicts (those injecting the drug themselves into their own veins) are prone to develop a staphylococcal bacteremia. An indwelling intravenous catheter should not remain in place longer than 3 to 4 days because of the risk of serious infection. Boils around the nose and lip are easily complicated by septicemia and for this reason should not be traumatized.

The "dangerous triangle" is a triangular area of the face with its base along the opening of the mouth and its apex in the region above the upper part of the nose (Fig. 23-4). A person's life is threatened if infection originating there spreads backward into the cranial vault. It is a peculiar area in that anatomic factors there favor just such a disaster. (A comparable lesion in another part of the body would be inconsequential.) Such factors include poor connective supports that provide no mechanical barriers, veins without valves connecting with veins that drain backward, and facial muscles that are more or less constantly in motion.

If piercing the ears (for cosmetic reasons) is done in an unsanitary setting, there is danger of secondary infection with potentially deadly staphylococci. They can easily enter such a wound that is kept open for several days, and in rare instances the spread of infection to the bloodstream has been fatal.

On occasion, implantation of staphylococci in the intestinal tract after antibiotic therapy results in enteritis or enterocolitis causing dysentery. Staphylococcal food poisoning is the most frequent type of food poisoning from ingestion of a bacterial toxin.

Staphylococci are an important cause of suppurative conditions in cattle and horses. Mastitis of staphylococcal origin in cows can be transmitted to other cows by the hands of the milker. Staphylococcal bacteremia may follow tick bites in lambs.

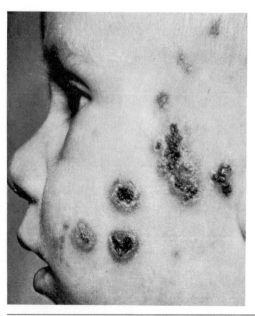

FIG. 23-3   Impetigo contagiosa, highly communicable skin disease. Note various-sized lesions, some dark and encrusted. (From Top, F.H., Sr., and Wehrle, P.F., editors: Communicable and infectious diseases, ed. 8, St. Louis, 1976, The C.V. Mosby Co.)

FIG. 23-4   Triangular area of face where staphylococcal infection is very dangerous.

Hospital-Acquired (Nosocomial) Staphylococcal Infection

Because of human beings' intimate contact with the ubiquitous staphylococcus, its human infections at any time or in any place receive serious consideration, and such is the case for hospital-acquired infections related to staphylococci. For a number of years, these infections have been troublesome in hospitals all over the world.

Consider the following background factors conducive to hospital-acquired staphylococcal infections. First, the problem is a complication of modern antibiotic therapy. Important antimicrobial agents are freely given, but antibiotics generally are bacteriostatic, *not* bactericidal. Staphylococci are well endowed for survival, and consequently antibiotic-resistant strains develop. Second, prepaid medical plans and medical advances favoring early detection of disease mean more persons are now treated in hospitals. Third, modern surgery is expanded. Complicated surgical technics effect a greater exposure of tissue for a longer period than ever before. Fourth, selected forms of modern therapy result in the immunosuppressed patient—that patient whose natural resistance mechanisms to infection have been intentionally depressed, such as the patient before a kidney transplant to whom drugs that depress the patient's resistance to infection are administered.

The hospital employee population is already large and complex, but it is growing with the increased demand for persons with specialized technical skills. Here, then, is a patient, sometimes seriously ill, confined within a large institution where contacts with all kinds of persons are many and varied. Consider these factors in light of the fact that staphylococci are everywhere. Is it any wonder they are such troublemakers!

Four major categories of disease caused by antibiotic-resistant staphylococci related to hospital-acquired infection are stressed:

1. Skin abscesses (pyoderma—an inclusive term for purulent, or pus-forming, lesions of skin and subcutaneous tissues) in newborn infants (Many of these infants develop breast abscess. Fatal staphylococcal pneumonia or septicemia is prevalent. The nursing mother can pick up a virulent organism from her baby. Abscess formation in her breast may result.)
2. Wound infections, especially of surgical wounds
3. Secondary staphylococcal infections in hospitalized elderly and debilitated persons
4. Gastroenteritis, as a result of a change in the bacterial flora of the intestinal tract

Toxic Shock Syndrome

Toxic shock syndrome (TSS) was first described and named in 1978, although it is now believed that this disorder can be traced as far back as the 1920s. Probably the disease at one time was termed *staphylococcal scarlet fever*. Up until the late 1970s, cases of TSS were not related to menstruation.

In 1980 an epidemic of TSS occurred, this time in menstruating women, and it was quickly linked to the use of high-absorbency tampons. In the first 9 months of 1980 299 cases of TSS with 25 deaths were reported to the Centers for Disease Control (CDC) in Atlanta.

Clinical features. The disease of the 1980s typically afflicts healthy young women during their menstrual period; one third of the cases reported are in adolescents, who seem especially vulnerable. Although most cases occur in menstruating women, some are seen in nonmenstruating women, men, and children under a variety of circumstances. The incubation period is short, about 12 to 24 hours.

The course of the disease is dramatic, with a sudden onset of high fever, vomiting, diarrhea, muscle pains, and rapid progression to hypotension (low blood pressure) and shock. A sunburnlike rash appears, which peels a week or so later, particularly on the palms of the hands and the soles of the feet. Manifestations of widespread involvement of the organ systems of the body are observed; for instance, the patient may be disoriented, exhibit respiratory distress, or cease to put out normal amounts of urine. The mortality is between 3% and 10%.

The agent. Certain strains of toxigenic S. *aureus* cause TSS. A preexisting soft tissue infection with these toxigenic strains or mucous membrane colonization is a prerequisite. In tampon-associated disease, colonization has occurred on the mucous membranes of the vagina or cervix.

Staphylococci that express large amounts of hemolysin, lipase, and nuclease usually inhabit the skin. Lipase facilitates their survival there. Staphylococci that elaborate these extracellular products only in small amounts may be relegated to mucous membranes. Indeed, such a staphylococcus is isolated from patients with menstrual toxic shock. Although it seems harmless, this staphylococcus can, under the right conditions, make significant amounts of an exotoxin that explains the manifestations of the disease.

The toxin. Interestingly, many manifestations of toxic shock are like those of scarlet fever, a disease in which a known toxin exerts a major effect. Two toxins (both proteins), pyrogenic exotoxin C and enterotoxin F, elaborated by S. *aureus* strains have been isolated from TSS patients. (Because of biochemical and immunologic similarities, some observers believe them to be the same.) Their combined action seems to induce the fever, enhance the skin reactivity, and depress the immune mechanisms of the body.

Relation to tampon use. The epidemic of 1980 was associated with the use of highly absorbent tampons introduced in the late 1970s (for example, Rely). Of cases reported to the CDC, 85% occurred in women menstruating at the time they got sick, and 98% of those women were using a high-risk tampon. The highly absorbent material of the

Superabsorbency of the tampon (not the brand) is the highest risk factor in menstrually associated toxic shock syndrome. At a recent meeting of the New York City branch of the American Society for Microbiology, Bruce A. Hanna, Ph.D., and Philip Tierno, Ph.D., focused on characteristics of Rely (tampon brand) that make it an ideal growth substrate for S. *aureus,* the causative agent in toxic shock syndrome. First of all, beta-glucosidase, a bacterial enzyme normally present in the vagina, breaks down Rely's superabsorbent material, carboxymethylcellulose, to produce glucose. In a laboratory experiment, Drs. Hanna and Tierno showed that glucose enhances the growth of toxigenic staphylococci in the tampon's superabsorbent material. According to these two microbiologists, the following features of the Rely tampon make it an ideal medium for bacterial growth and a substrate that resists normal body defenses. (1) The carboxymethylcellulose, as it absorbs the menstrual fluid, changes from a solid compressed chip state to a gelatinous mass. Bacteria are trapped inside the gel, where they are protected from the phagocytic action of the body's white blood cells. (2) The gelatinous state of the tampon encourages more efficient production of staphylococcal enterotoxin. (3) Sodium in carboxymethylcellulose increases the sodium concentration in menstrual fluid by about 50%—sodium chloride is a selective medium for staphylococci. (4) The physical construction of the Rely tampon, a "tea bag" with neither inside seams nor cross-stitching, allows components to float freely within it. Dr. Hanna concludes, "You have to remember that the vagina is heavily colonized with microorganisms. By occluding this space in the body you provide conditions that are . . .analogous to closing off an abscess."

Modified from Baer, K: J.A.M.A. 247:2339, 1982. Copyright 1982, American Medical Association.

tampon, carboxymethylcellulose in the case of Rely, is believed to create an environment in the vagina favoring the growth of toxigenic staphylococci and conducive to
both toxin production and absorption.

Pathology

The hallmark of staphylococcal infection is the **abscess,** the type lesion. It reflects the
excellent pus-forming ability of staphylococci and their generally limited capacity for
spread. Since the microbes reside on the skin, this is the area most frequently involved.
A boil is a skin abscess. Abscess formation is modified by anatomic location and degree
of involvement. Staphylococcal pneumonia means multiple abscesses in the lungs. Pyelonephritis means multiple abscesses throughout the excretory system of the kidney.
Staphylococcal septicemia is the development of multiple abscesses over the body; the
word *pyemia*, literally meaning "pus in the blood," is more appropriate. When staphylococci infect wounds, they induce pus formation therein. They are an important
cause of wound infections. Staphylococci are the cause of 80% of suppurative disorders
in humans (disease associated with pus formation).

Sources and Modes of Infection

Staphylococci are normal inhabitants of the human skin, mouth, throat, and nose.
They live in these areas without effect, but once past the barrier of the skin and mucous
membrane, they can cause extensive disease. Under certain conditions, they penetrate
the unbroken skin via the hair follicles and ducts of the sweat glands. The natural
invasive traits of staphylococci and the resistance of the body are so well balanced that
infection probably never occurs unless a highly virulent organism is encountered or
body resistance is lowered. As a rule, a localized process such as an abscess or a boil is
first formed. Healing without widespread dissemination of infection usually takes place,
but in some cases organisms do escape from the localized process, invade the bloodstream, and affect distant parts of the body.

Crucial to the spread of staphylococcal infections are direct, person-to-person contacts. For example, in the hospital the hands of the doctor or medical attendant may
carry the infection from one patient to another. Staphylococci are abundant in hospitals. The hospital staff generally has a high carrier rate, and cross-infection is significant. Nasal carriers are a source. Virulent organisms may also be passed to humans
from livestock and household pets.

Bacteriologic Diagnosis

A bacteriologic diagnosis of a staphylococcal infection is made easily by Gram-stained
smears and cultures. When blood cultures for staphylococci are made, special pains
must be taken to exclude those resident on the skin. The coagulase test (slide and test
tube methods) is performed to differentiate nonpathogenic from pathogenic staphylococci. A freshly isolated staphylococcus is most likely to be a virulent pathogen if it
produces a yellow pigment, hemolyzes blood, ferments mannitol, elaborates deoxyribonuclease (DNAse), and is coagulase positive. Demonstration of the enzyme catalase
distinguishes staphylococci from streptococci, which are catalase negative.

Phage typing. Bacteriophages (p. 691) are viruses that attack bacteria and in certain
instances lyse (dissolve) the bacterial cell parasitized. The action of phages is specific—
only a certain phage or group of phages affects the given strain of bacteria. For this
reason bacteriophage or phage typing can be carried out. To do so, a suspension of
known bacteriophage type is deposited on an agar plate freshly inoculated with the
microbe to be tested or typed. If that microbe is susceptible to the action of the particular phage, it is lysed—it does not grow on the agar plate. The indication is a clear
area on the plate, known as a **plaque** (p. 693).

The bacteriophage typing of staphylococci deserves special mention. It has been found that specific bacteriophages (staphylophages, Fig. 23-1) react with about 60% of coagulase-positive staphylococci. Coagulase-negative strains of staphylococci are not so susceptible. Because of this specificity, phage typing of S. *aureus* can be done. For convenience, bacteriophages have been given identification numbers, and the strains of staphylococci related to these particular phages are designated by the number of the bacteriophages or phages lysing them.

Because of an increasing number of staphylococci that are not typable with the standard international basic set of phages, in 1975 the International Subcommittee on Phage Typing of Staphylococci recommended a revised set of typing phages for classification of S. *aureus* of human origin. On this revised basis, S. *aureus* can be placed into groups lysed by staphylococcal typing phages, as shown in Table 23-1.

Phage typing is a valuable epidemiologic tool because it provides for an epidemiologic "fingerprinting" of staphylococci and certain other bacteria. By means of this specific kind of identification, their role in the transmission of epidemic disease is conveniently established.

Immunity

Humans possess considerable natural immunity to staphylococci. Specific serum antibodies to staphylococci can be demonstrated in most human beings, since they are the result of intermittent minor staphylococcal infections of the skin and mucous membranes. Acquired immunity of a protective nature may exist, but it is hardly a practical mechanism of defense against serious staphylococcal disease. Patients with debilitating disorders, particularly diabetes and sometimes viral infections, are especially at risk.

Prevention and Control of Staphylococcal Infection

The key measures in the control of staphylococcal infections in hospitals (or anywhere) are the maintenance of good housekeeping standards (see box below) and adherence to strict aseptic technics. There is strong evidence that currently pathogenic strains of staphylococci are as susceptible to chemical germicides as are the ordinary nonpatho-

Table 23-1 International Classification of Staphylococcal Phage Types

Revised Basic Set Phage Group	Individual Phages	Common Phage Types of Staphylococci
I	29, 52, 52A, 79, 80	29, 52/52A, 52/52A/80/81, 80
II	3A, 3B, 3C, 55, 71	3A/3B/3C/, 3C/55
III	6, 7, 42E, 47, 53, 54, 75, 77, 83A, 84, 85	6/7/47/53/54/75/77 and many others, usually complex
IV	42D	—
Miscellaneous	81, 187	—

Recommendation of the International Subcommittee on Phage Typing of Staphylococci.

Welton Taylor has said:
Perhaps the most important ally in the hospital's campaign against hospital-borne infection is the housekeeper, who merely has to employ the common-sense sanitation that any good housewife knows.

From Taylor, W.: Hosp. Tribune 7:3, Oct. 9, 1967.

genic strains. The detection of nasal carriers, especially in the nurseries for newborn infants, is critical.

Phage typing of coagulase-positive staphylococci has proved valuable in the epidemiologic study of hospital-acquired infections to trace the sources of infection.

If one strain of coagulase-positive staphylococcus is biologically anchored to a given site in the human body, another coagulase-positive strain cannot grow there. The colonization of the second strain is blocked by the **bacterial interference** between the two. Bacterial interference may result from the competition for some particular nutrient or from the action of an inhibitor agent. In hospital nurseries practical application of this phenomenon is sometimes made by careful and selective colonization of the newborn infants with a nonpathogenic staphylococcus to protect them from colonization and subsequent infection with a pathogenic one.

STREPTOCOCCUS SPECIES (THE STREPTOCOCCI)
General Considerations

The term *streptococcus* is morphologic and includes cocci that occur in pairs and chains. The combination of two Greek words, *strepto*, twisted, plus *kokkos*, berry, to name these organisms was first suggested by the famous surgeon Theodor Billroth in 1874. The genus *Streptococcus* is distinguished by a biochemical feature. The members ferment glucose by the hexose diphosphate pathway to yield mainly dextrorotatory lactic acid. In the genus are micoorganisms with significant variations in cultural characteristics and disease-producing capacities. Some produce deadly disease, others do so only under special conditions, and still others are nonpathogenic. As a whole, streptococci are probably responsible for more illness and cause more different kinds of disease than any other group of organisms. They attack any part of the body and can cause primary as easily as secondary disease. They attack both human beings and lower animals. Some occur as saprophytes in milk and other dairy products.

A great deal of research is concerned with the biology of the streptococcus. The individual coccus is being studied in great detail, as are the reactions induced on contact with a given host.

Characteristics

Streptococci of varying sizes (but usually about 1 μm in diameter) are arranged in long or short chains. Long chains contain 50 or more cocci; short chains contain as few as

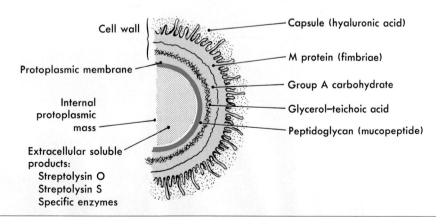

FIG. 23-5 Schematic representation of the gram-positive cell wall of group A streptococci. Note the cell wall constituents and source of extracellular soluble products. (Courtesy Dr. K.H. Johnston, The University of Texas Health Science Center at Dallas.)

four or six. Within the chains bacteria are found in pairs. Chains form when bacteria divide in one plane and still cling together. Streptococci are nonmotile, gram-positive organisms that do not form spores. Capsule (hyaluronic acid) formation is variable (Fig. 23-5). Some species form distinct capsules; many do not. Fig. 23-5 is a schematic representation of the gram-positive cell wall of streptococci. Specific components are discussed later.

Most streptococci grow best in the presence of oxygen, but they may also grow in its absence. A few species are strict anaerobes. Streptococci grow well on all fairly rich media, and visible growth usually appears within 24 to 48 hours. Growth is especially luxuriant on hormone media or media containing unheated serum, whole blood, or serous fluid. Streptococci also multiply in milk. They grow best at body temperature but may grow through a temperature range of 15° to 45° C. Most are *not* soluble in bile and do *not* ferment inulin (a fructose polymer). All are catalase negative and oxidase negative.

Streptococci may remain alive in sputum or other excreta for several weeks and in dried blood or pus for several months. They are destroyed at a temperature of 60° C in 30 to 60 minutes or in 15 minutes by 1:200 phenol solution, tincture of iodine, or 70% isopropyl alcohol. Penicillin is the most effective antibiotic against most types. Against beta-hemolytic streptococci it is the drug of choice, but against enterococci it is ineffective.

Classification

Streptococci may be classified on the basis of their action on blood agar, biochemical properties, or serologic behavior (agglutination and precipitation).

Corresponding to their action on 5% sheep blood agar plates (incubated at 35° C), streptococci are classified broadly as:

1. Alpha-hemolytic or viridans—the colony surrounded by a green halo; hemolysis slight or incomplete
2. Beta-hemolytic—the colony surrounded by a glass-clear, wide, colorless zone of hemolysis (Color plate 3, *B*).
3. Gamma (nonhemolytic)—the colony surrounded by neither hemolysis of blood agar nor a color change (Fig. 23-6)

As a general rule, the beta-hemolytic streptococcus is the most virulent and associ-

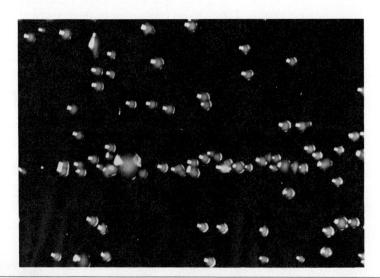

FIG. 23-6   Gamma-hemolytic streptococci with no hemolysis around their colonies.

ated with acute fulminating infections; the viridans type is associated with low-grade chronic infections, including nonlethal ones such as tooth abscesses and sinus infections. However, infections with alpha-hemolytic streptococci may be as serious as those with beta-hemolytic streptococci (for example, subacute bacterial endocarditis). Many strains of streptococci of the gamma type are nonpathogenic.

Another classification, related to biochemical properties, divides streptococci into the (1) pyogenic group, (2) viridans group, (3) lactic group, and (4) enterococcus group (Table 23-2). Enterococci, of which *Streptococcus faecalis* is representative, frequently infect the genitourinary tract and occasionally the respiratory tract; they are rare causes of subacute bacterial endocarditis. The lactic group, found in sour milk, includes *Streptococcus lactis* and *Streptococcus cremoris*. It is not pathogenic but is important in the dairy industry and in certain types of biologic assay. The representative of the viridans group is *Streptococcus mitis*, often referred to as *Streptococcus viridans*, and is responsible for 90% of the cases of subacute bacterial endocarditis. The type organism of the pyogenic group is *Streptococcus pyogenes*, a serious pathogen.

By serologic (precipitin) methods, streptococci fall into 13 groups (the classification of Lancefield and others), which correspond in a general way to their pathologic action (Table 23-3). In the 1920s Rebecca Lancefield studied the antigenic structure of the streptococcus and discovered that antigenic specificity could be based on **C substance** (*C carbohydrate*), a polysaccharide in the cell wall (Fig. 23-5). Using this group-specific substance, the serologic specificity of which is determined by an amino sugar, she arranged the streptococci into a number of antigenic groups, identified as Lancefield groups. Most streptococci can be placed into one of the following Lancefield groups: A, A-variant, B, C, D, E, F, G, H, K, L, M, O, and S. This classification has proved very reliable. The Lancefield capillary precipitin test is still the classic method for immunologic identification of streptococci. A white ring at the interface between extract of test organism and specific antiserum is a positive test.

Lancefield groups may be further subdivided on the basis of other cell surface antigens into serologic types given Arabic numbers. For more than 60 types in group A streptococci, the antigenic substance that determines type specificity is the **M protein** (a family of immunologically diverse proteins). Superficially attached to the bacterial cell surface where it is organized as fimbriae (a series of hairlike projections), M protein

Table 23-2 Biologic Properties of Streptococci

Division	Hemolytic Streptococcus	Viridans Streptococcus	Enterococcus	Lactic Streptococcus
Serologic group	A, B, C, E, F, G, H, K, L, M, O	None	D	N
Hemolysis on blood agar	Usually beta	Usually alpha	Alpha, beta, or gamma	Alpha or gamma
Growth in 0.1% methylene blue milk	−	−	+	+
Growth in 6.5% salt broth	−	−	+	−
Growth in presence of 40% bile	−	±	+	+
Antibiotic susceptibility (bacitracin)	Usually sensitive	May be resistant	May be resistant	(Nonpathogenic)

Table 23-3 Serologic Groups of Streptococci

Group	Species	Significance
A	*Streptococcus pyogenes*	Important human diseases; group sensitive to penicillin
B	*Streptococcus agalactiae*	Bovine mastitis
C	*Streptococcus equi*	Animal diseases; mild human respiratory infections
	Streptococcus zooepidemicus	
	Streptococcus equisimilis	
	Streptococcus dysgalactiae	
D	*Streptococcus faecalis*	Enterococci; genitourinary tract infections, endocarditis, human wound infections; found in dairy products
	Streptococcus faecalis subsp. *liquefaciens*	
	Streptococcus faecalis subsp. *zymogenes*	
	Streptococcus faecium	
E		Disease of swine; found in normal milk
F	*Streptococcus minutus*	Found in human respiratory tract
G	*Streptococcus anginosus*	Mild respiratory infections in humans; genital infections in dogs
H	*Streptococcus sanguis*	Found in human respiratory tract
K	*Streptococcus salivarius*	Found in human respiratory tract
L		Genital tract infections in dogs
M		Genital tract infections in dogs
N	*Streptococcus lactis*	Lactic group; found in dairy products
	Streptococcus cremoris	
O		Viridans group; subacute bacterial endocarditis; found in human upper respiratory tract
		Microaerophilic streptococci
		Anaerobic streptococci—13 species

constitutes a major virulence factor since it protects the virulent streptococci of this group from the phagocytes of the host (Fig. 15-8).

About 90% of human infections are caused by group A streptococci, which includes *Streptococcus pyogenes*. Streptococci of all other groups except those of N are indigenous to human beings and therefore potential pathogens. Certain members of groups B, C, D, H, K, and O and all members of group N are nonhemolytic.

Toxic Products

Streptococci elaborate many extracellular enzymes and poisons, some of which can be called exotoxins. Among these are (1) hemolysins, (2) leukocidins, (3) streptokinase, (4) streptodornase, (5) hyaluronidase, and (6) erythrogenic toxin.

Streptolysins produced by streptococci are of two types, S and O, and each is both hemolysin and leukocidin. *Streptolysin S* is produced primarily by members of group A: *streptolysin O* is elaborated by most members of group A, by the "human" members of group C, and by certain members of group G. The glass-clear zones of beta hemolysis around streptococcal colonies result from the combined action of these two hemolysins. Since streptolysin O is oxygen sensitive and soluble, it exerts its effect beneath the colonies where conditions for growth are anaerobic; streptolysin S is oxygen stable and accounts for surface hemolysis. Because streptolysin O (but not streptolysin S) is immunogenic, the antistreptolysin O titer may be applied to the diagnosis of streptococcal infection.

Streptokinase, or **fibrinolysin,** is notable in that it activates an enzyme that destroys fibrin (the framework of blood clots). Blood clots play an important part in wound healing and in blocking the spread of local infections. Streptokinase is a virulence factor in that its action favors the spread of the organisms.

Streptodornase (streptococcal deoxyribonuclease, DNAse B) acts to liquefy thick, tenacious exudates such as those seen in pneumonia. The enzymatic activity of this DNAse brings about the degradation of the deoxyribonucleoprotein in the exudate, the factor responsible for the viscosity. This nuclease is considered another virulence factor for streptococci.

Hyaluronidase (the spreading factor, invasin) by enzymatic action causes an increased permeability of host tissues to the cocci and their toxic products. It also breaks down hyaluronic acid in bacterial capsules. Thus capsules may be detectable only in young streptococci. The capsule itself (Fig. 23-5) may be a virulence factor, since it has been found in most epidemic strains of group A streptococci. Its role is unclear, however, in these organisms that also produce hyaluronidase.

Erythrogenic toxin produces erythema, or redness, when injected into the superficial layers of the skin, and with a large enough dose, a generalized rash. This is the scarlet fever toxin used in the Dick test (p. 495). It occurs in two immunologic types, A and B, and is phage mediated.

Pathogenicity

Streptococci may be responsible for a localized inflammatory reaction, such as an abscess, or a generalized reaction, such as septicemia. The nature of the lesion depends on the virulence of the streptococci, the number introduced into the body, the mode of introduction, the tissue invaded, and the resistance of the host.

Pathology

Pathologically, the type lesion of hemolytic streptococci is the diffuse, ill-defined spreading lesion of **cellulitis.** The exudate contains few cells and consists largely of fluid with little fibrin. The toxic products of the microbes greatly aid their extension through both natural tissue and inflammatory barriers, and they tend to infect lymphatic vessels at the site of invasion. Many well-known forms of streptococcal infection are an expression of cellulitis. Erysipelas is cellulitis with a specific anatomic pattern; septic sore throat is cellulitis of the throat.

Allergic manifestations follow certain streptococcal infections (usually of the throat)— for instance, acute rheumatic fever and one form of kidney disease (glomerulonephritis). Glomerulonephritis can follow streptococcal infections of the upper respiratory tract or of the skin, but rheumatic fever complicates group A streptococcal infections of the throat.

As a rule, the more virulent an infection, the more virulent are the causative streptococci for animals of the same species. Usually their virulence is lowered when they are introduced into animals of another species. When transferred from one animal to another, streptococci tend to produce the same type of lesion in the new host as in the original one. Rabbits and white mice are more susceptible to streptococci of human origin than are other laboratory animals.

Streptococci in human diseases. In addition to being the cause of erysipelas and two epidemic diseases, scarlet fever and epidemic sore throat, streptococci of group A (S. *pyogenes*) are the most common cause of acute endocarditis, septicemia, and puerperal sepsis. They may cause pneumonia, boils, abscesses, cellulitis, peritonitis, tonsillitis, cervical adenitis, lymphangitis, infection of surgical wounds, osteomyelitis, and empyema. From an infection of the middle ear (otitis media), streptococci may spread to the mastoid cells (the air spaces in the mastoid process of the temporal bone) and cause mastoiditis. From either the middle ear or the mastoid cells, spread of infection to the meninges results in streptococcal meningitis.

S. *pyogenes* is responsible for most bronchopneumonias complicating whooping cough, measles, and influenza. Such bronchopneumonias, often fatal, may reach epidemic proportions when outbreaks of these diseases occur in communities with a high carrier rate for S. *pyogenes*. Group A streptococcal pneumonia is often the terminal phase of chronic diseases such as tuberculosis and cancer.

Group A streptococci are the most common cause of the highly communicable skin disease, impetigo contagiosa (Fig. 23-3), possibly the single most common infection in children. Streptococci and staphylococci often act together in this disorder. In susceptible, usually small children, group A streptococci can colonize normal skin but not penetrate it until a break such as an insect bite allows entry. Local invasion leads to secondary infection of the superficial layers of the skin and the disease impetigo. Impetigo develops in a typical pattern. First a small blister filled with clear fluid forms and becomes well circumscribed. It displays a honey-colored crust at its center and is encircled by a reddened area. The skin of the arms and legs is most likely affected, and 20 lesions are average for a given patient. If staphylococci participate in the process, the lesions converge and become filled with pus. Impetigo resolves spontaneously within 6 to 8 days; if treated it resolves within 4 to 5 days. The serious hazard of impetigo is the likelihood of glomerulonephritis as a sequel

Group B streptococci, known as *Streptococcus agalactiae*, were linked to bovine mastitis as early as 1887, but they were not recognized as human pathogens until 1964, when their dramatic effects in newborn infants were demonstrated. In neonates group B streptococci can cause very grave disease, described as either early onset infection or late onset. Fulminant infection in the first 12 to 24 hours of neonatal life overwhelms the baby in the rapid extension to the bloodstream (septicemia) and all organs of the body. The baby seems to be a living culture medium. Mortality for early onset disease is close to 100%. Infection may first appear after the tenth day of life, at which time the affected infant presents a purulent meningitis. Infection of delayed onset is serious but carries a better prognosis. The principal reservoir for the group B streptococci has been shown to be the woman's genital tract, and 70% of offspring born to mothers carrying the organisms become so colonized. However, group B streptococci can also be spread to the babies from the nursery personnel. The overall risk to neonates of developing disease once they are colonized with group B streptococci appears to be 1 in 100. Because the organisms are recovered from sexual partners and harbored in the reproductive tract, group B streptococcal infection has been added to the expanding list of sexually transmitted diseases (p. 517).

Included in streptococci of group D (whose group antigen is a teichoic acid, not a polysaccharide) is *Streptococcus bovis*, an example of a sentinel microorganism, one with a peculiar and poorly understood predilection for patients with cancer. S. *bovis* is thus related to cancer of the large bowel.

Streptococcal infections in lower animals. The majority of streptococcal infections in lower animals are caused by streptococci in Lancefield groups B and C. (Group A streptococci possess a strict sort of specificity for humans.) Strangles, an acute communicable disease of the upper respiratory passage of horses, is caused by *Streptococcus equi*. Streptococcal mastitis, a serious disease of cows that renders milk unfit for use, is caused by S. *agalactiae* and most likely spread from cow to cow by the hands of the milker.

Sources and Modes of Infection

Streptococci are normal inhabitants of the mouth, nose, throat, and respiratory tract. They may be conveyed from person to person by direct contact or contaminated objects, hands, and surgical instruments. The hands are important conveyors of infection in puerperal sepsis and wound infections. Milk can be an important source. Streptococcal diseases, especially scarlet fever and septic sore throat, may be spread by milk that has been contaminated with the mouth and nose secretions of a carrier or by milk from a cow with mastitis caused by S. pyogenes.

Streptococci usually enter the body by the respiratory tract or through wounds of the skin. Only a minute abrasion is necessary, and steptococcal infections therein have many times in the past led to fatal septicemias in physicians and nurses. Streptococci leave the body by way of the mouth and nose and in the exudates from areas of infection. The nasal carrier is a source of infection. Enterococci as normal inhabitants of the intestinal canal are excreted in the feces.

Laboratory Diagnosis

Streptococci are detected by Gram-stained smears and cultures from the site of disease. In septicemias caused by beta-hemolytic streptococci the organisms can usually be detected by blood cultures. In subacute bacterial endocarditis, since the viridans organisms escape into the blood intermittently, cultures may have to be repeated.

A presumptive test of value in the identification of streptococci is the **bacitracin disk test.** Even in the low concentrations this antibiotic appears to be specifically active against members of Lancefield group A and without effect on other groups. A paper disk containing a known unit of bacitracin is placed on a blood agar plate previously streaked with the streptococcus in question. A zone of inhibition of growth found around the disk identifies it as being in group A.

ASO titer. When infection with streptococci that produce streptolysin O takes place, antibodies against the streptolysin O appear in the blood of the patient. The detection and quantitation of these antibodies is the **antistreptolysin O (ASO) titer.** It forms a useful procedure in diagnosis and management of streptococcal infections of a chronic and persistent nature, such as acute rheumatic fever and acute hemorrhagic glomerulonephritis (allergic reactions to beta-hemolytic streptococci). The course of these diseases relates to the titer (number of antibodies demonstrable), and the test helps to differentiate certain diseases similar to rheumatic fever where the streptococcus is not comparably implicated, since the titer in these diseases is not significant.

Determination of the ASO titer begins in the laboratory with a series of tube dilutions of a patient's serum to each of which the same volume of streptolysin O reagent is added. After an incubation period, an equal volume of group O human or rabbit red blood cells is added to each tube, and the mixtures are reincubated. The tubes are then examined for hemolysis, and the last tube (the highest dilution) of the series showing no hemolysis is taken as the ASO titer. The reciprocal of that dilution is expressed as the titer in Todd units. For example, if the highest dilution in the setup without any hemolysis is 1:250, then the ASO titer is expressed as 250 Todd units. Most normal adults show titers of up to 200 Todd units. An elevated titer usually appears 1 to 3 weeks after onset of infection, and a rising titer on repeated specified occasions can be very significant.

Group B streptococci. In the identification of S. agalactiae (group B streptococcus), hydrolysis of hippurate is a distinguishing biochemical feature. The capsular polysaccharide of S. agalactiae in the infected host diffuses through the tissues and is slowly excreted in the urine, where it is identified by counterimmunoelectrophoresis. A latex agglutination test for the detection of Group B carbohydrate antigens in cerebrospinal fluid, serum, and urine is now available.

Immunity

With the exception of scarlet fever, streptococcal infections are not followed by an immunity, and in scarlet fever the immunity is established only against scarlet fever toxin, not against the organisms.

Prevention

When caring for a patient with a streptococcal infection, medical attendants should remember that they are dealing with an infection that may be most virulent and easily spread. Medical personnel caring for patients with streptococcal infections such as erysipelas or scarlet fever should exercise great care that they do not carry the infection to the obstetric or surgical patient. The strictest adherence to aseptic technic is crucial here because the surgical wound and the recently pregnant uterus are extremely vulnerable.

Wounds and abrasions on the body should be thoroughly disinfected. The buccal and nasal secretions from a patient with streptococcal bronchopneumonia should be handled in the same manner as those from a patient with diphtheria. A patient with streptococcal bronchopneumonia should be isolated from patients with measles or influenza.

The fact that more than 60 types exist among group A streptococci, all immunologically specific, explains the delay in the development of a clinically feasible vaccine that centers on the highly antigenic M protein in the streptococcal cell membrane.

Scarlet Fever

Scarlet fever (scarlatina) is an acute infection of childhood characterized by sore throat, severe constitutional manifestations, and a distinct erythematous skin rash with massive desquamation of the outer layers of the skin (see Fig. 34-1, *B*). The nasopharynx and tonsils may be covered with a membrane, and the lymph nodes, especially those of the neck, are swollen and inflamed. The white blood cell count ranges from 15,000 to 30,000 cells per cubic millimeter, of which 85% to 95% are neutrophils. (A normal white blood cell count is 5000 to 9000 cells per cubic millimeter with 55% to 65% neutrophils.) Scarlet fever is caused by streptococci that produce erythrogenic toxin. They are lysogenic strains, since the erythrogenic toxin is coded for by a phage gene. In almost all cases the streptococci are of Lancefield group A. Only rarely is scarlet fever produced by streptococci in groups C and D.

Scarlet fever and streptococcal sore throat appear to be different manifestations of the same basic disease. If the streptococci causing the sore throat produce erythrogenic toxin and the recipient of the infection is not immune to the toxin, scarlet fever results. If the streptococci do not produce erythrogenic toxin or if the recipient is immune to the toxin, then only the sore throat is present.

In the past scarlet fever was a disease of great severity and often fatal. Now it is relatively benign with treatment but may be complicated by suppurative otitis media and peritonsillar abscess or followed by rheumatic fever and acute nephritis. It is much more prevalent in Europe than in North America.

Sources and Modes of Infection

The sources of infection are the nose and throat secretions of patients or carriers and pus from infected lymph nodes, ears, and other lesions. The organisms are present throughout the course of the illness and may persist in the nose and throat or in the purulent discharges from the lesions for weeks or months thereafter. As long as a person harbors the organism, that person is a dangerous source of infection. The desquamated flakes from the skin rash are not infectious.

The organisms usually enter the body through the mouth and nose, wounds (less often), burns, and the parturient uterus. Infection may be transmitted by direct contact or by contaminated objects, such as handkerchiefs, towels, pencils, toys, and dishes.

Erythrogenic Toxin

The erythrogenic toxin of scarlet fever streptococci, released by the organisms at the primary site and absorbed into the body, brings about the rash and other systemic effects. It may be found in the blood in a concentration as high as 300 units/ml and also in the urine. It is prepared artificially from a 5-day growth of scarlet fever streptococci in broth from which the bacteria have been removed by filtration. The filtrate contains the toxin in large amounts. Toxin prepared in this way causes scarlet fever. Injected in very small amounts into the skin of persons susceptible to scarlet fever, it gives rise to an inflammatory reaction, the basis of the Dick test. Scarlet fever toxin can induce an active immunity with the formation of antitoxin when injected into a suitable animal.

Immunity

Immunity to scarlet fever relates to scarlet fever antitoxin in the blood. Infants inherit an immunity from their mothers that is lost within a year, and susceptibility increases until the sixth year. After that, susceptibility decreases until adulthood, by which time most people are immune. An attack of scarlet fever is usually followed by permanent immunity. Remember that although an immune person will not be harmed by scarlet fever toxin, the streptococci themselves may invade the body and cause tonsillitis, abscesses, or otitis media. Immune persons may harbor the organisms for a long time and spread the infection widely.

Dick test. The Dick test is performed with the intradermal injection of one skin test dose (STD) of scarlet fever toxin. In immune persons the antitoxin in the blood neutralizes the injected toxin, and no reaction occurs. In susceptible persons the toxin injures the cells around the injection site, producing within 24 hours a zone of inflammation and redness at least 1 cm in diameter. The test is positive at the onset but becomes negative during the course of scarlet fever.

Schultz-Charlton phenomenon. If a small amount of the serum of a person convalescing from scarlet fever or of an animal immunized against scarlet fever is injected intradermally into an area of scarlet fever rash, the skin blanches at the injection site because the toxin in the area is neutralized by the injected antitoxin. This test differentiates scarlet fever from measles, rubella, and other skin diseases.

Prevention

A patient with scarlet fever should be isolated, and discharges from the mouth and nose as well as all contaminated articles should be disinfected. The patient should remain isolated until secretions from the mouth and nose are free of scarlet fever streptococci and all complications have cleared. Medical attendants for a patient should exercise every precaution to prevent the spread of infection to others, especially in obstetric and surgical wards. Remember that during epidemics of scarlet fever, persons with rhinitis and sinusitis may spread the disease just as effectively as those with a skin eruption. The disinfection procedures are the same as those for diphtheria (p. 592). Pasteurization prevents milkborne epidemics.

Erysipelas

Erysipelas (St. Anthony's fire) is an acute inflammation of the skin caused by hemolytic streptococci of Lancefield group A or, infrequently, of group C or D. It is a diffuse, intensely red, and rapidly progressive cellulitis of the skin, so sharply demarcated that its borders can be traced with a ballpoint pen. The streptococci grow at the periphery, not in the center, of the inflamed area, and they elaborate toxic substances effecting the toxic constitutional state.

Erysipelas should not be confused with erysipeloid, a localized infection of the skin caused by *Erysipelothrix rhusiopathiae* (gram-positive rod with a tendency to form long filaments) and found as an occupational disease in those who handle fish or meats.

Mode of Infection

For erysipelas, the portal of entry is a wound, fissure, or abrasion. A hard, red thickening of the skin beginning at the infection site extends peripherally. The streptococci grow almost exclusively in the lymph channels of the inflamed area, and as the disease progresses, they spread peripherally several centimeters beyond the line limiting the area of obvious inflammation. When streptococci enter the blood, the prognosis is grave. Erysipelas may be complicated by abscesses, pericarditis, arthritis, endocarditis, septicemia, and pneumonia. Patients with uncomplicated disease or without open wounds or superficial discharges will not transmit the infection to others. Erysipelas may be associated with the other group A streptococcal infections.

Immunity

Instead of inducing immunity, an attack of erysipelas seems to render the patient more vulnerable to future attacks.

Streptococcal Sore Throat (Septic Sore Throat)

Septic sore throat ("strep" throat) (Fig. 23-7) is an ulcerative inflammation of the throat associated with severe symptoms. It is caused by hemolytic streptococci of Lancefield group A or, in a small proportion of cases, group C. It may be transferred by direct contact or droplet infection. Streptococcal sore throat may be complicated by extension of the infection to the lungs with the development of streptococcal pneumonia, and like scarlet fever it may be followed by glomerulonephritis or rheumatic fever.

Although a variety of agents cause the familiar sore throat, the differential diagnosis usually narrows to either a viral infection or a strep throat, a distinction not readily made by clinical findings alone. Most of the time both are benign and self-limited, but the treatment for viral pharyngitis is supportive care, whereas for streptococcal infection treatment with penicillin is important in the prevention of rheumatic fever.

Rheumatic Fever

Acute rheumatic fever, the forerunner of rheumatic heart disease, has a clinical course similar to that of an acute infection. In the acute phase there is fever, an increased pulse rate (tachycardia), carditis (inflammation of the heart), and a characteristic type of polyarthritis. Rheumatic fever is more common in the northern than in the southern climates and is infrequent in the tropics. It is most common between the fifth and fifteenth years of life.

Attacks of acute rheumatic fever are preceded by tonsillitis or severe sore throat caused by any M type, group A streptococci. A diagnostic rise in the serum titer of antistreptolysin O occurs in more than 80% of the patients. Rheumatic fever tends to be a disease of long standing, with repeated attacks of the acute process. This favors the development of complications in the heart that add up to serious heart disease.

To prevent the harmful effects of rheumatic heart disease is to stop the recurrent attacks of acute rheumatic fever. Patients are protected as much as possible from group A streptococcal infections and are even given antibiotics prophylactically at times of the year when the incidence of such infections is high.

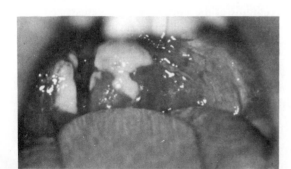

FIG. 23-7 "Strep" throat. In the throat of this child purulent exudate fills tonsillar crypts and extends over the surface of swollen tonsils like a membrane; pharyngeal mucosa in background is inflamed. (Courtesy Dr. J.D. Nelson, The University of Texas Health Science Center at Dallas.)

Puerperal Sepsis

Puerperal sepsis (puerperal septicemia) is usually caused by a hemolytic streptococcus from the nose and throat of the patient or those in close association with her. The streptococcus reaches the uterus by way of contaminated hands or instruments. Most streptococci that cause this condition belong to Lancefield group A of streptococci exogenous to the generative tract, but 20% to 25% of the cases come from anaerobic streptococci that are normal inhabitants of the vagina.

Dental Caries

Dental caries (tooth decay) is one of the most widespread disorders; it is estimated that every American has at least three cavities in his or her teeth. The factors implicated in the pathogenesis of caries and determining their severity are (1) individual susceptibility, (2) presence of bacteria capable of producing organic acids, and (3) presence of carbohydrates to support the cariogenic activities of acid-producing microorganisms. Some lactobacilli and certain acid-producing streptococci, notably *Streptococcus mutans*, are found in the mouth associated with tooth decay, and most observers state that they play a role, especially *S. mutans*.

S. mutans is a member of the viridans group of streptococci that resists classification by group-specific carbohydrates. As a group they elaborate extracellular polysaccharides that promote their adhesion to a surface. The cariogenic property of *S. mutans* is nicely shown in experimental animals (Fig. 23-8).

The first attack on tooth structure is postulated to come from the acid-producing *S. mutans*. It produces an enzyme, dextran-sucrase, that converts sucrose of food eaten to dextran. Dextran combines with salivary proteins to create on tooth surfaces a sticky, colorless film called a **plaque**. Plaque piles up continuously on teeth, forming again after removal at any given time. The plaque provides a haven for bacteria from which

FIG. 23-8 *Streptococcus mutans* in experimental cariogenesis, scanning electron micrograph. Note cocci attached to toothlike surface, "mesh" of which corresponds to salivary pellicle, the thin surface film of absorbed glycoproteins from saliva. After bacteria caught to pellicle were incubated in 1% glucose, they synthesized glucan, the extracellular glue of the micrograph. (Courtesy Dr. W.B. Clark, Boston.)

they can undermine and demineralize the enamel of a tooth. Subsequently, as tooth enamel breaks down, the typical cavity appears. Fragments of debris found therein support continued bacterial growth and thus tend to perpetuate the process of decay.

An anticavity vaccine against *S. mutans* is currently being tested in human beings. The immunoglobulin A antibodies it promotes inhibit the colonization of *S. mutans* and the development of caries in rats and monkeys. The vaccine is ingested or injected into the salivary glands.

STREPTOCOCCUS (DIPLOCOCCUS) PNEUMONIAE (THE PNEUMOCOCCI)

General Considerations

Characteristics

The organism *Streptococcus (Diplococcus) pneumoniae*, best known in relation to pneumonia, occurs typically in pairs of lancet-shaped diplococci with their broad ends apposed. The human being is the principal host, and within the human body or in respiratory secretions therefrom each pair is encapsulated. In vivo the cocci may also be found singly or in chains.

Pneumococci are pyogenic extracellular microbes that are gram positive, nonmotile, and nonspore-forming. Their optimum temperature is 37° C, and they grow best in a slightly alkaline milieu. Since the pneumococcus produces large amounts of lactic acid, the culture medium must be well buffered. Pneumococcal growth is most abundant on enriched media such as hormone agar and blood agar. Pneumococci grow equally well in the presence or absence of oxygen but, as facultative anaerobes, lack the enzymes catalase and peroxidase that prevent the accumulation of the hydrogen peroxide toxic to them. Catalase for their growth may be derived from the rich source in red blood cells in culture media. On the surface of blood agar, virulent, encapsulated strains grown aerobically form glistening, dome-shaped colonies, 0.5 to 3 mm in diameter (Fig. 23-9), that are soon surrounded by the diagnostic green zones of partial hemolysis (alpha hemolysis). The power to form capsules is lost when pneumococci are cultivated for a long time on artificial media.

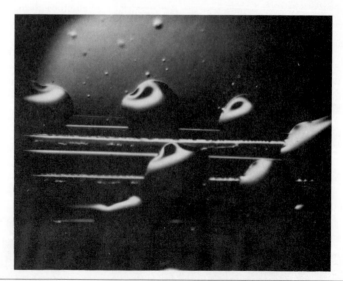

FIG. 23-9 Pneumococci. Greatly magnified, dome-shaped colonies are growing on routine blood agar plate. Young colonies are dome shaped; older ones flatten out in a distinctive way.

The pneumococcus is not very hardy and has no natural existence outside the animal body. In the finely divided spray thrown off from the nose and mouth, pneumococci live about 1½ hours in sunlight. In large masses of sputum they live for 1 month or more in the dark and about 2 weeks in sunlight. They do not resist drying at room temperature, are more susceptible to ordinary germicides than most other bacteria, and are destroyed by a temperature of 52° C in 10 minutes.

A remarkable capacity for autolysis (self-destruction) is displayed by the pneumococcus when its autolytic enzymes are activated. The bile solubility test that distinguishes pneumococci from alpha-streptococci applies the fact that surface agents such as bile salts readily trigger this reaction in pneumococci and only rarely in other cocci. Practically speaking, if an unidentified organism is bile insoluble, it is not a pneumococcus.

Types

Although all pneumococci are much alike microscopically and culturally, they show distinct differences immunologically. This was discovered in 1910 when different cultures of pneumococci were used to immunize animals and the serum of these animals was used to agglutinate pneumococci from various other sources.

Water-soluble polysaccharides in the capsule of pneumococci tend to slough off the bacteria as the grow. Spoken of as **specific soluble substances** (SSS), they give pneumococci their type characteristics. They can be detected by the precipitin reaction in broth cultures of pneumococci and in the blood and urine of patients with pneumonia. There are now at least 88 types or subtypes. All can cause pneumonia. Currently in the United States 80% of bacterial pneumonias are pneumococcal pneumonias. Most adult cases are caused by types 1, 2, 3, 4, 6, 7, 8, 14, 18, and 19. In children the primary types of pneumococci are 19, 23, 14, 3, 6, and 1 (in order of frequency).

Pneumococcus type 3 is the most virulent type and differs somewhat from other pneumococci, with its wide capsule and slimy growth on culture media. It is not lancet shaped. The direct relation between capsule development and virulence of pneumococci explains why type 3 infections have such a high mortality (more than 50% in most instances). Type 3 pneumonias menace the aged.

Formerly, therapeutic serums made from rabbits were available for the treatment of pneumonia. An antiserum prepared by immunization of an animal against pneumococci of one type agglutinates pneumococci of that type alone; therefore the determination of the pneumococcal type causing the infection was necessary before serum therapy could be instituted. Today serum therapy for pneumonia has been replaced for the most part by antimicrobial therapy.

Although classified in the genus *Streptococcus*, *S. pneumoniae* (the pneumococcus) does not belong to a Lancefield group. Somatic antigens demonstrated to be common to pneumococci include a C carbohydrate, an R antigen, and M antigens. **C carbohydrate**, derived from a cell wall constituent, can be precipitated by C reactive protein (CRP), a beta globulin (not an antibody) found in the serum of certain individuals with an inflammatory, neoplastic, or necrotizing disease (p. 398). **R antigen**, a protein derivative from rough strains (nonencapsulated and avirulent) of pneumococci, is found on or near the surface of the bacterial cell. **M antigens** are type-specific proteins analogous to the type-specific M proteins of group A hemolytic streptococci.

Toxic Products

The main virulence factor for pneumococci is the capsule. It is so considered because the presence of the capsule enables pneumococci to resist phagocytosis. Virulent organisms are smooth encapsulated strains; avirulent ones are rough nonencapsulated strains. Antiphagocytic properties stem from the highly acidic nature of the capsule combined with certain hydrophilic properties therein that render the encapsulated cocci very difficult for phagocytes to seize and ingest when suspended in fluid. On a surface, however, encapsulated pneumococci may be more easily attacked by the phagocytic cells. A preexisting disease in the lungs, such as a primary viral infection, associated with a hypersecretion of mucus may predispose to pneumococcal infection because the mucus protects the pneumococci from surface phagocytosis and thus allows for their free growth.

The pneumococcus has been cited as an example of a pathogen whose high order of invasiveness is combined with a minimal grade of toxigenicity. However, some of the clinical features of pneumococcal disease point to a toxemia, although a well-defined toxin has never been so implicated. Pneumococci do release hemolysins, leukocidins, certain necrotizing substances, and a neuraminidase that acts on cell membranes in the nasopharyngeal and bronchial mucosa. Many strains also release hyaluronidase to facilitate their spread in tissues.

Pathology

The pneumococcus is a major cause of three important diseases in the United States: pneumonia, meningitis (p. 523), and otitis media. The hallmark of the tissue lesion in pneumococcal infections is the presence of an adundance of fibrin in areas of inflammation. In lobar pneumonia there is much fibrin in the lungs; in pneumococcal meningitis much fibrin is deposited in the subarachnoid space.

Pneumonia. Pneumonia is an inflammatory condition of the air sacs (alveoli), bronchioles, and smaller bronchi of the lungs, in which these structures are filled with fibrinous exudate. **Consolidation** is the result whereby the air spaces of the lung are plugged up (Figs. 23-10 and 23-11). The pneumococcus is the usual cause of two kinds of pneumonia: (1) **lobar pneumonia,** involvement of one or more of the five major anatomic divisions (lobes) of the lungs, and (2) **bronchopneumonia** (bronchial pneumonia, lobular pneumonia), involvement of terminal bronchioles and adjacent lobules (small subdivisions of lobes). Confluent bronchopneumonia results from coalescence of areas of bronchopneumonia.

Sir William Osler wrote this description of lobar pneumonia before the turn of the century:

As a rule, the disease sets in abruptly with a severe chill, which lasts from 15 to 30 minutes or longer. In no acute disease is an initial chill so constant or so severe. The patient may be taken abruptly in the midst of his work, or may awaken out of a sound sleep in a rigor. The temperature taken during the chill shows that the fever has aready begun. If seen shortly after the onset, the patient has usually features of an acute fever, and complains of headache and general pains. Within a few hours there is pain in the side, often of an agonizing character; a short, dry, painful cough begins, and the respirations are increased in frequency. . . .

Toxaemia is the important prognostic feature in the disease, to which in a majority of the cases the degree of pyrexia and the extent of consolidation are entirely subsidiary. It is not proportionate to the degree of lung involved. A severe and fatal toxaemia may occur with the consolidation of only a small part of one lobe. On the other hand, a patient with complete solidification of one lung may have no signs of a general infection.

After the fever has persisted for from 5 to 9 or 10 days, there is an abrupt drop, known as the crisis, which is one of the most characteristic features of the disease.

Lobar pneumonia is a severely toxic disease. Rapid shallow breathing, increased pulse rate (tachycardia), cyanosis, and nausea with vomiting are manifestations, and the blood count usually shows a leukocytosis (30,000 to 40,000 per cubic millimeter) with 90% to 95% polymorphonuclear neutrophils. Pleurisy (inflammation of the pleura) is part of the disease. With recovery and resolution, the exudate in the lungs liquefies

FIG. 23-10   Pneumonia. Gross photograph of surface of the lung sliced through maximum area of disease. Air spaces on this surface are completely obliterated, being filled solidly with exudate—the wide, irregularly outlined, grayish white areas covering practically the entire surface.

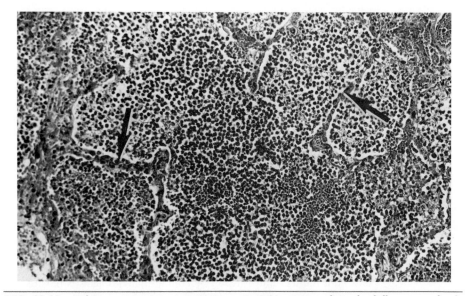

FIG. 23-11   Lobar pneumonia, microscopic section. Air sacs are plugged solidly as a result of inflammatory process. The many small irregular, closely packed, dark nuclei are polymorphonuclear neutrophils. Arrows point to alveolar walls about consolidated air spaces. (×800.) (From Smith, A.L.: Microbiology and pathology, ed. 12, St. Louis, 1980, The C.V. Mosby Co.)

and is removed partly by absorption and partly by expectoration. Air reenters the affected lobe or lobes, and the lung completely returns to its former efficiency. Occasionally, delayed resolution leads to abscess formation or chronic organizing (nonresolving) pneumonia. With delay in the liquefaction of the fibrinous exudate, the healing process brings about a replacement of the exudate with fibrous tissue. Thus air-containing spaces become solidified. The disease has yielded so dramatically to the use of the antimicrobials that the classic pathologic stages are rarely seen today.

Lobar pneumonia is a primary disease, and 95% of cases are pneumococcal.

Bronchopneumonia is usually pneumococcal, but it may be caused by any one of a number of microbes (see Fig. 31-3), including other streptococci, staphylococci, and influenza bacilli, operating singly or in variable combinations.

Bronchopneumonia, more often secondary than primary, is a serious complication to measles, influenza, whooping cough, and chronic diseases of the heart, blood vessels, lungs, and kidneys. It peaks in the early and late years of life and frequently is the terminal event in debilitating diseases of the very young or extremely old. (It has long been called the "old man's friend.") Bronchopneumonia may follow the administration of an anesthetic or the aspiration of infectious material into the lungs during an operation. In newborn infants it is often caused by aspiration of infected amniotic fluid.

Unlike lobar pneumonia, bronchopneumonia consists of scattered small inflammatory foci, usually more numerous at the lung bases. The exudate consists of leukocytes, fluid, and bacteria, but little or no fibrin and few red blood cells. Pleurisy and empyema are complications. This kind of pneumonia does not always resolve readily, and chronic pneumonia often persists. **Hypostatic pneumonia** is bronchopneumonia complicating the hypostatic congestion of heart failure.

Pneumonia with clinical manifestations of respiratory involvement and well-defined radiographic changes in the lungs may be caused by a variety of other microorganisms, such as the Friedländer bacillus *(Klebsiella pneumoniae)*, the influenza bacillus *(Haemophilus influenzae)*, and certain other agents (Table 23-4).

Otitis media. Otitis media, an inflammation of the middle ear associated with fever, earache, and abnormalities of hearing, is most commonly caused by the pneumococcus. It is said to be the diagnosis most often made by pediatricians. The disease-producing bacteria from the nasopharynx enter the middle ear through the eustachian tube, and in 40% of the cases S. *pneumoniae* can be cultured. *(Haemophilus influenzae* is also an important cause of this disorder.) Pneumococcal otitis media tends to spread to the meninges, with resultant pneumococcal meningitis.

Other diseases. Pneumococci cause other diseases, such as empyema, endocarditis, meningitis (most common cause of purulent meningitis after the age of 40), arthritis, peritonitis, and corneal ulcers. Some of these complicate pneumonia; others occur as primary conditions. Pneumococcal peritonitis is a primary disease in children. Sepsis after splenectomy is usually pneumococcal.

Sources and Modes of Infection

Lobar pneumonia is endemic in all centers of population. Epidemics seldom occur but may if conditions enhance exposure to infection with concomitant lowering of host resistance. The sources of infection are active cases and carriers.

Pneumococci enter and leave the body by the same route, the mouth and nose. Infection is usually transferred directly, most often by droplets from the mouth and nose, but indirect transmission by contaminated objects is possible.

Practically every person becomes a carrier of pneumococci for a short time during the year. Those who have come in contact with a patient often carry the organisms in their throats for a few days or weeks. Most carriers not in contact with pneumonia harbor comparatively avirulent pneumococci and are of little danger, although type 3

Table 23-4 Commonly Encountered Pneumonias

Clinical Setting	Usual Pathogen(s)
Community-Acquired Pneumonia (in Person Without Iatrogenic, Congenital, or Acquired Immune Deficiencies)	
"The big three" pneumonias (90%)	*Streptococcus pneumoniae*
	Virus, such as influenza or respiratory syncytial virus
	Mycoplasma pneumoniae
Legionnaires' disease	*Legionella pneumophilia*
Postinfluenzal pneumonia	*Staphylococcus aureus*
Chronic bronchitis with acute pneumonia	*Streptococcus pneumoniae*
	Haemophilus influenzae
Pneumonia in alcoholic or drug addict	*Streptococcus pneumoniae*
	Staphylococcus aureus
	Klebsiella pneumoniae
Aspiration pneumonia or lung abscess	Anaerobes
Pneumonia in immunocompromised host	*Staphylococcus aureus*
	Gram-negative enteric bacilli
	Fungus
	Pneumocystis carinii
	Virus
Hospital-Acquired Pneumonia	
Nosocomial pneumonia found in the intensive care units, the burn units, or in postoperative patients	*Staphylococcus aureus*
	Gram-negative enteric bacilli
Pneumonia in the immunocompromised host	Fungus
	Staphylococcus aureus
	Gram-negative enteric bacilli
	Pneumocystis carinii
	Virus

pneumococci may be found in them. That carriers of type 3 pneumococci are common whereas type 3 infections are comparatively rare is difficult to explain.

Laboratory Diagnosis

Pneumococci may be detected in sputum and other body fluids with some degree of certainty by direct microscopic examination of smears stained for capsules and by Gram's method. (Pneumococcocci and other streptococci bear a close microscopic resemblance to each other.) For rapid diagnosis in pneumococcal infections, counterimmunoelectrophoresis of body fluids is especially sensitive. Confirmatory methods are cultures and the inoculation of white mice. (Rabbits and mice are very susceptible to pneumococci.) In the mouse susceptibility test, death of the animal within 16 to 20 hours can result from the inoculation of only one virulent organism. Animal inoculation demonstrates pneumococci when direct smears fail to do so, and it always finds them more quickly than do cultures. Rough nonencapsulated strains of pneumococci do not kill the laboratory mouse.

Typing is done primarily to determine whether a highly virulent pneumococcus is present. There are several methods. All depend on the action of agglutinating and precipitating serums (typing serurms) prepared by immunizing animals against the different types of pneumococci. In the method devised by Neufeld flecks of sputum or other test material are mixed with the battery of type-specific serums. Where the type of pneumococcus matches that of the serum, the capsules of the pneumococci swell greatly. This is known as the **quellung** (German, swelling) reaction. It is an apparent enlargement of the capsule that may in part reflect a change in the refractive index of the capsule because of the antigen-antibody combination. If the sputum contains too few pneumococci or if the typing is otherwise unsatisfactory, some of the specimen may be injected into a white mouse, and in the course of a few hours, typing may be carried out on the yield from the peritoneal exudate.

Differentiation of pneumococci from other streptococci. Practical differences between pneumococci and other streptococci are: (1) on blood agar the colonies differ; (2) in animal tissues pneumococci have capsules, whereas other streptococci seldom do; (3) when 1 part of bile is added to 3 parts of liquid culture, pneumococci dissolve whereas streptococci do not; (4) pneumococci ferment inulin, whereas other streptococci do not; (5) pneumococci are inhibited by Optochin (ethylhydrocupreine hydrochloride), whereas other streptococci are not; and (6) pneumococci are more pathogenic for mice than are ordinary streptococci. Other differential aids are agglutination and precipitation tests with specific antiserums.

Immunity

Recovery from pneumococcal infection confers 6- to 12-month immunity to the type of pneumococcus causing the infection, but none to other types. (Instances have been reported in which a person has had pneumonia more than a dozen times.) Natural human resistance against the pneumococcus is comparatively high, and a person probably never contracts the disease unless the resistance is lowered. Blacks are more susceptible than whites, and men more so than women.

Prevention

The number of persons having contact with a patient with pneumonia should be restricted. The discharges from the mouth and nose of the patient should be burned or disinfected. The hands and all objects, such as spoons, cups, and other utensils, possibly contaminated by the patient should be disinfected. Measures should be taken to minimize droplet infection in the spray leaving the mouth of the patient when he talks or coughs.

Because of the antibiotic resistance that has been encountered in pneumococci, vaccine therapy is back. A polyvalent pneumococcal vaccine (Pneumovax) incorporates the purified capsular polysaccharides from the 14 types of pneumococci currently responsible for most pneumococcal disease. Protective type-specific antibody levels develop by the third week. Of the persons vaccinated, 60% have no reaction and 40% have only a mild local one (p. 428).

SUMMARY

1 The staphylococci, the streptococci, and the pneumococcus are the important gram-positive pyogenic cocci. The genera are *Staphylococcus* and *Streptococcus*. Formerly *Diplococcus pneumoniae*, the pneumococcus is now called *Streptococcus pneumoniae*.

2 Staphylococci are mostly aerobic and grow luxuriantly on all culture media, and some species produce catalase. On blood agar staphylococci produce pigments that range from gold to lemon yellow to white. Staphylococci are the bacteria most resistant to physical and chemical agents. They adapt quickly to antibiotics. About 80% are penicillin resistant.

3 The presence of the enzyme coagulase is the best evidence of pathogenicity with a staphylococcal species.

4 Examples of the many important metabolic products with toxic properties elaborated by staphylococci are hemolysins, enterotoxin, lipases, clumping factor, and nuclease.

5 The hallmark of staphylococcal infection is the abscess, and the best known staphylococcal infections are those of the skin and superficial tissues.

6 Toxic shock syndrome (TSS) in healthy young women is related to the action of an exotoxin elaborated by toxigenic staphylococci growing in the highly absorbent material of a menstrual tampon.

7 Phage typing is a valuable epidemiologic tool in tracing the spread of staphylococcal disease, particularly hospital-acquired infections.

8 Key measures in the control of hospital-acquired infections are the maintenance of good housekeeping standards, adherence to strict aseptic technic, and the detection of nasal carriers of staphylococci in the newborn nursery and the surgical suite.

9 Streptococci vary greatly in cultural characteristics and pathogenicity. They are responsible for more illness and cause more different kinds of disease than any other group of organisms. They can cause primary as easily as secondary disease in any part of the body and attack both humans and lower animals. Some are saprophytic in dairy products.

10 Penicillin is still the drug of choice against beta-hemolytic streptococci.

11 When grown on blood agar, streptococci are either hemolytic, incompletely hemolytic, or nonhemolytic. When classified according to their biochemical properties, streptococci may be placed in a pyogenes group, a viridans group, a lactic group, or an enterococcus group. The beta-hemolytic *Streptococcus pyogenes* is a serious pathogen.

12 By the most important classification of Lancefield, streptococci fall into 13 groups according to the antigenic specificity of their C substances. Some 90% of human infections are caused by Lancefield group A.

13 Important examples of the many extracellular enzymes and poisons elaborated by streptococci are hemolysins, leukocidins, hyaluronidase, and erythrogenic toxin.

14 The type lesion for hemolytic streptococci is the diffuse, ill-defined spreading lesion of cellulitis. Allergic reactions follow some streptococcal infections, usually of the throat (for example, acute rheumatic fever). Scarlet fever, epidemic sore throat, and impetigo contagiosa are serious streptococcal diseases.

15 Dental caries is related to individual susceptibility, to the presence of bacteria that produce organic acid in the mouth, and to the presence of carbohydrates to support the cariogenic activities of the acid-producers. *Streptococcus mutans* (a member of the viridans group) is implicated in the production of tooth decay.

16 The capsule of the pneumococcus is its main virulence factor, since the capsule enables it to resist phagocytosis by its principal host, the human being.

17 The pneumococcus lacks both catalase and peroxidase but grows equally well with or without oxygen on enriched media, producing large amounts of lactic acid. Unlike other streptococci it is soluble in bile.

18 Water-soluble polysaccharides in the pneumococcal capsule give type specificity. At least 88 types or subtypes of *Streptococcus pneumoniae* can cause disease. Pneumococcus type 3 is the most virulent type, with a wide capsule and slimy growth on culture media.

19 The pneumococcus causes three important diseases: pneumonia (both lobar pneumonia and bronchopneumonia), meningitis, and otitis media. The hallmark of pneumococcal infections is the abundance of fibrin present in the inflamed areas.

20 Pneumonia may be caused by a variety of bacteria and other microbes and may be either community acquired or hospital acquired.

21 A polyvalent pneumococcal vaccine incorporates the purified capsular polysaccharides from the 14 pneumococcal types currently responsible for pneumococcal disease.

QUESTIONS FOR REVIEW

1 Name important gram-positive cocci. Why are pyogenic cocci so called?
2 Characterize staphylococci.
3 List diseases caused by staphylococci.
4 Discuss the localization and invasion of the body by staphylococci.
5 How is pathogenicity of a staphylococcus established?
6 Comment on the background factors in hospital-acquired infections. How are they studied epidemiologically?
7 Outline the streptococci according to the following categories: morphology, general features, classification, and pathogenicity.
8 Name 10 human diseases and 2 animal diseases caused by streptococci. State the Lancefield groups responsible.
9 How are streptococci and pneumococci transmitted from person to person?
10 What is the Dick test? Neufeld reaction? Bacitracin disk test? Coagulase test? Schultz-Charlton phenomenon? Optochin test?
11 How is the antistreptolysin O titer used clinically?
12 Briefly discuss septic sore throat, puerperal sepsis, and erysipelas.
13 What is the relation between streptococcal disease and rheumatic fever?
14 Compare lobar pneumonia with bronchopneumonia.
15 Contrast the type lesion of staphylococcal and streptococcal infections. State the notable feature of pneumococcal injury.
16 What is the current status of vaccines against pyogenic cocci?
17 Explain what is meant by the scalded skin syndrome, otitis media, bacterial interference, group B streptococcal infection in newborns, the quellung reaction, a sentinel microorganism, the role of C substance in serogrouping, the role of lipase in staphylococcal infections, and virulence factors.
18 Give the salient features of toxic shock syndrome.
19 What appears to be the relation of acid-producing bacteria to dental caries? How is *Streptococcus mutans* implicated?

SUGGESTED READINGS

Abramson, C.: The serodiagnostic aspects of S. *aureus* infections, Lab. Manage. **21**:48, Nov. 1983.

Boerner, D.F., and Zwadyk, P.: The value of the sputum Gram's stain in community-acquired pneumonia, J.A.M.A. **247**:642, 1982.

Bannatyne, R.M., and others: Cumitech 10: laboratory diagnosis of upper respiratory tract infections, Washington, D.C., 1979, American Society for Microbiology.

Bartlett, J.G., and others: Cumitech 7: laboratory diagnosis of lower respiratory tract infections, Washington, D.C., 1978, American Society for Microbiology.

Kuhn, P.J.: Toxic shock syndrome, Diagn. Med. **3**:26, Nov.-Dec. 1980.

Marx, J.L.: New clue to the cause of toxic shock, Science **220**:290, 1983.

Mufson, M.A.: Pneumococcal infections, J.A.M.A. **246**:1942, 1981.

Philip, A.: Laboratory indentification of group B streptococci, Lab. Manage. **18**:33, April 1980.

Schutzer, S.E., and others: Toxic shock syndrome and lysogeny in *Staphylococcus aureus*, Science **220**:316, 1983.

Scott, J.R., and Fischetti, V.A.: Expression of streptococcal M protein in *Escherichia coli*, Science **221**:758, 1983.

The toxic shock syndrome: a conference held 20-22 November 1981, Ann. Intern. Med. **96**:831, 1982.

NEISSERIAE

Distinctive gram-negative cocci or plump coccobacilli, sometimes called neisseriae, are found in the family Neisseriaceae, which includes three parasitic genera—*Neisseria*, *Branhamella*, and *Moraxella*. Human beings are the only natural hosts for the most important ones, the gonococcus *(Neisseria gonorrhoeae)* and the meningococcus *(Neisseria meningitidis)*, two pathogenic neisseriae (also pyogenic cocci) of genus *Neisseria*. The members of this genus are aerobic or facultatively anaerobic parasites of human mucous membranes. All neisseriae produce the enzymes *indophenol oxidase* (that is, they are oxidase positive) and catalase. Members of the genera *Branhamella* and *Moraxella* and certain minor *Neisseria* species are important, not because of their pathogenicity but because of their habitat in the mouth and upper respiratory passages.

NEISSERIA GONORRHOEAE (THE GONOCOCCUS)

Neisseria gonorrhoeae causes gonorrhea, one of the most common sexually transmitted (venereal) diseases (STDs) and a widely prevalent disease of humans. It afflicts an estimated 65 million persons throughout the world and is a major communicable disease problem in the United States today. A disease known to the ancient Chinese, Egyptians, and Hebrews, it was termed *gonorrhea* (Greek *gonē*, seed or semen, plus *rhein*, to flow) by Galen about 130 AD. The gonococcus is sometimes called the diplococcus of Neisser after Albert Neisser (1855-1916), German physician, who in 1879 discovered it in purulent secretions from the urethra and cervix uteri and the conjunctiva of the eye. Infections caused by N. *gonorrhoeae* may be referred to as neisserian infections.

General Characteristics

The gonococcus is a gram-negative, nonmotile, nonencapsulated, nonspore-forming diplococcus. In smears the opposing sides of the two cocci are flattened, and the cocci are placed like two coffee beans with their flat sides together. The gonococcus possesses the characteristic gram-negative cell structure and is both an intracellular and extracellular parasite. There is only one defined strain.

Gonorrhea is a disease accompanied by a discharge from the genital tract that is at first thin and watery, then later purulent. The incubation period is short, 3 to 5 days. During the early stages gonococci are found free in the serous exudate or attached to epithelial cells, but when the exudate becomes purulent, phagocytosis takes place and gonococci are found within the cytoplasm of the pus cells (polymorphonuclear neutrophilic leukocytes). A single white blood cell may contain from 20 to 100 microorganisms, gonococci that are not dead but still infectious. In later stages of the disease they may be found outside the white blood cells, and when the disease becomes chronic, gonococci often cannot be found at all.

This fastidious microbe will not grow on ordinary culture media and is somewhat difficult to cultivate even on media prepared especially for it. Gonococci grow best at slightly below body temperature (35.5° C), in an atmosphere containing oxygen and carbon dioxide (10%). The appearance of colonies of gonococci is variable, and four types described are related to virulence. Colony types 1 and 2 come from infective organisms; colony types 3 and 4 from gonococci that are noninfective. The cocci of types 1 and 2 possess pili. Pili are attachment organelles, by means of which the bacteria are anchored to the columnar epithelial cells of a mucous membrane (urethral, cervical, or rectal) and can thus resist phagocytosis. The bacteria can then colonize the area. Attachment is considered a prerequisite for infection, and only gonococci with pili appear to be pathogenic.

Gonococci produce the enzyme indophenol oxidase as do other members of the genus *Neisseria*. Oxidases catalyze the reduction of molecular oxygen independently of hydrogen peroxide. The **oxidase reaction** is used to identify colonies of neisserian species in laboratory cultures (p. 174). The enzyme IgAase, an enzyme that cleaves the protective secretory IgA (immunoglobulin A) of mucous membranes, has been demonstrated for all four colony types of gonococci.

Although gonococci are difficult to destroy in the body, they possess little resistance outside it. They are killed in a very short time by sunlight and drying. In pus or on clothing in moist dark surroundings they may live from 18 to 24 hours. They are very susceptible to disinfectants, especially silver salts, and are killed by a temperature of 60° C in 10 minutes.

Although gonococci are susceptible to the action of the modern antibiotics, drug resistance is an increasingly serious problem, especially with strains of penicillinase-producing *N. gonorrhoeae* (PPNG), first recognized in the United States in early 1976, coming from Southeast Asia and the Philippine Islands. The gene for penicillinase (beta-lactamase) production in gonococci is carried by plasmids with similar nucleotide sequences to plasmids that mediate penicillin-resistance in certain gram-negative enteric organisms and may have been acquired from them (see Fig. 14-8). Certain gonococci can sexually transfer their beta-lactamase plasmids and probably do so rather freely in nature.

Pathogenicity

The gonococcus is parasitic specifically for humans. Nothing comparable to any of the clinical forms of gonorrhea had been produced artifically in lower animals up until the time gonococcal urethritis was produced experimentally in the chimpanzee and the male-to-female animal transmission was demonstrated.

In typical cases of gonorrhea the sites of primary infection in the human female are the urethra and cervix; in the human male the site is the urethra. Gonococci injure columnar epithelium, for which they have an affinity, like that lining the cervix uteri and the rectum, and the transitional (urothelial) epithelium lining the urinary tract.

Pelvic inflammatory disease (PID) refers to inflammation of the fallopian tubes (salpingitis) with or without inflammation of the ovary (oophoritis), localized pelvic peritonitis, and formation of an abscess (tuboovarian abscess). It is the most severe complication of the sexually transmitted diseases caused by *Neisseria gonorrhoeae* and *Chlamydia trachomatis*.

In this complex condition, microbes from the vagina and lower cervix uteri have crossed the normally sterile endometrial cavity and infected the fallopian tubes and adjoining structures. PID is commonly regarded as an infection caused by the gonococcus, but *Chlamydia trachomatis, Mycoplasma hominis,* and combinations of aerobes and anaerobes from the normal vagina are recognized etiologic agents.

Long-term changes in PID come from the scarring of the healing process. As the inflamed tubes heal, they may become completely or partially occluded by scar tissue. Complete occlusion results in involuntary infertility. Partial obstruction predisposes to ectopic pregnancy. (In an ectopic pregnancy, the products of conception are located away from the normal position in the uterine cavity.) Women who have had PID are 6 to 10 times as likely to have an ectopic pregnancy as are women who have not had this infection.

Adhesions that occur during the healing process result in chronic pelvic pain if the adhesive bands impinge on the ovaries, bowel, or other tender pelvic structure.

The economic costs of PID are staggering. The direct cost of treatment of acute PID in the United States has been estimated to be more than $600 million dollars annually—the total cost at upwards of $3 billion dollars a year.

Modified from McCormack, W.M.: J.A.M.A. **248**:177, 1982. Copyright 1982, American Medical Association.

Vaginal infection is not seen because the epithelium lining the vagina of an adult is a cornified stratified squamous type resistant to gonococci. Before the age of puberty, the vagina is lined with a softer, extremely susceptible epithelium. Gonorrheal vulvovaginitis in prepubertal girls may be epidemic and difficult to eradicate. The change in the epithelium with the onset of puberty usually eliminates this form of gonorrhea completely.

An important primary site of infection seen today is the conjunctiva of the eye, and the process (gonorrheal conjunctivitis and keratitis) is one that actively damages the eye (Fig. 24-1). **Ophthalmia neonatorum,** gonorrheal conjunctivitis in newborn infants, results when the eyes are infected during the birth passage. A profuse purulent discharge in the eyes of a neonate can build up a considerable pressure behind the lids. If the lids are forced apart, pus spurts out. The physician and attendants of these babies must be careful to protect their own eyes. In babies or adults such infection easily results in blindness or serious sight impairment because of the inflammatory distortion of the structures in the eye.

From the urethra of the human male, gonococcal infection may spread directly to the other parts of the male reproductive system. In the female it may likewise spread to other parts of the tract, especially to Bartholin's glands and the fallopian tubes. The lining of the uterus seems to resist the action of gonococci, but the presence of an intrauterine contraceptive device (IUD) may facilitate the spread of gonococci into the endometrium, thereby increasing the risk of complications in the fallopian tubes. Invasion of the fallopian tubes usually occurs with the first or second menstrual period after infection; however, in some cases it may not occur until later. Involvement of the

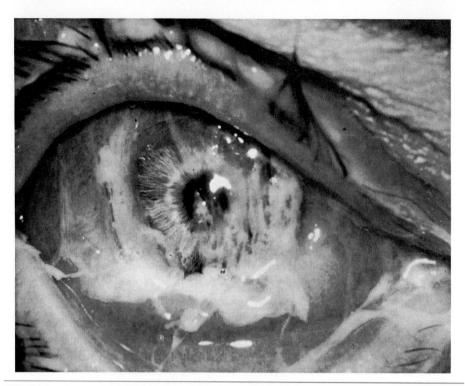

FIG. 24-1   Acute gonococcal conjunctivitis (gonorrheal ophthalmia of adults). Note inflammation, redness, irritation of conjunctivae, and copious pus. Smear of exudate showed pus cells with intracellular diplococci. Culture grew out gonococci. Patient also had gonococcal pelvic inflammatory disease. (From Donaldson, D.D.: Atlas of external diseases of the eye, vol. 1, St. Louis, 1966, The C.V. Mosby Co.)

fallopian tubes (salpingitis) is associated with considerable distortion and scarring if the disease becomes chronic. Scarring of the urethra in the male may lead to stricture or closure of the urethral lumen at one or more focal points.

Gonococci sometimes migrate from the genitourinary tract by the lymphatics or the bloodstream to set up distant foci of infection (for example, endocarditis, perihepatitis, and meningitis). Gonococcemia is associated with varied skin lesions from which organisms may be identified. An important manifestation of extragenital gonococcal infection is a purulent, destructive arthritis, noted especially between ages 15 and 35 years. As the overall incidence of gonorrhea increases, extragenital lesions become more prominent.

Sources and Modes of Infection

Gonococci are never found outside the human body unless they are on objects very recently contaminated with gonorrheal discharges, and there they live only for a short time. Therefore gonorrheal infections are practically always spread by direct contact, and mostly the mode of contact is sexual intercourse. One cannot deny that gonorrhea is sometimes transmitted indirectly by contaminated objects, but this is very rare.

Gonorrheal ophthalmia of adults is usually accidental. Infection from the genitourinary tract is inadvertently transferred to the eyes by the hands of the same or a different person. Vulvovaginitis in children is spread by the use of common bed linen, bathtubs, toilets, and such. It has been known to result from the use of contaminated rectal thermometers. It usually occurs where children live in crowded quarters.

Untreated gonococcal infections tend to become chronic. Females who are untreated or inadequately treated become infectious carriers for years after manifestations of disease have disappeared. An estimated 60% to 80% of females with the infection are asymptomatic. Infected males (perhaps as many as 40%) may also be asymptomatic.

Laboratory Diagnosis of Gonorrhea

Several microbiologic procedures apply to the diagnosis of gonorrhea. Smears, cultures, and the oxidase reaction are the *presumptive* tests. Fluorescent antibody technics and carbohydrate fermentation reactions are used to *confirm* the results of the presumptive tests and to establish the diagnosis of gonorrhea.

Direct smears of genital discharges may be stained with the Gram stain. There are rare exceptions to the rule that the finding of gram-negative diplococci in the pus cells of a genital exudate strongly *suggests* that they are gonococci. This is especially true if the exudate is from the male urethra, where, with typical acute purulent urethritis, the Gram-stained smear of the exudate containing the distinctive intracellular diplococci ordinarily makes the diagnosis. In females typical diplococci may be seen in smears of material from Skene's and Bartholin's glands early in the disease, but even a working diagnosis cannot be made on this basis alone. The reasons for this are several: (1) gram-negative diplococci other than gonococci occur outside cells; (2) gonococci occur outside cells, singly or in pairs; and (3) gram-positive organisms with the morphology of gonococci occur in cells. All that can be said about gram-negative diplococci found outside cells is that they *may be* gonococci. Very infrequently are gram-negative diplococci other than gonococci found within the pus cells of a genital exudate, but it is possible. The smear prepared from gonorrheal exudates should be quite thin; gonococci react to the Gram stain in an erratic way if the smear is thick and uneven. The microbes usually are not found in the exudate of chronic gonorrhea.

Cultural methods are of special value in the diagnosis of chronic disease and in the determination of a cure. Cultures are incubated under increased carbon dioxide tension (Fig. 24-2). Enriched media, such as chocolate agar, Hirschberg egg medium, and currently the Thayer-Martin (T-M) medium, are used to cultivate the delicate gonococcus. T-M medium contains hemoglobin, certain growth factors, and antimicrobial agents to inhibit selectively fungi, gram-positive organisms, and many gram-negative ones. (It is also called VCN medium for the three antibiotics it contains—vancomycin hydrochloride, colistimethate sodium, and nystatin. Recently trimethoprim [p. 235] has been incorporated to inhibit certain strains of *Proteus*.) The plates of T-M medium to be inoculated must be streaked in a specified way to spread the organisms out of the

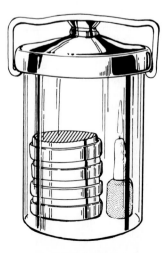

FIG. 24-2 Culture under increased carbon dioxide tension, simple method for partially anaerobic conditions. Cultures and lighted candle are placed in container, which is made airtight. Candle burns until oxygen is almost completely exhausted, then goes out.

associated mucus, which tends to lyse them (Fig. 24-3). A modification of T-M medium, Transgrow, has been developed so that suspect cultures can be sent into central laboratories. Transgrow comes in a screw-cap bottle containing a mixture of air and carbon dioxide that sustains growth of gonococci for 48 to 96 hours. After inoculation of the medium (Fig. 24-4) (and return of the screw cap), the bottle is ready for shipping (Fig. 24-4). The unopened bottle can be incubated directly at the receiving laboratory.

Public health authorities today recommend that suspect material for culture be obtained from the anorectal area and pharynx as well as from the urogenital tract (anterior urethra, endocervical canal). Rectal gonorrhea is easily overlooked. Carefully taken swabs are important in male homosexuals, and in females both cervical and rectal swabs are needed since half the women infected harbor the gonococcus in the rectum. Infection may persist there after it has been eliminated in the cervix. Sterile cotton swabs often contain fatty acids and other substances that inhibit the growth of the gonococcus; the use of noninhibitory swabs, such as calcium alginate swabs, is advised for specimen collection.

Gonococci can be recovered in urine from males if the first 10 ml of voided urine is

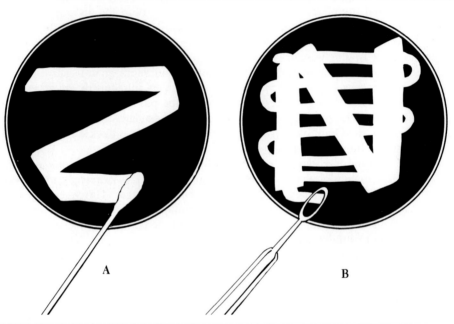

FIG. 24-3   Two-step method of streaking plate for culture of gonococci. **A,** Suspect secretions on swab rolled gently onto plate in **Z** pattern. **B,** Cross streaking of inoculum in **A** done with wire loop.

FIG. 24-4   Inoculation of Transgrow medium. Note neck of bottle elevated to prevent escape of carbon dioxide. The swab of test material is rolled over surface of medium as shown. (From Criteria and technics for diagnosis of gonorrhea, Atlanta, 1972, Centers for Disease Control.)

centrifuged and the sediment cultured. In a simpler screening method the first few drops of urine caught on a dry swab are immediately passed across T-M medium. Cultures of urine in screening programs have helped to define the reservoir existing in the asymptomatic human male.

The next steps in the bacteriologic study of the gonococcus are the determination of the biochemical reactions and identification of the organisms with fluorescein-labeled antiserums. The biochemical reactions of *Neisseria* species discussed in this chapter are given in Table 24-1. The fact that the gonococcus possesses a specific K-type antigen is the basis for the fluorescent antibody test to detect gonococci in direct smears of exudate or in smears made from cultures.

Several serologic tests, including an enzyme immunoassay, have been developed. If the major protein of the outer membrane of the gonococcus is used, 16 antigenically distinct serotypes may be demonstrated. With the protein antigen of the pili, six serologically distinct types are presented. However, a serologic test to catch the large number of persons in the silent, communicable reservoir, although highly desirable, is problematic. One problem is that the gonococcus seems to share its antigens with other, harmless neisseriae.

It is never within the province of the laboratory alone to say that a person is cured of gonorrhea, and in many cases the efforts of both laboratory and clinician do not determine whether the disease is eradicated. In dealing with a case of gonorrhea physicians and nurses should consider its medicolegal potential.

Social Importance of Gonorrhea

The medical, social, psychologic, and even medicolegal implications of gonorrhea are far-reaching. In the United States today gonorrhea is a disease of nearly pandemic proportions—the best estimates given are that approximately 2½ to 3 million new cases occur each year, about 1 every 15 seconds. More than half involve teenagers and young adults under 25 years of age. Gonorrhea not only is prevalent in teenagers but may be prevalent in prepubertal children as well. A vaginal or urethral discharge in any child must be suspect.

Although the ratio of the number of cases in males to females is approaching 1:1, generally three males are treated and reported for each female, since in males the manifestations may be sufficiently disagreeable to motivate them to seek medical attention. On the other hand, females are often asymptomatic or relatively so; many of them are reported and diagnosed only because of information from male consorts. The silent reservoir of asymptomatic females constitutes a primary obstacle to the control of the disease, and if widespread screening of the female population is not done, it may remain so. Some health authorities recommend that a culture for N. *gonorrhoeae* be considered an essential part of prenatal care and that a routine culture be taken more often at the time a pelvic examination is done.

The most destructive piece of misinformation handed down from generation to generation is that gonorrhea is no worse than a cold. Such a fallacy underestimates the danger of gonorrhea and creates the impression that colds are of no consequence.

Gonorrhea is said to be the most common cause of sterility in both sexes. In females sterility results from occlusion of the fallopian tubes by scar tissue formed during the healing of gonorrheal salpingitis. In males sterility results from occlusion of the vasa deferentia by a similar process of gonorrheal inflammation and healing with scarring.

Immunity

An attack of gonorrhea confers little if any immunity to subsequent attacks. An antibody response seems to be poor. The apparent lack of immunity may be a major factor as to why the disease remains endemic in human populations. A pronounced susceptibility

Table 24-1 Characteristics of Neisseriae

Organism	Growth on Thayer-Martin Medium	Growth on Nutrient Agar (No Blood) at 22° C	Oxidase	Fermentation of Sugars with Acid Only					Pigment Production	Habitat
				Glucose	Maltose	Sucrose	Fructose	Lactose		
Neisseria gonorrhoeae	+	–	+	+	–	–	–	–	–	Genital infections of humans
Neisseria meningitidis	+	–	+	+	+	–	–	–	–	Nasopharynx of humans
Neisseria sicca	–*	+	+	+	+	+	+	–	Variable	Nasopharynx of humans
Neisseria flavescens	–*	+	+	–	–	–	–	–	Golden yellow	Nasopharynx of humans
Neisseria subflava	–*	+	+	+	+	–	+	–	Greenish yellow	Nasopharynx of humans
Neisseria lactamica	–*	+	+	+	+	–	–	+ (slow)	Yellowish	Nasopharynx of humans

*Growth with heavy inoculum.

to reinfection does exist, which was amply documented by author James Boswell in the eighteenth century. Describing himself as "a compulsive patron of prostitutes," he recorded some 19 episodes of gonorrhea in the years between 1760 and 1790. He eventually died from complications of the disease.

It may be that specific acquired immunity with gonorrhea involves opsonins and is type specific.

Prevention

The general public should be warned of the dangers of gonorrhea and the difficulty of its cure. Unfortunately, the use of the "wonder drugs" has engendered an attitude that the disease is no problem. The dangers of quack doctors and folk remedies must be stressed. Patients must not allow their discharges to soil toilets or articles used by others, and they must be warned of the danger of transferring infectious material to their eyes by means of their hands.

Ophthalmia neonatorum can be prevented by the Credé method, a procedure so important that it is required by law in most of the 50 states.

The Credé method for prophylactic treatment in newborn babies is as follows. Immediately after birth, the eyelids of the baby are cleaned with sterile water. A different piece of cotton is used for each eye, and the lids are stroked (or irrigated) from the nose outward. Next, the lids are opened, and one or two drops of 1% silver nitrate solution are instilled into each eye, care being taken that the conjunctival sac is completely covered with the solution. After 2 minutes the eyes are irrigated with isotonic saline solution. (A mild irritation of the lining membranes of the eye may be produced, but it is short lived.)

Penicillin and other antibiotics are as effective as silver nitrate in the prevention of ophthalmia neonatorum, but where the law requires silver nitrate, such antibiotics are not used. A few states have passed laws allowing a suitable antibiotic to be used instead of silver nitrate. However, the National Society for the Prevention of Blindness in New York currently recommends that silver nitrate method be the standard and *preferred* procedure.

Vulvovaginitis in children may be prevented by proper care of bed linen, bathtubs, nightclothes, and wash water. All children should be examined for gonorrhea before admission to children's institutions or hospital wards with other children. Physicians and medical attendants wearing rubber or plastic gloves for palpation and inspection of cervix uteri and vagina must be careful to change them before a gloved finger is introduced into the rectum. Gloves contaminated with gonorrheal discharges can easily spread infection into the rectum.

Because of the mode of transfer, every person who has gonorrhea should be serologically tested for syphilis.

A gonococcal vaccine has been created by isolating and purifying a pilus protein. It is designed so that antibody produced in the recipient will coat the pili of any invading gonococcus. This action prevents adherence of the bacteria to host cells and thus prevents development of the disease.

The STD Pandemic

It is said that every 2 minutes somewhere in the United States a teenager contracts a sexually transmitted disease. Because of changing mores in our society,* forms of disease contracted mainly through intimate sexual contact and some that may also be

---

*Students of the social sciences, in correlating the change in social attitudes with the increased incidence of gonorrhea, define the problem in the context of seven P's: Promiscuity, Perversion, Pornography, Permissiveness, Pot, Pads, and the Pill.

Table 24-2 Sexually Transmitted Diseases (STDs)*

Disease	Agent
The "Classic Five"	**Bacterial**
Gonorrhea	*Neisseria gonorrhoeae*
Syphilis	*Treponema pallidum*
Chancroid (soft chancre)	*Haemophilus ducreyi*
Granuloma inguinale (Donovanosis)	*Calymmatobacterium granulomatis* (Donovan body)
Lymphogranuloma venereum	*Chlamydia trachomatis*
"Newcomers"	
	Viral
Herpes genitalis	*Herpesvirus hominis* types 1 and 2
Molluscum contagiosum	Poxvirus (molluscum body)
Condyloma acuminatum (genital wart)	Papovavirus
Heterophile-negative infectious mononucleosis	*Cytomegalovirus*
	Bacterial
Nongonococcal urethritis (NGU)	*Chlamydia trachomatis*
	Ureaplasma urealyticum (T strain mycoplasma)
	Mycoplasma hominis
Mycoplasmosis	*Genital mycoplasmas*
Oculogenital trachoma	*Chlamydia trachomatis*
Pelvic inflammatory disease	*Neisseria gonorrhoeae*
	Chlamydia trachomatis
	Mycoplasma hominis
Nonspecific vaginitis (bacterial vaginosis)	*Gardnerella vaginalis*
	Fungal
Candidiasis	*Candida albicans*
	Protozoan
Trichomoniasis	*Trichomonas vaginalis*
Amebiasis	*Entamoeba histolytica*
	Ectoparasitic
Pediculosis pubis	*Phthirus pubis* (crab louse)
Scabies	*Sarcoptes scabiei* (itch mite)
In the Neonate—Directly Related Complications	
	Viral
Neonatal herpes	*Herpesvirus hominis* types 1 and 2
Congenital abnormalities	*Cytomegalovirus*
	Bacterial
Inclusion conjunctivitis	*Chlamydia trachomatis* (TRIC agent)

*With the exception of gonorrhea, the ones caused by microbes are discussed in subsequent chapters.

Table 24-2 Sexually Transmitted Diseases (STDs)—cont'd

Disease	Agent
In the Neonate—Directly Related Complications—cont'd	
	Bacterial—cont'd
Pneumonia	*Chlamydia trachomatis*
Ophthalmia neonatorum	*Neisseria gonorrhoeae*
Congenital lues	*Treponema pallidum*
Sepsis, meningitis	Group B streptococcus
In the Homosexual Male—Special Problems	
	Viral
Hepatitis B	Hepatitis virus type B
Hepatitis A	Hepatitis virus type A
	Protozoan
Pneumocystosis (opportunistic infection)	*Pneumocystis carinii*
Acquired immunodeficiency syndrome (AIDS)	**Unknown**
Kaposi's sarcoma	
Lymphomas	
STDs with Fecal-Oral Transmission	
	Viral
Hepatitis A	Hepatitis virus type A
	Bacterial
Shigellosis	*Shigella* species
Salmonellosis	*Salmonella* species
Streptococcal proctitis	*Streptococcus* species
Enterocolitis	*Campylobacter* species
	Protozoan
Amebiasis	*Entamoeba histolytica*
Giardiasis	*Giardia lamblia*
	Metazoan
Pinworms	*Enterobius vermicularis*

contracted this way are on the rise. This is a matter of great concern to public health agencies. Table 24-2 gives a list of sexually transmitted diseases with causative agents. *

Treatment information for sexually transmitted diseases is available from a toll-free nationwide hot line, VD National Hotline, sponsored by the American Social Health Association. †

---

*With the exception of gonorrhea, the STDs in Table 24-2 are discussed in subsequent chapters.
†National Headquarters, 260 Sheridan Ave., Palo Alto, CA 94306. Phone 1-800-227-8922 (in California: 1-800-982-5883).

A report to be made to the council on Scientific Affairs, American Medical Association, at the annual meeting, June 1983, noted that in 1980, there were 7 million new or recurrent episodes of genital herpes, 3 million chlamydial infections, 3 million cases of trichomoniasis, 2 million cases of gonorrhea, 500,000 cases of venereal warts, and 100,000 cases of syphilis. These led to 1 million cases of pelvic inflammatory disease, 300,000 damaged infants, 200,000 infertile women, and $2.5 billion dollars in direct and indirect costs. The numbers represent only a fraction of the actual number, since only gonorrhea, syphilis, and hepatitis B are reportable diseases, and even then many private physicians do not report them. The report noted that whereas men are the major transmitters of sexually transmitted diseases, the greatest health and emotional consequences are borne by women and children.

Two highlights of the report are:

(1) It is believed that more than 2 million cases of gonorrhea occur annually, probably twice as many as are reported. Women are the primary victims, accounting for 90% of all complications. A woman has a 50% to 60% risk of contracting the disease after a single exposure to an infected man, whereas a man has a 20% risk after exposure to an infected woman.

(2) Studies indicate *Chlamydia trachomatis* may have overtaken gonorrhea as the number one sexually transmitted disease. An estimated 3 million Americans have chlamydial infections. Although chlamydia cause urethritis and epididymitis in men, women are its primary victims. The disease is responsible for at least 20% of all pelvic inflammatory disease, 11,000 cases of sterility, and 3,600 ectopic pregnancies annually. A majority of children born to infected mothers develop the disease.

Modified from Am. Med. News, p. 24, May 13, 1983.

NEISSERIA MENINGITIDIS (THE MENINGOCOCCUS)

The meningococcus *Neisseria meningitidis* was isolated in pure culture from the cerebrospinal fluid of a patient with meningitis and described in 1887 by Anton Weichselbaum (1845-1920), Austrian pathologist and military surgeon. It causes meningococcal septicemia with or without localization in the leptomeninges, which results in epidemic cerebrospinal meningitis. Epidemic cerebrospinal meningitis, known also as cerebrospinal fever, spotted fever, and meningococcal meningitis, is one form of acute bacterial meningitis. During World War II more soldiers in the Army of the United States died of meningococcal infection than of any other infectious disease.

General Characteristics

Meningococci are gram-negative diplococci strikingly similar to gonococci but more irregular in size and shape. They are nonmotile and do not form spores. They possess capsules and pili (fimbriae). Since the capsular material (polysaccharide) enables the meningococcus to resist phagocytosis, it is a virulence factor. Capsules are not usually seen in ordinary smears but may be demonstrated by special methods. Meningococci produce endotoxins (lipopolysaccharides) liberated when the cocci disintegrate; these toxins are partly responsible for the manifestations of meningitis. In the cerebrospinal fluid meningococci appear both inside and outside the polymorphonuclear neutrophilic leukocytes.

Meningococci grow best at body temperature in an atmosphere containing 10% carbon dioxide (Fig. 24-2). They do not grow at room temperature. Growth requires special media containing enrichments such as whole blood, serum, or ascitic fluid. Agar containing laked rabbit blood (red cells lysed) and dextrose supports the growth of meningococci especially well, as does enriched hormone agar. Different strains of meningococci vary considerably in the ease with which they grow on artificial culture media. Salt has a toxic effect on meningococci and should be left out of media on which they are to be grown. Like gonococci, they are oxidase positive and elaborate catalase and an IgAase.

Meningococci are such frail organisms that they survive only a short time outside the body. Sunlight and drying kill them within 24 hours. Away from the body they are easily killed by ordinary disinfectants, but in the nasopharynx they are very resistant. They are quite susceptible to heat and cold. Meningococci quickly lyse in cerebrospinal fluid removed from the body. Therefore specimens of suspect fluid should be examined as quickly as possible.

Meningococci and Gonococci Compared

Meningococci and gonococci are alike in that (1) both are strict parasites and cause disease only in humans, (2) they show little difference in resistance to injurious agents, (3) their distribution in the inflammatory exudate is the same, (4) they grow on artificial media with difficulty, and (5) their disease-producing properties are comparable. Skin lesions of gonococcemia are similar to those of meningococcemia, and gonococci have caused the Waterhouse-Friderichsen syndrome (below). It may be said that they are morphologically, physiologically, and immunologically much alike.

Groups of Meningococci

By serologic reactions, including a capsular swelling test similar to that used in typing pneumococci, meningococci are classified into serogroups—A, B, C, D, X, Y, Z, W-135, and 29-E.

Groups A and C are encapsulated and possess a specific capsular polysaccharide. There is a polysaccharide-polypeptide component in group B but usually no capsule. In the past group A meningococci have been responsible for 95% of cases of epidemic meningitis; group A was especially troublesome during World War II. Groups B and C have been the endemic organisms figuring in sporadic outbreaks between major epidemics. Until the mid-1960s, group B was the prevalent one. Since then, group C has been causing an increasing number of outbreaks at military bases. Groups Y and W-135 have figured in about one third of cases. In small populations such as households and day-care nurseries, all serogroups of N. *meningitidis* occasionally cause multiple cases. There are few meningococci of group D in this country.

Pathogenicity

The meningococcus is not very pathogenic for lower animals, and typical epidemic cerebrospinal meningitis occurs only in humans. Invasion of the body by meningococci occurs in three steps: (1) implantation in the nasopharynx, (2) entrance into the bloodstream with septicemia, and (3) localization in the meninges (cerebrospinal meningitis). For most patients the invasion ends with implantation in the nasopharynx (the carrier state).

About one third of meningococcal infections are septicemias of such severity that without treatment the patient dies before meningeal infection can occur. Meningococci proliferate as massively in blood and tissues as though growing in laboratory broth culture. The events of meningococcal septicemia come together under the designation **Waterhouse-Friderichsen syndrome** (Fig. 24-5), an acute fulminating condition characterized by many small areas of hemorrhage in the skin and, within a period of hours, death in peripheral circulatory failure. At autopsy large areas of hemorrhage are seen in

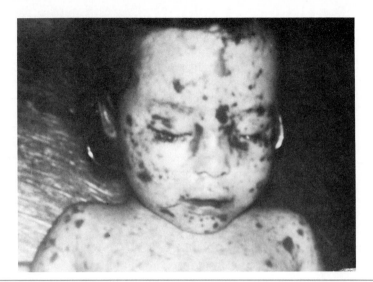

FIG. 24-5   Acute fatal meningococcemia in small child (Waterhouse-Friderichsen syndrome). Note hemorrhages over face and upper trunk. (From Smith, A.L.: Microbiology and pathology, ed. 12, St. Louis, 1980, The C.V. Mosby Co.)

the adrenal glands. The meningococcus possesses a unique capacity to induce fulminating illness and death within a matter of hours, and the rapid course of events possible with meningococcal disease—collapse and death within an hour—is terrifying to the social group. The disease can upset people to the point of mass hysteria.

In only a few patients meningococci in the bloodstream infect the meninges or other body tissues. When they do, purulent inflammation results. An exudate composed of leukocytes, fibrin, and meningococci forms in the subarachnoid space, the area between the two meningeal coverings, the arachnoid and the pia mater. Most prominent along the base of the brain, the exudate extends into the cerebral ventricles and down the spinal subarachnoid space. The cerebrospinal fluid may become so purulent that it scarcely flows through the lumen of the lumbar puncture needle.

Among the complications of epidemic cerebrospinal meningitis are arthritis, hydrocephalus, otitis media, retinitis, deafness from involvement of the eighth nerve, pericarditis, endocarditis, conjunctivitis, pneumonia, and blindness.

Sources and Modes of Infection

Since meningococci are such frail organisms and seldom found outside the human body, the source of infection must be a patient or a carrier. The organisms reside in the nasopharynx of both patients and carriers and leave the body in the nasal and buccal secretions (droplet infection). The mode of transfer is by close contact, and the organisms enter the body by the nose and mouth. Sexual transmission of meningococcal infection is reported. When meningococci reach a new host, they localize in the nasopharynx and multiply. Although the meningococcus colonizes the nasopharynx, it rarely causes a detectable nasopharyngeal infection and rarely causes primary pneumonia. Infection traced to articles recently contaminated by infected nasal and buccal secretions seldom occurs.

Meningococcal infections are not highly communicable, and there seems to be a high degree of natural immunity to them. Many persons exposed become carriers but few develop disease. It is estimated that about one carrier in 1000 develops meningococcemia or meningitis. In the absence of immunity the great number of carriers in the general population would maintain an epidemic at all times. Since meningococcal infection is endemic in all densely populated centers, it may become epidemic under overcrowded conditions, as in military camps. Overcrowding concentrates the carriers and promotes factors that lower individual resistance. When troops are mobilized, it is one serious disease to which the men seem to be exceedingly vulnerable.

Under ordinary conditions, from 5% to 30% of the general population are carriers. When epidemics occur, the number of carriers, especially of group A meningococci, increases. In military establishments 50% to 90% of the personnel may become carriers. About two thirds of those who contract meningitis harbor the organisms for a variable time after convalescence. The carrier state lasts about 6 months. However, healthy persons who have never had meningitis but nevertheless harbor meningococci in their nasopharynx play more of a part in the spread of meningitis than those who harbor the organisms after recovery from the disease.

At times the meningococcus has been recovered from nontraditional sites, such as the vagina and cervix, in situations where its relation to disease was not clear. This finding emphasizes the importance of adequately identifying gram-negative cocci and other organisms from the genitourinary tract and anus. Meningococci, although thought of as respiratory pathogens, on occasion may invade other parts of the body.

Laboratory Diagnosis

To give the patient the advantage of early treatment, which is so important in meningococcal infections, septicemia should be diagnosed clinically. The diagnosis is confirmed by blood cultures. A valuable diagnostic adjunct is the demonstration of meningococci by smears or cultures in the hemorrhagic skin lesions associated with septicemia (Fig. 24-5). When cultures for meningococci are made, the specimen must not be allowed to cool before inoculation of media.

In epidemic meningitis the cerebrospinal fluid is first turbid and then purulent. In smears of cerebrospinal fluid the typical arrangement of gram-negative diplococci within pus cells is sufficient for practical purposes to make a working diagnosis of epidemic meningitis. However, cultures must confirm that. Because of the tendency of meningococci to lyse, the specimen must be examined as quickly as possible after it is withdrawn.

Fluorescent antibody technics detect the organisms, and countercurrent immunoelectrophoresis (CIE), a very specific test, can be used to test for the presence of meningococcal antigen in cerebrospinal fluid and serum. Group typing is also done by immunoelectrophoresis.

Immunity

That epidemics of meningeal infection are not more common stems from the fact that whereas nasopharyngeal colonization is prevalent (low immunity to localization), resistance to invasion of the bloodstream or body tissues is high. Seven to ten days after N. meningitidis has colonized the nasopharynx or oropharynx, circulating bactericidal antibodies can be demonstrated. These are without noticeable effect on the nasopharyngeal resident meningococci but appear to prevent any of those entering the bloodstream from multiplying and disseminating to other tissues, particularly in the central nervous system. Therefore susceptibility or resistance to clinical meningococcal disease can be closely correlated with the absence or presence of serum bactericidal activity. Hemagglutinating antibodies are also present. Immunity seems to increase with age, because the disease is more prevalent in children than in adults.

Prevention

The patient with meningococcal infection should be isolated until cultures fail to show meningococci in the nasopharynx or until 24 hours have elapsed after the start of antibiotic therapy. All discharges from the mouth and nose and articles soiled therewith should be disinfected. The urine occasionally contains the organisms and therefore should be disinfected. The physician and nurse should use every precaution to avoid infection or the carrier state. Nurses should exercise care lest their hands convey infection. Meticulous hand washing after patient contact is mandatory. Dishes used by the patient should be properly sterilized. Persons who have been in close contact with a patient should not mingle with others until bacteriologic examination has proved them to be free of meningococci.

General preventive measures include the proper supervision of carriers. The wholesale isolation of carriers does not eliminate infection, and currently the trend is to isolate only those in immediate contact with a patient. Nose sprays are probably of little value. The antibiotic rifampin is used to eliminate the carrier state.

The control of epidemic meningitis is yet to be accomplished. For reasons that are not understood, it suddenly becomes virulent in a community, attains epidemic form, persists for a time, and then disappears.

Table 24-3 Pathogens Related to Meningitis

Agents of Primary Meningitis

1. Common and well-known ones
   Neisseria meningitidis—Epidemic cerebrospinal fever
   Haemophilus influenzae type b—Nonmeningococcal meningitis in infants and young children
   Streptococcus pneumoniae—Endemic meningitis in adults, less often in children
2. Less common and unusual ones
   Escherichia coli—Purulent meningitis of newborn infants
   NOTE: Any of the microbes in the following list may cause purulent or nonpurulent meningitis and the infection may be mixed.
   Mycobacterium tuberculosis
   Other *Neisseria* species
   Listeria monocytogenes
   Streptococcus pyogenes
   Klebsiella-Enterobacter-Serratia species
   Salmonella species
   Bacteroides species
   Pseudomonas species
   Proteus species
   Nocardia asteroides

Agents of Secondary Meningitis

NOTE: The following list gives microbes that may be introduced directly into meninges because of recent trauma, surgical procedure, lumbar puncture, or congenital malformation (for example, spina bifida).

1. Common ones
   Staphylococcus aureus
   Pseudomonas species
   Escherichia coli and other coliforms
   Proteus species
2. Less common ones
   Any pathogenic microbe with capacity to invade human tissue

---

Three meningococcal vaccines—monovalent A, monovalent C, and bivalent A-C vaccines—are now licensed for *selective* use in the United States. The U.S. Public Health Service Advisory Committee does not recommend *routine* vaccination against group A and C disease because not enough is known about the benefits thereof. Antigens used for meningococcal vaccines are purified bacterial cell wall polysaccharides (certain polymers of particular neuraminic acids). Vaccine is given as a single parenteral dose, and adverse reactions have been insignificant. Duration of immunity is unknown. Group C vaccine has been given to American military recruits for the last several years with good results. Group B vaccine is still under investigation.

Microbial Infection of Meninges

The term *meningitis* simply means inflammation of the lining membranes of the brain and spinal cord. However, the term is usually applied to involvement of two—the pia mater and the arachnoid mater—of the three membranes. In infections associated with the formation of an exudate, the exudate accumulates in the subarachnoid space (between the pia mater and the arachnoid mater) and changes the physical characteristics of the cerebrospinal fluid. For instance, the exudate with a pyogenic microorganism changes a water-clear cerebrospinal fluid to one that is visibly cloudy or creamy. The diagnosis of the kind of meningitis must be made microbiologically since a variety of

---

Table 24-3 Pathogens Related to Meningitis—cont'd

Agents of Meningitis as Part of Generalized Infection

1. Common and well-known ones
   Neisseria meningitidis—Part of fulminating meningococcal septicemia
   Streptococcus pneumoniae—Part of pneumococcal septicemia
   Mycobacterium tuberculosis—Part of miliary tuberculosis
2. Less common ones
   NOTE: Almost any generalized infection with bacteremia may be associated with meningitis. The following microbes have been so associated with meningitis.
   Neisseria gonorrhoeae
   Salmonella typhi
   Brucella species
   Staphylococcus aureus
   Bacillus anthracis
   Nocardia asteroides
   Cryptococcus neoformans (fungus)

Agents of Meningitis in a Defective (Immunocompromised) Host

1. With deficiency of immunoglobulin
   Streptococcus pneumoniae
   Haemophilus influenza type b
2. With T cell deficiency
   Listeria monocytogenes
   Mycobacterium tuberculosis
   Various fungi
3. With deficiency of complement
   Neisseria meningitidis
   Streptococcus pneumoniae

---

microbes with different antibiotic sensitivities may reach the meninges and cause an infection there. Unless the disease occurs in an epidemic pattern, in which case it is meningococcal, meningitis (nonepidemic) cannot be specifically related to a given causative agent.

Bacteria reach the meninges by (1) penetrating wounds; (2) passage from the nasopharyngeal mucosa through lymph spaces; (3) extension of regional infections in the middle ear, mastoid process, bony sinuses, or cranial bones; and (4) the bloodstream (bacteremia or septicemia).

Table 24-3 lists pathogens known to cause meningitis under the various circumstances indicated. It is to be emphasized that nonepidemic meningitis is a potential threat to the elderly, to alcoholics, to immunosuppressed patients, to patients on broad-spectrum antimicrobials, and to long-term survivors of trauma to the head. It also complicates neurosurgical procedures. Among the many notable etiologic agents are pneumococci, influenza bacilli, gram-positive pyogens, and gram-negative rods. Pneumococcal and influenzal meningitis have the highest incidence in patients between 3 months and 1 year of age. In the aged the main causes of meningitis are pneumococci and gram-negative rods. Staphylococcal meningitis is of concern; it occurs primarily in persons with chronic skin, bone, or sinus infections or after neurosurgical penetration of the subarachnoid space.

Diagnosis of meningitis (both epidemic and nonepidemic) is made by demonstration of the causal agent in smear and culture of cerebrospinal fluid. Countercurrent immunoelectrophoresis is advantageous in the diagnosis of meningitis caused by *Escherichia coli*, *Haemophilus influenzae* type b, and certain other gram-negative rods, especially when the patient has already received antibiotic drugs.

OTHER GRAM-NEGATIVE COCCI (AND COCCOBACILLI)
Moraxella (Branhamella) catarrhalis

The organism *Moraxella (Branhamella) catarrhalis* is a normal inhabitant of the mucous membranes, especially those of the respiratory tract. It may be confused with the meningococcus or the gonococcus. Like these organisms, it is a gram-negative, biscuit-shaped diplococcus. On rare occasions it may assume the intracellular position. Differentiation depends on agglutination tests and the ability of *Moraxella (Branhamella) catarrhalis* to grow on ordinary culture media at room temperature.

Other Neisseriae

Other gram-negative diplococci that may lead to some confusion in microbiologic diagnosis are *Neisseria subflava*, *Neisseria sicca*, and *Neisseria lactamica*, all normal inhabitants of the pharynx and nasopharynx (Table 24-1). *Neisseria flavescens* has been recovered from the cerebrospinal fluid of patients with meningitis, and its colonies look like those of *Neisseria meningitidis*, but it does produce a golden yellow pigment when first isolated.

Moraxella (Moraxella) lacunata (the Morax-Axenfeld Bacillus)

The oxidase-positive, gram-negative, strictly aerobic coccobacilli of the genus *Moraxella* are parasitic on mucous membranes of humans and animals. *Moraxella (Moraxella) lacunata* causes a subacute or chronic inflammation of the conjunctiva, eyelid, and cornea.

SUMMARY

1 Neisseriae are gram-negative, aerobic or facultatively anaerobic, oxidase- and catalase-positive cocci or coccobacilli that parasitize human mucous membranes.

2 The gonococcus, the cause of gonorrhea, is a nonencapsulated diplococcus that is found both inside and outside cells.

3 Neisseriae grow only in an atmosphere containing 10% carbon dioxide on media especially prepared for them. Four colony types are recognized.

4 Pili are attachment organelles enabling the gonococcus to anchor to the columnar epithelium of a mucous membrane.

5 From the initial focus of injury in the epithelium of the urethra, cervix uteri, or rectum, gonococcal infection spreads to other parts of the reproductive system in either sex. Repeated infections deform the fallopian tube, scarring and obstructing it. Scarring of the male urethra leads to stricture. Gonorrhea is reportedly the most common cause of sterility in both sexes.

6 An important primary site of gonococcal infection in the adult or neonate is the conjunctival lining of the eye. The purulent conjunctivitis destroys tissues of the eye, thus leading to blindness or serious sight impairment.

7 Ophthalmia neonatorum results when the baby's eyes are infected in the birth passage. The Credé maneuver—the instillation of a diluted solution of silver nitrate into each eye of the newborn—prevents this.

8 Gonorrhea, syphilis, chancroid, granuloma inguinale, and lymphogranuloma venereum, all sexually transmitted and bacterial in etiology, are the "classic five," but sexually transmitted diseases caused by viruses, fungi, and protozoa are of current interest. Examples include herpes genitalis (viral), candidiasis (fungal), and trichomoniasis (protozoan).

9 Examples of sexually transmitted diseases passed by the mother to her infant are neonatal herpes, inclusion conjunctivitis, and sepsis caused by group B streptococci.

10 An attack of gonorrhea confers little if any immunity to later attacks. A gonococcal vaccine has been made with pilus protein in an attempt to prevent the initial adherence of the bacteria to the host epithelium.

11 Meningococci are similar to gonococci but are more irregular in size and shape. They possess not only pili like gonococci but also a capsule. Their capsular material is an important virulent factor. Both microbes are strict parasites and cause disease only in humans. Only one strain of gonococci is defined. There are nine types of meningococci.

12 Meningococci are so frail that they survive, like gonococci, only briefly outside the human body.

13 Meningococci invade the body first by implanting in the nasopharynx. From this vantage point they enter the bloodstream, produce a septicemia, and finally may localize in the meninges, where a characteristic purulent inflammation is produced (meningitis). Droplet infection (patient or carrier) is the source of infection.

14 The Waterhouse-Friderichsen syndrome is the acute, fulminating meningococcal septicemia that results in peripheral circulatory failure and death within hours.

15 The diagnosis of epidemic or nonepidemic meningitis is made by the demonstration of the causative agent in the cerebrospinal fluid. Meningitis may be primary or secondary to a preexisting disease. Meningitis may complicate a congenital malformation such as spina bifida or be a part of a generalized infection such as miliary tuberculosis.

16 Bacteria causing nonepidemic meningitis reach the meninges by penetrating wounds; by passage from the nasopharynx through lymph spaces; by extension of regional infection in the middle ear, mastoid process, and bony sinuses; and by the bloodstream.

17 Important agents of meningitis in a defective host include the influenza bacillus when there is a deficiency of immunoglobulin, *Listeria monocytogenes* with a T cell deficiency, and the pneumococcus with a deficiency of complement.

QUESTIONS FOR REVIEW

1 Name and describe the microbe causing gonorrhea.
2 What is the social importance of gonorrhea?
3 Briefly discuss the STD pandemic.
4 List 10 diseases transmitted sexually and give the causative organism for each.
5 Outline the laboratory diagnosis of gonorrhea.
6 What are the sources and modes of infection in gonorrhea?
7 How is gonorrheal ophthalmia contracted in adults? In infants?
8 What is the Credé method?
9 Compare gonococci with meningococci.
10 Name and describe the microbe causing epidemic meningitis.
11 What are the sources and mode of infection in epidemic cerebrospinal meningitis?
12 What is the importance of the carrier in meningococcal infections?
13 Give the nursing precautions in epidemic meningitis.
14 How is the diagnosis of meningococcal infection made in the laboratory?
15 List 10 microorganisms causing meningitis.
16 What is the Waterhouse-Friderichsen syndrome?
17 Briefly explain the following: venereal, pili, oxidase reaction, ophthalmia neonatorum, keratitis, conjunctivitis, ectopic pregnancy, PID, proctitis, fimbriae, nasopharyngeal colonization, and immunocompromised host.

SUGGESTED READINGS

Counts, G.W., and Petersdorf, R.G.: The wheel within a wheel: meningococcal trends (editorial), J.A.M.A. **224**:2200, 1980.

Covey, R.C.: Love triangle: three unreported venereal diseases recognized, Lab. World **31**:16, April 1980.

Curran, J.W.: Economic consequences of pelvic inflammatory disease in the United States, Am. J. Obstet. Gynecol. **138**:848, 1980.

Eschenbach, D., and others: Cumitech 17: laboratory diagnosis of female genital tract infections, Washington, D.C., 1983, American Society for Microbiology.

Gilbaugh, J.H., Jr., and Fuchs, P.C.: The gonococcus and the toilet seat, N. Engl. J. Med. **301**:91, 1979.

Handsfield, H.H.: Sexually transmitted diseases, Hosp. Pract. **17**:99, Jan. 1982.

Hoffman, C.A.: Sexually transmitted diseases (editorial), J.A.M.A. **246**:1709, 1981.

Jacobs, N.F., Jr., and Kraus, S.J.: Gonococcal and nongonococcal urethritis in men, Ann. Intern. Med. **82**:7, 1975.

Kellogg, D.S., Jr., and others: Laboratory diagnosis of gonorrhea: Cumitech 4, Washington, D.C., 1976, American Society for Microbiology.

McCormack, W.M.: Sexually transmitted diseases: women as victims, J.A.M.A. **248**:177, 1982.

Potterat, J.J., and King, R.D.: A new approach to gonorrhea control: the asymptomatic man and incidence reduction, J.A.M.A. **245**:578, 1981.

Washington, A.E., and Wiesner, P.J.: The silent clap, J.A.M.A. **245**:609, 1981.

CHAPTER 25

ENTERIC BACILLI AND VIBRIOS

The enteric bacilli, a large, heterogeneous group in the family Enterobacteriaceae, include several closely related genera of short, straight, nonspore-forming, gram-negative rods, facultatively anaerobic, that inhabit or produce disease in the alimentary tract of warm-blooded animals, including human beings. The type genus is *Escherichia*; the type species, *Escherichia coli*. This family includes the nonpathogenic bacteria that normally inhabit the intestinal canal and the highly pathogenic bacteria that invade and injure it. Heading any list of enteric pathogens are the consistent troublemakers, the *Salmonella* and *Shigella* species. Normal residents include many species in a number of genera, and if given the right opportunity, many of these supposedly benign bacteria can be awesome in their behavior. Not all members of Enterobacteriaceae are intestinal parasites; they have been placed here because of other similarities.

Today members of this family of bacteria are notorious as causes of urinary tract infection and are recovered from a variety of clinical specimens taken from diseased foci other than in the gastrointestinal tract. The Enterobacteriaceae are probably responsible for more human misery than any other group.

The genus *Yersinia*, also in the family Enterobacteriaceae, is discussed with certain small, gram-negative bacilli in Chapter 26. But because of the nature of cholera, a discussion of certain vibrios in the family Vibrionaceae is included in this chapter.

CLASSIFICATION

Enteric microorganisms are defined here in the classification most widely used in microbiologic laboratories. The starting point is the family Enterobacteriaceae. The breakdown into tribes, as proposed by W.H. Ewing and approved by the subcommittee of the American Society for Microbiology, with certain modifications from the Centers for Disease Control,* is as follows:

	Genera	DNA Base Composition (% G + C)
Tribe I: Escherichieae	Escherichia (Escherichia coli, including Alkalescens-Dispar) Shigella	{50 to 53 moles %
Tribe II: Edwardsielleae	Edwardsiella	—
Tribe III: Salmonelleae	Salmonella Arizona† Citrobacter (including Bethesda-Ballerup)	{50 to 53 moles %
Tribe IV: Klebsielleae	Klebsiella Enterobacter (formerly Aerobacter, including Hafnia) Serratia	{52 to 59 moles %
Tribe V: Proteeae	Proteus Providencia	{39 to 42 moles %
Tribe VI: Yersinieae	Yersinia	45 to 47 moles %

*The federal facility for disease eradication, epidemiology, and education, located in Atlanta.
†The term *paracolon bacilli* was formerly used for enteric bacilli resembling *Escherichia coli* but fermenting lactose much more slowly. They were placed in the genus *Paracolobactrum* and separated into the Bethesda-Ballerup group, the Arizona group, the Providence group, and the Hafnia group. Here, paracolon bacilli are regrouped and the term *paracolon* discarded.

Table 25-1 A Profile of Enteric Pathogenicity*

	Pathogen			
	Salmonella		Shigella	
	S. typhi	Other Species	S. dysenteriae	Other Species
Disease	Typhoid fever	Septicemia Gastroenteritis	Bacillary dysentery (very disabling)	
Number of microbes for infectious dose	10^4 to 10^6	10^3 to 10^7	10^1 to 10^2	10^1 to 10^2
Virulence factors				
1. Attachment to bowel	—	—	Yes	Yes
2. Toxin production	No	No	Yes, exotoxin	No
3. Other	Survival in macrophages	Invasion of mucosa and bloodstream	Invasion and multiplication	
Invasion into bowel mucosa	Yes, spread beyond	Yes, limited	Yes, limited	
Control	Sanitation (emphasis on simple measures, such as careful hand washing)		Sanitation (emphasis on hand washing)	

*See text for discussions.

Within the genus *Salmonella*, Ewing and co-workers recognize only three species—*S. choleraesuis* as type species, *Salmonella typhi* as unique human pathogen, and *Salmonella enteritidis* to comprise all other serotypes and serobiotypes. By this arrangement (Edwards-Ewing classification), *S. enteritidis* would be *S. enteritidis* ser. *enteritidis*, and *S. typhimurium* would be *S. enteritidis* ser. *typhimurium*. Although this approach is employed by the CDC, it is otherwise not widely accepted. In this chapter traditional species and serotype names for members of the genus *Salmonella* are used.

GENERAL CHARACTERISTICS

The members of the family Enterobactericeae are diverse both biochemically and antigenically. Genera in the family have been defined primarily on a biochemical basis, whereas many of the original species, still with the same names, were established on both a biochemical and an ecologic basis. Most members of this family are motile by means of peritrichous flagella. Colonies grown on blood agar are fairly large, shiny, and gray; they may or may not be hemolytic. All species ferment glucose, although some do not produce gas, and nitrates are usually reduced to nitrites.

These bacteria are quite complex antigenically, and schemes that have been devised incorporate serotypes that, being biochemically similar, can only be distinguished by serologic typing. (The terms *serotype*, *biotype*, and *serovariant* are synonymous.) Serologic typing of the members of Enterobacteriaceae revolves around three main groups of antigens. (1) The O (German *Ohne Hauch*, nonspreading) antigens are somatic, heat-stable, polysaccharide antigens located primarily in the cell wall. The O complex determines the somatic subgroup of an organism (for example, in the genus *Salmonella* or *Escherichia*). (2) The K (German *Kapsel*, capsule) envelope antigens surround the cell. Mostly heat labile, they mask heat-stable, somatic antigens and block agglutination reactions in certain instances. An example is the Vi antigen of *Salmonella typhi*. (3) The H (German *Hauch*, spreading) antigens (flagellar antigens) are protein, heat

	Pathogen			
	Escherichia coli			
Vibrio cholerae	Enterotoxigenic	Enteroinvasive	Normal Flora†	Opportunistic Pathogens
Cholera (devastating) ("cholera kills")	Infectious diarrhea		Variable	Various diseases
10^4 to 10^8	10^6 to 10^9		Unknown	Unknown
Yes	Yes	—	—	—
Yes, classic one	Yes, enterotoxin	No	—	—
—	—	Invasion	Presence in gastrointestinal tract	Presence in gastrointestinal tract
No	No	Yes	—	—
Sanitation	Sanitation		Good hospital housekeeping practices	Good hospital housekeeping practices against nosocomial organisms

†Normal inhabitants of bowel represented here; they can produce extraintestinal infection from a steady supply.

labile, and located in the flagella. Serotypes within somatic groups in the genus *Salmonella* are determined by H antigens.

Enteric bacilli may be studied according to physiologic behavior, immunologic reactions, and disease production. Table 25-1 compares and focuses quickly on key points in the disease-producing capacity of the enteric pathogens discussed in this chapter. It is an overview, an introduction, and a summary tabulation.

In nature there is an enormous production of enteric organisms. It is said that some 10^{22} microbial cells are excreted per day by a human being.

LABORATORY STUDY

The practical identification of the enteric gram-negative bacilli, however, involves the use of an elaborate array of biochemical reactions. Demonstrating biologic activities of test organisms is usually a complex maneuver in the clinical laboratory. In a greatly simplified scheme Table 25-2 presents key reactions for bacteriologic separation of enteric organisms to be discussed.

Table 25-2 Simplified Scheme for the Separation of Some Gram-Negative Enteric Bacilli

Step 1. Fermentation of Lactose

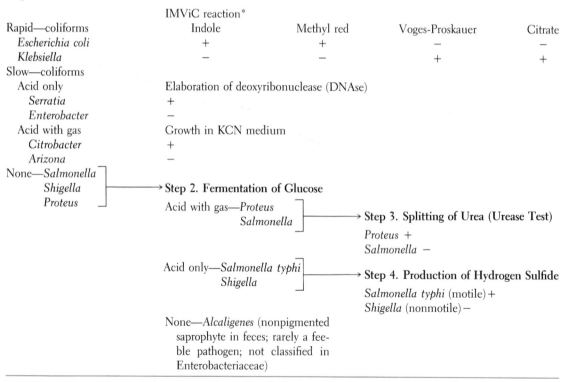

	IMViC reaction*			
Rapid—coliforms	Indole	Methyl red	Voges-Proskauer	Citrate
Escherichia coli	+	+	−	−
Klebsiella	−	−	+	+

Slow—coliforms
  Acid only   Elaboration of deoxyribonuclease (DNAse)
    Serratia   +
    Enterobacter   −
  Acid with gas   Growth in KCN medium
    Citrobacter   +
    Arizona   −
  None—*Salmonella*
      Shigella ——→ **Step 2. Fermentation of Glucose**
      Proteus
                     Acid with gas—*Proteus*
                              Salmonella ——→ **Step 3. Splitting of Urea (Urease Test)**
                                              Proteus +
                                              Salmonella −

                     Acid only—*Salmonella typhi*
                              Shigella ——→ **Step 4. Production of Hydrogen Sulfide**
                                              Salmonella typhi (motile) +
                                              Shigella (nonmotile) −

                     None—*Alcaligenes* (nonpigmented
                     saprophyte in feces; rarely a fee-
                     ble pathogen; not classified in
                     Enterobacteriaceae)

*IMViC reaction, useful in the differentiation of the coliform bacilli (normal residents of the enteron), refers to four biochemical color reactions. The first, "I," is the production of indole from tryptophan (pink or red color). The next two, the methyl red reaction, "M," and the Voges-Proskauer reaction, "Vi," indicate a difference in the fermentation of glucose in a special test medium after 2 to 4 days of incubation. If the test organism ferments glucose with the accumulation of acid end-products, the methyl red indicator gives a red color (pH below 4.5), a positive test. If glucose has been fermented with the accumulation of neutral end-products, the Voges-Proskauer test is positive; a red-orange color shows the presence of acetylmethylcarbinol. The fourth reaction, "C," refers to the utilization of Simmons citrate agar; if the test organism can use the citrate as its sole source of carbon, a blue color develops.

ESCHERICHIA COLI AND RELATED ORGANISMS
(THE COLIFORM BACILLI)
General Characteristics

Besides *E. coli*, coliform bacteria comprise members of the genera in the following groups: (1) *Klebsiella-Enterobacter-Serratia*, (2) *Arizona-Edwardsiella-Citrobacter*, and (3) *Providencia*.

In the colon (large bowel) 400 to 500 different types of bacteria are found. Forty percent of the fecal mass is bacterial.

As normal inhabitants of the intestinal tract, coliforms share certain traits. They are gram-negative, short rods that do not form spores. They grow at temperatures from 20° to 40° C but best at 37° C. A few species are surrounded by capsules. They are not so susceptible to the bacteriostatic action of dyes as other bacteria. Therefore media for the isolation of coliform bacilli can contain inhibitory dyes. Table 25-3 presents bacteriologic features of coliforms. (See also Color plate 1, C.)

In a sanitary water analysis the presence of coliform bacilli indicates fecal pollution of the water supply being tested.

Pathogenicity

E. coli and other coliform bacilli of the normal flora usually do not penetrate the intestinal wall to produce disease unless (1) the intestinal wall becomes diseased, (2) resistance of the host is lowered, or (3) virulence of the organisms is greatly increased. Under one of these conditions, coliforms may pass to the abdominal cavity or enter the bloodstream. Once outside the intestinal canal and in the tissues of the body, their virulence is remarkably enhanced. Out of its natural settings, *E. coli* is unabashed as an opportunist. Among the diseases coliforms may cause are pyelonephritis, cystitis, cholecystitis, abscesses, peritonitis, and meningitis (Fig. 25-1). Specifically, *E. coli* is the most common cause of pyelonephritis and urinary tract infections. The coliforms play a part in the formation of gallstones and are found in the cores of such stones. In peritonitis complicating intestinal perforation, the coliform group is joined by such organisms as streptococci and staphylococci. From any focus of inflammation they may enter the bloodstream and produce a septicemia.

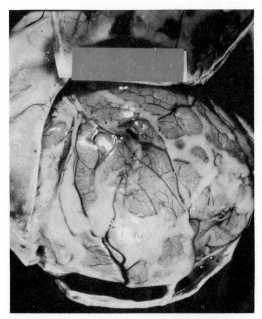

FIG. 25-1 *Escherichia coli* as cause of meningitis in 24-day-old premature infant. Note immature brain with coating of pus over outer surface and prominent vascular markings. Skull flaps are laid back; brain is seen in cranial cavity.

Table 25-3 Biochemical Reactions of Some Enteric Bacilli

Organism	Production of Indole	Methyl Red Test	Voges-Proskauer Reaction	Production of Hydrogen Sulfide	Liquefaction of Gelatin (22° C)
Coliforms					
Escherichia coli	+	+	−	−	−
Klebsiella species	−	−	+	−	−
Enterobacter species	−	Variable	Variable	−	Variable
Serratia marcescens	−	−	+	−	+
Arizona species	−	+	−	+	Slow
Edwardsiella tarda	+	+	−	+	−
Citrobacter species	−	+	−	+	−
Providencia species	+	+	−	−	−
Proteus Bacilli					
Proteus vulgaris	+	+	−	+	+
Morganella morganii	+	+	−	−	−

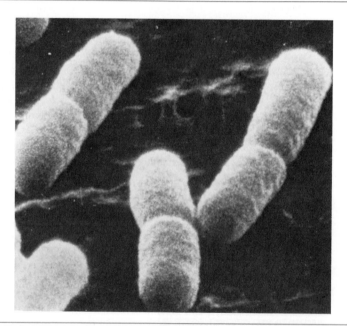

FIG. 25-2 *Escherichia coli,* the biologic workhorse (see box) in a scanning electron micrograph. (× 20,000). (Courtesy Lilly Research Laboratories, Indianapolis.)

Escherichia coli

The signal member of the coliform group is, of course, *E. coli,* first isolated in 1885 from feces by Theodor Escherich (1857-1911), a German pediatrician, and long known as the colon bacillus of humans and other vertebrates because of its natural habitat in the large bowel (Fig. 25-2).

The immunologic pattern for the genus *Escherichia* is as complex as that for the genus *Salmonella.* Immunologic subdivision of *E. coli* is made on the basis of O (somatic) antigens, H (flagellar) antigens, and K (envelope) antigens. More than 150 ser-

Use of Simmons Citrate	Splitting of Urea	Fermentation of Carbohydrates				
		Glucose with Gas	Lactose	Sucrose	Mannitol	Dulcitol
−	−	+	+	+	+	Variable
+	+	+	+	+	+	−
+	− (Variable)	+	Variable	+ (Variable)	+	−
+	Variable	+	Very slow	+	+	−
+	−	+	Slow	−	+	−
−	−	+	−	−	−	−
+	Variable	+	+	Variable	+	+
+	−	Variable	−	Variable	Variable	−
Variable	+	+	−	+	−	−
−	+	+	−	−	−	−

ologic types are known, of which 20 have been correlated with infantile diarrhea. It is said that 1000 to 2000 possible combinations of antigens exist for the genus *Escherichia*.

Most strains of *E. coli* are harmless, but some are pathogenic (Table 25-1). Normally organisms may be related to extraintestinal infections, as previously indicated, but the pathogenic strains are important specifically for their production of diarrheal disease; for instance, *E. coli* is a major cause of epidemic diarrhea in nurseries for newborn infants and the chief cause of traveler's diarrhea in North Americans returned from Mexico.

Diarrhea represents a change in the pattern of bowel movements—fecal material that is more of a liquid is passed more often. Many agents, both living and nonliving, cause it. Depending on the circumstances, diarrhea may be short term and inconsequential or long term and serious. It may be variably associated with manifestations of disease, such as fever, nausea and vomiting, abdominal cramps, and weight loss (Table 25-4). The consequence of prolonged diarrhea is the disturbance of fluid and electrolyte balance in the body that results from the loss of water and salts from the body. Infants and young children tolerate this physiologic disturbance very poorly.

Infectious Diarrhea

Infectious diarrhea, a very common problem, is tremendously significant worldwide. Although the disorder afflicts both adults and children, the implications are much more critical for the young individual than for the older one. In infants it represents an extremely serious threat to life; in the underdeveloped areas of the world 5 to 8 million infants succumb to it each year. The mortality is less in the United States by a factor of 25 or more, but even in this country much morbidity is associated with it. To emphasize that infectious diarrhea is caused by various agents, the box on p. 536, a list of the more important ones, is included.

Infantile, or summer, diarrhea is especially prevalent during hot weather among bottle-fed babies reared in unhygienic surroundings, but it is also a problem in modern hospital nurseries. In institutional outbreaks (hospitals, orphanages, and the like) half the cases are caused by *Shigella* (especially *S. sonnei*) and *Salmonella* species and pathogenic strains of *E. coli*, *Proteus*, *Pseudomonas*, and *Staphylococcus*. Viruses may also

Table 25-4 Clinicopathologic Correlations in Bacterial Diarrheas

Clinical Manifestation	Anatomic Site of Involvement	Mechanism of Action	Incubation	Example	Blood Cells in Feces
Vomiting (no fever)		Preformed toxin	Short (hours)	*Staphylococcus aureus*	No
Diarrhea (watery, no fever)		Enterotoxin (formed in vivo)	Hours to days	*Escherichia coli* *Vibrio cholerae*	No
"Enteric" fever		Invasion by pathogenic bacteria	Days	*Salmonella typhi* Other *Salmonella* species *Yersinia enterocolitica*	Yes
Dysentery (diarrhea plus blood and pus; fever)		Invasion by pathogenic bacteria	Days	*Shigella* species *Campylobacter* species *Vibrio parahaemolyticus* Invasive *Escherichia coli* *Clostridium difficile*	Yes

In 1885 Dr. Theodor Escherich (1857-1911) of Munich first isolated what is now known as *Escherichia coli.* In the hundred years or so that followed, *E. coli*'s meteoritic rise to prominence has been heavily punctuated with amazing discoveries.

Consider the potential of this tiny resident of the human intestine, one of the best known microbes. (A wag has said that if the ubiquitous *E. coli* were purple, we would live in a perpetual violet haze.) It thrives on a medium of inorganic salts and glucose, and it proliferates rapidly in 20 minutes under optimal conditions, 30 minutes, otherwise.

Because of its merits as a biologic workhorse, *E. coli* has indeed done yeoman services for research workers, both biologic and nonbiologic. Physicist, biochemist, and geneticist were first attracted to it because, with their limited knowledge of bacteriology, it seemed harmless. At first they did not bother to autoclave discarded cultures but poured them down the drain. The inadvisability of this break in technic was soon apparent in the recovery of *E. coli* from contaminated waters.

E. coli's contributions worldwide are reflected in a million or more published papers, including the prize winning works of Nobel laureates. The cracking of the genetic code was completed thanks to *E. coli*. Because *E. coli* was involved in the early work on bacterial viruses and first displayed sexual transfer of genetic material, it buttressed the new microbial genetics. It has been as crucial as the garden pea or the fruit fly. With recombinant DNA technology, *E. coli* produces human growth hormone so efficiently that the yield of 5 mg from only 2 gallons of culture matches that from nearly a half million sheep brains.

We acknowledge the Nobel laureates of our day, but as we salute them let us also give a well-earned toast to *E. coli.*

be implicated, with a significant number of the cases of summer diarrhea related to certain enteroviruses.

The *E. coli* recovered from infants in nursery outbreaks of epidemic diarrhea was first referred to as enteropathogenic *E. coli* a number of years ago. With time, special properties of strains similarly isolated were discovered, so that now it is known that diarrheal disease is produced by highly virulent types of *E. coli*, types not normally present in the bowel but brought there in food and water. At least two basic mechanisms for disease production have been studied, with the result that two types of diarrheagenic *E. coli* have been defined.

Enterotoxigenicity, or the ability to produce enterotoxin, is the first and more important mechanism for virulence, and **enterotoxigenic strains** are so designated. The second and less common mechanism relates to the quality of *invasiveness;* **enteroinvasive strains,** without known toxin formation, cause local inflammation with ulceration of the bowel mucosa when they invade the mucosal cells (Table 25-1).

When enterotoxigenic strains colonize, they first attach to small-bowel mucosa by means of fimbriae or pili (surface appendages), and at least 100 million organisms are required for infection. The attachment is necessary before the heat-labile enterotoxin can be formed by the infective types of *E. coli* possessing the plasmid that mediates toxin production. The heat-labile enterotoxin released acts on the intact intestinal wall in such a way as to move large amounts of fluid and electrolytes across the mucous membrane. The losses are massive, and a severe, watery diarrhea results. A heat-stable enterotoxin has also been demonstrated; its action is unknown.

Important agents of infectious diarrhea are as follows:

Bacteria
Invasive bacteria
  Shigella species
  Salmonella species
  Enteroinvasive *Escherichia coli*
  Staphylococcus species
Toxigenic bacteria
  Vibrio cholerae
  Vibrio parahaemolyticus
  Enterotoxigenic *Escherichia coli*
Microbes of toxin-mediated food poisoning
  Staphylococcus aureus
  Salmonella species
  Clostridium perfringens
  Bacillus cereus

Antibiotic-associated microbes
  Clostridium difficile with Clindamycin
    or Lincomycin therapy
  NOTE: Overgrowth of toxin producer
    results in pseudomembranous
    enterocolitis

Protozoa
Entamoeba histolytica
Giardia lamblia

Viruses
Rotaviruses
Parvoviruses
Certain of the adenoviruses, echoviruses,
  coxsackieviruses, and coronaviruses

In the hospital nursery, care in the preparation of infant formulas and the sterilization of bottles and nipples are essential to prevent this disorder, which is uncommon in the breast-fed baby. Fluorescent antibody technics detect the serotypes of *E. coli* in stools or on rectal swabs from the babies and help to trace carriers among personnel.

Klebsiella-Enterobacter-Serratia Species

The biologic qualities of the genus *Klebsiella* indicate a relationship to the coliforms; the disease-producing capacities indicate a kinship to respiratory tract pathogens. *Klebsiella* species are nonmotile, gram-negative, aerobic organisms surrounded by a broad, well-developed polysaccharide capsule. The large capsule is responsible for the large mucoid colonies of *Klebsiella* species seen on culture media and, because it resists phagocytosis, constitutes a primary virulence factor. *Klebsiella* species are differentiated from *E. coli* biochemically (Table 25-3). The type organism is *Klebsiella pneumoniae* (Friedländer's bacillus). It was discovered by Carl Friedländer (1847-1887) before the discovery of the pneumococcus; Friedländer considered it to be the sole cause of pneumonia. Actually, it is responsible for less than 10% of all cases.

A hazard to the chronic alcoholic, pneumonia caused by Friedländer's bacillus may be either lobar or lobular, is very severe, and is often fatal. Middle ear infections and meningitis may complicate it. Friedländer's bacillus is also responsible for septicemia.

The genus *Enterobacter* includes the microbes formerly known as *Aerobacter aerogenes*. These bacilli are found in the soil, water, dairy products, and the intestines of animals, including humans. As opportunists, they play a part in urinary tract infection and rarely cause more serious conditions—endocarditis, suppurative arthritis, and osteomyelitis. The type species is *Enterobacter cloacae*.

Serratia marcescens is a small, free-living, ubiquitous rod celebrated for the red pigment (prodigiosin) produced in cultures (Fig. 25-3). Since it has emerged as a sometimes formidable pathogen, only a small number of the organisms are seen to be chromogenic. The pigment-producing strains do so at room temperature, seldom at incubator temperatures. As is true for other coliforms, infections with this organism (serratiosis) complicate surgical procedures and antibiotic therapy, especially in elderly, debilitated patients in the hospital. The results may be urinary tract infections, pneumonia, empyema, meningitis, or wound infections, to mention a few.

Bacteriologically the organisms in the genera *Klebsiella*, *Enterobacter*, and *Serratia* are similar. However, when they are grown on a modified deoxyribonuclease (DNAse) agar containing toluidine blue as indicator, colonies of the genus *Serratia* may be distinguished from others. A bright pink zone around members of the genus *Serratia* indicates that it elaborates DNAse. The clear blue color of the other colonies indicates that these organisms do not.

These three closely related genera are important causes of nosocomial infections.

Arizona-Edwardsiella-Citrobacter Species

Formerly known as paracolon bacilli, the organisms of the genera *Arizona*, *Edwardsiella*, and *Citrobacter* ferment lactose very slowly, if at all.

Arizonae are short rods similar to the salmonellas, and in the eighth edition of *Bergey's Manual* they are classified in genus *Salmonella*, subgenus III. From 7 to 10 days may be required for them to ferment lactose (Table 25-3). The type species is *Arizona hinshawii* (or Bergey's *Salmonella arizonae*). Although usually found in animals, they are also found in human infections, such as gastroenteritis and cholecystitis, and occasionally cause systemic disease (septicemia).

Because of their pattern of biochemical reactivity, certain motile rods were separated into the genus *Edwardsiella* (Table 25-3). The type species *Edwardsiella tarda* was given a name implying a kind of biochemical *in*activity. *Edwardsiella tarda* has been isolated from humans and animals, especially snakes. In fact, snake meat is postulated to be a source of infection. Human infections are similar to the salmonelloses.

Citrobacter species are motile rods fermenting lactose and are similar to salmonellas (Table 25-3). The type species is *Citrobacter freundii*. They may be confused with *Salmonella* and *Arizona* organisms. Certain strains possess the Vi antigen found in *S. typhi*. *Citrobacter* species are differentiated from *Arizona* and *Salmonella* organisms by their ability to grow in the presence of potassium cyanide (positive KCN test). The significance of the genus *Citrobacter* in disease is undetermined, though it has been reported from serious urinary and respiratory tract infections in hospital patients.

FIG. 25-3   *Serratia marcescens*. Colonies of pigmented strain on nutrient agar plate. Its famous red color is suggested by the dark appearance.

Providencia ("Providence") Species

Providencia, or the "Providence," group of microbes (formerly known as paracolons) are free living, lactose negative, and biochemically similar to *Proteus* bacilli with the exception that members of the genus *Providencia* fail to degrade urea (Table 25-3). (In the eighth edition of *Bergey's Manual* they are classified in the genus *Proteus.*) The type species is *Providencia alcalifaciens.* Ordinarily nonpathogenic, they have been recovered from human diarrhea and urinary tract infections. They can also be nosocomial pathogens.

PROTEUS SPECIES (THE PROTEUS BACILLI)

Proteus bacilli comprise a genus of highly motile organisms that resemble and yet differ from other enteric bacilli in many ways (Table 25-3). They are gram negative, do not form spores, and are normal inhabitants of feces, water, soil, and sewage. They grow rapidly on ordinary media and, being very actively motile (Fig. 25-4), tend to "swarm" over the surface of cultures on solid media. As a consequence, colonies do not remain discrete. Growth from the colony edge in time extends over all available surface area as a thin, translucent, bluish film. The bacilli can be seen microscopically to break away, migrating over the agar surface. This property poses problems to the isolation of *Proteus* bacilli from mixed cultures, although generally with care it can be done.

Their primary pathogenicity is slight, but as secondary invaders they are vigorous. The type species is *Proteus vulgaris,* which is remarkably resistant to antimicrobial drugs. *Proteus* bacilli in the stool often increase in number after diarrhea caused by other organisms, especially when antibiotics have been given. *Proteus* bacilli are pathogenic for rabbits and guinea pigs.

Proteus bacilli rank next to *E. coli* in importance as a cause of urinary tract infections

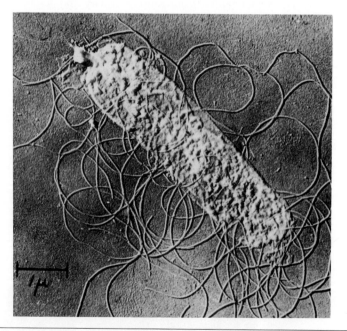

FIG. 25-4 *Proteus vulgaris.* Electron micrograph of actively motile bacterium. Note flagella all over bacterial cell. (Courtesy J.B. Roerig Division, Chas. Pfizer & Co., Inc., New York.)

and pyelonephritis. They are important in wound infections and are a rare cause of peritonitis. *Proteus morganii* is presumed to cause infectious diarrhea in infants.

It is a peculiar fact that although *Proteus* bacilli do not cause typhus fever nor act as secondary invaders, the serum of a patient with typhus fever agglutinates certain strains of *Proteus vulgaris*. This is the basis of the Weil-Felix reaction for typhus fever.

MORGANELLA MORGANII

The genus *Morganella* composed at this time of a single species, *M. morganii*, has recently been defined to contain the former *Proteus morganii* after certain differences between it and other *Proteus* species were demonstrated (Table 25-3). An important difference is in genetic relatedness. The guanine-plus-cytosine (G + C) percentage of total DNA for *M. morganii* is 50, whereas the percent G + C for *Proteus* species is 37 to 44. In addition, certain biochemical differences exist; for instance, *M. morganii* does not produce hydrogen sulfide nor does it liquefy gelatin. *M. morganii* is an occasional cause of urinary tract infections.

ENDOTOXIN SHOCK

Endotoxin shock (bacteremic shock, gram-negative shock, gram-negative sepsis) is a very serious complication of bacteremia with gram-negative enteric bacilli, notably *Pseudomonas aeruginosa* (p. 632) (85% mortality), *Proteus* species, *E. coli*, and coliforms of the genera *Klebsiella*, *Enterobacter*, and *Serratia*. (Bacterial shock rarely complicates bacteremia caused by gram-positive cocci or systemic infections from meningococci, clostridia, viruses, rickettsias, or fungi.) Mortality is over 50%. The dramatic clinical picture is drawn with the precipitous development of chills, fever, nausea, vomiting, diarrhea, and prostration. There is a sharp drop in blood pressure, and the state of shock is profound.

The combination of events results directly from the action of lipopolysaccharide endotoxin released into the circulation and activated there by certain components of the plasma. Endotoxin brings about constriction of small blood vessels, thus setting off a series of events leading to the shock state—increased peripheral vascular resistance, pooling of blood in the capillary circulation, hypotension, and shutdown of kidney function. The presence of the inciting gram-negative bacteria in the bloodstream also plays a part in pathogenesis.

This kind of bacteremia is especially prone to complicate infections where enteric bacilli are already entrenched, such as those of the genitourinary tract, lungs, intestinal tract, and biliary tract. Instrumentation or an operative procedure on the infected area (2 to 24 hours before) favors release of organisms into the bloodstream. In diseases such as leukemia, diabetes, and chronic liver disease, where host resistance is low, organisms flood the bloodstream from no apparent focus. Bacteremia is rarely complicated by shock in persons under the age of 40 years, except in pregnant women and neonates. The incidence is high in older men.

ENTERIC BACILLI IN HOSPITAL-ASSOCIATED INFECTIONS

Of all patients admitted to a modern hospital, 3% to 5% acquire an infection there. Since approximately 30 million people are admitted to hospitals annually, this means that about 1½ million patients develop an infection they did not bring into the hospital.

The predominating organism in hospital-associated (acquired) infections has been *Staphylococcus aureus*. It is still important, accounting for about one fifth of the infections. Today most observers agree that the prime offender is the gram-negative enteric bacilli as a group (*E. coli*, other coliforms, *Proteus* and *Pseudomonas* species). Such enteric bacilli easily account for two thirds of nosocomial infections. Other important agents are enterococci, pneumococci, and other streptococci.

Most hospital-acquired infections with enteric bacilli seem to occur in very young or very old patients. They tend to complicate chronic debilitating diseases, diseases that alter the patient's resistance to infection, and diseases treated with antibiotics to which the gram-negative bacilli have or easily acquire resistance. Most infections are endogenous (caused by organisms available from a plentiful supply within the gastrointestinal tract, respiratory tract, or urinary tract of a given patient). In about 90% of all cases there is a surgical wound infection (especially where the operation was done on the alimentary tract), a urinary tract infection, or respiratory tract infection such as bronchopneumonia.

Varied pathogens exist everywhere in the hospital environment, in foci one would least suspect—faucets, flower vases, soaps, and lotions, for example. The role of the more sophisticated items indigenous to a hospital is often overlooked—urethral catheters, intravenous catheters, respirators, reservoir nebulizers, and the like. Much current thought is directed to measures designed to reduce the hazards.

SALMONELLA SPECIES
General Characteristics

The genus *Salmonella* comprises the causative organisms of salmonellosis, some 250 species in four unnamed subgenera of motile enteric bacilli. The *Salmonella* genus was named for Daniel Elmer Salmon (1850-1914), American veterinary pathologist, who first isolated the organisms in 1885. Salmonellas are found everywhere that there are animals and humans. All can produce diseases in their natural hosts, and all are pathogenic for human beings. Salmonellosis is relatively common in the United States, where some 2 million cases occur annually.

Although the members of the different species are similar in morphology, staining reactions, and cultural characteristics, they can be separated by fermentation reactions (Table 25-5) and agglutination tests. In keeping with their role as intestinal pathogens, salmonellas do not ferment lactose. *Salmonella* species possess O, H, and K(Vi) antigens, and serologic typing is the ultimate step in classification. Some 1800 serotypes (Kaufmann-White classification) have been identified. Of these, 40 cause 90% of human infections. (Remember that in the Edwards-Ewing classification there are only three species.) The immunologic classification of *Salmonella* organisms is complex but important in tracing epidemiologic patterns of disease, as is phage typing. Salmonellas

Table 25-5 Biochemical Patterns for *Salmonella* Species

Organism	Production of Hydrogen Sulfide	Reduction of Nitrate	Production of Indole	Liquefaction of Gelatin
Salmonella typhi	+	+	−	−
Salmonella paratyphi-A	−	+	−	−
Salmonella schottmuelleri	+	+	−	−
Salmonella hirschfeldii	+	+	−	−
Salmonella typhimurium	+	+	−	−
Salmonella choleraesuis	Variable	+	−	−
Salmonella enteritidis	+	+	−	−
Salmonella gallinarum	Variable	+	−	−

are susceptible to the action of heat, disinfectants, and radiation. At refrigerator temperature they do not multiply but remain viable.

Sources and Modes of Infection

Since only a few species, such as *Salmonella typhi*, are indigenous to humans, most infections come from species in lower animals. Since, in nature, salmonellas occur in the intestinal tract of humans and animals, they must be spread by feces either directly or indirectly from one host to another. Excreta contain bacilli because the infected host has obvious disease or inapparent infection or is a carrier. The principal vehicle is water, milk, or food contaminated by feces. The most dangerous link in a chain of infection is the food handler who is a carrier. Flies are mechanical vectors (p. 304).

Typhoid Mary is a term sometimes used for a suspected or actual carrier of disease—typhoid fever or otherwise. The original Typhoid Mary, who lived and worked as a cook in the first part of this century, is said to be the first disease carrier discovered—the first such person apparently well and unaffected by the epidemics caused. A lifetime carrier, Typhoid Mary was linked to 10 outbreaks of typhoid fever for a total of 51 cases, 3 deaths. In 1907 she was arrested and held by the New York City Department of Health. It was then found that she regularly passed great numbers of typhoid bacilli in her stools. (Ironically, after her release she insisted on continuing her career as a cook.)

If the food ingested contains only a small number of salmonellas, acid in the stomach of the host destroys them readily. If food is heavily contaminated, some of the bacilli escape the effects of the gastric acid to enter and injure the small bowel.

Pathogenicity

Infection caused by *Salmonella* species is called salmonellosis, the major site of which is the lining of the intestinal tract (Tables 25-1 and 25-4). Because of their toxic properties, every known strain of *Salmonella* can cause any one of three types of salmonellosis: (1) acute gastroenteritis of the food infection type (10^3 to 10^7 organisms are said to be required for infection here), (2) septicemia or acute sepsis with localized complications similar to pyogenic infections, and (3) enteric fever such as typhoid or paratyphoid fevers (10^4 to 10^6 organisms are required for infection).

Most of the time, however, *Salmonella* infection is acute gastroenteritis, a condition marked by fever, nausea, vomiting, diarrhea, and abdominal cramps. Sometimes gastroenteritis is the forerunner of septicemia. Because salmonellas can survive and even grow within the phagocytic cells of the human body, they may thus be spread to cause a wide range of lesions in different anatomic areas. Examples are abscesses of various

Fermentation of Carbohydrates

Glucose	Lactose	Maltose	Sucrose	Mannitol	Dulcitol
Acid	—	Acid	—	Acid	—
Acid, gas	—	Acid, gas	—	Acid, gas	Variable
Acid, gas	—	Acid, gas	—	Acid, gas	Variable
Acid, gas	—	Acid, gas	—	Acid, gas	Acid, gas
Acid, gas	—	Acid, gas	—	Acid, gas	Variable
Acid, gas	—	Acid, gas	—	Acid, gas	Variable
Acid, gas	—	Acid, gas	—	Acid, gas	Variable
Acid	—	Acid	—	Acid	Acid

organs, arthritis, endocarditis, meningitis, pneumonia, and pyelonephritis. Salmonelloses may be severe, even fatal, or they may be mild and possibly inapparent. They are especially serious in the very young and the very old.

With regard to their natural host, salmonellas are divided into the following categories: (1) those that primarily affect humans, (2) those that primarily affect the lower animals but may cause human disease, and (3) those that affect lower animals only.

Salmonella typhi (the Typhoid Bacillus)

The organism *Salmonella typhi* causes typhoid fever, which as the natural infection occurs only in human beings. Although typhoid fever has been virtually eradicated in the United States, it is still endemic in parts of the world.

General Characteristics

A short, motile, nonencapsulated bacillus, *S. typhi*, also a facultative intracellular parasite, grows luxuriantly on all ordinary media. It grows best under aerobic conditions but may grow anaerobically. The temperature range for growth is from 4° to 40° C, the optimum, 37° C. Typhoid bacilli can survive outside the body, living about 1 week in sewage-contaminated water and not only living but multiplying in milk. They may be viable in fecal matter for 1 to 2 months.

Pathogenicity

Typhoid fever is an acute infectious disease with continuous fever, skin eruptions, bowel disturbances, and profound toxemia. Except in the first few days, leukopenia is always present in uncomplicated cases, probably because typhoid bacilli depress the bone marrow, where normal production of white blood cells occurs. Leukocytosis in the course of the disease signals a complication. The incubation period is from 7 to 21 days.

Ingested typhoid bacilli first penetrate the intestinal lining of the ileum to attack the lymphoid tissue there (p. 320). Peyer's patches in the small bowel bear the brunt of the attack. From there the bacilli taken up by macrophages pass to the regional mesenteric lymph nodes, which they colonize. Passing on to the thoracic duct, they enter the bloodstream, where many are lysed; the release of endotoxins brings about manifestations of the disease. The surviving bacilli in the bloodstream tend to localize in the gallbladder, bone marrow, and spleen, disappearing from the bloodstream beyond the first week of illness. The rose spots seen on the skin of the abdomen early in the illness contain many bacilli. As the disease progresses, the bacilli return to the intestinal tract and confine their activities there.

Pathology

Pathologically, typhoid fever is a granulomatous inflammation involving the lymphoid tissue of the body. In the intestinal mucosa hyperplastic lymphoid masses, including Peyer's patches, swell and form plateaus. When necrosis takes place with sloughing, an oval ulcer of varying depth is left. If ulceration extends deep enough, perforation of the intestinal wall results, a serious complication. If a blood vessel becomes necrotic, profuse hemorrhage may result, which may be life threatening. The mesenteric lymph nodes are swollen and may suppurate, and the hyperplastic lymphoid tissue in the spleen softens and enlarges it to two or three times normal size. The liver and kidneys show degenerative changes, and the gallbladder may be inflamed. Toxic injury to the heart muscle may precipitate acute heart failure and death. Cholecystitis may follow the disease, and typhoid bacilli may be found in the gallbladder years later. A focus of infection in the gallbladder generally complicates the control of the carrier state.

Spread

The fundamental basis of every typhoid infection is the same. *Typhoid bacilli from the feces or urine of a carrier or a person ill with typhoid fever have reached the mouth of the victim.* Any person contracting typhoid fever has swallowed typhoid bacilli. Bacilli entering the body by a route other than the alimentary tract do not infect.

Typhoid carriers are of two types: fecal and urinary. In the more common fecal carriers the bacilli multiply in the gallbladder and are excreted in the feces. (Infection of the gallbladder with stagnation of bile predisposes to the formation of gallstones.) In urinary carriers the organisms multiply in the kidney and are excreted in the urine. From 40% to 45% of patients become convalescent carriers for 3 to 10 weeks, but only 2% become permanent carriers. The average carrier is a woman over 40 years of age.

The most dangerous factor in the spread of typhoid fever is the carrier food handler who prepares foods that are served raw. It is hard to prevent the carrier state, and treatment for it can be difficult. Removal of the gallbladder cures selected cases.

With improvements in sanitation, widespread epidemics of typhoid fever from contamination of water supplies with sewage, once so common in large cities, are almost unknown. Strict laws regulate the cultivation of oysters and other shellfish eaten raw to eliminate contamination of these foods with sewage-laden water.

Laboratory Diagnosis

The laboratory offers four procedures for the diagnosis of typhoid fever: (1) blood culture, (2) stool culture, (3) urine culture, and (4) agglutination tests. (The determination of agglutinins in typhoid fever has been referred to as the Widal test, a term not now in general use.) Blood cultures are positive during the first 2 weeks in 95% of patients. The percentage falls to 10% by the fourth week. Isolation of the organism in blood culture is diagnostic of the disease.

Stool and urine cultures facilitate detection of carriers and indicate when a given patient ceases to be a source of infection. Typhoid bacilli may be cultivated from the stool in 85% of cases by the third week. Positive urine cultures are obtained after the second week in 25% of patients. Convalescent carriers excrete the bacilli in their feces or urine during convalescence only, but permanent carriers continue to excrete them.

When an animal is immunized with typhoid bacilli and other salmonellas, two major agglutinins are formed: one to somatic antigen, the O agglutinin (mainly IgM antibody), and the other to flagellar antigen, the H agglutinin (mainly IgM antibody). The heat-stable O, or somatic, antigen is lipopolysaccharide and part of the bacterial cell wall. The heat-labile H, or flagellar, antigen is protein. For epidemiologic studies, *Salmonella* species are grouped on the basis of a common O antigen (at least 140 species share the O antigen with *S. typhi*) and subdivided on the basis of the O complex of antigens and the H antigen (at least 61 species share this one with *S. typhi*). The O agglutinins appear earlier in disease and indicate actual infection with typhoid bacilli (or closely related salmonellas). If they are present in the serum of persons vaccinated, they occur in small amounts for only a short time. In addition to H and O antigens, typhoid bacilli responsible for active disease or carrier state contain the Vi antigen, which is also shared by some other *Salmonella* species. The Vi (for virulence) antigen is a specialized capsular or envelope K antigen found at the extreme periphery of the bacterial cell. Vi antibodies are not considered to be significant after vaccination.

After 7 to 10 days of infection, agglutinins against typhoid bacilli appear in the patient's blood, increasing during the second and third weeks of disease and often persisting for weeks, months, or years after recovery. These agglutinins are detectable in about 15% of patients during the first week and in about 90% or more of patients during the third week. Since agglutinins are present in the blood after typhoid vaccination and in the blood of carriers, serologic testing must be done on serum taken during the acute phase of disease as well as on serum from the convalescent phase. One serum only is of no value. (Narcotic addicts may have higher than average agglutinin titers to the O and H antigens of S. *typhi*.) A fourfold increase in titer between the acute specimen and the convalescent one is significant, but even this finding must be correlated with the clinical picture of disease.

Immunity

In about 98% of cases an attack of typhoid fever gives permanent immunity. The major component of the immunity associated with this disease is probably cell-meditated or cellular immunity. If the progressive activity of the salmonellas within macrophages could be contained, the disease could be restricted.

Prevention

The prevention of typhoid fever is twofold—community and personal. Community prevention refers to measures taken by the community as a whole to block the spread of disease among its members; personal prevention defines the measures taken to block the spread of infection from a person ill with the disease. The most important factors in the community program are (1) a supply of clean pasteurized milk, (2) a pure water supply, (3) efficient disposal of sewage, (4) proper sanitary control of food and eating places, (5) detection and isolation of carriers, especially food handlers, (6) destruction of flies, and (7) vaccination.

Personal prevention depends on isolation of the patient. Isolation does not mean merely putting the patient in a room and shutting the door; it means closing off all routes of spread.

Preferably the patient is hospitalized and kept there until he or she is no longer infectious. Nurses attending a patient with typhoid fever should consider every secretion and excretion of the patient to be a living culture of typhoid bacilli and should use every means to protect themselves, members of the patient's family, and people in the community. They should have nothing to do with the preparation of food, and their hands should be disinfected after each contact with the patient or anything that either the patient or the patient's secretions have touched. Feces and urine can be disinfected also (p. 274). Sputum should be received on disposable paper tissues and burned. Linen and bedclothes should be sterilized. Food remains should be burned and dishes boiled. The patient's bath water can be sterilized with chlorinated lime. The sickroom should be screened, and flies that accidentally gain access to it should be killed. All rugs, curtains, and similar fabric materials should be removed from the room. Pets should not be allowed to enter. Disinfection should continue through convalescence, and no patient should be discharged as cured until repeated cultures of feces and urine fail to show typhoid bacilli. After the patient has recovered, the sickroom should be thoroughly cleaned and sanitized.

The procedures just outlined together with vaccination have materially reduced the incidence of typhoid fever. This is proved by the following facts. In 1900 the death rate in the United States from typhoid fever was 39.5 per 100,000 persons; today approximately 400 cases are reported annually (0.18 cases per 100,000 persons reported in 1977). That universal vaccination alone will materially reduce the incidence of typhoid fever is proved by the experience of the U.S. Navy from 1911 to 1913, a period in which no revolutionary developments in sanitation took place. In 1911 there were 361 cases of typhoid fever per 100,000 men. In 1912 universal vaccination of naval personnel was instituted. In 1913 the rate fell to 34 per 100,000, a reduction of more than 90%.

The production of an artificial immunity by the administration of typhoid vaccine has been a most important factor in control, but currently the U.S. Public Health

Service is not recommending typhoid immunization in the United States. Its Advisory Committee does not indicate typhoid vaccination for persons going to summer camp or even for those surviving a flood disaster. Typhoid immunization is further discussed on pp. 428, 446, and 448.

Paratyphoid Bacilli

Paratyphoid bacilli are so called because they have been isolated from paratyphoid fever in humans, an enteric fever like typhoid but milder in its manifestations and shorter in duration. Although associated with enteric fever, each of the three species causes other forms of salmonellosis.

The three species of paratyphoid bacilli are *Salmonella paratyphi*-A, *Salmonella schottmuelleri*, and *Salmonella hirschfeldii*, formerly known as the paratyphoid bacilli A, B, and C, respectively. Infections with *S. paratyphi*-A occurs almost exclusively in humans; *S. schottmuelleri* occasionally infects lower animals. *S. hirschfeldii* is rarely found in the United States; infections are frequent in Eastern Europe, and mice are said to be its natural host. *S. paratyphi*-A resembles *S. typhi* more closely than does *S. schottmuelleri*, but infections with the latter are more prominent.

The mode of infection, sources, laboratory diagnosis, nursing precautions, and prevention are the same as those in typhoid fever. The immunity produced by a paratyphoid infection or vaccination is uncertain (p. 446).

Other Salmonellas

Among the many salmonellas that primarily affect lower animals but may infect humans are *Salmonella typhimurium* (cause of typhoid-like disease of mice and one cause of human gastroenteritis); *Salmonella choleraesuis* (once considered the cause of hog cholera and an important cause of *Salmonella* septicemia in humans); and *Salmonella enteritidis*, known also as Gärtner's bacillus (found in hogs, horses, mice, rats, and fowls; infectious for humans). Acute gastroenteritis of the food-infection type is usually attributable to *Salmonella enteritidis* or to one of its many serotypes. An organism primarily affecting lower animals is *Salmonella gallinarum* (cause of fowl typhoid); it differs from other *Salmonella* species in that it is not motile.

Attacks of enteric infection are on occasion traced to salmonellas from unusual sources, such as the small pet turtles given to children who subsequently developed salmonellosis. Contaminated baby chicks have been a similar source for small children.

SHIGELLA SPECIES (THE DYSENTERY BACILLI)

Dysentery (Greek *dys*, painful, plus *enteron*, intestine) is a painful diarrhea accompanied by the passage of blood, pus, and mucus and associated with abdominal pain and constitutional symptoms. It may be primary or secondary. When primary, it may be of protozoan, bacterial, or viral origin. Dysentery caused by members of the *Shigella* genus is known as bacillary dysentery. It is usually acute but may be chronic. Classic bacillary dysentery can be a very disabling disease. Bacillary dysentery may be endemic, epidemic, or pandemic. Pandemics still occur throughout the world; a 1970 pandemic in Central America killed 20,000 persons.

Bacillary dysentery is an important form of summer diarrhea in infants.

General Characteristics

Members of the genus *Shigella* are classified into four species, each of which is divided into serotypes: *Shigella dysenteriae* (10 types), isolated by Shiga in the Japanese epidemic in 1898; *Shigella flexneri* (six types), isolated by Flexner in the Philippine Islands in 1900; *Shigella boydii* (15 types); and *Shigella sonnei* (one type). All *Shigella* species are human pathogens. *S. flexneri* and *S. sonnei* are found worldwide, and in the United States *S. sonnei* accounts for 95% of the cases. Infection with *S. dysenteriae* in recent years has been limited to the Orient.

Dysentery caused by the Shiga bacillus (*S. dysenteriae*) is much more severe than that from the other organisms because this bacillus produces a powerful exotoxin in addition to its endotoxin. The exotoxin acts as a neurotoxin on the nervous system to paralyze the host. The endotoxin irritates the intestinal canal.

Dysentery bacilli are gram-negative, nonspore-bearing rods that grow on all ordinary media at temperatures from 10° to 42° C but best at 37° C (Color plate 1, C). They are aerobic and facultatively anaerobic. Unlike most other members of the enteric group, they are nonmotile and therefore do not have H antigens. All have typical O antigens; some have K antigens. Table 25-6 gives their biochemical reactions.

Pathogenicity

Dysentery is a human disease, and natural infections of lower animals do not exist. The incubation period is 1 to 7 days. The disease lasts 6 to 10 days and usually resolves. Epidemic dysentery is primarily an infection of the large intestine (Tables 25-1 and 25-5). The shigellas penetrate the bowel mucous membrane, but unlike typhoid bacilli, they do not invade the bloodsteam and are seldom if ever found in the internal organs or excreted in the urine. They are excreted in the feces.

Shigella species are highly infectious. Compared with that for other enteric pathogens, the number needed for infection is small; only 10 to 100 ingested organisms are required. (Remember, however, that viral disease may come from intake of a single viral particle.)

Pathology

Pathologically, bacillary dysentery is recognized as diffuse inflammation with ulceration of the large intestine and sometimes the lower portion of the small intestine. Early in the disease shallow ulcers or extensive raw surfaces that may be covered by a pseudomembrane form in the mucous membrane. In mild cases healing is complete, but in severe cases there is extensive scarring.

Mode of Infection

The mode of infection is practically the same as for typhoid fever; the bacilli enter the body of the victim by way of the mouth, having been transferred there from the feces of carriers or patients. Contaminated food, water, fingers, or other objects are the vehicles of spread. Contact with persons who have symptomless infections is especially significant in the spread of bacillary dysentery. Flies are a vector. Sexual transmission of enteric diseases, such as shigellosis, amebiasis, and viral hepatitis, is seen today in some large cities, especially among young men.

Since the human being is the only natural host and shigellosis is not found in nature, this disease could be eradicated, as smallpox has been.

Laboratory Diagnosis

The practical method for routine laboratory diagnosis is the cultivation of the organisms from the stool. Large numbers of shigellas are found in the feces. Unfortunately, this can be done only during the first 4 to 5 days of the disease.

Immunity

An attack of dysentery probably confers some degree of immunity, but two episodes in the same person do (rarely) occur in one season. Immunity may relate to stimulation of local immune mechanisms in the bowel to block penetration by the shigellas.

Table 25-6 Biochemical Reactions of *Shigella* Species

Organism	Production of Hydrogen Sulfide	Liquefaction of Gelatin	Reduction of Nitrate	Production of Indole	Fermentation of Carbohydrates				
					Glucose	Lactose	Sucrose	Mannitol	Dulcitol
Shigella dysenteriae	−	−	+	−	Acid	−	−	−	−
Shigella flexneri	−	−	+	+	Acid	−	−	Acid	−
Shigella boydii	−	−	+	Variable	Acid	−	−	Acid	Variable
Shigella sonnei	−	−	+	−	Acid	−	Acid	Acid	−

Prevention

Bacillary dysentery may be checked by the sanitary measures that control typhoid fever (p. 544) but it remains ready to rise in epidemic form when people are crowded together in unsanitary conditions. The feces and everything contaminated by patients should be handled exactly as for a case of typhoid fever. Food or milk should not be carried from the premises, and the medical attendants should not handle food for others. The patient should not be dismissed as cured until repeated feces cultures have failed to show the causative organism.

Preventive vaccination is not yet available. An oral attenuated vaccine against *S. flexneri* is being tested. One such vaccine is a "mutant hybrid," which is produced by "mating" *E. coli* with a nonvirulent strain of *S. flexneri*. There is no serum therapy.

VIBRIO SPECIES

The family Vibrionaceae contains rigid, gram-negative, straight or curved rods, facultatively anaerobic, found in fresh water and seawater. In its member genus *Vibrio* are the agents of cholera and vibriosis.

Vibrio cholerae (the Comma Bacillus)

The Classic Disease

Asiatic cholera, caused by *Vibrio cholerae*, is a specific infectious disease that affects the lower portion of the intestine and is described by violent purging, vomiting, burning thirst, muscular cramps, suppression of urine, and rapid collapse. Untreated, it can be a terrifying disease with a 70% or greater mortality. Only plague causes as much panic as Asiatic cholera. With massive diarrhea the patient's fluid losses are enormous— 10 to 20 L a day. There is severe, rapid dehydration, and death comes within hours. Four great pandemics spread throughout the world during the eighteenth century, and on two occasions in the nineteenth century the disease invaded the United States: in 1832 in New York City and in 1848 in New Orleans, whence it spread up the Mississippi Valley. The disease is endemic in India and China. The "scourge of antiquity," Asiatic cholera originated in India in the vicinity of Calcutta and on the delta of the Ganges River, where it was known at the time of Alexander the Great.

The Organism

V. *cholerae*, the comma bacillus, is a small, comma-shaped, gram-negative rod, motile by a polar flagellum (Fig. 25-5). The three immunologic types are the Inaba, Ogawa, and Hikojima. For the cholera vibrio to produce the devastating disease, it must attach to the epithelial lining of the small bowel (Tables 25-1 and 25-4 and box, p. 548). Virulent vibrios do so readily (Fig. 25-5). A mucinase, an enzyme they produce, facilitates penetration of the mucus, which forms a protective covering over the surface of the intestinal lining, so that an abundance of microorganisms can colonize. Once anchored on the mucous membrane, the vibrios multiply rapidly and produce the powerful enterotoxin that is responsible for the physiologic derangements of the disease.

The potent enterotoxin (exotoxin, cholera toxin, choleragen) is a polymeric molecule with two subunits. The B subunit binds toxin to a ganglioside receptor on the mucosal cells. The A subunit penetrates the cell membrane and activates adenyl cyclase, thus increasing intracellular levels of cyclic adenosine monophosphate (cAMP) and stimulating active secretion of water and electrolytes into the bowel lumen. Because of the action of toxin, reabsorption of the sodium ion stops. The outpouring of water into the bowel lumen becomes so great that the reabsorption mechanism of the large intestine is overwhelmed, and massive diarrhea begins. This explains why the fecal discharges are described as "rice water"—clear, not malodorous, with flecks of mucus. When the disease is at its height, the many stools are totally liquid. Vibrios abound in the stools.

The heat-labile enterotoxin of *E. coli* is quite like (if not identical to) the cholera toxin. The two toxins are alike in mode of action, subunit structure, and immunochemistry. However, the heat-labile enterotoxin of *E. coli* is encoded on a plasmid, whereas the cholera toxin is determined from the chromosome. Most of the cholera toxin is secreted; most of the *E. coli* enterotoxin remains cell associated.

Laboratory Diagnosis

In a suitable preparation the cholera vibrio possesses a very characteristic darting movement that is well displayed in the dark-field microscope. The vibrios may at times be demonstrated in direct smears of the mucus flecks in the rice water stools. V. *cholerae* grows at a very high pH (8.5 to 9.5) and is speedily killed by acid. Therefore standard media routine for enteric pathogens are not used, since other enterics present would produce acid, inhibiting its growth. Culture of the cholera vibrio may be made on the selective medium thiosulfate–citrate–bile salts–sucrose (TCBS) agar. Yellow colonies

Note the relative infectivity in terms of the infectious dose for certain important enteric pathogens in the following list:

	Number of organisms	Number of cysts
Bacteria		
Shigella species	10 to 100	—
Salmonella species	100,000	—
Escherichia coli	100,000,000	—
Vibrio cholerae	100,000,000	—
Protozoa		
Giardia lamblia	—	10 to 100
Entamoeba histolytica	—	10 to 100

result from the fermentation of the sucrose. If a few drops of sulfuric acid are added to a growth of cholera vibrios in nitrate-peptone broth, a red color develops—the cholera red reaction.

Mode of Infection

The disease is contracted by the ingestion of water or food contaminated by the excreta of persons harboring the bacilli. Since the cholera vibrio is exquisitely sensitive to gastric acid, at least 100 million vibrios are required for infection in a reasonably healthy person, but in a malnourished victim or someone with no gastric acidity, 10,000 to 1 million vibrios can produce disease. The human is the only host. The bacilli leave the body in the feces, urine, and secretions of the mouth. As a rule, the feces become free of bacilli during the last days of the disease, but some patients become short-term convalescent carriers. Permanent carriers do not occur.

Cholera today is rampant in the areas of the globe occupied by more than half the world's population, where, because of low standards of sanitation, human excrement

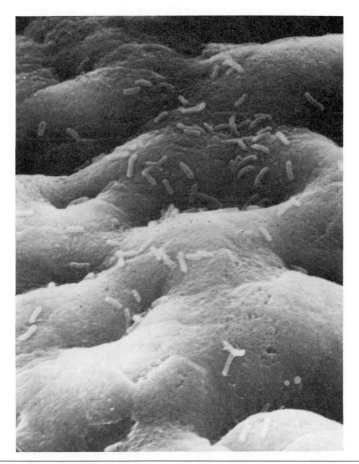

FIG. 25-5   Cholera vibrios adherent to mucosal lining of small bowel visualized in scanning electron micrograph. (Courtesy Dr. Richard Finkelstein, Dr. Edward T. Nelson, and Bob Fenley, Director, Office of Medical Information, The University of Texas Health Science Center at Dallas.)

easily contaminates the waterways and surface wells that are sources of drinking water. Vibrios may live in water for as long as 2 weeks; however, they have a limited tolerance for fresh water. Being halophilic (salt-loving), they survive for a long time in seawater.

As a disease, cholera should be only a historical note. Modern sanitation can eliminate waste-contaminated water supplies, the primary source of infection, and modern medical treatment can effectively deal with the disease.

Immunity

The immunity occurring after an attack of cholera is short-lived. The risk of reinfection is only slightly less than that of the initial infection. Although vaccination is done against cholera (pp. 427, 447, and 448), a truly effective vaccine is still not widely available today. An experimental "toxoid" has been put to field trial.

Vibrio cholerae biovar *eltor*

A biovar of V. *cholerae*, more resistant to physical and chemical agents and in itself quite virulent, is V. *cholerae* biovar *eltor* (V. *El Tor*). Unlike the classic one, the El Tor vibrio produces a hemolysin that lyses the red cells of sheep and goats (see Table 25-7 for comparison of the two). It causes a less devastating cholera and in recent times has been more significant than the classic vibrio as the cause of at least seven pandemics in South and Southeast Asia.

Vibrio parahaemolyticus

A vibrio like the classic one, *Vibrio parahaemolyticus* is a major cause of a gastroenteritis with manifestations like the classic disease. In countries such as Japan it is responsible for 50% to 60% of cases of summer diarrhea. Widely distributed as a marine microorganism, it is an important cause of food poisoning where seafood, especially raw or inadequately cooked, is consumed, and it can infect wounds incurred along the seashore. It is a small halophilic vibrio, urease positive, giving a negative cholera red reaction (Table 25-7). It also grows on TCBS agar, giving rise to blue-green colonies.

An infection with this vibrio is a *vibriosis*. Present vaccines are of no effect against it.

Table 25-7 A Comparison of Three Pathogenic Vibrios

Distinguishing Feature	Vibrio cholerae ("Classic" Vibrio)	Vibrio cholerae biovar *eltor*	Vibrio parahaemolyticus
Voges-Proskauer test	−	+	−
Production of indole	+	+	+
Liquefaction of gelatin	+	+	+
Production of hydrogen sulfide	−	−	−
Cholera red reaction	+	+	−
Fermentation of glucose	+	+	+
Fermentation of lactose	Slow	Slow	Slow
Growth in salt-free broth	+	+	−
Growth in broth containing 7% to 10% NaCl	−	−	+
Hemolysis of sheep (or goat) cells	−	+	−
Hemagglutination of chicken cells	−	+	−
Phage lysis	Susceptible	Resistant	
Polymyxin B susceptibility test	Susceptible	Resistant	
Agglutination in O group serum	+	+	−

SUMMARY

1 The enteric bacilli are a large, heterogeneous group of closely related genera in the family Enterobacteriaceae. They are short, straight, nonspore-forming, facultatively anaerobic, gram-negative rods that inhabit or produce disease in the alimentary tract of humans and other warm-blooded animals. The type species of the type genus is *Escherichia coli.*

2 Most enteric bacilli are motile. Their colonies on blood agar are fairly large, shiny, and gray and variably hemolytic. All species ferment glucose, but some do not form gas thereby. Media for their isolation can contain inhibitory dyes to advantage.

3 Serologic typing of these antigenically complex bacilli centers on (1) the O antigens, (2) the K antigens, and (3) the H antigens. O, or somatic, antigens are heat-stable polysaccharides of the cell wall. K, or envelope, antigens, which are mostly heat labile, surround the cell. H antigens are heat-labile proteins in the flagella.

4 Members of the following groups comprise the coliform bacteria: (1) *Klebsiella-Enterobacter-Serratia,* (2) *Arizona-Edwardsiella-Citrobacter,* and (3) *Providencia.* The signal coliform is *Escherichia coli.* More than 150 serotypes of *E. coli* are defined. Outside their normal habitat in the bowel and in the tissues of the host, coliforms become remarkably virulent. They may cause a variety of infections, important among which are pyelonephritis and peritonitis.

5 Infantile, or summer, diarrhea occurs during hot weather in bottle-fed babies reared in unhygienic environments. It is caused by species of *Salmonella, Shigella, Proteus, Pseudomonas,* and *Staphylococcus;* pathogenic strains of *E. coli;* and certain viruses.

6 Two virulent types of *E. coli,* both exogenous in origin, cause diarrhea. In one a heat-labile enterotoxin acts to move large amounts of fluid and electrolytes across the bowel mucosa. The attachment to the bowel lining of at least 100 million such *enterotoxigenic* organisms is required for the severe, watery diarrhea. The virulence of the other type is related to its property of invasiveness. *Enteroinvasive* strains of *E. coli* cause ulcerations.

7 *Proteus* bacilli tend to swarm over the surface of their cultures on solid media because they are so actively motile. Vigorous secondary invaders, *Proteus* bacilli cause urinary tract and wound infections. They are generally resistant to antimicrobial drugs.

8 Endotoxin shock is a grave complication to a gram-negative bacteremia. The events leading to the profound shock state are set off by lipopolysaccharide endotoxin in the circulation, which is activated there by certain components of plasma.

9 From 3% to 5% of patients admitted to a modern hospital develop an infection there. Ninety percent of the time a surgical wound, urinary tract infection, or respiratory tract infection is associated. Enteric bacilli account for two thirds of hospital-acquired infections.

10 Some 250 species of the genus *Salmonella* produce disease in lower animals, their natural hosts, from which infection is spread to humans by feces. Salmonellas are susceptible to heat, disinfectants, and radiation.

11 The three types of salmonellosis are (1) acute gastroenteritis of the food infection type, (2) septicemia with localized complications, and (3) enteric fever, such as typhoid.

12 All four species in the genus *Shigella* cause bacillary dysentery. The most severe form of disease results from the action of the neurotoxin produced by *S. dysenteriae.* The human is the sole host for the highly infectious, nonmotile shigellas. Only 10 to 100 ingested organisms are needed for infection.

13 In the family Vibronaceae are rigid, gram-negative, salt-loving, facultatively anaerobic, straight or curved rods found in fresh water and seawater.

14 The comma bacillus, *Vibrio cholerae*, causes Asiatic cholera, which is characterized by massive diarrhea with enormous fluid losses. In the "rice water" stools up to 10 or 20 L may be lost in a day. The powerful cholera enterotoxin is comparable in its action to the heat-labile enterotoxin of *E. coli*.

15 Cholera is contracted by the ingestion of fecally contaminated water or food. At least 100 million vibrios are required for disease, but if the victim is malnourished, as few as 10,000 may produce the infection.

QUESTIONS FOR REVIEW

1 Classify the enteric bacilli.
2 Comment on the various ways to identify enterics.
3 Briefly characterize organisms of the genus *Salmonella*.
4 Name the three major types of salmonellosis.
5 Name and describe the organism causing typhoid fever.
6 Draw a diagram tracing the spread of typhoid infection from person to person.
7 Discuss the nursing precautions in typhoid fever and other infections of the alimentary tract.
8 What is the laboratory diagnosis of salmonellosis?
9 Name and describe the organisms causing bacillary dysentery.
10 Which of the *Shigella* species causes the most severe disease and why?
11 Briefly comment on the pathogenicity of *Shigella* species.
12 Name the members of the coliform group of microorganisms.
13 Under what conditions can coliform bacilli invade the tissues of the body?
14 Discuss infantile diarrhea. Give causes.
15 What is the basis of the Weil-Felix reaction for typhus fever?
16 Compare *Proteus* bacilli with coliforms.
17 How does enteropathogenic *Escherichia coli* produce disease?
18 Briefly discuss hospital-acquired infections.
19 What is meant by gram-negative sepsis?
20 Make a list of opportunistic pathogens you have studied.
21 Compare the classic cholera vibrio with its biovar eltor.
22 How is Asiatic cholera transmitted? How is it prevented?
23 Explain the pathogenic mechanism for the diarrhea in cholera. State the serious consequences of this.
24 Briefly explain the following terms: nosocomial, enterotoxin, endotoxin, O antigen, H antigen, K antigen, Vi antigen, infectious diarrhea, heat-labile, cholera red reaction, opportunistic infection, "swarming," enteric fever, diarrhea, dysentery, rice water stool, mucinase, Weil-Felix test, choleragen, IMViC reaction, Peyer's patches, Widal test, summer diarrhea, and paracolon bacilli.

SUGGESTED READINGS

Barry, A.L., and others: Cumitech 2: laboratory diagnosis of urinary tract infections, Washington, D.C., 1975, American Society for Microbiology.

Blake, P.A.: *Vibrios* on the half shell: what the walrus and the carpenter didn't know (editorial), Ann. Intern. Med. **99**:558, 1983.

Burdash, N.M., and others: A comparison of four commercial systems for the identification of nonfermentative gram-negative bacilli, Am. J. Clin. Pathol. **73**:564, 1980.

Gleckman, R., and Esposito, A.: Gram-negative bacteremic shock, South. Med. J. **74**:335, 1981.

Gooch, W.M., III: Evaluation of a multitest system for rapid identification of *Salmonella* and *Shigella*, Am. J. Clin. Pathol. **73**:570, 1980.

McTighe, A.H., and others: The season for vibrios, Diagn. Med. **4**:33, May-June 1981.

Neill, R.J., and others: Synthesis and secretion of the plasmid-coded heat-labile enterotoxin of *Escherichia coli* in *Vibrio cholerae*, Science **221**:289, 1983.

Salmonellosis traced to marijuana—Ohio, Michigan, Morbid. Mortal. Weekly Rep. **30**:77, 1981.

Schuman, S.H.: Day-care–associated infection: more than meets the eye, J.A.M.A. **249**:76, 1983.

Scott, J.R., and Fischetti, V.A.: Expression of streptococcal M protein in *Escherichia coli*, Science **221**:758, 1983.

Soper, G.A.: Typhoid Mary, Military Surgeon **45**:1, 1919.

Taylor, D.N., and others: Salmonellosis associated with marijuana: a multistate outbreak traced by plasmid fingerprinting, N. Engl. J. Med. **306**:1249, 1982.

Vaughan, M. Cholera and cell regulation, Hosp. Pract. **17**:145, June 1982.

CHAPTER 26

SMALL GRAM-NEGATIVE RODS

Brucella Species (Agents of Brucellosis)
Hemophilic Bacteria
  Bordetella Species (Agents of
    Pertussis)
  Haemophilus influenzae (the
    Influenza Bacillus)
  Haemophilus ducreyi (Ducrey's
    Bacillus)
  Haemophilus aegyptius
  Haemophilus suis
  Haemophilus parainfluenzae

Yersinia Species
  Yersinia pestis (the Plague Bacillus)
  Other Yersinias
Francisella tularensis (the Agent of
  Tularemia)
Pasteurella Species
  Pasteurellas of Hemorrhagic
    Septicemia
Calymmatobacterium granulomatis
Gardnerella vaginalis

The microbes of this chapter are tiny gram-negative bacilli or coccobacilli, none measuring much more than 1 μm in greatest dimension. Three of the seven genera discussed are strict aerobes; the rest are facultative anaerobes.

BRUCELLA SPECIES (AGENTS OF BRUCELLOSIS)

The genus *Brucella* includes three important intracellular parasites that primarily attack lower animals, from which they are transmitted to humans to cause brucellosis. One, known as *Brucella melitensis*, produces Malta fever in goats and sheep, a disease with prolonged fever, arthritis, and a tendency to abort. Another, *Brucella abortus*, causes contagious abortion, or Bang's disease, in cattle. The features of contagious abortion of cattle are a tendency to abort, retention of placentas, sterility, and death of newborn offspring. The third, *Brucella suis*, causes contagious abortion in hogs. Male hogs are affected by brucellosis more often than female hogs, and in females the tendency to abort is less than in cows. Less well known is a fourth member of the genus, *Brucella canis*. It infects dogs and rarely humans. Brucellosis sometimes attacks animals other than the ones just mentioned. The disease known as *poll evil* in horses is caused by brucellas.

B. *melitensis* was discovered in Malta in 1887 by Sir David Bruce (1855-1931), a surgeon in the British army, and B. *abortus* was discovered in Denmark in 1897 by Bernhard L.F. Bang (1848-1932). In 1918 and 1925 Alice C. Evans (1881-1975) proved that these organisms are closely related. B. *suis*, discovered in infected hogs in the United States, was found to be closely related to the other two, and all three were placed in the genus *Brucella*, honoring Bruce. B. *melitensis* and B. *suis* probably represent variations of B. *abortus*, the result of its adaptation to the goat and hog, respectively.

General Characteristics

Brucellas are small, nonmotile, ovoid, gram-negative, nonspore-forming coccobacilli. Some form capsules. Brucellas elaborate an endotoxin, and as is true of other gram-negative organisms, their cell wall is made up of a lipoprotein-carbohydrate complex. Brucellas grow best, but slowly, on enriched media. Being so closely related to each other, they cannot be differentiated by ordinary cultural methods. For this purpose, special methods incorporate the dyes basic fuchsin and thionine into culture media. The predictable growth response for each of the three species separates them. Another property, variable among the three species and valuable in their differentiation, is the production of hydrogen sulfide. *B. melitensis* and *B. suis* grow in atmospheric oxygen, but for the primary isolation of strains of *B. abortus* the presence of 5% to 10% carbon dioxide in the atmosphere is necessary. The agglutination test separates *B. melitensis* from *B. abortus* and *B. suis* but not *B. abortus* from *B. suis*. Table 26-1 summarizes important differential features in the genus *Brucella*.

Brucellas are destroyed within 10 minutes by a temperature of 60° C and fortunately are quickly destroyed by pasteurization. They can remain alive and virulent for as long as 4 months in dark, damp surroundings.

Pathogenicity

B. melitensis is more pathogenic for humans than *B. suis*, which is more pathogenic than *B. abortus*. Animals may be infected with any one of the three. Pregnant domestic animals are more likely to be infected than young animals or nonpregnant adults, and abortion is probable when the infection is acquired during pregnancy. *B. melitensis* infections are common in the Mediterranean basin, from where imported goats brought the infection to the Southwestern United States.

The most frequently encountered form of human infection includes a long-continued fever or cycles of fever alternating with afebrile periods, pronounced weakness, and profuse sweating, but since brucellas can attack every organ and tissue of the human body, variants of this pattern exist. The incubation period varies from 5 to 30 days (sometimes longer).

Various names are applied to the disease in humans: Malta fever (because of its

---

Table 26-1 Differential Patterns of *Brucella* Species

	Species		
Characteristics	***Brucella abortus***	***Brucella melitensis***	***Brucella suis***
Culture in medium containing dyes			
1. Thionine	Growth inhibited	No effect	No effect
2. Basic fuchsin	No effect	No effect	Growth inhibited
Exposure to atmosphere of 10% carbon dioxide	Required for best growth	Not required	Not required
Production of hydrogen sulfide	For 2 to 4 days	None	For 6 to 10 days
Hydrolysis of urea (urease test)	None (or very slowly)	None (or very slowly)	Rapid
Lysis by phage	Yes	None	None at routine test dilution (RTD); yes at 10,000 times RTD
Agglutination reaction	Differentiates this organism from *Brucella melitensis* but not from *Brucella suis*	Differentiates this one from other two	Differentiates this one from *Brucella melitensis* but not from *Brucella abortus*
Biotypes	Nine	Three	Four
Host reservoir	Cattle	Goats (also sheep)	Pigs (also hares, reindeer)

prevalence on the island of Malta), undulant fever (because of the fever curve, Fig. 26-1), and brucellosis. In Southwest Texas it is called goat fever and Rio Grande fever.

Sources and Modes of Infection

Infection is widespread among goats, cattle, and hogs. In cattle, brucellosis ranks with tuberculosis as a source of economic loss. Infection in cattle is transmitted through food contaminated with infectious urine, feces, or lochia (the vaginal discharge of the first or second week after birth) and by contact with infected placentas or fetuses. Suckling animals may be infected by the milk of infected mothers. Cows become carriers and excrete the organisms in their milk for as long as 7 years.

Humans are very susceptible to infection with brucellas. Human infections, common but often overlooked, result from entry of the organisms through the gastrointestinal tract or skin. Humans can become infected by ingesting dairy products derived from infected cows or, for infection with *B. melitensis*, by consuming unpasteurized milk from infected goats. Dust may carry the organisms. Most of the time, however, infectious material is derived from the excreta of living animals or the blood and tissues of dead ones, and the virulent brucellas invade cuts or abrasions in the skin, probably even penetrating unbroken skin or mucous membrane. By handling infected cows or hogs and the meat of these animals, packinghouse workers, butchers, farmers, stockmen, and veterinarians are at risk. Transfer of infection directly from person to person is rare.

Laboratory Diagnosis

Laboratory aids in the diagnosis of undulant fever are an agglutination test, blood cultures (about one third positive), animal inoculation, feces cultures, urine cultures, cultures of bone marrow (taken by sternal puncture), and cultures of material aspirated from lymph nodes. The last two are important because, after the bacilli have spread through the body, they localize in the bone marrow, liver, spleen, and lymph nodes. Positive cultures confirm the diagnosis, but even with known disease brucellas often cannot be cultured. One must then rely on the agglutination test, done with a carefully standardized antigen. The one of choice is the tube agglutination test performed with the patient's serum and formalin-killed brucellas as antigen. More than 90% of patients

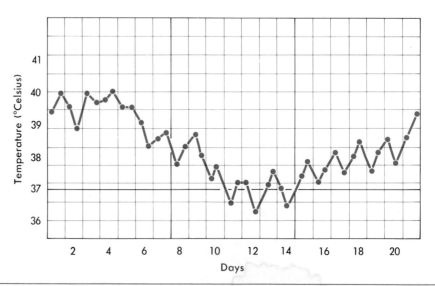

FIG. 26-1   Undulant fever curve.

with acute brucellosis demonstrate a positive agglutination test. Persons who have received cholera vaccine may develop agglutinins against brucellas, a point to be remembered in the investigation of suspected cases of undulant fever.

In human serum the pattern of antibody response in brucellosis is unique. In the acute phase the patient produces IgM antibodies (19S immunoglobulins, class M). These tend to persist. Subsequently, within days to weeks, IgG antibodies (7S immunoglobulins, class G) appear. The IgG titer decreases and disappears after treatment, usually within 6 months. Convalescents from one attack of brucellosis may continue to show agglutinins for months or years. If symptoms reappear and disease seems recurrent, the determination of the nature of that person's antibodies is significant. If they are IgG antibodies, even in low titers, then active disease is quite likely. If they are IgM antibodies in low titers, then previous contact with the organisms is indicated. Active disease is suggested only if IgM antibodies are elevated.

Also used in the laboratory diagnosis of brucellosis are a modified Coombs' test (helpful in chronic cases), a complement-fixation test, immunofluorescent technics, radioimmunoassay, and enzyme-linked immunosorbent assay (ELISA). Veterinarians use the milk ring test, an agglutination test with whole milk instead of serum as the source of antibodies. Guinea pigs may be inoculated with blood or cream from a suspect cow.

The skin test to detect hypersensitivity to brucellas is performed by the injection of **brucellergen,** a crystalline polypeptide extract of killed organisms, between the layers of the skin. If positive, the skin test indicates infection at some time past or present. A negative skin test eliminates that possibility. The skin test is comparable to the tuberculin skin test and as informative. However, commercial preparation of the antigen for intradermal testing is no longer feasible.

Immunity

Children under 10 years of age seldom contract brucellosis. Men are attacked more often than women (4:1), probably because of greater exposure. One attack may confer permanent immunity, although the disease is not self-limited. After an acute attack a person may be well, but there is reason to suppose that the organisms remain quiescent but alive in lymph nodes and spleen.

Prevention

Vaccines may provide temporary protection for humans but are still experimental. An avirulent live strain of *Brucella* is used to vaccinate cattle.

All milk (including goat's milk) and milk products should be pasteurized. All dairy cattle should be tested for brucellosis and infected animals removed from the herd. Those who work with animals must know how the disease is spread and what precautions to take to avoid contact with the flesh or excreta of infected animals. The strict application of sanitary and hygienic measures helps minimize the occupational hazards.

The excreta of a patient with undulant fever should be disposed of as though they were from a patient with typhoid fever.

B. suis, the most common agent of brucellosis in the United States today, replaces *B. abortus,* whose infection has been drastically reduced by three public health maneuvers: serologic surveys to detect the disease, destruction of infected herds, and vaccination of noninfected ones.

HEMOPHILIC BACTERIA

The hemophilic (blood-loving) bacteria include diverse species (see Table 26-2 for comparisons) with the following unifying features: (1) their growth depends on, or is aided by, hemoglobin, serum, ascitic fluid, or certain growth-accessory substances; (2) they are small; (3) they are gram negative and nonmotile; and (4) they are strict parasites.

Table 26-2 Biologic Features of Some Hemophilic Bacteria

Organism	Hemolysis	Capsule	Production of Indole	Catalase	Factor Required for Growth	
					X	V
Bordetella parapertussis	+	+	−	+	−	−
Bordetella pertussis	+	+	−	+	−	−
Haemophilus aegyptius	−	−	−	?	+	+
Haemophilus ducreyi	Slight	−	−	?	+	−
Haemophilus influenzae	−	+	+ or −	+	+	+
Haemophilus parainfluenzae	−	+	+ or −	+	−	+
Haemophilus suis	−	+	−	+	+	+
Haemophilus vaginalis	+ or −	−	−	−	−	−

Two factors in blood aiding their growth are X and V. Factor X is hemin (haemin), a complex of ferric iron with protoporphyrin IX, and factor V is nicotinamide adenine dinucleotide (NAD), a coenzyme. When culture media are prepared, factor X withstands the temperature of the autoclave. Since Factor V is heat labile, it must be supplied by potatoes or yeasts in culture media. Some hemophiles depend on both factors for continued multiplication, others rely on only one, and still others require neither. One genus in this category, *Haemophilus*, indicates by the name that its members are "blood-loving" bacteria, a property for which they are well known. Another genus name for some of these microbes is *Bordetella*. Because they can grow without blood, albeit *very* slowly, *Bordetella* species are sometimes classified separately.

Bordetella Species (Agents of Pertussis)

The Disease

Whooping cough (pertussis), a communicable disease of 1 to 2 months' duration, affects children with a catarrhal inflammation of the respiratory tract, productive of a paroxysmal cough that ends in a whoop. Why the victim whoops is unclear. At the time when the load of microorganisms is heaviest in the patient's lung, that person coughs the least. Also, the whoop may be absent in adults and in very small children. Severe paroxysms of coughing may result in rib fractures. The incubation period is 5 to 21 days. A widespread and dangerous disease, pertussis is one of the major causes of death in young children, and 90% of the deaths occur in children under 5 years of age. The danger lies in the frequency with which bronchopneumonia, malnutrition, or chronic diseases such as tuberculosis follow in its wake. In the 25% of children who have whooping cough before they are 1 year old (in 75% of cases the child is older), bronchopneumonia is the usual cause of the deaths.

The Agents

The causative organism in most cases of whooping cough was first observed in the sputum of patients with whooping cough by Jules Jean Baptiste Bordet and Octave Gengou in 1900. Hence it is often spoken of as the Bordet-Gengou bacillus, but it is properly known as *Bordetella pertussis*.

In possibly 5% of the patients a disease clinically indistinguishable from whooping cough has been related to *Bordetella parapertussis*, which is similar to *B. pertussis* in its growth patterns. One way to separate *B. parapertussis* from *B. pertussis* is biochemically. *B. pertussis* produces a brown pigment. It can also be identified serologically.

Bordetella pertussis

A small gram-negative bacillus, *B. pertussis (Haemophilus pertussis)* shows polar stain-ing, often forms capsules, is nonmotile, and does not form spores. It grows strictly in the presence of oxygen. When freshly isolated from the body, all pertussis bacilli are alike, but when cultivated, they resolve into four phases. Phase 1, or the S-form, represents the freshly isolated, virulent and encapsulated pathogen. Phases 2, 3, and 4, or R-forms, are nonencapsulated. Phase 4 is the completely nonpathogenic form, and phases 2 and 3 are intermediate. Only phase 1 organisms relate to disease in human beings.

Freshly isolated from the body, *B. pertussis* grows on special media containing blood. One especially suitable is glycerin-potato-blood agar (Bordet-Gengou medium), on which it grows slowly, with 2 or 3 days required for visible growth. If repeatedly cultured on media in which the amount of blood is decreased stepwise, *B. pertussis* is able to grow on blood-free media. With prolonged cultivation, *B. pertussis* loses its dependence on both X and V factors, and when this occurs, it loses its pathogenicity.

Pathogenicity

In the human respiratory system the bacilli aggregate in large masses among the cells lining the wall of the trachea and bronchi. They do not invade the bloodstream. The action of *B. pertussis* is, at least in part, the effect of toxic cell substances, including, in addition to the capsule, a heat-labile toxin and a lymphocytosis-promoting (or his-tamine-sensitizing) factor. The cell wall contains a thermostable endotoxin that is com-parable in many ways to the endotoxins of other gram-negative bacteria. With disrup-tion of the cell protoplasm, the thermolabile toxin is released. This substance is destroyed by a temperature of 56° C for 30 minutes. *B. pertussis* is also pathogenic for lower animals, especially rabbits and guinea pigs.

Modes of Infection

Whooping cough is one of the most highly communicable of all diseases. In geographic areas of the world where immunization is not practiced, fully 95% of the population contract the disease. It is usually transmitted by droplet infection and direct contact but may be spread by recently contaminated objects. The disease is communicable during any stage and often during convalescence. The child is most likely to spread the infec-tion just before the whoop appears and for about 3 weeks thereafter. The bacilli enter the body by the mouth and nose and are thrown off in the buccal and nasal secretions and in the sticky sputum that is coughed up. With the exception of those closely asso-ciated with the disease or patients convalescing from the disease, carriers do not exist. Convalescent carriers become noninfectious soon after the disease subsides.

Laboratory Diagnosis

Two culture methods for the bacteriologic diagnosis of whooping cough are the cough plate method and the nasal swab method. In the first an open Petri dish containing Bordet-Gengou medium is held in front of the child's mouth during a paroxysm of coughing. The organisms are sprayed on the medium in droplets from the mouth and nose. In the second a nasal swab is passed through the nose until it touches the posterior pharyngeal wall. After the swab is withdrawn, it is passed several times through a drop of penicillin solution on the surface of a plate of Bordet-Gengou medium. (The peni-cillin solution destroys many species of contaminating bacteria.) Then the material is spread over the surface of the medium with a wire loop. Cultures made by either method are incubated for about 3 days to allow visible growth. Identification of the bacilli may be made also in nasopharyngeal smears stained by fluorescent antibody technics.

A peculiar white blood count is found in pertussis (leukocyte count is 15,000 to

30,000 per cubic millimeter, of which 80% are lymphocytes), and in many cases it suggests the correct diagnosis.

Immunity

Humans possess no natural immunity to whooping cough. However, an attack of the disease confers lifelong immunity. Until recently vaccination was assumed to give permanent immunity to pertussis, but a waning immunity in previously immunized older persons is demonstrated in current epidemics of adult pertussis. Young adults, presumably immunized in early life, have been shown to serve as the primary reservoir for the spread of the disease to neonates and young infants.

Even though immune herself, a mother does not confer passive immunity on her offspring. Therefore a newborn child is susceptible. Some degree of immunity may be transferred to the baby, with no risk of intrauterine infection, if the mother is immunized during the latter months of pregnancy.

Prevention*

Whooping cough itself cannot be prevented completely by our present public health methods, but its death rate can be reduced to a remarkable degree. Unfortunately the current vaccine confers neither complete nor permanent immunity. An improved vaccine is needed. However, without any immunization program for pertussis, a 71-fold increase in the number of cases with a nearly fourfold increase in the number of deaths has been given as estimates in recent studies. Community vaccination was shown to reduce by 61% the costs related to the disease.

Mothers should be taught how deadly a disease whooping cough is among young children and the value of immunization. The unvaccinated child should be protected from exposure. The infected child should be isolated for several weeks after the whoop has disappeared. However, the child need not be kept in bed but may be allowed to play in the sunshine and fresh air.

Haemophilus influenzae (the Influenza Bacillus)

In 1892, Richard F.J. Pfeiffer (1858-1945) described in a one-page report to *Deutsche Medizinische Wochenschrift* the bacillus now known as Pfeiffer's bacillus, or *Haemophilus influenzae*, which he found in the sputum of patients with influenza and which he incorrectly believed to be the cause of the 1889-1891 flu pandemic. Until the 1918-1919 pandemic this small, pleomorphic bacillus was regarded as the sole cause of influenza. Now we know that influenza is caused by a virus. *H. influenzae* maintains an important position as secondary invader in influenza and other respiratory diseases as well. It may also be the primary cause of serious infections.

General Characteristics

The smallest known pathogenic bacillus, *H. influenzae* grows best in the presence of oxygen but may grow without it. Nonmotile and nonspore-forming, it does not occur outside the body and is very susceptible to destructive influences. There are two forms, the encapsulated and the nonencapsulated. All pathogenic influenza bacilli possess a polysaccharide capsule, which is highly antigenic, and the larger the capsule, the more virulent the organisms. On the basis of capsular antigens, the encapsulated organisms have been subdivided into six types: a, b, c, d, e, and f. Most infections in children and most systemic infections are related to the type b bacilli. The specific polyribosyl ribitol phosphate of the type b capsule appears to be the main reason for its virulence, and antibody to this substance is protective. Nonencapsulated bacilli are nontypable, and in adults two thirds of influenza bacilli related to disease are nontypable.

---

*For immunization, see pp. 428, 438, and 440.

The fastidious influenza bacilli grow only on special media such as chocolate agar or hemoglobin oleate agar. (Chocolate agar contains heat-lysed red blood cells, a source of hemin.) Both factors V and X are required for growth, which is often more luxuriant about an organism such as *Staphylococcus aureus* because some staphylococci and certain other bacteria provide the influenza bacilli with NAD, the required growth factor. The vigorous growth of bacteria in proximity to colonies of another organism is the **satellite phenomenon.**

Pathogenicity

H. influenzae is found in the throats of about a third of normal persons. For infection to occur, host defenses must be lowered. Adult humans show bactericidal substances for influenza bacilli in their blood related to the immunity probably derived from previous subclinical infections, and disease from *H. influenzae* is unusual in normal adults. Protective substances are not found in the serum of young children, and for them infection is an extremely serious matter. That systemic disease caused by influenza bacilli, almost always in children, is contagious can be shown in the secondary spread that occurs in families and day-care centers.

Since rifampin eradicates the nasopharyngeal carriage of these organisms, it can be given prophylactically to contacts of persons with disease caused by *H. influenzae* type b. This is the primary and the leading cause of purulent meningitis in young children, which is the most frequent form of nonepidemic meningitis. In 85% of cases the children are between 2 months and 3 years of age. A devastating disease, it is responsible for the deaths of one third of children affected and for serious mental deficits in a third of the survivors. Before modern methods of treatment, influenzal meningitis was uniformly fatal. Type b also causes acute epiglottitis, an obstructive inflammation that involves the soft tissues of the larynx, trachea, and bronchi. The process begins suddenly with great severity, blocks breathing, and may cause death shortly if obstruction is not relieved. Here one sees a blue, anoxic infant with a cherry red, gigantic epiglottis.

Influenza bacilli *not* of designated serotype and pneumococci are said by some observers to be the two major causes of otitis media (infection of the middle ear). Otitis produced by "untypable" influenza bacilli is the predominant kind in children.

A number of clinical infections may be caused by *H. influenzae*. These include cellulitis, pneumonia, pericarditis, subacute bacterial endocarditis, septic arthritis, osteomyelitis, and septicemia. This agent is moderately pathogenic for lower animals, especially the rabbit.

An immune anti-*Haemophilus* serum has been made in rabbits against type b bacilli. A vaccine against type b bacilli is available, but it is not effective where needed most— in babies 6 months to 1 year of age, since this age group does not yet have an immune system sufficiently mature to respond adequately to vaccination. Adults do not need this type of immunity because they are largely protected already by immunity derived from inapparent infections.

Laboratory Diagnosis

The laboratory diagnosis rests on the use of appropriate smear and culture technics to recover and demonstrate influenza bacilli in suitable specimens taken from the patient. The presence of satellite colonies on culture plates is a key point in identification. With immunofluorescent microscopy, influenza bacilli are easily found in cerebrospinal fluid taken from children with meningitis.

Immunodiagnostic tests available include complement fixation, latex agglutination, and countercurrent immunoelectrophoresis. This latter method is rapid, 90% accurate, and applicable to the diagnosis of meningitis in children who have already received antibiotics.

Haemophilus ducreyi (Ducrey's Bacillus)

Haemophilus ducreyi, or Ducrey's bacillus, an obligate parasite of human beings, causes **chancroid**, a localized, highly contagious, venereal ulcer with no relation to the chancre, the initial lesion of syphilis. It begins on the genitalia as a pustule that ruptures, exposing an ulcer with undermined edges and a gray base. Multiple ulcers are usually present that spread rapidly. These do not have the indurated edges of the syphilitic chancre; hence they are spoken of as *soft chancres*. Infection also may spread to the inguinal lymph nodes, where abscesses known as **buboes** are formed.

The chancroid, or soft chancre, must be differentiated from the "hard" chancre of syphilis, and the buboes from those of lymphogranuloma venereum (Table 26-3). It is not uncommon for a chancroid and a syphilitic chancre to occupy the same site. The chancroid appears first and heals, after which the chancre appears, or the chancre may appear before the chancroid heals.

Diagnosis

H. ducreyi may be detected in smears made directly from the edges of the lesions or cultivated (with difficulty) from the lesions if some of the exudate is inoculated into sterile rabbit blood. For the growth of this fastidious microbe a higher humidity and a slightly lower temperature are required than for most other bacteria. However, both smear and culture may fail to demonstrate the organisms.

A skin test for the diagnosis of chancroid, devised and used extensively in Europe, consists of the intradermal injection of a saline suspension of *H. ducreyi*. A positive result is an area of redness and induration at the injection site. The reaction reaches maximal intensity at the end of 48 hours.

Spread

Transmission of the infection is by direct contact only, usually by sexual intercourse, but it may be transmitted by surgical instruments or dressings. The disease is found more often in men than in women.

Table 26-3 Sexually Transmitted Diseases Associated with Ulcers

Disease	Cause	Ulcer Present Painful	Texture and Consistency	Regional Lymph Nodes Enlarged
Syphilis	Spirochete—*Treponema pallidum*	No	Chancre hard, indurated; ulceration shallow	Yes, firm
Chancroid (p. 562)	Bacterium—*Haemophilus ducreyi*	Yes, very painful	Soft ("soft chancre"); purulent component	Yes, fluctuant
Granuloma inguinale (p. 568)	Bacterium—*Calymmatobacterium granulomatis*	No	—	No
Genital herpes (p. 714)	Virus—*Herpesvirus hominis* type 2	Yes, exquisitely so	Soft	Yes, firm
Lymphogranuloma venereum (p. 661)	Bacterium—*Chlamydia trachomatis* L-1, L-2, L-3	Usually no ulcer		Yes, fluctuant

Haemophilus aegyptius

The Koch-Weeks bacillus, or *Haemophilus aegyptius*, causes pinkeye, a highly contagious conjunctivitis prone to epidemics. It is well named because of the intense inflammation of the conjunctival linings, which imparts a brilliant pink color to the white of the eyes. The itching associated with the infection is severe but is even further intensified if one rubs the eyes. A yellow discharge forms, dries, and crusts on the eyelids. The disease is transferred by hands, towels, handkerchiefs, and other objects that contact face and eyes. Other names for the bacillus are *Haemophilus Koch-Weeks* and *Haemophilus conjunctivitidis*.

Haemophilus suis

Haemophilus suis together with a virus causes swine influenza (p. 723).

Haemophilus parainfluenzae

Very similar to *H. influenzae* is *Haemophilus parainfluenzae*, which may cause subacute bacterial endocarditis, but usually it is a nonpathogenic microbe in the upper respiratory tract.

YERSINIA SPECIES

The genus *Yersinia* includes *Yersinia (Pasteurella) pestis*, the cause of plague, and two causes of yersiniosis, *Yersinia enterocolitica* and *Yersinia pseudotuberculosis*. Although it is discussed in this chapter with other small gram-negative bacilli, the genus *Yersinia* is classified in the family Enterobacteriaceae (p. 528).

Yersinia pestis (the Plague Bacillus)

Plague (a zoonosis) is an infectious disease of rodents, especially rats, and is transferred from them to humans. First described in Babylon, it has been a devastating pestilence for more than 3000 years. Pandemics in the past have swept over great areas of the world, with a terrifying mortality (50% or more). Hundreds of years ago the Chinese related the disease to rats by using the term *rat pestilence*. In these days of extensive travel and commerce, one must remember that this disease still exists over the world at all times. In the United States plague has been encountered in Texas, Louisiana, and many states in the far West. The disease is enzootic among the wild rodents of the Southwestern United States, especially ground squirrels and prairie dogs, and this focus of infection is an ever present and increasing threat. Among wild rodents it is known as **sylvatic plague** (Latin *silva*, forest).

General Characteristics

Yersinia (Pasteurella) pestis (discovered in 1894) is small, plump, sometimes pleomorphic, aerobic or facultatively anaerobic, gram negative, and nonspore-forming (Table 26-4). The plague bacillus tends to exhibit bipolar staining; the heavily stained ends of the bacillus contrast with the ligher central area. Plague bacilli grow on all ordinary culture media but, unlike other human pathogens, grow better at 25° to 30° C than at body temperature (37° C). Growth on agar containing from 3% to 5% salt is a specific feature for its identification.

Plague bacilli may live in the carcasses of dead rats, in the soil, and in sputum for some time. They retain their vitality for months in the presence of moisture and absence of light. Phenol, 5%, or Amphyl solution, 2%, destroys them in 20 minutes.

Plague bacilli are highly pathogenic for many animals: monkeys, rats, mice, guinea pigs, and rabbits. They owe their action to at least two toxic substances, one of which is an endotoxin, a lipopolysaccharide, similar to other bacterial endotoxins in both

Table 26-4 Biochemical Identity of Some Small Gram-Negative Bacilli

Organism	Optimal Temperature for Growth	Fermentation of Sugars				Production of Indole	Production of Hydrogen Sulfide	Oxidase Reaction
		Glucose	Maltose	Sucrose	Lactose			
Yersinia pestis	28° C	Acid	Acid	—	—	–	–	–
*Francisella tularensis**	37° C	Acid	Acid	—	—	–	+	–
Pasteurella multocida	37° C	Acid	—	Acid	—	+	+	+

*Note again the hazard of handling cultures of this organism. In routine identification biochemical differentiation is not necessary.

structure and physiologic effects. The other, a protein and part of the cell membrane, is like both endotoxins and exotoxins. It is released only after lysis of the bacterial cell.

Clinical Types

Plague occurs in three patterns: bubonic, septicemic, and pneumonic. Bubonic plague is the most frequent. The bacilli penetrate the skin and are carried by the lymphatics to the lymph nodes, draining the site of infection, commonly in the inguinal region. Here they multiply and form abscesses of the nodes (buboes). The bubo is extremely painful; a severe cellulitis surrounds it. Secondary buboes are formed in the nodes draining the primary buboes, and the bacilli finally move into the bloodstream, causing septicemia. The fatality rate with septicemia is 90% to 95%. The hemorrhages in bubonic plague that produce black splotches in the skin gave the name "Black Death" to plague during the Middle Ages. (From 1347 to 1349 AD the Black Death killed 25 to 40 million persons.)

Pneumonic plague, a highly virulent and rapidly fatal form, manifests as bronchopneumonia, and the bacilli are abundant in sputum. The mortality approaches 100%. Only a small percentage of cases in the average epidemic take the pneumonic form (or that of a meningitis), but epidemics strictly of pneumonic plague may occur. Septicemic plague is a rapidly progressive form in which the patient dies before buboes can develop.

Modes of Infection

Although many rodent species may be infected by Y. *pestis*, the most important source of primary infection is the rat, and, as a rule, epidemics of human plague closely follow epizootics of rat plague. Plague is transmitted from rat by the bite of the rat flea. (One species, the Indian rat flea, *Xenopsylla cheopis*, is considered the most effective vector.) A human contracts the bubonic form of plague by the bite of a flea that has previously fed on an infected rat or, rarely, on an infected person. The reason for epidemics of human plague occurring *after* epizootics of rat plague is that an epizootic destroys the rat flea's food source of first choice—the rat. The flea must then seek his food source of second choice—the human. At the time that it bites, if blood happens to be adherent to its mouth parts, the flea may transfer the infection mechanically. Otherwise the flea spreads the infection in a different way, 4 to 18 days after it has drawn the blood of an infected host. The plague bacilli that the flea ingests colonize the flea's intestine, where they multiply rapidly. The buildup of plague bacilli eventually produces an obstruction through which food cannot pass. As the hungry flea attempts to feed, the blood it sucks is mixed with plague bacilli. The mixture is regurgi-

Never handle a sick or dead wild animal. If an animal that is ordinarily thought of as wild lets you come near, consider it sick and avoid it altogether.

tated into the bite wound. The flea may die of the infection or become a carrier. The incubation period in humans is 2 to 6 days.

Pneumonic plague may result from involvement of the lungs during the course of bubonic plague (primary pneumonic plague) or from inhalation of particles of sputum thrown off by a person with pneumonic plague (secondary pneumonic plague). Epidemic pneumonic plague is the secondary type.

Dozens of wild and domestic rodents other than rats may contract plague naturally: ground squirrels, moles, prairie dogs, chipmunks, marmots, and guinea pigs. Infected rats carry the disease from one geographic locality to another. The black house rat is more dangerous to humans than other types because it lives in dwellings and therefore its fleas are more likely to bite a human being. The common gray sewer rat and the Egyptian rat can be infected. Bedbugs and the human flea may transmit plague from person to person. Plague may occasionally be acquired by the handling of infected rodents. Human carriers (without symptoms) have been found to harbor Y. pestis in their throats.

Today in Southeast Asia conditions are favorable for outbreaks of plague. A large infected rat population with the right number of rat fleas exists alongside a dense unimmunized human population in a humid, warm (not hot) climate. During the wet season the monsoons cancel out the effect of insecticides and force rats into human shelters. With their sticks the natives kill the rats and thus force the rat flea to seek a new host, this time a human one.

Laboratory Diagnosis

The important methods of laboratory diagnosis are the agglutination test on the patient's serum and the demonstration of the bacilli in lesions by smears, cultures, and animal inoculation. A fluorescent antibody test is used to identify the organisms in sputum.

Immunity

An attack of plague usually induces permanent immunity.

Prevention*

The prevention of plague depends on the eradication of rats and rat fleas. Patients with plague should be isolated in well-screened, vermin-free rooms. (The technic of quarantine was first used with plague.) Although patients with bubonic or septicemic plague seldom transmit the infection to others, no chance can be taken. They should be nursed with the same precautions as for patients with typhoid fever. The sputum from patients with pneumonic plague should receive special care. The sickroom should be dusted with a suitable insecticide to eliminate the vector flea. Attendants should be protected with plague vaccine.

Measures designed to block the transport of rats from infected to uninfected localities by trains, aircraft, and ships are imperative, and maritime rat-control measures are in force in most parts of the world. Rat-infested ships can be fumigated with a potent rodenticide. When rats can be trapped, fumigation may not be necessary. Fortunately relatively few rats are found on ships today.

---

*For immunization procedures, see pp. 428, 447, and 448.

Other Yersinias

Yersinia pseudotuberculosis and *Yersinia enterocolitica* cause yersiniosis, a febrile illness in humans and animals. Acute mesenteric lymphadenitis and enteritis with chronic arthritis are manifestations of infection with these two. Some species of *Yersinia* exhibit pectinolytic activity, a property formerly associated exclusively with bacteria pathogenic for plants.

FRANCISELLA TULARENSIS (THE AGENT OF TULAREMIA)

Francisella (Pasteurella) tularensis causes the zoonosis tularemia, or rabbit fever, an acute infectious disease of wild animals, especially rodents, transferred from animal to animal by the bite of an insect. It may be transmitted from animal to human by insect bite, but human infections come from contamination of the hands or conjunctivae by the tissues or body fluids of an infected animal or insect. This agent can penetrate intact skin.

General Characteristics

The organism *F. tularensis* derives its name from Tulare County, California, where the disease was first observed in 1911 in ground squirrels. *F. tularensis* is a small, highly pleomorphic, gram-negative, strictly aerobic, nonspore-forming, nonmotile organism varying considerably in size. It grows slowly on cystine agar (not on ordinary agar) and is easily demonstrated by animal inoculation. It behaves as an intracellular parasite, persisting for long periods in phagocytic and other body cells of the host. Enormous numbers fill these cells.

Clinical Types

Tularemia is a febrile disease accompanied by severe constitutional manifestations, pain, and prostration. The incubation period is usually 3 to 10 days. The clinical types are (1) ulceroglandular (an ulcer at the infection site, usually the hands or fingers, with regional lymph node involvement), (2) oculoglandular (pattern similar to that of the ulceroglandular type with conjunctiva as primary site), (3) glandular (lymph nodes involved but ulceration absent), and (4) typhoidal, also known as enteric or cryptogenic (neither ulceration nor lymph node involvement present). Death occurs in about 5% of the patients. Those who recover are incapacitated for weeks or months. (Clinically tularemia may be confused with cat-scratch disease.) The most prevalent and most readily recognized are the ulceroglandular and glandular types, which together account for over 85% of cases.

Modes of Infection

Tularemia is most prevalent in cottontail rabbits, jackrabbits, snowshoe rabbits, and ground squirrels. Certain birds have tularemia, and the disease is not unknown in cats and sheep. The water of streams inhabited by infected animals (beavers and muskrats), once contaminated, may long remain a source of infection. Epidemics caused by pollution of streams have been reported. In Russia human infections are often contracted from a species of fur-bearing water rat. More than 45 species of wild animals are known to be infected. Tame rabbits are susceptible but ordinarily do not contract the disease, since they do not harbor the small parasites that transmit the infection among animals.

The arthropods that transmit tularemia in animals are wood ticks, rabbit ticks, lice, horseflies, and squirrel fleas. Wood ticks, dog ticks, horseflies, and squirrel fleas sometimes may transmit the infection to humans. Rabbit ticks, rabbit lice, and mouse lice, important agents in the spread of the disease from rodent to rodent, do not bite humans.

In the United States a person usually contracts tularemia by handling infected rabbits; of these, about 70% are cottontails. Cold-storage rabbits remain infectious for 2 to 3 weeks, and if kept frozen, they may be a source of infection for 3½ years. The disease may be contracted if insufficiently cooked rabbit meat is eaten.

Inhalation of as few as 10 to 50 organisms can result in disease. (For infection to result from oral ingestion, the number of organisms required is 10^8.) The ocular type of tularemia usually results from a person rubbing his eye with contaminated fingers. Tularemia is not transmitted directly from person to person.

F. tularensis is the most easily communicable of all organisms. Not much exposure is necessary because human beings are highly susceptible. Many laboratory workers who have investigated tularemia have contracted the disease; some have died. Laboratory technologists are at risk for pulmonary tularemia, which is easily acquired through accidental inhalation of aerosols containing viable microbes.

Laboratory Diagnosis

The laboratory diagnosis of tularemia depends on the agglutination test, immunofluorescence (Color plate 3, C), Foshay's antiserum test, the tularemia skin test, and the inoculation of guinea pigs or rabbits with material from lesions. The earliest test to become positive is the skin test; the highly specific, tuberculin-type reaction appears within a few days of onset of illness. Currently this test is not available commercially. Cultures are too dangerous to be handled routinely.

Visitors, secretaries, various employees, and the like should be carefully segregated and barred from the specific laboratory area where work is carried on with *F. tularensis*.

Immunity

An attack of tularemia is followed by permanent immunity.

Prevention

The nature of the infection and its transfer dictate the preventive measures. Tularemia vaccine is a lyophilized, viable, attenuated strain of *F. tularensis*. Vaccination is indicated for those at risk.

PASTEURELLA SPECIES

The genus *Pasteurella* (named for Louis Pasteur) includes a group of organisms causing disease in animals but seldom attacking human beings.

Pasteurellas of Hemorrhagic Septicemia

Certain members of the genus *Pasteurella* cause hemorrhagic septicemia (pasteurellosis), a disease of cattle, horses, swine, bison, and poultry, characterized by septicemia, petechial hemorrhages of mucous membranes and internal organs, edema, and changes in the lungs. Mortality is high. The most important bacteria are *Pasteurella multocida* and *Pasteurella haemolytica*. *P. multocida* (meaning "many killing"), well

---

No animal, including the human being, is limitless in its love and good humor. A dog has only a few ways of expressing hostility, one of which is to bite and to claw. If sufficiently provoked, the dog cannot be blamed for obeying its instincts rather than its master or mistress. Let sleeping dogs lie and let strange dogs be!

From Goldwyn, R.M.: Man's best friend (editorial), Arch. Surg. 111:221, 1976.

known to veterinarians, can infect humans, in whom half of such infections result from an animal bite. The rest involve the respiratory tract in patients with preexisting pulmonary problems. The opportunistic infections it produces can be quite severe.

Microbiology of Dog and Cat Bites

Over half a million persons, principally children younger than 12 years, are bitten by animals each year in the United States. Most of the bites come from cats and dogs; dogs head the list. Generally, such wounds heal without complications, but they can be infected by microorganisms passed by the animal in biting.

Since it is normally found in the nose and throat of cats and dogs, *P. multocida* is an offender. Other organisms encountered, usually in mixed culture, include *Staphylococcus aureus* and alpha- and beta-hemolytic streptococci. The lesion produced may be a cellulitis, an abscess, or, about the extremities, a tenosynovitis (inflammation of tendon sheath). Osteomyelitis can complicate a deep-seated bite, and there is always the danger of bacteremic spread. Resolution of the tissue injury from the bite can be slow. Scars resulting from damage to tissue about the head and neck may be disfiguring.

CALYMMATOBACTERIUM GRANULOMATIS

Calymmatobacterium granulomatis (Greek *kalymma*, hood, plus *baktērion*, a little staff), the only species of its genus, causes granuloma inguinale, a venereal disease producing ulceration in the skin and subcutaneous tissues of the groin and genitalia. In the lesions, encapsulated, oval, rodlike bodies with a unique "safety pin" appearance are found within the cytoplasm of large mononuclear phagocytes (Fig. 26-2). These agents are also known as **Donovan bodies;** the disease, also as **donovanosis.**

The laboratory diagnosis of the disease rests on microscopic demonstration of the Donovan bodies in Wright-stained smears made of material taken from the lesions.

GARDNERELLA VAGINALIS

Gardnerella vaginalis (formerly *Haemophilus vaginalis, Corynebacterium vaginale*) is a small, pleomorphic, nonmotile, nonencapsulated, facultatively anaerobic, gram-negative rod. Although it was placed with other species in the genus *Haemophilus*, it is not comparable to them in its traits. Neither factor X nor factor V is required for

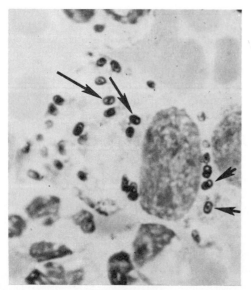

FIG. 26-2   Donovan bodies in cytoplasm of phagocytic cell. Note "safety pin" appearance of dark bipolar bodies *(arrows)*. Clear surrounding areas correspond to capsules about microorganisms. (From Davis, C.M.: J.A.M.A. **211**:632, 1970.)

growth. Since studies such as cell wall analysis, determination of guanine plus cytosine ratios, and electron microscopy show this microorganism to be distinct from *Coryne-bacterium* species and define its own cellular morphology and biochemical composition, it has been placed in a new genus, *Gardnerella*.

G. vaginalis is causally related to a nonspecific vaginitis, notable for the profuse malodorous discharge. Sexual transmission is indicated by the recovery of the organisms from the urethra and urine of male consorts.

The organism may be recovered in cultures taken from the genital tract of women with vaginitis. In Papanicolaou-stained smears taken from the cervix or made of the vaginal discharge, so-called clue cells are considered diagnostic by some observers. **Clue cells** are squamous epithelial cells heavily "dusted" with myriads of the small bacilli. An important diagnostic test in the laboratory is the addition of 10% potassium hydroxide to a portion of the vaginal discharge; a fishy, aminelike odor results when the test is positive.

SUMMARY

1 Three species in the genus *Brucella* cause brucellosis in humans. They are strict intracellular parasites that primarily attack animals, from which they are transmitted to humans. The most pathogenic species for humans is *Brucella melitensis*.

2 Characteristic fever cycles are observed in a patient with brucellosis, hence the term for the disease *undulant fever*. Since brucellas attack any anatomic area in the human body, the pattern of infection is variable.

3 Positive blood cultures confirm the diagnosis of brucellosis. When even a small amount of IgG is present, active disease is likely.

4 Pasteurization of dairy products prevents brucellosis. Infected dairy cattle should be removed from the herd.

5 The hemophilic bacteria are small (no larger than 1 μm), gram negative, nonmotile, and strict parasites. Their growth is facilitated by the heat-stable blood factor X and the heat-labile blood factor V. The genera *Bordetella* and *Haemophilus* of hemophilic bacteria contain important pathogens.

6 *Bordetella pertussis* causes whooping cough, or pertussis, one of the most communicable diseases of children. In unimmunized children under 5 years of age pertussis is a major cause of death.

7 All pathogenic influenza bacilli *(Haemophilus influenzae)* possess a highly antigenic capsule that is the basis for sorting the organisms into six types. Type b bacilli cause a purulent meningitis in young children.

8 In proximity to colonies of an organism that provides them with the needed growth factor NAD, influenza bacilli grow vigorously, thus demonstrating the satellite phenomenon.

9 The highly fastidious *Haemophilis ducreyi* (Ducrey's bacillus) causes chancroid, a localized sexually transmitted ulcer.

10 *Yersinia pestis* causes plague, which has been a devastating pestilence for more than 3000 years. The most frequent form, bubonic plague, is the same disease as the Black Death of the Middle Ages, which in 2 years killed from 25 to 40 million people.

11 The zoonosis plague is an infection of rodents and transferred from them to humans. Plague is transmitted from rat to rat by the bite of the rat flea, which is also a vector for the human. The prevention of plague depends on the eradication of rats and rat fleas. The technic of quarantine was first used with plague.

12 *Francisella tularensis* causes the zoonosis tularemia, or rabbit fever. In rodents and other wild animals it is transferred by an arthropod bite. In humans tularemia results from an arthropod bite or from contamination of the hands or eyes by tissues from an infected animal. In infected tissues, enormous numbers of the organisms fill the

macrophages and other body cells. Humans are very susceptible. Laboratory cultures of the organism are too dangerous for routine handling.

13 Members of the genus *Pasteurella* cause animal disease and rarely infect humans. Hemorrhagic septicemia is caused by certain *Pasteurella* species.

14 *Calymmatobacterium granulomatis* causes granuloma inguinale, a sexually transmitted disease producing ulceration in skin and subcutaneous tissues of the groin and genitalia. The causative organisms, known as Donovan bodies, are oval, encapsulated "safety pin" objects in the cytoplasm of macrophages.

15 *Gardnerella vaginalis* is causally related to a nonspecific vaginitis, notable for the profuse malodorous discharge and considered to be sexually transmitted.

QUESTIONS FOR REVIEW

1 Name the species of *Brucella*. How are they alike? How different?
2 Comment on the laboratory diagnosis of brucellosis, mentioning difficulties.
3 How is whooping cough transmitted? What are the dangers of this disease?
4 Give the laboratory diagnosis of pertussis.
5 List features shared by hemophilic bacteria.
6 What is the importance of factors V and X? What are they?
7 Briefly discuss pathogenicity of the influenza bacillus.
8 List five animals (other than humans) contracting tularemia. Name five vectors associated with it.
9 Give clinical types of plague and tularemia.
10 What is the reservoir of plague? How is it conveyed to humans?
11 What are Donovan bodies? Where are they found?
12 Briefly discuss the microbiology of dog and cat bites.
13 Describe the condition pinkeye. What organism causes it?
14 What is chancroid? Its cause? How does it differ from chancre?
15 Briefly explain yersiniosis, clue cells, the satellite phenomenon, an undulant fever curve, milk ring test, bubo, "Black Death," hemorrhagic septicemia, Malta fever, nontypable influenza bacilli, and zoonosis.
16 Why was *Gardnerella* set up as a new genus?

SUGGESTED READINGS

Barclay, W.R.: Brucellosis control, J.A.M.A. **244**:2315, 1980.

Beyt, B.E., Jr.: More from the pesky *Pasteurella*, J.A.M.A. **249**:516, 1983.

Cox, F., and others: Rifampin prophylaxis for contacts of *Haemophilus influenzae* type b disease, J.A.M.A. **245**:1043, 1981.

Finegold, S.M., and Martin, W.J.: Diagnostic microbiology, ed. 6, St. Louis, 1982, The C.V. Mosby Co.

Granoff, D.M., and others: Countercurrent immunoelectrophoresis in the diagnosis of *Haemophilus influenzae* type b infection, Am. J. Dis. Child. **131**:1357, 1977.

Koplan, J.P., and others: Pertussis vaccine—an analysis of benefits, risks and costs, N. Engl. J. Med. **301**:906, 1979.

Mertsola, J., and others: Serologic diagnosis of pertussis: comparison of enzyme-linked immunosorbent assay and bacterial agglutination, J. Infect. Dis. **147**:252, 1983.

Musher, D.M.: *Haemophilus influenzae* infections, Hosp. Pract. **18**:158, Aug. 1983.

Plague—South Carolina, Morbid. Mortal. Weekly Rep. **32**:417, 1983.

Wise, R.I.: Brucellosis in the United States, past, present, and future, J.A.M.A. **244**:2318, 1980.

ANAEROBES

OVERVIEW
Definition

Generally, anaerobes are bacteria that must have a low or zero oxygen tension for their continued growth and are sensitive to the action of oxidizing agents such as peroxide. (Most obligate anaerobes are bacteria. No fungi are known as such.) To anaerobes, oxygen is either inhibitory or toxic. Indeed, some are so fastidious that, for their manipulation in the laboratory, oxygen must be meticulously eliminated every step of the way. However, within such a large group of complex microorganisms there are varying degrees of tolerance, as would be expected.

The preceding definition suggests that anaerobes in their native habitats influence their surroundings. Studies show that most do synthesize regulatory substances such as certain relatively strong organic acids in large amounts or particular enzymes, which act to keep the oxidation-reduction potential of the milieu at the required low level.

Oxygen would be harmful to all bacteria were it not for protective intracellular enzyme systems possessed by certain ones. The complete reduction of a molecule of oxygen to water results in the appearance of intermediates too biochemically reactive to be well tolerated by living cells. One such intermediate is the superoxide radical. Aerobes and aerotolerant anaerobes possess the enzyme superoxide dismutase, which catalytically eliminates this intermediate by converting it to hydrogen peroxide. Hydrogen peroxide is also such a toxic intermediate, but the catalases and peroxidases present in these microorganisms remove it enzymatically, in turn, by converting it to water.

Various explanations are given as to why oxygen is toxic or inhibitory to anaerobes. With some exceptions, anaerobes do not contain catalase; therefore, with exposure to oxygen, peroxides harmful to them accumulate, In strict, highly oxygen-sensitive anaerobes, the enzyme superoxide dismutase is known to be absent, and the oxygen sensitivity of anaerobes does seem to be correlated with the lack of this enzyme or the presence of only subeffective amounts. The possibility also must be considered that oxygen may induce in some way oxidation of essential intracellular substances that cannot be easily regenerated by the anaerobe.

The upsurge of interest in anaerobic bacteriology is another example of progress following on the heels of advancements in technology. Now that we can view anaerobes realistically, and with the greatly increased concern about them, the confusion that fogs understanding of their biologic attributes is clearing.

Pertinence

Important to a consideration of anaerobes in disease is the knowledge of their position in health. It is amazing to realize that this poorly understood category must make a sizable contribution to human well-being, since in great abundance they are intimately bound to people. The vast majority are indigenous to humans (in space-age lingo, life on the human race is anaerobic). Anaerobes are the predominant population group of the resident microflora, and their myriads occupy the intestinal tract, oral and nasal cavities, genital tract (especially in the human female), respiratory tract, and skin of the human race. The colon, first, and the oral cavity, second, contain the most numerous and varied of the anaerobes. In most areas anaerobes outnumber aerobes 10:1; in the large intestine, 1000:1 (*Bacteroides* species, the anaerobe; *Escherichia coli*, the aerobe).

Anaerobes found in infection are commonly part of the normal flora. The more sophisticated technics of isolating, identifying, and studying anaerobes soon showed that they had capabilities for inducing severe disease and that under the right circumstances expressed them. The restrained one of great potential quickly became the unleashed opportunist of great vigor.

Many observers state that anaerobes cause practically any type of disease known, and currently, of specimens from cases of infectious disease submitted to clinical laboratories properly equipped to deal with them, anaerobes are found in 40% or more of specimens.

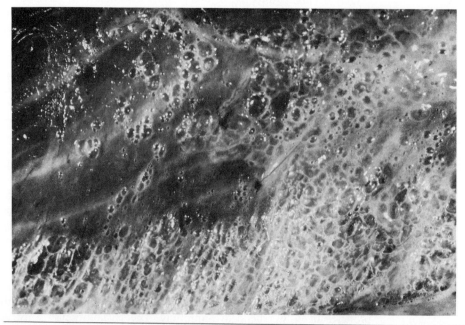

FIG. 27-1   Gas formation in soft tissues beneath the skin over belly of striated muscle, as seen in certain anaerobic infections. (From Smith, A.L.: Microbiology and pathology, ed. 12, St. Louis, 1980, The C.V. Mosby Co.)

Pathogenicity

The key to reversal of the biologic role of the normally saprophytic anaerobe is the level of tissue oxidation. Since this level is crucial for normal growth, it can, if lowered, further enhance growth. At the same time the level of tissue oxidation is also important to the function of body defense mechanisms such as that of certain bactericidal systems. What expands microbial growth also diminishes body resistance.

The most common reason for lower tissue oxygen potential is the pathologic state of tissue anoxia regularly accompanying many pathologic processes that impair the circulation to a given area. Other factors operating are (1) the presence of dead tissue from whatever cause, such as trauma or injection of drugs, and (2) the growth of aerobic microbes. When wounds are contaminated with well-tilled topsoil containing spore-forming anaerobes, the presence of ionized calcium salts aids in germination of their spores.

Most anaerobes are destructive. The type lesion is properly tissue necrosis. Two other, sometimes dramatic, features of their infections are the foul odor of discharges and purulent drainages and the manifestation of gas in soft tissues (crepitation) (Fig. 27-1).

The pattern for disease is obviously diffuse with a long list of possibilities. The more important are septicemia with metastatic suppuration; intraabdominal abscesses with peritonitis; infections of the female genital tract such as endometritis, tuboovarian abscess, pelvic cellulitis, and postsurgical wound infections; pleuropulmonary disease with empyema and abscess; chronic otitis media; and skin and soft tissue infections.

Practically speaking, only a limited number of specific anaerobes figure prominently in clinical infections. The following seven microorganisms, discussed in this chapter, probably account for two thirds of all such infections:

Bacteroides fragilis
Bacteroides thetaiotaomicron
Bacteroides melaninogenicus
Bacteroides asaccharolyticus
Fusobacterium nucleatum
Clostridium perfringens
Anaerobic cocci

Laboratory Diagnosis

Because of the ubiquity and profusion of anaerobic species, specimens for microbiologic study are taken directly only from a site normally sterile. Otherwise special collection technics are used (p. 187). Important sources for test material include wounds, abscesses, blood, abdominal discharges, and such. Getting the specimen or test materials to the laboratory may necessitate transport in a special carbon dioxide–filled container; the specimen requires immediate attention on arrival at its destination.

Because anaerobic infections tend to be mixed, and because many anaerobic species are fastidious, preliminary examination of a Gram-stained smear of specimen, particularly if it is exudative, is indicated.

For the isolation and identification of anaerobic species, three cultural methods are standard: (1) in the **roll-streak tube system**, PRAS (*p*rereduced, *a*naerobically *s*terilized) media are prepared in anaerobic culture tubes, stored, inoculated (rolled to distribute inoculum), and manipulated under a stream of oxygen-free gas; (2) the **anaerobic glove box** or chamber houses the cultures and cultural materials in an oxygen-free environment and yet it allows the technician access to the contents of the chamber through openings (ports) sealed off with gloves; and (3) in the **anaerobic jar system** (Fig. 27-2) plates are held during incubation in an atmosphere made appropriate by a specially devised elimination-replacement (of gases) scheme.

Colony growth is usually slow, with several days required for visible signs. For species identification, organisms recovered are Gram-stained to reveal their morphology

FIG. 27-2   GasPak 100 anaerobic system (self-contained, evacuation-replacement system). Anaerobic conditions are induced when water is added to GasPak envelope to form hydrogen, which reacts with atmospheric oxygen on palladium catalyst to form water. Carbon dioxide is also generated to produce growth of fastidious anaerobes. (Courtesy Bio-Quest, Division of Becton, Dickinson & Co., Cockeysville, Md.)

and subjected to selected biochemical reactions to show their metabolic patterns (all under oxygen-free gas, with the pH of reactions checked with a pH meter). Antibiotic susceptibility testing is done by a pour plate method.

Gas-liquid chromatography is valuable in the measurement of the acid and alcoholic by-products of anaerobic growth, and this technic has been a major asset in the identification of various anaerobic microbes. Only small amounts of bacterial growth are required for the emission of gaseous by-products, and the amounts and kinds of these gases register unique results like "fingerprints" that are diagnostic.

The Approach

A broad division of the total array of anaerobic microbes into two major categories on the basis of endospore formation is a convenient approach to the discussion of particular ones. The first such category would contain the spore-formers of a single genus, *Clostridium*, a fairly homogeneous group with several unifying features, a well-known and long-studied group, and one wherein the saprophytes exist in the external environment (although some are part of the intestinal flora of humans and animals). A fund of knowledge has accumulated on these, their patterns for clinical disease are classic for the most part, and in some instances a great deal has been accomplished in related areas of preventive medicine.

The second category—nonsporulating anaerobes—is not a straightforward one, gathering essentially all the rest. It is a more heterogeneous complex with few, if any, unifying features among the species other than their anaerobic specifications. No backlog of knowledge has accumulated on these; in fact, until recently information has been sketchy. Today technology makes the difference. (Technology, although designed for the demonstration of basic characteristics of both categories, has had a much greater impact for the nonspore-formers than the spore-formers.) The second category contains the microorganisms predominant in the internal environment of the human. They operate from a more strategic position in the body than do spore-formers. Their patterns for clinical disease are diffuse and diverse and range from minor superficial infections to septicemic dissemination.

Infections associated with anaerobes of whatever category are generally serious and carry a high mortality. Some of the more important genera and species now pass muster.

CLOSTRIDIUM SPECIES

Certain familiar sporulating anaerobes of the genus *Clostridium* (family Bacillaceae) are large bacilli peculiarly distorted by their heat-resistant spores and producing potent exotoxins and enzymes.

Clostridium tetani (the Bacillus of Tetanus)

The Disease

Tetanus, or lockjaw, is a disease worldwide in distribution, and globally, each year, an estimated 50,000 persons die of the disease. Manifestations in the central nervous system are produced by the potent exotoxin of *Clostridium tetani*. There are classic muscular spasms in tetanus involving the face, neck, and other parts of the body that are provoked by the slightest stimulation (noise, movement, or touch). These are very painful. Severe ones may compromise respiration. The name *lockjaw* comes from the muscular contractions that rigidly close or lock the jaws together. The corners of the mouth are turned up and the eyebrows peaked, producing a unique facial expression referred to as **risus sardonicus** (sardonic grin). Tetanus was known to Hippocrates and the ancients by the triad of wounding, lockjaw, and death (see box below).

Tetanus bacilli do not invade the body but remain at the site of infection, where they elaborate their powerful poison. The disease comes from infection of a wound and is not transmitted from person to person.

General Characteristics

The tetanus bacillus is an anaerobic spore-former. Biologically a saprophyte, it probably never infects a wound that does not contain some dead tissue. *C. tetani* stains with ordinary stains and is gram positive. The vegetative forms are slightly motile; the sporulating forms are nonmotile. The spore is situated in one end, giving the bacillus an appearance of a round-headed pin or drumstick. Growth is fairly luxuriant on all ordi-

---

Tetanus has been vividly described as follows:

[The spasms of tetanus] are characterized by a violent rigidity, usually sudden in onset but sometimes working up to a crescendo, with every single voluntary muscle in the body thrown into intense, painful tonic contraction. The eyes start, the jaw clenches, the tongue is bitten, the neck is retracted, the back arched, and opisthotonus is extreme. Often there is a muffled inspiratory cry, as the diaphragm contracts and draws air through the apposed vocal cords. Finally laryngeal spasm becomes complete, the chest fixed, and respiration ceases from muscle spasm. At the same time there is a gross outpouring of secretion, with profuse perspiration and foaming at the mouth.

Modified from Ablett, J.J.L.: Tetanus and the anaesthetist: review of symptomatology and recent advances in treatment, Br. J. Anaesth. 28:288, 1956.

Table 27-1 Bacteriologic Patterns of Clostridia

Organism	Growth in Litmus Milk	Growth in Cooked Meat	Gelatin Liquefaction	Synthesis of Lecithinase
Clostridium tetani	Soft clot	Gas, slow blackening	+ (blackening)	No
Clostridium perfringens	Stormy fermentation	Gas	+ (blackening)	Yes
Clostridium novyi	Acid, no clot	Gas	+	Yes
Clostridium septicum	Slow clot, gas	Gas	+ (gas)	Yes
Clostridium botulinum	Acid	Gas, blackening, digestion	+	Yes

nary media if anaerobic conditions are maintained (Table 27-1). The optimum temperature for growth is 37° C. In both cultures and infected wounds the presence of certain aerobic bacteria accelerates the growth of tetanus bacilli. Young cultures contain numerous vegetative bacilli, whereas old cultures are composed chiefly of sporulating organisms. The bacilli are seldom demonstrated in cultures from infected wounds because so few are present.

The vegetative bacilli are no more resistant to destructive influences than are other vegetative organisms. The spores, however, are very resistant. They withstand boiling for several minutes, pass through the intestinal canal unaffected, and when protected from the sunlight, remain infective for years.

Distribution

Tetanus bacilli are common residents of the superficial layers of soil. Since they normally inhabit the intestines of horses, cattle, and other herbivores, they are always found where manure is freely used as fertilizer, and barnyard soils are heavily contaminated with them. They are found in the intestinal canal of about 25% of human beings. Tetanus spores may be spread over a wide area by flies and high winds.

Extracellular Toxin

If it were not for their toxin, infection with tetanus bacilli would be without effect. Tetanus toxin is one of the most powerful poisons known, second only to that of botulism in potency. This explains why comparatively few bacilli at the infection site can induce such profound changes. **Tetanospasmin,** the toxin of tetanus, is a simple protein of a single antigenic type with a molecular weight of 68,000. It is a neurotoxin, that is, one with a specific affinity for the tissues of the central nervous system. From the infection site the toxin travels to the central nervous system along the axis cylinders of the motor nerves, where it is specifically and avidly bound to ganglioside in the spinal cord and brainstem. In the spinal cord, tetanospasmin interferes with inhibitory synaptic inputs on spinal motor neurons by blocking the release of the inhibitory neurotransmitter glycine. Since each motor neuron normally receives a balance of both stimulatory and inhibitory impulses from other neurons, loss of the inhibitory influences results in overstimulation of the neurons. The consequence is tonic spasms produced in the innervated muscles.

Besides tetanospasmin, tetanus bacilli elaborate small amounts of a toxic substance, **tetanolysin,** that destroys red blood cells and injures the heart. Taken orally, tetanus toxin is harmless.

Nitrate Reduction	Fermentation			Hemolytic	Spores	Motility
	Glucose	Lactose	Sucrose			
−	−	−	−	+	Round, terminal	+
+	+	+	+	+	Rare	−
−	+	−	−	+	Ovoid, eccentric	+
+	+	+	−	+	Ovoid, eccentric	+
−	+	−	−	±	Ovoid, eccentric	+

Pathogenicity

Tetanus in humans is fatal. Horses, cattle, sheep, and hogs may become infected.

Tetanus occurs in two clinical forms: one form is associated with a short incubation period (from 3 days to 3 weeks), an abrupt onset, severe manifestations, and a high mortality; the other is associated with a longer incubation period (from 4 to 5 weeks), less severe manifestations, and lower mortality. Rapidly fatal tetanus is more likely to follow wounds about the head and face, that is, near the brain.

Pathology

In tetanus pathologic changes are functional, *not* structural. The toxin of C. *tetani* acts directly on the central nervous system to originate the muscle spasms and convulsive seizures, and the impulses are transferred to the muscles by the motor nerves. No typical lesion exists, and even after death no organic lesions are seen at autopsy. The cerebrospinal fluid is normal.

Sources and Modes of Infection

Tetanus is practically always caused by spores introduced into a wound. Whether a given wound is complicated by tetanus depends on the type of wound, chance of contamination with tetanus spores, presence of dead tissue, and secondary infection. Deep puncture wounds are dangerous because they provide anaerobic conditions for growth of the bacilli. In lacerations, gunshot wounds, compound fractures, wounds containing foreign bodies, and infected wounds, the presence of dead tissue and certain other bacteria favors the growth of tetanus bacilli. If such wounds are contaminated by soil, especially heavily manured soil, the chances of tetanus infection are greatly increased. It is not surprising that war wounds are so easily complicated. Rusty nail wounds are a hazard, not because the nail is rusty but because rusty nails are usually dirty nails and tetanus spores are likely to be in the dirt.

In narcotic addicts, injection-related tetanus is a public health problem. The rapidly progressive and highly fatal disease in the drug addict is another, even more terrifying expression of tetanus. It is seen in the large metropolitan centers of the United States, especially in black women. The narcotic, usually heroin, has been given by the subcutaneous route of injection. Quinine, the adulterant, favors the growth of the bacilli by promoting anaerobic conditions in the tissues, enhancing the disease.

Puerperal tetanus exists in tropical countries. In these areas infection from bacilli inhabiting the intestinal canal sometimes complicates intestinal operations. Tetanus of the newborn, **tetanus neonatorum,** caused by infection through the navel is still seen in the Americas among black and Spanish-speaking groups with primitive medical care. Today in the United States most cases occur in persons with an average age over 50 years. Elderly persons are exposed to tetanus-prone injuries in their gardens. At the extremes of life, mortality is high.

Although tetanus is more likely to develop under conditions such as those enumerated, it may also follow wounds inflicted by apparently clean objects.

Prevention of Tetanus (pp. 422-424, 429, 438-440)

Any wound suggestive of the least danger from tetanus should be treated surgically. Puncture wounds should be widely opened and thoroughly cleaned to remove tetanus bacilli and other organisms and to allow access of oxygen, antagonistic to the growth of tetanus bacilli. Mangled wounds should be thoroughly cleaned and all dead tissue removed. Proper wound care cannot be overstressed, since tetanus organisms do not multiply in the surgically clean, aerobic wound with a good blood supply. Antibiotics help control associated pyogenic infections that damage tissues. *C. tetani* is sensitive to penicillin, but no amount of antibiotic has any effect on toxin released into the body from the wound site.

Tetanus is a *non*immunizing disease. No permanent or temporary active immunity develops in the patient who recovers. The amount of the very potent toxin needed to produce disease is too small to trigger the immune mechanisms that would prevent a second attack.

Active immunization for every member of the community cannot be too strongly recommended. In special need of protection are workers in industry or agriculture whose occupation predisposes them to wounds easily contaminated with tetanus bacilli, all allergy-prone individuals, and even recipients of a driver's license. All athletes must be fully immunized. (Football players are suffering more abrasions and more injuries from sliding and falling on artificial turf than on sod fields, although artificial turf is a less likely source of tetanus spores.)

Active immunization is practiced on a large scale in the armed forces of both the United States and England. Its effectiveness is proved by the fact that only 12 cases of tetanus appeared among 2,785,819 hospital admissions of military personnel for war wounds and other injuries during World War II. Of the 12 patients, 6 had not been adequately immunized and 2 failed to get the booster dose of toxoid at the time of injury.

Clostridium botulinum (the Bacillus of Botulism)

Nature of Botulism

Botulism is a specific intoxication caused by the toxin of *Clostridium botulinum*. The toxin, *not* the bacillus, produces disease, and like tetanus it is a poisoning, not an infection. The Centers for Disease Control classify botulism in four categories: (1) foodborne botulism, (2) infant botulism, (3) wound botulism, and (4) undetermined botulism.

The incubation period for all types of human botulism can be from a few hours to 8 days but is usually from 18 to 36 hours.

Foodborne botulism. Foodborne botulism, the most common form and the one usually referred to, is intoxication resulting from ingestion of food in which *C. botulinum* has grown and excreted its toxin. The foods most often linked to the disease are sausage (the disease derives its name from the Latin *botulus*, sausage), pork, and canned vegetables such as beans, peas, and asparagus. Cases have been traced to ripe olives, tuna, liver paste, salmon eggs, smoked whitefish products, tamales, spaghetti sauce, potato salad, and three-bean salad. Most of the foods incriminated have shared one feature: they were processed (improperly!) in the home by canning or pickling weeks or months before.

Foodborne disease develops as follows. Since the organism is widely distributed in nature, food is easily contaminated. If it is not canned or otherwise preserved properly, the spores are not destroyed. In the interval between preservation and use, the spores germinate and the bacilli multiply and excrete their toxin into the food. (Toxin remains potent in canned foods for 6 months or more.) If the contaminated food is not heated sufficiently for consumption, the toxin has not been inactivated. When such food is ingested, the preformed toxin is absorbed through the intestinal wall to exert its effects. Toxic signs usually appear within 24 to 48 hours and consist of generalized weakness, visual disturbances (often double vision), thickness of speech, nausea and vomiting, and difficulty in swallowing. There is no fever. Death from asphyxia usually occurs between the third and seventh days. The mortality ranges from 50% to 100%.

Natural food poisoning of this type exists among certain wild animals (if toxin is swallowed with their food). Examples are limberneck of chickens, fodder disease of horses, and duck sickness.

Infant botulism. Infant botulism, or the "floppy baby syndrome," was first recognized in 1976 as a form of botulism occurring in infants between 4 and 26 weeks of age. It is an intoxication caused *not* by ingestion of preformed toxin in food but by the absorption of toxin produced within the baby's own body. Sources for the disease are poorly understood; probably the baby swallows botulinal spores, which are everywhere in dust and dirt and are found on a variety of foods such as fresh fruits and vegetables, honey, and syrups. (The natural habitat of the microorganism is the soil.) The swallowed spores germinate, and the organisms colonize the infant's intestinal tract, multiply there, and form their toxin, which is absorbed. No one believes that spores or viable bacilli fed to adults induce botulism, but apparently the situation is different in babies. The first sign of disease is constipation. Lethargy, weakness, and feeding difficulty soon follow. Typically the infant shows an overall limpness or "floppiness." Progressive muscular involvement—a symmetric paralysis—may lead to respiratory failure and death.

Crucial to the prevention of this form of botulism is the thorough washing of anything a baby might put into or contact with the mouth—toys, food, surfaces of inanimate objects, even the baby's own fingers.

Sudden infant death syndrome (SIDS) (crib-death, cot-death, sudden unexpected death in infancy) designates the situation where an infant between 1 and 12 months dies suddenly without apparent reason. In the typical story, a well-cared for, healthy baby placed in bed at night is found dead the next morning. The baby died quietly during sleep with no outcry or signs of struggle. Autopsy demonstrates no cause of death. Infant botulism could be one important cause of this syndrome. Babies hospitalized for botulism rarely die, but if the intoxication were to develop rapidly, medical attention would not be obtained and the bacteria might produce enough toxin in the short interval of time to cause death.

Wound botulism. Wound botulism, the rarest form, comes from the absorption of botulinal toxin, usually type A or B, in an infected, traumatized wound. The characteristic picture of botulism without gastrointestinal manifestation develops.

Undetermined botulism. Undetermined botulism is the disease in persons older than 12 months in whom no food or wound source for toxin can be determined.

The Organism

The microbe C. *botulinum* is also an anaerobic, gram-positive, spore-forming bacillus (Table 27-1). A common inhabitant of the soil, the usual source from which foods are contaminated, C. *botulinum* is primarily a saprophyte. (Infections of laboratory animals have been caused experimentally.) Generally, the organism is unable to grow inside the body of a warm-blooded animal. C. *botulinum* and its toxin are destroyed in 10 minutes at the temperature of boiling water, but the "hard shell" spores must be held at a temperature of 120° C (249° F) for 15 minutes to be killed. Spores can withstand more than 2000 times the radiation lethal to a human being.

The Toxin

The toxin of C. *botulinum* is the most deadly of poisons, the mere tasting of food having been known to cause death. One ounce could exterminate all the people in the United States, and a mere half pound of botulinal toxin could wipe out the population of the world. (Botulinal toxin has been classified as an agent for chemical warfare.) It is 10,000 to 100,000 times more potent than diphtheria toxin or animal venoms. It differs from diphtheria and tetanus toxins in that it causes disease when swallowed and is more resistant to heat. A fast-acting neurotoxin, it affects the peripheral nervous system. The effect is specific at neuromuscular junctions. There, since toxin blocks the release of acetylcholine, neuronal activity halts and the innervated muscles are paralyzed. Muscles of vision, swallowing, and respiration are affected. With botulism the muscular paralysis is flaccid. Mental faculties are not impaired, and sensory disturbances do not coexist. Toxin also dilates blood vessels, and hemorrhages appear in different parts of the body.

There are six toxigenic types of C. *botulinum*: A, B, C, D, E, and F. Generally, one strain produces one type of toxin. All these toxins resist the intestinal juices, and the action of type E toxin can be increased 50 times by trypsin. Botulism almost always results from ingestion of toxin from types A, B, E, and F. C and D toxins cause disease in animals. Most A and B toxins are found in home-preserved vegetables and fruit (food of plant origin). Recent outbreaks of botulism have pointed to the presence of C. *botulinum*, type E, in water and marine wildlife. Therefore type E toxin may be found in fish and marine products. Type E spores can germinate at refrigerator temperatures and form toxin. Type F toxin has been reported in home-processed venison jerky. The pharmacologic action of the different specific toxins produced by the six toxigenic types of C. *botulinum* is the same in disease production.

There is good evidence that toxigenicity of clostridial species and even the type of toxin depend on the presence of specific bacteriophages, or bacterial viruses (lysogeny). For instance, if C. *botulinum*, type C, is "cured" of its phage, it loses its toxin. A startling fact is that experimentally the right phage can then convert it to a different clostridial species, also toxigenic and producing that specific toxin.

Laboratory Diagnosis

The laboratory diagnosis of botulism is made by (1) finding toxin in the patient's serum (toxemia may persist for prolonged periods), (2) isolating the microbes, and (3) identifying the toxin in food ingested. The mouse is a suitable test animal for inoculation. Being highly susceptible to the effects of the toxin, the mouse succumbs to even the small amounts of toxin in the blood sample received from the patient. Serum, gastric contents, and feces from a patient may be examined for toxin by mouse toxin–neutralization tests. Identification of the toxin as to type in suspected foods may be made by mouse tests. For type A toxin there is also a radioimmunoassay.

Prevention*

Prevention of botulism depends primarily on (1) heating foods to a temperature of 120° C for at least 15 minutes (at sea level) in the canning process (or steam cooking under adequate pressure, especially for low-acid foods) to destroy any spores present, (2) boiling canned foods for 15 minutes (at sea level) immediately before they are eaten to destroy the heat-labile toxins, and (3) properly refrigerating foods after cooking. In the canning of high-acid foods such as tomatoes and fruits, botulinal spores can be killed by a temperature of 100° C (boiling) for 15 minutes. For the processing of low-acid foods (beets, beans, corn, and meats) steam pressure methods are imperative. *Note the hazard for botulism at high altitudes—the boiling temperature is too low to destroy spores.*

---

*See p. 421 for passive immunization.

> *Do not so much as taste canned food until it has been fully heated. Remember: the poison is odorless and tasteless! The telltale signs of food spoilage (swollen container lid, cloudy or slimy appearance to food, putrid odor, and the like) may not be present!*

If fowls that have been eating discarded food develop limberneck, and if the responsible food can be traced, persons known to have eaten the same food should receive botulinal antitoxin. If a case develops in a group of people who have eaten the same food, the members should be given the antitoxin.

Round-the-clock emergency help—available antiserums, consultation, and laboratory assistance to establish the diagnosis—can be obtained from the U.S. Public Health Service Centers for Disease Control (Enteric Disease Branch) in Atlanta. (Call during the day 404-329-3753; at night, 404-329-3644.)

Clostridium perfringens, novyi, and *septicum* (the Clostridia of Gas Gangrene)

The Organisms

Gas gangrene (clostridial myonecrosis, clostridial myositis) is a highly fatal disease caused by the contamination of wounds with one or any combination of certain anaerobic, toxin-producing, spore-forming, gram-positive bacteria (Table 27-1). Almost invariably the infection is mixed. These microbes exist as saprophytes in the soil and as normal inhabitants of the intestinal canal of humans and animals. Infection ensues when soil contaminated by feces gets into a wound. Most important are *Clostridium perfringens, Clostridium novyi,* and *Clostridium septicum. Clostridium sporogenes,* sometimes considered nonpathogenic, *Clostridium histolyticum, Clostridium bifermentans,* and other clostridia are often present. The single outstanding one is *C. perfringens,* which is operative in 75% of all cases of gas gangrene and responsible for practically all those in civilian life.

C. perfringens is also known as Welch's bacillus, named for William Henry Welch (1850-1934), first dean of the medical school of Johns Hopkins University. Welch was well known for his original research in microbiology and pathology, especially for his studies (in 1892) of *C. perfringens* and its relation to gas gangrene.

In wartime gas gangrene complicates wounds 10 to 100 times more often than it does in peacetime and results from a variable combination of two or more species.

Aerobic pus-producing microbes and certain proteolytic organisms without effect on clean wounds often produce considerable destruction of tissue in wounds infected with gas bacilli. Aerobes in the wound facilitate the germination of clostridial spores and their vegetative growth by reducing the oxidation-reduction potential. In most cases of gas gangrene two or more clostridial anaerobes are associated with at least one predisposing aerobe. Generally, gas gangrene is brought about by aerobes and anaerobes working together.

The Disease

The organisms responsible for gas gangrene grow in the tissues of the wound, especially in muscle, releasing exotoxins and fermenting muscle sugars with such vigor that the pressure of accumulated gas tears the tissues apart. The air-filled (emphysematous) blebs of the wound give the name *gas gangrene* (Figs. 27-3 and 27-4). The exotoxins cause swelling and death of tissues locally, breakdown of red blood cells in the bloodstream, and damage to various organs over the body. The bacteria enter the blood just before death. Clinically there is a profound toxemia. The incubation period is 1 to 5 days.

C. perfringens elaborates several hemolysins and an extracellular proteolytic enzyme, collagenase, that facilitates spread of gas gangrene organisms through tissue spaces, since it is active against the fibrous protein supports of the body. *C. perfringens, novyi,* and *septicum* elaborate lecithinase; *C. histolyticum* and *sporogenes* do not. The potent lecithinase (the alpha-toxin) of *C. perfringens* is both hemolytic and necrotizing in its

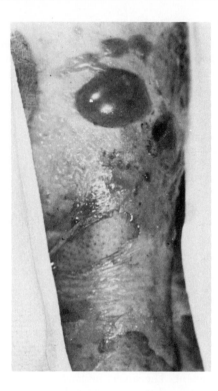

FIG. 27-3 Gas gangrene in 16-year-old boy, complicating compound fractures of leg sustained in motorcycle accident. The causative organism is *Clostridium perfringens*. Note blisters and discoloration. (From Altemeier, W.A., and Fullen, W.D.: J.A.M.A. **217**:806, 1972.)

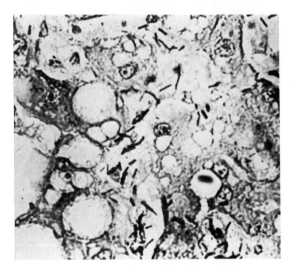

FIG. 27-4 Gas gangrene in Gram-stained tissue section of liver. Large air-filled spaces disrupt structure. Large gram-positive rods are present. The patient had leukemia. (Courtesy Dr. R.C. Reynolds, The University of Texas Health Science Center at Dallas.)

effects on tissues. The effect of no one single toxin is dominant in disease production; rather, the action of the different ones is additive.

Lacerated wounds, compound fractures, wounds with much dead tissue, war wounds, and injuries acquired in the vicinity of railroad tracks are likely sites of gas gangrene. Clostridial infection rarely complicates gangrenous appendicitis, strangulated hernia, and intestinal obstruction.

Skin samples taken from the thighs, groins, and buttocks of hospitalized patients sometimes yield a heavy growth of *C. perfringens*. To eliminate this kind of contamination, compresses of providone-iodine or 70% alcohol may be appplied. The presence of the organisms of gas gangrene in a wound does not necessarily mean gas gangrene.

Clostridial Infection and Malignancy

C. septicum is important in wartime gangrene but rarely complicates infection in otherwise healthy civilians. If the person's defenses are weakened, the situation may be reversed. This organism has acquired some notoriety by complicating cases of leukemia and intestinal cancer with clostridial septicemia.

Laboratory Diagnosis

Smears and cultures are used to advantage. When *C. perfringens* is responsible for a gaseous putrid discharge, a Gram-stained smear of material shows the presence of the large, gram-positive rods and the relative absence of neutrophils (Fig. 27-4). A differential medium such as lecithin-lactose agar containing neomycin and sodium azide to inhibit contaminating gram-negative bacteria allows species identification based on whether lactose is fermented or lecithinase is produced. If, at the same time, a plain blood agar plate is inoculated with test material, a significant nonclostridial organism will be picked up.

Prevention

Prevention of gas gangrene depends on proper surgical care of wounds. With established disease there must be a free incision to open the wound as widely as possible, all devitalized tissue must be excised, foreign bodies must be removed, and adequate drainage of the wound must be instituted. Gas gangrene polyvalent antitoxin has been prepared against the main organisms causing gas gangrene, but its efficacy today is challenged and many surgeons do not use it.

Clostridium difficile

Clostridium difficile, an obligate anaerobe and a typical spore-forming clostridial organism, is a normal constituent of the intestinal flora of infants up to 1 year of age but is rarely found in the adult. When the normal microbial flora is disturbed, as occurs with antibiotic therapy, this organism proliferates unchecked. It produces a cytopathic toxin that is directly implicated in the mediation of colitis secondary to antibiotic therapy. Counterimmunoelectrophoresis is an important tool for the identification of this toxigenic clostridium.

BACTEROIDES SPECIES

The family Bacteroidaceae contains the obligate anaerobes, the natural inhabitants of the natural cavities of humans, animals, and insects. (*Bacteroides fragilis* synthesizes vitamin K in the lower intestine.) The serious anaerobic infections in the hospital today come from the gram-negative rods of this family.

The Organisms

Bacteria of the genus *Bacteroides* are mostly nonmotile, nonspore-forming, usually gram-negative, very pleomorphic rods, with fastidious growth requirements in the laboratory. They reside in the colon, oral cavity, genital tract, and upper respiratory tract in humans. *Bacteroides* species make up 95% of the bacterial content of the stool and 20% by weight.

Pathogenicity

Outside their native haunts these bacteria are astounding pathogens, feared for the hemolytic and necrotizing (destructive) lesions of bacteroidosis. The expected portal of entry is a break in the lining of the gastrointestinal tract, the human female genital tract, an area of decubitus ulceration (bedsore), or a focus of gangrene. A notable feature of bacteroidosis is bacteremia. *Bacteroides* species have a peculiar affinity for venous channels, where they induce clotting (thrombosis), which in turn leads to tissue damage over an expanding front from impairment of the circulation. If the blood clots that are formed fragment and release infected bits of blood clot (emboli) into the circulation, abscesses form at locations remote from the primary process. Typical lesions of bacteroidosis are associated with the blood clots, foul-smelling gas, and liquefaction of the dead tissue.

Bacteroides species are often found in mucosal ulcers and foul-smelling abscesses of the lungs and other organs together with other microbes, and they are especially prone to complicate abdominal surgery, alcoholic liver disease, diabetes mellitus, and malignancy.

Specific Pathogens

Bacteroides fragilis and the related species, *Bacteroides thetaiotaomicron, ovatus, vulgatus,* and *distasonis,* represent the major bacterial component in the normal human intestinal tract. They cause most cases of bacteroidosis. A virulent capsular factor is postulated for *B. fragilis,* and the other species probably possess comparable material. *B. fragilis* is the type species of its genus and the single one most often isolated. *B. thetaiotaomicron* is next in frequency.

To a lesser extent *Bacteroides melaninogenicus* and the closely related *Bacteroides asaccharolyticus* are demonstrable human pathogens. *B. melaninogenicus* elaborates a melanin-like pigment that makes its colonies brown or black. Its colonies also give off a red fluorescence when viewed under ultraviolet light. *B. melaninogenicus* infections are less virulent and fulminant than those of *B. fragilis.*

FUSOBACTERIUM SPECIES

Members of the genus *Fusobacterium* (also in the family Bacteroidaceae), typically long, slender, spear-shaped bacilli with tapered ends, are incriminated in suppurative and gangrenous lesions. The species of note include *Fusobacterium nucleatum,* the type species, and *Fusobacterium necrophorum* (*necrophorum* means "necrosis producing").

OTHER ANAEROBES

The significance of many anaerobes is increasing in relation to disease. Mention is made of only a few being incriminated with greater regularity.

Gram-positive cocci of the genera *Peptococcus* (anaerobic staphylococci) and *Peptostreptococcus* (anaerobic streptococci) are important in anaerobic infection of the female genitalia. Small gram-negative, capnophilic cocci of the genus *Veillonella,* found ordinarily in the mouth and in dental abscesses, possess endotoxins. Species of gram-positive, nonspore-forming bacilli, notable for their output of organic acids, include *Propionibacterium acnes* and *Eubacterium limosum. Propionibacterium acnes,* based on the skin, fosters the development of acne and is a troublesome contaminant of blood cultures. *Eubacterium limosum* resides in the intestinal tract and is known to synthesize vitamin B_{12}.

SUMMARY

1 Anaerobes must have a low or zero oxygen tension for continued growth. To anaerobes, oxygen and oxidizing agents are either inhibitory or toxic.

2 The majority of anerobes are indigenous to human beings and are the predominant component of the resident microflora in the large bowel and the oral cavity. Although ordinarily restrained, these microbes can produce severe disease. Most are destructive pathogens; their type lesion is tissue necrosis.

3 Special collection and transport methods are necessary for laboratory specimens containing anaerobes, and cultural methods use oxygen-free environments, such as the anaerobic glove box or the anaerobic jar system. Gas-liquid chromatography "fingerprints" anaerobes, since the emission of their gaseous by-products is diagnostic.

4 Anaerobic spore-formers are found in a single genus, *Clostridium*. Members of the genus *Clostridium* exist in the external environment and also in the intestinal flora of humans and animals. They are large bacilli that are peculiarly distorted by their heat-resistant spores. Two clostridial exotoxins, the tetanus and the botulism exotoxins, are two of the most potent poisons known.

5 The fatal disease tetanus, or lockjaw, is caused by the effects of the potent neurotoxic exotoxin of *Clostridium tetani*. Tetanus is well known for the classic manifestations in the central nervous system. Tetanus is a nonimmunizing disease.

6 Tetanus bacilli are found where manure is freely used as fertilizer, and tetanus spores are very resistant. Tetanus bacilli probably never infect a wound that does not contain some dead tissue. Spores are introduced into a wound. Deep puncture wounds are dangerous, since they provide the anaerobic conditions for sporulation.

7 The effect of the potent botulinal exotoxin is botulism, a disease that is also characterized by central nervous system disturbances. Four kinds are recognized: (1) foodborne, the most common form, (2) infant, (3) wound, and (4) undetermined. Foodborne botulism results from ingestion of food in which bacilli have grown and excreted their toxin. Sources for infant botulism, or "floppy baby syndrome," are poorly understood. *Clostridium botulinum* is a common inhabitant of the soil, the usual source of contamination for food.

8 Phage infection, or lysogeny, of clostridial species determines toxigenicity.

9 Prevention of botulism depends primarily on heating foods adequately in the canning process and in properly refrigerating them.

10 Clostridial gas gangrene is a dramatic type of infection of a wound that has been contaminated by soil laden with feces. The pressure of gas accumulated in the infection tears the muscles and tissues apart. Gas gangrene, although 10 to 100 times more prevalent in wartime, may nevertheless complicate severe trauma in peacetime.

11 *Bacteroides* species, normally present in the natural cavities of humans, lower animals, and insects, are mostly nonmotile, pleomorphic rods with fastidious growth requirements in the laboratory. *Bacteroides* species have a peculiar affinity for veins, and the lesions of bacteroidosis are characterized by the formation of blood clots in veins, presence of foul-smelling gas, and liquefaction of dead tissue.

QUESTIONS FOR REVIEW

1 Define anaerobes.
2 Discuss anaerobes as organisms indigenous to humans.
3 How do anaerobes produce disease? Give four factors relating to their pathogenicity.
4 List diseases produced by anaerobes.
5 Outline briefly the laboratory diagnosis of anaerobic infections.
6 What has been the effect of technologic developments in anaerobic microbiology?
7 Name and describe the causative agents for tetanus, gas gangrene, and botulism.
8 What is the derivation of the word *botulism?*
9 What is the effect of tetanus toxin? Of diphtheria toxin? Of botulinal toxin? Compare the three as to potency.
10 List the types of wounds likely to be contaminated with *Clostridium tetani*.
11 What types of wounds are most likely infected with clostridia of gas gangrene?
12 What measures other than immunization help to prevent tetanus and gas gangrene?
13 Briefly describe tetanus.

14 What are the circumstances producing botulism? What are the main preventive measures?

15 Why is botulism a risk at high altitudes?

16 Briefly characterize *Bacteroides* species. What is their importance today in hospital infections?

SUGGESTED READINGS

Arnon, S.S.: Infant botulism, Ann. Rev. Med. **31**:541, 1980.

Bartlett, J.G.: Antibiotic-associated colitis, Clin. Gastroenterol. **8**:783, 1979.

Berry, P.L., and others: The use of an anaerobic incubator for the isolation of anaerobes from clinical samples, J. Clin. Pathol. **35**:1158, 1982.

Brook, I.: Anaerobic bacteria in pediatric respiratory infection: progress in diagnosis and treatment, South. Med. J. **74**:719, 1981.

Dellinger, E.P.: Severe necrotizing soft-tissue infections, multiple disease entities requiring a common approach, J.A.M.A. **246**:1717, 1981.

Dowell, V.R., Jr., Infant botulism: new guise for an old disease, Hospital Pract. **13**:67, Oct. 1978.

Finegold, S.M., and Sutter, V.L.; Anaerobic infections, a Scope publication, Kalamazoo, Mich., 1982, The Upjohn Co.

Finegold, S.M., and others: Cumitech 5: Practical anaerobic bacteriology, Washington, D.C., 1977, American Society for Microbiology.

Killgore, G.E., and others: Comparison of three anaerobic systems for the isolation of anaerobic bacteria from clinical specimens, Am. J. Clin. Pathol. **59**:552, 1973.

Kurzynski, T.A., and others: The use of CIE for the detection of *Clostridium difficile* toxin in stool filtrates: laboratory and clinical correlation, Am. J. Clin. Pathol. **79**:370, 1983.

Ladas, S., and others: Rapid diagnosis of anaerobic infections by gas-liquid chromatography, J. Clin. Pathol. **32**:1163, 1979.

Perry, J.L., and Plunkett, C.B.: An improved roll-tube culture system for anaerobes, Lab. Med. **10**:26, Jan. 1979.

Philip, A.: Infant botulism. Lab. Manage. **19**:53, Aug. 1981.

Reddy, D., and others: Nitrite inhibition of *Clostridium botulinum*: electron spin resonance detection of iron–nitric oxide complexes, Science **221**:769, 1983.

Watt, B., and others: Gas-liquid chromatography in the diagnosis of anaerobic infections: a 3 year experience, J. Clin. Pathol. **35**:709, 1982.

Wound botulism associated with parenteral cocaine abuse—New York City, Morbid. Mortal. Weekly Rep. **31**:87, 1982.

CHAPTER 28

CORYNEBACTERIA AND ACTINOMYCETES

Corynebacteria
    Corynebacterium diphtheriae (the
      Bacillus of Diphtheria)

Actinomycetes
    Actinomycosis
    Nocardiosis
    Mycetoma

CORYNEBACTERIA

The term **coryneform** is a working concept to sidestep for the time being unresolved problems in classification. The genus *Corynebacterium* (club bacteria) includes one dreaded pathogen—*Corynebacterium diphtheriae*, the type species.

Corynebacterium diphtheriae (the Bacillus of Diphtheria)

Many decades ago diphtheria, or "membranous croup" as it was called, was a major cause of death. The word *diphtheria* is derived from a Greek word meaning leather. The disease was so named because of the leathery consistency of the diphtheritic membrane. Since the causative organism, *Corynebacterium diphtheriae*, was first discovered by Edwin Klebs (1834-1913) in 1883, the epidemiology has become known, methods of producing a permanent immunity have been devised, and the mortality has been reduced from nearly 50% to a level such that death seldom occurs in patients who are adequately treated *early* in the disease. Few diseases have been so well studied and are today so well understood as diphtheria.

The genus *Corynebacterium* contains gram-positive, unevenly staining bacteria with clubbed or pointed ends. (*Korynē* signifies club.) *C. diphtheriae* is often called Klebs-Löffler bacillus because it was identified by Klebs and first grown in 1884 in pure culture by F.A.J. Löffler (1852-1915).

General Characteristics

C. diphtheriae is unique for its variation in size, shape, and appearance (pleomorphism). The bacteria may be straight or curved and swollen in the middle or clubbed at one or both ends. When division occurs, the bacteria remain attached at the point of separation in a V-shaped pattern (snapping). Stained organisms have a granular, solid, or barred appearance, and deeply staining granules (methachromatic granules) are distinctive. Granules at the ends of the bacilli are known as **polar bodies.**

C. diphtheriae is a gram-positive, nonmotile, strictly aerobic microbe that does not form spores. It grows best at 35° C on almost all ordinary media, but growth is luxuriant on Löffler (coagulated serum) or Pai (coagulated egg) media or on media containing potassium tellurite. Potassium tellurite inhibits the growth of many organisms found in the throat, and colonies of diphtheria bacilli assume a typical grayish black or gunmetal gray on media containing it. The morphologic features of the organisms are best preserved on Löffler's medium.

Based on cultural characteristics, fermentation tests, and immunologic reactions,

C. diphtheriae has been divided into three biochemical types: **gravis** (meaning severe), **intermedius** (meaning intermediate severity), and **mitis** (meaning mild). The existence of these three types is useful in epidemiologic studies. Epidemics are most often the *gravis* type, and its manifestations are the most severe.

For humans and some laboratory animals *C. diphtheriae* is highly pathogenic. In nature the disease is restricted to humans. An animal reservoir does not exist.

Diphtheria bacilli are fairly resistant to drying but easily destroyed by heat and by chemical disinfectants. Boiling destroys them in 1 minute. In bits of diphtheritic membrane they remain alive for several weeks.

Extracellular Toxin

Diphtheria bacilli owe their pathogenicity to their extracellular toxin, an exotoxin comparable in potency to the toxins produced by the sporulating anaerobes. Diphtheria bacilli are the only aerobes with this capability. In the production of disease, diphtheria bacilli do *not* invade tissues but grow superficially in restricted areas, usually on a mucous membrane. The exotoxin formed is released through an epithelial break at the site of infection into the bloodstream and circulated. As a result of the primary action of toxin, which is to block protein synthesis in and thereby injure all eukaryotic cells, systemic disturbances ensue and degenerative changes in organs of the body appear. Toxin affects certain nerves, the heart muscle, the kidneys, and the cortex of the adrenal gland. The features of diphtheria relate directly to the exotoxin; that is, diphtheria is a toxemia and a molecular disease.

In 1951 the remarkable discovery was made that exotoxin can be elaborated only by lysogenic strains of *C. diphtheriae*, that is, strains infected with certain bacteriophages (bacterial viruses) carrying the *tox* gene. The *tox* gene is essential neither for the phage carrying it nor for the bacterial host receiving it. If the specific phage is lost to the bacterium, the quality of toxigenicity disappears also, and conversely a nontoxigenic strain may be converted to a lysogenic and therefore toxigenic strain if treated (infected) with proper phage. The *tox* gene programs the structure of exotoxin, but biosynthesis is enacted by the bacterial cell. The amount of inorganic iron in the external and internal milieu is a critical factor regulating toxin production, which is depressed until that level of iron is critically reduced.

Soluble exotoxin released into media is concentrated to yield the diphtheria toxin of commerce. Such preparations contain 200 to 1000 MLD (minimum lethal dose) (p. 420) per milliliter. The toxin can be separated from the medium and purified. It deteriorates with age and is destroyed by a temperature of 60° C. Chemically exotoxin is a heat-labile, single polypeptide chain, the molecular weight of which is around 63,000.

Virulence Tests

Not all diphtheria bacilli produce toxin, and toxigenic organisms cannot be distinguished from nontoxigenic ones by microscopic appearance or cultural characteristics. Differentiation can be made by animal inoculation. A small amount of a liquid culture of the bacilli is injected either beneath or between the superficial layers of the skin of two guinea pigs, *only one* of which has been protected by a dose of diphtheria antitoxin. If the bacilli are toxigenic, a zone of inflammation appears at the inoculation site in the *un*protected pig, but there is no reaction at the site in the protected pig. If the bacilli do not produce toxin, neither pig shows a reaction. This is the **guinea pig virulence test.** Virulence tests may also be carried out on rabbits. The technic differs somewhat from that in guinea pigs, but the underlying principle is the same.

The **in vitro gel diffusion test** is a cultural method devised to determine the virulence of diphtheria bacilli. Diphtheria bacilli grown on an agar surface are brought into contact with antitoxin, which diffuses out from a well or trench in the medium or from a piece of filter paper soaked with it. Toxin, if produced, diffuses from the bacterial growth and forms a line of specific precipitate where it meets with antitoxin.

Pathogenicity

Diphtheria is an acute infectious disease of human beings induced by the extracellular toxin of *C. diphtheriae.* The recognition unit of the toxin relates to specific receptors found only on the cells of human beings. A characteristic inflammatory change takes place at the site of infection, and systemic disturbances accompany it. The incubation period is usually 2 to 6 days.

Pathologically the type lesion of diphtheria is the **pseudomembrane,** a superficial lesion occurring on mucous membranes (Fig. 28-1). The first stage in its formation is degeneration of the epithelial cells of the affected area. This is followed by an abundant fibrinous exudation onto the surface. As the fibrin precipitates, it entraps leukocytes, red blood cells, bacteria, and dead epithelial cells to form a thick, tough membranelike structure anchored to the underlying tissues. If the pseudomembrane is pulled off, a raw bleeding surface remains, but a new pseudomembrane soon forms. In the absence of antitoxin it persists for 7 to 10 days and then disappears. It may obstruct breathing, and in some patients tracheotomy or intubation is required to prevent suffocation. Organisms other than *C. diphtheriae* may produce pseudomembranes, and in a few patients with diphtheria, pseudomembranes do not form.

The most frequent sites for pseudomembrane formation are the tonsils, pharynx, larynx, and nasal passages. The diphtheritic membrane usually begins on one or both tonsils and spreads to the uvula and soft palate. Less often, diphtheria attacks the vulva, conjunctiva (ocular diphtheria), middle ear, umbilical cord (diphtheria neonatorum), and skin and infects wounds.

Diphtheria occurs in three clinical patterns: (1) **faucial,** in which the membrane appears on the tonsils and spreads to other parts of the pharynx; (2) **laryngeal,** in which the membrane may easily cause suffocation; and (3) **nasal,** in which the membrane is rarely associated with severe disease because toxin is poorly absorbed by the lining of the nose.

Bronchopneumonia is an important complication, but the really serious effects of diphtheria stem from the action of the toxin. The damage done to the heart may precipitate heart failure, even after other manifestations of the disease have subsided and sometimes in comparatively mild cases. Sudden death has been reported several weeks after apparent recovery. Degeneration of peripheral nerves induced by toxin leads to late paralysis, particularly of the soft palate.

Sources and Modes of Infection

The sources of infection in diphtheria are persons with clinical cases, persons with mild undetected cases, and carriers. Ordinarily the bacteria enter and leave the body by the same route (the mouth and nose). They may be transferred directly from person to person by droplets expelled from the mouth and nose or indirectly by cups, toys, pencils, dishes, and eating utensils contaminated by buccal or nasal secretions. The direct method of transfer is the more important.

Nasal diphtheria is an important source of infection because it is often overlooked. Diphtheria bacilli can be spread from skin infections appearing to be less serious ailments. In epidemics such skin sources figure significantly.

Skin infections were prevalent among soldiers during World War II, especially those in the South and Central Pacific areas. Today cutaneous diphtheria seems prevalent in children in rural areas and in both children and adults who live in crowded, unhygienic environments. In fact, cutaneous diphtheria in indigent adults ("skid-row" diphtheria)

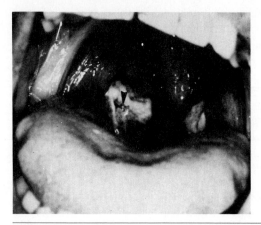

FIG. 28-1   Diphtheria in 43-year-old man; fourth day of disease. Arrow is over the membrane in tonsillar area of the throat. (Tongue is in the foreground.) (From McCloskey, R.V., and others: Ann. Intern. Med. 75:495, 1971.)

Table 28-1 Bacteriologic Diagnosis of Diphtheria

Organism	Morphology	Growth on Blood Tellurite	Hydrolysis of Urea	Reduction of Nitrate	Trehalose	Glucose	Maltose	Sucrose	Toxigenicity
					Fermentation of Carbohydrates				
Corynebacterium diphtheriae	Pleomorphic	Gray to black colonies	–	+	–	+	+	–	+
Corynebacterium pseudo-diphtheriticum (example of a diphtheroid)	More uniform	Opaque grayish	+	+	–	–	–	–	–

is an important reservoir. A few milkborne epidemics have been reported, usually caused by contamination of the milk by a person working in the dairy who either was a carrier or had a mild case of the disease. The hands transferred the buccal or nasal discharges. The notion that cats convey diphtheria is not true.

Over the last decade or so in this country diphtheria has continued to occur in the economically depressed areas of the South and Southwest, where living conditions are crowded, medical care lacking, hygiene poor, and immunization inadequate.

Diphtheria carriers. In about half of patients with diphtheria the bacilli leave the body within 3 days after the membrane disappears, and in four fifths of patients the bacilli have disappeared within 1 week. Sometimes they persist for a longer time, and the patient becomes a carrier. Also, certain contacts of a diphtheria patient or a carrier become carriers themselves without contracting the disease. An estimated 0.1% to 0.5% of the population carry virulent diphtheria bacilli. The percentage increases during epidemics and in crowded communities during cold weather.

Most carriers harbor the bacilli a short time only (from a few days to a few weeks), but a few harbor them permanently even with intensive treatment. Because a high percentage of organisms with the morphology and cultural features of diphtheria bacilli in normal throats are nontoxigenic and therefore not dangerous, virulence tests should always be done on diphtheria bacilli from suspected carriers. If a suitable antibiotic such as penicillin or erythromycin is given in conjunction with antitoxin during the acute stage of the disease and continued during convalescence, the carrier rate is reduced. In the treatment of diphtheria, antibiotics do not constitute the specific therapy but adjunctive or accessory measures, since penicillin does inhibit the growth of the bacilli, thereby influencing toxin production. Erythromycin is of value in the prevention or elimination of the carrier state.

Bacteriologic Diagnosis

The laboratory diagnosis of diphtheria is made by culture of the organism (Table 28-1). Some of the pseudomembrane is removed with a sterile swab; a Löffler slant or potassium tellurite medium is inoculated and incubated from 12 to 14 hours. Smears made from the culture are stained with methylene blue or one of the special stains for diphtheria bacilli. The bacilli sometimes may be found in smears made directly from the pseudomembrane, but failure to find them in no manner indicates that the patient does not have the disease. Antiseptics, gargles, or mouthwashes used before cultures are taken may prevent proper growth of bacteria, and antibiotics given 5 to 7 days before may inhibit bacterial growth. If possible, cultures should be taken to bring the swab in contact only with the site of disease, and cultures should be made from both throat and nose.

In a patient with ulcerative or membranous inflammation of the throat, both cultures for *C. diphtheriae* and smears for the organisms of Vincent's angina should be made. The two conditions may be easily confused. If only a culture is made, an infection with the organisms of Vincent's angina could be missed because these organisms do not grow in cultures, and diphtheria could be missed if smears alone are examined. Moreover the two diseases can coexist.

The finding of diphtheria bacilli in the throat does not necessarily mean that the patient has diphtheria; that person may be a carrier. Remember that membranous infections of the throat may be caused by organisms other than *C. diphtheriae* and that nonmembranous infections with *C. diphtheriae* occasionally occur. If streptococcal infection is suspected, a blood agar plate is inoculated.

Diphtheroid bacilli. The heterogeneous diphtheroid bacilli bear a close microscopic resemblance to *C. diphtheriae* and are sometimes confused with it on throat smears. They are abundant and have been isolated from varied sources such as the skin, nose, throat, urethra, bladder, vagina, and prostate gland. They reside in soil and water as saprophytes, do not produce toxins, and are nonpathogenic for humans except under exceptional circumstances. They have been reported as rare causes of wound infections, meningitis, osteomyelitis, and hepatitis. Like many other normally saprophytic microbes, they too can cause serious illness in the patient whose resistance mechanisms are impaired.

One member of the genus, *Corynebacterium parvum*, has been demonstrated to be an effective immunotherapeutic agent, stimulating the immune system to produce macrophages and augmenting the function of lymphocytes.

Immunity

Immunity to diphtheria results from the presence of diphtheria antitoxin in the blood. Newborn babies of immune mothers receive a passive immunity because of the transfer of antitoxin from the maternal to the fetal circulation across the placenta. In breast-fed infants this immunity is augmented by antibodies in the milk of the mother. This immunity is usually lost by the end of the first year, and from this time to the sixth year most children are susceptible. Minor infections reestablish an immunity during late childhood.

Possibly 50% of adults are immune, a figure once considerably higher. The reduction in the incidence of diphtheria, the decreased likelihood of minor infections, and the institution of vaccination during childhood (which does not give the permanent immunity that repeated minor infections do) have meant a decrease in adult immunity. The presence of from $\frac{1}{500}$ to $\frac{1}{250}$ unit of diphtheria antitoxin per milliliter of blood renders a person immune.

The **Schick test** is purported to determine whether a person has sufficient diphtheria antitoxin in his blood for immunity. One-fiftieth MLD of diphtheria toxin (0.1 ml of diluted toxin) is injected into the skin of one arm and a control dose of toxoid into the other. The skin sites on both arms are read 4 days later. In principle, if the subject's blood contains sufficient antitoxin for protection against diphtheria, no reaction occurs (negative test). An insufficient amount of antitoxin (positive test) is indicated by the appearance within 24 to 36 hours of a firm red area, 1 to 2 cm in diameter, persisting 4 or 5 days. In the past the Schick test has been used to detect persons lacking immunity and to determine the efficacy of active immunization once given. Since the highly purified, adult-type toxoids have become available, the need for routine Schick testing has been largely eliminated. Active immunization is desirable, and difficulties encountered with the Schick test can be bypassed.

An attack of diphtheria is usually followed by a fair degree of immunity, which may be temporary or permanent, but in a few patients immunity does not seem to develop. Most carriers of virulent diphtheria bacilli are immune to the toxin.

Prevention and Control of Diphtheria*

All persons ill with diphtheria should be isolated, and neither they nor their close contacts should be released until it is proved that they harbor no virulent diphtheria bacilli in their nose and throat. The mouth secretions and all objects so contaminated should be disinfected. The patient's eating utensils should be boiled. Nurses must be careful that they do not contaminate their hands with the mouth and nose secretions of patients so as to infect themselves. Persons known or presumed susceptible who contact a diphtheria patient should receive diphtheria toxoid (for active immunization) and an-

---

*For immunization in diphtheria, see pp. 421, 422, 429, and 438-440. For disinfection, see p. 273.

tibiotics. If they cannot be seen daily by the physician, they may be temporarily immunized with 10,000 units of diphtheria antitoxin intramuscularly. The administration of antitoxin can set the stage for an anaphylactic reaction should another agent containing horse serum be given to that person.

The general measures to reduce the incidence of diphtheria are (1) detection and treatment of carriers and (2) production of an active immunity in all susceptible persons—children under 6 years and all older children and adults, especially physicians and nurses not previously immunized.

ACTINOMYCETES

Actinomycetes are now classified as bacteria. For a long time these prokaryotic microbes, because of their similarities to both the true bacteria and the true fungi, were thought of as intermediates between the two, were sometimes called "higher bacteria," but were usually included in a discussion of medical mycology, mostly because actinomycetes are found in the same clinical specimens as fungi and produce similar diseases.

The order Actinomycetales contains gram-positive, sometimes acid-fast, mostly aerobic bacteria forming branching filaments that in some of the families develop into a mycelium. Some members are pathogens of humans, animals, and plants. In this classification of funguslike bacteria there are four families of note. The family Mycobacteriaceae comprises the genus *Mycobacterium* of acid-alcohol–fast microbes, of which *Mycobacterium tuberculosis* is the type species (Chapter 29). (We do not, however, refer to mycobacteria as actinomycetes.)

The family Streptomycetaceae is medically significant because the actinomycetes of genus *Streptomyces* provide many important antibiotics. Industrial microbiologists continue to show that streptomycetes elaborate a remarkable variety of metabolic by-products with varied and dramatic applications.

Two families of actinomycetes, each with an important genus, are Actinomycetaceae of nonacid-fast, diphtheroid bacteria with no mycelium and Nocardiaceae of variably acid-fast organisms with variable mycelial development. The genera are *Actinomyces* and *Nocardia*, respectively, and the diseases, actinomycosis and nocardiosis. Species of the genus *Actinomyces* are anaerobic or microaerophilic, whereas species of the genus *Nocardia* are aerobic. The acid-fast *Nocardia* species may be confused with *Mycobacterium tuberculosis*.

The actinomycetes (including streptomycetes) resemble fungi in that they may have a mycelium of masses of branched filaments, but their "hyphae" are much more slender than those of true fungi. The filaments of actinomycetes fragment into spherical or rod-shaped segments that function as spores and in turn develop into new hyphae. Spore formation comparable to that in bacteria is seen in the hyphae of the genus *Streptomyces*, wherein development of a mycelium is complete. Unlike fungi, actinomycetes lack a nuclear membrane.

Actinomycetes are widely distributed in nature and play a vital part in changes in organic material of the soil. This activity is more crucial to humans than any disease-producing capacity.

Actinomycosis

Actinomycosis is an infectious disease of lower animals (especially cattle) and humans caused by several species of bacteria belonging to the genus *Actinomyces*. Of these, *Actinomyces bovis* (bovine actinomycosis) and *Actinomyces israelii* (human actinomycosis) are the most important. Actinomycosis is typified clinically by formation of nodular swellings that soften and form abscesses, discharging a thin pus through multiple sinuses. The disease in cattle is known as *lumpy jaw*. Human disease follows three patterns: (1) cervicofacial, (2) thoracic, and (3) abdominal. The cervicofacial type is described by swelling and suppuration of the soft tissues of the face, jaw, and neck (Fig. 28-2). The usual type (about half the cases) and the least dangerous, it appears to have a special association with dental defects. The thoracic type is characterized by multiple small cavities and abscesses in the lungs. The abdominal type usually begins about the appendix or cecum. In advanced thoracic and abdominal disease, sinus tracts extend to the surface.

Pelvic inflammation as a complication to the presence of an intrauterine contraceptive device (IUD) may be produced by actinomycetes.

Laboratory Diagnosis

The etiologic agent is found in the pus or in the walls of abscesses as small, yellow **sulfur granules** about the size of a pinhead. When a sulfur granule is placed on a microslide and a cover glass pressed down on it, a distinct microscopic picture is seen—a central threadlike mass from which radiate many clublike structures (Fig. 28-3). For this reason *Actinomyces* species have been called ray fungi. The clubbed appearance is not pronounced in cultures.

The laboratory diagnosis is made by the demonstration of the organisms. They may be identified in the sulfur granules, which represent microcolonies in the exudate, or in sections of tissue taken from the lesions. Cultural methods help, but serologic and skin tests contribute little.

Sources and Modes of Infection

Pathogenic actinomycetes are normal inhabitants of the mouth of cattle and humans, probably existing in an attenuated state. No evidence exists that the organisms are saprophytic outside the animal body. Infection occurs in the event of injury to the mouth, tooth decay, or some other abnormal state. Such conditions favor invasion of the tissues. No evidence suggests direct transmission from animal to animal or from animal to human. People who pursue nonagricultural occupations are as likely to contract actinomycosis as are farmers and stockmen.

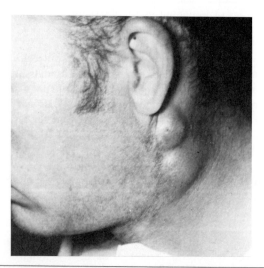

FIG. 28-2 Cerviofacial actinomycosis. Note two nodular swellings on side of neck; these represent actinomycotic abscesses. (Courtesy Dr. J.D. Nelson, The University of Texas Health Science Center at Dallas.)

Nocardiosis

Nocardia asteroides, an aerobic actinomycete found free in nature, is responsible for the human infection nocardiosis, which is usually a primary disease of the lungs (pulmonary nocardiosis) simulating tuberculosis and commonly found in the immunocompromised host. Bloodstream dissemination of infection to the rest of the body leads to abscesses in subcutaneous tissues and other internal organs. Brain abscess commonly complicates systemic infection. Nocardiosis (like cryptococcosis), with an affinity for the lungs, accompanies malignant diseases of the macrophage system— Hodgkin's disease, lymphosarcoma, and leukemia. Steroid hormones enhance the infection.

Mycetoma

Mycetoma is a generic name for a localized but destructive infection involving skin, subcutaneous tissue, bone, and fascia, usually of the foot and especially prevalent in the tropics. An alternate term, **Madura foot,** indicates the predilection of the disorder for the foot (Fig. 28-4). Mycetoma may be actinomycotic if caused by actinomycetes or maduromycotic if certain fungi are operative. *Nocardia asteroides* is one important cause of actinomycotic mycetoma. Eumycotic mycetoma, or maduromycosis, is caused by *Allescheria (Petriellidium) boydii* or certain other fungi. *Allescheria boydii* is a pathogenic ascomycete whose natural habitat is the soil. Mycetoma is related to the custom of walking barefoot and is an occupational hazard to farmers and field-workers. After initial injury the etiologic agents invade tissues to produce in time a network of interlocking abscesses and granulomas in the substance of the foot. End-stage disease may necessitate amputation. The organisms are identified in the granules or grains found in the purulent drainages from the sinus tracts.

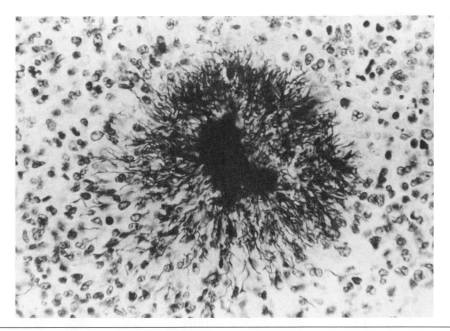

FIG. 28-3   *Actinomyces* sulfur granule. A filamentous colony (central tangled mass) of the microbes is surrounded by many pus cells (polymorphonuclear neutrophils) depicted by the many small irregular black nuclei in this microscopic field of pus. Note the stringy processes at the periphery of the colony.

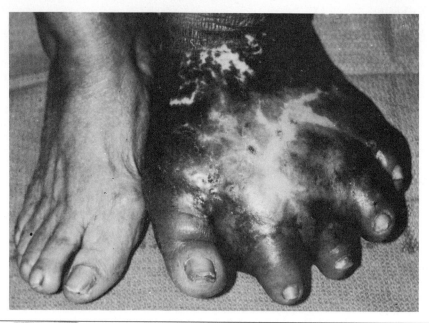

FIG. 28-4   Madura foot (mycetoma, maduromycosis of foot).

SUMMARY

1 The genus *Corynebacterium* of gram-positive, unevenly staining bacteria with clubbed or pointed ends contains the nonmotile, strictly aerobic *Corynebacterium diphtheriae*, the cause of the molecular disease diphtheria.

2 Phage-infected bacteria growing in the diphtheritic pseudomembrane on a mucous membrane release their potent exotoxin to be absorbed by the host. Since the primary action of the toxin is to block protein synthesis, all eukaryotic cells are vulnerable to its action. The serious effects of diphtheria toxin stem from the action of the toxin on the heart.

3 Not all diphtheria bacilli contain the *tox* gene, which structures the exotoxin. Virulence tests distinguish between those that do and those that do not.

4 Sources of infection are persons with diphtheria, either apparent or undetected, and carriers. No animal reservoir exists.

5 Actinomycetes, now classified as bacteria, are gram-positive, sometimes acid-fast, mostly aerobic bacteria forming branching filaments that in some families become a mycelium. Examples are the genera *Actinomyces* and *Nocardia*.

6 *Actinomyces* species are nonacid-fast, anaerobic or microaerophilic, diphtheroid bacteria with no mycelium. They cause actinomycosis, an infection of lower animals, particularly cattle, which may also occur in humans.

7 *Nocardia* species, the cause of nocardiosis, are variably acid-fast aerobic bacteria with variable mycelial development.

8 Nocardiosis is usually primary disease of the lungs, simulating tuberculosis and common in the compromised host.

9 Madura foot (mycetoma) is a generic name for a localized but destructive infection of the foot that is prevalent in the tropics. Several bacteria and certain fungi may cause this disorder.

QUESTIONS FOR REVIEW

1 What are actinomycetes? Their importance?
2 Briefly indicate what is meant by actinomycosis, toxigenicity, sulfur granule, Madura foot, maduromycosis, mycetoma, nocardiosis, mycelium, lumpy jaw, ray fungus, diphtheroid bacilli, *tox* gene, polar body, and pseudomembrane.
3 What does the word *diphtheria* mean?
4 Name and describe the bacterium causing diphtheria.
5 What are the clinical patterns in diphtheria? Characterize the pseudomembrane.
6 Explain the pathogenicity of the diphtheria bacillus. What is the mechanism for exotoxin production?
7 Briefly discuss virulence tests in diphtheria.
8 Outline the laboratory diagnosis of diphtheria.

SUGGESTED READINGS

Charnock, M., and Chambers, T.J.: Pelvic actinomycosis and intrauterine contraceptive devices, Lancet 1:1239, 1979.

Dobie, R.A., and Tobey, D.N.: Clinical features of diphtheria in the respiratory tract, J.A.M.A. 242:2197, 1979.

Gupta, P.K., and Woodruff, J.D.: Editorial: *Actinomyces* in vaginal smears, J.A.M.A. 247:1175, 1982.

Gupta, P.K., and others: Actinomycetes in cervicovaginal smears: an association with IUD usage, Acta Cytol. 20:295, 1976.

Lago, B.D., and Streicher, S.L.: New genetics of Streptomycetes, ASM News 48(2):53, Feb. 1982.

Singer, R.A., Lysogeny and toxinology in *Corynebacterium diphtheria*. In Bernheimer, A., editor: Mechanisms in bacterial toxinology, New York, 1976, John Wiley & Sons, Inc.

Valicenti, J.F., Jr., and others: Detection and prevalence of IUD-associated *Actinomyces* colonization and related morbidity, J.A.M.A. 247:1149, 1982.

CHAPTER 29

ACID-FAST MYCOBACTERIA

Tuberculosis
  Importance
  Etiologic Agents
  Mycobacterium tuberculosis (the
    Human Tubercle Bacillus)
  Mycobacterium bovis (the Bovine
    Tubercle Bacillus)

Leprosy (Hansen's Disease)
  Mycobacterium leprae (the Leprosy
    Bacillus)
Mycobacteriosis (Atypical Tuberculosis)
  Other Mycobacteria (Atypical,
    Anonymous, Unclassified
    Mycobacteria)

Acid-fast bacteria are classified in the genus *Mycobacterium*, the members of which are straight or curved rods that may show branching or irregular forms. Acid-fast mycobacteria possess the special property of being hydrophobic in nature because of their high lipid content; hence they are refractory to staining. Once dye has penetrated, however, they retain it, even though subjected to decolorization with acid or alcohol (p. 152). Those of greatest importance and pathogenic to humans are *Mycobacterium tuberculosis*, the cause of human tuberculosis; *Mycobacterium bovis*, the cause of tuberculosis in cattle and in humans; *Mycobacterium leprae*, the cause of leprosy; and certain species of atypical, formerly "unclassified" mycobacteria, the cause of mycobacteriosis and various other infections. In addition to these acid-fast organisms, 40 or more species exist. Most are saprophytes and nonpathogenic, but they may gain access to milk, butter, or other dairy products and be mistaken for *M. tuberculosis*. Some are pathogenic for lower animals; for example, one, *Mycobacterium paratuberculosis* (Johne's* bacillus), causes a granulomatous enteritis in cattle known as *Johne's disease*.

TUBERCULOSIS
Importance

In 1900 tuberculosis (a preventable disease) was the leading cause of death in the United States, as it was throughout the civilized world. (In the tropics it ranked second to malaria.) Since then it has dropped from first place in this country but still remains a major cause of chronic disability and ill health produced by a communicable disease (see box on p. 599). In the United States each year 23,000 or more new cases and slightly under 2000 deaths are reported.

Over the world tuberculosis is still a principal cause of death, ranking with malaria and malnutrition, and is the most frequent *infectious* cause of death. In the United States it is the third leading cause of death from infectious disease. Globally about 12 million persons have tuberculosis, and 3 million die annually. The mortality is the highest in the Orient, Africa, and Latin America.

Tuberculosis, a lifelong disease, is a health hazard of the lower socioeconomic groups, those in densely populated areas, the prison population, chronic alcoholics,

---

*Pronounced *Yo-nez* (German).

In just over 100 years, advances in diagnosis, prevention, and treatment have led to a dramatic decline in the severity of tuberculosis, especially in technically advanced countries.

Events making this possible began with the discovery of the tubercle bacillus and its acid-fast nature in 1882, the discovery of x rays in 1895, and the development of tuberculin skin testing in 1907 and 1908 and climaxed with the discovery of the first tuberculocidal drug in 1947. For many years nothing more than supportive measures were available for the treatment of tuberculosis—bed rest, special diets, and fresh air. Surgical procedures were in vogue for a short period of time in the midcentury. The diseased lung was collapsed or perhaps resected. When streptomycin was discovered in 1947, isoniazid in 1952, and more recently, rifampin in 1959, things completely changed. Sanitariums for the care of tuberculous patients closed. Today, when the proper combination of the modern drugs is properly administered, the disease tuberculosis is close to 100% curable.

persons over 50 and under 5 years of age, the chronically ill, and health-care professionals in contact with it. As a chronic disease, tuberculosis places the greatest impact on persons least able to sustain that impact.

The infectious nature of this ancient disease was suspected 5 centuries before the tubercle bacillus was discovered by Koch in 1882, and it had been produced by artificial inoculation 40 years before that time. Tuberculosis has claimed the lives of many great writers, painters, and musicians—Keats, Chopin, Goethe, Poe, Gauguin, Paganini, Molière, and many others. In fact, no other disease has had such a significant relation to literature and the arts.

Etiologic Agents

Three species of the genus *Mycobacterium* are important causes of tuberculosis. They are *M. tuberculosis* (the human bacillus; primary host, humans), *M. bovis* (the bovine bacillus; primary host, cattle), and *M. avium* (the avian bacillus; primary host, birds). The three resemble each other closely but may be differentiated by animal inoculation and cultural procedures. In 1898 Theobald Smith (1850-1934) of Albany, New York, differentiated the human and bovine forms of *M. tuberculosis*. America's foremost bacteriologist, he made several important contributions to microbiology.

Mycobacterium tuberculosis (the Human Tubercle Bacillus)

M. tuberculosis—the tubercle bacillus of common parlance—attacks all human races, other primates, and some domestic animals, such as swine and dogs. (A small percentage of hogs slaughtered for food shows evidence of tuberculosis; hogs are susceptible to the bovine, avian, and human bacilli.) Wild animals living in their natural surroundings do not get tuberculosis but may contract it in captivity.

General Characteristics

The organism *M. tuberculosis*, discovered by Robert Koch (1843-1910) in 1882, is a slender, rod-shaped, nonmotile, nonspore-forming bacillus, often beaded or granular in appearance on acid-fast smears. Tubercle bacilli are more resistant than other nonspore-forming organisms to the deleterious effects of drying and of acid or alkaline media. They remain alive in dried sputum or dust in a dark place for weeks or months and in moist sputum for 6 weeks or more. Direct sunlight kills them in 1 to 2 hours.

Sufficiently susceptible to heat, however, they are destroyed by the temperature of pasteurization. Phenol, 5%, kills them in sputum in 5 to 6 hours. Tubercle bacilli are not affected by routinely used antibiotics but are susceptible to streptomycin, dihydro-streptomycin, paraaminosalicylic acid (PAS), rifampin, and isoniazid (INH). *M. tuberculosis* tends to develop a resistance to these antimicrobial agents, and toxic effects are sometimes observed after their use. However, their value in treating and modifying the course of the disease has been considerable.

M. tuberculosis grows only on special media. Even then, growth is slow; 2 to 4 weeks or more often elapse before any growth is visible (ordinary bacteria display colonies on routinely used media within 24 to 48 hours). Compared with that for other bacteria, the generation time for mycobacteria is long, up to 12 hours during the logarithmic phase of their growth; most bacteria double their numbers in 20 to 30 minutes. Body temperature is best (37° C), but the tubercle bacillus may grow at a temperature as low as 29° or as high as 42° C. Although tubercle bacilli are aerobic, 5% to 10% carbon dioxide enhances their growth, and the presence of glycerin in the medium accelerates it for the human bacillus. A bacteriostatic dye such as malachite green added to the culture medium suppresses the more rapid growth of contaminants. (NOTE: The human tubercle bacillus is more readily cultivated than the bovine bacillus.)

Toxic Products

Tubercle bacilli do not produce exotoxins, hemolysins, or comparable substances, but poisonous products partly responsible for the clinical features of tuberculosis are liberated when the bacilli disintegrate.

When *M. tuberculosis* is grown artificially, the culture medium contains a product known as **tuberculin,** without effect on a nontuberculous animal (no history of contact with the tubercle bacillus) but with powerful effects in the body of a tuberculous animal. These effects come from surprisingly small doses, and if the dose is large enough, the results are disastrous.

There are more than 50 methods of preparing tuberculin, and the nature of each tuberculin depends to some extent on the method. The best known and, in the past, the most extensively used tuberculin is Koch's **original (or old) tuberculin,** often spoken of simply as OT. It is prepared from a culture of tubercle bacilli in 5% glycerin broth that is concentrated and filtered. The bacteria-free filtrate of tuberculin contains bacterial disintegration products, substances formed by the action of the bacilli on the culture medium, and concentrated culture medium.

When OT is chemically fractionated to remove impurities, a purified tuberculin, known as **purified protein derivative** (PPD), can be obtained and standardized. This is the tuberculin used today in tuberculin testing. It consists chiefly of the active principle of tuberculin without extraneous matter, is more stable than OT, and gives less variable results. Since PPD is relatively pure, a very small amount is required for a test reaction.

Sources and Modes of Infection

Human beings are the only reservoir for tuberculosis caused by *M. tuberculosis*. The tubercle bacilli have no natural existence outside the human body and are transmitted from source to destination by some form of direct or indirect contact. Important sources of infection in patients are the tuberculous sputum and discharges from tuberculosis foci other than the lungs.

Since the patient discharging tubercle bacilli in the sputum is by far the most important reservoir of infection, droplet infection is the most common mode of spread. Droplet infection refers to the inhalation of droplets from the mouth of the tuberculous patient or dust containing the partly dried but still viable bacilli. Size is crucial. If the inhaled particle is large, it lodges in the upper respiratory tract and is usually expelled.

Droplet nuclei, very small particles of 5 to 10 μm, are required. Particles so tiny can remain suspended in the environment indefinitely. When a recipient inhales these, they pass easily down the tracheobronchial tree to the alveoli, where disease is initiated. It is generally believed that only one or two bacilli are necessary to "inoculate" the alveoli.

The next most common mode is the transfer of the bacilli to the mouth by contaminated hands, handkerchiefs, or objects. The infant crawling on the floor may contract tuberculosis by contaminating the hands with tuberculous material and then placing them in the mouth. Unless properly washed and sterilized, the eating utensils used by a tuberculous patient may be a source of infection. The milk of a tuberculous mother may rarely convey the infection to her nursing child.

The avenues of exit for the tubercle bacilli depend on the part of the body infected. In pulmonary tuberculosis they are cast off in the sputum, although tubercle bacilli sometimes may be found in the feces of persons who swallow their sputum. In intestinal tuberculosis they are discharged in the feces; in tuberculosis of the genitourinary system they appear in the urine. Tubercle bacilli may be found in exudates from abscesses and in lesions of the lymph nodes, bones, and skin.

A week or two of drug therapy sharply reduces the likelihood of tuberculous patients being a source of infection, even though the bacilli are still found in their sputum.

Pathogenicity

Tuberculosis is the chronic granulomatous infection caused by *M. tuberculosis*. Although almost any tissue or organ of the body may be affected, the parts most often involved are the lungs, intestine, and kidneys in adults and the lungs, lymph nodes, bones, joints, and meninges in children.

Pathology

Tuberculosis is a granulomatous disease, since the macrophage system of the body provides the main line of cellular defense. The unit lesions, or **tubercles,** produced by *M. tuberculosis* are small nodules of macrophages (granulomas) reacting to the presence of the tubercle bacilli in the tissues of the body, and the term *tuberculosis* is derived from their presence.

Primary tuberculosis (childhood type). When tubercle bacilli first enter the body, the ensuing infection lasts several weeks, during which time the individual develops considerable immunity and a state of specific hypersensitivity to the tubercle bacillus. As a result, the tuberculin test becomes positive and remains so for life. The sequence of events in the first infection is termed the **primary complex,** and it may occur in the lungs (most commonly), intestinal tract, posterior pharynx, or skin (rarely). In the average person the primary complex is benign on the whole. The lesions in the lung heal or become latent without treatment. Most show residual calcification.

A feature of the primary complex is the extension of tubercle bacilli to the regional lymph nodes. With a primary complex in the lung the chest nodes are infected, as are the mesenteric nodes from the mucosal focus of the small bowel and the cervical lymph nodes from the mucosal focus in the tonsils, throat, or nasopharynx.

The lesions. As a consequence of the hypersensitive (allergic) state developed, lesions develop in sensitized tissues that are typical of tuberculosis and not exactly duplicated in any other disease.

The macrophage system responds to the presence of tubercle bacilli by sending out macrophages to engulf them and form restricting barriers around them. To function efficiently in this way, macrophages must have been activated by T lymphocytes at the onset of the infectious process. In their attack on the bacilli the activated macrophages form tubercles—firm, round or oval, white, gray, or yellow nodules from 1 to 3 mm in diameter. The merging of many macrophages results in a giant cell (Fig. 29-1).

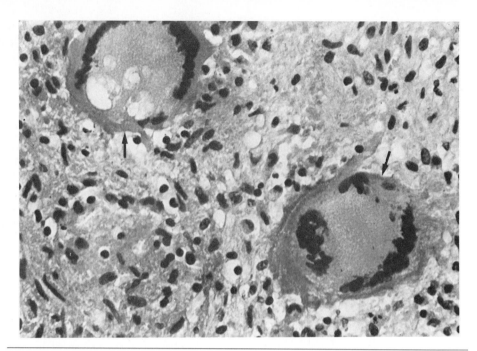

FIG. 29-1   Microsection of interior of a tubercle to show two multinucleated giant cells (*arrows*) and background of epithelioid cells (altered macrophages). The epithelioid cells are marked by the numerous, dispersed, small, ovoid, black bodies—the nuclei; their cytoplasmic outlines are blurred. Dark and dense cytoplasm of the giant cells defines their extent. Their many small nuclei lie in a half-moon pattern in the periphery. These are typical Langhans' giant cells as seen in tuberculosis.

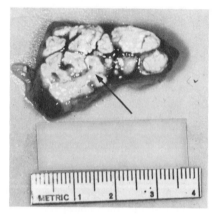

FIG. 29-2   Caseation in a lymph node. Note the cheesy quality. An arrow points to the white curdy material accumulating when this kind of tissue destruction occurs.

Microscopically a tubercle contains tubercle bacilli, epithelioid cells, and giant cells surrounded by a narrow band of lymphoid cells and encapsulated by fibrous tissue. Epithelioid cells are macrophages altered on contact with the fatty substances contained in the tubercle bacillus. They are usually prominent in tubercles, but their prime function is not clear. Since they tend to be found in close-set sheets that resemble a sheet of epithelium, they have long been designated as epithelium-like or epithelioid.

A tubercle may enlarge singly, or a number of small tubercles may coalesce to form a **conglomerate** tubercle. Because of the cellular reaction at the periphery of the tubercle, blood vessels are compressed and the circulation is impaired. This plus the lethal

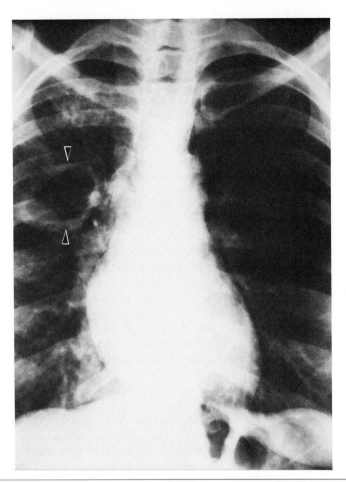

FIG. 29-3   X-ray film of the chest showing tuberculous cavity *(arrowheads)* in right upper lung field. (Compare the right lung field with the left one.)

action of the tubercle bacilli leads to a peculiar change in the center of the tubercle known as **caseation** (Fig. 29-2), a term derived from the dry, granular, cheesy quality of the dead tissue.

Such a caseous focus may long remain unchanged, or it may calcify. Calcification, the deposit of lime salts in dead tissue, is a reparative process. The caseous nodule may be organized, in which event the dead tissue is partially or completely replaced by connective tissue. The end stage of organization is a scar. On the other hand, the caseous focus may soften, enlarge, expand, and even dissect through the adjacent tissues. In the lung progression of the focus in time erodes the wall of a bronchus. The soft caseous material sloughs into its lumen, leaving an ovoid, scooped-out area termed a **cavity**. Near a body surface, such as the mucosal surface of the intestinal tract, a similar progression of a caseous focus with slough of its contents outward results in the formation of a tuberculous ulcer.

Chronic tuberculosis. Chronic tuberculosis (adult type) probably does not come from reinfection from the outside (exogenous) but from progression of the primary infection (primary complex). In some instances, after a period of latency following primary infection, there is reactivation of dormant foci.

Pulmonary tuberculosis (Fig. 29-3) is the usual form in the hypersensitive adult, involving the upper and posterior portion of the upper lobe of the lung, especially on the right side. This is **apical tuberculosis.** Tubercule bacilli prefer a high oxygen concentration, and the highest oxygen concentration in the body is found in this anatomic area. Accordingly, the bacilli grow better here than elsewhere in the body. As a part of apical tuberculosis, a cavity may form, enlarge, and become secondarily infected. Secondary infection, probably caused by streptococci and staphylococci, contributes to the hectic fever occurring in tuberculosis. Hemorrhage comes from erosion of a blood vessel in the wall of the cavity.

The body's attempts to check and heal a progressively destructive tuberculous focus in the lung (and elsewhere) are reflected in a buildup of fibrous and scar tissue. Scar tissue not only replaces and eliminates functioning tissue, but as it contracts, it also further distorts and damages often even relatively uninvolved tissue.

Tuberculous pneumonia results from the sudden spilling of tuberculous exudate into the sensitized air sacs of a large lung area. It is an acute dramatic manifestation of tuberculous allergy, clinically giving the picture of "galloping consumption."

Tuberculous pleurisy, or **pleuritis,** is secondary to tuberculosis of the lungs. It is an inflammation of the pleural membrane of the lung accompanied by a collection of fluid in the pleural cavity (pleural effusion). *All unexplained pleural effusions should be considered tuberculous until proved otherwise.*

Spread. Tubercle bacilli may be spread in the lymph or bloodstream to different parts of the body. This is the usual way. Tuberculous infection may also permeate adjacent tissues, move along natural passages (from kidneys to the bladder through the ureters), and expand over a surface. Occasionally material heavily laden with tubercle bacilli (for example, the liquefied center of a tubercle) is discharged into a blood vessel and infection seeded widely. Many small tubercles resembling millet seeds form in the lungs, spleen, liver, and various organs. This is **miliary tuberculosis.**

Tuberculosis of the intestine and regional mesenteric lymph nodes in adults is secondary to pulmonary tuberculosis, developing because human tubercle bacilli are swallowed in sputum. (In children intestinal tuberculosis is usually a primary infection caused by the ingestion of bovine bacilli in unpasteurized milk.) Tubercles and caseous foci form in the lymphoid tissue of the lower end of the ileum and cecum, leading to the formation of ulcers on the mucosal surface that tend to encircle the bowel wall. Perforation is uncommon, but adhesions form on the peritoneal surface of the bowel at the ulcer sites.

Tuberculous meningitis is a well-marked, acute allergic inflammation of the meninges associated with the formation of tubercles, especially concentrated at the base of the brain. It is usually a disease of childhood but may occur in adults. It may appear after generalized miliary tuberculosis or result from bacilli brought to the meninges by the bloodstream from distant foci. Tubercle bacilli may extend directly from adjacent tuberculous foci in the brain or the bones of the skull and spinal column.

Laboratory Diagnosis

Direct microscopic examination of suspect material after it has been stained with an acid-fast stain is the first procedure in the laboratory diagnosis of tuberculosis. The Kinyoun or Ziehl-Neelsen carbolfuchsin methods are suitable. *M. tuberculosis* is the only acid-fast organism consistently found in sputum. After staining with the two fluorescent dyes, auramine and rhodamine (Truant's method), tubercle bacilli glow with a bright yellow fluorescence at low-power magnification ($\times 100$) in a properly aligned fluorescent optical system. The rapid screening of smears possible with this method makes it widely applicable.

The failure to find tubercle bacilli in sputum does not rule out pulmonary disease, since the bacilli may not appear there until the disease is advanced. Moreover they may occur plentifully in one specimen and be scanty or absent in the next. In sputum a good chance of finding tubercle bacilli exists only if around 10,000 per milliliter are present. Tubercle bacilli are especially hard to find in smears of urine, cerebrospinal fluid, pleural fluid, joint fluid, and pus because of the relatively small number present, sometimes even with advanced disease. Various methods of concentration are applied to the test material, especially sputum, if direct smears fail to reveal the bacilli.

The second step in the laboratory diagnosis of tuberculosis is the culture of suspect material on special media devised for the growth of the tubercle bacillus, regardless of whether acid-fast organisms were demonstrated in the stained smear (Color plate 3, *D*). The best media are (1) the egg-containing ones, which are opaque, and (2) those made of oleic acid agar, which are translucent. Earlier detection of colonies is possible on translucent media. Cultural methods require a number of days or weeks, sometimes even 10 to 12 weeks, for the dry, crumbly, colorless colonies to form. The presence of a lipid substance called **cord factor** is responsible for serpentine cords seen microscopically in cultures of virulent tubercle bacilli on artificial media. Because of this factor, the proliferating organisms tend to line up in parallel rows that seem to curve in all directions, thus forming the singular pattern. Cord factor is correlated with the pathogenicity of the bacilli.

When bacilli are not demonstrated directly in suitably prepared smears, frequently they may be found in cultures or by animal inoculation, the third step in laboratory diagnosis. Some of the test material is injected subcutaneously into the groin of the extremely susceptible guinea pig. Guinea pigs are susceptible to both the human and bovine tubercle bacilli but are unaffected by the avian bacilli. If tubercle bacilli are present, the guinea pig becomes infected, the tissues about the site of inoculation thicken and may ulcerate, the inguinal lymph nodes enlarge, a generalized tuberculosis develops, and the animal dies about 6 weeks later. As a rule, the guinea pig is not allowed to die but is killed at a stated time and an autopsy performed when the disease is known to be far advanced.

Infection of a guinea pig with a given acid-fast organism is used to indicate the virulence of that organism; this is the **guinea pig virulence test.** If the organism produces disease in the animal, it is, for practical purposes, the pathogenic *M. tuberculosis*. If it fails, it is a nonpathogenic acid-fast organism, with certain important exceptions (p. 612).

NOTE: The bovine bacillus *(M. bovis)* is more pathogenic for ordinary laboratory animals than is the human bacillus. When inoculated into a rabbit, the bovine bacillus kills the animal in 2 to 5 weeks, whereas the human bacillus kills it in about 6 months or in some cases not at all.

There is presently a fluorescent antibody test in human tuberculosis that is applied to the serum (indirect fluorescent antibody), but practically speaking, serologic tests are not available for the laboratory diagnosis of tuberculosis.

Tuberculin Tests

The tuberculin test is based on the fact that persons infected with tubercle bacilli are allergic to the tubercle bacillus and its products (tuberculin). If a scratch is made on the arm of a person at some time infected and tuberculin is rubbed into the scratch (the **von Pirquet test**), or if some of the diluted tuberculin is injected *between the layers* of the skin (the **Mantoux test**), an area of redness and swelling appears at the site. If the person has never been infected, no reaction appears. In the patch test (the **Vollmer test**) a drop of ointment containing tuberculin or a small square of filter paper saturated with tuberculin is placed on the properly cleaned skin and held in place with adhesive tape for 48 hours. A positive test is indicated by redness and papule formation at the site.

The multiple-puncture technic (tuberculin **tine test**) applies tuberculin *transcutaneously*. Tuberculin four times the standard strength of OT is dried onto the tines of a small, specially constructed metal disk backed by a plastic holder, which is packaged commercially as a sterile disposable unit. After the skin has been cleaned, the disk is applied briefly to the test area with firm downward pressure, and the prongs ae allowed to pierce the skin. Reactions to this technic are seen in Fig. 29-4.

The Mantoux test, performed with serial dilutions of tuberculin beginning with a high dilution (less tuberculin) and descending to lower dilutions (more tuberculin), is considered the most reliable of the tuberculin tests. The dose used in the Mantoux test may be given by jet gun. The intradermal wheal produced should be 6 to 10 mm in diameter. In case-finding, multiple-puncture tests are practical and satisfactory for screening large groups, but a positive tine test, unless strongly reactive, should be confirmed with a Mantoux test.

The interval of time between the tuberculous infection and the appearance of a positive tuberculin reaction is around 6 to 15 weeks. Since many reach adulthood today in this country with a negative reaction, the circumstances under which conversion to a positive reaction is made may be known and the time of infection approximated. Timing is important because in the early stages tubercle bacilli proliferate actively. Adults known to have had previous negative tuberculin reactions but who now have positive reactions must be presumed infected with tubercle bacilli even though clinical, radiographic, and laboratory signs of the disease do not exist. Infection with tubercle bacilli is not synonymous with clinical tuberculosis, but persons harboring the organisms are at risk for overt disease and therefore are properly considered candidates for chemoprophylaxis. Since infection in the infant or young child is probably a recent event, a positive tuberculin reaction is quite serious. For all persons of any age with a positive tuberculin reaction, a course of antituberculous therapy might be advisable for at least 1 year. After a preliminary chest film the individual is given isoniazid as a single drug (p. 235). In adults where the presence of calcified lymph nodes in the chest indicates a process of long standing, such therapy is recommended because of the 5% hazard that they will have active tuberculosis during their life.

Positive reactions in the tests just described are known as **local reactions.** If tuberculin is given *beneath the skin*, two other reactions may occur: the focal and the constitutional. A **focal reaction** is an acute inflammation around a tuberculous focus in the body. This may have serious consequences; for instance, if the tuberculous focus is in the lung, a hemorrhage may result. A **constitutional reaction** is a systemic reaction with a sharp rise in temperature and a feeling of malaise lasting for several hours.

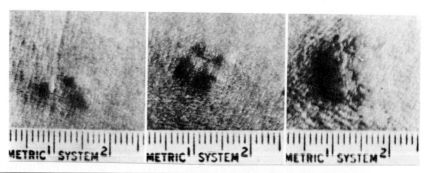

FIG. 29-4 Tine test—typical reactions with OT. In good light and with the subject's forearm slightly flexed, make readings at 48 to 72 hours. Inspect and gently palpate the site to determine induration. (Induration is an important feature of positive reaction.) Measure the diameter in millimeters of the largest single area. Results: 5 mm or more of induration—positive reaction; 2 to 4 mm—doubtful reaction; and 2 mm or no induration—negative reaction. (Courtesy Lederle Laboratories, Pearl River, N.Y.)

Immunity

White people possess considerable inherent resistance to tuberculosis but never complete immunity. The incidence in blacks is high, averaging (in proportion to population) eight cases to one in whites. In addition, when the black person contracts tuberculosis, the average life expectancy for that person without treatment is approximately one sixth that of the white person who contracts the disease. The American Indian and Mexican are also very susceptible.

Persons who live in isolated communities seem to be more vulnerable when exposed for the first time during adult life than those who have been reared in closely crowded and highly infected communities.

Not long ago it was assumed that tuberculosis was hereditary and ran in families, but tuberculosis runs in families because closely associated members of the family pass it to each other. Children of tuberculous parents may inherit certain predisposing factors, but a child born of tuberculous parents and at once removed to infection-free surroundings has a better chance of escaping the disease than one born of healthy parents but reared with tubercle bacilli.

The defensive factors that operate in a given infection depend on whether infection has occurred before. Most observers state that a degree of protection against subsequent infections is afforded the individual by the first infection. Whether active disease develops depends on (1) the number of bacilli, (2) their virulence, (3) the state of allergy, and (4) resistance of the subject. Progressive disease is most likely to occur when host defenses are weakened for whatever reason. Conditions that lower body resistance include malnutrition and predisposing diseases such as measles, whooping cough, and diabetes mellitus. One of the most vulnerable persons to tuberculosis is the chronic alcoholic.

Prevention*

Since tuberculous sputum is the chief source of infection with human bacilli, it should be disposed of carefully. It should not be allowed to dry, since then it could be blown from place to place and the germs spread over a wide area. Sputum should be received in suitable covered containers and burned. Promiscuous spitting should be taboo, and patients must cover their mouth when they cough. A tuberculous mother should not nurse her child. The woodwork of a room occupied by a tuberculous patient can be washed with soap and water and a suitable disinfectant, usually one of the phenolic derivatives, applied.

In communities with the best type of health supervision there has been a decided decrease in the incidence of tuberculosis because of several factors, among which are better living conditions and medical prophylaxis. Mass surveys of the population by chest x-ray examination have been of great value in the detection of pulmonary tuberculosis and other lung lesions. Radiographs of the chest at regular intervals are recommended for individuals whose work brings them into contact with active cases of tuberculosis.

Mycobacterium bovis (the Bovine Tubercle Bacillus)

M. bovis (*M. tuberculosis* var. *bovis*) is a bit shorter and plumper than *M. tuberculosis*. It grows more slowly in culture media than does the human bacillus, forms slightly smaller colonies, and is niacin test negative. However, its infection cannot be distinguished from that with the human bacillus by the tuberculin test.

M. bovis is highly virulent for humans. Human infections caused by bovine bacilli were once common in children and young people and affected the cervical lymph

---

*For BCG vaccination, see pp. 427 and 443.

nodes, intestines, mesenteric lymph nodes, and bones. **Scrofula**, tuberculosis of the lymph nodes of the neck, is typically part of the primary complex produced by the bovine bacillus in this area. Bovine bacilli cause pulmonary tuberculosis in cattle and in humans, but practically all pulmonary infections in both children and adults result from the human bacillus. Pulmonary tuberculosis is the same human disease whether caused by the bovine or human bacillus.

Infection is acquired from the milk of tuberculous cows. As a rule, the bacilli get into milk by fecal contamination. In the cow tuberculosis usually attacks the lungs, but since the cow swallows her sputum, the bacilli are excreted in the feces. The udder and flanks of the cow become contaminated, and the bacilli gain access to the milk, which, if consumed unpasteurized, is the source of infection. Sometimes the bacilli are excreted directly into the milk as it comes from an infected udder. This may occur without demonstrable tuberculous lesions in the udder.

The tuberculin reaction is of inestimable value in detecting tuberculous cows. Routine tuberculin testing of cattle by the U.S. Department of Agriculture and elimination of infected cattle from the herd, plus the pasteurization of milk, have practically eliminated bovine infection in the United States. In countries where no such safeguards exist, the bovine bacillus is responsible for 15% to 30% of the cases of tuberculosis in young persons.

LEPROSY (HANSEN'S DISEASE)

Leprosy is a geographically widespread, chronic, communicable disease caused by *Mycobacterium leprae* (Hansen's bacillus) that involves primarily skin, mucous membranes, and nerves. (Hansen's bacillus is the only known bacterium to invade selectively peripheral nerves.) Leprosy (Hansen's disease) has afflicted humans from the beginning of history, and descriptions of it occupy a prominent place in the Old Testament and other ancient writings. It is estimated that currently there are about 15 to 20 million leprosy patients in the world, more than at any time in history. They are mostly in the tropics and subtropics, with approximately 2000 in the United States. The International Leprosy Association has unanimously resolved that the term *leper* be abandoned and that the person suffering with the disease be designated the *leprosy patient*.

Mycobacterium leprae (the Leprosy Bacillus)

M. leprae is a long, slender, acid-fast rod closely resembling *M. tuberculosis. M. leprae*, discovered by Gerhard A. Hansen (1841-1912) in Norway in 1874, was actually the first bacterium identified as a cause of human disease. For nearly a century it remained the only one known to infect humans that could *not* be cultured in the laboratory and that did not produce progressive disease if inoculated into a test animal.

Today the bacillus of leprosy can be cultivated in the ears and rear footpads of white mice and hamsters, sites chosen as the largest and coolest parts of the experimental animals. (It grows best in the cooler tissues of human beings.) After many months mild changes detectable with the microscope occur at the sites of inoculation, but there is no gross advanced disease. The generation time for the leprosy bacillus is far greater than that of any known bacterium. The bacillus can be grown also in cell culture.

Studies of this bacillus show it to be closely related to *Corynebacterium* species because of its genome size and the guanine plus cytosine ratio of its DNA.

Mode of Infection

Human imagination has clothed Hansen's disease with many attributes it does not possess. One is that it is a highly communicable disease, which is not true. A human being contracts leprosy only after prolonged and intimate contact, and even then that person often escapes infection. Probably no more than 5% of spouses living with patients get the disease.

The bacilli leave the body in great numbers from degenerating lesions of the skin and mucous membranes and spread directly from person to person. Skin-to-skin transmission is important, and nose and mouth discharges are especially dangerous because lesions are very common in these locations. The portals of entry are probably skin and mucous membranes. Small breaks in the skin may admit organisms discharged from a patient with the disease. Fomites may have some effect, since the bacilli are viable in dried nasal mucus for up to 7 days. Although the bacteria prefer certain cool sites, they are spread widely over the body. They may be found in feces and urine. The disease is not hereditary, but since children are unusually susceptible, the percentage of infection in children associated with leprous parents is very high (30% to 40%). The highly variable incubation period is estimated to be from 5 to 15 years (sometimes 20 to 40 years).

Pathogenicity

Leprosy belongs with the infectious granulomas (diseases with a defensive multiplication of macrophages at the sites of infection). Although leprosy is generalized with a variety of changes, there are two conspicuous patterns: (1) the **lepromatous** or **nodular** form, characterized by tumorlike overgrowths of the skin and mucous membranes, and (2) **tuberculoid** or **anesthetic** leprosy, manifested by involvement of peripheral nerves with localized areas of skin anesthesia. Different forms of the disease reflect the host response to the causative agent. The end stages of the disease are associated with extensive defor-

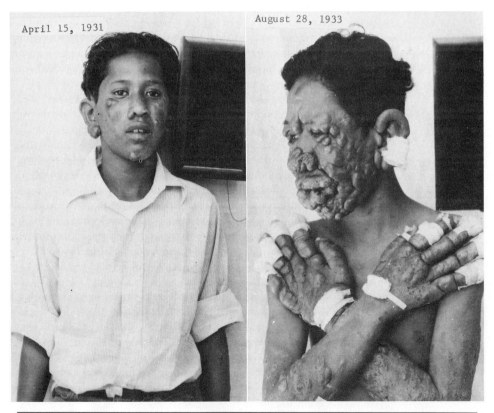

FIG. 29-5   Progression of leprosy in a teenaged Filipino. Note dates; this occurred before modern drugs for leprosy were available. (Courtesy Dr. C. Binford, Washington, D.C.)

FIG. 29-6   Armadillo (nine-banded), *Dasypus novemcinctus* Linn, shown here with its heavy coat of armor that incorporates bucklelike plates and movable transverse bands. It is a mammal closely related to sloths and anteaters that is found in the Western Hemisphere from Argentina to the northern part of Texas. An adult male averages 32 inches in length and 13 pounds in weight. The armadillo is omnivorous, mostly nocturnal, and harmless and on its short legs runs from danger. Its flesh is sometimes eaten. Some 5% of armadillos captured in Louisiana for experimental use are naturally infected with sylvatic leprosy. (Courtesy Dr. Eleanor E. Storrs, New Iberia, La.)

mity and destruction of tissue (Fig. 29-5). The leprosy patient may be left without nose, ears, hands, or feet and perhaps without eyesight.

The earliest lesion in leprosy may be diagnostic in that it contains acid-fast bacilli. Since it may not indicate the subsequent course, it is referred to as an **indeterminate** lesion. **Dimorphous leprosy** refers to the coexistence of the two forms—lepromatous and tuberculoid.

The armadillo is currently an experimental model for the study of leprosy. Since the body temperature of the armadillo (Fig. 29-6) is cool at 32° to 35° C, this animal was believed to be suitable for laboratory study, and indeed it was found to share a susceptibility to leprosy comparable to that of the human being. The discovery was made that this creature develops leprosy, analogous to lepromatous leprosy in humans, some 3 years after initial infection and it plays host to an enormous profusion of mycobacteria. The number of leprosy bacilli in its infections is 1000 to 10,000 times greater than that in advanced human disease. The experimental animals also succumb to a more severe disease, showing involvement of the central nervous system and lungs, areas spared in human beings. An African mangabey monkey (genus *Cercocebus*), also discovered to be susceptible to leprosy, may become another animal model.

Laboratory Diagnosis

M. leprae occurs abundantly in the lesions of leprosy. A diagnosis of leprosy is most often made with scrapings from the nasal septum or biopsies of the skin of the ear that have been stained for acid-fast bacilli. Large numbers of the bacilli are found within phagocytic cells packed together like packets of cigars. Bacilli also may be demonstrated in acid-fast stained material from lesions in other areas of skin and in lymph nodes. Fluorochrome staining can be used to demonstrate leprosy bacilli (Fig. 29-7). Leprosy and tubercle bacilli are differentiated by guinea pig inoculation, since leprosy bacilli have no effect on the animal.

Lepromin is a suspension of killed leprosy bacilli used in a skin test (the Mitsuda test) to detect susceptibility to this disease. The armadillo is a good source of lepromin.

FIG. 29-7 *Mycobacterium leprae*, fluorescent staining. Note the morphology. (×1000.) (Courtesy Dr. R.E. Mansfield, Canton, Ill.)

Prevention

Leprosy is best prevented by good living conditions. Prolonged and intimate contact with leprosy patients is hazardous. Strict isolation with segregation has not proved completely successful as a method of control, but most persons in the United States discharging *M. leprae* are referred to the national leprosarium, the U.S. Public Health Service Hospital at Carville, Louisiana.

A patient who has improved and has failed to discharge the bacilli for a period of 6 months may be paroled. Paroled patients should be examined twice yearly. Back home the patient should be semisegregated, with that person's own room, linens, and dishes. The patient's discharges, cooking utensils, and linens should be carefully disinfected. No children should live in the house. Children of leprosy patients should be separated from their parents at birth because the chances of infection are much greater in infants and young children. All persons who contact a patient should be examined every 6 months.

Results of clinical trials in eastern Uganda in Africa indicate that BCG vaccine protects children against leprosy. The possibility of such a surrogate vaccine strengthens the case for a strong cross-immunity among mycobacteria.

MYCOBACTERIOSIS (ATYPICAL TUBERCULOSIS)

Mycobacteriosis, or atypical tuberculosis, refers to any tuberculosis-like disease caused by nontuberculous mycobacteria (NTM), or mycobacteria other than *M. tuberculosis*. Mycobacteria other than *M. tuberculosis* are said to cause 10% of cases of clinical tuberculosis.

Other Mycobacteria (Atypical, Anonymous, Unclassified Mycobacteria)
The Organisms

In tuberculosis hospitals unusual acid-fast bacilli are sometimes found in the sputum of patients with a typical setting for tuberculosis and with cavitary disease seen in their x-ray films. Although they cause pulmonary disease mimicking tuberculosis (mycobacteriosis), these acid-fast organisms differ sharply from the tubercle bacillus. They also cause infection in lymph nodes and skin. Infections are usually chronic, variably destructive granulomas.

Such infections are caused by an assortment of mycobacteria, designated in various ways. In the past these mycobacteria were often called *MOTT* bacilli to indicate that they were *m*ycobacteria *o*ther *t*han *t*ubercle bacilli. Since, for a long time, most of them were not adequately identified and therefore not assigned to a species, they were referred to as anonymous or unclassified mycobacteria. The designation *atypical mycobacteria* is still generally acceptable, although not necessarily appropriate. Today classification of the atypical or nontuberculous mycobacteria is based on the results of careful study of their properties, and speciation is usually possible.

Classification

Atypical mycobacteria are readily differentiated from *M. tuberculosis* by their colonial characteristics, biochemical reactions, susceptibility to antituberculous drugs, and animal pathogenicity. Although species identification is important, these organisms are commonly considered in four categories (Table 29-1). In group I the photochromogens produce yellow-orange colonies in the light. *Mycobacterium kansasii* ("yellow bacillus"), the respresentative of this group, is a respiratory pathogen. *Mycobacterium marinum* causes "swimming pool" granuloma. Group I members are most closely associated with human pulmonary disease.

The scotochromogens of group II produce a yellow to orange pigment in the dark as well as in the light. *Mycobacterium scrofulaceum* ("tapwater bacillus"), the representative, causes a cervical lymphadenitis in children.

Not affected by light, the nonphotochromogens of group III form buff-colored colonies, soft in consistency in comparison with the rough ones of the tubercle bacillus. These include the Battey bacillus *(Mycobacterium intracellulare)*, the avian variety of tubercle bacilli *(M. avium)*, and other respiratory pathogens. The mycobacteria of the *M. avium–intracellulare* group or complex (MAC) cause chronic progressive pulmonary disease in middle-aged men and infect homosexual males with AIDS (p. 365).

The rapid growers of group IV form colonies on simple media within a short time, even at room temperature. Their nonpigmented colonies most closely resemble those of the tubercle bacillus. The pathogen of note here, among many saprophytes, is *Mycobacterium fortuitum*. Group IV mycobacteria produce subcutaneous abscesses after trauma to the skin.

Mycobacterium smegmatis (the smegma bacillus) is an acid-fast bacillus widespread in dust, soil, and water. It is inconsequential as a disease producer, but because it is a normal resident of the prepuce and vulva, it may gain access to urine and be mistaken for *M. tuberculosis* unless the specimen is properly collected. Smegma bacilli are sometimes found in feces. They are differentiated from tubercle bacilli by their growth pattern and by the fact that they do not produce disease in guinea pigs. They are now placed in group IV.

The pathogenicity of members of groups I, III, and IV is established; that of group

Table 29-1 Pathogenicity of Some Other Mycobacteria

Group and Common Name	Some Significant Members	Niacin Test*	Pathogenicity for		
			Human	Guinea Pig	Mouse
I Photochromogens	*Mycobacterium kansasii* *Mycobacterium marinum (balnei)*	Negative	+	−	+
II Scotochromogens	*Mycobacterium scrofulaceum* *Mycobacterium gordonae*	Negative	+	−	−
III Nonphotochromogens	*Mycobacterium intracellulare* ("avium-intracellulare group") *Mycobacterium xenopi*	Negative	+, −	−	+, −
IV Rapid growers	*Mycobacterium fortuitum* *Mycobacterium chelonei* *Mycobacterium smegmatis*	Negative	+, −	−	+, −

*Niacin test positive with *Mycobacterium tuberculosis*. It is the best test to separate tubercle bacilli from other mycobacteria. In the clinical setting, production of niacin by an acid-fast bacillus identifies it as the human tubercle bacillus.

II is equivocal. All the mycobacteria except those in group II cause disease in the mouse; none is pathogenic for the guinea pig.

The tuberculin test in mycobacteriosis is negative or weakly positive. A tuberculin reaction less than 10 mm in diameter may represent the cross-reaction to atypical mycobacterial sensitivity.

Transmission

Mycobacteriosis is not contagious, and humans are not the reservoir of infection, but much remains to be learned of the mode of spread and source of these mycobacteria. Their tolerance for adverse environments and their survival in sparse media are consistent with their wide dispersion in nature. They are found worldwide. Because they are found in soil, which is said to be the most common source, they do tend to form geographic patterns of distribution. Presumably human beings contact the organisms in the environment, and in this event the lung would likely be the portal of entry. A wide host range may exist in wild and domestic animals. Members of group I mycobacteria are found in raw milk and are known to survive the temperature of pasteurization. Some of these mycobacteria never produce infection, and others are opportunistic pathogens only on favorable occasions.

SUMMARY

1 The acid-fast mycobacteria of the genus *Mycobacterium* are straight or curved rods with branching and irregular forms. *Mycobacterium tuberculosis* causes tuberculosis, *Mycobacterium leprae* causes leprosy, and certain atypical mycobacteria cause mycobacteriosis. Saprophytic species of mycobacteria in dairy products may be mistaken for *M. tuberculosis*.

2 *M. tuberculosis*, the tubercle bacillus, is more resistant than other nonspore-forming microbes to the deleterious effects of drying and of acid or alkaline media. Tubercle bacilli grow slowly on special media only.

3 Tuberculosis is a chronic inflammatory process in which the macrophage system provides the main line of cellular defense. The unit lesions are called tubercles. The primary complex is the sequence of events of the first contact that a person has with the tubercle bacillus. It may occur in the lungs, intestinal tract, throat, or skin. Thereafter that person has a state of hypersensitivity to the tubercle bacillus and its product, tuberculin.

4 Tuberculin testing identifies the state of hypersensitivity. The tine test is used for the screening of large groups of people; the Mantoux test is the most reliable of the tuberculin tests.

5 Pulmonary tuberculosis is the usual manifestation of tuberculosis in the hypersensitive adult. All unexplained pleural effusions should be considered tuberculous until proved otherwise.

6 In the laboratory, diagnosis of tuberculosis is made by (1) the direct microscopic demonstration of the acid-fast bacilli in smears of suspect material, (2) the recovery of the bacilli in cultures, and (3) the demonstration of both the bacilli and the disease changes in a guinea pig inoculated with suspect material.

7 The bovine tubercle bacillus, the cause of tuberculosis in cows, is highly virulent for humans. Bovine tuberculosis can be prevented by pasteurization of milk and tuberculin testing of cows, with infected cows eliminated from the herd.

8 Leprosy, one of the oldest described diseases of human beings, is a chronic infectious granuloma that slowly deforms and destroys skin, mucous membranes, and nerves. The patient may be left without nose, ears, hands, feet, or even eyes.

9 The generation time for the leprosy bacillus is far greater than that of any known

bacterium. A human contracts leprosy only after prolonged and intimate contact with a leprosy patient. Children are highly susceptible.

10 Leprosy bacilli are readily identified in scrapings from the patient's nasal septum or in biopsies of the patient's ear.

11 The peculiar susceptibility of the armadillo has meant new developments in the study of leprosy. This animal hosts an enormous profusion of bacilli.

12 Atypical mycobacteria cause disease in the lung mimicking tuberculosis. Mycobacteriosis is usually a chronic, variably destructive granuloma.

13 Atypical mycobacteria are easily distinguished because of their properties. Three of the four groups defined are pathogenic.

QUESTIONS FOR REVIEW

1 What are the salient features of acid-fast mycobacteria?

2 Name and describe the acid-fast bacilli causing human tuberculosis.

3 Classify acid-fast bacteria and indicate their pathogenicity.

4 Discuss sources of infection in tuberculosis.

5 Outline the laboratory diagnosis of tuberculosis.

6 Give the technics for and the significance of the tuberculin test.

7 How should tuberculin tine tests be evaluated?

8 Compare the tubercle bacillus with mycobacteria of mycobacteriosis.

9 Compare the tubercle bacillus with Hansen's bacillus.

10 Briefly characterize leprosy.

11 What provision does the U.S. Public Health Service make for the care of patients with leprosy?

12 Define tuberculin, tubercle, scotochromogen, cavity, granuloma, infectious granuloma, scrofula, miliary, MOTT bacilli, smegma bacilli, primary complex, Battey bacilli, "yellow bacillus," tuberculoid leprosy, lepromatous leprosy, lepromin, cord factor, PPD, MAC, OT, and caseation.

SUGGESTED READINGS

Baillie, R.A., and Baillie, E.E.: Biblical leprosy as compared to present-day leprosy, South. Med. J. 75:855, 1982.

Chapman, J.S.: The atypical mycobacteria and human mycobacteriosis, New York, 1977, Plenum Medical Book Co.

Curtis, J., and Turk, J.L.: Mitsuda-type lepromin reactions as measure of host resistance in *Mycobacterium lepraemurium* infection, Infect. Immunol. 24:492, 1979.

Dillman, T.O., and Graham, D.Y.: Primary tuberculous enteritis: forgotten but not gone, Tex. Med. 75:48, April 1979.

Farer, L.S.: Tuberculosis in the United States today, part 2, Am. Lung Assoc. Bull. 65:9, May 1979.

Freedman, S.O.: Tuberculin testing and screening: a critical evaluation, Hosp. Pract. 7:63, May 1972.

Laurenzi, G.A.: Beethoven and tuberculosis, Forum Med. 2:802, 1979.

Meier, J.L., and others: Leprosy in wild armadillos (*Dasypus novemcinctus*) on the Texas Gulf Coast, ultrastructure of the liver and spleen, Lab. Invest. 49:281, 1983.

Sommers, H.M., and McClatchy, J.K.: Cumitech 16: Laboratory diagnosis of the mycobacterioses, Washington, D.C., 1983, American Society for Microbiology.

Stead, W.W.: Undetected tuberculosis in prison, source of infection for community at large, J.A.M.A. 240:2544, 1978.

Thompson, N.J., and others: The booster phenomenon in serial tuberculin testing, Am. Rev. Respir. Dis. 119:587, 1979.

Young, D.B., and Buchanan, T.M.: A serological test for leprosy with a glycolipid specific for *Mycobacterium leprae*, Science 221:1057, 1983.

SPIROCHETES AND SPIRALS

OVERVIEW

Spirochetes are actively motile, flexible, spiral bacteria found in contaminated water, sewage, soil, decaying organic matter, and the bodies of animals and humans (Fig. 30-1). They may be free living, commensal, or parasitic. Spiral microbes vary in length from only a few to 500 μm and move by rapidly rotating about their long axis, by bending, or by "snaking" along a corkscrew path. They are aerobic, facultatively anaerobic, or anaerobic, with no flagella, no rigid cell wall, and no endospores. Many of them are best visualized by phase-contrast and dark-field microscopy.

Slender spiral bacteria are classified in five genera that include nonpathogenic spirochetes often found on the mucous membranes of the mouth, about the teeth, and on the genitalia and three significant pathogens. The pathogenic spirochetes are found in the genus *Treponema*, containing the organisms responsible for syphilis and yaws; the genus *Borrelia*, with the organisms responsible for relapsing fever; and the genus *Leptospira*, with the organisms responsible for infectious jaundice (Weil's disease) and other forms of leptospirosis.

Other spiral-shaped bacteria with slightly different properties fall into the family Spirillaceae of spirally twisted rods. These are rigid, possess one flagellum or a tuft of flagella, and are actively motile, swimming in straight lines in corkscrew fashion. They may be saprophytes or pathogens. The genus *Spirillum* contains the microbes of one form of rat-bite fever. The genus *Campylobacter* contains microaerophilic, spirally curved rods that cause gastroenteritis.

TREPONEMA PALLIDUM (THE SPIROCHETE OF SYPHILIS)

Syphilis *(lues venerea)* is an infectious disease caused by *Treponema pallidum*. Clinically it may be either *acquired* or *congenital*, that is, incurred after or before birth; the former is more usual. The historic origin of syphilis is debated. Many observers believe that Columbus's sailors introduce it into Europe on their return from the New World. About that time the disease did spread over certain parts of Europe in virulent epidemic form.

T. pallidum (the pale spirochete) is an actively motile, slender, corkscrewlike organ-

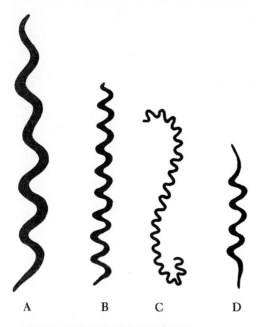

FIG. 30-1 Configurations of the major spiral microorganisms. **A**, *Borrelia*. **B**, *Treponema*. **C**, *Leptospira*. **D**, *Spirillum*.

A   B   C   D

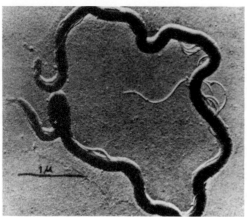

FIG. 30-2 *Treponema pallidum*, electron micrograph; a slender spiral microorganism, 5 to 15 μm in length. Axial fibrils arise from either pole of the spirochete, but in such a micrograph they appear collectively as a single, composite helical structure (intertwined about the spirochete), referred to as the *axial filament*. The typical movement of the spirochete is believed to be associated with the contractility of the fibrils. The cell wall contains peptidoglycan but is sufficiently pliable to allow the microbe to bend easily. (Courtesy Centers for Disease Control, Atlanta.)

ism that, when properly searched for, can be found in practically every syphilitic lesion (Fig. 30-2). It is especially abundant in chancres and mucous patches (lesions of the skin and mucous membranes found early in the disease). It is 6 to 15 μm long, and because it is only 0.25 μm in width, it cannot be seen with the ordinary light microscope. It has 6 to 14 spirals and rotates on its long axis. Usually rigid, it may bend on itself. In addition to its rotary motion, it has a slowly progressive one.

T. *pallidum* is difficult to stain with ordinarily used bacteriologic dyes and is best demonstrated by dark-field microscopy in the syphilitic chancres and mucous patches. Since the organisms are examined in the living state by this method, the diagnostic features of motility and shape are seen. Fritz Schaudinn (1871-1906), parasitologist, and Eric Hoffman (1868-1959), clinician, described the causative agent of syphilis, T. *pallidum*, in 1905. Their discovery of an almost invisible parasite was the result of incomparable skill in technic and staining methods.

T. *pallidum* can live outside the body under suitable conditions for 10 to 12 hours but is killed within 1 hour by drying. A fragile microbe, it may be destroyed by mild disinfectants or even soap and water and does not remain outside the body except briefly on objects contaminated with syphilitic secretions. It is sensitive to temperatures above 40° C. In whole blood or plasma at refrigerator temperature it remains viable for 24 hours but dies within 48 hours.

Under natural conditions *T. pallidum* infects humans only, but anthropoid apes, monkeys, rabbits, and guinea pigs may be infected artificially. The disease in rabbits and monkeys resembles the human disease in many but not all respects.

Acquired Syphilis

Modes of Infection

Syphilis is caused by a microbe that is not borne by food, air, water, or insect. Human beings are its only reservoir. To contract the disease, a susceptible person must be in close and intimate contact with an infectious person. Acquired syphilis therefore is contracted in most instances by sexual intercourse. At times it is contracted by other types of direct contact of skin or mucous membranes, such as kissing if the patient has lesions in the mouth. It is rarely spread by contaminated objects such as drinking cups or towels. Physicians, dentists, and nurses may become infected during the examination of a syphilitic patient.

With modern blood banks and blood banking methods the risk of transferring syphilis by blood transfusion is minimal. Blood in the modern blood bank is routinely tested by a standard serologic test for syphilis, usually by the VDRL test (p. 622). If the result is positive, that blood is not released.

Evolution of a Typical Case

The course of syphilis is outlined in Table 30-1.

Incubation stage. *T. pallidium* is a highly invasive organism. When introduced into the human body, it multiplies. Many of the treponemes migrate through the lymphatics to the regional lymph nodes. From the thoracic duct they enter the bloodstream and quickly spread over the body. Some continue to multiply at the original site. From 2 to 6 weeks later an inflammatory reaction at the inoculation site develops, and the primary lesion, the **chancre,** forms.

Primary stage. The chancre is the first clinical sign of infection. It usually appears on the genitalia, where in the human male it is readily observed. In the human female the chancre, if on the cervix, may escape detection. Some 10% or more of chancres are extragenital, being found on the face, lips, throat, tongue, tonsils, breasts, or fingers.

The chancre appears beneath the mucous membrane or skin as a small nodule that feels like a shot. It breaks down, forming a shallow ulcer with indurated edges and a hard, clean base. There is little pain or discharge unless secondary infection occurs. Chancres are usually single but may be multiple. They vary in size but are seldom more than ½ inch in diameter. After a few days the lymph nodes draining the site enlarge. Pain is absent, and the nodes do not suppurate.

After the chancre has persisted for 4 to 6 weeks, it heals spontaneously without scar or residual change. For the next 4 to 8 weeks (before the secondary stage) the patient shows no signs of disease. This is the **primary latent period.**

Secondary stage (stage of systemic involvement). The manifestations of secondary syphilis may recur over a period of 5 years. They are (1) skin lesions (Fig. 30-3), (2) mucosal lesions, (3) generalized rubbery lymphadenopathy, and (4) an influenza-like syndrome. The variable skin eruption is usually symmetrically arranged, macular, and copper colored and seldom itches or burns. A patchy loss of hair, even in the eyebrows, is associated. Known as **mucous patches** and most often found in the mouth, the mucosal lesions are painful superficial ulcers with a white raised surface and are swarming with treponemes. The secondary stage lesions heal spontaneously.

A basic contradiction exists between the cardinal sign and the expected symptoms in

Table 30-1 Evolution of Typical Case of Syphilis

Stage	Duration	Clinical Disease	Activity of *Treponema pallidum*	Diagnosis	Tissue Change
Incubation	2 to 6 weeks (most often 3 to 4 weeks)	None	Spirochetes actively proliferate at entry site, spread over body	Identification of *Treponema pallidum*: a. Dark-field microscopy b. Fluorescent antibody technic	Chancre appears at inoculation site
Primary	4 to 6 weeks	1. Chancre present at inoculation site 2. Regional lymphadenopathy	Chancre teeming with them	1. Dark-field microscopy of chancre 2. STS* become positive	Chancre present
Primary latent	4 to 8 weeks	None	Inconspicuous	STS positive	None demonstrable; chancre healed with little scarring

Secondary	Variable over period of 5 years (latent periods with recurrences)	1. Skin and mucosal lesions ("mucous patches")
		2. Generalized lymphadenopathy
			Skin and mucosal lesions rich in spirochetes (highly infectious)	1. Dark-field microscopy of lesions
				2. STS positive
					1. Infection active:
					  a. Vascular changes
					  b. Cuffs of inflammatory round cells about small blood vessels
					2. Resolution spontaneous—little scarring

Latent	Few months to a lifetime (average 6 to 7 years)	None	Inconspicuous	STS positive (can be negative)	
Tertiary	Variable—rest of patient's life	Related to organ system diseased and the incapacity thereof	Paucity of spirochetes in classic lesions	1. STS positive or negative 2. Special silver stains of tissue lesions may show spirochetes	1. Gumma 2. Definite predilection to heal in lesions 3. Scarring 4. Tissue distortion and abnormal function

*Serologic tests for syphilis.

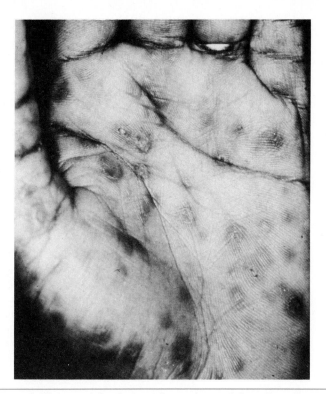

FIG. 30-3   A singular feature of skin lesions of secondary syphilis is their occurrence on the palm of the hand and sole of the foot.

both primary and secondary syphilis. The chancre looks painful but is not; the secondary skin rash looks pruritic (as though it should itch) but is not. (Of course exceptions are found.)

Latent stage. After the secondary stage is a period during which the patient shows no signs; syphilis is recognized only by serologic tests. This stage may last months or years, but in 25% to 50% of the patients it lasts a lifetime. The disease may become active at any time.

For some patients there is no latent period; the tertiary stage appears right after systemic manifestations have disappeared. In those patients in whom the chancre and secondary manifestations are not present or not detected, the whole course of the disease may be latent. Lesions of primary and secondary syphilis may be atypical and inconspicuous. Many patients are unaware or ignore the signs of the disease in these stages. Latent syphilis probably represents a biologic balance between the pathogenicity of *T. pallidium* and the defensive forces of the body.

Tertiary stage. Permanent tissue damage marks tertiary stage syphilis. The inflammatory response incorporates blood vessel involvement, cellular infiltrates, fibrosis for repair, and end-stage scarring. The hallmark of tertiary syphilis is the destructive **gumma,** a firm, yellowish white central focus surrounded by fibrous tissue. The tertiary lesions involve the deeper structures and organs of the body and interfere materially with their functions. Syphilis favors the cardiovascular and central nervous systems, but other organ systems are vulnerable. Treponemes are sparse in tertiary lesions, and the tissue reaction is usually attributed to some form of allergy.

Syphilis has been called the "great imitator." There is practically no organic disease whose manifestations it cannot copy, and no other disease can be clinically so diverse.

The two major expressions of tertiary syphilis are cardiovascular syphilis and neurosyphilis.

Cardiovascular syphilis is tertiary-stage syphilitic involvement of the heart and blood vessels. It results in inflammation of the aorta (syphilitic aortitis), aneurysmal dilation of the thoracic aorta, and distortion of the aortic valve to produce aortic insufficiency (one form of valvular disease of the heart). Syphilis also predisposes the person to arteriosclerosis in the aorta.

Neurosyphilis is syphilitic involvement of the central nervous system. When syphilis becomes generalized during its early stages, the central nervous system seldom escapes. In many patients the treponemes die without causing any damage there, but in 30% to 40% of patients they remain alive and initiate tertiary stage changes that come to light weeks, months, or years later. Anatomically neurosyphilis takes three forms: (1) syphilitic meningitis or meningovascular syphilis, (2) tabes dorsalis, and (3) general paresis. In addition, gummas occur as isolated lesions in various parts of the central nervous system.

A regular manifestation of neurosyphilis, *syphilitic meningitis,* more properly called **meningovascular syphilis,** is an inflammation of the meninges with or without involvement of the brain. The meninges are variably thickened, and typical changes occur in the meningeal blood vessels. Although not all patients with syphilis develop meningovascular disease, a good percentage do. It usually appears within 5 or 6 years after infection but may appear as early as 2 months or as late as 40 years.

Tabes dorsalis, or *locomotor ataxia,* is a degeneration of the posterior columns of the spinal cord and the posterior nerve roots and ganglia. Since the sensory pathways of the spinal cord are involved, the disease is chiefly one of muscular incoordination and sensory disturbances.

General paresis (general paralysis of the insane; paralytic dementia) is a diffuse meningoencephalitis expressed in progressive mental deterioration, insanity, and generalized paralysis, terminating in death. About 3% of patients with syphilis are affected. Paresis is more common in the highly civilized than in the primitive races and in better educated than in less educated persons and is five times more frequent in men than in women.

Neurosyphilis may simulate almost any disease of the central nervous system, but a careful history with a correct interpretation of laboratory tests on the blood and cerebrospinal fluid usually makes the distinction.

Immunity

Immune mechanisms are poorly understood in syphilis. The victim does develop some kind of resistance, since that person is not susceptible to superimposed syphilitic infection. After the patient is completely cured, however, susceptibility becomes as great as ever.

With syphilitic infection a heterogeneous group of antibodies results, but the exact nature of their contribution to immunity is unclear. Invading treponemes presumably interact with tissue cells of the host and release fatty substances that may combine with protein from treponemes to form antigens. These, in turn, stimulate formation of antibodies to both lipids and the organisms. Two broad categories of syphilitic antibodies are defined: (1) nonspecific, *nontreponemal* antibodies directed against lipid antigens and (2) specific *antitreponemal* ones against the spirochetes themselves.

It has been known for a long time that the serum of a patient with syphilis (cerebrospinal fluid in neurosyphilis) contains a nonspecific, nontreponemal antibody–like substance not found in normal blood or spinal fluid that can combine with certain lipid antigens, for example, cardiolipin extracted from the heart muscle of an ox. (Cardiolipin is one of the relatively few lipid antigens of importance.) On this phenomenon are based the serologic tests for syphilis (STS), since this combination fixes complement and aggregates antigen from colloidal suspensions (complement-fixation and flocculation tests). The antibody-like substance was originally called **reagin,** a term that reflected the lack of understanding of its nature. Still called reagin, it is now known to be an IgM antibody primarily directed against cardiolipin, and although an accidental discovery, it has remained through the years a mainstay in the serologic diagnosis of syphilis.

There is no method of artificially inducing immunity against syphilis. Inadequate early treatment may do more harm than good because it fails to cure and at the same time retards the establishment of any degree of immunity.

Laboratory Diagnosis

There are two important laboratory approaches to the diagnosis of syphilis: (1) the demonstration of the spirochetes in the lesions by dark-field microscopy or by immunofluorescence and (2) serologic testing for syphilis, notably the use of complement-fixation and flocculation tests (Table 30-2). Serologic tests include nontreponemal ones to demonstrate reagin in serum and cerebrospinal fluid and treponemal tests to detect the antigens of *T. pallidum* in serum.

Generally, nontreponemal or reagin tests are widely available, whereas treponemal tests are more likely to be done only in reference laboratories. Titers with the reagin tests decline with recovery from disease, whereas the treponemal tests remain positive for that person's life. Table 30-3 compares the reactivity rates of the two different kinds of serologic tests for syphilis.

The dark-field microscope is most practical in the investigation of suspected chancres

Table 30-2 Laboratory Diagnosis of Syphilis

Procedure	Primary Stage	Secondary Stage	Tertiary Stage	Syphilis Not Present
Morphologic: identification of spirochete by dark-field microscopy	+	+	−	−
Serologic				
Reagin tests (nontreponemal)	+ or −	+	Usually +	May be a false +
Treponemal tests (FTA-ABS or MHA-TP)*	+ or −	+	+	−

*FTA-ABS, Fluorescent treponemal antibody absorption test; MHA-TP, microhemagglutination *Treponema pallidum* test.

Table 30-3 Comparison of Reactivity of Serologic Tests in Stages of Syphilis

	Reactivity Rate	
Stage of Disease	Nontreponemal (Reagin) Test (VDRL or RPR) (%)	Treponemal Test (FTA-ABS) (%)
Primary		85 to 88
Early	50 to 70	
Late	75 to 80	
Secondary	98 to 100	99 to 100
Latent	60 to 75	97
Late	70	95
After treatment	Variable	Reactivity persists

and mucous patches. Fluorescent antibody technics also can be used on exudates therefrom. During the first few days of the chancre the spirochetes are found in almost all patients (Fig. 30-4). With time *T. pallidum* gradually disappears.

Serologic tests are seldom positive during the first few days of the chancre but usually become positive before the secondary stage appears. In neurosyphilis, serologic tests on both blood and cerebrospinal fluid are positive in most patients.

The complement-fixation (nontreponemal) test, regardless of the technic, is referred to as the **Wassermann test** because August von Wassermann (1866-1925) in 1906 first applied the principle of complement fixation to the diagnosis of syphilis. The test has been so much improved that, of the original, only the principle and the name remain. Two modifications of the Wassermann test in routine use are the Kolmer and the Eagle.

Precipitation (or flocculation) tests are the Kline, Kahn, Eagle, Hinton, Mazzini, RPR (rapid plasma reagin), and VDRL (Venereal Disease Research Laboratory) tests. Fast, inexpensive, and easy to do, the VDRL test is probably the most widely used. It is used almost exclusively for testing cerebrospinal fluid in the diagnosis of neurosyphilis and generally is the routine test performed on blood for transfusion.

Two terms applied to serologic tests for syphilis (STS), and other laboratory procedures as well, are **sensitivity** (*adj.* sensitive) and **specificity** (*adj.* specific). Sensitivity refers to the percentage of positive results in patients with the disease; specificity refers to the percentage of negative results among those who do *not* have the disease. The Kahn presumptive test and the Kline exclusion test are highly sensitive tests that can detect amounts of syphilitic reagin much less than the smallest amount demonstrated by the so-called standard or diagnostic tests. Sensitive tests have the disadvantage that positive results may be obtained in perfectly healthy persons or in persons who at least do not have syphilis. Therefore, more significant when negative than positive, they are used for *screening*. A positive result must be followed by other serologic tests, and for the laboratory diagnosis of syphilis, positive results are required from two or more additional tests less sensitive than the initial screening test.

Nontreponemal flocculation and complement-fixation tests for syphilis may give negative results in the presence of syphilis and sometimes give positive results in its absence. A serologic test for syphilis cannot diagnose the disease. It only gives immunologic information, and none devised so far is absolutely specific. Such results are called false negative and false positive reactions. False negatives may be caused by technical errors or by undetectably small amounts of syphilitic reagin. False positives come from technical error, or they may be biologic false positives (BFPs). Biologic false positives are occasionally or uniformly found in certain diseases such as yaws, infectious mononucleosis, malaria, leprosy, and rat-bite fever and in drug addicts. Partly to eliminate the biologic false positive, serologic research has focused on tests using as antigens either the organisms, dead or alive, or chemical extracts of them.

The **Treponema pallidum immobilization test** (TPI), a treponemal test in which *T. pallidum* itself is immobilized in the presence of guinea pig complement and syphilitic serum, is negative with serums giving BFP reactions. Its clinical usefulness depends on this fact. As specific as, and more sensitive than, the TPI test is the FTA-ABS (fluorescent treponemal antibody absorption) test, considered to be the best of the treponemal tests today. Some authorities believe it to be 99% accurate. As the name indicates, test serum is absorbed of nonspecific (confusing) antibodies. It is then brought into contact with *T. pallidum* and fluorescein-tagged antihuman globulin in a special way so that the combination may be viewed microscopically. Antibodies to the spirochetes, if present, attach to the organism, and the antihuman globulin in turn unites with them. When seen through the ultraviolet microscope, the result glows beautifully.

Other treponemal tests are the *T. pallidum* complement-fixation test, the Reiter protein complement-fixation test (RPCF), the microhemagglutination assay for *T. pal-*

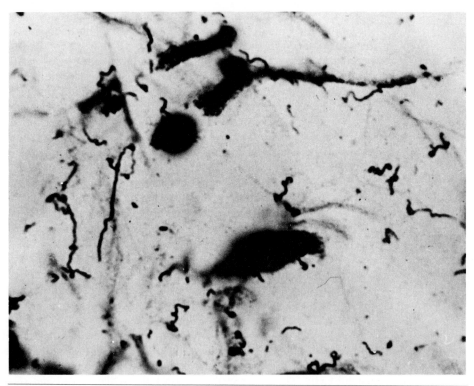

FIG. 30-4   Spirochetes of a primary chancre demonstrated by special silver stain. (Courtesy Dr. R.G. Freeman, The University of Texas Health Science Center at Dallas.)

lidum antibodies (MHA-TP), and the *automated* microhemagglutination assay (AMHA-TP). The antigen for the hemagglutination technics is the red blood cell sensitized with cell components of *T. pallidum*. The microhemagglutination *Treponema pallidum* test is coming to the forefront today, since it not only takes less time to perform but also is less expensive.

Remember that the diagnosis of syphilis must be made only after careful evaluation of both clinical features and laboratory findings.

Prevention

The patient is most likely to convey syphilis during the primary and secondary stages, since chancres and mucous patches are living cultures of syphilitic spirochetes. Medical attendants to patients with these lesions should be careful not to contact the patient's secretions. Certain manual examinations always must be made with gloves because a patient may show no evidence of syphilis but at the same time be infectious.

People should still be educated as to the universal prevalence of syphilis and the danger of syphilis not only to the person who has it but also to the sexual partner and the children. Today a growing public health problem, the disease is significant among young adults and especially teenagers, in whom there has been better than a 200% increase in the incidence of syphilis over the last decade.

Congenital (Prenatal) Syphilis

Congenital syphilis refers to the disease acquired before birth. As a result of the upsurge in the incidence of sexually transmitted diseases, including syphilis, several hundred babies will be born this year in the United States with congenital syphilis. For this form of syphilis to occur, the mother must be infected. The treponemes are bloodborne to the maternal side of the placenta and deposited there. Syphilitic foci develop in the placenta, and the spirochetes cross to the fetal circulation. A syphilitic father can transmit the infection to his child only indirectly, that is, by infecting the mother.

The shorter the time elapsing between infection of the mother and conception, the more likely it is that the unborn baby will be infected. If adequate treatment of the mother is instituted before the fifth month of pregnancy, the child should be born free of syphilis. The requirements by law in many states for premarital examination for syphilis and for prenatal serologic tests for syphilis have been important public health measures in the prevention of this form of the disease.

Congenital syphilis usually appears at birth or within a few weeks thereafter, but it may appear years later. The child of a syphilitic mother may be (1) born dead, (2) born alive with syphilis, (3) born in apparently good health but show evidence of syphilis several weeks or months later, or (4) entirely free of the disease.

The placenta of a syphilitic child is large for the weight of the child. In stillborn infants the lungs fill the entire thoracic cavity, are grayish white, and are incompletely developed, a condition known as white pneumonia, or **pneumonia alba,** and pathognomonic for congenital syphilis. Congenital neurosyphilis simulates the acquired form and may appear in early life or be delayed until adolescence.

Infants born with active syphilis are undersized and strikingly resemble an old man. A vesicular skin eruption and persistent nasal discharge (**snuffles**) are often present. The child may have linear scars at the angles of the mouth (**syphilitic rhagades**). Among the late manifestations of congenital syphilis are poorly developed, small, peg-shaped permanent teeth. The upper central incisors are wedge shaped and show a central notch (**Hutchinson's teeth**). Other late manifestations are interstitial keratitis, anterior bowing of the tibia (saber shin), dactylitis, and neurosyphilis.

To detect congenital syphilis in a newborn baby with no outward signs, a modified FTA-ABS test is applied. Specific antihuman immunoglobulin tagged with fluorescein indicates the presence of 19S immunoglobulins (IgM antibodies). Unlike the smaller 7S immunoglobulins, which are formed by the mother, these macroglobulins cannot cross the intact placenta and must be formed by the baby. Their presence indicates the baby's reaction to the baby's disease, not to the mother's.

Syphilis and Yaws

Yaws, or frambesia, a disease of unhygienic moist tropics and closely akin to syphilis, is caused by *Treponema pallidum* subsp. *pertenue,* a spirochete that cannot be differentiated serologically from *T. pallidum* and that, like *T. pallidum,* is susceptible to arsenic and penicillin. Yaws may be a special form of syphilis; however, most observers believe it is distinct.

Yaws is neither sexually transmitted nor congenital. The disease is spread through direct contact with the lesions, and flies may be mechanical vectors. Yaws is seen mostly in young persons, especially those between 4 and 10 years of age. The first sign of infection in infancy or childhood is a primary skin lesion that persists. From this focus treponemes move throughout the body, and in a secondary stage a generalized skin eruption appears, the lesions of which are swarming with spirochetes. When the yellowish crusts covering the large pustules in the skin are removed, the slightly bleeding surface looks exactly like a raspberry stuck to the skin—hence the name *frambesia,* or raspberry disease. This stage may last for months before it regresses. After a variable latent period the painful skin and the destructive, disfiguring bone lesions of the tertiary stage appear. In yaws, skin and bone lesions are prominent; visceral ones are rare. **Crab yaws** refers to the incapacitating and painful lesions on the soles of the feet.

BORRELIA SPECIES

Members of the genus *Borrelia* are helical cells with coarse, uneven coils. They are anaerobic parasites found on mucous membranes, and some cause disease in humans and animals (for example, avian and bovine spirochetosis).

Human Disease

The relapsing fevers, a group of clinically similar fevers, are acute infectious diseases caused by one or another of closely related species of spirochetes in genus *Borrelia* (Fig. 30-1). A relapsing fever is described clinically as alternating periods of severe febrile illness with apparent recovery (Fig. 30-5). (Relapsing fever has been called bilious typhoid, famine fever, vagabond fever, mianeh fever, carapateh fever, and kimputu.) The characteristic fever is abrupt in onset; the temperature climbs rapidly to 102° to 104° F (39° to 40° C) and remains elevated for 3 to 10 days; then, just as abruptly as it appeared, the fever clears. Two or three relapses are usual, with each successive episode of fever a little less severe in its effect. The incubation period is 3 to 10 days.

The spirochetes, found in the peripheral blood during the fever, are transferred from person to person by body lice *(Pediculus humanus* subsp. *humanus)* and from rodent to human by ticks of *Ornithodoros* species. (*Ornithodoros* ticks usually bite painlessly, take a short blood meal, and leave the host after 30 minutes to 1 hour. The host may be unaware of the tick bite.) Thus there are two main types of relapsing fever—louse-borne and tick-borne. The former occurs as epidemic relapsing fever in the general pattern of louse-borne disease, and the spirochete causing it is *Borrelia recurrentis.* Tick-borne relapsing fever does not occur in epidemics but is endemic relapsing fever caused by other species of *Borrelia.* It is the only type found in North America. Infected ticks may transmit the infection to their offspring for generations. Body lice do not do this. Rodents such as the ground squirrel and prairie dog are the main reservoirs of the tick-borne disease, but a wide variety of animals may be infected with borrelias, including armadillos, bats, dogs, foxes, horses, rabbits, and porcupines.

Laboratory Diagnosis

Actively motile spiral organisms, *Borrelia* species differ from *T. pallidum* in that they take the usual laboratory stains. They may be detected in the Wright-stained smears of peripheral blood or by dark-field illumination. Blood taken from a patient during the fever may be inoculated intraperitoneally into a white mouse or young guinea pig.

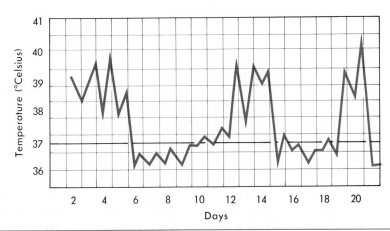

FIG. 30-5   Fever curve of tick-borne relapsing fever. Note fever-free intervals in relation to elevations of temperature.

Wright-stained films made from the tail blood 1 to 4 days later will show the borrelias. *Borrelia* species can be grown in the chick embryo.

LEPTOSPIRA SPECIES

Leptospires are tightly and finely coiled spiral bacteria with one or both ends bent typically to form a hook. Some are free living; others are parasitic or pathogenic in vertebrates. The type species of the genus *Leptospira* recognized in Bergey's classification (1984) is *Leptospira interrogans (Leptospira icterohaemorrhagiae)*. Taxonomic data in this genus are incomplete at present.

The Disease

Leptospirosis is an acute febrile disease of protean manifestations caused by serotypes of *Leptospira interrogans*. There are various names for this increasingly frequent condition, including swamp fever, swineherd's disease, infectious jaundice, spirochetal jaundice, and Weil's disease, the most severe form.

Clinically leptospirosis is manifested by high fever, muscular pains, redness of the conjunctivae, myalgia, jaundice (not invariable), and aseptic meningitis. The triad of nephritis, hepatitis, and meningitis marks Weil's disease. An attack is followed by a lasting immunity. Inapparent infections also occur.

Leptospires (Fig. 30-1) are found in wild and domestic animals all over the world. The disease is mainly a zoonosis; humans are only accidentally infected. In animals the spirochetes localize in the kidney and are excreted profusely in the urine. They can survive if urine is discharged to neutral or slightly alkaline water, sewage, or mud. A person probably acquires the infection through the skin from soil contaminated with the urine of infected rats or through the mouth by food or water that has been contaminated in the same way. The organisms enter the animal or human body through the abraded skin and mucous membranes of the eyes, nose, and mouth. Leptospirosis is important as an occupational disease. Especially vulnerable are workers in rat-infested mines, rice fields, and sewage disposal plants.

A hazard for leptospirosis exists when apparently healthy dogs, which can be carriers, are allowed into swimming pools. Today, in instances where the source of infection can be pinpointed, contact with dogs is the most frequent and contact with contaminated water is second. Dogs should be vaccinated against leptospirosis. For treatment of swimming pool water, 3 to 5 parts per million of residual chlorine kills *L. interrogans* (and probably its serotype in dogs). For continued protection 1 to 2 parts per million of residual chlorine in the pool is recommended.

Laboratory Diagnosis

The spirochetes are found in the blood early in the disease and in the urine after the seventh day. Leptospires of typical shape and motility are best seen by dark-field examination of blood and urine. They may be recovered from blood cultures and from the peritoneal cavity of a guinea pig inoculated with blood or urine. A microagglutination test using live antigen is diagnostic.

OTHER SPIROCHETES AND ASSOCIATED ORGANISMS

Treponema denticola is one of several saprophytic treponemes found in the mouth. Their presence may cause confusion in the examination of material from the mouth that might contain the spirochete of syphilis. *Treponema refringens* is part of the normal flora of the human male and female genitalia and may be associated with *T. pallidum* in various syphilitic lesions.

A medical mystery, first brought to researchers' attention by two Connecticut mothers, involves an inflammatory disorder that has come to be known as Lyme disease. Researchers are now reporting a possible cause—a spirochete isolated from ticks of the *Ixodes* genus.

Lyme, the town from which the disease draws its name, is a small community about 15 km north of Long Island Sound near the Connecticut River. In late 1975, a woman in Lyme, whose daughter was diagnosed as having juvenile rheumatoid arthritis, telephoned Connecticut's state health department in Hartford to point out that at least 11 other children in the locality had similar symptoms. Could there be an explanation other than arthritis? Similar inquiries triggered a clinical and epidemiologic investigation that continues to this day.

The clinical picture emerged as an illness of uncertain cause manifested typically by a distinctive expanding annular, erythematous skin lesion and recurrent bouts of arthritis. The culprit is now thought to be a penicillin-sensitive infectious agent transmitted by *Ixodes* spp ticks. Only one in five patients remembers a tick bite.

Investigators collected adult *Ixodes dammini* ticks from vegetation on Shelter Island, a small summer resort area at the eastern end of New York's Long Island and "a known endemic focus of Lyme disease." When they dissected 126 of the ticks they found spirochetes primarily in the midguts of 65 males and 12 females. Some ticks contained clumps of spirochetes. Electron microscopy indicated structural features of the spirochetes that were "similar to those reported for *Treponema* species." One indication that the spirochetes may be involved is that they bind antibodies in the blood of persons with clinically diagnosed Lyme disease. Finally in experiments with eight New Zealand white rabbits on which ticks harboring spirochetes were allowed to feed, the skin lesions developed on the back and lateral trunk of the animals 10 to 12 weeks later, and antibodies to spirochetes in significant titers were found in the blood of the animals 30 to 60 days later.

Modified from Gunby, P.: J.A.M.A. 248:813, 1982. Copyright 1982, American Medical Association.

Vincent's Angina (Fusospirochetal Disease)

In Vincent's infection (fusospirochetal disease) a grayish white pseudomembrane forms in the throat or mouth, beneath which ulceration occurs. When the gums and mouth are primarily involved, the disease is known as **trench mouth**. When the throat and tonsils are ulcerated, it is called *Vincent's angina* (Fig. 30-6). The disease is more properly called **acute necrotizing ulcerative gingivitis**. The disease is important by itself and because the membrane may be mistaken for a diphtheritic membrane. The extensive ulceration may cause the disease to be confused with syphilis.

Associated together in the lesions are two microbes: (1) a large, gram-negative, cigar-shaped, anaerobic fusiform bacillus and (2) a gram-negative spirochete, *Treponema vincentii*. Vincent's infection is probably caused by *Bacteroides melaninogenicus*, found in the mouth and pathogenic usually when associated with other kinds of organisms. Early the cigar-shaped bacilli are more numerous, whereas later on in the disease the spirochetes are. These two do not cause this condition, and most workers believe that they are only secondary invaders. They grow symbiotically here as opportunists, being

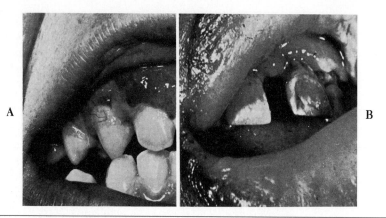

FIG. 30-6   Vincent's angina (necrotizing gingivitis.) **A,** Early stage. **B,** Late stage with marked destruction of tissues. (From Bhaskar, S.N.: Synopsis of oral pathology, ed. 5, St. Louis, 1977, The C.V. Mosby Co.)

already in the mouth. Their mere presence is not enough to cause disease. This must be triggered by some unusual circumstance, such as injury to the mouth or decreased oral resistance. Vincent's angina accompanies malnutrition, debilitating states, viral infections, and poor oral hygiene. Trench mouth is said to be associated with a high level of psychosocial stress.

Fusospirochetal organisms are identified directly on a crystal violet–stained smear of membranous exudate.

SPIRILLUM MINUS
Rat-Bite Fever (Sodoku)

Two clinical entities known as rat-bite fever are conveyed by the bite of a rat or other rodent. One, known in Japan as sodoku, is caused by *Spirillum minus*. This small and rigid spiral microbe (Fig. 30-1) is found primarily in wild rats and is spread from rat to rat and from rat to human by the bite of the rat (the healthy carrier). The features of the disease are ulceration of the bite, fever, and a skin eruption. Toes or fingers are likely sites of the rat bite, and regional lymph nodes enlarge as a consequence. In a typical incident an infant in bed is bitten in the night, and a massive lymph node appears 3 to 4 days later.

The spirilla may be seen in material from the ulcer examined in stained smears or by dark-field illumination, and a guinea pig may be inoculated with blood or tissue from lymph nodes to isolate the microbes. This type of rat-bite fever, although uncommon, is worldwide in distribution.

Streptobacillary Rat-Bite Fever

The other entity is rat-bite fever (Haverhill fever) caused by *Streptobacillus moniliformis,* an actinomycete-like, necklace-shaped bacterium found in the mouth and nasopharynx of normal rats, among which it causes widespread epidemics. A person contracts the infection from the bite of a rat. The disease is a febrile one with manifestations similar to those of rat-bite fever caused by *Spirillum minus.* Both resemble tularemia clinically. *Streptobacillus moniliformis* is identified by fluorescent antibody technics.

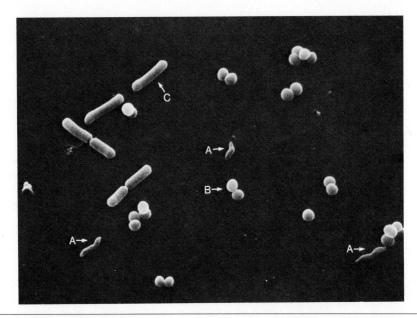

FIG. 30-7 *Campylobacter fetus (A)* in scanning electron micrograph to show characteristic "seagull" configuration and small size relative to two other microbes, *Staphylococcus aureus (B)* and *Bacillus subtilis (C). S. aureus* here is either a single coccus, a pair, or a small cluster of cocci (Chapter 23). The large rods are *B. subtilis* (p. 116). (×9500.) (Courtesy Lilly Research Laboratories, Indianapolis.)

CAMPYLOBACTER FETUS

Campylobacter (Greek *kampylos*, curved) *fetus* (formerly V*ibrio fetus*), the type species of a heterogeneous group noted for a flagellum three times longer than the bacterial body, is a fastidious, oxidase-positive, nonfermentative spiral (curved or corkscrew-shaped), gram-negative microbe found in the reproductive organs, intestinal tract, and oral cavity of animals and humans (Fig. 30-7). It is highly motile and moves in corkscrew fashion.

Human Disease

Long known as a pathogen in animals, *C. fetus* is gaining attention as a troublemaker in humans, causing a diversity of clinical infections: an undulating fever, thrombophlebitis of upper and lower extremities, and sepsis with localization in viscera. *Campylobacter* gastroenteritis is an acute mucosanguineous diarrhea lasting several days with or without fever; the incubation period is 3 days. An underlying condition such as alcoholism or cardiovascular disease may predispose the person to infection with *Campylobacter* species, but exactly how human beings become infected is unclear. Ingestion of contaminated food or water is implicated.

Laboratory Diagnosis

With dark-field or phase-contrast microscopy these microbes may be identified in stool specimens. They may be recovered in blood culture and from rectal swabs or stool specimens when special technics and selective media are used. Cultures are incubated at 42° C, since this microbe is thermophilic. A sensitive indirect hemagglutination test is available.

SUMMARY

1 Spirochetes are actively motile, flexible, spiral bacteria found in contaminated water, soil, decaying organic matter, and the bodies of animals and humans. They move by rapidly rotating about their long axis, by bending, or by snaking along a corkscrew path.

2 The genus *Treponema* contains the agents of syphilis and yaws. The genus *Borrelia* contains the agents of relapsing fever. In the genus *Leptospira* are found the microbes responsible for Weil's disease (infectious jaundice).

3 Slightly different spiral microbes are found in the genus *Spirillum*, which contains the twisted rods causing one form of rat-bite fever, and in the genus *Campylobacter*, which contains microaerophilic, curved rods causing gastroenteritis.

4 Syphilis is transmitted by close and intimate contact, usually sexual, and humans are the only reservoir. Syphilis may be acquired or congenital, that is, incurred around the time of birth.

5 The stages in the evolution of a typical case of syphilis include (1) the incubation stage, (2) the primary stage, (3) the primary latent stage, (4) the secondary stage, (5) the latent stage, and (6) the tertiary stage.

6 The chancre of syphilis appears from 2 to 6 weeks after infection, persists 4 to 6 weeks, and then heals without residual sign.

7 Manifestations of secondary syphilis include skin and mucosal lesions, rubbery lymph nodes, and influenza-like symptoms.

8 The hallmark of tertiary stage is the destructive gumma, a firm, yellowish white focus of disease walled off by scar tissue.

9 The antibody-like substance tested for in syphilitic patients is an IgM primarily directed against cardiolipin, a lipid antigen that can be extracted from the heart muscle of an ox.

10 In the laboratory serologic testing for syphilis is done with either nontreponemal or treponemal tests. Nontreponemal tests are designed to detect reagin in serum and cerebrospinal fluid. Treponemal tests detect the antigens of the causative spirochete. The fluorescent treponemal antibody adsorption test is considered to be 99% accurate.

11 A relapsing fever alternates periods of severe febrile illness with apparent recovery. Helical bacteria with coarse, uneven coils, the borrelias, cause relapsing fevers. They are transmitted by body lice and ticks; certain rodents are the main reservoirs.

12 Leptospires, tightly and finely coiled spiral bacteria with one or both ends hooked, cause leptospirosis. Leptospirosis is a zoonosis; humans are accidentally infected. The most severe form, Weil's disease, is characterized by kidney, liver, and central nervous system involvement.

13 Vincent's angina, also known as fusospirochetal disease, is an acute necrotizing inflammation of the gums associated with ulceration. A large, gram-negative, cigar-shaped anaerobic bacillus, probably *Bacteroides melaninogenicus*, and a gram-negative spiral microbe, *Treponema vincentii*, together, are implicated in this disease.

14 *Campylobacter fetus* is the type species of a heterogeneous group noted for a flagellum three times longer than the bacterial body. It is a fastidious, oxidase-positive, nonfermentative microbe that causes gastroenteritis.

QUESTIONS FOR REVIEW

1 Name and describe the causative agent of syphilis.
2 Outline the evolution of a typical case of syphilis.
3 Characterize congenital syphilis. What are its hazards?
4 How is syphilis transmitted?
5 Outline the laboratory diagnosis of syphilis. Briefly discuss the serologic tests.
6 Classify spiral bacteria. Present salient features.
7 Comment on the pathogenicity of *Treponema pallidum*. Why is syphilis called the "great imitator"? What is a gumma?
8 How does the causative agent of yaws compare with that of syphilis?
9 Name two vectors of relapsing fever and possible rodent reservoirs.
10 Characterize leptospirosis.
11 Give the names of the two microbes causing rat-bite fever.
12 How is Vincent's angina diagnosed bacteriologically?
13 Briefly explain what is meant by frambesia, reagin, STS, STD, latent disease, gumma, granuloma, congenital disease, cardiolipin, sensitivity of a test, specificity of a test, relapsing fever, trench mouth, Vincent's angina, sodoku, kimputu, and fusospirochetal disease.

SUGGESTED READINGS

Blaser, M.J., and others (The Collaborative Diarrheal Disease Study Group, Centers for Disease Control, Atlanta): *Campylobacter* enteritis in the United States, a multicenter study, Ann. Intern. Med. 98:360, 1983.

Bokkenheuser, V.D.: Human campylobacteriosis: laboratory and clinical aspects, Lab. Manage. 19:55, June 1981.

Chang, Y.W.: A guideline to serologic tests for syphilis, Diagn. Med. 6:51, March-April 1983.

Duel, W.: Leptospirosis risk from dogs in family swimming pools, J.A.M.A. 239:2177, 1978

Edelman, R., and Levine, M.M.: Acute diarrheal infections in infants. II. Bacterial and viral causes, Hosp. Pract. 15:97, Jan. 1980.

Goodhart, G.L.: Use and interpretation of serologic tests for the diagnosis of syphilis, South. Med. J. 76:373, 1983.

Kaplan, R.L., and others: *C. fetus* ssp. *jejuni*: a prime suspect in enteritis, Diagn. Med. 5:23, March-April, 1982.

Syphilis before birth, Emerg. Med. 12:126, June 15, 1980.

Ticks with other tricks, Emerg. Med. 11:63, Oct. 15, 1979.

Wadera, M., and Mukhopadhyay, A.K.: *Campylobacter* enteritis—an important cause of diarrhea, Tex. Med. 77:45, 1981.

Wang, W.L.L., and others: Effects of disinfectants on *Campylobacter jejuni*, Appl. Environ. Microbiol. 45:1202, 1983.

Waring, G.W., Jr.: False-positive tests for syphilis revisited, the intersection of Bayes' theorem and Wassermann's test, J.A.M.A. 243:2321, 1980.

CHAPTER 31

MISCELLANEOUS MICROBES

Pseudomonas Species
    Pseudomonas aeruginosa
    Pseudomonas pseudomallei
    Pseudomonas mallei
Bacillus anthracis (the Anthrax Bacillus)

Lactobacillus Species (the Lactobacilli)
Listeria monocytogenes
Mycoplasma Species
Legionella pneumophilia

PSEUDOMONAS SPECIES

The family Pseudomonadaceae of strictly aerobic, motile, straight or curved, gram-negative rods contains the genus *Pseudomonas* (Greek *pseudēs*, false, plus *monas*, unit) of soil and water bacteria (even seawater and heavy brine bacteria). They are noted for their water-soluble, fluorescent pigments and for their role in mineralization of organic matter.

Pseudomonas aeruginosa

The organism of blue pus, *Pseudomonas aeruginosa* (Latin *aerugo*, rust of copper), is an actively motile, nonspore-forming, oxidase-positive bacillus (Fig. 31-1). When stained with simple stains, it resembles *Corynebacterium diphtheriae*. It is unique for the diffusion of a blue-green pigment (pyocyanin, a blue pigment in a neutral or alkaline medium and red in an acid one, and pyoverdin, a yellowish green fluorescent pigment) through the medium on which it is grown; the same color is seen in the purulent discharges it produces in disease. Green pigment belongs to several species of bacteria, but the blue color of chloroform-soluble pyocyanin seems to be specific for *P. aeruginosa*. Pathogenic pseudomonads produce a great deal of fluorescent pigment. This pigment is readily detected with Wood's ultraviolet light, even under the black, silver-stained eschar of a burn wound that has been treated with silver nitrate.

    P. aeruginosa is nutritionally versatile, growing on almost any medium, and cultures have an odor like that of fermented grapes. It oxidizes, not ferments, carbohydrates. Aging, surface-grown cultures of *P. aeruginosa* display a diffuse bright iridescence (metallic sheen) over the growth. Varied explanations are given for the **iridescent phenomenon,** but the mechanism is not understood.

Its Potential

As a saprophyte, *P. aeruginosa* is ubiquitous in nature and in every part of the human habitat. As a pathogen, it has a wide, all-inclusive host range. Among human pathogens it is unique in that it also infects lower animals, both warm- and cold-blooded vertebrates, insects, and plants. It can destroy a tobacco or sugar cane crop, and hospital patients have even acquired infection from flowers growing in the room.

    Ordinarily it is but mildly pathogenic; it does not threaten the healthy laboratory technologist but causes primary infection only when the resistance of the host is lowered. Therefore it is a problem in the debilitated, the compromised host, young infants

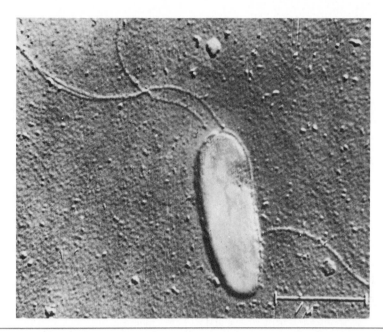

FIG. 31-1 *Pseudomonas aeruginosa*, electron micrograph. Note the flagella. (Courtesy J.B. Roerig Division, Chas. Pfizer & Co., Inc., New York, N.Y.)

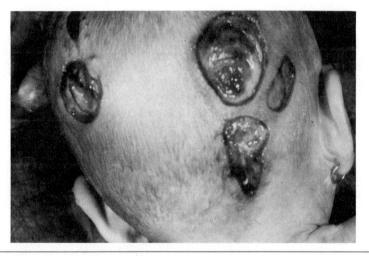

FIG. 31-2 Skin ulcers on the head of a young child with systemic *Pseudomonas aeruginosa* infection.

(Fig. 31-2), and the aged. The mortality from its infections is high. As a secondary invader, it often delays the repair of wounds. Among the diseases that it may cause are otitis media, abscesses, wound infections, sinus infections, and bronchopneumonia (Fig. 31-3). It is also an important cause of gram-negative shock (mortality, 80%), urinary tract infections, and the sepsis complicating severe burns. Infected burn wounds exude a blue pus. Mucoid strains of *P. aeruginosa* relate to the disease cystic fibrosis. They colonize the respiratory tract in such patients (rarely being found in sputum otherwise), persist, and tend to correlate with an unfavorable prognosis. *Pseudomonas*

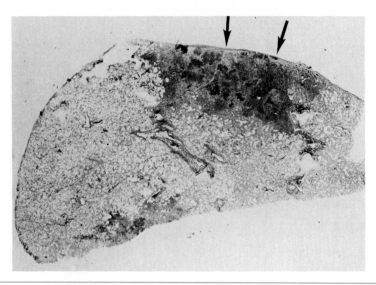

FIG. 31-3   Bronchopneumonia caused by *Pseudomonas aeruginosa*. A well-defined, solidified, darkened portion *(arrows)* is sharply distinct in a thin slice of air-containing lung. The focus of disease is hemorrhagic and partly necrotic.

pneumonia is the principal bacterial infection in these patients and often the cause of their death.

The way in which *P. aeruginosa* brings about disease is not fully understood. It is known to possess a weak endotoxin and to yield many extracellular products, indeed a battery of exotoxins. The *Pseudomonas* arsenal includes proteases, which enhance the invasiveness of the microbe; elastases, which injure blood vessel walls; collagenases, which degrade the collagen of the skin; phospholipase and hemolysins, which damage cell membranes; and an exotoxin (exotoxin A) comparable in its action to diphtheria exotoxin, which disrupts protein synthesis in eukaryotic cells. The necrosis of tissue, a striking feature of infection with this microbe, is related to its necrotizing proteolytic enzymes, the proteases. Microscopic examination of the areas of dead tissue reveals a massive overgrowth of the organisms therein.

This organism is very hardy. It is more resistant to the chemical disinfectants usually employed than are the other gram-negative bacilli and can actually grow in the dilute solutions. However, it can be killed by exposure to a temperature of 55° C for 1 hour. In the treatment of *Pseudomonas* infections, polymyxin and colistin, rather toxic drugs, are often the only antibiotics to which the organisms are susceptible. Since *P. aeruginosa* is resistant to the action of most antibiotics, it tends to become the dominant organism in a diseased area after extended antimicrobial therapy has eliminated the primary ones. Then it takes on a vigorous, destructive disease-producing capacity, especially with lowered host resistance.

The sources of *P. aeruginosa* in hospital-acquired infections are many, especially pieces of equipment that are hard to clean and sterilize, such as face masks, certain kinds of humid rubber tubing, water containers, nebulizers, and parts of the intermittent positive-pressure breathing machines. Psychrophilic strains that can grow at refrigerator temperatures can contaminate stored fluids or blood. A source easily overlooked is the hospital supply of distilled water, a serious hazard because of the widespread need for distilled water in preparation of various hospital solutions—detergents, disinfectants, and even parenteral medications. Pseudomonads have been demonstrated not only to survive but also to multiply rapidly in distilled water. Although pseudomonads are vir-

"Hot tub dermatitis," a new disease, is caused by a pseudomonad that thrives in a very warm, moist climate, as would be found in a hot tub, particularly when the tub is not properly chlorinated and when the filters are not changed often enough. The disease first appeared around 1975 at the time the hot tub was becoming popular. It is characterized by a rash and gastrointestinal distress that last for several days. The eyes and ears may be irritated. The ailment usually clears in a week.

tually impossible to eliminate, maximal aseptic technic is a must in the prevention of hospital infections caused by them.

Patients with 40% or more burned surface area are vulnerable to *Pseudomonas* sepsis. Administration of a heptavalent lipopolysaccharide *Pseudomonas* vaccine to these seriously burned persons is a great advance in the management of their injury.

Pseudomonas pseudomallei

Pseudomonas pseudomallei is a gram-negative, motile, nonpigmented bacillus that causes **melioidosis,** an uncommon tropical disease of humans and animals. Its pulmonary manifestations are easily confused with those of tuberculosis. There may be a septicemia, with widespread lesions over the body. Nicknamed the "Vietnamese time bomb," the disease can be dormant for a number of years only to appear suddenly, become active, and kill within days or weeks. The disease is endemic in Southeast Asia, where the organism is saprophytic in surface waters, soil (notably the rice fields), and organic matter. The epidemiology and transmission of the disease are not known. The organisms probably enter the body through the mouth and nose or open wounds in the skin.

Growth of the bacilli is typically wrinkled and crinkly in cultures, with a distinctive odor. A hemagglutination test is available for diagnostic study.

Pseudomonas mallei

Glanders (farcy) is chiefly a disease of horses, mules, and donkeys but may be transmitted to humans. Cattle are immune. Ulcerating, tubercle-like nodules form in the lungs, superficial lymph nodes, and mucous membranes. If lymph nodes are affected, the disease is known as **glanders;** if mucous membranes are affected, it is **farcy.** With the replacement of the horse by the automobile, this once prevalent disease is now rare.

The cause, *Pseudomonas (Actinobacillus) mallei*, is a narrow, sometimes slightly curved, small bacillus that is gram negative, nonmotile, nonencapsulated, and nonspore-forming, with little resistance to physical and chemical agents. A human being is easily infected in contact with tissues or excreta of diseased animals, since these contain virulent bacilli. The microbes enter the body by a wound, scratch, or abrasion of the skin.

Laboratory Diagnosis

Glanders is diagnosed in humans and animals by the isolation of the bacilli from the lesions or in blood cultures. There is a complement-fixation test. The skin test is performed with **mallein,** a product of *P. mallei*, injected subcutaneously in animals. Mallein tablets may be placed in the conjunctival sac and the reaction observed. When infectious material is injected intraperitoneally into a male guinea pig, the testicular swelling and generalized reaction within 3 or 4 days is known as the **Straus reaction.**

Prevention

No immunity exists to glanders. Control depends on destruction of animals with clinical or occult disease and on disinfection of stables, blankets, harnesses, and drinking troughs used by sick animals.

BACILLUS ANTHRACIS (THE ANTHRAX BACILLUS)

Known since antiquity, anthrax (charbon) is an acute infectious disease caused by *Bacillus anthracis*. Primarily a disease of lower animals, especially cattle and sheep, it is easily communicated to humans. Horses and hogs may become infected. Anthrax is found worldwide but, with the exception of certain restricted areas, is unusual in the United States.

In humans anthrax presents two forms: **external** (malignant pustule or carbuncle and anthrax edema) and **internal** (pulmonary anthrax or wool-sorter's disease and intestinal anthrax). External anthrax is the more common. Because of the fever and enlargement of the spleen that accompany the disease, anthrax is often called *splenic fever*. The incubation period for external anthrax is 1 to 5 days.

In the body of an animal that has died of acute disease, the dark blood and a swollen spleen are prominent.

General Characteristics

The gram-positive *B. anthracis* is quite large, in fact one of the largest bacterial pathogens. It forms central spores, and the swollen and concave ends of the bacilli give the chains of bacilli the appearance of bamboo rods. The organism grows in the presence of oxygen and also in its absence. It is the only spore-forming aerobic pathogenic bacterium and, unlike most spore-forming aerobic bacteria, is nonmotile. Since spores are not formed in the absence of oxygen, they are not formed in the animal body. In the blood and tissues of infected animals the anthrax bacillus is surrounded by a capsule. Growth is luxuriant on all ordinary culture media. Spores retain their vitality for years and are extremely resistant to heat and chemical disinfectants in the concentrations ordinarily used but are moderately susceptible to sunlight.

B. anthracis is of historical interest because it was the first pathogenic organism to be seen under the microscope, the first one *proved* to be the cause of a specific disease, the first one grown in pure culture, and the organism that Pasteur used in his classic experiments on artificial immunization.

Modes of Infection

Animals usually become infected by the intestinal route while grazing in infected pastures. Buzzards carry the infection long distances and contaminate soil and water with organisms on their feet and beak. Dogs discharge spores in their feces after eating the carcasses of infected animals. This disease is sometimes transmitted from animal to animal and from animal to human by greenhead flies and houseflies.

In human beings, anthrax is primarily an occupational disease confined to those who handle animals and animal products, such as hair and hides. Imported animal products are especially dangerous. The most common route of infection is through wounds or abrasions of the skin (cutaneous route). In numerous cases of anthrax, infection of the face by the bristles of cheap cow hair shaving brushes has been reported. The pulmonary form of anthrax is caused by inhalation of dust containing the spores (respiratory route) and endangers those who handle dry hides, wool, or hair. The intestinal form of anthrax is acquired by ingestion of infected milk or insufficiently cooked food (intestinal route). The bacilli leave the body in the exudate of the local lesion (malignant pustule) and in the sputum, feces, and urine.

Prevention

Patients with anthrax should be isolated. The dressings of external lesions should be burned, and the person doing the dressings should wear gloves. The feces, urine, sputum, and other excreta should be disinfected at once to prevent the formation of spores. The disinfectant should be strong and its application prolonged. Phenol, 5%, is probably the best one. The local lesions of anthrax should not be traumatized, as by squeezing, because this may cause the bacilli to invade the bloodstream.

The possibility of confusing a malignant pustule with an ordinary boil is so important that a description of the former is necessary. A malignant pustule begins as a small, hard, red area; a small vesicle soon develops in the center. The area increases in size, vesicles develop at the periphery, and the surrounding tissues become swollen. The regional lymph nodes draining the site enlarge. The center of the lesion softens, and a dark eschar forms. Pain is not present.

Infected animals should be separated from the herd. The bodies of animals that have died of anthrax should be completely burned to ashes. (Cremation prevents coyotes, bobcats, dogs, and other scavengers from catching the disease.) Their blood should not be allowed to escape, and their bodies should not be opened for autopsy except by an experienced veterinarian because the bacilli form spores when exposed to the air. If the body is buried, it should be packed with lime and buried at least 3 feet in the ground. When infection occurs in a herd, the pasture must be changed and the herd quarantined. A sharp watch should be kept for new cases, and the quarantine should not be lifted until 3 weeks after the last case. Cattle that have been vaccinated and that survive an epidemic must not be taken to slaughter for 42 days. Milk from infected herds should not be used. Hides and hair should be sterilized. If the soil becomes contaminated, it remains infectious for years. The best soil disinfectant is lye. Anthrax spores from buried animals have been brought to the surface by earthworms.

Prophylactic vaccination of animals against anthrax must be carried out with vigor. Cattle and sheep may be actively immunized with a vaccine that contains living but attenuated bacteria. Dead bacteria are without effect. A therapeutic serum has been prepared by immunization of horses against anthrax bacilli.

An anthrax vaccine prepared from a culture filtrate of an avirulent, nonencapsulated strain is available for laboratory workers at risk and for workers in occupations who are exposed to the disease.

LACTOBACILLUS SPECIES (THE LACTOBACILLI)

The lactobacilli are gram-positive, nonsporulating, microaerophilic rods that produce lactic acid from simple carbohydrates and preferentially grow in a more highly acid environment (pH 5) than most other bacteria. They are widely distributed in nature and out of character as disease producers. Many are normal inhabitants of the alimentary tract; most are nonmotile. Since they require many essential growth factors for their metabolism, they are used commercially in the bioassay of the B complex vitamins and certain amino acids. They are a more sensitive index than current chemical methods. They are also used extensively in the fermentation and dairy industries.

Some noteworthy members of the genus are *Lactobacillus delbrueckii, Lactobacillus acidophilus, Lactobacillus bulgaricus,* and *Lactobacillus casei. L. delbrueckii* is the type species. *L. acidophilus,* a normal inhabitant of the intestinal tract, is increased by a diet rich in milk or carbohydrates. The *Boas-Oppler bacillus,* found in the stomach in gastric cancer, is most likely the same organism. *L. bulgaricus* was isolated from Bulgarian fermented milk and is the organism that Metchnikoff believed would prevent intestinal putrefaction and thereby increase length of life. *L. casei* is most often used in the bioassay of vitamins and amino acids. *Lactobacillus* species make up the normal flora of the vagina during a woman's active reproductive years, where they are referred to collectively as *Döderlein's bacilli*.

Lactobacillus bifidus, found in the intestine of breast-fed infants, is now classified as *Bifidobacterium bifidum*, the type species of that genus.

LISTERIA MONOCYTOGENES

Listeria monocytogenes (monocyte plus Greek *gennan*, to produce) is a small (less than 2 μm long), gram-positive, aerobic to microaerophilic, nonspore-forming, motile coccobacillus. The genus name is for Lord Joseph Lister (1827-1912), an English Quaker physician and the founder of antiseptic surgery. Four serogroups with eleven serotypes exist based on O and H antigens.

L. monocytogenes affects humans and a great variety of animals. An intracellar pathogen, it causes listeriosis, a disease remarkable for the increase of large mononuclear leukocytes (monocytes) found in the bloodstream and one that is a prime example of disordered cell-mediated immunity. Because of this, listeriosis has proved to be a satisfactory disease model for the study of cell-mediated immunity, and much of what has been learned in this respect has come from the study of this disease.

Listeriosis

Listeriosis, especially in ruminant animals, is an encephalitis or an encephalomyelitis. In humans the most prominent manifestation of the disease is a purulent meningitis. Infection, however, can disseminate; listeriosis has been reported as conjunctivitis, endocarditis, urethritis, and septicemia with multiple abscesses. If untreated, it is fatal in 90% of cases. **Granulomatosis infantiseptica,** peculiar to the fetus and newborn, is an intrauterine listerial infection associated with widespread necrosis of the internal organs and hemorrhagic areas in the skin. The newborn infant usually dies within 2 or 3 days. Listeriosis represents a distinct hazard to the compromised host, but especially to the baby, the aged person, and the renal transplant recipient. In newborns and immuno*in*competent persons, listeral meningitis may not present the classic signs of meningeal irritation such as a stiff neck. Although human listeriosis is rare, the number of human cases is increasing, as is the number in poultry and livestock.

Laboratory Diagnosis

From specimens of blood, cerebrospinal fluid, amniotic fluid, genital secretions, and pus, the microorganism can be identified by cultural methods in the laboratory. Its colonies on blood-free agar display a blue-green iridescence if observed under obliquely reflected light. It is catalase-positive, produces beta-hemolysis, and is novel in that it can grow at the temperature of the refrigerator. In a highly specific test for identification of the organism, the Anton test, the everted eyelid of a rabbit is swabbed with the culture. Typically, the positive reaction to *Listeria* species is the development of purulent keratoconjunctivitis.

Spread

L. monocytogenes is widely distributed in nature and can be recovered from diverse sources such as human and animal feces, ferrets, insects, sewage, silage, and decaying vegetation. Listeriosis is a classic zoonosis: humans are accidental hosts. The transmission of listeriosis in humans and animals is not known.

MYCOPLASMA SPECIES

Mycoplasmas are out of the ordinary. They do not have a cell wall, and therefore they demonstrate a set of remarkable and unusual features. In the 1984 edition of *Bergey's*

Manual they are segregated into section 10. The classification begins with class I Mollicutes in division Tenericutes to encompass the delicate coccoid to filamentous prokaryotic microbes with a "pliable cell boundary." Order I, Mycoplasmatales, and family I, Mycoplasmataceae, gather the ones needing sterol for continued growth. Table 31-1 compares mycoplasmas with other microbes.

Gram-negative, nonmotile, and tiny, some mycoplasmas are so small as to be ultramicroscopic (about 200 nm). In fact, they are considered to be the smallest free-living self-replicating units, that is, forms of life capable of independent existence. Some of the spherical ones with a diameter of about 0.35 μm are so small that 8000 of them would fit inside a human red blood cell. The entire body of a mycoplasma consists of about 1 billion atoms.

Mycoplasmas are everywhere. They are saprophytes, parasites, and pathogens for a wide range of hosts.

General Characteristics

Mycoplasmas (historically pleuropneumonia-like organisms, or PPLOs) are smaller than ordinary bacteria and about the size of the larger viruses (Table 31-1); their cell volume is estimated to be one fifth that of *Escherichia coli*. They are the smallest microorganisms that can be cultivated on cell-free media (Fig. 31-4). They possess soft, fragile cell bodies, which are bounded by a trilaminar membrane but lack a rigid cell wall. Thus they are highly pleomorphic. Since they have no cell wall, they would also have to be gram negative and insensitive to penicillin, an antibiotic that inhibits cell wall synthesis. They contain both DNA and RNA and some metabolic enzymes. Unlike bacteria, they incorporate sterols in their unit membrane. Like viruses, they are filtrable and, in the animal body, intracellular parasites. Also, they are sensitive to ether and *in*sensitive to many antibiotics. Their ability to induce cytopathogenicity in cell cultures is a nuisance. The type species is *Mycoplasma mycoides*, the cause of pleuropneumonia in cattle.

Pathogenicity

In addition to bovine pleuropneumonia, mycoplasmas cause contagious mastitis in sheep and goats and respiratory disease in poultry. Dogs, rats, and mice may become infected with them. In humans they are linked to infections of the genitourinary tract and the upper and lower respiratory passages and rarely to other systemic disorders.

An important virulence mechanism in mycoplasma is the presence on the bacterial body of a specialized structure for attachment to epithelium. With this "tip" structure the microbes anchor themselves firmly to the lining epithelium of the respiratory tract, and thus secure, they resist the cleaning action of the respiratory cilia and mucus. The infectious dose is 10 to 100 organisms.

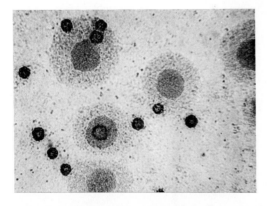

FIG. 31-4 Mycoplasmas. Mixture of classic *(Mycoplasma hominis)* colonies and smaller, very dark colonies of T (T strain) mycoplasmas. Direct test for urease (that of Shepard and Howard) has been applied to this preparation. Colonies of T strain mycoplasmas are urease positive, showing a dark color of reaction of product. Note the fried-egg appearance of classic colonies. (×300.) (From Shepard, M.C., and Howard, D.R.: Ann. N.Y. Acad. Sci. 174:809, 1970.)

Table 31-1 Comparison of Mycoplasmas with Other Microorganisms

Microbe	Size (nm)	Shape	Rigid Envelope	Extracellular Multiplication	Reproduction	Sensitivity to Antimicrobials	Nucleic Acids	Ribosomes	Metabolic Enzymes
Viruses	15 to 350	Symmetric	No	No	Eclipse-synthesis-assembly	No	DNA or RNA	No	No
Mycoplasmas (bacteria)	150 to 300	Varied	No	Yes	Fission	Tetracyclines only	DNA and RNA	Yes	Yes
Chlamydiae (bacteria)	200 to 400	Coccoid	Variable in life cycle	No	Fission; unique developmental cycle	Yes	DNA and RNA	Yes	Yes
Rickettsias (bacteria)	300 to 2000	Varied	Yes	No (most species)	Fission	Yes	DNA and RNA	Yes	Yes
Other bacteria	300 to 3000	Varied	Yes	Yes	Fission	Yes	DNA and RNA	Yes	Yes

Primary atypical pneumonia. *Mycoplasma pneumoniae,* the smallest known pathogen that can live outside of cells, causes primary atypical pneumonia. Formerly known
as the *Eaton agent,* this mycoplasma was discovered by Monroe Davis Eaton of Harvard University in 1944. Originally grown on chick embryo lung culture, it was considered a virus until it was cultured on artificial agar medium. Eaton suspected its relation
to primary atypical pneumonia, but only after fluorescent antibody technics were applied years later was it demonstrated that the Eaton agent caused this disease.

Primary atypical pneumonia, an acute, febrile, self-limited disease of humans, begins as an upper respiratory tract infection and spreads to the lungs. It was among the
most common respiratory infections of World War II and can still be recovered from a
large number of military recruits. The incubation period extends 2 to 3 weeks. The
manifestations of primary atypical pneumonia vary in severity from asymptomatic infection to serious pneumonia. With frank pneumonia, headache and malaise are fairly
severe, cough is paroxysmal, and scanty bits of sputum are raised. The lungs may be
extensively involved, although in most patients only one lobe is affected. The attack
lasts several weeks. Recovery is gradual. Probably most patients have a milder form or
even an inapparent infection—a "walking pneumonia."

The disease, although prevalent, does not occur in widespread epidemics. Epidemics are confined to persons living in crowded conditions and have been reported among
newborn infants. It is not seasonal, as are other pneumonias. Most cases are recorded
in persons 5 to 25 years old, and in children the peak incidence is between the ages of
of 5 and 10 years.

Primary atypical pneumonia is not very contagious, although apparently it is spread
directly from one person to another by oral or nasal secretions (droplet infection), and
the portal of entry is the upper respiratory tract. The disease is not always followed by
immunity.

The classic case of primary atypical pneumonia is associated with an increased number of cold agglutinins in the patient's serum. These are agglutinins that clump human
type O, Rh-negative erythrocytes at low temperatures (5° to 20° C) but not at body
temperature. They are found in several unrelated diseases.

In the serums of some convalescent patients, agglutinins develop to *Streptococcus
MG* (From the name *McGuinness* of the patient from whom it was originally recov-
ered), an alpha-hemolytic streptococcus now classified as *Streptococcus angionosus.*
There is also a complement-fixation test.

An alum vaccine prepared from formalin-killed organisms has been tested experimentally in animals and humans. An experimental intranasal vaccine against *M. pneumoniae* has been prepared from temperature-sensitive mutants of a virulent wild strain.

Ureaplasma urealyticum (Mycoplasma T Strains)

The microbes of the species *Ureaplasma urealyticum* (formerly the T strains of mycoplasmas, "tiny forms," or "T-form" PPLO colonies) represent human mycoplasmas
with a distinct growth pattern. This species produces very small colonies on its agar
media (only a few micrometers in diameter). Colonies of the "classic" organisms first
studied are much larger, many micrometers in diameter, even up to 0.75 mm (Fig.
31-4). (Both "tiny-form" and "large-form" colonies must be viewed with a microscope.)
Moreover, the mycoplasmas of *U. urealyticum* are unique among mycoplasmas in
being sensitive to therapeutic agents not affecting the classic ones and in possessing an
active urease system. Ten percent urea is required in culture medium for growth. Since
their ability to hydrolyze urea sets them apart, a reclassification of T strains created a
new genus called *Ureaplasma,* in family I, Mycoplasmataceae, with *U. urealyticum*
for the single human species and its serotypes.

Mycoplasmas are common parasites of the genital tract. Colonization is related
therein to sexual activity. In fact, they are considered part of the normal flora of sex-

ually active males and females. However, there is reason to state that genital mycoplasmas—both *U. urealyticum* and *Mycoplasma hominis*—are not necessarily benign but may be significant agents of sexually transmitted infection. *U. urealyticum* is recognized as an important cause of nongonococcal urethritis (p. 660). It also produces prostatitis in the human male and, in the human female, cervicitis, cystitis, and endometritis and contributes to infertility and premature births. Classic mycoplasmas are also implicated in so-called reproductive failure.

Laboratory Diagnosis

The laboratory diagnosis for mycoplasmas is made by cultural methods adapted to their size and peculiar growth requirements. Mycoplasmas tend to grow down into a solid medium and in a distinctive fashion produce a colony with a fried-egg appearance. They proliferate also in cell culture and in the developing chick embryo. Serologic identification is practical because mycoplasmas are vulnerable to neutralizing antibodies and induce hemagglutination and hemadsorption reactions. Since many normal persons demonstrate circulating antibodies to them, it is the *rising* titer that is of diagnostic significance in acute and convalescent serums.

LEGIONELLA PNEUMOPHILIA

Legionella pneumophilia is a newly identified bacterium that causes **legionellosis.** Two basic forms of this disease exist: (1) Legionnaires' disease and (2) Pontiac fever.

Legionnaires' Disease

Legionnaires' disease has generated considerable excitement and interest since first being recognized in July of the bicentennial year at the American Legion convention in Philadelphia. In a dramatic outbreak, nearly 200 Legionnaires were affected and 29 died. Since then the disease has been carefully studied in the epidemics and sporadic cases reported from other parts of the country.

After an incubation period of 2 to 10 days, an influenza-like illness appears with rapidly rising fever, chills, muscle aches, and headache; within a week pneumonia develops and progresses over several days. A dry cough, pleuritic chest pain, abdominal pain, or gastrointestinal symptoms may be present. Although is it now known to be an illness with multisystem involvement, the prime target of the infection is the lung and Legionnaires' disease is usually discussed as an acute bacterial pneumonia. Pathologically the air sacs in both lungs fill with an exudate composed of neutrophils, macrophages, and fibrin, in the pattern of a confluent bronchopneumonia. Sometimes abscesses form.

Most of the time the disease is self-limited, although the mortality may be 15% to 25%. Undiagnosed mild cases may exist. The infection is seasonal, with most cases in the late summer and fall. It occurs most often in middle-aged and older persons (mean age, early 50s) and affects more men than women.

Implicated in the pathogenesis of the disease is the vulnerability of the lung to infection resulting from the damage to alveolar macrophages normally resident in the air sacs of the lungs by known specific factors, such as smoking and breathing pollutants, and by less well-defined ones. As would be expected, in the immunosuppressed patient with known depression of resistance mechanisms in the respiratory tract, Legionnaires' disease is a significant illness, usually fatal.

Pontiac Fever

Pontiac fever (nonpneumonic legionellosis, steam-turbine cleaner's disease) is like Legionnaires' disease in its manifestations but is milder. It begins sooner after exposure to the causative agent than Legionnaires' disease, usually within a day or two. In both diseases the patient experiences headache, muscle aches, and fever, but those symptoms pointing to infection of the lung and other organs are not as conspicuous in Pontiac fever. Pneumonia does not supervene, and liver and kidney complications do not appear. Patients are sick for a shorter time, 2 to 5 days, and apparently all recover completely.

The Organism

The causative bacterium of Legionnaires' disease was isolated from lungs of victims at the Center for Disease Control, Atlanta, in 1976, about 6 months after the Philadelphia outbreak. It was given the name *Legionella pneumophilia*.

L. pneumophilia is a small, strictly aerobic, moderately pleomorphic, probably encapsulated, gram-negative rod with an elusive life cycle. Most strains have single polar flagella, and replication is by nonseptate constriction. Sugars are not fermented. There are six serotypes.

Special media and growth conditions are required for the exceedingly fastidious members of this species. They do grow, but slowly, on Mueller-Hinton agar containing hemoglobin and IsoVitaleX, on which they produce a soluble brown pigment. Buffered charcoal–yeast extract agar may be used, and experimentally selenium has been found to enhance growth. Their growth in the yolk sac of a fertilized hen's egg kills the embryo within a few days. After being inoculated with this organism, the highly susceptible guinea pig manifests fever, watery eyes, and prostration in a predictable fashion. Its spleen yields the agent. Freezing kills *L. pneumophilia*, as does formaldehyde.

Before this microbe was named provisionally, its characteristics were carefully examined and compared with those of other microbes. It does share a few traits with known bacterial species, but overall its profile of biologic and immunologic qualities is distinct. After certain genetic studies were made, the accumulated evidence became stronger that here indeed a new genus and species had been defined—a rare event since the golden age of microbiology.

Currently the family Legionellaceae has been established, and at least nine species in the single genus *Legionella* have been distinguished and named. The new species are phenotypically similar to *L. pneumophilia*. They are fastidious, gram-negative rods, sharing an aquatic habitat and producing comparable diseases. One species, *Legionella micdadei*, the Pittsburgh pneumonia agent, has attracted attention as a cause of nosocomial pneumonia in renal transplant patients and other immunocompromised hosts.

Laboratory Diagnosis

Various methods are designed to detect organisms or antibody. Gas-liquid chromatography is used to identify the organisms. Antibodies appear the third week after onset. Collection of paired serums—one from the acute and one from the convalescent state— allows for documentation of a significant (fourfold) rise in serum titer by the indirect fluorescent antibody test. Serodiagnosis may be made with a rising titer to 1:128 or more or, in the proper clinical setting, from a single specimen of convalescent serum with a titer of 1:256 or more.

Legionella species can be cultured from suitable specimens, such as blood, sputum, pleural fluid, transtracheal aspirates, bronchial washings, and lung tissue (from biopsy or autopsy), and it can also be recovered from a guinea pig.

The direct immunofluorescent technic may be applied to sputum, respiratory secretions, pleural fluid, or lung tissue, which may even be formalin-fixed. In this way the organisms can be demonstrated within a few hours. The Dieterle silver impregnation stain is an excellent tissue stain for *Legionella* species. The enzyme-linked immunosorbent assay (ELISA) test has been used with patient serum and urine.

Transmission

Much is unresolved regarding the life story of the Legionnaires' disease bacterium—its habitat in nature, transmission, entry into the lung, and disease-producing capacity there.

The major reservoirs of *Legionella* species are in moist soil and in natural bodies of water over wide geographic areas. In its natural aquatic habitat it demonstrates an interesting relationship to a common microbe, a blue-green alga, also found in natural waters and even in cooling towers. Members of the genus *Legionella* apparently require certain exotic growth factors supplied by the algae. Since *Legionella* organisms have been recovered from the water in air-conditioning cooling towers, they presumably become airborne when soil is disturbed and thus spread into evaporating pans and filters of large units. With the right temperature and humidity, they multiply there (with the blue-green algae) and can then be released in an infectious aerosol through fans and exhaust vents of the system.

Direct spread from person to person is unlikely. Surprisingly, the causative agent of legionellosis is not considered highly infectious; health care professionals do not seem to be at risk, even when an outbreak threatens the patients they care for. However, laboratory workers do well to take every indicated precaution until the organism is better understood.

Prevention

A vaccine has been tested successfully in laboratory animals, and vaccine development is feasible in human beings.

SUMMARY

1 Pseudomonads are strictly aerobic, motile, gram-negative rods that are noted for their water-soluble, fluorescent pigments. *Pseudomonas aeruginosa* is a well-known one. As a saprophyte, *Pseudomonas aeruginosa* is everywhere in nature. As a pathogen, it infects a wide host range of animals, insects, and plants. In humans it threatens the compromised host and young infants and the aged.

2 The sources of *Pseudomonas aeruginosa* in hospital-acquired infections are many. It is an extremely hardy microbe. It grows on almost any medium, even in dilute disinfectant solutions and distilled water. It is resistant to all antibiotics except for a few toxic ones.

3 The pathologic hallmark of *Pseudomonas* infection is the destruction of tissue, associated with a massive overgrowth of microorganisms. Two serious diseases typically produced by *Pseudomonas aeruginosa* are gram-negative shock and sepsis complicating burns.

4 *Pseudomonas pseudomallei* causes melioidosis, an endemic disease in Southeast Asia, where the organism is a saprophyte of the rice fields. *Pseudomonas mallei* causes glanders, a disease of lower animals that is transmissible to humans.

5 *Bacillus anthracis* is the first pathogen to have been seen under a microscope, the first one to have been proved to cause a specific disease, the first to have been grown in pure culture, and the one Pasteur used for artificial immunization. It causes anthrax, a disease of lower animals easily communicated to humans.

6 *Bacillus anthracis* is the only spore-forming aerobic pathogen. Central spores give it the shape of a bamboo rod. It is one of the largest bacterial pathogens and is not motile.

7 For humans anthrax is primarily an occupational disease. The common route of infection is through abrasions of the skin. Prophylactic vaccination of animals should be carried out with vigor.

8 Since the gram-positive, nonsporulating, microaerophilic lactobacilli require many essential growth factors, they are useful in the commercial bioassay of the B complex vitamins and certain amino acids. They are used in the fermentation and dairy industries.

9 The small, gram-positive, motile coccobacillus, *Listeria monocytogenes*, is novel in that it can grow in the refrigerator. It causes listeriosis, an inflammation of the central nervous system. The species name refers to the increase of monocytes in the bloodstream with listerial infection.

10 The disease listeriosis is a prime example of disordered cell-mediated immunity.

11 Mycoplasmas, the smallest, free-living, self-replicating units, are unique because they do not have a cell wall. Their soft fragile bodies are bounded by a trilaminar membrane. They are everywhere as saprophytes, parasites, and pathogens for a wide host range.

12 Mycoplasmas can be cultivated in cell-free media, but in their infected host they are intracellular parasites.

13 *Mycoplasma pneumoniae*, the smallest known extracellular pathogen, causes primary atypical pneumonia, a self-limited disease associated with a significant titer of cold agglutinins in the patient's serum.

14 *Ureaplasma urealyticum* causes nongonococcal urethritis in males and, in females, cervicitis and endometritis.

15 *Legionella pneumophilia* causes the mild Pontiac fever and the severe Legionnaires' disease. It is a small fastidious, strictly aerobic, gram-negative rod with an elusive life cycle. Major reservoirs of *Legionella* species appear to be in watery environments, such as in moist soil and in natural bodies of water.

16 Legionnaires' disease, first recognized at the American Legion convention in Philadelphia, is usually a self-limited influenza-like illness primarily affecting the lungs.

QUESTIONS FOR REVIEW

1 What is the significance of *Pseudomonas aeruginosa* in clinical medicine? Why is it so hard to deal with?

2 What is melioidosis? Name and describe the causative microbe.

3 Briefly characterize mycoplasmas. How is the laboratory diagnosis made for mycoplasmas?

4 List diseases caused by *Pseudomonas* species.

5 List five toxic products of *Pseudomonas* species.

6 State the practical importance of lactobacilli. List notable members of the genus *Lactobacillus*.

7 Define listeriosis. What forms are seen in humans?

8 Compare mycoplasmas with viruses.

9 Give the role of *Ureaplasma urealyticum* in human disease.

10 Outline salient features of primary atypical pneumonia.

11 What are Döderlein bacilli?

12 Give the method of spread for glanders. How may this disease be controlled?

13 Describe the organism causing anthrax.

14 Discuss the proper disposal of bodies of animals that have died of anthrax.

15 Why is the anthrax bacillus of historical interest?

16 Briefly explain "Vietnamese time bomb," "walking pneumonia," Straus reaction, T strains, the Eaton agent, pyoverdin, mallein, the iridescent phenomenon, farcy, cold agglutinins, pyocyanin, "blue pus," Pontiac fever, legionellosis, and the Anton test.

17 Describe the causative agent of Legionnaires' disease. What laboratory methods are used for its detection?

18 What is known about the source of infection in Legionnaires' disease? How is it presumed to be transmitted? Who is at greatest risk?

SUGGESTED READINGS

Edelstein, P.H.: Editorial: what to do about *Legionella?* J.A.M.A. **249**:3214, 1983.

Fraser, D.W., and McDade, J.E.: Legionelosis, Sci. Am. **241**:82, April 1979.

Hot tub *Pseudomonas*, Morbid. Mortal. Weekly Rep. **31**:541, 1982.

Kundsin, R.B., and others: *Ureaplasma urealyticum* incriminated in perinatal morbidity and mortality, Science **213**:474, 1981.

Martin, F.: *Legionella pneumophilia*, the microbe who came in from the cold, Diagn. Med. **2**:46, 1979.

Pseudomonas cepacia colonization—Minnesota, Morbid. Mortal. Weekly Rep. **30**:610, 1981.

Rinke, C.M.: Hot tub hygiene (editorial), J.A.M.A. **250**:2031, 1983.

Rose, H.D., and others: *Pseudomonas* pneumonia associated with use of a home whirlpool spa, J.A.M.A. **250**:2027, 1983.

Salmen, P., and others: Whirlpool-associated *Pseudomonas aeruginosa*, J.A.M.A. **250**:2027, 1983.

Wade, N.: Death at Sverdlovsk: a critical diagnosis, Science **209**:1501, 1980.

RICKETTSIAS AND CHLAMYDIAE

In Section 9 of *Bergey's Manual* (1984) the Rickettsias and Chlamydiae (or Chlamydias) are separated into two orders. The first includes microbes traditionally considered rickettsias; the second, those better known as chlamydiae or bedsoniae. The Rickettsias and Chlamydiae comprise about a dozen different microorganisms, which vary greatly in pathophysiology, and they produce an equal number of diseases, which are widely divergent in the clinical manifestations and epidemiologic patterns.

GENERAL DISCUSSION

Rickettsias are known for a distinct, selective type of parasitism of cells in disease and for a special relation to an arthropod, be it vector or host. Long relegated to an intermediate position between bacteria and viruses, these prokaryotic microbes are now recognized as bacteria.

Bergey's Manual lists order I, Rickettsiales, as having two families. The members of family I, Rickettsiaceae, parasitize tissue cells (except the red blood cells) of a vertebrate host; within this family, the microbes of tribe I, Rickettsieae, have adapted to existence in arthropods, and three genera are of medical importance—*Rickettsia*, *Rochalimaea*, and *Coxiella*. The members of family II, Bartonellaceae, affect both red blood cells and tissue cells; one genus, *Bartonella*, causes human disease. Only members of these two familes are discussed here.

Special Properties

Rickettsias of family I, small, pleomorphic, bacillary or coccobacillary forms, were named in honor of Dr. Howard T. Ricketts of Chicago, who lost his life in the scientific investigation of rickettsial disease. Dr. Ricketts (1871-1910) first observed rickettsia bodies in a case of Rocky Mountain spotted fever in 1909 and demonstrated that the disease was transmitted by the wood tick. Later he showed that tabardillo (Mexican typhus) was transmitted by the body louse.

Rickettsias occur singly or in pairs, chains, or irregular clusters. The electron microscope indicates that they have an internal structure much like that of bacteria (with a cell wall) and that they divide by binary fission. Unlike chlamydiae, they have no intracellular growth cycle but divide as bacteria do. Most rickettsias are held back by bacteria-retaining filters, but they are so small as to be close to the size of some of the larger viruses, being about 0.3 μm in diameter. They are barely visible in the light microscope. They are nonmotile, gram negative, and stain with difficulty. Since they stain faintly with the Gram stain, Giemsa or special stains are used to demonstrate them.

Obligate intracellular parasites, they do not multiply in the absence of living cells, with rare exception; the pathogenic forms grow only in the cells of infected animals. Some grow only in the cytoplasm (rickettsias of typhus fever), but others grow in both cytoplasm and nucleus (rickettsias of Rocky Mountain spotted fever) of the infected cell. They contain both DNA and RNA, and the fragments of enzyme systems that rickettsias possess allow them a range of metabolic activity. Their resistance to deleterious influences such as heat, drying, and chemicals is about the same as that of most bacteria (see Table 31-1).

The rickettsial diseases are transmitted to humans by arthropod vectors, the natural and primary hosts of the rickettsias.

Table 32-1 Rickettsial Infections in Humans

Group	Agent	Weil-Felix Reaction			Complement Fixation*
		OX-19	OX-2	OX-K	
Typhus Fevers					
Epidemic typhus	*Rickettsia prowazekii*	4+	+	−	+ (specific)
Murine typhus	*Rickettsia typhi (mooseri)*	4+	+	−	+ (specific)
Brill-Zinsser disease	*Rickettsia prowazekii*	−	−	−	
Spotted Fevers					
Rocky Mountain spotted fever	*Rickettsia rickettsii*	+	+	−	+
Rickettsialpox	*Rickettsia akari*	−	−	−	+
Fièvre boutonneuse	*Rickettsia conorii*	+	+	+	
Scrub Typhus	*Rickettsia tsutsugamushi*	−	−	3+	±
Trench Fever	*Rochalimaea quintana*				
Q Fever	*Coxiella burnetii*	−	−	−	+

*Cross-reactions occur between Rocky Mountain spotted fever and rickettsialpox.

Pathogenicity

When rickettsias invade humans, they attack the macrophage system, colonizing the endothelial lining cells of the walls of small blood vessels (small arteries, arterioles, and capillaries). Within these cells they induce a vasculitis (inflammation of the blood vessel) that is distributed all over the body, just as small blood vessels are. The consequence is disease in many anatomic areas in the body. A classic skin rash and a whole host of pathologic changes ensue. Rickettsial diseases can be quite serious and many times life-threatening in spite of the best available therapy.

An attack of rickettsial disease is usually followed by a lasting immunity.

Laboratory Diagnosis

Rickettsias can be cultivated in the yolk sac of the chick embryo (provided the hen did not receive antibiotics) or in cell cultures. Recovery of the microorganisms may be made when blood (rickettsemia present) and suitable specimens (sputum in the case of Q fever) from a patient are inoculated into laboratory animals, such as guinea pigs, mice, and rabbits.

Reaction in Guinea Pig	Reservoir in Nature	Vector	Geography
Fever, vasculitis, encephalitis within 4 to 8 days	Humans	Body lice (feces into broken skin)	Worldwide
Fever; tunica reaction (Neill-Mooser reaction) in males—scrotal swelling with adhesions between testis and scrotal sac	Rats	Rat fleas (feces into broken skin)	Worldwide
	Humans		Worldwide
Fever; widespread hemorrhagic necrosis, sloughing of footpads and ears; severe scrotal reaction—edema, reddening, necrosis, sloughing	Wild rodents	Ticks	Western hemisphere
Fever; scrotal reaction fifth day	House mice	Mites	United States, Russia, Korea
Fever fourth to sixth day; pronounced scrotal swelling but no slough	Small wild mammals	Ticks	Mediterranean coast, Middle East
	Wild rodents	Mites	Asia, Australia, Pacific Islands
	Humans	Body lice	Europe, Mexico, Africa
Splenomegaly 2 to 4 times normal	Ruminants, small mammals, domestic livestock	Ticks (to animals) Airborne animal products (to humans)	Worldwide

The most important rickettsial diseases are associated with a positive **Weil-Felix reaction,** an agglutination test similar to the Widal test for typhoid fever except that the test serum is mixed with different strains of *Proteus vulgaris*. The Weil-Felix reaction is a heterophil antibody reaction, since *Proteus* bacilli *do not* cause any of the rickettsial diseases. Complement-fixation and agglutination tests are of value in differentiating the rickettsial diseases, and immunofluorescent methods detect rickettsias in tissues of host and vector and demonstrate antibodies in serum.

Classification

Rickettsial diseases (rickettsioses) may be divided into the following groups:
1. Typhus fevers
2. Spotted fevers
3. Scrub typhus
4. Trench fever
5. Q fever

Table 32-1 indicates epidemiologic features of these groups. Rocky Mountain spotted fever, rickettsialpox, and Q fever are the three prevalent in North America.

RICKETTSIAL DISEASES
Typhus Fever Group

General Considerations

The typhus fever group includes epidemic typhus, murine typhus, and Brill-Zinsser disease. Clinically the different types of typhus fever closely resemble each other but vary in severity. They usually begin with severe headache, chills, and fever. (Headaches with rickettsial diseases are said to be the most severe of headaches.) A rash develops about the fourth day and persists throughout the course of the disease. Mentally the patient is dull and stuporous. This feature gave origin to the name *typhus*, which is derived from the Greek *typhos*, meaning vapor or smoke. Disease in this group extends from 3 to 5 weeks. The mortality varies from 5% to 70%.

Epidemic (classic, European, Old World, louse-borne) typhus. Epidemic typhus is caused by *Rickettsia prowazekii*. The species name is derived from Stanislas von Prowazek (1876-1915), an early investigator who also lost his life in the study of typhus. Epidemic typhus is spread from person to person by the body louse *(Pediculus humanus corporis)* (Fig. 32-1). Head lice *(Pediculus humanus capitis)* can transmit the disease but seldom do. No animal reservoir has been found; this is the only rickettsial disease for which human beings are the reservoir. Epidemic typhus is an acute, severe disease with a high mortality (10% to 40%) that has been known to spread over the world in devastating epidemics. Typhus fever and malaria are probably the greatest killers of human beings in recorded history. Epidemic typhus is a disease of overcrowding, famine, filth, and war. (It is also known as jail fever, war fever, and famine fever.) It is a disease of areas where heavy clothing is worn, since such garments readily harbor lice.

When a louse bites a person with epidemic typhus, the rickettsias are taken into the stomach of the louse and invade the cells lining the intestinal tract. They multiply to such an extent that the cells become greatly swollen, burst, and liberate the rickettsias into the feces of the louse. When a louse bites a person, it defecates at the same time. The site of the bite itches, and that person introduces the infectious fecal material of the louse into the skin by scratching. Infected lice die 8 to 10 days after infection from intestinal obstruction caused by parasitism of the lining cells.

Murine (endemic, New World, flea-borne) typhus. Murine typhus fever, caused by *Rickettsia typhi,* is relatively mild. A natural infection of rats, less often of mice, it is transmitted from rat to human by the rat flea (Fig. 32-2) and from rat to rat by the rat louse and the rat flea. Rats do not die of the disease, nor do the fleas or lice that infest them. Murine typhus may be transmitted from person to person by the body louse and become epidemic in louse-infested communities. The human mortality is about 2%. The endemic typhus fever of Mexico is known as *tabardillo* (from the Spanish word *tabardo,* meaning a colored cloak, to designate the mantlelike spotted rash of the disease).

Brill-Zinsser disease (recrudescent typhus). Brill-Zinsser disease, a mild form of typhus along the Atlantic coast, occurs in persons who have had classic typhus fever many years before. Recrudescence of infection with no part played by a louse is possible because even though a patient recovers from epidemic typhus, the causative rickettsias may remain latent in their host's cells for many years.

Nathan E. Brill (1860-1925) first observed this mild form of typhus in 1898 in immigrants to the United States from various countries of Eastern Europe. The exact nature of the disease was not known until 1934, when Hans Zinsser (1878-1940) advanced the hypothesis that the disease was a second but mild attack of classic typhus, an idea now amply confirmed.

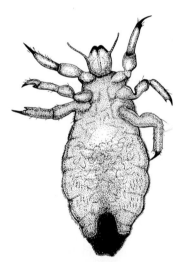

FIG. 32-1   Human body louse, vector in epidemic typhus fever, louse-borne relapsing fever, and trench fever. Lice are species specific and, within the given host, specific for a given anatomic site. This louse attacks specifically the body (not head) of its human host. With heavy infestation, one may recover 400 or 500 lice per person, and within the last decade or so lice infestation in Americans has increased 800%. (From Smith, A.L.: Microbiology and pathology, ed. 12, St. Louis, 1980, The C.V. Mosby Co.)

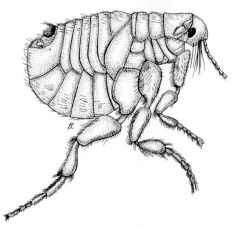

FIG. 32-2   Cosmopolitan rat flea (1500 species of rodent fleas). Vector in murine typhus and plague. (From Smith, A.L.: Microbiology and pathology, ed. 12, St. Louis, 1980, The C.V. Mosby Co.)

Laboratory Diagnosis

The typhus fevers all are identified by a positive Weil-Felix reaction (agglutination of the bacteria known as *Proteus* OX-19 in patient serum). Although the Weil-Felix reaction may be detected as early as the fourth day, it is strongly positive about the eighth day. The Weil-Felix reaction does not differentiate the typhus fevers of Rocky Mountain spotted fever, but it does separate them from certain other rickettsial infections. Fluorescent antibody technics can also be used in the diagnosis of the typhus fevers. The epidemic and endemic forms may be distinguished and the rickettsias demonstrated if blood from a patient is inoculated into a male guinea pig. The scrotal reaction (Table 32-1) is more likely to be seen if the patient has endemic typhus.

Immunity

If a person recovers from either epidemic or endemic typhus, that person is immune to both types, but a vaccine prepared against one type of the disease protects only against that one. Vaccines seem to be rather effective in preventing the typhus fevers, and when disease does occur in spite of vaccination, the severity is lessened and mortality reduced.* In only a few cases is recovery from epidemic or endemic typhus not followed by permanent immunity.

Prevention

Prevention of epidemic typhus fever depends on (1) isolation of the patient in a vermin-free room, (2) use of insecticides on clothing and bedding, with destruction of insect eggs (nits) attached to hair, (3) quarantine of all susceptible persons if many lice are in the vicinity of the patient, (4) systematic delousing and vaccination of susceptible persons, and (5) general improvement of living and sanitary conditions. The control of the endemic form depends on rat-proofing buildings and destroying rodents with their louse and flea populations.

Spotted Fever Group

Rocky Mountain Spotted Fever

Recognized as one of the most severe of all infectious diseases, Rocky Mountain spotted fever (mountain fever, tick fever†) is like typhus fever and is caused by a similar organism, *Rickettsia rickettsii*. In the Rocky Mountain states of Idaho and Montana many cases are found, but the disease is increasingly prevalent on the Atlantic seaboard, chiefly east of the Appalachian Mountains (mostly in the "tick-belt" states); two types of the disease are designated: the western type (the original Rocky Mountain spotted fever) and the eastern type. The difference is mainly in geographic distribution and mode of spread. The disease does show, however, great variation in severity. In Idaho the mortality varies from 3% to 10%, whereas in the Bitter Root Valley of Montana it may reach 90%. In the eastern states the disease is comparatively mild, and the mortality is about 6%.

The western variety is transmitted by the Rocky Mountain wood tick *(Dermacentor andersoni)*, and the eastern variety is transmitted by the American dog tick *(Dermacentor variabilis)*. In the southwestern United States the disease is transmitted by the Lone Star tick *(Amblyomma americanum)*. Several species of hard-shell ticks (Ixodidae) (Fig. 32-3) are infected. The male tick can infect the female, who is able to transmit the infection to her offspring. Fluorescent antibody technics show that infected female ticks may pass rickettsias to all their progeny, an example of **transovarian infection.** The infection may be retained in ticks through many generations. The tick is therefore able

---

*For immunization, see p. 428.

†Relapsing fever, a spirochetal disease, is also called tick fever.

to maintain the disease in nature without either rodent or human help. A reservoir of infection appears to exist among a variety of rodents, particularly jackrabbits and cottontails, among which the disease is transmitted from rabbit to rabbit by the rabbit tick, a tick that does not bite humans. Birds and large wild and domestic animals also maintain the disease.

The rickettsias of Rocky Mountain spotted fever differ from those of typhus fever in that they invade both the cytoplasm and nucleus of the infected cell. An attack renders a person immune for life. The serum from persons with Rocky Mountain spotted fever agglutinates *Proteus* OX-19 (positive Weil-Felix reaction), but the agglutination is not so strong as in typhus fever. *Proteus* OX-2 also may be agglutinated. A complement-fixation test is highly specific, and there is an indirect hemaggluntination test. The patient's blood may be inoculated into a male guinea pig and the rickettsias recovered from the animal.

Suspect ticks may be examined microscopically for the presence of rickettsias. A tick engorged from feeding is dissected from its shell, fixed in a suitable solution, and stained with Giemsa stain. If infectious, the tick will display rickettsias within the nuclei of cells of the salivary glands.

Health officials advise persons to wear protective clothing such as high boots in tick-infested areas. When persons leave such areas, they should examine their clothing and bodies carefully for ticks as soon as possible. Personal-use repellants are effective when applied to socks, trouser cuffs, and openings in clothing worn, such as neck, button areas, and the like. A tick may wander over a person's body for an hour or two before it feeds. It seems to favor a tender area of skin, but tick movements and bites are seldom felt. It is said that the time required for rickettsias to be transferred from the attached tick to the human being on whom it is feeding is at least 2 hours or more. When found, ticks should be gently removed with tweezers and plucked as close to the skin surface as possible. One must be sure to remove the tick with its mouth parts intact. The bite wound should be treated with a suitable germicide. Dogs and other pets should be inspected daily. Infective ticks are likely to be found in underbrush, tall grasses, weeds, and wet woods, especially in spring and early summer. It is advisable to clear out the underbrush from an area where children may be playing or else to keep the children away from that area. Parents should inspect their children, particularly their hair, for ticks after they have played in grassy or wooded areas or with pets.

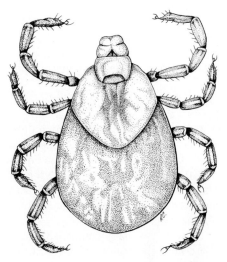

FIG. 32-3 Hard-shell tick, vector in Rocky Mountain spotted fever and tularemia. Hard-shell ticks have a hard, smooth skin and a head. Adult ticks are brown to reddish brown and about ⅛ inch long. With feeding, their bodies can swell to as much as ½ inch in length. The term "wood tick" is applied to several species of hard ticks. (From Smith, A.L.: Microbiology and pathology, ed. 12, St. Louis, 1980, The C.V. Mosby Co.)

Rickettsialpox

Rickettsialpox is an acute, mild febrile disease described by the appearance of a red papule at the site of inoculation followed later by a skin rash. The disease is caused by *Rickettsia akari*. The house mouse is the reservoir of infection, and the bite of the house mouse mite conveys the infection from mouse to mouse and to the human being. The serum of those ill with or convalescing from the disease does *not* give a positive Weil-Felix reaction. Clinically, rickettsialpox may be mistaken for chickenpox. The primary lesion at the site of inoculation often resembles the vesicle of a smallpox vaccination. The destruction of mice and their mites eliminates the disease.

Other Fevers

Included with the spotted fevers are tick-borne typhus fevers—fièvre boutonneuse, Queensland tick typhus, South African tick-bite fever, and tick-bite rickettsioses of India.

These illnesses are nonfatal and only moderately severe. They are caused by rickettsias related to the agent of spotted fever, and each can be traced to the bite of an ixodid tick. Fièvre boutonneuse, caused by *Rickettsia conorii*, is the most widespread and the prototype. In this disease there is an initial lesion, the *tache noire* (an ulcer covered by a black crust is the "black spot"), which is followed by a generalized rash. The Weil-Felix reaction is positive but not strongly so.

Scrub Typhus

Scrub typhus, known also as tsutsugamushi disease (Japanese, dangerous bug) and mite typhus, is so called because the mites that transmit the disease to humans are found in scrubland—land covered by a stunted growth of vegetation. The disease attacked thousands of soldiers in the Southwest Pacific area during World War II. It is an acute febrile disease, with a rash, generalized lymphadenopathy, and lymphocytosis, which may simulate infectious mononucleosis. The mortality ranges from 0.5% to 60%. The disease resembles Rocky Mountain spotted fever clinically, except that in scrub typhus there is a primary sore at the site of the mite bite. It is caused by *Rickettsia tsutsugamushi*.

Scrub typhus is a disease primarily of mites and wild rodents, transmitted from rodent to rodent and to humans by mites. Two types of mites are known to transmit the disease to humans. The rodents and mites that propagate the disease probably differ in various parts of the world. The mites inhabit short blades of grass and top layers of the soil. Persons who come in contact with the soil are therefore likely to contract the infection. There is no evidence that the disease spreads from person to person. The serum of persons who have the disease agglutinates *Proteus* OX-K but not OX-19. There is an indirect fluorescent antibody test, and inoculation of a mouse may be done.

Prevention depends on the destruction of the mites in the soil by insecticides and the application of insecticidal chemicals to clothing. There is as yet no effective preventive vaccine.

Trench Fever

Trench fever is a remittent or relapsing fever that affects soldiers on trench duty, but it may occur in civilian life when people live under conditions comparable to those of the trenches. It was prevalent during World War I but has since almost passed out of existence. It was infrequent during World War II. *Rochalimaea quintana*, the causative agent, is transmitted by the body louse. Unlike other rickettsias, a human being is its primary host, and the rickettsias of the genus *Rochalimaea* can be cultured in host cell–free media; they are not obligate intracellular parasites.

Although trench fever is never fatal and recovery is usually complete, it may recur, and in some cases it is followed by a state of chronic ill health with pain in the limbs, mental depression, and circulatory disorders. An attack does not confer immunity.

Q Fever

Q fever is a febrile disease of short duration caused by *Coxiella burnetii*. The pleomorphic rickettsias of the genus *Coxiella* are set apart from other rickettsias by their amazing resistance outside cells to heat, drying, and sunlight, agents that destroy most nonsporogenic bacteria. The agent of Q fever undergoes a complex developmental cycle and unlike other rickettsias is not an obligate intracellular parasite. It grows well in the placenta.

The usually mild respiratory symptoms of Q fever make it similar to primary atypical pneumonia or influenza. It is rarely fatal but can give rise to a severe form of endocarditis (inflammation of a heart valve). Q fever differs from most other rickettsial diseases in that it is not transmitted to humans by an arthropod vector, there is no rash, and the serum of an infected person does not agglutinate *Proteus* bacilli. The disease was first observed in Queensland, Australia, but the place of discovery did not give origin to the term *Q fever*; Q is for "query"—the etiologic agent was unknown.

Q fever is worldwide in distribution and prevalent in the United States, especially in certain areas along the west coast. It is a disease of animals (a zoonosis). Humans become infected by living or working in contact with infected livestock, such as cattle, sheep, and goats. Infection is especially common among packinghouse employees and laboratory workers. Most human infection is acquired by breathing contaminated air such as the dust from dairy barns and lambing sheds. Mass aerial infection of large groups has occurred. Such an instance happened to more than 1600 soldiers returning from Italy to the United States and who apparently were infected by the winds blowing at an airport. The microorganisms may survive and remain infective for several miles downwind from the source.

Infected cattle and sheep may shed infected placentas and fluids at the time of parturition. About half of the cows harboring this rickettsia discharge it in their milk, and the ingestion of the milk of infected cows or goats is a source of infection for humans. Although the rickettsias of Q fever are fairly resistant to heat, they are effectively destroyed in milk by the high-temperature, short-time method of pasteurization.

A capillary tube agglutination test detects antibodies in serum from human beings and animals. Blood from a patient may be inoculated into a male guinea pig.

BARTONELLA BACILLIFORMIS

The intracellular parasites of the genus *Bartonella* are small, gram-negative, flagellated, and very pleomorphic, hemotropic coccobacilli. They can be cultivated on cell-free media, unlike other rickettsias. All infect red blood cells as well as endothelial cells of liver, spleen, and lymph nodes. They are transmitted by the night bite of the sandfly *(Phlebotomus)*. The one species of the genus, *Bartonella bacilliformis*, is the cause of the South American disease known as Carrión's disease (Oroya fever), which is characterized by an acute anemia with fever followed by a verrucous, or wartlike, skin eruption (verruga peruana). *Salmonella* species are prone to complicate. Infection with *Bartonella*, a South American parasite, is endemoepidemic in the Andean valleys of the Peruvian sierra.

B. bacilliformis is easily recognized in Wright-stained films of peripheral blood and in material from skin lesions because of its unique microscopic appearance.

CHLAMYDIAE (BEDSONIAE)

In *Bergey's Manual* (1984), order II, Chlamydiales contains gram-negative, obligate intracellular pathogens of vertebrates with an intracellular pattern for reproduction peculiarly their own. They were given the name *chlamydiae* (Greek *chlamys*, cloak) because each organism is surrounded by a dense cell wall in part of its life cycle.

Description

Chlamydiae (bedsoniae) constitute a large group of micoorganisms (psittacosis group) with comparable design and makeup and are the cause of several important diseases.

Life Cycle

The life cycle of the chlamydiae is complex and unique, beginning when the infectious form of the chlamydial organism contacts surface receptors on a susceptible cell. The small (0.3 μm in diameter), thick-walled infectious particle is called an **elementary body** (Fig. 32-4). It is phagocytized by the cell under attack and incorporated into a membrane-bound vacuole in the cytoplasm. After this, a series of events ensues during the next 24 to 48 hours that brings about reorganization and growth of the elementary body within the vacuole. A large particle, the **reticulate or initial body,** appears, which is 0.8 to 1 μm in diameter, thin walled, and rich in RNA. It grows and divides repeatedly by binary fission so that eventually the vacuole is completely filled with a host of small particles; a microcolony is formed. The typical well-defined inclusion body is now demonstrable in host cell cytoplasm. The daughter cells of the microcolony differentiate into thick-walled infectious particles that, when the parasitized cell ruptures, repeat the growth cycle.

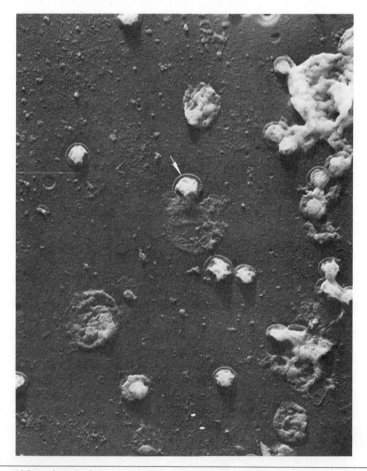

FIG. 32-4   Chlamydiae (bedsoniae), electron micrograph. Purified elementary bodies look like derby hats. Extensive studies were important to the conclusion that these organisms, now termed *chlamydiae,* were *not* viruses. (×28,000.)

Comparison with Viruses

For a long time chlamydiae were thought of as large viruses and were classified and referred to as basophilic viruses and as mantle viruses. Like viruses they are obligate intracellular parasites, but important differences exist (see Tables 33-1 and 33-2). Unlike viruses they possess enzymes and both DNA and RNA. They have no capsid symmetry and are susceptible to broad-spectrum antimicrobial drugs (very sensitive to tetracyclines). The patterns of their diseases differ in many ways from those of viruses.

Bases for Classification

Recent studies indicate that these parasites are better separated from true viruses and considered small bacteria. They are comparable to rickettsias but still distinct (see Table 31-1). The name *Bedsonia* was given in honor of Sir Samuel Bedson (1886-1969), who studied them extensively. However, many microbiologists believe that the genus name should be *Chlamydia*, and therefore these microbes are most often called chlamydiae. They are nonmotile, reproduce by binary fission, and have an outer cell wall chemically similar to that of gram-negative bacteria, from which they were probably derived.

Laboratory Diagnosis

Many procedures used in the laboratory diagnosis of viral infection apply to chlamydiae (p. 687). They grow in the fertile hen's eggs but better in cell cultures, where they express their distinct cytopathogenicity. They may be recovered from susceptible animals. They are easily stained with basophilic dyes (hence the term *basophilic viruses*), and because of their unique staining qualities, they can be visualized readily in suitable specimens by light or fluorescence microscopy.

Chlamydiae contain a common heat-stable group antigen in the cell wall, which induces complement-fixing antibodies. Although it demonstrates this group antigen, the complement-fixation test, nevertheless, is of value and has been widely used as a diagnostic serologic test for a given chlamydial disease. Chlamydiae also possess type-specific heat-labile antigens. These are usually best demonstrated by fluorescent antibody technics. Chlamydia have been typed by a specially designed microimmunofluorescence test that is based on their specific antigens. The well-known but unreliable skin test, the Frei test, is no longer used. (In the United States it is neither recommended nor available.)

Importance

The genus *Chlamydia* of family I, Chlamydiaceae, comprises only two known species, which are separated on the basis of inclusion type. The inclusions of *Chlamydia trachomatis* stain with iodine; those of *Chlamydia psittaci* do not. *C. trachomatis* is susceptible to sulfonamides; *C. psittaci* is not. Table 32-2 compares the two species in the genus *Chlamydia*.

Two biologically distinct subdivisions are recognized in the species *C. trachomatis*—the lymphogranuloma venereum (LGV) agent for one and the trachoma–inclusion conjunctivitis (TRIC) agents for the other. Each subdivision displays a number of serotypes, which align with the pattern of oculogenital diseases produced by these chlamydiae (Table 32-3).

As a result of the present classification of chlamydiae, organisms with widely diverse biologic and antigenic properties have been placed into each of two species; the greater diversity of biologic properties is seen within the species *psittaci*.

Chlamydial diseases sort out ecologically into three categories: (1) humans—oculogenital and respiratory, (2) birds—respiratory and generalized, and (3) nonprimate animals—respiratory, placental, arthritic, enteric, and so forth. Depending on the specific serotype, *C. trachomatis* is responsible for trachoma, lymphogranuloma venereum, inclusion conjunctivitis, pneumonia, cervicitis, salpingitis, urethritis, and proctitis in the human being, the natural host. *C. psittaci* causes human psittacosis (a zoonosis), psittacosis and ornithosis in birds, and various other diseases, such as abortion, arthritis, and systemic infections in sheep and cattle.

CHLAMYDIAL INFECTIONS
Trachoma

Although rare in the United States, trachoma, one of the oldest human diseases known, afflicts more than 500 million persons in the world (mostly in unindustrialized countries). The causative organisms, certain serotypes of *Chlamydia trachomatis*, specifically attack the lining cells of the cornea and the conjunctival membrane of the eye.

Table 32-2 Comparison of the Two Species in Genus *Chlamydia*

Distinguishing Feature	*Chlamydia psittaci*	*Chlamydia trachomatis*
Type of inclusion	Diffuse	Rigid, compact
Glycogen present	Negative	Positive
Iodine staining	Negative	Positive
Guanine and cytosine in DNA	41%	45%
Response to sulfonamides	Resistant	Sensitive
Hosts	Many animal species, humans (birds principal hosts)	Humans only
Serotypes	Many but not identified	15
Grade of disease	Usually severe	Mild, tend to be chronic

Table 32-3 Medical Importance of Chlamydial Species

Species	Serotypes*	Usual Mode of Spread	Major Clinical Syndromes†
Chlamydia psittaci	Unidentified (many)	Airborne (inhalation)	Psittacosis, ornithosis
TRIC Agents			
Chlamydia trachomatis	A, B, Ba, C	Flies, fomites, non-sexual person-to-person contact	Trachoma—classic, hyperendemic (genital infections rare)
Chlamydia trachomatis	D, E, F, G, H, I, J, K	Sexual, perinatal contact	Nongonococcal urethritis, cervicitis, pelvic inflammatory disease, epididymitis, inclusion conjunctivitis of newborns and adults, newborn pneumonia, proctitis
LGV Agent			
Chlamydia trachomatis	L-1, L-2, L-3	Sexual contact	Lymphogranuloma venereum

*Type of disease is related to serotype.
†Disease is matched to the serotype that is usually predominant.

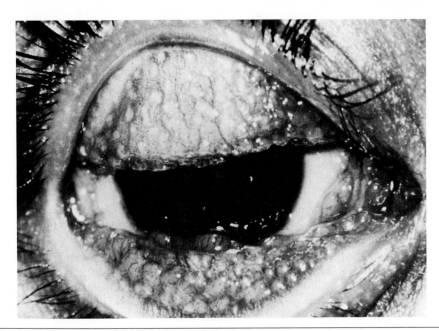

FIG. 32-5   Trachoma. Inflammatory nodules are spread over thickened conjunctiva of eye of Pima Indian (from tribe of southern Arizona and northern Mexico). Trachoma is prevalent among Indians of Southwest. (Courtesy of Dr. Phillips Thygeson, Los Altos, Calif.)

Growth within these cells produces very distinctive inclusion bodies. The disease process in trachoma is designated a keratoconjunctivitis. As is often the case with viral diseases, bacterial infection is superimposed to intensify the injury already done. The inflammatory changes are pronounced, and the scar tissue formed over the cornea, ordinarily transparent, results in impaired vision. The name *trachoma* is derived from a Greek word meaning "roughness" to describe the pebblelike appearance of the infected conjunctival membrane (Fig. 32-5). This disease is the world's leading cause of blindness (in some 20 million or more).

Spread

Trachoma is endemic in the large underprivileged areas of the tropics and subtropics, favored by poverty and filth; in some countries 75% of the population suffer. Human beings are the only sources of the agent. Transmission requires close personal contact, and for this reason the disease is often passed from mother to child. Chlamydiae may be spread by contaminated fingers and articles of clothing. Flies can carry the infection mechanically.

Laboratory Diagnosis

The diagnosis is made by identification of the typical inclusion bodies in scrapings from the conjunctival membranes and by the application of a complement-fixation text. Conjunctival scrapings may be stained with Giemsa stain to demonstrate the agent, but fluorescent antibody technics are more sensitive for this purpose.

Control

The agent has been grown in fertile eggs and an effective vaccine prepared. The prognosis in this disease has been greatly improved with the use of antibiotics alone. The incidence of the disease has been shown to drop simply with improvements in living conditions.

Oculogenital Infections

Knowledge of the range of oculogenital infections in both adults and infants caused by immunotypes of *C. trachomatis* is rapidly expanding. Varied chlamydial infections in the lungs, ears, and nasopharynx are being studied, as well as those in the eye and in the genital tract. In the latter areas such infections are now recognized as highly prevalent and important in the spread of sexually transmitted disease. A number of studies have shown certain strains of chlamydiae to be among the most frequently recovered pathogens in women attending venereal disease clinics. *C. trachomatis* may well be the most common sexually transmitted microorganism.

Genital trachoma is a term sometimes used for the sexually transmitted chlamydial infection that is seen as urethritis in men and cervicitis and salpingitis in women. Most chlamydial infections of the genital tract are probably asymptomatic; therefore carriers spread the infection.

Nongonococcal Urethritis

C. trachomatis is considered to be the major cause (found in 40% to 60% of cases) of an inflammation of the urethra (urethritis) in males. To emphasize that this condition, which is probably three times more common than gonorrhea, is a separate disease from gonococcal urethritis and that an etiologic agent other than the gonococcus is operative, it is designated *nongonococcal urethritis* (NGU) (p. 642). Gonorrheal urethritis often coexists with NGU, and if NGU is present after gonorrhea is adequately treated, it is referred to as *postgonococcal urethritis* (PGU); chlamydiae are responsible for 60% to 80% of these cases. Nongonococcal urethritis is said to be the most common sexually transmitted disease in the United States, as it is known to be in several European countries. NGU may be caused by agents other than chlamydiae, as follows:

Bacterial
  Chlamydia trachomatis (40% to 60% of cases)
  Ureaplasma urealyticum ("T strain mycoplasmas"—10% to 30% of cases)
  Mycoplasma hominis
  Gardnerella vaginalis
Protozoan
  Trichomonas vaginalis
Fungal
  Candidia albicans (rare)
Viral
  Herpesvirus hominis (?)
  Cytomegalovirus (?)

Males with NGU may have few or no manifestations of disease. Most of their female sexual partners are asymptomatic but do have a cervicitis from which chlamydiae are recoverable.

Serious complications in men include epididymitis, prostatitis, and in genetically susceptible persons Reiter's syndrome. Reiter's syndrome is a multisystem involvement of unknown etiology; prominent among its manifestations are arthritis and conjunctivitis. It is a rare sequel to NGU, and chlamydiae are associated with two thirds of the patients with arthritis.

Comparison of nongonococcal urethritis with gonorrhea. Nongonococcal urethritis tends to be a more slowly moving, milder, and more prolonged infection than gonorrheal urethritis, which is usually acute and fairly severe. The incubation period for nongonococcal inflammation is 8 to 21 days; for gonorrheal urethritis it is 3 to 7 days. The dysuria (painful urination) in NGU is mild, and the often scant urethral discharge is clear or white, thin, and mucoid. The profuse and spontaneous gonorrheal discharge is purulent.

Cervicitis

Chlamydial cervicitis has been documented in 30% of women examined in various venereal disease clinics and in up to 60% of such women with gonorrhea. Although her chlamydial infection often is asymptomatic carriage, at times in the woman it may be fairly severe.

The infected cervix may appear normal, or it may be reddened, edematous, and friable. A mucopurulent or purulent discharge may be present. Complications include the development of salpingitis (pelvic inflammatory disease), which is associated with increased likelihood of ectopic pregnancy (development of the fertilized ovum in an unfavorable site, such as the fallopian tube) and eventual sterility. In the pregnant woman chlamydial infection is especially serious because it is readily related to miscarriages and stillbirths and the infant is at risk for eye infection and pneumonia.

Adult Inclusion Conjunctivitis

Adult inclusion conjunctivitis is an acute eye infection acquired by exposure to infective genital tract discharges. Infected babies may transmit the disease to adults. Particularly prevalent in sexually active young adults, the disease is benign and self-limited most of the time but may be associated with serious sequelae.

Inclusion Conjunctivitis of the Newborn

Inclusion conjunctivitis of the newborn (inclusion blennorhea) is an acute mucopurulent chlamydial conjunctivitis. From the source of the organisms in the male urethritis, the female is infected sexually, and her infant is colonized with the same chlamydiae from her cervix during the passage through the birth canal. Of infants born to infected mothers, 33% to 50% will develop this condition. The incubation period is usually 5 to 14 days. The disease, usually benign and tending to resolve spontaneously within a week or two, may at times be severe with resultant scarring in the baby's eyes.

Inclusion conjunctivitis is caused by a serotype of *C. trachomatis* different from the one causing trachoma. It is remarkable that although produced by organisms of the same species, the two eye diseases—trachoma and inclusion conjunctivitis—have such different clinical features and epidemiologic patterns and pose different public health problems.

Neonatal Pneumonia

Chlamydiae picked up by the infant at birth may also colonize the baby's nasopharynx and possibly sites in the tracheobronchial tree. The nasopharyngeal infection, often the primary site, comes 1 or 2 weeks postnatally. This focus may be responsible subsequently for a pneumonia in that infant at 1 to 4 months of age, a pneumonia characterized by an afebrile course, chronic diffuse lung changes, and a bothersome staccato cough.

Lymphogranuloma Venereum

Lymphogranuloma venereum (venereal lymphogranuloma, lymphopathia venereum, lymphogranuloma inguinale), formerly known as climatic bubo, is a sexually transmitted disease caused by the LGV agent, a member of the species *C. trachomatis*. The LGV agent is biologically different from but immunologically close to other human pathogens in the species. Three serotypes are defined. The LGV agent, more highly invasive than other members of the species *trachomatis*, tends to cause systemic disease rather than a restricted process on a mucous membrane.

Formerly considered a disease of the lower echelon of society, it is appearing today in the affluent and seems to be increasing in incidence not only in this country but also throughout the world. It affects males more than females in a 20:1 ratio. The LGV agent is found only in human beings.

Pathology

In males the disease starts with a primary ulceration on the external genitalia. The infection extends to the inguinal lymph nodes, where buboes are formed. Inguinal lymphadenopathy is a salient feature. In females infection extends from the primary lesion along the lymphatic drainage to lymph nodes in the pelvis, where the chronic inflammation set up may in time lead to stricture of the rectum. The chlamydiae of this disease have an affinity for lymphoid cells; these are the cells that are parasitized. In both sexes the lymphatics are mainly involved. The severe scarring associated with long-standing lymphangitis results in a dramatic deformity of the external genitalia.

Lymphogranuloma venereum should not be confused with *granuloma inguinale*, also a sexually transmitted disease but caused by a different agent (p. 568).

Laboratory Diagnosis

For the laboratory diagnosis of lymphogranuloma venereum, inclusion bodies may be demonstrated in suitably stained smears of pus from the buboes.

The Frei test, discredited today, was once used to detect past or present infection in lymphogranuloma venereum. Material prepared by growth of the chlamydial agent in the yolk sac of a developing chick embryo was injected into the skin, and the appearance of a bright red papule was a positive test. The test was inconsistent in infection with the LGV agent—the one it was designed to detect.

A complement-fixation test and a more sensitive microimmunofluorescence test are diagnostic. Complement-fixing antibodies appear 2 to 4 weeks after onset of disease, and the significant finding is the rise in titer of antibodies between the serum sample of acute disease and that of convalescence.

Associated with the chronicity of lymphogranuloma venereum, profound changes occur in the serum globulin levels, often with reversal of the normal albumin/globulin ratio. Hyperglobulinemia is characteristic.

Psittacosis (Parrot Fever) and Ornithosis

Psittacosis (parrot fever, chlamydiosis) is a chlamydial infection so named because it affects psittacine birds, most often parrots and parakeets. Ornithosis is the corresponding disease in domestic fowl and birds other than those of the parrot family—chickens, turkeys, pigeons, and sea birds. Psittacosis is worldwide, but the number of human beings attacked is small.

Transmission

Psittacosis is widespread among the birds just named, the sources of most human infections. One contracts the disease from birds by (1) handling of sick birds or their feathers, (2) transmission of the agent through the air, (3) contact with material contaminated by infected birds, and (4) bites or wounds inflicted by sick birds. Sick birds show the chlamydiae in their nasal discharges and droppings and may become carriers. In birds the gastrointestinal tract is primarily affected. In humans the infection enters through the respiratory tract and localizes there. In a few cases infection is transmitted directly from person to person. The onset of human disease may be sudden, and the mortality is high. Explosive outbreaks sometimes occur in poultry-processing plants.

Laboratory Diagnosis

Diagnosis is made by the demonstration of the agent in a laboratory mouse that has been inoculated with blood or sputum from the patient. For diagnostic purposes, suspect material may also be injected into the yolk sac of the embryonated hen's egg.

A complement-fixation test is available, and as with the other chlamydial infections, it is the rising titer of antibodies obtained after 10 days between serum from the acute phase of disease and that of convalescence.

Prevention

The patient must be isolated and all discharges disinfected. A vaccine has been prepared against psittacosis, but its effectiveness is undetermined.

CAT-SCRATCH DISEASE

Cat-scratch disease (benign lymphoreticulosis, benign inoculation lymphoreticulosis, nonbacterial regional lymphadenitis) is an uncommon, febrile illness that may come after scratching, biting, or licking by a cat. It affects children primarily.

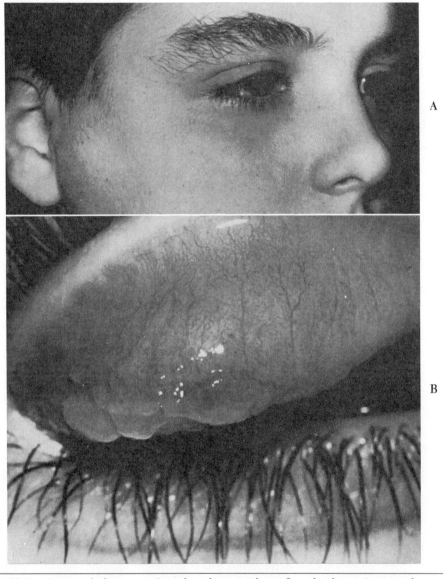

FIG. 32-6 Cat-scratch disease. **A,** Lymph node regional to inflamed right eye (preauricular node) is swollen and tender. **B,** Upper eyelid everted to show inflammatory swelling and congestion of blood vessels. Purulent exudate is present. Not only did this patient have a cat, but cat-scratch antigen test was positive. (From Donaldson, D.D.: Atlas of external diseases of the eye, vol. 1, St. Louis, 1966, The C.V. Mosby Co.)

The discussion of cat-scratch disease is placed in this chapter because for a long time the agent was believed to be one of the chlamydiae. The newly recognized bacterium is as yet unclassified.

A purulent lesion appears at the site of infection, and 2 or 3 weeks later the lymph nodes draining the area enlarge (Fig. 32-6). Incubation time from injury to illness is usually 3 to 10 days. As a rule, the disease is mild and self-limited. Cats are mechanical vectors and do not show related illness.

It is possible that animals other than cats carry the infection, including birds and insects, but the monkey is the only animal artificially infected.

Laboratory Diagnosis

Antigen for the skin test is prepared from the purulent material aspirated from the lesions or from a lymph node extract. It is evaluated in 48 hours. The skin test should be done at least 1 week after the onset of lymphadenitis. A positive test indicates previous or present infection.

Cause

A small, delicate, pleomorphic, gram-negative bacillus is currently considered the cause of cat-scratch disease. It was first demonstrated in tissue sections of lymph nodes that were stained with a special silver stain. Silver impregnation of the tissues makes the bacteria more readily seen microscopically. The lymph nodes have come from patients with a history of animal scratches and a positive skin test for cat-scratch disease. The bacillus does not seem to look like any other known microbe and may well be a newly recognized pathogenic agent.

Prevention

The only preventive measure known is the avoidance of cats.

SUMMARY

1 Rickettsias distinctively parasitize tissue cells and red blood cells of a vertebrate host. Arthropods are their natural and primary hosts. The three medically important genera are *Rickettisa*, *Rochalimaea*, and *Coxiella*.

2 Rickettsias are small, bacillary or coccobacillary forms, dividing by binary fission and containing both DNA and RNA.

3 The rickettsial diseases are grouped into the typhus fevers, the spotted fevers, scrub typhus, trench fever, and Q fever. Because rickettsias parasitize endothelial cells of small blood vessels, their anatomic lesions are widespread. Rocky Mountain spotted fever is one of the most severe of all infectious diseases.

4 Certain rickettsial diseases are associated with a positive Weil-Felix reaction, an agglutination test utilizing different strains of *Proteus*. It is a heterophil antibody reaction because *Proteus* bacilli bear no relation to rickettsial disease.

5 Q fever, a zoonosis, is caused by rickettsias in the genus *Coxiella*. Unlike the other rickettsias these are resistant to heat, drying, and sunlight; are not obligate intracellular parasites; do not produce a classic skin rash; and have no arthropod vector; and the serum of a patient does not agglutinate *Proteus* bacilli.

6 The small, gram-negative, flagellated, pleomorphic, hemotropic coccobacilli of the genus *Bartonella* can be cultivated on cell-free media. *Bartonella bacilliformis*, the cause of Carrión's disease, is transmitted by the sandfly. It is easily recognized in Wright-stained blood films.

7 Chlamydiae are gram-negative, obligate, intracellular pathogens with a distinctive life cycle. They possess both DNA and RNA, have no capsid symmetry as viruses do, and are susceptible to certain antimicrobial drugs.

8 The two species within the genus *Chlamydia* are *Chlamydia trachomatis* and *Chlamydia psittaci*. Two subdivisions in the species *trachomatis* are the lymphogranuloma venereum (LGV) agent (three serotypes) and the trachoma–inclusion conjunctivitis (TRIC) agents (12 serotypes).

9 Classic trachoma, the leading cause of blindness in the world, is caused by four specific serotypes, and oculogenital trachoma is caused by eight specific serotypes.

10 *Chlamydia trachomatis* is perhaps the most common sexually transmitted microbe. The infant infected at birth with chlamydiae from the mother may develop pneumonia or inclusion conjunctivitis as a consequence.

11 Chlamydial diseases sort out ecologically into (1) human diseases—oculogenital and respiratory; (2) bird diseases—respiratory and generalized; and (3) nonprimate animal diseases—respiratory, placental, arthritic, enteric, and the like.

12 *Chlamydia psittaci* causes psittacosis in humans and birds.

13 The agent of cat-scratch disease has been identified as a small, delicate, pleomorphic, gram-negative bacillus, first demonstrated in tissue sections of lymph nodes that were stained with a special silver stain.

QUESTIONS FOR REVIEW

1 What are rickettsias? How did they get their name? Give their characteristics.
2 How do rickettsias resemble bacteria? Viruses? How are they transmitted?
3 Into what five groups may rickettsial diseases be classified? Which ones are found in North America?
4 Give the laboratory diagnosis of rickettsial infections.
5 What is the Weil-Felix reaction? Its importance?
6 Briefly discuss Q fever and its causative agent.
7 Explain the pathologic changes in rickettsial diseases. (Consult outside references.)
8 State special properties of chlamydiae. Compare them with other microbes, including the rickettsias.
9 Give examples of chlamydial diseases with diagnostic inclusion bodies in parasitized cells. Outline the growth cycle for chlamydiae.
10 What disease is responsible for most of the world's blindness? Give its salient features.
11 What is cause of Carrión's disease?
12 Briefly explain Frei test, tabardillo, Ixodidae, transovarian infection, benign lymphoreticulosis, "Q" of Q fever, nongonococcal urethritis, inclusion conjunctivitis, blennorrhea, Neill-Mooser reaction, genital trachoma, and lymphogranuloma venereum.
13 Briefly discuss oculogenital infections caused by chlamydiae. What is their importance?
14 Give salient features of psittacosis and ornithosis.

SUGGESTED READINGS

Allibone, G.W., and others: Lymphogranuloma venereum revisited, Tex. Med. 78:54, March 1982.

Baca, O.G., and Paretsky, D.: Q fever and *Coxiella burnetii*: a model for host-parasite interactions, Microbiol. Rev. 47:127, 1983.

Books, J.B., and others: Rapid differentiation of Rocky Mountain spotted fever from chickenpox, measles, and enterovirus infections and bacterial meningitis by frequence-pulsed electron capture gas-liquid chromatographic analysis of sera, J. Clin. Microbiol. 14:165, 1981.

Ellis, R.E.: Chlamydial genital infections: manifestations and management, South. Med. J. 74:809, 1981.

Goldsmith, M.F.: Medical News: Has AFIP debugged the cat-scratch mystery? J.A.M.A. 250:2745, 1983.

Hechemy, K.E., and Michaelson, E.E.: Rocky Mountain spotted fever: a resurgent problem, Lab. Manage. 19:29, Oct. 1981.

McCaul, T.F., and Williams, J.C.: Developmental cycle of *Coxiella burnetii*: structure and morphogenesis of vegetative and sporogenic differentiations, J. Bacteriol. 147:1063, 1981.

Peter, J.B., and Lovett, M.A.: Genital infection from *Chlamydia trachomatis*, Diagn. Med. 5:17, Nov./ Dec. 1982.

Schachter, J., and Caldwell, H.: Chlamydiae, Ann. Rev. Microbiol. 34:285, 1980.

Schultz, M.G.: Daniel Carrión's experiment, N. Engl. J. Med. 278:1323, 1968.

Smith, T.F.: Chlamydial infections: clinical and laboratory features, Lab. Manage. 21:40, Aug. 1983.

Thomas, B.J., and others: Early detection of chlamydial inclusions combining the use of cyclohexamide-treated McCoy cells and immunofluorescence staining, J. Clin. Microbiol. 6:285, 1977.

Wear, D.J., and others: Cat-scratch disease: a bacterial infection, Science 221:1403, 1983.

Weiss, E.: The discovery of a sporogenic cycle in *Coxiella burnetii*, the agent of Q fever, resolves the enigma of its high stability in the environment, ASM News 47:455, 1981.

Winkler, H.: Rickettsiae: intracytoplasmic life, ASM News 48:184, 1982.

VIRUSES

GENERAL CONSIDERATIONS
Definition

The first and earliest working diagnosis of a virus (Latin *vīrus*, slime, poison) was of an agent that caused disease but one small enough to pass through a filter that would retain bacteria. For many years viruses were called *filtrable viruses*. Later, a virus was identified as the smallest infectious agent with a molecule of nucleic acid for its genome that can only be propagated in the presence of living cells (that is, it is also an obligate intracellular parasite). The viral nucleic acid the virus possesses contains the necessary information to enable it to synthesize in the host cell a number of virus-specific macromolecules it requires to produce progeny. Like other biologic units viruses reproduce not only with genetic continuity but also with the genetic potential for change.

If the least requirement for life is the ability of a living thing to duplicate itself, or reproduce, viruses are the smallest known living bodies. Most are so small that they cannot be seen with an ordinary light microscope, and some are so small that they approximate the size of the large protein molecules. Viruses seem to lie in a partially explored twilight zone between the cells the biologist studies and the molecules the chemist handles, being smaller than the smallest known bacterial cells and just larger than the largest macromolecules. (It would take 2500 poliovirus particles to span the point of a pin.)

Since the turn of the century and especially in the last few decades, our knowledge of the nature, structure, and natural history of viruses and of their disease-producing potential has increased greatly. Many viruses previously unknown have been isolated and identified—more than 100 in the last 30 years or so. The technologic advances that defined the modern dimensions of the science of virology have been the invention of the electron microscope; the introduction of the ultracentrifuge; the development of modern cell (tissue) culture technics; the applications of immunofluorescence and cytochemistry; the emergence of hybridoma technology for the production of monoclonal, monospecific antibodies; and the implementation of recombinant DNA technics.

Structure

Viruses are particulate and vary considerably in size (Table 33-1) and shape (Fig. 33-1). They can be very small and simple, about 20 nm in diameter with perhaps a dozen genes, or large and complex, about 200 to 300 nm in diameter with several hundred genes. Generally, the viruses of humans and animals are spherical, those of plants are rod shaped or many sided, and those of bacteria (bacterial viruses, or bacteriophages) are tadpole shaped. Once presumed to be fairly simple, viruses are now known to be highly complex structures in which viral components are fitted together into rigid geometric patterns with mathematic precision. In fact, a highly purified preparation of virus is referred to as a **virus crystal** (Fig. 33-2). There are three basic forms in viral structure—the icosahedral, helical, and complex. The **icosahedron**, a crystal, is a solid, many-sided geometric form with 20 triangular faces and 12 apexes. (A virus with the icosahedral pattern is said to have *cubic symmetry*.) The **helical** form indicates a spiral tubular structure bound up to make a compact, long rod. The **complex** form is enclosed by a loose covering envelope, and because the envelope is nonrigid, the virus is highly variable in size and shape.

In the simplest viruses there is a long coil of nucleic acid sufficient for a complement of genes (the **genome** of the virus), tightly folded and packed within a symmetric protein coat called the **capsid**. The capsid plus the contained nucleic acid is termed the **nu-**

Table 33-1 Relative Sizes of Animal Viruses

Biologic Unit	Approximate Diameter (or Diameter × Length in nm)*	Approximate Number of Genes
Red blood cell†	7500	—
Chlamydia species†‡	300 to 800	—
Poxvirus	230 × 300	400
Herpesvirus	100 to 200	180
Adenovirus	70 to 90	50
Retrovirus (leukovirus)	100	50
Reovirus	60 to 80	40
Coronavirus	80 to 130	30
Paramyxovirus	150 to 300	30
Rhabdovirus	70 to 175	20
Myxovirus	80 to 120	15
Arenavirus	50 to 300	15
Bunyavirus	90 to 100	15
Togavirus	40 to 70	15
Papovavirus	45 to 55	10
Picornavirus	20 to 30	12
Parvovirus (picodnavirus)	18 to 26	7
Bacterial virus (bacteriophage)	25 to 100	—
Plant virus	17 to 30	—
Serum albumin molecule†	5	—

*1 nm = 1/1000 μm = 1/1,000,000 mm.
†These are not viruses but are shown here for comparison.
‡See p. 655.

cleocapsid. The capsid is made up of subunits, or **capsomeres** (aggregates of oligopeptides), arranged in precise fashion around the nucleic acid core. Such a virus is a **naked virus.** In the larger viruses an envelope, a phospholipid bilayer membrane with "spikes" of glycoprotein projecting from the surface, links fatty substances and complex sugars to the protein coat. (The penetration of these fatty substances accounts for the ether sensitivity of a given virus.) The viral particle as a complete structural unit, either naked or enveloped, is the **virion** (Fig. 33-3). The virion is the extracellular, nonreplicating, and infectious form of virus, the special package in which the virus is transported from cell to cell. The envelope (if present) and the capsid aid in the attachment of the virion to and its penetration into a host cell.

Cells of bacteria, plants, and animals contain both deoxyribonucleic acid (DNA) and ribonucleic acid (RNA). True viruses, however, contain only one. DNA serves as the genetic material in complex viruses, such as vaccinia and bacteriophages, as it does for all other living things. On the other hand, RNA is the genetic material not only in some complex viruses, such as the influenza viruses, but also in some of the simplest and smallest, such as the polioviruses. The DNA and RNA may be either single stranded or double stranded.

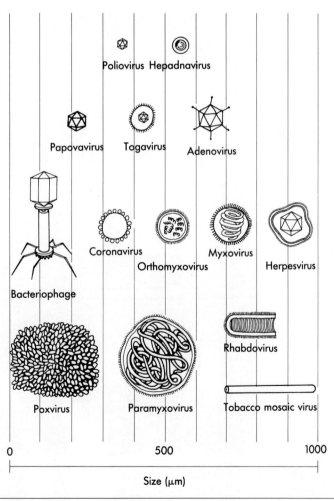

FIG. 33-1   Viruses, relative sizes and shapes. (A micrometer is enlarged approximately 125,000 times.)

The genetic material, or genome, of a virus refers to the sum of the specific genes that define the characteristics of the virus, order its replication, and function to preserve its genetic heritage intact. Details of the structure of the virus are remarkably adapted to the expression of the genetic strategy of that particular virus.

Life Cycle

Viruses, being obligate intracellular parasites, can only replicate inside given cells. The term *life cycle* encompasses the series of events that begins with the entry of a given virus into a specific cell, extends through intracellular sequences of replication and maturation, and ends with release of mature virions from the parasitized cell for passage through the extracellular medium (Fig. 33-4). The events of the life cycle are of considerable importance inasmuch as they bear directly on the cause and nature of viral injury to cells and thus may engender disease. Since genetic material varies among different viruses, the scheme or strategy of replication must also vary. Therefore in a general way only the life cycle of a virus may be sketched as follows:

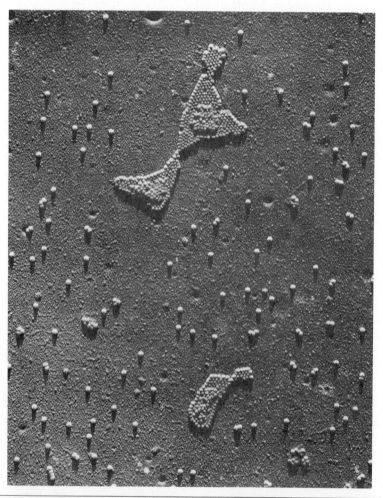

FIG. 33-2   Poliomyelitis virus crystal (type 1 strain purified and crystallized) in electron micrograph. Individual viral particles are 28 nm in diameter. In the absence of any immunity, one crystal of poliovirus contains enough particles to infect the population of the world. (×64,000.) (Courtesy Parke, Davis, & Co., Detroit.)

1. **Attachment and adsorption**—On contact with a susceptible cell the virion is able
to fasten to a specific site (a *viral receptor*) on the cell surface, and as a result of virus-
cell interaction, virus is adsorbed to the cell. In a virus without envelope a specific
protein complex on the capsid may be the mechanism for attachment. General factors,
such as age and inherent genetic susceptibility of the host, influence viral parasitism,
but it is also specific—only certain tissues and organs of the proper host species are
vulnerable. In an enveloped virus, host cell specificity comes from the envelope. The
interaction of specific molecules on the envelope with precise receptors on target cells
has been demonstrated in several instances, one such being the single, very precise
affinity of the mumps viral envelope for the sialic acid residues of human respiratory
epithelium. A viral envelope also contains powerful fusion factors that can enhance
viral invasion. Such factors aid the fusion of virus to cell and also cause one cell under
attack to merge with a second cell about to be attacked. Coalescence of cells facilitates
spread of virus. Going straight from one cell to the next, the viral invader can bypass
the extracellular area and is shielded from it, since it is in the extracellular milieu that

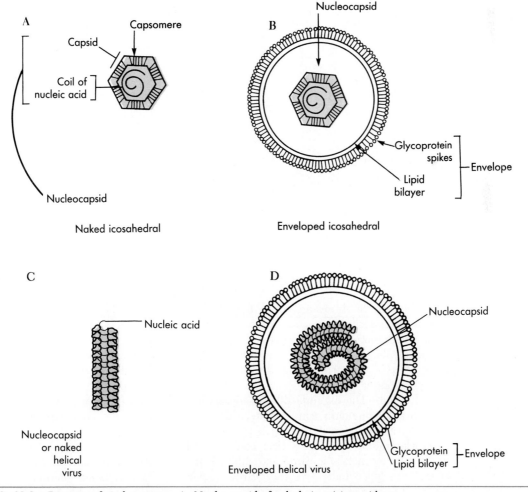

FIG. 33-3   Diagram of viral structure. **A,** Nucleocapsid of naked virus (virus without
envelope). The capsid contains the compacted genome (the coil of nucleic acid). Subunits of
capsid, the capsomeres, are indicated. **B,** Envelope around nucleocapsid of enveloped virus.
Components of envelope are indicated. **C,** Nucleocapsid of naked helical virus. **D,** Enveloped
helical virus. Components of envelope around nucleocapsid are indicated.

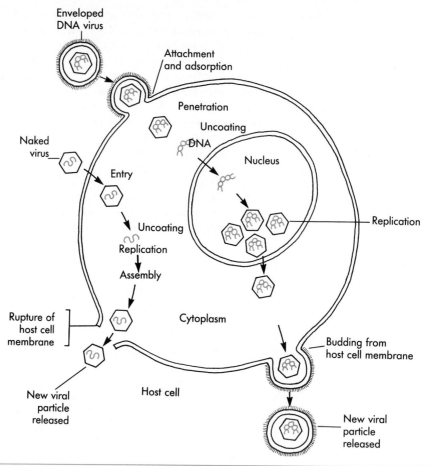

FIG. 33-4   Model of the events in the life cycle of a virus. Replication of RNA virus such as the poliovirus occurs in the host cell cytoplasm. Replication of DNA virus such as a herpesvirus occurs in the nucleus (see also Table 33-3).

antibodies against viruses are concentrated. Thus viruses may be spread *inter*cellularly as well as *extra*cellularly.

2. **Penetration**—Penetration comes right after attachment and may be the result of a kind of phagocytosis called *viropexis*. Viropexis is a process that is also similar to pinocytosis. The protein coat of the virus is stripped off, and viral nucleic acid enters the cytoplasm of the cell.

3. **Eclipse**—The dissembled virus appears to vanish into the protoplasmic maze of the cell and temporarily cannot be recovered in infectious form. The eclipse period is that interval of time so indicated.

4. **Maturation**—Because a virus contains such a small number of proteins, it can do little if any metabolic work, and although it carries its own genetic information, it cannot reproduce independently. (Viruses are said to act as independently existing genes.) It must seek out an environment in which chemical building blocks and energy for work are provided, and such an environment exists only inside a cell. Invasion of the cell by a virus is an act of piracy. A virus takes over, gains control of the genetic machinery of the parasitized cell, and redirects it to the synthesis of new viral particles. Replication is accomplished, and the new viral particles acquire protein coats (become encapsidated). The capsid protects the virus during its extracellular existence. Usually

the entry of viral nucleic acid into the host cell results in an infected cell, but one that is then immune—it cannot be reinfected by the same or related virus.

5. **Release**—A variable period elapses after the resources of the cell have been commandeered before masses of full-grown viruses appear. These mature progeny virions carrying their own genetic information are ready to be released from the cell. The variability of the life cycle for different viruses is reflected in different rates of production of infectious progeny. Some, such as poliovirus, produce many very quickly—in only a few hours a single, parasitized cell has produced 100,000 poliovirus particles.

Release of virions from the host cell is possible in at least two ways. One is by a process referred to as **budding.** In the case of an enveloped virus the encapsidated particle approaches the plasma membrane of the host cell, and a bud is formed in a distinct manner. It is then pinched off. With the exception of the glycoprotein spikes thereon, which are viral in origin, the envelope of an enveloped virus is derived from the host cell membrane. Another mechanism of release involves the virus-induced lysis of the cell. This frees the viral particles. The new generation of viruses is able to survive (not replicate) outside a cell until it can reach other suitable cells, but it must eventually find the new quarters.

Genetic Considerations

The genetic apparatus of a virus is much the same as for other living microorganisms but far less complex. Even large viruses have only a few hundred genes; smaller ones have considerably less. Genetic analysis is greatly simplified in viruses also, since, generally speaking, viruses are haploid (only one copy of each gene is present). (Retroviruses, being diploid, are the exception.)

Viruses display permanent genetic alterations that include mutations ranging from single base changes to sizable deletions, and they participate in certain genetic interactions that result in an altered genotypic or phenotypic expression (see outline below). An *intra*cellular milieu is required for genetic expression, and for the demonstration of certain genetic interactions the host cell must be infected by more than one virus. The following are genetic phenomena observable in viruses:

A. Alteration of genetic material—heritable
  1. Mutation—heritable change in genome of virus
  2. Gene rearrangements
     a. Recombination—exchange of genes between two viral genomes
     b. Reassortment—exchange of gene segments between two segmented genomes
     c. Transduction or rearrangement between viral genes and those of parasitized host cells
B. Interaction between gene products—nonheritable
  1. Complementation—one virus provides a missing gene product for replication of a second defective virus
  2. Phenotypic mixing—genotype from one virus but phenotype from two viruses

Mutations occur at about the same rate as in other microorganisms. In most viral infections with thousands of genome duplications, mutations are inevitable. Most are without effect (neutral) or are lethal to the virus. Deletions of genetic material are often lethal when they occur in a virus, usually a small one, with such a small amount of genetic material that none of it is expendable. With such factors operative it is easy to understand that most viruses are limited in their possible biologic variation. The notable exception is the influenza virus (p. 721).

Two rearrangements of genetic material of importance are **recombination** and **reassortment.** Recombination, the exchange of genes between two viral genomes, occurs during replication of most types of DNA viruses and is detected in the phenotypic characteristics of a large number of the offspring. RNA viruses display a wider range of nucleic acid interactions in recombination than do DNA viruses. Some RNA viruses recombine quite effectively (for example, retroviruses), whereas some indicate no ability to do so (for example, togaviruses, rhabdoviruses, and paramyxoviruses). Some are intermediate to the two (for example, picornaviruses). Reassortment, the exchange of gene segments between two segmented genomes, occurs readily with the influenza virus. In a segmented virus different genes are encoded in separate pieces of nucleic acid.

Rearrangements may take place between viral genes and those of the host cell. Instead of destroying its host, a virus may form a stable union with it. It is believed that in the induction of cancer by a virus, the genetic integration of all or more often part of a viral genome into a cell chromosome is of prime importance. This happens with retroviruses, papovaviruses, and adenoviruses. Herpesvirus may take up residence in the host cell as an unintegrated piece of genetic material. The persistent expression and effect of the viral genes in the cell are believed to stimulate cellular proliferation in such a way as to "transform" the cell. Transformation is the in vitro process whereby a normal cell phenotype is changed to one resembling a malignant neoplastic cell in its expression of uncontrolled and unrestricted growth.

Interactions between viral gene products are seen in complementation and phenotypic mixing. When two different mutants infect the same cell, complementation may occur if each mutant possesses a normal counterpart of a gene that is defective in its partner. Their normal gene products can then be directed to the replication of both parents. Phenotypic mixing results when complementation involves structural proteins. In the assembly process of the viral units, progeny is produced with structural features of both precursors.

Pathogenicity

In many instances little or no harm is done to the host in the life cycle of a virus, and many viral infections are silent and inapparent. For example, unrecognized infection with polioviruses is hundreds of times more common than the clinical disease. Some infections are latent, undetectable until activated by the proper stimulus (for example, infections with herpes simplex). However, if the cells are damaged or altered by the viral attack, disease exists and the signs of viral infection naturally reflect the anatomic location of the cells.

The pathology of viral diseases relates primarily to the visible effects of intracellular parasitism. When cells that have been specifically attacked by a virus show changes that directly reflect cell injury, the sum total of these is designated the **cytopathogenic, or cytopathic, effect** (CPE), an especially useful concept in cell (tissue) cultures.

In some cells small, round or oval bodies known as **inclusion bodies** will be formed as a manifestation of the cytopathic effect (Fig. 33-5). Inclusion bodies are distinct in

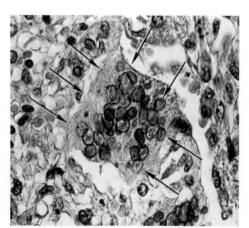

FIG. 33-5 Inclusion bodies of measles virus within nuclei of multinucleated giant cell (*arrows*), as seen in cell cultures and lymphoid tissues of patients with measles. (Microscopic section of lung from patient with rare measles giant cell pneumonia.) Note many nuclei piled on one another in murky cytoplasm. In each, chromatin is displaced to nuclear membrane by inclusion body. Viral particles in cytoplasm are not shown in routine hematoxylin-eosin stain. (×800.)

size, shape, and intracellular position for a given virus. They are emphasized by the pathologist, since they are usually easily visualized in the compound light microscope. In stained preparations inclusion bodies are commonly about the size of a red blood cell and rounded. Some viruses produce inclusion bodies in the cytoplasm of the parasitized cell and some in the nucleus; others produce them in both cytoplasm and nucleus. At times within the inclusion bodies still smaller units known as **elementary bodies** are seen. The elementary bodies are the virus particles, and an inclusion body with its elementary bodies may be likened to a colony of bacteria. Table 33-2 lists some infections where an intracellular parasite is related to an inclusion body. Note that inclusion bodies are seen with some bacterial infections.

Another reflection of cytopathogenic effect of virus is the formation of a giant cell, or syncytium, as is prominent in measles and cytomegalic inclusion disease (Fig. 33-5).

Mechanisms of viral injury to the infected cell vary with several possibilities. Viral interference with cell function may stem from the alteration or inhibition of protein synthesis or the induction of chromosomal changes, even to the incorporation of viral nucleic acid into the host cell genome. The injurious effect may be a direct one; for example, certain viruses, such as poliovirus, can activate and release enzymes contained in the lysosomes of the cell. In whatever way sustained progressive injury to a parasitized cell probably means that the cell is either partly or completely destroyed.

There is an opposite effect that viruses at times produce on parasitized cells, an effect not seen in relation to any other known living agent. Viruses can stimulate the parasitized cells to increase their numbers in a way the cells would not do if they did not contain virus. Such an increase in number of cells is referred to as **hyperplasia**. This "opposite effect" on biologic properties of cells parasitized by virus is called **conversion (transformation)**. Virus has not damaged the host cell so much as it has changed it. The cell continues to grow, synthesizes protein, and divides, but because of something added by the virus not present before, the cell behaves differently. For instance, the cell grows more rapidly but at the same time requires fewer nutrients to sustain the enhanced growth. Conversion (transformation) also affects cell morphology—details of size and shape deviate from the normal prototype. Probably the most singular trait thus conferred by virus on the transformed cell is mobility—the cell in its tissue environment can now move about, migrate, and even invade. It is this response of cells parasitized by certain viruses that so strongly implicates viruses as a cause of cancer in human beings.

Two factors influencing the response of cells to viral injury are (1) the rapidity with which a virus produces its effects—the more rapid its action, the more likely cells are to be damaged severely, and (2) the power of the cells injured to multiply. Nerve cells regulating sensory and motor function in the body cannot duplicate themselves (**regenerate**); if parasitized, they tend to be killed in the process. Cells lining the skin possess great powers of regeneration; therefore viral lesions in the lining cells of the skin may result in a piling up of increased numbers of epithelial cells at the site of injury (for example, the common wart). Sometimes viral effects are mixed, combining features of both destruction and hyperplasia.

There is often little in the way of measurable inflammatory response in viral infection. Where there is cell destruction, the tissues may show a minimal to moderate infiltration of inflammatory cells (lymphocytes and mononuclear cells, a few neutrophils) and vascular changes (edema and hyperemia) of limited extent. If not followed by any kind of cellular deterioration, virus-induced hyperplasia is not accompanied by detectable inflammation.

Viruses are responsible for more than 80 diseases of plants and many diseases of the lower animals. (In fact, they have been found to infect all known biounits from bacteria to human beings.) Practically all plants of commercial importance are attacked by them.

Table 33-2 Inclusion Bodies in Infections with Intracellular Parasites

Disease	Etiologic Agent	Location of Inclusion*	Name of Inclusion Body or Comment
Rabies	Virus	Cytoplasm	Negri bodies
Yellow fever	Virus	Cytoplasm	Councilman bodies—areas of necrosis in cell
Cytomegalic inclusion disease	Virus	Nucleus, also cytoplasm	Cytomegalic inclusions; prominent in enlarged cells
Molluscum contagiosum	Virus	Cytoplasm	Molluscum bodies; elementary bodies are Lipschütz bodies
Varicella-zoster	Virus	Nucleus	Prominent inclusions
Herpes simplex	Virus	Nucleus	Prominent inclusions
Vaccinia-variola	Virus	Cytoplasm	Guarnieri bodies; elementary bodies are Paschen bodies
Adenovirus pneumonia	Virus	Nucleus	Inclusions of rosette type
Measles giant cell pneumonia	Virus	Nucleus, also cytoplasm	Prominent within syncytial giant cells
Trachoma	Chlamydia (bacterium)	Cytoplasm	Prominent
Psittacosis	Chlamydia (bacterium)	Cytoplasm	Psittacosis bodies
Lymphogranuloma venereum	Chlamydia (bacterium)	Cytoplasm	Gamna-Favre bodies
Granuloma inguinale	Bacterium	Cytoplasm	Donovan bodies

*Note that what is called an inclusion body is a body visible with the *compound light microscope*. The electron microscope visualizes viral particles throughout the cell even in instances where there is a characteristic inclusion body in only one area.

Viruses have been known to attack the fungus from which penicillin is derived, thereby interfering with the manufacture of penicillin. Fungal viruses are common among the fungi. Viruses also attack insects. Of more than 550 viruses known, more than 200 produce over 50 viral diseases in humans, some of which are the most highly communicable and dangerous diseases known. Today the prevalent human infections (more than half) are viral. In addition a number of diseases of obscure etiology are now postulated to be viral in origin—multiple sclerosis, juvenile diabetes, systemic lupus erythematosus, and certain disorders of the central nervous system.

Classification

Currently most virologists favor a classification system based on major biologic properties such as nucleic acid present, size, structure (geometric pattern of capsid and number of capsomeres), presence or absence of an envelope, sensitivity to physical and chemical agents (ether sensitivity related to lipid envelope), immunologic aspects, epidemiologic features, and pathologic changes. Consistent with this, most animal viruses fall into 17 groups—6 have a DNA genome (deoxyriboviruses) and 10 have an RNA genome (riboviruses). The major groups are the following (see also Table 33-3):

1. Poxviruses (family Poxviridae)
2. Herpesviruses (family Herpetoviridae)
3. Adenoviruses (family Adenoviridae)
4. Papovaviruses (family Papovaviridae)
5. Parvoviruses (family Parvoviridae)
6. Myxoviruses (family Orthomyxoviridae)
7. Paramyxoviruses (family Paramyxoviridae)
8. Rhabdoviruses (family Rhabdoviridae)
9. Togaviruses (family Togaviridae)
10. Arenaviruses (family Arenaviridae)
11. Bunyaviruses (family Bunyaviridae)
12. Reoviruses (family Reoviridae)
13. Picornaviruses (family Picornaviridae)
14. Coronaviruses (family Coronaviridae)
15. Retroviruses (family Retroviridae)
16. Hepadnaviruses (family Hepadnaviridae)
17. Other viruses (unclassified)

Poxviruses

The largest and most complex of the animal viruses, the brick-shaped, enveloped, double-stranded DNA (200 to 300 genes present) poxviruses form characteristic inclusions in the cytoplasm of parasitized cells. Member viruses vary in sensitivity to ether. Poxviruses may be categorized into three groups: (1) poxviruses of mammals, (2) poxviruses of birds, and (3) oncogenic poxviruses (myxoma and fibroma poxviruses and Yaba monkey virus). There are 22 members whose action may either destroy cells or stimulate them to proliferate. The skin is a prime target, and viruses included are those of smallpox (variola), vaccinia, molluscum contagiosum, and cowpox.

Herpesviruses

Herpesviruses are medium-sized, ether-sensitive, enveloped, double-stranded DNA (100 to 200 genes) viruses, that, like poxviruses, often parasitize lining cells of skin and can pass from cell to cell without killing them. The characteristic viral inclusions are found in the nucleus. The protein shells of herpesviruses show cubic symmetry, with 162 capsomeres. Latent infections with these agents endure a lifetime. There are 20 members, including varicella-zoster virus, herpes simplex virus types 1 and 2, Epstein-

Table 33-3 General Properties of Viruses Related to Classification

Nucleic Acid Core	Viruses	Disease (Example)	Capsid Symmetry	Site of Capsid Assembly	Envelope Present
DNA					
Single stranded	Parvoviruses	Gastroenteritis	Icosahedral	Nucleus	No (naked virus)
Double stranded (circular)	Papovaviruses	Warts	Icosahedral	Nucleus	No (naked virus)
Double stranded	Adenoviruses	Respiratory disease	Icosahedral	Nucleus	No (naked virus)
Double stranded	Herpesviruses	Chickenpox	Icosahedral	Nucleus	Yes
Double stranded	Poxviruses	Smallpox	Complex	Cytoplasm	Complex coats
RNA					
Single stranded	Picornaviruses	Poliomyelitis	Icosahedral	Cytoplasm	No (naked virus)
Double stranded (segmented)	Reoviruses	Infantile diarrhea	Icosahedral	Cytoplasm	No (naked virus)
Single stranded	Togaviruses	Rubella	Icosahedral	Cytoplasm	Yes
Single stranded (segmented)	Arenaviruses	Hemorrhagic fever	Unknown or complex	Cytoplasm	Yes
Single stranded	Coronaviruses	Common cold	Unknown or complex	Cytoplasm	Yes
Single stranded (diploid)	Retroviruses	T cell leukemia	Unknown or complex	Cytoplasm	Yes
Single stranded (segmented)	Bunyaviruses	Encephalitis	Helical	Cytoplasm	Yes
Single stranded (segmented)	Myxoviruses	Influenza	Helical	Cytoplasm	Yes
Single stranded	Paramyxoviruses	Measles	Helical	Cytoplasm	Yes
Single stranded	Rhabdoviruses	Rabies	Helical	Cytoplasm	Yes

Barr virus (EBV), and cytomegaloviruses (some potential oncogenic agents are found here).

Adenoviruses

The medium-sized, ether-resistant, double-stranded DNA (30 to 50 genes) adenoviruses may persist for years as latent viruses in human lymphoid tissue. Adenoviruses have cubic symmetry, with 252 capsomeres. There are at least 33 human types, and distinct serotypes are known for many different animals. Adenoviruses produce typical cytopathic changes in cell culture, and their inclusion bodies form within the nucleus. These agents produce a range of diseases; some cause tumors in animals.

Papovaviruses

Papovaviruses are small, ether-resistant, nonenveloped, double-stranded DNA (circular) (5 to 8 genes) viruses that are noted for producing neoplasms in animals (oncogenic). Their growth cycles are relatively slow, and they replicate in the nucleus of the parasitized cell. Capsid symmetry is cubic. Papovaviruses include the *pa*pilloma viruses of humans, rabbits, cows, and dogs; the *po*lyoma virus of mice; and the *va*cuolating virus of monkeys (SV/40), all known oncogenic (tumor-producing) agents. There are at least 11 of them.

Parvoviruses (Picodnaviruses)

The parvoviruses are ether-resistant DNA viruses of limited genetic capacity and are *ultra*small (*parvo* means small). Capsid symmetry is cubic. Certain adeno-associated or adenosatellite viruses (defective viruses not able to replicate without adenovirus) and certain viruses of hamsters, rats, and mice are found here. They can produce latent infections in animals but do not seem to be associated with human disease.

Myxoviruses (Orthomyxoviruses)

The medium-sized, spherical, ether-sensitive, enveloped, single-stranded RNA myxoviruses with segmented genome have helical symmetry and replicate in the nucleus of the parasitized cell. Viral particles are pleomorphic, sometimes filamentous, and mature by budding of the cell surface. Most have a layer of prominent spikes in their outer wall. Members of the group agglutinate red blood cells of many mammals and birds. This hemagglutination is associated with the viral particle itself and inhibited by antibody acting against it. Some contain an enzyme, neuraminidase, that splits neuraminic acid from mucoproteins. (*Myxo* refers to mucus, for which these viruses have an affinity.) Myxoviruses include the influenza viruses A, B, and C, the virus of swine influenza, and that of fowl plague; myxoviruses are influenza viruses.

Paramyxoviruses

Paramyxoviruses are medium-sized, ether-sensitive, enveloped, single-stranded RNA viruses with helical symmetry. Unlike that of the myxoviruses, the genome of paramyxoviruses is *not* segmented. They are similar in appearance to but somewhat larger than myxoviruses. Paramyxoviruses appear to be antigenically stable, and certain ones hemagglutinate red blood cells, as do myxoviruses, with or without hemolysis. Replication in cell cultures occurs within cytoplasm. Some cause multinucleated giant cells to form in tissue cultures and sometimes in human tissues (for example, measles virus). Distinctive inclusion bodies are seen with certain ones. In this category are the parainfluenza viruses (four types), respiratory syncytial virus, and the viruses of measles and mumps in humans and of Newcastle disease and distemper in animals. The range of clinical disease problems caused by paramyxoviruses is a broad one.

Rhabdoviruses

Rhabdoviruses are ether-sensitive, single-stranded RNA viruses and have helical symmetry. Members of this group have an unusual appearance. (*Rhabdo* means rod.) Mature virions are shaped like bells or bullets. Intracytoplasmic inclusions, the Negri bodies, are seen with the rabies virus. Rhabdoviruses include the viruses of rabies and vesicular stomatitis of cattle, some insect viruses, and three important plant viruses.

Togaviruses

The small, ether-sensitive, enveloped, single-stranded RNA togaviruses mature by budding from the cell surface and are so named because of their envelope, or "toga."

Arboviruses (*ar*thropod-*bo*rne) is a general designation for an ecologic grouping of more than 350 viruses of diverse properties (most of which are togaviruses), sorted into several virus families. They all have, by definition, a complex life cycle involving biting (hemophagous) insects, especially mosquitoes and ticks. Arboviruses multiply in many species—humans, horses, birds, bats, snakes, and insects—and with the exception of the viruses of dengue fever and urban yellow fever, human beings are only accidental hosts. Arboviruses are most prevalent in the tropics, notably in rain forests. The ecologic hodgepodge of the total grouping is reflected by the exotic names for the diseases, such as Bwamba fever, Singapore splenic fever, Kyasanur Forest disease of India, Chikungunya encephalitis, and O'nyong-nyong infection in Uganda. Three patterns emerge in arboviral diseases: one is fairly mild; the other two are severe and often fatal. They are (1) fever (denguelike), with or without skin rash and polyarthralgia, (2) encephalitis, and (3) hemorrhagic fever with, as its name suggests, skin hemorrhages and visceral bleeding. Currently, on the basis of antigenicity, arboviruses are laid out in groups A and B, or in two genera. In group A (genus *Alphavirus*) are the viruses of western equine encephalitis, eastern equine encephalitis, and Venezuelan equine encephalitis; in group B (genus *Flavivirus*) are Japanese B encephalitis, St. Louis encephalitis, Murray Valley encephalitis, West Nile fever, dengue fever, and yellow fever.

Most of the togaviruses are arthropod-borne, but some are not. These include the rubella virus of humans (genus *Rubivirus*), the hog cholera virus, the border disease virus of sheep, the virus of equine arteritis, the virus of simian hemorrhagic fever, and the lactic dehydrogenase virus of mice.

Arenaviruses

Arenaviruses are enveloped, single-stranded RNA viruses, so named (Latin *arenaceus*, sandy) because of their unique appearance in electron micrographs, wherein the virions display a number of electron-dense, RNA-containing granules comparable to ribosomes. They include the arboviruses of the Tacaribe complex (South American hemorrhagic fevers) and the viruses of Lassa fever and lymphocytic choriomeningitis. Some of the group cause slow virus infections.

Bunyaviruses (Bunyamwera Supergroup)

The spherical, ether-sensitive, enveloped, single-stranded RNA bunyaviruses are all arboviruses. They replicate in the cytoplasm and mature by budding into the Golgi vesicles. The Bunyamwera supergroup comprises at least 124 viruses, including that of California encephalitis.

Reoviruses

Reoviruses are medium-sized, ether-resistant, uniquely double-stranded (segmented) RNA viruses. A double capsid or shell containing eight polypeptides surrounds this virus. Capsid symmetry is cubic (icosahedral), with 92 capsomeres. Reoviruses replicate within the cytoplasm. The three members in this group were originally so named because of their presence in the respiratory tract and in the enteric canal and because of their orphan status (*respiratory enteric orphans*). (Orphans are not known to produce disease.) Reoviruses are excellent inducers of interferon. A number of reovirus strains have been recovered from patients in Africa with Burkitt's lymphoma. *Rotavirus*, a new genus in the family Reoviridae, causes infantile gastroenteritis, a common illness and a major cause of infant death in the developing countries. The term *diplornavirus* has been suggested to take in the reoviruses of mammals, wound tumor virus of plants, Colorado tick fever virus, and certain arboviruses with double-stranded RNA.

Picornaviruses

The single-stranded RNA picornaviruses are the smallest known, simplest, and most readily crystallizable. The term *picorna* was coined to designate enteric and related viruses. *Pico* is for very small viruses, and *rna* indicates their nucleic acid. In addition, *p* is for polioviruses (four to five genes), the first known members of the group; *i* is for insensitivity to ether, a distinguishing feature of the group; *c* is for coxsackieviruses; *o* is for orphan or echoviruses; and *r* is for rhinoviruses, further indicating their membership. Capsid symmetry is cubic (icosahedral), with 32 capsomeres.

The nearly 200 members are subdivided into (1) enteroviruses, including polioviruses (3 types), coxsackieviruses (29 types), echoviruses (31 types), hepatitis type A virus, and unspecified enteroviruses (4 types), and (2) rhinoviruses (more than 100 types), the major causes of the common cold. A rhinovirus in cattle causes foot-and-mouth disease. Picornaviruses produce a wide range of diseases in many areas of the human body, the best known of which is poliomyelitis. By comparison with that for bacterial pathogens, the dose of enterovirus required for human disease is extremely small; for example, only one or two infectious polioviral particles would be needed for poliovirus infection.

Coronaviruses

Coronaviruses are ether-sensitive, enveloped, single-stranded RNA viruses that are similar to myxoviruses. Symmetry is helical, and the corona-like arrangement of the petal-shaped surface projections suggested the name. Replication is cytoplasmic, and the viral particles mature by budding into cytoplasmic vesicles. Included in this group are the viruses of avian infectious bronchitis, mouse hepatitis, and certain human upper respiratory tract infections.

Retroviruses (Oncoviruses, Leukoviruses, Oncogenic RNA Viruses)

Retroviruses are small, ether-sensitive, enveloped, unique, single-stranded RNA tumor viruses, budding at the cell membrane. They were the first viruses isolated from animal neoplasms. They are the only RNA viruses actually replicating in the nucleus of the parasitized cell. Their oncogenic (cancer-inducing) potential relates to a large, nonsegmented genome and to the possession of an RNA-directed (RNA-dependent) DNA polymerase (reverse transcriptase). Reverse transcriptase, discovered in 1970, mediates the synthesis of DNA from an RNA template and indicates a biochemical mechanism for perpetuation of the viral genome when the host cell divides. The genetic apparatus of the virus becomes a permanent part of that of the host cell, and thus a lifelong relationship is formed between the virus and the parasitized cell from which the virus cannot escape. With cell division the viral genome is carried as a gene that is replicated and passed on to daughter cells. (This is called *nuclear*, or *vertical*, spread of the virus.) These viruses can transform the parasitized cells and are generally associated with the appearance of tumor antigens.

Based on their differences, RNA tumor viruses are separated into A, B, and C types (the most important); included here are the mouse mammary tumor virus and oncogenic viruses falling into species-specific groups inducing either leukemia or sarcoma in mice, cats, fowls, and monkeys. The term *leukemia-sarcoma complex* is applied to such a group—for example, the hamster leukemia-sarcoma complex, feline leukemia-sarcoma complex, and murine leukemia-sarcoma complex.

Hepadnaviruses

The DNA hepadnaviruses, replicating in liver cells (hepatocytes), exhibit a characteristic ultrastructure. A double-shelled particle 40 to 50 nm in diameter is present with incomplete 22 nm spheres and filaments. Surface, core, and "e" antigens are found. The circular DNA contains a single-stranded region. This category includes the hepatitis B virus in humans and three similar viruses found in woodchucks, Peking ducks, and Beechey ground squirrels—viruses related to hepatitis, hepatocellular carcinoma, and certain immune complex diseases.

Other Viruses (Unclassified)

Some viruses remain unclassified because of the lack of pertinent information. Currently this category includes some slow viruses related to neurologic disorders.

Viroids

These infectious agents are the smallest ones known. A viroid is about one eightieth the size of the smallest known virus. A viroid, said to be like a virus, is a stable, self-replicating, infectious nucleic acid, remarkably insensitive to heat, organic solvents, and ionizing and ultraviolet radiation. It is nonencapsulated and does not possess nucleoprotein.

Viroids have been defined as single-stranded, covalently closed circular or linear RNA molecules of low molecular weight (75,000 to 100,000), about 360 nucleotides long. In electron micrographs they appear as double-stranded rods because of the self-complementary nature of the RNA sequence. Viroids replicate in the nuclei of susceptible plant species. If the viroid interferes with host metabolism, disease is present. Plant viroids are best known as causes of a certain number of economically devastating plant diseases—potato spindle tuber disease, citrus exocortis disease, cucumber pale fruit, chrysanthemum stunt, and coconut cadang/cadang. Other viroids may cause disease in animals (such as scrapie, a disease of sheep) or be the responsible agents for forms of cancer, but these implications for viroids are not substantiated.

Categories as to Source of Virus

When the sources of viruses are considered, the following three categories are named: (1) enteroviruses, or those isolated from the alimentary tract (at least 70 human ones known), (2) respiratory viruses, or those isolated from the respiratory tract, and (3) arboviruses (arthropod-borne), or those isolated from insects. (The term *arbovirus* emphasizes the biologic feature of insect transmission.)

Sometimes in the classification of viruses emphasis is given to the anatomic area in the body where the virus produces its dramatic effects. Viruses so distinguished include (1) those whose typical lesions appear on the skin and mucous membranes—**dermotropic viruses** of smallpox, measles, chickenpox, and herpes simplex; (2) those related to acute infection of the respiratory tract—**pneumotropic viruses** of the common cold, influenza, and viral pneumonia; (3) those that primarily affect the central nervous system—**neurotropic viruses** of rabies, poliomyelitis, and encephalitis; and (4) those in a miscellaneous group with no common organ system (each virus having its own special affinity for a given organ)—**viscerotropic viruses** of viral hepatitis (liver), mumps (salivary glands), and other diseases. This restricted scheme of classification, although of some interest to the pathologist, is far from rigid, since certain viruses or groups of viruses induce various disease processes. In some instances a virus may invade the body without necessarily attacking the area that it ordinarily does. Certain viruses once considered limited in their cellular affinities are now known to infect a number of different cells in humans and animals. For example, polioviruses, regarded as highly neurotropic for many years, can grow quite well in different kinds of nonneural tissues.

Viral diseases may be either generalized or localized. In **generalized infection** the virus is disseminated by the bloodstream (viremia) throughout the body but without significant localization, even though a skin rash may be present. Examples of generalized infections include smallpox, vaccinia, chickenpox, measles, yellow fever, rubella, dengue fever, and Colorado tick fever. In some viral diseases there is restriction of viral effect to a particular organ to which the virus travels by the bloodstream, peripheral nerves, or other body route. Examples of **localized viral infections** include poliomyelitis, the encephalitides, rabies, herpes simplex, warts, influenza, common cold, mumps, and hepatitis.

Time Frames in Viral Infections

Like other microbes, viruses cause infections that are, in the classic sense, short-term or acute and ones that are long-range or chronic. Like other *acute* microbial infections, those attributable to viruses may express all the disease-related events in only a few days to weeks. Comparably, in chronic viral infections, the disease-related processes are prolonged from months to years. However, because of the remarkable nature of the intracellular parasitism established by viruses, correlations of viral disease with specified time periods may be quite different from those of most other living agents.

The viral infections smallpox and influenza are examples of acute infections in the usual sense. A period of immediate disease follows the initial contact with infectious virus. Virus replicates in the specific anatomic area and disseminates predictably. After a short incubation period, the patient manifests an illness of several days' duration (no more than 20), after which the virus is cleared from the body, and the patient recovers. Thereafter that patient is free of that virus.

A variation to the acute pattern exists with herpes simplex, a disease in which an acute but persistent infection is present. Initial contact with virus results in a familiar sign of immediate disease—the fever blister—a lesion lasting for several days that heals as expected. With recovery, however, the host does not eliminate this virus but instead incorporates it into certain cells and inactivates it, in which condition the virus and the infection are said to be *latent*. Virus does not express itself in the latent state. From time to time, with the proper stimulus, latent virus does reactivate to produce another episode of acute infection quite like the primary one. Virus reappears, releases infectious particles, and is recoverable; but again, with healing of the lesion, virus disappears and becomes latent once more. Infection with herpes simplex virus can recur in this manner any number of times; infection with herpes zoster virus does not recur so frequently, but may occur once in the individual's lifetime.

Patterns unique to viruses appear with chronic infections. One such is the chronic persistent infection, of which cytomegalovirus infection (p. 716) is an example. Cytomegalovirus infection is initiated by a primary contact with virus that most likely occurs without immediate disease, but this contact does serve to establish a long-term, even lifelong, state of infection in the host that is also most likely without event, even though the virus continuously replicates and the host continuously sheds it. For some of these infected persons, later in life, when their resistance mechanisms are weakened, the clinical infection appears. Chronic persistent infections such as hepatitis type B, Colorado tick fever, and rubella are slightly different in that an episode of primary infection with viral replication is recognized from which the patient recovers. The former patient remains free of clinical illness but continues to shed infectious particles, even perhaps for a lifetime.

Slow virus infection is a variant of the pattern of chronic persistent infection. Between an initial virus-to-host contact and the end stage of cellular injury is a prolonged period during which the very slowly progressive viral activity builds up and becomes cumulative. The primary event may be known but is often obscure. An example of slow virus infection is subacute sclerosing panencephalitis (SSPE), for which a variant measles virus is strongly suspected to be the cause. Here the initial episode of measles is clear-cut. From the point of first-time contact with host, virus presumably spreads to the specific area where replication will occur. Sites selected by viruses of this category usually seem to be in the brain, an area where, as virus is released, it could only be detected locally. Then many years pass, sometimes 20 or more, with no detectable sign of viral effect. That so much time is required may be because of the slow rate of multiplication of these viruses in their target sites and also because only a limited number of specific sites may be available. Finally the devastation of the end stage is reached and manifested in clinical disease.

Cultivation

To repeat, viruses multiply only inside living cells. Viruses in virus-containing material may be artificially brought into contact with living cells by (1) inoculation of cell (tissue) culture systems, (2) inoculation of the membranes or cavities of the developing chick embryo (Fig. 33-6), and (3) inoculation of a susceptible animal, such as a weanling mouse (Fig. 33-7).

Animal Inoculation

Very young mice are exquisitely sensitive to many viral infections, although not all viruses have the mouse in their host range. Small animal colonies for diagnostic purposes are mostly found in large reference laboratories.

Chick Embryo

The fertile hen's egg as a growth medium for viruses is cheap, sterile, and easily manipulated and presents a selection of different types of suitable tissue. Positive signs of growth therein include death of tissue and the appearance of pocks—small scarlike foci (Fig. 33-8). Since many viruses have surface antigens that bind to red blood cell receptors, the demonstration of the phenomenon of hemagglutination is a useful indicator

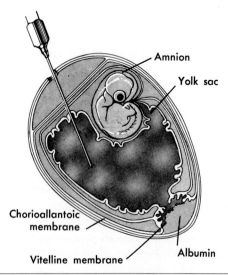

FIG. 33-6   Schematic drawing of injection of yolk sac in hen's egg containing embryo.

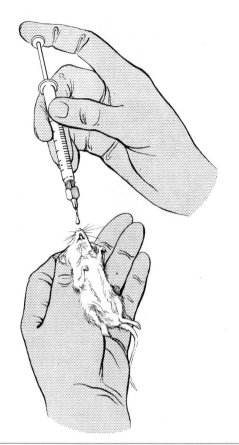

FIG. 33-7   Mouse inoculation with virus-containing material, intranasal route.

FIG. 33-8   Chorioallantoic membrane, showing whitish plaques (pocks) produced by smallpox virus. (From Hahon, N., and Ratner, M.: J. Bacteriol. 74:696, 1957.)

of viral growth in the fertile egg. The chick embryo method is carried out as follows. A fertile egg is incubated for 7 to 15 days. With a syringe and needle, the virus-containing material is injected into the membranes of the embryo or into the cavities connected with its development (yolk sac, amniotic cavity, or allantoic sac). This may be done through a window made in the shell of the egg over the embryo or through a hole drilled in the shell (Fig. 33-9). Of course, *aseptic technic is mandatory*. After inoculation, the egg is incubated for 48 to 72 hours, which is the time required for the virus to multiply. The material in which the virus has multiplied is then removed, used for the study of the virus, or sometimes purified, processed, and used as a vaccine.

Cell Culture

The most widely used laboratory method of growing viruses today is the cell (tissue) culture method. Almost every known animal virus has been propagated in cell culture. (An important exception is the hepatitis B virus.) This method is possible because of discoveries in basic technics, such as the preparation of improved culture media for living cells. Today commercially prepared media and cells for culture systems are read-

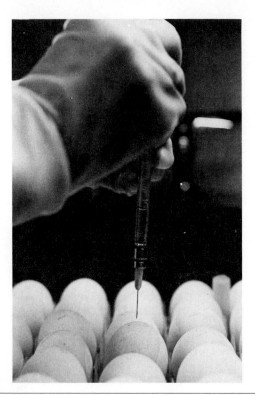

FIG. 33-9   Cultivation of influenza virus in fertile hen's eggs. Injection site on top of each embryo is disinfected. A needle and syringe deliver live virus. (Courtesy J.E. Bailey, Parke, Davis, & Co., Detroit.)

ily available, and the use of antimicrobials eliminates microbial contamination. The recognition of a cytopathogenic effect (CPE), the result of virus-cell interaction, furnishes a measure of viral injury in growing cells. Not all viruses induce such an effect, but for a CPE to appear, virus must be living and growing in the culture. Cytopathic changes induced in the culture medium include death and lysis of cells, vacuolization of cell cytoplasm, formation of giant cells with definite borders or the larger syncytial masses with ill-defined borders, and in certain instances alterations consistent with conversion or transformation of the cells.

Plaques (Fig. 33-13) are clear, precise areas of virus-induced cellular lysis seen in the cell culture under a layer of agar. The formation of plaques, the basis of standard assays, also reflects viral growth, as do the phenomena of hemadsorption and hemagglutination. Hemadsorption and hemagglutination are displayed under specified conditions when either guinea pig or chicken erythrocytes incubated in the culture system adhere in clumps to infected host cells. Assay of hemadsorption can be carried out without the cytopathic effect being obscured.

There are three standard cell culture systems, using (1) human fibroblasts from embryo lung, (2) a neoplastic cell line, or (3) primary monkey kidney. Cells that persist indefinitely in culture are spoken of as **cell lines.**

Both chick embryo and cell culture enable the laboratory worker to obtain large quantities of virus for study, vaccine production, or whatever purpose.

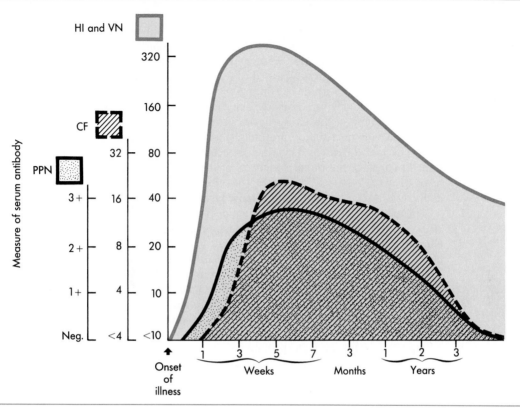

FIG. 33-10 Typical serum antibody responses in many viral infections. Note appearance in serum, curve of values, and duration for the four different kinds of antibodies in relation to time after onset of illness. *CF*, Complement-fixing antibodies; *HI*, hemagglutination-inhibiting antibodies; *PPN*, precipitins, or precipitating antibodies; *VN*, virus-neutralizing antibodies.

Laboratory Diagnosis of Viral Infections

Many procedures are available for the laboratory diagnosis of viral infections, although some are most efficiently carried out in specially equipped laboratories. These include (1) technics for isolation and identification of the virus, (2) serologic tests done serially during the course of infection, and other immunologic reactions, (3) direct microscopic examination of suitable specimens, and (4) skin tests.

Isolation

Test material may be inoculated into the embryonated hen's egg, cell culture, or a susceptible laboratory animal (for example, mouse, guinea pig, cotton rat, rabbit, or monkey), from which the virus is recovered and identified. The appearance of typical lesions or inclusion bodies may be diagnostic.

Serologic Procedures

Important in the identification of viruses are several serologic procedures of which the virus neutralization, hemagglutination-inhibition, complement-fixation, and precipitin reaction tests are especially valuable (Fig. 33-10). For the serologic evaluation of a patient's serum it is important that *paired samples* of blood be taken. The first is collected as early as possible in the *acute* stages of disease, and the second is drawn 2 to 4 weeks later during convalescence. The significant finding is at least a fourfold increase in the titer of antibodies between the specimens.

Viruses contain good antigens in their makeup and as a result stimulate antibody formation. The capsid is antigenic, as are the glycoprotein spikes of the envelope in those viruses possessing one. Since that glycoprotein is derived from the virus, unlike the envelope of which it is a part, it can carry specificity for the virus. One of the most important antibodies formed in humans and animals with viral infection is the virus-neutralizing antibody, an antibody that neutralizes or obliterates the infective and destructive capacity of the virus. To measure or titrate its activity, suitable mixtures of serum and virus preparation are inoculated into white mice susceptible to the pathogenic effect of the virus. Thus the protective quality of the test serum can be determined. To gauge the protective effects of the serum, either cell cultures or chick embryos may also be used. This is the **virus neutralization test.** (Virus-neutralizing antibody is a key constituent of immune serum given for viral infection.)

The phenomenon of hemagglutination, valuable in the identification of a given virus, is valid because most disease-producing viruses either directly or indirectly effect clumping of the red blood cells of humans and animals. Viral hemagglutination may be blocked by the action of specific antibodies found in immune or convalescent serum. This is the basis of the **hemagglutination-inhibition (HI) test.**

Many viruses induce the formation of complement-fixing antibodies. **Complement fixation** is the most widely used test. **Precipitin reactions** associated with several viruses are analogous to those observed with soluble products and toxins of bacteria. Precipitation is measured by immunodiffusion (p. 404).

The fluorescent antibody technic can be used to detect viral antigens directly in specimens from patients or diseased animals. A diagnosis may sometimes be made within a few hours. In rabies, for example, the fluorescent antibody is applied to thin sections or smears of brain tissue or salivary gland, and fluorescence is observed within hours if viral antigen is present.

Sophisticated and highly sensitive immunologic procedures, such as the immunoperoxidase technic, counterimmunoelectrophoresis, radioimmunoassay, and enzyme-linked immunosorbent assay (ELISA), are being widely applied to the diagnosis of viral disease.

Direct Examination

Microscopic examination may be made of suitable preparations, such as smears, imprints, scrapings, or sections of diseased tissue. In certain viral diseases inclusion bodies (viral aggregates) and other virus-induced cellular alterations are easily recognized with the compound light microscope in thin tissue sections stained with conventional stains or in cellular spreads stained with the Papanicolaou stain (Table 33-2). The electron microscope is a valuable tool to reveal viral particles both in body fluids and secretions and in *ultra*thin sections of tissue. Negative staining with an electron-dense material, such as phosphotungstic acid, is a technic used in the preparation of electron micrographs of viruses (see Fig. 8-8).

Skin Tests

Skin tests are available to aid in the diagnosis of some viral diseases. Vaccinia virus vaccine administered in a skin test indicates immunity to smallpox. Chick embryo antigens prepared for skin testing in mumps and herpes simplex infections indicate past or present infection.

Table 33-4 indicates how the laboratory is used in the diagnosis of viral disease.

Spread

The most significant of the viral diseases are caused by viruses that have as their natural host a human being, and spread from one person to another is either by direct or indirect contact, especially by means of nose and throat secretions, fecal material, and inanimate objects so contaminated. Droplet infection is very common. Insect vectors (flies, cockroaches, mosquitoes, or ticks) spread certain viral infections to humans from a reservoir of infection in either a lower animal or in the insect. Rabies is transmitted directly by the bite of the rabid animal. Water transmits diseases such as infectious hepatitis and enteroviral infections, and milk- or food-borne epidemics of hepatitis A, poliomyelitis, and enteroviral infections are occasionally seen. The asymptomatic human carrier state is not of consequence in the spread of viral disease, except with enteric disorders, rubella, and hepatitis.

Immunity

The mechanisms of natural resistance are poorly understood. Newborn mice are extremely susceptible to coxsackievirus and easily infected, whereas adult mice are quite

Table 33-4 Laboratory Diagnosis of Selected Viral Diseases in Humans

Virus	Test Specimen	Practical Diagnostic Tests							
		Animal Inoculation	Chick Embryo Culture	Cell Culture—Cytopathic Effect	Complement Fixation	Hemagglutination Inhibition	Virus Neutralization	Fluorescent Antibody	Inclusion Bodies
Viruses in Respiratory Disease									
Influenza viruses	Nose and throat secretions		X		X	X	X	X	
Adenoviruses	Throat secretions, fecal material, eye fluid, cerebrospinal fluid			X	X	X	X		
Viruses in Central Nervous System Disease									
Polioviruses	Throat secretions, rectal swabs, blood, urine, cerebrospinal fluid	X		X	X		X	X	
Encephalitis viruses	Blood (whole clotted or serum), throat swabs, cerebrospinal fluid, urine; brain, other tissues if illness fatal	X		X	X	X	X		
Rabies virus	Saliva, throat swabs, eye fluid, cerebrospinal fluid	X			X		X	X	X
Viruses in Skin Disease									
Variola virus	Material from vesicle, pustule, or scab		X	X	X	X	X	X	X
Measles virus	Blood, urine, throat swabs, eye fluid			X	X	X	X		
Rubella virus	Blood, urine, throat swabs	X	X		X	X	X	X	
Other Viruses									
Colorado tick fever virus	Blood, throat swabs, fecal material	X	X	X	X	X	X		

resistant under ordinary conditions. The virus of chickenpox affects only humans; other animals are resistant.

Specific antibody response in viral infection is comparable to that in other microbial infections except that in viral infection the antibody response may be a persistent one, lasting a long time—even a lifetime. Lifelong immunity is not seen in bacterial and fungal infections. Examples of viral diseases wherein one attack confers lifelong immunity are mumps, smallpox, and measles. However, in diseases such as the common cold and influenza no immunity of any appreciable duration results. Where immunity is short-lived, the incubation period has been short, viruses have not circulated in the bloodstream (viremia), and antibody-forming tissues have failed to receive adequate stimulation. In most persons in whom immunity lasts, antibodies in their serum may be demonstrated for many years.

Humoral immunity (the presence of neutralizing antibodies in serum) mediates a protective effect against viral infections in which viremia is a part of the disease. Cell-mediated immunity, also important in viral infection, operates to clear the virus and to facilitate recovery from the infection. In its absence serious manifestations of late disease are seen.

Viral Interference

An interference phenomenon is observed with viruses at the cellular level. A plant or animal cell exposed to a given virus subsequently develops a resistance to infection by a closely related strain of the same virus or another but similar one. This kind of interference exists between the viruses of yellow fever and dengue fever in the body of their insect vector, the mosquito *Aedes aegypti*. Such a mosquito cannot spread both of these diseases at the same time.

Interferon

Interferon or the interferon system might be considered a broad-spectrum, nonspecific type of antiviral agent. Interferons constitute a family of glycoprotein molecules with molecular weights of about 15,000 to 25,000 that are soluble, nontoxic, nonantigenic, and smaller in size than antibodies. Interferon is elaborated in small amounts and secreted by a normal body cell under attack from an invading virus. The molecules that are formed diffuse to neighboring cells, where they bind to receptors on cell surfaces and in so doing promote an *antiviral state* in the uninfected cells nearby. Related to a specific cell, interferon by its presence triggers a cellular reaction that halts synthesis of viral nucleic acid. With disruption of its life cycle, the effects of the invading virus are thereby blocked. The action of interferon is also to inhibit the proliferation of normal and transformed cells. Interferon is cell specific and species but *not* virus specific. Several animal viruses induce its formation. However, only a few are resistant to it. The action of interferon is a manifestation of viral interference and a most important part of the body's defense against viral infection—part of nature's first line of defense.

A great deal of interest is currently evident in interferon inducers, substances stimulating the endogenous production of interferon and thereby active against viral infection. All animal cells apparently produce interferon when appropriately stimulated with an inducer. A well-known one in the investigational field and one of the best is polyinosinic:polycytidylic acid (Poly I:C), a synthetic, double-stranded RNA.

Prevention of Viral Disease

Immunization procedures prevent viral diseases (Chapters 22 and 23), as do proper technics of sterilization (Chapter 16). Effective disinfectants to inactivate or destroy viruses are alkaline glutaraldehyde, formalin, dilute hydrochloric acid, organic iodine, and phenol, 1%. Roentgen rays and ultraviolet light destroy viruses, but the effective dose varies. Most viruses, except those of hepatitis, are destroyed in pasteurized milk. Influenza viruses are readily destroyed by soap and water.

Except to treat bacterial complications that are prone to complicate viral diseases, present-day antimicrobial drugs are generally of no value in the management of diseases caused by true viruses.

BACTERIOPHAGES (BACTERIAL VIRUSES)
General Characteristics

Two investigators independently discovered bacteriophage—F.W. Twort (1877-1950) in 1915 and F.H. d'Herelle (1873-1949) in 1917 Felix d'Herelle reported that a bacteria-free filtrate obtained from the stools of patients with bacillary dysentery contained an agent that when added to a liquid culture of dysentery bacilli dissolved the bacteria. If but a minute portion of the dissolved culture were added to another culture of dysentery bacilli, the bacteria in the second culture were likewise dissolved. This transfer from culture to culture could be kept up until the bacteria in hundreds of cultures were lysed, results that proved that the agent was not consumed in the process but apparently increased in amount. d'Herelle called this agent *bacteriophage*, meaning bacteria-eater. Today bacteriophages are recognized as viruses that attack bacteria—bacterial viruses (p. 92).

Viruses infecting many strains of bacteria have been isolated, and it has been shown that a given phage acts only on its own particular species or group of species. In fact, their highly specific nature makes them useful to the epidemiologist in the classification of bacteria; for example, phage typing of pathologic staphylococci is crucial to the epidemiologic study of hospital-acquired staphylococcal infections.

Tending to occur in nature with their specific hosts, bacteriophages are found most plentifully in the intestinal discharges of humans and the higher animals or in water and other materials contaminated with these discharges. They are also found in pus and even in soil. When phages are named, reference is made to the specific hosts, as with coliphages, staphylophages, cholera phages, and typhoid phages. Of all the viruses known, bacterial viruses are the most easily studied in the research laboratory, and the ones most thoroughly investigated have been those related to the enteric group of bacteria. As obligate intracellular parasites, bacteriophages closely resemble other viruses in their biologic properties.

Life Cycle

With the aid of the electron microscope, phages are seen to be tiny, tadpole-shaped units possessing a head, which is either rounded or many-sided, and a tail, which is a specialized structure for attachment. Most phages are composed of nuclear material (DNA) encased in a protein coat (Fig. 33-11). Some contain only RNA.

Phage attack on its specific bacterium occurs in a fantastic series of steps. Drawn to the susceptible bacterial cell, the virus fixes itself tail first to the cell at a specific receptor site (Fig. 33-12). By chemical action, it drills out a tiny hole in the cell wall, and the tail penetrates the cell membrane to the interior of the cell. The head changes shape, and soon the DNA of the virus is injected through the tail into the cell. The bacteriophage seems to work like the world's smallest syringe and needle when it injects its nuclear DNA into the bacterial cell. Once the injection is made, drastic changes occur. The DNA takes command of the vital forces of the microbe and in a matter of minutes imposes the synthesis of several hundred bacterial viruses exactly like the one originally invading the injured bacterial cell. The deranged cell swells and shatters, setting free a multitude of new viruses able to attach to a new host and repeat the cycle.

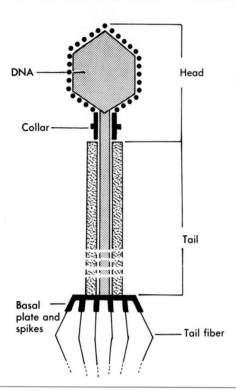

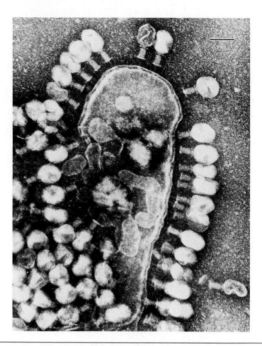

FIG. 33-11 Diagram of mycobacteriophage. Large dots around the head are protein coat. Measurements of this phage: head, about 95 nm; tail, about 300 nm; and tail fibers, 150 nm. (From Arnett, R.H., Jr., and Bazinet, G.F., Jr.: Plant biology: a concise introduction, ed.4, St. Louis, 1977, The C.V. Mosby Co.)

FIG. 33-12 Bacteriophage infection of bacteria. Numerous specific phages are adsorbed on *Escherichia coli*. Distance between baseplates of phage particles and cell wall is 30 to 40 nm. (Negative staining.) (From Simon, L.D., and Anderson, T.F.: Virology **32**:279, 1967.)

This rapid destruction of a bacterium is **lysis,** and the phage inducing it is referred to as a **virulent phage.**

Lysis is not the invariable result of bacteriophage action. When **temperate phages** infect, they do not destroy but seem to be able to establish a relatively stable symbiotic relation with the host they have parasitized. The host bacterium continues to grow and multiply, carrying the virus in its interior in a noninfective condition more or less indefinitely. The virus meantime replicates with the bacterial DNA replication. The phage genome replicating within the bacterial cell is referred to as a **prophage,** and this condition of mutual tolerance is termed **lysogeny** (p. 92) or **lysogenesis.** Lysogeny is widespread in nature. That the process is not without effect is seen in changes in physiologic characteristics in the infected bacteria. Phages may be responsible for the differences between an avirulent bacterial strain and a virulent one (the conversion of an avirulent strain to a virulent one). Diphtheria is caused by an exotoxin produced by toxigenic strains of *Corynebacterium diphtheriae*. Toxigenicity comes from phage activity in lysogenic strains of the diphtheria bacillus; uninfected, therefore nontoxigenic, strains are nonpathogenic.

Laboratory Diagnosis

Bacteriophages may be isolated and studied in the laboratory. If they are taken from a source in nature, they must be separated from bacteria by filtration. When they are added to a growing culture of specific bacteria, their action to block bacterial

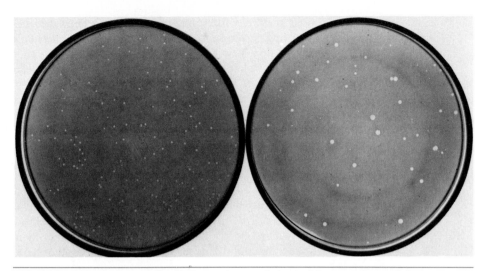

FIG. 33-13   Phage plaques. Small plaques (zones of clearing) in assay plate on left are *Staphylococcus aureus* phages; those on right are *Staphylococcus epidermidis* phages. In preparation of an assay plate, a mixture of test phages and microorganisms is poured over nutrient base. After plate has been incubated, plaques are evaluated and counted. (Courtesy Drs. E.D. Rosenblum and B. Minshew, Dallas.)

growth may be nicely shown. In a liquid culture medium, clearing indicates bacterial lysis. On a solid culture medium, lysis by virulent phages is seen in zones (plaques) (Fig. 33-13), usually circular, where bacterial growth has disappeared (clearing of bacterial growth). Each plaque contains many particles that in turn can form plaques. Pure preparations of phage may be obtained by picking material from well-isolated plaques.

Importance

Bacteriophages are of practical importance in many ways. In certain of the fermentation industries in which the commercial product depends on bacterial action (for example, the manufacture of streptomycin, acetone, and butyl alcohol), an "epidemic" of viral infection in the large vats used to grow the microorganisms can be a serious economic matter.

With the advent of recombinant DNA technology, bacteriophages have come into widespread use as vectors for molecular cloning (p. 100).

VIRUSES AND TERATOGENESIS

Currently the role of viruses in teratogenesis (the production of physical defects in the offspring in utero) is being carefully studied. When the pregnant woman contracts a viral infection accompanied by viremia, the infection often crosses the placenta to the susceptible embryo or fetus. Notable examples are measles, smallpox, vaccinia, western equine encephalitis, chickenpox, poliomyelitis, hepatitis, and coxsackievirus infection. Viral infections involving the offspring in the uterus may be more common than suspected, especially in lower socioeconomic groups. Although the possibilities for fetal infection are many, only three viruses fully qualify as teratogens. For rubella virus, cytomegalovirus, and herpesvirus, the evidence is clear-cut, with patterns of fetal malformation established.

Birth defects caused by rubella virus (p. 705) have been more thoroughly examined than those caused by other viruses, and most of what is known about viral teratogenesis comes from such investigations. In rubella the teratogenic mechanism seems to stem from a direct interaction between virus and parasitized cell that continues throughout the length of gestation. That plus viral damage to blood vessels disrupts normal organ development.

The cytomegalovirus (salivary gland virus) produces a mild infection in the mother; however, it can cause extensive and widespread damage in the neonate (p. 716). Herpesvirus has been known to cross the placenta to produce generalized infection with documented malformations in the central nervous system and eyes (p. 714).

VIRUSES AND CANCER

At the turn of the century the suggestion was made that viruses might cause cancer. In 1910 Peyton Rous (1879-1970), pathologist and medical researcher at the Rockefeller University, used material from a large tumor on the breast of a Plymouth Rock hen to show that cell-free filtrates of it induced the same tumor in other hens, that the induced tumor behaved as the original one did, and that the properties of the agent in his experiments were those of a filtrable virus. This important landmark in cancer research was years ahead of the times. To the early 1950s, nearly half a century, the viral etiology of cancer made slow headway, although substantiated by a number of significant observations, and for many years it took considerable courage to hold such a theory.

Today experimental evidence clearly indicates that filtrable viruses do cause several kinds of cancerous growths in lower animals—in rabbits (Fig. 33-14), mice, chickens,

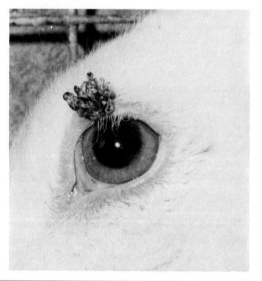

FIG. 33-14   Example of virus-induced neoplasm. Wartlike growth on eyelid of wild cottontail rabbit. Eyelids and ears are common sites for such tumors. An enzootic disease of wild rabbits, rabbit papillomatosis is also seen in domestic rabbits. Virus is transmitted in nature by rabbit tick and in laboratory by mosquito. (From Hagen, K.W.: Bull. Pathol. 8:308, 1967.)

hamsters, rats, cats, dogs, frogs, apes, monkeys, horses, squirrels, cattle, and deer (Table 33-5)—not only in the laboratory but also in their natural setting. Cancer is induced in all major groups of animals (including subhuman primates). Leukemia is an important form of virus-induced neoplasm in animals. Many different solid cancers in the skin, breast, lungs, gastrointestinal tract, and salivary glands of several species of laboratory animals are caused by the polyoma virus.

Tumor Viruses

Oncogenesis (Greek *onkos*, tumor, plus *genesis*, production) refers to the causation or induction of tumors ("cancers"). *Oncogenic*, the adjective, is used especially to refer to tumor-inducing viruses. *Oncogen*, the noun, refers to a given tumor virus. There are more than 100 known viral oncogens. Table 33-6 emphasizes certain patterns of viral oncogenesis.

Oncogenic viruses can be either DNA or RNA viruses, but with both kinds oncogenesis is generally associated with *integration of viral DNA into the host cell genome*. RNA tumor viruses comprising the family known as *retroviruses* satisfy this requirement by means of their virally encoded enzyme, reverse transcriptase, which converts viral RNA into DNA. As part of their normal replication cycle, retroviruses are able to insert (integrate) their genomes into cellular chromosomes. In this event they may become endogenous viruses, vertically transmissible. Although usually latent, they may at times be expressed as viruses.

Exogenous retroviruses exist, and they are horizontally transmitted like other animal viruses. In animals exogenous retroviruses are the ones that are primarily related to the production of naturally occurring cancers. In animals the principal two such DNA tumor viruses are a herpesvirus and a papovavirus (bovine papilloma virus).

Oncogenic tumor viruses are known to lie dormant for a long time in animals before being activated. It is postulated that either external or internal factors trigger the mech-

Table 33-5 Examples of Viral Neoplasms in Animals

Animal Species	Date Reported	Cancerous Growth (Neoplasm)*
Chicken	1908	Chicken leukemias
Chicken	1910	Rous sarcoma
Rabbit	1933	Shope papilloma-carcinoma
Mouse	1936	Breast carcinoma†
Frog	1938	Kidney carcinoma
Mouse	1951	Spontaneous leukemia
Mouse	1957	Salivary gland tumors
Hamster, rat, and rabbit	1958	Great variety of solid tumors (produced by polyoma virus)
Hamster	1962	Chest and liver tumors (adenovirus responsible isolated from human cancer)
Cat	1964	Leukemia
Monkey	1971	Woolly monkey fibrosarcoma
Gibbon ape	1972	Spontaneous lymphosarcoma

*See Glossary for definitions of terms used for cancerous growths.
†This most common form of cancer in the most commonly used laboratory animal was shown by John Bittner to be related to a virus, long referred to as Bittner's milk factor. Bittner's discovery stimulated greatly the study of the role of viruses in the induction of cancer.

Table 33-6 Patterns of Tumor Viruses*

Category	Nucleic Acid	Approximate Size (nm)	Member Viruses	Natural Hosts	Tumors Produced	Experimental Hosts
Hepadnaviruses	DNA	42	Hepatitis B	Human; woodchuck	Hepatocellular carcinoma (?)	Chimpanzee; woodchuck; Peking duck
Papovaviruses	DNA	45	Polyoma	Mouse	Solid tumors in many sites	Mouse, hamster, guinea pig, rat, rabbit, ferret
			SV/40	Rhesus monkey	Sarcomas (malignant tumors of connective tissue)	Hamster
			Papilloma Human	Human	Common warts (papillomas); cancer of uterine cervix (?); bowel cancer (?)	Human
			Rabbit	Rabbit	Papillomas	Rabbit
			Bovine	Cow	Papillomas; cancer alimentary tract	Cow, hamster, mouse
Poxviruses	DNA	250 × 200	Dog	Dog	Papillomas	Dog
			Yaba	Monkey	Benign histiocytomas	Monkey
			Fibroma	Rabbit, squirrel, deer	Fibromas, myxomas	Rabbit, squirrel, deer
Adenoviruses	DNA	80	Types 3, 7, 12, 18, 31	Human	Sarcomas, malignant lymphomas	Hamster, mouse, rat
Herpesviruses	DNA	100	Lucké	Leopard frog	Renal adenocarcinoma	Leopard frog
			Marek's disease	Chicken	Neurolymphoma (T cell lymphoproliferative tumor; Marek's disease)	Chicken
			Herpesvirus saimiri	Squirrel monkey	Meléndez' lymphoma of owl monkey	Owl monkey, marmoset

			Virus	Natural host	Cancer*	Experimental hosts
			Herpesvirus of rabbit	Rabbit	Hinze's lymphoma	Cottontail rabbit
			Epstein-Barr	Human	Burkitt's lymphoma (?), nasopharyngeal carcinoma (?)	
			Herpesvirus hominis type 2	Human	Carcinoma of cervix of uterus (?)	
Retro-viruses	RNA	70 to 110	Rous sarcoma	Chicken	Sarcomas, leukemias, adenocarcinoma of kidney	Chicken, turkey, rat, monkey, hamster, guinea pig
			Avian leukosis complex	Chicken	Leukemias, sarcomas	Chicken
			Murine leukemia complex: 14 strains, including Gross, Friend, Graffi, Rauscher, Moloney, Abelson	Mouse	Leukemias, malignant lymphoma	Mouse, rat, hamster
			Murine mammary tumor (Bittner)	Mouse	Mammary carcinoma	Mouse
			Bovine leukemia	Cattle	Lymphoid leukemia, lymphosarcoma	Sheep, cattle
			Feline leukemia	Cat	Leukemia-lymphoma	Cat
			Gibbon leukemia	Ape	Acute T cell leukemia, chronic myelogenous leukemia	Ape
			Human T cell leukemia-lymphoma	Human	Adult T cell leukemia	None

*See Glossary for definitions of terms used for cancerous growths.

anisms of neoplasia (new growth). External factors might be x rays and other forms of radiant energy. Internal factors might include the age at time of infection, dosage of virus, status of host immune response, and certain metabolic and hormonal states. (John Bittner demonstrated that breast cancer in mice is caused by an interplay of the virus [milk factor], hormones, and hereditary background.)

The role of viruses in human neoplasms is at this time undefined, but experimental reports suggest a relationship. For instance, viruslike particles, resembling those producing breast cancer in mice, have been isolated from human breast cancers and from the milk of breast cancer patients.

Much indirect evidence concerning DNA viruses suggests that certain ones are indeed involved. The DNA viruses presently linked to human cancer are the (1) Epstein-Barr virus—Burkitt's lymphoma (cancer of lymphoid tissue) and nasopharyngeal carcinoma; (2) *Herpesvirus hominis* type 2—carcinoma of uterine cervix; (3) papilloma viruses—a variety of warts (genital, common, plantar) and cancer of uterine cervix; and (4) hepatitis B virus—primary cancer of the liver.

Recently a retrovirus has been recovered from cells of adult human patients with a leukemia-lymphoma of mature T lymphocytes. Much attention is being focused on the human T cell leukemia-lymphoma virus (HTLV), since it is the first human retrovirus to be isolated, and the parallels between this retrovirus-induced neoplasm and those in animals are striking. The human T cell leukemia virus seems to be concentrated in certain geographic pockets, notably on the island of Kyushu in Japan and in the Caribbean, where it is endemic. It does not appear to be highly infectious; prolonged and intimate contact seems necessary for transmission. How it produces leukemia is not clear.

Table 33-7 presents viruses "indicted," not "convicted," as oncogens in human beings.

Oncogenes

Oncogenesis is a complex process influenced by a multiplicity of factors. In light of recent observations some of the mechanisms whereby certain viruses participate in the process have been studied. With the use of certain research technics, RNA sequences that differ from structural viral genes have been detected in rapidly transforming retroviruses. These RNA sequences are termed *oncogenes*, and the protein products encoded for by them appear to induce transformation (conversion) of a normal, nonneoplastic cell to a malignant, neoplastic one.

Investigation of the origin of these oncogenes revealed a startling observation—that

---

Table 33-7 Viruses "Indicted" for Oncogenicity in Humans*

Virus	Tumor
Retrovirus type C, human	T cell leukemia-lymphoma
Retrovirus type B	Breast cancer
Epstein-Barr virus (*Herpesvirus hominis* type 4)	Burkitt's lymphoma
	Hodgkin's disease
	Leukemia
	Lymphoma
Herpesvirus hominis type 2 (genital strain)	Carcinoma of uterine cervix
Hepatitis B virus	Hepatocellular carcinoma
Papilloma viruses	Carcinoma of uterine cervix
	Warts

*Relationship is not proved.

cellular counterparts (homologues) of these genes are present in normal cells. Some time earlier genes in the normal cell were usurped by virus and incorporated into the viral genome through a recombination mechanism similar to transduction. When these viruses, which have cell-derived oncogenes present in their genome, infect a cell, the oncogenes are expressed but obviously are no longer regulated by the host cell. Therefore an oncogene, a single gene with the potential to produce a protein that can of itself change the normal cell to a cancerous one, exists in a nontransformed normal cell. There it may cause no harm, since such a gene is tightly regulated under normal conditions, but when this gene comes under viral control, it plays a significant role in the induction of cancer.

Many of the retroviruses just described (with oncogenes) can cause acute (short latent period) neoplasms by bringing a gene that is necessary for tumor formation to the cell. By contrast, a second strategy for neoplasia induced by other types of retroviruses (those without oncogenes) is seen—for instance, in avian lymphomas, tumors with a long latent period. The mechanism here for tumor production involves the integration of the virus into the host chromosomal region adjacent to a specific cell oncogene, in this case the *c-myc* gene. This mechanism allows the virus to alter the transcriptional activity of the particular oncogene (the *c-myc* gene).

In addition, it is postulated that some oncogenes arise from damage to the genetic apparatus that may come, for instance, from known chemical or physical carcinogens (in no way referable to any transmissible viable agent). Therefore the study of oncogenes may be important in the understanding of chemical and physical carcinogenesis.

It has been shown recently that in a case of human cancer, a point mutation in the responsible oncogene represented a change of only 1 of 3 billion nucleotides making up the human genome.

Burkitt's Lymphoma

In 1958 Denis Burkitt defined the jaw tumor in central African children that now bears his name and mapped out its geographic distribution. He described it as a cancer but much about it impressed him as an infectious disease.

African Burkitt's lymphoma is a malignant disease of the jaw and abdomen affecting children between the ages of 2 and 14 years. The striking feature of the disease is its sharp geographic distribution. It is found in Central Africa limited to a malarious belt where the conditions of rainfall, temperature, altitude, vegetation, and humidity are the same (Fig. 33-15). It also appears to be a geographic variant of lymphomas of children in other parts of the world. But, unlike lymphomas elsewhere, it has a pattern for a specific infectious disease, strongly suggesting a viral etiology. Several viruses have been found in tumor tissue and in cell culture made of tumor. The Epstein-Barr virus (p. 718) was first isolated from such a cell culture, and this virus is significantly linked to this form of lymphoma.

At first Denis Burkitt postulated that the viral agent causing the disease was spread by a vector mosquito. This lymphoma is prevalent in areas where malaria is endemic and is rare outside tropical Africa. Now it is known that the mosquito is no more implicated than in the transmission of malaria, for it is believed that the damage from chronic malaria to the lymphoid system, especially in the spleen, is a crucial factor that determines in some way the oncogenic potential of the proposed virus and the course of the clinical disease. The current evidence suggests that the cause of African Burkitt's lymphoma is related to Epstein-Barr viral infection occurring in a host whose immunologic state is severely affected by constant and severe malarial infections.

Of importance also is a chromosomal abnormality, a translocation involving chromosome 8, consistently found in the tumor cells of Burkitt's lymphoma. This finding suggests that a genetic error appearing during the development of the disease may be a contributing factor to disease production. The part of the chromosome translocated

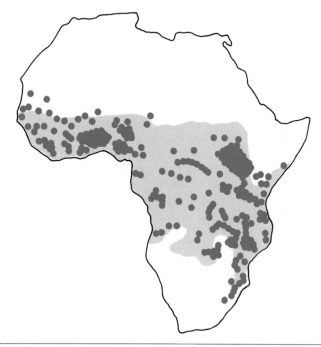

FIG. 33-15 Distribution of Burkitt's lymphoma within malarious belt of Central Africa *(colored area)*. Solid colored circles in this area indicate clusters of cases studied epidemiologically.

contains the oncogene known as *c-myc*. Experimental studies in chickens suggest that virally induced activation of *c-myc* influences malignant transformation of cells.

The Epstein-Barr virus strongly implicated here is also found elsewhere in the world as a cause of the nonneoplastic disease infectious mononucleosis (p. 717).

SUMMARY

1 A virus is the smallest known living body able to replicate itself and then only inside a host cell. Its nucleic acid, either DNA or RNA (but not both), contains the necessary information for viral synthesis within the host cell (the number of specific macromolecules required for viral progeny).

2 Viruses range from a particle 20 nm in diameter with a dozen genes or so to a large, complex one about 200 to 300 nm in diameter with several hundred genes.

3 The genome, the viral nucleic acid, is packed within a protein capsid. The capsid with the nucleic acid is the nucleocapsid. Subunits of the capsid, or capsomeres, are arranged precisely around the nucleic acid core in a naked, or nonenveloped, virus. Around the enveloped virus, there is a phospholipid bilayer membrane with spikes of glycoprotein. By a process of budding, an enveloped virus is pinched off from the host cell deriving its envelope from the host cell membrane.

4 The genetic apparatus of a virus is much the same as for other living organisms but much less complex. Viruses are haploid, except for the diploid retroviruses.

5 The events in the life cycle of a virus are (1) attachment and adsorption, (2) penetration, (3) eclipse, (4) maturation, and (5) release.

6 Viral infections may be silent and unapparent, with little or no harm coming to the host in the viral life cycle. Some viral infections are latent, undetectable until the virus is activated by the proper stimulus.

7 Certain viruses induce lysis of the host cell when the viral particles are released. A few viruses can stimulate the parasitized tissue cell to proliferate. The effect is a virus-induced hyperplasia (increase in number of cells).

8 The process of conversion or transformation is a viral effect whereby cells are induced (1) to grow more rapidly but at the same time require fewer nutrients, (2) to take on an appearance notably deviant from normal, and (3) to move about, migrate, and even invade other tissues.

9 Cytopathogenic, or cytopathic, effects are changes in cells, particularly in cell cultures, that directly reflect viral injury, for example, the formation of a giant cell or a syncytium.

10 Viruses are responsible for more than 80 diseases of plants and many diseases of lower animals. Of more than 550 viruses known, more than 200 produce over 50 viral infections in humans, some of which are the most dangerous diseases known.

11 Major biologic properties are used to classify viruses. Most animal viruses fall into 16 groups, 6 with a DNA genome and 10 with an RNA genome. An additional group is undefined because of lack of pertinent information.

12 The DNA viruses are the poxviruses, herpesviruses, adenoviruses, papovaviruses, parvoviruses, and hepadnaviruses. The RNA viruses are the myxoviruses, paramyxoviruses, rhabdoviruses, togaviruses, arenaviruses, bunyaviruses, reoviruses, picornaviruses, coronaviruses, and retroviruses.

13 Cytomegalovirus and hepatitis B virus produce chronic persistent infection, a long-term state of infection, most likely without manifestations even though the virus continuously replicates and the host continuously sheds it.

14 In the laboratory the diagnostic technics for viral infection include (1) cultivation, (2) serial serologic tests and other immunologic reactions, (3) direct microscopic examination of suitable specimens, and (4) skin tests.

15 Viruses may be artificially brought into contact with living cells by inoculation of (1) cell culture systems, (2) the membranes and cavities of the developing chick embryo, and (3) a susceptible animal, such as a weanling mouse.

16 A fourfold increase in the titer of antibodies between a serum sample taken early in the disease and one drawn 2 to 4 weeks later is a significant finding.

17 Unlike those in other microbial infections, the specific antibody responses in viral infections may last a lifetime.

18 Viral interference is the phenomenon whereby an animal cell exposed to a given virus subsequently develops a resistance to infection by a closely related strain of the same or another similar virus.

19 Bacteriophages, bacterial viruses, are used in the epidemiologic classification of bacteria and as vectors in molecular cloning.

20 Teratogenicity is demonstrated for rubella virus, cytomegalovirus, and herpesvirus.

21 Tumor-producing viruses can be either DNA or RNA viruses. Oncogenesis is generally associated with integration of viral DNA into the host cell genome. The RNA retrovirus satisfies this requirement by means of its viral encoded enzyme, reverse transcriptase, which converts viral RNA into DNA.

22 The role of viruses in human cancer is undefined. Hepatitis B virus is linked to human cancer of the liver. The human T cell leukemia-lymphoma virus is the first human retrovirus to be isolated from adult human patients with T cell leukemia-lymphoma, and strong parallels exist between this disease in humans and the retrovirus-induced neoplasm in animals.

23 An oncogene is a single gene in untransformed normal cells with the potential to produce a protein that can of itself change the normal cell into a cancerous one.

24 African Burkitt's lymphoma is a malignancy of the jaw and abdomen affecting children between the ages of 2 and 14, which is found in a sharply defined malarious belt in Central Africa. The Epstein-Barr virus is implicated in its etiology.

QUESTIONS FOR REVIEW

1 State briefly the salient features of viruses. Compare them with other microbes.
2 Sketch the life cycle of viruses (including bacterial viruses).
3 What are inclusion bodies? Their importance? Their specificity? Cite examples.
4 What is meant by cytopathic effect of viruses? How is this used in virology?
5 Give the two major pathologic effects viruses produce on cells they parasitize.
6 How may viruses be cultivated in the laboratory?
7 Discuss briefly the spread of viral diseases.
8 Outline the laboratory diagnosis of viral infections.
9 Comment on the nature and importance of interferon and viral interference.
10 Classify viruses. What is the basis for the most widely used system?
11 Discuss briefly the role of viruses in teratogenesis.
12 What is the importance of the nucleic acids in viruses?
13 State the case for viral oncogenesis. What is an oncogene?
14 Give an example of a DNA tumor virus and an RNA tumor virus. State the tumor produced by each.
15 Define or briefly explain bacteriophage, arbovirus, enterovirus, virion, viroid, viropexis, capsid, picorna, virology, lysogeny, papovaviruses, viral hemagglutination, plaque, icosahedron, virus-induced hyperplasia, dermotropic, prophage, capsomere, life cycle, helix, regeneration, virus neutralization, elementary body, Burkitt's lymphoma, tumor, neoplasm, neoplasia, and complementation.
16 What is the importance of reverse transcriptase in retroviruses?
17 Name the recently isolated human retrovirus.
18 Why is it difficult to demonstrate with absolute certainty the contribution of viruses to human cancers?

SUGGESTED READINGS

Chernesky, M.A., and others: Cumitech 15: laboratory diagnosis of viral infections, Washington, D.C., 1982, American Society for Microbiology.

Duckworth, D.H.: "Who discovered bacteriophage?" Bacteriol. Rev. **40**:793, 1976.

Fields, B.N.: How do viruses cause different diseases? J.A.M.A. **250**:1754, 1983.

Hsiung, G.D., and Fong, C.K.P., and August, M.J.: The use of electron microscopy for diagnosis of virus infections: an overview, Progr. Med. Virol. **25**:133, 1979.

Kaplan, M.M., and Koprowski, H.: Rabies, Sci. Am. **242**:120, Jan. 1980.

Kovi, J.: Viruses and human cancer, J. Natl. Med. Assoc. **71**:989, 1979.

Lennette, D.A., and others: Diagnosis of viral infections, Baltimore, 1979, University Park Press.

Luria, S.E., and others: General virology, ed. 3, New York, 1978, John Wiley & Sons, Inc.

Macek, C.: Medical News: Oncogenes: new evidence on link to cancer, J.A.M.A. **247**:1098, 1982.

Nahmias, A.J., and Hall, C.: Diagnosis of viral disease: today and tomorrow, Hosp. Pract. **16**:49, April 1981.

Oldstone, M.B.A.: Immunopathology of persistent viral infections, Hosp. Pract. **17**:61, Dec. 1982.

Popovic, M., and others: Isolation and transmission of human retrovirus (human T-cell leukemia virus), Science **219**:856, 1983.

Ray, C.G., and Minnich, L.L.: Regional diagnostic virology services, J.A.M.A. **247**:1309, 1982.

Reitz, M.S., Jr., and others: Human T-cell leukemia/lymphoma virus: the retrovirus of adult T-cell leukemia/lymphoma, J. Infect. Dis. **147**:399, 1983.

Schlesinger, R.W., editor: The togaviruses: biology, structure, replication, New York, 1980, Academic Press.

Storch, G., and others: Viral hepatitis associated with day-care centers, J.A.M.A. **242**:1514, 1979.

Tipple, M., and Saxon, E.: Diagnostic virology in clinical practice, Primary Care **6**:195, 1979.

Wackenhut, J.S., and Barnwell, R.A.: Burkitt's lymphoma, Am. J. Nurs. **79**:1766, 1979.

VIRAL DISEASES

SKIN DISEASES
Measles (Rubeola)*

Measles is an acute communicable disease associated with a catarrhal inflammation of the respiratory passages, fever, constitutional symptoms, a skin rash (Fig. 34-1, A) (that does not itch), and a distinct predilection for grave complications. The cause is the measles virus, a paramyxovirus, only one immunologic type of which is known. Measles is one of the most underrated of diseases. Bacterial infections such as group A streptococcal or pneumococcal bronchopneumonia, otitis media, and mastoiditis complicate 15% to 20% of cases, and the encephalitis that appears in 1 out of 1000 cases is responsible for permanent brain damage and even death. The incubation period is 10 to 12 days.

Measles depresses certain allergic conditions and immune processes. It renders the tuberculin test and agglutination tests less positive or even negative, and eczema and asthma often disappear during or after the attack.

The disease is especially virulent in populations native to tropics or in primitive races anywhere with no ethnic history of previous exposure.

---

*For a discussion of measles vaccines and immunization procedures, see pp. 424, 431, 439, and 446. For the laboratory diagnosis, see Table 33-4.

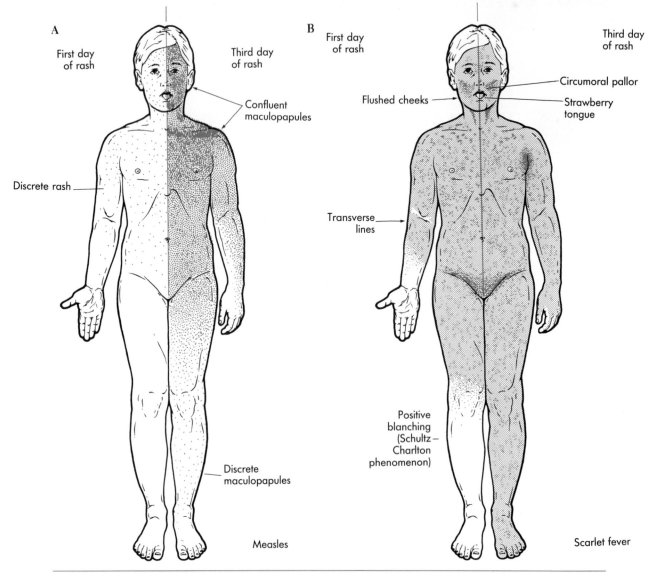

A

First day
of rash

Third day
of rash

Confluent
maculopapules

Discrete rash

Discrete
maculopapules

Measles

B

First day
of rash

Third day
of rash

Flushed cheeks

Circumoral pallor

Strawberry
tongue

Transverse
lines

Positive
blanching
(Schultz–
Charlton
phenomenon)

Scarlet fever

FIG. 34-1   Skin rashes in, **A**, measles and, **B**, scarlet fever (p. 494).

Spread

The virus of measles is thrown off in the lacrimal, nasal, and buccal secretions and enters the body by the mouth and nose. It is found in the blood, urine, secretions of the eyes, and discharges of the respiratory tract. Infection is usually transmitted directly from person to person. Healthy carriers are unknown. The time that an object contaminated by the secretions of a patient with measles remains infectious is short. Measles is not transferred by the scales from the skin. Measles virus may be spread a considerable distance through the air and is most highly communicable during the 3 or 4 days preceding the skin eruption. It is not transmitted after the fever has subsided. Children of mothers who have had measles are immune to the disease until they are about 6 months old. An attack of measles almost invariably produces a permanent immunity.

Since a nonhuman host does not exist for measles or for rubella, these two diseases, like smallpox, could be eradicated.

Prevention

The measles patient should be isolated and protected against streptococcal infections, staphylococcal infections, and common colds. Discharges from the nose, mouth, and eyes should be disinfected. When measles appears in a population group, such as armed forces personnel, daily inspections should be made, and persons having conjunctivitis, colds, or fever should be isolated.

In 1960, a representative year for the prevaccine era, around 400,000 cases of measles with about 400 deaths were reported in the United States. In 1963 measles vaccination was introduced, and by the mid-1970s the incidence of the disease had dropped sharply to around 30,000 cases with some 30 deaths annually. The efficacy of the immunization program and the safety of the vaccines are impressive. Severe reactions to the vaccine are rare—approximately 1 in 1 million children immunized—and severe illness or death is over 1000 times more likely to occur in a person with natural measles than in one who has just been vaccinated.

Atypical Measles

Children or young adults who have received inactivated measles vaccine can develop a sinister illness, an atypical form of measles, when they are exposed to natural measles several or many years later. Atypical measles is unlike the classic disease in that the lungs are involved, a measles pneumonitis appears with varied manifestations, and the skin rash evolves in an unusual pattern. A mixture of macules, papules, blisters, pinpoint hemorrhages, and wheals, the rash begins and concentrates on the extremities, especially on the wrists and ankles, and spreads toward the center and upper parts of the body. The distribution of the rash is much like that of Rocky Mountain spotted fever and is easily confused with it. The mechanisms responsible for this response to measles virus infection are not clear; probably this form of measles represents a hypersensitivity reaction. Fortunately it is self-limited.

Rubella (German Measles)*

German measles (3-day measles) is a mild but highly prevalent disease described by low fever, enlargement of lymph nodes (particularly the nodes behind the ears), milk catarrhal inflammation of the respiratory tract, joint pains, and a fine, pink skin rash similar to that of measles (rubeola) or scarlet fever (Fig. 34-2, A). The incubation period is 14 to 21 days. Transmission is airborne from person to person. A common source is someone with an inapparent infection.

The Agent

The cause is a medium-sized, spherical unit with an RNA core, an extremely pleomorphic virus that, isolated from throat washings and the blood of patients, has been grown in cell cultures. This peculiar virus is now classified as a togavirus. No insect vector is known, and only one antigen type exists.

Relation to Congenital Defects

Rubella is important because approximately one out of four children born to mothers contracting German measles during the first 4 months of pregnancy will have congenital defects. If rubella is contracted during the first 4 weeks of pregnancy, approximately half the children born will be deformed.

An Australian ophthalmologist, Norman McAllister Gregg, in 1941 first associated the congenital cataracts he was seeing at the time with the fact that the mothers of the affected offspring had had rubella in the recent epidemic in his country, and he described other malformations present as well. The epidemic had followed a 17-year pe-

---

*For immunization, see pp. 424, 433, and 441.

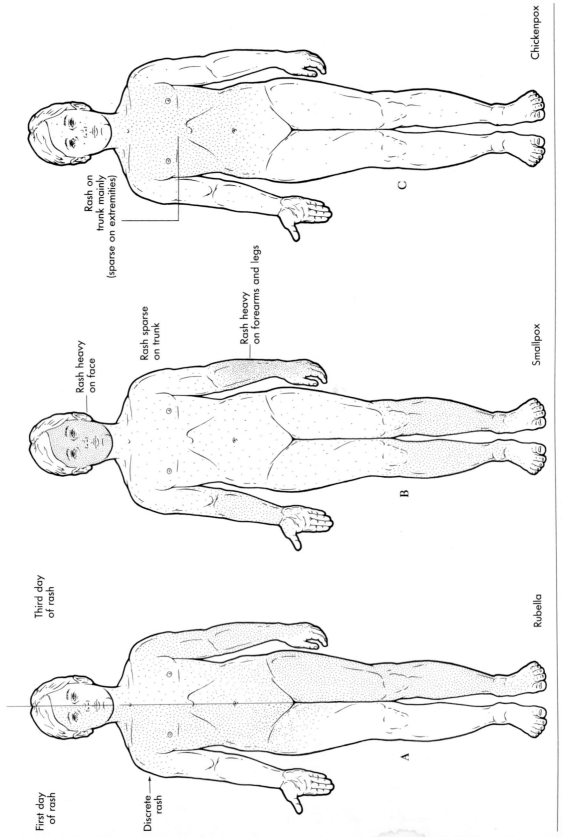

First day of rash

Third day of rash

Discrete rash

Rash heavy on face

Rash sparse on trunk

Rash heavy on forearms and legs

Rash on trunk mainly (sparse on extremities)

A

B

C

Rubella

Smallpox

Chickenpox

FIG. 34-2   Skin rashes in, **A**, rubella, **B**, smallpox, and, **C**, chickenpox.

riod of minimal incidence of rubella in Australia, and consequently many young adults were infected. Gregg noted that practically all mothers of the affected infants had had German measles early in gestation, some even before they realized they were pregnant. His correlations showed that malformations may result from extrinsic agents, even infectious ones such as viruses, rather than being exclusively bound to inherent genetic or developmental mechanisms, a concept that has had wide-ranging effects. As a result of the pandemic of 1964 in the United States, 30,000 infants were stillborn and 20,000 babies were born with congenital anomalies to mothers whose pregnancies were complicated early by rubella.

Congenital Rubella

From the mother's blood, the rubella virus crosses the placenta into the developing tissues of the new individual, where it can persist throughout gestation and into the neonatal period. The unborn child is infected at the same time as the mother, and since the viral injury leads to irregularities in the development of one or more organs, malformations result. The defects are common in the eyes, ears, heart, and brain. Examples are microcephaly (extremely small head), deaf-mutism, cardiac defects, and cataracts.

Viral injury in rubella is unique. In keeping with the mild nature of the infection, the virus neither destroys nor invigorates. It merely slows things down. The cells it parasitizes continue to grow and multiply but at a decreased rate. The embryo and fetus of the first 3 to 4 months of pregnancy are in a period of development during which the anatomic units define their shapes, take their places, and lay out their interconnections. Timing here is as critical as it is with a trapeze artist.

When infection with this virus persists after birth, the baby is born with congenital rubella, the manifestations of which may be mild or severe. There may be a single serious defect or multiple ones in a small undersized baby.

The **rubella syndrome** indicates active infection in the newborn, which is easily demonstrated by recovery of the rubella virus from nasopharyngeal washings, conjunctivae, urine, and cerebrospinal fluid.

Laboratory Diagnosis

As a routine, virus isolation is impractical. The following serologic procedures are available for the laboratory diagnosis of rubella infection: virus neutralization, complement fixation, hemagglutination inhibition (HI), immunofluorescence, and enzyme immunoassays specific for IgG or IgM antibody.

The hemagglutination-inhibition test is the most sensitive and most widely done. With a single specimen this test indicates immunity in the presence of HI antibody. With paired serums spaced 1 or 2 weeks apart and a fourfold increase in the HI titer (confirmed by the rubella complement-fixation test), it can make the diagnosis. If a woman is exposed to a possible case of rubella during her pregnancy, the HI test may be used to determine both in the pregnant woman and in the contact whether the exposure is to rubella and what the woman's immune status is to rubella. In fact *routine* hemagglutination-inhibition testing for rubella antibody is recommended in the woman planning to have children and in the pregnant woman early in gestation. For the diagnosis of congenital rubella there is the determination of rubella-specific IgM in the baby. (This is an antibody that does not cross the placenta and that therefore could not be derived from the baby's mother.) As with other viral diseases, IgM is the first antibody formed in response to infection, and the presence of rubella-specific IgM at birth indicates intrauterine infection.

Immunity

An attack of German measles confers permanent immunity. Girls should have German measles if possible before child-bearing years, and a pregnant woman should avoid exposure to the disease. This is true even in women supposed to have had an attack because an erroneous diagnosis is often made. Although the newborn baby possesses virus-neutralizing antibodies in high titer, the baby may continue to harbor virus and to shed it, even for several years, thus constituting an important reservoir and source of infection to susceptible persons in the environment.

Prevention

When rubella vaccine was licensed in 1969, its use was directed to preschool and young school-age children in an effort to decrease the overall number of cases and therefore the chances that a susceptible pregnant human female would be exposed. Vaccination was not stressed for susceptible pospubertal girls and women. Because of this approach, the characteristic pattern for rubella epidemics every 6 to 9 years was interrupted, and a reduction in the incidence of rubella occurred. In young school-age children the incidence decreased by 89% between the mid-1960s and the mid-1970s. However, control of rubella has not been completely achieved among adolescents and young adults in whom the overall susceptibility rate remains at 10% to 20%.

Smallpox (Variola)

Smallpox, the most infectious of diseases, was one of the most fearsome scourges in human history. Before the days of vaccination it spread over the world in devastating epidemics, in some of which 95% of the population were attacked and 25% died. Now it is only of historical interest.

Smallpox was characterized by severe constitutional manifestations and a rash (Fig. 34-2, *B*) that evolved in typical stages to become hemorrhagic with severe disease (Fig. 34-3). Healing of the skin lesions left the patient with many disfiguring scars.

Two types of smallpox virus (a poxvirus) are known. One causes a severe form, variola major (formerly seen in Asia, mortality of 20% to 40%); the other causes a mild form, variola minor or alastrim (formerly found in Africa, mortality of 1%). The elementary bodies of the smallpox virus may be seen in suitable preparations in the light and electron microscopes.

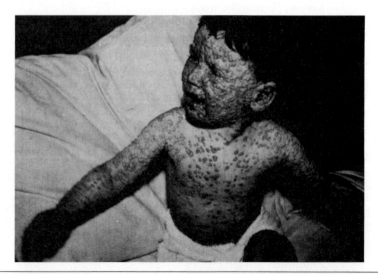

FIG. 34-3   Smallpox in unvaccinated 2½-year-old boy on eighth day of illness. Attack was severe, but boy recovered. (Courtesy Dr. Derrick Baxby, University of Liverpool, England.)

Eradication of Smallpox

In 1967 the World Health Organization (WHO) launched its "Smallpox Target Zero" campaign, which was to span a decade, cost $250 million, and integrate the effects of many nations before the goal of the campaign would be reached. The last known, naturally occurring case of smallpox affected a young hospital cook in Somalia (Africa) in October 1977. (Two cases of laboratory-acquired smallpox occurred in Birmingham, England, in 1978.)

A disease such as smallpox with obvious clinical features, virtually no asymptomatic carriers, no known animal or insect host or vector to serve as a reservoir (no nonhuman reservoir), and a short period of infectivity (3 to 4 weeks) can be eliminated readily. Since humans are the only hosts, prevention of spread to the point where no new cases are recognized means eradication. (Virus does not persist after clinical disease, and the limited antigenic variation in the virus means an effective vaccine can be produced.)

Necessary to the efficiency of the campaign were the identification and reporting of

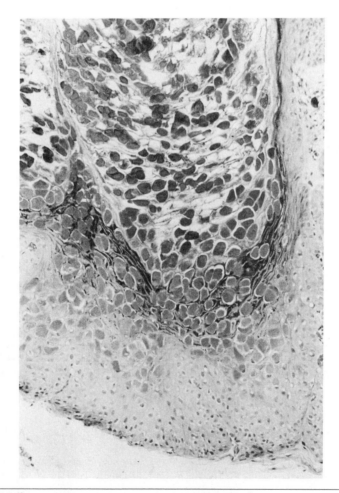

FIG. 34-4   Molluscum contagiosum in microscopic section to show numerous intracytoplasmic inclusions, the molluscum bodies. These rounded homogeneous masses form within squamous epithelial cells, enlarge, crowd nuclei aside, and soon fill the cells. Electron micrographs show molluscum bodies to be aggregations of viral particles.

all outbreaks of whatever size and the vaccination of all susceptibles in that area. By adherence to a surveillance and containment program, wherein cases were sought out and contacts vaccinated, it was possible to eliminate the disease in geographic areas where perhaps only 6% of the population were vaccinated. A long-lasting, potent, freeze-dried vaccine conferring a good level of protection was used, and a simple, bifurcated needle holding a drop of vaccine between two prongs made administration easy, even by untrained workers.

For the future, WHO recommends that stocks of smallpox viruses be confined to four designated reference laboratory centers, including the Centers for Disease Control and the American Type Culture Collection.

Vaccinia

Smallpox and vaccinia, or cowpox, are caused by pox viruses with similar biologic properties. The virus of vaccinia produces a mild disease either in humans or in cattle (its natural host), and its importance is that it can be used to produce immunity against the more severe disease, smallpox.

Molluscum Contagiosum

Molluscum contagiosum is a skin disease associated with small, pink, wartlike lesions on the face, extremities, and buttocks. It is spread from person to person by direct and indirect contact. The cause is a large poxvirus that produces very dramatic intracytoplasmic inclusions easily recognized in the squamous cells lining the affected skin site (Fig. 34-4).

HERPESVIRUS INFECTIONS

The many herpesviruses (Greek *herpein*, to creep) in the family Herpesviridae infect a wide variety of biologic species. In fact in every family of vertebrates there is a representative one. In human beings the five herpesviruses found are classified in Table 34-1. Although human herpesviruses are diverse in many ways, they are alike in certain properties that relate to disease: (1) the comparable size, (2) the unique, diagnostic appearance in an electron micrograph, (3) the cytopathic pattern of inclusion bodies in

Table 34-1 Human Herpesviruses

Classification	Common Name	In Single Human Host May Be Latent in	Diseases
Herpesvirus hominis 1 (HVH-1) Subfamily: Alphaherpesvirinae	Herpes simplex virus 1 (HSV-1)	Trigeminal ganglia	Herpes labialis, herpetic keratitis, herpetic whitlow, herpes simplex encephalitis
Herpesvirus hominis 2 (HVH-2) Subfamily: Alphaherpesvirinae	Herpes simplex virus 2 (HSV-2)	Sacral sensory root ganglia	Genital herpes
Herpesvirus hominis 3 (HVH-3) Subfamily: Alphaherpesvirinae	Varicella-zoster (VZ) virus	Dorsal root ganglia	Chickenpox, shingles
Herpesvirus hominis 4 (HVH-4) Subfamily: Gammaherpesvirinae	Epstein-Barr virus (EBV)	Lymphocytes and tonsillar epithelial cells	Infectious mononucleosis, Burkitt's lymphoma, nasopharyngeal carcinoma (?)
Herpesvirus hominis 5 (HVH-5) Subfamily: Betaherpesvirinae	Cytomegalovirus (CMV)	Renal tubular cells	Congenital and perinatal infections in infants, severe infections in immunosuppressed patients

infected cells, (4) the pattern of replication, and (5) the distinctive sequences of latency and reactivation in infection. All species of herpesviruses possess a remarkable ability for survival. After a herpesvirus has initiated the first, or primary, infection, it establishes itself in a latent state at an anatomic site. As a latent virus, it would seem to be dormant and inactive and actually is undetectable in the cells at the site of latency. From this site the virus may become reactivated, virus excretion may recur, and disease can result, a sequence of events that is apparently unaffected by the immunity of the host. This property of latency greatly enhances the possibility and ease of transmission in human populations that otherwise might not sustain such a common virus.

Chickenpox and Shingles (Varicella–Herpes Zoster)

The two diseases, chickenpox (varicella) and shingles (herpes zoster), represent two phases of activity of a single herpesvirus (*Varicellavirus* or *Herpesvirus hominis* type 3). It is also referred to as the varicella-zoster (VZ) virus. The first invasion of the body by the virus produces chickenpox, the generalized infection. Shingles, or zoster, is the localized infection in a partially immune host. It is assumed to be the reactivation of latent infection brought on by exogenous factors such as trauma, intercurrent disease, or drugs or by exposure to chickenpox. The localization of the latent virus is speculative, however.

Varicella, usually a mild but very contagious disease of childhood, presents a typical (teardrop) vesicular (blisterlike) rash (Fig. 34-5), which although generalized in the skin and mucous membranes, is concentrated on the trunk (Fig. 34-2, C). The incubation period of chickenpox is 14 to 16 days. Herpes zoster, on the other hand, is a disease of adults defined by the appearance of a vesicular eruption similar to that of varicella but with a quite different distribution. In shingles, vesicles usually occur on one side of the chest, following the course of the peripheral nerves supplying that part. The

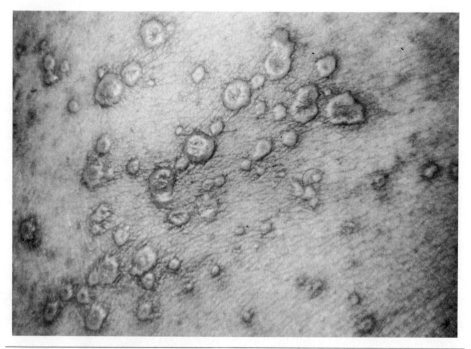

FIG. 34-5 Vesicles in chickenpox.

Latin word *cingulum*, "girdle," accounts for the name *shingles* and refers to the anatomic distribution of involvement. Shingles may occur in other parts of the body, but since the primary involvement is in the ganglion of the posterior nerve root, the vesicles always follow the course of the nerve or nerves supplying the affected part. Shingles may cause prolonged suffering because the skin eruption is associated with intense burning pain. The incubation period is from 10 to 23 days.

Diagnosis

In the vesicles of the skin in either disease there are typical reddish inclusion bodies in the nuclei of injured epithelial cells, and virus is present in the fluid (Fig. 34-6). Virus may be isolated. For serologic diagnosis, a complement-fixation test and a fluorescent antibody procedure are available.

Spread

The moist crusts of chickenpox are infectious, whereas the dry ones are not. Chickenpox is transmitted by direct contact with a patient. Herpes zoster in adults has served as a source of varicella in children.

No immunity for chickenpox is conferred by the mother on her newborn infant, and convalescent serum is of little value in preventing or modifying the disease. Immunosuppressed persons are at risk for serious infections with the varicella-zoster virus. No satisfactory animal model of varicella-zoster infection exists. A live, attenuated virus vaccine for chickenpox is still experimental.

Herpes Simplex Infections

Infection with herpes simplex virus *(Simplexvirus, Herpesvirus hominis)* is related to either of two recognized serotypes, each with unique biologic features.

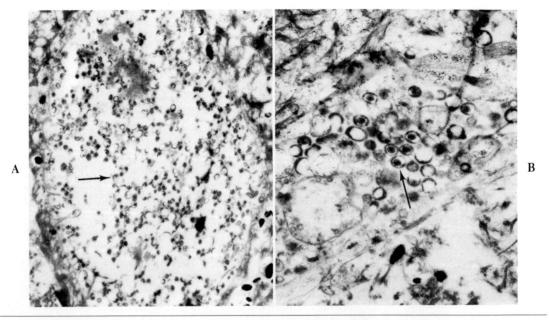

FIG. 34-6   Virus of chickenpox, electron micrograph. Numerous viral particles are seen in segment of squamous cell from skin. (**A,** ×15,000; **B,** ×30,000.)

Herpes simplex virus infection in humans is as follows:

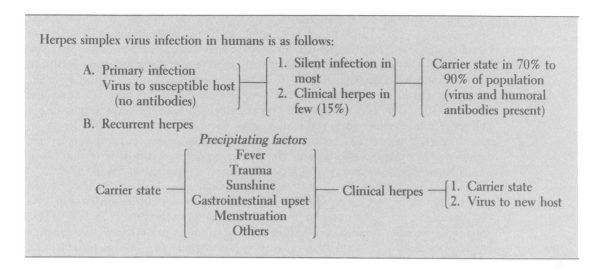

Type 1 Infections

The more common herpes simplex virus 1 is responsible for the familiar fever blisters (herpes simplex) and cold sores found around the mouth and lips (**herpes labialis**). It is said that 70 million people get cold sores each year. Primary *Herpesvirus hominis* type 1 infections are spread by direct contact with a lesion, although the high prevalence of such infections suggests other means of transmission. The primary infection with type 1 usually occurs in childhood.

In most instances the first infection with herpesvirus type 1 is a mild or inapparent one; clinical signs of disease occur in only about 10% to 15% of persons infected. The incubation period is 2 to 12 days. After the susceptible person has made primary contact with a lesion, the fever blister virus in the primary infection in that person travels up to the trigeminal ganglion, where it remains in a latent state for the lifetime of the infected person. When the body resistance is weakened from any one of a number of possible causes, the virus is reactivated. It travels from the ganglion by the peripheral nerve to the border of the lip to produce another fever blister. It descends along the peripheral nerves to erupt, as it were, over the nerve endings. Recurrent herpes simplex often accompanies febrile illness, exposure to cold or sunlight, fatigue, mental strain, or menstruation in women (see box above). Unlike other viral infections, the host's humoral antibodies apparently give no protection against subsequent lesions.

In addition to recurrent labialis, type 1 lesions include gingivostomatitis and corneal lesions in the eye (**herpetic keratitis**). Progressive involvement of the cornea of the eye with recurrent infection may lead to blindness. **Herpetic whitlow,** an occupational hazard to dentists, consists of small vesicles on the fingers that may rupture and leave slowly healing, painful ulcerations.

Herpesvirus infections usually are localized and transient, but occasionally serious complications ensue. Encephalitis in a newborn may result from delivery through an infected birth canal, but encephalitis can also occur in older children and adults who have repeated episodes of infection. **Herpes simplex encephalitis** is a severe involvement of the brain, initially localized to the temporal lobe, that is characterized by hemorrhage and tissue destruction. Clinically, the manifestations include fever, headache, a stiff neck, speech disturbances, and focal paralyses. The immunosuppressed patient is at high risk for dissemination of the herpesvirus with serious consequences.

Genital Herpes (Type 2 Infection)

Herpesvirus type 2 found around the genital organs causes **herpes genitalis,** the third most common sexually transmitted disease (STD). An estimated 20 million persons in the United States have it, with an estimated 500,000 new cases appearing each year. After an incubation period of 2 or 3 days, the exquisitely painful lesions of genital herpes appear as crops of clear vesicles on a reddened, elevated base. Vesicular fluid contains 1 to 10 million viral particles per milliliter. In time the vesicles fill with pus, ulcerate, and scab over. Regional lymph nodes enlarge. A viremia associated with primary infection may bring about systemic manifestations, more severe in women than in men. Since many viral particles are in the vesicles, the patient is very infectious from the time vesicles appear until scabs are completely gone, 2 to 3 weeks later.

In most women with the first episode of genital herpes, virus can be isolated from the urethra in 82% of cases, from the cervix in 88%, and from the pharynx in 13%. In men with the first infection, virus may be isolated from the urethra in 28% of cases and from the pharynx in 7%. Thereafter the virus travels to the sacral root ganglion, where it persists as a latent virus. It may be reactivated, and 75% of patients suffer repeated attacks.

A woman with a herpetic lesion confined to the cervix can be an asymptomatic carrier of the virus, and in some women the infection does not present any visible sign Because women with cancer of the mouth of the womb (cervix uteri) show a statistically significant correlation with previous type 2 infection, this herpesvirus is implicated in the causation of this form of cancer and also in the causation of carcinoma in situ of the vulva.

Neonatal Herpes

One very serious complication of herpes genitalis is infection of the newborn. By contacting a herpetic lesion in the birth canal, the baby acquires the infection. In 1 to 3 weeks the infant becomes gravely ill with generalized herpesvirus infection and usually succumbs. Herpetic lesions can be found in all the organs, including the brain (herpes encephalitis).

Anatomic Specificity

Until recently it was stated that herpesvirus type 2 infection as a sexually transmitted disease produced skin and mucosal lesions below the waist as compared with type 1 virus, which produced lesions above it. However, epidemiologic patterns today for herpesvirus types 1 and 2 show a less precise anatomic localization, presumably as a result of changes in sexual mores. Genital herpes is caused by type 1 in 15% or more of cases. Type 2 is found in the anal and oral areas, and herpesvirus type 1 is recovered from neonatal infection.

Laboratory Diagnosis

Herpes simplex virus forms characteristic intranuclear inclusion bodies in multinucleated giant cells present within the lesions. The intranuclear inclusions may be identified by the light or electron microscope in suitable preparations (smears, imprints, Pap smears of cervix uteri, and the like) of fluid taken from the superficial vesicles or of parasitized tissue cells (Figs. 34-7 and 34-8). The virus can be cultivated in cell cultures and recovered from newborn mice. Fluorescent antibody technics, as well as immunoenzyme assays and neutralization, hemagglutination, and complement-fixation tests, are serologic aids to the diagnosis.

Relation of Types 1 and 2

Herpesviruses are large DNA viruses coding for about 100 specific proteins. The DNA sequence homology between the two herpesvirus types is around 50%. Genetically the

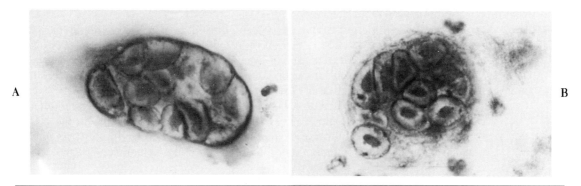

FIG. 34-7   Herpesvirus *(Herpesvirus hominis)* seen with light microscope, high-power photomicrograph. Inclusion bodies within nuclei of multinucleated squamous cell from skin. **A,** Early stage in formation of inclusion bodies. **B,** Inclusion bodies well defined.

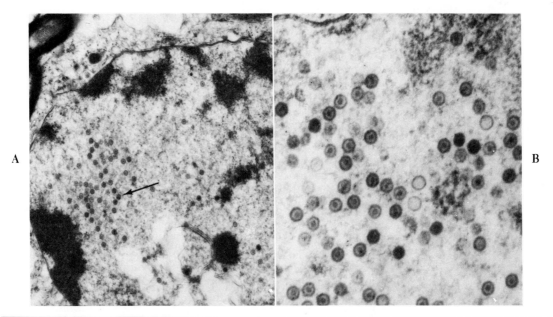

FIG. 34-8   Herpesvirus *(Herpesvirus hominis)*, electron micrograph. Numerous viral particles in one nucleus of squamous cell from skin. (**A,** ×20,000; **B,** ×56,000.)

two types are more closely related to each other than to any of the other herpesviruses, which are quite different in terms of genetic relatedness. Immunologically a strong cross-reactivity exists between the two. Of the antigens involved in routine serologic testing, 90% are possessed by both types. This makes it difficult to detect type 2–specific antibodies in persons with high antibody titers to type 1 herpesvirus. Some of the newer serologic methods are more sensitive. Some definite differences do exist. For instance, type 1 forms no plaques in chick embryo cells; type 2 does. Type 1 is highly neurovirulent for mice; type 2 is moderately so.

Work is being done on vaccines for genital herpes. One is genetically engineered, developed from live herpesvirus, and is currently being tested. The other is a killed virus vaccine being used in clinical trials in human beings.

Cytomegalic Inclusion Disease

Cytomegalic inclusion disease (salivary gland disease) is infection caused by the ubiquitous cytomegalovirus (CMV) or salivary gland virus *(Cytomegalovirus)*. It is also a herpesvirus (Table 34-1). The human cytomegalovirus is antigenically heterogeneous, and serotypes are described. Infection covers a wide range of possibilities from an inapparent process to lethal disease. Most infections are subclinical and benign.

Cytomegalovirus is widely distributed over the world and is especially common in the lower socioeconomic groups. It is said that 80% of the population in the United States have antibodies indicative of infection by the age of 35 or 40 years; yet not too much is known about person-to-person transmission under natural conditions.

The Inclusion Bodies

The pathologic lesions are striking in their nature. Large, well-defined viral inclusions are seen in the nucleus, and smaller ones are found in the cytoplasm of the injured cells, which are typically enlarged (**cytomegaly** means cell enlargement) (Fig. 34-9). In fatal cases cell changes are seen in the gastrointestinal tract, lung, liver, spleen, and other organs. In nonfatal cases inclusions may be found in epithelial cells from the kidney shed into the urinary sediment.

Spectrum of Infection

In an overt form cytomegalic inclusion disease is seen in the newborn as a congenital infection acquired from a mother who was probably asymptomatic (latent infection). Probably a tenth of newborn infections are acquired in utero; the rest result from the infant's passage through the birth canal. In the infected newborn cytomegalovirus causes major birth defects, especially in the central nervous system (for example, microcephaly, hydrocephaly, blindness, mental retardation, deafness). Brain damage is severe. Malformed babies exhibit the viral inclusions.

In healthy adult cytomegalovirus may produce, as a primary infection, an acute febrile illness not unlike infectious mononucleosis, with enlargement of the spleen and an increase in the number of circulating lymphocytes, many of which are atypical. Cytomegalovirus is the most common cause of heterophil antibody–negative infectious mononucleosis. In addition the virus may produce hepatitis or pneumonitis.

In the older person secondary infection from cytomegalovirus may complicate preexisting disease. The leukemic patient receiving cancer chemotherapy and the patient of any age receiving immunosuppressive therapy after organ transplant are vulnerable for generalized cytomegalovirus infection. In bone marrow recipients death usually comes from the interstitial pneumonitis induced by this virus.

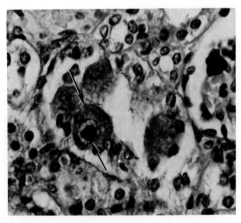

FIG. 34-9 Cytomegalic inclusion disease of kidney, photomicrograph. Arrows spot enlarged renal tubular cell with large intranuclear inclusion body of cytomegalovirus. (× 800.)

Cytomegalovirus infection is the most frequent viral complication of renal transplantation. The virus is found in 40% to 80% of patients postoperatively, and it is the pathogen most often recognized in the first 6 months after transplant. Clinical manifestations of viral infection such as pneumonia, hepatitis, or encephalitis are said to be more likely in seroimmune patients (antibodies present). In many instances infection is believed to be primary, since virus may be transplanted with the graft. In some studies primary disease (no antibodies demonstrated) has been shown to be much more severe for the graft recipient than secondary disease where antibodies are present.

Cytomegalic inclusion disease is associated with the postperfusion, or posttransfusion, syndrome, an infectious mononucleosis–like syndrome following perfusion of fresh blood in patients undergoing open-heart surgery. It appears 2 to 4 weeks after the transfusion of fresh blood containing cytomegalovirus and is characterized by splenomegaly, generalized lymphadenopathy, and an atypical lymphocytosis comparable to that found in infectious mononucleosis.

Human cytomegalovirus is known to depress both cell-mediated and humoral immune responses, a fact that may predispose certain patients to superinfection with bacterial and fungal pathogens. It is implicated in the acquired immunodeficiency syndrome of homosexual men and in Kaposi's sarcoma (p. 367), since the prevalence of past or present cytomegalovirus infection in homosexual men is very high.

Spread

Transmission of infection has been demonstrated by way of the respiratory, digestive, and genital tracts, by organ transplantation (especially kidneys and bone marrow), and by blood transfusion. Transmission may be vertical, that is, across the placenta, by contact with infected secretions of the birth canal, or by contact with saliva or milk. Transmission may occur by direct contact between one child and another, between child and adult, or between two adults (which is most likely sexual). Semen and cervical secretions are the most common specimens into which the virus is shed. Latent virus in a host may be inactivated by immunosuppression of that host. Cytomegalovirus has been isolated from human cancers.

Laboratory Diagnosis

The "owl-eye" viral inclusions are readily visualized and diagnostic, and virus retrieved from saliva, urine, and circulating lymphocytes can be propagated in cell culture. Serologic technics may be used to measure specific antibodies, including those of neutralization, complement fixation, hemagglutination, and immunofluorescence. If paired acute and convalescent serums show a fourfold increase in titer of antibodies to cytomegalovirus, a current or very recent infection is presumed. Counterimmunoelectrophoresis is useful. A monoclonal antibody against infected cells has been prepared for use in immunofluorescent assays applied to cell cultures. In the nuclear fluorescent urine test a human fibroblast culture is inoculated with urine from a patient. The cytomegalovirus antigen is demonstrated in nuclei from the cell culture by means of an immunofluorescent assay.

Prevention

An experimental live virus vaccine is being tested.

Infectious Mononucleosis

Infectious mononucleosis, or glandular fever, is an acute infectious viral disease with enlargement of the lymph nodes and spleen, sore throat, and mild fever. It is usually benign and self-limited, although subject to a variety of complications. Probably most of the time infection is not present as clinically overt disease but rather as a silent, inapparent infection or, at most, a mild upper respiratory tract ailment. It is rarely fatal. The disease may occur sporadically or in epidemics. Infectious mononucleosis, postulated to occur early in life, usually attacks children and young adults (up to 30 years of age). The incubation period is stated to be 4 to 14 days or longer (possibly up to 5 weeks). Most of the time the disease lasts 1 to 6 weeks.

The Agent

The cause is the Epstein-Barr virus (EBV) (*Lymphocryptovirus, Herpesvirus hominis* type 4). Drs. M.A. Epstein and Y.M. Barr first detected in 1964 this herpesvirus in cell culture lines derived from African Burkitt's lymphoma. It has also been consistently related to another neoplastic disease, a type of nasopharyngeal cancer seen mainly in Cantonese Chinese, natives of Alaska, and some of the population groups of northern and equatorial Africa.

Spread

The Epstein-Barr virus is found in the throats and saliva of patients and is spread by mouth-to-mouth contact, hence the designation the "kissing disease" for infectious mononucleosis. The disease is especially prevalent in colleges among adolescents and young adults. The Epstein-Barr virus is worldwide, and contact with it is widespread in underdeveloped countries and in better developed ones as well.

Laboratory Diagnosis

A dramatic feature of infectious mononucleosis is the Epstein-Barr virus–induced hematologic response of lymphocytes in the patient's blood. The Epstein-Barr virus is a lymphotrophic virus—cytotrophic for B lymphocytes—and one of its biologic attributes enables it to change lymphocytes into rapidly proliferating cells. Viral action therefore results in an increase in the total number of circulating white blood cells, among which will be found many lymphocytes (spoken of as virocytes) of an unusual appearance, described variously as atypical, reactive, stimulated, and even "turned on." (Hematologically, infectious mononucleosis is a lymphoproliferative disorder, though a self-limiting one.) An important part of the laboratory diagnosis of the disease is the morphologic identification of the atypical lymphocytes in a suitably stained peripheral blood smear.

During the course of clinical infection 50% to 80% of patients with infectious mononucleosis develop heterophil antibodies, which can be measured by their ability to clump sheep red blood cells. The demonstration of these under the prescribed conditions of the Paul-Bunnell (heterophil antibody) test is diagnostic. Since the antibodies to the Epstein-Barr virus capsid antigen appear early and remain the predominant antibodies during clinical illness, the immunofluorescent test that assays their level is a more reliable diagnostic approach. It is stated that infectious mononucleosis is a disease associated with the formation of more antibodies to different antigens than almost any other disease. It seems to make antigens, as it were, even more antigenic. Note that in infectious mononucleosis, both in acute and convalescent states, the nonsyphilitic patient may have a biologic false positive test for syphilis.

RESPIRATORY DISEASES
Influenza

Influenza* is a highly communicable disease occurring in epidemics that are described by explosive onset, rapid spread, involvement of a high percentage of the population, and frequency of serious secondary bronchopneumonia. Like many viral diseases, influenza seems to potentiate serious bacterial infection, mostly bronchopneumonia. The

---

*For immunization, see pp. 430, 443, and 447.

mortality of the 1918 pandemic was largely the result of the severe, complicating bronchopneumonia produced by virulent streptococci (or pneumococci).

Influenza causes thousands of deaths each year and costs the United States billions of dollars in medical costs and in lost earnings. The disease is described by fever, chills, malaise and muscular aches, sore throat, nasal discharge, headache, watery and burning eyes, cough, and respiratory congestion. Manifestations of infection appear after 24 to 48 hours' incubation and last from a few days to more than a week. Spread is by the respiratory route, and, as would be expected, influenza is most prevalent in the winter.

An influenza pandemic occurs about every 10 to 14 years. The name *influenza* comes from Italian astrologers of long ago who believed that the periodic appearance of the disease was in some way related to the *influence* of the heavenly bodies.

1918-1919 Pandemic. One of the world's greatest catastrophes was the pandemic of 1918-1919. The influenza of this pandemic was referred to as the white plague or Spanish flu. Although other epidemics have had higher death rates, on the basis of total number of deaths this was the worst pestilence that civilization had ever experienced. A guess is that the lives of a billion persons were affected, or half the world population at that time. Only two places on the globe allegedly escaped—St. Helena, an island in the South Atlantic, and Mauritius, one in the Indian Ocean. It is estimated that there were 200 million cases with 20 million deaths worldwide. In the United States alone there were 850,000 deaths, a figure greater than that for the combined battlefield losses of World War I, World War II, and the Korean War. The mortality of the influenza pandemic dwarfed the mortality of World War I. And this event occurred within the twentieth century!

The agent. Influenza is caused by a myxovirus (so tiny that 29 to 30 million of them could rest comfortably on the head of a pin). It was originally believed that *Haemophilus influenzae* was the cause of influenza. (This organism is an important *secondary* invader in this and other respiratory diseases.)

Based on its nucleocapsid proteins (those associated with the RNA core), the influenza virus is classified into three distinct antigenic types: A, B, and C. Of these, A and B have been best studied. They are alike in many ways but differ serologically.

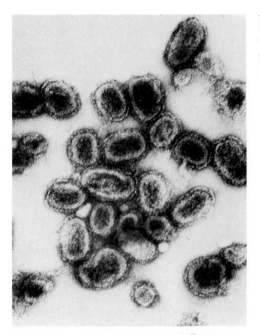

FIG. 34-10 Influenza virus negatively stained (A_2/Aichi/ 68 or "Hong Kong" virus), electron micrograph. ($\times 303,100$.) (Courtesy C.A. Baechler, Virus Research, Parke, Davis & Co., Detroit.)

Each includes numerous distinct strains. Influenza A strains (Figs. 34-10 and 34-11) have caused every known pandemic of flu and have figured much more frequently in epidemics than those of type B, the strains of which are considered to be less virulent than those of type A.

The envelope of the influenza A virus contains projections or spikes (Fig. 34-12) holding two viral antigens, **hemagglutinin (H)** and **neuraminidase (N)**, which designate subtypes for this flu virus. Protective antibodies resulting from immunization are directed against the hemagglutinin and to a lesser degree against the neuraminidase antigens. New H or N subtypes result from major changes in the H and N antigens that are referred to as **antigenic shifts.** Since the 1930s, clearly separate, medically important H (H1, H2, H3) and N (N1, N2) subtypes have been defined. (More are known.) Lesser changes in these antigens, termed **antigenic drifts,** result in a different strain of virus.

The name given to a particular influenza A virus includes the nucleocapsid type, the geographic site of isolation, the strain number, and the year of isolation with H and N subtypes.

INFLUENZA B INFECTION. B strains have recently been reported in scattered outbreaks. Influenza B virus infections have a predilection for schoolchildren and tend to recur every 4 to 6 years. They are also the most common forerunner of Reye's syndrome, the etiology of which is otherwise unknown.

REYE'S SYNDROME. An uncommon disorder first described in 1963, Reye's syndrome is seen in a child apparently recovering from a viral illness who suddenly becomes confused or agitated and begins to vomit. Most recover completely, although severe illness may lead to brain damage with resultant defects. In fatal cases the blood suger level is lowered, the blood ammonia level is elevated, the brain is dramatically swollen, and fatty change in the liver and kidneys is pronounced. The Centers for Disease Control warn that the use of salicylate-containing products to treat fevers in

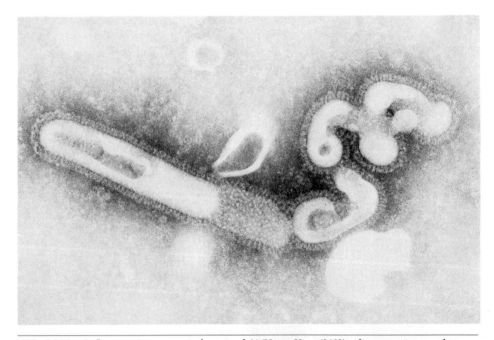

FIG. 34-11 Influenza virus, negatively stained (A/Hong Kong/1/68), electron micrograph. Note filamentous and pleomorphic particles. Virus isolated from throat of newborn baby. (×200,000.) (From Bauer, C.R., and others: J.A.M.A. **223:**1233, 1973. Copyright 1973, American Medical Association.)

children with varicella or influenza should be avoided if possible. This implies a strong link between the administration of aspirin and the development of the disease.

INFLUENZA C INFECTION. Influenza C virus is one cause of the common cold. Although this virus is widespread, infection is rarely associated with disease.

Epidemiology. The clinical manifestations of influenza remain fairly constant throughout the years although caused by an agent capable of remarkable changes. It has been said that it is an "*un*varying disease caused by a varying virus." The influenza virus enters the body by the mouth and nose and leaves by the same route. Recently influenza virus has been recovered from anal swabbings. Influenza has been produced artificially in chimpanzees, ferrets, and mice by injection of filtrates of nose and throat washings from known human cases.

The disease is spread by direct and indirect contact, including droplet infection. An epidemic is usually short lived and quickly subsides, to be followed after several weeks by a secondary wave, a free period, and a tertiary wave. Never does the infection spread faster than people travel. What happens to the virus between epidemics still remains a mystery, although the relationships of influenza viruses to animals are now apparent. There is good reason to assume that in the interval the virus resides in animals, multiplies, and changes genetically, possibly by genetic recombination of human strains with those from lower animals, thus becoming infective for humans again. Influenza A viruses are the only influenza viruses shown to have a natural host range; they have been isolated from pigs, horses, and birds (as well as from human beings). In most other viruses nucleic acid is a single strand. Most likely because its RNA is in pieces, the flu virus rearranges or shifts its genetic material from time to time.

An epidemic seems to depend on the emergence of the new strain. Since this organism possesses a continuing and dramatic tendency to change its chemical and genetic

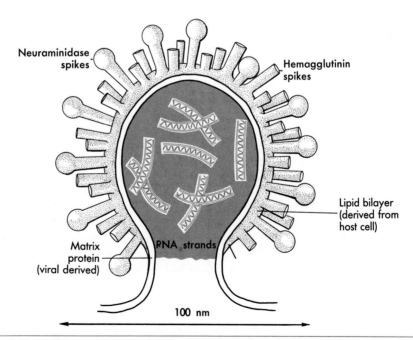

FIG. 34-12   Influenza virus emerging from cell (host cell not shown), schematic drawing. Note unusual nature of viral nucleic acid with discrete segments. In most other viruses nucleic acid is a single strand. Most likely because its RNA is in pieces, the flu virus rearranges or shifts its genetic material from time to time. Mutations result from alterations in biologically active surface proteins of hemagglutinin and neuraminidase spikes.

nature, that is, to mutate, new variants of the virus type arise constantly from the major antigenic shifts. Flu virus mutates to a magnitude not found for any other infectious agent. If the majority of persons do not possess the immune mechanisms to meet the new variation, then far-reaching epidemics and pandemics threaten (Fig. 34-12).

The explosive outbreak of an epidemic may be explained by the high communicability of the disease, the great number of susceptibles, and the fact that during the early days of the attack the patient is not confined to bed but mingles freely with other people. The average duration of an epidemic in a community is from 6 to 8 weeks. In secondary and tertiary outbreaks the number of people attacked is less than in a primary outbreak, but the disease tends to be severe, complications are more common, and the mortality is higher. Interepidemic cases are usually mild.

Recent pandemics and epidemics. In early 1957 influenza appeared out of northern China and within the next several months spread all over the world. There were millions of cases over a wide area of the Far East. Referred to as Asian flu, it quickly spanned the extent of the United States. The virus was found to comprise new strains of influenza virus A (H2N2), designated "Asian strains." On the whole, the mortality in this pandemic was low; in areas where notable it was associated with severe staphylococcal pneumonia.

In the summer of 1968 another pandemic of influenza came out of the Orient, starting in Hong Kong with close to half a million cases. Hong Kong influenza was caused by a virus not completely different from the Asian flu virus of 1957, a variant in the A group—A/Hong Kong/68 (H3N2), and the clinical disease was similar to that of the prior decade. During the last 3 months of 1968 it was estimated that 30 million persons were stricken. There were more deaths with Hong Kong influenza than with Asian influenza, especially among chronic invalids. *Staphylococcus aureus* and *Pseudomonas aeruginosa* were the complicating bacterial infections.

Since the 1968 pandemic, lesser epidemics have tended to occur annually. Several variants of the Hong Kong virus have emerged since 1968, and Asian strains of the flu virus may be important in the epidemiologic picture of influenza for some time, until the level of immunity in the general population has risen significantly.

The killer flu virus of the unprecedented 1918-1919 pandemic is postulated to have been an A/swine type of flu virus, a virus with a virulence unmatched in human history. (Influenza A virus was not isolated from a human being until 1933.) Flu in the United States at that time of pandemic was sometimes referred to as hog flu. Interest is renewed in such a virus because an A/swine type of flu virus, very similar to and assumed to be identical with that killer virus, has been isolated in the United States, an event suggesting its reappearance after over half a century. Serologic studies showed that antibodies to this swine type of flu virus were widespread in persons tested who were alive in 1918 and 1919. The virus, A/New Jersey/76 (H1N1), has peculiar properties. It is difficult to detect, growing poorly in hens' eggs and not at all in standard kidney cell cultures. Like its viral agent, the disease of 1918-1919 had unusual features; for instance, the highly susceptibles were persons 15 to 50 years of age, whereas in the flu epidemics of today they are the very old or the very young. The isolation of the swine type of flu virus was viewed with much alarm, but the anticipated epidemic did not materialize.

In the winter of 1977-1978 an epidemic of flu broke out in the Soviet Union and became widespread across Asia and Europe. Russian flu was mild and seen primarily in children and young adults under 24 years of age. In the United States mostly school-children and military recruits were affected. The Russian flu virus A/USSR/90/77 (H1N1) most closely resembles a strain prevalent in the early 1950s but not seen since, and presumably persons older than 25 years acquired some degree of immunity to it at that time.

Probably as far back as 1918 to the year 1957, influenza A viruses of the subtype

H1N1 circulated over the world. In 1957 this subtype was replaced by A strains of the H2N2 subtype. In 1968 this A subtype in turn gave way to an A subtype, H3N2, one that is still in circulation today. In 1977 strains of the A subtype H1N1 reappeared and thus began a period of time extending to the present in which strains of two subtypes of influenza virus type A have been prevalent and of epidemiologic importance. Some of the viruses of the subtype H1N1 that have been isolated and studied show genes derived from H3N2 strains, a finding that indicates that a reassortment of genes can occur in nature.

Immunity. Part of the population apparently possesses natural immunity to influenza because during an epidemic some persons do escape infection. Natural immunity acquired from an attack of the disease protects against a repeat attack by the same or very closely related strain, but it cannot protect against the new strains that are appearing. Little success has been obtained in producing passive immunity by means of convalescent or immune serum.

Laboratory diagnosis (see Table 33-4). The diagnosis of influenza can be confirmed only by isolation of the virus or by serologic studies. The virus can be grown in cell cultures or in the embryonated egg. The serologic tests include hemagglutination inhibition, neuraminidase inhibition, hemadsorption inhibition, neutralization, and complement fixation.

Prevention. When an epidemic of influenza strikes, all the methods known to preventive medicine fail to check it. Wholesale isolation seems to be of little value and almost impossible to achieve. Medical attendants to patients with influenza should disinfect the mouth and nasal secretions of the patient and should avoid exposing themselves to droplet infection.

Swine Influenza

Swine influenza is a severe respiratory infection of hogs, widely seeded in nature, caused by the combined action of a myxovirus and a bacterium. The bacterium, *Haemophilus suis*, is closely related to *Haemophilus influenzae*, and the swine flu virus (H1N1) is closely related to human influenza virus A. The human virus can produce the complete picture of influenza, whereas the virus of swine influenza alone produces only a mild form. A/swine (H1N1) can infect human beings.

Swine become infected with the strains of type A influenza virus prevalent in the human population at a given time. The Asian strains of the 1957 epidemic were recovered from hogs in Japan, and the evidence suggests that during the 1918 pandemic, the flu virus spread from humans to pigs. (Swine influenza was first recognized in 1918.) The swine flu virus commonly produces disease in swine every year. It can become established and remain latent in pigs (a potential reservoir) and may be present in half the swine herds in the United States.

Acute Respiratory Syndromes

Acute viral infection of the respiratory tract is one of the most common causes of human illness, responsible for approximately one third of all days lost from work and two thirds of days missed from school. Acute respiratory infection of viral type is not a single disease but a spectrum of entities such as rhinitis, pharyngitis, tonsillitis, laryngitis, and bronchitis (Table 34-2). These inflammatory processes can occur singly, but combinations are frequent. Remember that a syndrome is a set of clinical manifestations found together in a typical pattern; the cause, however, is variable.

Clinical Features

Viral infection of the lining of the nose results in a reddened mucous membrane and a hypersecretion of mucus—hence the well-known runny nose. A sensation of stuffiness comes from the nasal congestion and blockage. The throat is reddened and sore; tonsils and other lymphoid masses are enlarged. Hoarseness may result from inflammatory swelling of the larynx. Other clinical features associated in whole or in part with acute respiratory disease include low-grade fever, cough, pain and discomfort in the chest, headache, sensations of chilliness, and sometimes enlargement of cervical lymph nodes. Droplet infection accounts for spread; respiratory diseases are prevalent in cool months, and the epidemic pattern is a familiar one.

The disease influenza was originally considered to be part of the general complex of respiratory disorders. With a typical pattern for recurrent epidemics, influenza could be easily separated when the causative virus was discovered. Primary atypical pneumonia during World War II emerged in part from the respiratory assortment because of the characteristic findings in the lungs of the soldiers. A mycoplasma, *Mycoplasma pneumoniae* (not a virus), is responsible for a significant number of these cases (p. 641). In Great Britain certain viruses have been recovered from adults with the common cold (acute coryza) and the infection transferred to the chimpanzee, the only animal susceptible to the particular agents. The experimental studies indicate that the common cold has unique features.

Relation of Viruses

By the early 1950s cell (tissue) culture technology was developed to a high degree of efficiency. Viruses were recognized and recovered in such rapid-fire succession as to incur a kind of viral explosion. So many were identified that proper assimilation and definition of medical importance are as yet incomplete. Regarding respiratory infections, it soon became clear that an array of viral agents was associated with a variety of clinical infections with no fixed or specific relationship between the ailment and the

Table 34-2 Respiratory Illnesses Related to Known Respiratory Viruses

Illness	Virus
Influenza	Influenza viruses A, B, C
Common cold of adults	Rhinovirus (Salisbury)—110 types
	Coronavirus—3 types
Acute respiratory syndromes*	Influenza viruses A, B, C
Common cold syndrome in children, some adults;	Parainfluenza viruses 1, 2, 3, 4
rhinitis; pharyngitis and nasopharyngitis; tonsillitis;	Respiratory syncytial virus—1 type
laryngitis; tracheitis; bronchitis; bronchiolitis; bron-	Adenovirus—10 types
chopneumonia; pneumonitis; croup (acute laryngo-	Echovirus—6 types
tracheobronchitis)	Coxsackievirus A—23 types
	Coxsackievirus B—6 types
	Poliovirus, types 1, 2, 3
	Reovirus, types 1, 2, 3
Viral pneumonia (primary atypical pneumonia)	Influenza viruses A, B, C
	Parainfluenza viruses 1, 2, 3, 4
	Adenovirus—2 types
	Respiratory syncytial virus
	Measles virus
	Varicella virus
	Vaccinia virus
	Rubella virus
Pharyngitis (part of poliomyelitis)	Poliovirus, types 1, 2, 3
Pharyngitis (part of infectious mononucleosis)	Epstein-Barr virus

*No specific constant relation between given disorder and viral agent. See text.

virus. Viruses were recovered singly or together from the respiratory tract, both in health and in disease. Many observers prefer the term **syndrome,** the sum total of the clinical features of an illness, to emphasize the lack of specific correlation between any precise agent and clinical manifestations. In the light of present knowledge of viruses associated with respiratory disease, the best we can do is to present certain viruses as being found in respiratory secretions and tissues and to point out the variety of associated syndromes. Table 34-2 lists the viral agents recovered from acute respiratory syndromes.

Respiratory Viruses

From the known acute respiratory infections, viral agents have been recovered in most instances. Viruses that cause respiratory disease are spoken of collectively as respiratory viruses, and as an entity they encompass several major groups of viruses, including myxoviruses (influenza viruses), paramyxoviruses (parainfluenza viruses and the respiratory syncytial virus), picornaviruses (enteroviruses and rhinoviruses), adenoviruses, and reoviruses.

Parainfluenza viruses were first recognized in 1957. The four antigenic types are more stable than the influenza viruses; they do not mutate as often. Parainfluenza viruses are present in the community most of the year. The typically mild, first contact infection usually comes early in life. These viruses enter by way of the respiratory tract, and except in infants and young children, inflammation is usually confined to the upper part. They cause a number of illnesses, primarily in infants and young children, ranging in severity from mild upper respiratory tract infection to croup and pneumonia. (Croup, described by difficulty in breathing, a resonant barking cough, and hoarseness, results when inflammatory swelling acutely blocks the upper airway in the small child.) The clinical features of parainfluenza infections are not distinct; the majority are clinically inapparent.

Respiratory syncytial (RS) virus (an RNA paramyxovirus) infects a child before the age of 4 years. It is the most common cause of serious acute respiratory disease (including bronchiolitis and pneumonitis) in infants and very young children.

Spread of the viral infection is probably by direct contact with respiratory secretions, and prevalence peaks in late autumn and early winter. The incubation period is 3 to 7 days, and virus may be shed for as long as 27 days. Respiratory syncytial viral infection is more prevalent in babies who are not breast-fed; mother's milk seems to contain antibodies neutralizing the viral activity. Although the virus is not very antigenic—only one serotype exists—the appearance of virus-specific nasal IgA correlates with decreased nasal virus shedding and clinical recovery. Cellular immune mechanisms may both enhance disease and mediate recovery. Respiratory syncytial viral infections do not confer a solid kind of immunity. Therefore the infections, although prominent in young ages, may be seen throughout life, and aged persons and persons at high risk are reported to have very severe ones. This virus is a serious cause of nosocomial respiratory tract disease.

The cytopathic effect of respiratory syncytial virus in cell culture is truly remarkable. A wide extent of the monolayer fuses into a single cell with all the nuclei gathered together in one area of its cytoplasm. The syncytium so formed is not duplicated by any other virus.

A live, attenuated respiratory syncytial virus vaccine is under development, but there is no effective vaccine as yet. The use of the inactivated (formalin-killed) respiratory syncytial virus vaccine of the late 1960s has provoked in the recipients a paradoxical response to respiratory syncytial virus, a response comparable to that seen in atypical measles (p. 705) and also in some chlamydial infections. The disease resulting from reexposure to the virus in persons so vaccinated is worse than the native infection, partly because of the increased likelihood of serious lung involvement.

Among the **picornoviruses,** certain **enteroviruses** are respiratory pathogens. All the coxsackieviruses and some of the echoviruses (p. 727) are believed to cause respiratory disease. Coxsackieviruses have been isolated from nasal and pharyngeal secretions, and coxsackievirus A, type 20, has been studied in its relation to an acute upper respiratory tract coldlike illness. The respiratory agents among the echoviruses probably infect through the respiratory tract. They are known to multiply in the pharynx, producing a pharyngitis, and most likely are transmitted in respiratory secretions. Upper respiratory tract infections of echovirus are generally mild; the syndrome is many times that of the common cold. Echovirus types 11 and 20 have been specifically observed in such infections.

Adenoviruses,[*] also known as adenoidal-pharyngeal-conjunctival (APC) viruses, were first discovered in adenoids removed surgically from persons without any detectable clinical disease. To date, 33 types have been isolated from human beings. Adenoviruses are distinct for their size and stability in varied media such as sewage or swimming pool water, and since they are naked viruses, they are resistant to virucidal agents that are effective against viruses with envelopes. In fact adenoviruses may persist in certain disinfectant solutions. Although these viruses may be recovered from persons without disease, some do cause acute infection of respiratory mucous membranes and conjunctival membranes. Adenoviruses are related to a striking variety of clinical conditions such as acute respiratory disease, febrile pharyngitis or pharyngoconjunctival fever (especially in children), acute follicular conjunctivitis, epidemic keratoconjunctivitis (shipyard eye), tracheobronchitis, bronchiolitis, and pneumonitis. Types 3, 4, and 7 are related to epidemics of respiratory disease. Type 8 virus is the major cause of epidemic keratoconjunctivitis. Type 3, sometimes found in swimming pools, is the cause of pharyngoconjunctival fever. In the immunosuppressed patient adenoviruses can give rise to a severe pneumonitis.

In civilian groups only a small percentage of acute viral respiratory diseases is caused by adenoviruses, but this group, particularly types 3, 4, 7, 14, and 21, is of special importance to the armed forces. It incapacitates many of the recruits in the fall and winter about 8 weeks after induction. Immunization of military inductees therefore is highly desirable.

Humans are the only known reservoir of adenoviruses, and the virus is transmitted in respiratory and ocular secretions. The fact that the conjunctival inflammation often precedes the respiratory infection indicates the eye to be a significant portal of entry. Epidemics of pharyngoconjunctival fever and conjunctivitis have been traced to dissemination of the virus in swimming pools. Persons swimming in pools in the latter part of the day are more at risk for adenovirus infections. When the water is chlorinated once in the morning, the chlorine content of the water during the day may drop to a level at which adenoviruses are not affected. The respiratory and ocular involvement, often combined in adenovirus infection, is associated with enlargement (hyperplasia) of the lymphoid tissue in the respiratory passages and with lymphadenopathy in the neck.

Adenoviruses (see Fig. 39-6) cannot be studied by their effects in animals, since they do not produce disease in the routinely used laboratory animals, nor do they grow on the membranes of the fertile egg. However, they can be demonstrated nicely in cell cultures in which certain human and monkey cells are used. In virus-infected cells examined with the electron microscope, viral particles can be seen. This group shares a common antigen demonstrable in the serum of an infected person by means of a complement-fixation test. The breakdown of the group into the different types is done by means of the virus neutralization test (see Table 33-4).

Reoviruses are very stable viruses of undetermined pathogenicity. They are ubiquitous in nature and were named reoviruses from the designation "*respiratory enteric orphans*" to indicate that they were recovered from the respiratory and alimentary tract

---

*For immunization, see p. 430.

in humans in health and disease and were "orphans," that is, were not related to any other particular virus. Respiratory disease from all three types is usually low grade. Diagnosis of reovirus infections is by isolation of virus in cell culture and demonstration of hemagglutination inhibition.

Note that in the early stages of certain viral diseases—for example, poliomyelitis and infectious mononucleosis—involvement of the upper respiratory tract figures prominently.

Influence of Age

The patterns of respiratory disease in adults and children are basically similar in many respects. Influenza, for example, is much the same at any age. Yet there are points of contrast. These are presented in Table 34-3.

Common Cold (Acute Coryza)

The common cold, a major health problem, is said to be the most commonly occurring ailment of humans and to disable temporarily more people than any other infectious disease. The typical nasal discharge of only a few days' duration is spread by direct contact, and a cold is most communicable in the early stages. The incubation period is 2 to 3 days. The agent enters the body by way of the upper respiratory tract. In sneezing and coughing the affected persons spread not only their disease but also the bacteria they may be carrying in their throat. As with other viral diseases, the cold is often complicated by more serious bacterial infections such as pneumonia.

Some persons appear comparatively resistant to colds, whereas others are relatively susceptible. Factors said to predispose one to colds include exposure to chilling and dampness, sudden changes in temperature, dusty atmospheres, drafts, loss of sleep, overwork, and lowered body resistance. Most colds occur between October and May, and preschool children have the greatest number. Immunity after an attack is brief. A person may average two to four colds a year. The development of a vaccine is impractical at this time because of the number and diversity of causative agents.

Cold Viruses

In patients with the syndrome of the common cold, adenoviruses, certain enteroviruses, respiratory syncytial virus, influenza viruses, parainfluenza viruses, reoviruses, and even the Epstein-Barr virus have at times been incriminated specifically. A true cold may be caused by one or more than one virus. The designation "cold viruses," however, usually applies to rhinoviruses and coronaviruses.

Echovirus 28, reclassified as rhinovirus type 1 (p. 736), was the first to be implicated

---

Table 34-3 Comparison of Viral Infection of the Respiratory Tract in Children and Adults

	Children	Adults
Type of infection	Primary contact often	Inactivation of latent virus often; rarely, primary contact
Degree of involvement	Severe; infection may be fatal; mild cases also seen	Mild (except in aged and debilitated)
Clinical picture	Spectrum of respiratory disease	Common cold and related syndromes prominent
Causative viruses	Parainfluenza viruses, adenoviruses, and respiratory syncytial virus (viruses of childhood)	Cold viruses (coronaviruses and rhinoviruses) important; parainfluenza viruses, adenoviruses, and enteroviruses implicated

as a "cold virus." Later, pathogenic strains referred to as **rhinoviruses** or **Salisbury viruses** (110 rhinoviruses recognized, 80 characterized) were found in persons with colds and were grown in cell cultures containing cells from lungs of human embryos. Rhinoviruses do well in the human nose, where the temperature of 33° to 35° C is optimum for their replication. They rarely infect the deeper, warmer areas of the respiratory tract, and being acid labile, they cannot penetrate the alimentary tract. Rhinoviruses, first isolated in 1956, are responsible for more colds than any other known agent. According to some observers, they account for 70% of colds in adults and 35% of those in children. In the environment they persist, especially on the hands, which must be considered a major factor in the spread of the common cold.

Another group of cold viruses, the **coronaviruses,** so named because of a segmented ring around the virion, is made up of at least 10 serotypes. The cold caused by coronaviruses has a longer incubation period and a shorter sequence of illness than is seen with rhinoviruses.

Viral Pneumonia*

Viral pneumonia is a clinical syndrome—*not* a single, specific disease—acute, infectious, and self-limited. It is similar in its manifestations to primary atypical pneumonia, also a syndrome. Viruses implicated include the influenza viruses, the parainfluenza viruses, the adenoviruses, and in infants the respiratory syncytial virus.

CENTRAL NERVOUS SYSTEM DISEASES

Rabies (Hydrophobia)†

Rabies is an acute, paralytic, ordinarily fatal, infectious disease of warm-blooded animals, including humans, spread by the introduction of virus-laden saliva into a bite wound, sometimes by ingestion of infected material, and rarely by other means. In late 1978, secondary, human-to-human transmission of rabies was documented. Virus was inadvertently transferred in the cornea taken from an Oregon forester (who carried an eye donor card) who had died of an atypical case of rabies to the recipient, an Idaho housewife, who died of rabies 7 weeks later.

Rabies in Animals

Rabies has been recognized for 2000 years and has changed little in that time. It is primarily a disease of the lower animals, and dogs have been chiefly responsible for its propagation in civilized communities. Other domestic animals that contract rabies are cats, horses, cows, sheep, goats, and hogs. When the disease occurs in wildlife, it is referred to as sylvatic rabies, or wildlife rabies. Wild animals that often contract it are wolves, foxes, skunks, coyotes, raccoons, and hyenas. Raccoons can be a problem, since they come into the cities. Bats represent a primary source of rabies, since they are perpetual carriers of rabies virus. The disease may weaken a bat to the point that it falls to the ground, no longer able to fly. The bat may then be eaten by a skunk.

The rabid skunk is most dangerous to humans or other animals, since the saliva of the skunk carries 100 to 1000 times the amount of virus carried by the dog. Skunks are more susceptible than are dogs and spread the disease among themselves. The skunk is said to be the most frequent rabies carrier in the United States today. Skunks account for 51% of reported cases of rabies in the United States; dogs account for 4%.

In certain parts of the world (South America, West Indies, Central America, and Mexico) the vampire bat transmits the disease to horses and other livestock. This is a

---

*For immunization, see pp. 432 and 441-443.
†See the discussion on primary atypical pneumonia, p. 641.

serious veterinary problem because the vampire bat differs from other animals in that it does not succumb to the disease but becomes a symptomless carrier, remaining infective a long time. Bats are dangerous carriers of rabies because of their habits and close contact with human dwellings. They are more likely to be found in barns and in deteriorating houses than out in the wilds. They migrate constantly and thus evade control. The disease has been recognized in certain insect-eating bats in parts of the United States (Florida, Texas, Pennsylvania, California, and Montana), in the Caribbean area, and in Europe. Rabies in dogs is declining in incidence, but the disease in wild animals is definitely on the increase. It is rare in rodents (rats, mice, squirrels) and almost unheard of in laboratory animals (hamsters, gerbils, guinea pigs). Since such animals are small, they are not likely to survive the attack from a rabid animal. Rabies virus does not survive long on inanimate objects contaminated by secretions from a rabid animal. The virus is killed by sunlight and by heat. Therefore it is unlikely that rabies could be contracted from an inanimate object. Vaccines against rabies that are available are ineffective in wild animals.

Forms of Rabies

Rabies in animals occurs in two forms, the **furious** and the **dumb.** In the former a stage of increasing excitability is followed by a stage of paralysis ending in death. Absence of fear or of the instinct for self-preservation in a wild animal is abnormal. The rabid animal, however, fearlessly attacks anything that it encounters; it drools saliva because its throat is paralyzed. Such an animal has an uncontrollable urge to bite and chew. The skunk, ordinarily a shy and nocturnal animal, when rabid becomes a hard-biting, aggressive animal that attacks in broad daylight.

In dumb rabies, paralysis and death supervene without a preceding stage of excitement. The animal neither bites nor attacks. Except that its lower jaw droops, it may show little sign of illness. It often hides and may be found dead. In either form of rabies there is no fear of water, and the animals do attempt to drink. The human patient, however, avoids fluids because of painful spasms in the throat muscles induced with

On May 10, 1980, a rabid dog in Yuba County, California, bit three persons in a parking lot in the Olivehurst area. In the investigation by the Sutter-Yuba County Health Department, 70 persons were identified, who, because they had been exposed to the dog, were given antirabies prophylaxis. Since only 20% of the dogs and cats of the area were found to have up-to-date vaccinations, special clinics were held to vaccinate 2000 dogs. Over 300 stray dogs and cats were destroyed. Fortunately no cases of rabies developed in humans or animals.

Human rabies has become a rarity in the United States with only a few cases reported each year, but the cost of vigilance is a sizable one. Consider the breakdown of known costs generated by one rabid dog in Yuba County:

For human antirabies treatment	$92,650
For animal vaccination and veterinary services	$4,190
For health department and animal control programs	$8,950
TOTAL	$105,790

This figure represents over $1500 per person treated, and the total cost of the episode does not include lost work time, patient travel time, and costs of the 6-month quarantine imposed on animals exposed to the rabid dog.

Modified from The cost of one rabid dog—California, Morbid. Mortal. Weekly Rep. 30:527, 1981.

the act of swallowing. This explains the name **hydrophobia** for the disease. It has been said that the agony of the spasms of the victim of rabies possibly exceeds all other forms of human suffering.

The Virus

The bullet-shaped rabies virus (rhabdovirus) (Fig. 34-13) is found in the nervous system and saliva of infected animals. It is transmitted in the saliva introduced into the body through a wound that is usually the bite of a rabid animal. If a wound such as a cut or abrasion becomes accidentally contaminated with the saliva of a rabid animal, infection is as likely to occur as if the animal had inflicted the wound. Rabid animals can transmit the infection to others for several days before they show signs of disease. When a person or animal is infected, the infectious agent passes from the inoculation site along nerve trunks to the central nervous system. When the virus reaches the brain, the manifestations of rabies appear. About 30% to 40% of the people and 50% of the dogs bitten by rabid animals develop the disease. *But once rabies develops, it is almost invariably fatal.* In 1970, in Lima, Ohio, the first person with rabies to survive the ordeal was given extensive coverage by the news media.

Rabies virus in the brains of animals naturally infected is known as **street virus.** With serial passage through a certain number of rabbit brains, street virus becomes highly virulent for the rabbit but loses virulence for other species of animals. It is then **fixed virus,** which, treated with phenol, makes up the Semple vaccine. The Semple method of antirabies vaccination is no longer valid.

Incubation Period

The period of incubation of rabies is remarkable for its length. In humans it can vary from 10 days to 1 year, with an average of 2 to 6 weeks. In dogs it varies from 8 days to 1 year, with an average of 2 to 8 weeks. The nearer the site of inoculation to the brain, the shorter the period of incubation. It is also shorter in children than in adults.

Laboratory Diagnosis (see Table 33-4)

Adelchi Negri (1876-1912) discovered certain inclusion bodies in the brain cells of animals with rabies. The **Negri bodies** store ribonucleoprotein and virus antigen and are found in almost 100% of the patients with fully developed disease. By finding them,

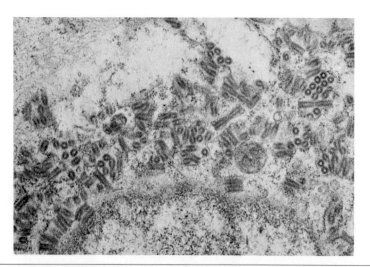

FIG. 34-13   Bullet-shaped rabies virus particles from hamster kidney cells shown in clusters, electron micrograph. (× 32,000.) (Courtesy Dr. Klaus Hummeler, Philadelphia.)

one makes the laboratory diagnosis of rabies quickly, but the fluorescent antibody examination of brain tissue is highly reliable and the preferred diagnostic method. When brain tissue is to be examined, the whole, intact brain should be submitted to the diagnostic laboratory. Fluorescent antibody examination may also be made on smears or imprints of the epithelial cells of the cornea of the eye.

Infant mice, susceptible to rabies virus, may be infected by intracerebral inoculation of suspect material; the virus is then identified by immunologic methods or by demonstration of Negri bodies in brain tissue.

Prevention

If a person is bitten by an animal suspected of having rabies, the animal should not be killed but should be placed in the hands of a competent veterinarian for observation. If the animal has rabies, the disease will be sufficiently developed within a few days for a definite diagnosis to be made. If the dog is well 10 days after biting a person, the person is in no danger from the bite. If, on the other hand, the animal is destroyed at the time of the bite, examination of the brain may fail to show Negri bodies because they are often absent in the early stages of the disease.

If examination of the brain of the animal fails to show Negri bodies, six or eight mice should be inoculated intracerebrally with an emulsion of the brain. If the animal had rabies, the mice may show Negri bodies in their brain tissue as early as 6 days and will exhibit signs of the disease on the seventh or eighth day. If the mice survive the twenty-first day, the animal may be considered not to have had rabies.

Local treatment of wounds inflicted by rabid animals is very important in prevention of the disease. Adequate first aid treatment is crucial. Immediate and thorough cleansing of superficial wounds with a tincture of green soap or benzalkonium chloride solution may inactivate the virus. (Even tap water has merit.) Remember that if soap is used before a quaternary ammonium compound such as benzalkonium chloride (Zephiran), thorough rinsing with water or saline to remove all traces of soap is imperative because soap inactivates the "quat." An antiseptic may be applied and the wound dressed. If a wound is deeply placed and washing by soap and water is not feasible, it is then considered advisable to cauterize the wound with fuming (concentrated) nitric acid. Antiserum may be injected into the base of the wound. Dogs, cats, and other pets bitten by rabid animals should be destroyed at once or, otherwise, vaccinated and kept in strict isolation for 6 months.

Persons who work with this disease are often confronted with the question of what course to take when a person drinks the milk or eats the flesh of a rabid animal. Eating or handling infected flesh or drinking contaminated milk can produce rabies if there is an open lesion on the skin or in the alimentary tract.

The nursing precautions in rabies are rather simple. All that is necessary is to sterilize the secretions from the mouth and nose of the patient and articles so contaminated. Rabies can be eradicated from civilized communities by measures designed for intensive control of stray dogs, immunization of resident dogs, and elimination of any reservoir of infection in the wild animals of the area.

Viral Encephalitis and Encephalomyelitis

The term *encephalitis* (*pl.* encephalitides) means inflammation of the brain, but by common usage it is applied to inflammatory conditions of the brain accompanied by degenerative changes instead of suppuration (pus formation). When the brain is involved, the spinal cord usually is also, hence the term *encephalomyelitis*.

Togaviral (Arboviral) Encephalitides

Animal viruses carried in the body of an insect vector are **arboviruses,** a simplification of "*ar*thropod-*bo*rne viruses." Arboviruses are the most numerous of the viruses infecting humans. Important forms of encephalitis, the togaviral (arboviral) encephalitides, are caused by neurotropic arboviruses, now classified as **togaviruses.** A noteworthy form of encephalitis, **postinfection encephalomyelitis,** neither caused by a neurotropic virus nor transmitted by an insect, resembles viral encephalitis clinically (not pathologically), and viral infection is usually implicated in its development.

Togaviral infections occur principally in mammals and birds, and for most, a human is only an accidental host. They are the largest group of zoonoses. The mosquito is an important vector, although togavirus is found in ticks and mites. Although the togaviral encephalitides are similar in many respects, the causative agents are immunologically distinct, and the geographic distribution of the diseases is clear-cut.

Equine encephalomyelitis, a disease of horses and mules, secondarily of humans, occurs in three types, each with its own virus: the eastern type (EEE) in the southern and eastern United States, the western type (WEE) in the western United States and Canada, and the Venezuelan type (VEE) in South America and Panama. **St. Louis encephalitis** (SLE), so named because it was first recognized in an epidemic in the vicinity of St. Louis in 1933, is a disease widespread in the United States. It occurs only in humans. **Japanese B encephalitis** is found in the Far East, and **Murray Valley encephalitis** is found in Australia.

The eastern type of equine encephalitis is a severe form with a mortality up to 70%. The death rate is much lower in the western form and in St. Louis encephalitis. Young persons, particularly infants, have an increased susceptibility to western equine encephalomyelitis, and St. Louis encephalitis has its greatest incidence in persons of middle age or older. The incubation period for the different encephalitides ranges from 5 to 15 days.

The epidemiologic pattern for the different forms of togaviral encephalitis is reasonably consistent, with but slight variations. Togavirus resides and multiplies in wild birds, its natural and primary hosts, and occasionally in domestic fowl. Venezuelan equine encephalomyelitis multiplies best in a reservoir of mammals; the eastern equine encephalomyelitis virus multiplies in both mammals and birds. From its natural reservoir, togavirus is transmitted from fowl to fowl (or mammal to mammal) and from bird to horse and human (terminal hosts) mainly by mosquitoes of the genus *Culex*. There is no known instance of direct person-to-person transmission of encephalitis.

Encephalitis is a disease of warm weather and the summertime. In nontropical areas snakes and other cold-blooded animals harbor the virus for the cold months of the year. When snakes come out of hibernation, they seek out areas that also shelter large numbers of wild birds. Virus circulates in the bloodstream of snakes, as it does in birds, so that in the early spring mosquitoes can pick up the virus from such an overwintering host and carry it straight to the wild bird reservoir. Once infected, a mosquito remains so for life.

For mosquito-borne encephalitis to reach epidemic proportions, a combination of factors is required, including a large wild bird population, favorable breeding sites for mosquitoes in stagnant pools and puddles, a high temperature, and susceptible terminal hosts.

The laboratory diagnosis of encephalitis (see Table 33-4) is made on recovery of the virus from suitable specimens and its proper identification. Young or newborn mice may be inoculated and observed.

Effective vaccines made by growing the equine viruses in a chick embryo confer immunity in horses lasting from 6 months to 1 year or longer. A vaccine prepared against eastern equine encephalitis is suitable for use in human beings, but the vaccine for the western form of the disease is given only in special circumstances to persons at high risk because of contact with the exotic and virulent form of the disease. The practical approach to the control of the disease is eradication of the mosquito vector.

Postinfection (Demyelinating) Encephalomyelitis

Postinfection encephalomyelitis (allergic encephalomyelitis) is an acute disorder of the central nervous system that occasionally arises during convalescence from infectious diseases, most often viral. It may occur after vaccination for such diseases. Notable among these are measles, rubella, smallpox, and influenza. Vaccination for smallpox and rabies can be so complicated.

The following theories are offered to explain postinfection encephalomyelitis: (1) it is caused by the virus of the primary disease; (2) vaccination activates some latent virus; and (3) the disease reflects an allergic reaction either to the virus of the preceding infection or to the patient's nervous tissue now altered by it. Currently this last, the autoimmune, approach is favored.

Slow Virus Infections

In some chronic diseases of animals and humans experimental evidence indicates that the end-stage changes in the tissues, particularly in the central nervous system, are the result of a slowly progressive and damaging proliferation of a virus or of a subviral unit such as a viroid (p. 682) or some unknown agent. The term *slow viruses* is applied to these agents because of their very long developmental cycles during which they are masked or inapparent. Some in this category are not viruses at all, but to date these have not been isolated.

In such diseases there is a long incubation period (years), slow start, protracted course, fatal outcome, no demonstrable formation of antibodies, no fever, and no inflammatory changes along the way. An example of a slow virus infection induced by a virus is **subacute sclerosing panencephalitis,** a chronic neurologic disorder for which the measles virus has been confirmed as causative agent. Recovery of the measles virus from affected brain tissue indicates its capacity for latency in infected humans and its ability to cause serious disease many years after the initial infection.

An example of a slow virus infection induced by a viruslike agent, perhaps a viroid, is **scrapie,** a slowly developing, brain-destroying, fatal disease of sheep. The human counterpart is **kuru,** a neurologic disease formerly of cannibals in New Guinea. It is transmitted in the undercooked brain (containing the agent) of the dead patient eaten at the funeral by the family as a mark of respect for the dead relative. (The word *kuru* in the Fore language of New Guinea means "trembling with fear.") When cannibalism disappeared in New Guinea, so did kuru.

Slow viruses may also be implicated in arthritic and rheumatic diseases and in the autoimmune diseases. The current interest in them stems from the suggestions they give as to the cause of many poorly understood central nervous system disorders in humans.

Poliomyelitis (Infantile Paralysis)*

Poliovirus Infection

Poliomyelitis is an acute infectious disease that in its severest form affects the brain, spinal cord, and certain nerves (Fig. 34-14).

Poliomyelitis occurs in four forms:

1. Silent or asymptomatic infection—any symptoms present are so mild as to be over-looked. Such a person is a healthy carrier. Silent infection exists among the members of a patient's family. Poliovirus infection of the human alimentary tract is exceedingly common all over the world. From the alimentary tract the virus at times enters the blood or lymph stream.
2. Abortive infection—findings referable to the nervous system are absent, although there may be a brief febrile illness, such as a mild respiratory tract infection or a simple gastrointestinal upset. Most cases are abortive.

---

*For immunization, see pp. 424, 432, 444, 445, and 447.

FIG. 34-14   Poliomyelitis, cross section of spinal cord from patient with paralytic poliomyelitis. (Paralysis results when motor neurons of anterior horns of spinal cord are destroyed by poliovirus.) Arrow points to anterior horn on left, a softened, depressed area of dead tissue, focally hemorrhagic. Anterior horn on right is similarly involved. (Courtesy Dr. B.D. Fallis, Dallas.)

3. Nonparalytic infection—findings point to the nervous system but without residual paralysis.
4. Paralytic infection—paralysis persists.

Polioviruses

Three serologic types of poliovirus are designated type 1 (Brunhilde), type 2 (Lansing), and type 3 (Leon). These viruses, among the smallest in size, may be grown in monkeys and chimpanzees. Cell cultures of monkey kidney sustain a generous growth, as do certain human cell cultures. Polioviruses are pathogenic for humans and other primates, but as far as we know the human is the only one having the disease; no reservoir of infection exists.

Polioviruses can be inactivated by ultraviolet radiation, by drying, and, if in a *watery* suspension, by a temperature of 50° to 55° C for 30 minutes. Inactivation is slow with disinfectant alcohol and unsatisfactory with many bacterial germicides in wide use.

Transmission

The incubation period ranges from 3 to 35 days. The infected person is most infectious during the latter part of the incubation period and the first week of the clinical illness, the time at which the virus is present in the throat. Before the onset of symptoms in infected persons, poliovirus is found in the secretions from the mouth and throat and in the feces. Poliovirus enters the body at the upper part of the alimentary tract and exits through either the upper or the lower end.

Poliovirus has been recovered in large amounts from sewage, and milkborne epidemics have been recorded. But the epidemiologic pattern for poliomyelitis is not that for the enteric infections. The favored mode of spread seems to be the direct one from person to person. Houseflies, filth flies, and cockroaches can be contaminated with virus, especially during an epidemic, but the role of these agents in transmission is undefined.

Epidemiology

Poliomyelitis occurs sporadically but tends to be epidemic. Infantile paralysis is not a good name for the disease because it occurs in adults and paralysis may be absent. In the early epidemics the disease chiefly attacked children less than 5 years of age, but then, in time, the children affected were older and the number of adult cases increased.

The disease is worldwide. All races and classes of people are affected. In temperate climates poliomyelitis appears in early summer. The number of patients and the severity of the disease increase, a peak is reached in late summer and early fall, and the disease subsides after the first frost. Occurrence in more than one member of a family is frequent. Pregnant women are more susceptible to the disease, and there have been cases of congenital poliomyelitis in which the mother had had the disease late in pregnancy.

Radical changes took place in the epidemiology of poliomyelitis with the use of the polio vaccines. In 1957, for the first time, all states and territories of the United States were free of epidemics. For the year 1957 just less than 6000 cases of poliomyelitis were reported to the U.S. Public Health Service, of which about 2500 were paralytic. A decade later in 1967 there were 44 cases (29 were paralytic), mostly in unimmunized or inadequately immunized children. Compare these figures with over 57,000 cases (21,000 paralytic) reported for the prevaccine year of 1952!

Immunity

Poliomyelitis confers lasting immunity *only* for the viral type responsible. A high percentage of adults have virus-neutralizing substances in their blood. Infants can inherit an immunity from their mothers by placental transfer.

Prevention

Preferably, the patient with poliomyelitis is isolated, although with widespread distribution of the virus, real insulation is impossible. Cross-infections between patients probably are inconsequential in hospitals where no attempt to isolate is made. However, the disease is sometimes spread from patient to medical attendant.

Contacts between children should be minimized during epidemics, and in an epidemic public health measures such as closing public swimming pools are advisable. Tonsillectomies should not be done during epidemics or in the season of the year when the incidence of the disease is highest. There is a greater risk for the severe form of the disease in persons who have recently had their tonsils removed. Healthy children may carry the virus in their throats, and it has been demonstrated in surgically removed tonsils.

OTHER ENTEROVIRUS DISEASES
Infections with Coxsackieviruses

The coxsackieviruses (picornaviruses) are worldwide in distribution and have been frequently associated with epidemics of poliomyelitis. They were named after the town in New York where the first such virus was identified as research work was being done on poliomyelitis. They resemble the viruses of poliomyelitis in many respects, including their epidemiology, and most strains of coxsackieviruses (and echoviruses as well) can cause a disease closely mimicking either nonparalytic or paralytic poliomyelitis. They are found in the nasopharynx and feces and at times in other parts of the body but *not* in the cerebrospinal fluid. They possess an unusual pathogenicity for infant mice and hamsters but none for the adult animals. Most illness produced by coxsackieviruses in humans appears in children.

They are divided into groups A and B. Within each group are a number of types, of which at least 29 are known. Group B is the more important in humans. Cocksackieviruses produce conditions such as aseptic meningitis, epidemic pleurodynia (group B), herpangina (group A), vesicular pharyngitis, encephalitis, hepatitis, orchitis, an influenza-like disease, "3-day fever," pericarditis, and a severe form of myocarditis in infants. Group A viruses usually are associated with infections of the mouth. Neutralizing antibodies to the virus are found in the convalescent serum of patients recovering from the diseases listed.

FIG. 34-15   *Aedes aegypti* mosquito, larval forms in Petri dish. (From Med. World News 6:168, Nov. 12, 1965; courtesy Eli Lilly & Co., Indianapolis.)

Infections with Echoviruses

Echoviruses (picornaviruses), found by accident during epidemiologic studies of poliomyelitis, were recovered from human fecal material in cell cultures. At first their destructive effect on tissue culture cells and the fact that they were not related to a known disease led to the designation of **echo**—*e*nteric, *c*ytopathogenic, *h*uman *o*rphan viruses. Some 31 members of this group have been classified so far. Many of these viruses may be harmless parasites, but some do cause epidemics of aseptic meningitis, summer diarrhea in infants and young children, and febrile illnesses, with or without rash. To date, the echoviruses have been found as etiologic agents only in clinical syndromes (not specific diseases) that may be produced by a number of other viruses (and bacteria) as well. For example, their chief central nervous system manifestation, aseptic meningitis, a well-defined entity, is caused by a great variety of agents.

OTHER ARTHROPOD-BORNE VIRAL DISEASES
Yellow Fever

Yellow fever,* a mosquito-borne hemorrhagic fever once known as yellow jack, is an acute infectious disease, the most striking pathologic feature of which is the rapid and extensive destruction of the liver. The onset is abrupt, the course rapid, the mortality high, and the clinical disease defined by jaundice (yellow discoloration of skin and mucous membranes), pronounced hematemesis (vomiting of blood), and hemorrhage from other body orifices.

Mention of yellow fever brings eight names to mind: Carlos Juan Finley (1833-1915), who first accused the mosquito of spreading the disease; Walter Reed (1851-1902); James Carroll, Jesse W. Lazear, and Aristides Agramonte, who, in 1900, formed the commission of the U.S. Army to study yellow fever in Cuba; Privates John

*For immunization, see pp. 433, 447, and 448.

R. Kissinger and John J. Moran, who permitted themselves to be inoculated with yellow fever; and William C. Gorgas, who applied the knowledge obtained to make the tropics more habitable for human beings. The members of the commission went to Cuba, lived in the tents of those who had had yellow fever, wore their clothes, and were bitten by infected mosquitoes. Carroll and Lazear contracted the disease; Lazear died. To these men, all who live in tropical and temperate zones owe a debt of gratitude.

The flavivirus (group B arbovirus) of yellow fever is transmitted by the mosquito *Aedes aegypti* (Fig. 34-15), which also transmits dengue fever. The mosquito bites a person during the first few days of illness and becomes infected. The virus multiplies in the body of the mosquito and reaches the salivary glands at the end of about 12 days; the mosquito is thus rendered infective the rest of *her* life. (Only the female transmits the disease.) When a nonimmune person is bitten by an infected mosquito, manifestations of yellow fever develop in 3 to 5 days. The virus is found in the bloodstream (viremia) during the first 3 days of the illness.

Yellow fever still remains endemic in many parts of the world. We must always guard against it because an epidemic requires but three things: a person ill of the disease, mosquitoes to be infected, and susceptible recipients. The likely transport of infected mosquitoes by airplanes and ships must be prevented.

A source of infection in tropical Africa and Latin America is jungle yellow fever found in monkeys. It is transmitted from animal to animal and from animal to human by forest mosquitoes (genus *Haemagogus*) living high in treetops. Epidemics occur. The causative virus seems to differ in no respect from that producing ordinary yellow fever, but if humans become infected with this virus, it is transmitted from them by *Aedes aegypti*.

An active immunity to yellow fever can be established. The value of convalescent serum is undetermined.

Dengue Fever

Dengue, or breakbone, fever, an acute disease lasting about 10 days, is depicted by sudden onset of paroxysmal fever (Fig. 34-16), intense joint pain, skin rash, and mental depression. The cause is a flavivirus (group B arbovirus) that is found in the bloodstream during the early days of the attack and is transmitted by the *Aedes aegypti* mosquito, which, once infected, remains so for the rest of *her* life. When dengue fever is introduced into a community, a high percentage of the population contracts the disease. An attack is followed by immunity that may persist for 1 to 2 years or the remainder of the patient's life.

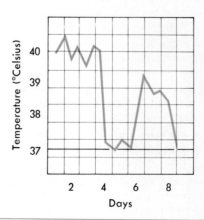

FIG. 34-16 "Saddle-back" fever curve seen in dengue.

There are four distinct immunologic types of the virus, of which types 1 and 2 are most often responsible for the disease in the western hemisphere. The dengue viruses can be grown in cell cultures. Type 1, live virus, weakened by repeated passages through mouse brain has been an effective vaccine in the clinical trials to date. Neutralizing and complement-fixing antibodies may be demonstrated in the serum of a patient after infection.

Colorado Tick Fever

An arboviral infection carried by an insect vector other than the mosquito is Colorado tick fever. The small arbovirus, classified as a reovirus, is transmitted by the wood tick, *Dermacentor andersoni*. A mild febrile, diphasic illness, it is found largely in the western states, the natural habitat of the tick vector. Patients usually have been in a tick-infested area 4 to 6 days before becoming ill, and ticks are even found still attached to their bodies. Because of the protracted meal taken by the vector tick, it is not easily overlooked. There is almost always a history of tick bite. The onset of disease is sudden, with chills, fever, and bodily aches that continue for 2 days or so. Then follows a period wherein the patient is essentially free of clinical manifestations. The fever returns to last for several days more before the disease has run its course. There are no complications. The disease has not been fatal, and the pathology is not known.

Laboratory Diagnosis (see Table 33-4)

Colorado tick fever virus grows in tissue culture and in the fertile hen's egg. When inoculated into young mice, it produces paralysis.

Immunity and Prevention

Infection probably produces lasting immunity. Prevention depends on the avoidance of tick-infested areas or the wearing of adequate protective clothing. The body should be regularly inspected for ticks, and any present should be detached. Effective vaccines have been prepared by propagation of the virus in chick embryos.

LIVER DISEASES (VIRAL HEPATITIS)

Hepatitis is an inflammation of the liver of variable severity producing a clinical illness characterized by fever, jaundice, and gastrointestinal manifestations such as nausea and vomiting. By definition viral hepatitis results from the inflammation that is caused by viral infection. Viruses that may cause a hepatitis in one way or another include myxovirus, adenovirus, herpesvirus, coxsackievirus, reovirus, cytomegalovirus, and the Epstein-Barr virus. Although hepatitis may be an important part of a systemic viral disease such as yellow fever or congenital rubella, the term *viral hepatitis* emphasizes those viral infections in which the injury to the liver determines the primary expression of disease. In systemic diseases in which hepatitis is a feature, the hepatitis, not being severe usually, is overshadowed by characteristic manifestations that are extrahepatic.

Three major diseases of the liver closely resemble each other in the clinical manifestations and in the pathologic findings that come from viral injury to the liver. One is viral hepatitis type A, caused by hepatitis A virus (HAV) (see p. 739 for acceptable terms). It has been known as infectious hepatitis, catarrhal jaundice, epidemic jaundice, and short-incubation hepatitis. Another, viral hepatitis type B, caused by hepatitis B virus (HBV), has been known as serum hepatitis, homologous serum jaundice, transfusion jaundice, tattoo hepatitis, posttransfusion jaundice, and long-incubation hepatitis. A third form of viral hepatitis exists, neither A nor B hepatitis. Although it is clinically indistinguishable from the other forms of viral hepatitis, though possibly somewhat milder in nature than the other two, it is antigenically separate and serologically cannot be linked to either hepatitis A or B. It is referred to as non-A, non-B (NANB) hepatitis. The three forms are acute or subacute infections of the liver found naturally only in human beings.

The true overall incidence of viral hepatitis is unknown, since subclinical infection occurs without jaundice. It has been estimated that 80% to 90% of cases go unrecognized. Immunity from an attack of one form of the disease results in lifelong immunity to that form only.

As a frame of reference, a traditional consideration of the three major forms is presented in Table 34-4, but with the impact of new discoveries related to viral hepatitis and its causation, concepts are changing.

Viral Hepatitis Type A

Viral hepatitis type A is usually spread directly from person to person, principally by the fecal-oral route. No evidence for chronic carriers exists; spread depends on person-to-person contact. Poor sanitation favors its spread, and the incidence of the disease increases with crowding of people. It is a hazard of the day-care center. Water, food, and milk can be sources of infection, and outbreaks have occurred in persons who have eaten raw shellfish contaminated by sewage. The cockroach and fly may be vectors. The disease occurs sporadically or in epidemics, which tend to recur cyclically every 7 years. Hepatitis type A is likely a commonplace infection, with clinical illness only occasionally seen. (It is not related to blood transfusion.)

The Virus

By technics of immune electron microscopy, hepatitis A virus and its antibody, anti-HAV, have been identified (p. 681). The A virus (*Enterovirus* type 72 in family Picornaviridae) is a single-stranded RNA virus possessing cubic symmetry but without an outer envelope. It is an exceedingly small enterovirus particle (27 nm, or one millionth of an inch in diameter) and serologically distinct. The virus is found in blood and feces

The World Health Organization Expert Committee on Viral Hepatitis (1976) has adopted the following terms with appropriate abbreviations for viral hepatitis:

For hepatitis A virus

HAV	Hepatitis A virus
HAAg	Hepatitis A antigen
Anti-HAV	Antibody to HAV (can be IgM or IgG)

For hepatitis B virus

HBV	Hepatitis B virus
HBsAg	Hepatitis B surface antigen (found on surface of virus and on the accompanying unattached, 22 nm sperical and tabular forms; particulate, viral coat protein; has major and minor antigenic subdeterminants; formerly the Australia antigen)
HBcAg	Hepatitis B core antigen (hepatitis B antigen found within the core of the virus; particulate)
HBeAg	The *e* antigen, which is closely associated with hepatitis B infectivity (soluble; present in viral core; several subtypes)
Anti-HBs	Antibody to HBsAg
Anti-HBe	Antibody to HBeAg
Anti-HBc	Antibody to HBcAg

Table 34-4 Comparison of Major Types of Viral Hepatitis

	A	B	NANB
Cause			
Virus	A (enterovirus)—27 nm, RNA	B—42 nm, DNA	Unknown (probably multiple viral agents)
Epidemiology			
Transmission	Fecal-oral; parenteral rarely; airborne (?)	From blood and blood products—parenteral; close personal contact with body fluids and secretions; by insect (?)	Parenteral; personal contact
Incubation	25 to 30 days (15 to 50 days)	30 to 180 days (60 days)	15 to 150 days (45 days)
Age preference	Young adults, 15 to 24 years; also children*	15 to 24 years (but all ages)†	Primarily adults, young men
Duration of infectivity	Short—virus in feces and blood 1 to 2 weeks before disease, remains 3 to 4 weeks longer	Long—virus in blood 3 months before disease; asymptomatic carrier for as long as 5 years	Long (?)
Carrier	No	Yes—5% to 10%	Yes—20% to 35% of patients become asymptomatic carriers
Virus present in	Feces, blood	Blood, body fluids, excretions	Blood
Distribution of cases	Point-source outbreaks, random cases	Prevalent in young adults and urban populations	Posttransfusion hepatitis and random cases
Duration of infectivity	Short	Long	Long
Clinical Features‡			
Onset	Acute	Slow, usually insidious	Insidious
Fever	Common before jaundice	Less common, low grade	Uncommon
Jaundice	Rare in children, more often in adults (50%)	Rare in children, more often in adults (30%)	Variable (20%)

*Hepatitis type A in children is commonly inapparent.
†Younger ages reflect drug abuse; transfusion-associated patients are 30 years and older.
‡Many clinical features are the same.

of infected persons, and virus recovery has been made from clinical disease. Hepatitis A virus has been shown to grow in fetal rhesus monkey kidney cells and in marmoset liver cells. It can be transmitted to marmosets and chimpanzees.

Laboratory Diagnosis
Serologic tests used to detect antibody against the hepatitis A virus demonstrate either seroconversion or at least a fourfold elevation in antibody titer between acute and con-

Table 34-4 Comparison of Major Types of Viral Hepatitis—cont'd

	A	B	NANB
Clinical Features—cont'd			
Severity of disease	Usually mild	Often severe (long period of morbidity)	Moderately severe
Chronic disease	Probably none	30% of cases of chronic hepatitis here	Yes—probably most cases of chronic hepatitis
Recovery	99%	85%	Variable
Mortality	0.1%	1% to 3%	1% to 2%
Prognosis	Good	Less favorable (variable)	20% cirrhosis; chronic active liver disease a good possibility
Laboratory Evaluation			
Elevated AST§	Transient, 1 to 3 weeks	Prolonged, 1 to 8 months or longer	Prolonged
Serologic markers	HAAg, anti-HAV	HBsAg, anti-HBs HBcAg, anti-HBc HBeAg, anti-HBe	?
Prevention and Control			
Prophylactic effect of gamma-globulin	Good	Present—recommended in specific instances	Effective in reducing incidence of disease and number of patients developing chronic liver disease
Nursing precautions and control	1. Isolation of patient 2. Sterilization of contaminated items 3. Meticulous hand washing; good personal hygiene 4. Wearing of protective clothing for patient contacts	1. No isolation needed 2. Sterilization of permanent medical apparatus in contact with blood 3. Use of disposables (needles, syringes, tubing) for all patients 4. Protective clothing for patient contacts 5. Meticulous hand washing; good personal hygiene	1. No isolation needed 2. Sterilization of permanent medical apparatus in contact with blood 3. Use of disposables 4. Meticulous hand washing; good personal hygiene

§The enzyme aspartate aminotransferase (AST), formerly known as glutamate oxaloacetate transaminase, is elevated with liver disease.

valescent sera (paired sera). Serologic tests used in identification include complement fixation, immune adherence, hemagglutination, and radioimmunoassay.

An IgM antibody against hepatitis A virus appears early in the serum of a patient with acute disease and remains detectable for 2 to 3 months. A direct radioimmunoassay test specific for this IgM-type antibody when positive is diagnostic of acute or very recent infection. An IgG against hepatitis A virus appears 2 to 4 weeks after onset of illness,

In day-care centers an emerging problem is the transmission of hepatitis A. From the outbreaks described, the epidemiologist knows that (1) hepatitis in children of day-care age is usually asymptomatic, (2) household contacts infected from the spread within the center constitute the majority of cases, and (3) most important of all, children who are in diapers up to 2 years are passing the infection (rarely older children). Officials at the center, parents, and health authorities must be alert to the fact that asymptomatic spread among children at the center is indicated when cases develop in their families.

The factors increasing the risk of an outbreak of hepatitis A at a day-care center are (1) an enrollment of more than 20 children, (2) a business day of greater than 9 hours, and (3) the presence of the babies 2 years old and younger in diapers. Prevention and control should focus on these children, since their presence in diapers facilitates the spread of disease.

Maintenance of appropriate hygienic standards begins with proper handwashing routines for staff and young children who cannot wash their hands themselves. Diapers should be changed only on impermeable surfaces that are cleansed and disinfected properly. One-half cup of household bleach in a gallon of tap water (a 1:32 dilution) prepared daily and dispensed in spray bottles is a satisfactory disinfectant. Accessory items, such as cans of baby powder, jars of petroleum jelly, and the like, should also be disinfected daily, since they are easily soiled during a diaper change.

In the event of an outbreak in the day-care center, new admissions may be suspended, or they may be required to receive prophylactic human normal immune globulin (HNIG).

Modified from Hepatitis A outbreak in a day-care center—Texas, Morbid. Mortal. Weekly Rep. 29:565, 1980.

and this one persists for life. A radioimmunoassay test is available that measures total antibody, both the IgM and the IgG types, but this test does not differentiate between the two types. Here a positive test is not specific for acute hepatitis, since it indicates well persons who have had the disease in the past.

Prevention and Control

True epidemic hepatitis is usually type A, the mortality of which is low. Widespread immunity does exist, probably gained from inapparent childhood infections. There is little evidence for the existence of chronic hepatitis with this virus. A carrier state is rare. Gamma-globulin given as late as 6 days before onset of illness may protect those exposed for as long as 6 to 8 weeks.

The patient should be isolated. Diligent hand washing, wearing of protective gowns, and autoclaving of articles contaminated by the patient are necessary.

Viral Hepatitis Type B

Infection with hepatitis B virus is worldwide. Globally the number of persistent human carriers of hepatitis B virus is estimated at 200 million. Although infection may be transmitted in a variety of ways, transmission through blood and blood products has special significance in clinical medicine.

Spread

Since viral hepatitis type B is carried by human serum (or plasma), it may complicate blood transfusion or the administration of convalescent serum, vaccines, and other biologic products containing human serum. Needles, syringes, and tubing sets for transfusions and stylets for finger punctures are important conveyors of the infection when soiled by blood or blood products.* *As little as 0.000025 ml of blood contaminated with B virus (0.01 ml with A virus) has been shown to cause disease.* For this reason, commercially available disposable needles, disposable stylets for finger puncture, and disposable syringes are strongly recommended. Disposable units of plastic tubing suitable for blood transfusion are widely used.

Hepatitis B represents an occupational hazard among personnel working with blood or serum, including laboratory technologists, blood bank workers, physicians, dentists, and nurses. Surgeons and pathologists are at high risk because of the possibility of accidental self-inoculation. WHO estimates that hepatitis B infection among hospital personnel is three to six times greater than in employees in other occupations. Persons so exposed should observe rigid asepsis. The attack rate is high in the renal dialysis unit, and paradoxically, the hepatitis is much more severe in the medical attendants than it is in the patients.

The disease is also prevalent among drug addicts who share unsterilized and contaminated hypodermic needles. Heroin addicts are especially notorious for "passing the needle." Epidemiologic changes in viral hepatitis within the last decade reflect the increase of illicit drug use and its patterns. The changes have been shifts in seasonal and age incidences, a trend from rural to urban cases, and more cases of hepatitis B than of hepatitis A.

Hepatitis B virus may spread directly from person to person. It has been identified in a variety of body fluids and excretions, including saliva, throat washings, serous fluid, urine, semen, menstrual blood, vaginal secretions, feces, sweat, bile, cord blood, and breast milk. Demonstration of the viral particle in body fluid or excretion does not necessarily correlate with infectivity, however. Hepatitis B virus has been recovered from mosquitoes, but the role of an insect vector is unclear. Carriers of B virus may transmit the virus even though they are without symptoms, and a chronic carrier state does exist. The hepatitis B virus is the only well-characterized extracellular human virus with a prolonged carrier state. (Over 20% of the population of the third world harbor the virus.)

The Virus

Hepatitis B virus (Fig. 34-17) is a hepadnavirus, a spherical, parvovirus-like, double-stranded DNA, double-shelled, 42 nm particle, formerly referred to as the **Dane particle** (first defined in 1970). The virion consists of two parts, an outer lipoprotein envelope and a DNA-containing core, a 27 nm nucleocapsid that contains a molecule of DNA. The dense core has antigenic specificity, for which the term **core antigen** (HBcAg) is used (p. 739). It is covered by the lipoprotein coat, which contains the hepatitis surface antigen (HBsAg), formerly called the **Australia antigen.** Infected plasma contains varying numbers of particles of varying sizes and forms. Commonly seen spherical and filamentous particles with a mean diameter of 22 nm are devoid of DNA and represent free envelopes of the virus (Fig. 34-17).

Four major antigenic subtypes of the hepatitis surface antigen exist. The core of the virus also contains the e antigen, HBeAg, which is associated with active liver disease and with infectivity of the virus. The e antigen is a protein found only in HBsAg-positive serum. It is important to identify those HBsAg-positive persons with e antigen,

---

*For sterilization technics to prevent the spread of the hepatitis viruses, see pp. 228, 266, and 270.

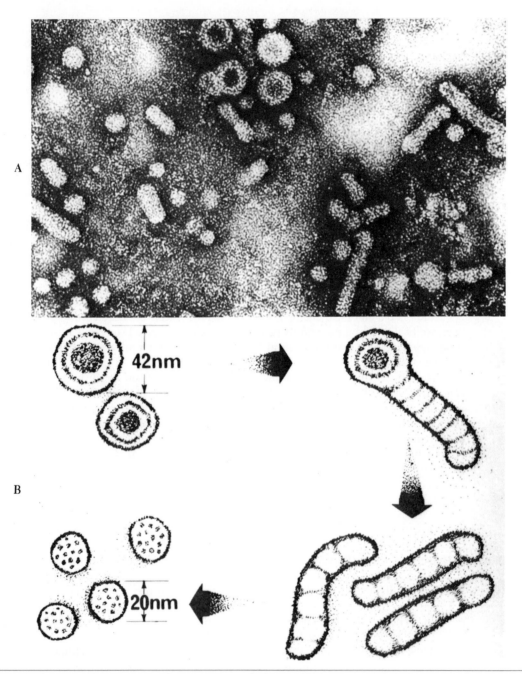

FIG. 34-17   Hepatitis B virus. **A,** Electron micrograph showing the intact virions (the large, round, 42 nm Dane particles) and other virus-associated particles (the small round and elongate forms). All particles contain surface antigen (HBsAg). In addition, an intact virion has a core with its own antigenic specificity (HBcAg). **B,** Diagrammatic representation of the three distinct forms visualized in **A.** The large round 42 nm particle is the virus. Although tubular or filamentous forms are observed, the small spherical particles are the more plentiful in sera of patients by a factor of more than 1000. These particles are found in concentrations as high as 10^{12} particles per milliliter of serum. (Courtesy Dr. Burton Combes, The University of Texas Health Science Center at Dallas.)

since it is a part of the hepatitis B virus particle and persons with HBeAg are the most infective.

Antibodies to the antigens of hepatitis B virus are anti-HBc, anti-HBs, and anti-HBe.

Laboratory Diagnosis

Identification of the hepatitis B virus may be made by technics of counter-immunoelec-trophoresis, complement fixation, and passive hemagglutination, but the more sensitive methods of reverse passive hemagglutination and radioimmunoassay are preferred. Radioimmunoassays (sensitive and specific) exist for HBsAg, anti-HBs, HBcAg, HBeAg, and anti-HBe (see box on p. 739). The demonstration of HBsAg indicates acute or chronic type B hepatitis or an asymptomatic carrier state. HBsAg may be detected several weeks before evidence of infection; it persists during clinical disease and usually disappears when liver function tests return to normal. Acute infection is indicated when a rising titer of anti-HBs is demonstrated between an acute serum paired with a convalescent serum taken 2 to 3 months later. The presence of **antibody** to HBsAg indicates previous exposure to B virus with or without history of hepatitis. Anti-HBs is the protective antibody against hepatitis B virus, and with sensitive radioimmunoassay technics, it usually is detectable for the life of the individual. The duration of immunity to the hepatitis B virus is considered to be lifelong following natural disease.

Implications for Blood Banking

The immediate application of the knowledge gained about HBsAg and the hepatitis B virus is in the blood bank to screen donors. The incidence of viral hepatitis among patients receiving blood containing HBsAg is five times greater than it is among those patients given HBsAg-negative blood. Many tests have been designed to detect the surface antigen in blood for transfusion, and the very serious risk involved has been minimized to the extent that presently only about one tenth of cases of posttransfusion hepatitis are attributable to the agent of hepatitis B. The carrier state for HBsAg in the United States exists in an estimated 0.05% to 0.5% of blood donors. (The high-risk carriers of the hepatitis B virus are most likely to be the commercial donors from the skid row, drug addict, and prison inmate populations.)

Prevention and Control

Measures for control include (1) technics for detection of HBsAg-positive blood units with elimination of such units, (2) the use of environmental barrier methods in specific areas where workers are at risk because of their association with blood or blood products, (3) administration of immune globulin to persons at risk and to infants born to HBsAg-positive mothers, and (4) immunization (pp. 424–425, 430, 444).

Hepatitis B Virus and Liver Cancer

The sequence of events leading to the formation of hepatocellular carcinoma (primary liver cancer) is not known, but evidence is mounting that the hepatitis B virus is implicated. Infection with this virus does precede the development of the liver cancer, and the hepatitis B virus may well be a primary tumor-producing (oncogenic) agent for this form of cancer of the liver. Some of the clinical and epidemiologic findings that strongly suggest such a correlation include the following: (1) the high incidence of hepatocellular carcinoma in areas of prevalent hepatitis B virus infection, especially in the part of Africa below the Sahara Desert and in Southeast Asia; (2) the clusters of hepatocellular carcinoma in families whose members carry hepatitis B virus; and (3) the presence of markers for this virus in 80% to 90% of patients with this form of liver cancer. It is said that the relative risk of primary liver cancer is 273 times greater in carriers than in noncarriers of the virus.

HBsAg has reportedly been recovered from a culture system of human malignant liver cells.

Non-A, Non-B Viral Hepatitis

A serious concern to blood banks is posttransfusion hepatitis, a complication of blood transfusion (not a transfusion reaction). In persons receiving multiple transfusions, even where careful screening of would-be donors has been carried out, the incidence of posttransfusion hepatitis is still high. Sensitive immunologic tests that are used to identify infection with hepatitis A virus or hepatitis B virus indicate with certainty that the hepatitis, comparable to the other forms of viral hepatitis, is *not* caused by either the A or B virus. In this frame of reference the disorder is regularly termed *non-A, non-B (NANB) viral hepatitis,* and currently in the United States 80% to 90% of cases of posttransfusion hepatitis fall into this category (around 240,000 cases annually of clinical and subclinical posttransfusion hepatitis). The serious consequences of non-A, non-B hepatitis are like those of viral hepatitis type B. It may at times be seen as acute and fulminant disease, and it also demonstrates a propensity, perhaps even a greater one, for progression to a chronic state. The incubation period is intermediate to that for the other two types of viral hepatitis, being about 7 weeks.

Spread

The disease is spread parenterally by blood or blood products, a manner of spread consistent with the existence of a chronic carrier state in the asymptomatic persons who serve as blood donors. The risk of non-A, non-B hepatitis is said to be greater from donors with an elevated serum level of alanine aminotransferase. Since sporadic disease is reported where no prior blood transfusion was given, there must also be a type of exposure other than by an injection.

Laboratory Diagnosis

No specific laboratory procedure is presently available to pinpoint non-A, non-B hepatitis. The diagnosis, largely one of exclusion, is made after laboratory tests have eliminated the possibility of infection with hepatitis A or B virus and other viruses known to cause liver damage.

The Cause

Viruslike particles associated with this human disease can be recovered from experimentally infected chimpanzees and visualized by immune electron microscopy in the injured liver tissue. A virus similar to hepatitis B virus has been isolated that is implicated as a causative agent. It is a 35 to 40 nm DNA virus, intermediate in size to the A and B viruses. Good reasons exist, however, to indicate that more than one virus may be capable of producing what is presently called non-A, non-B viral hepatitis.

MISCELLANEOUS VIRAL INFECTIONS

Mumps (Epidemic Parotitis)*

The Infection

Mumps, or epidemic parotitis, an acute contagious disease prevalent in winter and early spring, is distinct for a painful inflammatory swelling of one or both parotid glands. It is easily recognized because of the typical appearance of the patient. A child may look like a chipmunk with nuts in its cheeks (Fig. 34-18). Caused by a paramyxo-

---

*For immunization, see pp. 432, 439 and 441.

virus, it occurs most often between the fifth and fifteenth years. Adult epidemics may erupt in military organizations and are extremely difficult to control.

The period of contagion begins before the glandular swelling and persists until it subsides. The incubation period is usually 14 to 21 days. Mumps is mostly transferred directly from person to person by droplets of saliva, but indirect transfer by contaminated hands or inanimate objects may occur. Generally, the virus does not survive long outside the warm, moist environment of the body of its only host, a human being. An estimated 30% to 40% of persons infected with the mumps virus have a silent infection followed by permanent immunity. During the course of the inapparent infection, these persons may pass the virus to others.

The virus invades the body through the mouth and throat and during the initial, asymptomatic phase of mumps actively replicates in the respiratory epithelium. Entry of the virus into the bloodstream enables it to seed the parotid gland and other target sites as well—kidney, pancreas, ovary, testis, heart, and central nervous system. Viremia is thus responsible for such important complications as orchitis, oophoritis, encephalitis, and pancreatitis. Inflammation of the testicle (orchitis), usually one sided, occurs in 20% of men with this disease. If, as is rarely the case, both testicles are involved, sterility may result. The mumps virus is an important cause of unilateral deafness and also of the aseptic meningitis syndrome. Aseptic meningitis is a mild, self-limited disorder with manifestations such as fever, headache, stiff neck, and muscular weakness consistent with inflammatory involvement of the meninges. The disease may be caused by a variety of viruses and some bacteria as well. Mumps virus inflammation in the central nervous system can be present with or without parotitis.

Laboratory Diagnosis

Mumps virus may be recovered from the urine, saliva, or blood; in cases of central nervous system disease it is found in the cerebrospinal fluid. The chick embryo and cell cultures are used for its identification. Only one serotype is known for this virus, which agglutinates red blood cells of the chicken and of the human blood group O. A skin test, not widely used, is available. The antigen for the test is killed virus from fertile eggs. Inflammation at the site of injection 18 to 36 hours later indicates the presence of immunity (in the absence of sensitivity to egg protein).

Immunity

A passive immunity is given by an immune mother to her baby and persists for about 6 months. Convalescent serum given from 7 to 10 days after exposure protects a high

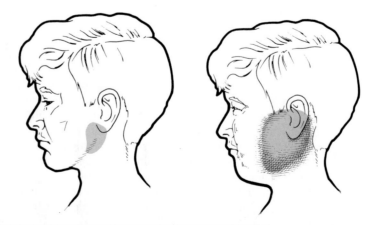

FIG. 34-18   Mumps, showing extent of swelling of parotid gland.

proportion of children from infection. An attack of mumps confers permanent immunity, and, contrary to lay belief, the same immunity follows either unilateral or bilateral involvement. Mumps is another example of a disease that could easily be eradicated. Vaccination is practical, and no host reservoir other than human beings exists.

Infantile Gastroenteritis (Infection with Rotavirus)

Diarrheal disease—infantile gastroenteritis, acute enteritis, infantile diarrhea—is one of the commonest disorders of childhood in all parts of the world. Mortality is high in the underdeveloped areas of the world, and morbidity is substantial even in the industrialized nations. Rotaviruses closely relate to enteric infection in infants and young children in whatever geographic area. Worldwide they have been recovered from 40% of infants with diarrheal disease and in some countries from as many as 88% of the infant patients. They are a frequent and important cause of diarrheal disease not only in human infants but also in the young (calves, piglets, lambs, and so on) of a variety of animals.

Overall, not much is known about factors that influence this disease process. Diarrhea, headache, and fever, the clinical manifestations of rotaviral infantile gastroenteritis, come on after an incubation period of about 48 hours. Most cases occur in children less than 6 years of age, and in temperate climates a seasonal distribution in the cooler months of the year is evident. The names *winter diarrhea* and *winter vomiting disease* indicate this. Rotavirus infection can vary in severity; it can be mild and sometimes asymptomatic in young infants less than 2 months of age.

In the United States infection is widespread. Few children have escaped it within the first few years of life, and more than half have been infected during the first few months of their life. Spread is most likely along the fecal-oral pathway. In some instances the infant acquires infection from an older person with no sign of disease. Parents of infected children have antibodies to the virus yet rarely secrete it. Passive immunity may be acquired by the newborn through the placenta and the mother's milk.

The Agent

Rotavirus, now classified in the family Reoviridae, has been termed orbivirus, duovirus, reovirus-like agent, and infantile gastroenteritis agent. Two serotypes infect humans; four infect animals. Newborn animals are susceptible to infection with human virus, but the role of animal types in human disease is unknown.

Rotavirus is a naked viral particle, 70 nm in diameter, with a core 38 nm in diameter. The core is encircled by an inner layer of capsomeres to which a sharply defined, circular outer layer of capsomeres is attached. The name of the virus, from the Latin word *rota*, meaning wheel, is derived from the distinctive structure so produced. The outer capsid resembles the rim of a wheel to which short spokes radiate from a wide hub.

Diagnosis

Rotavirus does not replicate readily in simple cell culture systems, even when exposed to a variety of cells, but does do so when placed in organ cultures of fetal intestine. By technics of electron microscopy, the virus has been demonstrated in duodenal aspirates or biopsies and in fecal specimens (stool and rectal swabs) taken from sick infants. Virions appear in stool filtrates throughout the course of illness but disappear in conspicuous fashion just before the diarrhea ceases. Viral particles abound in the stools with infection; as many as 10^9 to 10^{10} per gram of feces have been demonstrated. ELISA distinguishes the members of the rotavirus group. Serologic tests for clinical diagnosis include complement fixation and indirect immunofluorescence.

Condyloma Acuminatum

Condyloma acuminatum is a cauliflower-like mass of coalescent warty excrescences formed on the external genitalia and around the anal region in either sex (Fig. 34-19). The growths, often multiple, vary considerably in size. The lesion is called a venereal wart and is transmitted sexually. The incubation period is 1 to 6 months. Large and bulky warts cause considerable discomfort and are associated with a disagreeable odor. The cause is a human papilloma virus (a papovavirus), an allied strain of the one causing the human wart so familiar in its extragenital setting.

Foot-and-Mouth Disease

Foot-and-mouth disease, or aphthous fever, worldwide in distribution, is an acute, infectious, severely debilitating process in animals described by the formation of vesicles (the aphthae) in the mouth, on the udder, and on the skin around the hoofs. One of the most contagious of all diseases, it primarily affects cloven-footed animals (cattle, sheep, swine, and goats). Horses are immune, and humans seldom contract the disease. The cause is a picornavirus, termed an **aphthovirus** (genus *Aphthovirus*). Several serotypes are known.

Spread

This dread disease is readily transmitted by direct contact or indirectly by contamination of fodder with infectious discharges. It has been transferred to humans by contact with infected animals, by the milk of infected animals, and by contaminated material.

Laboratory Diagnosis

A diagnosis may be established with the injection of suspicious material into the foot-pads of a guinea pig. Typical aphthous lesions will be produced if the rhinovirus is present.

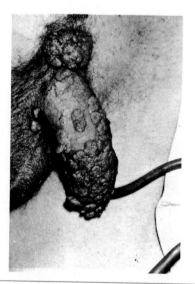

FIG. 34-19 Condylomata acuminata of skin of penis. Note confluence of wartlike growths. A giant condyloma acuminatum is present at penile-abdominal junction. (From Winter, C.C.: Practical urology, St. Louis, 1969, The C.V. Mosby Co.)

Prevention

Control of the disease depends on the slaughter of all exposed animals, with burial or cremation of their bodies, disinfection of pens, and proper quarantine.

Currently this disease has been eradicated in the United States and a number of other countries, but it does remain endemic in wide areas over the world. There is no specific treatment. A vaccine is forthcoming through recombinant DNA technology. (A protective antigen has been cloned and expressed in bacteria.)

Distemper of Dogs

Canine distemper, an important disease of young dogs, is caused by a paramyxovirus in the same genus as the measles virus. Distemper of dogs affects primarily the respiratory and nervous systems and is spread by food and water and, to a lesser degree, respiratory tract secretions contaminated with virus. Dogs and related species are susceptible; humans are not. A highly successful vaccine has been prepared against it.

Additional Diseases Caused by Viruses

In the following lists are viral diseases that affect humans, animals, plants, and insects, respectively.

Humans
Bwamba fever
Hemorrhagic fevers (Omsk, Crimean, Bolivian, Indian, and Korean types, Kyasanur Forest disease, Lassa fever, hemorrhagic nephrosonephritis)
Louping ill
Lymphocytic choriomeningitis
Phlebotomus (sandfly) fever
Rift Valley fever
Russian spring-summer encephalitis
Semliki Forest disease
Warts (verrucae)
West Nile fever

Animals
Aleutian disease of mink
Duck virus enteritis
Ectromelia
Fowl plague
Fowlpox
Hog cholera
Infectious canine hepatitis
Infectious pancreatic necrosis of trout and Atlantic salmon
Louping ill of sheep
Mengo fever
Newcastle disease of chickens
Orf
Pox of horses, sheep, goats, swine, buffalo, and camels

Animals—cont'd
Rift Valley of sheep
Rinderpest
Vesicular stomatitis
Visna in sheep

Plants
Alfalfa mosaic
Barley stripe mosaic
Curly top of sugar beet
Lettuce necrotic yellows
Peach yellows
Potato X
Striate mosaic of wheat
Swollen shoot
Tobacco mosaic
Tobacco necrosis
Tobacco rattle
Tomato bushy stunt
Tomato spotted wilt
Turnip yellow mosaic
Wound tumor

Insects (Mostly the Larval Stage Actively Diseased)
Chronic bee paralysis
Granuloses
Polyhedrosis of spruce sawfly and other polyhedroses
Silkworm jaundice
Tipula iridescence

SUMMARY

1 Certain viruses consistently favor certain anatomic areas for their diseases: dermotropic viruses, the skin; pneumotropic viruses, the respiratory tract; neurotropic viruses, the central nervous system; and viscerotropic viruses, unrelated systems such as the liver or salivary glands.

2 One of four children born to mothers contracting rubella (German measles) during the first 4 months of pregnancy will have congenital defects. If rubella is contracted during the first 4 weeks of pregnancy, about half the children born will be deformed.

3 In 1967 WHO began its "Smallpox Target Zero" campaign to eradicate smallpox. Success came after a decade, the expenditure of $250 million, and the combined efforts of many nations. The last known, naturally occurring case was found in Africa in October 1977.

4 The diverse human herpesviruses share properties such as (1) a unique, diagnostic appearance in the electron micrograph; (2) a comparable size; (3) a cytopathic effect in infected cells; (4) a pattern of replication; and (5) a unique sequence of latency and reactivation.

5 Herpes genitalis is the third most common sexually transmitted disease.

6 Cytomegalovirus infection is usually subclinical and benign, except for the newborn, who is at risk for birth defects, and the immunosuppressed patient, who is at risk for generalized viral infection. Renal transplant patients are especially vulnerable.

7 Infectious mononucleosis, the "kissing disease," is caused by a lymphotrophic virus, the Epstein-Barr virus, one of whose biologic attributes is to change lymphocytes into rapidly proliferating cells.

8 Influenza, a highly communicable disease occurring in epidemics and pandemics, seems to potentiate serious bacterial infection, mostly bronchopneumonia.

9 The nucleic acid of the flu virus, unlike the single thread of most other viruses, is segmented. Its RNA, because it is in segments, enables the flu virus to rearrange or shift its genetic material. Thus the flu virus constantly mutates to new variants of virus type to which no immunity exists in the general population.

10 The bullet-shaped rhabdovirus causes rabies, which is an acute, paralytic, fatal infectious disease of warm-blooded animals, spread by the introduction of virus-laden saliva into a bite wound. The rabid skunk is especially dangerous.

11 The most numerous of the viruses infecting humans are the arboviruses, animal viruses carried in the body of an arthropod vector (*ar*thropod-*bo*rne viruses). Arboviral or togaviral infections occur mainly in mammals and birds; humans are only accidental hosts. The mosquito is often the vector. These infections, constituting the largest group of zoonoses, are similar yet are caused by immunologically distinct viruses and are distributed in clear-cut geographic patterns. These viruses cause important encephalitides.

12 The pattern of infection for poliomyelitis varies from the silent to the severe, paralytic one.

13 Viral hepatitis type A, viral hepatitis type B, and non-A, non-B hepatitis closely resemble each other, but the agents are antigenically separate. These forms of hepatitis are found naturally in humans. Hepatitis type A is spread by the fecal-oral route; hepatitis type B, through blood and blood products; and non-A, non-B hepatitis, through blood transfusions.

14 The cause of hepatitis A is an enterovirus; the agent of hepatitis B, a hepadnavirus; and the agent of non-A, non-B hepatitis, as yet unidentified.

15 A paramyxovirus causes mumps, or epidemic parotitis, a painful, inflammatory swelling of one or both parotid glands (salivary glands).

16 A correlation is emerging between hepatitis B and primary liver cancer.

17 Rotavirus infection—infantile gastroenteritis—is the most common disorder of childhood in all parts of the world, and the mortality is high.

18 Condyloma acuminatum, caused by a human papilloma virus (a papovavirus), is a highly prevalent, sexually transmitted disease. Venereal warts are produced.

19 The CDC has warned against the administration of aspirin or a salicylate-containing product to children with flu or chickenpox. Their action strongly suggests a link between the two factors in the production of Reye's syndrome.

QUESTIONS FOR REVIEW

1 Make a chart showing the causative agent, clinical features, laboratory diagnosis, and prevention of measles, rubella, mumps, influenza, poliomyelitis, rabies, yellow fever, dengue fever, Colorado tick fever, and infectious mononucleosis.

2 Discuss briefly the relation of chickenpox to shingles.

3 Outline the logical procedure to follow when a person is bitten by an animal suspected of having rabies.

4 Briefly compare hepatitis A with hepatitis B.

5 Briefly characterize herpesviruses. What diseases do they cause? List types.

6 Briefly characterize adenoviruses, coxsackieviruses, echoviruses, arboviruses, rhinoviruses, rotaviruses, reoviruses, enteroviruses, coronaviruses, and cold viruses.

7 What is the most common form of posttransfusion hepatitis today? Why is this so?

8 What is meant by a syndrome? Briefly discuss acute respiratory syndromes.

9 Define or briefly explain alastrim, variola major, variola minor, microcephaly, pustule, coryza, rhinitis, encephalomyelitis, yellow jack, silent infection, AST, vesicle, virocyte, latency of a virus, whitlow, keratitis, and hepatitis.

10 Why was eradication of a disease such as smallpox possible?

11 What serious effect may rubella have in a pregnant woman? Explain congenital rubella.

12 It is superstition among laymen that if the eruption of shingles encircles the body, death occurs. Explain why the eruption does not do this.

13 Discuss sylvatic or wildlife rabies and its implications for the spread of the disease.

14 What is meant by slow virus infection? What is the implication?

SUGGESTED READINGS

Ahtone, J., and Maynard, J.E.: Laboratory diagnosis of hepatitis B, J.A.M.A. **249**:2067, 1983.

Chernesdy, M.A., and others: Cumitech 15: laboratory diagnosis of viral infections, Washington, D.C., 1982, American Society for Microbiology.

Cherry, J.D.: The "new" epidemiology of measles and rubella, Hosp. Pract. **15**:49, July 1980.

De Vivo, D.C., and Nicholson, J.F.: Pathologic and clinical findings in Reye syndrome, Lab. Manage. **19**:41, March 1981.

Drew, W.L., and others: Cytomegalovirus and Kaposi's sarcoma in young homosexual men, Lancet **2**:125, 1982.

Favero, M.S., and others: Guidelines for the care of patients hospitalized with viral hepatitis, Ann. Intern. Med. **91**:872, 1979.

Gitnick, G.: Non-A, non-B hepatitis: etiology and clinical course, Lab. Med. **14**:721, 1983.

Gunby, P.: Medical news: How to contain respiratory syncytial virus, J.A.M.A. **244**:225, 1980.

Hinshaw, V.S., and others: Swine influenza-like viruses in turkeys: potential source of virus for humans? Science **220**:206, 1983.

Hoofnagle, J.H.: Type A and type B hepatitis, Lab. Med. **14**:705, 1983.

Kaplan, M.M., and Koprowski, H.: Rabies, Sci. Am. **242**:120, Jan. 1980.

King, J.W.: A clinical approach to non-A, non-B hepatitis, South. Med. J. **76**:1017, 1983.

Knowles, B.B., and others: Human hepatocellular carcinoma cell lines secrete the major plasma proteins and hepatitis B surface antigen, Science **209**:497, 1980.

Krugman, S.: Hepatitis B immunoprophylaxis, Lab. Med. **14**:727, 1983.

Medical News: Reye's syndrome—aspirin link: a bit stronger, J.A.M.A. **247**:1534, 1982.

Melnick, J.L.: Classification of hepatitis A virus as *Enterovirus* type 72 and of hepatitis B virus as Hepadnavirus type 1, Intervirology **18**:105, 1982.

Morgan, D.G., and others: Site of Epstein-Barr virus replication in the oropharynx, Lancet **2**:1154, 1979.

Nevalainen, D.E.: Editorial: Viral hepatitis: yesterday, today, and tomorrow, Lab. Med. **14**:698, 1983.

Palese, P., and Young, J.F.: Variation of influenza A, B, and C viruses, Science **215**:1468, 1982.

Peter, J.B.: Genital herpes: urgent questions, elusive answers, Diagn. Med. **60**:70, March/April 1983.

Polesky, H.F., and Hanson, M.: Transfusion-associated hepatitis: a dilemma, Lab. Med. **14**:717, 1983.

Robinson, W.S.: The enigma of non-A, non-B hepatitis, J. Infect. Dis. **145**:387, 1982.

Waldman, A.A., and Pindyck, J.: Blood donor alanine aminotransferase levels: an automated screening system, Lab. Manage. **21**:53, 1983.

FUNGI
Medical Mycology

FUNGI IN PROFILE

Molds, yeasts, and certain related forms constitute the organisms known as fungi. Their science is **mycology.** From this group of microbes come those whose presence is a common sight on stale bread, rotten fruit, or damp leather. Be it fuzzy or sooty, green, black, or white, the growth on moldy food and clothing is familiar to everyone. Fungi (100,000 species or more) are among the most plentiful forms of life—they powder the earth and dust the atmosphere. About 100 species of fungi affect human beings and animals; of these, at least 20 species cause potentially fatal disease, 35 causes less severe systemic involvement, and 45 result in minor superficial infections of skin or mucous membranes.

Fungi, being unable to make their own food by photosynthesis, as higher plants do, must either exist as parasites on other living organisms or as saprophytes avail themselves of the dead remains. Within the protoplasm of saprophytic fungi are elaborated chemical substances and enzymes that diffuse into the environment, changing what complex substances are there (wood, leather, clothing, bread, and dead organic plant or animal matter) to simpler substances that can be used for their food. The chemical processes of digestion are completed outside the organism, and the end-products are then absorbed by the fungus.

Structure

Certain differences exist between fungi and bacteria. Bacteria are prokaryotes; fungi are eukaryotes and therefore, unlike bacteria, possess a well-defined nucleus, a nuclear membrane, and other eukaryotic organelles. Their cell wall contains chitin, and their plasma membrane, sterols, which the comparable structures in bacteria do not.

Fungi vary in size but are larger than bacteria. Some, the mushrooms and toadstools, are large and easily visible to the naked eye, but most of the medically important ones are microscopic or at least so small that the microscope is necessary for their complete investigation. A given fungus may be a single cell, or it may be composed of many cells laid out in a definite pattern. This distinction is not always sharp, for the two forms may represent different phases of fungal growth. For example, certain important pathogenic fungi produce disease in the body tissues as single cells (in vivo at 37° C) but, when cultured in the laboratory (in vitro at 25° C), present a complex multicellular arrangement. Most disease-producing fungi are dimorphic, with both a yeast and a mold phase, and the capacity of a fungus, largely dependent on the environment, to grow in either form is referred to as **dimorphism**. In this chapter, to describe fungi, we shall separate them loosely on this basis, designating unicellular forms as **yeasts** and **yeastlike** organisms and multicellular ones as **molds** and **moldlike** forms.

The unicellular fungi, or yeasts, are nonmotile, round or oval organisms most of which reproduce by a characteristic process of budding. They vary considerably in size, depending on age and species, but all are microscopic. Nuclei are demonstrated by a suitable stain.

The multicellular fungi, the molds and related forms, present typical structures associated with nutrition and reproduction. Most molds are made up of a **mycelium,** a network or matlike growth of branched threads bearing fruiting bodies. A rudimentary plant known as a **thallus** (no root, stem, or leaf) is formed. The individual threads of the mycelium are known as **hyphae.** In some fungi nonseptate hyphae consist of single threads containing many nuclei spaced along the thread. In most, septate hyphae are divided by cross walls, or **septa,** into distinct cells, each containing a nucleus. The hyphae have thin walls to allow for ready absorption of food and water. This fact helps account for the rapid growth of fungi. The portion of the mycelium concerned with nutrition is referred to as the **vegetative** mycelium. The part that usually projects into the air is the **aerial** or **reproductive** mycelium.

Reproduction

Multicellular fungi reproduce by the conversion of a spore into a vegetative fungus. Spores of fungi are not to be confused with those of bacteria, which are the resistant bodies formed by bacteria for survival, *not* for reproduction. Only one spore forms from one bacterial cell.

Spores are formed in a great variety of ways from the reproductive mycelium, depending on the species. Asexual spores called **conidia** are borne on morphologically distinctive hyphal stalks, the **conidiophores,** from which they are pinched free. In some molds the asexual spores are simply attached to the ends or sides of the hyphae. In some the hyphae bear little pods or sacs called **sporangia,** in which the spores, or **sporangiospores,** rest (Fig. 35-1, A). The sporangium develops at the tip of an aseptate hyphal stalk, the **sporangiophore.** In other molds the hyphae are branched to form brushlike processes or branched conidiophores, each of which bears conidia (Fig. 35-1, B). In still other molds the hyphae or conidiophores produce heads from which radiate fine chains of conidia (Fig. 35-1, C). In most cases the spores separate from the hyphae before vegetation.

If a spore is formed after the nuclei of two hyphae have contacted and fused or after an association of a specialized structure on the mycelium with the nucleus of another specialized structure developed close by, that spore is designated a **sexual spore.** When there is no fusion of nuclei and the spore is simply formed as a swollen body at the end of the hypha, that is an **asexual** spore. Asexual spores are formed in great numbers, sexual spores only occasionally.

Fungi producing sexual spores are "perfect fungi." The stage in fungal growth involving sexual recombination is the "perfect stage." **Fungi Imperfecti** ("imperfect fungi") represent a category of fungi long considered to produce only asexual spores, possibly because of the presence of mutant genes influencing sexual expression, and distinct also because of the concentration therein of pathogenic fungi. Recently, sexual phases of reproduction have been described for certain ones of these medically important imperfect fungi.

Spores are resistant to drying, cold, and moderate heat and may maintain their

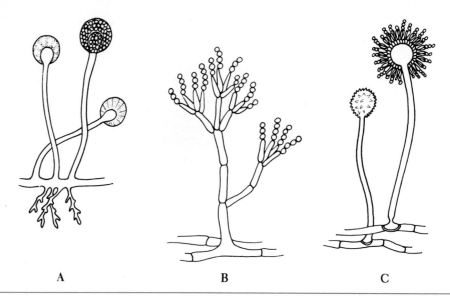

A   B   C

FIG. 35-1 Well-known saprophytic fungi. Note fruiting bodies in which asexual reproductive spores form. **A,** Genus *Rhizopus,* black mold, showing a sporangium containing sporangiospores at end of a sporangiophore. **B,** Genus *Penicillium,* green mold, showing conidia on branched conidiophores. **C,** Genus *Aspergillus,* gray-green mold, showing conidia on unbranched conidiophores.

vitality for a long time. They are constantly present in the air. When they contact food or other material supplying the necessary elements for growth, they develop into new fungi.

Unicellular fungi thought of as yeasts may reproduce by spore formation. The cell enlarges slightly, and the nucleus becomes converted into a definite number of spores. However, **budding,** considered by some a simple type of spore formation, is the usual way. In the process of budding the nucleus moves toward the edge of the cell and divides into two daughter nuclei. A knoblike protrusion of the cytoplasm forms at this point, and one of the nuclei passes into it. The protrusion increases in size and becomes constricted at its base until there is only a narrow connection between the protrusion and the parent cell. Finally, the two separate, and the protrusion or bud, now a small yeast cell, continues to grow until it reaches full size; then the process is repeated.

Multiplication by simple fission (amitotic splitting), characteristic of some yeasts, resembles the process in bacteria. The yeast develops to its full size, and a membrane forms across the middle of the cell. A dividing wall forms, and the two parts separate.

Conditions Affecting Growth

Fungi grow best under much the same conditions as do bacteria, that is, in warm, moist surroundings, but fungi grow more slowly. They grow in the presence of much acid and large amounts of sugar, which bacteria cannot do. Most do not grow in the absence of free oxygen, and large amounts of carbon dioxide are harmful to them. Only a few are anaerobic. Many grow best at temperatures somewhat lower than body temperature. At low temperatures metabolic activities may be slowed, but the organisms do not die. In fact they are rather resistant to cold; many of the commonly encountered ones survive freezing temperatures for long periods of time (months or years). Some can even grow at temperatures below freezing. To prevent growth of mold, meats and certain food products must be refrigerated at temperatures less than $-6.67°$ C ($20°$ F). On the other hand, fungi are quite susceptible to heat, being easily killed at high temperatures. Most species are little affected by light, being able to grow either in the light or in the dark.

Patterns of antibiotic susceptibility in fungi differ sharply from those of bacteria.

Classification

The division Eumycota contains the true fungi (eumycetes), the organisms without chlorophyl ordinarily referred to as yeasts, molds, and related forms.

Fungi are classified on the basis of morphology—the appearance of the colony, spores, and mycelium. Four major subdivisions, pertinent to our study, follow:

1. *Zygomycotina*—Zygomycetes constitutes an important class in this subdivision. The term *Phycomycetes* for a formal class is no longer generally recognized, and the reassignment of certain of its members to the class Zygomycetes is a more workable system. Zygomycetes, or water molds, reproduce in a morphologically simple way either sexually by fusion of two compatible gametes to form a **zygospore** or asexually by production of conidia or sporangiospores. Included here are the ubiquitous bread molds of genera *Rhizopus* (Fig. 35-2) and *Mucor* and other widespread contaminants. Under certain conditions some of these primitive molds (for example, *Mucor*) may be human pathogens. Note that the hyphae of the water molds are nonseptate (coenocytic) (Fig. 35-3); in the other subdivisions the hyphae of the mycelium, where present, are septate.

2. *Ascomycotina*—Ascomycetes, or sac fungi, make up a large grouping. **Ascospores** (sexual spores) are formed within a specially developed sac, or **ascus.** Included here are single-celled yeasts of the genus *Saccharomyces* in class Hemiascomycetes, crucial to the brewing, baking, and wine industries. For some human pathogens formerly considered as imperfect, the perfect stage is now demonstrable. In reclassification and assignment to this subdivision these fungi have been given new names. However, the former designations are still widely used and may be standard for some time to come. The genus name *Arthroderma* reclassifies *Trichophyton*, and the genus name *Nannizzia*, *Microsporum*. *Emmonsiella capsulata* refers to *Histoplasma capsulatum* and *Ajellomyces dermatitidis* to *Blastomyces dermatitidis*.

3. *Basidiomycotina*—Basidiomycetes, or club fungi, form **basidiospores** (sexual spores) on the surface of a specialized structure, a **basidium.** These club fungi encompass the large fleshy toadstools, puffballs, and mushrooms, as well as the small plant

FIG. 35-2 Black bread mold, *Rhizopus stolonifer*, on nutrient agar plate. Each tiny dot is a fruiting body. (From Noland, G.B.: General biology, ed. 11, St. Louis, 1983, The C.V. Mosby Co.)

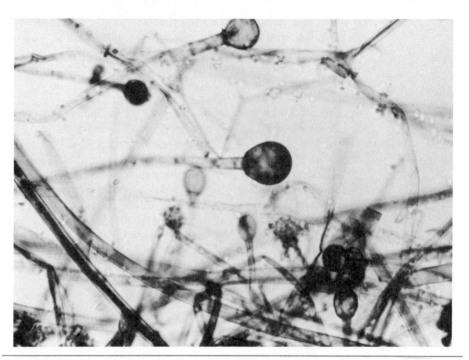

FIG. 35-3 Zygomycetes (Phycomycetes), wet mount, showing sporangia at ends of sporangiophores. Note that hyphae are nonseptate. (Courtesy Dr. R.C. Reynolds, The University of Texas Health Science Center at Dallas.)

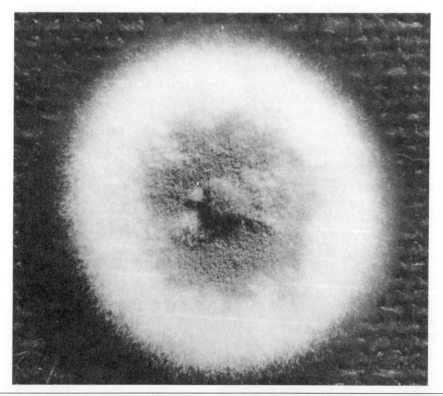

FIG. 35-4 *Penicillium chrysogenium*, mold growth; mutant form seen in Fig. 14-9, p. 246.

smuts and rusts. Certain members produce poisons toxic to humans. *Filobasidiella neoformans*, assigned to this subdivision, indicates the perfect stage of *Cryptococcus neoformans*.

4. *Deuteromycotina*—Deuteromycetes, Fungi Imperfecti, or adelomycetes encompass a taxonomic wastebasket of sorts created to accommodate those fungi for which the sexual state is still unknown but which form asexual spores in a typical way at the ends and sides of their hyphae. Practically, this category also includes fungi in which the asexual stage is the dominant one. Even after the sexual stage has been discovered and the fungus placed in another subdivision, the name of the deuteromycetic genus may persist. Now classified here are the multicellular molds *Aspergillus* and *Penicillium*. The perennial contaminant *Aspergillus* is a sometime pathogen, and *Penicillium* species (Figs. 35-4 and 35-5), the source of the antibiotic penicillin, are rarely so. A number of important fungal pathogens found in this subdivision are to be discussed in their role as disease producers.

Laboratory Study

Fungi may be studied and identified in a number of ways.

Direct Visualization

Stained or unstained material may be observed directly with the microscope. Usually a wet mount is prepared. The specimen—a bit of fungal growth, sputum, pus, skin scrapings, or infected hairs—is placed in a drop of mounting fluid on a glass slide, covered with a cover glass, and examined under the microscope. Because of the translucent quality of fungal hyphae, the intensity of light in the compound light microscope must be lowered to reduce the illumination of the specimen and enhance constrast, or a phase-contrast microscope may be used.

Yeasts may simply be suspended in water. Material containing molds is routinely mounted in a 10% to 20% solution of potassium hydroxide and left for 10 to 15 minutes in order that the specimen may be cleared, that is, become more transparent and more easily defined microscopically. At times special stains are useful. All fungi are gram positive because of the chitin in the cell wall.

Microscopic identification of a given fungus rests on the study of its structure, especially the kind of spores and their relation to the hyphae. Only four fungal elements are identifiable ordinarily in medical specimens: the hyphae, yeast forms, pseudohyphae (elongated budding cells that have failed to detach), and spherules.

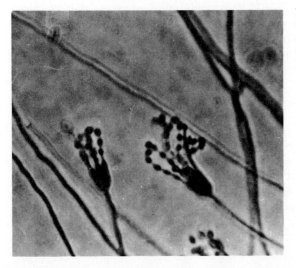

FIG. 35-5 Blue-green mold *Penicillium* (wet mount examined microscopically). Note conidia (asexual spores) and conidiophores (specialized hyphae). (From Noland, G.B.: General biology, ed. 11, St. Louis, 1983, The C.V. Mosby Co.)

The phenomenon of fluorescence is important in the study of superficial fungal infections. For example, infected hairs often fluoresce under a filtered source of ultraviolet light.

Culture

Both pathogenic and saprophytic fungi are highly resistant to acid environments, and both prefer large amounts of sugar in their food supply. Consistent with these requirements, the French mycologist Sabouraud, around the turn of the century, devised a culture medium of maltose (or dextrose), peptone, and agar (Sabouraud agar) still widely used today. Regularly used with it is Sabouraud agar to which the antibiotics chloramphenicol and cycloheximide have been added. Antibiotics added to the various culture media used for fungi suppress bacterial growth in the specimen.

A fungal culture on Sabouraud agar is incubated at room temperature (20° C). Blood agar may also be inoculated but is incubated at body or incubator temperature (37° C). Littman oxgall agar is frequently used. Since fungi do particularly well when portions of raw or cooked vegetables are added to nutrient media, potato and carrot combinations with cornmeal agar are valuable. Pathogenic fungi grow slowly (unlike bacteria), so cultures must be kept for at least 2 or 3 weeks, sometimes longer.

If set up as a slide culture, fungal growth may be observed daily under the microscope (see Fig. 9-11). In a liquid culture medium, bacterial growth is dispersed; few bacteria form the characteristic sheet or pellicle on the surface of the liquid that is seen with the growth of fungi. If the very center of the agar medium in a Petri dish is inoculated, a giant colony grows out to the edge, covering the surface of the dish. The gross appearance of the colony can be diagnostic for a specific organism.

As with bacteria, fermentation reactions are important in identification of fungi, and, rarely, animal inoculations are performed.

Mycoserology

Immunologic reactions in the laboratory evaluation of fungal diseases include agglutination, precipitation, and complement-fixation tests and immunofluorescent and immunoelectrophoretic technics (Table 35-1). These reactions are generally positive with active diease. Since fungi share common antigens, a battery of serologic tests will more likely demonstrate the higher titer of antibodies resulting from the presence of the infectious agent. When serum samples are taken at 2- to 3-week intervals, a rise in antibody titer implicates the particular fungus as the causative one.

Skin testing, which measures cell-mediated immunity, is of great value. Remember, however, that some skin test antigens are cross-reactive. Skin tests are read like the tuberculin test—an area of induration greater than 5 mm in diameter is a positive reaction. Skin testing must be done cautiously, because the exposure to antigen associated with the procedure can convert that person's serologic reaction from negative to a low titer value.

Pathology of Fungal Disease

Fungi are important causes of disease in humans and animals and one of the chief causes in plants. In humans, **mycoses** (fungal infections) are of three types: superficial, subcutaneous, and systemic (or deep). The three types differ not only as to anatomic site but also as to the seriousness of the situation. Superficial disorders are cosmetically disgusting but mild. The gravity of disease increases with the subcutaneous and the systemic mycoses; the latter are dangerous and even potentially life-threatening infections.

Superficial fungi in the skin, hair, and nails cause the **dermatomycoses.** They

Table 35-1 **Immunodiagnostic Tests of Value in Systemic Mycoses**

	Aspergillosis	Blasto-mycosis	Candidiasis	Coccidio-idomycosis	Crypto-coccosis	Histo-plasmosis
Serologic Tests						
Immunodiffusion	X*	X	X	X		X
Complement fixation	X	X		X		X
Precipitation			X	X		X
Latex agglutination			X	X	X	X
Immunofluorescence	X		X	X	X	X
Counterimmuno-electrophoresis			X	X	X	
Intradermal Test		X		X	X	X

*X indicates an accepted test.

spread from animal to human or from one human being to another; they even cause epidemics, but they do not invade.

Subcutaneous and **systemic** fungi contact people from the environment—from the soil, the vegetation, bird droppings, and such. For a subcutaneous mycosis to develop, the agent must be introduced into the tissue beneath the skin, as by trauma. For systemic mycosis, inhalation of infectious material is considered the important route of infection. Ordinarily such fungi, very insidious in their approach, gain a foothold in the human body but progress rather leisurely. Regression also seems slow moving with this category of disease.

Systemic mycoses fall into three categories: (1) **primary** infections, usually with a geographic pattern; (2) **secondary** infections, or "superinfections"; and (3) **complicating** infections. Most of the fungal diseases discussed in this chapter are systemic, including the important primary mycoses. The four major ones are blastomycosis, coccidioidomycosis, cryptococcosis, and histoplasmosis.

Secondary mycoses develop during the course of a bacterial or viral disease for which antibiotics are being given. (Bacteria help to control fungi in nature.) The single most important example is candidiasis. Superinfection of this kind, often hospital acquired, is produced by both fungi and a number of bacteria—many gram-negative bacilli, including the enterics, and staphylococci.

Complicating infections appear after special therapeutic procedures such as peritoneal dialysis or the prolonged use of a catheter indwelling a blood vessel. They are a distinct hazard in the management of the compromised host, that patient with severe, chronic, debilitating disease for which antibiotics, steroid hormones, or immunosuppressive agents are required. They also follow close on disturbances of the immune mechanisms such as are seen in cancer of the lymphoid system and bone marrow failure.

Certain fungi (for example, *Cryptococcus*) are readily pathogenic either as primary or secondary invaders. Others (for example, *Candida*, *Aspergillus*, *Mucor*, and *Rhizopus*), benign in nature, seize the "opportunity" with inordinate vigor.

Skin and mucous membranes provide natural barriers to fungal invasion of the human body. The T cell–dependent system and antibody-complement–mediated phagocytosis by circulating and fixed phagocytes are important in the body's defense. In the study of tissues the body's response to fungal intrusion is seen to be granulomatous inflammation, that is, the reaction in which macrophages are the main participating cells. Microscopically they are numerous and conspicuous in distinctive arrangements. Tissue damage comes after the state of allergy has been set up in the host to the proteins of the fungus. Usually the etiologic agent can be demonstrated in sections of infected tissues or in fluids therefrom; *its presence makes the diagnosis.*

Importance

Fungi are important in the processes of nature, agriculture, manufacturing, and medicine. Commonly encountered molds can injure woodwork and fabrics and spoil food. They destroy food during its growth, in the process of manufacture, and after it has reached the consumer. Foods most vulnerable are bread, vegetables, fruits, and preserves. Molds are an important factor in the decay of dead animal and vegetable matter; complex organic compounds broken down into simple ones are returned to the soil to be used as food by green plants. Some fungi, such as certain mushrooms, are human food sources. Molds are used commercially in the manufacture of beverages and to give flavor to cheeses (for example, Roquefort, Camembert). Penicillin, an effective antibacterial substance, is derived from a common mold, *Penicillium notatum.*

Yeasts are economically important because they ferment sugars. The fact that they convert sugars by enzymatic action into alcohol and carbon dioxide is practically applied in the manufacture of alcoholic beverages and in baking. In the manufacture of alcoholic beverages, carbon dioxide is a by-product, whereas in baking it is the essential factor (p. 891). In the preparation of commercial yeast the cells are grown in a suitable liquid medium, separated out from the liquid portion by centrifugation, mixed with starch or vegetable oil, and then molded and cut into cakes. Yeast is a source of a vitamin B and of ergosterol, from which vitamin D is obtained.

DISEASES CAUSED BY FUNGI

Medical mycology deals with the fungi that bring about disease.

Superficial Mycoses

Dermatomycoses

Superficial fungal infections of the skin, hair, and nails, generally called **ringworm** or **tinea** (Latin, a worm), are *dermatomycoses* or *dermatophytoses.* Fungi causing dermatomycoses and showing no tendency to invade the deeper structures of the body are called **dermatophytes.** In the skin, hair, and nails, dermatophytes affect, grow, and form their hyphae in the keratin, the hard protein that is the principal constituent of these structures. Most of these fungi are distributed worldwide, but some species favor a particular geographic area. Many domestic and other animals harbor these infections and may transmit the causative agent to humans. The three important genera, which are closely related botanically, are *Microsporum, Trichophyton,* and *Epidermophyton.*

The genus *Microsporum* (*Nannizzia*) (Fig. 35-6) is the most frequent cause of ringworm of the scalp and may give rise to ringworm in other parts of the body. Hairs removed from the affected regions are surrounded by a coat of spores, and scales of skin show many branched mycelial threads. Infected hairs fluoresce. *Trichophyton* (*Arthroderma*) causes ringworm of the scalp, beard, other areas of skin, and nails (Fig. 35-7). The organisms are found as chains of spores, inside or on the surface of affected hairs,

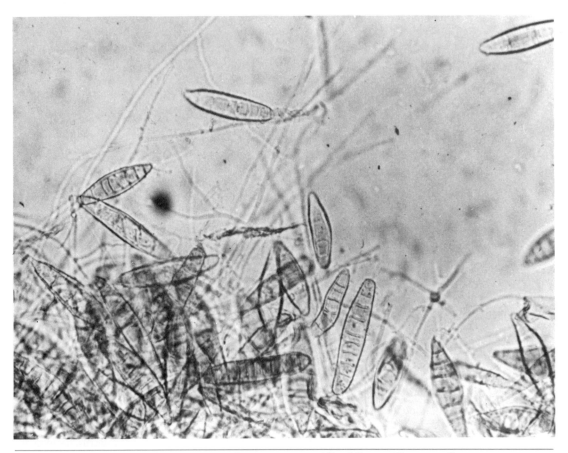

FIG. 35-6 *Microsporum canis*, wet mount, showing macroconidia. *Micro*conidia are small and single celled, whereas *macro*conidia are large and usually multicelled. (Courtesy Dr. R.C. Reynolds, The University of Texas Health Science Center at Dallas.)

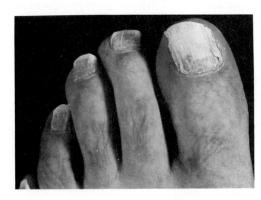

FIG. 35-7 Onychomycosis (tinea unguium) of nail of great toe caused by *Trichophyton rubrum*. Nail is white and unattached to nail bed over most of its extent. (AFIP Neg. 74-19321.) (From Anderson, W.A.D., and Kissane, J.M., editors: Pathology, ed. 7, St. Louis, 1977, The C.V. Mosby Co.)

or as hyphae and characteristic spores in skin scrapings. *Trichophyton schoenleini* is the cause of almost all cases of favus. The spores and mycelial threads are found in the favus crusts. Hairs in the affected areas are filled with vesicles and channels from which the mycelia have disappeared. *Epidermophyton* is largely responsible for ringworm of the skin, body, hands, and feet. Epidermophyta appear as interlacing threads in the skin and do not invade the hairs.

Ringworm of the scalp (**tinea capitis**), seen most often in children, is a common and highly communicable disease. It may be spread directly from person to person or by articles of wearing apparel. It occurs in domestic animals, from which it may be transmitted to humans. From 10% to 30% of ringworm infections occurring in cities and 80% of those in rural areas are thus transmitted, either by direct or indirect contact with the animal. Pets (dogs and cats) readily pass the ringworm fungi on to their human masters. **Juvenile type scalp ringworm** is associated with a patchy loss of hair. It resolves spontaneously at puberty because at this time secretion of fatty acids from the scalp is increased. This form of ringworm causes no permanent damage.

The lesions of ringworm are circular scaly patches at the affected site. Ringworm of the beard (**tinea barbae**) is known as barber's itch. Ringworm of the groin is known as **tinea cruris** or **dhobie itch.** The most prevalent dermatophytosis is ringworm of the feet, known as **tinea pedis** or **athlete's foot.** For years athlete's foot was assumed to be contracted from footwear, lockers, and floors, but experiments indicate that exposure to the pathogenic dermatophytes in public swimming pools or shower stalls plays a minor role. These fungi are everywhere, even on the feet of uninfected persons. The lesions of athlete's foot most probably appear because of a factor of decreased resistance in the skin to contact with the causative fungi. Uncomplicated tinea pedis is a dry scaly lesion of the skin with few symptoms. However, the feet sweat, and in such a moist environment bacteria normally resident on the feet invade the fungal lesion. Athlete's foot (the white, soggy, malodorous, itching changes between the toes) results. When these changes are present, the infection is more properly considered both fungal and bacterial.

Dermatophytids, or "ids," are secondary vesicular lesions found most often on the hands of a person with a dermatophytosis such as athlete's foot. They develop as a manifestation of that person's hypersensitivity to the products of the dermatophyte responsible. The particular fungus cannot be cultured from the blisters, but the lesions clear with effective treatment of the basic disorder.

Favus usually affects the scalp, with the formation of yellowish, cup-shaped crusts, or **scutula,** around the mouths of the hair follicles. These crusts consist of masses of spores and mycelial threads mixed with leukocytes and epithelial cells. Permanent damage may be done to the scalp. Favus may be transmitted directly or indirectly from person to person and tends to run in families. The disease is mostly found in the Mediterranean area of Europe and North Africa.

Table 35-2 compares the most common dermatophytoses and relates them to the causative fungi.

Laboratory diagnosis of the dermatomycoses relies on demonstration of fungi in hair and skin scrapings by direct microscopy or by cultural methods.

Control of the dermatomycoses is very difficult. It consists of the proper sterilization of clothing, bathing suits, and objects subjected to frequent handling. Hygiene of the feet is extremely important.

Subcutaneous Mycoses

Sporotrichosis

Sporotrichosis, caused by the dimorphic *Sporothrix schenckii*, may affect humans, lower animals, or plants. *Sporothrix* is widely distributed in nature as a saprophyte on

Table 35-2 Comparison of Common Dermatophytoses

Disorder	Anatomic Site (or Structure)	Clinical Features	Usual Dermatophytes Responsible
Ringworm of scalp (tinea capitis)	Scalp hair	Circular bald spots with hair stubs; infected hairs may fluoresce	*Microsporum canis* *Trichophyton tonsurans*
Ringworm of beard (tinea barbae)	Beard hair	Reddened, soggy areas	*Trichophyton rubrum* *Trichophyton mentagrophytes*
Jock itch, dhobie itch (tinea cruris)	Skin of groin	Red, scaling, pruritic changes in intertriginous area	*Trichophyton rubrum* *Trichophyton mentagrophytes* *Epidermophyton floccosum*
Athlete's foot (tinea pedis)	Webs of toes	Itching, scaling lesions with fissures	*Trichophyton rubrum* *Trichophyton mentagrophytes* *Epidermophyton floccosum*
Ringworm of body (tinea corporis)	Nonhairy, smooth body skin	Circular patches with central scaling; red vesicular border	*Microsporum canis* *Trichophyton mentagrophytes*
Onychomycosis (tinea unguium)	Nails	Thickened, crumbly, discolored, lusterless fingernail or toenail	*Trichophyton rubrum* *Trichophyton mentagrophytes* *Epidermophyton floccosum*

vegetation. The infection is usually acquired from plants, especially barberry shrubs, roses, and certain mosses, which seem to harbor the fungus. Sphagnum moss may be contaminated, and no method is known to decontaminate it. The fungi are introduced into wounds (inoculation infection) with penetration of the subcutaneous tissue by infected plants or vegetable matter. The disease is a hazard to florists, alcoholic rose gardeners, forestry workers, and persons with occupations providing comparable exposure. The agent may be a normal inhabitant of the human alimentary and respiratory tracts, and in a few instances the disease has been transferred from person to person. Transmission of the infection from lower animals to humans by bites or indirect routes has been noted. Recently sporotrichosis was reported as a complication of catfish stings. Animals most often affected are horses, mules, dogs, rats, and mice. The majority of cases occur in the United States, especially in the Missouri and Mississippi valleys.

Sporotrichosis is a chronic infection usually limited to the skin and underlying tissues and is accompanied by the formation of nodular masses that slowly undergo softening and ulceration. In typical cases the first evidence of disease is seen at the site of some trivial injury, usually on the fingers. The wound does not heal, and an ulcer (Fig. 35-8) appears, to be followed by nodular corklike swellings in chains up the forearm. The disease seldom extends farther than to the regional lymph nodes, but secondary foci sometimes crop up in other parts of the body such as the lungs, spleen, liver, and other organs. In rare cases sporotrichosis disseminates to sites other than the skin (such as bone marrow and brain) in persons with compromised body defenses.

The oval or cigar-shaped fungus of sporotrichosis, resident within mononuclear cells, is rarely found in smears made from the pus of a skin lesion or in sections of tissue taken from the lesions. As a rule, it is demonstrated only in cultures (Fig. 35-9). At room temperature *Sporothrix* produces a dirty white, grayish, or black filamentous growth, but at 37° C on enriched agar the growth is a creamy yeastlike colony. Fluorescent antibody technics detect the microorganisms in exudates from the lesions.

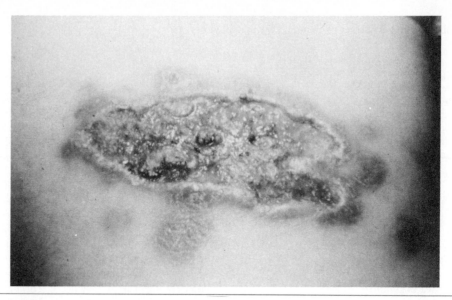

FIG. 35-8   Skin ulcer in sporotrichosis. (Courtesy Dr. J.D. Nelson, The University of Texas Health Science Center at Dallas.)

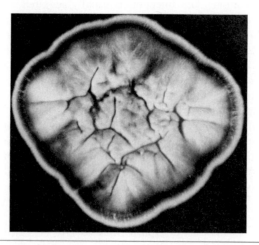

FIG. 35-9   *Sporothrix schenckii*, giant colony. (Courtesy Dr. R.H. Musgnug, Haddonfield, N.J.)

Mycetoma (Madura Foot)

Mycetoma (p. 595) is caused by a variety of fungi, as well as by the filamentous actinomycetes now classified as bacteria. The term **maduromycosis** is often used when the cause is a fungus, but the clinical disease is essentially the same regardless of the etiologic agent. As would be expected, therapy is different.

Systemic Mycoses

Aspergillosis

Aspergillosis is an infection most often produced by *Aspergillus fumigatus*, a gray-green mold growing in the soil. Dust is an important source. This fungus may cause various

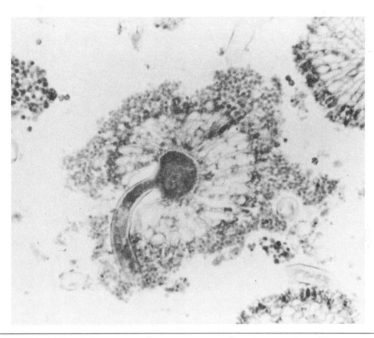

FIG. 35-10   *Aspergillus* in specimen of sputum prepared as tissue section. Note structure of fruiting head. Patient had pulmonary aspergillosis with intracavitary fungus ball (mass of mycelial growth) in an upper lobe of lung. (Courtesy Dr. R.C. Reynolds, The University of Texas Health Science Center at Dallas.)

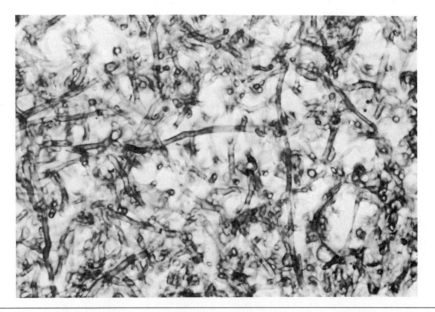

FIG. 35-11   *Aspergillus* species in special stain of microsection of lung. Abundant growth of this opportunist occurred in diabetic patient. Note compact clusters of branching septate hyphae (4 to 6 μm in diameter). (From Zugibe, F.T.: Diagnostic histochemistry, St. Louis, 1970, The C.V. Mosby Co.)

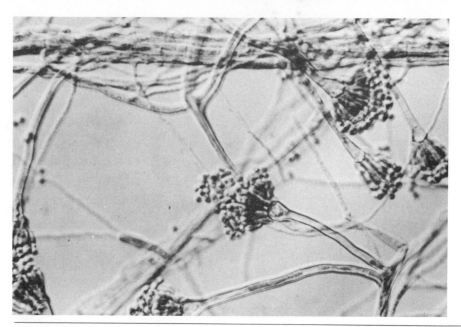

FIG. 35-12 *Aspergillus nidulans*, wet mount. Note unbranched conidiophores and conidia. (Courtesy Dr. R.C. Reynolds, The University of Texas Health Science Center at Dallas.)

types of infection in chickens, ducks, pigeons, cattle, sheep, and horses. Animals usually contract the disease from moldy feed.

In humans the disease generally occurs as an infection of the external ear (otomycosis). The infection may be superficial and mild, or it may cause ulceration and perforation of the tympanic membrane. Infection most likely comes from fungi living saprophytically on the earwax. Other types of human infections are pulmonary infections (Fig. 35-10), sinus infections, and infections of the subcutaneous tissues. Aspergillosis as a superinfection complicating antibiotic therapy affects the lungs (Fig. 35-11). Aspergillosis is a common opportunistic infection in the immunosuppressed patient, with involvement of lungs, gastrointestinal tract, and brain. Aspergilli can no longer be passed off simply as "weeds" in the laboratory, likely to contaminate any culture.

Aspergillosis may be caused by aspergilli other than *Aspergillus fumigatus*. Among these are *Aspergillus nidulans* (Fig. 35-12), *Aspergillus niger*, and *Aspergillus flavus*. Certain strains of *Aspergillus flavus* (and also *Penicillium puberulum*) elaborate **aflatoxins,** toxic substances that can cause severe damage or even cancer in the liver of animals ingesting them. Aflatoxins are exceedingly potent in this respect, cancer formation requiring an amount no more than 0.05 ppm, and they are also natural mutagens. Animals contact these mycotoxins (fungal toxins) in the mold produced by *Aspergillus* on peanuts and other harvest feeds.

The aflatoxin content of such foods as peanut butter is greatly reduced by careful removal of infected peanuts before processing. Currently the Food and Drug Administration has established the safe level of aflatoxin in peanuts as 20 parts per billion (ppb), but it may lower this value to 15 ppb aflatoxin.* An FDA survey has demonstrated that 93% of peanut butter samples are below 20 ppb. There is no direct evidence that aflatoxins are cancer producing in human beings.

Serologic aids to the diagnosis of aspergillosis include a complement-fixation test, an immunodiffusion test, and the indirect fluorescent antibody determination.

---

*Federal Register **39:**42751, 45299, Dec. 6, 1974.

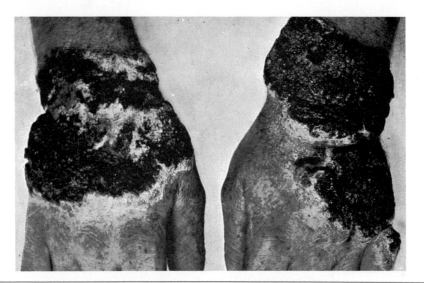

FIG. 35-13   Blastomycosis in Minnesota farmer. Note verrucous or warty nature and sharp margins.

FIG. 35-14   *Blastomyces dermatitidis*, yeast phase culture at 37° C. Note wrinkled, waxy growth. (Courtesy Dr. R.C. Reynolds, The University of Texas Health Science Center at Dallas.)

Blastomycosis

There are two kinds of blastomycosis—the **North American** and the **South American** (paracoccidioidomycosis). North American blastomycosis, known as **Gilchrist's disease** after its discoverer, is a granulomatous and suppurative inflammation. Multiple abscesses form in the skin and subcutaneous tissues (blastomycetic dermatitis) or in the internal organs of the body (systemic blastomycosis). The cutaneous form is more often seen than the systemic (Fig. 35-13). The lesions of blastomycetic dermatitis may be mistaken for cancer or tuberculosis. With systemic disease, pulmonary blastomycosis is most likely present, closely resembling pulmonary tuberculosis.

North American blastomycosis, confined almost exclusively to the United States and Canada, but concentrated largely in the central and southeastern United States, is caused by *Blastomyces dermatitidis*. The organisms are demonstrated by direct microscopic examination of pus from the lesions or by cultural methods (Figs. 35-14 and 35-15). In pus they are round or oval granular yeast forms, varying from 8 to 15 μm in diameter. They are surrounded by a thick refractile wall, which makes them doubly contoured, and single budding forms are present (Fig. 35-16). The organisms grow typically in the mycelial phase (Fig. 35-17) on all media, but isolation is often complicated by overgrowth of bacterial contaminants from the lesion.

Accumulated evidence indicates that cutaneous blastomycosis (Fig. 35-13) results from infection through wounds and that the portal of entry for systemic blastomycosis is the lung. The organisms spread to other organs in the bloodstream. In some cases of systemic blastomycosis, infection may come from the skin.

Although fungi are weak antigens, their infections are associated with a degree of allergy useful in laboratory diagnosis. For North American blastomycosis there are a complement-fixation test, an immunodiffusion test, and a skin test (the blastomycin test). The first test is specific and the most widely used. Cross-reactions may occur with the skin test for blastomycosis and that for histoplasmosis. (A person with histoplasmosis may appear to give a positive test for blastomycosis.) Skin testing may be done with *Blastomyces dermatitidis* vaccine, considered a more effective antigen than blastomycin.

Paracoccidiodomycosis, or South American blastomycosis, is similar to the North American type. Caused by *Paracoccidioides (Blastomyces) brasiliensis*, it is known also as **paracoccidioidal granuloma.** The fungus enters the body by way of the mouth, where it localizes and causes ulcers and granulomas. From these lesions the fungi spread to the lungs and other parts of the body. The disease may terminate fatally. Most cases have been reported in Brazil.

Candidiasis (Moniliasis)

Candida albicans is a budding yeastlike organism, worldwide in distribution. It is found on the mucous membranes of the mouth, intestinal tract, and vagina and on the skin of normal persons with no disease. Most fungi belong to the environment; only a few are normal inhabitants of the human body, such as *Candida albicans*, which is rarely found outside its natural habitat. Although the presence of *Candida albicans* in the intestinal tract is accepted as normal, the reservoir for infection is also considered to be there.

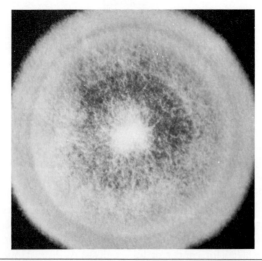

FIG. 35-15 *Blastomyces dermatitidis,* giant colony, showing abundant fully mycelial growth. (From Musgnug, R.H.: Med. Tribune **3:**16, May 28, 1962.)

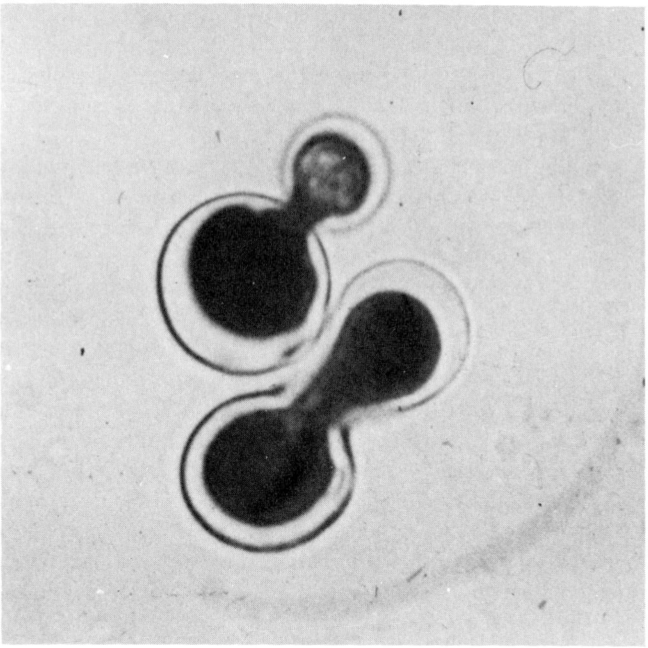

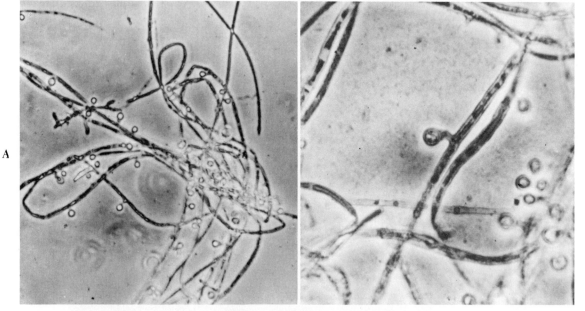

FIG. 35-16 *Blastomyces dermatitidis*, microscopic appearance of budding yeast forms. (Courtesy Dr. R.C. Reynolds, The University of Texas Health Science Center at Dallas.)

A   B

FIG. 35-17 *Blastomyces dermatitidis*, microscopic appearance of mycelial phase. Note filamentous growth and numerous conidia attached to hyphae near septations. **A,** Low-power magnification. **B,** Higher magnification. (Courtesy Dr. R.C. Reynolds, The University of Texas Health Science Center at Dallas.)

The relationship of the genus *Candida* to disease may be hard to determine. It is found in association with known pathogens in persons with known illness but in whom there is no reason to suspect its pathogenicity. For the diagnosis of candidiasis, *Candida albicans*, the chief pathogen of the genus, must be repeatedly isolated from the lesions to the exclusion of better-defined agents.

Two infections of *Candida albicans* have been with us for a long time. One, on the mucous membranes of the mouth, is known as **thrush** (oral candidiasis). The other, involving the mucous membranes of the female genitalia, is **candidal vulvovaginitis,** or **vaginal thrush.**

Thrush is seen as many small milklike flecks that may coalesce and cover the entire lining of the mouth. Beneath these patches on the inside of the lips, on the hard palate, and on the tips and edges of the tongue are areas of catarrhal inflammation. Thrush is an especially troublesome infection in newborn infants in hospital nurseries. The baby usually acquires the organism from the vagina of the mother during the birth process. Maternal infection is the primary source, but the organisms may be spread from person to person by contaminated fingers, utensils, and nipples. Thrush tends to be rare after the newborn period in healthy subjects of any age but is prevalent in poorly nourished children and in chronically ill and aged adults.

Vulvovaginitis is a thrushlike infection associated with a typical vaginal discharge and represents a spread of microorganisms to a mucocutaneous surface in contact with the anus. It tends to recur and can be troublesome. The sugar content of the urine in pregnancy and uncontrolled diabetes may be contributing factors, and vaginal thrush is associated with oral contraceptives. Today it has taken on the proportions of sexually transmitted disease.

Candidal skin infections affect mainly the moist warm parts of the body, such as the axilla, intergluteal folds, groin, or inframammary folds, and are found in the interdigital webs of the hands in those persons, such as fruit canners and vegetable and fish handlers, who by occupation must keep their hands constantly soaked in water. Infection of the nail thickens, damages, and eventually destroys it. **Chronic mucocutaneous**

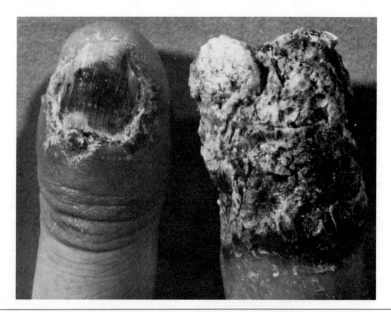

FIG. 35-18   Candidiasis, long standing, in patient with immunologic disorder. Thumbs show destruction of nails with accumulation of horny material. (From Kirkpatrick, C.H., and others: Ann. Intern. Med. 74:955, 1971.)

candidiasis is a singular disorder that bedevils the patient with deficiency of cellular immunity (Fig. 35-18).

Candida albicans may cause disease in lungs, kidneys, and various other organs of the body in persons in whom there is a primary predisposing condition. Chest disease is frequent. Bronchopulmonary or pulmonary candidiasis varies from a mild inflam-

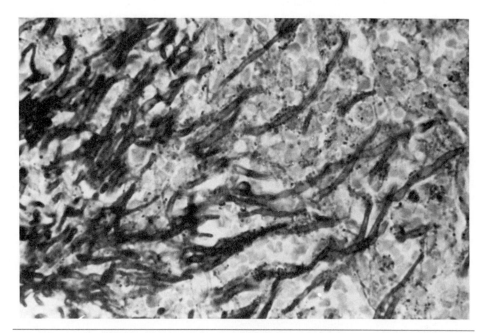

FIG. 35-19 Candidiasis of spleen in disseminated disease, microsection. Note abundant mycelial filaments of *Candida albicans*. (×800.)

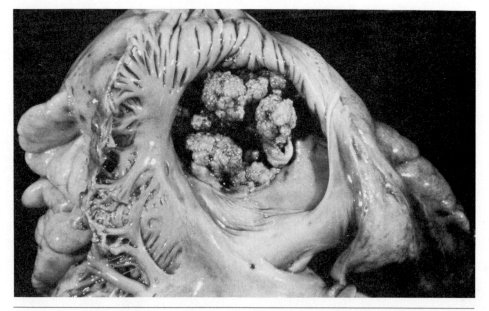

FIG. 35-20 Candidiasis of heart valve. Atrium of heart has been opened to expose valvular opening. Note large cauliflower-like excrescences (vegetations) on leaflets of tricuspid valve. Patient was a drug addict and, as such, a person prone to involvement of tricuspid valve.

mation to a severe infection such as tuberculosis. Systemic and invasive candidiasis is seen more often today. It is prone to follow persistent skin or mucosal lesions in a person with lowered resistance or altered immunologic mechanisms. *Candida* easily gains the ascendancy when dosage with a broad-spectrum antibiotic is prolonged, since the normal bacterial flora is thereby depressed. It is considered the major hospital-acquired, pathogenic fungus.

Overgrowth of *Candida* for any reason is an ominous event. If the fungi circulate in the bloodstream (a fungemia), they induce a serious toxic reaction and are widely disseminated in the body (Fig. 35-19). The patient at greatest risk is the one with leukemia, some kind of bone marrow failure, or an organ transplant. Other pathologic conditions over which the threat of candidiasis hangs are diabetes, chronic alcoholism, endocrine disorders, malnutrition, and certain kinds of cancers, especially hematologic malignancies. Candidiasis is the most frequent fungal infection in cancer patients. Drug addicts are vulnerable to candidal endocarditis. Under the right circumstances, *Candida* strikes as a formidable pathogen (Fig. 35-20). *Candida tropicalis* (less often a colonizer) is said to be the major pathogen in the immunocompromised host.

The laboratory demonstration of *Candida albicans* is easily made either by direct microscopic examination of unstained or stained material or by culture (Figs. 35-21 to 35-23). In the exudate from a lesion or in sputum the organisms appear as oval, bud-

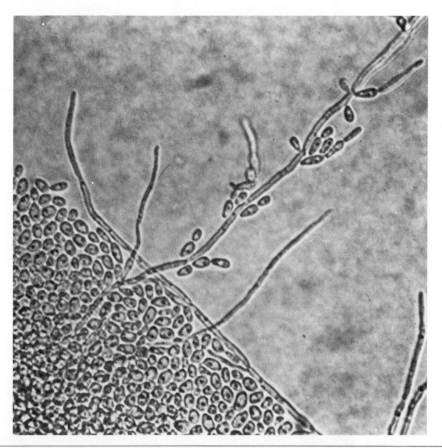

FIG. 35-21 *Candida albicans*, microscopic appearance. Note small oval budding yeastlike cells attached to hyphae (pseudohyphae) at points of constriction. Mycelial fragments vary in thickness and length. (Courtesy Dr. R.C. Reynolds, The University of Texas Health Science Center at Dallas.)

ding yeasts, 2 to 4 μm, with scattered pseudohyphal segments of varying dimensions. A biologic test of value not only for the identification of *Candida albicans* but also for the differentiation of *albicans* from *tropicalis* is the germ tube test (Fig. 35-23). *Candida* is strongly gram positive.

Serologically there are the agglutination tests, the precipitin reaction, an indirect immunofluorescent technic, an immunodiffusion test, and counterimmunoelectrophoresis. Inoculation of the chorioallantoic membrane of the chick embryo, with visible lesions appearing 72 to 96 hours later, is a good test for the pathogenicity of *Candida*. Since candidal skin testing is almost invariably positive in normal adults, it is applicable as an indicator of competent cellular immunity rather than of candidal infection, past or present.

Coccidioidomycosis (Coccidioidal Granuloma)

Coccidioidomycosis (valley fever, desert rheumatism), one of the most infectious of the fungal diseases, exists in two forms—the primary (usually self-limited) and the progressive. In the primary form the lesions are confined to the lungs, giving pulmonary symptoms of varying severity, sometimes with cavitation (Fig. 35-24). As a rule, the infection ends in recovery, but in a small percentage of cases the process spreads from

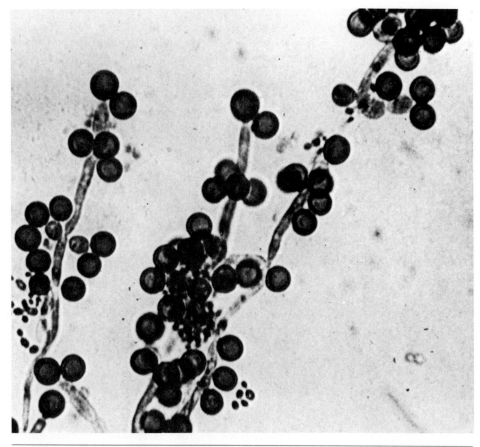

FIG. 35-22   *Candida albicans*, microscopic appearance of chlamydospores from fungal growth on corn meal agar. Note ball-like clusters of characteristic thick-walled, rounded, resting spores designated chlamydospores. (Courtesy Dr. R.C. Reynolds, The University of Texas Health Science Center at Dallas.)

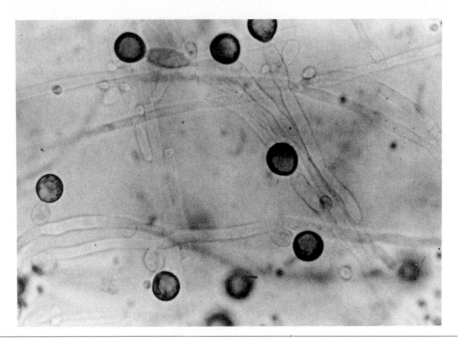

FIG. 35-23 *Candida albicans,* microscopic appearance of germ tubes (3 to 4 μm wide, up to 20 μm long) formed when 24-hour-old culture of this fungus was placed in serum at 37° C for 2 to 3 hours. With germination of a fungal spore, one or more tubelike processes (germ tubes) elongate into filaments (hyphae) and eventually branch. (Courtesy Dr. R.C. Reynolds, The University of Texas Health Science Center at Dallas.)

the lungs to produce the progressive form. This happens more in blacks and the more darkly pigmented races than in whites.

In the progressive form the disease spreads to the skin, subcutaneous tissues, bones, meninges, and internal organs. The lesions in the skin resemble those of blastomycosis. In other parts of the body they resemble those of tuberculosis. The mortality is high. This form of the disease is designated coccidioidal granuloma.

Coccidioidomycosis is endemic in the desert valleys of California and the dry dusty areas of the southwestern United States, especially in Arizona, New Mexico, Texas, and parts of Mexico. Humans and animals are most likely infected by the inhalation of spore-bearing dust. The infection is found in cattle, sheep, dogs, and certain wild rodents. A reservoir of infection may exist in small wild rodents, these animals passing the spores in their feces to contaminate the soil, after which the spores would be spread by the wind. The rather frequent dust storms of the Southwest could carry these organisms long distances.

The cause is *Coccidioides immitis.* Diagnosis is made when it is found in the disease. The appearance of the dimorphic fungus in lesions differs from its appearance on culture media. In body tissues and exudates one sees yeastlike forms—large thick-walled, nonbudding spherules, 20 to 60 μm in diameter, filled with endospores, 2 to 5 μm in diameter (reproduction by endosporulation) (Fig. 35-25). As many as 1000 endospores may be released from a single spherule. In laboratory cultures growth is that of a mold (Fig. 35-26), and one sees a fluffy, cottony white colony.

The **coccidioidin test,** a test of sensitivity to an extract of the organism, is of value. (Cross-reactions may occur with skin tests for histoplasmosis and blastomycosis, however.) **Spherulin,** an extract of the spherule phase of growth, is also used in sensitivity

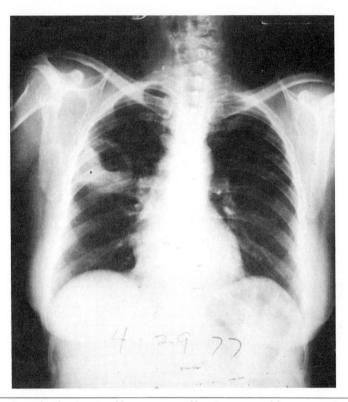

FIG. 35-24   Coccidioidomycosis of lung on x-ray film. Large round lesion in right midlung field represents a cavity caused by *Coccidioides immitis*. (Courtesy Dr. Jack Reynolds, The University of Texas Health Science Center at Dallas.)

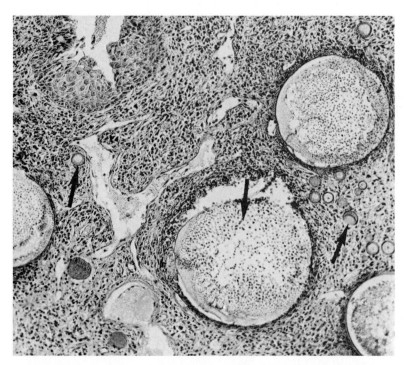

FIG. 35-25   *Coccidioides immitis* spherules *(arrows)* in tissue microsection; reproduction by endosporulation. Large balloonlike structures *(central arrow)* contain myriad endospores that they release into tissue spaces. Doubly refractile capsule seen about varying-sized spherules.

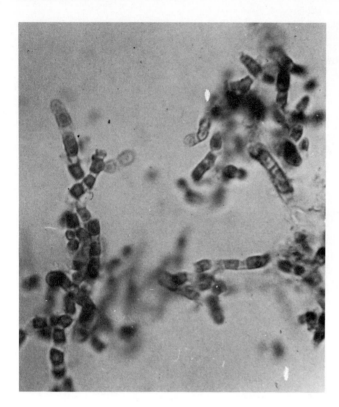

FIG. 35-26 *Coccidioides immitis,* microscopic appearance of highly infectious arthrospores from mold growth of this fungus. Numerous chains of arthrospores alternate with empty shells. Arthrospores result when hyphae segment to cut off rectangular, somewhat thick-walled cells, and this pattern of spore formation is typical for *Coccidioides.* (Courtesy Dr. R.C. Reynolds, The University of Texas Health Science Center at Dallas.)

testing and may prove to be a superior skin test agent to coccidioidin. Immunofluorescence can be used to identify the infection, as can precipitation, latex agglutination, and complement-fixation (most widely used) tests and quantitative immunodiffusion. These serologic aids are used to follow the course of the infection in a given patient. Currently a coccidioidal vaccine made from formalin-inactivated spherules has been successful in trial runs on human beings.

Cryptococcosis (Torulosis)

Cryptococcus neoformans (Torula histolytica) is a yeastlike organism that usually infects the lungs and central nervous system but may attack other parts of the body. *Cryptococcus* is the only encapsulated yeast to invade the central nervous system. With infection, multiple small nodules form, with the gross and microscopic appearance of tubercles. In the central nervous system the meninges are thickened and matted together and the brain is invaded (Fig. 35-27). In the patient with preexisting malignancy of the macrophage system, as well as in the debilitated or compromised host, this agent is an important opportunist.

Humans become infected through the skin, mouth, nose, and throat. Transmission from person to person or from animal to human has not been recorded. This fungus, saprophytic in nature, has been found in cattle, horses, dogs, and cats. Birds are not its hosts, but it is a saprophyte in pigeon droppings, and cases of cryptococcal meningitis have been traced to the vast pigeon populations found in many large cities. Pigeons are mechanical vectors, carrying the fungi on their feet and beaks. The birds are not affected, probably because of their high body temperature. *Cryptococcus neoformans* has a definite predilection for pigeon droppings, which are rich in creatinine. Creatinine is assimilated only by this organism and not by other species of cryptococci or other fungi.

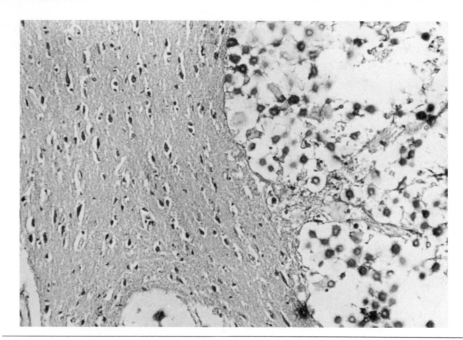

FIG. 35-27   Cryptococcosis of brain. To the right, a large mass of microorganisms with wide and prominent capsules occupies a cystic space in the brain. Considerable loss of brain tissue has resulted from their presence, and no inflammatory response is indicated, as is often the case with *Cryptococcus* in the central nervous system.

Since *Cryptococcus neoformans* is known also as *Torula histolytica*, infections with it may be referred to as either cryptococcosis or torulosis. Torulosis or cryptococcosis can be diagnosed with certainty only when the budding organisms are found in the affected tissues, pus, sputum, or cerebrospinal fluid (Figs. 35-28 and 35-29). In wet mounts (prepared with nigrosin) cryptococci are ovoid to spherical, budding yeast forms, 5 to 15 μm in diameter. Colonies of this fungus growing on culture media containing a substrate for phenol oxidase produce the brown pigment melanin. With fluorescent antibody technics the diagnosis may be made within hours. Agglutination tests are sensitive and specific, and counterimmunoelectrophoresis is useful. Four serotypes of *Cryptococcus neoformans*, A, B, C, and D, are demonstrated. Most human disease comes from type A. There is a **cryptococcin** skin test.

Histoplasmosis

Histoplasmosis, sometimes called **Darling's disease,** is caused by *Histoplasma capsulatum*, a diphasic (dimorphic) fungus—a single budding yeast at body temperature and a mold at room temperature and in nature. The fungus attacks primarily the macrophage system, where it parasitizes the component cells. Like coccidioidomycosis, the disease exists in the primary and progressive forms. The primary form involves the lungs (Fig. 35-30) but usually heals, leaving many small calcified foci in the lungs and lymph nodes of the chest. Most of the time the disease is the primary form, benign and self-limited. In the progressive disseminated form, ulcerating lesions are found in the nose and mouth, and there is enlargement of the spleen, liver, and lymph nodes. The progressive form is generally fatal. Histoplasmosis has generally not been considered an opportunistic fungal infection, but in the susceptible patient it can be a widely disseminated and serious one.

FIG. 35-28   *Cryptococcus neoformans* in culture. Moist, slimy, mucoid, cream-colored growth shown is typical for this yeastlike fungus. (Courtesy Dr. R.C. Reynolds, The University of Texas Health Science Center at Dallas.)

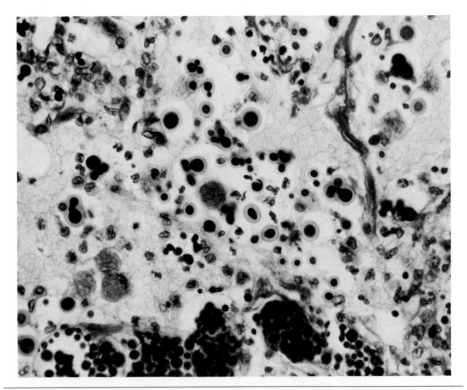

FIG. 35-29   *Cryptococcus neoformans (Torula histolytica)* among partially hemolyzed red cells in air sacs of lung. Wide gray capsules encompass budding yeast forms in this specially stained microsection. (From Kent, T.H., and Layton, J.M.: Am. J. Clin. Pathol. **38:**596, 1962.)

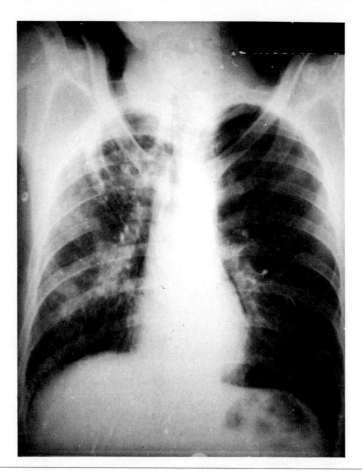

FIG. 35-30   Histoplasmosis of lung on x-ray film. An extensive strandlike and mottled infiltrate (whitish areas) involving upper two thirds of right lung field represents infection with *Histoplasma capsulatum.* (Air-containing, uninvolved lung tissue on the left is black.) (Courtesy Dr. Jack Reynolds, The University of Texas Health Science Center at Dallas.)

Humans probably contract the disease by inhalation of spores from fungi growing in the soil. Infection of the soil comes from the excreta of a variety of birds and bats in which the microbes have been found. No intermediate host is identified. Spores may be carried by prevailing winds and even by tornadoes. Outbreaks of the disease have been traced to inhalation of dust from caves. Histoplasmosis is also referred to as cave sickness or **speleonosis.** Victims of histoplasmosis and blastomycosis as well tend to be outdoor types—construction workers, farmers, spelunkers, and the like. The disease is encountered in the central Mississippi Valley and the Ohio Valley and in widespread areas of the world. The danger of working with soil contaminated with *Histoplasma capsulatum* can be minimized if only workers with a positive histoplasmin skin test are exposed and if the soil is decontaminated with a 3% formalin spray.

For identification of the fungi, stained smears and imprints, as well as cultures (Figs. 35-31 and 35-32), are made of peripheral blood, bone marrow (Fig. 35-33), aspirated material from lymph nodes, and sputum. To aid in the diagnosis and follow-up of histoplasmosis, the laboratory offers a skin test (the histoplasmin test or the histoplasmin tine test) and five serologic tests—precipitation, latex agglutination, complement fixa-

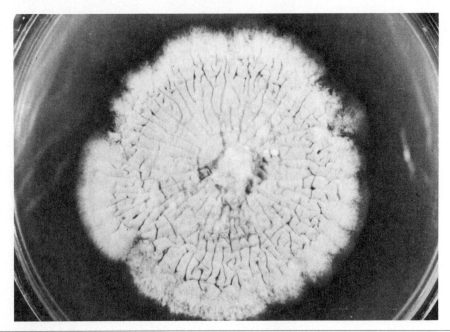

FIG. 35-31   *Histoplasma capsulatum* in culture. Note convoluted cerebriform yeastlike colony. (Courtesy Dr. R.C. Reynolds, The University of Texas Health Science Center at Dallas.)

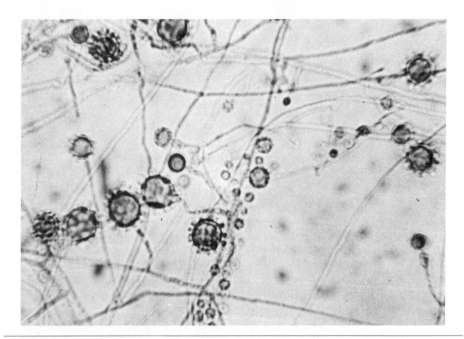

FIG. 35-32   *Histoplasma capsulatum*, microscopic appearance of large, thick-walled, tuberculate macroconidia unique among pathogenic fungi for *Histoplasma capsulatum*. (Courtesy Dr. R.C. Reynolds, The University of Texas Health Science Center at Dallas.)

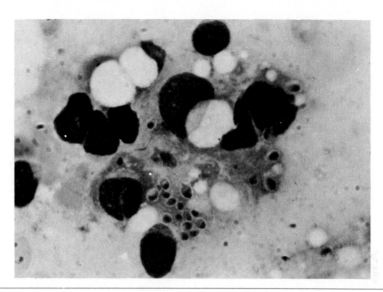

FIG. 35-33   Histoplasmosis of bone marrow. Small oval organisms 2 to 3 μm by 3 to 4 μm are seen within cytoplasm of microphages. Wright-stained smear examined microscopically. (×950.) (From Anderson, W.A.D., and Kissane, J.M., editors: Pathology, ed. 7, St. Louis, 1977, The C.V. Mosby Co.)

tion, fluorescent antibody detection, and immunodiffusion. Perhaps the most reliable is the complement-fixation test, the method of choice being the Laboratory Branch Complement-Fixation Test of the Centers for Disease Control.

Mucormycosis

Mucormycosis (**zygomycosis, phycomycosis**) can be an overwhelmingly acute and fatal infection caused by species of *Mucor* and *Rhizopus* and other normally harmless zygomycetes of the soil and decaying organic matter. The organisms are also ubiquitous in the hospital environment.

Diabetes, untreated and out of control, is the most important forerunner; the ketoacidosis rather than the hyperglycemia triggers the process. Along with disease in the lungs and central nervous system, an intraorbital cellulitis is a prominent feature. In the tissues the hyphae abound, very broad, 10 to 30 μm wide, and branching, especially within walls and lumen of blood vessels. No spores are seen, and little if any inflammation is present. In tissue section, hyphae are easily identified as belonging to Zygomycetes because they are nonseptate and coenocytic; that is, they contain many nuclei within a continuous mass of cytoplasm. The term *mucormycosis* is often used simply to indicate infections in which such hyphae are seen, and the demonstration of hyphae in infected tissues is essential for the diagnosis.

Other Fungal Diseases of Humans

Other fungal infections, infrequent in humans, follow:
1. Rhinosporidiosis (caused by *Rhinosporidium seeberi*)
2. Geotrichosis (caused by one or more species of *Geotrichum*)
3. Chromoblastomycosis (caused by three different fungi)
4. Penicilliosis (caused by certain species of *Penicillium*)
5. Piedra (caused by two different species of fungi)
6. Otomycosis (caused by species of *Penicillium* and other fungi)

Fungal Diseases of Lower Animals

Fungal infections are very important to the veterinary microbiologist. Some common ones follow:

1. Ringworm of the horse (caused by *Trichophyton equinum*); may occur in humans
2. Ringworm of horses, cattle, dogs, and possibly sheep and hogs (caused by *Trichophyton mentagrophytes*); human infection possible
3. Ringworm of cats and dogs (caused by *Microsporum canis*)
4. Epizootic lymphangitis of horses (caused by *Blastomyces farciminosus*)
5. Favus of chickens (caused by *Trichophyton gallinae*)
6. Coccidioidomycosis of cattle, dogs, horses, and sheep (caused by *Coccidioides immitis*); same organism causes human disease
7. Aspergillosis of wild and domestic fowls (caused by *Aspergillus fumigatus*)
8. Candidiasis of poultry (caused by *Candida albicans*)
9. Histoplasmosis of dogs, cats, cattle, sheep, swine, poultry, and horses (caused by *Histoplasma capsulatum*)
10. Cryptococcosis (mastitis) of cattle (caused by *Cryptococcus neoformans*)

Fungal Diseases of Plants

Molds and yeasts are economically important, for they cause so many diseases of plants. Plants imported from other countries are rigidly inspected on arrival in American seaports and points of entry. If a fungal infection is found, the plant or plant product is not allowed entry. Some important plant diseases caused by molds follow:

1. Brown rot of peaches and plums
2. Chestnut blight

Professor Matossian, writing in the *American Scientist*, states that the witchcraft trials of 1692 in Massachusetts and Connecticut, the worst outbreak of witch persecution in American history, may have been part of an unrecognized American health problem of the time.

The aberrant behavior of the men and women hanged as witches in Salem in 1692 may well have occurred because they had eaten the poisonous fungus ergot. The afflicted adults and children were described as behaving as though "Satan were loosed in Salem." When ergot is ingested, especially by children, it causes behavior reactions easily attributable to witches, such as fits, strange visions, and the like. Animals with convulsive ergotism may become wild, make loud distressed noises, and die. Several cows died during the 1692 outbreak.

Ergot grows on rye in cool, damp weather. Occasional cool periods in coastal lowlands, Matossian believes, could have caused the rye to become infected with ergot, and in the American colonies at that time rye bread was still a dietary staple. From information gathered from such records as court transcripts, climate indicators, and diaries, it can be determined that the growing season in eastern New England was abnormally cool in 1690, 1691, and 1692. Diaries kept in Boston recorded the intervening winters as "very cold." The households chiefly striken by the bewitchment were those closest to the marshy land. "New Englanders believed in witchcraft both before and after 1692, yet in no other year was there such severe persecution."

Modified from Matossian, M.K.: Ergot and the Salem witchcraft affair, Am. Scientist 70:355, 1982.

3. Mildew of grapes
4. White pine blister rust
5. Rust of oats, wheat, and barley
6. Smuts of various grains
7. Potato rot
8. Corn leaf blight

Ergotism

Ergot, a drug whose derivatives are widely used to check hemorrhage from childbirth, is composed of several alkaloidal poisons (mycotoxins) produced by the growth of a mold, *Claviceps purpurea*, in the grains of rye, wheat, and barley. The fungus (also referred to as ergot) grows as a purple-black, slightly curved mass that replaces the infected grain and converts it to a black sclerotium, from which the drug is extracted. An enzyme secreted by the hyphae is contained in a thick honeydew that attracts insects and helps to spread the fungus. When bread made from infected grain is eaten, the condition known as ergotism develops. Ergotism is described by gangrene of the extremities, abortion, and convulsions. It was at one time prevalent in Central Europe.

A specific component of the pharmacologically potent alkaloids produced by the fungus ergot is lysergic acid. A well-known derivative is the hallucinogen lysergic acid diethylamide, or LSD.

SUMMARY

1 Fungi abound in nature—100,000 species exist either as parasites on other living organisms or as saprophytes on their dead remains. About 100 species affect humans. Fungi are larger than bacteria and are eukaryotes.

2 Dimorphic fungi grow as single cells in the living host (the yeast phase) and as multicellular structures in the laboratory (the mold phase).

3 Multicellular fungi reproduce by the conversion of a spore from the reproductive mycelium into a vegetative fungus. Spores are asexual, or sexual if there is fusion of nuclei. Fungi producing sexual spores are "perfect fungi."

4 Fungi, mostly aerobic, grow best in warm, moist surroundings but much more slowly than bacteria. They are more tolerant to acid and are more resistant to cold than are bacteria.

5 Eumycota, the division of true fungi, contains (1) zygomycetes, or water molds, (2) ascomycetes, or sac fungi (here are single-celled yeasts crucial to baking and brewing), (3) basidiomycetes, or club fungi, and (4) deuteromycetes, a taxonomic wastebasket. The sexual state for deuteromycetes, or so-called Fungi Imperfecti, is still unknown; some are pathogens.

6 Fungal infection is insidious and progresses slowly; it also regresses slowly.

7 Fungi are studied in the laboratory by (1) direct microscopic examination, (2) cultural methods, and (3) standard immunologic reactions.

8 Superficial mycoses (ringworm, tinea), cosmetically disgusting but mild, involve the skin, hair, and nails and are caused by the dermatophytes *Microsporum*, *Trichophyton*, and *Epidermophyton*.

9 Sporotrichosis is a subcutaneous mycosis caused by *Sporothrix*, an agent widespread in nature.

10 Systemic mycoses may be primary, secondary, and complicating infections. Secondary or so-called superinfections are often hospital acquired. They are secondary to a disease for which antimicrobial drugs are being given.

11 Complicating infections come after special therapeutic procedures, such as peritoneal dialysis, and constitute a distinct hazard to the compromised host.

12 The best-known systemic mycoses are aspergillosis, blastomycosis, cryptococcosis, coccidioidomycosis, and histoplasmosis. Disease patterns for these overlap; therefore in every instance diagnosis *must be made* with the morphologic identification of the causative fungus.

13 The serious disease ergotism results from the consumption of cereal grains supporting the growth of the mold *Claviceps purpurea*.

QUESTIONS FOR REVIEW

1 Give the general characteristics of fungi.
2 Name the science that deals with fungi.
3 What is a fungal infection called? Cite five examples.
4 Make the distinction between molds and yeasts. Is it a clear one?
5 How do fungi perpetuate themselves?
6 What are dermatomycoses? Name three major dermatophytes.
7 Outline the laboratory diagnosis of fungal disease.
8 Name and briefly describe the chief systemic diseases caused by fungi.
9 What is ergotism? What is the pharmacologic nature of LSD?
10 Classify fungi.
11 Make pertinent comments regarding the pathology of fungal disease.
12 What threat does fungal infection pose to the "compromised host"? Cite the common offenders.
13 Briefly define speleonosis, torulosis, mycotoxin, maduromycosis, tinea, superinfection, complicating infection (in fungal disease), mycelium, aflatoxins, favus, otomycosis, thallus, cave sickness, germ tubes, "ids," dimorphism, conidia, pseudohyphae, mycoserology, and thrush.

SUGGESTED READINGS

Ainsworth, G.C.: Introduction to history of mycology, New York, 1976, Cambridge University Press.

Ainsworth, G.C., and others: The fungi, New York, 1973, Academic Press, Inc.

Balows, A.: Serodiagnosis of mycotic disease. Springfield, Ill., 1978, Charles C Thomas, Publisher.

Beneke, E.S.: Human mycoses, a Scope publication, Kalamazoo, Mich., 1979, The Upjohn Company.

Cooper, B.H.: Rapid methods for laboratory diagnosis of fungal disease, Lab. Manage. 19:37, April 1981.

Delacrétaz, J., and others: Color atlas of medical mycology, Chicago, 1976, Year Book Medical Publishers, Inc.

Denys, G.A., and others: Evaluation of a commercial exoantigen test system for the rapid identification of systemic fungal pathogens, Am. J. Clin. Pathol. 79:379, 1983.

Graham, A.R.: Fungal autofluorescence with ultraviolet illumination, Am. J. Clin. Pathol. 79:231, 1983.

Gregory, D.W.: Saturday conference: *Candida* infections, South. Med. J. 75:339, 1982.

Grotte, M., and Younger, B.: Sporotrichosis associated with sphagnum moss exposure, Arch. Pathol. Lab. Med. 105:50, 1981.

Haley, L.D., and others: Cumitech 11: practical methods for culture and identification of fungi in the clinical microbiology laboratory, Washington, D.C., 1980, American Society for Microbiology.

Kauffman, C.A., and others: Detection of cryptococcal antigen, comparison of two latex agglutination tests, Am. J. Clin. Pathol. 75:106, 1981.

Land, G.A., and others: Rapid identification of medically important yeasts, Lab. Med. 10:533, 1979.

Larson, D.M., and others: Primary cutaneous (inoculation) blastomycosis: an occupational hazard to pathologists, Am. J. Clin. Pathol. 79:253, 1983.

Lawrence, R.M.: Acute pulmonary blastomycosis acquired in West Texas, Tex. Med. 77:50, 1981.

MacDonald, F., and Odds, F.C.: Purified *Candida albicans* proteinase in the serological diagnosis of systemic candidosis, J.A.M.A. **243**:2409, 1980.

Mann, J.L.: Autofluorescence of fungi: an aid to detection in tissue sections, Am. J. Clin. Pathol. **79**:587, 1983.

Martin, M.V., and White, F.H.: A microbiologic and ultrastructural investigation of germ-tube formation by oral strains of *Candida tropicalis*, Am. J. Clin. Pathol. **75**:671, 1981.

Schwarz, J.: Histoplasmosis, New York, 1981, Praeger Publishers.

Singleton, P., and Sainsbury, D.: A dictionary of microbiology, New York, 1978, John Wiley & Sons, Inc.

Snell, W.H., and Dick, E.A.: A glossary of mycology, ed. 2, Cambridge, Mass., 1971, Harvard University Press.

PROTOZOA
Medical Parasitology

Parasites are generally defined as organisms that require living matter for their nourishment; that is, they must live within or on the bodies of other living organisms. Parasites are the weaker or dependent organisms, deriving all the benefit from the relation. The science dealing with such organisms is **parasitology.** According to this definition, a parasite may be a bacterium, virus, rickettsia, protozoon, plant (for example, mistletoe), or animal. However, by common usage, *medical parasitology* refers to certain parasites of medical interest—protozoa, helminths, and arthropods—and their diseases, some of the most important and prevalent diseases known. Table 36-1 lists the important protozoan diseases with the pertinent features of each. Although parasitic infections are commonly treated as exotic diseases of another world and the parasitic burden of North Americans is not great, it is nonetheless true that practically every parasitic disease known to human beings has been recognized in recent times in the United States.

The animal within or on which a parasite lives is the **host.** All stages of the parasite's development may take place in the same animal host. On the other hand, a parasite may have one or more hosts. It undergoes its larval or asexual stage in the **intermediate** host and its adult or sexual stage in the **definitive** host. A parasite that lives within the body of the host is known as an **endoparasite;** one that lives on the outside of the body is an **ectoparasite.** A tapeworm is an example of an endoparasite; a louse is an ectoparasite.

GENERAL CHARACTERISTICS

Protozoa have traditionally been classified in the animal kingdom, wherein, as unicellular microorganisms, they have represented the lowest form of animal life. In many instances this classification still holds. Current classification places them within the kingdom Protista, in the subkingdom or branch designated Protozoa. This approach has not yet been universally accepted. Protozoa are more complex in their functional activities than bacteria or the average cell of a multicellular organism. Each is a complete, self-contained unit, with special structures known as **organelles** to carry out such functions as nutrition, locomotion, respiration, excretion, and attachment to objects. The

Table 36-1 Overview of Protozoan Parasitic Diseases

Disease	Causative Agent	Infective Cyst	Expected Habitat in Humans	Usual Portal of Entry	Common Source of Infection	Name of Arthropod Vector	Practical Laboratory Diagnosis Morphologic	Practical Laboratory Diagnosis Immunologic
Amebiasis	*Entamoeba histolytica*	Yes	Large bowel lumen and wall	Mouth	Fecally contaminated water, food	None	X	X
African sleeping sickness	*Trypanosoma gambiense, Trypanosoma rhodesiense*	No	Blood, lymph, tissue fluids	Skin	Insect bite	Tsetse fly (*Glossina*)	X	X
South American trypanosomiasis	*Trypanosoma cruzi*	No	Blood, tissue cells	Skin	Insect bite	Triatomid bug (*Triatoma*)	X	X
Leishmaniasis (visceral, cutaneous)	*Leishmania donovani, Leishmania tropica, Leishmania braziliensis*	No	Macrophages of skin, mucous membranes	Skin	Insect bite	Sandfly (*Phlebotomus*)	X	X
Trichomoniasis	*Trichomonas vaginalis*	No	Vagina—female; urethra, prostate—male	Genitalia	Trophozoites in genital secretions	None	X	
Giardiasis	*Giardia lamblia*	Yes	Duodenum, upper jejunum	Mouth	Fecally contaminated water	None	X	
Malaria	*Plasmodium vivax, Plasmodium falciparum, Plasmodium ovale, Plasmodium malariae*	No	Liver cells, red blood cells	Skin	Insect bite	*Anopheles* mosquito	X	
Toxoplasmosis	*Toxoplasma gondii*	Yes	All organs	Mouth	Food contaminated with cat feces	None	X	X
Pneumocystosis	*Pneumocystis carinii*	Yes	Lungs	Respiratory tract	Respiratory secretions (?)	None	X	
Balantidiasis	*Balantidium coli*	Yes	Large bowel	Mouth	Fecally contaminated food	None	X	

majority are of microscopic size. Protozoa range in size from 1 μm to 50 mm or more; most are from 5 to 250 μm. As a rule, the pathogenic ones are smaller than the nonpathogenic ones. They may be spherical, spindle, spiral, or cup shaped.

In medical parasitology identification of a given animal parasite is of paramount importance. Practically, this is done by the recognition of specific structural (morphologic) details in the makeup of the given parasite. There are many species, but only about 30 affect humans.

Structure

Protozoa are units of protoplasm differentiated into cytoplasm circumscribed by the cell or plasma membrane and nucleus encased by the nuclear envelope. Most have just one vesicular nucleus. The cytoplasm is separated into a homogeneous **ectoplasm** and a granular **endoplasm**. The ectoplasm helps to form the various organs of locomotion, contraction, and prehension, such as pseudopods, flagella, cilia, and suctorial tubes, and in certain species the ectoplasm contains a definite opening or portal for intake of food. The endoplasm digests food materials and surrounds the nucleus.

Many protozoa, particularly pathogenic ones, absorb fluid directly through the plasma membrane. Most take in solid particles, such as small animal or vegetable organisms, and digest them enzymatically. Because their food consists chiefly of bacteria, protozoa may be important in limiting the bacterial population of the universe. Waste material is excreted through the cell membrane or, in some cases, through an ejection pore. Parasitic protozoa can utilize atmospheric oxygen and release carbon dioxide, but since free oxygen is rare in the intestinal lumen and in tissues, most have an anaerobic metabolism.

Locomotion

All protozoa possess some type of motility. It may be by pseudopod formation (Fig. 36-1) or by the action of flagella or cilia. For locomotion by **pseudopod** (false foot) formation a sharp or blunt ectoplasmic process flows forward, pulling the rest of the organism after it. **Flagella** are whiplike prolongations of protoplasm that propel the organism by their lashing motions. Some protozoa have only one flagellum; others have several. Some of the flagellate protozoa also have an **undulating membrane** to help in locomotion. This is a fluted membranous process attached to one side of the organism. **Cilia** are similar to flagella except that they are shorter, more delicate, more plentiful, and attached to the entire outer surface of the microbe. Individually they are less powerful than flagella, but the synchronous action of the many cilia accomplishes the most rapid motion of which unicellular organisms are capable.

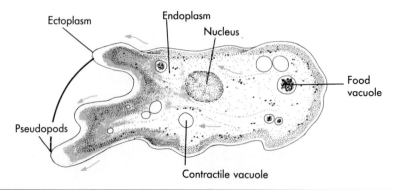

FIG. 36-1   Diagram of active locomotion in an ameba. Arrows indicate direction.

Cyst Formation

When protozoa are subjected to adverse conditions, they become inactive, assume a more or less rounded form, and secrete and surround themselves with a resistant membranous wall within which they may live for a long time and resist various destructive agents in their environment. This is **cyst** formation. When conditions suitable for growth are reestablished, the cyst imbibes water, and the protozoon returns to the vegetative state. Sometimes cyst formation precedes reproduction. Since vegetative protozoa, or **trophozoites**, are very susceptible to deleterious influences and cysts are very resistant, it is the cysts that are primarily responsible for the spread of protozoan infections.

Reproduction

Protozoa, unlike worms or helminths, multiply within their hosts. A few taken into the host soon give rise to many. In protozoa reproduction may be either sexual or asexual (simple fission). In some (for example, *Plasmodium* of malaria) the sexual cycle occurs in one species of animal and the asexual cycle in another. The sexual cycle occurs in the **definitive** host, the asexual cycle in the **intermediate** host. Protozoan cells capable of sexual reproduction are known as **gametes**. The cell formed by the union of two gametes is a **zygote**. Asexual reproduction occurs in amebas and flagellates. Lengthwise or crosswise division of the protozoon yields two new members of the species.

Classification

The designation Protozoa contains four important categories of organisms of medical interest in human beings. In each the method of locomotion varies. These are variously classified. The names used in this text are those used when these categories were considered classes.

Rhizopodea

This category includes the pathogenic amebas and some nonpathogenic ones (Fig. 36-2). Locomotion is characterized by pseudopod formation, and the cytoplasm is divided into ectoplasm and endoplasm. One nucleus is usual. (In a recent classification phylum I is Sarcomastigophora, subphylum III is Sarcodina, superclass 1 is Rhizopoda, and class 1 containing *Entamoeba* is Lobosea.)

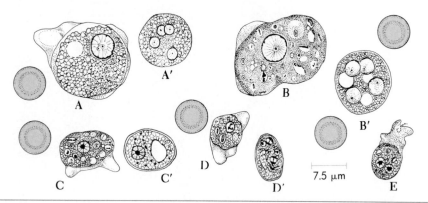

FIG. 36-2   Diagram of protozoa: amebas—pathogenic and nonpathogenic. Note rounded cyst to right of trophozoite. **A** and **A′**, *Entamoeba histolytica* (the pathogen, the rest commensal). **B** and **B′**, *Entamoeba coli*. **C** and **C′**, *Iodamoeba bütschlii (williamsi)*. **D** and **D′**, *Endolimax nana*. **E**, *Dientamoeba fragilis*, now classified as a trichomonad (p. 798). (Red blood cells, 7.5 to 8 μm in diameter, are included as references for size of protozoa.)

Zoomastigophorea

These protozoa are commonly called flagellates because movement is by means of flagella and an undulating membrane. Flagellates have two nuclei, and the cytoplasm is not differentiated into endoplasm and ectoplasm. Cell bodies are often pear shaped and fixed in outline. The most important flagellates medically are in the genera *Trypanosoma, Leishmania, Trichomonas, Giardia, Chilomastix* (Fig. 36-3), and *Dientamoeba* (Fig. 36-2, *E*). (In a recent classification phylum I is Sarcomastigophora, subphylum I is Mastigophora, and class 2 is Zoomastigophorea.)

Sporozoa

There are no external organs of locomotion here. These organisms living within the cells, tissues, cavities, and fluids of the body are represented by *Plasmodium* of malaria. Also placed here are *Toxoplasma gondii*, which moves by bending and gliding movements of its body, and as advocated by some authors, *Pneumocystis carinii*, which may form pseudopods. *Pneumocystis carinii* is still in an uncertain taxonomic position. (In a recent classification phylum III is Acicomplexa, class 2 is Sporozoea, and subclass 2 is Coccidia.)

Ciliata

Cilia are present for locomotion. The only pathogenic member in this category is *Balantidium coli* (Fig. 36-3, *B* and *B'*). (In a recent classification phylum VII is Ciliophora, and class I is Kinetofragminophorea.)

Laboratory Diagnosis

The structure of many protozoan parasites makes it easy for them to be identified microscopically in suitably prepared clinical specimens such as blood or stool. At times, however, a morphologic diagnosis is not possible in parasitic infections.

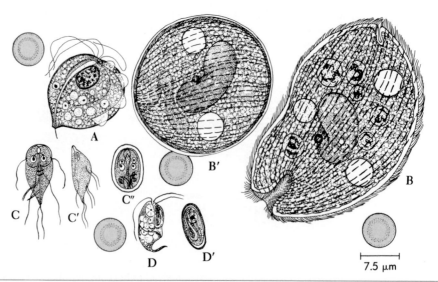

FIG. 36-3 Diagram of protozoa, flagellates and the ciliate. Note rounded or ovoid cyst to the side of the trophozoite. **A,** *Trichomonas vaginalis* (no cyst). **B** and **B',** *Balantidium coli* (**B'** the cyst). **C** to **C'',** *Giardia lamblia* (**C'** the side view). **D** and **D',** *Chilomastix mesnili.* (Red blood cells, 7.5 to 8 μm in diameter, are included as references for size of protozoa.)

Table 36-2   Immunodiagnostic Tests of Value in Protozoan Diseases

	Amebiasis	Chagas' Disease	African Trypanosomiasis	Leishmaniasis	Malaria	Toxoplasmosis	Pneumocystosis	Giardiasis
Serologic Test								
Complement fixation	X*	X	X	X	X	X	X	
Indirect hemagglutination	X	X	X	X	X	X		
Indirect immunofluorescence	X	X	X	X	X	X	X	X
Precipitin	X	X	X	X	X	X		X
Immunoelectrophoresis	X	X		X	X			
Double diffusion	X	X	X	X	X			X
Enzyme-linked immunosorbent assay (ELISA)	X	X	X	X	X	X		X
Intradermal Test				X		X		

From Kagan, I.G.: Hosp. Pract. **9:**157, Sept. 1974.
*X indicates an accepted test for routine use.

Immunodiagnosis

Although incompletely defined, immunity in parasitic infections is both humoral and cell mediated. Fortunately parasites possess a variety of antigens with distinct components within their makeup and therefore lend themselves nicely to serologic testing. Antibodies produced may be composed of different immunoglobulin classes (M, G, or E), and elevations, for instance, of total IgM or IgE may be helpful in diagnosis. When present, skin sensitizing antibodies of IgE or of a subclass of IgG form the basis of the immediate reaction of the skin test.

The immunodiagnostic tests that have been standardized for the identification of the protozoa of this chapter and the metazoa of the next are applications of complement-fixation, indirect hemagglutination (most sensitive), latex agglutination, precipitation, indirect fluorescent antibody detection (indirect immunofluorescence), counterimmunoelectrophoresis, bentonite flocculation, double diffusion, enzyme-linked immunosorbent assay (ELISA), and intradermal testing. For most situations there is not necessarily one best test; two or more may have to be used for clinical evaluation. Tables 36-2 and 37-6 show the applications of various ones to diseases for which they are useful. Within the last few years at least 30 reagents and tests have been packaged commercially for the serodiagnosis of major parasitic diseases. Well-known ones are the kits and reagents for the protozoan diseases amebiasis and toxoplasmosis and for the metazoan diseases echinococcosis and trichinosis.

Prevention and Control

Measures for the control of protozoan infections include (1) the eradication of sources of infection in human beings by treatment and by dissemination of information designed to prevent spread; (2) the sanitary control of water, food and food handlers, living and working conditions, and waste disposal; and (3) the institution of measures that interfere with the biologic activity of the parasite, such as the destruction or control of reservoir hosts and vectors. To date, effective immunization against parasitic diseases has not been accomplished. One of the few exceptions is with the disease cutaneous leishmaniasis. For a long time in the Middle East, vaccination of a child has been practiced (away from the face) using material from an active leishmanial sore because children, particularly little girls, are so protected not just from the disease but more especially from the ugly scars resulting from it.

PROTOZOAN DISEASES

Amebiasis

Amebiasis means infection with *Entamoeba histolytica*. The disease occurs in two forms: acute amebiasis (amebic dysentery), characterized by an intense dysentery with bloody, mucus-filled stools, and chronic or latent amebiasis, described by vague intestinal disturbances, muscular aching, loss of weight, even constipation, or no manifestations at all. Latent amebiasis is a long-term process with few symptoms that may suddenly flare up to acute serious disease. An estimated 3% to 5% of persons in the United States are affected by amebiasis. Worldwide an estimated 400 million persons have this parasitic infection.

The Organism

The organism *Entamoeba histolytica* exists as a vegetative ameba, or trophozoite, and as a cyst (Fig. 36-2, A and A'). The trophozoite represents the feeding stage; the cyst, the protective stage of this protozoon. Vegetative trophozoites possess an active type of ameboid motion on a warm microscopic stage. Microscopically one sees the pseudopods, a distinctive nucleus, and red blood cells within the cytoplasm of the trophozoite. Vegetative amebas are very susceptible to injurious agents. In an unfavorable environment they quickly succumb; therefore they do little to transmit the disease. The cysts are nonmotile; smaller (average diameter 12 μm) than vegetative amebas, which are 8 to 25 μm in diameter; and surrounded by a resistant wall.

Life History

The life cycle of *Entamoeba histolytica* begins with the cysts by which the disease is transmitted from person to person (Fig. 36-4). After the cysts are passed in the feces, they remain infectious for several days if not destroyed by heat and drying. When the cysts are swallowed by a new host, they pass through the stomach unchanged. The shells are dissolved by juices of the small intestine in the ileocecal area, and the vegetative forms are liberated. The trophozoites pass to the large intestine to attack the mucous membrane and produce ulceration. Since amebas do not possess mitochondria, they depend on resident bacteria for help when they establish themselves in the bowel. The vegetative amebas multiply in the ulcers; some escape into the lumen of the intestine. If diarrhea is present, they are swept out of the intestinal tract. If diarrhea is not present, they multiply one or more times and then encyst. Encystment occurs in the large bowel, not outside the body. Cysts are excreted in the feces.

Sources and Modes of Infection

The life cycle of *Entamoeba histolytica* reveals three facts: (1) infection can be acquired only by swallowing cysts, (2) infection comes from the feces of a person excreting cysts, and (3) acute cases are of little danger. The feces of patients with acute amebiasis contain largely vegetative parasites that die quickly; they could not survive the acid gastric juice should they accidentally be swallowed. Infection is usually acquired by ingestion of uncooked food contaminated with feces containing cysts. The most important single source of infection is the food handler with chronic amebiasis, especially the one preparing uncooked foods. Other sources of infection are vegetables fertilized with human

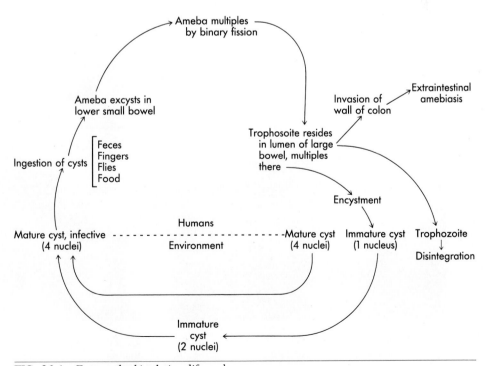

FIG. 36-4   *Entamoeba histolytica*, life cycle.

excreta and drinking water contaminated with sewage. Apparently the latter was the
cause of the Chicago epidemic of 1933, in which there were 1409 cases with 98 deaths.
The water in two hotels had been contaminated by sewage. Flies and other insects may
spread the cysts mechanically, and spread may come from direct hand-to-mouth contact.

Lesions

In the majority (85% to 95%) of cases there seems to be a state of balance between the
amebas and the host. The patient experiences mild disturbances or none at all and is
able to repair the ulcers almost as fast as they are formed. This is chronic or latent
amebiasis.

If the resistance of the host is lowered or massive infection occurs, the host is unable
to repair the ulcers as fast as they are formed, and the increasing ulceration causes a
violent dysentery in which the stools consist entirely of blood and mucus. This is acute
amebiasis or amebic dysentery. Occasionally intestinal perforation occurs. Amebic ulcers are deep, ragged ulcers with undermined edges, described as flask-shaped or water
bottle lesions.

Extraintestinal amebiasis. A serious concern with intestinal amebiasis is that the
complications outside the intestinal lumen are life-threatening ones. Sometimes amebas
penetrate deeper in the intestinal wall and enter tributaries of the portal vein to be
carried to the liver. Here they produce amebic hepatitis or a liver abscess (Fig. 36-5).
The "abscess" is actually not a true one, since it contains necrotic liver tissue that is
not associated with an inflammatory or tissue reaction at the borders. The liver abscess
forms with lytic destruction of liver tissue by the amebas, and classically the fluid contents have always been referred to as "anchovy paste." (A liver abscess can develop

FIG. 36-5   Amebic abscess of liver, cross section of the organ. On the surface of this slice of
liver tissue may be seen a large rounded mass of soft necrotic material that represents the
abscess.

without being clinically conspicuous.) Amebiasis in the liver is not walled off; thus it can extend either to perforate the diaphragm and rupture into the chest cavity and lung or, occasionally, to rupture into the abdominal cavity. Amebic abscesses, metastatic from an extraintestinal site in the liver, can form in the lung, brain, and various other organs of the body (Fig. 36-6).

Laboratory Diagnosis

The laboratory diagnosis of amebiasis necessitates the examination of *fresh warm* stools for vegetative amebas and the examination of ordinary specimens for cysts. The examination of iron hematoxylin–stained smears of specimens is helpful. Some authorities state that *Entamoeba histolytica* should never be identified without careful study of nuclear and cytoplasmic detail in a permanently stained preparation. A stool specimen positive for the presence of *Entamoeba histolytica* is indisputable proof of amebiasis.

Some medications—antimicrobial drugs, barium, soap, hypertonic enema solu-

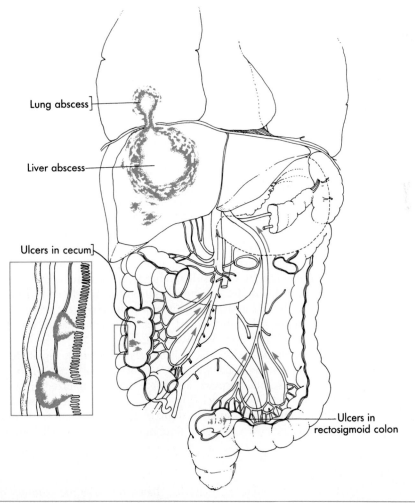

Lung abscess

Liver abscess

Ulcers in cecum

Ulcers in rectosigmoid colon

FIG. 36-6   Lesions in amebiasis. Amebic ulcers in the large bowel are called "water bottle" ulcers because of the characteristic shape *(inset)*. From these ulcers amebas penetrate the intestinal wall to enter the portal venous system and be carried to the liver. (*Arrows* indicate the flow of blood in the veins to the liver.) An abscess in the liver or hepatitis may result, and at times the trophozoites move across the diaphragm to form an abscess in the lung.

tions, and the like—if given even 2 weeks before laboratory examination of feces, may cause the amebas to disappear. Because the trophozoites are lysed in water, stool specimens are better collected on paper or in suitable cartons.

Artifacts in feces may confuse or obscure the presence of amebas in the specimen. This is so because of the cellular material in feces that may be mistaken for cysts, such as partly digested plant or animal cells, epithelial cells, macrophages, and pus cells.

Although morphologic recognition of *Entamoeba histolytica* under the microscope is the prime concern of the laboratory, serologic tests are used to advantage to identify this parasite. These include an indirect hemagglutination test, an indirect fluorescent antibody test, and an agar gel double diffusion technic. The ELISA test (p. 408) has been adapted to a commercial kit for the detection of amebiasis in the routine laboratory. A positive test reaction exhibits a red dye.

Prevention

The prevention of amebiasis depends on the proper control of carriers, proper sanitary supervision of foods, and general cleanliness. Protective immunity has not been demonstrated in humans. Note that routine chlorination of water does *not* destroy cysts of *Entamoeba histolytica* (p. 228).

Other Amebas

Besides *Entamoeba histolytica*, other amebas may be found in the intestinal canal, but *Entamoeba histolytica* is the only one that causes disease. The harmless, nondisease-producing ones are referred to as **commensals.**

A well-known commensal is *Entamoeba coli* (Fig. 36-2, *B* and *B'*). It is considered to be nonpathogenic because it does not invade tissues, but it must be distinguished from *Entamoeba histolytica*, which it closely resembles. *Entamoeba coli* looks much like *Entamoeba histolytica* except that its trophozoites are usually larger (12 to 30 μm in diameter), their movements in fresh stool are distinctive, and they do not normally ingest red blood cells. The cysts average 18 μm in diameter and possess a much thicker cyst wall than do those of *Entamoeba histolytica*.

Endolimax nana (Fig. 36-2, *D* and *D'*) is a comparable nonpathogenic ameba except that it is smaller. Trophozoites range from 6 to 12 μm in diameter. Cysts are oval, usually contain four nuclei, and average 8 μm in diameter.

Iodamoeba bütschlii is still another nonpathogenic ameba with a vegetative and a cyst phase in its life cycle. Trophozoites are similar to those of *Entamoeba coli* but smaller, averaging 10 μm in diameter. One or more glycogen masses detectable with iodine are present and persist into the cyst, hence the name *Iodamoeba*. Cysts, 10 μm in diameter, vary in shape (Fig. 36-2, *C* and *C'*).

Dientamoeba fragilis, like *Entamoeba histolytica*, ingests red blood cells. Its habitat is the cecum and appendix, and it is occasionally related to low-grade disease in humans. Unlike other amebas discussed, only a trophozoite (no cyst) is found, 10 μm in diameter, with two nuclei (Fig. 36-2, *E*). This microorganism has long been placed with the amebas although it displays certain differences. It is now classified with the trichomonads, even though it does not move with flagella but with pseudopods.

Infections with Hemoflagellates

Hemoflagellates

As the name indicates, these flagellated protozoa are parasites of the peripheral blood (in the plasma, not in blood cells) and known as medically important members of the genera *Trypanosoma* and *Leishmania*. Hemoflagellates have both a vertebrate and an invertebrate host. At one time in their life cycle they live in blood and fixed tissues of vertebrates (all classes), and at another time they live in the intestines of bloodsucking insects. Hemoflagellates are either elongate with a single, whiplike flagellum or rounded with a very short one. Not only does the flagellum propel the organism, but it also facilitates attachment of the parasite in the intestine and salivary gland of the insect host. Unique to these protozoa is the sausage-shaped kinetoplast containing mitochondrial DNA; mitochondria arise from it. All hemoflagellates possess a single nucleus.

Morphologic forms. Hemoflagellates vary in their appearance according to the phase of their life cycle and the host parasitized (Fig. 36-7). Differences in body shape, position of kinetoplast, and development of the flagellum define four morphologic stages—trypomastigote, amastigote, promastigote, and epimastigote (Table 36-3). (Adult, larval, or sexual forms cannot be differentiated with hemoflagellates.)

The definitive stage of the genus *Trypanosoma* is the **trypomastigote.** In this form the kinetoplast is near the posterior end of the body with an undulating membrane extending the full length of the body. The flagellum runs along the surface usually extending out as a free whip anterior to the body. The flagellar membrane is closely apposed to the body surface so that beating of the flagellum pulls this area into a fold, which together with the flagellum constitutes the undulating membrane.

The **amastigote** is a small oval parasite with a very short flagellum that projects only slightly beyond the flagellar pocket. It is seen in some of the hemoflagellate life cycles and is the definitive stage in the genus *Leishmania.*

The **promastigote** form is found in insect hosts as an elongate, slender parasite with an anterior kinetoplast and a free flagellum. The **epimastigote,** still another stage en-

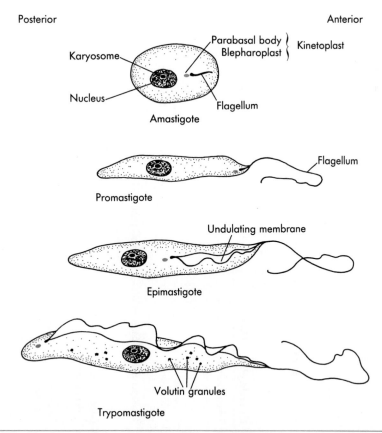

FIG. 36-7 The four developmental stages of hemoflagellates in hosts parasitized. Note changes in body shape, position of kinetoplast, and presence of a flagellum and undulating membrane.

Table 36-3 Expression of Morphologic Forms of Hemoflagellates

Hemoflagellate	Amastigote		Promastigote		Epimastigote		Trypomastigote	
	Human	Vector	Human	Vector	Human	Vector	Human	Vector
Trypanosoma gambiense	—	—	—	Salivary glands—tsetse fly	—	Salivary glands—tsetse fly	Blood, lymph nodes, central nervous system	Proboscis—tsetse fly
Trypanosoma rhodesiense	—	—	—	Salivary glands—tsetse fly	—	Salivary glands—tsetse fly	Blood, lymph nodes, central nervous system	Proboscis—tsetse fly
Trypanosoma cruzi	Skin, multiple sites	Gut—triatomid bug	—	Hindgut—triatomid bug	—	Hindgut—triatomid bug	Blood	Feces—triatomid bug
Leishmania donovani	Liver, spleen, lymph nodes	—	—	Midgut—sandfly	—	—	—	—
Leishmania tropica	Skin	—	—	Midgut—sandfly	—	—	—	—
Leishmania brasiliense	Skin, mucosae	—	—	Midgut—sandfly	—	—	—	—

countered in some of the life cycles, is similar to the promastigote. In this stage the kinetoplast has shifted in front of the nucleus, and a short undulating membrane has developed, which becomes a free flagellum.

Trypanosomes

Trypanosomes (Fig. 36-8), of which there are many species, are hemoflagellates, and their infection is trypanosomiasis. The two types important to humans are African trypanosomiasis, or African sleeping sickness, and South African trypanosomiasis, or Chagas' disease.

Trypanosomes are spindle-shaped protozoa, 14 to 33 μm long, that enter the bloodstream of many different species of animals. In their vertebrate hosts most trypanosomes live in blood, in tissue spaces (lymph nodes and spleen, especially), and in the cerebrospinal fluid. With the exception of *Trypanosoma cruzi*, they do not invade cells. They are especially abundant in lymph vessels and in the intercellular spaces of the brain.

Life cycle. The 15- to 35-day cycle in the insect begins when the biting insect ingests trypanosomes along with the blood meal. In the insect the parasites are first found in the midgut, where they multiply as trypomastigotes for about 10 days. Then the parasites move forward into the foregut, finally reaching the salivary glands. Here they become epimastigotes in form and either attach to epithelial cells or remain free. Again they multiply. After several generations the cycle in the insect is complete with the appearance of small, stumpy forms lacking a free flagellum, called metacyclic trypomastigotes.

The metacyclic trypomastigotes are the infective forms for the vertebrate host, and an infected insect can inject several thousand with a single bite. Once in the new host, the trypanosomes multiply as trypomastigotes in blood and lymph and, after a period of time, invade the central nervous system.

African Trypanosomiasis

African trypanosomiasis is seen in two forms: Gambian trypanosomiasis (agent, *Trypanosoma gambiense*) and the more virulent Rhodesian trypanosomiasis (agent, *Trypanosoma rhodesiense*). Each is transmitted by a species of the tsetse fly, a fly found only in Africa. The fly becomes infected by ingesting the blood of a person with the disease, and the parasite undergoes a cycle of development in its body. From the salivary glands of the fly, they are transferred to persons bitten. (The skin is the portal of entry.) The incubation period is 1 to 2 weeks. Cattle, swine, and wild game animals, especially antelope, may harbor the parasites and be a source of human infection (Fig. 36-9).

Early in the course of either form of trypanosomiasis there are acute episodes of fever and inflammation of lymph nodes as the trypanosomes multiply in the bloodstream and invade lymph nodes, liver, and spleen. The Rhodesian form is usually fatal within a

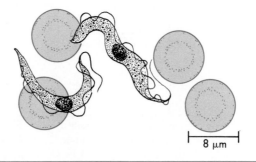

8 μm

FIG. 36-8   *Trypanosoma gambiense* sketched in blood smear.

mattter of months and rarely progresses to the chronic stages of the Gambian form. In the end stages of the disease invasion of the brain and its coverings produces the celebrated and uncontrollable sleepiness. Emaciation of the victim is extreme.

The old Lingala curse "Owa na ntolo" expresses the African native's fear of sleeping sickness. Translated, it reads, "May you die of sleeping sickness!"

South American Trypanosomiasis

South American trypanosomiasis, or Chagas' disease, is caused by *Trypanosoma cruzi*, a thin, undulating organism about 20 μm long with one flagellum. The insect vectors here are triatomid bugs (Reduviidae), kissing or cone-nosed bugs, related to bedbugs. These small, bloodsucking insects transfer the disease from a reservoir in humans and in domestic and wild animals such as dogs, cats, rats, armadillos, and opossums. Infection results from the contamination of the skin with insect feces and is not transferred by the actual insect bite. South American trypanosomiasis differs from African trypanosomiasis in that the parasites multiply in the tissue cells rather than in the blood. They reappear in the blood to be picked up by the vector. If the patient survives the acute stage, the disease becomes chronic, with the trypanosomes localized in various organs.

Laboratory diagnosis. During the fever trypanosomes of the African disease may be demonstrated in Giemsa-stained films of peripheral blood. Concentration technics for peripheral blood facilitate the search for parasites. Smears and imprints of lymph nodes may contain them. *Trypanosoma cruzi* is also identified in aspirated material from spleen, liver, lymph nodes, and bone marrow.

Xenodiagnosis refers to the demonstration of infective organisms in the feces of clean laboratory-bred bugs that are allowed to feed on a given patient. It is a method applicable to the diagnosis of South American trypanosomiasis. From a colony of triatomid

FIG. 36-9 Economic impact of parasitic disease. Shaded area on map of Africa represents about 7 million km of land rendered unproductive because of trypanosomiasis, land that could otherwise be grazed. Trypanosomiasis kills 3 million cattle annually; this area could support an estimated 120 million head of cattle. Map of British Isles is a scale reference.

bugs reared on pigeon's blood, 10 hungry ones are allowed access to a given patient. These are kept for 10 days, at the end of which time their excreta are examined for the causative trypanosomes.

Serologic methods useful in diagnosis are given in Table 36-2.

Prevention. Any design for a vaccine against trypanosomiasis must provide for frequent changes in the antigens of the African parasites and for a multiplicity of antigens in each strain of the South American one.

An experimental vaccine has been prepared, with the microorganisms of a culture being killed by physical means—subjection of the trypanosomes to high-frequency sound waves, to pressure, or to mechanical forces evoked when the culture is shaken with glass beads. It has been used only in mice.

Leishmaniasis

Leishmaniasis is a protozoan disease caused by what is probably human beings' most ancient parasite. The disease exists in two forms: the visceral (agent, *Leishmania donovani*) and the cutaneous (agents, *Leishmania tropica* and *Leishmania braziliensis*). The incubation period extends from 2 weeks to many months or even several years.

In the vertebrate host, the hemoflagellate, the small, oval amastigote, 2 to 5 by 1 to 3 μm, resides and multiplies within macrophages and even ruptures them. Some species live in endothelial cells. In the insect the longer promastigote, 14 to 20 by 1.5 to 4 μm, is found.

Visceral leishmaniasis (kala-azar, dumdum fever). The visceral form of leishmaniasis is characterized by fever, enlargement of the spleen and liver, progressive emaciation, weakness, and in untreated patients, death. The agent, *Leishmania donovani*, is transmitted from human to human by the bite of sandflies of the genus *Phlebotomus* (Fig. 36-10). Sandflies are tiny, primitive flies that can pass through ordinary household screens. The disease is endemic among dogs, which may be a source of infection. It occurs in the countries bordering the Mediterranean Sea, in India, in the Middle East, in China, and in parts of Africa.

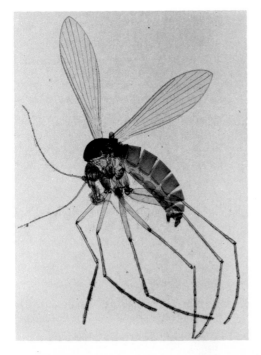

FIG. 36-10   The sandfly *Phlebotomus*, vector of *Leishmania* species. (From Schmidt, G.D., and Roberts, L.S.: Foundations of parasitology, ed. 3, St. Louis, 1985, The C.V. Mosby Co.; photograph by Jay Georgi.)

Cutaneous leishmaniasis. Cutaneous leishmaniasis is described by the presence of nodular and ulcerating lesions in the skin. There are two types. One, known as Oriental sore, Aleppo button, or Delhi boil, is caused by *Leishmania tropica*. The other, known as American leishmaniasis or espundia, is caused by *Leishmania braziliensis*. Like the visceral disease, cutaneous leishmaniasis is transmitted by sandflies. The individual lesions on the skin represent the bites of insects or the mechanical transfer of infection by scratching or some form of abrasion. This disease is seen in the same parts of the world as visceral leishmaniasis, but the two forms of leishmaniasis are said not to occur in exactly the same localities. Cutaneous leishmaniasis is occasionally seen in the United States in persons coming from endemic areas.

Laboratory diagnosis. In all forms of leishmaniasis the diagnosis is best made microscopically under oil immersion by demonstration of the intracellular, nonflagellated organisms (Leishman-Donovan bodies) in Giemsa-stained smears from lesions or in biopsies of involved tissues. Diagnostic material obtained by puncture of the enlarged, firm spleen is particularly valuable, and liver puncture may also be done. These parasites can also be cultured in special media, and the positive Montenegro skin test indicates past or present infection. Serologic tests of value are given in Table 36-2.

Trichomoniasis

Trichomonads

Flagellate protozoa in the genus *Trichomonas* look alike, are designated for specific anatomic sites in human beings, and are unique in that only the trophozoite is evident. No cysts are found. Trichomonads are oval, tapering to a point, and possess three to five flagella, an undulating membrane, and an axostyle. (An **axostyle** is a stiff, supporting rod coursing through the length of the parasite and protruding posteriorly as a tail-like spine.) The three human species are *Trichomonas vaginalis* (10 to 30 μm long, found in the vagina), *Trichomonas (Pentatrichomonas) hominis* (8 to 20 μm long, found in the intestine), and *Trichomonas tenax* (6 to 8 μm long, found in the mouth). *Trichomonas vaginalis* is the largest, the most vigorous, and the only definitely pathogenic one.

Trichomonas vaginalis (Fig. 36-3, A) is a typical trichomonad, although quite variable in size. It resides in the vagina of the human female and in the urethra, epididymis, and prostate gland of the human male. It moves with a certain jerky motion, with its short undulating membrane viewed microscopically as coming in and out of focus, contributing to a cogwheel effect.

The Disease

Trichomoniasis occurs as a widespread infection of the genitourinary tract caused by the atrial flagellate *Trichomonas vaginalis*. In human females it is an intractable vaginitis associated with itching, burning, and profuse, frothy, cream-colored, foul-smelling discharge in which trichomonads abound. The physiologic state of the vagina, including the pH, are among factors that predispose to infection.

In human males the parasites are found in the prepuce and prostatic urethra, but symptoms seldom occur. The human male is usually considered the reservoir of infection. The infection, largely transmitted by sexual intercourse, is a disease of generally unrecognized significance. The source of infection are the trophozoites in vaginal and prostatic fluids.

Laboratory Diagnosis

The leaflike trichomonads are readily identified microscopically in females in vaginal discharges prepared as wet mounts or in Papanicolaou-stained vaginal and cervical smears and, in males, in cellular mounts of urine and prostatic discharges.

Infection with Intestinal Flagellates

Important intestinal flagellates found in stool specimens are *Giardia lamblia, Trichomonas hominis* (p. 804), and *Chilomastix mesnili* (Fig. 36-3). The latter two are not considered pathogenic by most protozoologists. *Giardia lambli* causes giardiasis (Table 36-1).

Intestinal Flagellates

Most intestinal flagellates have a trophozoite and a cyst phase. The cyst, if present, is the infective agent. If not, as with trichomonads, the trophozoite passes the infection. Reproduction, however, comes from the longitudinal division of the vegetative trophozoite.

Since these parasites reside in the small and large intestine, they must withstand the peristaltic action of the intestines. The presence of specialized structures, such as the sucking disk, the axostyle, and the undulating membrane, enable them to do so. Flagella may be single or multiple, and one or more nuclei may be found.

Spread of intestinal flagellates is mainly by contaminated food and drink and also by hand-to-mouth contact. Their distribution is worldwide.

Nonpathogenic parasites (commensals) in this category (as well as in other categories) must be properly identified and distinguished from the pathogens with which they are likely to be confused. Because of their distinctive morphology and motility, the two commensals, *Trichomonas hominis* and *Chilomastix mesnili*, are readily differentiated from the pathogens in clinical specimens. *Trichomonas hominis* is a typical trichomonad, as described previously. The cone-shaped *Chilomastix mesnili* (average length 12 μm) possesses four flagella, one nucleus, and a lemon-shaped cyst (6 μm in diameter) (Fig. 36-3, *D* and *D'*).

Giardiasis

The residence of *Giardia lamblia* in the lumen of the upper small intestine (duodenum and upper jejunum) with a sometime invasion of the mucosa is regularly associated with bowel disturbances—persistent diarrhea, flatulence, cramping pain, and epigastric tenderness—and inflammatory changes in the small bowel, which usually are benign and self-limited, but not invariably so (Fig. 36-11). Manifestations of disease are often chronic. The incubation period is 6 to 22 days.

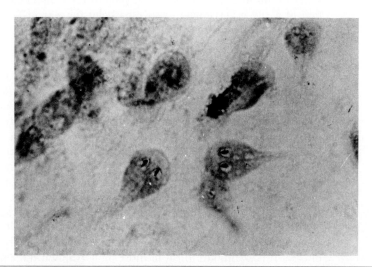

FIG. 36-11   Photomicrograph of *Giardia lamblia* in gastric aspirate processed for cytologic examination. Many are present.

The organism. *Giardia lamblia* is a heart-shaped parasite, 9 to 16 μm wide and 2 to 4 μm thick, with four pairs of flagella and two nuclei. On its ventral surface is a slightly concave, circular sucking disk. Cysts (the infective stage) are 8 to 12 μm long and 7 to 10 μm wide (Fig. 36-3, *C* to *C"*).

Nourished from the intestinal surface, they sit on top of the intestinal epithelium. If they thus blanket the mucosa, their presence mechanically blocks the penetration and absorption of nutrients and fat-soluble vitamins. The action of the powerful sucking disk to attach and detach the parasite injures the mucosal epithelium, thereby impairing digestive functions and absorption. A heavy infection easily results in a malabsorption state with which light yellowish, clay-colored, bubbly, frothy stools are associated.

Spread. Spread through contaminated water (also food), giardiasis is common in many areas of the world and endemic in the United States. With international travel expanding, more travelers are at risk. Giardiasis is a reasonable explanation for "traveler's diarrhea," with the likely source of infection being contaminated tap water or ice and iced beverages consumed on the trip. Recent epidemics have been described in Leningrad in the Soviet Union—nearly a fourth of the tourists to Leningrad returned home with the disease—and in certain resorts of the Colorado Rockies, where skiers and visitors drank untreated mountain water.

Although giardiasis is usually waterborne, recent evidence indicates a sexually transmitted spread (along with the other enteric diseases, shigellosis and amebiasis), in homosexual males. Cysts may be passed from hand to hand. Outbreaks have been reported in day-care centers. Vaginal giardiasis in young girls results from the migration of the parasites from the anus to the vaginal area. Because *Giardia lamblia* may be an opportunist, persons with an immunodeficiency state are at risk. Reportedly the most frequently identified organism in human stool examinations performed in state and territorial public health laboratories is *Giardia lamblia*.

Laboratory diagnosis. Diagnosis is made by identification of the protozoa (trophozoites or cysts) in the feces, either in direct smears or in saline mounts prepared after fecal concentration. The pattern of excretion is variable so that several stools may need to be examined over several days before the parasites are found. Better yet, the parasite may be identified in material aspirated from the duodenum or biopsied from the jejunum. An indirect immunofluorescence test is available for serologic evaluation.

Control. Boiling or freezing kills *Giardia lamblia*, but it can remain viable in cold water for up to 2 months. Efficient systems for sedimentation, flocculation, and filtration of a city water supply remove particles the size of *Giardia* cysts, but the accompanying chlorination at dosages used may not.

Malaria

Malaria is an acute febrile disease caused by the malarial parasite, a protozoon of the genus *Plasmodium*. The word *malaria* is derived from the Italian for "bad air," and the disease, although well described by Hippocrates in the late fifth century BC, got its name in the eighteenth century AD because of its association with the ill-smelling vapors from the marshes around Rome. The parasite of malaria was discovered in 1880 by Charles Louis Alphonse Laveran (1845-1922), French army surgeon, and within the next 20 years transmission of malaria by the mosquito was worked out by Sir Ronald Ross (1857-1932), an English pathologist and parasitologist, who also devised suitable methods for the elimination of mosquitoes over wide areas.

Malaria, one of the most widely prevalent diseases in the world, remains the number one public health problem globally. Worldwide it results in a greater morbidity and mortality than any other infectious disease. More people have malaria than any other disease. It is a constant threat to more than 1 billion human beings. Almost the entire adult population in Africa and India is infected, and a million deaths annually occur in Africa alone in children under 5 years of age.

Before America's civil and military involvement in Southeast Asia, malaria was infrequently seen in the United States (60 cases reported in 1961). In 1970 there were over 3500 cases, mostly in persons returning from Southeast Asia. During the 1970s, malaria (60% vivax) was imported from the major malarious areas of the world by travelers, such as tourists, students, teachers, missionaries, and business people, and at least 12 such persons died of the disease after returning home. Several hundred cases are reported annually to the Centers for Disease Control in Atlanta. In 1982 there were 1056 cases.

Malaria has been a scourge of war and of humanity throughout the ages. The Army Medical Corps states that this disease can put more soldiers out of action than battle casualties. In World War II there were over 490,000 cases of malaria with 8 million man-days lost.

Types

Malaria exists in three distinct types (Table 36-4) (each caused by its own species of *Plasmodium*):

1. **Tertian**—a paroxysm of chill and fever every 48 hours; cause, *Plasmodium vivax*; the most common type and the least severe
2. **Quartan**—a paroxysm of chill and fever every 72 hours; cause, *Plasmodium malariae*; the least common type
3. **Falciparum, estivoautumnal, or malignant**—irregular paroxysms; cause, *Plasmodium falciparum*; the second most common type

The first two types are known as **regular intermittent types.** After the paroxysm of chill and fever, temperature returns to normal and the patient is fairly comfortable until the next paroxysm. In the third, or **remittent,** type the fever varies in intensity, but the patient does not become completely afebrile. Falciparum (also called pernicious)

Table 36-4 Distinguishing Features of the Malarias

Feature	Disease Caused by			
	Plasmodium falciparum	*Plasmodium vivax*	*Plasmodium malariae*	*Plasmodium ovale*
Usual incubation period	8 to 25 days	8 to 27 days	15 to 30 days	9 to 17 days
Hours in erythrocytic cycle	48	48	72	50
Multiple infection in red blood cell	Yes	Rare	No	No
All forms parasitic in blood	Rare	Yes	Yes	Yes
Maximum number parasites per cubic millimeter blood	Up to 2,500,000 (no limit)	50,000	20,000	30,000
Schüffner's dots on red blood cell	No	Yes	No	Yes
Large coarse rings on parasites	No	Yes	Yes	Yes
Sausage-shaped gametocytes	Yes	No	No	No
Duration of untreated infection	6 months to 2 years	1½ to 4 years	1 to 30 years	1½ to 4 years

malaria is the most severe in its consequences and is the treacherous form. Its clinical picture is diverse, sometimes obscure, and often dramatic, and it can be rapidly fatal. Death, it is said, may come within a matter of hours. Deaths associated with malaria are usually attributable to this form, and if mixed infection is present, falciparum is usually dominant. Relapses do not tend to occur with falciparum malaria.

Ninety percent of malaria in Southeast Asia is caused by *Plasmodium falciparum*. Although this is true, 85% of malaria in returnees has been caused by *Plasmodium vivax*. In the Korean conflict malaria was almost entirely vivax, a less severe form clinically but one that is more likely to lie dormant only to recur many months after the initial episodes. Falciparum malaria is the most common type in tropical Africa.

A fourth parasite in human disease is *Plasmodium ovale*, whose appearance and life cycle are much like those of *Plasmodium vivax*. Infection with this parasite, tertian malaria, is usually mild and not widely distributed over the world.

Modes of Infection

The different species of malarial parasites are closely related and transmitted in the same way—by the bite of a female mosquito of the genus *Anopheles*. Of this genus, nearly 100 species may transmit the infection naturally. Human beings are the main reservoir of infection. In malaria-infected countries many of the inhabitants become asymptomatic carriers. They harbor the parasites in their blood (parasitemia) and tissues without manifesting the disease. Repeated attacks seem to give the patient some immunity.

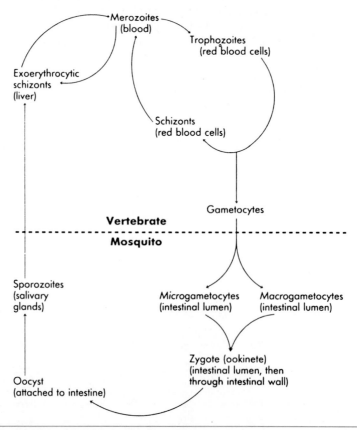

FIG. 36-12   The life cycle of the malarial parasite.

Unrecognized infections and insufficient treatment lead to the carrier state, so important in the spread of malaria.

The disease is occasionally transferred by the use of contaminated hypodermic syringes, as is common among heroin addicts, or by blood or platelet transfusion. There is an increased awareness of this hazard in blood banking partly because of the likelihood of the source of infection being a blood donor who is a drug addict.

A reservoir of malaria exists in monkeys, from which the infection is transmitted to humans and other primates by certain forest species of *Anopheles*.

Life Story of the Parasite

Two major events occur in the complicated life story of the malarial parasite (Fig. 36-12).

Asexual development (schizogony) in humans. The **preerythrocytic phase** begins when young malarial parasites (**sporozoites**) introduced into the bloodstream by the mosquito bite localize in the cells of the liver, therein multiply, and mature. (Some 500,000 parasites must be injected by the mosquito for human infection.) After 6 to 9 days young parasites (**merozoites**) are released into the bloodstream. Merozoites recognize specific receptors on the surface of red blood cells that allow them to be taken into these cells. The Duffy blood group antigen (p. 414) is associated with the receptor site on the red cell for *Plasmodium vivax*. Persons lacking this blood group antigen are resistant to vivax malaria.

The **erythrocytic phase** begins when a merozoite enters a red blood cell to grow and develop therein. The parasite, not fully utilizing the hemoglobin of the red cell, accumulates pigment granules of hemozoin, an iron porphyrin hematin, within its cytoplasm. This is "malarial pigment" and not a normal breakdown product of hemoglobin. When the merozoite matures in the red cell, it becomes a **schizont**. Still within the red cell, the mature schizont, by the process of **segmentation,** forms a number of segments that suddenly separate, releasing a generation of merozoites (about 20) and thereby destroying the red cell (Fig. 36-13). Some of the merozoites are destroyed by white blood cells, but the majority enter red blood cells to repeat the process of schizogony (Color plate 4). In falciparum malaria two or even three or four parasites invade a single red cell (Fig. 36-14).

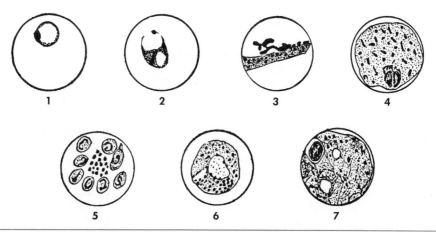

FIG. 36-13   *Plasmodium malariae* (parasite of quartan malaria), development in red blood cell. *1,* Young ring form (trophozoite); *2* to *4,* successive stages; *5,* mature schizont (made up of merozoites); *6,* microgametocyte; *7,* macrogametocyte.

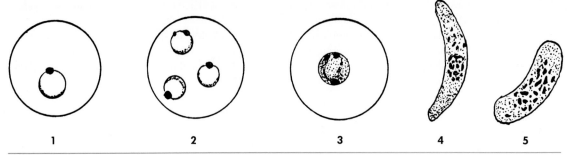

FIG. 36-14   *Plasmodium falciparum* (parasite of estivoautumnal malaria), development in red blood cell. *1*, Trophozoite; *2*, three trophozoites (multiple infection with *Plasmodium falciparum*) in red blood cell; *3*, schizont, early formation and production of pigment (seldom found in peripheral blood); *4*, macrogametocyte; *5*, microgametocyte.

About 2 weeks (sometimes longer) after the infecting mosquito bite, enough parasites are present for red blood cell destruction to cause trouble. The *incubation period* covers the initial preerythrocytic phase and the first 2 weeks or so of the erythrocytic phase. The time elapsing between the entrance of a parasite into a red blood cell and its segmentation is the **periodicity** of the parasite. For *Plasmodium vivax* it is 48 hours and for *Plasmodium malariae*, 72 hours. For *Plasmodium falciparum* it is usually 48 hours, but not regularly so.

The paroxysms of chills and fever in malaria stem from the liberation of metabolic by-products from the parasite and toxic breakdown products from the disrupted blood cell. Sharp paroxysms occur in tertian and quartan malaria because all parasitized cells rupture at about the same time. In falciparum infections some parasitized cells rupture ahead of time, and some rupture behind time, so that several hours are required for the whole brood of parasites to be released. Thus chills are usually absent, and fever may be continuous.

Some parasites do not repeat the asexual phase of development but produce male and female sexual forms, or *gametocytes* (Figs. 36-13 and 36-14 and Color plate 4), the sexual development of which is completed within the stomach of an *Anopheles* mosquito. Sexual forms do not appear in the blood until infection is of 2 to 3 weeks' duration.

Sexual development (sporogony) in the mosquito. When a female *Anopheles* mosquito ingests male and female sexual parasites from the blood of an infected person, a rather complicated sexual cycle begins in its stomach. Within 10 to 20 minutes initially the red cell membrane about each gametocyte disintegrates, freeing a macrogamete (female) or a microgamete*cyte* (male), whose continued development by the dramatic process of **exflagellation** releases eight highly motile, flagellated **microgametes** (male). Fertilization, the fusion of an entire microgamete with a macrogamete, follows quickly. Microgametes correspond to spermatozoa in the higher forms of life. The fertilized parasite, now the **ookinete**, penetrates the stomach wall and encysts, or transforms to the **oocyst**, within which are formed myriad small, spindle-shaped parasites, the **sporozoites**. The cyst ruptures, and the sporozoites are carried by the lymphatic system to the salivary gland of the mosquito, which is so constructed that the parasites are ejected in the saliva when the mosquito bites.

Pathology

The chief pathologic changes in malaria relate to destruction of red blood cells (hemolysis). Hemolysis leads to different degrees of anemia and jaundice. The parasitic infection seems to make the blood more viscous, and the sticky parasitized cells plug and obstruct small blood vessels. This is likely to occur with falciparum malaria and accounts for its worst complication, cerebral malaria. The liver and spleen enlarge (hepatosplenomegaly), partly because the presence of the malarial pigment stimulates the macrophage system to activity. Macrophages ingest the pigment and deposit it in liver, spleen, and bone marrow. In acute malaria the spleen is moderately enlarged, soft, and friable. In chronic malaria it is enlarged and fibrotic. Such a spleen easily ruptures as the result of a blow or fall or even spontaneously.

A dreaded complication, blackwater fever, develops with acute and massive hemolysis. Large amounts of hemoglobin are released into the plasma (hemoglobinemia), spilling over into the urine (hemoglobinuria). A severe kind of acute renal failure is associated with the passage of reddish black urine.

Laboratory Diagnosis

Malarial parasites are easily seen in the red cells of properly prepared (and stained) thin smears of peripheral blood taken just before or at the peak of the paroxysm. Wright's and Giemsa stains are most often used. The young parasites, seen as blue rings with a red chromatin dot attached, have a signet-ring appearance. The quartan ring is thicker than the tertian, and the falciparum ring is thin and hairlike. The latter often has two chromatin dots. Full-grown malarial parasites almost completely fill the red cell and contain numerous red granules. Just before red cell ruptures, the parasites assume a segmented or rosette pattern. So-called malarial crescents are sausage-shaped gametocytes with a chromatin mass near their center and are the sexual parasites of falciparum malaria. They are frequently seen in the blood, whereas the fully developed parasite, or schizont, is seldom seen in this form of malaria.

At times, when organisms are too few to be seen in thin smear, a sample of blood may be smeared thickly on a glass slide, treated to remove hemoglobin from the erythrocytes, and then stained. The disadvantage of the thick smear is that the shape and appearance of the parasites are altered in preparation. Its advantage is that a much greater volume of blood may be examined within a given length of time. *To be reliable, it must be prepared meticulously.*

Practically speaking, both thick and thin smears are needed for the diagnosis of malaria (stained preferably with Giemsa stain). In routine laboratory work at least 100 oil immersion fields on a thin smear and 200 similar fields on a thick smear must be examined before a negative report is given. Every 6 to 12 hours for the following 24 to 36 hours is appropriate timing for repeat thick and thin films, as indicated. In falciparum malaria thick smears may not show parasites for several days after onset.

The laboratory diagnosis of malaria *not only consists of finding the parasites but also includes the determination of the species.* Identification of the parasite makes the clinical diagnosis.

Fluorescent antibody technics are also used to stain the malarial parasite specifically. A soluble antigen fluorescent antibody (SAFA) diagnostic test fills a need for the blood bank in the screening of donors. *Plasmodium malariae* is feared as the cause of the rare transfusion-induced malaria because it tends to produce a prolonged low-level asymptomatic parasitemia and is almost impossible to identify in smears from such a donor's blood. Serologic testing for antimalarial antibody is therefore important in the investigation of the febrile transfusion reaction.

A small dose of quinine or other antimalarial drug drives the parasites out of the peripheral blood; therefore it is practically useless to examine blood for malaria right after such drugs are given.

Mosquitoes Transmitting Malaria

Malaria is transmitted by various species of *Anopheles* mosquitoes and only by the female because the male lives on fruits and vegetables, not on blood. Since the common house mosquito (*Culex*) is not a vector, one must distinguish it from *Anopheles*. The *Culex* mosquito bites during the daytime; the *Anopheles* at night or near dusk. The wings of *Anopheles* are spotted, whereas those of *Culex* are not. When *Culex* is resting on a wall, its body is almost parallel to the wall; the body of *Anopheles* stands at an acute angle (see Fig. 16-2).

Immunity

Most of what is known about immunity in malaria has come out of studies on human volunteers infected with human plasmodia. Both humoral and cellular factors exist, but cellular defenses seem to be the prominent ones.

Active immunity may be acquired with present or past infection. In hyperendemic areas a mother confers passive immunity that is effective for the first 3 months of her baby's life. After that her child develops active immunity for active infections, usually within the next 3 years. That adults of the area are immune is reflected in only mild or no manifestations of disease when plasmodia are found circulating in the bloodstream.

Abnormalities of red blood cells are correlated with resistance to malaria. Persons with the sickle cell trait, that is, persons whose red blood cells contain the abnormal hemoglobin S and therefore assume bizarre, crescent shapes, and persons whose red blood cells exhibit a glucose-6-phosphate dehydrogenase deficiency, have lower infection rates and lower parasite densities than do individuals with normal red cells. Plasmodia of malaria do not seem to penetrate sickled or otherwise abnormal red cells.

Prevention

Prevention of malaria depends on blocking the transfer of infection from person to person by mosquitoes. Recommended for this purpose are (1) screening of houses, (2) draining and oiling of ponds of water to prevent mosquito breeding and the use of minnows to destroy the larvae, (3) proper treatment of patients with antimalarial drugs, and (4) detection and cure of carriers. The presence of an animal reservoir greatly complicates the problem of malarial control in those countries where jungles swarm with monkeys and other primates. The countless numbers of insects in Africa and India preclude the possibility of massive spraying of insecticides being effective for mosquito (or disease) control in those areas.

Development of a vaccine for malaria is still experimental. Partially purified material from the malarial plasmodium is used.

Travelers can reduce the risk of exposure by wearing adequate protective clothing, applying mosquito repellent to exposed skin, remaining in well-screened areas between dusk and dawn (the feeding time for *Anopheles*), and sleeping under mosquito netting (16 to 18 mesh). Chemoprophylaxis is advised in areas of endemic malaria (p. 448).

Toxoplasmosis

Toxoplasma gondii, the cause of toxoplasmosis, is a delicate, banana-shaped, obligate intracellular parasite, 3 by 6 μm, that is easily killed by physical agents. It is a cosmopolitan sporozoan, being found in animals and birds all over the world. Its infection is a zoonosis.

The Disease

The organism can invade practically any tissue cell and is especially likely to affect cells of the macrophage system, including the lining cells of blood vessels. Within a cell the organisms rapidly proliferate in typical fashion to form a rosette that may become a cyst containing some 3000 parasites. Inflammation may be present in the lungs, lymph nodes, eyes, and brain. Characteristic are the focal deposits of calcium in nervous tissue (especially in the fetus).

In humans the disease occurs in three forms: acquired, congenital, and recrudescent. The acquired, or adult, form often passes unnoticed, since most cases are asymptomatic and benign. One fourth to one half of adults in the United States have been infected at some time. The congenital form is acquired in utero from a mother who may have no history of previous disease but who probably had an asymptomatic infection during the pregnancy. Infection of the mother in the early months of pregnancy means her child is more likely to be severely affected, since the embryo is most vulnerable at this time. The newborn baby becomes very ill, develops a skin rash, turns yellow, and may convulse. Because of the brain damage associated with the intracerebral calcium deposits, the baby who survives the disease may be born with microcephaly, hydrocephalus, and mental retardation. *Toxoplasma gondii* also produces chorioretinitis, an inflammatory process in the choroid and retina of the eye. The resultant changes in the infant's eyes can easily lead to blindness.

The recrudescent form is seen in a person who has become immunodeficient. No effective form of treatment for the disease is known.

Toxoplasmosis is associated with the formation of tumors in birds; in humans it is seen with neoplasms of the central nervous system.

Sources and Mode of Infection

This sporozoan has a life cycle in cats similar to that of the malarial parasite in mosquitoes. Cats, the definitive hosts, pick up the parasites when they consume the intermediate hosts—infected birds and mice (Fig. 36-15). In the cat's intestine, which provides a peculiarly suitable habitat for the parasite, the organisms go through asexual (schizogony) and sexual (gametogony) stages of development in the epithelium, and the oocysts are passed in the feces. After being passed into soil, sand, or litterbox, oocysts become infectious after 3 or 4 days in warm, moist surroundings. They can be recovered after many months from water or wet soil and are generally resistant to many chemical agents, including ordinary disinfectants. However, drying and heat kill them. The common house cat is implicated as the human reservoir.

The parasite is universal. Many animals harbor it, and it may persist in the raw flesh of the slaughtered animal until killed by heat, drying, or freezing ($-20°$ C). Eating raw or undercooked meat is a principal source of human infection. Pork and lamb are prime sources; beef can be also. The National Livestock and Meat Board recommends that all meat be heated to at least $140°$ F ($60°$ C) throughout to kill the toxoplasmas.

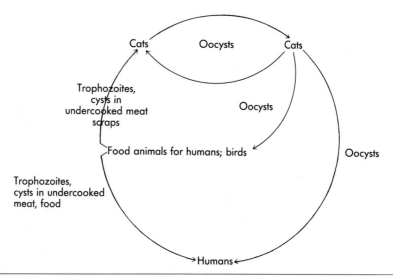

FIG. 36-15   *Toxoplasma gondii*, life cycle.

(This is the "rare" reading on meat thermometers designed for home use.) Pregnant women should be very careful in handling cats, cat feces, or articles contaminated with cat feces and should avoid altogether any contact with a strange cat or one newly brought into the household. If kitty litter is changed daily, the eggs of the parasite cannot mature and become infective. In the prevention of the disease meticulous hand washing is an important measure.

Toxoplasmas are taken into the human body by way of the mouth because the person has either contacted an infected cat or consumed infected meat. The toxoplasmas are released from the oocysts, invade the intestinal lining, migrate into body tissues and fluids, and spread by the bloodstream throughout the body, multiplying within parenchymal cells of many organs. Cyst formation occurs especially in the brain, heart, and skeletal muscle.

For the congenital form to develop, toxoplasmas must get into the bloodstream of the mother sometime after the first trimester of pregnancy. They establish a focus of infection in the placenta that enables them to penetrate the fetal circulation and so infect the fetus. Congenital infection is said to occur as often as 1 per 500 or 1000 pregnancies.

Toxoplasmosis can be an opportunistic pathogen in patients receiving organ transplants. Since the heart muscle is known to be affected during acute infection, a donor heart might be a source of a rapidly fatal condition to the immunosuppressed recipient. Such a transmission of toxoplasmosis has been described in two heart transplant patients.

Laboratory Diagnosis

These microorganisms (Fig. 36-16), possessing a delicate crescent-shaped body with tapered ends, are easily stained and identified in smears made from body fluids, exudates, and even diseased tissues, such as lymph nodes. Toxoplasmas can be cultured

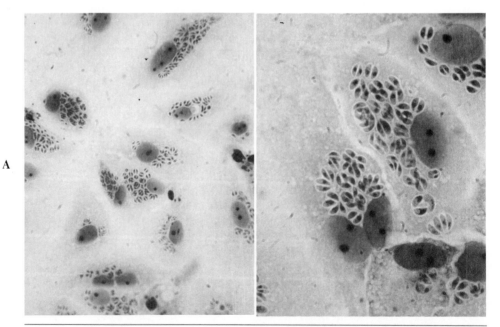

A   B

FIG. 36-16 *Toxoplasma gondii* in monkey kidney cell culture. Many crescent-shaped microorganisms within cell cytoplasm. (**A,** ×400; **B,** ×1000.) (Courtesy Dr. H.G. Sheffield, Laboratory of Parasitic Diseases, National Institute of Allergy and Infectious Diseases, Bethesda, Md.)

in cells or in a fertile hen's egg. Practically speaking, however, the diagnosis of toxoplasmosis usually must be made serologically, and the examinations available include an indirect hemagglutination test, a complement-fixation test, a neutralization test, an immunofluorescence test (the most widely applied), and ELISA. The antibodies detected by the Sabin-Feldman dye test are those preventing parasites of a laboratory culture from taking up methylene blue dye. *Rising* titers with serologic tests are necessary for diagnosis. A toxoplasmin skin test is available. A mouse can be inoculated with suspect material so that the organisms can be recovered and identified from that animal.

Toxoplasmas may be detected in the feces of a cat. Direct smears are not adequate; the fecal flotation method must be used. Forty-one percent of cats in the United States are reported to have antibodies as measured by a positive Sabin-Feldman dye test.

Pneumocystosis

Pneumocystis carinii, first recognized in the United States in 1956, is a universal saprophyte widespread in nature in the lungs of many animals, including human beings. Little is known about it—its life cycle, mode of reproduction, host specificity, mode of infection, or epidemiology. It was not recognized as a pathogen until the advent of immunosuppression. Although considered an organism of low pathogenicity, in the susceptible person its effects are devastating. Infection is now reported from all parts of the world. *Pneumocystis carinii* causes pneumocystosis.

The Disease

Pneumocystosis (diffuse interstitial pneumonitis, interstitial plasma cell pneumonia) is an inflammatory process unique in the lungs. In pneumocystosis the air sacs of the lungs are filled with a foamy, semiliquid, lightly staining material representing clustered masses of oval, minute organisms (trophozoites) about 5 to 7 μm in diameter, each surrounded by a thin homogeneous capsule, and cysts, 3 to 5 μm in diameter, with a thicker wall. The walls of the air sacs are permeated by inflammatory cells. The incubation period is said to be 2 to 6 weeks.

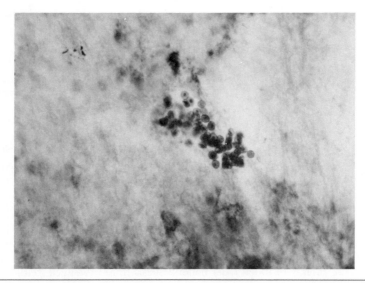

FIG. 36-17 *Pneumocystis carinii* in direct smear of pulmonary lavage sediment (methenamine silver stain). Note staining of wall of "cystic" forms and their typical target-like pattern. Within each would be up to eight smaller forms, the "sporozoites" (1 to 2 μm in diameter). (×500.) (From Drew, W.L., and others: J.A.M.A. **230**:713, 1974.)

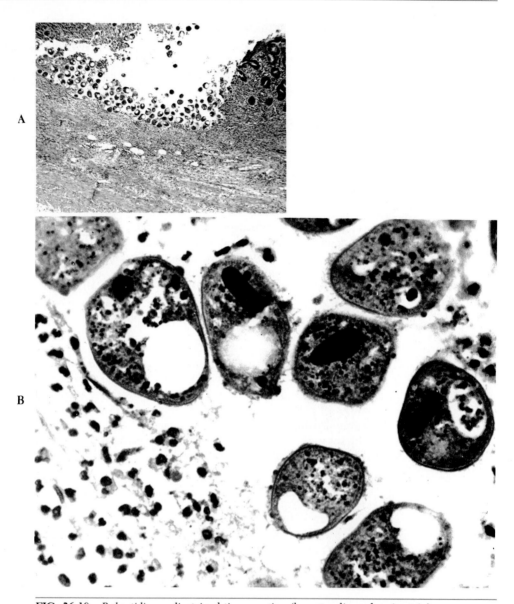

FIG. 36-18   *Balantidium coli,* stained tissue section (hematoxylin and eosin stain).
A, Balantidiasis of appendix. In crater of an ulcer of appendiceal wall are numerous ciliates cut
in cross section. (×35.) **B,** Cross-sectional areas of *Balantidium coli* as seen with higher
microscopic magnification. (×430.) (From Anderson, W.A.D., and Kissane, J.M.: Pathology,
ed. 7, St. Louis, 1977, The C.V. Mosby Co.)

Pneumocystosis is another not uncommon disorder of the compromised host, particularly the immunosuppressed one. First encountered in debilitated infants, it is said to
be one of the worst, fastest moving, and cruelest killers of such children. The vulnerability of much older patients stems from prolonged steroid hormone therapy, extended
immunosuppression, or the presence of leukemia or other cancer of the lymphoid system. Cortisone seems to enhance the growth of these organisms. Pneumocystosis is
important in homosexual males, in some of whom a defect in cell-mediated immunity
may be ascribed to factors such as multiple concurrent infections or the use of certain
drugs.

Laboratory Diagnosis

The diagnosis is a morphologic one; the organisms are identified and studied in smears of aspirates and imprints from the diseased lungs stained by a suitable stain, such as the Papanicolaou technic, and also in tissue sections of lung obtained from either ante-mortem biopsy or autopsy, wherein special silver stains demonstrate them nicely (Fig. 36-17). *Pneumocystis carinii* can be cultured in suitable cell cultures, such as the chick lung epithelial cell culture, and it is studied with the scanning electron microscope.

Balantidiasis

Balantidium coli, the most important intestinal ciliate, the only pathogenic ciliate, and the largest protozoon to invade humans, causes balantidiasis.

The Organism

Balantidium coli is shaped like a sac; *Balantidium* means "little bag." It is seen in two life stages: the motile trophozoite, averaging 60 by 45 μm, and the cyst, the infective form, averaging 52 to 55 μm in diameter (Fig. 36-3, *B* and *B'*). The life cycle is somewhat similar to that of *Entamoeba histolytica* except that multiplication of the organism does not occur in the cyst. The trophozoite is found in the cecum (large intestine) and terminal ileum (small intestine), primarily in the lumen but also in the mucosal lining.

The Disease

In some cases the trophozoite seems to be a harmless inhabitant of the large intestine, but usually its presence is associated with diarrhea. The multiplying parasites may invade the intestinal wall, form nests, and produce small abscesses that break down into irregular ulcers with undermined edges (Fig. 36-18). An intense dysentery that results may even cause death. The mushy feces smell like a pigsty.

Spread

Balantidium coli is a normal inhabitant of the large intestine of the domestic hog. Humans are infected when they ingest cysts passed by the host. Spread may be hand to mouth or through contaminated food.

Laboratory Diagnosis

The laboratory identification of cysts and trophozoites in the stool or in the exudate from intestinal ulcers makes the diagnosis.

SUMMARY

1 Protozoa are single-celled, complete, self-contained units that require living sustenance. The dependent parasite derives all the benefit from its host. Only about 30 protozoa affect humans.

2 Protozoa develop in a larval or asexual stage in an intermediate host and in a sexual or adult stage in a definitive host.

3 With adverse conditions most protozoa form cysts; they secrete and enclose themselves within a resistant wall, where they may remain for a long time. When favorable conditions for growth return, the cyst imbibes water, and the microorganisms become trophozoites. The trophozoite is susceptible to deleterious influences; the cyst is resistant. Hence the cysts spread disease.

4 Protozoa may be classified according to their mode of locomotion. Motility may be by (1) pseudopod formation (for example, in amebas); (2) flagella and an undulating membrane (for example, in flagellates); and (3) cilia (for example, in ciliates). No external organs of locomotion are identified for the plasmodia of malaria.

5 The identification of protozoan parasites invariably involves the microscopic recog-

nition of specific structural details in suitable preparations of body fluids. Serologic testing may be done by standard immunodiagnostic methods.

6 Measures for control of protozoan infections include (1) the eradication of sources of human infection; (2) the sanitary control of water, food, and waste disposal; and (3) the institution of measures that interfere with the biologic activity of the parasite.

7 *Entamoeba histolytica* causes amebiasis. From the amebic "water bottle" ulcers in the large intestine, the amebas may disseminate through the portal venous system to the liver and then over the body.

8 Hemoflagellates are either spindle shaped with a single whiplike flagellum or rounded with a short one. They have an invertebrate host—a bloodsucking insect— and a vertebrate host, in which they are found in blood, tissue spaces, and cerebrospinal fluid.

9 Members of the genus *Trypanosoma* cause African trypanosomiasis, or African sleeping sickness, and South American trypanosomiasis, or Chagas' disease.

10 Three species in the genus *Leishmania* cause two forms of leishmaniasis—the cutaneous and the visceral. This hemoflagellate multiples in the macrophages of its vertebrate host. The tiny, primitive sandfly vector can penetrate ordinary household screens.

11 *Trichomonas vaginalis*, the largest flagellate in its genus, causes the sexually transmitted infection trichomoniasis. The reservoir is the genitourinary tract of the human male.

12 Intestinal flagellates possess specialized structures that enable them to withstand the peristaltic action of their intestinal habitat.

13 *Giardia lamblia* causes the widespread giardiasis. This intestinal flagellate blankets the duodenal mucosa, thereby mechanically blocking digestive function.

14 Malaria, the number one public health problem globally, is caused by several *Plasmodium* species, each with its distinct disease pattern. The female *Anopheles* mosquito transmits the parasite, and humans are the main reservoir.

15 The manifestations of malaria come from the destruction (hemolysis) of the parasitized red blood cells; splenomegaly is a constant feature.

16 Travelers to malarious areas are advised to take every precaution, including chemoprophylaxis.

17 *Toxoplasma gondii*, an obligate intracellular parasite, causes toxoplasmosis, a disease occurring in an acquired and a congenital form. The congenital form is the more serious, since it is likely associated with severe malformations in the infant.

18 Since cats are the definitive hosts, pregnant women should carefully handle cats, cat feces, or articles so contaminated and should avoid contact with a strange cat.

19 *Pneumocystis carinii*, a universal saprophyte in the lungs of many animals, causes pneumocystosis, a unique pulmonary infection, especially in immunocompromised hosts, notably debilitated infants, and AIDS patients.

20 *Balantidium coli*, the only pathogenic ciliate and the largest protozoon to invade humans, causes balantidiasis. It is a normal inhabitant of the domestic hog.

QUESTIONS FOR REVIEW

1 Describe protozoa. What are their salient features?

2 Delineate the four classes of protozoa of medical interest.

3 Define intermediate host, definitive host, organelle, pseudopod, cyst, flagella, cilia, trophozoite, parasite, ectoplasm, endoplasm, commensal, hemoflagellate, kinetoplast, amastigote, promastigote, epimastigote, merozoite, recrudescent, atrial, flagellate, endoparasite, ectoparasite, schizogony, sporogony, zoonosis, and medical parasitology.

4 How is amebiasis spread? Give the life history of *Entamoeba histolytica*.

5 Outline the development of the malarial parasite in (a) the human and (b) the mosquito.
6 Discuss the prevention of malaria.
7 Compare the *Anopheles* mosquito with the *Culex*.
8 Give the laboratory diagnosis for:
   a. Malaria
   b. Amebiasis
   c. Trypanosomiasis
   d. Leishmaniasis
   e. Toxoplasmosis
   f. Pneumocystosis
   g. Balantidiasis
   h. Giardiasis
   i. Trichomoniasis
9 Characterize briefly Chagas' disease, African trypanosomiasis, leishmaniasis, trichomoniasis, giardiasis, balantidiasis, toxoplasmosis, and pneumocystosis.
10 Compare the life cycle of *Plasmodium* with that of *Toxoplasma*.
11 Discuss prevention and control of parasitic diseases. (Consult outside sources.)
12 Name six commensal organisms that must be differentiated from parasitic protozoa in stool specimens.
13 What is the role of the arthropod vector in protozoan diseases? Name three important arthropod vectors.
14 Name protozoan diseases that are also important as sexually transmitted diseases.
15 Do protozoan diseases pose a threat to the day-care nursery? (Consult outside sources.)

SUGGESTED READINGS

Beck, J.W., and Davies, J.E.: Medical parasitology, ed. 3, St. Louis, 1981, The C.V. Mosby Co.

Bloom, B.R.: Games parasites play: how parasites evade immune surveillance, Nature **279**:21, 1979.

Campbell, C.C., and Chin, W.: Diagnosing and monitoring malaria, Diagn. Med. **4**:46, May-June 1981.

Carswell, H.: Medical News: drug-resistant malaria raises concern, controversy, J.A.M.A. **249**:2291, 1983.

Committee on Systematics and Evolution of the Society of Protozoologists: A newly revised classification of the Protozoa, J. Protozool. **27**:37, 1980.

Cooper, K.F., III, and others: Giardiasis: a common, sexually transmissible parasitic diarrhea with pitfalls in diagnosis, South. Med. J. **76**:863, 1983.

Farmer, J.N.: The protozoa, introduction to protozoology, St. Louis, 1978, The C.V. Mosby Co.

Frenkel, J.K., and others: *Toxoplasma gondii* in cats: fecal stages identified as coccidian oocysts, Science **167**:893, 1970.

Gillin, F.D., and Reiner, D.S.: Human milk kills parasitic intestinal protozoa, Science **221**:1290, 1983.

Harrison, G.: Mosquitoes, malaria, and man: a history of the hostilities since 1880, New York, 1978, E.P. Dutton.

Lewin, R.: Nairobi laboratory fights more than disease, Science **216**:500, 1982.

Schmidt, G.D., and Roberts, L.J.: Foundations of parasitology, ed. 3, St. Louis, 1985, The C.V. Mosby Co.

Visvesvara, G.S.: *Giardia lamblia*: America's no. 1 intestinal parasite, Diagn. Med. **4**:24, March-April 1981.

Weller, T.H.: Health crusades and tropical diseases, Hosp. Pract. **15**:191, March 1980.

METAZOA
Medical Helminthology

PERSPECTIVES

Among multicellular animal parasites, or Metazoa, three phyla are medically noteworthy in humans: (1) **Platyhelminthes,** or flatworms, which includes the two classes **Trematoda** (flukes) and **Cestoidea** (tapeworms); (2) **Nematoda** or **Nemathelminthes** (roundworms), and (3) **Arthropoda.** The first two phyla listed contain worms, or helminths, elongate, invertebrate animals without appendages or bilateral symmetry. The study of the pathogenic ones is medical helminthology. The third phylum contains mites, spiders, ticks, flies, lice, and fleas, which are important vectors in the transmission of disease.

Helminths: General Considerations

Comparison with Protozoa

Whereas protozoa are unicellular and *micro*scopic, worms are multicellular and *macro*scopic. Like protozoa, they take up residence in a characteristic anatomic habitat. Unlike protozoa, which continually multiply within the host like bacteria, the infectious unit of the worm, the egg or larva, matures into but a single adult. The number of adult helminths, that is, the worm burden, therefore varies and considerably influences the development of disease. Like protozoa, the disease-producing capacity of worms relates to their location, numbers, and toxic products, as well as to the resistance mechanisms and immune responses in the host.

Like protozoa, the geographic distribution of worms varies. Some worms are widespread over the world, but most prefer a warm climate and thrive under conditions of poor sanitation. Like higher forms of life, worms have a functional anatomy. Most have a digestive tract, an excretory tract, a reproductive system, and a nervous system. Most worms demonstrate comparable events in their life story.

Basic Life Pattern

Parasitism of a human or animal host begins when eggs and/or larval forms of the worm are taken in by that host. Therein they become adult worms and initiate the sexual cycle. After copulation, fertilized adult females lay eggs, which in turn develop into larvae. With few exceptions, the eggs or their larvae must pass outside the body of the definitive host (in which the sexual cycle occurs) to complete their growth and asexual development into larval forms infective for the next definitive host. This may come about in a sequence of events that is often quite complicated. Intermediate hosts may be required for the asexual development of the larval stages. A cycle is completed when the infective larvae again reach a definitive host.

Spread of Infections

Eggs and/or larvae of worms enter the body of the definitive host in several ways. They may be ingested in contaminated food or water, larvae may penetrate the skin in contact

Table 37-1 Overview of Arthropoda in Spread of Disease

Examples	Vectors	
	Common Names	Genus Names
Metazoan Diseases		
Filariasis	Mosquito	*Culex*
River blindness	Black fly	*Simulium*
Ascariasis	Housefly	*Musca*
Protozoan Diseases		
Malaria	Mosquito	*Anopheles*
Sleeping sickness	Tsetse fly	*Glossina*
Leishmaniasis	Sandfly	*Phlebotomus*
Amebiasis	Housefly	*Musca*
Chagas' disease	Triatomid bug	*Triatoma, Rhodnius*
Bacterial Diseases		
Tularemia	Tick	*Dermacentor, Ambylomma, Rhipicephalus*
Relapsing fever	Louse	*Pediculus*
	Tick	*Ornithodoros*
Plague	Flea	*Xenopsylla, Pulex*
Salmonellosis	Housefly	*Musca*
Bacillary dysentery	Housefly	*Musca*
Cholera	Housefly	*Musca*
Rickettsial Diseases*		
Spotted fevers	Tick	*Dermacentor, Ambylomma, Rhipicephalus, Ornithodorus*
Scrub typhus	Mite	*Trombicula*
Rickettsialpox	Mite	*Liponyssoides*
Typhus fever	Louse	*Pediculus*
Murine typhus	Flea	*Xenopsylla*
Bartonellosis	Sandfly	*Phlebotomus*
Trachoma†	Housefly	*Musca*
Viral Diseases		
Yellow fever	Mosquito	*Aedes*
Encephalitis	Mosquito	*Culex*
Poliomyelitis	Housefly	*Musca*

*Rickettsias are now classified as bacteria.
†Chlamydiae, the cause of trachoma, are now classified with rickettsias as bacteria.

with infective soil, or larvae may be injected through the skin with the bite of an insect. Once inside the host, the eggs and/or larvae reach their characteristic habitat directly with little movement, or indirectly after following a circuitous route, in the circulation, tissue spaces, and the alimentary tract. Eggs and/or larvae leave the body of the definitive host through either end of the alimentary tract or through the skin, as when larvae are picked up from the blood by a bloodsucking insect.

Eosinophilia

An increased number of eosinophils (one of the white blood cells) in peripheral blood and in tissues is a characteristic feature of parasitic infections, particularly with worms. A response requiring T cell participation, it indicates that the parasite has invaded tissues. Eosinophils are considered to be phagocytic cells and at times show cytotoxic activity on the surface of helminths.

Arthropods

Arthropods are widely distributed and make up the most highly diversified of all the phyla of the animal kingdom. Arthropods may directly injure human beings in various ways, for instance, by poisoning, blistering the skin, sucking blood, and penetrating body tissues. They are presented briefly at this point to emphasize their importance as vectors in the transmission of disease. Since the phylum Arthropoda is not discussed further in this chapter or text, Table 37-1 shows the extent to which arthropod agents participate in the production of disease. (Some diseases listed have been previously discussed in the text.) The table gives the spectrum of etiologic agents so transmitted, indicates the arthropods most often encountered, and lists common names alongside scientific ones. Remember that if the parasite is carried unchanged in the body of the arthropod vector, the vector is a **mechanical** one. The parasite or pathogenic agent undergoes a series of developmental changes in the body of the **biologic** vector.

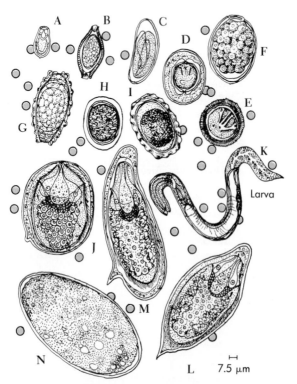

FIG. 37-1 Ova of metazoa. Note one larval form included. **A,** *Opisthorchis (Clonorchis) sinensis.* **B,** *Trichuris trichiura.* **C,** *Enterobius vermicularis.* **D,** *Hymenolepis nana.* **E,** *Taenia solium.* **F,** *Diphyllobothrium latum.* **G,** *Ascaris lumbricoides,* unfertilized ovum. **H** and **I,** *Ascaris lumbricoides,* fertilized ova. **J,** *Schistosoma japonicum.* **K,** *Strongyloides stercoralis* larva (embryonated ovum rare in feces, the rhabditiform larva usually found). **L,** *Schistosoma haematobium.* **M,** *Schistosoma mansoni.* **N,** *Fasciolopsis buski.* (Red blood cells, 7.5 to 8 μm in diameter, are included in the sketch as references for size of the ova.)

REGISTRY OF PATHOGENS
Trematodes (Flukes)

Trematodes, or flukes, are usually flat, leaflike, nonsegmented parasites provided with suckers for attachment to the host. All species, except those that inhabit the bloodstream (the blood flukes, or schistosomes), are hermaphrodites and have operculate eggs (Fig. 37-1, *A*, *J*, *L* to *N*). Some have the most complicated life histories in the animal kingdom. Trematode infections are prevalent in the Orient and in the tropics, where contamination of fresh water by human feces is widespread.

Flukes vary in size from less than 1 mm to several centimeters. The life span varies with the species but usually covers several years, even up to 30 in some. Respiration in the adult worms is largely anaerobic (glycogen is split into carbon dioxide and fatty acid), whereas the larval forms require oxygen.

Classification

Flukes are classified according to the area of the body in which their development into adults is completed and their eggs are deposited. From the standpoint of habitat there are (1) intestinal, (2) liver, (3) lung, and (4) blood flukes.

Intestinal flukes. An important example of an intestinal fluke is the largest human trematode, *Fasciolopsis buski*, which averages up to 7 or 8 cm in length (Fig. 37-2). In residence in the small intestine, each adult worm deposits 25,000 eggs a day. It is widespread in pigs in Southeast Asia.

Liver flukes. Two important liver flukes are *Opisthorchis (Clonorchis) sinensis*, the Chinese or Oriental liver fluke, and *Fasciola hepatica*, the sheep liver fluke. Liver flukes reside in the bile ducts of the liver. *Opisthorchis sinensis*, with a life span up to 20 years, parasitizes fish-eating animals. *Fasciola hepatica*, a leaflike fluke with a dis-

FIG. 37-2 *Fasciolopsis buski.* (From Sonnenwirth, A.C., and Jarett, L., editors: Gradwohl's clinical laboratory methods and diagnosis, ed. 8, St. Louis, 1980, The C.V. Mosby Co.)

Table 37-2 Some Differential Features of Parasitic Flukes in Humans

Fluke	Disease	Geographic Distribution	Dimensions		Characteristics of Egg	Laboratory Identification
			Hermaphrodites	Eggs (μm)		
Intestinal						
Fasciolopsis buski	Fasciolopsiasis; intestinal distomiasis	East and Southeast Asia	7 to 8 cm long	140 by 80	Operculate	Eggs in stool
Liver						
Opisthorchis sinensis	Clonorchiasis; hepatic distomiasis	China, Korea, Indochina, Japan, Taiwan	10 to 25 by 3 to 5 mm	29 by 16	Yellowish brown; elevation around rim of operculum; contains miracidium	Eggs in stool or in biliary drainage
Fasciola hepatica	Fascioliasis ("liver rot"); hepatic distomiasis	Worldwide (in sheep-raising areas)	25 to 30 by 13 mm	130 to 150 by 63 to 90	Operculate; much like egg of *Fasciolopsis*	Eggs in feces or bile or material from duodenal intubation
Lung						
Paragonimus westermani	Paragonomiasis; endemic hemoptysis (bloody sputum); pulmonary distomiasis	East and Southeast Asia, North Central Africa, South America, North America (in animals)	5 to 10 mm	85 by 55	Ovoid, yellowish brown with flattened operculum; thick shelled	Eggs in sputum and feces, in material aspirated from lesions, or in pleural fluid

Blood

			Separate sexes				
			Male	Female			
Schistosoma mansoni	Schistosomiasis; intestinal bilharziasis	Africa to Near East, South America, Caribbean	10 by 1.1 mm	14 by 0.16 mm	115 to 175 by 45 to 70	Yellowish brown, ovoid with sharp *lateral* spine	Eggs in stool; rectal or liver biopsy
Schistosoma haematobium	Vesical schistosomiasis	Africa, Arabia to Lebanon	10 to 15 by 0.25 mm	20 by 0.25 mm	115 to 170 by 40 to 65	Elliptical with sharp *terminal* spine	Eggs in urine (last few drops voided or passed at noon, after exercise, after prostatic massage)
Schistosoma japonicum	Oriental schistosomiasis; Katayama disease	China, Japan, Philippines	12 to 20 by 0.5 mm	15 to 30 by 0.3 mm	70 to 100 by 50 to 65	Round with rudimentary lateral spine; laid in clusters; well mixed with feces; sticky outer coat covered with debris (smallest schistosome egg)	Eggs in stool, liver biopsy

tinct anterior cone, infects sheep and cattle, which in turn seed soil and water with viable eggs.

Lung flukes. The well-known lung fluke *Paragonimus westermani* (Oriental lung fluke) is a small, reddish brown worm, residing, as its name indicates, in the lung. Its natural definitive hosts are a wide variety of fur-bearing animals. The disease produced by this fluke is primarily a zoonosis.

Blood flukes. The important blood flukes are three species in the genus *Schistosoma*—*Schistosoma mansoni*, *Schistosoma haematobium*, and *Schistosoma japonicum*. *Schistosoma mansoni* resides in the veins of the large intestine (the inferior mesenteric veins, which are in the portal venous system draining blood to the liver). This is the blood fluke most often seen in the United States. New York City is said to contain 100,000 cases, with probably that many also spread over the rest of the country. *Schistosoma haematobium* inhabits the vesical veins, which drain the urinary bladder. In some parts of Africa this schistosomal infection is so prevalent that hematuria, or blood in the urine, is considered a sign of manhood in adolescent boys. *Schistosoma japonicum* resides in the veins of the small intestine (superior mesenteric veins). This blood fluke is found in many animal hosts, as well as in human beings. The adult female of this species is the most prolific egg producer among the schistosomes, depositing 3000 eggs daily.

Differential features of the aforementioned flukes are presented in Table 37-2.

Life History

Events in the life cycle of trematodes (Table 37-3) may be sketched as follows. In the definitive host sexual reproduction occurs with production of eggs. The name *Schistosoma* (Greek *schisto*, split, plus Greek *sōma*, body) is derived from the presence of a

---

Table 37-3 Contrasting Events in Life Cycles of Flukes

Fluke	Intermediate Host(s)	Source of Infection	Portal of Entry in Human	Habitat in Definitive Human Host
Fasciolopsis buski (intestinal)	Snail; water plants (water nuts)	Eating water nuts and contaminated vegetation	Mouth	Small intestine (duodenum and upper jejunum)
Opisthorchis sinensis (liver)	Snail; fish of family Cyprinidae	Eating raw or improperly cooked freshwater fish	Mouth	Bile ducts and gallbladder
Fasciola hepatica (liver)	Snail; water plants	Eating fresh watercress	Mouth	Bile ducts
Paragonimus westermani (lung)	Snail; crabs and crayfish	Eating freshwater crustaceans (crabs)	Mouth	Lungs
Schistosoma mansoni (blood)	Snail	Cercariae in fresh water	Skin	Veins of large intestine (inferior mesenteric)
Schistosoma haematobium (blood)	Snail	Cercariae in fresh water	Skin	Veins of urinary bladder (vesical)
Schistosoma japonicum (blood)	Snail	Cercariae in fresh water	Skin	Veins of small intestine (superior mesenteric)

ventral slit in the body of the short, stout male schistosome. In the slit, or gynecophoral groove, of the monogamous male, the female worm lies in a continual state of copulation and daily deposits eggs. (Adult male and female worms may persist for 30 years.)

Depending on the anatomic residence of the adult flukes, eggs may escape through the intestinal canal, urinary tract, or pulmonary passages. The embryo within the egg is the **miracidium** (*pl.* miracidia), similar in many ways to a ciliate. Eggs hatch in fresh water. In eggs from the hermaphrodites the operculum opens up like a lid to allow the larval form to get out, but in the nonoperculated egg of a schistosome the larval form must crack the shell to escape.

The miracidium. In fresh water the miracidium swims (about 5 to 6 hours usually) until it reaches an intermediate host in certain species of freshwater snails, such as would be prevalent in rice paddies and irrigation ditches. Of the 100,000 known species of snails, only about 70 serve as intermediate hosts to the miracidia (Fig. 37-3). Species and even strains of flukes have adapted to but a single or at most only a few species of snails. The miracidium has only a matter of hours in which to find its appropriate molluscan host if it is to survive. Hatching miracidia are phototrophic, that is, they swim toward light. (This phenomenon can sometimes be demonstrated in fecal or urinary specimens.) On contact the miracidium penetrates the host snail. In some instances the snail host may ingest unhatched eggs, which hatch in its intestine.

The asexual development and multiplication of flukes require an intermediate host

FIG. 37-3 *Biomphalaria glabrata (Australorbis glabratus)*, snail vector for schistosomiasis in western hemisphere. (From Med. World News **6:**35, Dec. 3, 1965; photograph by Pete Peters.)

or hosts. The schistosomes complete the asexual phase of their life cycle in but one intermediate host—the snail (Fig. 37-4). The other trematodes require a secondary intermediate host, which may be animal or plant (Fig. 37-5 and Table 37-3).

Cercariae. In the snail the miracidium undergoes a series of developmental changes as a result of which cercariae are formed. Cercariae are elliptic, elongate, minute worms with a tail for swimming. So extensive is the process within the snail that thousands of cercariae may come from one miracidium. It is said that one infected snail alone can release 169 million cercariae in 1 month's time. The viability of the cercariae outside a host is limited, being seldom more than 1 day.

Since a secondary intermediate host is not necessary, the cercariae of the schisto-

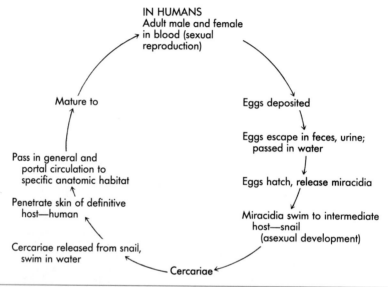

FIG. 37-4   Blood flukes, life cycle.

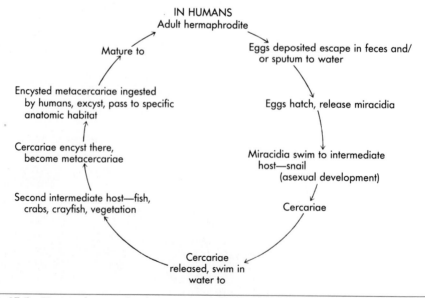

FIG. 37-5   Trematodes passed to humans through vegetation, raw fish, crabs, and crayfish: life cycle.

somes move directly to their definitive host, the human. Being phototrophic, cercariae (Fig. 37-6) emerge during the hours of sunlight; the greatest number is at high noon (hence the most likely time for a human to get the disease). After the cercariae have penetrated the wet skin of a human, especially likely between the toes, and shed their forked tail, they make their way in the general circulation and through the portal venous system to their specific anatomic habitat in the human, where they mature into adult worms.

Metacercariae. If a secondary intermediate host is required, as for flukes other than schistosomes, the cercariae swim around until they can attach to blades of grass or other aquatic vegetation, where they encyst and become metacercariae. Sometimes the cercariae enter aquatic animals, such as certain freshwater fish, crabs, and crayfish, to encyst. When the encysted organisms in raw or improperly cooked food are swallowed by a human being (the definitive host), the metacercariae excyst, migrate to their specific anatomic site, and develop into adult flukes.

Pathogenicity

The major pathologic change in the human body as a consequence of fluke infection centers primarily on the eggs that cannot escape to the outside in feces, urine, or sputum but are trapped in the tissues. The presence of eggs at or near the anatomic residence of the adult worms incites an inflammatory reaction that is designed to wall off the eggs but is one that may result in considerable distortion of that area because of fibrosis and scarring. Depending on the habitat of the worm, lesions may be found in a variety of areas, such as the large and small intestines, liver, mesentery, peritoneum, and urinary bladder. If the inflammatory response occurs in or close to a body surface

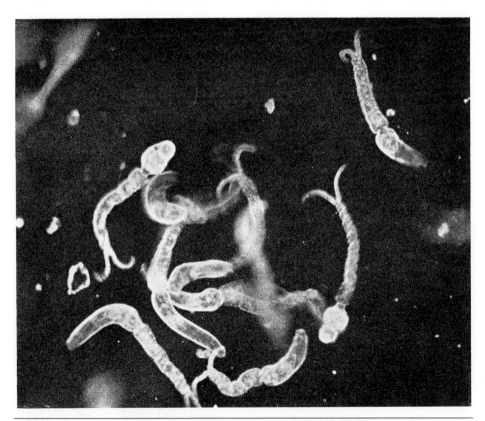

FIG. 37-6   Schistosome cercariae photographed by dark-field illumination.

lined by a mucous membrane, a focus develops that, if it is a destructive one, leads to ulceration. If the eggs by their presence have an irritative effect on the epithelial cells of a surface, a secondary buildup of the lining epithelial cells produces polypoid outgrowths or papillomatous thickenings and folds that project into the lumen of the organ.

The large intestinal fluke causes ulceration in its small bowel habitat. The liver flukes from their position in bile ducts within the liver induce hepatitis (inflammation of the liver), deformities of the bile ducts with proliferation of lining cells, and dilatation of duct lumens. A heavy fluke infection can severely damage or even destroy the liver. In the lung the specific fluke becomes encapsulated and encysts, forming up to 20 cysts, 1 cm in diameter. If the capsule of a cyst ruptures, eggs are released and inflammation ensues in the adjacent lung tissue. Blood flukes inhabiting intestinal and vesical veins seed small veins with eggs, which can break through to the extravascular spaces. Here many are indeed walled off by the tissue response. Many ova die, become surrounded by a foreign body reaction, and are calcified. The small fibrous nodules formed around schistosomal eggs are called **pseudotubercles,** since they resemble the tubercles of tuberculosis (p. 601). Pseudotubercles may be found in the large and small intestines, liver, and urinary bladder. Because of the greater volume of eggs, schistosomiasis caused by *Schistosoma japonicum* is considered the most severe form.

Manifestations of fluke infection are variable in nature and severity, as would be expected from the multiplicity of factors that influence the course of infection. Digestive disturbances, fever, pain, allergic reactions, and the like are commonly associated.

The period of time between initial infection and the appearance of fluke eggs is

Table 37-4 Differential Features of Some Tapeworms

Worm	Disease	Geographic Distribution	Portal of Entry	Habitat in Human
Taenia saginata	Taeniasis; beef tapeworm infection	Cosmopolitan where beef eaten raw or improperly cooked	Mouth	Upper jejunum (small bowel)
Taenia solium	Pork tapeworm infection; cysticercosis (if human intermediate host)	Latin America, Africa, India	Mouth	Upper jejunum
Dipylidium caninum	Dipylidiasis; dog tapeworm infection	Worldwide	Mouth	Small intestine
Hymenolepsis nana	Dwarf tapeworm infection	Worldwide	Mouth	Ileum (small bowel)
Diphyllobothrium latum	Fish or broad tapeworm infection	Temperate zones where freshwater fish are in diet (Baltic countries, parts of Asia, South America, Great Lakes region in United States)	Mouth	Ileum, jejunum
Echinococcus granulosus	Echinococcosis; hydatid cyst	Worldwide where humans live close to dogs (Australia, Middle East, South America)	Mouth	Liver, lungs, bones, brain

generally 6 to 10 weeks. Ova from all types of flukes may be found in stool specimens, regardless of the habitat of adult worms.

Infection with blood flukes (schistosomiasis or bilharziasis) affects 250 million persons. One of the oldest human diseases, it is documented over 4000 years of human history. Calcified ova have been found in Egyptian mummies. Also called snail fever, schistosomiasis is one of the world's most important medical problems. As a global disease, it is second only to malaria in the geographic extent of the incapacity and morbidity produced.

As an aside, schistosomes recovered from human beings have been found to contain *Salmonella* species in what is believed to be a synergistic relationship between the two. The association and interaction may explain why chronic salmonellosis often accompanies schistosomiasis.

Prevention and Control

Measures of control have been largely directed toward the elimination of the intermediate host, the snail, the single most effective measure being the use of molluscides. Travelers or residents in an infected geographic area must avoid all contact with infected water for whatever purpose—drinking, bathing, and swimming. Water for personal use must be boiled or chlorinated. Waterproof boots, plastic trousers, or clothing (also skin) treated with certain chemicals can be made cercariae repellent. Ointments containing 20% benzyl benzoate or an insect repellent such as dibutyl phthalate can be used to protect the skin. Water should be chlorinated, and feces should be properly treated and disposed of.

Source of Infection	Definitive Host(s)	Intermediate Host(s)	Length of Worm	Laboratory Diagnosis
Ingestion of meat with larval cysts	Human	Cattle, other herbivores	To 12 m (1000 to 2000 segments)	Segments and ova in stool; Scotch tape swab for segments
Ingestion of measly pork	Human	Hog, human (at times), wild boars, sheep, and others	To 7 m (800 to 1000 segments)	Segments and ova in stool; Scotch tape swab for segments
Ingestion of arthropod vector	Dogs, cats, humans (accidental)	Larval fleas of dog, cat, human; dog louse	To 70 cm (60 to 75 segments)	Egg sacks and segments in stool
Ingestion of feces from soil; autoinfection	Mice, rats, humans	Larvae of certain arthropods, humans may be	To 4 cm (200 segments)	Eggs in stool
Ingestion of larval form in fish	Human, dog, other mammals	Freshwater copepods (*Cyclops*), freshwater pike, salmon, trout, whitefish	To 10 m (3000 segments)	Operculate eggs or evacuated proglottids in stool
Ingestion of eggs in feces of infected dog or carnivore	Dogs, wolves, jackals, coyotes, other carnivores	Humans may be, sheep, cattle, other herbivores	To 9 mm (4 segments)	Serologic tests; tissue examination

Cestodes (Tapeworms)

Tapeworms *(Taenia)*, elongate, ribbonlike worms, are intestinal parasites in all classes of vertebrates and produce chronic digestive disturbances of variable degree. Table 37-4 presents an overview of tapeworms important in humans. Generally, tapeworms do not compete with the host for food substances; therefore their presence neither increases the appetite nor causes the host to lose weight.

Anatomy

Adult tapeworms have a small head, which buries itself in the intestinal mucosa and anchors the worm, and a nonsegmented neck. The head and neck together make up the **scolex,** or holdfast, behind which is the segmented part of the worm, or the **strobila.** The strobila is made up of a variable number of segments, or **proglottids,** which are attached in a line to the scolex. New ones are formed by a process of segmentation from the scolex—the youngest segment is the one joining the scolex; the last segment, the oldest. Up to a certain point the farther the segment from the scolex, the larger it is. There is no alimentary canal. Each segment obtains its nourishment from the host's intestinal juices by osmosis.

The head is extremely small in comparison with the remainder of the body, often the size of a pinhead (Fig. 37-7). It is provided with hooklets, suckers, or both for attachment to the intestinal wall. The hooklets are arranged in one or more rows around a small prominence (**rostellum**) situated on the head. In at least one species attachment is accomplished by suctorial grooves on the sides of the head. Treatment that fails to recover the head, regardless of the number of segments removed, is valueless because the head immediately replaces the lost segments. The peculiar shape, arrangement, and deeply embedded position of the hooklets often make removal of the head from the intestine difficult.

Each fully developed segment is a sexually complete hermaphrodite. From the scolex to the other end of the worm, there are the following:
1. Undeveloped segments: **immature proglottids**
2. Segments with both male and female elements: well-developed, **mature proglottids**
3. Segments filled by the egg-laden uterus: **gravid proglottids**
4. Degenerating gravid proglottids

In a few species the ova are extruded from the segment through a birth pore, but in most species no birth pore is present and the ova escape from a proglottid that has

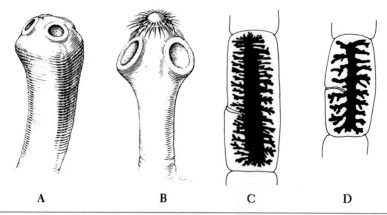

| A | B | C | D |

FIG. 37-7 *Taenia saginata* (beef tapeworm) compared with *Taenia solium* (pork tapeworm). **A,** Scolex of *Taenia saginata* with suckers only. **B,** Scolex of *Taenia solium* with apical hooks and suckers. **C,** Uterus of *Taenia saginata* with more than 14 primary lateral branches. **D,** Uterus of *Taenia solium* with fewer than 12 primary lateral branches.

disintegrated and is either still attached to the strobila or broken off from it. The gravid proglottids toward the end of the tapeworm separate and may be passed in the feces. A person can harbor a tapeworm without ova appearing in the stool because the segments may be expelled before the ova are liberated.

Life Cycle

Tapeworms have a larval and an adult cycle of existence (Figs. 37-8 and 37-9). As a rule, the cycles take place in different species of animals, although autoinfection occurs. The adult cycle occurs within the intestinal canal, the larval cycle within the tissues of the host.

The egg develops into an adult tapeworm as follows. Through a series of changes, an embryo is formed within it. After the eggs are swallowed by a susceptible intermediate host, larvae are set free. By means of their hooklets the larvae penetrate the intestinal wall and pass into the tissues, where the hooklets are lost. They then reach the

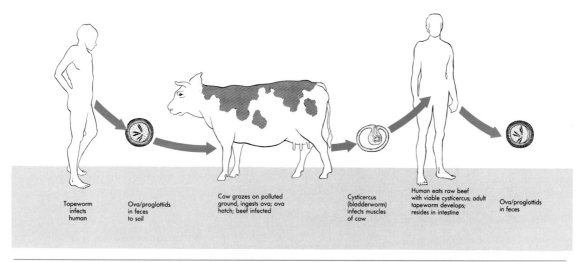

| Tapeworm infects human | Ova/proglottids in feces to soil | Cow grazes on polluted ground, ingests ova; ova hatch; beef infected | Cysticercus (bladderworm) infects muscles of cow | Human eats raw beef with viable cysticercus; adult tapeworm develops; resides in intestine | Ova/proglottids in feces |

FIG. 37-8   *Taenia saginata*, life cycle.

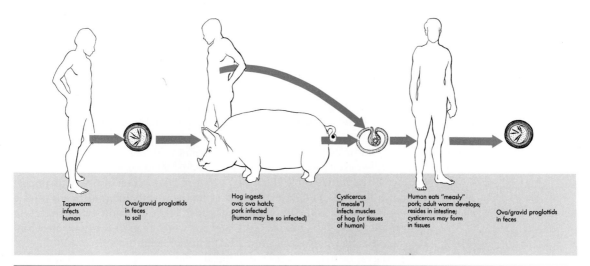

| Tapeworm infects human | Ova/gravid proglottids in feces to soil | Hog ingests ova; ova hatch; pork infected (human may be so infected) | Cysticercus ("measle") infects muscles of hog (or tissues of human) | Human eats "measly" pork; adult worm develops; resides in intestine; cysticercus may form in tissues | Ova/gravid proglottids in feces |

FIG. 37-9   *Taenia solium*, life cycle. When eggs are ingested by a human being, cysticercosis develops in the human host.

bloodstream to be carried to different parts of the body to lodge and each to develop into a scolex. The irritation caused by formation of the scolex sets up a tissue reaction, and a cyst wall is formed around it. The scolex encased in the cyst is termed a **cysticer-cus.** When the raw or insufficiently cooked flesh of the animal containing the cyst is eaten by a susceptible host, the cyst wall is digested, the scolex attaches itself to the intestinal wall of the new host, and an adult tapeworm develops.

In worms (for example, *Diphyllobothrium*) whose eggs (Fig. 37-1, *F*) escape through a birth pore, ciliated embryos escape from the egg after it is passed. They take up an aquatic existence and swim until they gain access to certain species of freshwater fish, which they parasitize with the help of *Cyclops* species, or water fleas, to act as trans-ferring hosts. Consumption of the raw or poorly cooked fish transfers the parasite to humans.

Taenia saginata (Beef Tapeworm, Unarmed Tapeworm)

Infection by *Taenia saginata* is common. The human being is the definitive host; the cow and giraffe are the intermediate hosts. Since the eggs of this worm do not hatch in human beings, only one saginatum typically is present per person. Its favored site is the upper jejunum, where it can exist for up to 25 years.

Anatomy. The organism *Taenia saginata* ranges in length from 4 to 12 m, with a small (1.5 mm in diameter), pear-shaped, somewhat quadrangular head with four suck-ers and a rather long and slender neck. The absence of hooklets gives it the name *unarmed tapeworm*. The strobila is made up of 1000 to 2000 proglottids, the mature ones of which measure 5 to 7 mm by 18 to 20 mm. The segments possess independent motility and may escape through the anal canal. Gravid proglottids are highly motile across the feces and may be mistaken for other worms. The uterus extends along the midline of the gravid proglottid and gives off 20 to 30 delicate branches on each side. Within the gravid uterus are 100,000 eggs. Eggs are spherical or ovoid, yellow or brown, and measure 20 to 30 μm by 30 to 40 μm.

Taenia solium (Pork Tapeworm)

Infection with *Taenia solium*, rare in the United States and Canada, is acquired by eating measly pork. As a rule only one worm is present, but occasionally two or more are found in the usual habitat, the upper jejunum, where they may live for up to 25 years. Humans are the only definitive hosts. The eggs of *Taenia solium*, unlike those of *Taenia saginata*, are also infectious for human beings, because when swallowed, the eggs hatch, and it is possible for both larval and adult cycles of development of the pork tapeworm to occur in the human being. (The human can serve as intermediate host for the larval cycle here.)

Biologic features. Of the two worms, *Taenia solium* is shorter than *Taenia saginata*, measuring 2 to 8 m in length. The small head (1 mm in diameter) is very dark and globular or quadrangular. It is provided with four suckers and two rows of hooklets (a double crown of 25 to 30 hooklets) projecting from the rostellum, the cushioned pro-tuberance of the scolex. The neck is threadlike. The mature segments measure 5 to 6 mm by 10 to 12 mm, and a gravid proglottid liberates 30,000 to 50,000 eggs when it ruptures. The fully developed tapeworm is made up of 800 to 1000 segments.

The ova (Fig. 37-1, *E*) resemble those of *Taenia saginata*. Practically, it is impos-sible to distinguish the two. The differentiation of the worms therefore depends on the characteristics of the uterus in the terminal proglottids. There are more branches (20 to 30) coming from the sides of the uterus in the beef tapeworm than in the pork tapeworm (Fig. 37-7), where the uterus gives off only 7 to 12 thick lateral branches.

Cysticercosis (larval taeniasis). A fully developed, mature cysticercus (known as *Cys-ticercus cellulosae*) is a small, translucent, thin-walled bladder filled with fluid. It is approximately 5 to 10 mm, and from the inner wall, an invaginated scolex equipped with sucker and hooks is visible as a white spot. The cysticercus may be present in muscle or in a viscus where it can persist for 3 to 5 years.

When the intermediate host—the pig or human—takes in embryonated eggs of *Taenia solium*, hatching occurs in the small intestine. From the small intestine the liberated larval forms burrow through to the circulation and disseminate over the body of the intermediate host to various muscles and viscera.

The human may be infected through ingestion of contaminated food or drink, by hand-to-mouth contact, or by displacement of eggs to the stomach through reverse peristalsis. (Because of the likelihood of such reverse peristalsis, vomiting may be an unfortunate event in a person with a pork tapeworm.) When infected pork (measly pork) is eaten, the cyst is dissolved by the action of the digestive juices, the now evaginated scolex attaches to the jejunal mucosa, and, in several months, it develops into an adult worm.

Unfortunately thorough cooking of pork does not guarantee protection against cysticercosis; it only prevents the adult tapeworm from colonizing the intestine.

Hymenolepis nana (Dwarf Tapeworm)

The dwarf tapeworm, *Hymenolepis nana*, the smallest and the one most frequently found in humans, measures from 1 to 4 cm in length. The head is round and provided with four suckers and a single row of 24 to 30 small hooklets. Ova are distinctive (Fig. 37-1, *D*). As a rule, many (maybe as many as 2000) worms and ova are present, but the worms are so degenerated as to be unrecognizable.

Infection is spread directly from one person to another with no intermediate host. Eggs containing a fully developed embryo are released from a disintegrating end segment and passed in the feces. Within the new host the eggs hatch in the stomach or small intestine, and the resultant larval forms move in the lumen to attach to the intestinal wall at a lower site, usually in the upper two thirds of the ileum, where they develop in the mucous membrane. In about 2 weeks adult worms appear.

Dipylidium caninum (Dog Tapeworm)

Dipylidium caninum, the most common of the dog tapeworms, parasitizes both dogs and cats. It is 15 to 70 cm in length with a small head displaying four suckers and about 60 hooklets arranged in four rows. It resides in the small intestine. Hooked ova remain within the ripe proglottid until it has broken loose and migrated from the anus. Egg sacks are found in the stool.

The intermediate host is the flea; young children and babies especially are likely to be occasional hosts. They may acquire the infection if a dog that has just been crushing, eating, or nipping its fleas licks them in the face. The dog tapeworm is not found in anyone over 1 year of age. This is an example of an age-related immunity.

Diphyllobothrium latum (Fish Tapeworm or Broad Russian Tapeworm)

Diphyllobothrium latum, the largest human tapeworm, usually measures 3 to 10 m in length, but occasionally a length of 12 m is reached. The small head, flattened and almond shaped, is provided with two lateral sucking grooves for attachment to the intestinal mucosa of the host. The name *latum* indicates that the mature segments, of which there may be over 3000, are broader than they are long. The uterus filled with ova is the reason for the characteristic brown or black rosette displayed by a segment. One worm may discharge 1 million eggs daily from its gravid uteri (Fig. 37-1, *F*).

The ileum or sometimes the jejunum is the habitat of this worm, where it may live for up to 20 years. Its presence gives rise to irregular fever and digestive disturbances, and because this worm absorbs all available vitamin B_{12}, its presence is associated with a blood picture in the host identical to that of pernicious anemia. Manifestations of the blood disorder subside promptly after the worm is eliminated.

This parasite is destroyed if infected fish is cooked to a temperature of 56° C or frozen at −18° C for 24 hours (at −10° C, 12 hours).

Echinococcus granulosus

Echinococcus granulosus is a tiny worm 3 to 9 mm long with a pear-shaped head, four suckers, 30 to 36 hooklets, but only one each of the immature, mature, and gravid proglottids. The adult host is the dog and other canines. By ingesting the eggs, a human being or a sheep can become the larval or intermediate host, and the unique development of the scolex within a cystic cavity can occur in human organs. The liver is a favored site, and the result is termed the **hydatid cyst.** Such a cyst is unilocular and from 1 to 7 cm in diameter. In cyst fluid free-floating scolices and thousands of brood capsules (structures within which scolices are formed) are known as hydatid sand (Fig. 37-10). The gravity of the infection in humans is determined by the location and the progression of the cyst with time. Rupture of the cyst can produce acute anaphylaxis.

Nematodes (Roundworms)

Nematodes are unsegmented worms with a tapered cylindric body tapering toward both ends. The mouth is frequently surrounded by thick lips or papillae, and there is a complete digestive tract. The sexes are distinct, and the male is shorter and more slender than the female. The adults inhabit the intestinal tract of humans, and as a rule, there is no intermediate host. The nematode life cycle passes through a series of stages from the larval forms to the adult worm.

Ancylostoma duodenale and *Necator americanus*

Ancylostoma duodenale and *Necator americanus* are, respectively, the Old World and New World hookworms. They resemble each other closely but differ in several important details (Fig. 37-11).

The severity of the digestive disorders related to nematode infection depends on the load of worms carried. Table 37-5 emphasizes roundworms important in humans.

Anatomy. Both species are pale red and pointed at both ends. As a rule, the adult *Ancylostoma duodenale* (1.3 cm long) is larger than *Necator americanus* (1.1 cm long).

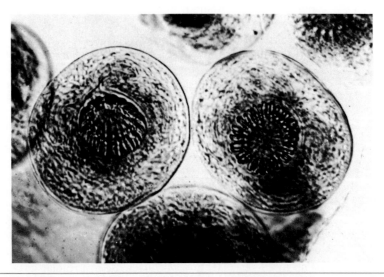

FIG. 37-10 Hydatid sand *(Echinococcus granulosus)*. (From Sonnenwirth, A.C., and Jarett, L., editors: Gradwohl's clinical laboratory methods and diagnosis, ed. 8, St. Louis, 1980, The C.V. Mosby Co.)

Table 37-5   Differential Features of Some Roundworms

Worm	Disease	Geographic Distribution	Portal of Entry	Source of Infection	Habitat in Host	Length of Worm	Laboratory Diagnosis
Necator americanus Ancylostoma duodenale	Hookworm infection; uncinariasis	Tropical and subtropical zones between 45° N and 30° S latitude	Skin (of feet)	Filariform larvae in soil	Small intestine	To 1.3 cm	Eggs in stool; egg count per gram stool an estimate of worm burden
Ascaris lumbricoides (large roundworm of human)	Ascariasis	Temperate and tropical zones (worldwide)	Mouth	Eggs in soil or on vegetables	Small intestine	To 35 cm	Eggs in stool
Toxocara canis	Toxocariasis; visceral larva migrans	Worldwide	Mouth	Eggs in soil	Liver, lung, brain, eye	Larva to 0.3 mm	Serologic tests
Enterobius vermicularis (pinworm)	Enterobiasis	Widest of any helminth (worldwide)	Mouth	Eggs in environment; autoinfection	Large bowel, appendix	To 1.3 cm (size of a pin)	Eggs from perianal skin on Scotch tape swab
Trichuris trichiura (whipworm)	Trichuriasis; whipworm infection	Tropics and subtropics (southern United States)	Mouth	Eggs in soil or on vegetables	Cecum, other areas of large bowel, ileum	To 5 cm	Eggs in stool
Trichinella spiralis	Trichinosis	United States, worldwide (except Asia, Australia, Pacific Islands, Puerto Rico), arctic regions	Mouth	Larvae encysted in pork	Adult—small intestine; larva—encysted in skeletal muscle	To 0.4 cm	Skin test; serologic tests; muscle biopsy
Strongyloides stercoralis	Cochin China diarrhea; Vietnam diarrhea	Tropics and subtropics	Skin	Larvae in soil; autoinfection	Small intestine	To 0.2 cm	Larvae in stool
Wuchereria bancrofti (filarial worm)	Filariasis; elephantiasis	Tropics and subtropics—Africa, Near and Far East, South America	Skin	Bite of *Culex* mosquito (intermediate host)	Lymphatics (human the definitive host)	To 10 cm	Microfilaria in blood smear taken at night
Onchocerca volvulus	River blindness	Tropical zone of Africa and Latin America (near fast-running streams in mountainous areas)	Skin	Bite of *Simulium* fly (intermediate host)	Subcutaneous area (human the definitive host)	To 50 cm	Skin biopsy; aspirate of nodule

Its mouth has a pair of ventral hooks on each side of the midline and a pair of dorsal hooks. The mouth of *Necator americanus* is provided with plates instead of ventral hooks and has a distinct dorsal conical toothlike structure.

The ova of the two species are practically identical, except that those of *Necator americanus* are larger. They are oval or oblong but in certain positions appear spherical. They have three distinct parts—the shell, the yolk, and a clear space between the yolk and shell (Fig. 37-12). The thin, smooth shell appears as a distinct line. Eggs that have been passed for 24 hours or more show well-developed embryos. The ova stick to glass or other surfaces. Advantage is taken of this feature in certain diagnostic procedures, but it makes thorough washing of laboratory glassware imperative.

Habitat. The adult worms live in the upper small intestine, attached to the mucous membrane, where their presence produces a unique chain of events. Great numbers are usually present, tearing the tissues to get to small blood vessels, from which they pump blood into their intestines. They also produce an anticoagulant to facilitate their bloodsucking at the place of attachment. However, they extravasate wastefully much of the blood into the lumen of the intestinal tract so that a profound anemia may be secondary to their bloodsucking activities. Hookworms parasitize an estimated 700 million persons, and the blood loss from such a parasitic burden is possibly 7 million L/day, approximately the total blood volume of 1 million persons. Hookworm disease is sometimes referred to as **uncinariasis** or ancylostomiasis.

Life history. The life history of the hookworm is as follows (Fig. 37-13). After the ova are passed, development begins with the proper environmental temperature and moisture. (A mature female worm releases 10,000 to 20,000 eggs per day.) The larvae hatch and undergo certain developmental changes whereby they can infect a new host.

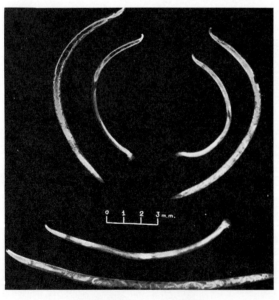

FIG. 37-11   Hookworms. Above, *Necator americanus* (male and female), two pairs; below, *Ancylostoma duodenale* (male and female), one pair. Note millimeter scale. (From Sonnenwirth, A.C., and Jarett, L., editors: Gradwohl's clinical laboratory methods and diagnosis, ed. 8, St. Louis, 1980, The C.V. Mosby Co.; original figure of P. Kouri.)

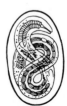

FIG. 37-12   Hookworm eggs, sketch. Development from earliest stage to embryo.

The first-stage larvae emerging are the free-living ones of distinct shape, the **rhabditoid** larvae. Rhabditoid larvae can metamorphose into long, delicate, threadlike forms, the **filariform** larvae of the infective stage. When the filariform larvae contact the skin of a human being, they penetrate it, producing a dermatitis (ground itch). They travel in the lymph and bloodstream to the lungs. Here they gain access to the bronchi and are carried by the bronchial secretions to the pharynx, where they are swallowed. After they reach the small intestine, they develop into adult worms (Fig. 37-11). Around 10 days are required for the migration from skin to bowel. The adult worm is seldom found in the feces in the absence of antihelminthic treatment.

Toxocariasis

Two roundworms nearly universal in their definitive hosts are the dog roundworm, *Toxocara canis*, and the cat roundworm, *Toxocara catis*. *Toxocara canis* is the main one in temperate climates, and toxocariasis, its infection, is a more common parasitic ailment in the United States than is generally recognized, partly because of a dog population now numbered around 80 million. The infection rate of *Toxocara canis* in puppies is nearly 100%, and in older dogs, around 20%.

Life cycle. The life cycle of *Toxocara* species in its definitive host is identical to that of *Ascaris lumbricoides* in its host with this variation. In the dog the larval form of *Toxocara canis* is transmitted prenatally across the placenta to puppies because encysted larvae are activated in the pregnant bitch. In the puppies, larvae pass to the lungs and then to the small intestine, where they mature to adults. Adult worms are 10 to 12 cm long, and females may shed at least 200,000 eggs per day. These do not become infective for 2 to 3 weeks. The spheroidal eggs, 85 by 75 μm, are covered by a thick shell and may survive in humid soil for months.

Human beings usually contact the eggs in soil (in parks, playgrounds, schoolyards), sandboxes, and like areas contaminated by dog feces. Transmission usually results if a person ingests such eggs either directly from soil or on contaminated hands and fomites. Children are vulnerable, especially those 1 to 4 years of age who eat dirt. However, adults with poor sanitary habits can easily contract the infection. After the egg is ingested, it usually hatches in the small intestine and enters the larval stage. Because the human is not the definitive host, larvae do not develop further. But like *Ascaris*, the larval worm penetrates the bowel wall and migrates in the portal circulation to the liver. Most larvae are stopped here, but some pass in venous channels to the lungs and

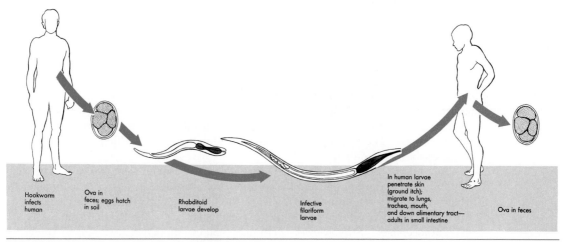

Hookworm infects human

Ova in feces; eggs hatch in soil

Rhabditoid larvae develop

Infective filariform larvae

In human larvae penetrate skin (ground itch); migrate to lungs, trachea, mouth, and down alimentary tract—adults in small intestine

Ova in feces

FIG. 37-13   *Ancylostoma duodenale,* life cycle.

eventually enter the systemic circulation to be widely disseminated. If the larval dimensions of 350 by 75 μm exceed the diameter of a systemic blood vessel, larvae bore through the vascular wall into the extravascular tissues, where they may survive indefinitely. Once in tissues, they begin an aimless migration that stimulates an inflammatory response on the part of the host.

The disease. Toxocariasis is the disease consequent to the mechanical damage done by the migrating larvae and the resultant inflammatory host response, often a severe one. The term **visceral larva migrans** denotes the syndrome caused by the prolonged migration of larvae in the tissues. Clinical manifestations depend on the number and distribution of larvae in the host, whether there is reinfection, and the host's immunologic response. With no reinfection the disease may be self-limited and even asymptomatic. But it can be severe, particularly with ocular involvement. Larva-induced encephalitis or myocarditis can cause death.

Laboratory diagnosis. Since the parasite does not mature in humans, it cannot be demonstrated morphologically in their feces, and the demonstration of larvae in tissue biopsies is not easy. Serologic evidence is helpful. A *Toxocara canis* larval antigen preparation has been adapted for use in an ELISA test.

Prevention. For the prevention of this disease, public health recommendations would be to (1) keep down the number of stray dogs (or cats), (2) prevent fouling of public places with dog excreta (some communities already require that dog owners immediately remove solid wastes deposited in public places by their pets), (3) protect children by covering sandboxes when not in use and excluding dogs from children's playgrounds, (4) eliminate roundworms in dogs with appropriate drugs, and (5) educate the public to the risks of the disease.

Strongyloides stercoralis

Anatomy. One of the smallest of human nematodes, the adult *Strongyloides stercoralis* is only about 2 mm long, with a four-lipped mouth and an esophagus that extends through the anterior fourth of its body. Male worms reportedly have not been found in humans. The slender adult females live deep in the intestinal mucosa (duodenum and upper jejunum), where the ova are deposited. Eggs are 50 by 32 μm, similar to hookworm eggs, with a thin clear shell. Neither adult worms nor ova appear in feces unless active purgation is present.

The larvae known as rhabditiform larvae are hatched in the intestine and passed in the stool (Fig. 37-1, *K*). They are 250 to 500 μm long and actively motile. In fresh specimens they constantly wiggle and bend but do not move progressively. The disturbance produced in the laboratory preparation is often noticed under the microscope before the larvae come into view.

Life cycle. Rhabditiform larvae passed in the stool molt within 24 hours and in the environment change into filariform larvae, which are the infectious organisms (Fig. 37-14). Infection is acquired by a human being when the filariform larvae penetrate the skin or are accidentally swallowed. From the skin the larvae pass into the circulation, move to the lungs, escape into the air sacs, pass up the tracheobronchial tree to the pharynx, and are swallowed there by the host. In the small intestine they mature into adults. Filariform larvae may develop in the soil into free-living adults that mate and produce eggs. The eggs hatch to release rhabditiform larvae, which in turn molt and thus establish a free-living cycle.

Autoinfection occurs in a unique way with this parasite in a human host. The rhabditiform larvae can transform to the infective filariform larvae in a way that enables the filariform larvae to continuously reinvade the host's intestinal mucosa or penetrate the perineal skin. The infection can thus be perpetuated for a long period of time.

Pathogenicity. *Strongyloides stercoralis* causes strongyloidiasis (also known as Cochin China diarrhea or Vietnam diarrhea). It parasitizes an estimated 35 million persons, for many of whom the host-parasite relationship is an innocuous one. In certain instances, however, autoinfection can lead to a massive infection, and this parasite can be responsible for serious disease in the immunocompromised host. It is primarily a human parasite.

Ascaris lumbricoides (Eelworm or Roundworm)

Ascaris lumbricoides, the largest intestinal nematode, is harbored by approximately 1 billion persons and is responsible for the most common helminth infection in human beings. Infection is mostly in children under 10 years of age. Most persons with this worm are asymptomatic.

Biologic features. *Ascaris lumbricoides* is fusiform in shape and yellow or reddish. The male measures 15 to 20 cm in length; the female, 20 to 40 cm. The head is relatively small, and the oral cavity has three serrated lips. This worm looks like the ordinary earthworm but is not as red (Fig. 37-15).

The habitat of *Ascaris lumbricoides* is the upper end of the small intestine, but it may be found in any part of the intestinal tract, free in the peritoneal cavity, or in the trachea and bronchi. Its life span is from 12 to 18 months. Several are usually

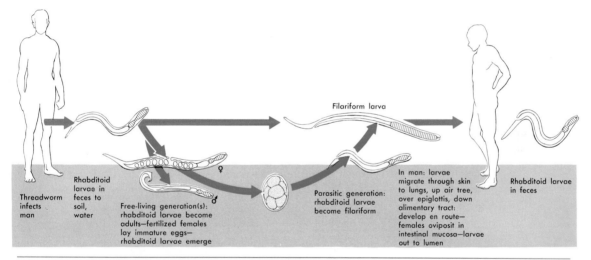

Filariform larva

Threadworm infects man

Rhabditoid larvae in feces to soil, water

Free-living generation(s): rhabditoid larvae become adults—fertilized females lay immature eggs— rhabditoid larvae emerge

Parasitic generation: rhabditoid larvae become filariform

In man: larvae migrate through skin to lungs, up air tree, over epiglottis, down alimentary tract: develop en route— females oviposit in intestinal mucosa—larvae out to lumen

Rhabditoid larvae in feces

FIG. 37-14   *Strongyloides stercoralis*, life cycle. (Autoinfection occurs.)

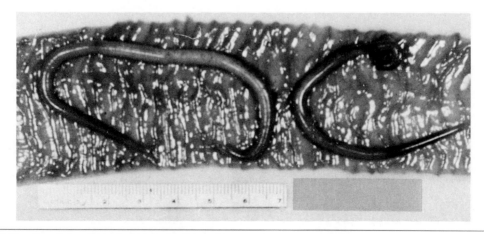

FIG. 37-15   *Ascaris lumbricoides* in lumen of intestine.

present (as many as 100 can be), clinging together to form palpable masses or even cause intestinal obstruction. Adult worms appear in the feces only with active purgation.

The daily output of a female worm is said to be 200,000 eggs. The oval fertilized eggs (Fig. 37-1, *H* and *I*) average 48 μm in diameter and 62 μm in length. If only females worms are present, unfertilized eggs (Fig. 37-1, *G*) will be found. These are elongated, irregular in shape, and bear little resemblance to the fertilized egg. They frequently escape detection. After the eggs are passed, segmentation takes place, and in 3 weeks an infective larva has developed within the eggshell. If these mature eggs are swallowed, the larvae escape from the eggs and travel to the lungs by the bloodstream. They reach the intestines in the same manner as do hookworm larvae (Fig. 37-16) and mature into adults there. Some 5 to 8 weeks are required for completion of the life cycle. Adult worms persist only a year at most before they are passed from the intestinal tract.

Enterobius vermicularis (Oxyuris vermicularis, Threadworm, Pinworm, or Seatworm)

Infection with *Enterobius vermicularis* (enterobiasis) is the most prevalent worm infection of children and adults in the United States. Over the world some 208 million persons are affected. This is the worm with the widest geographic distribution of any of the medically important helminths.

Biologic features. The males of this species measure 3 to 5 mm in length; the females, about 10 mm. The adult female worms migrate through the anus and deposit their eggs on the perianal region, most frequently at night. Because of the peculiar laying habits of the female, the eggs (Fig. 37-1, *C*) seldom occur in the feces but are present around the anal region. They are best found in skin scrapings taken from this region. Eggs may be picked up from the perianal skin by means of swabs made of cellulose adhesive tape, if the sticky side is applied to the skin. The eggs are removed from the tape by toluene and identified under the microscope. An enema may be given and the adult worms identified in the stool that results.

Deposition of eggs on the perianal skin results in intense itching. In children, pinworms should be suspected from this finding alone. Small children may maintain the infection from ova collected under their fingernails when they scratch themselves (Fig. 37-17).

Infection comes from swallowing the eggs, after which male and female parasites

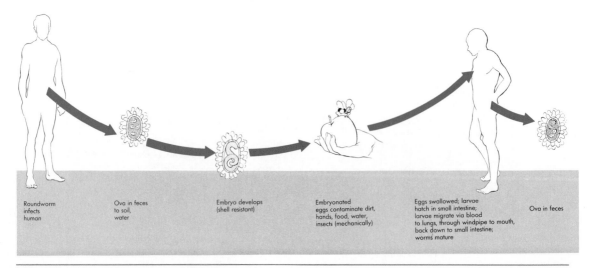

Roundworm infects human | Ova in feces to soil, water | Embryo develops (shell resistant) | Embryonated eggs contaminate dirt, hands, food, water, insects (mechanically) | Eggs swallowed; larvae hatch in small intestine; larvae migrate via blood to lungs, through windpipe to mouth, back down to small intestine; worms mature | Ova in feces

FIG. 37-16 *Ascaris lumbricoides,* life cycle.

hatch out at the lower end of the small intestine. After fertilizing the females, the males die, and the females migrate to the colon and rectum. Patients continually reinfect themselves, and parents may acquire the infection from their children.

Control. Eggs may be widely disseminated in a household or in an institution—in the dust, clothing, and bedding and on furniture, doorknobs, and the like. To eliminate them is an exasperating and almost hopeless task.

The U.S. Public Health Service recommends that families with pinworms pay careful attention not only to general household cleanliness but also to personal hygiene among the members. It stresses the following health measures in the control of the parasites: (1) regular morning showers should be taken; (2) hands should be washed often, especially before food is eaten or prepared; (3) fingernails should be kept short and clean; (4) household surfaces and floors should be vacuumed thoroughly and often; and (5) bed linens should be washed two or three times a week in a machine in which the temperature exposure is at least 150° C for several minutes or more.

Trichuris trichiura (Trichocephalus trichiurus, Trichocephalus dispar, **Whipworm)**

Trichuris trichiura is characterized by a long threadlike neck that makes up about half the length of the body. The male is 30 to 45 mm in length; the female is somewhat longer, 45 to 50 mm.

Biologic features. The worms live in the cecum and large intestine, with the slender end of the worm embedded in the mucosa. The adult worm with the aid of a small "spear" fastens itself into the mucosa much like a needle and thread in a piece of cloth. The name *Trichuris* means "threadtail."

Eggs are discharged in feces, and within the eggshell, the embryo therein contained develops into the infective larval stage in 3 to 6 weeks. Eggs are typically barrel shaped, average 50 by 20 μm, and are stained with bile pigments from the feces. Two shell membranes and translucent polar plugs make these eggs distinctive (Fig. 37-1, *B*).

The portal of entry is the mouth. Infection results from ingestion of the eggs or from hand-to-mouth contact. The fully developed larva hatches in the small intestine and penetrates the mucous membrane for a 3- to 10-day period of development. An adolescent worm then moves into the cecum, where it matures fully to an adult. The life cycle is completed in approximately 90 days. The live span of the adult worms is usually 4 to 6 years.

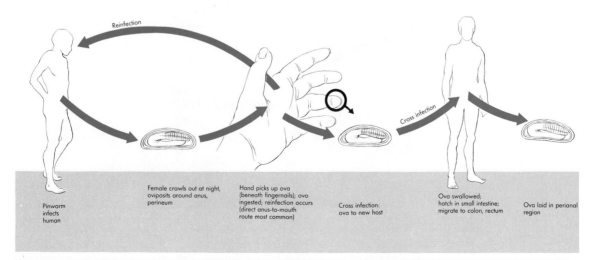

Reinfection

Pinworm infects human

Female crawls out at night, oviposits around anus, perineum

Hand picks up ova (beneath fingernails); ova ingested; reinfection occurs (direct anus-to-mouth route most common)

Cross infection: ova to new host

Cross infection

Ova swallowed; hatch in small intestine; migrate to colon, rectum

Ova laid in perianal region

FIG. 37-17   *Enterobius vermicularis,* life cycle.

The worms themselves are rare in feces, and eggs are not abundant in stool specimens.

Generally, manifestations of disease are related to the number of worms in the bowel, or the worm burden (Fig. 37-18). Humans are the principal hosts, and an estimated 500 million persons are parasitized.

Trichinella spiralis

Trichinella spiralis is the cause of trichinosis (trichiniasis, trichinellosis). It is the smallest worm, with the exception of *Strongyloides stercoralis*, found in the intestinal canal and is barely visible with the unaided eye. The males are about 1.5 mm in length; the females, 3 to 4 mm. The posterior end of the male is bifid and has two tonguelike appendages.

The infection is primarily one of rats, propagated because rats eat their dead. Hogs acquire the infection from rats, and a human becomes infected by eating insufficiently cooked pork. German, Italian, and Polish ethnic groups run the risk of infection because of their fondness for foods containing pork, especially sausage, prepared in traditional ways, which do not necessarily destroy the organisms. Pork sausage is the major source of trichinosis in the United States, but pork is not the only source. The infection has been acquired from the ingestion of bear and walrus meat. Polar bears are said to be heavily infected. The fact that it is found in the arctic region indicates that the parasite possesses a unique tolerance to cold.

Life history. The cycle of development of *Trichinella spiralis* (Fig. 37-19) is practically the same in humans as in other animals. Pork containing the encysted larvae is eaten and the cyst capsule digested away. The larvae pass to the small intestine, where

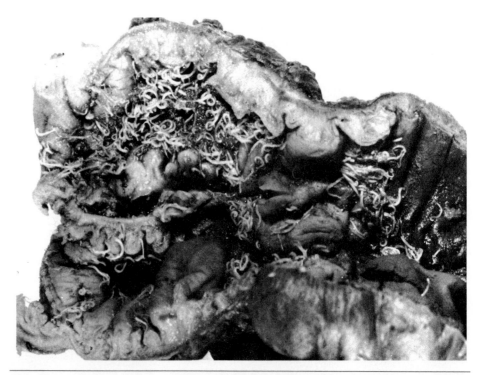

FIG. 37-18   *Trichuris trichiura,* massive infestation in young child producing severe hemorrhagic diarrhea and death. Segment of bowel is opened to show the many whipworms on mucosal surface. (From Anderson, W.A.D., and Kissane, J.M., editors: Pathology, ed. 7, St. Louis, 1977, The C.V. Mosby Co.)

they mature. After copulation the males die, and the females embed in the mucous membrane, where they give birth to as many as 1000 to 1500 larvae. These larvae migrate by the lymph and bloodstream to the skeletal muscles to encyst, become encapsulated, and subsequently calcify (Fig. 37-20).

The free larvae measure 90 to 100 μm in length and 6 μm in diameter. They may be found in the blood and cerebrospinal fluid during the period of migration (6 to 22 days after infection).

After encystment the coiled embryos may be found with the low power of the microscope in a teased portion of muscle and the diagnosis made. The lemon-shaped cysts

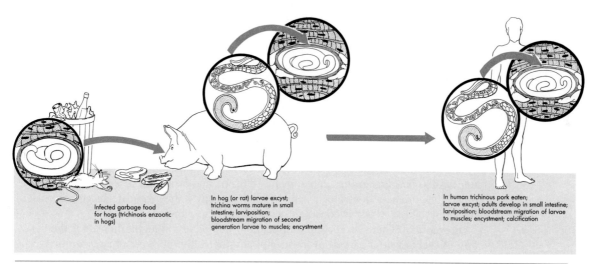

Infected garbage food
for hogs (trichinosis enzootic
in hogs)

In hog (or rat) larvae excyst;
trichina worms mature in small
intestine; larviposition;
bloodstream migration of second
generation larvae to muscles; encystment

In human trichinous pork eaten;
larvae excyst; adults develop in small intestine;
larviposition; bloodstream migration of larvae
to muscles; encystment; calcification

FIG. 37-19 *Trichinella spiralis*, life cycle.

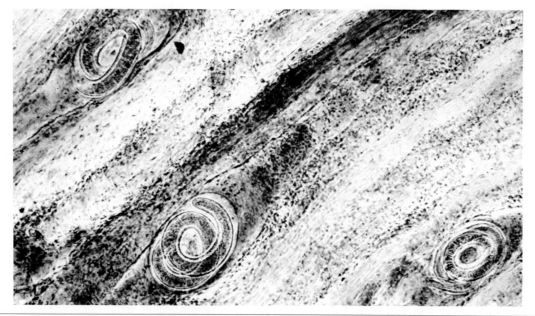

FIG. 37-20 *Trichinella spiralis* encysted in muscle, microscopic section. Larvae may live 10 to 20 years in these cysts. (From Hickman, C.P., and others: Integrated principles of zoology, ed. 7, St. Louis, 1984, The C.V. Mosby Co.)

with coiled larvae therein are most frequently found at the tendinous insertions of the muscle, and the muscles most frequently infected are the pectoralis major, the outer head of the gastrocnemius, the deltoid, and the lower portion of the biceps. The cysts appear as white specks, measuring 250 to 400 μm, and their long axis extends in the same general direction as the fibers of the muscle. Muscle biopsy is the surest method in diagnosis.

Trichinosis. When the parasites are developing in the intestines, gastrointestinal disturbances are prominent. These appear 2 to 3 days after ingestion of the contaminated pork. During this time the adult worms may be found in the feces. When the larvae migrate, fever, delirium, rheumatic pains, and labored respiration are present. This period begins at the end of 1 week after infection and lasts 1 to 2 weeks. When encystment begins, edema and skin eruptions appear. This period lasts about 1 week. After the disease becomes chronic, muscular pains of a rheumatic character may be present for months.

Laboratory diagnosis. Intense eosinophilia, commonly over 500 eosinophilic leukocytes per cubic millimeter of blood, is a feature of all stages of the disease. A skin test is available. A flocculation test, the Sussenguth-Kline test, becomes positive 2 to 3 weeks after infection and remains so for 10 months or longer. There are other serologic tests including complement fixation, latex agglutination, hemagglutination, fluorescent antibody technics, and counterimmunoelectrophoresis.

Identification of the worms in feces or larvae in other clinical specimens is usually impractical because of the course of the disease, and biopsy of a tender muscle poses difficulties. Generally serologic tests are helpful, since they indicate recent or current infection. The skin test is consistent with more remote infection.

Prevention. There is no simple inspection method at a slaughterhouse for the detection of trichinas in a carcass of meat, and commercially, pork is not so examined. The elimination of the disease rests practically with *adequate cooking of pork*. For example, one should cook a pork roast at an oven temperature of at least 350° F (176° C), allowing 35 to 50 minutes per pound. An internal temperature of at least 137° F (58° C) is required to destroy the larvae. Cooking pork to an internal temperature of 165° to 170° F (74° to 77° C) is said to bring out the best in flavor and juiciness. Practically speaking, pork should be cooked until the center is no longer pink. Smoking, pickling, heavy seasoning, or spicing does not make uncooked pork products safe. Freezing meat at −15° C for 30 days or at −28.9° C for 6 to 12 days eliminates the larvae.

Wuchereria bancrofti (Bancroft's Filaria Worm)

Filariasis is infection with the filarial worms, which incorporate certain unique features in their life cycle. The best known is *Wuchereria bancrofti*. A human being is its reservoir of infection and the mosquito its vector. Of historical note is the fact that the first demonstration of the mosquito as a vector of disease was in connection with Bancroft's filaria worm. In 1878 Sir Patrick Manson, in Amoy, China, showed *Culex* mosquitoes to be the natural transmitters of filaria worms. The work suggested to Sir Ronald Ross the possibility that malaria might be similarly transmitted. In Great Britain Manson is known as the father of tropical medicine.

Life cycle. The life cycle of *Wuchereria bancrofti* is sketched as follows (Fig. 37-21). When a human, the only definitive host, is infected, slender white adult male and female worms reside in the lymphatic system. The females are 80 to 100 mm in length and 0.24 to 0.3 mm in diameter; the males are 40 mm long and 0.1 mm in diameter. Within the uterus of the adult female, embryos develop as tightly coiled threads within eggshells. About the time an egg is laid, the embryo uncoils into a tiny, delicate, eel-like form. The eggshell remains applied around the elongate embryo, which is said to be **sheathed.** In some species of filaria worms a naked, or **unsheathed,** embryo is discharged. The embryos in the lymph and bloodstream become **microfilariae** (Fig. 37-22) with remarkable habits. One is their migration into the peripheral blood at night at a time that coincides with the feeding time of the vector mosquito. During the day they are hidden away in an undetermined site.

Taken into the body of the mosquito, the microfilariae lose their sheath and further develop in a larval series. In time the infective larvae escape from the mosquito as it takes a blood meal, returning to the definitive host in whose lymphatic channels they continue their growth. Adolescent worms gather within sinuses of lymph nodes of the groin and pelvis, where they mature and mate to repeat the cycle.

Disease. In localizing within the human lymphatic system, the adult worms start an inflammatory process that progresses from acute to chronic stages. It is associated with tissue changes causing obstruction of lymphatic vessels, stagnation of lymph flow, and proliferation of connective tissue. The skin of the lower extremities and external geni-

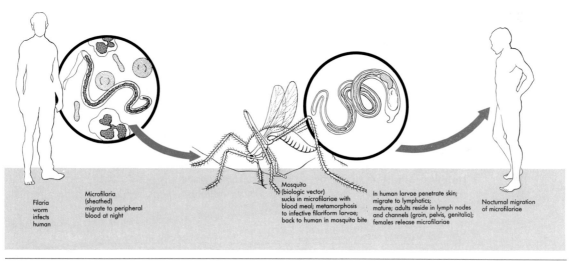

| Filaria worm infects human | Microfilaria (sheathed) migrate to peripheral blood at night | Mosquito (biologic vector) sucks in microfilariae with blood meal; metamorphosis to infective filariform larvae; back to human in mosquito bite | In human larvae penetrate skin; migrate to lymphatics; mature; adults reside in lymph nodes and channels (groin, pelvis, genitalia); females release microfilariae | Nocturnal migration of microfilariae |

FIG. 37-21   *Wuchereria bancrofti,* life cycle.

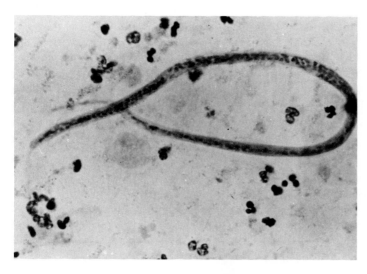

FIG. 37-22   Microfilaria in blood smear, photomicrograph.

In tropical and subtropical areas of the world, where the hazard of contacting so many human parasites is great, travelers are urged to do the following:
1. Carry out faithfully the prescribed regimen for malaria prophylaxis. The medications for prevention of this disease are *considerably* cheaper than hospital costs for treatment.
2. Use insect repellents (physical, chemical, and mechanical).
3. Avoid ingestion of unboiled water, fresh vegetables, and raw or rare meat. Remember that ice is only as safe as the water used for making it.
4. Avoid bathing and swimming in freshwater streams and pools.

talia becomes thickened, coarse, and redundant. The tremendous enlargement of the affected part becomes a great burden to the victim. The result of long years of infection, the end-stage lesion, is referred to as **elephantiasis.** This disorder was well known to the ancient Hindu physicians around 600 BC.

Filariasis is a widely distributed disease in tropical areas of the world and extends into some of the subtropical areas. Some 300 million persons have filariasis. The fact that the appropriate mosquitoes breed close by the dwelling places of human beings with microfilariae circulating in their bloodstream favors the continuous spread of this disease. Therefore methods of prevention and control must be designed for treatment of the human carrier and elimination of the insect vector.

Laboratory diagnosis. The diagnosis of filariasis lies in the identification of the microfilariae found in blood films, especially thick blood films stained with Giemsa stain. In a wet mount of blood they are seen to move gracefully, pushing the red blood cells gently aside. For best results with *Wuchereria bancrofti*, peripheral blood films are made between the hours of 10 PM and 2 AM.

Onchocerca volvulus

Onchocerciasis, or river blindness, a disease afflicting an estimated 40 million persons in Africa and Latin America, is produced by the filaria worm *Onchocerca volvulus*. A hardy black gnat of the genus *Simulium*, which breeds in fast-flowing, aerated mountain streams, transmits the infection as intermediate host. In the bite of the fly, larvae of the threadlike worm are deposited beneath the skin, the site of residence of the adult worms. The female worm gives birth to **unsheathed** microfilariae that pass into the lymphatic system. They are rarely found in peripheral blood but migrate throughout the body in the lymphatics. They die on reaching the eyes but not until they have induced serious changes in the ocular tissues that can in time lead to complete blindness. They can also produce subcutaneous nodules on the head or trunk, wherein a group of adult worms is walled off by fibrous connective tissue.

LABORATORY DIAGNOSIS

The laboratory diagnosis of metazoan, as with protozoan, infections depends most of the time on morphologic (gross and microscopic) identification of the parasite (adult or larva) or its ova (Fig. 37-1). In the morphologic evaluation of eggs, emphasis is given to size, shape, color, nature of eggshell, and stage of development of contained embryo. Specimens usually examined include feces (the most important), urine, blood, body tissues, duodenal aspirates, and the like (Tables 37-2, 37-4, and 37-5).

Examination of Feces for Parasites

Macroscopic Inspection

Sometimes the nurse or medical attendant must examine feces macroscopically for parasites. Proceed as follows.

If need be, improvise a sieve by removing both ends of a can and tying one or two layers of gauze over one opening. Allow water to run slowly over the feces, and break up the masses with a glass rod or wooden applicator. Note the presence of worms. If necessary, remove worms and examine with a hand lens or the eyepiece of the microscope. Small worms may be mounted in a drop of water for macroscopic examination. If a tapeworm is present, be sure to search carefully for the head. Speciation of tapeworms is based on the morphology of the gravid proglottids and the scolex. CAUTION: The handling of unpreserved tapeworm proglottids is extremely hazardous! After tapeworm treatment the entire quantity of feces passed should be saved so that the search for the head may be thorough. If it is suspected that the patient is suffering from typhoid fever or any other condition in which pathogenic bacteria appear in the feces, the washings should be received in a vessel containing disinfectant. Needless to say, the seive, glassware, and utensils used should be properly sterilized or carefully discarded.

Objects likely to be mistaken for intestinal worms. Segmented strands of mucus may be mistaken for tapeworms. The same is true of banana fibers because of their segmented structure and oval cells. Fibers of celery and green vegetables may be mistaken for roundworms and orange fibers for pinworms.

Microscopic Evaluation

Collection of feces. Careful handling of the stool specimen is essential to the laboratory diagnosis of both protozoan and metazoan parasites and their infections. *Fresh, unpreserved* specimens should be examined *within 1 hour after passage* to detect motility of protozoa. Exceptionally, formed stools may be refrigerated at 4° C for 1 or 2 days. At room or incubator temperatures, bacteria in feces overgrow the sample, producing toxic products that kill off trophozoites. Care must be taken to avoid contamination of stool specimens with urine or water. Fecal samples must be collected before diagnostic radiographic studies with barium are performed. At least 7 days are necessary for barium to be cleared from the bowel, and the presence of any amount of it interferes with the identification of parasites.

Generally, a minimum of three specimens is examined in the laboratory—one collected every other day, or a series of three, within 10 days. Cysts are produced cyclically so that more cysts may be found in a person's stools on one day than on another. Therefore more than one stool may be needed to improve recovery of the parasite.

If worms of any kind are to be sent to the laboratory, do not let them dry out. Keep them moist either in saline or in a fixative such as alcohol.

Fixatives. Preservatives used to fix portions of a specimen include 5% or 10% formalin (aqueous or saline), polyvinyl alcohol fixative (PVA), merthiolate-iodine-formaldehyde (MIF), and Schaudinn's fixative. Formalin fixation is good for wet mounts; iodine can be added. PVA fixative works well for trophozoites, cysts, helminth eggs, and larvae in the permanent preparation.

Methods. Feces are examined for parasites in a standard way by (1) direct wet (saline) mounts, (2) mounts prepared with the yield from a concentration technic, and (3) stained smears. With a calibrated microscope, the size of the parasite, a diagnostic feature of some species, can be measured.

The direct mount made speedily from several portions of the stool and *always from any bloody mucus* is useful in the identification of helminth eggs and larvae, motile trophozoites, and some protozoan cysts, but it is limited if there are cells, such as macrophages and neutrophils, that may be confused with protozoa. Lugol's iodine (not

Table 37-6 Immunodiagnostic Tests of Value in Metazoan Diseases

	Schisto-somiasis*	Cysticercosis	Echino-coccosis	Ancylo-stomiasis	Ascariasis	Toxo-cariasis	Trichinosis	Filariasis	Stron-gyloidi-asis
Serologic Test									
Complement fixation	X†	X	X	X	X	X	X	X	
Bentonite flocculation	X	X	X		X	X	X	X	
Indirect hemagglutination	X	X	X	X	X	X	X	X	X
Latex agglutination	X	X	X	X			X	X	
Immunofluorescence	X	X	X	X	X	X	X	X	X
Precipitin	X	X	X		X	X	X		
Enzyme-linked immuno-sorbent assay (ELISA)	X	X	X		X	X	X	X	X
Intradermal Test	X		X			X	X		

Modified from Kagan, I.G.: Hosp. Pract. 9:157, Sept. 1974.

*A highly sensitive and specific radioimmunoassay has been developed for the detection of human schistosomiasis with a purified *Schistosoma mansoni* egg antigen labeled with iodine 125.

†X indicates an accepted test for routine use. Commercial reagents are available.

Gram stain) added to the mount enhances detail in some cysts, but it also stops movement, distorts trophozoites, and may overstain eggs. The diagnosis of protozoan trophozoites and some cysts found on wet mounts should be confirmed with the stained
smear.

Concentration technics, usually flotation and sedimentation procedures, recover
from light infection small numbers of eggs, larvae, or cysts in fresh or preserved specimens, but parasites tend to be mixed in with and obscured by the debris resultant from
the technic. In the flotation method, some operculate and heavy eggs do not float.

Stained smears as permanent preparations should be made on all fresh or PVA-fixed
stools, since this maneuver is the most important means of identifying intestinal protozoa. Good stains are available.

Immunodiagnostic Tests

When the morphologic diagnosis of a parasite cannot be made readily, serologic testing
may be of value. Table 37-6 presents immunodiagnostic tests in relation to the disease
for which they are applicable. (See also p. 793.)

The ELISA test has been adapted to commercial kits wherein specific parasitic antigens are used for testing with patient serum. With a positive test, the end point is read
out in a color reaction.

SUMMARY

1 *Platyhelminthes*, or flatworms, *Nemathelminthes*, or roundworms, and *Arthropoda*
   (vectors of disease) are phyla of metazoa that cause disease in humans. The two
   classes of flatworms are *Trematoda*, or flukes, and *Cestoidea*, or tapeworms.
2 Worms (helminths), elongate, invertebrate animals without appendages or bilateral
   symmetry, demonstrate functional anatomy and characteristic cycles of development—adult (sexual) in the definitive host and larval (asexual) in the intermediate
   host. Most worms prefer a warm climate and poor sanitary conditions.
3 Unlike protozoa, worms are multicellular and macroscopic; the infectious unit matures into only one adult. Like protozoa, their anatomic habitat influences the pattern of their infections.
4 Worm eggs and/or larvae usually leave the definitive host through either end of the
   alimentary tract; some are picked up through the skin by a bloodsucking arthropod.
5 Trematodes, or flukes, are leaflike, unsegmented parasites provided with suckers
   for attachment to the host. Some have the most complicated life cycles in the animal kingdom. Flukes are classified as intestinal, liver, lung, or blood, according
   to habitat.
6 Three species of blood flukes are (1) *Schistosoma mansoni*, residing in the veins of
   the large intestine; (2) *Schistosoma haematobium*, residing in the veins draining the
   urinary bladder; and (3) *Schistosoma japonicum*, residing in the veins of the small
   intestine (the adult of this species deposits 3000 eggs daily).
7 Fluke eggs trapped in tissues incite a reaction there to their presence that results in
   considerable scarring. Schistosomiasis afflicts 250 million persons globally.
8 Tapeworms, elongate, ribbonlike worms found as intestinal parasites in all classes
   of vertebrates, produce chronic digestive disturbances.
9 The definitive host of the beef tapeworm, *Taenia saginata*, is the human. Humans
   are the only definitive host for the pork tapeworm, *Taenia solium*, and sometimes
   its intermediate host, an unfortunate event because the cysticercus of the larval
   cycle may develop in a muscle or in one of the visceral organs of the human affected.
10 Nematodes are unsegmented roundworms with tapered cylindric bodies, a com-

plete digestive tract, and distinct sexes. The adults reside in the human intestinal tract. Usually there is no intermediate host.

11 Hookworms live in the upper small intestine attached to the mucous membrane where they tear the tissues to reach small blood vessels. The blood loss globally from the hookworm burden in 700 million persons parasitized could well amount to 7 million L/day, approximately the total blood volume of 1 million persons.

12 *Strongyloides stercoralis* is passed in the stool as the actively motile larval form.

13 The largest intestinal nematode, *Ascaris lumbricoides*, is harbored by approximately 1 billion persons. (The population of the United States is approximately 250 million.)

14 The pinworm *Enterobius vermicularis*, responsible for the most prevalent worm infection in the United States, is recovered from the perianal skin on Scotch tape swabs.

15 *Trichinella spiralis*, a small roundworm that is the cause of trichinosis, is an infection primarily of rats, which infect hogs. A human becomes infected by eating insufficiently cooked pork. Larval forms of this worm encyst in the skeletal muscles and eventually calcify there.

16 The adult sexes of the best-known filarial worm, *Wuchereria bancrofti*, reside in the lymphatic system where they obstruct lymph flow and induce proliferation of scar tissue. Elephantiasis is the grotesque end stage. Microfilariae of this worm migrate in the peripheral blood at night at a time that coincides with the feeding period of the vector mosquito.

17 The laboratory diagnosis of protozoan and metazoan infections depends on the gross and microscopic identification of the parasite (adult or larva) or its ova. Careful handling of suspect stool specimens is essential to the laboratory evaluation, as well as to personal safety. Serologic testing is also of value.

QUESTIONS FOR REVIEW

1 Define microfilariae, hydatid sand, mechanical vector, biologic vector, ova, scolex, proglottid, ground itch, definitive host, cysticercus, hermaphrodite, operculate, phototrophic, autoinfection, and eosinophilia.

2 List five arthropod vectors and the diseases spread by them.

3 Briefly describe the three phyla of metazoa concerned in human disease production.

4 Outline the life cycle of the following:
   a. Blood flukes
   b. Liver flukes
   c. Beef tapeworm
   d. Pork tapeworm
   e. Fish tapeworm
   f. Hookworm
   g. Pinworm
   h. Roundworm
   i. Filaria worms

5 Discuss the pathogenicity of flukes.

6 What is elephantiasis? River blindness?

7 Give the laboratory diagnosis for:
   a. Schistosomiasis
   b. Trichinosis
   c. Filariasis
   d. Onchocerciasis
   e. Enterobiasis
   f. Tapeworm infections
   g. Hookworm infection
   h. Vietnam diarrhea

8 List objects mistaken for worms in a *macroscopic* stool examination.

9 Comment on the control and prevention of metazoan diseases.

10 Name two fixatives used with stool specimens to be examined for parasites.

11 Give the three standard methods for examination of stool specimens for parasites.

12 Compare multicellular metazoan parasites with unicellular protozoan parasites, em-

phasizing morphology, anatomic habitat, life cycle, geographic distribution, spread of infection, and laboratory diagnosis.

13 List ways in which roundworms differ from flatworms.

14 List precautions to be taken by the traveler in an underdeveloped country.

15 Why is infection with *Taenia solium* more serious than with *Taenia saginata?*

SUGGESTED READINGS

Barclay, W.R.: Diphyllobothriasis, editorial, J.A.M.A. **246:**2483, 1981.

Brown, H.W., and Neva, F.A.: Basic clinical parasitology, ed. 5, East Norwalk, Conn., 1983, Appleton-Century-Croft.

Finegold, S.M., and Martin, W.: Diagnostic microbiology, ed. 6, St. Louis, 1982, The C.V. Mosby Co.

Greenblatt, C.L.: Molecular mimicry and the carbohydrate language of parasitism, ASM News **49:**488, 1983.

James, M.T., and Harwood, R.F.: Herms' medical entomology, New York, 1969, Macmillan, Inc.

Micozzi, M.S., and Eveland, V.L.K.: Lab approaches to parasitic disease in the United States, Lab. Manage. **18:**42, Feb. 1980.

Nash, T.E., and others: Schistosome infections in humans; perspectives and recent findings, NIH conference, Ann. Intern. Med. **97:**740, 1982.

LABORATORY SURVEY OF
MODULE FIVE *

PROJECT   Pyogenic Cocci

Part A—Study of Morphology of Pyogenic Cocci

1. Examine microscopically prepared Gram-stained smears of the following:
   a. *Staphylococcus aureus*
   b. *Streptococcus* species
   c. *Streptococcus pneumoniae*
   d. *Neisseria gonorrhoeae*
2. Examine microscopically the capsule stain of *Streptococcus pneumoniae*. (A suitable specimen of sputum may be used if available.)
3. Note details of morphologic arrangement and Gram-staining reactions. Make comparisons.

Part B—Study of Cultural Characteristics

1. Make cultures of a hemolytic *Staphylococcus aureus* on the following:
   a. Nutrient (beef extract) agar—note pigment.
   b. Blood agar—note hemolysis.
   c. Mannitol salt agar—note growth on medium with a high concentration of salt and fermentation of mannitol.
2. Make culture of *Streptococcus pneumoniae* on blood agar. Note alpha-hemolysis and characteristic colonies.
3. Examine a culture of *Neisseria gonorrhoeae* on chocolate agar.
4. Examine a blood agar plate showing colonies of *Streptococcus pyogenes*. Note beta-hemolysis.

Part C—Identification of Colonies of Cocci

1. Make a culture of a nasal swab from the nose.
   a. Inoculate a blood agar plate and mannitol salt agar. Incubate.
   b. Note characteristics of colonies. Use Gram-stained bacterial smears to study morphology of bacteria in colonies present.
   NOTE: The differential characteristics of the organisms growing on blood agar are discussed by the instructor.
2. Make a culture of a throat swab rubbed over your tonsils (if present) and the back part of your throat.
   a. Inoculate a blood agar plate.
   b. Note colonies. Make Gram-stained smears.

Part D—Typing of Pneumococci with Demonstration by Instructor and Discussion

1. Discussion of methods of typing pneumococci (*Streptococcus pneumoniae*)
2. Comparison of pneumococci from cultures and from body fluids as sputum or peri-

---

*Since it would be impossible to investigate thoroughly all of the organisms in this unit, we will present only a few preliminary tests indicating the nature of some of the microbes and the diagnosis of their infections.

toneal fluid of the mouse. What important structure is not so well developed in pneumococci from cultures as in organisms from the body?

3. Demonstration of method of taking cultures to detect meningococcal carriers
4. Distribution of prepared smears
   a. Pus containing gonococci stained with methylene blue and by Gram's method
   b. Purulent cerebrospinal fluid containing meningococci stained with methylene blue and by Gram's method
5. Note the similarity of meningococci and gonococci.

PROJECT   Enteric Bacilli

Part A—Study of Morphology of Gram-Negative Bacilli

1. Examine microscopically Gram-stained smears of the following:
   a. *Escherichia coli*
   b. *Salmonella typhi* (or other *Salmonella* species)
   c. *Shigella flexneri*
   d. *Klebsiella pneumoniae*
   e. *Proteus vulgaris*
   f. *Vibrio cholerae**
2. Note detail and Gram-staining reaction.

Part B—Motility in Hanging Drop Preparation with Demonstration by Instructor and Discussion

1. Emphasis on technic of making hanging drop and precautions in handling a preparation of living infectious organisms
2. Demonstration of motility of *Proteus vulgaris* or one of the salmonellas

Part C—Study of Cultural Characteristics

1. Make cultures of *Escherichia coli* on the following:
   a. Blood agar
   b. MacConkey agar
   c. *Salmonella-Shigella* (SS) agar
   d. Deoxycholate citrate agar
   e. Eosin–methylene blue agar
   f. Triple sugar iron agar
   g. Bismuth sulfite agar
2. Select a pathogen such as one of the salmonellas and make corresponding cultures on the media just mentioned for culture of *Escherichia coli*.
3. Compare the growth obtained on these media for the pathogen with that of the nonpathogen.
   NOTE: The instructor discusses the varied appearance of the colonies obtained and the reasons. The student identifies the lactose and nonlactose fermenters. The importance of this biochemical test is stressed.
4. Make culture of *Pseudomonas aeruginosa*† on blood agar. Note pigment produced. Examine smears of *Pseudomonas aeruginosa*.

Part D—Demonstration of a Serologic Reaction by Instructor with Discussion

1. Demonstration of the macroscopic slide agglutination test using growth of *Salmonella typhi* (or other *Salmonella* species) and *Salmonella* polyvalent diagnostic serum
2. Discussion of bacterial agglutination in type-specific serum

---

*Not one of the enteric bacilli, this vibrio causes an important disease of the intestine.
†Although not discussed as an enteric bacillus, it is found in the intestinal tract.

3. Review of the significance of the agglutination test as a serologic reaction and diagnostic role of the Widal test in typhoid fever

PROJECT   Small Gram-Negative Rods

Part A—*Brucella* and *Francisella* Species

1. Demonstration by the instructor
   a. Cultures of *Brucella* species
   CAUTION: Because of danger of infection, cultures of *Francisella tularensis* should *not* be handled.
   b. Prepared stained smears of small gram-negative rods
   c. Slide agglutination test for undulant fever
   d. Preserved liver of a guinea pig that has died of tularemia, if available
2. Discussion by instructor: diagnostic tests for tularemia and brucellosis

Part B—The Hemophilic Bacteria

Demonstration by instructor with brief discussion of the cough plate method of detecting the presence of *Bordetella pertussis*

PROJECT   Gram-Positive Bacilli

Part A—Study of Morphology of the Gram-Positive Bacilli—Both Aerobic and Anaerobic

1. Examine microscopically prepared smears.
   a. *Corynebacterium diphtheriae*—in smears stained with:
      (1) Methylene blue stain
      (2) Albert stain
   b. *Clostridium tetani*—in Gram-stained smears
   c. *Bacillus anthracis*—in Gram-stained smears
2. Note morphology of the bacteria, Gram-staining reaction, arrangements, presence and position of spores. Make comparisons.
3. Compare prepared slides of the diphtheria bacillus with those of diphtheroid bacilli. What is the value of differential stains?

Part B—Study of Cultural Characteristics

1. Study prepared cultures of *Corynebacterium diphtheriae*.
   a. Examine a 24-hour culture on Löffler serum medium.
   b. Examine a culture on cystine tellurite agar.
2. Study prepared cultures of diphtheroid bacilli.
3. Note morphology of colonies, especially on the differential media.
4. Compare organisms grown on one medium when examined in stained smears with those grown on the other.
5. Compare the growth of the diphtheria bacilli with that of the diphtheroid bacilli.

Part C—Introduction to Anaerobic Culture Methods

1. Demonstration by instructor with discussion
   a. Selected technics for culture of strict anaerobes
      (1) Inoculation of thioglycollate broth
      (2) Manipulation of stab or solid tube cultures to contain growth within depths of tube
      (3) Use of layering technic to cover surface of medium with airtight substance—petroleum jelly or paraffin
      (4) Use of Brewer anaerobic jar
      (5) Use of GasPak anaerobic jar system

      b. Distinctions between anaerobes and microaerophiles

      c. Technics for culture of capnophiles—use of candle jar

  2. Student participation

      a. Study of anaerobic cultures prepared by some of the technics demonstrated

      b. Comparison of simple technics with more elaborate ones

      c. Evaluation of results obtained with various technics

Part D—Study of a Biochemical Reaction

1. Demonstrate the stormy fermentation of milk.
   a. Boil a tube of skim milk for 10 minutes to drive off oxygen.
   b. Use sterile technic to inoculate heavily the tube of milk from an aerobic culture of *Clostridium perfringens*. Allow the milk to cool to around 50° C before the inoculation is made.
   c. Obtain a tube of sterile mineral oil.
   d. Pour the oil over the surface of the milk to form a surface layer about ½ inch thick.
   e. Incubate the tube of milk at 37° C.
2. Observe nature of bacterial growth in the milk under anaerobic conditions. What method for anaerobiosis was used?

Part E—Demonstration of Animal Inoculation by Instructor with Discussion

1. Determination of the toxigenicity of a strain of suspected *Corynebacterium diphtheriae* in the guinea pig by the intracutaneous virulence test
2. Emphasis on applications of animal virulence tests
3. Comparison of a positive and a negative test result

Part F—Demonstration of the Technic of Throat Culture by Instructor with Brief Discussion

1. Emphasis on proper technic, precautions, and reasons for taking throat cultures
2. Use of differential stains for throat smears
   Students may stain suitable smears with:
   a. Albert stain
   b. Methylene blue stain
   c. Gram I stain with Gram II stain as a mordant
3. Discussion of differential features noted microscopically

PROJECT Acid-Fast Bacteria

Part A—Study of Morphology of Acid-Fast Bacteria

1. Use a specimen of tuberculous sputum that has been autoclaved. Why is this necessary?
2. Prepare smears of the sputum and make acid-fast stains. Make a drawing in color (red and blue) of the organisms present.
3. Make Gram-stained smears of sputum. Compare the two stained preparations.
4. Examine microscopically prepared smears of:
   a. *Mycobacterium tuberculosis*
   b. *Mycobacterium leprae*
   c. *Mycobacterium kansasii* (atypical mycobacterium)
   d. *Mycobacterium smegmatis* (atypical mycobacterium)
   e. *Nocardia asteroides* (actinomycete)

Part B—Study of Cultural Characteristics

1. Examine prepared cultures of *Mycobacterium tuberculosis* on Lowenstein-Jensen medium.

2. Examine prepared cultures of the following atypical mycobacteria. Note salient features.
    a. *Mycobacterium kansasii* (group I)
    b. *Mycobacterium scrofulaceum* (group II)
    c. *Mycobacterium intracellulare* (or *xenopi*) (group III)
    d. *Mycobacterium smegmatis* (group IV)
3. Compare the cultural characteristics of atypical mycobacteria with those of the mycobacteria of tuberculosis.
4. Why do we not examine cultures of *Mycobacterium leprae*? Can this microbe be cultured?

Part C—Animal Inoculation in the Laboratory Diagnosis of Tuberculosis with Demonstration by Instructor and Discussion

1. Study of organs of a guinea pig that has died from tuberculosis
   The instructor can carefully perform an autopsy on a tuberculous guinea pig. The effect of the bacilli on different organs should be demonstrated and discussed. How do the intestinal ulcers of tuberculosis differ from those of typhoid fever?
   NOTE: Organs must be fixed and noninfectious.
2. Demonstration of fixed tuberculous specimens from a human being, for example, a tuberculous lung with cavity formation
3. Presentation of microscopic tissue sections of tubercles
4. Demonstration of technic of inoculating a guinea pig to detect the presence of *Mycobacterium tuberculosis*

Part D—Skin Testing in Tuberculosis with Demonstration by Instructor and Discussion

1. Technic of doing a Mantoux test
2. Technic of the tine (multiple-puncture) test
3. Interpretation of test results (where positive and where negative)

PROJECT   Spirochetes

Part A—Study of Morphology of Spirochetes

1. Examine microscopically prepared smears of:
    a. *Borrelia recurrentis*—Wright's stain of peripheral blood film
    b. *Treponema pallidum*—silver stain of infected tissue
    c. Fusospirochetal organisms of Vincent's angina—stained with Gram stains I and II
2. Compare.

Part B—Study of Spirochetes from the Mouth

1. Take scrapings from between the teeth with a toothpick.
2. Make a thin smear. Let dry and fix in flame.
3. Stain with carbolfuchsin for 1 minute.
4. Note morphology of organisms present.

PROJECT   Fungi

Part A—Study of Unicellular Fungi

1. Observe the morphology of yeasts in a wet mount.
    a. Rub a small portion of yeast cake in about 5 to 10 ml of distilled water.
    b. Place a drop of the suspension on a microslide.
    c. Cover with a cover glass carefully to avoid bubbles.
    d. Examine with the low- and high-power objectives of the microscope.

2. Observe the morphology of yeasts in a wet mount admixed with Gram's iodine. NOTE: The starch granules from the yeast cake are stained dark blue. By contrast, the yeast cells appear yellowish.

Part B—Study of the Morphology of Multicellular Fungi

1. Observe the morphology of the common bread mold *Rhizopus.*
   a. Obtain a piece of bread containing a black moldy growth.
      (1) Examine the natural growth macroscopically, that is, with the naked eye.
      (2) Examine the growth using a strong hand lens.
      NOTE: Observe the color, consistency, texture, and general appearance of the black mold growing on the bread.
   b. Observe the black mold microscopically.
      (1) Carefully tease off some of the mold, and transfer it to a microslide. Place in a drop of water. Carefully position a cover glass.
      (2) Use the low- and high-power objectives of the microscope to examine the mycelial fragments of this mold. Reduce the opening of the iris diaphragm to enhance contrast in the specimen.
      (3) Observe the appearance of the hyphae, and note whether they possess partitions. Observe the nuclei and the appearance of the fruiting bodies.
2. Observe the morphology of the common powdery green mold on citrus fruit. Proceed as in procedure 1 above.
3. Make drawings. Note the presence of hyphae in test specimens of mold. Are they septate or nonseptate? Are nuclei present?

Part C—Study of Morphology and Cultural Characteristics of *Candida albicans*

1. Observe the growth of *Candida albicans* on:
   a. Blood agar (Incubate at 37° C.)
   b. Sabouraud agar (Incubate one plate at 37° C and the other at room temperature, 20° C.)
2. Prepare smears from colonies on blood and Sabouraud agar. Stain by Gram's method.
3. Prepare wet mounts of fungal growth. Examine microscopically.

Part D—Study of a Biochemical Reaction of Yeast

1. Place about 20 ml of 2% lactose in one large test tube (1 × 8 inches).
2. Place about 20 ml of 2% dextrose in another.
3. Rub a cake of yeast in a few milliliters of water, and add half the substance to each of the tubes.
4. Place either test tube at an angle of about 45 degrees in a vessel of water warmed to 42° C. What happens?
NOTE: This test is used to determine whether sugar found in the urine of a nursing mother is lactose or dextrose.

PROJECT   Protozoa
Part A—Study of Morphology and Biologic Functions of Protozoa in Hay Infusion
NOTE: Look for a protozoon of the genus *Paramecium*. Note its motility, feeding pattern, and expulsion of waste material from the body of the parasite. Observe cell detail.
1. Procure a few drops of hay infusion containing paramecia; place on a clean slide and add a few grains of sand to hold cover glass away from the organisms.
2. Gently apply cover glass.
3. Observe microscopically. Note movement of paramecia. Do they seem to have purposeful or aimless movement?

NOTE: In the microscopic study of the wet mount, use the low-power objective of the scope first. Then study the organisms under the high-power objective. Reduce the amount of light coming through the microscope by closing the iris diaphragm. What is the effect of this?

4. Sketch a quiescent organism.
5. Place a drop of india ink on a microslide and mix with a drop of hay infusion (containing paramecia).
6. Examine microscopically. Note activities of protozoa in regard to particles of carbon in the ink. Note currents in oral groove. Note ingestion of particles, attempt at digestion (india ink is indigestible), and expulsion of particles.

Part B—Study of Morphology of an Intestinal Ciliate

1. Prepare a wet mount of a specimen of fecal material from a guinea pig.
   a. Mix a small portion of the contents of the cecum of the animal with a drop of isotonic saline solution on a microslide.
   b. Superimpose a cover glass.
2. Examine microscopically using both low-power and high-power objectives.
3. Look especially for the large ciliate *(Balantidium coli)* that often inhabits the cecum.
4. Note the presence of various flagellates.
5. Observe morphology and movements of protozoa present.
6. Note comparative size of protozoa and bacteria. Make a drawing of each type of protozoon found.

Part C—Study of Morphology of Important Pathogenic Protozoa

1. Obtain prepared slides for study of the following:
   a. Malarial parasites in stained blood films
      (1) Thick films
      (2) Thin films
   b. Trypanosomes in stained blood film
   c. *Entamoeba histolytica* in iron hematoxylin–stained fecal smear
   d. *Giardia lamblia* in iron hematoxylin–stained fecal smear
   e. *Trichomonas vaginalis* in a Papanicolaou-stained smear from cervix of uterus
   f. *Toxoplasma gondii* in hematoxylin and eosin–stained tissue section
2. Examine the protozoa microscopically. Note morphologic features and location in the body where found. Make drawings.
3. Compare the vegetative and cyst forms of *Entamoeba histolytica* in stained fecal smears.
   NOTE: The vegetative forms of *Entamoeba histolytica* may be demonstrated by the instructor in a fresh specimen of feces, if available.

EVALUATION FOR MODULE FIVE

PART I

In the following exercises, indicate the one number on the right that completes the statement or best answers the question. Please circle that number.

1. A culture made from an infected burn wound grows out a microorganism that produces a soluble blue-green pigment on meat extract agar. This microbe is probably:
   (a) *Staphylococcus aureus*
   (b) *Citrobacter freundii*
   (c) *Pseudomonas mallei*
   (d) *Pseudomonas aeruginosa*
   (e) *Pseudomonas pseudomallei*

   1. a
   2. b
   3. c
   4. e
   5. d

2. Which of the following diseases may be caused by *Streptococcus pyogenes?*
   (a) Puerperal sepsis
   (b) Osteomyelitis
   (c) Scarlet fever
   (d) Erysipelas
   (e) Septic sore throat

   1. a and b
   2. a, b, and d
   3. c
   4. none of these
   5. all of these

3. *Coxiella burnetii* is different from the rickettsias causing typhus fever and Rocky Mountain spotted fever in that:
   (a) It can produce a skin rash in the patient.
   (b) It does not require an arthropod vector.
   (c) It is found only in the United States.
   (d) It is more resistant to heat.

   1. a
   2. b
   3. b and d
   4. none of these
   5. c

4. Which of the following statements apply to the enteric bacilli?
   (a) They are all gram negative.
   (b) They are all airborne infectious agents.
   (c) They are all strictly anaerobic.
   (d) Their mode of invasion is through the alimentary tract.
   (e) They all invade the bloodstream during the course of disease.

   1. b
   2. e
   3. d and e
   4. a and d
   5. all but c

5. The diagnosis of hookworm infection depends mainly on:
   (a) The identification of the skin lesions
   (b) The presence of cell inclusions
   (c) Skin tests
   (d) An increase in antibody titer
   (e) Detection of the typical eggs in the stool specimen

   1. a and d
   2. b only
   3. d only
   4. e only
   5. a and e

6. What characteristic do members of the genus *Haemophilus* have in common?
   (a) They destroy red blood cells by dissolving them.
   (b) They are found only in the red blood cells of an infected person.
   (c) They cannot be grown in the laboratory except in living animals.
   (d) They grow in the laboratory only in a medium that contains hemoglobin.
   (e) They all cause diseases of the central nervous system only.

   1. a
   2. b
   3. c
   4. d
   5. d and e

7. Which of the following answers this description: anaerobic, spore-forming organisms that produce exotoxins?
   (a) *Corynebacterium diphtheriae*
   (b) *Clostridium tetani*
   (c) *Clostridium perfringens*
   (d) *Clostridium botulinum*
   (e) *Streptococcus pyogenes*

   1. a
   2. all but e
   3. all but a
   4. b and d
   5. all but a and e

8. As a medical attendant in the emergency room of the city-county hospital, you are standing before a 35-year-old drug addict. You quickly observe the trismus (inability to open the mouth) and the muscular twitching that this patient demonstrates. When a history is taken from a companion, you learn that this person had diphtheria as a 4-year-old child, which was treated with antitoxin. In the evaluation of this clinical problem, one would need to:
   (a) Do a throat culture.
   (b) Test the patient's serum for antitoxic antibodies to tetanus.
   (c) Test the patient's serum for antitoxic antibodies to diphtheria.
   (d) Inject tetanus toxoid at once.
   (e) Inject horse serum antitoxin *intravenously* at once.

   1. a and c
   2. b only
   3. d and e
   4. b and e
   5. none of these

9. The viability of anthrax spores in clinical specimens that might contain *Bacillus anthracis* is about:
   (a) 48 hours
   (b) 7 days
   (c) 6 months
   (d) 1 year
   (e) 40 years

   1. a
   2. b
   3. c
   4. d
   5. e

10. In testing sputum from a patient suspected of having tuberculosis, the bacteriologist can do the following for the laboratory identification of the organism:
    (a) Culture the specimen in the absence of oxygen.
    (b) Use a special medium for the culture of the specimen.
    (c) Attempt to identify the causative organism in smears by means of the Gram stain.
    (d) Inoculate a guinea pig with the suspect material.
    (e) Attempt to identify the causative organism in smears by means of the acid-fast stain.

    1. a
    2. b
    3. c and d
    4. b and d
    5. b, d, and e

11. Which of the following diseases are associated with eosinophilia?

    (a) Hookworm disease
    (b) Ascariasis
    (c) Amebiasis
    (d) Trichomoniasis
    (e) Schistosomiasis

1. a, b, and d
2. a, b, and e
3. c and d
4. a
5. all but b

12. We are aware of the following facts concerning pathogenic rickettsias:

    (a) They grow only within living cells.
    (b) They may be grown within the embryonated hen's egg.
    (c) They are gram positive.
    (d) They are always found within the nucleus of the cell.
    (e) They are transmitted by arthropod vectors.

1. a
2. a, b, and e
3. c
4. d
5. none of these

13. We are aware of the following facts concerning viruses:

    (a) They cannot be seen with an ordinary light microscope.
    (b) They grow only within living cells.
    (c) They cannot be separated from media by filtration.
    (d) They cause only one important communicable disease—influenza.
    (e) They confer no immunity with disease.

1. a
2. a and b
3. a, b, and c
4. c
5. none of these

14. The rickettsial diseases are spread primarily by the following routine:

    (a) Droplet infection
    (b) Contaminated water
    (c) Contaminated food
    (d) Arthropods and rodents
    (e) Direct contact

1. a
2. b
3. c
4. d
5. e

15. The presence of an infection caused by *Treponema pallidum* can be demonstrated by three laboratory procedures as follows:

    (a) Gram stain
    (b) Complete-fixation test
    (c) Dark-field examination
    (d) Growth on special culture media
    (e) Precipitation tests

1. b, c, and d
2. b, c, and e
3. c, d, and e
4. a, d, and e
5. a, b, and c

16. A refugee has malaria. If it is characterized by chills and fever that occur regularly once every 3 days, which of the following statements apply?

    (a) Disease is the estivoautumnal type of malaria.
    (b) Infection is with *Plasmodium vivax*.
    (c) Disease is the tertian type.
    (d) Disease is the quartan type.
    (e) Infection is with *Plasmodium malariae*.

1. a and e
2. a and b
3. c and e
4. b and d
5. d and e

17. What sort of specimen should be collected from a patient to ascertain the presence of *Entamoeba histolytica*?

    (a) Early morning sputum placed in refrigerator until examined
    (b) Catheterized urine specimen kept at room temperature until examined
    (c) Stool specimen kept warm until examined
    (d) Voided urine specimen incubated until examined
    (e) Stool specimen refrigerated until examined

1. c
2. b and c
3. d and c
4. b
5. e

18. In meningococcal meningitis the following are true:
   (a) The Credé technic is useful in prophylaxis.   1. a and b
   (b) An exotoxin is prominent.   2. a and e
   (c) The Waterhouse-Friderichsen syndrome may be a   3. c and d
      feature.   4. c, d, and e
   (d) The causative organism is found in the cerebro-   5. none of these
      spinal fluid.
   (e) There are never skin lesions.

19. Rubella (German measles) is a serious disease because:
   (a) The virus is a togavirus.   1. a
   (b) There is a mild skin rash.   2. c and d
   (c) The virus is teratogenic.   3. b, c, and d
   (d) The virus is oncogenic.   4. c and e
   (e) The virus persists.   5. e

20. Hospital strains of coagulase-positive staphylococci:
   (a) Are a threat to newborn infants   1. a
   (b) Are maintained by carriers among personnel   2. a and b
   (c) Are distinguished by phage-typing patterns   3. b and c
   (d) Never become resistant to antibiotics   4. a, b, and c
   (e) Do not cause infections of surgical wounds   5. d and e

21. The substance elaborated by streptococci that initiates the dissolution of fibrin clots
   is:
   (a) Spreading factor   1. a
   (b) Leukocidin   2. b
   (c) Streptodornase   3. c
   (d) Streptolysin   4. d
   (e) Streptokinase   5. e

22. Which of the following is *not* typical for cholera?
   (a) Explosive outbreaks   1. a, b, c, and d
   (b) Severe dehydration   2. all of these
   (c) Transmission by water   3. a and c
   (d) Caused by a vibrio   4. a, b, and c
   (e) Enterotoxin important   5. none of these

23. Pulmonary tuberculosis:
   (a) Invariably develops when tubercle bacilli enter the   1. e
      respiratory tract   2. d
   (b) Is usually caused by the bovine bacillus in the   3. c
      United States   4. b
   (c) Results in immediate-type allergy   5. a
   (d) Is more likely to develop in a tuberculin-positive
      than in a tuberculin-negative person
   (e) Involves the lymphatics in the primary complex

24. Inclusion conjunctivitis (blenorrhea) of the newborn:
   (a) Is caused by *Neisseria gonorrhoeae*   1. a and b
   (b) May be acquired from the genital tract of the   2. c
      mother   3. d and e
   (c) Is caused by a virus unrelated to trachoma virus   4. b
   (d) Is caused by type 8 adenovirus   5. b and c
   (e) Is transmitted from the respiratory tract of the
      medical attendants

25. *Mycobacterium leprae:*
    (a) Tends to appear in large numbers in nasal mucosa   1. a and b
    (b) Causes a highly contagious disease   2. c and d
    (c) Is found in patients in the United States   3. b, c, and d
    (d) Grows well on media used for the tubercle bacil-   4. a, c, and e
       lus   5. e
    (e) Causes leprosy
26. Urinary tract infection is rarely, if ever, caused by:
    (a) *Proteus vulgaris*   1. a
    (b) *Escherichia coli*   2. b
    (c) *Pseudomonas aeruginosa*   3. c
    (d) *Mycobacterium tuberculosis*   4. d
    (e) *Mycobacterium smegmatis*   5. e
27. Human beings are "accidental hosts" in infection with:
    (a) *Pasteurella tularensis*   1. a
    (b) *Bordetella pertussis*   2. b
    (c) Poliomyelitis virus   3. c
    (d) *Mycobacterium tuberculosis*   4. d
    (e) *Treponema pallidum*   5. e
28. The manifestations of disease in a farmer are chills, fever, and headache; diagnosis of an atypical pneumonia is made. About 2 weeks ago a large number of turkeys on his farm became sick and died; the reason for the illness was not determined. Probably the farmer's ailment is:
    (a) Anthrax   1. b
    (b) Rabies   2. a
    (c) Relapsing fever   3. c
    (d) Leptospirosis   4. e
    (e) Ornithosis   5. d
29. All the following are notable traits of *Mycobacterium tuberculosis* except:
    (a) Grows very slowly   1. a
    (b) Highly resistant to drying   2. b and c
    (c) Spore-former   3. c
    (d) Strict aerobe   4. d and e
    (e) Rich in lipids   5. e
30. Which one of the following parasites most commonly causes blockage of the lumen of the intestinal tract?
    (a) *Taenia solium*   1. a
    (b) *Taenia saginata*   2. b
    (c) *Giardia lamblia*   3. c
    (d) *Ascaris lumbricoides*   4. d
    (e) *Necator americanus*   5. e
31. All the following statements relate to bacteriophages except:
    (a) Phages are viruses.   1. b
    (b) Phages are prepared as attenuated vaccines.   2. a and c
    (c) Phages can produce genetic changes in their bac-   3. e
       terial hosts.   4. b and d
    (d) Phages can confer virulence on certain bacteria.   5. b and e
    (e) Phages can cause bacteriolysis.

32. When antibiotic therapy is given to a patient in such a way as to suppress the normal microflora of the bowel, one of the microbes listed below may proliferate as a result, secrete an enterotoxin that damages the intestinal mucosa, and produce antibiotic-associated colitis. The microbe is:
    (a) *Aerobacter aerogenes*   1. a
    (b) *Escherichia coli*   2. b
    (c) *Clostridium difficile*   3. c
    (d) *Proteus vulgaris*   4. d
    (e) *Borrelia vincentii*   5. e

33. A parasite seen often in the urine, especially in men, is:
    (a) *Trichomonas vaginalis*   1. a
    (b) *Enterobius vermicularis*   2. b
    (c) *Giardia lamblia*   3. c
    (d) *Chilomastix mesnili*   4. d
    (e) *Trichuris trichiura*   5. e

34. A parasite prevalent in the Arctic is:
    (a) *Loa loa*   1. b
    (b) *Fasciolopsis buski*   2. c
    (c) *Onchocerca volvulus*   3. a
    (d) *Trichinella spiralis*   4. e
    (e) *Opisthorchis sinensis*   5. d

35. The majority of respiratory tract infections are caused by:
    (a) *Staphylococcus aureus*   1. a and b
    (b) *Staphylococcus albus*   2. c
    (c) Hemolytic streptococcus   3. d and e
    (d) Friedländer's bacillus   4. c, d, and e
    (e) A virus   5. e

36. Pneumonia in a meat packer may be caused by:
    (a) Q fever   1. a
    (b) Tularemia   2. b
    (c) Brill's disease   3. c
    (d) Rickettsialpox   4. d
    (e) Psittacosis   5. e

37. In tropical Africa 50% to 70% of the malignant tumors occurring in children are represented by:
    (a) Burkitt's lymphoma   1. a
    (b) Acute leukemia   2. b
    (c) Kaposi's sarcoma   3. c
    (d) Hodgkin's disease   4. d
    (e) Neuroblastoma   5. e

38. Echovirus is easily isolated from all these except:
    (a) Cerebrospinal fluid   1. a
    (b) Nose   2. a and b
    (c) Stools   3. c
    (d) Throat   4. d and e
    (e) Fingernail scrapings   5. e

39. A 20-year-old man has a hard, indurated, penile ulcer. He believes it is healing with applications of petrolatum and denies that it was ever painful. The most useful diagnostic procedure would be to take scrapings for:
    (a) Gram stain
    (b) Acid-fast stain
    (c) Tissue culture
    (d) Dark-field microscopy
    (e) Fluorescent antibody study

    1. a and b
    2. c
    3. d
    4. d and e
    5. e

40. Nosocomial infections:
    (a) Must be reported to local health department
    (b) Are acquired in the hospital
    (c) Usually are viral
    (d) Are invariably airborne
    (e) Are diseases of unknown origin

    1. a, b, c, and d
    2. all of these
    3. b
    4. c
    5. b and e

41. The presence of sulfur granules in an exudate from a draining sinus is diagnostic for:
    (a) Nocardiosis
    (b) Actinomycosis
    (c) Coccidioidomycosis
    (d) Dermatophytosis
    (e) Histoplasmosis

    1. a
    2. a and b
    3. b
    4. c and d
    5. e

42. Disease in the central nervous system may result from:
    (a) Epidemic typhus
    (b) Rabies immunization
    (c) Mumps
    (d) Cryptococcosis
    (e) Torulosis

    1. none of these
    2. all of these
    3. a
    4. a and b
    5. a, b, c, and d

43. The agent that causes psittacosis-ornithosis is classified as:
    (a) Virus
    (b) Protozoa
    (c) Fungus
    (d) Mycoplasma
    (e) Rickettsia

    1. a
    2. b
    3. c
    4. d
    5. e

44. Mycoplasmas differ from other bacteria in that:
    (a) Their colonial morphology is different.
    (b) They lack cell walls.
    (c) They do not grow on cell-free media.
    (d) They produce disease only in animals other than humans.
    (e) Their colonies are usually quite large.

    1. all of these
    2. a, b, c, and d
    3. a, b, and c
    4. a and b
    5. e

45. The "Black Death," or plague, which is said to have claimed over 1 billion lives in the last 2000 years, is:
    (a) A natural disease of wild rodents
    (b) Transmitted from rodent to rodent by fleas
    (c) Transmitted from animal to human by fleas
    (d) Endemic in the southwest United States
    (e) Caused by *Yersinia pestis*

    1. d and e
    2. c, d, and e
    3. a, b, c, and e
    4. all of these
    5. none of these

46. Spontaneous mutation of the virus is now known to be a problem in the epidemiology and control of:
    (a) Influenza   1. all of these
    (b) Rabies   2. none of these
    (c) Rubeola   3. a, b, and c
    (d) Rubella   4. a
    (e) Yellow fever   5. d and e

47. The patient with leukemia or bone marrow failure is vulnerable to infection with which of the following fungi?
    (a) *Histoplasma*   1. a
    (b) *Coccidioides*   2. b
    (c) *Candida*   3. c
    (d) *Blastomyces*   4. d and e
    (e) *Actinomyces*   5. d

48. The following are important fungi and ordinarily benign but they may be serious opportunistic pathogens:
    (a) *Candida*   1. all of these
    (b) *Aspergillus*   2. none of these
    (c) *Mucor*   3. all of these except b
    (d) *Rhizopus*   4. all of these except e
    (e) *Cryptococcus*   5. c

49. The three most important genera of dermatophytes are:
    (a) *Taenia*   1. a, b, and c
    (b) *Trichophyton*   2. b, c, and d
    (c) *Epidermophyton*   3. c, d, and e
    (d) *Microsporum*   4. b, d, and e
    (e) *Candida*   5. a, c, and d

50. Starlings may play a role in the spread of:
    (a) Nocardiosis   1. e
    (b) Actinomycosis   2. d
    (c) Histoplasmosis   3. c
    (d) Brucellosis   4. b
    (e) Mucormycosis   5. a

51. A primary lesion at the site that the infectious agent enters the body is an important feature of:
    (a) Syphilis   1. all of these
    (b) Sporotrichosis   2. none of these
    (c) Tularemia   3. b, c, d, and e
    (d) Boutonneuse fever   4. c, d, and e
    (e) Tsutsugamushi disease   5. a and e

52. In which of the following diseases would you be most likely to detect the pathogenic microbe by direct microscopic (bright-field or dark-field) examination of a specimen of blood from a patient?
    (a) Malaria   1. a, b, and c
    (b) Relapsing fever   2. a, b, c, and d
    (c) Trypanosomiasis   3. a, b, and d
    (d) Syphilis   4. all of these
    (e) Cholera   5. none of these

53. Strains of staphylococci can vary in:
    (a) Phage-typing patterns   1. a and c
    (b) Sensitivity to an antibiotic   2. b and d
    (c) Antigenic structure   3. d and e
    (d) Food poisoning potential   4. all of these
    (e) Preferred anatomic site for disease   5. none of these
54. The kind of infectious process most often caused by the members of the genus
    Salmonella is:
    (a) Septicemia   1. a
    (b) Gastroenteritis   2. b
    (c) Meningitis   3. c
    (d) Nephritis   4. e
    (e) Peritonitis   5. d
55. The antigenic type designations A, B, and C for influenza viruses derive from:
    (a) Nucleocapsid proteins of distinct antigenic types   1. b
    (b) Hemagglutinin glycoproteins of distinct antigenic   2. a
        types   3. d
    (c) Neuraminidase glycoproteins of distinct types 4. c, d, and e
    (d) Geographic sites of isolation   5. c
    (e) Years of isolation
56. For coccidioidomycosis:
    (a) The primary infection is usually benign and self-   1. b, c, and d
        limited.   2. a
    (b) The geographic distribution falls in the Mississippi   3. c and d
        River basin in the United States.   4. e
    (c) The disease is primarily one of birds.   5. a and e
    (d) A superficial skin infection usually is present.
    (e) The etiologic agent is a dimorphic fungus.
57. A positive tuberculin reaction:
    (a) Appears within 2 hours in most persons   1. a
    (b) Indicates that person's hypersensitivity to tuber-   2. b
        culosis   3. c
    (c) Is evidence of an active disease process   4. c and e
    (d) Rules out the possibility of active disease   5. b and e
    (e) Also indicates a hypersensitivity to mallein
58. The most effective control measure for hookworm and roundworm infection is:
    (a) Chlorination of the water supply   1. e
    (b) Isolation of infected persons   2. d
    (c) Chemical treatment of contaminated soil   3. c
    (d) Sanitary disposal of feces   4. a
    (e) Control of all arthopod vectors   5. b
59. *Herpesvirus hominis* type 2 is a:
    (a) DNA virus   1. a
    (b) RNA virus   2. b
    (c) Complex DNA-RNA bacterium   3. c
    (d) Defective virus   4. d
    (e) Rickettsia   5. e

60. A 17-year-old college freshman is found to have an enlarged spleen, leukocytosis, heterophil antibodies (positive Paul-Bunnell test), and atypical lymphocytes on the peripheral blood smear. This student probably has:

(a) Gonorrhea   1. a
(b) Typhus fever   2. b
(c) Infectious mononucleosis   3. c
(d) Lymphoma, unclassified   4. d
(e) Burkitt's lymphoma   5. e

61. Not nearly as many cases of tetanus, botulism, or diphtheria are diagnosed each year, but the overall prognosis for one of these, once full-blown clinical manifestations appear, has not changed appreciably in over 50 years. This is because:

(a) The causative agents are changing, and their tox-   1. b
    ins are becoming more severe in effect.   2. a and c
(b) With fully developed exotoxic disease, toxin is   3. b and d
    bound irreversibly to tissues and medical interven-   4. b and e
    tion cannot break this union.   5. e
(c) The organisms are resistant to even the newer an-
    tibiotics.
(d) Antitoxin is not widely available today—much de-
    lay occurs between the time of diagnosis and the
    administration of hyperimmune serum.
(e) The diseases are unusual; therefore diagnosis is
    not made until cultures are obtained.

62. Syphilis may be transmitted by contact with:

(a) The primary chancre   1. a
(b) Blood from a syphilitic person   2. a and b
(c) Condyloma acuminatum   3. a, b, and c
(d) Mucous patches   4. a, b, and d
(e) Tertiary gummas   5. a, b, and e

63. One would not expect to find the enzyme superoxide dismutase in:

(a) Strict aerobes   1. e
(b) Strict anaerobes   2. d
(c) Facultative microbes   3. b
(d) Aerotolerant anaerobes   4. c
(e) Microaerophilic microbes   5. a

64. A traveler, recently returned from Africa, has been having paroxysms of chills, fever, and sweating that last for 1 or 2 days and recur at 36- to 48-hour intervals. Examination of a stained blood film reveals ring and crescent forms within the red blood cells. The infecting microorganism is:

(a) *Plasmodium falciparum*   1. a
(b) *Trypanosoma gambiense*   2. b
(c) *Wuchereria bancrofti*   3. c
(d) *Schistosoma mansoni*   4. d
(e) *Onchocera volvulus*   5. e

65. The leading cause of treatment failures in the management of gonorrheal infection is:

(a) Reexposure of the patient   1. a
(b) Appearance of antibiotic-resistant strains of gono-   2. b and c
    cocci   3. b, c, and d
(c) Biochemical antagonism to the antibiotic used   4. d
(d) Presence of an intercurrent infection of a nonve-   5. b and e
    nereal nature
(e) Therapy not continued for a sufficient period of
    time

PART II

Comparisons: Match the item in column Beta to the phrase (or word) in column Alpha best describing it. Please place the letter in the appropriate space.

COLUMN ALPHA　　　　　　　　　　　　　　　　**COLUMN BETA**

Staphylococcus aureus with *Streptococcus pyogenes*

_____ 1. Can produce hyaluronidase　　　　　(a) *Staphylococcus aureus*
_____ 2. Phage typing in epidemiologic study　(b) *Streptococcus pyogenes*
_____ 3. Serologic typing in epidemiologic study　(c) Both
_____ 4. Can produce greenish discoloration on　(d) Neither
　　　　　　blood agar
_____ 5. Healthy carrier state possible
_____ 6. Nonpathogenic in humans
_____ 7. Septicemia a dangerous event
_____ 8. Produces localized abscesses
_____ 9. Produces cellulitis

Beef Tapeworm with Pork Tapeworm

_____ 10. Mosquito vector　　　　　　　　　(a) Beef tapeworm
_____ 11. Scolex possesses suckers only (unarmed　(b) Pork tapeworm
　　　　　　tapeworm)　　　　　　　　　　(c) Both
_____ 12. Scolex possesses suckers and apical hooks　(d) Neither
_____ 13. Usually only one worm present
_____ 14. Eggs infectious for human beings
_____ 15. Filariform larvae produce ground itch
_____ 16. Hydatid cyst formed in liver of host

Tetanus with Gas Gangrene

_____ 17. Antibiotic therapy highly effective　　(a) Tetanus
_____ 18. Air-filled blebs associated with wound in-　(b) Gas gangrene
　　　　　　fection　　　　　　　　　　　　(c) Both
_____ 19. Active immunity advised　　　　　　(d) Neither
_____ 20. Organisms found in human and animal
　　　　　　feces
_____ 21. Dirt of wound enhances growth of orga-
　　　　　　nisms
_____ 22. Organisms produce a most potent exo-
　　　　　　toxin
_____ 23. Disease result of wound infection
_____ 24. Blood culture positive early in disease

Arboviruses with Picornaviruses

_____ 25. Transmitted by arthropods　　　　　(a) Arboviruses
_____ 26. Includes polioviruses　　　　　　　(b) Picornaviruses
_____ 27. Includes viruses of equine encephalomy-　(c) Both
　　　　　　elitis　　　　　　　　　　　　　(d) Neither
_____ 28. A number of immunologically distinct vi-
　　　　　　ruses
_____ 29. Birds important in cycle in nature
_____ 30. Includes rabies virus
_____ 32. Includes herpesvirus
_____ 32. Humans an accidental host usually
_____ 33. Important diseases in humans
_____ 34. Ecologic hodgepodge
_____ 35. Known oncogenic viruses included

Coccidioidomycosis with Histoplasmosis

_____ 36. Disease occurs in primary and progressive forms

_____ 37. Organisms reproduce by endosporulation

_____ 38. Found in southwestern part of United States

_____ 39. Found in central Mississippi Valley

_____ 40. Causes paracoccidioidal granuloma

_____ 41. Organisms parasitize cells of macrophage system

_____ 42. Also known as speleonosis

_____ 43. Disease contracted from inhalation of spore-bearing dust

_____ 44. Causes disease in lungs

_____ 45. Organism grows in the laboratory as a mold

(a) Coccidioidomycosis
(b) Histoplasmosis
(c) Both
(d) Neither

Pathogenic Bacteria with Pathogenic Fungi

_____ 46. Eukaryotes

_____ 47. Prokaryotes

_____ 48. Chitinous cell wall present

_____ 49. Peptidoglycan in cell wall

_____ 50. Sterols normally present in cytoplasmic membrane

_____ 51. Slide cultures useful

_____ 52. Culture on blood agar routine

_____ 53. Culture on Sabouraud agar routine

_____ 54. Fermentation reactions useful for identification

_____ 55. Phenomenon of fluorescence applicable to laboratory diagnosis

_____ 56. Cultures incubated at body temperature

_____ 57. Cultures incubated at room temperature

_____ 58. More time needed for microbial growth

_____ 59. More complex nutrients may be required

_____ 60. Dimorphism exhibited

_____ 61. Important species grow in absence of free oxygen

_____ 62. Morphologic detection of microbes in diseased tissues pertinent to diagnosis

_____ 63. Nucleic acids not present

(a) Pathogenic bacteria
(b) Pathogenic fungi
(c) Both
(d) Neither

Hepatitis A with Hepatitis B

_____ 64. Inflammation of the liver
_____ 65. Disease in which jaundice not a feature
_____ 66. Viruses are causative agents
_____ 67. Autoclave sterilization inactivates causative agents
_____ 68. Long incubation period (months)
_____ 69. Short incubation period (days)
_____ 70. Serologic tests for diagnosis
_____ 71. Carrier state
_____ 72. Parenteral route of infection of concern in health care
_____ 73. Oral-anal route of infection important
_____ 74. Predilection for children
_____ 75. Predilection for adults

(a) Hepatitis A
(b) Hepatitis B
(c) Both
(d) Neither

Protozoa with Metazoa

_____ 76. Unicellular
_____ 77. Multicellular
_____ 78. Eosinophilia with infections
_____ 79. Arthropod vectors with some infections
_____ 80. Microscopic
_____ 81. Macroscopic
_____ 82. Characteristic anatomic habitat
_____ 83. Multiplication continued within host
_____ 84. Egg or larva matures to a single adult
_____ 85. Variable geographic distribution
_____ 86. Possess a functional anatomy like that of higher forms of life
_____ 87. Immune defenses in infections well defined

(a) Protozoa
(b) Metazoa
(c) Both
(d) Neither

Falciparum Malaria with Vivax Malaria

_____ 88. Infected red blood cell enlarged
_____ 89. Tertian malaria
_____ 90. Malignant malaria
_____ 91. Paroxysms irregular
_____ 92. Paroxysms every 48 hours
_____ 93. Paroxysms every 72 hours
_____ 94. Multiple infection in red blood cell
_____ 95. Parasitemia maximal (no limit as to number)
_____ 96. Parasitemia around 50,000 parasites per cubic millimeter blood
_____ 97. Sausage-shaped gametocytes
_____ 98. _Anopheles_ the mosquito vector
_____ 99. Chemoprophylaxis advised to travelers
_____ 100. Cerebral malaria

(a) Falciparum malaria
(b) Vivax malaria
(c) Both
(d) Neither

PART III

1 Please place the letter indicating the bacterium from column Beta in front of the statement regarding it in column Alpha.

COLUMN ALPHA—Identity

_____ 1. The organism infecting cattle and causing undulant fever in humans
_____ 2. The organism resembling the enteric group but causing respiratory tract infection in humans
_____ 3. The aerobic, gram-negative organism that caused epidemics of the Black Death during the Middle Ages
_____ 4. An organism attacking persons who handle infected animal products such as hides, wool, or hair
_____ 5. The organism causing rabbit fever
_____ 6. An organism losing its importance as a human pathogen since the advent of the automobile

COLUMN BETA—Microbe

(a) *Bacillus anthracis*
(b) *Brucella abortus*
(c) *Francisella tularensis*
(d) *Klebsiella pneumoniae*
(e) *Pseudomonas mallei*
(f) *Vibrio cholerae*
(g) *Yersinia pestis*

2 Please place the letter indicating the protozoan or metazoan organism from column Beta in front of the statement regarding it in column Alpha.

COLUMN ALPHA—Identity

_____ 1. A sporozoan with a life cycle in cats causing chorioretinitis in newborn humans
_____ 2. The small waterborne flagellate whose residence atop the duodenal mucosa may result in a malabsorption state
_____ 3. The protozoan causing intestinal disease with liver abscess
_____ 4. The lung fluke
_____ 5. The largest intestinal nematode for humans
_____ 6. The roundworm causing river blindness in Africa and Latin America
_____ 7. A roundworm with characteristically sheathed microfilaria
_____ 8. The dwarf tapeworm and the one most often seen in humans
_____ 9. The large ciliated protozoon causing one kind of dysentery in humans
_____ 10. The blood fluke whose egg possesses a lateral spine
_____ 11. A blood flagellate transmitted by triatomid bugs
_____ 12. A roundworm, whose larval form is transmitted to the bitch's litter, causing visceral larva migrans

COLUMN BETA—Microbe

(a) *Ascaris lumbricoides*
(b) *Balantidium coli*
(c) *Diphyllobothrium latum*
(d) *Entamoeba histolytica*
(e) *Enterobius vermicularis*
(f) *Giardia lamblia*
(g) *Hymenolepis nana*
(h) *Onchocerca volvulus*
(i) *Paragonimus westermani*
(j) *Schistosoma mansoni*
(k) *Taenia solium*
(l) *Toxocara canis*
(m) *Toxoplasma gondii*
(n) *Trichinella spiralis*
(o) *Trypanosoma cruzi*
(p) *Wuchereria bancrofti*

_____ 13. The cestode causing cysticercosis

_____ 14. The roundworm causing the most prevalent worm infection of children and adults in the continental United States

_____ 15. The roundworm associated with an intense eosinophilia

_____ 16. The fish tapeworm absorbing available vitamin B_{12} in its small intestinal habitat

3 Please place the letter indicating the virus(es) from columm Beta in front of the statement regarding it in column Alpha.

COLUMN ALPHA—Identity	COLUMN BETA—Virus(es)
_____ 1. Six groups of important animal viruses	(a) Arboviruses
_____ 2. Ten groups of important animal viruses	(b) Bacteriophage
_____ 3. Bullet-shaped virus	(c) Cold viruses
_____ 4. Active immunization possible after exposure	(d) Cytomegalovirus
_____ 5. Tadpole-shaped virus	(e) DNA viruses
_____ 6. Inactivation by ether	(f) Enveloped viruses
_____ 7. Insensitivity to ether	(g) Epstein-Barr virus
_____ 8. Group important for oncogenesis	(h) Hepatitis viruses
_____ 9. Type of transmission incorporated in group name	(i) Herpesvirus
_____ 10. Enlargement of parasitized cell	(j) Measles virus
_____ 11. Smallest, simplest, most readily crystallized viruses	(k) Mumps virus
_____ 12. Syncytium formed in cell culture	(l) Myxoviruses
_____ 13. Cause of an important sexually transmitted disease	(m) Naked viruses
_____ 14. Group encompassing myxoviruses, paramyxoviruses, picornaviruses, adenoviruses, and reoviruses	(n) Papovavirus
_____ 15. Rhinoviruses and coronaviruses	(o) Picornaviruses
_____ 16. Postulated causes of scrapie and kuru	(p) Rabies virus
_____ 17. From 80% to 90% of their infections may be inapparent	(q) Respiratory syncytial virus
_____ 18. Cytotrophism for B lymphocytes results in appearance of atypical or reactive lymphocytes	(r) Respiratory viruses
_____ 19. Cause of venereal warts	(s) Retroviruses
_____ 20. Cause of diarrheal disease in human infants	(t) RNA viruses
_____ 21. Members of group agglutinate red blood cells of many mammals and birds	(u) Rotaviruses
	(v) Slow viruses

_____ 22. Skin hypersensitivity to virus demonstrable

_____ 23. Latency a notable feature of infection

_____ 24. Release from parasitized host cell by process of budding

_____ 25. Release from parasitized host cell after lysis of host cell

_____ 26. Negri bodies

_____ 27. Prevalence in rain forests of tropics

_____ 28. Part of their structure derived from parasitized host cell

_____ 29. Overt disease preceded by prolonged period, even many years, of slowly progressive viral activity

_____ 30. Serious atypical variant of natural disease may complicate administration of this paramyxovirus in inactivated vaccine

4 When the paired statements are compared quantitatively, one of them may be found to exert or imply a greater effect or value. If so, please indicate which is the stronger statement as follows:

   A if the first statement is the greater
   B if the second statement represents the greater effect
   C if the two statements are equal or nearly so

_____ 1. The residual oxygen in a candle jar
          The residual oxygen in an anaerobic jar system

_____ 2. The pathogenicity of *Staphylococcus epidermidis*
          The pathogenicity of *Staphylococcus aureus*

_____ 3. The salt tolerance of beta-hemolytic streptococci
          The salt tolerance of coagulase-positive staphylococci

_____ 4. The number of antigenically distinct staphylococcal enterotoxins
          The number of distinct staphylococcal phage types

_____ 5. The size of the egg of *Taenia solium*
          The size of the egg of *Taenia saginata*

_____ 6. The possibility of the transmission of *Trypanosoma cruzi* in the Southwest United States
          The possibility of the transmission of *Trypanosoma gambiense* in the Southwest United States

_____ 7. The number of antigenic types of rhinoviruses
          The number of antigenic types of poliomyelitis viruses

_____ 8. The average incubation period for viral hepatitis type A
          The average incubation period for viral hepatitis type B

_____ 9. The incidence of inapparent histoplasmosis
          The incidence of overt acute histoplasmosis

_____ 10. The probability of infection with *Salmonella typhimurium* in North America
          The probability of infection with *Salmonella typhi* in North America

_____ 11. The tendency of staphylococci to acquire resistance to penicillin
          The tendency of group A streptococci to acquire resistance to penicillin

_____ 12. The number of well persons who carry viridans streptococci in their throat

The number of well persons who carry coliform bacilli in their intestinal tract

_____ 13. The size of the egg of *Necator americanus*

The size of the egg of *Ancylostoma duodenale*

_____ 14. Sensitivity of spores to lethal irradiation

Sensitivity of vegetative cells to lethal irradiation

_____ 15. The spectrum of microbes sensitive to streptomycin

The spectrum of microbes sensitive to tetracycline

_____ 16. The number of branches from the sides of the uterus in terminal proglottids of the beef tapeworm

The number of branches from the sides of the uterus in terminal proglottids of the pork tapeworm

_____ 17. The specificity of immune tolerance

The specificity of an antibody response

_____ 18. The duration of passive immunity

The duration of active immunity

_____ 19. The inhibition of antigen-antibody precipitation by an excess of antigen

The inhibition of antigen-antibody precipitation by an excess of antibody

_____ 20. Biologic abilities of an autotroph

Biologic abilities of an obligate parasite

_____ 21. The size of the liver fluke *Opisthorchis sinensis*

The size of the intestinal fluke *Fasciolopsis buski*

_____ 22. The likelihood of subacute bacterial endocarditis developing in a person with a prosthesis

The likelihood of subacute bacterial endocarditis developing in a person without a prosthesis

_____ 23. The incidence in the United States of brucellosis caused by *Brucella melitensis*

The incidence in the United States of brucellosis caused by *Brucella abortus*

_____ 24. The chance of a biologic false positive reaction with the VDRL flocculation test

The chance of a biologic false positive reaction with the fluorescent treponemal antibody absorption test

5 Match the given disease in the following list with its causative agent, with a specific diagnostic test, and, if indicated, with a culture medium suitable for isolation of the pathogenic agent. From column Alpha, please select the appropriate causative agent and place its corresponding number opposite the disease in the first blank to the right. From column Beta, please select the appropriate diagnostic test and place its corresponding number opposite the disease in the second blank to the right. Match the disease to the suitable culture medium by placing the proper number from column Gamma in the blank adjacent to the right margin.

DISEASE	PATHOGENIC AGENT	DIAGNOSTIC TEST	CULTURE MEDIUM SUITABLE FOR ISOLATION
Boils	_____	_____	_____
Diphtheria	_____	_____	_____
Epidemic cerebral meningitis	_____	_____	_____
Geotrichosis	_____		_____
Herpes simplex	_____	_____	
Hookworm dis-ease	_____	_____	
Infectious mono-nucleosis	_____	_____	
Lobar pneumonia	_____	_____	_____
Malaria	_____	_____	
Pinworm infec-tion	_____	_____	
Primary atypical pneumonia	_____	_____	
Rabies	_____	_____	
Ringworm of scalp	_____	_____	
Scarlet fever	_____	_____	_____
Syphilis	_____	_____	
Trachoma	_____	_____	
Tuberculosis	_____	_____	_____
Typhoid fever	_____	_____	_____
Typhus fever	_____	_____	
Whooping cough	_____	_____	_____

COLUMN ALPHA—
Pathogenic Agents

1. *Bordetella pertussis*
2. *Chlamydia trachomatis*
3. *Corynebacterium diphtheriae*
4. *Enterobius vermicularis*
5. Epstein-Barr virus
6. *Geotrichum* species
7. *Herpesvirus hominis* type 1
8. *Herpesvirus hominis* type 2
9. *Microsporum* species
10. *Mycobacterium tuberculosis*
11. *Mycoplasma pneumoniae*
12. *Necator americanus*
13. *Neisseria meningitidis*
14. *Plasmodium vivax*
15. Rhabdovirus
16. *Rickettsia prowazekii*
17. *Salmonella typhi*
18. *Staphylococcus aureus*
19. *Streptococcus pneumoniae*
20. *Streptococcus pyogenes*
21. *Treponema pallidum*

COLUMN BETA—
Diagnostic Tests

1. Anal swabs taken at night
2. Cerebrospinal fluid culture on chocolate agar
3. Cold agglutination test
4. Cough plate method
5. Demonstration of Negri bodies
6. Heterophil antibody test
7. Inclusions in stained conjunctival scrapings
8. Inoculation of Löffler blood serum culture medium
9. Intranuclear inclusion bodies in skin scrapings
10. Mantoux test
11. Neufeld typing method
12. Phage typing
13. Schultz-Charlton phenomenon
14. Stool examination for ova and parasites
15. Thick blood smears
16. Use of 10% potassium hydroxide in wet mount
17. Wassermann test
18. Weil-Felix reaction
19. Widal agglutination test

COLUMN GAMMA—Culture Medium Suitable for Isolation

1. Blood agar
2. Bordet-Gengou agar
3. Chocolate agar
4. MacConkey agar
5. Oleic acid–albumin medium
6. Potassium tellurite medium
7. Sabouraud agar

6 Match the species name in column Beta to the genus name in column Alpha to identify properly the microorganism.

COLUMN ALPHA—Genus Name

_____ 1. *Allescheria*
_____ 2. *Bacteroides*
_____ 3. *Balantidium*
_____ 4. *Bartonella*
_____ 5. *Bordetella*
_____ 6. *Branhamella*
_____ 7. *Brucella*
_____ 8. *Campylobacter*
_____ 9. *Clostridium*
_____ 10. *Corynebacterium*
_____ 11. *Cryptococcus*
_____ 12. *Entamoeba*
_____ 13. *Enterobacter*
_____ 14. *Escherichia*
_____ 15. *Gardnerella*
_____ 16. *Haemophilus*
_____ 17. *Morganella*
_____ 18. *Mycobacterium*
_____ 19. *Pasteurella*
_____ 20. *Propionibacterium*
_____ 21. *Shigella*
_____ 22. *Staphylococcus*
_____ 23. *Trichomonas*
_____ 24. *Vibrio*

COLUMN BETA—Species Name

(a) *acnes*
(b) *bacilliformis*
(c) *boydii*
(d) *catarrhalis*
(e) *cloacae*
(f) *coli*
(g) *fetus*
(h) *fortuitum*
(i) *fragilis*
(j) *histolytica*
(k) *histolyticum*
(l) *morganii*
(m) *multocida*
(n) *neoformans*
(o) *parahaemolyticus*
(p) *parapertussis*
(q) *parvum*
(r) *pertenue*
(s) *saprophyticus*
(t) *suis*
(u) *vaginalis*

PART IV

After the initial comment or question, please indicate by circling which item or items in the associated listing apply. If none applies, do not circle anything.

1. Giant cell pneumonia, a disease of debilitated children, is caused by the measles virus. Measles virus is also implicated in:
   (a) Infectious mononucleosis
   (b) Kuru
   (c) Burkitt's lymphoma
   (d) Subacute sclerosing panencephalitis
2. The property most often related to disease production by staphylococci is:
   (a) Production of a golden pigment
   (b) Coagulase production
   (c) Fermentation of mannitol
   (d) Ability to hemolyze human erythrocytes
3. Which of the following procedures in the laboratory would most likely yield a *Bacteroides* species out of a contaminated specimen?
   (a) Culture on Sabouraud agar
   (b) Culture in broth containing 30% serous fluid
   (c) Culture on a MacConkey agar plate
   (d) Culture on serum-tellurite agar
   (e) Culture on an anaerobic blood agar plate

4. *Bacteroides* species that produce black pigment may be isolated from human anaerobic infection of:
   (a) The oral cavity
   (b) The intestinal tract
   (c) The respiratory tract

5. Which of the following should *not* be used in preparing direct fecal smears for detection of parasites?
   (a) A wooden applicator
   (b) Iodine solution
   (c) Tap water
   (d) Physiologic saline
   (e) Glass microslides

6. A 14-year-old American boy is seen with a small scar on his left wrist, inflammation of the lymphatics of the left forearm and arm, and a slightly tender lymph node in the left axilla. He has fever and also severe pain in several joints. His hobby is raising small rodents and cats. Most likely this boy has:
   (a) Murine typhus
   (b) Cat-scratch disease
   (c) Rat-bite fever *(Streptobacillus moniliformis)*
   (d) Tularemia

7. Fungal infections may possibly be present in which of the following persons?
   (a) A college student with flu 2 weeks after a visit to a cave in South Texas
   (b) A 45-year-old alcoholic rose gardener with an ulcer on his forearm
   (c) A 30-year-old service station attendant from Phoenix who develops a cough and lung disease in July
   (d) A 40-year-old man with a lymphoid cancer who develops fever, a stiff neck, and signs of lung disease

8. The most accurate diagnosis of fungal disease must be:
   (a) The demonstration of the appropriate delayed hypersensitivity
   (b) The detection of fungi in the lesions of the disease
   (c) The detection of spores in sputum

9. A 2-year-old Pakistani girl, staying with relatives in the United States, suddenly becomes ill with a fever of 40° C (104° F). Clear, fragile vesicles (blisters) erupt on the trunk of her body. During the 4 days thereafter, new eruptions involve the face and extremities and the older vesicles become pustules. The fever is unabated, and the child remains ill. She most likely has:
   (a) Shingles
   (b) Chickenpox
   (c) Cat-scratch fever
   (d) Smallpox

10. Aflatoxin is implicated in the development of which of the following cancers?
    (a) Breast carcinoma
    (b) Acute leukemia
    (c) Lung carcinoma
    (d) Skin cancer
    (e) Liver cancer

11. All oncogenic (tumor) viruses:
    (a) Contain double-stranded DNA
    (b) Produce only sarcomas (one kind of cancer)
    (c) Change the growth properties of infected cells
    (d) Exist as episomes in the cytoplasm of infected cells
    (e) Are transmitted by arthropod vectors

12. Which of the following viral diseases is positively correlated with a malformation in the baby and infection in the baby's mother?
    (a) Rubella
    (b) Mumps
    (c) Cytomegalovirus
    (d) Influenza

13. A 3-year-old child is seen with the typical manifestations of bacterial meningitis. The organism most likely responsible is:
    (a) *Haemophilus influenzae*
    (b) *Staphylococcus aureus*
    (c) *Staphylococcus epidermidis*
    (d) *Streptococcus pneumoniae*

14. Five surgical patients developed postoperative wound infections at about the same time. The hospital administrator, concerned about a possible break in aseptic technic in the surgical suite, institutes an epidemiologic survey. An important part of the survey is the determination of the possible cause of the postoperative infections. One microbe that is likely to have caused the wound infections in clean surgical rooms belongs to the genus:
    (a) *Pseudomonas*
    (b) *Staphylococcus*
    (c) *Diplococcus*
    (d) *Proteus*
    (e) *Candida*

15. All hospital personnel involved in each of the surgical procedures carried out on the five patients with the postoperative infections must have nasopharyngeal cultures made:
    (a) To determine if one such employee is a carrier of the pathogen and, if so, to treat that person to eliminate that pathogen
    (b) Because this routine procedure is part of an epidemiologic survey
    (c) To provide a specimen to send to the Centers for Disease Control for bacteriophage typing if cultures prove positive for the microbe under suspicion

16. Sampling of the area of the surgical suite is carried out:
    (a) To locate a possible reservoir of microbes that might be responsible for the contamination
    (b) To determine the bacterial air count
    (c) To spot-check the adequacy of the routine cleaning procedures

17. Dimorphism in fungi means:
    (a) The development of a yeast phase at body temperature (37° C) and a mold phase at room temperature (20° C)
    (b) The ability to grow at both room and body temperatures
    (c) The expression of two different temperature-dependent colors

18. Which of the following statements applies to herpes genitalis?
    (a) Spread is most often venereal.
    (b) The same virus type causes both cold sores and fever blisters as causes herpes genitalis.
    (c) The virus has been implicated in carcinogenesis in the uterine cervix.
    (d) Herpes genitalis in the pregnant woman at term places her baby at risk for a serious herpetic infection with a high mortality.

19. The following represent appropriate ways for the management of herpes genitalis:
    (a) Ether
    (b) BCG vaccine
    (c) Neutral red dye plus light
    (d) Idoxuridine
    (e) Counseling
20. The number of known etiologic agents responsible for sexually transmitted diseases is:
    (a) 5
    (b) 10 to 15
    (c) More than 20
21. Both *Proteus* and *Pseudomonas* species:
    (a) May be isolated from stool cultures in the absence of disease
    (b) Often cause urinary tract infection
    (c) Produce colorless colonies on MacConkey agar
    (d) Produce water-soluble pigments
22. *Mycobacterium* species other than *Mycobacterium tuberculosis*:
    (a) Are usually nonvirulent for guinea pigs
    (b) Are acid fast
    (c) Are not important causes of pulmonary disease in humans
    (d) Grow on plain nutrient agar
23. Intranuclear inclusions are found in all these except:
    (a) Cytomegalovirus infection
    (b) Molluscum contagiosum
    (c) Herpes genitalis
    (d) Adenovirus infection
24. A positive Mantoux test in a 1-year-old child suggests:
    (a) Infection in utero
    (b) Contact with a person with active tuberculosis
    (c) That BCG immunization is advised
    (d) That the child is susceptible to primary infection
25. Gonorrhea:
    (a) Produces more serious disease with more serious consequences in human females than in males
    (b) Produces in males urethral infection that is always symptomatic
    (c) Is passed more often from an infected male to an uninfected female than from an infected female to an uninfected male
    (d) Is associated with an infection of the fallopian tubes that can lead to sterility in females

MICROBES
PUBLIC WELFARE

MICROBES EVERYWHERE

Persons in daily contact with disease are likely to look on microbes as agents of harm only, but such is not the case. *The majority of bacteria and other microbes are helpful to humans, animals, and plants, all of which depend on microbes for their very existence.* In a broad sense microorganisms producing disease form a small and inconspicuous group. About 50,000 species of microbes have been defined; of this number, probably less than 100 cause disease. Nothing gives us a better idea of the broad scope of microbial activity than a consideration of how bacteria and other microbes affect our daily lives.

MICROBES IN THE PROCESSES OF NATURE

The microbiology of nature defines the role that microbes play in the various processes of nature. Bacteria and other microbes have been said to be nature's garbage disposal system and fertilizer factory, for they are the active agents in the decomposition of dead organic matter of animal origin, releasing elements needed for growth of plants and returning them to the soil. Bacteria purify sewage by living on the impurities in it and converting these to inoffensive substances that serve as food material for plants. As water trickles through the soil, bacteria help to filter it. Though contaminated when it enters the earth, water may trickle out pure and clean. In short, microbes represent a major natural resource.

Participation in the Cycle of an Element

The chief elements entering the bodies of animals and plants are nitrogen, carbon, hydrogen, and oxygen, which are combined with other less common elements to form complex proteins. Animals depend on plants for these elements. Plants receive their hydrogen and oxygen from water and must obtain their carbon and nitrogen from an inorganic compound containing them. The assimilation of an element from inorganic compounds, its conversion into organic compounds to form the bodies of plants and animals, and its subsequent reappearance in the inorganic state to be used again constitute the *cycle* of the element.

Carbon Cycle

Plants can assimilate carbon only from carbon dioxide. Carbon dioxide is present in the air in small quantities, but the supply must be constantly renewed. This comes about as carbon dioxide, a waste product of metabolism, is constantly eliminated in the breath of all animals. It is also released from the respiration of microorganisms (Fig. 38-1).

If animals were sufficiently numerous to eat all plants, converting them into carbon dioxide and excreting the gas, the supply of plant food for animals and of carbon dioxide for plants would be perfectly balanced. However, tissues of certain plants such as wood and cellulose, which contain much carbon, pass through the intestinal tracts of animals unchanged. Therefore the carbon in them is not converted into carbon dioxide and released. Other plants die with much carbon bound in their bodies. Carbon is also present in the body tissues of all animals when they die. If there were not some method of recovering this carbon in a form available for plant growth (CO_2), plant life would soon cease and animal life would shortly thereafter. To recover the carbon, fungi and bacteria attack animal excreta and dead plants and animals to initiate processes of decay and decomposition. The complex carbon compounds are broken down by microbes into carbon dioxide, which floats away in the air to become plant food.

Nitrogen Cycle

Nitrogen, one of the most important elements in the composition of the plant body, is abundant in the air but not in a form suitable for plant (or animal) use. To be suitable, it must be in the form of nitrates. Nitrates present in the soil in very small amounts must constantly be renewed. Microbes participate by (1) decomposing organic matter and (2) converting the nitrogen of the air into nitrates (Fig. 38-2). Organic matter is broken down by different microorganisms with various actions to form simpler com-

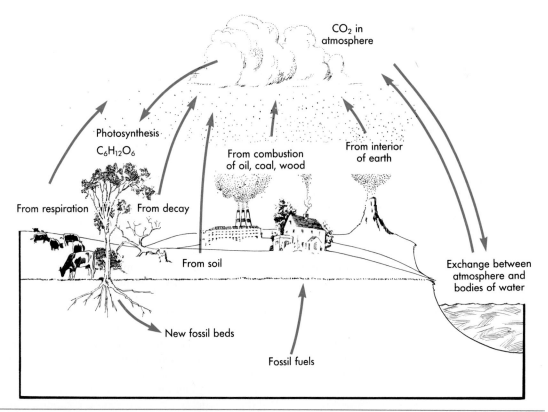

FIG. 38-1   Carbon cycle. Sources of carbon dioxide from the earth's surface and within its depths; assimilation of carbon dioxide from atmosphere by plant life through photosynthesis.

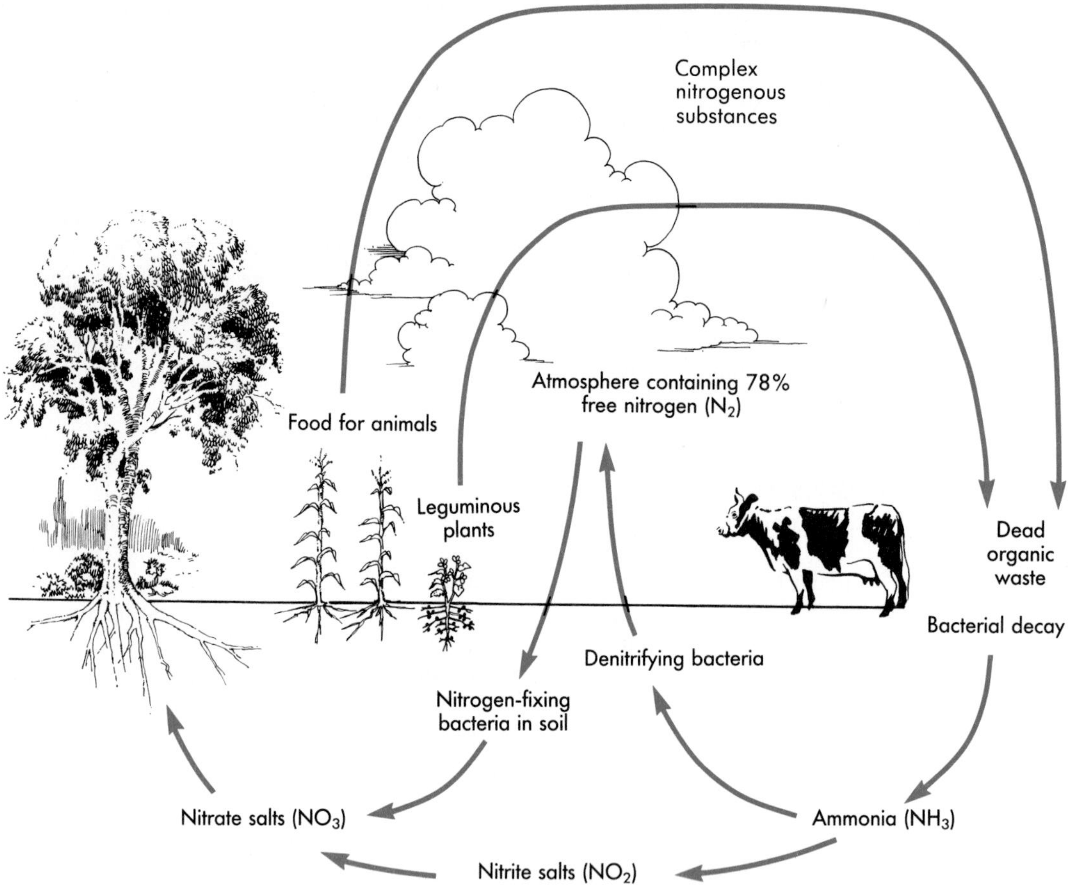

FIG. 38-2   Nitrogen cycle. Various sources indicated for nitrogen needed by plants and animals for protein synthesis. Note that microbes of soil convert some nitrogen stepwise into ammonia, nitrite, and nitrate. Nitrate is also derived from nitrogen of air by bacterial action and through oxidizing action of lightning. Nitrate from these sources is taken up by roots of plants.

pounds, among which is ammonia. Ammonia is attacked by **nitrifying** bacteria (genus *Nitrosomonas*), which convert it into nitrites. Other nitrifying bacteria (genus *Nitrobacter*) then convert the nitrites into nitrates, which are absorbed by plant roots and built into complex nitrogen compounds like those of the original organic matter.

The term **nitrogen fixation** applies to the recovery of free nitrogen from the air and its synthesis into nitrates for use by plants and by animals eating the plants. The first step in nitrogen fixation, the formation of ammonia, is accomplished by two groups of microorganisms. One (genus *Azotobacter*) lives in the soil. The other (genus *Rhizobium*) penetrates the root hairs and develops symbiotically in the cells of certain legumes such as clover. It receives its nourishment from the plant and forms nodules filled with rapidly multiplying bacteria. These bind the nitrogen of the air into a compound incorporated into the bacterial body, which is removed and used by the plant. This symbiotic relation between plant and microbes is a necessary and mutually beneficial arrangement known as **mutualism.** When the plant dies and decays, the nitrogenous compounds are set free in the soil. Soils in which such products of decay are present are fertile. A large amount of nitrogen may also be converted to nitrates and carried to the soil and living organisms during severe electrical rainstorms.

Within the upper several centimeters of well-tilled soil, there are 300 to 3000 kg of living microorganisms per hectare. An acre of alfalfa well nodulated by nitrogen-fixing rhizobia has the capacity to fix in the ground 200 kg of nitrogen from the air. The microbes in the earth's crust fix annually 10^8 metric tons of nitrogen from the air, six times more than can be accomplished artificially in industry.

Microbiology of the soil deals with the many aspects of the population of microorganisms on which fertility depends.

Other Cycles

Other elemental cycles in which microbes participate are those of phosphorus and sulfur in fresh and marine waters.

MICROBES IN ANIMAL NUTRITION

In 1861 Pasteur expressed the idea that animal life would not be possible were it not for the microbial inhabitants of the intestinal tract:

> If microscopic beings were to disappear from our globe, the surface of the earth would be encumbered with dead organic matter and corpses of all kinds, animal and vegetable. . . . Without them, life would become impossible because death would be incomplete.

Bacteria and protozoa take an active part in the digestive processes of both lower animals and humans, and in the large bowel they form many vitamins essential to animal and human nutrition. Large amounts of vitamins in the B complex group are so produced, and this is the major source of vitamin K, important in the prevention of hemorrhage. Vitamin K is primarily supplied to the body from the intestinal lumen, where about 1000 units of vitamin K per gram weight of dry bacteria are synthesized by *Escherichia coli*.

Administration of the modern broad-spectrum antibiotics (for example, the tetracyclines) may so sharply reduce the microbial population of the intestinal tract as to cause a sizable reduction in the production of vitamin substances there. Depletion of vitamin K is especially serious before surgery, and the vitamin may have to be administered to the patient as a drug to prevent hemorrhage. Microbes may have even much greater importance in the physical well-being of lower animals and humans than is apparent with our present state of knowledge.

MICROBES IN INDUSTRY

Microbiology of industry involves the roles that microbes play in processes of commerce and industry.

Manufacture of Dairy Products

Microbiology of agriculture is a broad subject encompassing the relations of microbes to domestic animals and to the soil. One important subdivision is **dairy microbiology,** the manufacture and processing of milk products.

When milk is churned, the fat globules coalesce, taking up a small amount of milk solids and water to form butter and leaving a residue of buttermilk. When milk is allowed to stand, it sours and curdles because bacteria convert the milk sugar, lactose, into lactic acid, a process of **ripening.** Normal milk bacteria (lactic acid bacteria primarily in genera *Streptococcus, Leuconostoc, Pediococcus,* and *Lactobacillus;* see also p. 911) play the key role in ripening.

Originally, ripening served to preserve the milk until enough to churn had accumulated, since the acidity of sour milk inhibits growth of other microbes. Ripening is often hastened by adding cultures of these lactic acid–producers known as **starters,** and today ripening is done to give flavor to butter. Butter is usually prepared from cream instead of whole milk, but for cultured buttermilk ordinary butter starters are added to fresh sweet milk.

Cheese consists of the curd separated from the liquid portion of milk either by bacterial fermentation of the lactose to form lactic acid or by the action of the milk-curdling enzyme **rennin,** which is obtained from the stomach of a calf. Today, however, in the United States a **microbial rennet** (from *Mucor*) is largely replacing the calf enzyme in the cheese industry.

Cheese prepared by the action of lactic acid is cottage cheese. Green cheese is produced when rennin is the curdling agent. The water is drained away and the curd pressed into a block and partly dried. When green cheese is set aside for a time, it ripens; that is, it undergoes changes in color, flavor, and odor that are induced by enzymes in the curd as well as by certain molds, yeasts, and bacteria. The kind of cheese resulting depends to a great extent on the kind of organisms causing it to ripen. Swiss cheese is ripened by *Streptococcus thermophilus* and *Lactobacillus bulgaricus* as starters, after which a gas-producing colony of *Propionibacterium freudenreichii* supbsp. *shermanii* literally blows out the familiar holes. In the preparation of blue cheese (Roquefort) the curd is sprinkled with spores of *Penicillium roqueforti* or similar mold to produce the blue-green mottling. The creamy consistency of Camembert is largely due to the growth of *Penicillium camemberti*, and the strong smell of Limburger comes from the growth of certain yeasts and bacteria.

Cheese is usually named after the locality in which it is produced, and a certain type of cheese may have several different names, depending on where it is produced. Cheddar (after Cheddar, England), Wisconsin, American, and brick cheeses are basically the same.

In the United States milk used to make cheese must either be pasteurized (heated to 161° F [72° C] for 15 seconds) or heat treated (150° F [65° C] for 15 seconds). When cheese is made from heat-treated milk, it must also be aged for at least 60 days at cool temperatures. In pasteurized milk, unlike heat-treated milk, both pathogens and bacteria responsible for cheese flavors are killed. Generally, milder cheeses such as processed cheese are made of pasteurized milk, whereas stronger cheeses such as cheddar are made from heat-treated milk. Cheeses made in the United States are produced in air-conditioned factories, and carefully selected, pure cultures of microorganisms are handled with strict asepsis.

Manufacture of Alcohol and Alcoholic Beverages

Alcoholic beverages are of two classes: those manufactured by fermentation alone and those in which the alcohol is distilled off the fermented mixture. The distillate has a higher alcohol content than the beverage manufactured by fermentation alone. Beverages manufactured by fermentation alone are wine, beer, and cider. Those manufactured by distillation include commercial alcohol, whiskey, rum, and brandy. The manufacture of both fermented and distilled beverages depends primarily on the conversion of sugar into alcohol by yeasts. Commercial fermentations rely mainly on enzymes of yeasts of *Saccharomyces* species.

Wines are made by the fermentation of grape juice; cider, by the fermentation of apple juice. Beer is made from grains, most often barley. Grains do not contain sugar at first, but when the grain is soaked in water, the seeds germinate and the starch-splitting enzyme known as *diastase* (amylase) is produced. It converts the starch of the grain into the sugars maltose and glucose. Grain starch that has been converted into sugar is known as *malt*. The maltose and glucose in the malt are converted into alcohol by yeasts. *Sake* is Japanese "rice beer," in the manufacture of which the mold *Aspergillus oryzae* is used to convert the starch of the rice to sugars. If foreign organisms grow in beer, they give it an undesirable taste. This condition is "disease" of beer and may be prevented by pasteurization.

Purification of alcoholic beverages by distillation depends on the fact that alcohol has a much lower boiling point than water. When a fermented mixture is heated to a temperature considerably below the boiling point of water, the alcohol distills over, leaving the water behind. Commercial alcohol and whiskey are usually made by distillation of fermented grains. Brandy is made by distillation of fermented fruits. Rum comes from the distillation of fermented sugar cane juice. Vodka is the distillate from fermented potatoes. The mixture of fermenting grains used in the manufacture of alcohol and alcoholic beverages is known as *mash*. The mash from which rye whiskey is made contains 51% or more rye, whereas that used for the manufacture of bourbon contains 51% or more corn. The mash used in the production of alcohol is usually a mixture of corn (88.5%), barley (9.75%), and rye (1.75%), known as *spirit mash*.

Industrial alcohol (ethyl) for years has been manufactured from molasses. Alcohol may be produced by the action of bacteria on the cellulose of plants. In this method by-products of the lumber industry, such as sawdust, trimmings, and remains of trees, are used. A large amount of alcohol is obtained as a by-product when petroleum is cracked to form gasoline.

Baking

An essential step in baking is the same as that for the manufacture of alcoholic beverages: conversion of sugar into alcohol and carbon dioxide by yeast. It is the carbon dioxide, not the alcohol, that plays the important role in baking. When yeasts are mixed with dough, their enzymes, notably zymase, ferment the sugar present; the carbon dioxide thus produced riddles the dough with small holes, causing it to rise. When the dough is baked, the heat volatilizes the alcohol and causes the carbon dioxide to expand and be driven off, leaving the bread light and spongy. Some sugar must be present or fermentation will not take place. The dough may contain a small amount of diastase, which converts some of the starch of the dough into sugar. If white flour in which the amylase has been destroyed in the refining process is used, sugar must be added.

Manufacture of Vinegar

There are several methods of making vinegar. It may be made from wine or cider or from grains that have undergone alcoholic fermentation. In most methods the first step is the production of alcohol by fermentation, and the second step is the conversion of the alcohol into acetic acid. The first step differs in no manner from the production of alcohol for other purposes. When the alcoholic liquor is exposed to the air, a film known as mother of vinegar and containing acetic acid bacteria (genera *Acetobacter* and *Gluconobacter* [*Acetomonas*]) forms on the surface, and these bacteria convert alcohol into acetic acid. If the process continues too long, the bacteria decompose the acetic acid into carbon dioxide and water, and the vinegar loses some of its strength.

Production of Sauerkraut

Sauerkraut is finely shredded cabbage that has been allowed to ferment in brine formed by salt and cabbage juice. The finely shredded cabbage is placed in layers in a cask, and salt is sprinkled over each layer. The layers are packed closely together by a weight placed on them. The salt extracts juice from the cabbage, and bacteria present convert the sugar of the juice into lactic acid, which contributes to sauerkraut's peculiar flavor.

Tanning

The recently removed hide or skin of an animal must be preserved in some way, or its microbial population will destroy it. The leather is soaked and softened in numerous changes of cool water to remove dirt, blood, and nonspecific debris. Cool water retards bacterial growth. The excess flesh is trimmed from the hide or skin and the hair is loosened. One method of loosening the hair is to soak the hide in a solution of lime. It has been found that old solutions, which contain many more bacteria, remove the hair more effectively than fresh ones, which contain few. During the process the bacteria partially decompose and soften the leather.

After the loosened hair is scraped away, minerals are removed by washing or chemical action. Before they are tanned, hides of some animals are pickled. Pickling is done by treating the hide with sulfuric acid and salt. Tanning is accomplished by the use of vegetable tannins or chemicals such as alum.

Curing Tobacco

When tobacco leaves are cured, they change in texture and flavor and take on a brown color. These changes are accompanied by a decrease in sugar, nicotine, and water. The process is apparently one of oxidation induced by fermentation.

Retting Flax and Hemp

In flax and hemp plants the fibers are closely bound to the wood and bark of the plant by a gluelike pectin. When these fibers are separated from the bark and wood, the commercial products linen and hemp are obtained. The dissolution of the pectin to effect separation is accomplished by the action of certain aerobes and the anaerobic butyric acid bacteria (organisms in the genus *Clostridium*).

Manufacture of Antibiotics

The manufacture of antibiotics is a major pharmaceutical activity. More than 100 are available today for plants, animals, and humans. The basic sources are cultures of molds or soil bacteria. Penicillin is produced by certain molds of the genus *Penicillium*, and antibiotics such as streptomycin are obtained from bacteria in genus *Streptomyces*. The development of penicillin and streptomycin fermentations into industrial processes was the beginning of the field of bioengineering (Fig. 38-3). At present, on an annual basis, the industrial yield of antibiotics from microbes is measured in the millions of pounds. Currently a little more than one third of antibiotics used are cephalosporins, for which worldwide sales total about $1 billion.

Oil Prospecting

Certain types of bacteria are known to be quite plentiful in the soil overlying deposits of oil. Their growth is supported by gases that seep from the oil to the surface, and their presence assists in the identification of such deposits. Microbes have also been used to recover copper, zinc, and uranium from low-grade ores and slag of mines.

Other Industrial Processes

The theoretical study of fermentation has been constantly stimulated by the industrial values of the many end-products obtained. To the practical applications of microbes just given, the following are added:
1. Manufacture of acetone (fermentation of sugars—*Clostridium acetobutylicum*)
2. Manufacture of lactic acid (fermentation of molasses—lactic acid bacteria)
3. Manufacture of dextran, a plasma expander (fermentation of cane sugar—*Leuconostoc mesenteroides*)
4. Manufacture of butyl alcohol (fermentation of sugars—butyric acid bacteria)

5. Manufacture of vitamins (for example, cyanocobalamin, or vitamin B_{12} [at least 2000 pounds annually], riboflavin, or vitamin B_2, and vitamin C)
6. Microbiologic assay—determination of amount of amino acids and vitamins in tissues and body fluids (for example, assay of vitamin B_{12})
7. Manufacture of steroid hormones, cortisone, and hydrocortisone (action of certain molds on plant steroids)
8. Manufacture of citric acid (more than 350 million pounds annually) used in lemon flavoring in the food, beverage, and pharmaceutical industries (oxidation of sugar—*Aspergillus niger*)
9. Manufacture of enzyme detergents (subtilisins, alkaline proteases from *Bacillus subtilis*, added to final detergent product)
10. Manufacture of insecticides (microbial insecticides or bioinsecticides—protein of *Bacillus thuringiensis* lethal to larvae [caterpillars] of serious food crop pests in order Lepidoptera; possible application of known insect pathogens as some fungi in genera *Beauvera* and *Coelomomyces*, protozoa in genus *Nosema*, and insect viruses such as baculoviruses or granulosis viruses)

    Microbes live in symbiosis with insects, furnishing essential vitamins and amino acids needed by the insect and probably nitrogen through nitrogen fixation. The herbivorous (wood-eating) termite most likely depends on the nitrogen-fixing bacteria of its intestinal flora as a significant source of its nitrogen.
11. Production of food in seas and oceans (photosynthesis carried on by certain marine microbes, role similar to that of plants on land)

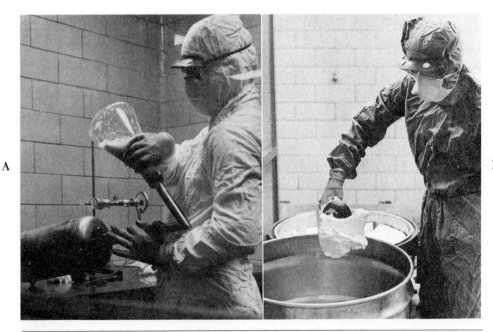

FIG. 38-3   Production of antibiotics is a complex process involving fermentation, filtration, extraction, and other intricate operations, many of which must be performed under sterile conditions. Antibiotic is grown in large fermentation tanks under controlled conditions, filtered to remove impurities, and finally recovered in crystalline form. **A,** Recharging seed tank with microbial culture in fermentation process. **B,** Handling bulk antibiotic under sterile conditions. (Courtesy Eli Lilly & Co., Indianapolis.)

12. Production of ensilage for animal feed (fermentation of sugars in shredded green plants by lactic acid bacteria)
13. Manufacture of plant growth factors (for example, gibberellins from *Gibberella fujikuroi*)
14. Manufacture of enzymes (for example, amylases, proteases, pectinases, amyloglucosidase, glucose isomerase, and rennet)

    The fruit, wine, baking, milling, dairy, and distilling industries use microbial enzymes—prime users are those dealing with starch degradation and the manufacture of detergents.
15. Manufacture of amino acid (for example, glutamate, lysine; 640 million pounds of glutamate and 60 million pounds of lysine produced annually in the United States)
16. Manufacture of flavor nucleotides (for example, inosinate, guanylate; 6 million pounds annually)
17. Manufacture of polysaccharides (for example, xanthan polymer, a viscosity agent important in the food and oil industries)
18. Disposal of sewage (Chapter 39)

Concepts and technics of genetic engineering employing microbes will revolutionize industrial processing of products derived from microbes (p. 103). Considerably larger quantities will be available at considerably less expense.

SUMMARY

1 The majority of bacteria and other microbes are helpful to humans, animals, and plants, all of which depend on microbes for their very existence.
2 Microbes producing disease are actually only a small group.
3 Bacteria and other microbes are called nature's garbage disposal system and fertilizer factory, for they are the active agents in the decomposition of dead organic matter of animal origin, releasing elements needed for growth of plants and returning them to the soil.
4 Within the upper several centimeters of well-tilled soil live 300 to 3000 kg of living microorganisms per hectare.
5 Microbes play vital roles in the elemental cycles in nature, notably in the carbon and nitrogen cycles.
6 Nitrogen, an important element in the composition of plants, is abundant in the air. Microbes render that nitrogen into a form suitable for plant or animal use by converting it to nitrate.
7 Nitrogen fixation, the recovery of free nitrogen from the air and its synthesis into nitrate for use by plants and animals eating plants, is accomplished by bacteria.
8 Bacteria and protozoa participate actively in digestive processes of animals and humans; in the large bowel they form substances essential for nutrition (for example, large amounts of vitamins in the B complex group).
9 Microbes are vital to processes of commerce and industry, such as the manufacture of dairy products.
10 The organism responsible for the ripening process of cheese to a large extent determines the kind of cheese. The variety of cheeses available shows that a number of microbes can be used.
11 Commercial fermentations applicable to the manufacture of alcohol and alcoholic beverages rely mainly on enzymes from yeasts of *Saccharomyces* species.
12 The basic sources for the manufacture of antibiotics are cultures of molds or soil bacteria; for example, penicillin is produced by certain molds of the genus *Penicillium*.

13 The field of bioengineering began with the industrial development of penicillin and streptomycin fermentations.

14 The theoretical study of fermentation has been constantly stimulated by the industrial values of the many end-products obtained.

15 Microbes figure in the production of a varied and almost inexhaustible assortment of substances—for example, acetone, lactic acid, dextran, steroid hormones, citric acid, insecticides, amino acids, and enzyme detergents.

QUESTIONS FOR REVIEW

1 What is meant by the cycle of an element?
2 Briefly outline the nitrogen cycle and the carbon cycle.
3 What is nitrogen fixation? Its importance?
4 What determines fertility of the soil?
5 What causes root nodules on clover plants? What purpose do they serve?
6 What organism is important in the synthesis of vitamin K in the intestine?
7 List 10 practical applications made of microorganisms. Name some microbes used in industry.
8 Explain how cheese is made. What is dairy microbiology?
9 What is diastase? Its role in fermentation?
10 Briefly tell what is meant by starters, rennin, hectare, ripening, mother of vinegar, bioinsecticide, nitrifying bacteria, fertility of soil, mutualism, and green cheese.

SUGGESTED READINGS

Abelson, P.: Biotechnology: an overview, Science 219:611, 1983.

Aharonowitz, Y., and Cohen, G.: The microbiological production of pharmaceuticals, Sci. Am. 245:140, Sept. 1981.

Anagnostakis, S.L.: Biological control of chestnut blight, Science 215:466, 1982.

Brill, W.J.: Agricultural microbiology, Sci. Am. 245:198, Sept. 1981.

Burg, R.W.: Fermentation products in animal health, ASM News 48:460, 1982.

Capone, C.G., and Carpenter, E.J.: Nitrogen fixation in the marine environment, Science 217:1140, 1982.

Colwell, R.R.: Biotechnology in the marine sciences, Science 222:19, 1983.

Demain, A.L.: Industrial microbiology, Science 214:987, 1981.

Demain, A.L.: New applications of microbial products, Science 219:709, 1983.

Demain, A.L., and Solomon, N.A.: Industrial microbiology, Sci. Am. 245:66, Sept. 1981.

Dorn, R.I., and Oberlander, T.M.: Microbial origin of desert varnish, Science 213:1245, 1981.

Eveleigh, D.E.: The microbiological production of industrial chemicals, Sci. Am. 245:154, Sept. 1981.

Gaden, E.L., Jr.: Production methods in industrial microbiology, Sci. Am. 245:180, Sept. 1981.

Geesey, G.G.: Microbial exopolymers: ecological and economic considerations, ASM News 48:9, 1982.

Fox, J.L.: Soil microbes pose problems for pesticides, Science 221:1029, 1983.

Hartline, B.K.: Fighting the spreading chestnut blight, Science 209:892, 1980.

Hopwood, D.A.: The genetic programming of industrial microorganisms, Sci. Am. 245:90, Sept. 1981.

Keyser, H.H., and others: Fast-growing *Rhizobia* isolated from root nodules of soybeans, Science 215:1631, 1982.

Miller, L.K., and others: Bacterial, viral and fungal insecticides, Science 219:715, 1983.

Phaff, H.J.: Industrial microorganisms, Sci. Am. 245:76, Sept. 1981.

Rose, A.H.: The microbiological production of food and drink, Sci. Am. 245:126, Sept. 1981.

Schroth, M.N., and Hancock, J.G.: Disease-suppressive soil and root-colonizing bacteria, Science 216:1376, 1982.

Thatcher, R.C., and Weaver, T.L.: Carbon-nitrogen cycling through microbial formamide metabolism, Science 192:1234, 1976.

Woodruff, H.B.: Natural products from microorganisms, Science 208:1225, 1980.

CHAPTER 39

MICROBIOLOGY OF WATER

GENERAL CONSIDERATIONS

Practically all waters under natural conditions contain microbes, including protozoa, bacteria, fungi, and viruses. Some contain many; others, only a few. The number and kind of microbes present depend on the source of water, the addition of the excreta from humans and animals, and the addition of other contaminated material.

Since sewage holds the pooled excreta from both the sick and well, it necessarily contains pathogenic organisms, especially those that leave the body by feces or urine. Sewage must be properly disposed of to avoid contamination of water supplies and the spread of disease by flies or other vectors.

Sanitary microbiology (the microbiology of sanitation) pertains to the treatment of drinking water supplies and sewage disposal.

Sanitary Classification

Waters may be classified as potable, contaminated, or polluted. **Potable** water is free of injurious agents and pleasant to the taste—it is a satisfactory drinking water. **Contaminated** water contains dangerous microbial or chemical agents. Contaminated water may be of pleasing taste, odor, and appearance. **Polluted** water has an unpleasant appearance, taste, or odor. Because of its content, it is unclear and unfit for use. It may or may not be contaminated with disease-producing agents.

The pollution of water supplies is a major health problem today. Not only sanitary sewage but also complex wastes from industry and agriculture may be discharged into water. In increasing volume substances such as plastics, detergents, pesticides, herbicides, insecticides, oils, animal and vegetable matter, chemical fertilizers, and even radioactive wastes are released from factories, canneries, poultry-processing plants, oil fields, and farms. The persistence of synthetic detergents in wastewater and sewage reflects a failure of microbial action. Bacteria do not possess the enzymes to degrade them chemically.

Fish provide a measure of the purity of a water supply, since they do not thrive in the water of a river or lake that is heavily polluted.

Sources of Water

Water, the most abundant of substances on the earth, singularly is found in nature as gas, liquid, or solid. It is surprisingly difficult to obtain or keep as a pure substance and contaminates easily or recontaminates after purification.

A water supply may come from (1) rain or snow, (2) surface water (shallow wells, rivers, ponds, lakes, and wastewater), and (3) groundwater (deep wells and springs) (Fig. 39-1). Generally, surface water contains more microbes than either groundwater or rainwater, since the majority of soil microorganisms are found in the upper 6 inches of the earth's crust. Surface water contains many harmless microbes from the soil, and in the vicinity of cities it is often contaminated with sewage bacteria. Surface water in sparsely settled localities may not be dangerous, but *the only safe rule is not to use surface water without purification.*

Unless properly constructed, shallow wells may become contaminated with the drainage from outhouses, stables, and other outbuildings. Improperly constructed shallow wells have been responsible for many outbreaks of typhoid fever and dysentery in rural communities. For a shallow well to be safe, its upper portion must be lined with an impervious material so that water from the surface does not seep into it, and it should be located so that outhouses drain away from it. Of course, it must be kept tightly closed. Shallow springs may be just as dangerous as shallow wells.

Deep-well water and deep-spring water usually contain few microorganisms because

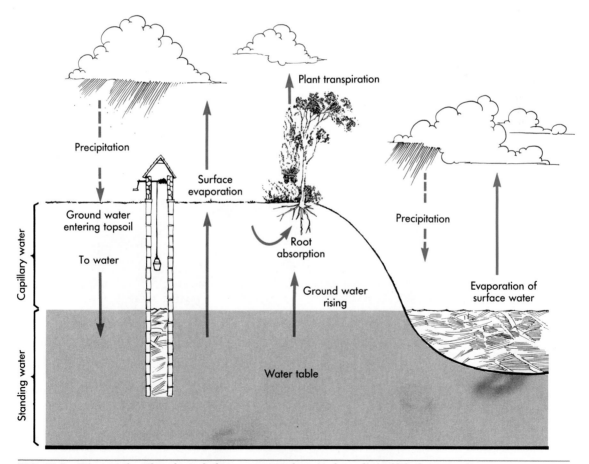

FIG. 39-1   Water cycle. This physical phenomenon is almost independent of life, but water is necessary for metabolism in all organisms.

microbes are filtered out as the water trickles through the layers of the earth. But, like shallow wells, deep wells must be protected against pollution by surface water. Microbial contamination may occur when a well is situated within 200 feet of the source of contamination, and chemical contamination may occur when the well is within 400 feet. Contamination is more likely during wet weather.

Waterborne Diseases

The sources for microbes in water are many—soil, air, the water itself, and the decaying bodies and excreta of humans and animals. As a result, many different kinds are found there—well-known pathogens as well as harmless water bacteria. There is increased concern in the United States about the presence of viruses in water, since most human viruses multiply in and are excreted by the alimentary canal (Fig. 39-2). Human enteric viruses have been recovered from a third of surface water samples examined. Groundwaters may also transmit virus.

The varied diseases spread by water are notably those whose agents leave the body by way of the alimentary or urinary tracts, and waterborne infections are mostly contracted by the ingestion of contaminated water. When a waterborne epidemic occurs, most cases appear within a few days, an indication that all were infected about the same time. Important waterborne diseases are typhoid fever, bacillary dysentery, amebic dysentery, salmonellosis, viral hepatitis, cholera, and giardiasis. Diseases that on occasion are transmitted by water are tularemia, anthrax, leptospirosis, schistosomiasis, ascariasis, rotavirus infection, poliomyelitis, and other enterovirus (coxsackievirus and echovirus) infections. Pathogenic bacteria do not live long in water, and they do not multiply there; but they may persist for a time in water that is cool and contains considerable organic matter.

Certain opportunistic pathogens, including many species of *Pseudomonas*, are referred to as water organisms because they survive and propagate in a wet, stagnant environment. They can grow in sterile water on traces of phosphorus and sulfur and are generally resistant to antibiotics and disinfectants. In hospitals they lurk as an unsuspected source of infection in sinks, drains, faucets, and the oxygen therapy apparatus. They can gain a foothold in the air ducts, mist generators, and humidifying pans in incubators of newborn and premature nurseries. Most susceptible to these water orga-

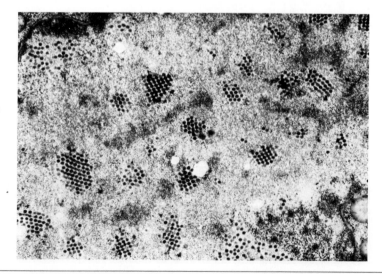

FIG. 39-2 Adenovirus crystals identified in sewage. (Courtesy Division of Microbiology and Infectious Diseases, Southwest Foundation for Research and Education, San Antonio.)

nisms are newborn infants, patients with surgical wounds, and chronic invalids. Removal of this source of infection lies in heat sterilization, absolute cleanliness, and removal of the moist, stagnant foci in hospital equipment and treatment areas.

Water Analysis: Bacteriologic Examination of Water

Experience shows that it is almost impossible to isolate from water the organisms responsible for the most important waterborne disease; few are present to begin with, and they do not multiply in water. Sanitarians and public health workers therefore have concluded that the only safe method to prevent waterborne diseases is to condemn any fecally polluted water as being unfit for human use. It *might* contain harmful organisms. Fecal pollution can be determined by examination of the water for colon bacilli *(Escherichia coli)*, abundant in feces and *not* found outside the intestinal tract in nature. Also found in water are certain bacteria that resemble colon bacilli but that may or may not be of fecal origin. These bacteria, like the colon bacillus, also ferment lactose with the formation of gas. In practical water analysis therefore testing treats such microorganisms collectively as a group, the **coliform** bacteria. The procedure recommended by the American Public Health Association, the American Water Works Association, and the Water Pollution Control Federation, as outlined in *Standard Methods for the Examination of Water and Wastewater,** consists of three tests: presumptive, confirmed, and completed (Fig. 39-3).

Included in *Standard Methods for the Examination of Water and Wastewater* is the **membrane filter procedure** for detection of coliform bacteria. In this method, developed in Germany during World War II, the water is filtered through a thin disk of bacteria-retaining material. After filtation the disk is inverted over an absorbent pad saturated with modified Endo agar and incubated for 20 ± 2 hours at $35° \pm 0.5°$ C. All organisms that produce a dark purplish green colony with a metallic sheen are considered to be coliforms. Colonies of coliforms are counted, and an estimation of the number per 100 ml of water sample is made.

European bacteriologists have devised systems whereby water is also examined for *Streptococcus faecalis* and *Clostridium perfringens*, both of which are just as constantly present in feces as coliforms but not so plentiful. Also, the use of phages as accurate indicators of fecal pollution of water is a possibility.

Purification

Drinking Water Standards

Drinking water must contain no impurity that would offend sight, taste, or smell, and substances with deleterious physiologic effects must be eliminated or not introduced.

The U.S. Public Health Service has set standards for drinking water used in interstate commerce. Originally designed to protect the traveling public, to a great extent these standards have been applied throughout the country. Specifications relate to (1) source and protection of the water, (2) physical and chemical properties, (3) bacteriologic content, and (4) limits on concentrations of radioactivity. At the end of 1974 Congress passed the Safe Drinking Water Act, which calls for stringent federal standards to protect consumers to "maximum feasible extent" from contaminants in water. The Environmental Protection Agency (EPA) was charged to prepare the recommended standards for safe drinking water. The standards establish the maximum allowable number of coliforms as well as the maximum bacterial content in terms of a standard plate count in samples taken at stated intervals from public drinking water supplies. The maximum allowable limit for coliforms is one coliform colony per milliliter of water, as deter-

---

*Rand, M.C., editor: Standard methods for the examination of water and wastewater, Washington, D.C., 1976, the American Public Health Association, the American Water Works Association, and the Water Pollution Control Federation.

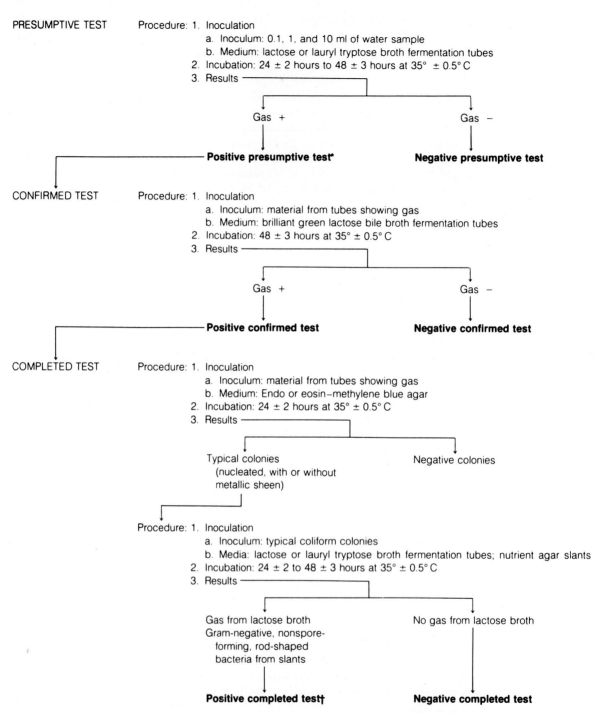

PRESUMPTIVE TEST   Procedure: 1. Inoculation
                               a. Inoculum: 0.1, 1, and 10 ml of water sample
                               b. Medium: lactose or lauryl tryptose broth fermentation tubes
                            2. Incubation: 24 ± 2 hours to 48 ± 3 hours at 35° ± 0.5°C
                            3. Results

                                    Gas +   Gas −

                      Positive presumptive test*   Negative presumptive test

CONFIRMED TEST   Procedure: 1. Inoculation
                               a. Inoculum: material from tubes showing gas
                               b. Medium: brilliant green lactose bile broth fermentation tubes
                            2. Incubation: 48 ± 3 hours at 35° ± 0.5°C
                            3. Results

                                    Gas +   Gas −

                      Positive confirmed test   Negative confirmed test

COMPLETED TEST   Procedure: 1. Inoculation
                               a. Inoculum: material from tubes showing gas
                               b. Medium: Endo or eosin–methylene blue agar
                            2. Incubation: 24 ± 2 hours at 35° ± 0.5°C
                            3. Results

                        Typical colonies   Negative colonies
                        (nucleated, with or without
                        metallic sheen)

                    Procedure: 1. Inoculation
                               a. Inoculum: typical coliform colonies
                               b. Media: lactose or lauryl tryptose broth fermentation tubes; nutrient agar slants
                            2. Incubation: 24 ± 2 to 48 ± 3 hours at 35° ± 0.5°C
                            3. Results

                        Gas from lactose broth   No gas from lactose broth
                        Gram-negative, nonspore-
                          forming, rod-shaped
                          bacteria from slants

                      Positive completed test†   Negative completed test

*The basis of the presumptive test is that coliforms ferment lactose with the production of gas. A positive result is presumptive evidence therefore that coliforms are present and it suggests fecal contamination. The smallest amount of sample in which the fermentation of lactose occurs is an index of the number of organisms present.
†A positive completed test indicates that coliforms are present and completes their identification. Further elucidation of bacteria identified in the completed test is not done, since sanitarians in the United States believe that satisfactory drinking water should be free of both fecal and nonfecal organisms.

FIG. 39-3   Flow chart for standard water analysis.

mined by the membrane filter procedure; for bacterial content, it is 500 microorganisms per milliliter of water, as determined by the standard plate count. Revised standards will follow further studies of the health effects of the various drinking water contaminants.

Investigation of the problem of viruses in water is accelerating. Since minimal amounts of virus can produce infection, total removal of viruses from a water supply for human consumption seems indicated.

Water Purification

Water may be purified by natural means. As water trickles through the earth, microbes are filtered out. Water standing in lakes and ponds undergoes some degree of purification because of the combined action of sunlight, sedimentation, dilution of impurities, and destruction of bacteria by protozoa. Streams may be purified in a comparable way as they flow along their course. However, the fact that all the foregoing methods of natural purification are slow and uncertain in their action and the possibility being great that contaminated material will enter the flowing stream along its course mean that one cannot and must not depend on self-purification of water. *All surface waters must be regarded as dangerous unless subjected to artificial purification.*

Emergency small-scale water purification.* In an emergency or in unusual circumstances water (such as surface water) can be made safe for consumption by (1) boiling or (2) chemical disinfection with 2% tincture of iodine (from medicine chest or first-aid kit) or tetraglycine hydroperiodide tablets. Cloudy water should first be strained through a clean cloth into a suitable container to remove sediment and floating matter before it is treated with either heat or iodine.

In the first method the water is brought to a vigorous boil. At high altitudes boiling should be continued for several minutes. A pinch of salt improves the taste of the boiled water. Note that, if tap water is uncomfortably hot to the touch, it is probably safe to use after it has cooled in a thoroughly clean container.

In chemical disinfection small quantities of water can be made safe by the addition of 5 drops (1 drop equals 0.05 ml) of 2% tincture of iodine to 1 L (quart†) of clear water. The mixture is then allowed to stand for 30 minutes. Ten drops of 2% tincture of iodine should be added to cloudy or very cold water, and the mixture should stand for several hours before it is used. Tetraglycine hydroperiodide tablets (Globaline, Potable-Aqua, Coghlan) are applied according to the manufacturer's specifications. The number of tablets is doubled when added to cloudy water. Very cold water is warmed to room temperature, and the recommended contact time is increased. Chemical disinfection is recommended only when it is not possible to boil the water.

Large-scale (municipal) water purification. Generally, large-scale artificial water purification (Fig. 39-4) is carried out by a combination of physical and chemical methods—sedimentation, filtration, coagulation, and chlorination (p. 227). **Sedimentation** is the process whereby water is held in settling basins for a period to allow mud solids and large particles of organic matter to settle out. **Coagulation** refers to the addition of chemicals to water to soften and clarify it. A chemical such as alum or ferric sulfate coagulates the suspended organic matter in the water to which microbes are attached. The flocculent precipitate that results carries down most of the suspended material, and in this way many bacteria are readily filtered out. Some clarification of water occurs with sedimentation and coagulation, but **filtration** is the process that puts the final sparkle in.

In water treatment filtration is done by either the **slow-sand** or **rapid-sand** process.

---

*See section "Health hints for the traveler" in Health information for international travel 1983, Morbid. Mortal. Weekly Rep. **32** (suppl.):100, Aug. 1983, HEW Publication No. (CDC) 83-8280.
†One liter equals 1.065 quarts.

FIG. 39-4 East side water purification plant in Dallas. Purification process extends from chemical plant in the foreground, where ferrous sulfate and lime are added to water, to pump house in background, from where water is pumped to reservoir. Between these points are rapid mixers, flocculators, settling basins, and sand filters. Water is chlorinated several times at points along the way. (Courtesy Dallas Water Utilities, Dallas.)

In slow-sand filtration water is passed directly into beds of fine sand, 1 to 5 feet thick, supported by graded layers of gravel, underneath which is a drainage system. Bacteria are removed not so much by mechanical straining as through biologic mechanisms related to the activity of protozoa present. Slow-sand filtration is one of the oldest and most effective methods of water purification, but few of these filtres are constructed today because the relatively large land area required for the layout is no longer available in the large population centers of the United States.

Rapid-sand filters, usually made of silica and crushed quartz or crushed anthracite coal, allow water to flow through larger grain sand at much faster rates. For rapid filters to be effective, prior treatment of the water by coagulation and sedimentation is necessary. Rapid-sand filtration can filter more than 200 million gallons per day per acre of filter area, with the removal of 90% to 95% of all bacteria.

Efficient filtration eliminates all pathogenic and most saprophytic microorganisms. Both types of filters must be cleaned periodically. A rapid-sand filter is made in units so that only one part of it is closed for cleaning at any one time.

After filtration, water is universally chlorinated. There are three methods in common use:

1. **Simple chlorination**—This method is the addition of a standard effective amount of chlorine gas to water. Chlorine is usually added from cylinders connected with regulating equipment to control the exact amount.
2. **Ammonia-chlorine treatment**—This treatment results in compounds called chloramines, formed when chlorine is added to water containing ammonia. Chloramines are less active as germicidal agents but more stable than free chlorine.
3. **Superchlorination**—This process involves the addition of a much larger dose of chlorine than with simple chlorination and the subsequent removal of the excess.

Granular, activated carbon, mixed with raw, untreated water, absorbs odors and tastes resulting from decayed vegetation and other organic matter present. Softening, the removal of dissolved limestone from hard water, is accomplished either by chemical precipitation of the calcium carbonate and magnesium hydroxide or by a process of ion exchange.

SPECIAL MEASURES
Fluoride Content of Water and Tooth Decay

It is well known that the fluoride content of the drinking water of infants and children profoundly influences the development of their teeth. If too much fluoride is present (4 to 5 ppm water), the condition known as **fluorosis,** or mottled enamel, develops; if too little, the incidence of dental caries is greatly increased. Adult teeth do not seem to be affected. The optimum ratio of fluoride to drinking water for infants and young children is 1 ppm. When this is present, there is sufficient fluoride to retard tooth decay significantly but not enough to stain or mottle the teeth. Reduction of the incidence of tooth decay is about 60% to 70%. In nature the fluoride content of water varies from much less to much more than 1 ppm, but the proper level may be maintained by the addition of fluoride to water whose content is low or by chemical removal of fluoride from water whose content is too high.

Ice

When pathogenic bacteria gain access to ice, the majority die. Perhaps 10% survive for more than a few hours. Ice that has been stored for 6 months is practically bacteria free. The use of water of high sanitary quality in the manufacture of ice has lessened the possibility that ice itself will convey disease. However, the handling of ice in the home, eating places, and fast food restaurants may be a source of infection.

Purification of Sewage

Sewage must be processed in treatment plants to render it harmless before it is discharged into any body of water (river or lake) (Fig. 39-5). Many methods of sewage purification are used. The usual sequence is screening and sedimentation to remove larger particles, chemical treatment to remove small ones, microbial action to digest organic matter, aeration to induce oxidation, filtration through sand, and chlorination. Breakdown of human waste in sewage by microbes is essentially decay of organic material, and the sanitary engineer designs the disposal plant to facilitate and speed up microbial activity and ensure its completion. Water reclamation is practical if purification of sewage can be accomplished efficiently and economically (Fig. 39-6).

No known way exists to remove all viruses from sewage (Fig. 39-2); they are not regularly inactivated by primary and secondary treatment and easily pass into the effluent. Specifically, enteroviruses, reoviruses, and adenoviruses have been demonstrated to come through standard treatment. Parasites readily disseminated in sewage

FIG. 39-5   Aerial survey of wastewater treatment plant in Dallas. Note layout of settling basins, microbial digestors, and trickling filters. In primary settling basins organic wastes are oxidized by bacteria and algae growing on rocks. Sludge from primary settling basins goes to a separate group of heated tanks called digestors, where, by a process of anaerobic decomposition, solids are broken down by microbes. Effluent from primary settling basins goes to 29 circular trickling filters, each with a half-acre rock bed. (Courtesy Dallas Water Utilities, Dallas.)

FIG. 39-6   Water reclamation. An "activated sludge" pilot plant in Dallas of a type used by many cities in the nation; capacity, 1 million gallons a day. In activated sludge, microbes are mixed with sewage to consume its organic products. Here is one large tank maintained in an aerobic state to facilitate growth of sewage microbes. It is seeded by sludge (primarily microorganisms) settled from a second large tank, the secondary clarifying tank. Input of raw sewage or sewage already settled into aeration tank provides substrate for microbial action. (Courtesy Dallas Water Utilities, Dallas.)

include *Entamoeba histolytica*, *Giardia lamblia*, roundworms, hookworms, and tapeworms.

Swimming Pool Sanitation

Unless kept sanitary, swimming pools convey conjunctivitis, ear infections, skin diseases, and intestinal infections. The source of infection is a person using the pool, and
fresh discharges from that person contaminating the water are more likely to be highly
infectious to other bathers in the pool. Swimming pool water should be kept as pure as
drinking water, according to standards set by the American Public Health Association.
It recommends that under specified conditions the mean bacterial count in such water
be no more than 200 organisms per 100 ml, with coliforms not being demonstrated by
positive confirmed tests for them. Swimming pool sanitation can be maintained with
frequent changes of water and the use of disinfectants. Chlorine is the most widely used
one, and the highest possible concentration should be maintained—about 1 ppm (p.
228). A higher concentration irritates the eyes. Before entering the pool, each bather
should take a shower using soap. Persons with infections of any kind should not be
allowed entry to the pool.

Some authorities prefer iodine to chlorine as a disinfectant for swimming pools. Its
disinfectant action is less hindered by organic matter, and there is less eye and skin
irritation than with chlorine. Bromine has also been used as a swimming pool disinfectant.

SUMMARY

1 Practically all waters under natural conditions contain a variable number of microbes—protozoa, bacteria, fungi, and viruses.

2 In sanitary microbiology water is classified as potable, contaminated, and polluted.
   Only potable water is satisfactory to drink.

3 Fish provide a measure of the purity of a water supply, since they do not thrive in
   heavily polluted water.

4 Water contaminates easily and recontaminates after purification.

5 Since the majority of soil microbes are found in the upper 6 inches of the earth's
   crust, surface water contains many microbes from the soil and is often contaminated
   with sewage bacteria. The only safe rule is not to use surface water without purification.

6 Important waterborne diseases are typhoid fever, bacillary dysentery, amebic dysentery, salmonellosis, viral hepatitis, and cholera.

7 When a waterborne epidemic occurs, most cases appear within a few days, indicating that all persons were infected at about the same time.

8 Pathogenic bacteria do not live long in water, do not multiply there, and generally
   do not persist there unless the water is cool and contains much organic matter.

9 Sanitary water analysis is based on the premise that fecally polluted water is unfit
   for human use because it might contain harmful microbes excreted from the alimentary tract.

10 Water is examined in a sanitary water analysis for the presence of coliform bacteria.
   The membrane filter procedure for detection of coliform bacteria is one method
   widely used.

11 The U.S. Public Health Service has set standards for drinking water defining specifications for physical and chemical properties, the bacteriologic content, concentration of radioactivity, and source and protection of the water.

12 The Environmental Protection Agency (EPA) is charged with the recommended
   standards for safe drinking water.

13 Water may be purified by natural means (generally unreliable) and by artificial

means that use a combination of physical and chemical methods, notably sedimentation, filtration, coagulation, and chlorination.

14 Rapid-sand filters allow for filtration of more than 200 million gallons of water per day per acre of filter area, with removal of 90% to 95% of all bacteria.

15 Three methods for chlorination of water are simple chlorination, ammonia-chlorine treatment, and superchlorination.

16 Reduction of the incidence of tooth decay with the judicious use of fluoride is about 60% to 70%.

17 Purification of sewage involves screening and sedimentation of human waste to remove larger particles, chemical treatment to remove small ones, microbial action to digest organic matter, aeration to induce oxidation, filtration through sand, and chlorination.

QUESTIONS FOR REVIEW

1 Name the diseases spread by water. Give their causal agents.
2 Discuss the bacteriologic examination of water.
3 Describe the purification of water by (a) natural means and (b) artificial means.
4 List sources of water. From a public health standpoint how do waters from these sources differ? State the sources of the microbes in water.
5 Why is sewage of such sanitary significance?
6 What specifications are made for drinking water standards?
7 Discuss water pollution today as a public health problem. (Consult references.)
8 Give the sanitary classification of water.
9 How are swimming pools made sanitary?
10 Briefly explain water cycle, fluorosis, activated sludge, water reclamation, harmless water bacteria, Endo agar, coliforms, and practical water analysis.
11 What is the relation of the fluoride content of water to tooth decay?

SUGGESTED READINGS

Andrus, P.L.: The role of fluoride in the prevention of dental caries, Tex. Med. 78:57, 1982.

Bitton, G., and Farrah, S.R.: Viral aspects of sludge application to land, ASM News 46:622, 1980.

Gerba, C.P., and McNabb, J.F.: Microbial aspects of groundwater pollution, ASM News 47:326, 1981.

Gunnison, D.: Microbial processes in recently impounded reservoirs, ASM News 47:527, 1981.

Hartenstin; R.: Sludge decomposition and stabilization, Science 212:743, 1981.

Hetrick, F.M.: Survival of human pathogenic viruses in estuarine and marine waters, ASM News 44:300, 1978.

Leverett, D.H.: Fluorides and the changing prevalence of dental caries, Science 217:26, 1983.

Newbrun, E.: Sugar and dental caries: a review of human studies, Science 217:418, 1982.

Palmer, S.R., and others: Water-borne outbreak of Campylobacter gastroenteritis, Lancet 1:287, 1983.

Pye, V.I., and Patrick, R.: Ground water contamination in the United States, Science 221:713, 1983.

Reinheimer, G.: Aquatic microbiology, ed. 2, London, 1980, John Wiley & Sons, Inc.

Water-related disease outbreaks in the United States— 1980, Morbid. Mortal. Weekly Rep. 30:623, 1982.

MICROBIOLOGY OF FOOD

MILK

If obtained in a pure state and kept pure, milk is our best single food; if improperly handled, it is our most dangerous one. Milk conveys infection because it is an excellent culture medium and is consumed uncooked. The microbiology of dairy products deals with (1) the microbes that make milk and milk products unfit for human consumption, (2) prevention of disease in cattle, and (3) manufacture of milk products.

Bacteria in Milk

Bacteria gain access to milk in different ways. Milk as secreted by the mammary glands of perfectly healthy cows is usually sterile, but as the cow is milked, bacteria from the teats and milk ducts get in, and by the time the milk enters the receptacle, contamination has taken place.

Under the best conditions the bacterial content of milk is relatively high. Some bacteria, even though they come from external sources, are so commonly present as to be regarded as normal milk bacteria. Normal milk bacteria sour milk; they also may destroy its food value.

The two main sources for pathogenic bacteria found in milk are the persons who handle it and infected cows. Bacteria may be present within the udders of diseased cows. Unclean and unsanitary milking utensils, pasteurizing tanks, and milk containers, as well as dust and manure, are important sources of contamination. The pathogens most often present in milk are staphylococci, streptococci, lactic acid bacteria, and enteric organisms. The presence of coliform bacteria points to fecal contamination.

Although it is not the only criterion, the number of bacteria in milk is the best index of its sanitary quality. The number present depends on the number originally introduced and the temperature at which the milk is kept. Many species of bacteria begin to multiply as soon as they gain access to milk. This can be largely prevented if the milk is rapidly chilled to 7° C (45 ° F) as soon as it is obtained and kept cold. If the milk is allowed to warm, there is a rapid rise in the number of bacteria.

Milk of high quality may contain only a few hundred bacteria per milliliter; bad milk, millions per milliliter. All authorities agree that a single high bacterial count does not invariably mean poor-quality milk and that, to be significant, a high bacterial count in milk from a specific dairy or other source must recur daily.

Practical methods applicable to the routine detection of viruses in milk and other foods are not available for universal use at this time.

Milkborne Diseases

The diseases spread by milk may be separated into (1) diseases of human origin and (2) diseases of milk-producing animals transmitted in their milk. Most notable of those that primarily affect humans and may be transmitted by milk are typhoid fever, salmonellosis, bacillary and amebic dysentery, scarlet fever, infantile diarrhea, diphtheria, poliomyelitis, infectious hepatitis, and septic sore throat. The diseases that primarily affect milk-producing animals and may be transmitted to humans in infected milk are undulant fever, Q fever, and bovine tuberculosis. Undulant fever may be contracted from the milk of infected cows or goats.

The spread of milk of human diseases usually stems from contamination of the milk by the discharges of a handler who is ill with the disease or is a carrier. Milkborne epidemics appear among the patrons of a given dairy, and the source of infection usually is traced to a food handler there.

Bovine tuberculosis, especially in children under 5 years of age, usually comes from ingestion of the milk of tuberculous cows. The milk of a single cow excreting *Mycobacterium tuberculosis* may render infectious the mixed milk of the whole herd. For humans the tuberculin testing of cows and the pasteurization of milk eliminate the threat of bovine tuberculosis.

A disease not of bacterial origin but of historical interest is **milk sickness** (called milksick by the pioneers). This disease cost the life of many Midwestern pioneers, including the mother of Abraham Lincoln. It is caused by the ingestion of milk from cows that have been poisoned when they have eaten certain species of goldenrod and the white snakeroot. The poisonous principles in these plants are excreted in the milk of the affected cow.

Pasteurized Milk

Pasteurization is the process of heating milk to a temperature high enough to kill all *nonspore-bearing pathogenic* bacteria but not so high as to affect the chemical composition. There are two methods of pasteurization. One is known as the **holding method**; the other as the **continuous flow, high-temperature, short-time, or flash method.** In the former the milk is held in tanks or vats, where it is subjected to a comparatively low temperature for a comparatively long time while being constantly agitated. In the latter the milk is subjected to a higher temperature for a short time as it flows through the pasteurizing equipment (Fig. 40-1). After both methods the milk is chilled as it is transferred through pipes to the packaging machines, where it is put in clean sanitary containers (glass, cardboard, or plastic).

The **Grade A Pasteurized Milk Ordinance** (1978) recommendations of the U.S. Public Health Service/Food and Drug Administration are that milk be subjected in the holding method to a temperature of *at least* 63° C (145° F) for *at least* 30 minutes; in the continuous flow method, to a temperature of *at least* 72° C (161° F) for *at least* 15 seconds (high temperature, short time); or in a continuous flow, to a temperature of at least 89° C (191° F) for *1 second or less* (higher temperature, shorter time).

It is important to note that pasteurization does *not* completely sterilize milk and that it should never be used to cover gross negligence in sanitation and the sanitary handling of milk.

Pasteurized Versus Raw Milk

Pasteurization is the single most crucial factor in the maintenance of a safe milk supply. Objections to pasteurization might be that (1) it destroys vitamins, (2) dairy workers may not be as careful when they know that the milk is to be pasteurized, and (3) pasteurization may be done carelessly and ineffectively. The latter two objections can be eliminated by thorough inspection systems. Vitamin C, which is destroyed, can easily be replaced in the diet or given as ascorbic acid.

The temperature of pasteurization is carefully controlled, since heating milk to high temperatures can drastically change it. Boiling decomposes milk proteins, changes the content of phosphorus, precipitates calcium and magnesium, drives off carbon dioxide, combusts lactose (milk sugar), and destroys enzymes in the milk.

Requirements for a Safe Milk Supply

To ensure a safe milk supply, the following requirements must be met:

1. Cows must be healthy, well fed, and free of diseases such as tuberculosis, brucellosis, and mastitis.
2. All workers handling milk must be free of infectious organisms, and their hands and person must be clean.

FIG. 40-1   High-temperature short-time (HTST) method of pasteurization. Milk is received from dairy farm in tank trucks and placed in holding tanks. Milk is pasteurized at 79° C (175° F) for 16 seconds through a plate heat-exchanger with holding tube and timing pump. It is also clarified, standardized, homogenized, and steam heated to 88° to 90° C (190° to 194° F) for 30 seconds, vacuum-cooled back to 79° C (175° F), and cooled at 1° C (33° F) through a plate heat-exchanger containing cooling medium. All is done automatically from a central control panel, using sanitary air-operated valves for routing the milk. Cleaning and sanitizing is also done by automatic circulation of cleaning and sanitizing compounds. Note the Vac-Heat unit in foreground. This unit ensures the same milk flavor year-round by removing various volatile feed-off flavors, which vary during the year's four seasons. (Courtesy CP Division, St. Regis, Dallas.)

3. The premises must be kept clean.
4. The udders and flanks of the cows must be washed before milking.
5. Milking utensils and machinery that contain the milk must be kept sterile and should be constructed to keep out dust and flies. (Practically all milking today in commercial dairies is done mechanically.)
6. Milk must be chilled immediately to 7° C (45° F) and kept cold.
7. Milk must be pasteurized under carefully controlled conditions and again chilled.
8. Directly after pasteurization, milk must be placed into the container reaching the customer.
9. Milk must be delivered cold (7° C) in refrigerated vehicles and immediately refrigerated at its destination.
10. A uniform statewide sanitary milk code must provide for proper inspection of dairies and pasteurization plants.
11. Local laboratory services must check on the purity, cleanliness, and safety of the milk.

Milk Grading

Any system of grading the milk supply is based generally on the sanitation of the dairy, health of the cows, methods of handling the milk, and bacterial content.

The bacterial count is obtained by either the **plate count** (p. 164) or **direct microscopic clump count.** In the direct microscopic clump count an accurately measured amount of milk (0.01 ml) is placed on a slide and spread into a thin smear of uniform thickness. The dried smear is stained. With a microscope calibrated so that the area of the field is known, the bacteria in representative fields are counted and the number of bacteria in the whole specimen calculated. In this clump count both individual bacteria and unseparated groups of bacteria are counted as units. In addition to the bacterial count of milk, testing for coliforms is done to detect contamination after pasteurization. Milk is also checked for possible adulteration.

The phosphatase test is used in grading to check whether milk has been adequately pasteurized, since the enzyme phosphatase is mostly destroyed at the temperature of pasteurization. A significant amount detectable in milk indicates incomplete pasteurization or the presence of raw milk mixed in with the sample of pasteurized milk.

According to the 1978 *Milk Ordinance and Code,* Grade A pasteurized milk (the only grade allowed in interstate commerce for retail sale) is milk that has been pasteurized, cooled, and packaged in accord with precise U.S. Public Health Service/Food and Drug Administration specifications. Its bacterial content has been reduced by pasteurization from no more than 300,000 per milliliter to no more than 20,000 per milliliter, with coliforms no more than 10 per milliliter.

The general improvement in milk sanitation has been such that most authorities state that there is no longer any practical necessity for special grades of milk. One standard, high-quality, safe, pure milk should be readily available to people in all walks of life at a reasonable cost.

Antibiotics in Milk

Antibiotics administered to a dairy animal may well be residual in the milk of that animal. An antibiotic residual in milk is a serious concern. For one reason, it causes difficulties in the manufacture of milk products, where processes of curdling, ripening, flavor production, and so on depend directly on microbes.

Penicillin is mainly responsible for this kind of milk contamination, since it is widely used in the treatment of bovine mastitis. A residual of this antibiotic also poses a hazard to the consumer who might be allergic to it. Unlike penicillin, other antibiotics residual in milk generally are poor allergens. Testing for the presence of penicillin and nonpenicillin antibiotics in milk is required in some geographic areas.

Milk Formulas for Hospital Nurseries

The preparation of infant milk formulas in hospitals should conform to standards set by the hospital medical staff or appropriate health agency. For proper microbiologic control the American Academy of Pediatrics, in its Standards and Recommendations for the Hospital Care of Newborn Infants, recommends that technics of formula preparation be checked at least once a week. As part of their surveillance plan, random samples of the ready-to-use formula are sent to the laboratory to be cultured. The limit for the bacterial plate count of a formula sample is 25 organisms per milliliter. In the identification of organisms present there should be none other than spore-formers. Otherwise a break in technic is indicated, and responsible authorities should be notified at once.

Bacterial Action on Milk

Milk may be decomposed by the action of bacteria on its carbohydrates (fermentation of milk) or proteins (putrefaction of milk). Souring, or normal milk decomposition, comes from lactic acid fermentation, the result of the formation of insoluble casein by the action of lactic acid on caseinogen. Lactic acid is formed by bacterial action on lactose, and the organisms responsible are *Streptococcus lactis, Lactobacillus acidophilus, Bifidobacterium bifidum (Lactobacillus bifidus), Lactobacillus casei, Lactobacillus bulgaricus, Escherichia coli,* and others.

Putrefaction of milk seldom occurs, but it is dangerous when it does. Putrefactive changes are caused by the action of spore-bearing and anaerobic bacilli on the milk proteins, which converts the milk into a bitter liquid having little resemblance to fresh milk.

Alcoholic fermentation of milk is rarely spontaneous under natural conditions, but artificially it is a feature of the preparation of certain alcoholic beverages, among which are koumiss, kefir, leben, and matzoon. Koumiss, a drink of the Tartars, is the alcoholic fermentation of mare's milk, the bacteriology of which is unclear. Both a bacterium and a yeast appear to be required for kefir, which is made by inhabitants of the Caucasus from the milk of cows, goats, and sheep. Leben is a drink of Egypt made from the milk of cows, buffaloes, and goats; matzoon, which is similar to yogurt, is a drink of Asia Minor.

Slimy or ropy milk fermentation usually results from the action of *Alcaligenes viscolactis* or a similar organism, but it may be attributable to streptococci and the lactic acid bacteria. Such an unsavory fermentation resulting from the action of a particular species of streptococcus is a deliberate step in the manufacture of Edam cheese in Holland.

Yogurt is the milk to which Metchnikoff referred in his book on the prolonged life of the Bulgarian tribes and is the forerunner of present-day Bulgarian buttermilk. The chief characteristic of yogurt is its high acidity, produced by the growth of *Lactobacillus bulgaricus. Streptococcus thermophilus* gives it a custardlike consistency. Because of its high acidity, it is considered to be one of the safest of all foods. Archaeologists working in underdeveloped areas of the world believe that natural yogurt, a major food item in the diet of many people, is always nutritious and safe to eat. Yogurt and buttermilk are at times used in the nursing care of patients with advanced cancerous growths, secondarily infected and malodorous. The alkaline medium of the growth allows bacterial growth and fermentation activities to flourish. The application of yogurt or buttermilk changes the tissue environment of the cancerous area to an acid one, thus inhibiting the growth of the odor-forming microbes.

Bacterial growth may color milk red, yellow, or blue, when the corresponding chromogens are present.

Ice Cream

Although ice cream is frozen, not all bacteria are destroyed. In fact some pathogens live longer in ice cream than in milk. Typhoid bacilli have been known to live as long as 2 years in ice cream. Outbreaks of typhoid fever, septic sore throat, diphtheria, food poisoning, and scarlet fever have been traced to ice cream. To prevent it from spreading disease, ice cream must be made with pasteurized milk and handled properly thereafter. Proper handling means cleanliness of utensils and factory and proper health supervision of employees.

Butter

Although derived from milk, butter poorly supports microbial growth because it is chiefly fat and water with little protein or sugar. However, certain bacteria do grow in it and render it rancid. The ordinary saprophytic organisms found in butter are the various types of lactic acid bacteria, especially streptococci. Improperly prepared butters may contain *Escherichia coli* and various yeasts and molds. The presence of an excessive quantity of yeasts in butter indicates that proper sanitary precautions were not exercised in preparation. *Mycobacterium tuberculosis* and the atypical mycobacteria have also been found in butter. Typhoid bacilli have been found, but they tend to die therein. Contamination of butter with pathogenic organisms may be eliminated by pasteurizing the cream used in making the butter, properly sterilizing utensils, and strictly supervising personnel.

FOOD

The microbiology of food is studied from many standpoints. The soil and agricultural microbiologist studies the microbes (free living and symbiotic) in the soil. These are utilized by food plants in growth and development. Microbes aid in the return of waste products of food utilization to the soil, where the waste products are used again by microbes and food plants. The food microbiologist studies the part that microbes play in (1) the manufacture of bread, butter, cheese, and many other foods and (2) the spoilage of foods and how this may be prevented. Public health microbiologists, primarily interested in the health of the community, study food in regard to the diseases it may convey, the harmful products of its spoilage, and how both foodborne disease and food spoilage may be prevented. The diseases conveyed by food are discussed here.

Food Poisoning

When foods are related to disease, they are considered first as carriers of infection, as shown on p. 303 (for example, poliomyelitis and hepatitis). Another group of illnesses is spoken of as food poisoning. By definition, food poisoning is an acute illness resulting from the ingestion of some injurious agent in food. It is classified as follows:
1. Nonmicrobial type
    a. Individual idiosyncrasy
    b. Chemical factor from
        (1) Foods naturally poisonous (toadstools and the like)
        (2) Poisons accidentally or intentionally added (plant sprays and the like)
2. Microbial type
    a. Food intoxications—effect of ingested toxin (for example, botulism)
    b. Food infections—multiplication of ingested microorganisms
        (1) Bacterial (for example, salmonellosis)
        (2) Parasitic (for example, trichinosis)

A consistent feature of food infections and intoxications is that many persons eat the offending food and most develop the disease. In food infection the diagnosis is made when the offending organism is cultured from the feces of the patient and from the food eaten. In food intoxication diagnostic efforts are directed to finding the offending toxin in the food eaten. Examination of the feces of the patient is not of consistent value.

Food poisoning, especially that caused by staphylococci, was formerly known as "ptomaine" poisoning. This is incorrect because ptomaines play no part in food poisoning.

Staphylococci, *Clostridium botulinum*, and, less often, certain other organisms produce food intoxication. *Salmonella* species and rarely *Streptococcus faecalis* (enterococcus) cause food infection. Table 40-1 lists bacterial types of food poisoning and the prominent features of each.

Food Intoxication

Staphylococcal food intoxication, the most common type of food poisoning, is caused by the action of an enterotoxin liberated by some strains of staphylococci. The staphylococci multiply in the offending food before it is eaten and elaborate their toxin there. (They do not multiply in the intestinal tract.) Manifestations come from the ingested toxin. Staphylococci require a period of not less than 8 hours to elaborate enough toxin in food to cause symptoms. After the food is eaten, the disorder appears within 6 hours. Recovery occurs in 24 to 48 hours. Staphylococcal food intoxication usually follows the ingestion of starchy foods, especially potato salad, custards, and pies. When the offending food is meat, it is usually ham. Many outbreaks have been traced to chicken salad.

Botulism is a very serious type of food intoxication produced by the ingestion of food containing the potent exotoxin of *Clostridium botulinum* (p. 580). The major source of botulism is improperly processed home-canned vegetables of low acid content. *Before they are eaten,* such foods should always be cooked *at the temperature of boiling water for at least 15 minutes (20 minutes for corn and spinach)* and thoroughly stirred and mixed. Contaminated foods may have unpleasant odors somewhat characteristic for a given food, but these off-odors are not easily recognized and do not suggest spoiled food. The cans or containers holding the food do not necessarily bulge, a conventional sign of food spoilage.

Food intoxication caused by staphylococci and agents other than *Clostridium botulinum* differs from botulism in that (1) the total time of illness is short and (2) cranial nerves are not involved.

Bacterial Food Infections

Food infection is caused by the multiplication of bacteria that have been taken into the intestinal canal. The responsible bacteria are most often salmonellas. Of special note in food infection are *Salmonella enteritidis* (Gärtner's bacillus), *Salmonella typhimurium*, and *Salmonella choleraesuis*. Once inside the intestine, they multiply rapidly and induce widespread inflammation with severe manifestations—nausea, vomiting, diarrhea, and fever.

Food infection usually occurs as an explosive outbreak following a meal attended by a large number of persons. Food cooked in large quantities is more often the source of infection than is food cooked in small quantities, since heat is not so likely to penetrate and destroy the organisms in a large quantity. Most outbreaks occur during the warm months. The incubation period for *Salmonella* food infection is 7 to 72 hours, and recovery occurs usually within 3 or 4 days; but in severe infections death may occur within 24 hours.

The insufficiently cooked meat of infected animals may convey the disease to humans, but usually salmonellas reach the food (often but not necessarily meat) from outside sources. Such sources are the intestinal contents of the slaughtered animals, the intestinal contents of animals (especially rats and mice) that have contacted the food, and (probably most important) human carriers of the bacteria. *Salmonella* infections are common in rats, and these animals often become carriers. Cockroaches have been shown to convey *Salmonella* food poisoning. Salmonellas are abundant in the intestinal tract of poultry, a very important source, and may be present on the shell of an egg.

Table 40-1 Bacterial Types of Food Poisoning

Organism	Foods	Onset (After Food Eaten)	Clinical Features	Toxin	Outcome
Food Intoxication					
Staphylococci	Potato salad, chicken salad, cream fillings, milk, meats, cheese	1 to 6 hours	Vomiting, diarrhea, abdominal cramps, prostration	Entero-toxin	Severe symptoms of short duration (death rare)
Clostridium botulinum	Canned string beans, corn, beets, ripe olives	2 hours to 8 days	Vomiting, diarrhea, double vision, difficulty in swallowing and speaking, respiratory paralysis	Neurotoxin	Mortality high
Food Infection					
Salmonellas	Eggs, poultry, meats	7 to 72 hours	Diarrhea, severe abdominal pain, fever, prostration	None	Recovery in a few days (death rare)
Streptococci	Sausage, poulty dressing, custard	2 to 18 hours	Nausea, pain, diarrhea	None	Recovery in 1 or 2 days

Contaminated eggs broken commercially may be a source of organisms in egg products. Foods such as the meringue on pies contaminated with *Salmonella* may have no abnormal odor, taste, or appearance.

Prevention of *Salmonella* food poisoning depends on cleanliness in handling food, proper cooking of food, proper refrigeration of food that has been cooked, and detection of carriers. Frequent and careful hand washing by food handlers is mandatory. Food that remains on their skin for a time provides nourishment for microbial contaminants, some foodborne, and thus contributes to microbial survival in the environment.

Under the provisions of the Egg Products Inspection Act of 1970, all commercial eggs broken out of the shell for manufacturing use must be pasteurized. The egg or egg product is heated to 60° to 61° C (140° to 143° F) and held there for 3½ to 4 minutes. The final product must be free of salmonellas. The U.S. Department of Agriculture currently requires that in the commercial processing of precooked roast beef it be heated to an internal temperature of 63° C to destroy any salmonellas present.

Rarely, streptococci cause food poisoning of the infection type. At times the addition of large numbers of organisms to the already resident enterococci in the bowel triggers gastroenteritis. The food usually responsible contains meat that has stood at warm temperatures for several hours, and the organism usually recovered from the feces is *Streptococcus faecalis*. The incubation period is 2 to 18 hours, and the illness lasts a short time.

Other organisms sometimes infecting food are *Clostridium perfringens*, *Bacillus cereus*, *Proteus* bacilli, and *Escherichia coli*. In outbreaks traced to these it has been assumed that the microbes were able to multiply in food, especially meat, that had stood at room temperature overnight or longer.

Parasitic Food Infection

For a discussion of trichinosis, see p. 844.

Measures to Safeguard Food

To safeguard a food supply, remember that:

1. Appearance, taste, and smell are not always reliable indicators that a food is safe for consumption.
2. Any unusual change in color, consistency, or odor or the production of gas in food means that the food should be discarded *without being eaten*. NOTE: If *suspect* food has been tasted or eaten, it should not be discarded but kept for 48 hours in the event that it might have to be examined for the presence of *Clostridium botulinum* exotoxin.
3. Dishes, platters, cutlery, can openers, utensils, cutting boards, kitchen counters, and equipment of all kinds used to prepare, serve, and store food must be clean and sanitary.
4. Hands that prepare, serve, and store food should be clean. Preferably, food should be handled as little as possible.
5. Foods served raw should be washed carefully and thoroughly.
6. Bacteria grow fastest in the temperature range of 40° to 140° F (Fig. 40-2). Therefore:
   a. Hot food should be kept hot (above 140° F).
   b. Cold food should be kept cold (below 40° F). All dairy foods should be refrigerated.
   c. Cooked food that is to be refrigerated should be cooled quickly and refrigerated promptly. If cooked food is to be kept longer than a few days, it should be frozen.
   d. Perishable foods should be kept chilled if taken on a trip or picnic. Foods may spoil easily if exposed to a warmer temperature for no longer than ½ hour before they are eaten.

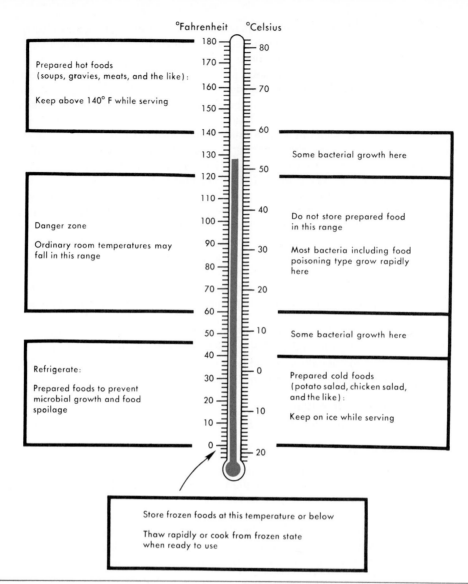

°Fahrenheit   °Celsius

Prepared hot foods
(soups, gravies, meats, and the like):

Keep above 140° F while serving

Some bacterial growth here

Danger zone

Ordinary room temperatures may
fall in this range

Do not store prepared food
in this range

Most bacteria including food
poisoning type grow rapidly
here

Some bacterial growth here

Refrigerate:

Prepared foods to prevent
microbial growth and food
spoilage

Prepared cold foods
(potato salad, chicken salad,
and the like):

Keep on ice while serving

Store frozen foods at this temperature or below

Thaw rapidly or cook from frozen state
when ready to use

FIG. 40-2   Temperature guide for proper food handling. *Remember to keep cold foods cold and hot foods hot!* Temperatures above 45° F and below 140° F provide breeding grounds for many foodborne illnesses.

      e. Extra care should be taken with foods easily contaminated by microbial growth—meats, poultry, stuffing, gravy, salads, eggs, custards, and cream pies.
   7. Feeding dishes, bedding, or toys used by pets should be kept out of the kitchen and away from items in contact with the family's food.

Food Preservation

There are numerous methods of preserving food. The following are a few.

Refrigerating

Refrigeration, one of the best and most universally used methods of food preservation, has distinct advantages: (1) there is little change in the composition of food; therefore taste, odor, and appearance are preserved; (2) the nutritive value of the food is not reduced; (3) there is no decrease in digestibility; and (4) there is little effect on vitamin content. During refrigeration pathogenic bacteria are inhibited and many are destroyed. When refrigeration is discontinued, those that remain viable begin to multiply. Some saprophytes can multiply at refrigeration temperatures.

Freezing

Freezing as a way of preserving food is in widespread use today because of its application in the preparation of "convenience" food items. Quick freezing is carried out at very low temperatures ($-35°$ F [$2°$ C] with a holding temperature of $-10°$ F [$-23°$ C]). A preliminary step to freezing is blanching. The food is heated quickly to inactivate enzymes that would break down proteins and change the texture, flavor, and appearance of the food. Blanching is followed by instant cooling in ice water. Food that is frozen retains its nutrients and is palatable. Freezing meat kills encysted larvae therein. Since cold-storage foods decompose rapidly when thawed or warmed to room temperature, they should be consumed right away.

Drying

When food is preserved by drying, microbes are unable to multiply because of a lack of water.

Smoking

When food is preserved by smoking, it is dried and preservatives from the smoke are added to it.

Pickling

In food preservation by pickling, the high-acid content of the medium prevents microbial multiplication.

Salting and Preserving in Brine

Salting and preserving in brine is an effective method of food preservation because the osmotic pressure of the medium is so changed that microbes will not multiply therein; sometimes water is extracted from the microbial cell.

Canning

The heat of canning destroys microorganisms. The container, with its content of food, is hermetically sealed to keep out contaminants (see box on p. 918). The sealing also excludes free oxygen, which would aid the growth of molds, yeasts, and most species of pathogenic bacteria.

Commercial canning is a very safe method of food preservation. It is done scientifically and under supervision of the indicated health agency. Home canning, on the other hand, may be done under "rough-and-ready" conditions by unskilled hands. Since there is always a possibility for food contamination, home canned foods, especially meats and vegetables, should always be heated before consumption.

Preserving

With preserving the food is heated and sugar added. A large amount of sugar retards bacterial multiplication in the same manner as does salt. However, it does not retard the multiplication of fungi. Heating and sealing have the same effect as in canning.

Cooking

In the cooking process microbes in food are killed by heat, the composition of the food is changed, and water may be recovered.

Toward the end of the eighteenth century the French government offered a prize of 12,000 francs to anyone who could devise a method for "preserving fresh food for transport during military and naval movements of great distance and duration." In 1795 Nicolas Appert (1750-1841), confectioner and distiller in Paris, began to work on the problem. Assuming that air was responsible for food spoilage, he placed various foods in widemouthed bottles sealed with airtight stoppers and cooked them in boiling water. He soon found his method to be successful. He could store food for prolonged periods without it deteriorating. The "founder of applied microbiology" personally received the prize in 1810 from the Emperor Napoleon Bonaparte himself. About the same time the tin can was patented. A decade later, in 1819, the first American canneries were opened in Boston and New York City.

Use of Chemicals

The addition of chemicals such as benzoic acid in the form of the sodium salt, sodium and calcium propionates, sodium and potassium nitrites, and esters of parahydroxybenzoic acid (parabens) to foods as preservatives is regulated by the Food and Drug Administration (FDA) and by the U.S. Department of Agriculture.

Ionizing Radiation

Ultraviolet irradiation has already been used successfully by food industries to treat the air in storage and processing rooms, to prevent the growth of fungi on shelves and walls of preparation rooms, and to destroy parasites in meat, but the FDA does not sanction irradiation of the food itself.

Cold sterilization of foods by means of ionizing radiation has been under investigation by the U.S. armed forces, and one of the major obstacles encountered has been the terrific resistance of the spores of *Clostridium botulinum* to the effects of radiation. This is especially a problem with seafoods.

The U.S. Army is pressing for sterilization of meat by irradiation. In the Army's process, meat is vacuum packed, heated to 160° F (71° C) to inactivate enzymes, frozen, irradiated, and then thawed. With a shelf life equivalent to that of conventionally canned food, the Army considers their finished product to be more nutritious and palatable. Irradiated meat has been consumed since 1975 by astronauts in flight.

Food in Vending Machines

Vending machines today dispense many foods that support the rapid growth of infectious or toxigenic microbes. To be safe for consumption, the food must be prepared under sanitary conditions and transported in properly refrigerated vehicles. In the vending machine food items must be maintained at proper temperatures and replaced often.

SUMMARY

1 Milk conveys infection because it is an excellent culture medium and is consumed uncooked. Under the best conditions the bacterial content of milk is relatively high.

2 The two main sources for pathogenic bacteria in milk are the persons who handle it and infected cows.

3 The number of bacteria in milk is the best index of its sanitary quality.

4 Diseases spread by milk include typhoid fever, salmonellosis, bacillary and amebic

dysentery, scarlet fever, infantile diarrhea, infectious hepatitis, and septic sore throat.

5 Diseases of milk-producing animals transmissible to humans in infected milk are undulant fever, Q fever, and bovine tuberculosis.

6 Pasteurization is the process of heating milk (or other liquids) to a temperature high enough to kill all nonspore-bearing pathogenic bacteria but not so high as to affect the chemical composition.

7 Pasteurization is the single most crucial factor in the maintenance of a safe milk supply.

8 Souring of milk comes from the formation of insoluble casein from caseinogen as a result of the action of lactic acid produced by bacterial action on milk sugar, or lactose.

9 Food may be a carrier of disease, and food poisoning results from the ingestion of some injurious agent in the food.

10 Food poisoning may be of the nonmicrobial type (for example, food poisoning from the ingestion of a naturally poisonous food such as certain mushrooms) and of the microbial type.

11 Microbial food poisoning resulting from the ingestion of a preformed toxin is *food intoxication* (for example, botulism), and that coming from the proliferation of microbes ingested is *food infection* (for example, salmonellosis).

12 Remember in the safeguarding of food that cold foods must be kept cold and hot ones hot, since temperatures above 45° F and below 140° F provide breeding conditions for many foodborne illnesses.

13 The numerous methods of preserving food include refrigerating, freezing, drying, smoking, pickling, canning, and applying certain chemicals.

QUESTIONS FOR REVIEW

1 Name diseases spread by contaminated milk and food.
2 Give differential characteristics of milkborne and waterborne epidemics.
3 What is meant by Grade A pasteurized milk?
4 What is flash pasteurization?
5 Give the requirements for a safe milk supply.
6 Make an outline showing the different kinds of food poisoning. Name causal agents.
7 Differentiate food intoxication from food infection. Give examples.
8 List the different methods of food preservation. Name five chemical food preservatives.
9 List practical measures to safeguard food.
10 Explain briefly lactic acid bacteria, normal milk bacteria, yogurt, buttermilk, holding method of pasteurization, milk sickness, phosphatase test, souring of milk, food idiosyncrasy, low-acid foods, high-acid foods, and blanching of foods.
11 Indicate how a bacterial count of a milk sample is done.
12 State the possible consequences of an antibiotic residual in milk.

SUGGESTED READINGS

Bauman, H.E.: Food microbiology, ASM News **38**:312, 1972.

Boer, H.R., and others: Bacterial colonization of human milk, South. Med. J. **74**:716, 1981.

Grade A pasteurized milk ordinance, 1978 recommendations, Public Health Service/Food and Drug Administration, Washington, D.C., 1978, U.S. Government Printing Office.

Kuhn, P.J.: "It didn't look quite right, but I ate it anyway," Diagn. Med. **3**:46, 1980.

Mackey, B.M., and others: The recovery of sublethally injured *Esherichia coli* from frozen meat, J. Appl. Bacteriol. **48**:315, 1980.

Porter, J.R.: Microbiology and the food and energy crisis, ASM News **40**:813, 1974.

Price, J.F., and Schweigert, B.S., editors: The science of meat and meat products, ed. 2, San Francisco, 1971, W.H. Freeman & Co., Publishers.

Reddy, D., and others: Nitrite inhibition of *Clostridium botulinum*: electron spin resonance detection of iron–nitric oxide complexes, Science **221**:769, 1983.

LABORATORY SURVEY
OF MODULE SIX

PROJECT   Bacteriology of Water
Part A—Bacterial Plate Count of Water
1. Obtain from the instructor the following:
   a. Sample of water—may be of low bacterial count (tap water) or high bacterial count (river water)
   b. Sterile pipettes
   c. Three tubes, each of which contains 9 ml of sterile water
   d. Three sterile Petri dishes
   e. Three tubes of nutrient agar
2. Prepare dilutions of water sample.
   a. To one tube of sterile water, add 1 ml of water sample and mix.
   b. Transfer 1 ml from this first tube with a new sterile pipette to a second tube of 9 ml of sterile water and mix.
   c. Repeat procedure with a new pipette—transfer 1 ml from the second to third tube and mix.
3. Calculate dilution of water in each tube. Why is a new pipette used for each procedure?
4. Mix diluted water samples with a suitable culture medium.
   a. With a new pipette, transfer 1 ml from each tube to a Petri dish, beginning with highest dilution and proceeding to lowest. Why do you go from the highest to the lowest dilution?
   b. Melt tubes of agar in water bath, let cool to 42° to 45° C, and add one tube to each Petri dish.
   c. Mix agar with water by rotating, let the agar solidify, invert plates, and incubate at 37° C for 24 hours.
5. Select a plate on which colonies are distinctly separated and count the colonies.
6. Calculate from the dilution the bacterial content of the water per milliliter. Record your results below.

Plate	Dilution	Number of Bacterial Colonies Present	Estimated Number of Bacteria per ml in Water Sample
1			
2			
3			

Part B—Presumptive Test for Coliforms in Water
1. Obtain from the instructor the following:
   a. Sample of contaminated water
   b. Sterile pipettes
   c. Nine fermentation tubes containing lactose (or lauryl tryptose) broth
2. Inoculate a sample of contaminated water as follows:
   a. To each of two fermentation tubes, add 0.1 ml of water sample.

   b. To two more tubes, add 1 ml of water sample.
   c. To remaining five fermentation tubes, add 10 ml portions of water sample.
3. Mix water sample with contents of each tube by gently rolling the tube between the palms of the hands.
   NOTE: Do not let the contents of a tube come in contact with the cotton plug or screw cap.
3. Mix water sample with contents of each tube by gently rolling the tube between the palms of the hands.
   NOTE: Do not let the contents of a tube come in contact with the cotton plug.
4. Place tubes in incubator at 37° C and examine at 24 and 48 hours. For each period note percentage of gas formed in each tube as indicated by portion of collecting tube that is filled with gas.
5. Record your findings below.

Tube	Amount of Water Sample (ml)	Percentage gas Formed in Fermentation Tube in		Presumptive Test	
		24 hrs	48 hrs	Positive	Negative
1	0.1				
2	0.1				
3	1				
4	1				
5	10				
6	10				
7	10				
8	10				
9	10				

Part C—Demonstration of a Positive Confirmed Test by Instructor with Discussion

1. Selection of one of the fermentation tubes from the presumptive test in which gas formation is nicely shown
2. Demonstration of plates (Endo medium or eosin–methylene blue agar) made from tube selected
3. Observation of characteristics that differentiate coliform colonies from those of other organisms on the medium

PROJECT   Bacteriology of Milk

Part A—Bacterial Plate Count of Milk

1. Use same method here that was used in obtaining bacterial plate count of water (p. 921). Use milk samples instead of water. Adapt technic so that dilution bottles contain 99 ml instead of 9 ml of sterile water.
2. Plate milk sample dilutions and incubate at 37° C for 48 hours.
3. Count the colonies and calculate the number of bacteria per milliliter of original milk sample. Are colonies on the plates all alike or are there varieties?
4. Chart results.

Plate	Dilution	Number of Bacterial Colonies Present	Estimated Number of Bacteria per ml of Milk
1			
2			
3			

Part B—Effect of Pasteurization on Milk

1. Obtain from the instructor the following:
   a. Two tubes of raw milk
   b. Two 99 ml dilution bottles
   c. Two tubes of nutrient agar
   d. Pipettes
2. Pasteurize one raw milk sample as follows:
   a. Place one tube of milk in a water bath and raise temperature of the water to 65° C.
   b. Hold at this temperature for 30 minutes.
3. Prepare dilutions of both milk samples.
   a. Make a 1:100 dilution of this milk and the unheated milk.
   b. With separate sterile pipettes, place 1 ml of each dilution in a Petri dish and mix with melted agar as outlined in steps 4 (b) and (c) in the experiment on counting bacteria in water (p. 921).
   c. Observe. Which plate contains more bacteria? Why?
   d. Chart results.

Sample	Bacterial Plate Count	Number of Bacteria per ml in Milk Sample
Pasteurized milk (heated to 65° C for 30 min)		
Raw milk (not heated)		

Part C—Phosphatase Test*

NOTE: This test depends on the fact that raw milk contains an enzyme known as phosphatase, which is more resistant to pasteurization than any known pathogenic bacterium. Its inactivation therefore is a reliable index that the milk has been properly pasteurized and is safe.

---

*The method of preparing the reagents needed in the test may be found in *Standard Methods for the Examination of Dairy Products*, published by the American Public Health Association.

1. Test the two samples of milk with which you have previously worked—one a sample of raw milk and one a sample of raw milk that you have pasteurized. Proceed as follows for a given sample:
   a. Pipette 5 ml of buffered substrate into a test tube.
   b. Add 0.5 ml of milk sample and shake. Label tube.
   c. Place the test tube in a 37° C incubator or water bath for 30 minutes. Remove.
   d. Add 6 drops of CQC reagent and 2 drops of catalyst to the tube. Stopper test tube, mix contents, and incubate for 5 minutes.
      NOTE: If milk sample has not been properly pasteurized, a blue color will appear after about 5 minutes. If milk has been properly pasteurized, the color will be gray or light brown. No attempt is made in this exercise to extract the pigment and quantitate it.
2. Compare results obtained for the two samples of milk—one pasteurized and one not.

Part D—Methylene Blue Reduction Test

1. Work in pairs.
2. Obtain two milk samples from the instructor—one of low bacterial count and one that is heavily contaminated.
3. On each sample, carry out the test as follows:
   a. Place 10 ml of each sample of milk in a test tube. Label tube.
   b. Add 1 ml of methylene blue thiocyanate solution* to each test tube.
   c. Place rubber stoppers in the two tubes and mix thoroughly by inverting.
   d. Place in water bath at 37° C.
   e. Observe tubes occasionally to determine decolorization.
      NOTE: The more bacteria present, the more rapid the methylene blue is decolorized. On this basis, test milk may be classified as follows:
      Class 1. Excellent quality milk (not decolorized in 8 hours)
      Class 2. Milk of good quality (decolorized in less than 8 but not less than 6 hours)
      Class 3. Milk of fair quality (decolorized in less than 6 but not less than 2 hours)
      Class 4. Poor quality milk (decolorized in less than 2 hours)
4. Classify the two milk samples that you and your partner have tested. Record findings.

*Prepared from methylene blue thiocyanate tablets (about 8.8 g) certified by the Commission on Standardization of Biological Stains. Tablets may be obtained from laboratory supply houses. Dissolve one tablet in 200 ml hot sterile distilled water and allow solution to stand overnight. Store in amber glass bottles away from light.

REFERENCES

Standard methods for the examination of water and wastewater, Washington, D.C., 1976, American Public Health Association, Inc.

Standard methods for the examination of dairy products: microbiological and chemical, Washington, D.C., 1978, American Public Health Association, Inc.

EVALUATION FOR MODULE SIX

1 Please circle the *one* number from the column on the right that indicates the *best* answer to the question or that correctly completes the statement.

1. Why does the bacteriologist measure the safety of a water supply by testing for the comparatively nonpathogenic colon bacillus?
   (a) Although the colon bacillus is normal in the intestines, if ingested it usually causes disease.
   (b) The colon bacillus indicates that the water supply is probably contaminated with fecal material.
   (c) Pathogenic organisms from infected fecal material are hard to isolate in water.
   (d) Although fecal material from some sources may be safe, that from other sources may carry disease germs.
   (e) Presence of the colon bacillus means that the typhoid bacillus is also present.

   1. a
   2. b and c
   3. b, c, and d
   4. c, d, and e
   5. all of these

2. Which of the following tests for the pasteurization of milk is most practicable to detect raw milk mixed with pasteurized milk or to indicate milk incompletely heated in the pasteurization process?
   (a) Acidity test
   (b) Phosphatase test
   (c) Catalase test
   (d) Test for coliforms
   (e) Methylene blue test

   1. a, b, and c
   2. a, b, and d
   3. b
   4. c
   5. c and e

3. The most effective method of minimizing the spread of infectious diseases from food handlers to consumers in fast-food restaurants is:
   (a) To enforce legal penalties for violators of the sanitary code
   (b) To take periodic cultures from utensils and equipment in contact with food
   (c) To require periodic health examinations of food handlers
   (d) To require annual chest radiographic examinations
   (e) To institute programs wherein food handlers are instructed in principles of hygiene

   1. a
   2. b
   3. c
   4. d
   5. e

4. The presence of 1 ppm of fluoride in drinking water may lead to:
   (a) Mottled tooth enamel
   (b) Decreased number of dental caries
   (c) Impaired renal function
   (d) Softening of the bones
   (e) Fluorosis

   1. a and b
   2. a and e
   3. b
   4. c
   5. d

5. Of the following statements, which hold(s) equally true for staphylococcal food poisoning, shigellosis, and amebiasis?

    (a) Humans are the natural reservoir of infection.

    (b) The infectious agent dies quickly outside the human body.

    (c) The refrigeration of foods would successfully block spread.

    (d) The etiologic diagnosis can be made by stool culture or microscopic examination of the stool specimen.

    (e) Each is caused by a soluble enterotoxin.

    1. a
    2. a, d, and e
    3. b and c
    4. all of these
    5. none of these

6. A gourmand developed food poisoning shortly after a trip to the Orient. In one country this person dined on sashimi (raw fish). If the food poisoning is the result of the ingestion of the sashimi, the likely infectious agent is:

    (a) *Vibrio cholerae*

    (b) Enterogenic *Escherichia coli*

    (c) *Shigella sonnei*

    (d) *Vibrio parahaemolyticus*

    (e) *Salmonella typhi*

    1. a
    2. b
    3. c
    4. d
    5. e

7. Staphylococcal food poisoning is:

    (a) A food poisoning that does not produce any manifestations for 24 hours

    (b) An intoxication

    (c) The direct consequence of ingestion of live staphylococci that multiply subsequently in the colon

    (d) Associated with an enterotoxin that induces severe manifestations of disease

    1. a, b, and c
    2. a and c
    3. b, c, and d
    4. b and d
    5. all of these

8. John Snow is remembered because:

    (a) He was interested in the transmission of cholera.

    (b) He investigated the epidemic of measles on the Faroe Islands.

    (c) He was the founder of the American Trudeau Society for tuberculosis.

    (d) He was the sponsor of the Social Security Law.

    1. a
    2. a and b
    3. b
    4. c
    5. d

9. Which of the following vitamins are produced in the large bowel of humans?

    (a) Vitamin A

    (b) Vitamin B complex

    (c) Vitamin C

    (d) Vitamin D

    (e) Vitamin K

    1. a, b, and c
    2. b
    3. b and c
    4. b and e
    5. c, d, and e

10. Which of the following statements are true?

    (a) Nitrogen fixation is a stepwise process whereby atmospheric nitrogen is converted into nitrates.

    (b) Only one or two species of bacteria participate in nitrogen fixation.

    (c) Nitrifying bacteria convert ammonia to nitrate.

    (d) Symbiotic relationships among bacteria are important in the nitrogen cycle.

    (e) Nitrates in the soil must be constantly renewed.

    1. a, b, c, and d
    2. a, c, and d
    3. a, c, d, and e
    4. all of these
    5. none of these

11. Small amounts of water can be made safe for drinking by:
    (a) Vigorous boiling for several minutes
    (b) Filtration through gauze
    (c) Allowing the water to stand for an hour, then filtering it through gauze
    (d) Chemical disinfection with 2% tincture of iodine (5 drops to 1 L)
    (e) Addition of a pinch of table salt

    1. a, b, and d
    2. a and d
    3. a and e
    4. b and c
    5. b, c, d, and e

12. An important characteristic of yogurt is its high acidity. Which of the following statements is (are) true?
    (a) The high acidity results from the growth of *Streptococcus thermophilus*.
    (b) The high acidity inhibits microbial growth, thus making yogurt a safe food to eat.
    (c) The custardlike consistency of yogurt comes from the growth of *Lactobacillus bulgaricus*.
    (d) Yogurt as a food is found only in the better developed countries.
    (e) Yogurt is produced from the slimy fermentation of milk by *Alcaligenes viscolactis*.

    1. a
    2. b
    3. a, b, c, and d
    4. all of these
    5. none of these

13. Botulism is a serious type of food poisoning. We know that:
    (a) It results from the ingestion of the potent exotoxin of *Clostridium difficile*.
    (b) The exclusive source is improperly processed home canned vegetables of low acid content.
    (c) It is a true food intoxication but also a food infection.
    (d) The suspect food emits a characteristic and easily detected odor.
    (e) The mortality from the disease is high.

    1. b, c, d, and e
    2. c, d, and e
    3. e
    4. all of these
    5. none of these

2 Match the item in column Beta to the most descriptive word or phrase in column Alpha. Please place the related letter in the appropriate space.

COLUMN ALPHA—Bacterial Food Poisoning	COLUMN BETA
_____ 1. The result of ingestion of certain bacteria normally present in intestine	(a) Food infection
_____ 2. The result of ingestion of preformed toxin	(b) Food intoxication
_____ 3. The result of multiplication of microbes ingested	(c) Neither
_____ 4. Type of food poisoning often involving a large number of persons who have eaten the same offending food	(d) Both
_____ 5. Important examples are bacterial in origin, not protozoan	
_____ 6. Offending microbe should be cultured from patient's feces as well as from food consumed	
_____ 7. Kind of "ptomaine" poisoning	
_____ 8. Salmonellas the major causative agents	

PART II

1 Most of the cowhands who ate the potato salad at the roundup became ill with nausea and vomiting 2 to 6 hours after chowtime. The cowboys who did *not* eat the potato salad did *not* become ill. Please circle one response for each of the statements.

1. The most likely organism causing the illness would be:
   (a) Group A streptococcus
   (b) Enterococcus
   (c) Coagulase-positive staphylococcus
   (d) Coagulase-negative staphylococcus
   (e) None of these

2. The symptoms were probably caused by:
   (a) An infection of the alimentary canal
   (b) A toxin produced by the organisms growing in the food before it was eaten
   (c) A toxin produced by organisms growing in the intestinal tract
   (d) A substance resulting from bacterial decomposition of the food
   (e) None of these

3. The organism associated with the illness would most likely be identified by:
   (a) Culture of the potato salad
   (b) Culture of the vomitus from a patient
   (c) Culture of stool specimens
   (d) Microscopic examination of the remaining potato salad
   (e) None of these

4. Epidemiologic investigation of the episode of food poisoning revealed that the range cook who prepared the salad had an infection of the hand. An organism was cultured from the sample of the salad dressing used in preparation of the potato salad. To determine the source of the food contamination, one should compare this organism with any organisms recovered from other cultures taken. If organisms are of the same species, they are evaluated epidemiologically by means of:
   (a) A precipitin test to determine serologic type
   (b) The fluorescent antibody technic
   (c) A test to determine bacteriophage type
   (d) A slide agglutination test
   (e) Animal inoculations

2 Consider the following list of diseases. Are they spread by ingestion of contaminated milk or water? If the disease is spread by contaminated water, please place a W in the blank space in front of the disease; if it is spread by contaminated milk, then place an M in the blank space. Some of these diseases may be spread by both milk and water; in this case write both M and W in the blank space. If the disease is not spread by either medium, leave the space blank.

_____ Typhoid fever	_____ Rabies	_____ Cholera
_____ Bacillary dysentery	_____ Amebic dysentery	_____ Botulism
_____ Septic sore throat	_____ Pertussis	_____ Rocky Mountain
_____ Syphilis	_____ Q fever	spotted fever
_____ Bovine tuberculosis	_____ Relapsing fever	_____ Brucellosis
_____ Actinomycosis	_____ Salmonellosis	_____ Diphtheria
_____ Scarlet fever	_____ Viral hepatitis	_____ Tetanus
_____ Infantile diarrhea		

3 Please circle either the *T* or the *F* to indicate whether the statement is true or false.

T   F   1. Bacteria are able to obtain carbon and nitrogen from the air because they contain chlorophyl.

T   F   2. Pathogenic bacteria that grow in the soil never form spores.

T   F   3. Bacteria are seldom found below 6 feet in the soil.

T   F   4. Diseases may be transmitted by means of an intermediate host.

T   F   5. Drinking water always contains a large number of bacteria.

T   F   6. A silo depends on bacteria for its action.

T   F   7. In food poisoning from *Salmonella* species digestive disturbances are prominent.

T   F   8. *Salmonella* food poisoning has been chiefly associated with canned vegetables.

T   F   9. Ultraviolet radiation is bactericidal because the heat that results from it coagulates the bacterial proteins.

T   F   10. Bacterial counts of pasteurized milk cannot be expected to yield any information useful in the prevention of disease because pasteurization kills all pathogenic organisms except spore forms.

T   F   11. Most cases of pulmonary tuberculosis are acquired by drinking infected milk.

T   F   12. Food poisoning is a hazard of picnics because the food served may not have been properly refrigerated for several hours.

T   F   13. Food is an important vehicle in the transmission of syphilis.

T   F   14. Most microbes in nature are not harmful to humans.

T   F   15. The term *nitrogen fixation* refers to the recovery of carbon dioxide from the air by unicellular animals.

T   F   16. Lactic acid bacteria are normal milk inhabitants.

T   F   17. The chief milk-curdling enzyme is diastase.

T   F   18. *Saccharomyces* species are important in commercial fermentations.

T   F   19. *Bacillus subtilis* provides alkaline proteases for the manufacture of enzyme detergents.

T   F   20. *Pseudomonas* species can survive in a wet, stagnant environment by utilizing even traces of phosphorus and sulfur.

T   F   21. Some saprophytes can multiply at refrigeration temperatures.

T   F   22. The spores of *Clostridium botulinum* are quite resistant to the effects of radiation.

T   F   23. Food in vending machines does not present any hazards to the consumer even if it is carelessly handled and improperly refrigerated.

T   F   24. Bacteria grow fastest in the temperature range 40° to 140° F (4° to 60° C).

T   F   25. Viruses are regularly inactivated by primary and secondary standard treatment of sewage; therefore they are not found in the effluent.

T   F   26. Pasteurization of milk has great public health value because bacterial spores do not survive pasteurization.

T   F   27. Botulism has not resulted from commercially prepared foods in the past 50 years.

GLOSSARY

abscess Circumscribed collection of pus.

acquired immunity Immunity acquired after birth.

activated lymphocyte Any lymphocyte in active state of differentiation, such as lymphocyte proliferating on first contact with antigen, lymphocyte engaging in cell-mediated immune reaction, or one developing to produce antibody.

activated macrophage ("angry" macrophage) Macrophage more efficient than nonactivated macrophage at killing bacteria or other target cells; macrophage from antigen-sensitized animal.

active carrier Person or animal who becomes carrier after recovery from given disease.

active immunity Immunity brought about by activity of certain body cells of person becoming immune on direct exposure to antigen.

acute disease Disease that runs a rapid course with more or less severe manifestations.

adjuvant Substance mixed with antigen to enhance antigenicity and antibody response.

aerobe Organism whose growth requires the presence of oxygen.

afferent Bringing to or into.

affinity Attraction.

agammaglobulinemia Deficiency or absence of gamma-globulin in blood, usually associated with increased susceptibility to infection.

agar Gelatinous substance prepared from Japanese seaweed used as a base for solid culture media.

agglutination Visible clumping of cells suspended in a fluid.

agglutinins Antibodies that cause agglutination.

agglutinogen Any substance that, acting as an antigen, stimulates production of agglutinins.

algid Cold; an algid fever is one in which the patient goes into a state of collapse.

allergen Substance that induces allergic state when introduced into body of susceptible person.

allergy (hypersensitivity) State in which the affected person exhibits unusual manifestations on contacting an allergen.

alloantigens Different forms of an antigen in individuals of a species.

allogenic Genetically dissimilar within the same species.

allogenic (allogeneic) disease Graft-versus-host reaction in immunosuppressed subjects who have been given allogenic lymphocytes.

allograft (homograft) Graft exchanged between two genetically dissimilar individuals of same species.

alum toxoid Toxoid treated with an aluminum compound.

amboceptor Substance that combines with cells and complement to dissolve cells (for example, hemolysin).

ameba, amoeba (*pl.* amebas, amoebae) Protozoon that moves by extruding fingerlike processes (pseudopods).

ameboid Resembling an ameba.

amino acids Organic chemical compounds containing an amino (NH_2) group and a carboxyl (COOH) group; they form the chief structure of proteins.

anabolism Phase of intermediary metabolism concerned with the energy-requiring biosynthesis of cell components from smaller precursor molecules.

anaerobe Organism that grows only or best in the absence of atmospheric oxygen.

anaphylactoid Like anaphylaxis.

anaphylaxis State of hypersusceptibility to a protein resulting from a previous introduction of the protein into the body.

antibacterial serum Antiserum that destroys or prevents the growth of bacteria.

antibiotic Agent produced by one organism that destroys or inhibits another organism

antibody Agent in the body that destroys or inactivates certain foreign substances that gain access to body, particularly microbes and their products.

anticoagulant Agent that prevents coagulation.

anticodon Sequence of three nucleotides in transfer RNA that, in process of protein synthesis, binds to a specific codon in messenger RNA by complementary base pairing.

antigen Substance that, when introduced into the body, causes production of antibodies.

antiseptic Substance that prevents the growth of bacteria.

antiserum Immune serum.

antitoxin Immune serum that neutralizes the action of a toxin.

antivenin Antitoxic serum for snake venom.

aphthous Characterized by the presence of small ulcers.

ascites Abnormal collection of fluid in the peritoneal cavity.

aseptic Free from living microorganisms.

ataxia Lack of muscular coordination.

atopy Human allergy with hereditary background.

attenuated Weakened.

auto Self (combining form).

autoantibody Antibody formed against autoantigens.

autoantigen Antigens present in same individual as antibody-producing cells and not originally "foreign" to body.

autoclave Apparatus for sterilizing by steam under pressure; pressure steam sterilizer.

autogenous vaccine Vaccine made from culture of bacteria obtained from given patient.

autograft Graft derived from and applied to same individual.

autoimmune disease Disease wherein autoimmunization against certain body proteins is pathogenetic mechanism.

autoimmunity (autoallergy) Unusual state resulting from production (*autoimmunization*) by body of antibodies against its own proteins.

autoinfection Infection of one part of body by bacteria derived from some other part.

autotrophic Organisms that can form their proteins and carbohydrates out of inorganic salts and carbon dioxide.

B lymphocyte (B cell) Thymus-*in*dependent lymphocyte.

bacteremia Condition in which bacteria are in bloodstream but do not multiply there.

bactericide (*adj.* **bactericidal**) Agent lethal to bacteria.

bacteriology Science that deals with bacteria.

bacteriolysins Antibodies that lyse bacteria in presence of complement.

bacteriophages (phages) Bacterial viruses.

bacteriostasis (*adj.* **bacteriostatic**) Inhibition of bacterial growth; bacteria, however, not directly killed.

Bang's disease Contagious abortion of cattle.

base pairing Pairing of nucleotide nitrogenous bases: adenine with thymine and guanine with cytosine; the phenomenon that permits accurate copying of genetic material by cells.

BCG (bacillus of Calmette-Guérin) Vaccine against tuberculosis made from bovine strain of tubercle bacilli attenuated through long culturing; name derived from two French scientists who developed the strain.

biologic transfer of infection Mode of transfer of infection from host to host by an animal or insect in which agent causing disease undergoes a cycle of development.

biotherapy Treatment of disease with a living agent or its products.

blastogenesis Morphologic transformation of small lymphocytes into larger cells appearing like blast cells when exposed to antigens to which they are sensitized.

blood serum *see* Serum.

boil Abscess of the skin and subcutaneous tissue.

broad spectrum Term used to indicate that an antibiotic is effective against large array of micro-organisms.

bronchiolitis Inflammation of bronchiole, a finer subdivision of respiratory tract.

bronchopneumonia Small focal areas of inflammatory consolidation in the lungs.

bubo Inflammatory enlargement of a lymph node, often with formation of pus.

cancer Lay term for neoplasm.

capnophilic Growing best in presence of carbon dioxide.

capsule Envelope that surrounds certain bacteria.

carcinogenesis Induction of cancer.

carcinogenic Applied to agent capable of inducing cancer under suitable circumstances.

carditis Inflammation of heart.

caries Caries in teeth (dental caries) are layman's cavities.

carrier Person in apparent health who harbors pathogenic agent.

caseous Cheeselike.

catabolism Metabolic breakdown (energy-yielding degradation) of complex substances into products of simpler makeup.

catarrhal Characterized by an outpouring of mucus.

CDC Centers for Disease Control, Atlanta.

cell Minute protoplasmic structure, anatomic and physiologic unit of all animals and plants; that is, all animals and plants are made up of one or more cells, and their activities depend on combined activities of those cells; the simplest unit that can exist as an independent living system.

cell culture Cultivation of tissue cells away from human or animal body.

cellulitis Diffuse inflammation of connective tissues.

Celsius Inventor of the temperature scale setting 100° between the freezing point at zero and the boiling point of water.

chain reaction Series of successive reactions in which each succeeding reaction depends on preceding one and which, when once begun, continues until one or more of chemicals taking part in reaction are exhausted.

chancre Initial lesion of syphilis.

chemotaxis Reaction to a chemical whereby cells are attracted (*positive chemotaxis*) or repelled (*negative chemotaxis*) by the chemical.

chemotherapy Treatment of disease by administration of drugs that destroy causative organism of disease but do not injure patient.

chromatography Method of chemical analysis whereby certain compounds of a mixture are separated by use of their solubility and adsorptive properties; a solution containing a mixture of closely related compounds is allowed to seep through an adsorbent medium, such as a gel, so that each compound becomes adsorbed in a separate layer.

chromogenic Color-producing.

chromosomes Rod-shaped masses of chromatin that appear in the cell nucleus during mitosis; play important part in cell division and transmit the hereditary characteristics of the cell.

chronic Long-continued.

cilia (*sing. cilium*) Hairlike processes that spring from certain cells and by their action create currents in liquids; if the cells are fixed, the liquid is made to flow, but if the cells are unicellular organisms suspended in liquid, the cells move.

ciliates Unicellular organisms that move by means of cilia.

clinical Founded on actual observation.

clinical case Person ill and showing signs of a disease.

clone Progeny of a single cell; term used as either noun or verb; to clone is to establish strain of cells with identical properties of particular parent cell.

cloning Process of selectively reproducing a cell or given entity that results in production of a clone.

coagulase Enzyme that hastens coagulation.

coagulation Formation of a blood clot.

codon A triplet of nucleotides that specifies particular amino acid in protein or stops protein synthesis.

coliform bacteria Group of bacteria consisting of *Escherichia coli* and related intestinal inhabitants.

colon bacillus *Escherichia coli.*

colony Visible growth of bacteria on culture medium; all progeny of single preexisting bacterium.

commensalism Stable, close physical association of two species of organism in which one species is benefited and the other not affected.

communicable Capable of being transmitted from one person to another.

complement fixation Activation of complement system; that is, promotion of lytic action of complement by combination of antigen, antibody, and complement; the basis of complement-fixation tests for syphilis and other diseases.

complication Disease state concurrent with another disease.

condyloma Wartlike growth.

congenital Existing at the time of birth or shortly thereafter.

consolidation Solidification, as of the lung in pneumonia.

consumption Wasting away; a lay term, now rarely used, for pulmonary tuberculosis.

contagious Highly communicable; in common parlance, a disease easily "caught."

contamination Soiling with infectious material.

convalescent carrier Carrier who harbors organisms of a disease during recovery from disease.

convalescent serum Serum of person recently recovered from a disease; in a few cases injection of convalescent serum seems to be of value in treating or preventing the disease in others.

Coombs' test Test to detect presence of globulin antibodies on surface of red blood cells; used to detect sensitized red blood cells in hemolytic diseases.

coproantibodies Antibodies formed in the colon.

coprophilic Affinity for feces and filth.

coprozoic Living in or found in feces.

counterstain Second stain of different color applied to a smear to make effects of first stain more distinct.

covalent bonds A nonionic chemical bond formed by shared electrons, like those formed by reaction between an acid and base.

Credé method Instillation of 2% silver nitrate solution into each eye of newborn infant to prevent ophthalmia neonatorum.

crisis Sudden change in course of a disease; diseases that terminate by sudden change for better are said to end by crisis.

cryobiology Science that deals with effects of low temperatures on biologic systems.

culture Growth of microorganisms on nutrient medium; to grow microorganisms on such a medium.

culture media (*sing.* **medium**) Artificial food material on which microbes are grown (cultured).

cutaneous Pertaining to the skin; cutaneous inoculation is done by rubbing the infectious material on the abraded skin.

cycle Series of changes leading back to starting point.

cyst Abnormal saccular structure containing liquid, air, or solid material; stage in history of certain protozoa during which time encysted organism is protected by surrounding wall.

cysticercus Larval form of tapeworm in tissues of intermediate host; develops into adult worm on entry into intestinal canal of definitive host.

cytology Science dealing with study of cells, their origin, structure, and function.

cytolysin Antibody that dissolves cells.

cytopathogenic (cytopathic) effect Pathologic changes in cells of given cell (tissue) culture referable to action of some injurious agent, especially those caused by viruses grown in the cell culture.

Dakin's solution Neutral solution of sodium hypochlorite once used in disinfection of wounds.

defibrinate To remove the fibrin of blood to prevent clotting.

degenerate To undergo progressive deterioration.

degeneration Deterioration; change from higher to lower form.

degerm To remove bacteria from skin by mechanical cleaning or application of antiseptics.

dental plaque Deposit of material on surface of a tooth that may serve as a nidus for disease (for example, dental caries).

deoxyribonucleic acid (DNA) One of two nucleic acids that have been identified; essential for biologic inheritance.

dermatitis Inflammation of skin.

dermatomycoses Superficial fungal infections of skin and appendages.

dermatophyte Fungus parasitic for skin.

dermotropic Affinity for the skin.

desensitization Condition wherein an organism does not react to a specific antigen to which it previously did; process of bringing about this state.

desquamation Shedding of the superficial layer of the skin in scales or shreds.

determinant Factor establishing the nature of an entity or event.

diarrhea Abnormal frequency and/or fluidity of bowel movements.

diatomaceous earth Earth made up of petrified bodies of diatoms, or unicellular algae.

Dick test Skin test to determine susceptibility to scarlet fever.

differential stain Stain distinguishing between different groups of organisms.

direct contact Spread of a disease more or less directly from person to person.

disinfectant Substance that disinfects.

disinfection Destruction of all disease-producing organisms and their products.

dissociation Separation; dissolution of relations.

disulfide bridge Chemical bond using two sulfur atoms combined with another element, frequently used to bind protein chains together.

DNA *see* Deoxyribonucleic acid.

droplet infection Infection conveyed by spray thrown off from mouth and nose during talking, coughing, etc.

drug fast Resistance to the action of drugs; microbes able to withstand action of given drug.

DTP Diphtheria-tetanus-pertussis vaccine.

dysentery Diarrhea plus blood and mucus in stool; associated with inflammation of alimentary tract.

EBV Epstein-Barr virus.

ecology Science of organisms as affected by factors of their environments.

ecosystem Basic unit in ecology derived from interaction of living and nonliving elements in given area.

ectoenzyme Enzyme excreted into surrounding medium through plasma membrane of cell forming it.

ectoplasm Outer clear zone of the cytoplasm of unicellular organism.

efferent Bearing away.

electrophoresis Application of an electric current to separate substances that move faster in an electric field in a fluid or gel from those that move more slowly.

Elek plate Gel diffusion test for the demonstration of toxigenicity of *Corynebacterium diphtheriae*.

elementary bodies Virus particles (also used with chlamydiae).

emporiatrics Study of diseases of travelers.

empyema Collection of pus in a cavity; when used without qualification, collection of pus in pleural cavity.

encephalitis Inflammation of brain.

encephalomyelitis Inflammation of brain and spinal cord.

endemic More or less continuously present in a community.

endocarditis Inflammation of endocardium, or lining membrane of heart, including that of the valves.

endoenzyme Enzyme liberated only when cell that produces it disintegrates.

endogenous Originating within the organism.

endoplasm Zone of granular cytoplasm found near nucleus of many unicellular organisms.

endothelium Flattened cell lining of heart, blood vessels, and lymph channels.

endotoxin Toxin liberated only when cell producing it disintegrates.

enteric bacteria Bacteria isolated from gastrointestinal tract.

enterotoxin Toxin bringing about diarrhea and vomiting usually succeeding ingestion of contaminated food.

enzyme Catalytic substance secreted by living cell capable of changing other substances without undergoing any change itself.

eosinophil Granular leukocyte whose granules stain with eosin.

epidemic Disease that attacks large number of persons in a community at same time.

epidemiology Science that deals with epidemics.

essential (disease) Of unknown cause; idiopathic.

etiology Cause.

eukaryote Protist with true nucleus.

exacerbation Increase in severity of a disease.

exanthem Febrile disease accompanied by skin eruption.

excision (genetic) Enzymatic removal of a polynucleotide segment from a nucleic acid molecule.

exfoliate To come off in strips or sheets, particularly stripping of skin after certain exanthematous diseases.

exfoliative cytology Study of cells cast off from body surface.

exogenous Coming from the outside of body.

exotoxin Toxin secreted by microorganism into surrounding medium.

expression (genetic) Process by which information in DNA or RNA is transferred and ultimately translated into protein in the cell.

exudate Fluid and formed elements of blood extravasated into tissues or cavities of body as part of inflammatory reaction.

facultative Able to do a thing although not ordinarily doing it; for example, a facultative anaerobe is an organism that can live in absence of oxygen but does not ordinarily do so.

familial Affecting several members of the same family.

FDA Food and Drug Administration; part of U.S. Department of Health and Human Services.

feedback Return of part of output of a system as input.

ferment Enzyme.

fever Abnormally high body temperature usually related to a disease process.

FIAC Federal Interagency Advisory Committee (recombinant DNA regulation working group).

fibrinolysin Substance that dissolves or destroys fibrin.

filaria Long, threadlike roundworm that lives in circulatory or lymphatic system.

filariform Resembling filariae.

fixed virus (rabies) Street or wild virus adapted to rabbit and less virulent for human and dog.

flagella (sing. flagellum) Long, hairlike processes that by their lashing activity cause organism to move; one or more flagella may be attached to one or both ends of organism or completely around it.

flagellates Organisms that move by means of flagella.

flocculation test Test dependent on coalescence of finely divided or colloidal particles into larger visible particles.

fluid toxoid Toxoid not further treated with aluminum compounds.

fluorescence Property of emitting light after exposure to light.

fluorescent antibodies Antibodies that fluoresce.

focal infection Localized site of more or less chronic infection from which bacteria or their products are spread to other parts of body.

fomites Substances other than food that may transmit infectious organisms.

fulminating Sudden, severe, and overwhelming.

fumigation Exposure to fumes of a gas that destroys bacteria or vermin.
fungicide Agent destructive to fungi.
furuncle A boil.
furunculosis Presence of a number of boils.

gamete One of two cells (either male or female) whose union is necessary for sexual reproduction.
gamma globulin Fraction of globulin of blood with which antibodies are associated.
gene Biologic unit of heredity, self-reproducing and located in definite position (locus) on a particular chromosome; segment of DNA that transmits a given bit of genetic information by indicating the structure of a protein.
genetic determinants Carriers of genetic information.
genetics Branch of biology concerned with the mechanisms of heredity.
genome Complete set of hereditary factors.
germ Pathogenic microbe.
germ cell Cell specialized for reproduction.
germicide Agent that destroys germs.
Giemsa stain Stain (azure and eosin dyes) used to demonstrate protozoa, viral inclusion bodies, and rickettsias.
globulin Class of proteins characterized by being insoluble in water but soluble in weak solutions of various salts; one of important proteins of plasma of blood.
gnotobiosis Science of rearing and keeping animals either born germ free or with limited known microbial flora.
gram negative Bacteria decolorized by Gram's method but stained with counterstain (red or brown).
gram positive Bacteria not decolorized by Gram's method; they retain original violet color of Gram stain and are not stained by the counterstain.
Gram stain, Gram's method Method of differential staining devised by Hans Christian Gram, Danish bacteriologist.
granulocyte Cell containing granules within its cytoplasm; granular leukocyte (polymorphonuclear neutrophil, eosinophil, or basophil).
granuloma Circumscribed collection of macrophages surrounding point of irritation.
granulomatous Inflammation with pronounced response of macrophage system.
grouping Classification; *see also* Typing.
gumma Granuloma of late stages of syphilis.

habitat Place where plant or animal is found in nature.
halophile Salt loving.
hectare Ten thousand square meters.
hectic fever Daily recurring fever characterized by chills, sweating, and flushed countenance.
hematogenous Originating in blood or borne by blood.
hematology Study of the blood and its diseases.
hemoglobinophilic Pertaining to organisms that grow especially well in culture media containing hemoglobin.
hemolysin Antibody dissolving red blood cells in presence of complement.
hemolysis Lysis (dissolution) of red blood cells.
hemolytic Causing hemolysis.
hemophilic Blood loving; hemoglobinophilic.
hepatitis Inflammation of liver.
hepatogenous Originating in the liver.
herd immunity Resistance to disease related to immunity of high proportion of members of group.
hereditary Transmitted through members of family from generation to generation.
heterologous Derived from animal of another species.
heterophil Having affinity for antigens or antibodies other than one for which it is specific.

heterotrophic Refers to organisms requiring simple form of carbon for metabolism.

HEW U.S. Department of Health, Education, and Welfare, reorganized and renamed HHS (Health and Human Services) in 1979.

HHS *see* HEW

Hinton test Precipitation test for syphilis.

homeostasis Tendency to stability within internal environment or fluid matrix of an organism.

homo Same (combining form).

homograft Graft from one individual to another of same species to include *allogenic* grafts, or *allografts*, between genetically dissimilar individuals and *syngeneic* grafts between genetically identical individuals (identical twins).

host Animal or plant on which a parasite lives.

hydrogen bonds Linkage between a hydrogen atom and electronegative atoms, such as nitrogen, with one side of the bond being covalent and the other electrostatic.

hydrolysis Decomposition from incorporation and splitting of water.

hyperimmune Quality of possessing degree of immunity greater than that usually found under similar circumstances.

hyperplasia Increase in size related to increase in number of component units.

hyperpyrexia Very high fever, usually more than 41° C (106° F).

hypersensitivity *see* Allergy.

hypha (*pl.* **hyphae**) One of filaments composing a fungus.

iatrogenic Applied to adverse condition resulting from physician's treatment.

icterus Jaundice.

idiopathic Of unknown cause.

idiosyncrasy Individual and peculiar susceptibility or sensitivity to a drug, protein, food, or other agent.

"immortalize" Capacity of a virus or other agent to alter a normal cell so that the cell will reproduce indefinitely in appropriate media.

immune Exempt from a given infection.

immune bodies Antibodies.

immune globulin Sterile preparation of globulin from blood; antibodies normally associated with gamma-globulin fraction of blood proteins.

immune serum Serum containing immune bodies.

immunity Natural or acquired resistance to a disease.

immunochemistry Study of the complex chemical reactions in immunity.

immunocompetent cell Any cell that can be stimulated by antigen either to form antibodies or to give rise to cells that form antibodies.

immunoelectrophoresis Technic in which proteins are first separated by electrophoresis and then allowed to react with antiserum so that a pattern of precipitation arcs is developed.

immunofluorescence Technic in which antigen or antibody is conjugated to a fluorochrome and then allowed to react with corresponding antibody or antigen in a suitable specimen; microscopic observation of pattern of fluorescence locates antibody or antigen.

immunoglobulins Antibodies.

immunohematology Branch of hematology dealing with immune bodies in blood.

immunologic enhancement Enhanced survival of grafts of tumor or normal tissue because of specific humoral antibodies.

immunologic tolerance Failure of antibody response to a potential antigen after exposure to the antigen.

immunologist One versed in immunity.

immunology Science that deals with immunity.

impetigo contagiosa Infectious vesicular and pustular eruption most often seen on face and other body areas of children.

in vitro In the laboratory.

in vivo In the living body.

inclusion bodies Round, oval, or irregularly shaped particles in cytoplasm or nucleus of cells parasitized by viruses; colonies of viruses.

incompatible Not capable of being mixed without undergoing destructive chemical changes or acting antagonistically.

incubation period Period between time of infection or inoculation and appearance of manifestations.

incubator Cabinet in which constant temperature is maintained for growing cultures of bacteria.

indicator A substance that renders visible completion of a chemical reaction.

indirect contact Transfer of infection by means of inanimate objects, contaminated fingers, water, food, and the like.

infection Invasion of body by pathogenic agents with their subsequent multiplication and production of disease; inflammation from living agents.

infectious Having qualities that may transmit disease.

infectious granuloma Granuloma from a specific microbe.

infestation Invasion of body by macroscopic parasites such as insects; refers particularly to parasites on surface of body.

inflammation Protective reaction on part of tissues brought about by presence of an irritant—reaction to injury.

inhibition Diminution or arrest of function.

inoculate To implant microbes or infectious material onto culture media; to introduce artificially biologic product or disease-producing agent into body.

insecticide Substance that destroys insects.

intercellular Between cells.

intercurrent infection Infection that attacks a person already ill with another disease.

intermediates of metabolism Substances formed in a chemical process that is essential to the formation of a desired end-product.

intermittent Periods of activity separated by periods of quietude.

intoxication Poisoning.

intracellular Within a cell.

intracutaneous *see* Intradermal.

intradermal Within substance of skin.

intraperitoneal Within peritoneal cavity.

intraspinal Within vertebral canal.

intravenous Within a vein.

intrinsic From within; inherent.

involution forms Abnormal forms assumed by microorganisms growing under unfavorable conditions.

iodophors Disinfectants with germ-killing iodine carried by surface-active solvent.

isoantigens (alloantigens) Antigens from individuals of same species.

isolate To close all avenues by which a person may spread infection to others; to separate from others.

isologous Pertaining to same species.

isotonic Solution having the same osmotic pressure as that of standard reference solution.

isotopes Atoms of same elements having different atomic weights but generally same chemical behavior.

jaundice Deposition of bilirubin in skin and tissues with resultant yellowish discoloration in a patient with hyperbilirubinemia (excess of bilirubin in the blood).

Kahn test Precipitation test for syphilis.

keratitis Inflammation of cornea.

keratoconjunctivitis Inflammation of cornea and conjunctiva.

kernicterus Cerebral manifestations of severe jaundice in newborn infant; degeneration of nerve cells caused by accumulation of bilirubin in brain.

killer cell Cell that in vitro is cytotoxic to cell cultures.

Kline test Precipitation test for syphilis.

Koch's postulates Certain requirements that must be met before a given microorganism can be considered the cause of a certain disease.

lac operon Cluster of genes controlling synthesis of enzymes involved in the metabolism of lactose in *Escherichia coli*.

lag phase Period of time between stimulus and resultant reaction.

larva (*pl.* **larvae**) Young of any animal differing in form from its parent.

latent Seemingly inactive, potential, or concealed.

latex particle test Passive agglutination test utilizing soluble antigen adsorbed to minute particles of polystyrene latex; in presence of antibody specific for bound antigen, latex spheres are visibly clumped.

lawn Solid confluent growth of microorganisms of one species or strain on solid medium.

lesion Specific pathologic structural or functional (or both) change brought about by disease.

leukocidin Substance that destroys leukocytes.

leukocyte White blood cell.

leukocytosis Transient protective increase in leukocytes in blood in response to injury.

leukopenia Abnormal decrease in leukocytes in blood.

local infection One that is confined to a restricted area.

locus Position.

lues venerea Syphilis.

lumbar Pertaining to that part of the back between the ribs and pelvis.

lymphadenitis Inflammation of a lymph node.

lymphadenopathy Disease of a lymph node (or nodes).

lymphocyte Nongranular white blood cell of lymphoid origin.

lymphocytosis Increase in lymphocytes in blood.

lymphoid tissue Delicate connective tissue lattice with lymphocytes and related cells in its meshes.

lymphoma Tumor of lymphoid tissue.

lyophilization Creation of stable product by rapid freezing and drying of frozen product under high vacuum.

macromolecule Very large molecule having polymeric chain structure.

macrophage Large mononuclear wandering phagocytic cell originating in macrophage system.

macrophage system System of phagocytic cells scattered through various organs and tissues, particularly spleen, liver, bone marrow, and lymph nodes, playing an important part in immunity.

macroscopic Visible to naked eye.

macular Consisting of small, flat, reddish spots in skin.

malignant Virulent; going from bad to worse.

Mantoux test Tuberculin skin test.

marker Something that identifies or is used to identify.

Mazzini test Precipitation test for syphilis.

meatus Opening.

mechanical transfer of infection Transfer of infection by arthropods in which infectious agent undergoes no cycle of development in body of particular insect or other arthropod.

medulla Center of a part or organ.

meiosis Special type of cell division during maturation of sex cells by which normal number of chromosomes is halved.

membranous croup Lay term for diphtheria.

metabolism Sum total of chemical changes whereby nutrition and functional activities of body are maintained.

metachromatic granules Granules of deeply staining material found in certain bacteria.

metazoa (*sing.* **metazoon**) Multicellular animals.

microaerophilic Applied to microorganisms that require free oxygen for their growth but in an amount less than that of oxygen of atmosphere.

microbe Microscopic unicellular organism.

microgram One thousandth of a milligram, or 1/1,000,000 of a gram.

micrometer One thousandth of a millimeter, or 1/25,000 of an inch; a micron.

micron *see* Micrometer.

microorganism Organism of microscopic size.

microscopy Study of objects by means of the microscope.

miliary Small, resembling a millet seed in size.

minimum lethal dose (MLD) Smallest dose that will cause death.

mitochondrion (*pl.* **mitochondria**) Small spherical or rod-shaped cytoplasmic organelle in eukaryotic cell.

mitosis Indirect cell division.

mixed culture Culture containing two or more kinds of organisms.

mixed infection Infection with two or more kinds of organisms.

mixed vaccine Vaccine containing two or more kinds of organisms.

molds Multicellular fungi.

molt Act of shedding outer body covering (skin, cuticle, feathers).

monocyte A white blood cell, 12 to 30 μm in diameter; precursor to macrophage.

morbid Pertaining to or affected with disease.

morbidity Departure (subjective or objective) from state of physiologic well-being (WHO definition).

mordant Chemical added to a dye to make it stain more intensely.

morphologic Pertaining to shape or form.

mutation Change or alteration in form or qualities; permanent transmissible change in characters of an offspring from those of its parents.

mutualism Stable arrangement by which two organisms of different species live closely, with each deriving some benefit from the association.

mycelium Vegetative part of a fungus, consisting of many hyphae.

mycobacteriosis Disease produced by certain members of the genus *Mycobacterium*.

mycohemia Presence of fungi in bloodstream.

mycology Science that deals with fungi.

mycophage Fungal virus.

mycosis Disease caused by fungi.

mycotoxin Fungal toxin.

nanometer (**millimicron**) One millionth of a millimeter; 10 angstrom (Å).

nasopharynx Portion of pharynx above palate.

natural immunity Immunity with which a person or animal is born.

natural selection Natural process tending to eliminate organisms poorly suited to their environment and perpetuate those with genetic qualities that enhance their survival.

necrosis Death of tissue while yet part of living body.

negative staining Staining the background, not the organism.

Negri bodies Diagnostic inclusion bodies found in certain brain cells of animal with rabies.

neoplasm (**neoplasia**) New growth of abnormal cells of autonomous nature (scientific term for layman's "cancer"); the word *tumor* is used synonymously, although its true meaning is mass.

neuritis Inflammation of a nerve.

neurotropic Affinity for central nervous system or nervous tissue.

NIH National Institutes of Health.

NIOSH National Institute for Occupational Safety and Health, part of HHS.

normal flora Bacterial content of given area during health; reasonably constant as to quantity and proportions.

nosocomial Pertaining to hospital or infirmary.

nosology Science of classification of disease.

nuclear medicine Use of radioisotopes in medicine.

nucleic acids Complex chemical substances closely associated with transmission of genetic characteristics of cells; the two identified are ribonucleic acid and deoxyribonucleic acid.

nucleoprotein Simple basic protein combined with nucleic acid.

nucleoside Purine or pyrimidine linked to a pentose.

nucleotide Compound consisting of a nitrogenous base (a purine or pyrimidine derivative), a sugar, and one or more phosphate groups (a phosphate ester of a nucleoside).

nucleus (*pl.* **nuclei**) Central, compact portion of cell, the functional center of a eukaryotic cell.

old tuberculin (OT) Special type of tuberculin.

oncogenic Tumor producing.

oncology Study of neoplasms.

operon Cluster of structural genes whose coordinated expression is controlled by regulator gene.

opportunists Microbes that produce infection only under especially favorable conditions.

opsonins Substances in blood that render microorganisms more susceptible to phagocytosis.

organelle(s) Specialized part of protozoon that performs special function; specific particles of organized living substance present in almost all cells.

osteomyelitis Inflammation of bone marrow.

otitis media Inflammation of middle ear.

pandemic Very widespread epidemic, even of worldwide extent.

parasite Animal or plant organism that lives on another organism at whose expense it obtains some advantage.

parasitology Science that deals with parasites and their effects on other living organisms.

parenchyma Specialized functioning tissue or cells of an organ.

parenterally In some manner other than gastrointestinal tract; by injection.

paroxysm Sudden attack of disease or acceleration of manifestations of existing disease.

passive carrier Carrier who harbors causative agent of disease without having had the disease.

passive immunity Immunity produced without body of person or animal becoming immune taking any part in its production.

pasteurization Heating of milk or other liquids for a short time to a temperature that will destroy pathogenic bacteria but will not affect its food properties and flavor.

pathogenesis Sequence of events in development of given disease state.

pathogenicity Disease-producing quality.

pathognomonic Specifically characteristic or diagnostic of a disease.

pathology Branch of medicine dealing with nature of disease, especially with reference to structural and functional changes in tissues and organs of human body.

Paul-Bunnell test Heterophil antibody test on serum; important in diagnosis of infectious mononucleosis.

peptide Two or more amino acids covalently joined by peptide bonds; a two–amino acid linkage is a *dipeptide*; three or more amino acids so linked form a *polypeptide*.

peptide linkage Covalent bond formed between the amino terminal nitrogen of one amino acid or peptide and the carboxyl terminal carbon of another amino acid or peptide.

peritonitis Inflammation of peritoneum.

permanent carrier Carrier who harbors disease-producing agent for months or years.

petechia (*pl.* **petechiae**) Pinpoint hemorrhage.

Petri dish Round glass dish with cover used for growing bacterial cultures.

Peyer's patches Collection of lymphoid nodules packed together to form oblong elevations of mucous membrane of the small intestine, their long axis corresponding to that of intestine.

phage typing Use of bacteriophages and their lytic properties to classify bacteria.

phagocyte Cell capable of ingesting bacteria or other foreign particles.

phagocytic Related to phagocytes or phagocytosis.

phagocytosis Process of ingestion by phagocytes.

phenol coefficient Measure (ratio) of disinfecting property of a chemical compared with that of phenol (carbolic acid).

phlegmon Acute diffuse inflammation of subcutaneous connective tissue.

pinocytosis Cell drinking, or imbibition of liquids by cells; minute invaginations on surface of cells close to form fluid-filled vacuoles.

Pirquet's test (von Pirquet) One of tuberculin skin tests.

plaques Visible areas of cellular damage caused by virus inoculated into susceptible cell culture, analogous to colonies of bacteria on agar plate; *see also* Dental plaque.

plasma Fluid portion of circulating blood; fluid portion of *clotted* blood is *serum*.

plasma membrane Outer membrane encasing the protoplasm of a cell.

plasmid Circular DNA with the capacity to use replicative mechanisms of the cell for its replication; it can pass into or out of a cell without killing the cell; it can also code for protein production by host cell; plasmids are one type of vector used in gene cloning.

plasmolysis Shrinking of cell suspended in hypertonic solution.

plasmoptysis Swelling and bursting of cell suspended in hypotonic solution.

platelet (thrombocyte) Small, nonnucleate structure, 3 μm in diameter, that is found in blood and derived from megakaryocytes of bone marrow and is important in coagulation of blood; it is essential for hemostasis, and since it contains histamine, it may play role in hypersensitivity reactions.

pleomorphism Existence of different forms in the same species.

pleurisy Inflammation of pleura.

pleuropneumonia Infectious pneumonia and pleurisy of cattle.

pneumonia Inflammatory consolidation or solidification of lung tissue; presence of exudate blots out air-containing spaces.

pneumonitis Inflammation of supporting framework of lung.

pneumotropic Having affinity for lungs.

polar bodies Deeply staining bodies found in one or both ends of certain species of bacteria.

pollution State of being unclean; as used in bacteriology, containing harmful substances other than bacteria.

precipitation Clumping of proteins in solution by addition of specific precipitin.

precipitins Antibodies that cause precipitation.

prehension Act of taking hold; grasping.

preservative Substance added to product to prevent bacterial growth and consequent spoiling.

primary First (first focus of disease).

primary infection First of two infections, one occurring during course of other.

prognosis Forecast of outcome of a disease.

prokaryote Protist without true nucleus.

properdin Protein component of globulin fraction of blood playing role in immunity.

prophylaxis Prevention of disease.

proprietary Referring to fact that a given item is a commercial one.

proteins Naturally occurring complex polymers of amino acids that are the essential constituents of life.

proteolytic Bringing about digestion or liquefaction of proteins.

protist Member of kingdom Protista, which includes all single-celled organisms (bacteria, algae, slime molds, fungi, protozoa), some plantlike, some animal-like, some neither.

protoplasm Living material of which cells are composed.

protozoology Science that deals with protozoa.

protozoon (*pl.* protozoa; *adj.* protozoan) Unicellular animal organism.

provocative dose Dose stimulating appearance of given effect.

pseudomembrane Fibrinous exudate forming a tough membranous structure on surface of skin or mucous membrane.

pseudopod Temporary protoplasmic process put forth by protozoon for purposes of locomotion or obtaining food.

psychrophile Cold-loving organism, growing best at low temperatures.

ptomaines Basic substances resembling alkaloids formed during decomposition of dead organic matter.

pure culture Culture containing only one species of organism.

purulent Containing pus.

pus Fluid product of inflammation consisting of leukocytes, bacteria, dead tissue cells, foreign elements, and fluid from blood.

pustule Circumscribed elevation on skin containing pus.

putrefaction Decomposition of proteins.

pyelitis Inflammation of renal pelvis.

pyelonephritis Inflammation of kidney parenchyma and pelvis.

pyemia Form of septicemia in which organisms in bloodstream lodge in organs and tissues and set up secondary abscesses.

pyogenic Pus forming.

quarantine Limitation of freedom of movement of well persons or animals that have been exposed to a communicable disease for a time equal to the longest incubation period.

quiescent Not active.

quinsy Peritonsillar abscess

racial immunity Immunity peculiar to a race.

radiant energy Energy from radioactive source.

radioisotope Isotope form of element that is radioactive.

raw milk Unpasteurized milk.

reading (genetic) Process whereby the sequence information in one polymer is used to produce a defined sequence in another polymer; replication, transcription, and translation are processes that entail reading.

receptor site Specific area on cell membrane that senses the environment in which cell lives.

receptors Precise chemical groupings on surface of target cell; those on immunologically competent cell can combine specifically with antigen; term used by Ehrlich in his side-chain theory of immunity to denote specialized portions of cell that combine with foreign substances.

recombinant DNA technology Technics for cutting apart and splicing together pieces of DNA from different organisms; when segments of human or animal DNA are transferred into another organism (for example, bacterium), these segments replicate and may be expressed together with bacterium's own genetic material.

recrudescent Recurrence of symptoms of disease after days or weeks.

recurrent Reappearance of symptoms after an intermission.

reduction Removal of oxygen from or addition of hydrogen to a compound.

remission (*adj.* **remittent**) Temporary cessation of manifestations of a disease.

rennin Milk-curdling enzyme.

replication Process by which genetic determinants are duplicated during cell multiplication so that identical genetic characters are passed to next generation.

resident bacteria Bacteria normally occurring at given anatomic site.

resistance Inherent power of body to ward off disease.

resolution Subsidence of inflammation.

respiration An ATP-generating process in which an inorganic compound, such as oxygen, serves as the ultimate electron acceptor; the electron donor can be either an organic or an inorganic compound.

Rh factor Blood factor (agglutinogen) found on red blood cells of most persons; so named because of its occurrence on red blood cells of rhesus monkeys.

rhinitis Inflammation of nose.

ribonucleic acid (RNA) One of two nucleic acids; *see also* Deoxyribonucleic acid.

ribosomes Ribonucleoprotein granules in cell cytoplasm.

ringworm Fungal disease of skin.

RNA *see* Ribonucleic acid.

rodent Gnawing mammal (rats, mice, etc.)

rose spots Characteristic spots in skin over lower portion of trunk and abdomen in typhoid fever.

sanitary Conducive to health.

sanitize To reduce number of bacteria to safe level as judged by public health standards.

sapremia Condition in which products of action of saprophytic bacteria on dead tissues are absorbed into body and produce disease.

saprophyte Organism that normally grows on dead matter.

Schick test Skin test to detect susceptibility to diphtheria.

scrofula Tuberculosis of lymph nodes, particularly those of neck.

secondary infection Infection occurring in host suffering from primary infection.

sedimentation Settling of solid matter to bottom of a liquid.

selective action Tendency on part of disease-producing agents to attack certain parts of body.

Semple vaccine Rabies vaccine prepared from rabbit brain treated with phenol.

sepsis Poisoning by microbes or their products.

septic Relating to presence of pathogenic organisms or their poisonous products.

septicemia Systemic disease caused by invasion of bloodstream by pathogenic organisms, with their subsequent multiplication therein.

serology Branch of science that deals with serums, especially immune serums.

serotype Subtype within a bacterial species identified by serologic technics.

serum Fluid that exudes when blood coagulates; portion of plasma left after plasma protein fibrinogen is removed.

signs Objective disturbances produced by disease; observed by physician, nurse, or other medical attendant.

simian viruses Viral contaminants in cell cultures or normal monkey cells.

simple stain Stain using only one dye.

sinus Tract leading from an area of disease to body surface.

skin test dose (STD) Unit of measurement of scarlet fever toxin, amount required to produce positive reaction on skin of person susceptible to scarlet fever.

smear Very thin layer of material spread on glass microslide.

species immunity Immunity peculiar to a species.

sporadic disease Disease that occurs in neither an endemic nor an epidemic but occasionally.

spore Highly resistant form assumed by certain species of bacteria when grown under adverse influences; reproductive cells of certain types of organisms.

sporulation Production of spores or division into spores.

sterile Free of living microorganisms and their products.

sterilization Process of making sterile.

stock vaccine Vaccine made from cultures other than those from patient who is to receive the vaccine.

strain Subdivision of species.

streptolysin O and S Hemolysins produced by streptococci.

stroma Supporting connective tissue framework of organ or gland.

subacute Between acute and chronic in time.

subcutaneous Under the skin.

substrate Substance on which an enzyme acts.

sulfur granules Small yellow granules present in pus from lesions of actinomycosis.

superinfection New (superimposed) infection with drug-resistant microbes as complication of antimicrobial therapy for preexisting infection.

suppressor cell Lymphocyte capable of regulating extent of antibody production by other cells; best exemplified by suppressor T lymphocytes, subset population of lymphocytes.

suppuration Formation of pus.

Svedberg unit Unit in which sedimentation coefficients are measured; symbol S; one Svedberg unit equals 10^{-13} seconds; after Svedberg, the Swedish inventor of the ultracentrifuge.

sycosis barbae Folliculitis of beard.

symbiosis Mutually advantageous association of two or more organisms.

symptoms Subjective disturbances of disease, felt or experienced by patient but not directly measurable; for example, pain—the patient feels it definitely but it cannot be seen, heard, or touched.

syndrome Set of symptoms and signs occurring together in a pattern; etiologic agents variable.

T lymphocyte (T cell) Thymus-dependent lymphocyte.

taxon Particular group into which related organisms are classified.

taxonomy Branch of biology dealing with arrangement and classification of biologic organisms.

template Pattern, guide, mold, or blueprint.

terminal disinfection Disinfection of room after it has been vacated by patient.

terminal infection Infection that occurs during course of chronic disease and causes death.

thermal death point Degree of heat necessary to kill liquid culture of given species of bacteria in 10 minutes.

tinea A dermatomycosis.

tissue culture *see* Cell culture.

titer Measure of minimal amount of given substance needed for precise results in titration; standard strength of a solution established by titration.

titration Determination of concentration of contained substance by measurement of least amount of another substance that, when added to test solution, produces precise reaction.

tolerance Ability to endure without ill effect.

toxemia Presence of toxins in the blood.

toxin Poisonous substance elaborated during growth of pathogenic bacteria.

toxoid Toxin treated in such a manner that its toxic properties are destroyed without affecting its antibody-producing properties.

transcription Synthesis of messenger RNA or DNA template.

transduction Transmission of genetic factor from one bacterial cell to another by viral agent.

transformation (bacterial) Artificial conversion of bacterial types within a species by transfer of DNA from one bacterium to another.

transformation (cellular) Conversion of normal cells in culture to cells that exhibit properties of cancer cells.

transformation (lymphocytic) Morphologic and other changes in lymphocytes resulting from exposure to antigens to which they are specifically reactive.

trophozoite Active, motile, feeding stage of protozoan organism.

tubercle Granuloma that forms the unit specific lesion of tuberculosis.

tuberculin Toxic protein extract obtained from tubercle bacilli.

tumor Mass; by common usage, word designates neoplasm or a lesion to be differentiated from neoplasm.

typing (classification) Determination of category to which an individual, object, microbe, and the like belong with respect to given standard of reference.

ulcer Circumscribed area of inflammatory necrosis of epithelial lining of a surface.

ulceration Process of ulcer formation.

ultracentrifuge Powerful centrifuge similar in principle to a cream separator; used to separate many of the smaller and lighter components of a cell.

unit Standard of measurement.

vaccination Introduction of a vaccine into body.

vaccine Causative agent of disease so modified that it is incapable of producing disease while retaining its power to cause antibody formation.

vaccinia Cowpox.

variation Deviation from parent form.

VDRL (Venereal Disease Research Laboratory) test Precipitation test for syphilis.

vector Carrier of disease-producing agents from one host to another, especially an arthropod (fly, mosquito, flea, louse, or other insect); also, a DNA structure, such as a virus, plasmid, or bacteriophage, that can transmit additional DNA sequences into a cell, primarily conferring the ability to replicate the additional DNA.

vegetative bacteria Nonspore-forming bacteria or spore-forming bacteria in their nonsporulating state.

venereal Applied to a transmission of disease by intimate sexual contact.

vesicle (blister) Small circumscribed elevation of skin containing thin nonpurulent fluid.

viremia Presence of virus in bloodstream.

virion Complete mature infective virus particle; in some viruses it is identical to the nucleocapsid; in more complex viruses it includes the nucleocapsid plus a surrounding envelope.

virology Science that deals with viruses and viral diseases.

virucide Agent destroying or inactivating viruses.

virulence Ability of organism to produce disease.

virus-neutralizing antibodies Antibodies that inactivate viruses.

viscerotropic Affinity for internal organs of chest or abdomen.

viscus (*pl.* **viscera**) Internal organ, expecially one of abdominal organs.

vital functions Functions necessary for maintenance of life.

vitamins Certain little-understood food substances whose presence in very small amounts is necessary for normal functioning of body cells.

Wassermann test Complement-fixation test for syphilis devised by August von Wassermann.

Weil-Felix reaction Nonspecific but highly valuable agglutination test for typhus fever; uses a member of *Proteus* group as organism agglutinated.

wheal Circumscribed elevation of skin caused by edema of underlying connective tissue.

WHO World Health Organization

Widal test Agglutination test for typhoid fever.

wild virus Virus found in nature.

Wright's stain Mixture of eosin and methylene blue used to demonstrate blood cells and malarial parasites.

X rays (roentgen rays) Highly penetrating form of ionizing radiation produced in special high-voltage equipment.

xenogenic (xenogeneic) Preferred term for graft derived from species different from that of recipient.

xenograft (heterograft) Preferred term for graft from donor of dissimilar species.

yeasts Unicellular fungi.

zoonosis (*pl.* **zoonoses**) Disease of animals that may be secondarily transmitted to humans.

INDEX

Page numbers in *italics* indicate illustrations.
Page numbers followed by *t* indicate tables.